AAPG Treatise of Petroleum Geology Reprint Series

The American Association of Petroleum Geologists
gratefully acknowledges and appreciates the leadership and support
of the AAPG Foundation in the development of the
Treatise of Petroleum Geology.

GEOCHEMISTRY

Compiled by
Edward A. Beaumont
and
Norman H. Foster

Treatise of Petroleum Geology
Reprint Series, No. 8

Published by
The American Association of Petroleum Geologists
Tulsa, Oklahoma 74101, U.S.A.

Copyright © 1988
The American Association of Petroleum Geologists
All Rights Reserved

Library of Congress Cataloging-in-Publication Data

Geochemistry.

(Treatise of petroleum geology reprint series ; no. 8)
Includes bibliographies.
1. Geochemistry.   I. Beaumont, E. A. (Edward A.)
II. Foster, Norman H.   III. American Association of
Petroleum Geologists.   IV. Series.
QE515.G38526   1988   551.9   88-16671
ISBN 0-89181-407-8

INTRODUCTION

This reprint volume belongs to a series that is part of the *Treatise of Petroleum Geology*. The *Treatise of Petroleum Geology* was conceived during a discussion we had at the 1984 AAPG Annual Meeting in San Antonio, Texas. When our discussions ended, we had decided to write a state-of-the-art textbook in petroleum geology, directed not at the student, but at the practicing petroleum geologist. The project to put together one textbook gradually evolved into a series of three different publications: the Reprint Series, the Atlas of Oil and Gas Fields, and the Handbook of Petroleum Geology; collectively these publications are known as the *Treatise of Petroleum Geology*. With the help of the Treatise of Petroleum Geology Advisory Board (whose names are listed on the next page) we designed this set of publications to represent the cutting edge in petroleum exploration knowledge and application. The Reprint Series provides previously published landmark literature; the Atlas collects detailed field studies to illustrate the various ways oil and gas are trapped; and the Handbook is a professional explorationist's guide to the latest knowledge in the various areas of petroleum geology and related fields.

The papers in the volumes of the Reprint Series complement the chapters of the Handbook. Papers were selected on the basis of their usefulness today in petroleum exploration and development. Many "classic papers" that led to our present state of knowledge have not been included because of space limitations.

We divided the topic of geochemistry into four sections: (1) Petroleum Formation, (2) Source Rock Evaluation, (3) Migration, and (4) Surface Geochemistry. Section one, Petroleum Formation, contains papers that discuss accumulation and preservation of organic matter, conversion of organic matter to kerogen, conversion of kerogen to petroleum, and effects that different types of kerogen have on the types of petroleum generated.

Section two, Source Rock Evaluation, contains papers that examine the application of geochemical techniques to evaluation of source rocks. These papers review methods for estimating total organic content of source rocks, potential amount of petroleum a source rock can expel, and quantity of petroleum expelled from a source rock. The papers also discuss ways to match oil in a reservoir to its source rock, procedures to measure maturity of source rocks, and techniques for estimating time of expulsion of petroleum from source rocks.

Papers in section three, Migration, deal with mechanisms of expulsion of petroleum from source rocks and with migration of petroleum in carrier beds.

Section four, Surface Geochemistry, has two papers that discuss the potential and problems of applying surface geochemical techniques to petroleum exploration.

Edward A. Beaumont
Tulsa, Oklahoma

Norman H. Foster
Denver, Colorado

Treatise of Petroleum Geology
Advisory Board

W.O. Abbott
Robert S. Agaston
John J. Amoruso
J.D. Armstrong
George B. Asquith
Colin Barker
Ted L. Bear
Edward A. Beaumont
Robert R. Berg
Richard R. Bloomer
Louis C. Bortz
Donald R. Boyd
Robert L. Brenner
Raymond Buchanan
Daniel A. Busch
David G. Campbell
J. Ben Carsey
Duncan M. Chisholm
H. Victor Church
Don Clutterbuck
Robert J. Cordell
Robert D. Cowdery
William H. Curry, III
Doris M. Curtis
Graham R. Curtis
Clint A. Darnall
Patrick Daugherty
Herbert G. Davis
James R. Davis
Gerard J. Demaison
Parke A. Dickey
F.A. Dix, Jr.
Charles F. Dodge
Edward D. Dolly
Ben Donegan
Robert H. Dott, Sr.
John H. Doveton
Marlan W. Downey
John G. Drake
Richard J. Ebens
William L. Fisher
Norman H. Foster
Lawrence W. Funkhouser
William E. Galloway
Lee C. Gerhard
James A. Gibbs
Arthur R. Green
Robert D. Gunn
Merrill W. Haas
Robert N. Hacker
J. Bill Hailey
Michel T. Halbouty
Bernold M. Hanson
Tod P. Harding
Donald G. Harris
Frank W. Harrison, Jr.

Ronald L. Hart
Dan J. Hartmann
John D. Haun
Hollis D. Hedberg
James A. Helwig
Thomas B. Henderson, Jr.
Francis E. Heritier
Paul D. Hess
Mason L. Hill
David K. Hobday
David S. Holland
Myron K. Horn
Michael E. Hriskevich
J.J.C. Ingels
Michael S. Johnson
Bradley B. Jones
R.W. Jones
John E. Kilkenny
H. Douglas Klemme
Allan J. Koch
Raden P. Koesoemadinate
Hans H. Krause
Naresh Kumar
Rolf Magne Larsen
Jay E. Leonard
Ray Leonard
Howard H. Lester
Detlev Leythaeuser
John P. Lockridge
Tony Lomando
John M. Long
Susan A. Longacre
James D. Lowell
Peter T. Lucas
Harold C. MacDonald
Andrew S. Mackenzie
Jack P. Martin
Michael E. Mathy
Vincent Matthews, III
James A. McCaleb
Dean A. McGee
Philip J. McKenna
Robert E. Megill
Fred F. Meissner
Robert K. Merrill
David L. Mikesh
Marcus E. Milling
George Mirkin
Richard J. Moiola
D. Keith Murray
Norman S. Neidell
Ronald A. Nelson
Charles R. Noll
Clifton J. Nolte

Susan E. Palmer
Arthur J. Pansze
John M. Parker
Alain Perrodon
James A. Peterson
R. Michael Peterson
David E. Powley
A. Pulunggono
Donald L. Rasmussen
R. Randolf Ray
Dudley D. Rice
Edward C. Roy, Jr.
Eric A. Rudd
Floyd F. Sabins, Jr.
Nahum Schneidermann
Peter A. Scholle
George L. Scott, Jr.
Robert T. Sellars, Jr.
John W. Shelton
Robert M. Sneider
Stephen A. Sonnenberg
William E. Speer
Ernest J. Spradlin
Bill St. John
Philip H. Stark
Richard Steinmetz
Per R. Stokke
Donald S. Stone
Doug K. Strickland
James V. Taranik
Harry TerBest, Jr.
Bruce K. Thatcher, Jr.
M. Raymond Thomasson
Bernard Tissot
Donald Todd
M.O. Turner
Peter R. Vail
Arthur M. Van Tyne
Harry K. Veal
Richard R. Vincelette
Fred J. Wagner, Jr.
Anthony Walton
Douglas W. Waples
Harry W. Wassall, III
W. Lynn Watney
N.L. Watts
Koenradd J. Weber
Robert J. Weimer
Dietrich H. Welte
Alun H. Whittaker
James E. Wilson, Jr.
Martha O. Withjack
P.W.J. Wood
Homer O. Woodbury
Mehmet A. Yukler
Zhai Guangming

TABLE OF CONTENTS

GEOCHEMISTRY

PETROLEUM FORMATION

Hydrocarbons. A. Perrodon...3
Physical-chemical models for oil generation. D.W. Waples 27
The generative basin concept. G. Demaison...43
Time, temperature and organic maturation—the evolution of rank within a sedimentary pile. N.J.R. Wright 57
Influence of nature and diagenesis of organic matter in formation of petroleum. B. Tissot, B. Durand,
　J. Espitalié, and A. Combaz..73
Sedimentary organic matter and kerogen. Definition and quantitative importance of kerogen. B. Durand 81
Relationship between petroleum composition and depositional environment of petroleum source rocks.
　J.M. Moldowan, W.K. Seifert, and E.J. Gallegos...103
Comparison of carbonate and shale source rocks. R.W. Jones....................................117
Anoxic environments and oil source bed genesis. G.J. Demaison and G.T. Moore.................135
Generation, accumulation, and resource potential of biogenic gas. D.D. Rice and G.E. Claypool..................167
Source rocks and hydrocarbons of the North Sea. C. Cornford 189

SOURCE ROCK EVALUATION

Principles of geochemical prospect appraisal. A.S. Mackenzie and T.M. Quigley...............................231
Locating petroleum prospects: application of principle of petroleum generation and migration—geological
　modeling. B. Tissot and D.H. Welte...249
Geochemical modeling: a quantitative approach to the evaluation of oil and gas prospects. B.P. Tissot
　and D.H. Welte 261
Recent advances in petroleum geochemistry applied to hydrocarbon exploration. B.P. Tissot....................287
Modern approaches in source-rock evaluation. D.W. Waples 307
Predictive source bed stratigraphy; a guide to regional petroleum occurrence—North Sea basin and eastern
　North American continental margin. G. Demaison, A.J.J. Holck, R.W. Jones, and G.T. Moore 323
Biological markers in fossil fuel production. R.P. Philp......................................337
Guidelines for evaluating petroleum source rock using programmed pyrolysis. K.E. Peters 392
Identification of source rocks on wireline logs by density/resistivity and sonic transit time/resistivity
　crossplots. B.L. Meyer and M.H. Nederlof..405
A geochemical reconstruction of oil generation in the Barrow sub-basin of Western Australia.
　J.K. Volkman, R. Alexander, R.I. Kagi, R.A. Noble, and G.W. Woodhouse 415
Compilation and correlation of major termal maturation indicators. Y. Héroux, A. Chagnon, and R. Bertrand........430

Migration

Primary migrations and the source rock concept. A. Perrodon 449

Present trends in organic geochemistry in research on migration of hydrocarbons. B. Durand 461

Detecting migration phenomena in a geological series by means of C_1—C_{35} hydrocarbon amounts and distributions. M. Vandenbroucke, B. Durand, and J.L. Oudin 473

Some mass balance and geological constraints on migration mechanisms. R.W. Jones 483

Some factors in oil accumulation. V.C. Illing 503

Stress fields, a key to oil migration. J. du Rouchet 528

A novel approach for recognition and quantification of hydrocarbon migration effects in shale-sandstone sequences. D. Leythaeuser, A. Mackenzie, R.G. Schaefer, and M. Bjoroy 540

Petroleum geology of the Bakken Formation, Williston basin, North Dakota and Montana. F.F. Meissner 565

Geochemical exploration in the Powder River basin. J.A. Momper and J.A. Williams 587

Gas generation and migration in the Deep Basin of western Canada. D.H. Welte, R.G. Schaefer, W. Stoessinger, and M. Radke 599

Focused gas migration and concentration of deep-gas accumulations. J-C. Pratsch 613

Surface Geochemistry

Near-surface hydrocarbon surveys in oil and gas exploration. R.D. McIver 623

Surface geochemical methods used for oil and gas prospecting—a review. R.P. Philp and P.T. Crisp 627

PETROLEUM FORMATION

From
Dynamics of Oil and Gas Accumulations
by Alain Perrodon

1. — HYDROCARBONS

« ... Au-delà des éléments, il y a le fait qu'ils forment un tout. »
'... And beyond the elements lies the fact that they form a whole.'

Georges DUMEZIL.

1.1. CHARACTERISTICS OF CRUDE OILS AND NATURAL GASES

As opposed to kerogen, which is insoluble in organic solvents, oils and gases form the great family of bitumens (from the Latin *bitus*, resinous wood). The word asphalt — from the Greek *asphaltos* — corresponds more generally to thick or solid hydrocarbon substances. The word 'naphtha', of Mesopotamian origin, derives from one of the oldest known languages, from *napata,* to burn, more specifically applies to liquid petroleum products, and is found in the Arabic word '*al naft*'.

In technical language, bitumen is a derivative of petroleum. It exists in the natural state, in which case it is an alteration product of crude oil. It is also an industrial product, the heavy fraction produced by the distillation of certain crude oils. In the petroleum world, asphalt also designates a sedimentary rock impregnated with bitumen. The words *petroleum* and *Erdöl* are neologisms which mean stone oil, and which may appear inaccurate today in the light of our knowledge of the biological origin of crude oils. But the rocks sometimes have the same lineage.

Crude oils and natural gases are fluids which generally occur in the liquid state or the gaseous state in a given set of conditions. They consist essentially or chiefly of hydrocarbons, namely substances basically made up of carbon and hydrogen. One of the essential properties of these substances is to dissolve intimately in each other. In subsurface conditions, a crude oil may thus contain light alkanes that can be separated into a gas phase and heavy products which, if isolated, would normally occur in the solid state. However, these mutual solution possibilities are not unlimited, and vary in accordance with the hydrocarbon species, while benzene and toluene are the best solvents.

Conversely, resins and asphaltenes can 'precipitate' by the dissolution in water of the most soluble hydrocarbons, particularly the aromatics. This leads to the process known as crude oil 'deasphalting'. The same process can be triggered by the addition of methane, as the increase in the proportion of this gas causes asphaltenes to precipitate. Similarly, a drop in temperature and pressure can cause the precipitation of heavy paraffins in paraffinic crudes, as observed in Nigeria and Indonesia.

Due to these 'associative' properties, crude oils and natural gases are mixtures, in various proportions, of different hydrocarbon compounds and a number of other chemical compounds. They are complex substances which vary in accordance with the characteristics of the environment in which they are found, which are born, transformed, degraded, disappear, and consequently to some extent, live, like all natural objects. They represent a transition stage in the vast carbon cycle. Their physical and chemical properties provide an image of these mixtures, which are themselves consequences of the geological framework and history.

Nearly 500 chemical substances have been isolated in the different crude oils analyzed throughout the world, not including organometallic compounds. However, only 150 account for more than half of the components of crude oils. About 200 of these compounds are not hydrocarbons, but these are generally present in low concentrations. Exceptionally, certain heteroatomic 'nonhydrocarbon' compounds form the bulk of some heavy crudes, which have generally evolved very little if at all. Hence the Venezuelan 'Boscan', a Tertiary crude oil, known for its poor quality, only contains 35 to 38 % hydrocarbons.

As a rule, oils and gases can be grouped together under the term of hydrocarbons, which make up the bulk of their content.

1.1.1. Composition of hydrocarbons

1.1.1.1. *Chemical composition*

From the chemical standpoint, hydrocarbons are divided in practice into three major families :
— saturated hydrocarbons : alkanes or paraffins,
— unsaturated hydrocarbons : alkenes, naphtheno-aromatics and aromatics,
— resins and asphaltenes.

Saturated hydrocarbons

As a rule these are quantitatively the most important, with 50 to 60 % of all components. They are distributed in crude oils in three main families.

N-alkanes or n-paraffins (from the Latin *parum affinis*, little affinity) include saturated hydrocarbons with a linear chain and the general formula C_nH_{2n+2}.

They constitute the more or less long chains from C_1 to C_{40}, and the C_5 to C_7 group is generally most widespread. These hydrocarbons occur in three states :
— gas from C_1 to C_4,
— liquid from C_5 to C_{15},
— solid (natural paraffins) from C_{16}.

The n-alkanes with an odd number of carbon atoms, preferentially synthesized by living organisms, are true biological markers. The predominance of low molecular weight n-alkanes (C_{15}, C_{17}, C_{19}) is mainly observed in microscopic algae, whereas in the higher plants, n-alkanes with higher molecular weights (C_{21}, C_{23}, C_{25}, etc.) are encountered exclusively. On the other hand, light alkanes from C_2 to C_8, which are not present in living organisms, but abundant in crude oils, are the transformation products of organic matter after its deposition (Silverman, 1971).

The n-alkanes account for about 15 to 20 % of crude oils. They are especially important in crude oils derived from the organic matter of the higher plants. Their proportion tends to increase during diagenesis. Biodegradation reduces them, as certain bacteria feed preferentially on these substances.

Isoalkanes exhibit branches (methyl groups) in the C_6-C_8 families. Some compounds, alkanes with an isoprenoid chain such as pristane (C_{19}) and phytane (C_{70}), are derived from the molecules present in living organic matter (phytil chain of chlorophyll). Like some n-alkanes, they may be considered as true biochemical fossils.

Cycloalkanes or naphthenes are formed of cyclic hydrocarbons. Their general formula is C_nH_{2n}. Their proportion in crude oils ranges from 20 to 40 %, with an average of around 30 %.

Unsaturated hydrocarbons

This family essentially includes the aromatics, so-called because of their pleasant odor, and the naphtheno-aromatics. The latter are often combined with sulfur-bearing compounds characterized by very unpleasant odors. They account for an average of 20 to 45 % by weight of crude oils.
— True aromatics are molecules formed exclusively by aromatic rings — generally four or five : the simplest is benzene C_6H_6.
— Naphtheno-aromatics consist of one or more condensed aromatic rings, combined with naphthenic hydrocarbons and alkyl chains. They are extremely abundant in young and immature crudes, and also in oils that have undergone strong catagenesis. They are also often combined with sulfur-bearing compounds.
— Alkenes or unsaturated acyclic hydrocarbons are very unstable and therefore extremely rare, but not totally absent from natural crude oils.

Resins and asphaltenes

These are complex compounds with high molecular weight, relatively rich in N, S and O, and also nickel and vanadium. They form the heaviest fraction of crude oils. They appear to be formed chiefly of polyaromatic or naphtheno-aromatic nuclei and chains, partly heteroatomic, including compounds containing sulfur, nitrogen and oxygen. The resins have a molecular weight of 500 to 1200, the asphaltenes up to and above 50,000, and their diameter is about 40 to 50 Å on average.

Resins are relatively more soluble than asphaltenes in organic solvents and light alkanes. The former include acids and esters, and are generally slightly less aromatic than the latter. Resins are unstable in air and light and tend towards the structure of asphaltenes.

These substances account for 0 to 40 % of crude oils, with an average around 20 %. They are encountered abundantly in immature crudes and also altered and biodegraded crude oils, particularly tar sands. They account for 50 % or more of bitumens disseminated in sediments. Note that asphaltenes do not have any equivalent in living matter, and are a product of diagenesis/catagenesis.

1.1.1.2. *Isotopic composition*

The carbon in hydrocarbons is represented by two isotopes ^{12}C and ^{13}C, whose ratio $\delta^{13}C$, in relation to that of a standard sample, is given by the formula:

$$\delta^{13}C = \frac{(^{13}C/^{12}C) - (^{13}C/^{12}C)\ \text{standard}}{(^{13}C/^{12}C)\ \text{standard}} \times 1000$$

If $\delta > 0$, the sample is enriched in heavy isotope, and if $\delta < 0$, it is depleted. In terms of the value of δ, the accuracy of measurements of the isotopes of C is around ± 0.1 to 0.2‰, or 1 to 2 ppm of ^{13}C in relation to total C. In marine carbonates, the δ value oscillates some 3‰ about zero, and this is explained by the fact that the international standard is a belemnite (Figure 16). Non-marine carbonates are lighter, around -5 ± 3‰/PDB.

Organic substances are always lighter than carbonates. The δ of marine organisms are statistically higher than those of continental organisms (by about 5‰). This difference is explained by the fact that marine plants use the carbonate and bicarbonate ions of sea-water by photosynthesis, while land plants utilize atmospheric CO_2 ($\delta \cong 7$‰/PDB).

This difference of 5‰ is also found in oils of marine origin and those of continental origin. The δ value for all oils and kerogens generally ranges from -32 to -2‰. For methane the range is -85 to -30‰, and in rare cases -20‰. The carbonates are always heavier, and their δ value varies according to their origin.

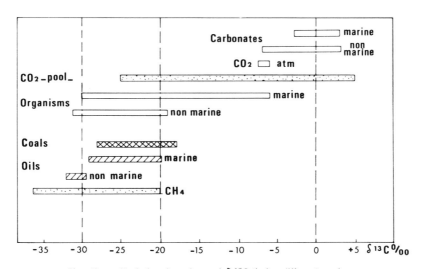

Fig. 16. — Variation in values of $\delta^{13}C$‰ for different rocks.

1.1.2. **Principal hydrocarbon families**

In order of greater complexity, we shall examine the following in succession:
— natural gases,
— oils or liquid products,
— tar sands, oil shales and hydrates, or solid products.

These subdivisions are actually approximate and somewhat vague, since hydrocarbons form a continuous chain in nature. Pressure and temperature conditions prevailing deep in the earth superimpose the light oil and gas areas, and in these conditions the latter may be found in a liquid phase.

More accurately, it is customary to draw a distinction between dry gases and wet gases among the gaseous materials, and light oils (and condensates) and heavy oils among the liquid hydrocarbons. Dry gases consist essentially of methane — at least 97 % — and very small amounts of C_2, C_3, etc. Wet gases display high proportions of 'condensable' gases C_4, C_5, C_6, etc., and light liquid fractions (condensate).

Light oils consist chiefly of the light fractions of liquid hydrocarbons and gases, and heavy oils of the heavy fractions of liquid hydrocarbons, which keep the solid substances in solution. These heavy components may result from the absence of light products:

— by non-formation: immature oils,

— by elimination, alteration and biodegradation: degraded oils.

More or less solid materials are designated by the all-purpose term of bitumens or asphalts, while the term paraffin designates paraffinic hydrocarbons with C_{17} and higher, and clathrates are the hydrates of methane.

1.1.2.1. *Natural gases*

The natural gases found in sedimentary basins often consist mainly of hydrocarbons, chiefly methane. They occur:

— in individual gas pools,

— associated with oil pools, saturated with gas and exhibiting a gas cap, or unsaturated and essentially in solution in the liquid phase.

Chemical composition

Methane is generally the essential constituent of natural gas, as much as 99 % or even 100 % in some pools. The remaining alkanes, ethane, propane, butane etc., account for decreasing proportions in the range of a few percent. Boiling points are — 165 °C for methane, — 42 °C for propane (dry gases), and 0.6 °C for butane for example (wet gas).

The isotopic composition of methane varies approximately from — 85 to — 20 ‰. These differences appear to be related essentially to the degree of maturation of the organic matter and hydrocarbons, the gas becoming increasingly 'heavier' as the diagenesis process is more complete (Figure 17). Hence methane of bacterial origin exhibits a δ of about — 90 to — 70 ‰, and gases resulting from weak diagenesis have a δ of — 70 to — 40 ‰. The gases formed concurrently with liquid hydrocarbons generally range from — 50 to — 40 ‰. Finally, the so-called diagenetic gases, or preferably catagenetic gases, usually exhibit values lower than 45 ‰, and possibly as low as — 30 to — 20 ‰.

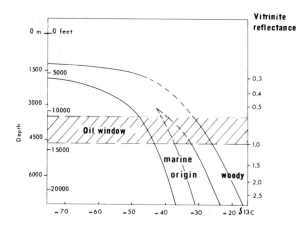

Fig. 17. — Diagram showing the variation in δ ^{13}C of methane versus vitrinite reflectance (Stahl, 1977).

Other constituents

Natural gases may contain variable amounts of other non-hydrocarbon gases, especially H_2S, CO_2 and N_2.

Together with carbon dioxide, hydrogen sulfide is a standard product of the diagenesis and catagenesis of organic matter, especially at great depths, by the action of methane on sulfates, and also due to magmatic activity. It is also produced by the bacterial reduction of sulfates, as observed in some lakes and in the Black Sea. It appears routinely in many provinces, especially in the Canadian Rockies. The proportion of H_2S is up to 15 % in the Lacq gas field, and as high as 88 % in a Devonian pool in Alberta. Gases containing H_2S, even in traces, are called 'sour' gases, as opposed to 'sweet' gases, which contain no H_2S.

Carbon dioxide appears initially at different stages in the sedimentation of organic matter, by oxidation, fermentation or decarboxylation of its constituents. It is also a product of metamorphism, especially of carbonates. Some regions, like the eastern part of West Germany, the Pannonian Basin, and New Mexico, are particularly rich in CO_2; some pools containing a gas consisting of 100 % CO_2. Carbon dioxide may also be formed by the oxidation of hydrocarbons by sulfates. In some cases, as in Hungary, the Gulf of Gabès, and in Sicily in the Ragusa Field, CO_2 may be associated with oil pools. The isotopic composition of CO_2 resulting from the oxidation of organic matter is close to that of the kerogens, around -25 to $-15\,\%_o$. That of carbon dioxide of volcanic origin is heavier, around -6 to $-4\,\%_o$. In many cases, the δ value is close to that of marine carbonates, $+3$ to $-3\,\%_o$.

Nitrogen, produced by diagenesis, or of crustal origin, is found in some fields, especially in North Germany and the Netherlands and in the adjacent part of the North Sea, in Oklahoma, and in Texas. The great gas fields of Slochteren in Holland and in the Texas Panhandle in the United States contain 14 and 11 % respectively. This percentage may be as high as 100 % in some pools. Nitrogen appears to exist more frequently in carbonate and evaporitic series than in shaly environments, where it is absorbed by the clays (Welte, Panel Discussion: Time and Temperature Relation, 9th World Petroleum Conference, 1975, **2**, 203). In the case of coal formations, it appears that the proportion of nitrogen in comparison with methane increases rapidly above an R_v of 3.2. In Holland, this value corresponds to a burial depth of about 5000 m (van Wijhe *et al.*, 1980).

All these incompatible gases generally exhibit more drawbacks than advantages, especially from the economic standpoint, if only because they supplant hydrocarbons in the traps. Whereas H_2S may offer a large source of sulfur, it is viciously poisonous and a cause of pollution and degradation, requiring its removal.

Helium and argon, often associated with nitrogen, are generally of crustal origin, and an increase in the percentage of the former is normally observed from the Tertiary to the Paleozoic, in a ratio of 1 to 2 and more. These two rare gases obviously have considerable economic value, even if they only occur in very small amounts.

Physical and economic characteristics

The physical properties of natural gases faithfully reflect their chemical compositions.

The heating value in particular, which accounts for practically their entire economic value, varies in accordance with the hydrocarbons in the composition and the other gas constituents. It ranges on the average from 38 to 40 MJ/kg or 9500 to 10,000 cal/g. Some examples are given by the gases from the following fields:
— Ripalta (Po Plain) (00 % C_1) : 8600 to 9500 cal/g or 34.4 to 38 MJ kg^{-1},
— Cortemaggiore (Po Plain) (92 % C_1) : 8150 to 9050 cal g^{-1} or 32.6 to 36.2 MJ kg^{-1},
— Frigg (North Sea) (95 % C_1, 4 % C_2, 0.8 % $N_2 + CO_2$) : 38 MJ kg^{-1},
— Hassi Rmel (Algerian Sahara) (75.6 % C_1, 9.3 % C_2) : 39 MJ kg^{-1},
— Groningen (Holland) (81 % C_1, 14 % N_2) : 8000 cal g^{-1} or 32 MJ kg^{-1},
— Lacq (France) (69 % C_1, 9.65 % CO_2, 15.3 % H_2S) : 38 MJ kg^{-1},
— Ekofisk (North Sea) (83 % C_1, 9 % C_2, 2.3 % $N_2 + CO_2$) : 11,500 cal g^{-1} or 46 MJ kg^{-1}.

The solubility of methane in water is relatively high, about 5 to 10 g dm^{-3} at 25 °C. It decreases with increasing temperature (up to 80 °C) and salinity, rises with increasing pressure to a value of around 70 bar, when the appearance of methane hydrate considerably decreases the dissolved volumes (Figure 18). Solubility decreases with an increase in the number of carbons (McAuliffe, 1979). The solubility of gaseous hydrocarbons in liquid hydrocarbons is much higher.

In relation to air, the density varies from 0.55, the density of methane, for gases consisting essentially of this hydrocarbon, to values of about 1 for wet gases rich in condensate or in H_2S and CO_2. The density of alkanes increases rapidly with atomic number:

$$C_1 = 0.55 \qquad C_2 = 1.04 \qquad C_3 = 1.52 \qquad C_4 = 2.00$$

The density in relation to water, or in g/cm^3, ranges from 0.00073 for methane at atmospheric pressure to 0.5 for a standard mixture at high pressure, between 350 and 700 bar, or 5000 to 10,000 psi (Schowalter, 1979).

Compressibility is another important property of natural gases, and in the conditions prevailing in the pool, their volume may be reduced by compression by a factor of 200 to 300.

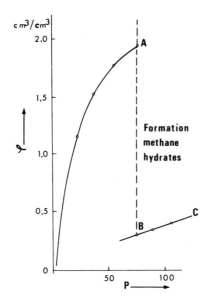

Fig. 18. — Diagram showing the variation in solubility of methane versus pressure (Makogon, Tsare & Cherski, 1971).

1.1.2.2. Crude oils

Chemical composition

Crude oils generally consist mainly of hydrocarbons and polar compounds (resins, asphaltenes, etc.).

Hydrocarbons

Crude oils are formed mainly of liquid hydrocarbons, but the latter also contain gaseous and solid hydrocarbons in solution, in variable amounts. Light crude oils display a high proportion of the former, while heavy crudes contain large amounts of the latter. These different substances recover their individual identity in the distillation process. The isotopic composition of the oils generally ranges from — 30 to — 20‰, similar to the kerogens. Crude oils of continental origin, which are rich in waxes, are lighter, around — 32 to — 28‰, than the majority, which are of marine origin (Figure 19).

The δ value for oils is often closely comparable or slightly lighter, by 1 or 2‰, than that of the associated kerogens likely to be the parent formation and ranging from — 32 to — 21‰. As a rule, the isotopic composition becomes steadily lighter from the kerogens towards the saturates, and including the asphaltenes, resins and aromatics (Figure 19).

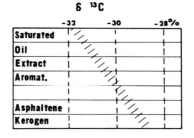

Fig. 19. — Variation in δ ^{13}C versus hydrocarbon type (Stahl, 1971).

Other constituents

Sulfur compounds are the most frequently found, especially in heavy crude oils, in which sulfur is the third most important element after C and H. It accounts for an average of 0.65 % by weight of the crude oils, and often more than 1 % (up to 5 to 6 %, for example, in the heavy oils of the Aquitaine Basin). Sulfur is generally present in the form of thiols or mercaptans, sulfides and thiophene derivatives, more rarely in the form of free sulfur. These thiophenic compounds are very abundant in crude oils which are rich in aromatics, resins and asphaltenes, where their proportion may be as high as 20 % in some cases. Sulfur compounds are often related to the depositional environments of evaporitic and carbonate source rocks. They show a tendency to decrease slightly with increasing depth. On the other hand, they increase with alterations, and the S content may be as high as 10 %, as in the Tertiary oils of the Camargue.

Nitrogen compounds are found less frequently and account for less than 0.1 % on the average. They are found in complex molecules with polyaromatic structures such as organometallic compounds, including the porphyrins which are also very rich in nickel and vanadium, and also in pyridine and quinoline.

Oxygenated compounds are chiefly associated with the acid function — COOH, particularly with the fatty acids and isoprenoid and naphthenic acids. They are found in low concentrations, mainly in immature crude oils.

Physical and economic characteristics

The chemical compositions of crude oils display certain correlations between constituents and condition their physical and economic properties. A close correlation appears between sulfur content and the proportion of aromatic constituents (thiophenic derivatives), resins and asphaltenes.

It is well known that the sulfur found in the constituents and in aromatic crude oils combines with the unsaturated hydrocarbons and facilitates the formation of cycles, particularly benzene rings (Tissot & Welte, 1978). Moreover, crude oils which are relatively rich in sulfur are generally heavy and viscous.

The chief economic property of crude oils is their heating value, for which they are used as an energy source. This heating value varies according to the density and chemical composition, ranging from :

47 to 44.5 MJ kg^{-1} (11,700 to 11,100 cal g^{-1}) for a crude of 0.7 to 0.8; 70 to 45 °API,
44.5 to 43 MJ kg^{-1} (11,100 to 10,675 cal g^{-1}) for a crude of 0.8 to 0.9; 45 to 25 °API,
43 to 42 MJ kg^{-1} (10,675 to 10,500 cal g^{-1}) for a crude of 0.9 to 0.95; 25 to 16 °API.

In comparison, a bituminous coal has a heating value of 22 MJ (5600 kcal) and 32 MJ (8000 kcal) per kg, yielding the approximate equivalence :

$$\boxed{1.5 \text{ t coal} \cong 1 \text{ t oil} \cong 1000 \text{ m}^3 \text{ gas}}$$

Crude oils also have the following uses :
— feedstock for the petrochemical industry,
— lubricant in oils and greases,
— road surfacing material with heavy fractions (for example, natural asphalt from the Val de Travers Mine in Switzerland).

All these uses depend closely on the chemical composition.

The chief remaining physical properties of crude oils are their density, viscosity and solubility. Added to these are secondary properties, which are often of great practical or theoretical value, such as fluorescence and the rotational effect on polarized light (see Perrodon, 1966).

The density of a crude oil reflects its chemical composition. It increases with the percentage of hydrocarbons and heavy products, including resins and asphaltenes. Hence heavy crude oils are often rich in sulfur. It decreases with the rise in temperature level at which the pool is found or has developed, namely with burial depth. Density is normally expressed in g/ml or degrees API, which varies inversely.

It ranges approximately :
— from 10 °API (1 g/ml or 1) for heavy crude oils (density equal to that of fresh water),
— to 50 °API (0.77) for light crude oils.

Density is a very representative indicator of the economic quality of the crudes and serves to set their prices.

Viscosity, the opposite of fluidity, represents the ability of a crude oil to flow. It depends on chemical composition, increasing with the percentage of heavy constituents and consequently with the density of the crude. However, some crudes which are relatively light but rich in paraffins of C_{30} and higher, display high viscosities (paraffinic crudes). Viscosity decreases sharply with an increase in dissolved gases and with rising temperature. It is expressed in centipoises and varies from less than 1 cP to tens of thousands of cP.

This property may be associated with the pour point of a crude oil, a property which is also very sensitive to the heavy paraffins content. Light and fluid crudes solidify at very low temperatures. High viscosity crudes, generally very paraffinic, may become pasty and solid at positive temperatures around 10 degrees Celsius (50 °F). Paraffinic crudes have a relatively high pour point. Bacterially modified heavy crudes, namely crudes which are very poor in paraffins, exhibit a relatively low pour point, which often suffices to identify them as a degraded product, considering their density.

The solubility of hydrocarbons is their ability to dissolve mutually in each other. Heavy fractions and solid phases are dissolved in the light hydrocarbons, gases dissolve the light liquid fractions and are dissolved in large quantities in the liquid fractions. This dissolution of the heavy fractions by the light hydrocarbons could facilitate their migration, as postulated for the Handil Field in East Kalimantan (Durand & Oudin, 1979).

Distillation is merely the separation of these gaseous, liquid and solid compounds gathered together in complex combinations in the crude oils. The most remarkable property of these mutual solubilities is that of the gas dissolved in the liquid phase, which can reach very high proportions. The ratio of gas dissolved in the oil, or the gas/oil ratio (GOR) may be as high as 1000 m^3 per ton (6000 $cfbl^{-1}$) for very light crude oils and condensates. This ratio varies inversely with the oil density, increases with rising pressure, in other words with burial depth, until it reaches the saturation pressure (the equivalent of the bubble point), above which a differentiated gas phase occurs and forms a gas cap. Conversely, it decreases with pressure and with the production of a field.

The gas/oil ratio is traditionally expressed in cubic meters of gas per ton ($m^3 t^{-1}$) or cubic meter of oil ($m^3 m^{-3}$), or in cubic feet per barrel ($ft^3 bl^{-1}$) or standard cubic feet per barrel ($Scf bl^{-1}$), both expressions being related as follows :

$$1 \ m^3 \ m^{-3} = 5.65 \ ft^3 \ bl^{-1} \ \text{or Scft per bl or}$$
$$100 \ ft^3 \ bl^{-1} \ \text{or Scft per bl} = 17.7 \ m^3 \ m^{-3}$$

This dissolved gas increases the volume of the liquid phase by a quantity called the volume factor. The accurate knowledge of this factor is important in calculating reserves, since the volume of crude oil in a pool is in this case greater than the same quantity deprived of its gas in surface conditions. The volume shrinkage factor increases as the quantity of gas in solution is greater, i.e. as the crude oil is lighter, as shown by the following table.

volume factor	*GOR*	
	$m^3 \ m^{-3}$	$ft^3 \ bl^{-1}$
1.10	17.5	100
1.20	52.5	300
1.31	87.5	500
1.59	175.0	1000
1.85	260.0	1500
2.00	350.0	2000

The solubility of liquid hydrocarbons in water is generally very low and decreases sharply with rising molecular weight, but varies substantially in accordance with their chemical properties. Hence aromatics are far more soluble than alkanes with an equivalent number of C atoms. For instance hexane, cyclohexane and benzene exhibit solubilities of 9.5, 60 and 1750 $mg \ l^{-1}$ respectively. Benzene and toluene are by far the most soluble liquid hydrocarbons. Solubility in water increases with temperature at constant pressure, especially for aromatics. It declines with salinity and with pressure (McAuliffe, 1979).

The fluorescence of crude oils and more specifically of bitumens, under the action of ultraviolet radiation, varies from pale yellow to greenish blue. It is a characteristic property of aromatic hydrocarbons, and is associated with the presence of double bonds and aliphatic radicals. This property is extremely important for the discovery at the wellsite of small quantities of hydrocarbons, and also in the field. It generally makes it possible to distinguish them from coals.

Optical activity, or rotational effect on polarized light, is a general property of oils that testifies to their biological origin and particularly to their relations with lipids. It decreases gradually with geological time and is altered by the diagenesis/catagenesis process. Some authors claim that it is related to the production of organic compounds by microbes (Philippi, 1977).

Among the other physical properties are :
— resistivity, ranging from 10^{11} to 10^{18} ohms cm^{-1},
— dielectric constant, equal to 2, compared with 6 to 11 for rocks, 94 for ice,
— index of refraction, ranging from 1.7 to 1.48, which decreases slightly with increasing diagenesis.

Chief types of crude oil

These chemical properties, and especially the variations of the different components, give rise to the classification of crude oils, as made by Tissot & Welte (1978), according to the distribution of hydrocarbons that they contain.

Paraffinic crude oils, containing more than 50 % saturated hydrocarbons and over 40 % paraffinic hydrocarbons (iso- and n-paraffins), are light crude oils with density of about 0.85, sometimes highly viscous, containing less than 10 % resins and asphaltenes, and less than 1 % sulfur. This class includes the crude oils of the Paleozoic in the Sahara and the United States, the Cretaceous in Gabon and Congo, the Tertiary in Libya and Indonesia.

Naphtheno-paraffinic crude oils also contain more than 50 % saturated hydrocarbons, and less than 40 % paraffinic hydrocarbons (iso- and n-paraffins) and naphthenic hydrocarbons. They are also poor in sulfur, and may contain 5 to 15 % resins and asphaltenes and 25 to 40 % aromatics. This category includes a number of crudes from Alberta, the Paris Basin, North Aquitaine and the North Sea.

Naphthenic crude oils contain less than 50 % saturated hydrocarbons and over 40 % naphthenic hydrocarbons. This proportion sometimes results from the elimination of alkanes by the biodegradation of paraffinic or naphtheno-paraffinic oils. Naphthenic crude oils are found in particular in the Gulf Coast and the North Sea

Aromatic crude oils contain less than 50 % saturated hydrocarbons, over 50 % aromatics, resins and asphaltenes. Their sulfur content is often greater than 1 % and they frequently correspond to degraded products. These oils are heavy and viscous, often containing more than 25 % resins and asphaltenes.

According to the proportion of alkanes, a distinction is drawn between :
— asphaltic aromatic oils, containing less than 25 % naphthenes, and generally rich in sulfur; this category includes the Athabasca tar sands and they heavy oils of Aquitaine and Venezuela, constituting extremely large volumes of oil in place,
— naphthenic aromatics containing more than 25 % naphthenes, which are poorer in sulfur ($S < 1$ %); this group includes the Cretaceous oils of Gabon and Congo (Emeraude field).

Between the naphthenic and aromatic crudes, it is possible to identify a class of intermediate aromatics, characterized by 40 to 70 % aromatics, 10 to 30 % resins and asphaltenes. These crudes are generally rich in sulfur ($S > 1$ %) and rather heavy ($d > 0.85$). They include a large share of the Jurassic and Cretaceous oils of the Middle East, representing considerable reserves, oils of the Permian and Carboniferous of the Permian Basin of West Texas, and the Cretaceous/Jurassic in Aquitaine, Spain, and Sicily.

The large majority of crudes actually can be divided approximately into two main groups :
— intermediate aromatics corresponding to reducing marine environments,
— naphtheno-paraffinics and paraffinics, generally associated with deltaic or non-marine environments.

In each province, it is also necessary to consider the geothermal conditions and alterations which more or less obliterate the imprint of the depositional environment and leave a stamp of their own characteristics. Thus immature crudes are generally heavy and rich in nonhydrocarbon compounds and in asphaltenes.

Crude oils subjected to strong catagenesis are often relatively rich in paraffins due to the decrease in aromatics and heavy products. They become increasingly lighter with greater depth (or more precisely with rising temperature).

Alterations and biodegradations also modify the composition of crude oils :
— by favoring the relative concentration of complex heavy polar compounds, and especially resins and asphaltenes, by the degradation of alkanes and aromatics, to yield naphthenic and asphaltic aromatic crudes : this is the case in particular of 'oils' from tar sands,
— by eliminating a more or less broad range of paraffins by bacterial destruction, resulting in a pattern of naphthenic crudes.

Many of these effects often occur in combination, as in the case of shallow crude oils, which are immature and accordingly altered and degraded. These different effects are often difficult to identify individually. However, it appears that catagenesis cannot obliterate the marks of prior degradation.

Heavy crude oils

It is customary to apply this term to crude oils exhibiting a specific gravity between 0.93 and 1 or 20 to 10º API, and high viscosity in subsurface conditions, greater than 10,000 cP. Their sulfur content is often high, greater than 5 % and sometimes 10 %, as well as the proportion of metals (V, Ni) which may be as high as 600 ppm. The composition of the hydrocarbons is distinguished by a high proportion of C_{15}^+ and a very low proportion, if any, of light products. The share of these hydrocarbons in the composition of these crudes may be relatively small.

Heavy crude oils may be of various origins. They are often normal oils which have been degraded by the departure of light components, leaching by fresh water, oxidation, or microbial degradation. They may belong to varieties of crude oil which are especially rich in heavy compounds, such as paraffins and asphaltenes, giving them their very high viscosity.

Crudes of this type appear to be more specifically associated with carbonate source rocks. They may also be immature crudes whose composition, relatively poor in hydrocarbons, includes many polar and slightly evolved substances such as asphaltenes and resins. It is sometimes difficult to distinguish between different origins, as alteration affects mature as well as immature oils, especially since these two occurrences characterize shallow zones, together with the fact that maturation criteria are often obliterated by the destructive action of alterations.

Heavy crude oils represent considerable reserves in place, around 280 to 350 Gt, largely concentrated in the great tar sands accumulations in Canada and Venezuela.

1.1.2.3. *Solid hydrocarbons*

This category includes natural hydrocarbons which are relatively simple from the chemical standpoint, such as methane hydrates, or complex and often pasty, like the vast family of bitumens and asphalts, together with the sedimentary formations in which they are most often encountered, tar sands and oil shales. We shall deal with the following briefly :
— hydrates,
— the bitumen family.

1.2. HYDROCARBON GENESIS

Hydrocarbons are formed in sedimentary formations, rich in organic matter, and subjected subsequently to a degree of diagenesis/catagenesis. These formations are qualified as source rocks. In certain conditions, and especially above a certain concentration, these products escape from the beds in which they were formed into porous and permeable rocks, called reservoirs, characterized in particular by lower internal pressure. This process is called 'primary migration'.

For a clearer understanding of the mechanisms that give rise to hydrocarbons and the characteristics of their source rocks, we shall examine the following in succession :
— the main constituents of living organic matter,
— the environments and mechanisms of deposition, in which the kerogen is formed,
— subsequent transformations during the diagenesis and catagenesis phases.

A sufficient number of facts, universally recognized, today allow us to assert the biological and sedimentary origin of petroleum, and inorganic theories are irrelevant except historically. In rare cases, however, methane may be of inorganic origin.

Throughout this chapter reference is frequently made to the recent book by Tissot & Welte, "Petroleum Formation and Occurrence" (1978), which offers the most complete review of the subject (see also Hunt, 1979).

1.2.1. **The major constituents of living organic matter**

All living organisms consist essentially today, as they did in the past, of a very small number of chemical substances, proteins, carbohydrates, lipids and lignin, and their proportions often vary substantially according to the groups and families of the animal and vegetable kingdoms.

— Proteins, which generally form a large share of organic matter, are polymers featuring a high degree of organization, formed of amino acids. They are present in the composition of animal tissues and enzymes. The amino acids are their monomer representatives, soluble in water.

— Carbohydrates include glucides and their polymers. They account for a large part of the tissues, especially with cellulose, chitin and sporopollenin. Cellulose is chiefly produced by higher plants and is a major constituent of wood. Chitin is mainly found in certain animals, and sporopollenin in spores and pollens.

— Carbohydrates are also a source of energy for living organisms.

— Lipids include all organic compounds which are insoluble in water and soluble in organic solvents (e.g. chloroform), and also unsaturated fatty acids, frequently found in vegetable oils. Lipids are abundant in algae, especially the botryococcus family, and in diatoms, in which they form up to 70 % of the dry weight, as well as pigments, grains and spores which are rich in C_{34} (Moldowan & Seifert, 1980), where they concentrate high energy potential.

— Lipids also contain fats, which are esters of fatty acids. Saturated fatty acids are represented by the simplified formula:

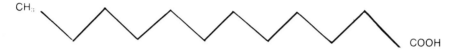

where each corner corresponds to CH_2. Unsaturated fatty acids are also frequently found. Lipids also include terpenic compounds consisting of isoprenoids, which can be polymerized into long chains or form cycles. Lipids are a constituent of a number of plant materials, such as gums, and also of sporopollenin, cutin, chorophyll and carotenoids.

— Lipids, which are insoluble in water and often more chemically resistant than the other components, play a predominant role in hydrocarbon genesis. Fine chemical analyses, and especially the $^{13}C/^{12}C$ isotopic ratio, show that crude oils are mainly derived from the lipid fractions of organic matter, while coals are produced from the remains of plants, and particularly from cellulose matter which is their major non-lipid constituent.

— Close chemical relationships have also been observed between hydrocarbons and lipids, especially through certain constituents called 'biochemical fossils' (Silverman, 1971). Lipids account for the bulk of the organic constituents capable of being converted into oil.

In addition to these basic materials, lignin and tannins deserve to be mentioned. Lignin is a polyphenol with high molecular weight aromatic structures that forms the skeleton of the tissue of the higher plants. Tannins are less condensed polyphenols that impregnate the cortex tissues.

Organic properties of some families making up the biomass

The biomass consists of a small number of large groups of living organisms: marine plankton, bacteria, algae and the higher plants.

— Marine and lacustrine plankton, especially microscopic algae, consist mainly of proteins for more than half of their mass, carbohydrates, and 25 to 5 % lipids. Certain microorganisms, such as diatoms, are especially rich in lipids, the latter consisting largely of fatty acids with 3 to 5 % hydrocarbons, particularly n-alkanes.

— Bacteria, among the earliest organisms, apart from water, consist mainly of proteins and may contain up to 10 % lipids, particularly hydrocarbons and mostly C_{10} to C_{30}. They feed largely on fatty acids.

— The higher land plants are formed mainly of cellulose and lignin, 30 to 50 % and 15 to 25 % respectively, although certain parts, such as seeds, pollens, waxes and resins contain significant fractions of lipids, especially n-alkanes with C_{27}, C_{29} and C_{31}.

The organic matter produced by plankton, algae and bacteria is generally deposited in place along the same vertical: this is called autochtonous. The organic matter from the higher land plants may be conveyed into sedimentary basins by air or river currents, and is qualified as allochtonous or inherited.

1.2.2. Depositional environments and mechanisms - kerogen

After the death of different living organisms, organic matter is found in their remains, disseminated or dissolved in water, and especially in combination with clay particles. To be able to settle, this organic matter must be degraded before deposition, and must be protected at the time of its deposition and subsequently, initially implying an aquatic environment. Protection before deposition requires an anaerobic environment, with water devoid of oxygen, i.e. perfectly calm, confined, and often stratified. As a rule, this preservation is facilitated by a slow time of fall for planktonic organic matter, which forms the bulk of marine organic sediments.

Most of the organic matter of planktonic origin is destroyed as it falls through the body of water at a speed of about 100 m per week. It is estimated that on the average:
— only 2 % reaches the shallow floors,
— about 0.02 % reaches the deep ocean floors.

Another part is destroyed by benthic organisms living on the sea bed. Hence the quantities that deposit finally depend on:
— the biomass produced, namely, the organic productivity of the environment,
— the physicochemical conditions of the depositional environment.

In other words, the characteristics of the basin directly condition the deposition of future source rocks (Curtis, 1980).

1.2.2.1. *Depositional environments*

Environments with high organic productivity, which generally create an oxygen shortage by their reactivity, are governed by the combination of an external energy, the sun, which furthers photosynthesis, and nutritive elements, such as phosphates and nitrates. These favorable and often exceptional environments are found at sea and on land.

Solar energy is limited to the surface layer of the oceans and land masses, but it extends to the entire surface of the Earth. It offers the driving force of the mechanism of photosynthesis, which has formed the basis of all biological activity for some two billion years. The nutritive elements, consisting mainly of nitrogen, phosphorus, silica and iron, are derived on land from the soil, and at sea from the continents by the drainage of rivers and waterways, or from deep waters, by upwellings which, especially in the trade wind zones, push the cold and more mineralized waters from the depths to the surface.

The most favorable environments, all of them aquatic, are thus found on the continental margins, shelves and slopes, in coastal waters, and also in lagoons, landlocked or silled seas, and lakes. Favorable environments may develop in the great deltaic units marked by high subsidence and sedimentation, like the deltaic zones and basins of the Niger, the Mississippi, the Ganges, and the Mahakam in Indonesia. They generally correspond to a warm, humid climate and high sedimentation, predominantly clay. They are rich in inherited organic matter, of the humic or lignitous type.

In the upwelling zones, such as those found off the coast of southwest Africa, the area of Walvis Bay, and the coasts of Chile and California in particular, the oyxgen deficiency created by the intense biological activity at the surface produces an anaerobic state in the underlying layers. The planktonic populations are also regularly poisoned by the proliferation of certain microalgae, causing water-blooming and the massive deposition of an organic mass. These environments may be considered as especially favorable to the production of source rocks. They often correspond to arid climates exhibiting low sedimentation and predominantly planktonic or sapropelic organic matter.

Similar mechanisms may also be found in lakes and lagoons, where certain algae proliferate, producing eutrophic conditions. It may also be considered that certain lacustrine basins, which are poor in detrital inflows, and display a high pH, constitute the best carbon traps. Certain lacustrine algae, like the botryococcus family, are particularly important constituents of source rocks. The wealth of the lacustrine basins of Reconcavo in Brazil, and Green River in Utah, provides an illustration of this (see Chapter 2 of Part 3). Some of the constituents of these algae, especially $C_{34}H_{70}$, have been found in the oils from Minas and Duri, Indonesia (Moldowan & Seifert, 1980).

Landlocked or semi-landlocked seas, like the Black, Baltic and Caspian Seas, the Gulf of California, and lagoons, like that of Maracaibo, also represent excellent models of environments for organic sedimentation. In these exceptional conditions, the percentage of organic matter in the sediments, normally less than 0.5 % (lower than 0.9 % in clays, 0.2 % in sandstones and carbonates) may reach values of 2 to 10 % and more. Hence the levels showing a high concentration of organic matter can be considered as the result of exceptional preservation conditions as well as the productivity of these conditions, and the alternations often found in these deposits correspond to variations in the sedimentation environment on the floor. It is these exceptional environments that form the cradle for the particular sediments that subsequently become oil source rocks.

The sedimentation rate plays a definite role in the dispersion and preservation of the organic matrix. Rapid sedimentation disseminates the organic matter in detrital materials. On the other hand, it guarantees rapid burial and effective protection of these delicate substances. By conveying these layers rapidly to greater depths and higher temperatures, it limits the period of bacterial destruction of organic matter. On the other hand, slow sedimentation favors the concentration of organic matter in certain beds. In this case preservation must be guaranteed by a rigorously confined environment. Thus the study of these environments is especially interesting from this point of view.

Clay minerals appear to play a more or less important role in the fixation and preservation of organic matter. They seem to absorb certain polar organic compounds and transform dissolved organic matter, reducing the risks of alteration in water (Dow, 1978). Two main types of depositional environment emerge from these considerations, corresponding to two clearly distinct biological families :
— a restricted marine or lacustrine environment, characterized by a predominance of planktonic or algal organic matter, qualified as sapropelic, or oilprone,
— an environment with a brackish tendency, and the rapid sedimentation of clay materials especially rich in ligneous or humic organic matter, also called gasprone.

This naturally includes a whole sequence of intermediate environments. We shall see below that the former provide the origin of oil provinces, often naphthenic, while the latter give rise to provinces which are rich in gas and paraffinic oils.

1.2.2.2. *Sedimentation mechanisms*

This section considers the initial and important transformations occurring in a freshly deposited sediment, especially in the mud. These are the first steps of the transformation of organic matter into hydrocarbons, corresponding to biological (or biochemical) diagenesis.

This initial stage of diagenesis begins immediately after deposition under the water/sediment interface, and continues while declining gradually but irregularly down to depths of a few hundred meters in cold zones, corresponding to temperatures below 60 to 70 °C (Philippi, 1977). During this phase, the sediments are subject to the energetic action of the microorganisms in the mud, and particularly the bacteria, in the first decameters of the deposit, especially in a shallow environment.

This biological activity is closely related to the situation of the depositional environment, and primarily its depth. Activity is intense in shallow environments, appears to decline fairly rapidly with the depth of the water, and is very low on the deep ocean floor (Pelet, 1974). During these reactions, the pH rises slightly and the environment becomes strongly reducing. Calcium carbonate and silica are partly dissolved. Simultaneously, CO_2, H_2S and CH_4 produced by the degradation of organic matter are evolved. The bacteria split the organic molecules to yield carbon dioxide and methane primarily, and a complex insoluble residue that forms part of the composition of kerogen. These transformations are also marked by the elimination of certain functions, and this results in a slight enrichment in the light ^{12}C isotopes.

The bacteria, particularly the sulfate-reducing strains, generate H_2S which is fixed by the clay minerals in the presence of iron to yield pyrite. In the case of carbonate source rocks, it is only fixed in small proportions and combines with the organic matter, passing subsequently into the hydrocarbons generated. In some conditions, as in fairly deep lakes without detrital inflows, such as Kivu Lake, the fermentation of deposited organic matter, chiefly cellulose, may yield considerable amounts of methane. This is one of the formation processes of what is called 'early gas'.

But the most important product of these early transformations is undeniably kerogen, the starting point of a long evolution which may culminate in hydrocarbons. More specifically, the true sources of the formation of hydrocarbons are represented by organic molecules, especially lipids, which resist the intense activity of the biological diagenesis of the upper layers of the deposit. This phase of biological diagenesis may be considered to correspond in fact to a stage of degradation of organic matter. It generates methane, apparently at the expense of liquid hydrocarbons.

Hence the rate of burial may play an important role : if rapid, it reduces the period of microbial action and degradation, halted by the rise in temperature, and allows the preservation of a larger fraction of the stock or organic matter; if slow, it will only allow the preservation of very little source rock materials (Coleman *et al.*, 1979).

Organic facies

In accordance with the characteristics of the organic matter, two main types of organic facies may be distinguished :
— humic or detrital facies,
— sapropelic or authigenic facies.

The former may be related to aquatic or eolian transport : plant tissues, spores, pollens, and particularly vitrinite (coal beds form the boundary of these two modes of transport); the latter are essentially aquatic, marine, brackish or lacustrine, and largely consist of algae and planktonic materials.

The sapropelic facies is generally represented by the presence of recognizable algae, and by the accumulation of undifferentiated algal materials exhibiting a fluffy appearance. These environments often appear to be devoid of any detrital inputs, especially humic, and particularly vitrinite. They are associated with evaporitic facies. In this case, they correspond to sub-desert climates. The humic material, largely formed of debris of higher plants, spores and pollens generally associated with active sand/shale sedimentation, usually corresponds to humid climates (Robert, unpublished document, 1978). These two facies are in some way merged with the chemical properties of the kerogens, defined in three categories I, II and III.

1.2.2.3. Kerogen

The term kerogen is applied to the organic constituents of sedimentary rocks which are insoluble in organic or alkaline solvents. Kerogen is thus different from the organic constituents which are soluble in these solvents, and which are designated by the term bitumens. It is the most widespread form of organic matter on Earth, a thousand times more than coal and liquid petroleum, and fifty times more than the bitumens dispersed in non-reservoir rocks (Hunt, 1972; Tissot & Welte, 1978; Durand, 1980) (Figure 22).

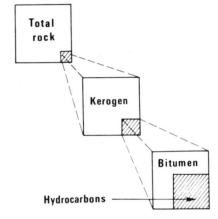

Fig. 22. — Ratios of hydrocarbons to bitumens, kerogen and total rock (Tissot & Welte, 1978).

From the chemical standpoint, kerogen is a macro-molecular complex composed of condensed benzene rings, connected by heteroatomic or aliphatic links. It may also contain lipids.

Its structure is disordered and only assumes a degree of organization during catagenesis. Under detailed observation, it appears to be composed of recognizable plant particles, of diffuse materials resulting from the degradation of previous constituents, and of secondary bituminous materials (the flaky amorphous material observed in palynological preparations is a concentrated form of diffuse materials, after the chemical destruction of the mineral support).

Certain specific organic compounds, synthesized by plants or 'animals', are also found in trace amounts in and outside the kerogen. These compounds are alkanes, fatty acids, terpenes, steroids, porphyrins and linear hydrocarbons which survive the ordeal of organic decomposition and diagenesis unscathed, and which are called geochemical fossils. These substances undoubtedly play a secondary role in the genesis of petroleum. On the other hand, they are of considerable importance in the identification of these mechanisms. Their structures reflect the architecture of organic substances present in the original organisms, and also the reactions to which these substances were subjected during their burial. Among these geochemical fossils is hopane, $C_{30}H_{52}$, of which many derivatives have been identified.

From the isotopic standpoint, the nature and the depositional environment of organic matter are reflected by certain variations in $\delta\ ^{13}C$. Kerogen, which is rich in aliphatic structures, often corresponds to more marine environments, and appears to be poorer in ^{13}C isotopes than the organic matter containing aromatic structures, often related to the higher plants and humic matter. The $\delta\ ^{13}C$ value appears to range from -23 to $-30‰$ in marine formations, and more often between -28 and $-32‰$ in organic matter and oils of non-marine origin (Figures 16 and 19). The isotopic composition also varies during burial:
— fairly rapidly, at the very outset of burial, the δ goes from -32 to $-5‰$ for the organic matter of living matter, to -32 to $-20‰$ for the organic matter of recent sediments,
— very slightly in the diagenesis and catagenesis stages,
— more sharply at the approach of metamorphism, when values of $-10‰$ are reached.

From the chronostratigraphic standpoint, a very slight enrichment in ^{13}C is observed with age. Hence Cenozoic oils exhibit a $\delta\ ^{13}C$ of around $-25‰$, and those of the near Paleozoic a $\delta\ ^{13}C$ of $-30‰$ (Stahl, 1977).

1.2.3. Diagenetic and catagenetic transformations

As a rule, organic substances are more fragile and more unstable than mineral compounds, and they undergo considerable transformations from the outset of the diagenesis stage. It is during these mutations subsequent to deposition that kerogen gives birth to oil and gas (Kübler, 1980). However, the presence of hydrocarbons closely comparable to crude oils in composition has been noted in recent sediments, as in the Santa Barbara Channel off the coast of California. This observation has led some investigators to postulate that the hydrocarbon generation process begins very early at the start of the sedimentary cycle (Copelin, 1979).

If nevertheless remains true that oil and, to a lesser extent, gas are the result of chemical transformations during burial. Hence the rapidity and intensity of burial, in other words, of subsidence, exert a major effect on the maturation of organic matter. These mutations and geneses fall within the general framework of diagenetic and catagenetic processes examined in Section 1.2. These effects, which lead to the formation of hydrocarbons, may be compared closely with coalification, which yields different categories of coal and methane, from organic materials derived from the higher plants (Robert, 1980).

Table I

Principal parameters of the diagenetic transformation of hydrocarbons and coals.

Stage	Vitrinite reflectance %	LOM	TAI	Hydrocarbons	Coal	% v.m
Diagenesis	0.30	2- 4- 6-		Biogenic and early diagenesis gas (Immature zone)	Peat	
					Lignite	
					Sub-bituminous	
Catagenesis	0.5 1 1.5	7- 8- 10- 11- 12-	2.5 3 3.5 4	Oil (Oil window) Wet gas (Condensate)	high volatile	45 40 30
					medium vol. low vol.	20
Metagenesis	2 2.5 4	14- 16- 18-	5	Methane ?	semi-anthrac.	10 5
					Anthracite	
Metamorph.		20-			meta-anthrac.	

In both cases, these rearrangements tend towards a higher degree of order and a more stable product, as the most disordered structures, particularly the functional groups, are gradually eliminated by the breakage of the weakest bonds. The separated radicals thus yield CO_2, H_2S and water. From this standpoint, the genesis of hydrocarbons is only one aspect of a general redox reaction scheme, or of the disproportionation of organic matter, simultaneously reflected by the genesis of a reduced substance enriched in H, and an oxidized compound depleted of H and assuming an aromatic structure, according to the following pattern:

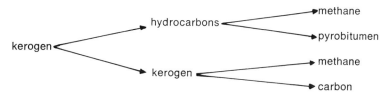

The hydrocarbons thus appear as a necessary step in the conversion of organic matter under the action of temperature, reaching towards a higher degree of order and greater stability.

The chemical properties of the early kerogens are grouped under three types defined by the 'evolutionary paths' and qualified I, II, III.

— Type I is characterized by a high H/C ratio (1.5 or more) and by a low O/C ratio (less than 1). It consists largely of lipids, particularly aliphatic chains, resulting from the accumulation of algal materials, for example of the botryococcus family, or from the elimination of the other constituents by biodegradation. Type I is especially prolific in oil, or oilprone. It is found in large quantities in the Green River oil shales It may be compared with the alginites (van Krevelen, 1961) and the sapropelites (Combaz, 1975 and 1980).

— Type II is frequent, especially in the marine sediments of restricted environments. It is characterized by relatively high H/C ratios and fairly low O/C ratios. It is fairly rich in aliphatic chains and naphthenic and benzene rings, and often in sulfur. Type II may be compared with the exinite and liptobiolit of the previous authors. It is also called oilprone.

— Type III, derived chiefly from the constituents of the higher continental plants, but also from marine planktonic organsims, mainly contains condensed and oxygenated polyaromatic groups, with a few aliphatic chains. The H/C ratio is low, but the O/C ratio is relatively high. The petroleum potential of this type of kerogen has a low yield, about five times less productive than that of Type II (Tissot, 1973). It only produces mediocre source rocks, chiefly containing gas, as well as carbon dioxide and water and is gasprone. It is comparable to the vitrinite of coals.

This classification does not cover all the organic matter existing in sedimentary rocks. Organic matter also exists with an H/C ratio lower than that of Type III kerogen, including oxidized organic matter, called inertinite, consisting of 'dead carbon', which is practically unaffected by rising temperature, and is consequently incapable of generating gas.

These different types are related to the characteristics of the depositional environment and the materials that have deposited. Hence types I and II appear to be predominant in carbonate facies, while type III can be found in a marine sand/shale environment.

Type III kerogen is opposed to both bituminous materials and sapropelic coals (bogheads and cannelcoals). It corresponds to the humic part (vitrinite) of normal coals, which include two other groups of materials, inertinite (fusinite) and liptinite (exinite).

*
**

Following the same plan as above, we shall now analyze the great phases of diagenesis and catagenesis.

1.2.3.1. *Diagenesis*

The transformations of organic matter and hydrocarbons can be observed in detail in the fine sediments where they have good chances of being produced, and where errors related to migration can be discarded (Hunt, 1977).

These transformations are the result of biological action (bacteria) and thermo-catalytic action. Bacterial action appears to be limited to the first hundred meters of sediments in the total sulfate-reducing zone. As the pressure and temperature increase, their activity declines due to the development of toxic materials or the depletion of the stock of organic matter. Hence they diminish in number, but very irregularly, with depth. However, in certain low temperature conditions, below 30 °C, microorganism activity could extend down to 1000 or 2000 meters (Claypool & Kaplan, 1974; Rice & Claypool, 1981). The bacteria degrade the organic matter and sometimes certain categories more specifically, such as carbonaceous materials, and possibly sapropelic kerogens. This could be the cause of a large proportion of the early gases, but also of the alteration and destruction of potential source rocks. These biogenic gases could account for about 20 % of the total gas reserves (Rice & Claypool, 1981).

In this initial stage corresponding to an R_v lower than 0.5 % and to average temperatures below 50 to 60 °C, the first chemical reactions, undoubtedly favored by the catalytic action of certain minerals, including smectite and dolomite, cause the breakage of the weakest links of the organic structure, in other words, chiefly the splitting of the acid and ether functions (decarboxylation). This is reflected on a van Krevelen diagram by a rapid decrease in the O/C ratio for type III, and to a lesser extent for type II, and a smaller decrease in the H/C ratio (Figure 23). Especially for kerogen type III, this results in the formation of H_2O, CO_2, CH_4 and heavy heteroatomic compounds, such as resins and asphaltenes. Hence it is partly from organic matter qualified as immature that the stage of genesis occurs of methane and small amounts of ethane, which constitute the 'early diagenetic gases' (Weber & Maximov, 1976).

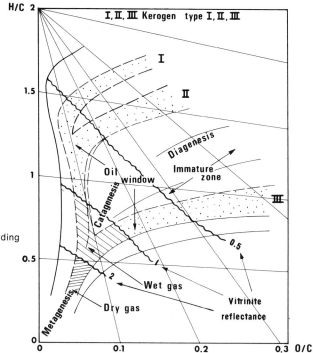

Fig. 23. — General scheme of kerogen evolution according to van Krevelen diagram (Tissot & Welte, 1978).

These early gases are differentiated on the isotopic level from the later gases by a $\delta\ ^{13}C$ less than $-60\ ‰$ (-60 to $-95‰$), close to the value of biogenic CO_2, and the remaining gases generally display values above $-50‰$ (Rice & Claypool, 1981). Also encountered in this stage are 'geochemical fossils', in other words, molecules synthesized by living organisms at the time of deposition and preserved without any notable change. Their value is more theoretical than practical, because they never represent more than a small part of the hydrocarbons formed.

1.2.3.2. Catagenesis

It is above temperatures of around 50 to 60 °C, but variable depending on the characteristics of the kerogen, that the organic matter 'matures' and 'cooks', to give birth to oil and gas. Two main phases can be distinguished depending on temperature : the first corresponds to the principal phase of the genesis of liquid hydrocarbons, and is often qualified as the 'oil window', while the second is characterized chiefly by the formation of wet gases.

Weak catagenesis stage or 'oil window'

It is in the temperature range from 50 to 120/150 °C, and mainly above around 100 °C, that the transformations of kerogen into hydrocarbons, and particularly the C_{15}, C_{40} fractions, appear to be the most extensive (Hunt, 1977) (Figure 24).

This zone of formation and preservation of liquid hydrocarbons, called an 'oil window', is characterized by the following organic markers (see Chapter 2 of Part 1, p. 42), while time and the composition of the kerogen cause these limits to vary slightly within the following ranges :

— R_v vitrinite   0.5 to 1.0/1.2
— TAI   2.5 to about 3.7 to 4
— LOM   7 to about 11 to 12.

For an average geothermal gradient, this corresponds to approximate burial depths of 1.5 to 4 km, or for Tertiary formations to maximum paleotemperatures of 60 to 130 °C. These temperatures may be lowered for pre-Tertiary series and for type I and type II kerogens (or sapropelic organic matter). From the practical standpoint, note that the TAI is more sensitive in areas of weak diagenesis, while the R_v is sensitive above the value 0.5.

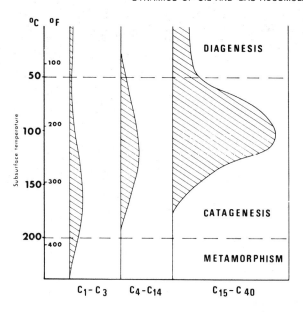

Fig. 24. — Diagram of variation in quantities of hydrocarbons contained in fine-grained sediments versus temperature (Hunt, 1977).

The 'paths' of this transformation have been clarified by van Krevelen on a diagram, according to the composition of the kerogen, and are often designated by his name (Figure 23).

From the chemical standpoint, most of these transformations are reflected by a faster drop in the H/C ratio than the O/C ratio. This pattern varies substantially from one type of kerogen to another, and is distinguished by the decrease in :
— the O/C ratio, more rapid in the case of wet kerogen, which chiefly yields gas,
— the H/C ratio from 1.25 to about 0.5 in the case of a type II kerogen.

If the chemical analysis of the hydrocarbons formed proceeds further, as catagenesis becomes more intense, a shortening of the chains is observed as well as a reduction in molecular weight, an increase in the content of n-alkanes in comparison with iso-alkanes and cyclanes, and a progressive increase in the benzene ring/saturated ring ratio. In the n-alkane family, the disappearance of the predominance of individuals with an odd number of C is usually observed simultaneously (although parity is found in some

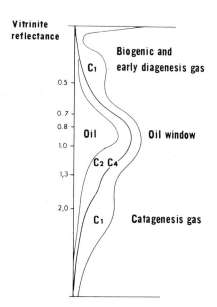

Fig. 25. — Diagram showing the intensity of hydrocarbon formation as a function of diagenesis/catagenesis (Chilingarian & Yen, 1978).

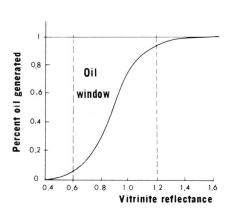

Fig. 26. — Variation in percentage of oil generated versus vitrinite reflectance (Waples, 1979).

sediments and organisms, and even sometimes the predominance of an even number of C) (Connan & van der Weide, 1978).

The rate of these transformations does not appear to be constant as a function of temperature, and accelerates exponentially (Philippi, 1965). This is shown by the curve depicting the transformation of kerogen or percentage of oil generated as a function of the increase in R_r, for example for a medium-grade kerogen, consisting of 50 % alginite and liptinite (exinite). In this case, the graph shows that half of the oil was formed when the R_r of this genesis occurred in the reflectance range between 0.6 and 1.2 %, or the 'oil window' (Waples, 1979) (Figures 25 and 26).

The presence of aromatics, more stable and more resistant to thermal cracking, appears to consolidate resistance to evolution. These characteristics seem to be offset by the inherent features of the carbonate source rocks, especially their high content of lipids, which finally guarantees a high yield in oil genesis.

From the mechanical standpoint, this birth of hydrocarbons is reflected by an increase in volume, and consequently, if the rock is only slightly permeable or impermeable, by a rise in pressure. It also introduces certain changes in the properties of the source rock, including increased resistivity, as observed in the Bakken formation in the Williston Basin (Meissner, 1978) (Figure 36).

Advanced catagenesis stage

Above temperatures in the neighborhood of 110 to 120 °C and up to around 200 °C, in the geological series corresponding on the average to a vitrinite reflectance of 2 %, new splits occur in the C-C bonds of the remaining kerogen and the hydrocarbons already formed. The latter become increasingly lighter and are eventually only represented by gases, and finally methane. The H/C ratio falls steadily in the kerogen and reaches zero when only insoluble organic matter is left. It increases in the hydrocarbons formed to a value of 4 when only methane remains.

From the isotopic standpoint, the $\delta\ ^{13}C$ of these gases, which rises during the maturation process, generally ranges from -60 to $-30\ ‰$. The value for kerogen also increases to $-10\ ‰$ at the limit of metamorphism boundary (Figure 27). In these conditions, the carbonates form CO_2, which can be added to the hydrocarbons, and may even supplant them. Large volumes of H_2S may also be formed from the sulfur remaining in the kerogen, by the reduction of sulfates by organic compounds, and of N_2 from benzene rings (Tissot & Bessereau, 1982). These gases may play an important role by increasing the permeability of the carbonates and hence facilitating hydrocarbon migration (Erdman, 1975).

Changes in fluorescence

At the outset, the fluorescence of the organic matter varies according to the type of compound involved — it is associated with the existence of the C=C bond (or benzene rings in the hydrocarbons) — and it also varies with the degree of catagenesis.

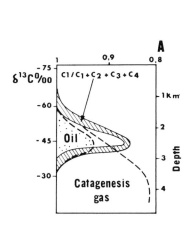

Fig. 27. — Diagram showing the variation in $\delta\ ^{13}C$ as a function of the ratio $C_1/(C_1 + C_2 + C_3 + C_4)$ and of burial depth (Stahl, 1977).

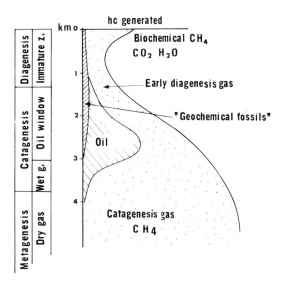

Fig. 28. — General scheme of hydrocarbon genesis versus depth for an average geothermal gradient.

Lacustrine algae, particularly the botryococcus family, display the following changes in their fluorescence colors :
— green in immature zones ($R_v < 0.5\%$),
— yellow in the zones at the beginning of the oil window,
— orange in the zones at the end of the oil window,
— reddish in zones with R_v 1 to 1.2 %
(P. Robert, 1978, unpublished document).

1.2.3.3. Evolutionary pattern

This series of transformations with increasing pressure, and especially temperature, may be summarized by a graph showing the burial depths on the y axis, and assuming an average geothermal gradient, with the quantities of hydrocarbons formed on the x axis. This diagram shows the following (Figure 28) :
— at shallow depth, an immature zone characterized by a vitrinite reflectance lower than 0.6, the area of the formation of H_2O, CO_2 and some early hydrocarbons, including CH_4, sometimes with the presence of the same quantities of natural gasoline (Connan et al., 1978),
— below, an oil genesis zone, initially heavy and rich in polar compounds, then in naphthenic resins and finally paraffinic, up to an R_v of 1 to 1.3, corresponding to a temperature of about 150 °C, a zone of wet gas genesis up to an R_v of 2,
— deeper still, a zone of dry gas genesis (methane) beyond an R_v of 2, but with a lower R_v for type III kerogens; the destruction of the benzene rings at temperatures above 150 °C may be accompanied by the formation of H_2S and N_2.

These transformations appear to depend on the composition of the organic matter. Type III kerogens give liquid hydrocarbons at higher temperatures than a type II kerogen. In these conditions, the oil window concept must be interpreted according to the type of kerogen, especially if type III predominates.

A vitrinite reflectance greater than 2 % marks the entry into the metagenesis stage, distinguished by the exclusive presence of methane, and a stable residue consisting chiefly of stacks of aromatic layers. As a rule, the methane remains stable at high temperatures, above 550 °C. However, the presence of water, a permanent subsurface constituent, causes a degree of dissociation. At depths around 9000 m, depending on the geothermal gradient, 5 to 40 % of the methane is destroyed. The presence of sulfur or sulfur compounds significantly decreases its stability (Barker & Kemp, 1979).

In the case of carbonaceous source rocks, the formation of nitrogen predominates above an R_v of 3.2 (about 5000 m), and some authors use this figure to define a 'gas window' (R_v between 1.2 and 3.2) (van Wijhe et al., 1980). The analysis of the hydrocarbons imprisoned in the fine-grained source rocks, which may be considered as a first approximation to be in place, confirms this statistical scheme and shows that 82 % of the gas, 91 % of the liquids and 60 % of the asphaltic compounds are found in the catagenetic conditions, corresponding to a temperature interval lying between 50 and 200 °C (Hunt, 1977).

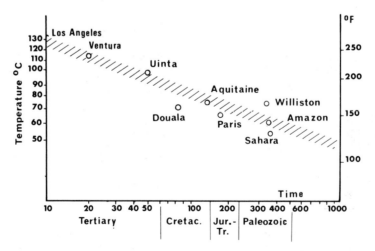

Fig. 29. — Variation of oil genesis temperature versus time.

HYDROCARBONS

Importance of the time factor

Time plays a major role in hydrocarbon genesis, which is known to require even higher temperatures for more recent source formations (see Chapter 2 of Part 1). Simultaneously, laboratory experiments have shown that temperatures around 300 °C must be obtained to 'degrade' organic matter into hydrocarbons in the space of weeks or months. In this respect, the investigation not only introduced an acceleration of the time factor, but altered the quality of the reactions, and is not a faithful reflection of the transformations occurring in nature with the passage of geological time. The study of different basins provides the following approximate 'critical' temperatures (Figure 29) :
— 50 to 60 °C for the Silurian/Devonian in the Sahara,
— 60 to 70 °C for the Toarcian in the Paris Basin,
— 100 °C for the Eocene in the Utah Basin,
— 115 °C for the Mio-Pliocene in the Los Angeles and Ventura Basins.

This genesis may extend over a more or less long period, possibly several million years, in very subsident basins, and over a hundred million years in the case of platform provinces. This more or less long 'manufacture' probably passes through intermittent phases in case of accelerated subsidence or the increase of heat flows. Everything occurs as if this transformation takes place from a sort of threshold concentration, as Zieglar & Spotts (1978) pointed out for the Great Valley of California (Figure 30).

These graphs, which are evidently based on a number of assumptions, show the difference in historical behavior between the northwest and southeast zones of the basin. In the northwest, where subsidence is relatively weak, the diagram suggests that hydrocarbon genesis proceeded during less than five million years, and that only the pre-Miocene formations reached the 'oil window'. In the southeast part, which is more subsident, the Miocene has been brought to the 'cooking' depth required for more than five to six

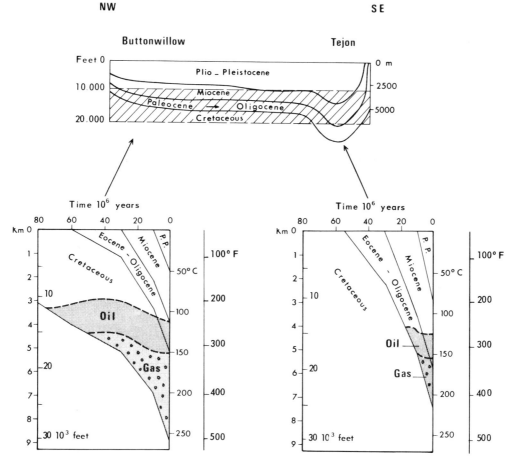

Fig. 30. — Longitudinal section of San Joaquin Basin in California, and time/depth diagrams / temperature of two zones of Buttonwillow and Tejon (Zieglar & Spotts, 1978).

million years, and the older formations for more than ten million years. This southern region turned out to be richer, with a cumulative oil output of more than 600 Mm³ (see Chapter 3 of Part 3). The process is more complex and somewhat different for gases, and gas formed in the Plio-Pleistocene, in other words less than one million years ago, is found in the Po Plain and in the Gulf of Mexico.

In all cases, the thermal history must be reconstructed, particularly by the analysis of vitrinite reflectance values, of the basin or the zone concerned. Like Lopatin (1971), Waples (1980) recommends the consideration of a time/temperature maturity index.

Evolution of coals and oil shales

This evolution of bitumens occurs simultaneously with that of coals during their 'coalification':
— the immature phase of the organic matter corresponds approximately to peats, and to soft and dull lignites,
— the oil window corresponds to high-volatile bituminous coals and gas coals,
— the gas zone of catagenesis corresponds to semibituminous and nongaseous coals and to anthracites.

This 'coalification' also corresponds to a progressive concentration of carbon and a concomitant increase in the reflectance of vitrinite, the main constituent. Volatile compounds are liberated throughout this catagenesis, carbon dioxide and water. Liquid hydrocarbons may also be generated from coal, but normally in very small amounts, possibly due to the absence of drains for their migration. No commercial oil accumulation appears to have originated in this way (Tissot & Welte, 1978).

Oil shales form another large family of carbonaceous rocks, characterized by high kerogen content, and especially type I kerogen. These rocks, particularly tasmanites, correspond to lagoonal or epicontinental sea deposits, and the matrix is argillaceous or carbonate. The amorphous organic matter is essentially of algal origin, and the *Botryococcus* family plays an important role. The H/C ratio is often higher than 1.5 %. They contain more than 2.5 % organic matter and often 5 to 10 %. Two-thirds of the organic compounds are susceptible to conversion to bitumen, with proportions reaching 50 to 100 and even 200 liters of oil per ton of rock (25 to 50 US gallons per short ton).

An oil shale subjected to sufficient catagenesis can generate liquid hydrocarbons.

*
* *

The evolution of kerogen with depth and temperature, and its transformation into hydrocarbons, may be reflected by the diagram in Figure 28, which shows that gases, especially methane, originate more clearly (apart from the eventuality of gas of deeper origin):
— at the time of deposition, gas of bacterial origin,
— in the first phases of diagenesis, partly early gas,
— in the advanced catagenesis phases, so-called catagenesis gas, produced from both sapropelic and humic organic matter, as well as coals.

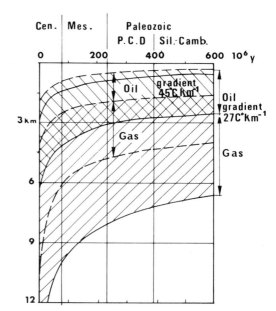

Fig. 31. — Graph showing the depth of oil and gas genesis zones versus time, with respective geothermal gradients of 45 °C/km at the center and 27 °C/km on the right (Magara, 1978).

Gas, particularly methane, which is more resistant and hence more omnipresent, is thus the culmination of the different evolutionary pathways.

The isotopic composition of the carbon in the methane offers direct help in identifying the origin of a natural gas. Hence gases in immature zones, up to depths around 1100 m, exhibit a $\delta^{13}C$ ranging from -95 to $-55‰$ on average. In the intermediate zone, around 1000 to 1700 m, the $\delta^{13}C$ values are roughly between -55 and $-65‰$. Below approximately 1700 m, in the catagenesis zone, the $\delta^{13}C$ average is between -58 and $-36‰$. At extreme depths, below 6000 to 7000 m, $\delta^{13}C$ reaches values around $-10‰$ (Alekseyev, 1974).

In the central area, liquid hydrocarbons occur essentially in the approximate temperature interval of 60 to 120 °C, called the 'oil window'. As shown above, the depths of this interval vary according to time and geothermal gradients. They appear to be shallower in older series and with higher temperatures, as shown by Figure 31.

Roughly it may be stated, as by Tissot et al., (1975), that periods of hydrocarbon genesis correspond substantially to subsidence eras, in other words the downwarping periods of the basins.

1.2.4. Relationships between crude oils and depositional environments

From the sedimentological standpoint, the different depositional environments favorable to the formation of source rocks or kerogen rich rocks may be related to two factors: sand/shale sedimentation of the paralic type, and carbonate/shale sedimentation with an evaporitic tendency. However, some continental environments of the lacustrine type may exhibit remarkable abundance and quality. These environments often strongly condition the characteristics of crude oils, especially if the action of catagenesis is weaker.

1.2.4.1. *Paralic sand / shale environments*

These environments are characterized by thick pyritic black shale series, often with coal seams and sand intervals, in variable proportions, often lens-shaped, and of lacustrine, brackish or marine facies. They correspond to high subsidence areas, with intense alluvial activity and rapid burial, generally situated in the neighborhood of the mouths of the great rivers. The organic matter, of continental origin, consists of debris of the higher plants, rich in cellulose and lignin, conveyed directly to the lakes, lagoons and the sea by currents, with the detrital materials, sands and clays. Rapid burial or a deep environment pose obstacles to reworking on the floor.

In areas of slower sedimentation, these materials are attacked in the mud by microorganisms. Cellulose and part of the lignin are attacked, and very little remains but the waxes and long chains of n-alkanes. The kerogen resulting from these transformations, usually type II, yields paraffinic crudes, and rarely naphtheno-paraffinic oils. Saturated hydrocarbons account for 60 to 90 % of these crudes, aromatics 10 to 30 % of total hydrocarbons, and resins and asphaltenes generally less than 10 %. Sulfur is normally absent or in proportions less than 0.5 %.

Some of these crudes of non-marine origin, which are remarkable for their high proportion of waxes, display high viscosities. They are encountered in particular in rift type basins, especially in the pre-drift Cretaceous coastal basins of Brazil, Magellan and Gabon, as well as the Tertiary Basins of the Orinoco, Indonesia and the Rocky Mountains (Uinta Basin in particular) (Tissot & Welte, 1978).

Another closely studied example of this type of crude is offered by the oils of the Handil and Bekapai Fields in the Mahakam Delta (East Kalimantan) (Combaz & de Matharel, 1978) (see Chapter 2 of Part 3). These crudes, most probably originating in a type III kerogen, are highly paraffinic, particularly rich in high molecular weight paraffins, and poor in S (around 0.06 %). Their specific gravity oscillates between 0.82 and 0.87 — 41 and 31 °API —, decreasing with depth. Their pour point is + 30 °C for Handil crude.

The crude oil of the Gamba Field in Gabon, which also has a high paraffin content (13 %), and a specific gravity of 0.86, has a viscosity at 25 °C of 165 cP and a pour point of + 23 °C (see Chapter 2 of Part 3).

1.2.4.2. *Carbonate and evaporitic environments*

The organic matter in purely marine environments is rich in proteins, carbohydrates and lipids derived from plankton and algae. It deposits to yield a type II kerogen that evolves normally towards naphtheno-paraffinic or intermediate aromatic crude oils. Saturated hydrocarbons account for 40 to 70 % and aromatics 25 to 60 % of total hydrocarbons. Resins and asphaltenes may be relatively abundant, together with sulfur, especially in poorly evolved crudes. Good examples of these oils are found in the Jurassic and Cretaceous formations of the Middle East, the Devonian formations of Alberta, and the Upper Cretaceous formations of Equatorial Africa.

Environments with an evaporitic tendency are generally represented by black shales and dark limestones, often bituminous, phosphatic and radioactive, and sometimes siliceous. They frequently correspond to older continental margins or marginal depressions in 'upwelling' and eutrophic zones in a generally arid climate. Detrital deposits are meager, stratification is very finely bedded, and the environment often highly saline. Sedimentation may be relatively slow and the proportion of organic matter high. Benthic organisms are generally absent and planktonic forms predominate, characteristic of a eutrophic environment. The sedimentation is formed of very fine beds, very thin alternations of various types, denoting the absence of currents and organisms on the floor.

The kerogen of carbonate source rocks has the reputation of being of excellent quality. It normally consists of amorphous organic matter, presumably of algal or sapropelic origin. The absence of ligneous or reworked components goes hand in hand with the limited clastic deposits, and these factors contribute to the development of the high quality of this kerogen. Its H/C ratio is high, giving it high evolutive potential (Elf Aquitaine, 1977).

While the carbonate muds are rapidly lithified, it appears that they subsequently develop very dense fissures due to their lack of plasticity. These microfractures are probably the reason for the effective expulsion of the hydrocarbons formed, reaching a clearly higher level than in shale formations. The crudes formed in these environments are usually heavy, because they are rich in polar compounds, resins and asphaltenes, and in sulfur, which may reach 5 % and even 10 % in the altered aromatic base crudes. They are rich in trace metals, Cu, Ni, Co, Mn and U. The branched chains of these hydrocarbons are more resistant to alterations and produce most of the shows (Weeks, 1968).

Crude oils of this type are found in the Middle East, especially in Syria and Iraq, and also in Aquitaine, Venezuela and Mexico. The Wyoming Basin in the Rocky Mountains offers an interesting example of the superposition of two hydrocarbon families. Sulfur-rich naphtheno-aromatic crudes are found in the carbonate and evaporitic formations of the Paleozoic, while gasoline-rich paraffinic and naphtheno-paraffinic crudes are found in the clastic shale series of the Cretaceous. In these conditions, the deepest crudes are the heaviest.

Reef environments are very prolific, but the organic matter they contain is largely degraded and mineralized. It can only be preserved and accumulate in the sheltered depressions which are often formed behind the reef barriers. Highly saline environments, often displaying a very high organic matter content, derived from a few rare species of living organisms, slight elastic deposition, and evident preservation characteristics, can provide excellent source rocks (Kirkland & Evans, 1981).

1.2.4.3. *Lacustrine environments*

While most source rocks correspond to marine environments, this is not indispensable, and in different basins, gas and oil clearly appear to be derived from lacustrine formations. Among these are the following :
— Tertiary of the Cook Inlet Basin, the Middle Magdalena (Colombia), and the Pakistan and Chinese basins,
— Carboniferous in the British Midlands,
— above all, the oil shales and oils of the Eocene Green River formations in the Green River and Uinta Basins (Wyoming, Utah, Colorado).

Lakes and some lagoons may offer especially favorable eutrophic conditions for the development of algae, particularly the botrycoccus and cyanophyceae algae as in Lake Maracaibo, which are capable of using atmospheric nitrogen directly. In these environments, the density of plankton and other organisms rapidly stops the penetration of light, creating a contrast between the surface waters which are very rich in organic matter and the deeper levels which are stratified, stagnant and anaerobic. Thus life develops intensely at the surface, and the organic matter of dead organisms finds ideal conditions on the floor for putrefaction and preservation. These eutrophic environments thus constitute true 'kerogen factories' (du Rouchet), on highly variable scales, ranging from the simple pond to inland seas, like the Black Sea.

PHYSICAL-CHEMICAL MODELS FOR OIL GENERATION

Douglas W. Waples

ABSTRACT

Modern physical-chemical models for petroleum-forming processes have evolved from work on coals and oil shales that began in the early twentieth century. By extrapolating the results of laboratory pyrolysis studies of fossil organic matter to the lower temperatures and longer times available in geologic settings, it has been found that oil is generated from kerogen by first-order thermal decomposition reactions. The "pseudo" kinetic parameters measured for these reactions are artifacts of the complexity of the system, and should not be confused with true kinetic parameters.

Over the years several models that attempt to predict thermal maturation of kerogens as a function of time and temperature have been developed. The most effective models are those which take into account the complete thermal history of the source rock. At the present time several models appear to give adequate correspondences between measured and predicted maturities. Our ability to select the best of these models will be dependent upon finding test cases involving unusually long or short times, or high or low temperatures. The present models are being widely applied by explorationists in many organizations and are of great value in both frontier areas and maturely explored regions. Rapid advances in computerization indicate that the future of thermal modeling of oil generation is very bright.

INTRODUCTION

Although men have been interested in the origin of coal, petroleum, and oil shale for millenia, scientific approaches to solving these problems have only emerged within the last two centuries. The genetic relationships among coal, oil shale, and petroleum were recognized early by a few workers, but until the 1960s research on each of these fossil fuels proceeded more or less independently and oblivious of studies on the others. Coal was the subject of earliest study, both because of its long-standing economic importance and because of its obvious affinity to plant material. Oil shale has also been known and utilized for a long time, but its more limited distribution made early investigations less popular. In fact, most work on oil shale in the early twentieth century concentrated not on determining how it was formed in nature, but rather on how it could be transformed by man into a more useful material, shale oil.

Petroleum, the fossil fuel of greatest current importance, was the last to be studied. Not until the middle of the twentieth century

Douglas W. Waples, a Beckham grant recipient, received a B.A. in chemistry from DePauw University, Indiana, and a Ph.D. in physical organic chemistry from Stanford University. He held a postdoctoral fellowship with the Alexander-von-Humboldt Foundation, University of Göttingen, West Germany, and a Latin American Teaching Fellowship at Valparaiso, Chile. From 1976 to 1980 he served on the faculty of the Chemistry and Geochemistry Department at Colorado School of Mines, being promoted to associate professor in 1979. Dr. Waples has recently become a geochemical consultant to the petroleum industry and is located in Dallas, Texas. His research interests include thermal modeling of petroleum generation, organic facies analysis, phosphorites, and the application of organic geochemistry to petroleum exploration. In 1982 he received the J.C. Sproule award from the American Association of Petroleum Geologists for the best paper published by an author under 35 years of age, based on a paper he authored in 1980. Dr. Waples is the author of *Organic Geochemistry for Exploration Geologists*, and is currently associate editor of the Bulletin of the American Association of Petroleum Geologists. His present business address is: 1717 Place One Lane, Garland, Texas 75042.

0163-9153/83/7804-0015

© 1983 Colorado School of Mines

did an understanding of the mechanisms by which organic matter preserved in fine-grained sediments is transformed into petroleum begin to emerge. In the last two decades, the great progress in our knowledge of petroleum formation has been a direct result of the synthesis and application to oil generation of preceding work on coalification and oil-shale pyrolysis.

Our understanding of transformations of organic matter in sediments and rocks has evolved very slowly, as numerous theories emerged, were eventually proven wrong or found to be limited in applicability, and became extinct. Although many of these now seem naive, each has served as a work point and has focused attention on the essential requirements for coalification and petroleum-forming processes. A brief historical review of earlier theories is therefore useful.

COALIFICATION

The presence in coals of macroscopic plant remains has made it apparent for a long time that at least some portions of coal are of organic origin. It was not immediately obvious, however, that coals of all ranks were derived from basically the same organic sources. The suggestion in 1778 by von Beroldingen that peat, lignite, bituminous coal, and anthracite all represent different stages in the transformation of organic matter buried in the earth's crust did not meet with universal or unconditional acceptance. Various other theories cited additional factors in the formation of coals of different ranks: climate, geologic age, initial composition of organic matter, and microbial activity. It was not until the early part of the twentieth century that the concepts of rank and type were distinguished clearly (White 1908, 1913).

One early theory that both survived and flourished, however, was the hypothesis that burial is the decisive factor in coalification. The relationship between burial and coal rank was first enunciated by Carl Hilt in 1873, who noted that fixed carbon content in coals increased regularly with depth of burial. The utility and general validity of "Hilt's Law" led naturally to attempts to explain the dependence of coal rank on burial depth (e.g., Reeves 1928; Heck 1943; Bottcher, and others 1949). The numerous coalification models that derived from these efforts focused on three factors that also are functions of burial depth: temperature, static pressure, and time.

Temperature

Early workers who supported the dominant influence of temperature on coalification generally believed that extremely high temperatures were required. Erdmann (1924) and Roberts (1924) claimed that temperatures in excess of 325° and 550°C were necessary to produce bituminous and anthracite coals, respectively. Sources for such immense amounts of heat were a problem, however, for temperatures of 500°C could only be reached by igneous activity or by burial to immense depths. Although some early workers attempted to correlate coalification with igneous activity, neither extreme burial or igneous activity was defensible globally as the primary cause of coalification.

In the first part of the twentieth century White (1913) and Bergius (1913) finally recognized that, given the vastness of geologic time, reasonably low temperatures were adequate even for formation of anthracite. There were problems, however, with coalification schemes that relied solely on temperature as the agent of transformation, because high-quality coals could not be produced by heating either plant material or low-rank coals in the laboratory.

White (1908, 1913, 1935) therefore provided a very attractive alternative with his thrust-pressure hypothesis. The empirical correlation between coal rank and orogenic folding first noted by Rogers (1843) led White to propose that frictional heating during orogenic movements had caused the coalification.

White's ideas were adopted by many other workers over the next half century, but were never universally accepted. For example, those coal workers whose favorite areas were not associated with thrusting strongly opposed the thrust-pressure hypothesis. The thrust-pressure hypothesis thus persisted until the middle of the twentieth century before finally being discounted as the dominant factor in coalification (Heck 1943; Teichmüller and Teichmüller 1951; Huck and Karweil 1955).

Static Pressure

Static pressure associated with overburden was also suggested as the most important factor in coalification (e.g., Reeves 1928), but because of its inability to provide a source of heat never gained the general acceptance that White's dynamic-pressure hypothesis enjoyed. Despite the numerous strong arguments against it as an important influence in the chemical transformation of coals (e.g., Bergius 1913; White 1935; Huck and Karweil 1955; Karweil 1956; Lopatin and Bostick 1973), the static-pressure hypothesis has persisted with some workers almost to the present day. Most modern workers believe that although static pressure is apparently required for the formation of true coals, it affects physical properties rather than rates of chemical reactions.

Time

The role of geologic time has taken on a gradually increasing importance in our understanding of coalification since Bergius (1913) and White (1913) first recognized its power, but time alone has seldom been considered seriously as the dominant factor in coalification. Most theories that allocate an important role to time give an even more vital one to temperature.

Combined Time and Temperature

Early discussions of the effects of time in coalification processes were generally qualitative; the only quantitative measurements which showed that time and temperature could to some extent be substituted for each other came from experimental work carried out in chemistry laboratories. Bergius (1913) was probably the first to suggest that time and temperature enjoy an exponential interrelationship in coalification by proposing that the rates of coalification reactions double with each 10°C rise in temperature. In the first half of this century, however, there were very few such studies.

There were also some difficulties in using laboratory data to understand natural coalification processes. Van Krevelen (1952) showed that coal pyrolysis and normal coalification are chemically distinct processes, and suggested that laboratory data may not be applicable to coals in natural settings. Hanbaba and Jüntgen (1969), however, found that such extrapolations were valid if made with care. Numerous workers (e.g. Fuchs and Sandhoff 1942; van Krevelen, and others 1951; Stone, and others 1954; Huck and Karweil 1955; Hanbaba and Jüntgen 1969) showed that pyrolysis of humic coals (and thus presumably coalification) is a much more complex chemical process than had been previously appreciated, and that kinetic analyses therefore are also much more difficult to interpret. For example, activation energies measured for coal pyrolysis by various workers varied between 4 and 59 kcal/mole, depending upon experimental design. Many of these activation energies and the corresponding A factors in the Arrhenius equation (1) are far too low to represent true kinetic parameters for chemical reactions occurring during coalification:

$$k = Ae^{-E_a/RT} \qquad (1)$$

These results indicated that coal-pyrolysis kinetics may be controlled by the rates of physical processes, such as diffusion or removal of steric hindrance, rather than by rates of chemical reactions (Stone, and others 1954; Huck and Karweil 1955). The unusual kinetic parameters often found for coal-pyrolysis reactions suggest that the term "pseudoactivation energy" is therefore appropriate for measured E_a values.

In spite of the relatively abundant work on the kinetics of coal pyrolysis, kinetic calculations were not utilized by coal geologists until Karweil, with his classic (1956) nomograph (figure 1), established the first means of predicting the rank of a coal knowing only its thermal and burial history. Karweil's method of utilizing time and temperature represented a quantum leap forward in quantifying the process of thermal transformation of organic matter in sediments, and it is a testimony to his insight that essentially no improvement was made on his method over the next decade and a half.

OIL-SHALE RETORTING

It was realized long ago that heating oil shale in the absence of oxygen produces an organic liquid somewhat similar in chemical composition to crude oil. Crum-Brown originally defined "kerogen" as the organic material in oil shale that produces oil during retorting. Engler (1913) proposed that kerogen decomposes thermally in a two-step process:

Kerogen → Bitumen → Oil

The "bitumen" consists of fragments that have been broken off from the larger kerogen molecules but which are different in composition from oil molecules. Engler observed, as Bergius did simultaneously for coal pyrolysis, that time and temperature could be substituted for each other, and that the effect of temperature was greater than that of pressure. The studies of McKee and Lyder (1921) and Franks and Goodier (1922) confirmed these results.

Maier and Zimmerley (1924) showed that the decomposition of Green River Shale kerogen was an irreversible first-order reaction. Together with Franks and Goodier (1922), they also made

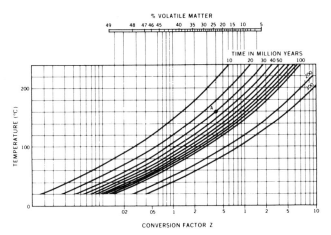

Figure 1.—Karweil's 1956 nomograph relating coal rank (determined by volatile-matter content) to length of time spent at a certain paleotemperature. Each line represents a different paleotemperature. The x coordinate of any point along any time line represents the rank of a coal seam whose maximum paleotemperature is given by the y coordinate of the same point. Example: Point A has spent 30 million years at 160°C and contains 24 percent volatile matter (corresponding to the medium-volatile bituminous rank). Reprinted by permission in modified form.

the important statement that there was no definite temperature threshold for the pyrolysis, but rather a gradual increase in rate with increasing temperature.

Hubbard and Robinson (1950) also proposed a slight modification in Engler's earlier mechanism: gas is formed along with bitumen in the initial decomposition reaction.

Kerogen → Bitumen → Oil + Gas + Carbon residue
　　　　　　　　　+
　　　　　　　　　Gas

Hill and Dougan (1967) found the mechanism of oil-shale pyrolysis at low temperatures to be simpler than at the higher temperatures employed by most earlier workers. Their reactions were first order in the earlier stages, gradually giving over to zero-order processes in the later stages. At high temperatures bitumen production apparently occurred so rapidly that diffusional escape from the rock matrix was not possible; polymerization and kinetic complexities thus resulted. Weitkamp and Gutberlet (1970) also emphasized the importance of diffusion on both the mechanism and energetics of oil-shale pyrolysis. It became clear from these studies that extrapolation of pyrolysis data on oil shales to reactions in natural settings would be difficult.

In 1972 Cummins and Robinson helped solve this problem by carrying out oil-shale-pyrolysis experiments at much lower temperatures (150-350°C) than previous workers had. At those temperatures the pyrolysis was adequately described by a first-order kinetic expression having a pseudoactivation energy of 19 kcal/mole.

Braun and Rothman (1975) simplified the mechanism of oil-shale pyrolysis even futher by reinterpreting Hubbard and Robinson's data, concluding that oil-shale pyrolysis could indeed be represented by the two sequential reactions originally proposed by Engler (1913). The pseudoactivation energy for bitumen formation was found to be about 10 kcal/mole, and that for subsequent bitumen decomposition, 42.5 kcal/mole.

By the mid-1970s, therefore, work on the kinetics of oil-shale pyrolysis had essentially confirmed Engler's early analysis. Pyrolysis experiments carried out at low to moderate temperatures are adequately described by first-order kinetics. As van Krevelen, and others (1951) had observed for coal pyrolysis, pseudoactivation energies increase as oil-shale retorting proceeds (Weitkamp and Gutberlet 1970), but such changes are probably linked to physical, rather than chemical, changes in the oil shale during retorting.

No one has ever attempted to compare natural or artificial coalification processes with oil-shale retorting. Neither the relationship of algal coals to oil-shale kerogens nor rank changes in algal coals were studied until fairly recently (e.g., Schopf 1949; Cane 1950). Only in the last few years has there developed a concept of thermal maturation of oil-shale kerogen analogous to coalification.

Because of the poor empirical correlation between oil-shale occurrences and petroleum deposits, it was thought until recently that oil-shale kerogen could not be transformed into bitumen or hydrocarbons in natural settings. Kinetic data compiled from laboratory retorting experiments were therefore not applied to the behavior of oil-shale kerogen under natural conditions. Instead, as our understanding of the similarities among coalification, oil-shale retorting, and petroleum generation has grown, data from artificial-coalification and retorting studies have been applied directly to understand the chemistry and energetics of petroleum formation in the subsurface.

PETROLEUM FORMATION

The origin of petroleum was poorly understood prior to the 1960s. An organic origin had been suggested by early chemical analyses of petroleum constituents (e.g., Marcusson 1908; Treibs 1936), but there was no consensus about how petroleum forms from organic matter. Although microbial activity at shallow depths of burial was proposed by some workers to transform the various functionalized organic molecules into hydrocarbons, it was not until the 1950s that indigenous hydrocarbons were finally identified for the first time in recent sediments (Smith 1954). These mixtures were soon shown to be quite distinct from petroleum hydrocarbons, however. In contrast, bitumens similar in composition to petroleum were found in all lithified shales and limestones studied by Hunt and Jamieson (1956). It was thus apparent that petroleum does not form immediately after sedimentation, but rather at greater depths of burial.

Oil Generation at Depth

The mechanism by which oil is generated remained obscure through the 1950s, however, because the relationships among coal, oil-shale kerogen, petroleum, and disseminated organic matter (kerogen) in ordinary sedimentary rocks were not fully appreciated. Early workers were often on the right track, but never quite solved the problem in detail. Rogers (1860) had proposed long ago that oil and gas in Pennsylvania were derived from coal-bearing strata at a certain coal rank. Potonié (1910) and Anderson and Pack (1915) suggested that "sapropelites" and diatoms might be sources of oil when subjected to elevated pressures, temperatures, or both. Several other early workers had specifically proposed that kerogen is somehow converted to bitumen, which then yields petroleum (e.g., Engler 1913; McKee and Lyder 1921).

Karrick (1926) had suggested that virtually all shales could, like oil shales, generate some bitumen upon retorting, but the significance of this idea for oil generation in natural settings went unappreciated until much later. The mechanism by which oil forms in shales was poorly understood even in the late 1950s, despite the proposal by McNab, and others (1952) that petroleum is produced by thermal decomposition of kerogen.

An important advance occurred when Forsman and Hunt (1958) distinguished between coal-type kerogens and oil-shale-type kerogens. Subsequent work has shown that oil is principally derived from kerogens that are similar to oil-shale kerogens in origin and chemical composition. There had previously been a great reluctance among coal workers and petroleum geochemists alike to see the close similarities among coals, oil shales, and finely disseminated organic matter in sediments, a failure which no doubt retarded the development of petroleum geochemistry for many years.

Role of Kerogen

By the 1960s evidence had begun to accumulate that some dispersed kerogen in fine-grained sedimentary rocks is indeed transformed into bitumen by thermal processes. Philippi (1965) provided particularly strong empirical evidence for the role of kerogen in petroleum formation in his classic paper on generation of petroleum in Southern California.

As the analytical methods commonly employed by coal scientists were gradually adapted for studying disseminated kerogens, the kerogen-transformation model was gradually confirmed. Spores and pollen grains in shales were shown to darken with increasing thermal exposure. Vitrinite reflectance measurements were applied to disseminated kerogens as well as to coals (Teichmüller 1971), and it became common to assume that increases in vitrinite reflectance values were valid indicators of the extent to which oil generation had occurred. Elemental analyses confirmed that chemical transformations accompany thermal maturation of kerogens as well as coals (Hlauschek 1950; Forsman and Hunt 1958; Durand and Espitalié 1973). The coal maceral concept (White 1908; Potonié 1910) was finally applied to kerogens (Forsman and Hunt 1958), and coal-evolution diagrams (figure 2) (van Krevelen 1950) were eventually used to describe kerogen transformations (Durand and Espitalié 1973).

Pyrolysis studies have proven that virtually all kerogens in sedimentary rocks are capable of generating hydrocarbons by thermal decomposition, the nature and quantity of product being functions of kerogen type (Dow 1977). As earlier workers had for coal pyrolysis and oil-shale retorting, Barker (1974) and Peters, and others (1977) established for kerogen pyrolysis that pseudoactivation energy increases and remaining hydrocarbon-generative capacity decreases with increasing rank.

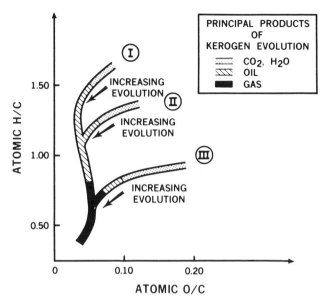

Figure 2.—Van Krevelen diagram showing elemental compositional differences among kerogen types (I, II, and III), and changes in H/C and O/C ratios with maturation for each kerogen type. After Tissot, and others (1974); reprinted by permission.

KINETICS

Philippi's (1965) work showed that both time and temperature are important in oil generation, and that they are at least to some degree interchangeable. Oil was formed at lower temperatures in the Los Angeles Basin than in the Ventura Basin because the Los Angeles Basin source rocks were cooked longer. Philippi's approach was qualitative, however; he did not attempt to describe precisely the relationship between time and temperature.

Soon thereafter workers at the French Petroleum Institute began to develop mathematical models for petroleum-generative processes (Tissot 1969). Their models assumed essentially the mechanism of Engler (1913) for petroleum formation, and employed pseudoactivation energies similar to those found by Cummins and Robinson (1972) for low-temperature pyrolysis of Green River oil shale: 10 kcal/mole for bitumen formation and 15 to 20 kcal/mole for hydrocarbon formation. A factors were 10^2 to 10^8 my^{-1}.

Because these pseudoactivation energies and A factors were far lower than could be explained on purely chemical grounds, it was proposed that catalysis by minerals, particularly clays, was responsible for the low measured pseudoactivation energies (e.g., Brooks 1948; Louis and Tissot 1967; Tissot 1969; Shimoyama and Johns 1971). Geocatalysis models have always been beset by two difficulties, however: the improbability of appreciable surface interactions between large, relatively immobile kerogen molecules and catalytic mineral surfaces, and the inability of known catalysts to lower activation energies sufficiently to explain the empirical and experimental data on oil generation. Recent data (e.g., Horsfield and Douglas 1980) indicate that if geocatalysis is a factor in oil generation, it is much more important in cracking the initial bitumen than in the breakdown of kerogen.

A much more plausible explanation was finally offered in 1975 for the unusual kinetic parameters obtained from laboratory pyrolysis experiments of coals, oil shales, and kerogens and from empirical data on oil generation in natural settings (Jüntgen and Klein 1975; Tissot and Espitalié 1975). Both groups observed that pseudoactivation energies representing a group of parallel reactions can be very different than the true activation energy for any one of the individual reactions. Jüntgen and Klein showed (figure 3) that the pyrolysis behavior of a group of parallel reactions having normal kinetic parameters (activation energies from 48 to 62 kcal/mole; A factor = 10^{15} min^{-1}) can be described quite accurately as though they were actually a single reaction having kinetic parameters (E_a = 20 kcal/mole; A = 10^4 min^{-1}) that are physically impossible for the first-order thermal decomposition of organic matter. Jüntgen and Klein's contribution therefore offered a much more plausible alternative to geocatalysis in explaining the unusual kinetic parameters that were often reported in pyrolysis work on fossil fuels and from empirical studies on petroleum formation.

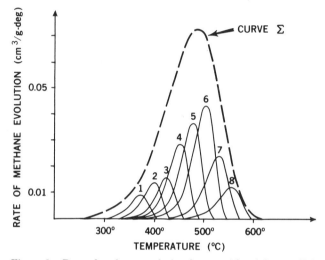

Figure 3.—Rate of methane evolution from coal for eight parallel reactions as a function of pyrolysis temperature during programmed-temperature pyrolysis. Each numbered curve represents one of the parallel reactions. A Gaussian distribution is assumed for the initial concentrations of the eight starting materials. Activation energies assumed (kcal/mole) are:

(1) = 48 (2) = 50 (3) = 52 (4) = 54
(5) = 56 (6) = 58 (7) = 60 (8) = 62.

A factors are 10^{15} min^{-1} for each reaction. Curve Σ, representing the sum of the eight parallel reactions, has a pseudoactivation energy of 20 kcal/mole and an A factor of 10^4 min^{-1}. After Jüntgen and Klein (1975); reprinted by permission.

KEROGEN TYPE

Although earlier workers had suggested that the kinetics of oil generation are different for different kerogen types, Louis and Tissot (1967) were perhaps the first to attribute the different temperature ranges for oil generation in different basins at least partly to different types of organic matter. In 1975 Tissot and

Espitalié described each of the three principal kerogen types by a distribution of activation energies ranging from 10 to 80 kcal/mole. Type I kerogen has the highest "average" pseudoactivation energy because it contains mainly cross-linked aliphatic chains that require cleavage of strong carbon-carbon bonds in order to produce bitumen. Type II kerogen has a lower average pseudo-activation energy, because it comprises a substantial proportion of weaker C-N, C-S, and C-O bonds.

According to Tissot and Espitalié, Types I and III kerogens begin to generate hydrocarbons even earlier than does Type II kerogen, but reach their maximum generation rates later (figure 4). Because all the experiments that led to these conclusions were carried out at elevated temperatures during programmed-temperature pyrolysis, the validity of their application to petroleum generative processes in natural settings has not been established.

Other investigators have come to opposite conclusions about oil-generation thresholds for different kerogen types. Powell, and others (1978) showed that amorphous and resinous (Type II) kerogens begin to generate oil at R_o values of 0.5 percent, whereas terrestrial (Type III) kerogens commence later, at R_o = 0.7 percent. This apparent contradiction of the results of Tissot and Espitalié (1975) could be explained by different definitions of "threshold." Although figure 4 shows that in an absolute sense generation begins in Type III kerogen earlier than in Type II, if the threshold for "effective" or "measurable" generation is raised slightly (to value G in figure 4, for example), this latter threshold is attained earlier in Type II kerogen.

On the basis of changes in fluorescence intensity, Leythaeuser, and others (1980) claimed that the threshold for the beginning of oil generation increases in going from Type I to II to III kerogen. The discrepancy between their results and those of other workers may be partly a result of different methodologies, but may also in fact represent an important advance in our understanding, because fluorescence increases are probably more directly related to petroleum formation than are changes in reflectance of vitrinite particles (Radke, and others 1980).

Despite the accumulation of evidence suggesting that the kinetic parameters of the various kerogen types are distinct, most workers (e.g., Dow 1977; Waples 1980) have ignored these differences. This approach is undoubtedly an oversimplification that should be eliminated by further research.

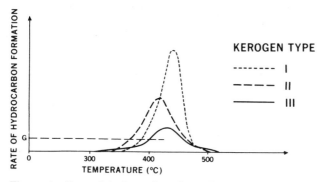

Figure 4.—Rate of hydrocarbon formation by programmed-temperature pyrolysis of kerogens of Types I, II, and III. Time and temperature both increase to the right. From Tissot and Espitalié (1975); reprinted by permission.

PRESSURE

The role of pressure in oil generation has never been carefully examined. Early discussions often took coalification as a model, and assumed that if pressure aided the chemical reactions involved in coalification (an assumption that was probably incorrect), it should also facilitate oil generation. This simplistic approach was superseded by more sophisticated considerations, in which it was assumed that petroleum is formed by thermal decomposition of large kerogen molecules. The most important factor in such a model was retardation of kerogen decomposition by the buildup of internal pressure from volatile light hydrocarbons. Low pressures were therefore believed to favor kerogen decomposition, but were not thought to be nearly as important as temperature.

As microfracturing has come into favor as an important mechanism for primary migration, buildup of internal pressure during oil generation has come to be recognized as a necessity. The question of possible retardation of oil generation by high pressures has thus become moot. The remaining interests in pressure influences on petroleum formation seem to be the effect of shale overpressuring on oil generation, how pressure influences subsurface temperatures, and what precise conditions are required for initiation of microfracturing.

DESTRUCTION OF HYDROCARBONS

It has always been more important to explorationists to know where oil might be found than to understand its origin. In 1915, in an effort to avoid drilling in areas that could not possibly yield liquid hydrocarbons, White proposed his famous "Carbon-Ratio Theory," based on empirical studies of liquid-hydrocarbon occurrences, which stated that commercial oil deposits would not be found where fixed-carbon contents of coals exceeded 65 or 70 percent.

It was subsequently hypothesized by many geologists (e.g., Fuller 1919; Pratt 1934; Barton 1934; White 1935) that light crude oils are formed by thermal cracking of initially heavy oils. Critics noted, however, that the correlation between fixed-carbon content and oil gravity was very imperfect in some areas (e.g., McCoy 1921; Russell 1927; Jones 1928).

These objections were eventually overcome and in the intervening years additional evidence has supported the carbon-ratio theory. For example, many workers have shown that liquid bitumens and petroleums change in composition and gravity as a result of thermal-cracking reactions in the subsurface, and can eventually be destroyed in extreme cases.

Adoption of new methods of determining coal rank, especially vitrinite reflectance, has permitted extension of the carbon-ratio theory to sediments that lack coal beds. Various workers have proposed oil-preservation deadlines at vitrinite-reflectance (oil immersion) values from 1.0 to 1.35 percent. The dry-gas preservation deadline probably lies at vitrinite reflectance values of about 3.0 to 3.2 percent, although the exact value is poorly known. Dry-gas preservation may, however, be more dependent upon the presence of oxidizing agents like elemental sulfur than on temperature alone (Baker and Kemp 1982).

Hydrocarbon-preservation deadlines have often been fixed in terms of subsurface temperature (e.g., Hunt 1975). Such schemes have the virtue of simplicity, because subsurface temperatures are readily predictable from measured or estimated geothermal gradients, but these models, which ignore the effects of time, are subject to considerable error, when reservoirs of greatly different ages are compared. Because destruction of hydrocarbons by cracking reactions follows first-order kinetics with measured pseudoactivation energies of about 50 kcal/mole (McNab, and others 1952), it is necessary to consider both time and temperature effects in establishing hydrocarbon-preservation deadlines.

MODELS FOR THERMAL GENERATION OF PETROLEUM

It has become possible, through the use of spore coloration, vitrinite reflectance, and a host of other less precise or less common techniques, to estimate the degree to which petroleum-generative processes have progressed in sedimentary rocks and whether preservation deadlines for liquid or gaseous hydrocarbons have been reached. Whether or not these various techniques are really adequate measures of hydrocarbon generation and destruction is being debated. Most of the above-mentioned techniques measure changes that are assumed to occur simultaneously with petroleum formation, but which are distinct processes chemically. Their kinetic dependences may therefore be very different from those of petroleum formation and destruction. Although these accomplishments have been very important to petroleum explorationists working in areas where samples are readily available, they have been of limited value in frontier areas. Furthermore, measured maturity values give no direct information about timing of oil generation, a question of critical importance in many cases.

It is therefore desirable to be able to predict the progress of thermal evolution of sedimentary rocks through time. The earliest model for thermal maturation, developed in 1956 by Karweil, was designed to predict coal rank. Models for petroleum generation were first published a decade later (Louis and Tissot 1967). The sophistication and value of subsequent models have gradually increased, although progress has been somewhat erratic. Models for coalification and petroleum generation are fully compatible with each other, because each assumes that time, temperature, or both are important factors, whereas effects of pressure are negligible. Some models have also taken into consideration variations in the type of organic matter present. Time-temperature models can be divided into several groups based on the ways they take time and temperature into consideration.

TIME EFFECTS

Negligible Effects of Time

The simplest models (e.g., Neruchev and Parparova 1972a, 1972b) assume that time is not a significant factor in oil generation. This assumption allows the establishment of strict minimum temperatures below which petroleum formation does not occur. Most workers have assumed an oil-generation-threshold temperature of 50 to 70°C, although some have also admitted that time does play a role. It has also been claimed, most recently by Price (1982), that coal rank (or kerogen maturity) is a maximum-reading thermometer, a statement tantamount to saying that only temperature affects maturity, and that any temperature below the maximum is completely inconsequential.

Statements about absolute temperature thresholds are of some value in establishing an approximate boundary for oil generation, but are of little use as predictive tools because there are so many exceptions. For example, one of the earliest studies of oil generation (Philippi 1965) showed clearly that young petroliferous basins generate oil at much higher temperatures (115-140°C) than predicted by the above rule of thumb. It is apparent from comparing the temperatures of oil generation in California from Philippi's study with those from the much older Paris Basin (60-65°C: Louis and Tissot 1967) that time must be considered in developing models for oil generation. If generation and destruction thresholds could in fact be determined merely from maximum paleotemperatures, our problems would be greatly simplified.

Geologic Age as Cooking Time

Connan's (1974) empirical study on the kinetics of oil generation used formation ages as their cooking time and the maximum paleotemperature as the cooking temperature. He was well aware that this assumption introduced errors into his method, but hoped that the errors were systematic and would thus cancel each other.

Connan's model is satisfactory for comparing basins whose histories are simple and similar. In complex tectonic regimes or where basins of different ages must be compared, however, the deficiencies of the model become apparent.

Effective Heating Time

Lopatin (1969a, 1969b) showed through the use of burial-history plots (figure 5) that heating effects are not constant throughout a rock's history. For example, the oldest rocks depicted in figure 5 (125 my age: lowermost burial history line) spent their first 85 my at temperatures significantly below their maximum paleotemperature. The cooking that occurred during the first 85 my is therefore much less important than that which took place later. Lopatin (1969b) was thus the first to attempt to define an "effective heating time" for oil generation by observing a good correlation between measured thermal maturity and the length of time spent at temperatures above 100°C.

Hood, and co-workers (1975), in contrast, did not use a predetermined temperature to delineate the onset of their "effective heating time," but instead defined it as that period spent within 15°C of the rock's maximum paleotemperature. They created a scale of thermal maturity called "level of organic metamorphism" (LOM), values of which they calculated using only the effective heating time and maximum paleotemperature. Their choice of a 15°C range for effective heating was an acknowledgment of the very rapid increase in oil-generation rate with increasing temperature. Because of the simplifying assumptions employed in calculating effective heating time, Hood and co-workers cautioned against putting too much faith in calculated LOM values. They preferred to limit their predictions to extrapolations within sections where some measured data already existed. Tests of their

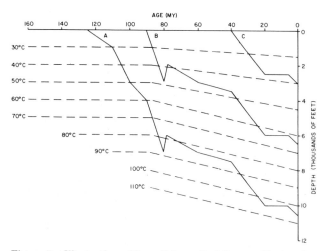

Figure 5.—Illustration of Lopatin's method for considering the complete thermal history of sedimentary rocks. Burial-history curves for three horizons (A, B, C) in a hypothetical sedimentary sequence are solid lines; superimposed temperature grid is represented by dashed lines. From Waples (1980); reprinted with permission.

model in both Paleozoic and Mesozoic basins gave reasonably good correspondences with measured data.

The main weakness with the "effective heating time" model lies in its inability to consider a rock's complete thermal history. In many cases, particularly in old rocks, important thermal maturation may have occurred at temperatures more than 15°C lower than the maximum paleotemperature. The next generation of total-thermal-history models was able to take a rock's entire thermal history into account.

Total-Thermal-History Models

Karweil's (1956) model was capable of taking the complete burial and thermal history of a coal into account in calculating its maturity. He explained how one can sum the effects of several time-temperature regimes and calculate a coal's cumulative maturity using the conversion factor Z (figure 1). Karweil's scheme was developed long before thermal conversion of kerogen to oil was understood, however, and was intended only to predict coal rank. It therefore has never been popularly applied to predicting oil generation. Karweil's method has been criticized for overestimating the effect of time (Nagornyi and Nagornyi 1974; Karweil 1975; Kettel 1981).

The first mathematical model for oil generation that used the entire thermal history of a sediment in predicting its oil-generative history was developed by Tissot (1969). His model was quite sophisticated, employing several reactions, linked both in parallel and in series, to define the oil-generative process. Kinetic parameters were selected to give the best fit with empirical data.

Tissot's work gave an excellent correspondence between measured and predicted values for 23 wells from the Paris Basin. However, because the mathematical apparatus he used was rather formidable and because little effort was expended to make his model available to geologists as a predictive tool, Tissot's model has been used mainly by his colleagues in France (e.g.,

Deroo, and others 1969; Tissot, and others 1975; Tissot and Espitalié 1975; Du Rouchet 1980).

Deroo and co-workers (1969) modeled bitumen formation in the Paris Basin as a function of time. Application of Tissot's model in other basins showed clearly how strongly the thermal history of sedimentary rocks influences the timing and quantity of hydrocarbons generated (figure 6).

Subsequent models developed by Lopatin (1971) and Bostick (1973) sacrificed some of Tissot's rigor but more than compensated by developing better geologic frameworks. Lopatin's (1971) idea, like Karweil's (1956) original nomograph for predicting coal rank, was revolutionary because it provided a much more convenient method for combining the maturation effects of time and temperature. Lopatin's method, which grew out of Karweil's work, has been explained in detail elsewhere (Waples 1980), but a brief summary is appropriate here.

Starting with a burial-history diagram and a superimposed grid of isotherms that represent the temperature history of the sedimentary section (figure 5), one can calculate the thermal maturity of any rock layer at the present or at any time in the past by merely summing the amount of thermal exposure attained in discrete time intervals in the past. In Lopatin's original (1971) scheme each 10°C temperature interval is assigned a temperature factor (γ) based on consideration of the kinetics of oil generation. This γ value is then multiplied by that temperature interval's time factor, which is the length of time that the rock spent within that temperature interval. The sum of these n products for all relevant temperature and time intervals gives the "Time-Temperature Index" of maturity (TTI) (Lopatin 1971), or "Sum Heat Impulse" (Lopatin and Bostick 1973). Mathematically stated,

$$\text{TTI} = \sum_{i=1}^{n} (t_i)(T_i) \qquad (2)$$

where t_i is the length of time in million years (my) spent by the rock in the ith temperature interval, and T_i is the temperature factor for the ith temperature interval.

The value of T (given by T_{i+1}/T_i) determines the relative importances of time and temperature in oil generation. Following an old chemical approximation, Lopatin assumed T = 2, which means that the rate of oil generation doubles with every 10°C rise in temperature. To avoid dealing with large numbers he arbitrarily assigned the 100-110°C temperature interval a T value of 1. The 80-90°C temperature interval thus has T = ¼, whereas 120-130°C has T = 4.

Lopatin tested and calibrated his model on a very difficult well, Münsterland-1, drilled in West Germany in Paleozoic rocks that have had a very complicated tectonic and thermal history. The results of his testing showed a very high internal consistency, but the absolute accuracy of his calibration was questioned by several workers on the basis of errors in his geologic reconstruction and thermal history. Recalibration of Lopatin's method with larger and more reliable data sets (Waples 1980; Kettel 1981) has verified the general validity of the model itself, but has modified Lopatin's original TTI-vitrinite reflectance correlation.

Lopatin and Bostick (1973) and Lopatin (1976) himself later suggested some improvements in the original scheme. Lopatin (1976) used fewer and larger temperature intervals (15 to 30°C instead of 10°C), and created a nomograph to minimize mathematical manipulations. The logic in making this change is discussed in the next section. Lopatin's nomograph is probably unnecessary, because both construction of the burial and thermal history of a sedimentary section and calculation of TTI values can easily be computerized. In any case, using larger temperature

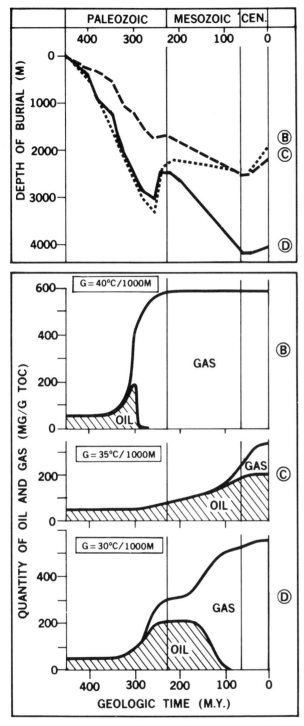

Figure 6.—Burial history and hydrocarbon generation in Paleozoic rocks at three locations in the Illizi Basin, Algeria. Hydrocarbon generation predicted using the model of Tissot (1969). From Tissot, and others (1975); reprinted by permission.

intervals reduces accuracy. The approach of Gretener and Curtis (1982) is essentially that of Lopatin (1971), although cosmetic changes have been made.

One weakness of Lopatin's and Karweil's total-thermal-history models in application to the problem of oil generation is that they were developed and calibrated for transformations of solid organic matter (coal and vitrinitic kerogens) rather than for generation of liquid or gaseous hydrocarbons. Waples (1980) tried to extend this methodology to hydrocarbon generation, but further work is necessary to establish more securely the correlation of TTI values with the various stages of hydrocarbon generation and destruction.

TEMPERATURE EFFECTS

There are several methods of interrelating time and temperature in total-thermal-history models of oil generation. One way to consider the effect of temperature on reaction rates is to assume that the pseudoactivation energy remains constant (e.g., Du Rouchet 1980). As temperature increases with increasing burial, the ratio γ of the oil-generation rate at the new, higher temperature (T_h) to that at the previous, lower temperature (T_l) decreases, as is shown by dividing the Arrhenius equation (1) at the higher temperature by that at the lower:

$$\gamma = \frac{k_h}{k_l} = \frac{Ae^{-E_a/RT_h}}{Ae^{-E_a/RT_l}} \quad (3)$$

Simplification yields

$$\gamma = \exp\left[\frac{E_a(\Delta T)}{RT_l T_h}\right] \quad (4)$$

If the activation energy and temperature increment ($\Delta T = T_h - T_l$) are held constant as T_l and T_h both increase, there is a gradual decrease in γ. Because γ depends on ΔT, it is important to specify the size of the temperature intervals used. A 10°C interval has generally been employed in the past, and, except as noted, all discussions of γ in this paper assume $\Delta T = 10°C$.

Golitsyn (1973) proposed that γ decreases with increasing temperature: he used $\gamma = 1.4$ up to 60°C, $\gamma = 1.3$ to 100°C, and $\gamma = 1.2$ to 150°C. The pseudoactivation energy required by his model would be about 7 kcal/mole, a value much lower than that determined experimentally or empirically for oil generation by any worker, but justified by Golitsyn in the belief that oil-generation kinetics are diffusion controlled. Tissot (1969), in contrast, proposed that a single pseudoactivation energy of 50 kcal/mole was satisfactory for modeling oil generation. The much higher γ values (2.3 at 60°C, 1.8 at 100°C, and 1.4 at 150°C) implied by Tissot's higher pseudoactivation energy indicate that Tissot's model for oil generation is much more sensitive to temperature than is Golitsyn's.

Lopatin (1976) favored an approximately constant pseudo-activation energy of 10 kcal/mole for petroleum formation, but elected to maintain a constant $\gamma = 2$. As Equation (4) shows, under these conditions ΔT must increase as the reaction temperature increases. Lopatin used 15°C ΔT intervals from 50 to 80°C, gradually increasing to 30°C intervals from 170 to 230°C. There is no fundamental difference between Lopatin's and Golitsyn's approaches; in both cases E_a is constant. Whether γ or ΔT is also maintained constant is a matter of computational convenience.

Other investigators have chosen to maintain a constant γ, rather than E_a. If γ and ΔT remain constant, Equation (4) shows that as temperature increases, the pseudoactivation energy must decrease. After testing several γ values against empirical data on thermal maturity of kerogens, Waples (1980) concluded that no better value than Lopatin's original one ($\gamma = 2$) was available. A γ value of 2 implies that pseudoactivation energies increase from approximately 12 kcal/mole at 25°C to 20 kcal/mole at 100°C to 26 kcal/mole at 160°C. Kettel (1981) favored $\gamma = 1.6$ on the basis of data from four wells near the Bramsche Massif in northwest Germany, but his study lacks an adequate statistical base.

On the basis of empirical and experimental evidence from coal and oil-shale pyrolysis, Golitsyn's (1973) and Du Rouchet's (1980) hypothesis that E_a remains constant during oil generation appears untenable. The models of Lopatin (1971) and Waples (1980) are more consistent with experimental evidence, but suggest that the increase in activation energy is a function of temperature alone. However, all experimental evidence from pyrolysis work indicates that pseudoactivation energy increases with increasing maturity rather than with temperature.

Hood, and others (1975) anticipated this difficulty by relating the increase in pseudoactivation energy to LOM values rather than to temperature. Their pseudoactivation energies range from about 15 to 26 kcal/mole within the oil-generation window. To be even more rigorous, we should consider γ to be a function of kerogen type as well as of maturity.

One curious aspect even of sophisticated models of oil generation is that many of them assume an absolute minimum temperature for oil generation, even though their kinetic foundations predict that generation should proceed at finite (albeit very modest) rates even at low temperatures. From a theoretical point of view such absolute temperature thresholds have little to recommend them, but from a practical standpoint they probably are valuable: the amount of maturation that occurs at low temperatures in reasonable spans of geologic time is of little or no consequence.

MODERN MODELS: WEAKNESSES AND POSSIBLE IMPROVEMENTS

The most sophisticated published model presently in widespread use is that developed by Lopatin (1971). His model could be improved in several ways, however, either by adapting positive features of other models, by refinement of basic parameters already in the model, or by further research. One complaint is that errors in Lopatin's original reconstruction invalidated his original calibration of TTI values to measured maturity. Later work (Waples 1980) has improved the status of the calibration, but more work is needed, particularly on hydrocarbon deadlines.

The success of any calibration of Lopatin's model will be critically dependent upon choosing a satisfactory mathematical relationship between the effects of time and temperature. On the basis of pyrolysis studies, most workers believe that oil generation is adequately described by a first-order kinetic expression, with the Arrhenius equation (1) defining the rate constant k. More research is needed to determine pseudoactivation energies and A factors of the different kerogen types. An effort should be made to integrate the model of Hood, and others (1975), in which E_a values change with maturity, into Lopatin's model. The main difficulty will be to make E_a dependent upon maturity rather than on temperature.

All the time-temperature models cited earlier seem to give good correspondences between measured and predicted maturity values. A plausible explanation for this success is that the temperature and time ranges over which oil generation occurs are too narrow to define clearly its temperature dependence (Snowdon 1979). The maximum range for oil generation in known examples is only about 80°C (60 to 140°C), with the majority lying between 80 and 110°C.

Much of the world's oil has been generated from rocks of Mesozoic age. In order to separate the effects of time and temperature, extreme cases containing very young or very old sediments or unusually high or low temperatures must be studied carefully. The Miocene-age source rocks of California and Paleozoic source rocks of the Williston Basin, for example, would be very useful in this regard.

The models must then be forced to conform to the data from such examples. Extreme examples have often been discarded in the past because results from them did not conform with expectations. Paleozoic sediments are a good example; poor correlations between predicted and measured values have often been explained by postulating changes in heat flow through time. Facile explanations for poor correlations may actually obscure fundamental problems with some of the models. The apparent success of all the models in most instances is comforting; we can generally get close to the right answer even if our understanding is not perfect. These modest accomplishments should not lull us into complacency about the ultimate accuracy of our models, however.

Uplift and erosion retard or stop the chemical reactions involved in organic maturation by lowering subsurface temperatures. A precise knowledge of the timing, duration, and amount of erosional removal is thus critical to the success of total-thermal-history models. Tectonic events that occurred when the source rocks were at their maximum paleotemperatures are most important; in most settings of interest to petroleum explorationists these are also the most recent and best-documented events. In cases where the exact magnitude of even recent tectonic events is poorly known, however, modeling may be problematic. It is then necessary to obtain measured maturity data to check the geologic reconstructions. When such data are not available Lopatin's method should be used with caution.

Compaction introduces a further complication. Although most workers have ignored compaction effects, Du Rouchet (1980) and Falvey and Deighton (1982) have published thermal maturation studies in which sediment compaction was considered. If compaction curves were identical in all basins, and if essentially all compaction occurred prior to the onset of oil generation, neglecting compaction would be of little consequence. In comparing areas of greatly different lithologies, however, particularly those in which overpressuring (undercompaction) occurs, it is conceivable that ignoring compaction might introduce a significant error in burial history.

Even where precise burial-history curves can be constructed, paleotemperatures can present problems. In many areas it may be reasonable to extrapolate modern gradients into the past on the basis of stable heat-flow models. In other areas, however, gradients certainly have changed through time. Provinces in which salt diapirism, permafrost formation, or intrusive volcanism have occurred are particularly difficult to model. Issler (1982) believes that the differences between present-day and past temperatures are often very large and essentially impossible to assess accurately. He therefore concludes that there are no TTI scales that in all cases define the oil-generation window, and that each basin must be calibrated individually. This view seems to me to be unduly pessimistic, because historical models for heat flow in sedimentary basins can be developed. Although accurate modeling of such changes requires a more thorough understanding of the geophysics of basin development than we are generally capable of at the present time, recent' work (e.g., McKenzie 1981; Ho and Sahai 1982; Cohen 1982) is encouraging.

Modern geothermal gradients are themselves often poorly known. Most measured gradients are interpolated from only a few subsurface temperatures, which in turn are normally unequilibrated measurements corrected by standard methods that may not always be satisfactory. In general, temperature data are by far the weakest part of any time-temperature model. Our greatest effort should be expanded to improve our knowledge of modern and ancient temperatures.

Waples' (1980) application of Lopatin's method to oil- and gas-preservation deadlines has created another important problem. His implicit assumption that destruction of oil and gas follows first-order kinetics with pseudoactivation parameters similar to those for oil generation is almost certainly not true. Oil destruction, in contrast to oil generation, involves numerous reactions having true activation energies that are rather similar to each other, because they are mainly hydrocarbon-cracking reactions. The pseudoactivation energy and A factors for the entire cracking process are therefore much closer to the true activation energies and A factors than are the kinetic parameters for oil generation. The high pseudoactivation energy for oil destruction compared to that for oil generation means that the kinetics of generation and destruction may have very different temperature dependences. The accurate application of Lopatin's method to oil and gas preservation deadlines thus requires a more sophisticated investigation than has yet been published.

APPLICATION TO HYDROCARBON EXPLORATION

Lopatin's method is now being rapidly adopted by many petroleum explorationists. The model is conceptually simple, yet reasonably satisfying from a theoretical point of view. It lends itself well to computerization, and can be applied in both frontier and heavily explored areas where deeper new plays are being considered. Required input data are minimal (time stratigraphy, amounts and timing of erosional removal, and geothermal history), and can be based either on data obtained from careful analyses or on pure speculation. The quality of the predictions from the model is, of course, a direct function of the quality of input data, but even highly speculative models can be extremely valuable in frontier areas. Varying the input parameters, such as geothermal gradient or amount of erosional removal, allows one to create both optimistic and pessimistic scenarios.

There have been several published applications of Lopatin's method, most of which have utilized Waples' (1980) recalibration and elaboration of generation and preservation deadlines. Zieglar and Spotts (1978), for example, used Waples' data prior to publication to explain variability of oil and gas occurrences in California's Great Valley. The large amount of Cretaceous sediment that accumulated in the Delta depocenter near Sacramento started hydrocarbon generation there in Late Cretaceous time (figure 7). Because the organic matter in the Cretaceous rocks is principally gas-prone kerogen, Cretaceous beds have been significant sources only for the gaseous hydrocarbons reservoired in Paleogene rocks. In contrast, the two depocenters in the southern Great Valley, which contain oil-prone kerogen in the Tertiary-age source beds, have generated liquid hydrocarbons, but only since the Miocene. The Tejon depocenter began to generate oil from Paleogene source beds nearly 20 my ago (figure 8), whereas in the Buttonwillow depocenter slower sediment accumulation retarded maturation, restricting oil generation to the Plio-Pleistocene. The authors suggested that the smaller oil reserves adjacent to the Buttonwillow depocenter may be a result of less-complete maturation than in the Tejon area.

Moshier and Waples (1982) superimposed TTI lines on cross sections through the drainage area of Alberta's Lower Cretaceous Mannville Group shales, which were thought to be possible source rocks for Alberta's vast heavy oil deposits (figure 9). By making burial-history profiles for several sites instead of a single point like Zieglar and Spotts (1978), they calculated the volume of sediment at various levels of thermal maturity more accurately, and showed that the Mannville shales alone were inadequate sources for the heavy-oil deposits.

Magoon and Claypool (in press) applied Lopitan's method to the Inigok-1 well on Alaska's North Slope in an effort to determine timing of oil and gas generation. The accuracy of their geologic reconstructions was verified by comparing measured

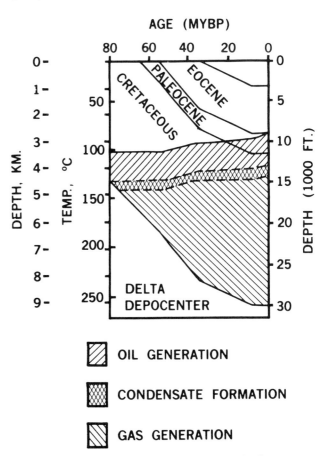

Figure 7.—Hydrocarbon-generation history in the deepest part of the Delta depocenter south of Sacramento, California. From Zieglar and Spotts (1978); reprinted by permission.

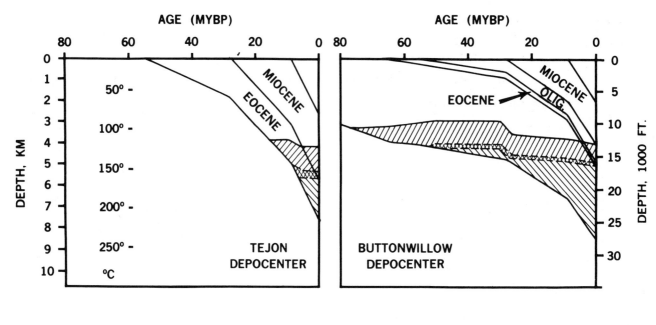

Figure 8.—Hydrocarbon-generation history in the deepest part of the Tejon and Buttonwillow depocenters, southern Great Valley, California. Horizontal and vertical scales and subsurface temperatures are the same in both diagrams. From Zieglar and Spotts (1978); reprinted by permission.

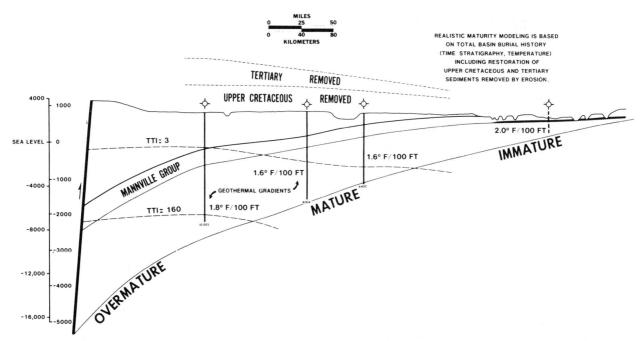

Figure 9.—Cross section from foothills to tar sands, Alberta, showing present-day TTI values through the basin.

maturity data with predicted maturities using a slight modification of Waples' (1980) recalibration of Lopatin's TTI values. From three possible temperature histories for the sediments they selected the one which best fitted the measured data.

Using the best temperature history and a burial-history diagram (figure 10), Magoon and Claypool showed that at Inigok, maturation commenced in the lowermost potential source rock (Shublik Fm.) about 80 mybp, and is still continuing today in the uppermost potential source rock (Pebble Shale Unit). The Shublik Fm. at Inigok is now in the early gas-generative phase.

Lopatin's method has also been used as one small part of comprehensive models of basin evolution. In Welte and Yukler's (1981) model the entire sedimentary history of a basin is modeled, including heat flow, sediment accumulation rates, physical properties of sediments, and organic matter type and quantity. Lopatin's method is used to predict the timing of hydrocarbon formation. Sluijk and Nederlof's (in press) model focuses on both generation and preservation of hydrocarbons, using Lopatin's method to predict both kerogen maturity and gas/oil ratios as cracking proceeds.

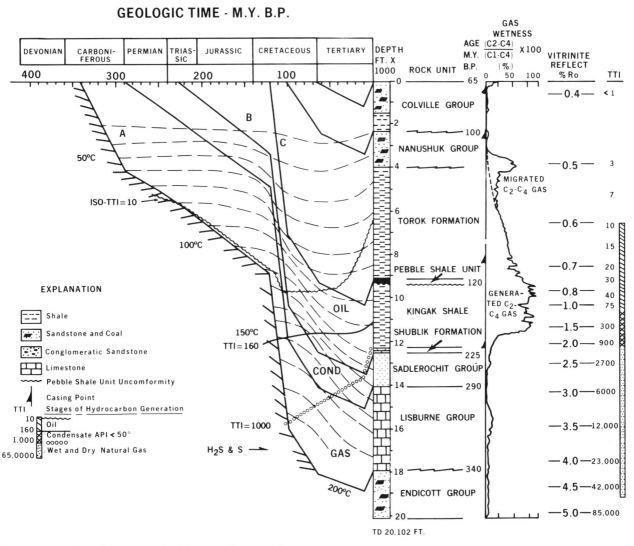

Figure 10.—Hydrocarbon-generation history and measured geochemical data at Inigok-1, North Slope, Alaska. From Magoon and Claypool, to be published.

THE FUTURE OF THERMAL MODELING

Over the last few years time-temperature models for oil generation have become accepted exploration tools within many oil companies. This trend is likely to continue at an accelerated pace in the future, but there will probably be some important changes in both the models and their applications.

REFINEMENT OF PRESENT MODELS

Lopatin's method is probably good enough and workable enough to serve as the basis for future modeling work. Several possible improvements in Lopatin's scheme have already been mentioned. Among the refinements most likely to be brought to fruition in the next few years are:

1. Relatively minor recalibrations will be made in the correlation between measured maturity parameters and TTI values.
2. Mathematical expressions interrelating the effects of time and temperature on the cracking of bitumen and petroleum will be developed. Integration of these formulas with those for oil generation will give a much firmer basis for estimating hydrocarbon-preservation deadlines.
3. Recalibration of hydrocarbon deadlines will probably occur as more data become available.
4. Better methods of predicting paleotemperatures from geophysical models will increase confidence in our knowledge of the temperature histories of rocks. Calibration of thermal models in areas where tectonic histories are well understood will permit evaluation of paleotemperatures in complex areas by forcing predicted maturity data to correspond to measured data.
5. Better methods will be developed for correcting measured, unequilibrated borehole temperatures.
6. Other mathematical expressions relating time and temperature during oil generation will be developed. Without improvements in our knowledge of thermal histories, however, these new models are unlikely to be significantly more accurate than Lopatin's original one.
7. Computerization will facilitate TTI calculations and will also permit the use of much smaller temperature intervals.

NOVEL APPLICATIONS

Zieglar and Spotts (1978), Waples (1980), and Magoon and Claypool (in press) have already acknowledged that Lopatin's model is extremely useful in determining the history of hydrocarbon generation. An understanding of this history is valuable in comparing timing of oil generation with timing of trap formation, trap tilting, trap breaching, and reservoir diagenesis.

Recently Moshier and Waples (1982) and Sluijk and Nederlof (in press) have placed calculated isomaturity lines on a cross section to facilitate making regional predictions about maturity levels. In figure 9, for example, a now-overmature pile of sediment near the center of the basin has generated oil. In following the most natural migration path (in this case laterally continuous sands of the Mannville Group), the oil has moved out of the mature region into a cooler zone along the structurally higher basin margin. This rapid and efficient movement out of the zone of generation permitted preservation of oil, for had it remained behind it would have quickly been cracked to gas. It is thus necessary to have good lines of migrational communication between the generation zone and the preservation zone. Any factors that hinder removal of the oil from the generative zone (such as sealing faults or laterally discontinuous conduits) will greatly reduce the oil-source potential of an area. Regional analyses integrating maturation studies with stratigraphy and structural geology are likely to become common and very important in the future.

It is also illuminating to follow basin development and maturation through time by means of isomaturity lines superimposed on cross sections at different times in the past. It is likely that in the near future it will be possible to model basin development through time on interactive computer terminals and watch maturation occur on the CRT. Three-dimensional modeling may even become feasible. It may also be possible to show the competition between generation and cracking and thus to gain a better understanding of the delicate balance that petroleum reservoirs strike between oil generation and destruction.

The future of total-thermal-history models is bright. Aided by computers and computer-graphics capabilities, Lopatin's method (or some slightly modified version) will become an important, routinely used tool among explorationists.

REFERENCES

Anderson, R., and Pack, R.W., 1915, Geology and oil resources of the west border of the San Joaquin Valley north of Coalinga, California: U.S. Geological Survey Bulletin 603.

Barker, C., 1974, Pyrolysis techniques for source-rock evaluation: American Association of Petroleum Geologists Bulletin, volume 58, pages 2349-2361.

Barker, C., and Kemp, M.K., 1982, Stability of natural gas at high temperatures, deep subsurface: American Association of Petroleum Geologists Bulletin, volume 66, page 545.

Barton, D.C., 1934, Natural history of the Gulf Coast crude oil, in Problems of Petroleum Geology, Wrather, W.E., and Lahee, F.H., eds., American Association of Petroleum Geologists, Tulsa, pages 109-155.

Bergius, F., 1913, Production of hydrogen from water and coal from cellulose at high temperatures and pressure: Journal of the Society of Chemical Industry, London, volume 32, pages 462-467.

Bostick, N.H., 1973, Time as a factor in thermal metamorphism of phytoclasts (coaly particles): Congres International de Stratigraphie et de Geologie du Carbonifere, Septieme, Krefeld, Compte Rendu, volume 2, Illinois State Geological Survey reprint series 1974-H, pages 183-193.

Bottcher, H., Teichmüller, M., and Teichmuller, R., 1949, Hilt's Law in the Bochum Trough of the Ruhr: Gluckauf, volume 85, pages 81-92 (in German).

Braun, R.L., and Rothman, A.J., 1975, Oil-shale pyrolysis: kinetics and mechanism of oil production: Fuel, volume 54, pages 129-131.

Brooks, B.T., 1948, Active-surface catalysts in formation of petroleum: American Association of Petroleum Geologists Bulletin, volume 32, pages 2269-2286.

Cane, R.F., 1950, The mechanism of the pyrolysis of torbanite: Proceedings of the Second Oil Shale and Cannel Coal Conference, Glasgow, pages 592-604.

Cohen, C.R., 1982, Model for a passive to active continental margin transition: implications for hydrocarbon exploration: American Association of Petroleum Geologists Bulletin, volume 66, pages 708-718.

Connan, J., 1974, Time-temperature relation in oil genesis: American Association of Petroleum Geologists Bulletin, volume 58, pages 2516-2521.

Cummins, J.J., and Robinson, W.E., 1972, Thermal degradation of Green River kerogen at 150° to 350°C: U.S. Bureau of Mines Report of Investigations 7620.

Deroo, G., Durand, B., Espitalié, J., Pelet, R., and Tissot, B., 1969, Possible application of mathematical models of petroleum formation in sedimentary basins, in Advances in Organic Geochemistry 1968, Schenck, P.A., and Havenaar, I., eds., Pergamon Press, pages 345-354 (in French).

Dow, W.G., 1977, Kerogen studies and geological interpretations: Journal of Geochemical Exploration, volume 7, pages 79-99.

Du Rouchet, J., 1980, The program DIAGEN, two methods for considering the chemical evolution of organic matter: Bulletin Centre Recherche Exploration-Production Elf-Aquitaine, volume 4, pages 813-831 (in French).

Durand, B., and Espitalié, J., 1973, Evolution of organic matter in the course of burial of sediments: Academie des Sciences Paris, Comptes Rendus, Serie D, volume 276, pages 2253-2256 (in French).

Engler, K.O.V., 1913, Die Chemie und Physik des Erdols, volume 1, S. Hirzel, Leipzig (in German).

Erdmann, E., 1924, The genetic connection of brown coal and hard coal, based on new studies: Brennstoff-Chemie, volume 5, pages 177-186 (in German).

Falvey, D.A., and Deighton, I., 1982, Recent advances in burial and thermal geohistory analysis: Journal of the Australian Petroleum Exploration Association, volume 22, pages 65-81.

Forsman, J.P., and Hunt, J.M., 1958, Insoluble organic matter (kerogen) in sedimentary rocks of marine origin, in Habitat of Oil, Weeks, L.G., ed., American Association of Petroleum Geologists, Tulsa, pages 747-778.

Franks, A.J., and Goodier, B.D., 1922, Preliminary study of the organic matter of Colorado oil shales: Colorado School of Mines Quarterly, volume 17, Supplement A, pages 3-16.

Fuchs, W., and Sandhoff, A.G., 1942, Theory of coal pyrolysis: Industrial and Engineering Chemistry, volume 34, pages 567-571.

Fuller, M.L., 1919, Relation of oil to carbon ratios of Pennsylvanian coals in north Texas: Economic Geology, volume 14, pages 536-542.

Golitsyn, M.V., 1973, The duration of the process of coal metamorphism: Izvestiya Akademii Nauk USSR, Seriya Geologicheskaya, number 8, pages 90-97 (in Russian).

Gretener, P.E., and Curtis, C.D., 1982, On the role of temperature and time in organic metamorphism: American Association of Petroleum Geologists Bulletin, volume 66, pages 1124-1129.

Hanbaba, P., and Jüntgen, H., 1969, The applicability of laboratory investigations of the geochemical processes of gas formation from hard coal, and the influence of oxygen on gas formation, in Advances in Organic Geochemistry 1968, Schenck, P.A., and Havenaar, I., eds., Pergamon Press, pages 459-471 (in German).

Heck, E.T., 1943, Regional metamorphism of coal in southeastern West Virginia: American Association of Petroleum Geologists Bulletin, volume 27, pages 1194-1227.

Hill, G.R., and Dougan, P., 1967, The characteristics of a low temperature in situ shale oil: Colorado School of Mines Quarterly, volume 62, number 3, pages 75-90.

Hilt, C., 1973, The relationships between the composition and technical properties of hard coals: Zeitschrift des Vereins Deutscher Ingenieure, volume 17, pages 193-202 (in German).

Hlauschek, H., 1950, Roumanian crude oils: American Association of Petroleum Geologists Bulletin, volume 34, pages 755-781.

Ho, T.T.Y., and Sahai, S.K., 1982, Estimation of organic maturation from seismic and heat-flow data: American Association of Petroleum Geologists Bulletin, volume 66, pages 581-582.

Hood, A., Gutjahr, C.C.M., and Heacock, R.L., 1975, Organic metamorphism and the generation of petroleum: American Association of Petroleum Geologists Bulletin, volume 59, pages 986-996.

Horsfield, B., and Douglas, A.G., 1980, The influence of minerals on the pyrolysis of kerogens: Geochimica et Cosmochimica Acta, volume 44, pages 1119-1131.

Hubbard, A.B., and Robinson, W.E., 1950, A thermal decomposition study of Colorado oil shale: U.S. Bureau of Mines Report of Investigations 4744.

Huck, G., and Karweil, J., 1955, Physical chemical problems of coalification: Brennstoff-Chemie, volume 36, pages 1-11 (in German).

Hunt, J.M., 1975, Is there a geochemical depth limit for hydrocarbons?: Petroleum Engineer, volume 47, number 3, pages 112, 116, 118, 120, 123, and 127.

Hunt, J.M., and Jamieson, G.W., 1956, Oil and organic matter in source rocks of petroleum: American Association of Petroleum Geologists Bulletin, volume 40, pages 477-488.

Issler, D., 1982, Calculation of organic maturation levels from downhole temperature/burial history curves: unpublished Bachelor of Science Honors thesis, Department of Earth Sciences, University of Waterloo, Canada.

Jones, I.W., 1928, Carbon ratios as an index of oil and gas in western Canada: Economic Geology, volume 23, pages 353-380.

Jüntgen, H., and Klein, J., 1975, Formation of natural gas from coaly sediments: Erdol und Kohle-Erdgas-Petrochemie, volume 28, pages 65-73 (in German).

Karrick, L.C., 1926, Manual of testing methods for oil shale and shale oil: U.S. Bureau of Mines Bulletin 249.

Karweil, J., 1956, The metamorphosis of coals from the standpoint of physical chemistry: Zeitschrift der Geologischen Gesellschaft, volume 107, pages 132-139 (in German).

Kettel, D., 1981, Maturity calculations for the Upper Carboniferous of northwest Germany—a test of several methods: Erdol Erdgas Zeitschrift, volume 97, pages 395-404 (in German).

Leythaeuser, D., Hagemann, H.W., Hollerbach, A., and Schaefer, R.G., 1980, Hydrocarbon generation in source beds as function of type and maturation of their organic matter: a mass balance approach: Proceedings Tenth World Petroleum Congress, Bucharest, volume 2, pages 31-41.

Lopatin, N.V., 1969a, The main phase of oil generation, Izvestiya Akademii Nauk USSR, Seriya Geologicheskaya, number 5, pages 69-76 (in Russian).

―――― 1969b, The role of geologic time in carbonization processes of coals: Vestnik Moscow University, Seriya 4, Geology, number 1, pages 95-98 (in Russian).

―――― 1971, Temperature and geologic time as factors in coalification: Izvestiya Akademii Nauk USSR, Seriya Geologicheskaya, number 3, pages 95-106 (in Russian).

―――― 1976, Determination of the influence of temperature and geologic time on the catagenetic processes of coalification and oil-gas generation, in Issledovaniya Organicheskogo Veshchestva Sovremennykh i Iskopaemykh Osadkov, Otdel'nye Ottiski, Nauka, Moscow, pages 361-366 (in Russian).

Lopatin, N.V., and Bostick, N.H., 1973, The geologic factors in coalification, in Priroda Organicheskogo Veshchestva Sovremennykh i Iskopaemykh Osadkov, Nauka, Moscow, pages 79-90 (in Russian).

Louis, M., and Tissot, B., 1967, Influence of temperature and pressure on hydrocarbon formation from kerogen in shales: Proceedings Seventh World Petroleum Congress, Mexico, volume 2, pages 47-60 (in French).

Magoon, L.B., and Claypool, G.E., in press, Petroleum geochemistry of the North Slope of Alaska—time and degree of thermal maturity, in Advances in Organic Geochemistry 1981, Bjoroy, M., ed., Pergamon Press.

Maier, C.G., and Zimmerley, S.R., 1924, The chemical dynamics of the transformation of the organic matter to bitumen in oil shale: University of Utah Bulletin number 14, pages 62-81.

Marcusson, J., 1908, The optically active components of petroleum:

Chemiker-Zeitung, number 30, pages 377-378 (in German).

McCoy, A.W., 1921, A short sketch of the paleogeography and historical geology of the mid-continent oil district and its importance to petroleum geology: American Association of Petroleum Geologists Bulletin, volume 6, pages 541-584.

McKee, R.H., and Lyder, E.E., 1921, Thermal decomposition of shales. I—Heat effects: Industrial and Engineering Chemistry Journal, volume 13, pages 613-618.

McKenzie, D., 1981, The variation of temperature with time and hydrocarbon maturation in sedimentary basins formed by extension: Earth and Planetary Science Letters, volume 55, pages 87-98.

McNab, J.G., Smith, P.V., Jr., and Betts, R.L., 1952, The evolution of petroleum: Industrial and Engineering Chemistry, volume 44, pages 2556-2563.

Moshier, S.O., and Waples, D.W., 1982, Was the Mannville Group the source for Alberta's heavy oils?: American Association of Petroleum Geologists Bulletin, volume 66, page 610.

Nagornyi, V.N., and Nagornyi, Yu.N., 1974, The question of the quantitative evaluation of the role of time in processes of regional metamorphism of coals: Solid Fuel Chemistry, volume 8, number 4, pages 30-36.

Neruchev, S.G., and Parparova, G.M., 1972a, The role of geologic time in the process of metamorphism of coal and dispersed organic matter in rocks: Akademiya Nauk USSR, Sibirskoe Otdelenie, Geologiya i Geofizika, number 10, pages 3-10 (in Russian).

——— 1972b, Depth zonality of the metamorphism of coals and organic matter in rocks: Akademiya Nauk USSR, Sibirskoe Otdelenie, Geologiya i Geofizika, number 9, pages 28-36 (in Russian).

Peters, K.E., Ishiwatari, R., and Kaplan, I.R., 1977, Color of kerogen as index of organic maturity: American Association of Petroleum Geologists Bulletin, volume 61, pages 504-510.

Philippi, G.T., 1965, On the depth, time, and mechanism of petroleum generation: Geochimica et Cosmochimica Acta, volume 29, pages 1021-1049.

Potonié, H., 1910, Die Entstehung der Steinkohle, Gebruder Borntrager, Berlin (in German).

Powell, T.G., Foscolos, A.E., Gunther, P.R., and Snowdon, L.R., 1978, Diagenesis of organic matter and fine clay minerals: a comparative study: Geochimica et Cosmochimica Acta, volume 42, pages 1181-1197.

Pratt, W.E., 1934, Hydrogenation and the origin of oil, in Problems of Petroleum Geology, Wrather, W.E., and Lahee, F.H., eds., American Association of Petroleum Geologists, Tulsa, pages 235-245.

Price, L.C., 1982, Time as factor in organic metamorphism and use of vitrinite reflectance as an absolute paleogeothermometer: American Association of Petroleum Geologists Bulletin, volume 66, pages 619-620.

Radke, M., Schaefer, R.G., Leythaeuser, D., and Teichmüller, M., 1980, Composition of soluble organic matter in coals: relation to rank and liptinite fluorescence: Geochimica et Cosmochimica Acta, volume 44, pages 1787-1800.

Reeves, F., 1928, The carbon-ratio theory in the light of Hilt's Law: American Association of Petroleum Geologists Bulletin, volume 12, pages 795-823.

Roberts, J., 1924, The origin of anthracite: Proceedings of the South Wales Institute of Engineering, volume 40, pages 97-138.

Rogers, H.D., 1843, An inquiry into the origin of the Appalachian coal strata, bituminous and anthracitic: Proceedings and Transactions of the Association of American Geologists and Naturalists, Gould, Kendall, and Lincoln, Boston, pages 433-474.

——— 1860, On the distribution and probable origin of the petroleum, or rock oil, of western Pennsylvania: Proceedings of the Royal Philosophical Society of Glasgow, volume 4, pages 355-359.

Russell, W.L., 1927, The proofs of the carbon-ratio theory: American Association of Petroleum Geologists Bulletin, volume 11, pages 977-989.

Schopf, J.M., 1949, Cannel, boghead, torbanite, oil shale: Economic Geology, volume 44, pages 68-71.

Shimoyama, A., and Johns, W.D., 1971, Catalytic conversion of fatty acids to petroleumlike paraffins and their maturation: Nature Physical Science, volume 232, pages 140-144.

Sluijk, D., and Nederlof, M.H., in press, Geochemical evaluation of sedimentary basins, in American Association of Petroleum Geologists Memoir, Demaison, G.J, and Murris, R.J., eds.

Smith, P.V., Jr., 1954, Studies on origin of petroleum: occurrence of hydrocarbons in recent sediments: American Association of Petroleum Geologists Bulletin, volume 38, pages 377-404.

Snowdon, L.R., 1979, Errors in extrapolation of experimental kinetic parameters to organic geochemical systems: American Association of Petroleum Geologists Bulletin, volume 63, pages 1128-1138.

Stone, H.N., Batchelor, J.D., and Johnstone, H.F., 1954, Low temperature carbonization rates in a fluidized bed: Industrial and Engineering Chemistry, Engineering and Process Development, volume 46, pages 274-278.

Teichmüller, M., 1971, Application of coal-petrographic methods in petroleum and natural-gas prospecting: Erdol und Kohle-Erdgas-Petrochemie, volume 24, pages 69-76 (in German).

Teichmüller, M., and Teichmüller, R., 1951, Questions on coalification in the Osnabruck region: Neues Jahrbuch fur Geologie und Palaeontologie, volume 93, pages 69-85 (in German).

Tissot, B., 1969, First data on the mechanisms and kinetics of the formation of petroleum in sediments: Revue de l'Institut Francais du Petrole, volume 24, pages 470-501 (in French).

Tissot, B., and Espitalié, J., 1975, Thermal evolution of organic material in sediments: applications of a mathematical simulation: Revue de l'Institut Francais du Petrole, volume 30, pages 743-777 (in French).

Tissot, B., Durand, B., Espitalié, J., and Combaz, A., 1974, Influence of nature and diagenesis of organic matter in formation of petroleum: American Association of Petroleum Geologists Bulletin, volume 258, pages 499-506.

Tissot, B., Deroo, G., and Espitalié, J., 1975, Comparative study of the duration of formation and expulsion of oil in several geological provinces: Proceedings Ninth World Petroleum Congress, Tokyo, volume 2, pages 159-169 (in French).

Treibs, A., 1936, Chlorophyll and hemin derivatives in organic mineral substances: Angewandte Chemie, volume 49, pages 682-686 (in German); translation in Geochemistry of Organic Molecules, Kvenvolden, K.A., ed., Dowden, Hutchinson, and Ross, Stroudsburg, Pennsylvania, pages 17-25 (1980).

van Krevelen, D.W., 1950, Graphical-statistical method for the study of structure and reaction processes of coal: Fuel, volume 29, pages 269-284.

——— 1952, Some chemical aspects of coal genesis and coal structure, in Troisieme Congres pour l'Avancement des Etudes de Stratigraphie et du Geologie du Carbonifere, Compte Rendu, Heerlen, volume 1, pages 359-368.

van Krevelen, D.W., van Heerden, C., and Hüntjens, F.J., 1951, Physicochemical aspects of the pyrolysis of coal and related organic compounds: Fuel, volume 30, pages 253-259.

Waples, D.W., 1980, Time and temperature in petroleum formation: application of Lopatin's method to petroleum exploration: American Association of Petroleum Geologists Bulletin, volume 64, pages 916-926.

Weitkamp, A.W., and Gutberlet, L.C., 1970, Application of a microretort to problems in shale pyrolysis: Industrial and Engineering Chemistry Process Design Development, volume 9, pages 386-395.

Welte, D.H., and Yukler, M.A., 1981, Petroleum origin and accumulation in basin evolution—a quantitative model: American Association of Petroleum Geologists Bulletin, volume 65, pages 1387-1396.

White, D., 1908, Some problems of the formation of coal: Economic Geology, volume 3, pages 292-318.

——— 1913, Regional metamorphism of coal, in The Origin of Coal, White, D., and Thiessen, R., eds., U.S. Bureau of Mines Bulletin 38, pages 91-130.

——— 1915, Some relations in origin between coal and petroleum: Journal of the Washington Academy of Sciences, volume 5, pages 189-212.

——— 1935, Metamorphism of organic sediments and derived oils: American Association of Petroleum Geologists Bulletin, volume 19, pages 589-617.

Zieglar, D.L., and Spotts, J.H., 1978, Reservoir and source bed history of Great Valley, California: American Association of Petroleum Geologists Bulletin, volume 62, pages 813-826.

The Generative Basin Concept

Gerard Demaison
*Chevron Overseas Petroleum Inc.
San Francisco, California*

Recent progress in organic geochemical and other fields of earth sciences has made feasible the development of methods for evaluating the likeliness of hydrocarbon occurrence in an undrilled trap. A fundamental step in hydrocarbon charge prediction is to determine whether an undrilled trap has had access to hydrocarbon migration from mature source rocks.

Areas underlain by mature source rocks are called "petroleum generative depressions" or "hydrocarbon kitchens." A "generative basin" is defined as a sedimentary basin that contains one or more petroleum generative depressions. Mapping generative depressions is achieved by integrating geochemical data relevant to maturation and organic facies with structural and stratigraphic information derived from seismic surveys and deep wells.

Locales of high success ratios in finding petroleum are called "areas of high potential," "plays," or "petroleum zones." A rapid worldwide review of 12 sedimentary basins, described in order of geotectonic style, reveals the following regularities:

1. The zones of concentrated petroleum occurrence ("areas of high potential") and high success ratios are genetically related to oil generative depressions or basins. These depressions are mappable by integrated methods (geology, geophysics, and geochemistry).

2. The largest petroleum accumulations tend to be located close to the center of the generative basins or on structurally high trends neighboring deep generative depressions.

3. Migration distances commonly range in tens rather than hundreds of miles and are limited by the drainage areas of individual structures. Thus the outlines of generative depressions commonly include most of the producible hydrocarbon accumulations and the largest fields. Unusual cases of long distance migration are documented on certain foreland basin plates where stratigraphy and structure permitted uninterrupted updip movement of oil.

These three regularities provide powerful analogs for forecasting areas of high petroleum potential in undrilled or sparsely drilled basins.

It is important to keep in mind, however, that the predictive accuracy of geochemical mapping systems is dictated by the available data base. Hence, at early stages of exploration, geologists should guard from overtaxing geological and geochemical knowledge, and remain alert to the potentialities of the unknown.

INTRODUCTION

Since the turn of the century, one of the basic rules for finding petroleum has been to drill anticlinal structures in sedimentary basins. Straightforward application of the anticlinal theory has been totally successful: over 95 of the world's producible petroleum has been found stored in structural traps. Moreover, since the 1930s, most oil-bearing structural traps have been detected by geophysical methods. Seismic applications, have now reached a level of sophistication permitting the detection of traps in unusually difficult geologic conditions, thus opening new frontiers to exploration.

However, most explored anticlinal structures are found barren of hydrocarbons when drilled. Furthermore it has been observed that in some basins or portions of basins virtually all anticlinal traps are found barren of hydrocarbons although reservoirs and seals are present.

Lastly, the producible petroleum reserves of explored sedimentary basins follow an areal distribution which obeys temporal and paleolatitudinal considerations more than geotectonic style (Bois, Bouche, and Pelet, 1982). Bally (1980) even reached the earlier conclusion that "the classification of basins does little to improve our hydrocarbon volume forecasting ability." These observations point to regional inequalities in petroleum distribution caused by geologic factors other than just structure and reservoirs.

Geologic rationalizations heard in the 1950s and 1960s to explain the occurrence of barren traps and petroleum-poor basins were: 1) The oil and gas were once entrapped but have escaped through the seals, given geologic time; and 2) The oil has been "flushed out" by water movement.

Underlying these rationalizations were also the opinions that: 1) Entrapped oil can dissipate from the subsurface without leaving any residual traces; 2) Any dark shale is, a source rock; 3) Oil generation can occur at very shallow depth; and 4) There are neither limits nor restraints to migration distances in sedimentary basins.

These opinions supported the working concept that petroleum was ubiquitous in the subsurface, either because it was generated everywhere or because it would migrate anywhere. This optimistic outlook was necessary and useful in view of our lack of knowledge of petroleum formation mechanisms. Given such uncertainties, there was no alternative to systematic assumption of hydrocarbon presence in undrilled traps. This positive stand was an essential safeguard against the risk of prematurely condemning of viable exploration plays.

Since the mid-1960s, scientific breakthroughs in understanding petroleum formation and destruction processes have reduced many areas of uncertainty in petroleum geology. We owe the new perspectives to the post-war emergence of analytical techniques such as gas chromatography and mass spectrometry: they made possible the observation of geologic phenomena that, hitherto had been only a matter of speculation and controversy.

Because of these recent post-war technological advances, the following concepts are now recognized as useful in petroleum geology:

1. Evaporites are most efficient seals mainly because they offer very little or no pore space; however, the long term sealing properties of very fine grained, water-wet, porous rocks such as shales are also remarkably efficient *in the absence of open fractures.* This is due to the displacement pressure barrier effect created by the capillary pressure between oil and water in rock pores (Berg, 1975, Schowalter, 1976). Long-term sealing properties of very fine grained water-wet rocks are demonstrated by the excellent preservation of light oil and gas reserves in some very old sedimentary basins. For instance, shallow Paleozoic oil and gas in the Illinois, Michigan and Appalachian basins, major reserves in the Paleozoic Volga-Ural Basin (USSR) and giant Devonian and Ordovician fields in the southern Algerian Sahara demonstrate the sealing efficiency of very low permeability rocks, provided geologic history following entrapment has remained quiescent. All the above basins feature stable tectonic conditions and a lack of adverse thermal history since petroleum generation took place (during Late Paleozoic time, U.S.A. and Volga-Ural; and Cretaceous time, Algeria).

Rationalizations for barren traps and petroleum-poor basins by diffusion of *oil* through apparently tight seals, are difficult to reconcile with these geological observations, particularly where rocks have not suffered excessive thermal or tectonic stresses. These rationalizations also contradict theoretical considerations and laboratory experiments demonstrating that entry pressures are independent of time (Schowalter, 1976). Because of high solubility in water, *natural gas* diffusion through undisturbed seals is possible, and has been observed in certain geologic settings (Leythauser et al, 1982).

Leakage from an oil accumulation through disturbed or poor quality seals, or in-situ degradation of entrapped oil, leaves residual traces in a reservoir because oil-wetting of the mineral grains does occur when oil saturations reach a level capable of sustaining water-free production (Schowalter and Hess, 1982). These traces can be observed by geochemical techniques or, in most cases, by modern hydrocarbon show detection methods as used in mud logging. *Hence, it is an inescapable conclusion that truly barren traps are found empty because no oil ever filled the reservoirs to begin with.*

2. Prolific oil producing basins, when geochemically evaluated, are shown to contain at least one adequately mature, deeply buried source rock system. It is often stratigraphically widespread and was deposited in an oxygen-depleted environment. Conversely, the bulk of dark, fine-grained sediments, in the sedimentary record, are *not* source rocks, and were deposited under oxygen-rich water, as in most of today's world oceans and lakes (Demaison et al, 1983).

3. Petroleum generation results from the transformation of kerogen (a solid organic substance) in the deep subsurface under the influence of both subsurface temperature and geologic time (Tissot and Welte, 1978).

4. Except in unusual cases of very long-range migration typically encountered on foreland basin plates, most entrapped oil in sedimentary basins originates from synclinal drainage areas that surround the trap itself. Thus migration distances commonly range in tens rather than hundreds of miles, particularly in strongly structured and/or faulted basins.

GEOLOGIC RISK REDUCTION BY GEOCHEMICAL METHODS

Exploration risk, being defined as the probability of spending exploration funds without economic success, has always been at the heart of the oil business. Geologic risk, which is a part of overall exploration risk, is fueled by uncertainties in subsurface geologic conditions, prior to drilling. It can also be expressed in terms of the probability of simultaneous occurrence of the key factors that determine the habitat of oil and gas in the subsurface.

Successful exploration for producible hydrocarbons in the subsurface depends on satisfying the following probabilities: 1) probability of existence of a trap (structure × reservoir × seal); 2) probability that the trap has received and physically retained a petroleum charge (source × maturation × migration paths × timing); and 3) probability that the entrapped petroleum has been preserved from the effects of thermal or bacterial degradation (temperature regime × meteoric water ingress).

Since these three main probabilities are independent of each other, the overall probability of discovering producible hydrocarbons at a given location is the product (not the sum) of the probabilities of these individual factors, that is, if any one of these three main factors is 0, the overall probability of success is 0, regardless of how favorable the other two remaining factors are.

Common sense agrees with probability concepts: no geologist will recommend drilling a syncline or a section void of reservoirs. However equaliy high risks, in a mathematical sense, are often taken by explorationists with respect to oil charge and oil preservation probabilities.

This is commonly the case if geochemical information is absent, poorly integrated, or ignored. In the face of uncertainty or skepticism regarding source and degradation regimes, explorationists tend to rely mainly on structural information to evaluate geologic risk prior to drilling. This exploration approach is justifiable when applied to proven petroliferous basins or plays. However, it can be improved by also taking into account measurable parameters relevant to oil and gas generation, migration, and destruction. This integrated approach is particularly effective in frontier basins, deep gas plays, and offshore environments, where exploration costs are inordinately high.

THE GENERATIVE BASIN CONCEPT

A fundamental step in the prediction of petroleum occurrence by geochemical methods is evaluation of whether an undrilled trap has had access to hydrocarbon migration from mature petroleum source beds. This statement holds the essence of the "generative basin concept."

Areas underlain by hydrocarbon generative source beds are called "petroleum generative depressions" or "hydrocarbon kitchens." The term "generative basin" describes a sedimentary basin containing one or more petroleum generating depression.

Recognition of generative depressions is achieved by overlaying organic facies maps and maturation maps of each key petroleum source horizon. Maturation maps are compiled from seismic depth maps, near the potential source horizons, and from maturation gradients derived from well data and calibrated time-temperature models (Waples, 1983). Organic facies maps reflect the stratigraphic distribution of organic matter types within a given source rock unit. They are compiled by integrating kerogen type data into the known paleogeographic and paleo-oceanographic context (Demaison et al, 1983).

The geochemical approach, in prospect appraisal, begins by investigating whether mature source beds are present in the drainage area of a trap. A further step consists of mapping areas of mature source beds and calculating both mature source rock volumes and petroleum yield. Lastly, migration pathways can be modeled between the mature source-rocks and the trap. This type of geologic exercise permits a ranking of prospects by the criterion of degree-of-access to mature source rocks.

The geochemical approach to basin evaluation consists of mapping oil generative depressions or basins and erecting a matrix of drilling success ratios, volumes of discovered hydrocarbons and "kitchen" potential. When these correlations have been established they may be used for comparative purposes and for future evaluation of geologic risk. Application of the "generative basin concept," leading to recognition and prediction of areas of high potential, is the object of this contribution.

RECOGNITION AND PREDICTION OF AREAS OF HIGH POTENTIAL

Historically recognized zones of high success ratios in finding petroleum in sedimentary basins are called "areas of high potential," "plays," "fair-ways," or "petroleum zones." Geochemical data, when mapped together with geology and structure derived from seismic information, permit early geographic and geologic delineation of these areas of high potential. Moreover, the generative depression mapping approach leads to a more realistic evaluation of geologic risk prior to drilling, than the blanket application of worldwide success ratios or arbitrary risk factors.

The following illustrations are meant to demonstrate that basin-wide geochemical mapping can help identify areas of high potential in sedimentary basins. The subject basins are rapidly reviewed in order of tectonic style.

RIFT BASINS

North Sea Basin

The heavily explored portion of the North Sea between the 55th and 62nd parallels is a major oil and gas province with estimated reserves of over 24 billion barrels of oil and over 55 Tcf of gas (Figure 1). Reservoirs range in age from Devonian to Eocene typically producing on tilted fault blocks and, to a lesser degree, on salt swells.

The principal oil sourcing system is of Late Jurassic Kimmeridgian-Volgian age. It is thermally mature and actively generating and expelling oil at formation temperatures higher than 200°F (93°C) corresponding to depths in excess of 10,000 ft (3,048 m). The generated oil moves to the nearest available reservoirs: older, fault-juxtaposed Middle Jurassic sands in the northern North Sea (Viking Graben); and Upper Cretaceous chalk in the southern North Sea (Ekofisk area and "Tail-end" graben).

A maturation map of the Kimmeridgian shale permits one to delineate both immature areas where temperatures are lower than about 200°F (93 °C) and no significant oil generation has taken place, and mature oil generative depressions (where present-day earth subsurface temperature exceeds 200°F [93°C]). This corresponds approximately to a vitrinite reflectance level of 0.6.

A statistical count of all dry holes and successful holes in the North Sea shows that: 1) Virtually all the oil and gas fields lie within or very near the oil generative depression containing mature Kimmeridgian source rocks; 2) The historical success ratio in the mature source fairway, or generative basin, is in the order of 1 in 3; 3) Outside the Kimmeridgian generative depression the historical success ratio is in the order of 1 in 30. The oil fields found thus far in this higher risk area (Beatrice, Briesling and Bream) are probably sourced from Middle or Lower Jurassic beds. The crude oils are different in composition from Kimmeridgian-sourced crudes; 4) Migration distances in the North Sea are commonly short and limited by the drainage areas of individual structures; and 5) The fields with the largest reserves tend to be close to the center of the generative depression (Statfjord, Piper, Forties, Ekofisk), in close proximity to the most thermally mature and thickest

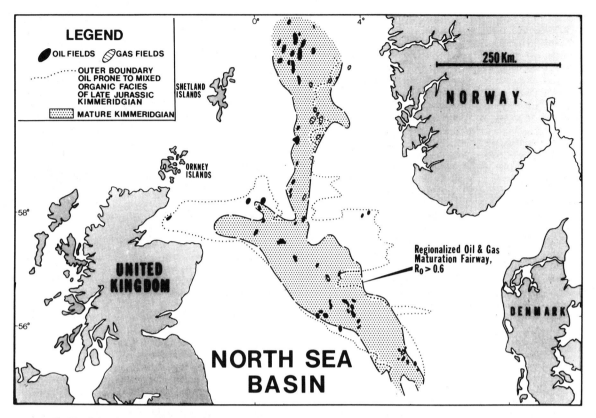

Figure 1. The North Sea Basin (Ziegler, 1980; Demaison et al, 1984).

Kimmeridgian shale depocenters.

Syrte Basin

The formline structure map at the base of the Upper Cretaceous in the Syrte Basin and Western Desert indicates that the oil fields are closely associated with regional structural depressions where Cretaceous and Paleogene source rocks are buried below about 2,500 m (8,202 ft) (Figure 2). Most of the largest fields (Sarir, Amal, Nafoora, Gialo) tend to occur close to the deepest parts of the generative depressions. Other large fields (Zelten, Waha, Defa, Dahra-Hofra) are on high horst blocks but in close proximity to deeply buried generative depressions. Oil occurrence in the western desert of Egypt relates to generative depressions involving Jurassic and lowermost Cretaceous source rocks to the north of the Quattara Ridge and Upper Cretaceous source rocks in the Abu Gharadig Basin (Parker, 1982).

Gippsland Basin

In the Gippsland Basin (Figure 3) only a relatively limited volume of source beds (Eocene, Paleocene and perhaps Upper Cretaceous paralic coal measures) is within the generative window, hence major oil and gas generation is confined to the central part of the basin. The oil fields (3 of them giants) are located close to the thickest and deepest hydrocarbon kitchens, within the oil window (Ro 0.6). The gas fields are located on the outer edge of the generative depression, in a fairway comprised between 0.45 and 0.6 vitrinite reflectance. However, carbon isotope data for the methanes of these gas fields are not available, therefore it is not known whether the gas is early or late thermal. Success ratios are spectacularly high in the oil and gas fairways which clearly coincide with source maturation patterns.

INTRACRATONIC DOWNWARPS

Western Siberian Basin

This mega-basin hosts the major share of Soviet oil reserves and about 6% of the world's reserves. It is a site of active exploration and production (Figure 4). The main source system is the Bhazenov Formation, a widespread laminated black shale of Late Jurassic age which is nearly synchronous with the Late Jurassic source beds of the North Sea. They are closely related to the same Late Jurassic "oceanic anoxic event."

It should be noted that the supergiant oil fields (Samotlor, Ust-Balisk, Surgut) are all located close to the center of this very large generative basin. The origin of the

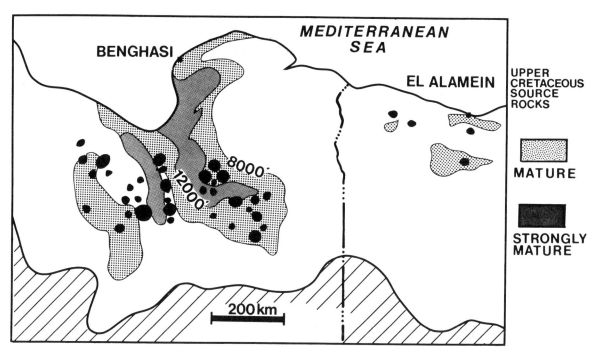

Figure 2. Syrte Basin, North Africa (Parsons et al, 1980; Parker, 1982).

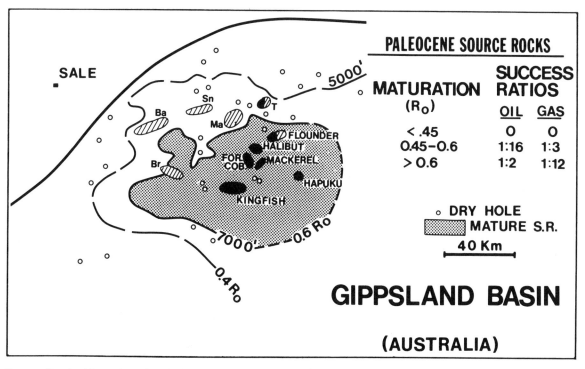

Figure 3. Gippsland Basin, Australia (Threllfall, Brown, and Griffith, 1976; Kantsler et al, 1978; Shibaoka, Saxby, and Taylor, 1978).

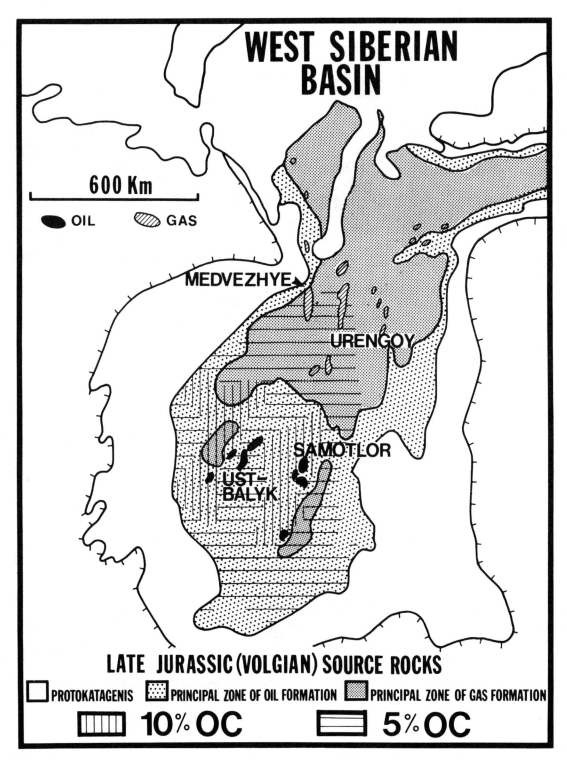

Figure 4. Western Siberia, USSR (Kontorovich, 1971, 1983).

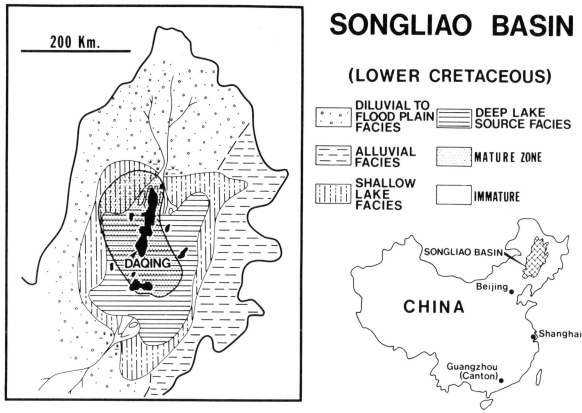

Figure 5. Songliao Basin, Peoples Republic of China (Wan Shangwen, Hu Wenhai, and Tan Shidian, 1982).

gas in the giant fields (for example, Urengoy) in the north is still controversial and could be partly "biogenic."

Songliao Basin

The Songliao Basin (Figure 5) is a large intracratonic rift-style basin in eastern China. It is filled with non-marine fluviatile to lacustrine sediments of Cretaceous to Tertiary age. The source beds are Early Cretaceous lacustrine, highly oil-prone, black shales. They are thermally mature and oil generative in the central part of the basin. The combination of deep lake facies, in Lower Cretaceous time, and favorable thermal maturation levels, geographically outlines that area containing the bulk of the oil reserves in the Songliao Basin. Geologists from the PRC have recently expressed the view that the oil generating depressions contain over 80 percent of the oil reserves of Songliao, Domying and other depressions of East China (Hu, 1983).

The largest oil-bearing structural complex is the giant Daqing oil field. It is located immediately updip from the center of the generative basin, the Pijia-Gulong generative depression. The giant Daqing field complex is believed to be the largest oil accumulation in existence in a nonmarine basin, anywhere in the world. It accounts for 80 percent of the total discovered oil in place in the Songliao Basin. The maximum distance of hydrocarbon migration is less than 25 mi (40 km).

Illinois Basin

A generalized isoreflectance (maturation) contour map on top of the Lower Carboniferous (New Albany) oil source shale sequence in the Illinois Basin (U.S.A.) shows more than 3/4 of the basin's oil occurring in the Lower Carboniferous (Mississippian) above the New Albany shale (Figure 6). Also, the bulk of Devonian and Silurian oil yet found in the Illinois Basin occurs within a few hundred feet of the base of the New Albany source shale. Note that over 90 percent of the oil produced so far from the Illinois Basin lies in a fairway comprised between 0.7 Ro and 30 mi (48 km) updip from 0.6 Ro (approximately 0.55 Ro) as measured in the New Albany shale. Migration distance does not exceed 30 mi (48 km) and most of the large accumulations, such as the La Salle anticline fields, are immediately updip from the center of the generative depression.

Cooper Basin

Figure 7 is a simplified and schematized isoreflectance

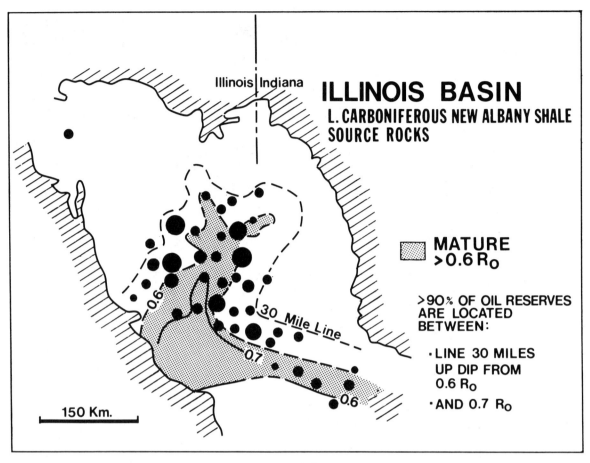

Figure 6. Illinois Basin (U.S.A.) (Swann and Bell, 1958; Barrows and Cluff, 1983).

maturation contour map on top of the Permian coal measures in the Cooper Basin of Central Australia. Giant gas fields sourced by the Permian coals and coaly shales occur in or near the zone of optimum maturation (reflectance 0.9 to 2.0). Permian sands on anticlinal structures which are both in the immature zone, and not within drainage reach from the gas generative zone, have been found to be water wet.

Anticlines with Permian sands in the post-mature zone have yielded natural gas high in carbon dioxide: sandstone porosities in these postmature zones are very low due to extensive silicification. The largest gas field (Moomba) lies immediately updip of the main gas generative depression.

Paris Basin

The Paris Basin (Figure 8) has minor oil production, mainly from Middle Jurassic limestones. Reportedly the co-sources for most of these oil accumulations are black shales at the base of the Middle Jurassic, and the Liassic (Toarcian) shales. However, the latter are isolated by a thick nonsource shale from the mid-Jurassic reservoirs. As earlier recognized by Tissot et al (*in* Tissot and Welte, 1978, p. 510), producing fields are in, or very close to, the zone where the source shales are most deeply buried and oil generative in the central part of the basin. The Paris Basin contains many anticlines outside of the generative depression, some large, with adequate well-sealed reservoirs. So far, they have been found barren. Short migration distances can be explained by a combination of mediocre carrier continuity and rather limited oil volumes available for migration.

PASSIVE ATLANTIC-TYPE MARGINS

Barrow-Dampier Basin

Apparently the major source unit in the stratigraphic section of the northwest shelf of Australia (Figure 9) is in marine shales of Late Jurassic to Lower Cretaceous age which tend to be overpressured. Although marine, these rocks are mainly gas-prone, with, however, subordinate, sporadic potential for oil. The outline of the mature Jurassic generative depression delineates virtually all the commercial

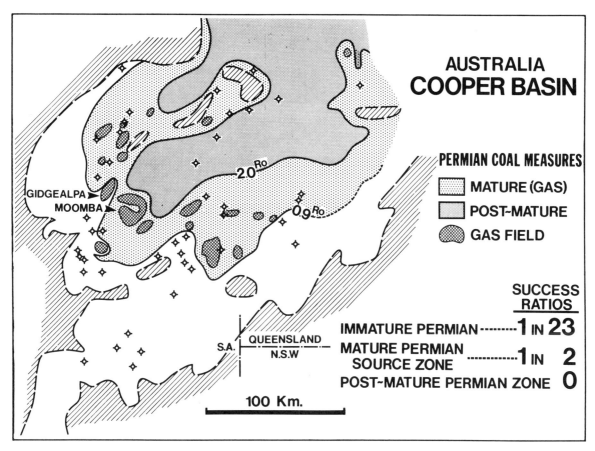

Figure 7. Cooper Basin, Australia (Kantsler et al, 1983).

gas and oil accumulations. Migration distances are short. Lower Jurassic and Triassic coals are rich in inertinite and give mediocre hydrocarbon yields. The suggestion that they are only a secondary co-source is supported by the observation that there are no commercial gas accumulations in Triassic reservoirs, except where these are within drainage reach from the Jurassic generative depression.

COMPLEX MULTICYCLE BASINS

Illizi and Ahnet Basins

The regional patterns of oil and gas occurrence in the Illizi and Ahnet Basins (Figure 10) were recognized as far back as the late 1960s. Lack of oil occurrence in the Devonian and Ordovician of the Ahnet Basin relates to post-maturity of the Silurian source shales, reflectances being in excess of 2%. In the Illizi Basin the high success ratio oil fairway, which contains the four largest oil fields (Zarzatine, Tin Fouye, Edjeleh and Tiguentourine) is located in the zone of more moderate maturation (0.6 to 1.0 Ro) of the Paleozoic source shales.

FORELAND BASINS

Middle East Petroleum Zones

Figure 11 shows the clear geographic association of Cretaceous fields in the Middle East with zones of deep burial of Cretaceous rocks, which include several source horizons. Several large oil fields, in the shallow Cretaceous burial area on the south side of the Gulf, seem to have been sourced from the Upper Jurassic (Murris, personal communication). Gas fields coincide with those areas where Cretaceous rocks are buried below 5,000 m (16,404 ft) in the post mature, metagenetic stage. It is interesting to note that Cretaceous organic-rich rocks are widespread in the Levant (Syria, Lebanon, Jordan, Israel) where they often reach oil-shale levels of richness. With the exception of parts of Syria, and perhaps under or near the Dead Sea, the general lack of maturity due to insufficient burial of the source beds precludes the occurrence of major petroleum producing areas.

Williston Basin

About half of the discovered oil in the Williston Basin

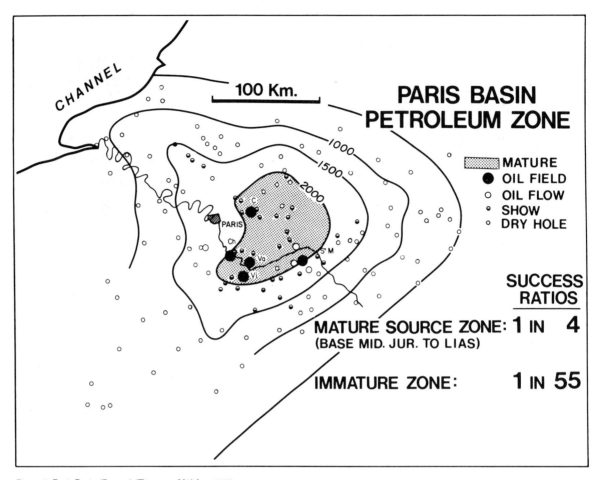

Figure 8. Paris Basin (France) (Tissot and Welte, 1978).

(Figure 12) has been entrapped within the generative depression, and the other half has migrated a rather long distance updip (in excess of 60 mi [97 km]) on the northeastern flank. This has been made possible by adequate lateral reservoir continuity, enhanced by "shunting" from one permeable carrier to another. Moreover the structural style which is homoclinal, with little interruptions, has allowed extensive lateral drainage to take place. The largest accumulation in the Williston Basin is on the Nesson Anticline. It contains about 500 million barrels of oil. It is noteworthy that this major anticlinal trend is plunging into the heart of the Bakken generative depression, and drains its deepest and most mature portion.

CONCLUSIONS

A rapid review of the petroleum generation aspects of 12 basins around the world, for which stratigraphic source bed occurrence and maturation gradients are known, reveals the following regularities:

1. In all reviewed cases, regardless of sediment age, basin size, and tectonic style, the zones of preferred petroleum occurrence ("areas of high potential") are genetically correlated to geochemically identified oil or gas generative depressions. The latter are geographically outlined by the favorable thermal maturation fairways relevant to key petroleum source horizons. The success ratios within, or close to, the generative depressions are in the order of 1 in 3, wherever such ratios have been statistically evaluated.

2. The largest petroleum accumulations tend to be located close to the center of the generative depressions. One common denominator to the giant accumulations, besides structural size, is the large volumes of hydrocarbon-generating sediments, drained for long spans of geologic time.

3. Migration distances for most reviewed cases commonly range in tens rather than hundreds of miles and are limited to the drainage areas of individual structures. Thus the outlines of the mature hydrocarbon generative depressions commonly include most of the basin's hydrocarbon reserves and the largest fields. These conclusions apply mainly to rift basins, intra-cratonic

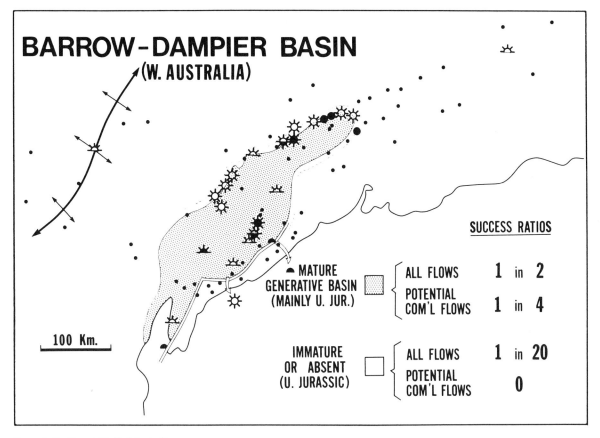

Figure 9. Northwest Shelf of Australia.

platform-type basins, and passive oceanic margins, including deltas, and wherever migration distances are limited by structural style. Cases of long-distance migration are documented in certain foreland basins of North America (for example, the Denver Basin and Williston Basin), and in the peri-Andean basins of Ecuador and Peru, or for the heavy oils of Western Canada and Eastern Venezuela (Demaison, 1977). In all cases, both stratigraphy and structure favored uninterrupted updip movements of oil.

In conclusion, these observations demonstrate the fundamental reason for geographic patterns of regional petroleum occurrence: that being their association with widespread, petroleum generative, source beds in the subsurface. The generative basin concept may also be applied in a predictive mode, since the three regularities, listed earlier, provide powerful analogs in forecasting areas of high potential in undrilled or sparsely drilled basins.

It is important to keep in mind, that the implementation of generative basin mapping methods, requires a sufficient data bank: regional seismic compilations are needed, with geochemical profiles from existing deep wells: together they permit realistic basinwide reconstruction of generation, migration, and accumulation processes. Furthermore, oil-to-oil and oil-to-source correlation studies, by biological markers and other methods, are often indispensable.

In summary, the predictive accuracy of geochemical mapping systems is dictated by the available data base, while final interpretation is as dependent as ever on geologic skills and experience. Thus at early stages of exploration, when data are sparse, geologists should guard from overtaxing geochemical knowledge. The words of Wallace Pratt: "The oil-finder...need always be always alert to the potentialities of what he does not know," still apply today, as they did some 30 years ago.

ACKNOWLEDGMENTS

The writer thanks the management of Chevron Overseas Petroleum Inc., and Standard Oil Company of California, for permission to publish this paper. The writer is also indebted to R.G. Alexander Jr., A.J.J. Holck, R.J. Murris, and M.H. Nederlof for critical reviews of the manuscript.

REFERENCES

Bally, A.W., and S. Snelson, 1980, Realms of subsidence, *in* Facts and principles of world petroleum occurrence:

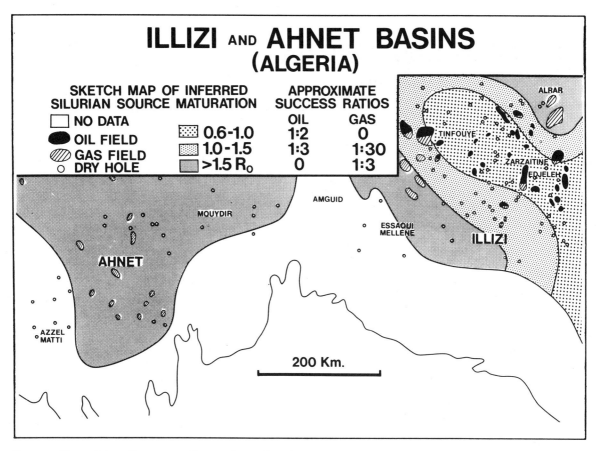

Figure 10. Algerian Sahara (Correia, 1967; Tissot et al, 1973; Perrodon, 1980).

Canadian Society of Petroleum Geologists Memoir 6, p. 9-94.

Barrows, M.H., and R.M. Cluff, 1983, New Albany shale group (Devonian-Mississippian) source rocks and hydrocarbon occurrence in the Illinois Basin, *in* G. Demaison and R.J. Murris, eds., Geochemistry and basin evaluation: AAPG Memoir 35, this volume.

Berg, R.R., 1975, Capillary pressure in stratigraphic traps: AAPG Bulletin, v. 59, p. 939-956.

Bois, C., P. Bouche, and R. Pelet, 1982, Global geologic history and distribution of hydrocarbon reserves: AAPG Bulletin, v. 66, p. 1248-1270.

Correia, M., 1967, Relations possibles entre l'etat de conservation des elements figures de la matiere organique et l'existence de gisements d'hydrocarbures: Revue Institute Francais de Petrole 22, n. 9, p. 1285-1306.

Crostella, A., and M.A. Chaney, 1978, The petroleum geology of the outer Dampier Sub-basin: Australian Petroleum Association Journal, v. 18, pt. 1, p. 13-33.

Demaison, G.J., 1977, Tar sands and supergiant oil fields: AAPG Bulletin, v. 61, p. 1950-1961.

——, et al, 1984, Predictive source bed stratigraphy; a guide to regional petroleum occurrence: London, Proceedings 11th World Petroleum Congress, John Wiley and Sons.

Dow, W.G., 1974, Application of oil-correlation and source rock data to exploration in Williston Basin: AAPG Bulletin, v. 58, p. 1252-1262.

Hu, 1983, Panel discussion 1: London, Proceedings 11th World Petroleum Congress, John Wiley and Sons.

Kantsler, A.J., et al, 1983, Hydrocarbon habitat of the Cooper-Eromanga Basin, Australia, *in* G.J. Demaison and R.J. Murris, eds., Geochemistry and basin evaluation: AAPG Memoir 35, this volume.

Kontorovich, A.E., ed., 1971, Regional geochemistry of petroleum bearing Mesozoic formations in Siberian basins: Ministry of Geology of the USSR, Memoir 118 (in Russian).

——, 1983, Geochemical methods for quantitative evaluation of petroleum potential in sedimentary basins; examples from Siberian basins (USSR), *in* G.J. Demaison and R.J. Murris, eds., Geochemistry and basin evaluation: AAPG Memoir 35, this volume.

Leythaeuser, D., R.G. Schaefer, and A. Yukler, 1983, Role of diffusion in primary migration of hydrocarbons: AAPG Bulletin, v. 66, p. 408-429.

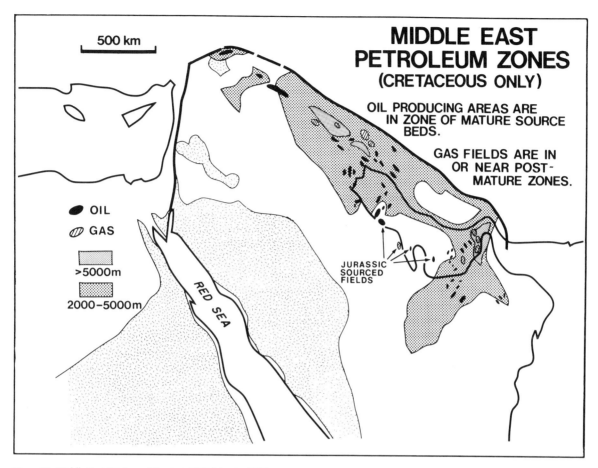

Figure 11. Middle East (Riche and Prestat, 1978; Murris, 1983).

Meissner, F.F., 1983, Petroleum geology of the Bakken Formation, Williston Basin, North Dakota and Montana, in G.J. Demaison and R.J. Murris, eds., Geochemistry and basin evaluation: AAPG Memoir 35, this volume.

Murris, R.J., 1983, Middle-East stratigraphic evolution and oil habitat, in G.J. Demaison and R.J. Murris, eds., Geochemistry and basin evaluation: AAPG Memoir 35, this volume.

Parker, J.R., 1982, Hydrocarbon habitat of the western desert of Egypt: Abstract, EGPG Sixth Exploration Seminar, Cairo.

Parsons, M.G., A.M. Zagaar, and J.J. Curry, 1980, Hydrocarbon occurrence in the Syrte Basin, Libya, in Facts and principles of world petroleum occurrence: Canadian Society of Petroleum Geologists Memoir 6, p. 723-732.

Perrodon, A., 1980, Geodynamique petroliere Genese et repartition des gisements d'hydrocarbures: Paris, New York, Masson.

Riche, P.H., and B. Prestat, 1979, Paleogeographie du Cretace de proche et moyen orient et sa signification petroliere: Proceedings 10th World Petroleum Congress, Heyden and Son Ltd., London.

Schowalter, T.T., 1976, The mechanics of secondary migration and entrapment: Wyoming Geological Association Earth Science Bulletin, v. 9, no. 4, p. 1-43.

———— and P.D. Hess, 1982, Interpretation of hydrocarbon subsurface shows: AAPG Bulletin, v. 66, p. 1302-1327.

Swann, D.H., and A.H. Bell, 1958, Habitat of oil in the Illinois Basin, in L.G. Weeks, ed., Habitat of Oil: AAPG Special Publication.

Shibaoka, M., J.D. Saxby, and G.H. Taylor, 1978, Hydrocarbon generation in Gippsland Basin, Australia; comparison with Cooper Basin, Australia: AAPG Bulletin, v. 62, p. 1159-1170.

Threlfall, W.F., B.R. Brown, and B.R. Griffith, 1976, Gippsland Basin offshore, in Economic geology of Papua-New Guinea: Australasian Institute of Mining and Metallurgy Monograph 7, p. 41-67.

Tissot, B., et al, 1973, Origine et migration des hydrocarbures dans le Sahara oriental (Algerie), in Advances in organic geochemistry: Paris, Technip.

————, and D.H. Welte, 1978, Petroleum formation and occurrence: New York, Springer-Verlag.

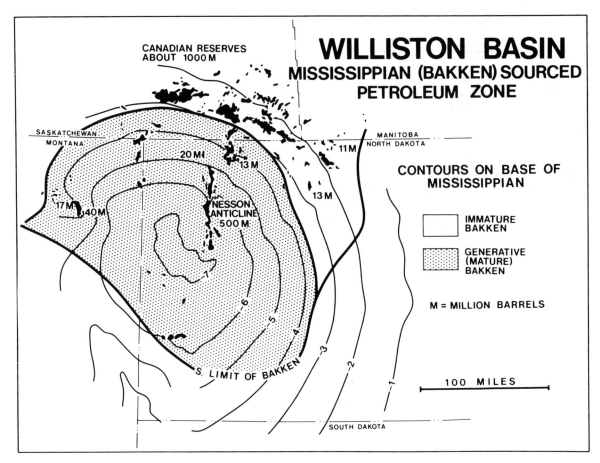

Figure 12. Williston Basin (Dow, 1974; Meissner, 1983).

Wang, Shangwen, Hu Wenhai and Tan Shidian, 1982, Habitat of oil and gas fields in China: Oil and Gas Journal, v. 80, n. 24, p. 119-128.

Waples, D.W., 1983, Physical-chemical models for oil generation: Colorado School of Mines Quarterly, v. 78, no. 4, p. 15-30

Ziegler, P.A., 1980, Northwest European Basin geology and hydrocarbon provinces, in Facts and principles of world petroleum occurrence: Canadian Society of Petroleum Geologists Memoir 6, p. 653-706.

TIME, TEMPERATURE AND ORGANIC MATURATION —
the evolution of rank within a sedimentary pile

N. J. R. Wright*

Through using the current published literature and attempting to gain insights into the maturation history of sedimentary basins, in particular the time of potential oil generation, the author has become aware of certain problems associated with the diagrammatic representation of burial histories and rank evolution. This note does not represent the product of any new experimental work or any rethinking of the thermodynamics of the complex organic reactions involved in maturation of kerogen; I have simply taken published "models" of organic maturation and attempted to apply them to a simple and, I think, intuitively understandable pictorial representation of the progressive evolution of rank within a sedimentary column.

INTRODUCTION

Organic maturation and the generation of oil

Excellent reviews of the various empirical methods of determining maturation levels in organic matter are given by Tissot & Welte (1978, pp 449-471) and by Bostick (1979). Methods include pyrolysis, vitrinite reflectance, palynomorph colouration, fluorescence of liptinites, elemental and electron spin resonance (ESR) analysis of kerogen, bitumen abundance, and methods based on hydrocarbon composition such as measurement of the carbon preference index (CPI). Hood, Gutjahr and Peacock (1975) proposed a continuous numerical scale of maturation, the LOM (Level of Organic Metamorphism) scale; this and a number of the other parameters mentioned above are compared in Figs. 1 and 2, from Tissot & Welte (1978).

The quantity of liquid hydrocarbons generated from Kerogen is most directly defined by the "Bitumen ratio", the ratio of extractable bitumen (milligrams) to the total organic carbon (grams) in a "tight" sample. The clearly defined maximum in the bitumen ratio shown in Fig. 2 for the main types of kerogen, has long been known and constitutes the "oil generation zone". This corresponds to vitrinite reflectance $R_o = 0.5\%$ to $R_o = 1.3\%$ (LOM 7 to LOM 11), although Hood et al., restricted their zone of generation to LOM 9 to LOM 11.

The concept of the oil generation zone related to vitrinite reflectance fits well with regional data for the distribution of oil and gas pools as Bostick's (1979) recent compilation of global data shows. He points out that, on average, pools might be expected at shallower depths (lower reflectances) than those of the generation zone, due to upwards migration into reservoirs, and this is reflected in the data.

Coalification, temperature and time

The processes of organic maturation were first examined by coal geologists who had long noted increased ranks at deeper levels in a coal basin. At first the role of pressure was thought to be of primary importance, but although pressure is still thought to be influential at low levels of maturation (Damberger 1968) it is

---

* *Phillips Petroleum Co., Future Ventures Secn., 266 FPB, Bartlesville, Oklahoma, USA.*

now recognised that temperature is the single most important parameter affecting rank. The close dependence of these parameters has been amply demonstrated in laboratory heating experiments (e.g. Bostick 1973, 1979) and in the field. However, laboratory heated samples always show lower ranks than natural samples heated to the same temperature but for geological periods of time; similarly, samples from older (Palaeozoic, Mesozoic) basins tend to be higher ranks than comparably buried Cenozoic samples. Thus, the role of time is also recognised as critical in the process of maturation.

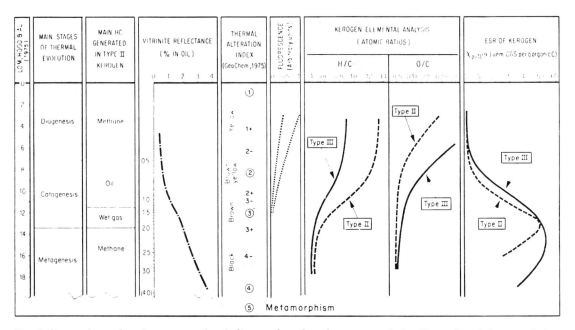

Fig. 1 Comparison of various maturation indicators based on kerogen analysis. (Reproduced, by permission of the authors, from B. P. Tissot and D. H. Welte, "Petroleum Formation and Occurrence", Springer Verlag, Berlin, 1978.)

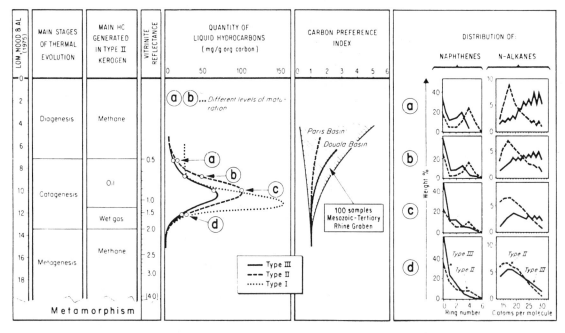

Fig. 2 Comparison of various maturation indicators based on abundance and composition of the extractable bitumen. (Adapted from an original sketch by B. Durand). (Reproduced, by permission of the authors, from B. P. Tissot and D. H. Welte, "Petroleum Formation and Occurrence", Springer Verlag, Berlin, 1978.)

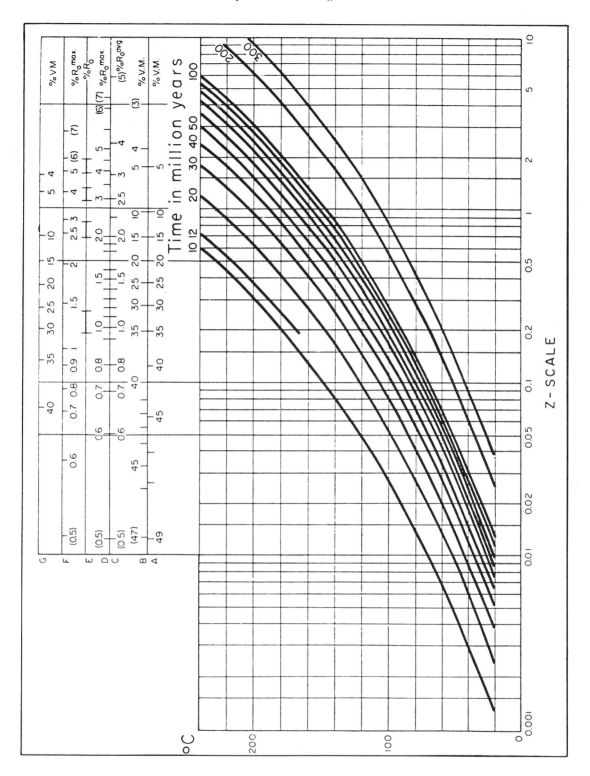

Fig. 3 Relations between temperature, burial time, and vitrinite properties used to determine rank of coal and phytoclasts. The diagram and scale A are after Karweil (1956). Other scales are: B — Correction of the relation between relative depth and volatile matter, from Teichmüller and Teichmüller (1968, Fig. 10). C — Average vitrinite reflectance in oil; converted from scale B by Teichmüller (1971, Fig. 2). D — Maximum vitrinite reflectance in oil; from scale B by Alpern and Lemos de Sousa (1970, Fig. 1) and De Vries *et al.* (1968, Figs. 8 and 11). E — Estimated uncertainty in the volatile matter-maximum reflectance conversion; from sources in C and D, plus Kotter, (1960) and Lensch (1963). F — Maximum vitrinite reflectance in oil; from scale D by conversion with the regression line on Figure 7, this paper. G — Volatile matter; from scale F by using the sources given in D. (Reproduced by permission of the author, from Bostick, 1973.)

MODELS OF ORGANIC MATURATION

A number of differing time–temperature–maturation models are now used in order to predict palaeotemperatures from rank values or to predict maturation levels from burial and palaeotemperature assumptions. They derive in the main from coalification reaction equations proposed by Huck & Karweil (1955); Karweil (1956) synthesised empirical rank data and theoretical modelling into a convenient "nomogram", relating temperature, heating time and coal rank. Karweil's diagram is shown in Fig. 3 (from Bostick 1973), with various modifications to the rank scale proposed by subsequent workers. In general, Bostick notes that the reaction rates predicted by Huck & Karweil, assuming an activation energy of 8.4 Kcal/mole over the whole range of coalification, are too slow for high levels of maturation (above $R_o = 0.6\%$) and too fast for lower levels.

Bostick 1973

Bostick's adaptation of the Karweil nomogram for prediction of coal rank is shown in Fig. 4: thermal histories are plotted on the diagram, and the ratio of the area enclosed by the thermal pathway to the maximum temperature maximum time rectangle is taken as an estimated proportion of the maximum possible rank attained by the sample. Bostick (1973) showed that this method gave predictions close to the actual vitrinite reflectances observed in a number of borehole samples.

In a later paper (1979), Bostick used a simplified maturation diagram in which the actual vitrinite reflectance predictions were plotted; this has been re-drawn in Fig. 5.

These maturation diagrams are informative and give a graphic portrayal of possible phases of oil generation. However, they do suffer from limitations. Firstly, each potential source horizon has to be plotted separately on the diagram, as the rank curves are related to time since initial deposition and apply only to the sediment deposited at time zero. This may not

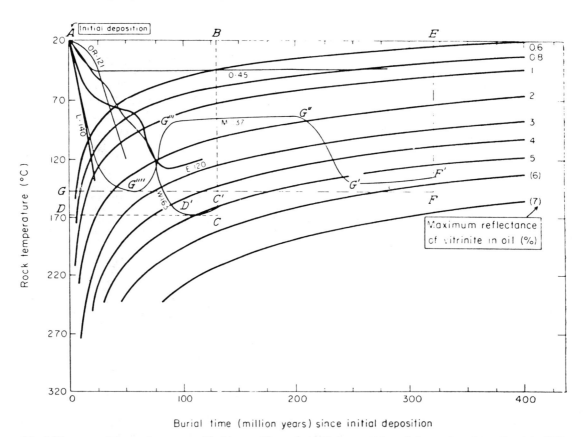

Fig. 4 Diagram of thermal metamorphic history. (Reproduced by permission of the author, from Bostick 1973) Thermal histories are shown as curves such as AG''''G'''G''G'F'. Maximum metamorphism would be achieved by the immediate attainment of the highest temperature, sustained for the longest time, i.e. pathway AGF, and this maximum vitrinite reflectance is predicted by the point F on the diagram. The ratio of actual metamorphism to greatest potential metamorphism is estimated by the ratio of the area above the actual heating curve (AEF'G'G''G'''G'''') to the area of the maximum rectangle (AEFG). In this case the ratio is about 3/4, and the predicted vitrinite reflectance is $3/4 \times 6.2\% = 4.6\%$.

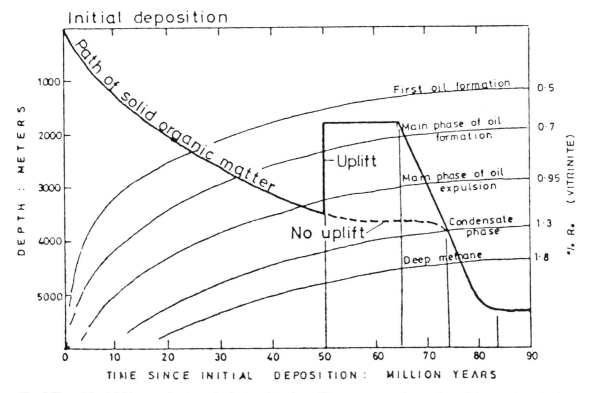

Fig. 5 Plot of burial history of a hypothetical rock unit, to illustrate generation and possible trapping of oil. (Redrawn from Bostick, 1979).

be a drawback when studying maturation in a well-known petroleum basin where the source horizon is unequivocally defined, but in many less well understood basins maturation information of a more general kind is useful.

Secondly, although the diagrams seem to predict quite accurately the final rank of a sample, there must be reservations about the veracity of the implied rank pathway, particularly where uplift and cooling is involved. For example, the uplift pathway in Fig. 5 at 50 m.y. seems to imply a reduction of rank from $R_o = 1.0\%$ to $R_o = 0.6\%$. This is of course impossible, as coalification reactions are not reversible, and rank will continue to increase even though some cooling has taken place.

Hood, Gutjahr and Peacock, 1975

These authors have proposed an alternative method of predicting maturation levels in a sediment. This is based on two parameters, the maximum temperature reached (T_{max}) and an "effective heating time" (t_{eff}) defined as the time during which a rock has been within 15°C of its maximum temperature (T_{max}). The relationship of these parameters to rank, defined by their Level of Organic Metamorphism (LOM) scale, is shown in Fig. 6; this is based on observed LOM values for 40 fine-grained rocks of varied and well known thermal histories and on the rank/depth relationships of 13 separate deep wells, and seems to provide a convenient and quick method of estimating potential maturation levels. The authors noted that the time/temperature relationship implied by the slope of the iso-rank lines in Fig. 6 is equivalent to a doubling of the reaction rate for every increase of 10°C, a relationship previously noted by a number of other authors (e.g. Lopatin 1971 — see below).

Lopatin 1971 and 1976

Lopatin showed, from a consideration of the Arrhenius (reaction rate) formula and empirical data for activation energies observed in natural coalification processes, that the ratio of reaction rates at T and T+10° was approximately 2 over the temperatures commonly considered for coalification. On this basis, Lopatin defined an arbitrary time/temperature index of coalification, τ such that

$$\tau = \Delta_1 \gamma_1 + \Delta_2 \gamma_2 + \Delta_3 \gamma_3 + \ldots + \Delta_{18} \gamma_{18} \quad (1)$$

where

Δ_1 = time in m.y. at 50-60°C
Δ_2 = time in m.y. at 60-70°C
Δ_3 = time in m.y. at 70-80°C
Δ_{18} = time in m.y. at 220-230°C

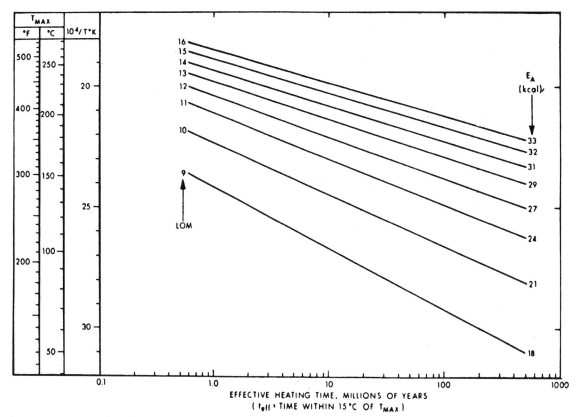

Fig. 6 Relation of LOM to maximum temperature and effective heating time. (Reproduced from Hood *et al.*, 1975).

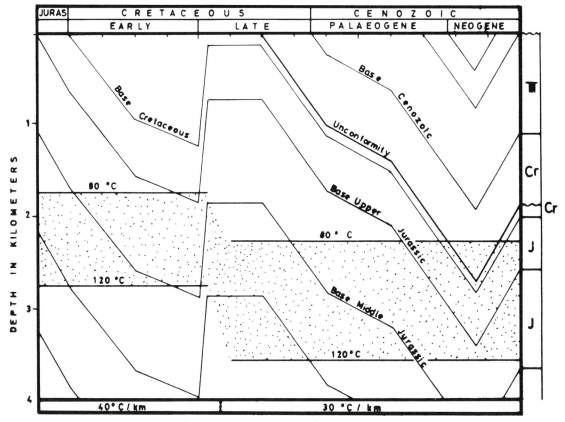

Fig. 7 Hypothetical burial history diagram for a Jurassic to Tertiary stratigraphic column. Possible "temperature windows" for two assumed geothermal gradients are shown.

Time, Temperature and Organic Maturation

INTERNATIONAL COAL PETROLOGY HANDBOOK	ASTM COAL CLASS		SOVIET COAL CLASS	RANK PARAMETERS		
				LOM (Hood et al)	$R_0\%$	τ (Lopatin)
PEAT	LIGNITE			—1		
SOFT BROWN COAL				—2		
				—3		
					0.3	
				—4		
HARD BROWN COAL			BROWN	—5		
	SUB-BITUMINOUS COAL	C		—6	0.4	
		B				
		A	BROWN/ FLAME	—7	0.5	
	HIGH VOLATILE BITUMINOUS COAL	C		—8	0.6	
HARD COAL / BITUMINOUS		B	FLAME	—9	0.7	8
					0.8	
				—10		
		A	GAS		0.9	
				—11	1.0	
	MEDIUM VOLATILE		FAT			25
				—12	1.5	55
	LOW VOLATILE BITUMINOUS COAL		COKING			
			CAKING/ SINTERING	—13		
	SEMI-ANTHRACITE		LEAN		2.0	88
				—14		110
			SEMI-ANTHRACITE	—15		
					2.5	220
ANTHRACITE	ANTHRACITE		ANTHRACITE	—16		

Rank parameter annotations: "Fat" (≈25–55), "Low volatile" (≈55–110), "Lean" (≈110–220).

Table 1. Approximate correlation of coal classifications and Rank parameters.

and γ is a continuously doubling multiplier based arbitrarily on γ_6 as unity—thus $\gamma_1=2^{-5}$, $\gamma_2=2^{-4}$, $\gamma_3=2^{-3}$ etc. (see Table 2).

Thus, by a simple arithmetic process of time/temperature integration the additive τ index can be calculated for any sediment from its thermal history. Lopatin tested this model against the vitrinite reflectance profiles in a number of deep boreholes, and showed that there was a very close correlation between τ and R_0. The relationship is given by

$$\lg\tau = 0.77 R_0 + 0.4062 \qquad (II)$$

and τ values in relation to rank stages are shown on Table 1.

ORGANIC MATURATION IN GEOLOGICAL HISTORY

The coal geologist is primarily interested in final rank attained at the present day and in explaining any variations or anomalies; indeed, present day maturation levels are the only direct test for any of the coalification models and their rank predictions. However, the oil geologist must also be concerned with the maturation history of a sedimentary basin if he is to predict possible time of oil generation and relate it to other critical factors such as porosity development and formation of structures.

Given reasonable palaeontologic or other time control of a stratigraphic column, a diagram of burial history can be constructed, as shown in Fig. 7, plotting the evolution with respect to depth and time of any part of the stratigraphic sequence. Uncertainties exist, of course, where periods of uplift and erosion occur, but regional geological information is normally sufficient to provide constraints on these.

If the time factor in organic maturation is ignored, the oil generation zone is defined simply as a "temperature window" in present day sections. The threshold of oil maturity varies from 60°C to 100°C in different basins, while the upper limit varies from 100°C to 150°C; a reasonable average for the window is sometimes taken as 80°C to 120°C. The simplest form of oil maturation modelling is therefore to reconstruct this temperature window with respect to depth and time, and plot this on the burial history diagram as shown in Fig. 7.

This approach, however, tells nothing of the evolving state of organic maturation in the sedimentary column. Each of the coalification models described above has been used to predict the heating time required for a particular sediment to have generated oil; for example Bostick's diagrams (Fig. 4, Fig. 5) and similar maturation nomograms are quite widely used. As has been said, this is ideal when a particular horizon is being focussed upon, but does not give an overall picture of the evolving maturation in the column. However, each of the maturation analyses is quite capable of modelling maturation at any point in geological history, and therefore of mapping iso-rank lines through time. These should then provide a complete picture of potential sourcing.

Rate of heating

It is convenient to combine rate of subsidence and geothermal gradient into a single parameter, the Rate of Heating (ROH). The organic reactions concerned are generally assumed independent of any pressure effects, and are simply controlled by temperature and time.

In Lopatin's analysis, ROH is expressed by the Δ factors and for constant heating $ROH = {}^{10}/\Delta$. In Hood et al., (op cit) the heating rate is expressed by t_{eff}; for constant heating $ROH = {}^{15}/t_{eff}$.

For many basins, ROH values vary typically from 0.3 to 1.0 °C/m.y., but may reach 10 °C/m.y. or more in rapidly subsiding hot basins. Values as high as 50° or 100°C m.y. may occur in unusual cases.

Predictions of the models

Maturation, being temperature time dependent, depends critically on changes in the Rate of Heating (ROH); the three maturation models described above give similar predictions as to the nature of this dependence and therefore as to the form of iso-rank lines on a burial diagram. However, the numerical analysis of Lopatin is most useful for predicting the precise shape of these lines under variable heating conditions.

Constant ROH

(i.e. steady subsidence through a constant geothermal gradient.) Obviously, under these conditions any maturation level will be achieved at a constant depth with time; in Fig. 8, a, b and c have identical thermal histories and hence identical rank. Lines of equal R_0, LOM, τ are therefore at constant depth/temperature and the critical maturation level is equivalent to a temperature window.

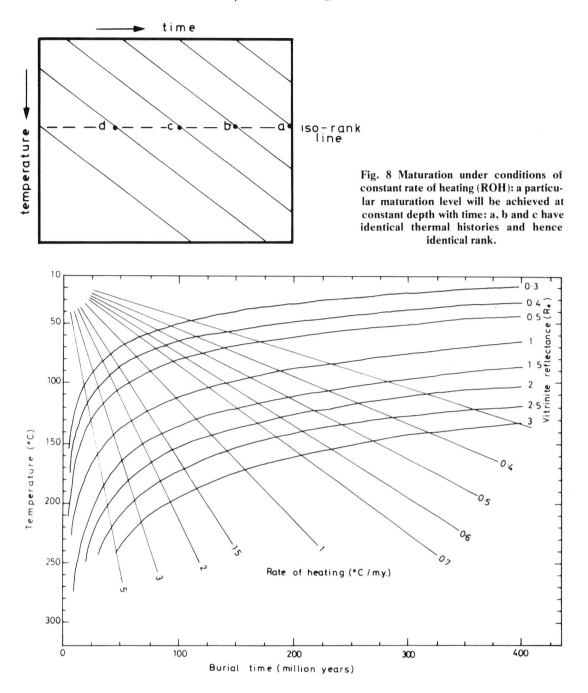

Fig. 8 Maturation under conditions of constant rate of heating (ROH): a particular maturation level will be achieved at constant depth with time: a, b and c have identical thermal histories and hence identical rank.

Fig. 9 Diagram of vitrinite metamorphism for various constant Rates of Heating. The curves are redrawn from Bostick's (1973) diagram (see Fig. 4), but under constant heating conditions the vitrinite reflectances can be read directly from the curves, which represent *half* the maximum values of Fig. 4.

In Bostick's graph, Fig. 4, rank attained under constant ROH is exactly half the maximum possible value. This figure has been redrawn to give vitrinite reflectance directly for various constant ROH curves which are plotted as gradients on the nomogram (Fig. 9). In the model of Hood et al., for constant heating ROH is simply represented by the t_{eff} scale ($t_{eff} = {}^{15}/ROH$) in Fig. 6. Temperature values for a particular rank (R_0, LOM) can be read directly from these figures.

In Lopatin's model, ROH is represented by Δ factors in the serial equation (1), which will be equal for constant ROH ($\Delta = {}^{10}/ROH$). $\Delta \times \gamma$ increments for two ROH values are tabulated in Table 2. Temperatures for a particular rank (τ) can be derived by progressive summation of these increments.

Temperature interval °C	γ	ROH = 0.33 °C/m.y.		ROH = 1 °C/m.y.	
		$\gamma \times \Delta$ increment	$\Sigma \gamma \Delta = \tau$	$\gamma \times \Delta$ increment	$\Sigma \gamma \Delta = \tau$
50-60	$\gamma_1 = 2^{-5}$	0.9	0.9	0.3	0.3
60-70	$\gamma_2 = 2^{-4}$	1.8	2.7	0.6	0.9
70-80	$\gamma_3 = 2^{-3}$	3.7	6.4	1.2	2.1
80-90	$\gamma_4 = 2^{-2}$	7.5	13.9	2.5	4.6
90-100	$\gamma_5 = 2^{-1}$	15	29	5	10
100-110	$\gamma_6 = 2^0$	$\Delta = 30$	59	$\Delta = 10$	20
110-120	$\gamma_7 = 2^1$	60	119	20	40
120-130	$\gamma_8 = 2^2$	120	239	40	80
130-140	$\gamma_9 = 2^3$	240	479	80	160
140-150	$\gamma_{10} = 2^4$	480	959	160	320
150-160	$\gamma_{11} = 2^5$			320	640
160-170	$\gamma_{12} = 2^6$			640	1280
170-180	$\gamma_{13} = 2^7$				
180-190	$\gamma_{14} = 2^8$				
190-200	$\gamma_{15} = 2^9$				

Table 2 Calculation of the time/temperature index τ according to Lopatin's (1971) analysis, for two constant Rates of Heating — 0.33 and 1 °C/m.y. Left hand columns give the time/temperature increment for each temperature interval; right hand columns give the summation up to, hence the τ value at, each temperature.

		Hood et al		Bostick		Lopatin	
	ROH, °C/m.y.	1	0.33	1	0.33	1	0.33
LOM 9 $R_0 = 0.65$	$\tau = 8$	95°C	80°C	95°C	72°C	96°C	82°C
LOM 11 $R_0 = 1.15$	$\tau = 20$	155°C	130°C	115°C	90°C	110°C	94°C

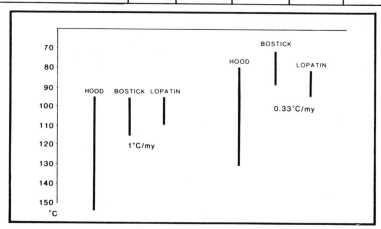

Table 3 Comparison of the "oil generation zone" temperatures corresponding to LOM 9 to 11, $R_0 = 0.65$ to 1.15 and $\tau = 8$ to 20 for the three maturation models of Hood et al (1975), Bostick (1973) and Lopatin (1971).

Table 3 shows the temperature values for the "oil generation zone" (LOM 9 to 11, $R_0 = 0.65$ to 1.15, $\tau = 8$ to 20) for two typical ROH values. The lower temperature threshold shows good agreement between the models, but the upper, LOM 11 temperature is much higher in the Hood et al., model than for the other two. This discrepency is too large to be ignored, and implies that further experimental work is required.

Increased ROH

As is clear from Table 3 and particularly from Fig. 9, increased ROH implies a higher temperature for a particular rank. If the increase is due to a higher geothermal gradient, higher temperatures will be attained at comparable depths. If the increase is due to a faster rate of subsidence, one can perhaps view the iso-rank curves being depressed with the sediment pile until a new equilibrium level is achieved.

The numerical Lopatin analysis has been used to illustrate the form of the transition predicted by the models, and is shown in Fig. 10. Appropriate terms in the serial equation are derived from Table 2. The new equilibrium is achieved by the first rock with no "memory" of the previous thermal regime, i.e. the rock at 50°C at the time of change of ROH. The two equilibrium levels are linked by a gentle curve; the precise form at this curve is unimportant, but is related to the "double reaction rate with 10°C" principle which underlies the thermodynamic interpretation of all three models.

Decreased ROH

Iso-rank lines are raised to a new equilibrium level, as shown in Fig. 11. Increments are again derived from Table 2.

ROH becomes zero

In the extreme case, sedimentation ceases and the sediment pile is held at an even temperature. Increasing rank under these

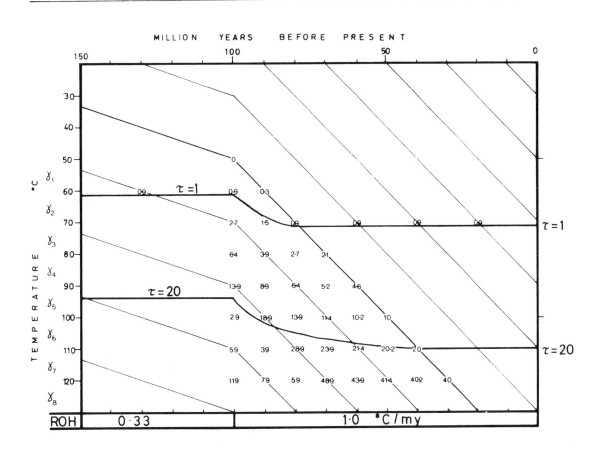

Fig. 10 Diagram showing calculation of theoretical τ values according to Lopatin's (1971) analysis, to show predicted form of iso-rank lines following an *increase* in the Rate of Heating (ROH). Time/temperature increments along individual heating (burial) curves are derived from Table 2.

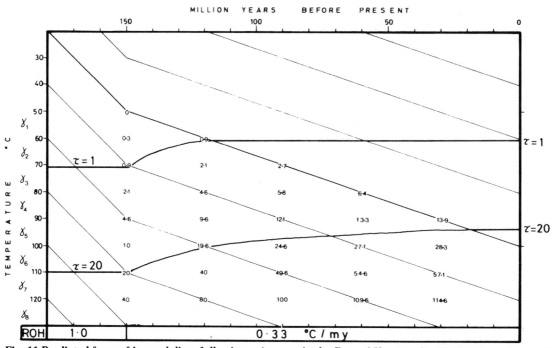

Fig. 11 Predicted form of iso-rank lines following a *decrease* in the Rate of Heating. See Table 2 for values and increments.

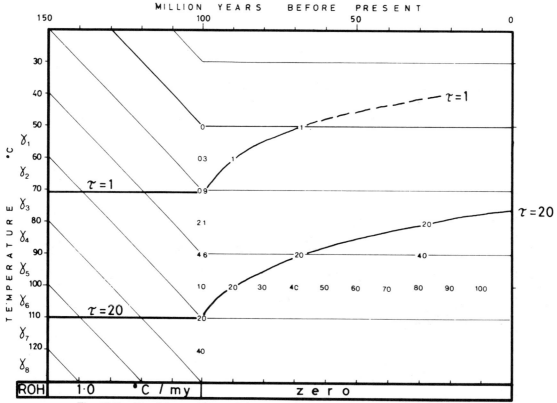

Fig. 12 Predicted form of iso-rank lines on Rate of Heating becoming zero.

conditions is represented by moving along horizontal lines in Hood *et al.*, or Bostick's nomograms (Fig. 4 or Fig. 6); the form of the rising iso-rank curves can be directly plotted from the "double reaction rate" relationship (see Fig. 12).

ROH becomes negative
(i.e. cooling on uplift and erosion.)

Rank parameters are properties of rock, not of space occupied by rock, and since the coalification reactions are treated as non-regressive, iso-rank lines are uplifted with the stratigraphic column. Reactions continue, and therefore burial lines continue to cross iso-rank lines, though at an ever decreasing rate as cooling progresses. Lopatin's analysis allows some qualification of this process, shown in Fig. 13. In Fig. 13, a total cooling at 80°C is effected over 20 m.y., and it can be seen that the rank increments during cooling are relatively small. Since uplift and cooling is commonly more rapid than this, these increments might be ignored.

No actual discontinuity in rank is likely to be registered at the unconformity unless uplift has been sufficient to bring significantly matured rocks to the surface; in Lopatin's model, rocks that have been heated to at least 50°C. However, as burial and maturation continues after the uplifted episode, rank discontinuities tend to become obscured, as is shown in Fig. 13 by the calculated τ-values above and below the unconformity.

APPLICATION TO BASIN STUDIES

The above observations on the predicted response of maturation levels in a subsiding basin to variations in the Rate of Heating can be applied in a generalised form to any sedimentary basin. Having constructed a burial diagram from stratigraphic data, and making a considered assumption as to the past geothermal gradient, it is usually possible to recognise distinct episodes for which the ROH can be averaged. Equilibrium temperatures for the maturation levels of interest (e.g. the oil generation zone), can then be deduced from Fig. 9 (Bostick), Fig. 6 (Hood et al.), or by calculation from Equation I (Lopatin) for each of the separate ROH phases. The general form of the curves linking these levels can be estimated from Figs. 9, 10, 11 and 13.

Temperature threshold

Although the "temperature window" discussed above is not sufficient for determining oil generation in geological history, it cannot be simply replaced by an "oil generation zone" defined entirely in terms of rank stages. Temperature remains critical; a source shale with $R_0 = 0.8\%$ can remain at that rank for many millions of years and never generate a drop of oil. The vital effect is the *progression* through the critical rank stages—through $R_0 = 0.65$ to 1.15%, LOH 9 to LOH 11, or $\tau = 8$ to $\tau = 20$—and this rate of reaction is a straightforward function of temperature.

The definition of a "threshold temperature" below which oil generation is unlikely even if maturation levels are favourable is obviously arbitrary, as examination of Figs. 6 or 9 will

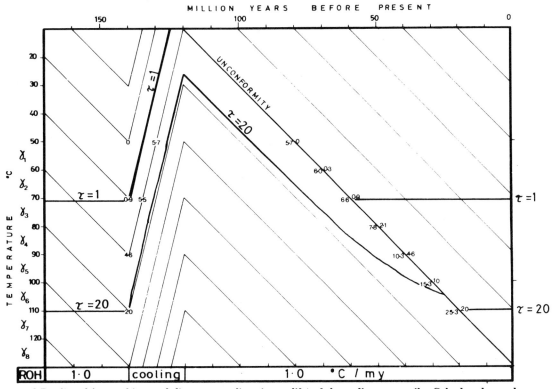

Fig. 13 Predicted form of iso-rank lines on cooling (i.e. uplift) of the sedimentary pile. Calculated τ values on either side of the resulting unconformity are shown.

show. In Lopatin's numerical analysis, no rank increments are considered for temperatures below 50°C, and examination of Table 2 shows that for typical ROH values increments are very low (less than 2) for the first few terms. Since the observed temperature threshold varies from 60°C to 100°C and is often taken as 80°C, a threshold of 70°C has been assumed.

Thus the oil generation zone is defined only as the LOM 9 to LOM 11 interval at temperatures above 70°C. Under normal geological conditions it is only through uplift and erosion that rocks of such rank would be cooled to below the threshold temperature, although special conditions involving rapidly reduced geothermal gradients might be imagined.

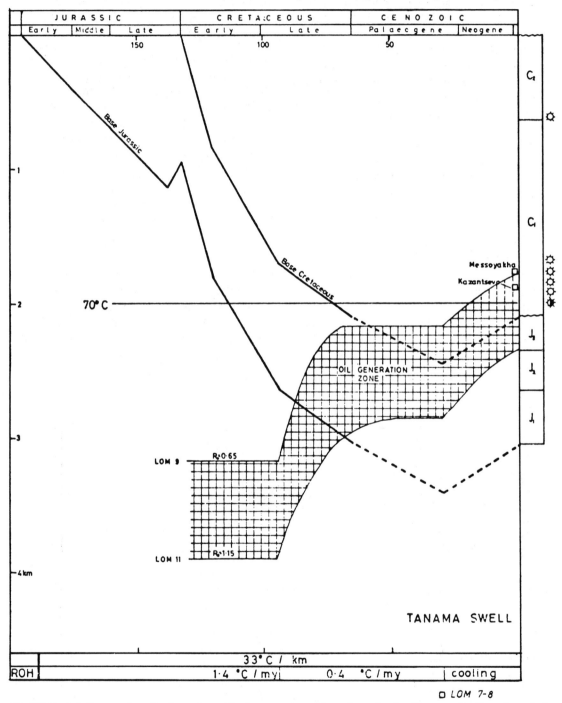

Fig. 14 Burial diagram for the Jurassic to Tertiary stratigraphic column in the Tanama Swell region of the Yenisey-Khatanga trough, northern Central Siberia. Rates of Heating assuming a constant geothermal gradient (33°C/km) are calculated and used to predict the level of a possible "Oil Generation Zone" (LOM 9-11, R = 0.65 − 1.15) within the evolving sedimentary pile. Present-day gas and oil occurrences are shown against the stratigraphic column.

An example

Unfortunately, the author has not had an opportunity to work in detail on vitrinite data from deep wells and to use this method directly. Instead it has been applied in a generalised way to a number of Soviet petroleum basins which the author was studying, and indeed it is in situations where detailed well data is somewhat lacking where the method is thought to be most useful.

Fig. 14 is a burial diagram for Jurassic, Cretaceous and Cenozoic strata in the Tanama Swell region of the Yenisey Khatanga Trough. The assumed geothermal gradient and the averaged ROH values are recorded at the bottom, and the estimated LOM 9 to LOM 11 field is shaded. There is some vitrinite data available; this is rather scattered and individual boreholes cannot be identified but in general terms the coal rank stages are consistent with the predictions:—

Brown coal
  (LOM 4–7?) occurs no deeper than 1460 m
Brown/flame coal
  (LOM 7–8?) occurs no deeper than 1550 m
Flame coal
  (LOM 8–9?) occurs no deeper than 2300 m
Gas coal
  (LOM 9–11) occurs no deeper than 3020 m

Soviet geochemists commonly use the disappearance of humic acid as an indicator of the Brown coal to Flame coal transition, corresponding to LOM 7–8. The depth of this parameter for two oilfields in the area is shown on Fig. 14. Gas and oil pools are also shown.

The 70°C temperature threshold is plotted, and in this case is constant at 2 km depth.

As can be seen, the reconstructed oil generation zone rises sharply through the section by over 1.5 km from Mid-Cretaceous to recent time, first as a response to lower ROH and then to Late Cenozoic uplift. A number of implications follow from the model:—

(i) All pre-Jurassic rocks are post-mature, but might contain gas.
(ii) No Mesozoic oil was generated prior to Late Cretaceous time.
(iii) Lower and Middle Jurassic rocks have been potentially oil generating since the Late Cretaceous; they may now be post mature (oil expulsion) although vitrinite data indicates that they still lie within the LOM 9–11 zone.
(iv) The Upper Jurassic shales, which are a possible source horizon in parts of the basin, have been within the oil generation zone for 55 to 70 m.y., since latest Cretaceous time.
(v) The basal Cretaceous (which also contains possible source shales) was potentially generating during much of Palaeogene time, but was cooled to below the temperature threshold due to uplift during the Neogene.
(vi) Overlying Cretaceous strata are immature.

Acknowledgements

This work developed from studies undertaken while the author was with the Cambridge Arctic Shelf Programme, directed by W. B. Harland. Input from various colleagues and Mr. Harland's encouragement to publish are gratefully acknowledged.

REFERENCES

BOSTICK, N. H., 1973. Time as a factor in thermal metamorphism of phytoclasts. Congrès International de Stratigraphie et de Géologie du Carbonifère Septième, Krefeld, August 23-28 1971. *Compte Rendu*, **2**, 183-193.

BOSTICK, N. H., 1979. Microscopic measurement of the level of catagenesis of solid organic matter in sedimentary rocks—a review. SEPM Special Publication No. 26, 17-43.

DAMBERGER, H. H., 1968. *In:* Nachweis der Abhängigkeit der Inkohlung von der Temperatur *Brennstoft-Chemie*. **49** (3), 73-77.

DOW, W. G., 1977. Kerogen studies and geological interpretations. *Journal of Geochemical Exploration*, **7**, (1977), 79-99.

HOOD, H., GUTJAHR, C. C. M., and PEACOCK, R. L., 1975. Organic Metamorphism and the Generation of Petroleum. *Amer. Ass. Pet. Geol. Bull.*, **59** (6), 986-996.

HUCK, G., and KARWEIL, J., 1955. Physikalisch-chemische Probleme der Inkohlung *Brennstoft-Chemie*. **36** (1), 1-11.

KARWEIL, J., 1956. Die Metamorphose der Kohlen vom Standpunkt der physikalischem Chemie. *Deut. geol. Ges. Z.*, **107**, 132-139.

LOPATIN, N. V., 1971. Temperature and geologic time as factors in coalification. *An SSSR Izvestiya*, Ser. Geol. No. 3, 95-106 [In Russian].

—, 1976. Historico-genetic analysis of petroleum generation: Application of a model of uniform continuous subsidence of the oil-source bed. *An SSSR Izvestiya*, Ser. Geol. No. 8, 93-101 [in Russian].

TISSOT, B. P., and WELTE, D. H., 1978. Petroleum formation and occurence. Springer-Verlag-Berlin, Heidelberg, New York, 1978, 538 p.

Influence of Nature and Diagenesis of Organic Matter in Formation of Petroleum[1]

B. TISSOT,[2] B. DURAND,[2] J. ESPITALIÉ,[2] and A. COMBAZ[3]

Rueil-Malmaison, France, and Paris, France

Abstract Elevation of temperature and pressure during burial of sediments results in a physiochemical transformation of the kerogen, and particularly in changes of elementary composition, infrared spectra, and thermal properties of kerogen. The combined variation of these characters defines an evolution path for each kerogen type (organic matter from similar environments of deposition group along the same evolution path). The degree of evolution of individual samples can be evaluated by using these techniques.

An experimental evolution test, using thermogravimetric, infrared, and associated techniques, can reproduce the natural degradation of kerogen during its burial in a sedimentary basin. The test allows the petroleum geologist to make an evaluation of the petroleum potential of a given formation in the buried parts of a sedimentary basin, by using shallow or surface samples and simulating their degradation in the laboratory.

The major products generated during the three successive steps of kerogen evolution are carbon dioxide and water; oil; gas. The relative abundance of these products depends on the composition of the original kerogen. Kerogen from a "high" evolution path generates abundant oil, whereas that from a "low" evolution path produces methane at depth, but little or no oil.

GENERAL SCHEME OF HYDROCARBON GENERATION

Many studies have shown that most commercial hydrocarbons are generated by thermal transformation of kerogen during burial of the source rocks (Larskaya and Zhabrev, 1964; Louis, 1964; Philippi, 1965; Louis and Tissot, 1967; Albrecht and Ourisson, 1969; Vassoyevich et al., 1969; Tissot et al., 1971).

The general scheme of hydrocarbon generation is summarized on Figure 1, by gathering the results of several studies carried out in different sedimentary basins from France, north and west Africa, Canada, and South America. The diagram shows the relative amount and nature of hydrocarbons generated from sedimentary organic matter as a function of depth of burial. The depth scale is only indicative and may vary according to the nature of the original organic matter and to the geothermal gradient.

In young sediments, freshly deposited, little hydrocarbon is present. With the exception of methane of biochemical origin (marsh gas),

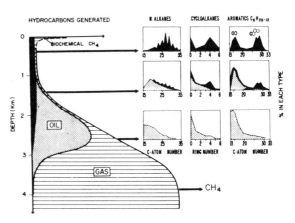

FIG. 1—General scheme of hydrocarbon generation. Depth scale represented is based on examples of Mesozoic and Paleozoic source rocks. It is only indicative and may vary according to nature of original organic matter, burial history, and geothermal gradient.

these hydrocarbons are inherited more or less directly from the organisms incorporated in the sediment. They have very peculiar structures, characteristic of molecules synthesized by living organisms. Some are incorporated in the sediment without any change, like n-alkanes from higher plants or from algae; others, like steroids or pentacyclic triterpenoids, result from biogenic molecules through an early transformation (for example, loss of a functional group) that does not affect the hydrocarbon skeleton. These molecules can be considered as "geochemical fossils," in the same sense as

---

© 1974. The American Association of Petroleum Geologists. All rights reserved.

[1] Manuscript received, July 5, 1973; accepted, September 14, 1973. Read at the AAPG Research Symposium, Anaheim, California, May 15, 1973.

[2] Institut Francais du Petrole.

[3] Compagnie Francaise des Petroles.

The writers are indebted to Compagnie Francaise des Petroles and Compagnie ELF-Re for their assistance in supplying samples. The writers also thank B. Alpern, P. Robin, J. C. Roussel, C. Souron, A. Marchand, and M. F. Achard for their participation in reflectance, infrared, thermogravimetric, and ESR studies.

ammonites or foraminifers, are macro- or microfossils.

During a certain time and depth span, the hydrocarbon content remains low, as shown in Figure 1, to about 1,000 m. Then, beyond a certain threshold, new hydrocarbons are generated, first gradually, then more rapidly, by thermal breakdown of the insoluble organic matter ("kerogen"). These hydrocarbons have no characteristic structure, and the previous hydrocarbons (the geochemical fossils) are diluted progressively by the new ones. This is the principal phase of oil formation, as designated by Vassoyevich et al. (1969).

At greater depth, increased cracking of carbon-carbon bonds occurs and generates light hydrocarbons from kerogen and from previous oil as well. This is the phase of condensate and gas formation. Later, under epimetamorphic conditions, methane, carbon dioxide, and a carbon-rich residue would be the end point of organic matter.

Kerogen Evolution Paths

Although this is a general scheme of the evolution of organic matter during burial, the total amount of hydrocarbons that can be generated and the relative importance of oil and gas depend on the composition of the original organic matter; some formations may provide oil and gas, and others generate mostly gas. The chemical and physical study of the insoluble organic matter, or kerogen, is a way to characterize the various types of organic matter and thus to evaluate the oil and gas potential of the formations.

The elemental composition (atomic H/C and O/C ratios) of a number of kerogens is shown on Figure 2. This presentation allows a first approach to the classification of the kerogens. Kerogens taken at different depths from the same formation normally group along a curve that we propose to call the "evolution path," for example, the lower Toarcian shales of the Paris basin, the Silurian shales from the Sahara in Algeria and Libya, the Upper Cretaceous shales of the Douala basin, or the Mannville shales from the Lower Cretaceous of Alberta, which have been studied by McIver (1967).

Furthermore, rocks from closely related environments of deposition, like shales containing Tasmanites algae from the Silurian of the Sahara and from the lower Toarcian of the Paris basin, result in the same evolution path. The different paths start from rather different hydrogen and oxygen original contents, according to the nature of the organic material and to the conditions of deposition. All curves tend to join each other at great depths and approach pure carbon as shown in Figure 2 (Durand and Espitalié, 1973).

Kerogens rich in aliphatic structures, able to generate oil, have an original high H/C ratio and a relatively low O/C ratio (evolution path I on Fig. 2). This group includes algal-rich sediments, in particular those derived from lacustrine Botryococcus (like Autun and Campine bogheads), and their marine equivalents (like tasmanite from Tasmania). It also includes normal disseminated kerogen from the most prolific source rocks (including some source rocks from the Middle East).

Kerogens with a predominant aromatic structure result from abundant contributions from higher plants and terrestrial humic material accumulated in nonmarine or paralic environments. They have a high original oxygen content, with an O/C ratio that may reach 0.2 or 0.3 in some samples, and a relatively low H/C ratio (evolution path III on Fig. 2). This group includes the Mannville shales from the Lower Cretaceous of Alberta and the Upper Cretaceous of the Douala basin, Cameroon.

Most of the kerogens appear on Figure 2 between these two extreme types. Some have relatively high values of H/C ratio and low values of O/C ratio, like the Toarcian shales of the Paris basin, the Silurian shales of the Sahara, and the upper Paleozoic–Triassic shales of Spitzbergen (evolution path II on Fig. 2). Some others have a comparatively lower H/C ratio, like the Viking shales from the Cretaceous of Alberta (McIver, 1967, evolution path not represented here), resulting from the participation of plant material from the continent.

The global composition of the kerogen can be compared with the composition of the coal macerals by using the same diagram (Fig. 3). The evolution path of type I resembles the carbonization path of alginites. The kerogen type III, rich in material from higher plants, has an evolution path comparable to that of vitrinite. The other evolution paths, like type II, occupy an intermediate position, as exinite does, but the H/C ratio may be higher or lower according to the particular formation.

Transformation of Kerogen during Burial

The burial of sediments at depth results in a temperature and pressure increase that in-

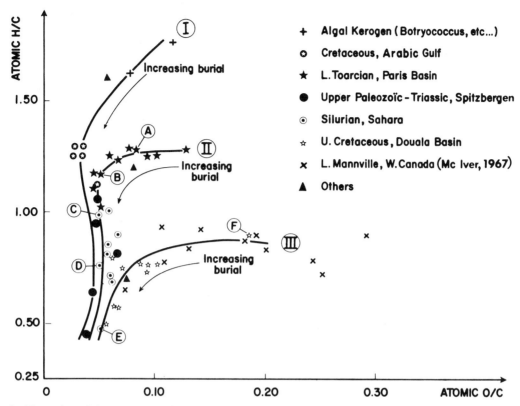

FIG. 2—Examples of kerogen evolution paths. Path I includes algal kerogen and excellent source rocks from Middle East; path II includes good source rocks from North Africa and other basins; path III corresponds to less oil-productive organic matter, but may include gas source rocks. Evolution of kerogen composition with depth is marked by arrow along each particular path.

duces a progressive transformation of kerogen (McIver, 1967). The successive steps of increasing thermal evolution can be characterized to locate them along an evolution path, like

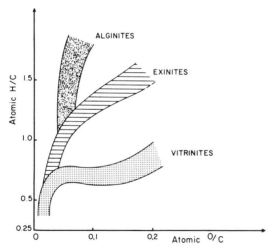

FIG. 3—Evolution of coal macerals (after Van Krevelen, 1961).

path II, that comprises the lower Toarcian shales of the Paris basin and the Silurian shales from the Sahara. In the same basin, for example the Paris basin, subsurface surveys enable geologists to reconstruct the burial history and arrange the samples in order of increasing burial depth (Tissot et al., 1971). To compare samples from different basins, like the Paris basin and Sahara, burial depths cannot be used readily, as the age of the formations, the duration of burial, and the geothermal gradients may be different. However, the order of increasing evolution still can be evaluated by simultaneous use of vitrinite reflectance, microfossil preservation and color, and electron-spin resonance (ESR) measurements, together with available burial and geothermal data.

When the samples along an evolution path, like path II on Figure 2, are considered in this manner the shallow samples, with a low grade of evolution, are at the beginning of the curve with relatively high H/C and O/C ratios. A progressive evolution results first in a partial

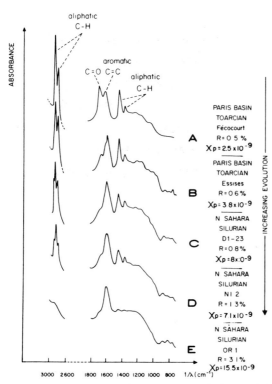

FIG. 4—Evolution of kerogen structure during burial. Samples are taken along evolution path II and marked A-E on Figure 2. Infrared spectra show progressive elimination of C = O (carbon dioxide and water formation) and aliphatic CH_2, CH_3 bands (hydrocarbons formation). Reflectance of vitrinite (R) and ESR data (paramagnetic susceptibility Xp, C G S electromagnetic units) are shown for comparison.

elimination of oxygen, marked by a decrease of the O/C ratio (sample B). Later the elimination of hydrogen, especially in the form of hydrocarbons, makes the H/C ratio decrease (samples C, D). In very deep samples, the H/C ratio is lowered to about 0.5, and the O/C ratio is only 0.05.

Infrared analysis of the same kerogen samples (Fig. 4) shows that the diminution of the oxygen content during the first step is related to the progressive elimination of the C = O group. The following stage, that of hydrocarbon generation, is linked with a decrease of the H/C ratio and of aliphatic bands. In very deep samples, aliphatic and carbonyl bands have nearly disappeared. The remaining oxygen probably is restricted to the nonlabile heterocyclic structures (Espitalié et al., 1973).

The thermogravimetric analysis (TGA) of the same set of kerogen samples, all derived from a similar environment, allows a classification of these samples along their particular evolution path. On heating kerogens in an inert nitrogen atmosphere, the weight loss from shallow samples is important for this type of organic matter and may reach 70 percent. With increasing burial and increasing level of organic-matter transformation, the weight loss decreases progressively down to 10 percent in very deep samples from the western Sahara (Fig. 5).

COMPARATIVE EVALUATION OF PETROLEUM POTENTIAL BY SIMULATED EVOLUTION OF KEROGEN

In an attempt to simulate in the laboratory the natural evolution of kerogen during burial, a shallow and immature sample of the same sequence (type II) was used. Several aliquots of the sample were prepared and heated in a nitrogen atmosphere, with the same thermogravimetric equipment. The temperature was increased at a constant 4°C/minute; one sample was heated to 350°C, the second sample to 400°C, etc. . . . up to 600°C. The weight loss was recorded. The heated sample then was subjected to optical examination, reflectance measurement, elementary, and infrared analysis. The higher temperatures used are not found in a typical sedimentary basin, but they are used to achieve results in the short laboratory times that are comparable on a relative basis to the longer times and lower temperatures in nature.

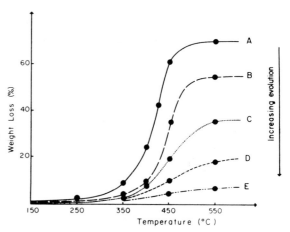

FIG. 5—Thermogravimetric analysis of kerogen. Samples are taken along evolution path II, and are marked A-E on Figure 2. Samples were heated at temperature increase of 4°C/min and weight loss was measured. Weight loss is highest in shallow samples taken along evolution paths I or II (here). Samples taken along path III would show least weight loss.

Heating to 350°C results in a rather small weight loss. Although the organic matter is somewhat darkened, its reflectance is still low (Fig. 6). From 350 to 470°C, kerogen shows a major change; the weight loss is maximal per time or temperature unit. The crystalline organization of carbon increases; reflectance goes up to 1.9 percent. Beyond 470°C the weight loss is small and very slow. But carbonization of kerogen goes on and is marked by a continuous increase of reflectance (2.9 percent at 600°C).

The elementary composition of the samples subjected to these experimental evolution tests varies continuously in relation to the temperature. It first is denoted by an oxygen loss, followed by a hydrogen loss. Furthermore, the succession of the different values of the H/C and O/C ratios reproduces exactly the natural evolution path that we have described previously (compare Figs. 7, 2).

The same observation applies to the infrared spectra; samples submitted to a thermal evolution in the laboratory show first a decrease of the carbonyl $C = O$ bond followed by a progressive loss of both aliphatic and carbonyl groups. Again the succession of IR spectra resulting from the experimental evolution tests is nearly identical with the natural evolution with depth in a sedimentary basin that we have seen previously (compare Figs. 8, 4).

The step-by-step analysis of the chemical composition and the physical properties of the kerogen that has undergone either a simulated degradation in the laboratory or natural degradation during the geologic history of a sedimentary basin is an important comparison. Their coincidence allows us to make an evalua-

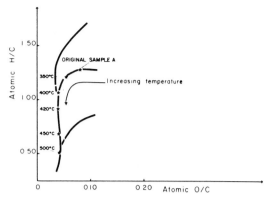

Fig. 7—Experimental evolution path of shallow kerogen. Aliquots of original sample (Fecocourt, lower Toarcian shales of Paris basin, marked A on Fig. 2) are respectively heated to temperatures shown on diagram (temperature increase 4°C/min). Succession of elementary compositions reproduces natural evolution path II shown on Figure 2.

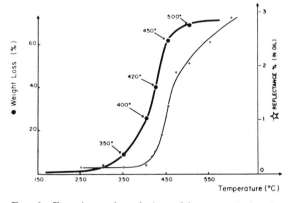

Fig. 6—Experimental evolution of kerogen A showing progressive weight loss and variation of reflectance of kerogen (temperature increase 4°C/min).

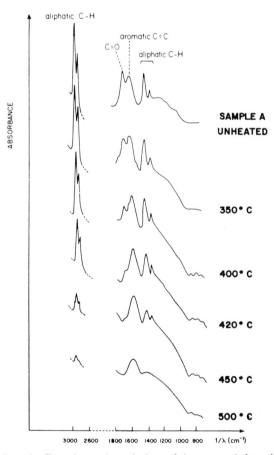

Fig. 8—Experimental evolution of kerogen (infrared analysis). Same aliquots of lower Toarcian shales from Fecocourt are heated to temperatures shown on diagram. With increasing temperature of experiments, succession of infrared spectra reproduces natural evolution during burial, described on Figure 4.

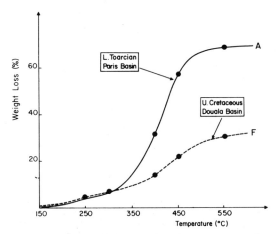

Fig. 9—Comparative experimental assay on two kerogens by using thermogravimetric analysis. Lower Toarcian shales of Paris basin (A) are two times more productive than Upper Cretaceous shales of Douala basin (F). Original samples A and F are marked on Figure 2.

tion of the petroleum potential of a given formation in the buried parts of a sedimentary basin, by using shallow or outcrop samples of the same formation and simulating their degradation in the laboratory.

As an example, two formations are compared here, one belonging to a relatively "high," hydrogen-rich, evolution path of type II (the lower Toarcian shales of the Paris basin), the other to a "low" evolution path of type III with a smaller hydrogen content (Upper Cretaceous shales of the Douala basin, studied by Albrecht and Ourisson, 1969). Two shallow kerogens, one from each of these formations, were heated gradually in the laboratory to simulate the evolutionary changes that would have occurred in nature if these shallow kerogens had been buried to greater depth.

On thermogravimetric analysis, the lower Toarcian shale kerogen yields twice as much volatile material as the Douala kerogen; the weight loss is between 60 and 70 percent, compared to 30 percent (Fig. 9). In addition, the infrared analysis of the two natural unheated samples shows that the lower Toarcian shales contain more aliphatic groups, capable of producing hydrocarbons, compared to the Upper Cretaceous of the Douala basin, where the aliphatic bands are much smaller and the C=O and aromatic bands are comparatively more important (Fig. 10). The relative abundances of these groups result in products of different composition from the thermal degradation of kerogen during the thermogravimetric analysis; CO_2 and H_2O amount to more than 50 percent of these products in the kerogen from the Douala basin, against only 25 percent in the kerogen of the lower Toarcian shales from the Paris basin. Altogether, the amount of petroleum resulting from the transformation of one gram of kerogen is much greater in the lower Toarcian shales. This is in agreement with the amount of hydrocarbons, resin, and asphaltenes effectively generated during burial at various depths, of these two types of kerogens, as measured directly by extraction of cores from the two basins. The amounts of natural petroleum, as extracted from cores, are plotted on Figure 11 as a function of the effective temperature corresponding to their natural burial depth. For a given temperature, the amount of petroleum generated in the lower Toarcian shales of the Paris basin is higher, as expected, than in the Upper Cretaceous from Douala basin.

Conclusion

The general structure usually proposed for kerogen is made of cyclic polycondensed nuclei, bearing alkyl chains and functions, and linked by heteroatomic bonds comprising in particular oxygen. As burial depth and tem-

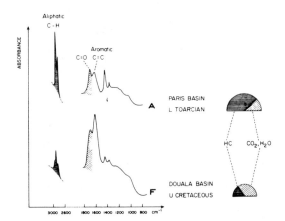

Fig. 10—Comparative infrared analysis of two shallow kerogens having maximum depth of burial of 800 m (left), and composition of products generated by simulated evolution (right). Lower Toarcian shales of Paris basin (A) are rich in aliphatic structures and generate mostly hydrocarbons upon simulated evolution. Upper Cretaceous shales of Douala basin (F) contain more C=O and aromatic C=C groups and generate much less hydrocarbons, but more carbon dioxide and water (surfaces on right are proportional to abundance of various products per gram of kerogen).

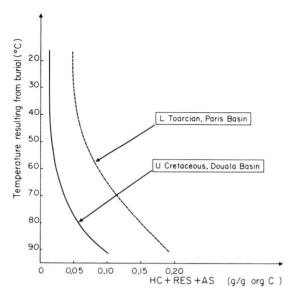

Fig. 11—Compared formation of petroleum in Paris and Douala basins, as function of temperature of source rocks. Transformation of organic matter into petroleum is measured on cores by ratio of chloroform extract (hydrocarbon + resins + asphaltenes) to total organic carbon (Tissot et al., 1971). Comparison is restricted to interval 20–95°C, as deeper samples are available from Douala, but not from Paris basin.

perature increase, heteroatomic bonds are broken successively, starting with some labile carbonyl and carboxyl groups, and roughly in order of ascending rupture energy. Heteroatoms, especially oxygen, are removed partly as CO_2 and H_2O; that step corresponds to a comparative greater decrease of O/C than H/C ratio, and is denoted on Figure 12 by a flattening of the evolution paths. As temperature continues to increase, hydrocarbon structures and particularly aliphatic groups are freed; this is the principal phase of oil formation. That step corresponds to a comparatively greater decrease of H/C than O/C ratio and is denoted on Figure 12 by a "steep" part of the evolution paths. In the final stage, breaking of C—C bonds becomes very important, and cracking generates gas from both kerogen and petroleum already formed (Fig. 12). All along the transformation the carbon content increases in the residual kerogen which becomes more condensed.

The relative importance of these three principal steps of the evolution is primarily dependent on the composition of the original kerogen; a "high" evolution path (like type I or II) corresponds to kerogen types that generate abundant oil, whereas a "low" evolution

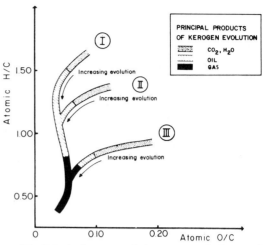

Fig. 12—Principal steps of kerogen evolution and hydrocarbon formation. Flat parts of curves indicate mostly oxygen loss as CO_2 and H_2O. Steep parts show loss of hydrogen and hydrocarbon generation. Evolution paths I or II correspond to kerogens able to generate abundant oil, whereas evolution path III is representative of hydrogen-poor kerogens, producing mostly methane at depth.

path (like type III) is representative of kerogen types that generate abundant CO_2 and H_2O, and produce methane at great depth, but little or no oil.

Chemical and physical analysis of kerogen, associated with simulated evolution in the laboratory, permits an evaluation of the comparative petroleum potential of the various kerogen types.

REFERENCES CITED

Albrecht, P., and G. Ourisson, 1969, Diagenese des hydrocarbures dans une serie sedimentaire epaisse (Douala, Cameroun): Geochim. et Cosmochim. Acta, v. 33, p. 138–142.

Durand, B., J. Espitalié, G. Nicaise, and A. Combaz, 1972, Etude de la matiere organique insoluble (kerogene) des argiles du Toarcien du Bassin de Paris: I, Etude par les procedes optiques, analyse elementaire, etude en microscopie et diffraction electroniques: Inst. Francais Petrole Rev., v. 27, no. 6, p. 865–884.

——— and J. Espitalié, 1973, Evolution de la matiere organique au cours de l'enfouissement des sediments: Acad. Sci. Comptes Rendus, t. 276, p. 2253–2256.

Espitalié, J., B. Durand, J. C. Roussel, C. Souron, 1973, Etude de la matiere organique insoluble (kerogene) des argiles du Toarcien du bassin de Paris: II, Etudes en spectroscopie infrarouge, en analyse thermique differentielle et en analyse thermogravimetrique: Inst. Francais Petrole Rev., v. 28, no. 1, p. 37–66.

Larskava, Ye.S., and D. V. Zhabrev, 1964, Effect of stratal temperature and pressure on the composition of dispersed organic matter: Akad. Nauk SSSR

Doklady, v. 157, no. 4, p. 897-900 (in Russian); English translation: v. 157, no. 1-6, 1965, p. 135-138.

Louis, M., 1964, Etudes geochimiques sur les schistes cartons du Toarcien du bassin de Paris, in G. D. Hobson and M. C. Louis, eds., Advances in organic geochemistry 1964: Oxford, Pergamon Press, p. 85-94.

──── and B. Tissot, 1967, Influence de la temperature et de la pression sur la formation des hydrocarbures dans les argiles a kerogene: 7th World Petroleum Cong. Proc., v. 2, p. 47-60.

Marchand, A., P. Libert, and A. Combaz, 1969, Essai de caracterisation physico-chimique de la diagenese de quelques roches biologiquement homogenes: Inst. Francais Petrole Rev., v. 24, p. 3-20.

McIver, R. D., 1967, Composition of kerogen; clue to its role in the origin of petroleum: 7th World Petroleum Cong. Proc., v. 2, p. 25-36.

Philippi, G. T., 1965, On the depth, time, and mechanism of petroleum generation: Geochim. et Cosmochim. Acta, v. 29, p. 1021-1049.

Tissot, B., Y. Califet-Debyser, G. Deroo, and J. L. Oudin, 1971, Origin and evolution of hydrocarbons in early Toarcian shales, Paris basin, France: Am. Assoc. Petroleum Geologists Bull., v. 55, p. 2177-2193.

Van Krevelen, D. W., 1961, Coal: Amsterdam, Elsevier Publ. Co.

Vassoyevich, N. B., Yu. I. Korchagina, N. V. Lopatin, and V. V. Chernyshev, 1969, Principal phase of oil formation: Moskov. Univ. Vestnik, no. 6, p. 3-27 (in Russian); English translation: Internat. Geology Rev., 1970, v. 12, no. 11, p. 1276-1296.

Sedimentary organic matter and kerogen. Definition and quantitative importance of kerogen

B. DURAND*

I. INTRODUCTION

The object of this Chapter is to define the term kerogen as used in this work, to show the importance of the study of this subject and to indicate a quantitative evaluation.

Since the term kerogen is now used in many different ways, it seemed of interest to inventory these uses and thus to supply the historical background for the concept of kerogen before giving the definition we use here. In order to do so, it is first necessary to describe the main features of the formation and evolution processes for sedimentary organic matter (OM) and the analysis procedures used. This is the only way to bring out the links between various uses of the word kerogen and to clearly show the meaning of the term.

This presentation will also contribute to the continuity between the various Chapters of this work since they introduce several subjects which are later further developed. At the same time a general conception of sedimentary OM is brought out which is essentially that of the co-authors of this work.

II. ORIGIN AND FORMATION OF SEDIMENTARY ORGANIC MATTER

Sedimentary OM is derived from living organic matter (and the products of its metabolism).

After the death of organisms the component substances such as carbohydrates, proteins, lipids, lignin, etc. are subjected to decomposition to various degrees depending on the sedimentation medium (especially its redox properties). Part of the products of this decomposition is recycled by other organisms which use them as energy sources. Simple molecules such

---

*Institut Français du Pétrole. Rueil-Malmaison, France.

as CO_2, H_2O, CH_4, NH_3, N_2, SH_2, etc. are formed as the products of metabolic processes. Another part is transformed into simple molecules (CO_2, H_2O, etc.) by physico-chemical processes (e.g. oxidation). The rest, which in most cases represents only a very small fraction of the initial quantity of living mater, escapes complete biological recycling or physico-chemical decomposition and is incorporated into sediments. This fraction is therefore the primary source of sedimentary OM.

This material is preserved in various ways (Chapter 14). As examples we can mention such possibilities as that the toxicity of the products or their incapacity for use as energy sources keeps them from being reused by organisms or that rapid structural modifications, polycondensation ("polymerization"), associations with minerals make enzymatic deterioration impossible or that the sedimentation medium contains neither organisms nor oxygen, etc.

Some of the preserved products have clearly and sometimes completely preserved the chemical structures which they had in living matter. These are, for example, substances with low chemical reactivity which in the organisms often had a protective role against modifications of the external medium (some saturated hydrocarbons, vegetable waxes, resins, sporopollenin, chitin, etc.) or substances derived from pigments (isoprenoids, porphyrins, etc.) or other metabolites (steroids, terpenoids, etc.).

These molecules, whose structural filiation with living matter is evident, are said to be **inherited** or may be called geochemical **fossils** or **geochemical markers.**

Other complex molecules (not inherited) are said to be **neoformations**. This signifies that they theoretically have no equivalent in living matter. The idea of neoformation is less clear than that of inheritance since:

(a) Neoformed structures can incorporate inherited elements.
(b) There are still many unknown structures in living matter.
(c) Many structures in living matter as well as in preserved products are much too complex to be analyzed by existing methods.

It is thus difficult to evaluate the share of inheritance and that of neoformation. However the very complexity of a large part of the preserved products supports the importance of the role of neoformation.

Organic products incorporated in sediment can be derived from organisms living in the sedimentation basin and are then called **authochtonous**. The material may also come from organisms living outside of the sedimentation medium such as that, for example, which is introduced by rivers, wind action, etc. This material is then called **allochthonous**. To these are added materials from organisms contemporary with more ancient sedimentary strata and organic products from metamorphic and crystalline rock or from the leaching of soils which have been transported by erosion, transgressions or other less important geological modifications. These products are called **reworked** and are generally allochtonous.

All these processes, decomposition of living material and incorporation into sediments of the fraction preserved, introduction of the various other organic materials, etc. are part of what is more generally called the process of formation of sedimentary OM.

III. GENERAL PROCEDURES FOR THE ANALYSIS OF SEDIMENTARY ORGANIC MATTER

The organic content of a sediment is therefore a complex and heterogenous group which is also, as will be seen below, in continuous evolution. The analysis generally proposes to define criteria for origin (progenitors, physico-chemistry of the initial materials, etc.); homogeneity (distribution of the various materials: autochthonous, allochthonous,, contemporary, reworked etc., physico-chemical variety of these materials, etc.); distribution (concentration, relations between organic and mineral phases) and evolution (nature of evolutionary processes, stage reached by these processes, etc.) It is then possible to begin the study of the main geological and geochemical problems (for example, the origin of oil, paleoenvironment of sedimentary deposits, etc.) and to make classifications.

At the present time there are two main approaches:

(a) Firstly there is the approach wich essentially uses optical analysis methods for the study of the **petrography of OM.**

(b) Secondly there is the approach using chemical analysis methods for **coals physico-chemistry, petroleum geochemistry, pedology** (soil study), **environmental geochemistry**, etc.

A. Optical methods.

Petrographers work on thin slices and polished sections made from sediments or concentrates of OM which are generally obtained by physical dressing (grinding and dense fluid separation, flotation, etc. — *cf.* Chapters 3, 11, and 12). On the basis of experience obtained in coal petrography, they identify and classify debris from organisms and microfossils, but their main task is to define the petrographic objects called macerals (the term maceral was created by Mary Stopes [67], by analogy with mineral), i.e. groups which are characterized by the same optical properties (color, reflectance, fluorescence, etc.) so that they can be identified and classified.

In coal petrography three main families of macerals are usually distinguished ([63], Chapter 11) which are differentiated within the same sample by different reflectances. Thus in increasing order of reflectivity we have the following :

(a) Macerals of the exinite family (also called liptinite) generally containing spore cases, leaf cuticles, resin globules, single-cell algae, etc. and also, more generally, debris derived from lipid — rich parts of higher plants.

(b) Macerals of the vitrinite family. They are gels derived especially from the lignin-cellulose walls of cells from higher plants.

(c) Macerals of the inertinite family. These elements are generally of the same origin as macerals of the vitrinite family but have been altered or oxidized for various reasons such as reworking, changes in redox conditions, action of microorganisms, forest fires, etc.

The optical properties of macerals also make it possible to study coal evolution processes. It is conventional to use the reflectance of macerals of the vitrinite family to classify coals according to their **rank** in the natural process of coalification. This process progressively transforms peat into anthracite during burial mainly, it is believed, because of the effects of increased temperatures accompanying burial (geothermal gradient) with a resulting increase in the reflectance of vitrinite.

The study of coals on the basis of maceral analysis has made it possible to utilize coal petrography as an instrument for the analysis of coal genesis and evolution processes. There is, in effect, a good correlation between the petrographic and morphological characteristics of macerals, their origin and physico-chemistry and the stage which they have reached in the evolutionary processes [78]. The application of maceral analysis principles to the study of dispersed OM has not yet, however, made it possible to etablish this coherence (Chapter 11). This is true because of the variety of deposit media and progenitors so that the relations between petrographic characteristics, genesis and physico-chemistry are more difficult to establish.

Petrographers also work with transmitted light on OM collected after the destruction of minerals by hydrochloric and hydrofluoric acid, according to methods inherited from palynology.

They then distinguish between recognizable and amorphous OM. Recognizable organic matter is "the sum of the fragments of identifiable biological origin" [52]. In them can be observed fragments of vegetable tissue, ligneous fibers, epidermises, cuticles, resins, etc. and microfossils especially pollens, which are one of the first objects of study for palynology, but also spores and algae: *Dinoflagellata, Tasmanaceae, Botryococcaceae,* etc. Important groups such as *Diatomaceae, Foraminifera, Radiolaria* and *Coccolithophoridae,* whose tests are destroyed by acids, cannot be identified.

Amorphous OM consists of flakes, powder and agglomerates without definite shapes. The proportion of the various amorphous and figured constituents makes it possible to define the palynofacies ([12], Chapter 3).

By means of these techniques it is possible to follow the evolution of OM, especially that which takes place during burial. During burial, spores, pollen, algae and in most cases amorphous OM as well, which are transparent and light yellow to light brown in color in the case of slightly or moderately evolved samples, become increasingly black and opaque [15, 64, 30].

Amorphous OM and at the same time OM which is very rich in amorphous fractions are often called colloidal, algal, sapropelic or marine. For many people the term sapropelic is a synonym for being derived from algae, although according to the definition of R. Potonie (in [63]) it designates a gel derived from organic debris which has putrified in an anaerobic medium.

The terms sapropelic, algal and marine should be used prudently when qualifying this amorphous material since the algal origin cannot be demonstrated by the amorphous characteristic alone even if much algal debris is observed at the same time, nor can the marine origin (i.e. the amorphous OM originates in marine organisms).

OM which is rich in opaque debris and especially in ligneous debris is often called ligneous, coaly, humic or terrestrial (i.e. originating in terrestrial organisms). To some extent the same criticisms which are made about the terms sapropelic, algal and marine can also be made about the use of these terms. In theory they mean that the OM is derived from coals, soils or at least from a continental sedimentation environment. This is not always clear on the basis of optical examination alone. In this case, however, the origin of the OM can be shown best through this examination.

B. Chemical methods.

In a relatively arbitrary way we will distinguish between the chemical methods intended to isolate fractions of OM without damaging the structures and those whose object is to degrade these structures in a controlled way.

The first consist of solvent extraction, gas entrainment techniques or gas phase extraction for the volatile compounds.

At the present time the most used of these methods are the following:

(a) Extraction with organic solvents at moderate temperatures ($< 80°$): chloroform, benzene, methanol-benzene mixture, etc. This method extracts hydrocarbons and more complex products which are heteroatomic and/or high atomic weight. The latter products will be called resins and asphaltenes hereunder. The organic fraction extracted in this way is often called **bitumen,** but this bitumen should not be confused with road bitumen or with the bitumens of petrographers. The latter are organic substances filling rock pores and under the microscope appear as homogeneous. These bitumens may or may not be soluble in organic solvents. Therefore the term bitumoid has been proposed (Alpern, *in:* Chapter 11) for calling the part of OM which is soluble in organic solvents. The insoluble fraction left in the rock is generally called **kerogen.** This is the conventional fractionation in petroleum geochemistry.

(b) Cold diluted base solutions possibly associated with dechelating agents (for example a NaOH 1N, 1% sodium pyrophosphate mixture) for the extraction of the products called fulvic and humic acids. The fraction remaining in the rock is called **humin.** This fractionation is classic in pedology (soil science).

The second group consist of the use of specific reagents (Chapter 10) for characterizing and quantifying chemical functions and bonds; or controlled hydrolysis, oxidation and hydrogenation methods or thermal degradation (pyrolysis), etc. The latter are less specific but can modify or fractionate structures which are highly complex into simpler elements.

Among these methods one of the most used is hydrolysis with strong hot concentrated mineral acids (for example HCl 6N at boiling point) which liberate amino acids and carbohydrates and other substances. The fraction remaining in the sediment is called stable residue. This fractionation is widely used for the study of recent sediments.

In all cases, the degradation or fractionation products are then refractionated and/or analyzed in detail by means of current techniques of physico-chemical analysis, i.e. elemental analysis, various spectroscopic techniques [visible , infrared (IR), ultraviolet (UV), nuclear magnetic resonance (NMR), electron spin resonance (ESR)], chromatography (liquid and gas phases), mass spectrometry (MS), etc.

Study of the organic compounds left in the sediment presents special difficulties, and for a long time this was not done. If the analysis is made directly on the rock, interferences with minerals hamper correct determination of the properties of OM, since rocks in which OM predominates are rare and the most frequent contents are about 1% (see below).

Modifications, often of a considerable extent, occur when these fractions are to be isolated from their mineral matrix (*cf.* Chapter 2).

C. Complementary nature of optical and chemical methods.

Each of these methods gives a partial picture of sedimentary OM. Theoretically optical methods give more comprehensive results. Used alone, however, these methods have the following disadvantages:

(a) It is not possible to see objects with a size below the power of resolution of the microscope, so that most of the organic content of the sediment cannot be seen. This tendancy is often further increased by the preparation methods, which favor elements of large size. The use of fluorescence techiques, which are becoming wide-spread at the present time, overcomes this drawback to a certain extent.

(b) Some specialists have a tendancy to systematize on the basis of morphological criteria from transmitted light study alone. Thus abundance of algae is considered a single chemism while amorphous matter is called algal, sapropelic or marine.

(c) The usual preparation techniques result in the loss of much organic matter.

The geochemical method used alone has, among other, the following disadvantages:

(a) Too often OM is considered a single entity which is to be fractionated, while in fact it is an assemblage of many entities. Thus this method does not give direct access to the different constituents (allochthonous, autochthonous, reworked).

(b) Some specialists have a tendancy to systematize on the basis of the compounds which they know how to analyze best, for example non volatile saturated hydrocarbons, while these are often not very representative of the OM as a whole.

(c) The method does not sufficiently take into account the fact that the same chemism can originate in rather different circumstances.

It is becoming increasingly clear that these two methods are both necessary for good comprehensive understanding of sedimentary OM and more and more laboratories are using them jointly.

IV. EVOLUTION AT DEPTH

A. General

The biochemical and physico-chemical processes mentioned above are replaced here by processes which we will call geochemical and geodynamic:

(a) Weathering phenomena due to readjustements or changes in the redox conditions. (These phenomena are often difficult to distinguish from formation processes). These phenomena are of relatively great importance in soils which are not subjected to subsidence.

(b) Phenomena due to temperature and pressure increases accompanying burial and possibly to the action of mineral catalysts and are called diagenetic and catagenetic phenomena.

(c) Phenomena due to the displacement of a part of the sedimentary organic content called migration (for example migration of hydrocarbons from source rocks to reservoir rocks).

These phenomena finally lead to a further deterioration of the organic matter which was already relatively degraded at the formation stage, i.e. to produce simple, thermodynamically stable molecules (CO_2, CH_4, H_2, H_2O, etc.) and a residue very rich in aromatic carbon from complex structures inherited from living or neoformed matter. This stage is generally reached before the zone of metamorphism where the sediments will be transformed into crystalline rocks and where the carbon residue may be transformed into graphite.

Although there is a tendency to move towards thermodynamically stable compounds, experience shows that these processes can be described only in kinetic terms since the carbon compounds formed are mainly metastable under subsurface conditions even on the geological time scale.

In addition, the existence of migration phenomena makes the sedimentary system an open system in which equilibria, even if they had the time to establish themselves, would be continually displaced. No doubt exceptions exist for closed systems such as fluid inclusions in which partial equilibria have the time to establish themselves between compounds with low molecular weights. In addition, since reaction rates increase as temperatures rise, it is possible that the OM system and its degradation products may be increasingly well described in thermodynamic terms as the zone of metamorphism are approached. However the zones of metamorphism have not been studied from this point of view.

These phenomena are indicated by progressive modifications in the properties of organic matter which can be brought out by many methods. We have already spoken of the increased reflectance of vitrinite which accompanies coalification and the variations in spore and pollen color and transparency. There are, therefore, a large number of properties whose evolution can be followed. Further details on this subject will be given in other Chapters of the present work.

It should be noted that one of the general effects of this evolution is the convergence of the properties of OM of different origins since the result is always the same relatively simple system. In particular, there is convergence of optical properties.

B. Evolution of soluble, insoluble and volatile fractions of organic matter (Fig. 1.1.).

Study of many sedimentary basins has little made by little made it possible to bring out the following main trends:

Recently sedimented OM (recent sediments, peat, soils) is not very soluble in organic solvents. In addition, this soluble fraction contains few hydrocarbons. On the other hand, it is partly soluble in bases (humic and fulvic acids) and in acids (hydrolyzable fractions).

There may be a volatile fraction which is the result of physico-chemical and/or biochemical degradation, and in certain cases there may be so-called biogenic methane (marsh gas) and hydrogen sulphide.

The beginnings of burial (Zone A in Fig. 1.1.) are marked by:

(a) The disappearance of soluble fractions in acids and bases.
(b) The formation of CO_2 and H_2O.
(c) A slight increase in the size of the fraction soluble in organic solvents which consists essentially of resins and asphaltenes and a small quantity of hydrocarbons.

(d) A relative decrease in the amount of oxygen in the fraction insoluble in organic solvents which is accompanied by various modifications in its physico-chemical characteristics, in particular, the oxygenated functions disappear (Chapter 6).

Beyond a certain depth (transition A > B in Fig. 1.1), which is generally about 1 km, there is an increase in the relative size of the fraction soluble in organic solvents and the volatile fraction. This phenomenon is principally due to the formation of hydrocarbons. The distribution of these hydrocarbons changes with depth through the progressive reduction in the average molecular weight. Thus in successive stages we can observe an oil zone, a wet gas zone and a methane zone. Thus less and less resins and asphaltenes are recovered. The frac-

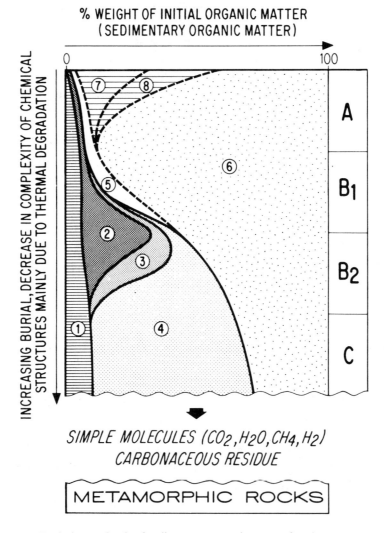

Fig. 1.1. — Evolution at depth of sedimentary organic matter fractions:
1. CO_2 + H_2O. 2. Oil. 3. Wet gas. 4. Methane. 5. Resins + Asphaltenes. 6. Insoluble matter in organic solvents (kerogen). 7. Soluble matter in alkalis. 8. Soluble matter in acids.
Dotted lines between 5, 6, 7 and 8 mean possible overlapping of these fractions.

tion soluble in organic solvents, and which can be recovered after evaporation of the solvent, thus ends by disappearing. Zones B (B_1 + B_2) and C in Fig. 1.1 have been located in reference to what Vassoyevitch [79] called the main oil formation zone. B_1 and B_2 correspond to the presence of oil with the boundary between the two corresponding to the maximum formation, while C indicates the absence of oil and wet gas and the presence of dry gas (methane).

The fraction insoluble in organic solvents which had lost oxygen during the preceding stage (A) now becomes poor in hydrogen. At the same time many of its physico-chemical properties are modified (disappearance of saturated CH functions, cf. Chapter 6), increase in reflectance (Chapters 11 and 12), variation in the number of free radicals (Chapter 8), etc.).

Transformation of organic material into lighter molecules is never complete. Light molecules such as CH_4, CO_2, H_2O, H_2, etc. coexist with an insoluble residue very rich in aromatic carbon (carbonaceous residue) which could produce graphite.

It is supposed that the evolution which we have just described is principally due to increased temperatures [42, 43, 44, 51, 79] and the observed phenomena are called diagenetic and/or catagenetic. There is a great deal of variation in the definitions and names of the characteristic zones depending upon the authors (Chapter 4), and the zones which we have called A, B, C are not necessarily those used. However the disagreement concerns the criteria to be used and the position of the zone boundaries rather than the description of the entire group of phenomena. In any case it should be noted that the location of the zones which may be used changes a great deal with the nature of the initial OM [73] and that exact boundaries cannot be assigned unless the organic content in the entire sedimentary column is very well known.

It is possible to introduce a quantitative element into the preceding sheme. As a first approach, the relative extent of hydrocarbon formation rises to the extent to which the OM (at the recent sedimentary stage) is rich in hydrogen and therefore poor in oxygen. Inversely, the extent of CO_2 and H_2O formation will be greater if the OM is initially highly oxygenated.

If we refer to the petrographic patterns, the first case (high hydrogen contents) will result from OM rich in macerals of the exinite family and/or with a significant amorphous fraction, while the second case (high initial oxygen content) will result from OM rich in macerals of the vitrinite family and/or with a large coaly fraction. However this correspondence between maceral analysis or the palynofacies and the capacity to form hydrocarbons has only a statistical character, and there are many exceptions. Indeed we shall see later on (Chapter 4) that optical properties are only an approximate indicator of hydrogen content.

C. Comparison between evolution at depth and pyrolysis.

The evolutionary process which has just been described is analogous in general to that which is observed during laboratory pyrolysis of sedimentary OM. This analogy was noted as early as 1915 by D. White [84, 85]. Indeed coalification and carbonization are both carbon enrichment processes of the solid phase of OM where fractions are progressively formed which are volatile or soluble in organic solvents and rich in hydrocarbons, while the insoluble residue loses first oxygen and then hydrogen and thus becomes richer in carbon. In both cases the final result is the formation of light molecules including methane and a residue which is very rich in carbon. The hydrocarbon yield of pyrolysis depends, as in the case of natural evolution, on the chemical structure of the original substance and especially on the H/C ratio.

Pyrolysis is used industrially to make coke from coal and graphite from organic substances, so-called graphitizable substances (for instance tars from coals or petroleum). It has also been used and may possibly be used again for producing petroleum products called shale oil from sediments called oil shales. From what has been said it can be understood that oil shale is rock which, under present technical and economic conditions, has the following favourable characteristics:

(a) Sufficient OM.
(b) The OM has a sufficiently high hydrogen content.

It should be noted that effects analogous to those obtained by burial of sediments at temperatures of about 20 to 200° C can be obtained in the laboratory only at significantly higher temperatures ranging from about 200 to 600° C, depending on the stage of natural degradation which it is intended to reproduce.

However the natural modifications observed in sedimentary OM are also due to the temperatures, and relatively low temperatures are compensated for by the very long geological periods involved.

It is not necessary to invent the existence of catalytic effects in order to explain temperature differentials between the artifical and natural processes. A calculation explains it more simply on the basis of a description of the phenomena in kinetic terms. By means of the same kinetic formulation based on first-order kinetics and the same range of activation energies (20-80 kcal/mol) [71, 72, 73], it is possible to correctly simulate both the laboratory phenomena (time scale = 1 hr, temperature range = 200 to 600° C) and those observed underground (time scale = millions of years, temperature range = 20 to 200° C).

There are however important differences between pyrolysis and natural evolution (Chapter 4). In general, the lower the activation energy of the phenomena, the more difficult it is to simulate them on the laboratory scale by increasing the temperature. This is true because secondary reactions are also activated at the same time which would be much less extensive at low temperatures.

V. DEFINITION OF KEROGEN

A. History and different meanings of the word kerogen.

According to Steuart [66], the word was proposed by A. Crum-Brown to describe the organic content of the oil shale in the Lothians (Scotland) which produce, by distillation, oils with a waxy (paraffinic) consistency (from Greek Keros = wax). However the term kerogen covered and still covers a variety of concepts which historically originated with the exploitation of oil shale and are closely linked to research on the origin of petroleum and especially with the development of theories on its organic origin. When we inventory the principal meanings of the word kerogen we find the following:

1. Organic matter in oil shale.

Historically, this definition is the result of the first extension of the definition of Crum-Brown. Since the OM in a variety of oil shales showed, when examined with the methods of the time, the same general characteristics as the Lothian shales, the conception was formed that this was a relatively well defined substance. For Cunningham Craig [16] it was the result of the inspissating of petroleum through argillaceous beds (and was consequently an asphaltoid substance). For most observers [2, 3, 53, 75] however, it was a syngenetic product. Under the microscope it consisted of a light yellow to brownish gel containing sometimes abundant "yellow-bodies" which were finally recognized as being derived from algae. This kerogen was therefore formed from algae and from products probably resulting from their decomposition, so that the terms algal and sapropelic were frequently used.

The very concept of OM in oil shale now seems very approximate since the notion of oil shale is essentially economic. Capacity for distillation under economically profitable conditions at a given time can be attained for a relatively diverse range of petrographic and chemical compositions and generating media [2, 3, 22, 31, 53]. In itself, therefore, it is not a sufficient basis for defining a certain type of OM.

2. *Sedimentary organic matter with a distillation yield of at least 50%.*

Historically, this concept was formulated by Himus and his coworkers [22, 31]. It defines a type of OM with a high capacity for distillation. At the same time, Himus describes another type of OM with little capacity for producing oil by distillation (10% at the most) called coaly OM. He supposes that intermediate situations can be obtained mainly by the mixture of these two types. Rocks whose organic content is composed essentially of kerogen are called kerogen rocks. When the concentration of OM is very high, the substance is called kerogen coal. Coal sensu stricto, also called humic coal, is the concentrated form of the type of organic matter with little capacity for distillation.

Under the microscope, this kerogen is characterized by the great abundance of macerals of the exinite family and/or of the amorphous fraction. It is often called sapropelic. The fraction of the OM which is not kerogen is characterized by the great abundance of macerals of the vitrinite family and/or of ligneous debris. This fraction is often called coaly or humic. We thus find here the mixture which we have mentioned above of optical and genetic criteria.

3. *Sedimentary organic matter capable of producing oil through artificial distillation (pyrolysis) or natural distillation (evolution during burial).*

This concept is historically derived from the parallel already indicated in 1915 by White [84, 85], between pyrolysis and evolution due to burial. It was then formulated by Trager [75] and applied concretely by Takahashi [69] to research on petroleum source rocks in the oil-bearing sedimentary basins of Japan.

This concept indicates a more comprehensive view than the preceding ones concerning the group of sedimentary organic substances since according to it the OM which produces oil through distillation is of the same nature as that producing oil during burial.

In addition, the characteristics of this sapropelic kerogen are those of the preceding definition. It is thus here again opposed to a coaly OM whith little capacity for producing oil by natural or artificial distillation.

We note that, in the three preceding concepts, kerogen loses its distinctive properties during burial. For (see above) as hydrocarbon formation proceeds, the remaining OM loses hydrogen and therefore its capacity for the distillation and formation of hydrocarbons. The optical characteristics also change, and there is a convergence of the optical properties of kerogen and of coaly OM.

4. *Insoluble organic matter resulting from the condensation of lipids.*

This concept was defined by Breger [6, 7, 8, 9] and is based on the following considerations:

Of the main chemical families consituting living OM, i.e. lipids, carbohydrates, proteins, lignin, etc., only lipids and lignin would be preserved to a significant degree during the formation of sedimentary OM. Part of the lipids would produce oil during burial. Another part would be condensed (polymerization) into an insoluble product, kerogen, which is also capable of producing oil but only under exceptional conditions (for example extensive burial). For structural reasons lignin derivatives are not able to form oil.

Erdman [26] gives a definition of kerogen which is close to that of Breger but in more general terms. In recently formed sedimentary OM he distinguishes two main classes of constituents. Only one of these will produce oil and kerogen during burial. The opposition mentioned above between a type of OM capable of generating hydrocarbons and a type unable to do so is thus found again in Breger and Erdman, this time at the level of the chemical structures. Kerogen, however, which elsewhere was a potential source of hydrocarbons, is here rather a by-product of their formation.

5. *The fraction of organic matter dispersed in sediments which is insoluble in organic solvents.*

Historically, the creators of this concept seem to have been Forsman and Hunt [28] who defined it on the basis of the OM disseminated in ancient sediments. Today this is the most widespread meaning of the word kerogen.

Originally this was a petroleum geochemistry concept since organic solvent extraction is used for examining the relations between hydrocarbons and the rest of the OM. In this sense a synonym of the word kerogen is **kerabitumen** which was recommended by the nomenclature committee of the *4th World Petroleum Congress*. However the term kerabitumen is still not in general use.

The definition depends on the nature and the conditions of use of the solvent. In practice (*cf.* paragr. III.B) the solvents are not very polar (chloroform, benzene, methanol-benzene) and are used at temperatures below 80° C. Under these conditions (Chapter 2), the quantities of OM dissolved (bitumen) still vary considerably, depending upon the solvent and the extraction conditions. However there is little variation in the quantities of hydrocarbons dissolved.

6. *Miscellaneous.*

The following meanings can also be mentioned:

(a) Total sedimentary OM [46].
(b) The insoluble OM in any kind of rock [61].
(c) The sedimentary OM insoluble in organic solvents and alkalis.
(d) The sedimentary OM non hydrolyzable by hot mineral acids.

This variety of concepts reflects the complexity of the sedimentary OM, and a clearer idea of this complexity can be formed on the basis of attempts at classification [74, 8, 14, 68]. This also indicates the diversity of the interpretations of the transformation processes in sediments, especially as concerns petroleum formation.

B. The definition used here: sedimentary organic matter insoluble in the usual organic solvents. Why this fraction should be studied.

In the present book we **call kerogen sensu stricto the fraction of sedimentary OM which is insoluble in the usual organic solvents** (under the conditions indicated in paragr. V.A.5) as opposed to the soluble fraction called **bitumen**. However we do not limit our use of the word to organic matter disseminated in ancient sediments as is the case of current uses of the term, and we apply it to **all sedimentary OM**. Thus we include in sedimentary OM the humic coals of various ranks (peat, lignite, bituminous coal, anthracite), boghead coal, cannel coal, asphaltoid substances (asphalts, bitumens, tar in tar sands), OM in recent sediments and soils.

The use of a single term for designating the insoluble fraction of sedimentary OM in general is justified by the progressive convergence now observed between the methods and philosophies of the various disciplines dealing with sedimentary OM. To an increasing degree a general conception of sedimentary OM is being developed which we are presenting here.

According to this view, the organic materials found in soils, recent and ancient sediments, coals, etc. are considered as special cases. Formation and evolutionary processes account for the transitions from one special case to the next. Thus the types of OM generating hydrocarbons are thus less and less contrasted with those which are not capable of doing so. There does not seem to be any criterion whether of a genetic or physico-chemical kind which seems to be sufficient to establish such sharply defined categories in spite of technical developments, and there are transitions between the various evolutionary stages of the OM and between the types of OM. (We will, however, see that some types are better represented, and this phenomenon is the basis for classifications which are too sharply defined).

The term used for designating this insoluble fraction is finally of secondary importance. It is more important to see the significance of its study.

The main reason is that examination of the relations between the soluble and insoluble fractions in standard organic solvents is a necessary stage in the understanding of the origin and formation of petroleum in sediments.

The soluble fraction contains "free" hydrocarbons which are formed during geological processes, whereas the insoluble fraction is where the largest proportion of compounds can be found that are liable to form hydrocarbons (resins and asphaltenes recovered from the soluble fraction are also capable of such formation).

Secondly, let us note that the insoluble nature is linked to a condensation state of OM, denoting a certain degree of stability under geological conditions (and even cosmic, *cf.* Chapter 15) which may justify, as is done in several of the texts presented, the use of the word **kerogen** to designate, in a more general way, a "polycondensed" or "polymerized" state of OM not belonging to the living realm (geopolymers).

From a more practical standpoint, it must be pointed out that what we call kerogen must be analyzed to achieve a physico-chemical description of sedimentary OM. Indeed, if the structures are not to be degraded, the only procedure now possible for as complete a physico-chemical analysis as possible is extraction by a solvent not having any solvolysis effect and the analysis of the products thus extracted, followed by an analysis of the insoluble residue. Yet this residue is by far the largest quantitative proportion, and so we cannot pretend to describe the sedimentary OM without having examined it.

Because of this quantitative importance, the physico-chemical description of kerogen is thus a good approximation of that of the sedimentary OM. This is why its analysis, as will be seen later on, serves to go well beyond an examination of oil-forming phenomena. (This quantitative importance also justifies the procedure that was used to deal with the optical analysis of kerogen (Chapters 3, 11 and 12). The litterature contains almost no optical description of kerogen, sensu stricto, because petrographers only very rarely practice solvent extraction before analysis. Their observations and conclusions are however applicable to kerogen).

Insolubility requires the development and use of methods which are not generally used for analyzing soluble fractions. It is thus a federating element from the analytical standpoint which by itself justifies the assembling of most of the texts presented in this book.

There are thus technical, theoretical and finally historical reasons for analyzing the insoluble fraction of sedimentary organic matter per se. However, the disadvantages of making it into a separate category must be looked at from the methodological standpoint.

This fraction is a heterogeneous mixture in which a priori neither the origin of the constituents nor the stages they have attained in the processes of oil formation are distinguished. As a result, kerogen contains:

(a) OM incapable of forming oil because, for example, it has undergone an oxidation process at the same time as it was incorporated into the sediment or before it was thus incorporated, such as fusinite, or again because it has undergone a previous sedimentary cycle, such as various reworked products.

(b) OM at varying stages of the oil formation process, ranging from little evolved OM (peat, insoluble organic contents of recent sediments) to insoluble residues that are very rich in carbon and are created by this formation;

(c) The insoluble fraction of "bitumens" as defined by petrographers, which are probably themselves the result of the migration of part of the hydrocarbons formed.

The fact of creating this category "kerogen" thus has, a priori, no genetic significance from either the standpoint of the origin of the constituents of the sedimentary OM or even from the standpoint of the origin of oil (hence we are a long way off from the etymological standpoint). It is only by a more thorough analysis, with emphasis being placed on optical methods for evaluating the heterogeneity of the organic content, that affiliations can be established between the different constituents.

The creation of this category thus has a highly pragmatic nature and so cannot satisfy naturalists (*cf.* Chapter 11) and even less philosophers. Unfortunately this is the case of most categories that we now know how to distinguish in sedimentary OM, and a great deal of progresses will have to be made in our knowledge of such matter before entirely satisfactory classifications are established.

VI. QUANTITATIVE IMPORTANCE

A. Quantities of kerogen imbedded in sediments.

Sediments almost always contain OM. Sandstone contains the least while some rocks (coal, boghead coal and veins of bitumen or asphaltite, etc.) consist almost entirely of it.

Kerogen, defined as the fraction of this sedimentary OM (including coal, asphalt and bitumen, the OM in soils and recent sediments, etc.) which is insoluble in organic solvents, makes up the largest part. It generally represents more than 95% of the weight in the case of recent OM. This proportion progressively falls during burial because of the formation of soluble and/or volatile products, especially hydrocarbons. However the extent of this formation varies considerably depending on the nature of the origin of the OM (10-80% of the initial OM depending on the case). In view of the variety of the situations, it is hardly possible to estimate the average proportion of kerogen in the total sedimentary OM.

The content by weight of kerogen for a sediment is generally estimated on the basis of its organic carbon content. This method leads to excessive underestimates for slightly evolved sediments (recent or shallow buried). This is true because the carbon content of slightly evolved kerogens is about 60-70% and, as we have seen (paragr. IV.B), at this stage kerogen constitutes almost all the OM. In order to obtain the kerogen content it is therefore necessary to multiply the organic carbon content by 1.5 or 1.6. During evolution due to burial, the organic carbon content of kerogens increases and can exceed 90% for very evolved sediments. At the same time the proportion of kerogen in the OM falls, but part of the products formed is no longer in the rock at the time of analysis, either because these products have left the rock in situ (migration) or because they have been lost during handling. Finally, at these stages, the remaining soluble and/or volatile products rarely form more than 20% of the total OM, and the organic carbon content in the end gives an estimate which is in the case quite close of kerogen content.

For all sediments, the organic carbon content of a sediment is an underestimate of its kerogen content, and it is generally admitted that the result must be multiplied by a coefficient of 1.2 to 1.3. However in view of the approximate nature of the calculations below, we will evaluate kerogen content on the basis of the organic carbon content.

The distribution of carbon in the earth's crust has recently been estimated [36, 49, 80]. We shall use here the estimate by Hunt [36], based on the work of Ronov et al. [56]. According to him, the total mass of organic carbon contained is sediments is 1.2×10^{16} t. Other studies, most of which are older [20, 34, 45, 81, 83] indicate quantities of between 10^{15} and 3.5×10^{16} t. The calculations of Hunt are partly based on a more recent evaluation of the total mass of sediments existing in the oceanic domain now thought to constitute about 8% of the total sediment mass, and also on a new estimate of the average organic carbon contents about which more data is now available. These calculations are still, however, largely subject to revision, especially as concerns the estimates for average organic carbon content. The latter figure is till distorted by the sampling since there is more data for the present continental domain than for the present subsea area, and there is more for shallow sediments than for deep ones. The figure of 10^{16} t will therefore be considered as indicating only an order of magnitude for the quantities of organic carbon and therefore of kerogen which are present in sediments. This figure represents [36] 10-15% of the total carbon (organic and inorganic) contained in the earth's crust (carbonates in sedimentary rock 50-60%, carbonates in non-sedimentary rock 25-30%), while the rest is found in the form of elementary carbon (graphite, diamonds, carbonaceous particles, etc.) in non-sedimentary rocks, i.e. eruptive, crystalline and metamorphic rocks. The quantities of carbon contained in the biomass, atmospheric CO_2 and that dissolved in the oceans are smaller by several orders of magnitude.

Table 1.1 compares quantities for the different forms of carbon. Most of the figures given are from Hunt [36] but have been rounded off. For the biomass, a range has been given on the basis of various estimates [82, 36, 13]. It is difficult to attain a high degree of accuracy for this figure.

TABLE 1.1
QUANTITIES OF THE DIFFERENT FORMS OF CARBON IN THE EARTH'S CRUST
(from Hunt [36])

Forms of carbon	Quantities (t)
Kerogen	$1 \cdot 10^{16}$
Carbonates	
in sedimentary rocks	$6 \cdot 10^{16}$
in non sedimentary rocks	$1 \cdot 10^{16}$
dissolved in oceans	$5 \cdot 10^{13}$
«Elemental» carbon (mainly in basaltic, granitic and metamorphic rocks)	$1 \cdot 10^{16}$
Dissolved in oceans (CO_2, organics)	$1 \cdot 10^{12}$
Atmosphere (CO_2)	$1 \cdot 10^{12}$
Biomass (living organisms)	$0.3 - 3 \cdot 10^{12}$

According to various sources [1, 4, 18, 21, 24, 41, 58, 62, 65, 86, 87, 88] the mass of carbon entering synthesized organic products each year through photosynthesis processes (primary organic productivity) is 1.5 to $7 \cdot 10^{10}$ t for the oceans and 1.5 to $8 \cdot 10^{10}$ t for the continental areas. A small fraction estimated at 0.01% to 10% of this quantity, depending on the sedimentation media [38, 48, 54, 82], escapes the biological cycle or surface alterations and is incorporated in sediments. The ratio between the quantities of kerogen present in sediments and the quantities incorporated annually through the water-sediment interface might possibly give an idea of the length in years of the "long" organic carbon cycle (as opposed to the "short" biological cycle). This estimate is, however, very doubtful since the estimate for the rate of incorporation by sediments is very inaccurate while this rate varies a great deal depending on the sedimentation media. It is also probable that this rate has varied at various geological epochs. The calculation gives about 10^8 years for the length of this long cycle.

The elementary carbon in crystalline rocks or the carbon in carbonates is finally partly of organic origin. The former is partially the result of the degradation of kerogen during burial ("carbonaceous residue", *cf.* Chapter 4 and above) while the second is often a by-product of biological activity. Theoretically, the reincorporation of these forms of carbon into the biological cycle takes place through their transformation into CO_2 or CO_3^{--} by the action of surface alteration, degradation in sediments and magmatic and volcanic phenomena. The broadest estimate for this long carbon cycle should therefore take into account the other forms of sedimentary carbon, so that the preceding length of time must be multiplied almost by 10.

The length of time obtained in this way is so great that the validity of this kind of calculation becomes very doubtful.

Figure 1.2 situates the total quantity of kerogen in relation to the ultimate resources for the main categories of fossil fuels. By ultimate resources we mean all tonnage recoverable from deposits, whether these are proven or to be discovered and whether or not this tonnage is economically workable under present conditions but which might become workable in the future according to reasonable projection. The above tonnage is opposed to reserves which constitute a small part of the resources which are both proven and recoverable at the present time. For the ultimate resources for oil, gas and asphalt we have selected the figures of 4×10^{11} t, 2×10^{11} t and 3×10^{11} t, taking into account the fact that in this field the margin of error is large [5].

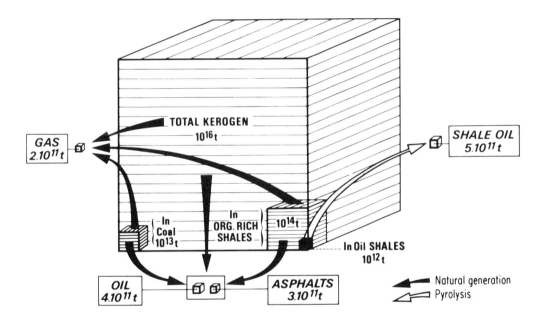

Fig. 1.2. — Comparison between the quantity of kerogen and the ultimate resources of fossil fuels.

Resources for asphalts may appear very high in the light of current figures. This is based on Demaison's recent estimate [19] which gives the figure of about 1.5×10^{11} t for the in situ asphalt in the three major asphaltic provinces: Orinoco (Venezuela), Athabasca (Canada) and Melekess (USSR).

If we suppose that most of the hydrocarbons found in oil, gas or asphaltic fields come from the transformation of kerogen, a comparison of the quantities of kerogen and quantities of oil, gas and asphalt resources expresses the average final yield of natural transformations (petroleum and gas formation, migration) and them of recovery, leading from kerogen to recoverable hydrocarbons. This yield is considered as between 10^{-5} and 10^{-4}.

Hunt [36] estimates the quantities of oil (liquid hydrocarbons) disseminated in sediments (the micropetroleum of Russian authors) at about 2×10^{14} t. The quantities of petroleum *in situ* in exploitable fields which are either proven or to be discovered would amount to 10^{12} t (if we multiply ultimate oil resources by 2.5, i.e. fixing at 40% the average rate of recovery which can finally be expected). If we consider on the other hand that asphalt also derives from this micropetroleum the average yield of the concentration processes leading from disseminated oil to exploitable fields would then be about 1%.

A similar calculation gives an average yield of about 2% for the transformation of kerogen into oil (liquid hydrocarbons). This is a mean value for the whole sedimentary column; therefore the mean maximum rate of this transformation is higher.

However these figures are very questionable. Firstly they do not reflect the diversity of geological situations, and secondly they are based on rough estimates.

The coal ressources indicated of 10^{13} t are from estimates from *World Coal* 1965 [89]. According to Feys [27] this estimate varies from 7 to 17 × 10^{12} t. The present reserves are about 0.5 × 10^{12} t [89]. The estimate of oil shale resources is very uncertain. We have therefore selected two values, one for oil shale in the broad sense, i.e. shale whose organic carbon content is more than 5%. According to Duncan *et al.* [23] their organic content amounts to 10^{15} t. This estimate seems very high, and based on the experience gained at *Institut Français du Pétrole (IFP)* we would be inclined to reduce this value by an order of magnitude. This figure would also be more in agreement with the figure by Vassoyevitch *et al.* [80], who gives 2.10^{14} t for the organic content of shales having an organic carbon content of between 4 and 10% and that by Nesterov *et al.* [49], which gives about 10^{13} t for shales having an organic content of more than 3%. We would thus have 10^{14} t of kerogen which could give 3 × 10^{13} t of oil by pyrolysis, i.e. 75 times the petroleum resources. The other value was obtained by multiplying by 2 an estimate of oil quantities obtainable through the pyrolysis of the presently inventoried oil shales on the basis of current technology (oil shale reserves), i.e. by fixing at 50% the average rate of transformation of kerogen from these shales into oil. This estimate [5, 10, 23] is about 0.5 × 10^{12} t which therefore gives 10^{12} t for the kerogen in these shales.

B. Distribution of kerogen in sediments.

Our knowledge of this distribution is based principally on the work of Trask *et al.* [76], who have compiled organic carbon contents in sediments in the US on a large scale, and on the work of Ronov [55] and Ronov *et al.* [56] who have done the same thing for the Russian platform, and also, for some aspects, on the work of Hunt [34, 35, 36] and of Gersanovic *et al.* [*in:* 50], Vassoyevitch *et al.* [80], Nesterov *et al.* [49].

The distribution of organic carbon, depending upon geological age since the Cambrian, shows two minimums. The first was in the Silurian and the second in the Triassic (Fig. 1.3). These classic results can be interpreted [25] as a variation in primary organic productivity which was due to variations in the climate or in the quantities of atmospheric CO_2. It is also possible to suppose a relative reduction, due to regressions of great amplitude which seem to have marked these epochs, in the surface of the continental margins which appear to be especially favorable for the accumulation and preservation of organic matter [25, 50].

The distribution of organic carbon, depending upon the main types of sediments, has been the subject of statistics for the present continental and oceanic domains (Table 1.2). The continental domain includes not only the continents but also the continental shelves and slopes. Average organic carbon contents are low. On the average clays and shales are richer than carbonates which are in turn richer than sandstone. Within each of these categories, the finest grained sediments are usually richest in organic carbon. The published literature does not

Sedimentary organic matter and kerogen

TABLE 1.2
ORGANIC CARBON IN SEDIMENTS
(from Hunt [36])

Location	Mean Values (Wt. %)	Mass (10^{16} t)
Continents, Shelf and Slope		
Clays and Shales	0.99	0.82
Carbonates	0.33	0.08
Sands	0.28	0.09
Oceanic		
Clays and Shales	0.22	0.07
Carbonates	0.28	0.10
Siliceous	0.26	0.04

give distributions of values for each sedimentary category. It is, however, possible to deduce on the basis of what has been said above that more than 95% of the kerogen is found in sediments with less than 5% organic carbon.

Kerogen seems to be linked to the finest grain size fractions of the sediments. For example, Hunt [35] found the following distribution in Viking Shale (Alberta) in terms of grain size:

Grain Size	Average OM Wt. %
Siltstone	1.79
Clay 2-4 μ	2.08
Clay less than 2 μ	6.50

However it should be observed that the method by which the fractions are recovered, generally centrifuging or decantation, can lead to the enrichment of fine fractions with kerogen although there may not be a real bond.

Examination of the distribution between the oceanic and continental domains snows that the largest quantities of kerogen are to be found in the clay and shale of the continental margins. This is confirmed by the work of Gersanovic *et al.* [*in:* 50] according to which more than 85% of the OM sedimented in the Holocene is to be found on the continental slopes.

Observations on the distribution of kerogen in terms of geological age, type of sedimentation and sedimentation domain are important for petroleum exploration since there is a clear correlation between the capacity of a sedimentary medium to fossilize OM and its petroleum possibilities. Thus Ronov [55] observes that the average organic carbon content of sediments is three times higher in petroliferous basins than in non-petroliferous basins. There is thus a correlation between the distribution of petroleum reserves and that of the average carbon content of sediments in relation to the geological age (Fig. 1.3).

While these observations are very interesting, much more fruitful observations could no doubt be made within the framework of studies on sedimentary models. In previous research, a very small number of sedimentary categories have been chosen and they are therefore badly defined. Only average organic carbon content values have actually been used, whereas it would also be interesting to study the distributions.

The nature of kerogen has not been taken into consideration. Moreover, when more detailed studies are available, they concern anomalous accumulations such as coals, oil shale, etc. which, as seen above, represent a very small portion of sedimentary OM. Therefore a true sedimentology of OM has still not been created. The work will be difficult since the phenomena determining sedimentation of OM have still not yet been well understood and weighted (Chapter 14).

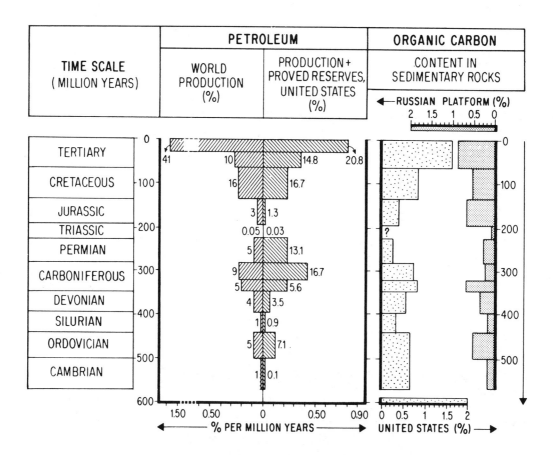

Fig. 1.3. — Geological distribution of oil and organic carbon (from J. Debyser and G. Deroo, (1969), *Rev. Inst. Franç. du Pétrole*, **XXIV**, 1, 21).

Little work in this direction has been published. As an example we can cite an interesting study by Cazes *et al.* [11] on the distribution of kerogen in the Oxfordian in the E of the Paris Basin which is a marly-carbonate medium with reefal limestone. This statistical study takes into consideration a large number of sedimentological and petrological variables. The fossilization of kerogen in this medium finally appears to be linked to a very calm type of sedimentation, in most cases accompanied by an argillaceous element facilitating the preservation of the OM. We can also cite the publications of the *Orgon* Mission in which *Centre National pour l'Exploitation des Océans (CNEXO), IFP* and French oil companies are associated [17].

REFERENCES

1. Basilevich, N.I., Rodin, L.E. and Rozov, N.N. (1971), "How Much Does the Living Matter of Our Planet Weight?" (in Russian), *Prinoda*, No. 1.
2. Bertrand, C.E., (1893), *Bull. Soc. Belge de Géologie*, **7**, 45.
3. Bertrand, C.E., (1897), *2nd Lecture to the Belgian Society of Geology, Paleontology and Hydrology*, 19 Oct. 1897.
4. Bogorov, V.G., (1971), in: *Organic Matter of Recent and Ancient Sediments*, (in Russian), N.B. Vassoyevitch ed. Akad. Nauk SSSR, Moscow.
5. Boy de la Tour, X., Brasseur, R. and Le Duigou, J., (1976), *Rev. Inst. Franç. du Pétrole*, **XXXI**, 741.
6. Breger, I.A., (1960), *Geochim. et Cosmochim. Acta*, **19**, 297.
7. Breger, I.A., (1961), "Kerogen", in: *McGraw Hill Encyclopedia of Science and Technology*, **9**, 337.
8. Breger, I.A., (1963), in: *Organic Geochemistry*, Pergamon Press, Oxford, **3**, 50.
9. Breger, I.A., (1976), 2nd IIASA Conference, *Future Supply of Nature-made Oil and Gas*, **51**, 913.
10. Burger J., (1973), *Rev. Inst. Franç. du Pétrole*, **XXVIII**, 315.
11. Cazes, P. and Reyre, Y., (1976), *Bull. BRGM* (2), IV, 2-1976, 85.
12. Combaz, A., (1964), *Rev. Micropaléont.*, 7, 4, 205.
13. Combaz, A., (1973), in: *Advances in Organic Geochemistry 1973*, B. Tissot and F. Bienner ed., Editions Technip, Paris, 1974, 423.
14. Combaz, A., (1973), in: *Pétrographie de la Matière Organique des Sédiments — Relations avec la Paléotempérature et le Potentiel Pétrolier*, B. Alpern ed., CNRS, Paris, 1975, 93.
15. Correia, M., (1967), *Rev. Inst. Franç. du Pétrole*, **XXIV**, 12, 1417.
16. Cunningham. Craig E. H., (1916), *Journ. Inst. Petr. Tech.*, 1916, 2, 246.
17. CEPM-CNEXO, (1974), *Géochimie Organique des Sédiments Marins Profonds — Orgon I — Mer de Norvège*, CNRS, Paris, 1977.
18. Dalton, L.V., (1909), *Econ. Geol.*, 4, 603.
19. Demaison, G.J., (1977), *AAPG Bull.*, **61**, 11, 1950.
20. Dietrich, G., (1963), *General Oceanography — An Introduction*, Interscience Pub., New York, 229.
21. Douce, R. and Joyard, J., (1977), *La Recherche*, **79**, 527.
22. Down, A. L. and Himus, G. W., (1940), *J. Inst. Petroleum*, **26**, 329.
23. Duncan, D. C. and Swanson, V. E., (1965), *Geological Survey Circular*, **523**.
24. Duvigneaud, P., in: Dajoz, (1970), *Précis d'Ecologie*, Dunod, Paris.
25. Debyser, J., (1975), *Rev. Inst. Franç. du Pétrole*, **XXIV**, 22.
26. Erdman, J.G., (1975), *Proceedings of the 9th World Petroleum Congress*, **2**, 139.
27. Feys, R., (1976), *Annales des Mines*, January 1976, 9.
28. Forsmann, J.P., and Hunt, J.M., (1958), in: *Habitat of Oil*, L.G. Weeks ed., AAPG, 747.
29. Forsmann, J.P., (1963), "Geochemistry of kerogen," in: *Organic Geochemistry*, I.A. Breger ed., Pergamon Press, 1963, 148.
30. Gutjahr, C.C.M., (1966), *Leidse Geol. Meded.*, **38**, 1.
31. Himus, H.W. (1941), in: *Oil Shale and Cannel Coal*, **2**, Institute of Petroleum, London, 1951.
32. Hunt, J.M., Steward, F., and Dickey, P.A., (1954), *AAPG Bull.*, **38**,8, 1671.
33. Hunt, J.M., and Jamieson, G.W., (1958), in: *Habitat of Oil*, L.G. Weeks ed., AAPG, 735.
34. Hunt, J.M., (1961), *Geochim. et Cosmochim. Acta*, **22**, 37.
35. Hunt, J.M., (1963), in: V. Bese, *Vortrage der III int. viss. Konf. für Geochemie, Microbiologie und Erdölchemie*, Budapest, 8-13/10/1962, Budapest, 1963.
36. Hunt, J.M., (1972), *AAPG Bull.*, **38**, 377.
37. Hunt. J.M., (1974), in: *Advances in Organic Geochemistry 1973*, B. Tissot and F. Bienner ed., Editions Technip, Paris, 1974, 593.
38. Ivanenkov, V.N., (1977), "Oxidation Rate of Organic Matter in the World Ocean". *Proceedings of the 8th International Meeting of Organic Geochemistry*, Moscow, 1977.
39. Kinney, J.W., (1923), PhD Thesis, McGill University.
40. Kinney, C.R., and Schwartz, D., (1957), *Ind. Eng. Chem.*, **49**, 1125.
41. Koblenz-Mishke, O.M., Volkovinsky, V.V., and Kabanova, J.G., (1968), in: *Symp. Sci. Explos. South Pacific*, N.S. Wooster ed., Scripps Institute of Oceanography, La Jolla, California, 183.

42. Larskaya, Y.S., and Zhabrev, V.D., (1964), *Dokl. Akad. Nauk SSR, 157,* N° **4** 897 (in Russian).
43. Louis, M.C., (1964), *in: Advances in Organic Geochemistry 1964,* G.D. Hobson and M.C. Louis ed., Pergamon Press, Oxford, 85.
44. Louis, A., and Tissot, B., (1967), *Proceedings of the 7th World Petroleum Congress,* **2**, 47.
45. Mason, B., (1966), *Principles of Geochemistry,* 3rd ed., John Wiley, New York.
46. Manskaya, S.M., Kodina, L.A., and Generalova, V.N., (1976), *Geokhimiya,* **1**, 3.
47. Menzel, D.W. and Ryther, J.H., (1970), *in: Organic Matter in Natural Waters,* Alaska Univ. Marine Sci. Inst. Publ, **1**, 31.
48. Menzel, D.W., (1974), *in: The Sea,* E. Goldberg ed., John Wiley, New York **5**, 659.
49. Nesterov, I.I., and Salmanov, F.K., (1976), 2nd IIASA Conference, *Future Supply of Nature-Made Oil and Gas,* **51**, 913.
50. Pelet, R., Cauwet G. and Combaz, A., (1975), *Rev. Inst. Franç. du Pétrole,* **XXX**, 18.
51. Philippi, G.T., (1965), *Geochim. et Cosmochim. Acta,* **29**, 1021.
52. Raynaud, J.F., and Robert, P., (1976) Bull. Centre Rech. Pau, SNPA, **10**, 1, 109.
53. Renault, B., (1899), "Sur Quelques Microorganismes des Combustibles Fossiles,", *Bull. Soc. Industries Min.,* 3ᵉ Ser. **13**, 865; 1900, 3ᵉ Ser. **14**, 5.
54. Romankevitch, E.A., Artemiev, V.E. and Belyaeva, A.N., (1977), "The Geochemistry of Organic Matter in Bottom Sediments of Seas and Oceans," *Proceedings of the 9th International Meeting on Organic Geochemistry,* Moscow, 1977.
55. Ronov, A.B., (1958), *Geochemistry,* **5**, 510.
56. Ronov, A.B., and Yaroshevkiy, A.A., (1969), *in: The Earth's Crust and Upper Mantle,* Am. Geophys. Union Geophys. Mon. Ser., **13**, 37.
57. Rubey, W.W., (1951), *Geol. Soc. America Bull.,* **62**, 1111.
58. Ryther, J.H., (1963), *in: The Sea,* **2**, M.N. Hill ed., Interscience Pub., New York.
59. Ryther, J.H., (1969), *Science,* **166**, 72.
60. Ryther, J.H., (1970), *Nature,* **227, 5276,** 374.
61. Saxby, J.D., (1976), *in: Oil Shale,* T.F. Yen and G.V. Chilingarian ed., Elsevier, 104.
62. Skopintsev, B.A., (1961), *in: Recent Sediments of Seas and Oceans,* Akad. Nauk. SSR. Moscow, 285 (in Russian).
63. *Stach's Textbook of Coal Petrology,* 2nd ed., Borntraeger, 1975.
64. Staplin, F., (1969), *Bull. Canad. Petrol. Geol.,* **17**, 1, 47.
65. Steeman, Nielsen E. and Jensen, J.A., (1957), *Galathea Rep.,* **1**, 49.
66. Steuart, D.R., (1912), *in: The Oil Shales of the Lothians — Part III: The Chemistry of the Oil Shales,* 2nd ed., Memoirs of the Geological Survey, Scotland, 143.
67. Stopes, M., (1935), *Fuel,* **14**, 4.
68. Subbota, M.I., Khodjakuliev, I.A., and Romaniuk, A.F., (1976), *in: Research on the Organic Matter in Contemporary and Fossil Deposits* (in Russian), Nauk ed., Moscow, **65**.
69. Takahashi, J., (1935), *Sci. Rpts. Tohoku Univ., 1*, 3, 63.
70. Teichmüller, M., (1958), *Rev. Industrie Minérale,* spec. issue, 15 July 1958, 99.
71. Tissot, B., (1969), *Rev. Inst. Franç. du Pétrole,* **XXIV**, 4, 470.
72. Tissot, B., and Pelet, R., (1971), *Proceedings of the 8th World Petroleum Congress,* **2**, 35.
73. Tissot, B., and Espitalié, J., (1975), *Rev. Inst. Franç. du Pétrole,* **XXX**, 5, 743.
74. Tomkeieff, S.I., (1954), *Coals and Bitumens,* Pergamon Press London.
75. Trager, E.A., (1924), *AAPG Bull.,* **8**, 301.
76. Trask, P.B., and Patnode, H.W., (1942), *Source Beds of Petroleum,* AAPG, Tulsa.
77. Trask, P.B., (1932), *Origin and Environment of Source Sediments of Petroleum* Gulf Publ. Col, Houston, Texas.
78. Van Krevelen, D.M., (1961), *Coal (Typology, Chemistry, Physics, Constitution),* Elsevier.
79. Vassoyevitch, N.B., Korghagina, Y.I., Lopatin, N.V. and Chernyshev, V.V., (1969), *Vestnik Mosk. Univ. 1969,* **6**, 3 (in Russian).
80. Vassoyevitch, N.B., Koniukov, A.I. and Lopatin, N.V., (1976), International Geological Congress, 25th Session, *Report of Soviet Geologists on Combustible Mineral Resources,* 7.
81. Weeks, L.G., (1958), *in: Habitat of Oil,* AAPG 1958, Tulsa, 1.
82. Welte, D.H., (1970), *Naturwissenschaften,* **57**, 17.
83. Wickman, F.E., (1956), *Geochim. et Cosmochim. Acta,* **9**, 136.
84. White, D., (1915), *J. Wash. Acad. Sc.,* **6**, 189.
85. White, D., (1916), *Geol. Soc. America Bull.,* **28**, 1917.
86. Whittle, K.J., (1976), *in: concepts in Marine Organic Chemistry,* Symposium, Edinburgh.
87. Williams, P.J., (1975), *in: Chemical Oceanography,* 2nd ed., J.P. Riley and G. Skinow ed., **2**, 301.
88. Winberg, G.G., (1960), *Primary Productivity of Aquatoria,* Akad, Nauk. SSSR. Minsk.
89. *World Coal,* (1975), Miller Freeman Publication, San Francisco.

Relationship Between Petroleum Composition and Depositional Environment of Petroleum Source Rocks[1]

J. MICHAEL MOLDOWAN,[2] WOLFGANG K. SEIFERT,[2] and EMILIO J. GALLEGOS[3]

ABSTRACT

Crude oils of nonmarine source can be distinguished from those of marine shale source and from oils originating in marine carbonate sequences by using a battery of geochemical parameters, as demonstrated with a sample suite of nearly 40 oils. A novel parameter based on the presence of C_{30} steranes in the oil was found to be a definitive indication of a contribution to the source from marine-derived organic matter. A second novel parameter based on monoaromatized steroid distributions was effective in helping to distinguish nonmarine from marine crudes and can be used to gauge relative amounts of higher plant input to oils within a given basin. Sterane distributions were similarly useful for detecting higher plant input but were less effective than monoaromatized steroid distributions for making marine versus nonmarine distinctions. Concentrations of high molecular-weight paraffin can also be effective nonmarine indicators but are influenced by maturation and biodegradation processes. Certain algal-derived nonmarine oils may show little high molecular-weight paraffin response. Oils from carbonate sources (with a few exceptions) can be distinguished by having low pristane-phytane ratios, low carbon preference indexes, and high sulfur contents. Gammacerane indexes and carbon isotope ratios of the whole crude are not effective in distinguishing these types of environmental differences on a global basis.

INTRODUCTION

The origin of the organic material in sedimentary rocks and crude oil has received much attention (e.g., Tissot and Welte, 1978; Demaison and Moore, 1980). To identify the origin in sediments, the most direct method is macroscopic or microscopic examination of the remnants of organisms. However, some kerogens, particularly types I and II, commonly do not contain identifiable organic remains and can be entirely amorphous. Moreover, in thermally mature rocks (which are often of interest in petroleum exploration), characteristic structures of the organisms, if present, become increasingly difficult to identify, and no such structures exist in petroleum. In these cases, the approach has been to relate the chemical composition of the geolipid to that expected to be derived from dominant organisms, both precursors and reworking microbes, in a given ecosystem.

Numerous geochemical parameters to relate crude oil composition with source rock origins have been proposed. This study addresses the question of general applicability of certain parameters, particularly regarding marine versus nonmarine environments of petroleum source rock deposition. In addition, novel parameters based on sterane and monoaromatized steroid hydrocarbon distributions are introduced.

CRUDE OILS STUDIED

The crude oils analyzed in this study are listed in Tables 1 and 2. No more than two crudes from a single source are listed, in most cases, only one. From an individual state or country, several oils may be listed that represent different source depositional environments or ages. The source type designations for the oils were developed with the help of Chevron geologists and are based on nearly certain geologic assumptions. In many cases, the geologic assumptions are supported by geochemical correlations (see references listed in Tables 1, 2). Oils whose source designation is uncertain were omitted from the study as were those that appeared to have both marine and nonmarine source rocks (such as some oils from epicontinental source rocks).

The oils selected for this study were generally the least mature and least biodegraded available, in order to minimize the effects of maturation and biodegradation. However, some mature oils were used in the study. The few biodegraded oils in Tables 1 and 2 were omitted from consideration when the level of biodegradation would have greatly affected the parameter being considered.

This screening of oils resulted in a deceptively small group of samples. However, most of the samples represent a large group of oils, each with a similar or identical source, and thus similar or identical geochemical-source parameters.

NOVEL PARAMETERS

C_{30} Steranes

The presence of C_{30} steranes in crude oils derived from marine sources (marine crudes) was detected using metastable scanning gas chromatography–mass spectrometry (GC-MS/MS) (for methods see Haddon, 1979; Moldowan et al, 1983). The first application to the analysis of steranes by

---

©Copyright 1985. The American Association of Petroleum Geologists. All rights reserved.
[1]Manuscript received, November 30, 1984; accepted, April 19, 1985.
[2]Chevron Oil Field Research Company, P.O. Box 1627, Richmond, California 94802.
[3]Chevron Research Company, 576 Standard Avenue, Richmond, California 94802.

The writers are grateful to G. J. Demaison, R. W. Jones, and L. W. Slentz for providing geologic interpretations of oil sources, and to P. Albrecht, G. J. Demaison, and R. W. Jones for helpful discussions. We thank the following for providing samples: J. Connan for New Zealand oil, A. A. Petrov for USSR oils, Towner Petroleum for Ohio oils, D. F. Huang for Central Hebei oil of China, and C. Djerassi for a sample of gorgostane. Sample and data handling were by L. A. Wraxall, F. J. Fago, C. Y. Y. Lee, D. A. Dyke, and P. Novotny. L. A. Wraxall provided editorial assistance. Appreciation is extended to the management of Chevron Oil Field Research Company for permission to publish.

Composition and Depositional Environment of Petroleum Source Rocks

Table 1. Marine Oils Environmental Indicators

Source Type	Location	Identity	GC-MS Run No.	Steranes (%) C_{30}/C_{27}–C_{30}	Biomarkers Steranes 17α-Hopanes	Gammacerane[a] Index	nC_{31}[b] nC_{19}	Acyclics CPI[c]	Pristane[d] Phytane	δC^{13}[e] C^{12}	Sulfur (%)	References[f]
Shale	California	Carneros (late Miocene)	257	0.027	1.20	0.7	0.35	1.00	1.5	+5.4	0.36	1
Shale	California	Oceanic (Oligocene)	263	0.03	0.72	0.4	0.23	1.04	1.8	+2.1	0.43	1
Shale	California	Buena Vista (Miocene)	36	0.016	1.12	0.5	0.35	1.12	1.4	+6.2	0.5	2
Shale	Louisiana	Bay Marchand (Miocene)	277	0.074	0.53	0.6	0.12	0.99	1.9	+3.5	0.25	2
Shale	Louisiana	Ship Shoal (Miocene)	235	0.051	0.12	0.5	0.11	1.03	2.6	+3.2	0.24	3
Shale	Wyoming	Spring Valley (Cretaceous)	161	0.064	0.36	0.6	0.11	1.00	2.0	+2.7	0.06	4
Shale	Alaska	Kingak (Jurassic)	219	0.088	0.24	0.3	0.11	1.00	2.0	−2.0	0.32	5
Shale	Alaska	Sadlerochit (early Mesozoic)	202	0.11	0.27	0.4	0.18	1.05	1.5	−0.3	0.75	5
Shale	Alaska	Cook Inlet (Jurassic)	47	0.10	0.44	0.4	0.26	0.97	3.3	+2.6	0.23	—
Shale	Australia	Windalia (Late Cretaceous)	90	0.086	0.28	2.6	0.12	1.01	2.5	+3.3	0.01	6
Shale	Australia	Barrow (Jurassic)	94	0.085	0.26	2.3	h	h	2.5	+3.6	0.07	6
Marl	Canada	Kee Scarp (Devonian)	23	0.036	0.47	0.9	0.06	1.03	1.2	+0.10	0.34	3
Marl	Canada	Leduc (Devonian)	28	0.036	0.19	0.6	0.13	0.92	1.4	+1.50	0.26	—
Marl	Spain	Casablanca (Miocene)	434	0.032	0.44	0.2	0.28	1.03	1.7	+1.8	0.1	7
Carbonate	Spain	Amposta-Marino (Cretaceous)	435	0.016	0.38	3.8	0.27	1.00	0.7	+3.2	4.3	7
Carbonate	Italy	Gela (Triassic)	386	0.0	0.20	2.9	0.53	0.89	1.2	+1.3	6.0	—
Carbonate	Greece	Prinos	472	0.012	1.35	11.0	g	g	0.3	+7.4	3.6	8
Carbonate	Greece	Dragopsa (Paleozoic)	446	0.043	0.15	1.3	0.31	0.71	h	+0.1	4.1	8
Carbonate	Florida	Sunniland (Cretaceous)	494	0.029	0.13	0.8	h	h	0.5	+3.3	2.8	9
Carbonate	Wyoming	Manderson (Permian)	359	0.043	0.22	2.5	0.32	0.91	0.68	+1.7	2.0	3
Carbonate	A	Jurassic	404	0.038	0.09	0.4	0.24	0.96	0.65	+3.0	1.8	—
Carbonate	A	Cretaceous	405	0.032	0.42	0.4	0.17	0.96	1.3	+3.3	0.8	—
Carbonate	A	Cretaceous	406	0.024	1.00	0.6	0.21	0.93	1.6	+2.6	0.3	—
Carbonate	B	Cretaceous	478	0.043	0.07	0.7	0.11	0.99	0.68	+1.4	2.4	—
?	Ohio	Knox Dolomite (Cambrian)	425	0	0.22	1.0	0.18	0.97	1.5	−0.3	0.10	10
?	Ohio	Chepultepec Dolomite (Late Cambrian)	428	0	0.79	1.4	0.24	1.02	1.4	−0.8	0.03	10
?	USSR	Siva (Precambrian)	259-1	0	0.21	1.0	0.22	1.03	1.1	−2.3	0.07	11
?	USSR	East Siberia (Cambrian)	317	0	0.19	1.7	0[i]	1.0	0.96	−4.9	0.17	12

[a] (Gammacerane/17α, 21β(H)-hopane) × 100.
[b] Normal paraffins by capillary column gas chromatography.
[c] $2(n-C_{23} + n-C_{25} + n-C_{27} + n-C_{29})/2(n-C_{24} + n-C_{26} + n-C_{28}) + n-C_{22} + n-C_{30}$.
[d] Isoprenoids C_{19}/C_{20} by capillary column gas chromatography.
[e] Relative to petroleum standard NBS-22, ‰.
[f] References: (1) Seifert and Moldowan (1978), (2) Seifert and Moldowan (1979), (3) Seifert and Moldowan (1981), (4) Orr (1974), (5) Seifert et al (1980), (6) Volkman et al (1983b), (7) Seifert et al (1983), (8) Seifert et al (in press), (9) J. G. Palacas (1984), private communication, (10) Towner Petroleum (1981), private communication, (11) Seifert (1980), (12) A. A. Petrov (1979), private communication.
[g] Low peaks.
[h] Biodegradation.
[i] nC_{31} too low to measure.

Table 2. Nonmarine Oils Environmental Indicators

Source Type	Location	Identity	GC-MS Run No.	Biomarkers				Acyclics			δC^{13e} C^{12}	Sulfur (%)	References[f]
				Steranes (%) C_{30}/C_{27}–C_{30}	Steranes 17α-Hopanes	Gammacerane[a] Index	$\frac{nC_{31}}{nC_{19}}$[b]	CPI[c]	Pristane[d] Phytane				
Shale	Sumatra	Damar	307	0	<0.1	0.6	1.8	1.07	6.8		+1.4	0.07	1
Shale	Sumatra	Southeast Balam	351	0	<0.1	0.6	1.4	1.11	1.8		+7.8	0.14	1
Shale	Sumatra	Minas	305	0	<0.1	0.9	0.85	1.04	2.4		+4.5	0.13	1
Shale	Sumatra	Petapahan	306	0	<0.1	0.9	0.78	1.06	2.3		+1.3	0.18	1
Shale	Sudan	Unity	396	0	0.16	1.5	0.25	1.04	1.6		+0.1	0.08	—
Shale	Sudan	Abu Gabra	382	0	0.02	1.9	0.51	1.05	2.2		−1.1	0.04	—
Shale	Chad	Mangara	374	0	<0.1	3.5	0.56	1.05	2.4		+1.1	0.2	—
Shale	Chad	Tega	376	0	0.06	4.1	0.18	1.06	2.0		+2.1	0.2	—
Shale	Chad	Sedigi	372	0	<0.1	0.6	0.50	1.03	4.4		+2.8	0.03	—
Shale	Brazil	Baracica	438	0	0.02	1.8	0.71	1.07	1.9		+1.3	0.12	2
Shale	China	Taching	369	0	0.08	2.8	0.37	1.07	1.3		−1.1	0.09	3
Shale	China	Central Hebei	532	0	0.30	0.6	g	g	g		+2.6	0.38	—
Shale	New Zealand		479	0	0.09	0.2	0.36	1.06	9.4		+3.0	0.12	4
Marl	Utah	Green River	1	0	0.12	8.0	0.74	1.03	1.0		−1.2	0.04	5

[a][Gammacerane/17α, 21β(H)-hopane] × 100.
[b]Normal paraffins by capillary column gas chromatography.
[c]2(n-C$_{25}$ + n-C$_{27}$ + n-C$_{29}$)/2(n-C$_{24}$ + n-C$_{26}$ + n-C$_{28}$) + n-C$_{22}$ + n-C$_{30}$
[d]Isoprenoids C$_{19}$/C$_{20}$ by capillary column gas chromatography.
[e]Relative to petroleum standard NBS-22, o/oo.
[f]References: (1) Seifert and Moldowan (1981). (2) cf Demaison and Moore (1980). (3) Shangwen et al (1982). (4) Connan and Cassou (1980). (5) Seifert (1977).
[g]Biodegradation.

Warburton and Zumberge (1983) demonstrated the power of GC-MS/MS for quantitative and selective sterane analysis. As illustrated in Figure 1, GC-MS/MS allows detection and quantitation of steranes by monitoring direct precursors to the mass to charge ratio (m/z) 217 fragment ion. In effect, those molecular ions that decompose giving m/z 217 fragments in the field-free region of the mass spectrometer are recorded. Thus, the C_{30} sterane group, which is common in relatively minor amounts in marine crudes (Table 1), can be observed in Sadlerochit oil (Figure 1).

The internal distribution of the C_{30} sterane isomers invariably was observed to mimic that of the C_{29} steranes (Seifert and Moldowan, 1979). Thus, a pseudodiagnostic fingerprint for C_{30} steranes is available internally in a given GC-MS/MS run. Confirmatory evidence for the C_{30} steranes is shown in the GC-MS/MS scan of m/z 414 daughter ions of Sadlerochit oil (Figure 2). The fingerprint of C_{30} steranes obtained on the trace representing m/z 414→m/z 217 in Figure 2 is identical with that in the parent-ion scan (Figure 1). In the same run shown on Figure 2, we also observed the fingerprint of C_{30} methyl steranes (methyl group on ring A, B, or C) on the chromatographic trace m/z 414 → 231 and of C_{30} terpanes on m/z 412 → 191. The m/z 412 parents appear on the same chromatographic trace as m/z 414 parents in the GC-MS/MS runs, and daughters of m/z 412 were observed along with those of m/z 414 in the m/z 414 daughter-ion scans because of the relatively low resolution obtained using the electrostatic sector. A diagnostic C_{30} sterane pattern (Figures 1, 2) was absent from oils generated from nonmarine sources (nonmarine crudes). Instead, terpane parents are generally observed. This can be confirmed by the GC-MS/MS m/z 414 daughter-ion scan (shown in Figure 3), where a weak C_{30} methyl sterane pattern is detectable in the m/z 414 → 231 daughter-ion chromatogram.

A comparison of Tables 1 and 2 shows without exception that C_{30} steranes are present in oils from post-Devonian marine source rocks and absent in oils derived solely from lacustrine or nonmarine rocks. Thus, C_{30} steranes are highly specific compounds that can be used to detect a marine depositional history of crude oil source rocks. The application of the C_{30} sterane parameter may be prevented by intervening circumstances pertaining to high maturation levels, such as those found in some condensates where steranes are severely reduced in concentration, and also in heavily biodegraded oils where the steranes are removed. C_{30} steranes might also be difficult to detect in oils from some deltaic sources (such as the Mahakam delta) with a high relative abundance of higher plant input to the source (Hoffmann et al, 1984). Interestingly, C_{30} steranes are absent in four oils, two from Ohio and two from USSR (Table 1), suspected of being derived from Cambrian marine sources.

The origin of C_{30} steranes cannot be given with certainty due to our lack of evidence for their structure. However, it seems possible that they arise from the marine sterols identified mainly through the research of C. Djerassi and coworkers (Djerassi, 1981). These sterols occurring in marine invertebrates and marine algae have been found with variously alkylated side chains and range in size from 26 to 31 carbon atoms.

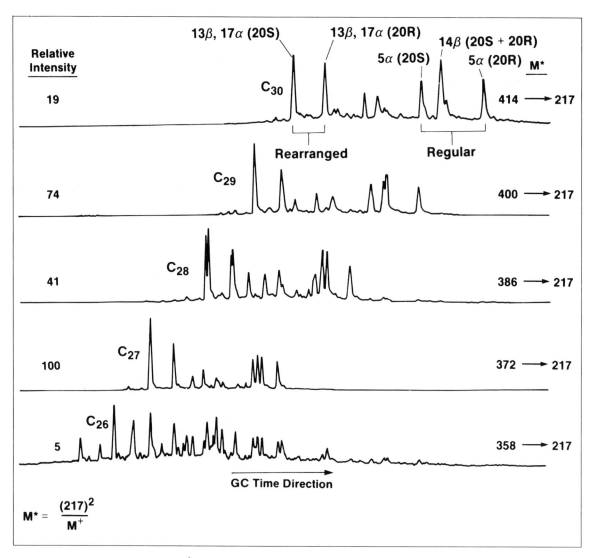

Figure 1—Parent-ion scans (B^2/E linked scans) for steranes, $M^+ \rightarrow 217$, in Sadlerochit, Alaska (marine) crude.

Present speculation on the major contributor or contributors to the C_{30} steranes in marine crude oils centers on 24-propylidenecholesterol (Figure 4A), a C_{30} sterol that is widely distributed in marine organisms in two isomeric forms: 24Z-propylidenecholesterol was found in a scallop (Idler et al, 1971) as well as other organisms (Sheikh and Djerassi, 1974), and 24E-propylidenecholesterol was determined to be the principal constituent of a chrysophyte alga (Rohmer et al, 1980). Both isomers were identified (Brassell and Eglinton, 1983) as environmental indicators in a diatomaceous ooze from the Japan Trench. Reduction of either isomer of 24-propylidenecholesterol during early diagenesis would result in 24-propylcholestane (Figure 4B) with a nearly 1:1 mixture of 24R and 24S epimers, similar to that found for ergostane (Mackenzie et al, 1980) in thermally mature geolipids.

The C_{30} sterols from marine organisms display a great diversity of structure with respect to side-chain alkylation, and any or all of these isomers may possibly contribute as precursors to the C_{30} sterane population. One possible precursor, gorgosterol (Figure 4C), an abundant constituent in marine invertebrates and symbionant algae (Withers et al, 1979), has been identified in a recent sediment from the Namibian shelf (Smith et al, 1982). An authentic sample of gorgostane (Figure 4D) did not match any of the C_{30} steranes by MS or GC retention time. Like the pentacyclic hopane, gorgostane (Figure 4D) has an m/z 412 molecular ion, but it still shows up on the metastable chromatogram due to the low resolution of the method (see C_{30} terpanes on Figures 2, 3). It is unlikely that the side-chain cyclopropyl of gorgosterol would survive diagenesis intact, and the cyclopropane ring-opened products (Proudfoot and Djerassi,

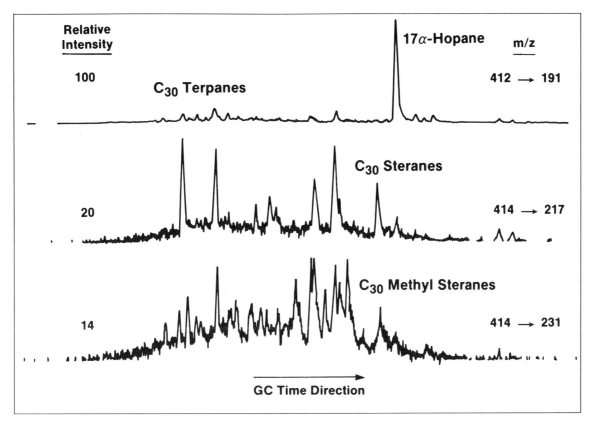

Figure 2—Daughter-ion scans (B/E linked scans) for C_{30} steranes in Sadlerochit, Alaska (marine) crude.

1984) have to be regarded as possible C_{30} sterane contributors.

Despite their great age, the Precambrian and Cambrian (age > 500 m.y.) crudes (Table 1) contain good concentrations of steranes, indicating relatively mild thermal histories. The absence of C_{30} steranes in these very old crudes suggests an evolutionary lag in the appearance of C_{30} sterols in marine organisms. Alternatively, organism distributions contributing to the sources of these oils may have been dominated by a few species that did not contain C_{30} sterols. Our next youngest entries, Kee Scarp and Leduc, from Devonian (age < 400 m.y.) rocks in Canada, already show abundant C_{30} steranes. Perhaps we can fill this data gap by future analyses of additional lower Paleozoic samples.

The C_{28}-C_{30} methyl steranes were found in both marine and nonmarine crudes. The presence of 4α-methylsteroids (4α-methylsterols, 4α-methylstanones, etc) in marine and lacustrine dinoflagellates and in recent sediments associated with them (Boon et al, 1979; Gagosian et al, 1980; Brassell and Eglinton, 1983; de Leeuw et al, 1983; Robinson et al, 1984) has been used to infer the contribution of dinoflagellates wherever such 4α-methylsteroids occur in sediments. By extension of this reasoning, we suggest that the widespread occurrence of methylsteranes (probably 4-methylsteranes) in crude oils may reflect the participation of dinoflagellates in ecosystems that eventually contribute to petroleum genesis.

The C_{26} steranes were also detected in our GC-MS/MS analyses in several crude oils both from marine and nonmarine sources (see Figure 1). Although suitable marine sterol precursors with 26 carbon atoms exist (Djerassi, 1981), these C_{26} steranes could also be derived solely or in part from the catagenic degradation of higher carbon numbered homologs. These alternatives are currently under study.

Sterane and Monoaromatic Steroid Hydrocarbon Distributions

The distributions of C_{27}, C_{28}, and C_{29} steranes and monoaromatic steroid hydrocarbons (MA-steroids) for each of the oils in Tables 1 and 2 have been plotted in triangular graphs. This method, used by Huang and Meinschein (1979) in plotting sterol distributions, allows the overview of a large data set, as shown in Figures 5 and 6. GC-MS/MS was used in obtaining the sterane data, using the sum of 5α,14α,17α (20S) + (20R) plus 5α, 14β,17β(20S) + (20R) steranes for C_{27}, C_{28}, and C_{29}. The quantitation method for MA-steroids has been previously published (Seifert et al, 1983).

It is clear from Figures 5 and 6 that the distributions of the two types of sterol-derived biomarkers are different,

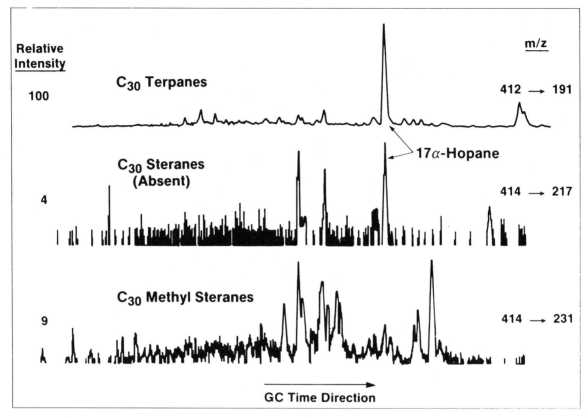

Figure 3—Daughter-ion scans (B/E linked scans) for C_{30} steranes in Green River, Utah (lacustrine) crude.

probably because of differing origins for the steranes and MA-steroids. Steranes can arise through complete defunctionalization and hydrogenation of any sterol with or without incorporation into the kerogen of the source rock, whereas MA-steroids are likely to be derived mainly from sterols containing more than one double bond (Mackenzie et al, 1982) and possibly without kerogen incorporation (Mackenzie et al, 1981).

There is so much overlap between the different oil source types in the sterane summary plot (Figure 5) that the analysis would seem of little use in obtaining the type of environmental designations we are trying to measure. For example, in Figure 5, area A for nonmarine oils falls in the middle of the marine oil distribution combined areas for marine shale plus marine carbonate-derived oils. The MA-steroid distributions, however, are more spread out and can clearly be used as environmental indicators in more cases. For example, most of the marine oils contain a higher proportion of C_{28} MA-steroids than the nonmarine oils in Figure 6.

The most apparent use of the MA-steroid triangle comes in distinguishing oils derived from nonmarine shale sources from those derived from marine shale sources. The main

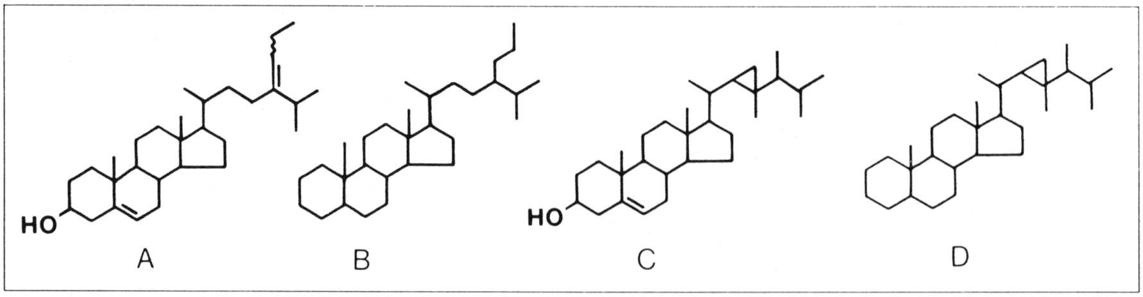

Figure 4—Two common marine sterols and potentially derived sedimentary steranes.

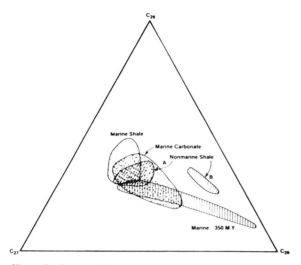

Figure 5—Sterane hydrocarbon distribution, composite for all oils. See text for explanation of A and B.

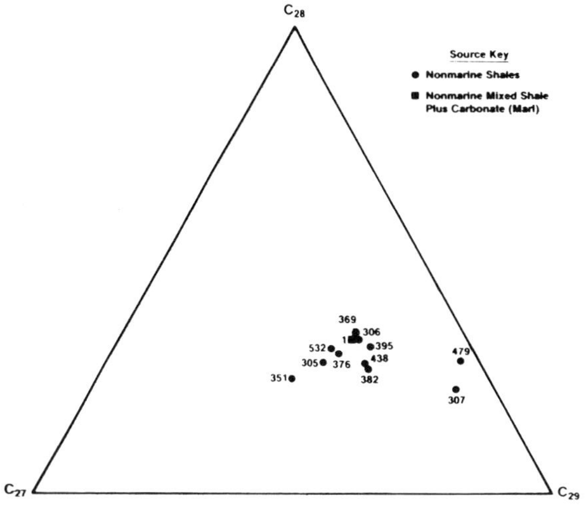

Figure 7—Monoaromatized steroid hydrocarbon distribution, nonmarine oils. GC-MS numbers are from Table 2.

area in Figure 6 occupied by the nonmarine oils (area A) is relatively small, although two outlying samples are circled separately (area B). These samples are from New Zealand (479) and the Damar oil from Sumatra (307) (see Figure 7). The samples from marine shale sources generally contain relatively less C_{29} MA-steroids than do the nonmarine oils (Figure 6). This result supports the original sterol work of Huang and Meinschein (1979). In theory, nonmarine samples had the greatest access to terrigenous material derived from higher plants, whose sterol content is largely C_{29} sterols. The distributions of biomarkers in nonmarine-derived petroleums should therefore be richer in C_{29}-steroids than in marine-derived petroleums.

The five oils from middle to lower Paleozoic sources (age > 350 m.y.) (Figures 5, 6) have high concentrations of C_{29} steranes, particularly C_{29} MA-steroids. Relying on comparison with an authentic standard, the C_{29} steranes are assumed to have the stigmastane carbon skeleton. It is not clear from this limited data set whether a general phenomenon is being observed. Widespread use of C_{29} sterols could have preceded use of C_{27} and C_{28} sterols in the evolution of certain microorganisms responsible for petroleum. However, this would be difficult to explain chemically, as the pathway to the common C_{29} sterols involves dialkylation of a C_{27} sterol precursor with a $\Delta 24$ double bond (Nes and Nes, 1980) and would represent a more advanced biosynthetic sequence than, for example, cholesterol (C. Djerassi, 1985, personal communication).

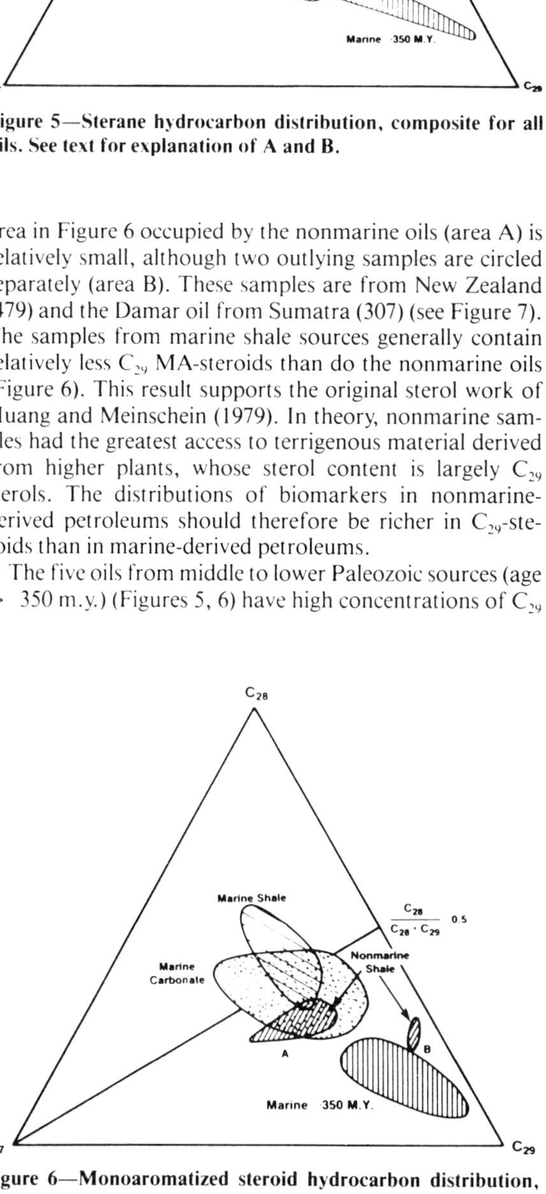

The spread in data for the marine shale oils in Figure 6 may be due to the incorporation of varying amounts of nonmarine organic matter into the marine sediments. This concept has been used recently (Volkman et al, 1983b) in conjunction with sterane ratios to support a source input variation between Windalia oil (oil 90 on Table 1) and the other Barrow subbasin oils from Australia, represented in this study by Barrow (oil 94 on Table 1). These Australian oils were correlated with the marine deltaic Dingo claystone (Volkman et al, 1983b). The Barrow oil is the only marine shale oil in this study to plot within the nonmarine oil group (compare Figures 6 and 8), and Windalia (oil) plots near the nonmarine group. This would support a significant nonmarine organic component in the source of these oils, such a component being most pronounced in Barrow (oil).

The MA-steroid distribution pattern for oils of marine carbonate sources (Figure 6) nearly encompasses the nonmarine shale group (A in Figure 6) and overlaps markedly with the marine shale group. The overlap between the marine carbonate and marine shale groups is strong enough that distinction between the two types in an unknown oil on the basis of MA-steroid distribution would seem tenuous. However, some oils of marine carbonate source have higher relative concentrations of C_{29} MA-steroids than do the marine shale oils. This situation can be explained on the

Figure 6—Monoaromatized steroid hydrocarbon distribution, composite for all oils. See text for explanation of A and B.

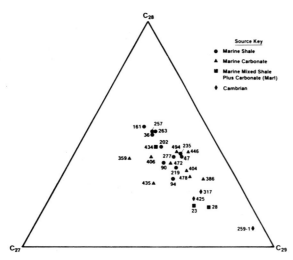

Figure 8—Monoaromatized steroid hydrocarbon distribution, marine oils. GC-MS numbers are from Table 1.

basis of environmental differences at the time of sedimentation. Environmental differences would affect the biotic distribution and, hence, the distribution of MA-steroids.

For example, brown algae contain predominantly fucosterol, a C_{29} sterol (Goodwin, 1973), and in many species of green algae the major sterols are reported to be C_{29} (Patterson, 1974; Paoletti et al, 1976; Dickson et al, 1979; Fattorusso, 1980).

On the other hand, the red algae that contain only C_{27} sterols (Goodwin, 1973) are abundant in marine environments and may make a substantial contribution to the microscopic algal biomass (plankton). If conditions during deposition of certain carbonate source rocks would favor the growth of brown or certain green algae or other C_{29} sterol-rich organisms, a C_{29} sterol predominance in the deposited organic matter could develop, possibly leading to a C_{29} MA-steroid preference in some carbonate-derived crude oils.

By comparison, nonmarine red and brown algae are uncommon, and their contribution to nonmarine ecosystems and biomass is minor (Bold and Wynne, 1978). Thus, fewer types of abundant algal precursors in nonmarine environments may lead to the narrower band of MA-steroid values (Figure 6) for the nonmarine oils, whose steroids appear to be derived principally from green algae and some higher plants.

Whereas further sampling will lead to expansion of MA-steroid distributions from both marine and nonmarine oils, this will be especially true for marine shale and carbonate-derived oils because of the greater diversity of biota available in the marine ecosystem.

Sterols also have also been sporadically reported from prokaryotes (e.g., Kohl et al, 1983); but many of the sterols may have been contributed from the bacterial food source. Often the amounts reported are miniscule. The prominent bacteria that contribute to petroleum source rocks (i.e., the archaebacteria) use other types of polycyclic and acyclic triterpenoids as cell-wall lipids (see Ourisson et al, 1982), but not sterols. Hence, the bacterial contribution to petroleum steroid distributions is probably insignificant.

This reasoning, based on algal sterol distributions, offers an alternative explanation to terrestrial input as an influence on sterane or MA-steroid ratios. The sterols reported from terrestrial soils (Huang and Meinschein, 1979) showed the virtual absence of C_{27} sterols, as is the case with higher plants. In contrast, most samples of the nonmarine oils studied have sterane and MA-steroid distributions with nearly equal C_{27}/C_{29} contents, and the distributions (Figures 5, 6) closely resemble those of many marine carbonate oils that may have largely marine planktonic and bacterial origins. The area occupied by marine shale oils in Figure 6 represents a shift toward C_{28} from the main band (A in Figure 6) of lacustrine oils, as opposed to the results of a previous study (Huang and Meinschein, 1979) that predicted a shift toward C_{27}. Organisms rich in C_{28} sterols, therefore, apparently shift the distribution in this selection of marine shale-derived oils, not the contribution of higher plants to nonmarine sediments (which would shift lacustrine oils toward C_{29}).

However, oils 307 and 479 (Figure 7) plot near the C_{29} apex of the triangle, close to the 0% C_{27} line. Oils from the Mahakam delta, Indonesia, were reported to have similar distributions with a predominance of C_{29} steranes and C_{29} MA-steroids (Hoffmann et al, 1984). Oils plotting in this way would appear to have an input of sterol precursors almost entirely derived from higher plants (Huang and Meinschein, 1979).

Connection of areas A and B in Figure 6 can be viewed as a band whose major axis lies parallel with the 0% C_{28} axis. In fact, all of the nonmarine oils in this study have a $C_{28}/(C_{28} + C_{29})$ MA-steroid ratio less than 0.5. Position of the points within this nonmarine band might be used as an indicator of the relative distributions of land plant-derived and algal-derived sterol inputs. For example, by virtue of decreasing C_{29} Ma-steroid contents, the hypothesis might be developed that the Sumatran oils plotted in Figure 7 (see Table 2) are derived increasingly from algal lipids and decreasingly from land plant organic matter in the order 307, 306, 305, and 351. This sequence is supported in part by the previous report (Seifert and Moldowan, 1981) that oils 305 and 351 contain high concentrations of botryococcane (Moldowan and Seifert, 1980) and compounds similar to botryococcane that are attributable to the nonmarine (but sometimes estuarian or lagoonal) alga *Botryococcus braunii*. Oils 306 and 307 lack those compounds, but the gas chromatogram of oil 307 was shown (Seifert and Moldowan, 1981, their Figure 6) to be more heavily skewed toward the high molecular-weight paraffins, which may reflect higher plant wax input. The $\delta C^{13}/C^{12}$ carbon isotope ratios (Table 2) are also significantly different among these oils. The oils with probably the greatest algal component (351 and 305) also have the highest δC^{13} values (+7.8 and +4.5, respectively), whereas oils 306 and 307, which are most likely (based on other geochemical parameters) to be influenced by land plant-derived organic matter, have the lower δC^{13} values (+1.3 and +1.4, respectively).

Data for sterane/hopane ratios are given in Tables 1 and 2 and are plotted in Figure 9a. Although no clear-cut divi-

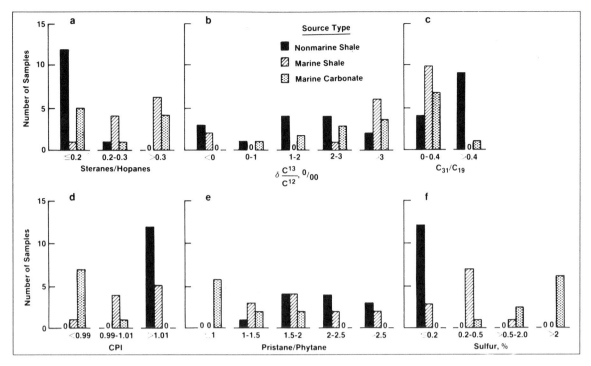

Figure 9—Evaluation of environmental indicators from Tables 1 and 2. CPI = carbon preference index (see Table 1).

sion exists between oils from marine and nonmarine sources with respect to this ratio, the nonmarine oils tend to have low sterane/hopane ratios and in general low sterane concentrations. We have also observed the general trend toward lower concentrations of MA-steroids in oils generated from nonmarine sources.

CLASSICAL PARAMETERS

Carbon Isotope Ratios

An early parameter used for classification of crude oils in marine and nonmarine was the C^{13}/C^{12} carbon isotopic ratio of the whole crude. This ratio was proposed to be lower in oils derived from nonmarine (terrestrial) organic matter than in marine-derived crudes (Silverman and Epstein, 1958; Silverman, 1964). Although this concept was originally supported by others (Hunt, 1970; Tissot and Welte, 1978; Rogers, 1980), doubts began to surface as to its general applicability (Deines, 1980; Yeh and Epstein, 1981). A recent study by Sofer (1984) failed to support the original hypothesis, but a refinement based on the isotopic relationship between saturated and aromatic hydrocarbon fractions was deemed successful in making the terrigenous/marine distinction. Whether this differentiation can be applied in general has been discussed by Schoell (in press), who suggests that plotting carbon versus hydrogen isotopes may be more successful in differentiating oils from carbonate sources from oils from shale sources.

This paper does not take issue with these applications of isotope chemistry. However, Figure 9b is a demonstration of the lack of general applicability of the whole-oil carbon isotopic ($\delta C^{13}/C^{12}$) ratio as an indicator of marine/nonmarine sources of crude oil. Many nonmarine crudes with high isotopic ratios were encountered in this study (Table 2), whereas low isotopic-ratio crudes derived from marine shales were also represented. The marine crude from Kingak, Alaska, which has been shown to correlate with Kingak Shale (Seifert et al, 1980, 1983), has the lowest carbon isotope ratio in this study at -2.0 (NBS-22 standard) (Table 1). It may have significant amounts of nonmarine organic matter in its source rock, as evidenced by its relatively high percentage of C_{29} MA-steroids compared with most of the other marine shale-sourced oils (Figure 8), a percentage that almost places it among the nonmarine group in Figure 7. The Sadlerochit oil from Alaska probably was generated from marine shale sources, including post-Neocomian and Triassic (Shublik Formation) shales as well as Jurassic (Kingak formation) shales (Seifert et al, 1980). Its higher carbon isotope ratio of -0.3 and relatively lower percentage of C_{29} MA-steroids are evidence of a greater contribution of marine-derived organic matter to the sources of Sadlerochit oil than to the sole source of Kingak oil. Thus, the carbon isotope ratio may, at best, be used cautiously as a weak indicator of source-derived contributions to crude oils within a given basin in certain cases (note also the Sumatra oils discussed previously), but it does not appear to be applicable when comparing oils from different basins.

Table 3. Specificity Summary Crude Oil Source-Type Indicators

	Nonmarine vs. Marine	Nonmarine vs. Marine Shale	Nonmarine vs. Marine Carbonate	Marine Shale vs. Carbonate
C_{30} steranes	+ + + +	+ + + +	+ + + +	−
Sulfur (%)	+ +	+ +	+ + +	+ +
MA-steranes	+ +	+ +	+ +	+
High molecular-weight paraffin	+ +	+ +	+ +	−
Carbon preference index	−	−	+ + +	+
Pristane/phytane	−	−	+ +	+
Steranes/hopanes	+	+	+	−
Carbon isotope	−	−	−	−
Gammacerane index	−	−	−	?

+ + + + Definitive.
 + + + Strong indicator (may be affected by secondary processes).
   + Weak indicator.
   − Not indicator.

Normal Paraffin Distribution Parameters

The standard method of categorizing the amount of land-derived organic material in an oil is to determine its degree of waxiness. This method assumes that terrigenous material contributes a high molecular-weight normal paraffin component to the oil (Hedberg, 1968; Albrecht and Ourisson, 1971). Thus, recent studies into oil classification by source input have relied heavily on waxiness as an environmental source input parameter (e.g., Connan and Cassou, 1980; Sofer, 1984). Furthermore, the preference for odd or even carbon numbered paraffins can reflect the environment of source rock deposition. An odd numbered preference in the high molecular-weight paraffins derived directly from terrestrial higher plants (Albrecht and Ourisson, 1971) or from decarboxylation of even numbered fatty acids present in terrestrial plant waxes (Cooper and Bray, 1963) is generally considered a marker of terrestrial input. Predominance of even numbered normal paraffins signals direct reduction of fatty acids (Kvenvolden, 1967) in an extremely anoxic, highly reducing, and hypersaline environment. Heavy crude oils from carbonate source rocks in evaporitic sequences exemplify oils with a predominance of even carbon numbered paraffins (Connan et al, in press).

Secondary processes in crude oil diagenesis present serious obstacles to using normal paraffins as environmental source indicators. Such processes as biodegradation, water washing, and weathering can selectively remove portions of the normal paraffins. Maturation progressively converts high molecular-weight paraffins into lower molecular-weight paraffins without odd-even preference, and subsurface migration appears to discriminate against high molecular-weight compounds in general. Also, it has been shown that primary migration can favor short-chain normal paraffins (Leythaeuser et al, 1984).

Another consideration is that varying amounts of terrigenous higher plant material may reach the anoxic zone of an aqueous system and be incorporated into a future petroleum source rock. However, this material generally has undergone, to various degrees, subaerial oxidation in the soil prior to transport (Demaison and Moore, 1980). It can range from refractory type IV kerogen to type III kerogen, which, while still having potential for liquids, is essentially gas-prone (Tissot and Welte, 1978). Therefore, the presence of terrestrial plant remains in the source rock does not necessarily mean that such remains have contributed significantly to petroleum produced from that rock, particularly when algal/bacterial remains are also abundant.

For example, there is no real need to invoke large contributions of higher plant material to explain normal paraffin distributions typically found in nonmarine crude oils. A study of nonmarine algae (Gelpi et al, 1970) demonstrated that several common species (not only *Botryococcus braunii*) contain two groups of odd carbon numbered n-alkanes and n-alkenes, one group ranging from C_{15} to C_{19} with the most abundant member at C_{17}, and the other, ranging from C_{23} to C_{33}, with abundant C_{23}, C_{27}, and C_{29} compounds. In fact, the alga *Botryococcus braunii* has more than 98% of the higher molecular-weight group. Various combinations of these algae are, therefore, capable of producing bimodal (high and low molecular weight) or single modal crude oil normal paraffin distributions. Algae and bacteria are generally considered to be the principal organisms to form oil-prone types I and II kerogens (Tissot and Welte, 1978). Many of the crudes plotting in nonmarine group A (Figure 6) have a bimodal n-paraffin distribution, as expected from a nonmarine algal mixture.

As a measure of crude oil waxiness, we have tabulated nC_{31}/nC_{19} ratios of the oils (Tables 1, 2) and plotted them in Figure 9c. We have already discriminated in favor of this parameter and the carbon preference index (CPI) parameter (discussed below) by selecting the oils with the highest biomarker concentrations, which tend to be the less mature oils from each basin, and avoiding those oils that show signs of biodegradation or weathering. The nonmarine oils display a tendency toward nC_{31}/nC_{19} ratios greater than 0.4, overlapping only in one case with a marine carbonate oil (Figure 9c). All types of marine and nonmarine oils can be found in the 0-0.4 nC_{31}/nC_{19} category. The parameter, therefore, is selective for nonmarine oils but is not definitive.

The CPI (Tables 1, 2), plotted in Figure 9d as a measure of odd-to-even n-paraffin predominance, appears to be well suited, within the constraints of this (not overly mature) oil selection, to aid in distinguishing nonmarine oils (CPI commonly > 1; odd predominance) from marine carbonate oils (CPI commonly < 1; even predominance). This distinction is thought to occur through even carbon numbered n-alkyl fatty acid reduction in the highly anoxic environment of carbonate source rock deposition to yield even numbered n-alkanes (CPI < 1). These same fatty acids yield odd numbered n-alkane predominance (CPI > 1) under somewhat more oxic circumstances pertaining to many other marine oils. Thus, the CPI values for marine shale-derived oils overlap the others (Figure 9d) but are less pronounced, reflecting a less extreme or moderately anoxic environment commonly encountered in that type of marine deposition. Odd numbered high molecular-weight normal alkanes and alkenes from nonmarine algae as well as higher plants undoubtedly contribute to the > 1 CPI in nonmarine oils.

Pristane/Phytane and Sulfur Concentrations

The pristane/phytane ratios and sulfur concentrations in Tables 1 and 2, like the CPI parameter, are useful to categorize oils from carbonate sources. The differences in carbonate oils also appear to be derived from the highly anoxic reducing environment characteristic of the selected oils, rather than from organic input. Distributions of pristane/phytane in Figure 9e demonstrate no chance of distinguishing marine shale oils from nonmarine oils because they have nearly identical distributions. However, carbonate oils have a tendency to show pristane/phytane ratios less than 1; if the ratio is much less than 1, it could act as a selective indicator for a carbonate source. Phytane predominates through favored reduction of phytol residues (Albaigés and Torradas, 1977) from chlorophyll (Seifert, 1975) or bacterial cell walls (Chappe et al, 1980) in highly anoxic environments. In less reducing environments, oxidative carbon-loss mechanisms allow pristane predominance. Thus, a parallelism between even carbon numbered paraffin predominance and phytane over pristane predominance (Seifert et al, 1983) is observed with few exceptions—namely, the Cretaceous oil from location A (oil 405) and the Sadlerochit, Alaska, oil (oil 202). The mixing of at least three marine shale sources in Sadlerochit (Seifert et al, 1980, 1983) may be responsible for its lack of parallelism in those parameters.

Sulfur concentrations, like the paraffin-based parameters, are susceptible to modification by secondary processes of diagenesis, including maturation (Orr, 1974; Seifert and Moldowan, 1978), biodegradation, and water washing (Bailey et al, 1973; Orr, 1978; Palmer, 1984). Any oils where biodegradation is indicated in the tables have been omitted in preparing Figure 9f. Again, as in the case of CPI and pristane/phytane, a strong difference between oils from marine carbonates and those from shaly sources is apparent (Figure 9f). The high sulfur in carbonate oils is opposed by low sulfur in nonmarine oils, but the marine shale oils overlap with both groups.

The tendency toward high sulfur concentrations in oils originating from carbonate source rocks is not too surprising since, in the highly reducing environments found during this type of deposition (Demaison and Bourgeois, 1985), reduced sulfur compounds are abundantly provided by microbial sulfate reduction (Barbat, 1967). However, the principal reason for high sulfur in carbonate oils is the lack of an inorganic acceptor for H_2S. In shales, iron scavenges most of the sulfur. A more detailed consideration of factors influencing sulfur concentrations in oils is presented by Tissot and Welte (1978). Based on the evidence in Figure 9f, sulfur concentrations in oils can be a useful piece of corroborative evidence to identify their source rock depositional environments.

Terpanes

Use of the diterpanes and triterpanes as environmental markers has been selective and restricted to special cases. The hopanes (Van Dorsselaer et al, 1977), tricyclic and tetracyclic terpanes (Aquino Neto et al, 1983; Moldowan et al, 1983), higher molecular-weight regular isoprenoids (Albaigés et al, 1978; Albaigés, 1980), and head-to-head linked isoprenoids (Moldowan and Seifert, 1979; Chappe et al, 1980) are considered to be derived from bacterial cell-wall lipids (Ourisson et al, 1982). As such, the wide distribution of these compounds in crude oils and sedimentary rocks attests to the importance of bacterial reworking of organic matter during early diagenesis.

Gammacerane (Hills et al, 1966) is a terpane that is biogenetically related to hopane, differing only in the manner of cyclization of the E-ring. Its origin was suggested from gammacerane-3β-ol, which occurs in ferns, and a nonmarine protozoan *Tetrahymena pyriformis* (Hills et al, 1966; Whitehead, 1974). It was originally isolated from the lacustrine Green River shale, making it a potential nonmarine marker. However, the present data (Tables 1, 2) demonstrate a complete lack of correlation between gammacerane content and the depositional environment of the oil. Its presence is detectable, although sometimes in small traces, in nearly every sample where hopanes occur. Therefore, like the hopanes, it may originate from precursors that are bacterial cell-wall constituents. It has been suggested (G. Demaison, 1984, personal communication) that exceedingly high gammacerane indices as found in Green River (oil 1 on Table 2) and Prinos (oil 472 on Table 1) oils may signal hypersaline episodes of source rock deposition, which may occur in alkaline lakes as well as in lagoonal carbonate-evaporite environments.

Unusual compounds sometimes occur as markers of special environments or organisms. The structure of 18α(H)-oleanane has been examined (Smith et al, 1970), and its presence in Nigerian crudes (Whitehead, 1974; Ekweozor et al, 1979) has been attributed to higher plant contribution to the source. Perhaps perhydro-β-carotene, found in Green River shale (Murphy et al, 1967; Gallegos, 1971) and oil (W. K. Seifert and J. M. Moldowan, unpublished research) is a terrestrial marker since it has not been reported from sources of marine origin. Botryococcane found in certain Sumatran crude oils (Moldowan and Seifert, 1980; Seifert and Moldowan, 1981) is apparently derived from botryococcene, which is known in only one organism, *Botryococcus braunii*, a freshwater or brackish-water alga. Bicyclic sesquiterpanes and diterpanes (Philp et al, 1981) and bridged tetracyclic diterpanes (Noble et al, 1985), found in crude oils from Australia, have been proposed as products of higher plant resins, and several other pentacyclic triterpanes of unknown structure may also be resin derived (Grantham et al, 1983). The origins of the curious 28,30-bisnorhopane (Seifert et al, 1978; Moldowan et al, 1984) and its companion 25,28,30-trisnorhopane (Bjorøy and Rullkötter, 1980; Volkman et al, 1983a; Moldowan et al, 1984) have been attributed to various possible environments, one of which is bacterial input involving free lipids as opposed to the familiar mechanism of binding into a kerogen polymer. These compounds appear to be associated with highly anoxic ecosystems (Katz and Elrod, 1983), but the relationship is not well defined.

Numerous other terpane-derived continental markers have been found in recent or immature sediments and coals, but not identified in petroleum (Corbet et al, 1980; Barrick and Hedges, 1981; Rullkötter et al, 1982; Chaffee and Johns, 1983). This may be caused by several factors, including the inherently poor quality of terrestrial organic matter as a petroleum source. Contributing factors may also be the lability of many continental marker molecular structures in the aquatic environments in which petroleum source rocks are deposited (i.e., Corbet et al, 1980) and the inability of those markers to survive intact incorporation into and liberation from the kerogen polymer.

SUMMARY AND CONCLUSIONS

Various crude oil composition parameters have been described in detail, and their potential use as indicators for depositional environments of petroleum source rocks has been examined. The results summarized in Table 3 may be used as a guide to a multiparameter assessment of crude oil and, by analogy, source rock origins. It is apparent that no one parameter can deliver all of the answers.

The most powerful parameter for identifying the input of marine organic matter to the source rock is the presence of C_{30} steranes. Oil from marine carbonate sources can best be characterized by a combination of percentage of sulfur, CPI, and pristane/phytane ratio, although none of these parameters appears to be definitive by itself. Distributions of monoaromatized steroid hydrocarbons and high molecular-weight normal paraffins are applicable to making a nonmarine/marine distinction and may help to delineate the proportion of nonmarine organic matter in the marine source rock. Among nonmarine oils, MA-steroid and sterane distributions may provide a means of assessing the relative contributions of organic matter derived from land plants compared with aquatic organisms. Several special terpanes are useful as indicators of specific nonmarine environments, and rare occurrences of resinite-derived oils can be documented. However, the contribution of terrestrially derived lipids to both marine and nonmarine crude oils may be commonly overstated. With a few exceptions, an explanation based on distributions of algal types and bacterial reworking can explain the majority of observed biologic marker distributions in crude oils.

SELECTED REFERENCES

Albaigés, J., 1980, Identification and geochemical significance of long chain acyclic isoprenoid hydrocarbons in crude oils, in A. G. Douglas and J. R. Maxwell, eds., Advances in organic geochemistry, 1979: Oxford, England, Pergamon Press, p. 265-274.

────── and J. Torradas, 1977, Geochemical characterization of Spanish crude oils, in R. Campos and J. Goni, eds., Advances in organic geochemistry, 1975: Madrid, Enadimsa, p. 99-115.

────── J. Borbón, and P. Salagre, 1978, Identification of a series of C_{25}-C_{40} acyclic isoprenoid hydrocarbons in crude oils: Tetrahedron Letters, p. 595-598.

Albrecht, P., and G. Ourisson, 1971, Biogenic substances in sediments and fossils: Angewandte Chemie (International Edition), v. 10, p. 209-225.

Aquino Neto, F. R., J. M. Trendel, A. Restle, J. Connan, and P. Albrecht, 1983, Occurrence and formation of tricyclic and tetracyclic terpanes in sediments and petroleums, in M. Bjorøy, C. Albrecht, C. Cornford, K. de Groot, G. Eglinton, E. Galimov, D. Leythaeuser, R. Pelet, J. Rullkötter, and G. Speers, eds., Advances in organic geochemistry, 1981: New York, John Wiley and Sons, p. 659-667.

Bailey, N. J. L., A. M. Jobson, and M. A. Rogers, 1973, Bacterial degradation of crude oil; comparison of field and experimental data: Chemical Geology, v. 11, p. 203-221.

Barbat, W. N., 1967, Crude-oil correlations and their role in exploration: AAPG Bulletin, v. 51, p. 1255-1292.

Barrick, R. C., and J. I. Hedges, 1981, Hydrocarbon geochemistry of the Puget Sound region; II, sedimentary diterpenoid, steroid and triterpenoid hydrocarbons: Geochimica et Cosmochimica Acta, v. 45, p. 381-392.

Bjorøy, M., and J. Rullkötter, 1980, An unusual C_{27}-triterpane: 25,28,30-trisnormoretane: Chemical Geology, v. 30, p. 27-34.

Bold, H. C., and M. J. Wynne, 1978, Introduction to the algae: Englewood Cliffs, New Jersey, Prentice-Hall, 706 p.

Boon, J. J., W. I. C. Rijpstra, F. de Lange, J. W. de Leeuw, M. Yoshioka, and Y. Shimizu, 1979, Black Sea sterol; a molecular fossil for dinoflagellate blooms: Nature, v. 277, p. 125-127.

Brassell, S. C., and G. Eglinton, 1983, Steroids and triterpenoids in deep sea sediments as environmental and diagenetic indicators, in M. Bjorøy, C. Albrecht, C. Cornford, K. de Groot, G. Eglinton, E. Galimov, D. Leythaeuser, R. Pelet, J. Rullkötter, and G. Speers, eds., Advances in organic geochemistry, 1981: New York, John Wiley and Sons, p. 684-697.

Chaffee, A. L., and R. B. Johns, 1983, Polycyclic aromatic hydrocarbons in Australian coals; angularly fused pentacyclic tri- and tetra-aromatic components of Victorian brown coal: Geochimica et Cosmochimica Acta, v. 47, p. 2141-2155.

Chappe, B., W. Michaelis, and P. Albrecht, 1980, Molecular fossils of archaebacteria as selective degradation products of kerogen, in A. G. Douglas and J. R. Maxwell, eds., Advances in organic geochemistry, 1979: Oxford, England, Pergamon Press, p. 265-274.

Connan, J., and A. M. Cassou, 1980, Properties of gases and petroleum liquids derived from terrestrial kerogen at various maturation levels: Geochimica et Cosmochimica Acta, v. 44, p. 1-23.

────── J.-L. Grondin, J.-P. Colin, G. Hussler, and P. Albrecht, in press, Nonbiodegraded heavy oils in some carbonate basins, in P. A. Schenck, J. W. de Leeuw, and G. W. M. Lijmbach, eds., Advances in organic geochemistry, 1983: Oxford, England, Pergamon Press.

Cooper, J. E., and E. E. Bray, 1963, A postulated role of fatty acids in petroleum formation: Geochimica et Cosmochimica Acta, v. 27, p. 1113-1127.

Corbet, B., P. Albrecht, and G. Ourisson, 1980, Photochemical or photomimetic fossil triterpenoids in sediments and petroleum: Journal of the American Chemical Society, v. 102, p. 1171-1173.

Deines, P., 1980, The isotopic composition of reduced organic carbon, in P. Fritz and C. H. Fontes, eds., Handbook of environmental isotope geochemistry 1, the terrestrial environment: Amsterdam, Elsevier, p. 329-406.

de Leeuw, J. W., W. I. C. Rijpstra, P. A. Schenck, and J. K. Volkman, 1983, Free, esterified and residual bound sterols in Black Sea Unit I sediments: Geochimica et Cosmochimica Acta, v. 47, p. 455-465.

Demaison, G., and F. T. Bourgeois, 1985, Environment of deposition of middle Miocene, Alcanar, carbonate source beds, Casablanca field, Tarragona basin, Spain, in Petroleum geochemistry and source rock potential of carbonate rocks: AAPG Studies in Geology 18, p. 151-161.

────── G. T. Moore, 1980, Anoxic environments and oil source bed genesis: AAPG Bulletin, v. 64, p. 1179-1209.

Dickson, L. G., G. W. Patterson, and B. A. Knights, 1979, Distribution of sterols in the marine Chlorophyceae: Proceedings of the International Seaweed Symposium, 1977, v. 9, p. 413-420.

Djerassi, C., 1981, Recent studies in the marine sterol field: Pure and Applied Chemistry, v. 53, p. 873-890.

Ekweozor, C. M., J. I. Okogun, D. E. U. Ekong, and J. R. Maxwell, 1979, Preliminary organic geochemical studies of samples from the Niger delta, Nigeria; I, analyses of crude oils for triterpanes: Chemical Geology, v. 27, p. 11-28.

Fattorusso, E., S. Magno, and L. Mayol, 1980, Sterols of Mediterranean Chlorophyceae: Experientia, v. 36, p. 1137-1138.

Gagosian, R. B., S. O. Smith, C. Lee, J. W. Farrington, and N. M. Frew, 1980, Steroid transformations in recent marine sediments, in A. G. Douglas and J. R. Maxwell, eds., Advances in organic geochemistry, 1979: Oxford, England, Pergamon Press, p. 407-419.

Gallegos, E. J., 1971, Identification of new steranes, terpanes, and branched paraffins in Green River shale by combined capillary gas chromatography and mass spectrometry: Analytical Chemistry, v. 43, p. 1151-1160.

Gelpi, E., H. Schneider, J. Mann, and J. Oró, 1970, Hydrocarbons of geochemical significance in microscopic algae: Phytochemistry v. 9, p. 603-612.

Goodwin, T. W., 1973, Comparative biochemistry of sterols in eukaryotic microorganisms, in J. A. Erwin, ed., Lipids and biomembranes of eukaryotic microorganisms: New York, Academic Press, p. 1-40.

Grantham, P. J., J. Posthuma, and A. Baak, 1983, Triterpanes in a number of Far-Eastern crude oils, in M. Bjorøy et al, eds., Advances in organic geochemistry, 1981: New York, John Wiley and Sons, p. 675-683.

Haddon, W. F., 1979, Computerized mass spectrometry linked scan system for recording metastable ions: Analytical Chemistry, v. 51, p. 983-988.

Hedberg, H. D., 1968, Significance of high-wax oils with respect to genesis of petroleum: AAPG Bulletin, v. 52, p. 736-750.

Hills, I. R., E. V. Whitehead, D. E. Anders, J. J. Cummins, and W. E. Robinson, 1966, An optically active triterpane, gammacerane in Green River, Colorado, oil shale bitumen: Journal of the Chemical Society, Chemical Communications, p. 752-754.

Hoffmann, C. F., A. S. Mackenzie, C. A. Lewis, J. R. Maxwell, J. L. Oudin, B. Durand, and M. Vandenbroucke, 1984, A biological marker study of coals, shales and oils from the Mahakam delta, Kalimantan, Indonesia: Chemical Geology, v. 42, p. 1-23.

Huang, W.-Y., and W. G. Meinschein, 1979, Sterols as ecological indicators: Geochimica et Cosmochimica Acta, v. 43, p. 739-745.

Hunt, J. M., 1970, The significance of carbon isotope variations in marine sediments, in G. D. Hobson and G. C. Speers, eds., Advances in organic geochemistry, 1966: New York, Pergamon Press, p. 27-35.

Idler, D. R., L. M. Safe, and E. F. MacDonald, 1971, A new C_{30} sterol, (Z)-24-propylidene-cholest-5-en-3β-ol (29-methylisofucosterol): Steroids, v. 18, p. 545-553.

Katz, B. J., and L. W. Elrod, 1983, organic geochemistry of DSDP Site 467, offshore California, middle Miocene to lower Pliocene strata: Geochimica et Cosmochimica Acta, v. 47, p. 389-396.

Kohl, W., A. Gloc, and H. Reichenbach, 1983, Steroids from the Myxobacterium Nannocystis exedens: Journal of General Microbiology, v. 129, p. 1624-1635.

Kvenvolden, K. A., 1967, Normal fatty acids in sediments: Journal of the American Oil Chemists' Society, v. 44, p. 628-636.

Leythaeuser, D., A. S. Mackenzie, R. G. Schaefer, and M. Bjorøy, 1984, A novel approach for recognition and quantification of hydrocarbon migration effects in shale-sandstone sequences: AAPG Bulletin, v. 68, p. 196-219.

Mackenzie, A. S., N. A. Lamb, and J. R. Maxwell, 1982, Steroid hydrocarbons and the thermal history of sediments: Nature, v. 295, p. 223-226.

——— C. A. Lewis, and J. R. Maxwell, 1981, Molecular parameters of maturation in the Toarcian shales, Paris Basin, France; IV, laboratory thermal alteration studies: Geochimica et Cosmochimica Acta, v. 45, p. 2369-2376.

——— R. L. Patience, J. R. Maxwell, M. Vandenbroucke, and B. Durand, 1980, Molecular parameters of maturation in the Toarcian shales, Paris basin, France; I, changes in the configurations of acyclic isoprenoid alkanes, steranes and triterpanes: Geochimica et Cosmochimica Acta, v. 44, p. 1709-1722.

Moldowan, J. M., and W. K. Seifert, 1979, Head-to-head linked isoprenoid hydrocarbons in petroleum: Science, v. 204, p. 169-171.

——— ——— 1980, First discovery of botryococcane in petroleum: Journal of the Chemical Society, Chemical Communications, p. 912-914.

——— ——— and E. J. Gallegos, 1983, Identification of an extended series of tricyclic terpanes in petroleum: Geochimica et Cosmochimica Acta, v. 47, p. 1531-1534.

——— ——— E. Arnold, and J. Clardy, 1984, Structure proof and significance of stereoisomeric 28,30-bisnorhopanes in petroleum and petroleum source rocks: Geochimica et Cosmochimica Acta, v. 48, p. 1651-1661.

Murphy, S. M. T. J., A. McCormick, and G. Eglinton, 1967, Perhydro-β-carotene in Green River shale: Science, v. 157, p. 1040-1042.

Nes, W. R., and W. D. Nes, 1980, Lipids in evolution: New York, Plenum Press, 244 p.

Noble, R., J. Knox, R. Alexander, and R. Kagi, 1985, Identification of tetracyclic diterpanes in Australian crude oils and sediments: Journal of the Chemical Society, Chemical Communications, p. 32-33.

Orr, W. L., 1974, Changes in sulfur content and isotopic ratios of sulfur during petroleum maturation—study of Big Horn basin Paleozoic oils: AAPG Bulletin, v. 58, p. 2295-2318.

——— 1978, Sulfur in heavy oils, oil sands, and oil shales, in O. P. Strausz and E. M. Lown, eds., Oil sand and oil shale chemistry: New York, Verlag Chemie, p. 223-241.

Ourisson, G., P. Albrecht, and M. Rohmer, 1982, Predictive microbial biochemistry from molecular fossils to procaryotic membranes: Trends in Biochemical Sciences, v. 7, p. 236-239.

Palmer, S. E., 1984, Effect of water washing on C_{15+} hydrocarbon fraction of crude oils from Northwest Palawan, Philippines: AAPG Bulletin, v. 68, p. 137-149.

Paoletti, C., B. Pushparaj, G. Florenzano, P. Capella, and G. Lercker, 1976, Unsaponifiable matter of green and blue-green algal lipids as a factor of biochemical differentiation of their biomasses: II, terpenic alcohol and sterol fractions: Lipids, v. 11, p. 266-271.

Patterson, G. W., 1974, Sterols of some green algae: Comparative Biochemistry and Physiology, v. 47B, p. 453-457.

Philp, R. P., T. D. Gilbert, and J. Friedrich, 1981, Bicyclic sesquiterpenoids and diterpenoids in Australian crude oils: Geochimica et Cosmochimica Acta, v. 45, p. 1173-1180.

Proudfoot, J. R., and C. Djerassi, 1984, Stereochemical effects in cyclopropane ring openings: Journal of American Chemical Society, v. 106, p. 5613-5622.

Robinson, N., G. Eglinton, S. C. Brassell, and P. A. Cranwell, 1984, Dinoflagellate origin for sedimentary 4α-methylsteroids and 5α(H)-stanols: Nature, v. 308, p. 439-442.

Rogers, M. A., 1980, Application of organic facies concepts to hydrocarbon source rock evaluation, in Proceedings of the 10th World Petroleum Congress, 1979, v. 2: London, Heyden and Son, p. 23-30.

Rohmer, M., W. C. M. C. Kokke, W. Fenical, and C. Djerassi, 1980, Isolation of two new C_{30} sterols, (24E)-24-n-propylidenecholesterol and (24ξ)-24-n-propylcholesterol from a cultured marine chrysophyte: Steroids, v. 35, p. 219-231.

Rullkötter, J., D. Leythaeuser, and D. Wendisch, 1982, Novel 23,28-bisnorlupanes in Tertiary sediments; widespread occurrence of nuclear demethylated triterpanes: Geochimica et Cosmochimica Acta, v. 46, p. 2501-2509.

Schoell, M., in press, Recent advances in petroleum isotope geochemistry, in P. A. Schenck, J. W. de Leeuw, and G. W. M. Lijmbach, eds., Advances in organic geochemistry, 1983: Oxford, England, Pergamon Press.

Seifert, W. K., 1975, Carboxylic acids in petroleum and sediments, in W. Herz, H. Grisebach, and G. W. Kirby, eds., Progress in the chemistry of organic natural products, v. 32: New York, Springer-Verlag, p. 1-49.

——— 1977, Source rock/oil correlations by C_{27}-C_{30} biological marker hydrocarbons, in R. Campos and J. Goni, eds., Advances in organic geochemistry, 1975: Madrid, Enadimsa, p. 21-44.

——— 1980, Impact of Treibs' discovery of porphyrins on present-day biological marker organic geochemistry, in A. A. Prashnowsky, ed., Proceedings of Treibs International Symposium: Würzburg, Halbigdruck, p. 13-35.

——— and J. M. Moldowan, 1978, Applications of steranes, terpanes, and monoaromatics to the maturation, migration, and source of crude oils: Geochimica et Cosmochimica Acta, v. 42, p. 77-95.

——— ——— 1979, The effect of biodegradation on steranes and terpanes in crude oils: Geochimica et Cosmochimica Acta, v. 43, p. 111-126.

——— ——— 1981, Paleoreconstruction by biological markers: Geochimica et Cosmochimica Acta, v. 45, p. 783-794.

——— R. M. K. Carlson, and J. M. Moldowan, 1983, Geomimetic synthesis, structure assignment and geochemical correlation application of monoaromatized petroleum steroids, in M. Bjorøy, C. Albrecht, C. Cornford, K. de Groot, G. Eglinton, E. Galimov, D. Leythaeuser, R. Pelet, J. Rullkötter, and G. Speers, eds., Advances in organic geochemistry, 1981: New York, John Wiley and Sons, p. 710-724.

———— J. M. Moldowan, and G. J. Demaison, in press, Source correlation of biodegraded oils, *in* P. A. Schenck, J. W. de Leeuw, and G. W. M. Lijmbach, eds., Advances in organic geochemistry, 1983: Oxford, England, Pergamon Press.

———— ———— and R. W. Jones, 1980, Application of biological marker chemistry to petroleum exploration, *in* Exploration, supply and demand: Proceedings of the 10th World Petroleum Congress, Bucharest, Romania, September 1979, London, Heyden and Sons, paper SP8, p. 425-438.

———— ———— G. W. Smith, and E. V. Whitehead, 1978, First proof of structure of a C_{28}-pentacyclic triterpane in petroleum: Nature, v. 271, p. 436-437.

Shangwen, W., H. Wenhai, and T. Shidian, 1982, Habitat of oil and gas fields in China: Oil & Gas Journal, v. 80 (June 14), p. 119-128.

Sheikh, Y. M., and C. Djerassi, 1974, Steroids from sponges: Tetrahedron, v. 30, p. 4095-4103.

Silverman, S. R., 1964, Investigations of petroleum origin and evolution mechanisms by carbon isotope studies, *in* H. Craig et al, eds., Isotope and cosmic chemistry: Amsterdam, North-Holland, p. 92-102.

———— and S. Epstein, 1958, Carbon isotopic compositions of petroleum and other sedimentary organic materials: AAPG Bulletin, v. 42, p. 998-1012.

Smith, D. J., G. Eglinton, R. J. Morris, and E. L. Poutanen, 1982, Aspects of steroid geochemistry of a recent diatomaceous sediment from the Namibian Shelf: Oceanologica Acta, v. 5, p. 365-378.

Smith, G. W., D. T. Foweil, and B. G. Melson, 1970, Crystal structure of 18αH-oleanane: Nature, v. 228, p. 355-356.

Sofer, Z., 1984, Stable carbon isotope compositions of crude oils: application to source depositional environments and petroleum alteration: AAPG Bulletin, v. 68, p. 31-49.

Tissot, B. P., and D. H. Welte, 1978, Petroleum formation and occurrence; a new approach to oil and gas exploration: New York, Springer-Verlag, 521 p.

van Dorsselaer, A., P. Albrecht, and G. Ourisson, 1977, Identification of novel, (17αH)-hopanes in shales, coals, lignites, sediments, and petroleum: Bulletin du Societe Chimique de France, p. 165-170.

Volkman, J. K., R. Alexander, R. I. Kagi, and J. Rullkötter, 1983a, GC-MS characterisation of C_{27} and C_{28} triterpanes in sediments and petroleum: Geochimica et Cosmochimica Acta, v. 47, p. 1033-1040.

———— ———— ———— R. A. Noble, and G. W. Woodhouse, 1983b, A geochemical reconstruction of oil generation in the Barrow sub-basin of Western Australia: Geochimica et Cosmochimica Acta, v. 47, p. 2091-2105.

Warburton, G. A., and J. E. Zumberge, 1983, Determination of petroleum sterane distributions by mass spectrometry with selective metastable ion monitoring: Analytical Chemistry, v. 55, p. 123-126.

Whitehead, E. V., 1974, The structure of petroleum pentacyclanes, *in* B. Tissot and F. Bienner, eds., Advances in organic geochemistry, 1973: Paris, Technip, p. 225-243.

Withers, N. W., W. C. M. C. Kokke, M. Rohmer, W. H. Fenical, and C. Djerassi, 1979, Isolation of sterols with cyclopropyl-containing side chains from the cultured marine alga *Peridinium foliaceum*: Tetrahedron Letters, p. 3605-3608.

Yeh, H. W., and S. Epstein, 1981, Hydrogen and carbon isotopes of petroleum and related organic matter: Geochimica et Cosmochimica Acta, v. 45, p. 753-762.

Comparison of Carbonate and Shale Source Rocks

R. W. Jones
Chevron Oil Field Research Company, La Habra, California

> As with shales, the source potential of carbonate rocks depends primarily upon the organic facies rather than the mineral matrix. Where the depositional and early diagenetic environment is highly oxygenated, the total-organic-carbon (TOC) content is low. The remaining kerogen is highly oxygenated, with a negligible generative capacity for hydrocarbons, despite a relatively high hydrocarbon/TOC ratio in the immature state. An anoxic depositional early diagenetic environment can result in the deposition of organic-rich, fine-grained carbonate sediments that are excellent potential source rocks.
>
> Excellent oil-prone source rocks, whether with carbonate- or clay-mineral matrices, have many characteristics in common. Both form in anoxic environments, are generally laminated and heterogeneous, have moderate to high TOC, and contain high-quality organic matter (OM). The latter is exemplified by atomic H/C ratios ≥ 1.2 near a vitrinite reflectance of 0.50% R_o. Although they constitute a small percentage of all carbonate rocks, organic-rich, fine-grained carbonate rocks are widespread in both time and space and are the probable source of 30–40% or more of the petroleum reserves of the world.
>
> Gas-prone organic facies are rare in carbonate rocks because they are usually dominated by terrestrial organic matter deposited in a dominantly clay matrix. However, gas-prone organic facies may occur in carbonate rocks as a result of turbidite deposition or by a mixture of kerogen types II and IV. Most carbonate rocks contain nongenerative organic facies, as do most siliceous rocks. Oxygen-rich depositional environments for carbonates are found from sea level (reefs) to the ocean depths (*Globigerina* ooze).
>
> Despite the basic relationship between organic-rich oil-prone carbonate and shale source rocks, some significant differences exist. Oils derived from carbonate rocks are often richer in cyclic hydrocarbons and sulfur compounds than oils derived from shales, owing to the dearth of terrestrial-plant waxes in the OM and less iron in the pore water. In addition, the generally earlier decrease of porosity and permeability and the greater contrast between the physical properties of the OM and the rock matrix in carbonate source rocks often result in different primary migration characteristics.

INTRODUCTION

The literature unfortunately has placed carbonates and shales in antagonistic positions regarding source-rock evaluation. Different criteria have been used to evaluate their source potential (Tissot and Welte, 1978; Hunt, 1967, 1979). This approach requires discussion. Why not also compare carbonates and siltstones, or carbonates and sandstones? Clearly, one missing ingredient here is the original grain size of the rock. Veber and Gorskaya (1965) noted that the variations in organic content of carbonate rocks with grain size are similar to those in terrigenous facies. Because the density of oil-prone organic matter (OM) is very close to that of water, oil-prone source rocks are almost invariably fine grained. Thus, to arrive at a meaningful comparison of carbonate and shale source rocks, we must compare the shales with fine-grained carbonates (e.g., micrites) deposited in quiet water. Most carbonates are not pertinent except to understand why they are not source rocks, just as siliceous siltstones and sandstones are not.

The source potential of any sediment depends primarily on its organic facies (Jones and Demaison, 1982). The latter is determined by the amount and type of the original OM input, the time the OM spends in an oxygenated water column, and the oxygen content of the water at and near the sediment–water interface during and shortly after deposition. The hydrologic factors that control the grain size of the rock matrix also influence the organic facies by their effect on the size and density of the organic particles deposited, by the rate of oxygen diffusion, and by the rate of replacement of the bottom water and, thereby, the input of oxygen to the bottom waters. Other important secondary factors that influence the organic facies preserved in the sediment include the sedimentation rate, the thickness of the overlying oxygenated water column, the near-surface organic productivity, and the varying amounts of terrestrial OM input (Jones, 1983).

This paper stresses the similarities between oil-prone source rocks that range in composition from nearly all clay and organic matter to nearly all carbonate and organic matter. Emphasis is on the independence of organic facies and overall petroleum-generating capacity from the composition of the enclosing rock matrix. Thus, carbonates, marls, and shales are placed in a continuum, with each term occupying one-third of the distance from 100 to 0 percent carbonate. Where additional refinement is available or necessary to make a point, numbers or modifiers are used. Also included is a discussion of TOC requirements and catalytic effects, two areas in which the literature suggests substantial differences between carbonate and shale source rocks; but I disagree. The paper concludes with a discussion of one area of major differences: the relative efficiency of different migration mechanisms in carbonate and shale source rocks.

ORGANIC FACIES

General

In this paper an organic facies is defined as a mappable rock unit, distinguishable from adjacent rock units by the character of its OM *without regard to the inorganic aspects of the sediment.* Thus, organic facies are defined similarly to other facies such as biofacies. All organic facies occur in both carbonates and shales, although in different proportions.

Previous Work

The origin of the concept of organic facies is obscure, but it clearly lies with the coal petrographers. They first recognized that the different coal macerals (types of OM) commonly occur in specific groupings with unique chemical and physical properties. They also recognized that the different microlithotypes, as organic facies are called in coal terminology, were determined by both the types of OM available in the coal swamp and the early diagenetic environments (Stach et al, 1982).

In the early 1960s, Krejci–Graf (1963, 1964a, 1964b) distinguished four main groups of organic substances in the earth's crust that were determined by the oxygen content of the depositional environment. The less oxygen in the depositional environment, the more oil-prone the OM. At the same time, Bitterli (1963a, 1963b) made an extensive study of the bituminous rock sequences in Europe and noted with moderate surprise that a major controlling factor in their deposition was not water depth but the anoxia of the depositional environment. The two authors mentioned carbonate content only in passing and clearly did not consider it to be a controlling parameter. Other studies relating depositional environment and/or organic-matter type to oil-generating capacity were made by Bass (1963), Breger and Brown (1963), Tissot et al (1974), Parparova and Neruchev (1977), Maksimov et al (1976), Cornelius (1978), Larskaia (1978), and many others. With the exception of Cornelius, who used carbonate content as a parameter to complete his Eh–pH diagrams, the carbonate content of the rock did not enter into their evaluations.

In the American literature the emphasis has been on differentiating a dominantly gas-generative, "terrestrial" organic facies from a dominantly oil-generative "marine organic facies" (Breger and Brown, 1963; Rogers and Koons, 1971; Dow, 1977; Barker, 1978, 1979; and Rogers, 1980). This dichotomy has not been conspicuously successful, despite its wide use. The distribution of gas and oil in reservoirs depends not only on the original organic facies of the source rock but also on the maturity level of the OM and on the relative ease of migration of gas and oil in different geologic situations (Spillers, 1965; Weber and Daukow, 1975; Tissot and Welte, 1978; Durand and Oudin, 1980). In addition, the degree of oxidation of both the terrestrial- and marine-derived OM and initial variations within each group have a profound effect on the amounts and types of generated products. For example, waxy terrestrial-plant debris is probably a major source contributor to some oils (Hedberg, 1968); plant resins have been touted as a source for some condensates (Snowdon and Powell, 1982); vitrinite yields dominantly gas to reservoirs despite the generation of abundant liquids in vitrinitic coals (Tissot and Welte, 1978); and highly oxidized terrestrial OM is essentially inert (Tissot and Welte, 1978). Although marine OM initially has a higher, more homogeneous oil-generating capacity than terrestrial OM, it has the same problems of varying degrees of oxidation in the water column and during early diagenesis as the terrestrial OM. Thus, its generating capacity also runs the full spectrum from highly oil prone to inert. Clearly some tie between generating capacity and comparatively easily measured chemical parameters has been needed.

Without explicitly using the concept of organic facies, Tissot et al (1974) used the atomic H/C ratio of kerogen to define three kerogen types with different hydrocarbon-generating capacities. The kerogen types were subsequently related to Rock-Eval pyrolysis data (Espitalié et al, 1977). Building on the concepts developed by the French investigators, Jones and Demaison (1982) explicitly defined different organic facies[1] in terms of atomic H/C ratios and Rock-Eval data at a maturation defined by a vitrinite reflectance of $\cong 0.50\%$ R_o—i.e., near the initiation of significant generation of liquid hydrocarbons. It is important to recognize that, as defined by nature, the different organic facies are completely gradational, and the assigned boundaries arbitrary. The basic effort of Jones and Demaison (1982) tied generating capacity, organic chemistry, early diagenetic environment, sedimentology, and original OM input into a concept understandable to and usable by a petroleum geologist (Tables 1, 2). The independence of organic facies and rock matrix is stressed in the following discussion of four different organic facies that occur in both carbonate-rich and carbonate-poor rocks.

Organic Facies A

In both carbonates and shales, organic facies A is characterized by deposition in a highly anoxic environment. The anoxic environment can be either preexisting or created by the high input of OM (Demaison and Moore, 1980; Calvert, 1983). The associated rock is usually laminated and heterogeneous, the TOC contents are moderate to very high, and atomic H/C ratios are ≥ 1.4 at $\cong 0.5\%$ R_o (Tables 1, 2). The Rock-Eval S_2 response curve is usually rather narrow. T_{max}, the temperature of maximum hydrocarbon generation from the kerogen, is unusually high for a given maturation. The narrow S_2 and high T_{max} probably reflect the high degree of polymerization and homogenization of the OM.

Chevron data indicate that organic facies A occurs more frequently in carbonate rocks than in shales. This association is probably due to the dearth of terrestrial and previously degraded or thermally altered OM in most carbonate sequences of organic facies A. Microscopically, the OM is usually amorphous, clumpy, and quite homogeneous. It is presumed to be of algal/bacterial origin (Tissot and Welte, 1978). However, a strong case for hydrogenation and chemical homogenization by high alkalinity has been made for the Eocene Green River Formation (Smith and Lee, 1982; Smith, 1983). A preferred depositional environment for documented cases is shallow-water saline lakes, as exemplified by the Green River Formation and some other Tertiary lakes in the Great Basin. However, in some marine carbonate sequences, most notably some carbonate source rocks in the Middle East, the input of clay and terrestrially derived OM is negligible, the atomic H/C ratios of immature OM slightly exceed 1.4, and the existence of organic facies A can be inferred (Tissot and Welte, 1978). In

---

[1] The concepts of kerogen type and organic facies overlap and seem similar, but they are quite different. For example, organic facies D, commercially nongenerative, can be composed of a variety of kerogen types, including (1) highly oxidized terrestrial OM oxidized above the water table (e.g., terrestrial red beds), (2) OM from a variety of sources oxidized by oxygen-rich water at any water depth, and (3) redeposited kerogen of any type that was overmatured in a previous thermal cycle.

Table 1

Organic Facies	Products	Depositional Environment	Internal Structures	Organic Matter
A	Oil	Anoxic (saline), lacustrine; rare marine	Finely laminated	Algal; amorphous; rare terrestrial
B	Oil	Anoxic; marine	Laminated, well bedded	Algal; amorphous; common terrestrial
B-C	Oil-gas	Variable; deltaic	Poorly bedded	Mixed marine, terrestrial
C	Gas	Mildly oxic; shelf/slope; coals	Poorly bedded; bioturbated	Terrestrial, mostly "vitrinites"; partially degraded algal
D	Dry gas	Highly oxic; anywhere	Massive; bioturbated	Highly oxidized; reworked

Table 1 — Some generalized sedimentary characteristics of organic facies A–D.

Table 2

Organic Facies	Dominant Products	Atomic H/C at %$R_o \cong 0.5$	Pyrolysis Yield[a]	
			HI	OI
A	Oil	≥ 1.4	700–1000+	10–40
B	Oil	1.2–1.4	350–700	20–60
B-C	Oil-gas	1.0–1.2	200–350	40–80
C	Gas	0.7–1.0	50–200	50–150
D	Dry gas	0.4–0.7	<50	20–200

[a]Derived from Rock-Eval pyrolysis data, where HI = hydrogen index = mg hydrocarbons generated/g TOC; OI = oxygen index = mg CO_2 generated/g TOC.

Table 2 — Some generalized geochemical characteristics of organic facies A–D.

carbonate-poor sequences containing organic facies A, discrete algae (e.g., *Botryococcus*, *Tasmanites*) usually dominate (Hutton, Kantsler, and Cook, 1980).

Organic facies A is likely to be more laterally and vertically extensive in carbonate rocks than in noncarbonates because of the greater stability of the depositional environment and the dearth of terrestrial input. Unfortunately, these factors, conducive to development of excellent potential source rocks, often carry with them a limited accessibility to reservoir rocks unless substantial tectonic activity has severely fractured the carbonates (Fouch, 1983b). Because everything has to be just right to form organic facies A, it is rare in the geologic record and has not been a major source of oil, except perhaps locally in the Middle East, whether it was deposited in a carbonate or a clay matrix.

Organic Facies B

Organic facies B is the source of most of the world's oil. The carbonate content of the rock matrix ranges from <1 to >99%. Irrespective of carbonate content, the TOC contents are usually >1.5 wt%, and the OM dominantly amorphous with an atomic H/C ratio ≥ 1.2 at %$R_o \cong 0.5$. The sediments are usually laminated or banded and either were deposited under anoxic water or else the pore water became anoxic soon after deposition. The Rock-Eval pyrograms of carbonate and shale organic-rich source rocks of organic facies B are similar. However, the S_2 peak is usually broader in shales and the hydrogen-index averages somewhat lower. The differences reflect a greater diversity of OM in the shales.

Conditions favorable for the deposition of organic facies B can be found in several scenarios, including the following:

1. Rapid deposition of OM can create local anoxia in a generally oxic environment, as occurs on the shelf off southwestern Africa (Calvert, 1983; Calvert and Price, 1971).

2. Deposition of OM within an oxygen minimum layer (OML) where it impinges on the upper slope can create local anoxia, as suggested by Summerhayes (1981) for the Monterey Formation in southern California.

3. Thermal or salinity layering can cause a persistent density stratification that precludes replenishment of the oxygen supply, even if oxygen is only slowly consumed by the oxidation of the incoming OM. In this situation, anoxia first develops in the basin's topographic lows and gradually expands upward until oxygen insertion by water mixing produces an upper limit.

Despite increasing recognition of the occurrences of scenarios 1 and 2 and their combination, most carbonate source rocks were deposited within the constraints of scenario 3.

Density stratification owing to salinity differences in a developing evaporitic regime is often a prelude to the deposition of carbonate-rich source rocks in a situation in which the anoxia builds upward from the sediment–water interface. Examples include (1) the basinal source rocks of Silurian pinnacle reefs in the Michigan basin (Gardner and Bray, 1984); (2) the lagoonal source rocks for the Lower Cretaceous rudist mounds in south Florida (Palacas, 1983); (3) the deep-water organic-rich Permian Lamar and Bone Spring limestones underlying the Castile evaporites in the Delaware basin of west Texas–New Mexico (Fig. 1); (4) the inferred interreef carbonate source beds in the Devonian of the Rainbow–Zama area in western Canada (McCames and Griffith, 1967); (5) the deep-water Rudeis Formation marls in the Miocene of the Gulf of Suez (Heybroek, 1965; Hassan and El-Dashlouty, 1970); (6) parts of the middle Paleozoic of the Russian platform (Zhuze, 1972); (7) the Cambrian of the southeastern Siberian platform (Bakhturov, 1981); (8) parts of the Tethys Ocean in the Mesozoic, particularly the Upper Jurassic of the Middle East (Murris, 1980); (9) the Jurassic of the Gulf Coast (Oehler, 1984); and (10) the precursors of the main deposition of the Permian Zechstein evaporites in western Europe (Fig. 2). There may be a source contribution from the "Stinkdolomit" and "Stinkkalk," but the primary source is probably the Stinkschiefer, which is a laminated carbonate-rich (70–90%) rock that contains the highest TOC contents (Botz and Müller, 1981; Botz et al, 1981).

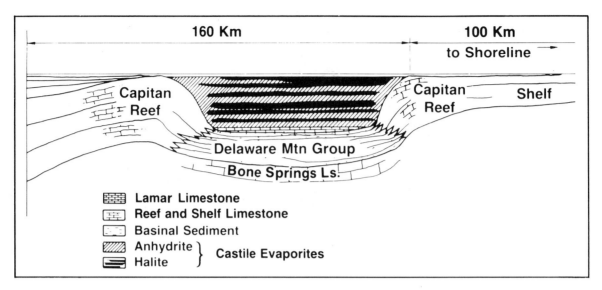

Figure 1 — Diagrammatic stratigraphic cross section, Late Permian time, Delaware basin, west Texas and southeastern New Mexico, showing the propitious geometrical relationships between the source rocks of the Lamar and Bone Springs Limestones, deposited in the basin center, and the adjacent reservoirs in the reef and on the shelf.

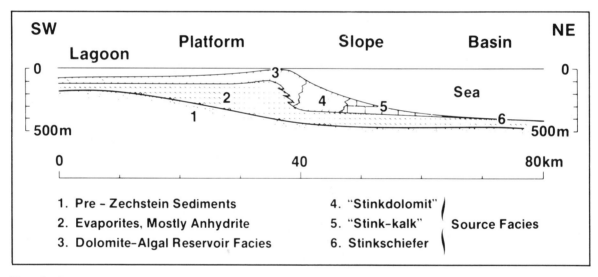

Figure 2 — Diagrammatic stratigraphic cross section of a part of East Germany showing the relationships between source and reservoir rocks in the lower part of the Permian Zechstein Formation prior to deposition of the succeeding evaporitic sequence. Cross section modified from Botz and Müller (1981).

Where carbonate rocks with good source potential are interbedded in evaporite sequences, migration problems often have precluded effective drainage. Any thermally generated hydrocarbons may remain where they originated or very close by. Thus, accumulations are likely to be small, as are those having a source in the Cretaceous Binga Formation carbonates in the Cuanza basin of Angola (Brognon and Verrier, 1966). The organic-rich calcareous shales interbedded with evaporite deposits in the Pennsylvanian Paradox Formation of the Paradox basin, Utah, still retain almost all of their hydrocarbons. Where they were deposited around the porous algal mounds at the southern end of the basin, they have been the source of locally prolific production (Peterson and Hite, 1969).

Many organic-rich carbonate source rocks have been deposited in lake systems. On the average they have been

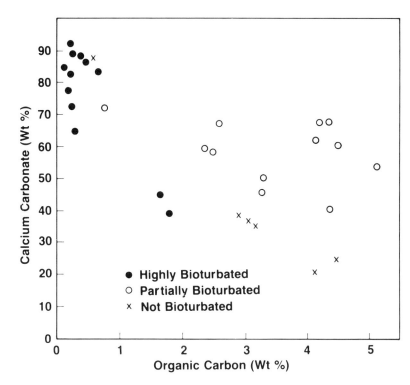

Figure 3 — Correlation of calcium carbonate and organic-carbon contents in a 30-m (98-ft) core from the middle Cretaceous Greenhorn Formation, Colorado. Data from Pratt (1982).

responsible for less commercial oil than their often less organic-rich shale counterparts because of poorer access to reservoirs (Fouch, 1983a, 1983b).

Density stratification caused by salinity differences also can be created by an ingress of fresh water at the top of the water column. Documented cases in the geologic record in which carbonate source rocks resulted from this type of density stratification are rare and difficult to identify. However, the flow of fresh water through deltaic systems into Lake Uinta must have contributed to the density stratification there that led to the formation of organic-rich dolomitic marls in the Green River Formation. Ozimic (1982) attributed formation of the carbonate-rich oil shales of the Lower Cretaceous in the Eromanga basin, Australia, to anoxia created by the input of fresh water into a silled marine basin.

Perhaps the best documented illustration of how density stratification created by fresh-water input into a marine environment can affect the carbonate and organic content of the underlying sediments is provided by an excellent study by Pratt (1982) of a 30-m (98-ft) core from the middle Cretaceous Greenhorn Formation near Pueblo, Colorado. Figure 3, generated from Pratt's tabular data, shows two distinct groupings of sediments: one that is carbonate-rich and organic-carbon-poor, and the other with a moderate carbonate content and a moderate to high organic-carbon content. Pratt explained the two sediment types as products of variable terrestrial and fresh-water input from highlands to the west. During times of high fresh-water input, enhanced density stratification and bottom-water anoxia resulted in inhibited growth of calcareous algae and preservation of organic-rich sediment.

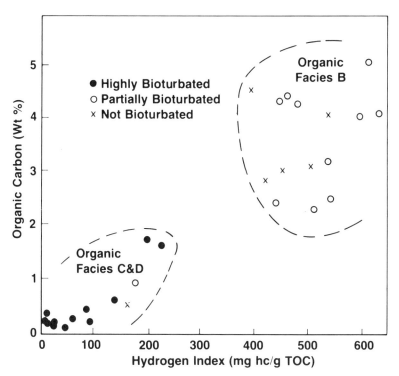

Figure 4 — Correlation of organic carbon and hydrogen index in a 30-m (98-ft) core from the middle Cretaceous Greenhorn Formation, Colorado. Data from Pratt (1982). The very low hydrogen indices of some of the high carbonate samples may be due in part to mineral-matrix effects during pyrolysis (Espitalié, Madec, and Tissot, 1980; Dembicki, Horsfield, and Ho, 1983).

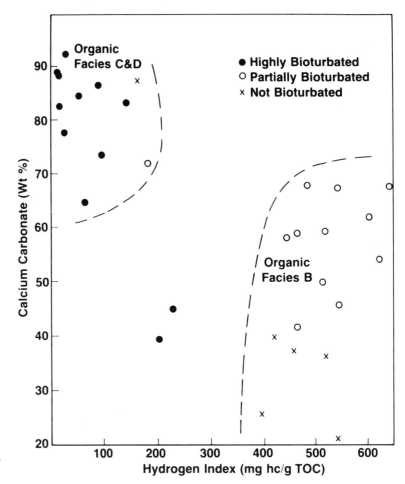

Figure 5 — Correlation of calcium carbonate and hydrogen index in a 30-m (98-ft) core from the middle Cretaceous Greenhorn Formation, Colorado. Data from Pratt (1982).

Pyrolysis and chromatographic data indicate that the OM in the Greenhorn Formation is at the beginning of the oil window. Therefore, the different organic facies can be recognized by using the hydrogen index (Jones and Demaison, 1982). Figure 4 shows a well-defined organic facies B, and a gradation from organic facies C to organic facies D, as defined by the hydrogen-index data of Pratt (1982). As the TOC decreases, the OM becomes less oil prone, reflecting the increased proportion of bacteria-resistant, residual, nongenerative OM. The plot of hydrogen index versus percentage of carbonate (Fig. 5) indicates the same separation of organic facies as shown in Figure 4. These results show the effects of increased bioturbation recorded by Pratt (1982) in the high-carbonate rocks.

Three other less clearly defined points can be made:

1. Five of the six nonbioturbated samples have a lower carbonate percentage and a slightly lower hydrogen index than the higher carbonate samples of organic facies B. This may reflect the increased input of terrestrial OM that Pratt (1982) observed in the more clay-rich samples.

2. Two highly bioturbated samples with 40–50% carbonate content occupy a distinct niche in Figure 5. Stratigraphically, they occur as thin marls within a limestone sequence and probably reflect an influx of fresh water and associated clay insufficient to create complete anoxia. Their hydrogen indices are at the extreme upper end of organic facies C and demonstrate that marls, despite a dominant marine-algal input, can, if partly degraded, contain organic facies C.

3. The anomalous, nonbioturbated sample with a high carbonate content is from a thin limestone bed within a marl. Thus, it may be surmised that the length of time of normal marine salinity at the surface necessary to create the limestone was insufficient to disturb the anoxia at the water–sediment interface. In addition, the modest TOC of the sample, albeit the highest of the high carbonate samples, suggests that most of the original OM was not derived from calcareous algae. The carbonate was a diluent.

Dilution is common to many potential carbonate source rocks. The Holocene sediments of the Black Sea are a case in point. Anoxia was created by density stratification, this time caused by influx of saline marine water into a fresh-water lake. However, the most organic-rich sediments usually contain the least carbonate, a fact in part caused by the com-

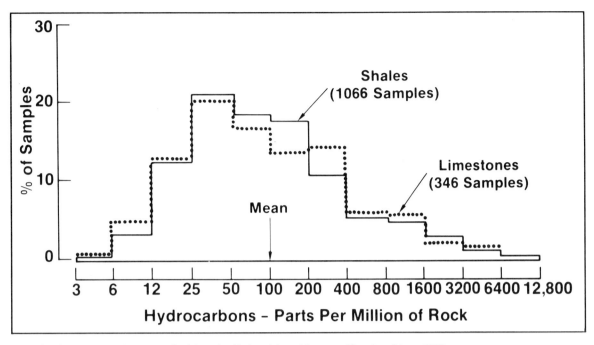

Figure 6 — Distribution of hydrocarbons in Cambrian to Late Tertiary shales and limestones. Figure from Gehman (1962).

bination of a roughly constant input of OM superimposed on a varying rate of coccolith deposition (Degens, 1974; Degens et al, 1978; Jones, 1983).

Carbonate dissolution caused by a low pH at or near the sediment–water interface can also increase the OM/carbonate ratio. This process also contributed to the strong inverse correlation between OM and carbonate content in the Black Sea (Degens and Stoffers, 1976). Dean, Gardner, and Cepek (1981) stress variations in the carbonate compensation depth and the resulting periodic dissolution of carbonate in Cretaceous–Eocene sediments of the northwest Atlantic as a major control on varying organic and carbonate content. Similar effects were postulated as one of the controls in the deposition of organic-rich, middle Cretaceous limestones in the equatorial Pacific (Dean, Claypool, and Thiede, 1981).

Sometimes the transition from carbonate-poor to carbonate-rich organic facies B can be very gradual. An excellent example is the Eagle Ford Shale to Austin Chalk transition in the Cretaceous of the East Texas basin. Grabowski (1981, 1984) showed that the organic facies in this interval remained essentially constant as the Cretaceous transgression proceeded. As the locus of deposition became farther removed from terrigenous input, the carbonate content of the sediments increased. Only when the water became shallower and the anoxia at the water–sediment interface was destroyed did the organic facies become less oil prone, although the carbonate content remained high.

Many of the world's prolific oil source rocks were deposited under conditions in which the density stratification necessary for anoxia to develop and persist at the ocean bottom was created by temperature gradients in widespread warm, temperate climates. Water turnover owing to temperature-induced density inversions did not occur, and oxygen injection near the water–sediment interface was minimal because of a lack of strong mixing. Such periods typically developed oxygen-deficient bottom water worldwide, occurring during quiet but persistent worldwide transgressions (Tissot, 1979), and extended the oxygen-deficient conditions well onto the shelf, particularly into broad shelf basins. These conditions led to deposition of carbonate-, siliceous-, and clay-rich organic facies B source rocks.

The most widespread transgressive source rocks are usually clay rich. Examples include the Upper Devonian–Lower Mississippian Bakken Shale–Woodford Shale–Chattanooga Shale unit in the Mid-Continent province of the United States and the Kimmeridgian source rocks of the North Sea and Grand Banks. Deeper water Woodford equivalents (Arkansas Novaculite) are also organic rich but are less so than the Woodford, whose organic content was not diluted by major siliceous input. The siliceous Lower Cretaceous Mowry Shale is widespread in the Rocky Mountain area and is a major oil source. The Upper Cretaceous Niobrara carbonate and equivalents of organic-rich organic facies B are also widespread in the Rocky Mountains; however, their contribution to oil production has been limited by maturity and reservoir-access problems. The very prolific Upper Jurassic source rocks in the Middle East also were deposited in relatively shallow water during a widespread transgression in a mild climate (Murris, 1980; Ayres et al, 1982). These rocks are virtually all carbonate and OM. The climate-induced density stratification was probably aided by salinity stratification, as indicated by associated evaporites in the Gotnia basin to the north and pseudomorph structures after halite in the broad lagoonal areas on the Arabian Jurassic shelf (Ayres et al, 1982).

Many shallow-water carbonate depositional environments can include a variety of local conditions that permit the accumulation and preservation of a significant amount of OM (Malek-Aslani, 1980). Local vertical salinity gradients can exist on a carbonate shelf, particularly in areas of irregular topography. Because of the low iron content of many carbonate environments, carbonate sediments tend both to generate and to preserve H_2S. The H_2S can create an anoxic environment at lesser sediment thicknesses than for clay muds (Lindbloom and Lupton, 1961). The combination of high organic productivity, high carbonate input, and low clay input can create an H_2S-O_2 interface at the water–sediment interface in shallow water, as postulated by Abed and Amireh (1983) for the Upper Cretaceous oil shales of Jordan. Anoxia can develop in shallow-water sediments at/near the sediment–water interface in fine-grained sediments of any type of sediment matrix if the input of organic matter is sufficiently rapid to deplete the available oxygen. A large amount of organic matter is destroyed to permit the preservation of the remainder (Jones, 1983). Examples include certain areas of noncarbonate muds in the Baltic Sea (P. J. Müller, written communication), the siliceous sediments in areas of upwelling off Peru and southwest Africa (Müller and Suess, 1979; Calvert and Price, 1971), and thin organic-rich layers interbedded with Ordovician shelf carbonates and evaporites in the Williston basin (Kohm and Louden, 1979). Blue-green algal (stromatolitic) mats containing thin, organic-rich laminae have existed throughout much of geologic time. However, their restricted lateral and vertical extent at any particular time has limited their significance as source rocks of organic facies B. The best documented exception to this generalization is the probable algal-mat source for the substantial hydrocarbon accumulations in the Upper Jurassic Smackover Formation of the Gulf Coast (Oehler, 1984).

Despite the overlapping and intertwined nature of carbonate-, siliceous-, and clay-dominated source rocks of organic facies B demonstrated above, some differences exist. Although most OM in most organic facies B rocks is marine derived in both clay- and carbonate-rich rocks, the clay-rich rocks usually contain more terrestrially derived OM. Fortunately the more waxy and hydrogen-rich macerals of terrestrial plants (e.g., spores, pollen, cuticle, and the more hydrogen-rich varieties of vitrinite) are the most resistant to bacterial attack. Thus, if the terrestrial input is minor, it is likely to be dominated by hydrogen-rich OM similar in oil-generating capacity to the marine OM. This is true for the Kimmeridgian of the North Sea and the better Cretaceous source rocks of the Rocky Mountain region. Of course, as the percentage of terrestrial OM increases, the overall quality decreases, resulting in organic facies B-C (e.g., Tertiary of the Gulf Coast) and ultimately the primarily gas-generative organic facies C, characterized by a dominant terrestrial input.

Statistical differences exist between the oils generated in carbonate- and clay-dominated rocks of organic facies B (Gransch and Posthuma, 1974; Tissot and Welte, 1978; Breger, 1980; Tissot and Roucaché, 1980). Although they both yield a naphthenic–paraffinic oil, the paraffin content increases with maturity and terrestrial OM content. Oils derived from source rocks with iron-rich clays are usually low in sulfur; conversely, carbonate and siliceous rocks of organic facies B typically yield moderate- to high-sulfur oils. In addition, the high sulfur content of the kerogens in some carbonate/siliceous oil-prone source rocks can result in the formation of low-gravity, sulfur-rich oils formed at less than normal thermal stress (unpublished Chevron data). Such oils primarily reflect the lower activation energy required to break sulfur–carbon than carbon–carbon bonds. The higher sulfur content can persist into the condensate stage of maturation. For example, in western Wyoming, condensates derived from the Phosphoria Formation (Permian) are distinctly higher in sulfur than those of the same API gravity derived from essentially carbonate-free Cretaceous sources.

Typically, giant oil fields derived from organic facies B in either carbonate-rich (e.g., Arabia) or clay-rich (e.g., North Sea) rocks have low gas–oil ratios unless the maturity of the OM is high. This reflects both the high TOC contents necessary for prolific oil production from either rock type and oil migration prior to reaching the main stage of gas generation.

Other differences exist, particularly in dominant migration mechanisms, as discussed later.

Organic Facies C

The gas-prone organic facies C is typically composed of partially oxidized terrestrial OM deposited in a carbonate-poor clay matrix. Tremendous thicknesses of this rock–OM association were deposited on continental margins during extensive episodes of the Mesozoic and Cenozoic Eras, as the oil industry discovered to its consternation during exploration on continental shelves during the past two decades.

Organic facies C can occur in carbonate rocks. However, it is volumetrically unimportant and usually distributed in thin layers of limited areal extent. Several processes can be involved. Limited amounts of terrestrially derived OM can be introduced into a carbonate environment, but because the terrestrial OM is almost always associated with clay the carbonate environment will usually be overwhelmed if the influx of terrestrial OM becomes dominant. The OM that initially formed in carbonate environments is dominated by the potential for organic facies A–B; however, the variation in oxygen content near the water–sediment interface can sometimes stabilize the degrading OM in the geochemical characteristics of organic facies C (Fig. 4; Dean, Gardner, and Cepek, 1981; Pratt, 1982). Another possible scenario for the development of organic facies C in carbonate rocks is the side-by-side existence of highly degraded OM and higher quality OM protected from degradation by inclusion in carbonate skeletal remains. In such cases, the TOC is usually quite low, and a small amount of late-stage gas migration is the best that can be hoped for. Basically, organic facies C is economically unimportant in carbonate rocks.

Organic Facies D

Unfortunately, most carbonate rocks contain nongenerative organic facies D, a fact that has undoubtedly postponed the recognition of some carbonates as excellent oil-source rocks. The primary reason for the dominance of organic facies D in carbonate rocks is straightforward: most carbonates were deposited in highly oxygenated environments (Larskaia, 1977). This environment existed in carbonate banks and reefs, in the shelf and ramp carbonates that are important volu-

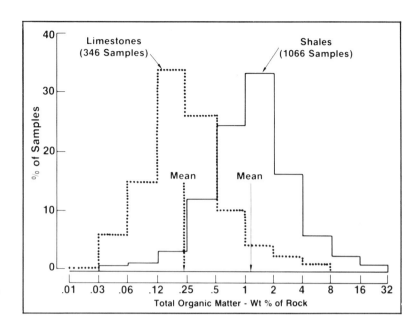

Figure 7 — Distribution of total organic matter in Cambrian to Late Tertiary shales and limestones. Figure from Gehman (1962).

Table 3

Total Organic Matter (wt%)	Carbonate (wt%)	Bitumen/TOM	Number of Samples
0–0.1	95.7	23.8	63
0.1–0.2	89.8	14.4	77
0.2–0.4	84.5	12.1	45
0.4–0.6	75.4	6.8	23
0.6–0.8	42.1	6.6	16
0.8–1.0	33.9	5.5	18
>1.0	27.2	2.2	37

[a]Data from Uspenskii and Chernysheva (1951)

Table 3 — Relationships between total-organic-matter (TOM), bitumen, and carbonate contents for 279 samples from the USSR.[a]

metrically, in most of the chalks deposited in intermediate water depths, and in deep-sea oozes deposited above the carbonate compensation depth. These are the rocks with low TOC contents but high hydrocarbon/TOC ratios that have occasionally been touted as possible source rocks (Hunt, 1967). I consider their generating capacity negligible. The epitome of organic facies D in carbonate rocks was reached in most of the Jurassic back-reef carbonates of the East Coast of the United States. There, a highly oxidizing environment was also subjected to an input of residual nongenerative OM derived from erosion of the strongly thermally altered Paleozoic rocks of the Appalachian Mountains.

Of course, organic facies D is also dominant in carbonate-poor rocks. Nonmarine rocks, except for most coals and some lake sediments, are composed of organic facies D, as are most shelf sediments, particularly the siltstones and sandstones. Red beds contain organic facies D, whether their highly oxidized kerogen was originally derived from structured terrestrial OM deposited in flood-plain deposits or nonstructured marine OM deposited in the ocean depths.

THE TOC PROBLEM

The repeated suggestions in the literature that carbonates may be effective oil source rocks at lower TOC contents than shales are largely based on interpretations by others of Gehman's (1962) excellent pioneer work. Thus, Gehman's study is worth reexamining despite the lack of both thermal-maturation data and the pairing of TOC and hydrocarbon data on the same samples. Gehman's study is summarized in Figures 6 and 7. For the 346 limestone and 1,066 shale samples analyzed, the data show that (1) the average total-organic-matter (TOM) content of the shales $\cong$ 4 times that of the limestones; (2) the absolute amounts and distribution of hydrocarbons in limestones and shales are nearly identical; and, consequently, (3) the ratio of hydrocarbons to TOM $\cong$ 4 times greater in limestones than in shales. Gehman's data, particularly item 3, have been used to argue that carbonates do not require as much organic carbon as shales to be source rocks. A minimum figure of 0.3 wt% TOC for carbonates has been suggested (Hunt, 1967, 1979; Tissot and Welte, 1978).

Other interpretations of Gehman's data are possible. The lower TOM and higher hydrocarbon/TOM ratio in carbonates could, in part, simply reflect the greater preferential destruction of nonhydrocarbon OM in highly oxygenated and bioturbated sediments. Although primarily concerned with the effects of oxygenated meteoric waters, Gehman (1962) suggested this interpretation. Unfortunately, Gehman did not break out the hydrocarbon/TOM ratios as a function of both TOM and carbonate percentage. However, others have.

Although using bitumen rather than

Table 4

Total Organic Matter (wt%)	Carbonate Rich (CaCO$_3$ ≥ 92 wt%)			Relatively Carbonate Poor (CaCO$_3$ ≤ 71.5 wt%)		
	Carbonate (wt%)	Bitumen/TOM	Number of Samples	Carbonate (wt%)	Bitumen/TOM	Number of Samples
0.1	98.7	24.2	56	71.5	21.4	7
0.1–0.2	98.1	14.3	61	51.7	14.7	16
0.2–0.4	95.8	13.2	34	49.4	9.6	11
0.4–0.6	92.7	5.8	14	48.6	8.5	9
0.6–0.8	95.9	5.6	6	9.7	7.2	10
0.8–1.0	93.1	6.1	5	11.2	5.2	13
>1.0	92.1	3.9	2	23.5	2.3	35

[a]Data from Uspenskii and Chernysheva (1951).

Table 4 — Correlation of total organic matter (TOM) with bitumen/TOM for 178 carbonate-rich (>90% CaCO$_3$) and 101 carbonate-poor (<72% CaCO$_3$) samples from the USSR.[a]

hydrocarbon data on 279 samples from Paleozoic–Tertiary carbonates of the USSR, Uspenskii and Chernysheva (1951) showed conclusively that the basic correlation is between low TOM and high bitumen/TOM ratios, not between carbonates and high bitumen/TOM ratios (Tables 3, 4). Table 3 shows a strong positive correlation between carbonate percentage and the bitumen/TOM ratio. The data also show an equally strong inverse correlation between the carbonate percentage and the TOM and between the TOM and the bitumen/TOM ratio. What underlying relationship controls these correlations? Uspenskii and Chernysheva (1951) resolved this problem by dividing each TOM group into high and low (relatively) carbonate subgroups (Table 4). These data clearly show that the primary controlling factor of the bitumen/TOM ratio is the amount of TOM, not the carbonate percentage. Uspenskii and Chernysheva (1951, p. 104) explicitly attributed the increased bitumen/TOM ratio in carbonate rocks to a faster rate of decomposition of the OM. Their explanation is consistent with the dominant control of both the TOM percentage and the bitumen/TOM ratio being the oxygen content of the depositional and early diagenetic environment rather than the chemical composition of the rock matrix.

Bordovskiy et al (1974) provided an unusual example of the preferential destruction of nonhydrocarbon OM in the bottom sediments of the Kuril–Kamchatka trench. They showed that as OM moved through the alimentary canal of a bottom feeder, the total amount of oils (≅ hydrocarbons) remained essentially constant, whereas the TOC dropped by ≅50% and the bitumen content by nearly an order of magnitude. Thus, the hydrocarbons were preferentially preserved, and the hydrocarbon/TOC ratio went up as the TOC dropped.

Despite the overriding inverse correlation between TOM and the bitumen/TOM ratio for all rock matrices, there is a definite tendency, particularly at very low TOMs, for the bitumen/TOM and hydrocarbon/TOM ratios to be higher in carbonate rocks. Bordovskiy and Takh (1978) compared the bitumen/TOM ratios for Holocene sediments defined as carbonate and carbonate-free sediments in the Caspian Sea (Fig. 8). The shapes of the two curves are essentially identical, confirming the basic TOM to bitumen/TOM correlation. However, the two curves show a persistent separation between the carbonate and carbonate-free sediments, a separation that clearly increases with decreased TOC. Do the substantially higher hydrocarbon/TOC ratios at low TOCs for carbonate rocks really matter when we are considering rocks that are potentially the source of significant amounts of petroleum? I think not. The total amount of bitumen–hydrocarbons in these low TOC rocks is negligible compared to what is necessary to produce significant accumulations. The total generating capacity is what is of interest, and the ratio of bitumen–hydrocarbons/TOC in immature low-TOC rocks, are not pertinent. For example, if one believes that in immature sediments, such as those described by Bordovskiy and Takh (1978), the higher the bitumen/TOC ratio the better the source potential, one is forced to prefer an infinitesimal TOC, a *reductio ad absurdum*.

Figure 9 shows TOC and hydrocarbon/TOC data compiled from the literature by Tissot and Welte (1978, p. 95) for 20 areas of Holocene clastic sedimentation in five different depositional environments. They also included analyses from two areas of carbonate muds. Although many unknown factors are operating, three conclusions pertinent to this discussion are apparent from Figure 9: (1) overall, an inverse correlation exists between the TOC and the hydrocarbon/TOC ratio; (2) the values for the carbonate samples are surrounded by samples from a variety of clastic environments; and (3) the highest hydrocarbon/TOC ratios are from clastic environments with low TOCs.

At least two effects might contribute to the increasing spread between the bitumen/TOC ratio for carbonates and shales that occur with decreasing TOC (Fig. 8). In rocks that are almost all carbonate and have a low TOC, the ratio may be inflated by loss of a substantial portion of the OM in the necessarily severe oxidation process. In shales with low TOC, a substantial portion of the OM may be land-derived inertinite with negligible bitumen content. Thus the bitumen/TOC ratio would be lower than for shales with a higher TOC but the same background inertinite content (Hood and Castaño, 1974; Leythaeuser et al, 1980).

Various reasons have been advanced to explain the differences (and similarities) that often exist in hydrocarbon/TOC and bitumen/TOC ratios and average TOC between carbonate-rich and carbonate-poor rocks at all values of TOC reported by Gehman (1962), Bordovskiy and Takh (1978), Tissot and Welte (1978), and others. None of the possible explanations has been accompanied by a suffic-

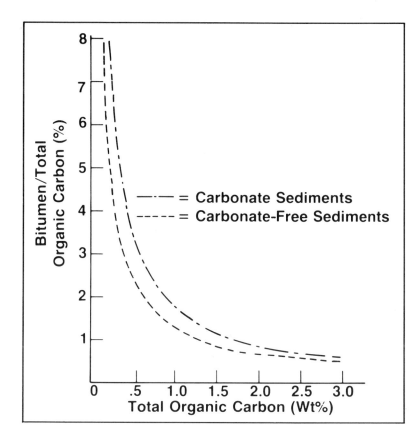

Figure 8 — Correlation of bitumen/total organic carbon with total organic carbon for carbonate and carbonate-free bottom sediments of the Caspian Sea, USSR. Figure from Bordovskiy and Takh (1978).

ient amount of data on the many factors that could influence the numbers to be completely convincing.

The quality (type) of the OM could be significant (Hunt, 1979), as suggested by an average greater algal input to carbonate rocks and a corresponding slightly higher H/C ratio. However, the H/C ratio difference in oil-prone source rocks is small, and the sulfur often incorporated in kerogens in carbonate rocks probably decreases the hydrocarbon-generating capacity of a given H/C ratio. Also, Figure 9, although containing no explicit data regarding kerogen quality, gives no support to the thesis that the types of OM in Holocene carbonate muds contain higher percentages of hydrocarbons than do their siliciclastic counterparts.

Differences in adsorption capacities between clays and carbonates for hydrocarbons and bitumens are a possible explanation. However, despite the excellent observations and experimental work on carbonates by Shearman and Skipwith (1965), Chave (1965), and Suess (1970, 1973), and many studies on clays (summarized in Hunt, 1979), the quantitative significance of adsorption effects of carbonates and shales on the hydrocarbon/TOC ratio at different maturation levels is not known. For example, most clays presumably have greater catalytic effects regarding conversion of various bitumen components to hydrocarbons; yet carbonates, statistically, have a higher hydrocarbon/TOC ratio.

I suggest (also with no proof) that the hydrocarbon–bitumen–TOC relationships being discussed are primarily caused by the overall longer exposure of OM in carbonate rocks to an oxygenated environment. At least four separate effects might contribute within such a framework: (1) higher average organic productivity in carbonates accompanied by a preferential destruction of nonhydro-

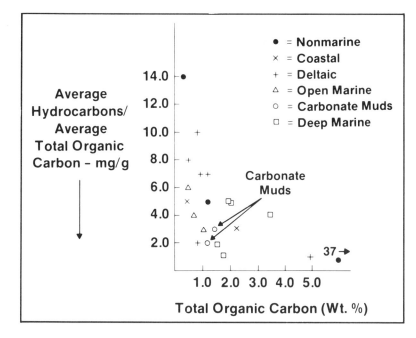

Figure 9 — Correlation of hydrocarbon/total organic carbon with total organic carbon for Holocene sediments from five siliciclastic environments and two carbonate-mud environments. Data from Tissot and Welte (1978, p. 95).

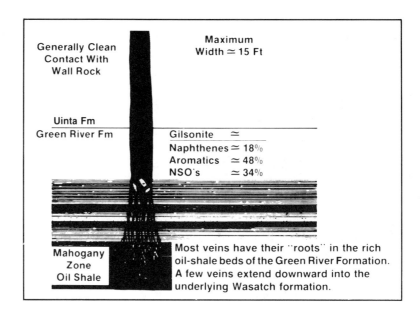

Figure 10 — Generalized cross section of a gilsonite mine, Uinta basin, Utah. After Eldridge (1901).

carbon OM could result in a higher hydrocarbon/TOC ratio than for shales with a similar TOC; (2) as a group, carbonates are deposited in a shallower, more agitated, and more oxygenated environment than shales and, everything else being equal, lower TOC contents should result; (3) most carbonates have a larger average grain size than shales, and diffusion of oxygen into such carbonates should be greater than for shales; (4) lower sedimentation rates and a larger average grain size for carbonates than for shales could result in the preservation of an oxic early diagenetic environment for a longer period of time for carbonates than for shales. Again, everything else being equal, lower TOC contents should result from the lower sedimentation rates (Toth and Lerman, 1977; Müller and Suess, 1979; Müller and Mangini, 1980).

In summary, the data presented and discussed above do not indicate that carbonate-rich rocks have a lower minimum TOC requirement to be source rocks than carbonate-poor rocks. This relationship simply reflects the lower TOC contents and more oxygenated depositional and early diagenetic environment of most carbonates analyzed. Only one of the carbonate source rocks reported on in this volume contains TOC values close to the 0.3 wt% minimum suggested by Tissot and Welte (1978) and Hunt (1979). Gardner and Bray (1984) report an average TOC of 0.35 wt% for the Silurian source rocks of the Michigan basin. However, many individual values are higher, and it is possible to speculate that the deeper and sparsely sampled parts of the depositional basin contain higher TOC contents. In all the other examples discussed in this volume the TOC contents are 1.5+ wt%, indistinguishable from the TOC contents of shales that were the source of similar amounts of oil.

EFFECTS OF VARIATIONS IN ROCK MATRIX ON GENERATION

Catalytic Effects

Arguments about the relative catalytic effects of carbonates and clays on hydrocarbon generation have been around for a long time. In my opinion, their longevity is due to the following reasons: (1) most experiments have been performed on hydrocarbons or organic acids (Shimoyama and Johns, 1972) with extremely high surface-to-volume ratios, not on kerogens with a much smaller surface-to-volume ratio; (2) any catalytic effects in the subsurface during the generative process in organic-rich rocks are minor and so are difficult to demonstrate conclusively as a significant subsurface effect; and (3) effects that have been demonstrated (Dembicki, Horsfield, and Ho, 1983) are strongest in low-TOC rocks with comparatively low surface-to-volume ratios of the kerogen. Such demonstrated alterations are also more directly related to catalytic effects as hydrocarbons move through the source rock than to the initial generation process. The lack of a rock-matrix effect in the more popular of the thermal-maturation models (Hood, Gutjahr, and Heacock, 1975; Waples, 1980) is a clear indication that geochemists in the oil industry have not perceived a significant difference between the catalytic effects of clay- and carbonate-rock matrices.

In excellent oil-prone source rocks, only a small percentage of the potentially generative OM is in contact with either clays or carbonates during generation, and generation is primarily a temperature effect. This problem is being increasingly recognized (Johns, 1982; Halpern and Kaplan, 1982). Undoubtedly, some differential catalytic effects occur as the bitumen moves through and out of an organic-rich source rock. However, if the movement is primarily through fractures (Momper, 1978), this effect is probably also small. Saturation occurs quickly. A possible exception to this generalization is clay-catalytic effects where hydrocarbons move substantial distances in overpressured high-porosity shales (Davis, 1982; Davis and Stanley, 1983). Even this possibility as a quantitatively significant effect can be questioned. Savin (1982) used isotopic studies to show that most water moving out of and through shales moves through cracks.

Thus the arguments go on. However, as stated previously, the surface-to-volume ratio of the kerogen particles in organic-rich oil-prone rocks strongly suggests that the catalytic effects of different rock matrices regarding generation are insignificant.

Delay in Generation

Nevertheless, there is some evidence that generation is delayed in fine-grained, tight, oil-prone carbonate source rocks (Palacas, 1984). Differences in generation timing within both carbonate and clay rocks of organic facies B having identical thermal histories probably reflect whether the system is open or closed. For example, in pyrolysis bomb experiments, generation reactions reach equilibrium at a constant pressure and temperature if the generation products are not removed from the bomb (S. R. Silverman, personal communication). In most stratigraphic sections, local pressure buildups tend to dissipate into the regional pressure gradient. A normal generation versus depth (temperature) profile results, as in the Douala basin, Cameroons (Albrecht, Vandenbroucke, and Mandengue, 1976).

However, in a relatively fine-grained carbonate sequence, particularly if it is argillaceous or organic rich, even a moderate amount of hydrocarbon generation is likely to build up and maintain the local pressure sufficiently to retard the chemical reactions necessary for additional generation. The retardation can also occur in tight or overpressured shales but is probably more common in fine-grained carbonates owing to their lower porosity and permeability for a given thermal history. This effect has not been discussed adequately in the source-rock literature, although it has been observed occasionally. It is probably the explanation for the liquid hydrocarbons observed by Price, Clayton, and Rumen (1981) in the subsurface at high temperatures; the higher percentage of saturate hydrocarbons in open pores than in closed pores in a Triassic limestone sequence in Hungary (Brukner and Vetö, 1983); the dramatic retardation in maturation of the OM in a Triassic calcareous shale overlying a reservoir at 200°C (392°F) in the AGIP Canoica well in Italy (Neglia, 1980); and an anomalous immaturity in overpressured shale in the Miesbach 1 well in West Germany (Jacob and Kuckelhorn, 1977). As expected, fine-grained carbonate rocks dominate the above examples.

In some carbonate source rocks the retardation of generation by local pressure buildup creates a bomb that is ready to explode if the system is driven to high enough temperatures or is tectonically fractured. This observation leads into the next section: migration similarities and contrasts in carbonate-rich and carbonate-poor, oil-prone source rocks.

MIGRATION

The same suite of possible migration mechanisms exists in all rocks. However, its relative quantitative significance depends not only on the organic characteristics but also on the rocks' chemical and, particularly, physical properties. Although the chemical differences between carbonate and shale source rocks are substantial, their effects are more evident on oil characteristics than on the efficiency or type of migration. On the other hand, effective shale and carbonate source rocks usually are quite different in their physical properties. In particular, carbonate source rocks are usually more brittle, more heterogeneous, and tighter than shale source rocks. Consequently they are more prone to fracture and emit hydrocarbons more readily than shales in a variety of geologic settings.

Fracturing on various scales has recently achieved considerable recognition as a principal cause of primary migration from organic-rich, oil-prone source rocks (Momper, 1978; Tissot and Welte, 1978; Hunt, 1979). However, Momper (1978), in an otherwise perspicacious paper, denigrated the significance of carbonate source rocks. He believed that their general dearth of clay would not permit the episodic buildup of pressure from hydrocarbon generation and its subsequent release by microfracturing that he considered the *sine qua non* of commercially significant expulsion and migration of hydrocarbons (Momper, 1978, p. B22–B24). In support of this view, Momper cited the lack of documented high pressures in carbonate sequences and a generally poor documentation of carbonate source rocks. The question of documentation has been answered by several papers in this volume and by Murris (1980) and Ayres et al (1982) on the Jurassic of the Arabian Peninsula.

The pressure-buildup question is more involved. Certainly, carbonate sections are leaky where they contain little clay or organic matter. For example, there is strong empirical support for not drilling below the top of massive carbonate sequences, such as the Ellenburger Formation in the southern Mid-Continent, if tests at the top are water wet (Cardwell, 1977). However, fine-grained carbonate rocks that contain considerable clay and/or organic matter are not leaky to fluids beneath them. In the process of normal compaction, such source rocks become overpressured with respect to both the immediately overlying and underlying sections. Thus, they can emit petroleum upward and downward. They can be a source for overlying accumulations with a separate cap, as with the Hanifa–Hadriya carbonate source rock in the Middle East (Murris, 1980; Ayres et al, 1982), or be both a source and cap for underlying accumulations, as with the Oxfordian carbonate source rocks in the Middle East and the Miocene marls that cap and act as the source for reservoirs in offshore southern Spain (Demaison and Bourgeois, 1984).

Organic-rich carbonate source rocks are typically deposited in an anoxic, salinity-stratified environment that is the prelude to evaporite deposition. If such a cycle is carried to completion, it is immaterial whether the emission of hydrocarbons is episodic or continuous; the evaporite cap is available at all times to channel the hydrocarbons into any available porosity. An example of this situation is the Jurassic Smackover section along the northeastern coast of the Gulf of Mexico, where the source and reservoir directly underlie an anhydritic caprock. There, it is easy to visualize a heterogeneous, finely laminated, shallow-water, stromatolitic source that underwent continuous compression, microfracturing, and hydrocarbon migration as it moved to greater depths and higher temperatures (Oehler, 1984). Additional overpressuring owing to hydrocarbon generation only improved an already felicitous situation.

The possibility for "explosive" migration from carbonate source rocks exists in certain geologic situations. If carbonate source rocks are dominantly composed of fine-grained carbonate and organic matter, they will, upon adequate burial, become impermeable, and almost all the noncarbonate space in the rock will be occupied by organic matter. Increasing burial will place the organic matter under near-lithostatic stress, and only a slight change in the regional stress system can result in the injection of large amounts of the more mobile fraction of the organic matter into the adjacent rocks. A dramatic illustration of this situation is given in Figure 10. The ejection of organic matter from the Mahogany Zone of the dolomitic Green River Formation (Eocene) and the injection of the thermally immature gilsonite into both the overlying and underlying section took place when the immature organic matter under lithostatic stress was buried $\cong 2,000$ m (6,560 ft) and the Uinta basin area in Utah was subjected to relatively mild regional warping that created tensile fracturing (Jones, 1981). Similar injections of other types of bitumens from different parts of the Green River Formation also occurred (Hunt, Stewart, and Dickey, 1954; Hunt, 1963). The most paraffinic, and presumably most mature, bitumens originated in the lower parts of the Green River Formation, which presumably were under the greatest thermal stress at the time of migration.

If an organic-rich carbonate source rock is subjected to high temperatures in a passive tectonic regime, petroleum migration may be limited. However, when tectonic fracturing finally occurs, the ejected and injected OM may be producible oil. Perhaps the Ghawar field in Arabia, the world's largest producible-petroleum accumulation, was partially charged with hydrocarbons by this mechanism.

The recent intensive study of the Miocene Monterey Formation in southern California has revealed many examples of "explosive" migration. The Monterey Formation is a heterogeneous, brittle rock whose rock matrix is composed of various silica phases, limestone, dolomite, a variable clay content, and OM (Isaacs, 1980, 1983). Although the siliceous aspects of the Monterey have received the most attention in the literature (e.g., Garrison and Douglas, 1981), the OM content shows an inverse correlation with silica wt%, whereas a positive correlation exists with calcite wt% (Isaacs, 1983). Organic contents reach 24 wt% and average 13 wt% in the phosphatic carbonaceous-marl member of the Monterey Formation along the Santa Barbara coast (Isaacs, 1983). After being subjected to tectonism, the Monterey was highly fractured locally, and the fractures were filled with its own oil, apparently in an explosive manner (Redwine, 1981; Surdam and Stanley, 1981; Roehl, 1981; Belfield et al, 1983). This type of fracturing commonly creates regions of lower pressure. The pressure gradients thus formed can draw the oil from the rock matrix to the fractures. Similar effects have been described in the Russian literature (Veber and Gorskaya, 1965).

Many, but by no means all, of the oils having the Monterey as their source are low gravity and are immature by geochemical criteria. They contain unusually high contents of sulfur and asphaltenes. Such oils originated under thermal conditions less severe than those usually associated with "normal" generation and migration, a possibility not recognized early on by geochemists and managers of all oil companies.

The migration of immature oils within and from carbonate-rich source rocks is not confined to the Monterey Formation. De Groot (1980) suggested that oils in the Eastern Mediterranean are immature, early-expulsion products. Palmer and Zumberge (1981) showed that petroliferous marls in the upper Miocene evaporites of Sicily emitted asphaltic-seep oils at low levels of maturation. Powell (1984) described heavy immature oils and bitumens in the Devonian of Canada. These occurrences have been tied to the existence of a low-molecular-weight kerogen rich in nitrogen, sulfur, and oxygen (NSO) compounds that was deposited under highly euxinic conditions (Powell and Snowdon, 1983). The source rocks are organic rich and interlayered with other rocks of quite different physical properties. When tectonically deformed, these rocks tend to mobilize large quantities of heavy oil.

One possible reason for the delay in recognition of carbonates as important source rocks is that, until recently, migration hypotheses had been dominated by those involving water. Continuous-phase oil migration from carbonate source rocks is clearly a preferred mechanism. Various concepts of gas flushing that have, in part, replaced the water-dominated mechanisms originally proposed for thick Tertiary delta systems cannot explain migration from carbonate source rocks. It is likely that the low porosity and permeability of carbonate source rocks would preclude gas from having access to significant amounts of liquid hydrocarbons. There is no need to invoke the solution of oil in either water or gas as the primary migration mechanism responsible for any of the carbonate source–oil correlations reported in this volume. Continuous-phase primary oil migration is the most reasonable explanation for the oil accumulations discussed herein that originated in the Austin Chalk (Grabowski, 1984), Sunniland carbonates (Palacas, Anders, and King, 1984), Alcanar Formation (Demaison and Bourgeois, 1984), Salina A-1 carbonates (Gardner and Bray, 1984), Smackover alginites (Oehler, 1984), and the La Luna Formation (Zumberge, 1984).

SUMMARY AND CONCLUSIONS

Whether dominated by a carbonate or shale rock matrix, excellent oil-prone source rocks share many characteristics. Both (1) were formed in anoxic environments, (2) are generally finely laminated, (3) contain moderate to high TOC contents ($>1.5-20+$ wt%), and (4) contain high-quality OM (atomic H/C ≥ 1.2 at %$R_o \cong 0.5$).

All of the organic facies exist in both carbonate and shale rocks, although they occur in different proportions and with slightly different organic characteristics in some cases.

Organic facies A is preferentially found in carbonate rocks primarily owing to the dearth of terrestrial organic input to marine carbonates, but perhaps also owing to inorganic hydrogenization in lakes under highly alkaline conditions (e.g., Green River Formation).

Organic facies B has been the source of hundreds of billions of barrels of oil from both carbonate and shale source rocks. Differences in oils derived from organic facies B in carbonate and shale source rocks primarily reflect the dearth of iron in carbonate systems and the variations in algal and terrestrial organic input to the OM.

Organic facies C is more important in shales, owing to the dominance of terrestrial input descending through an oxic water column, but this facies does occur in carbonates where partially degraded OM is preserved.

Organic facies D is the dominant facies in carbonate rocks, simply because most carbonates were deposited slowly in shallow, oxic water.

There is no convincing evidence that TOC requirements are less for carbonates than for shales, or that catalytic effects are more pronounced for shales with respect to subsurface generation from kerogen.

Substantial amounts of generally nonproducible bitumen (dominantly NSO compounds with a small percentage of hydrocarbons) can be mobilized early in organic-rich, laminated, heterogeneous carbonate rocks after they have been tectonically fractured.

There is no convincing evidence that shale and carbonate source rocks have different thermal requirements to produce moderate- to high-gravity oils.

ACKNOWLEDGMENTS

Thanks are expressed to D. Leythaeuser, G. T. Moore, P. Müller, J. G. Palacas, K. E. Peters, and an anonymous reviewer for helpful reviews of the manuscript at various stages. The responsibility for the occasional controversial opinions expressed in the manuscript is solely that of the author.

REFERENCES CITED

Abed, A.M., and B.S. Amireh, 1983, Petrography and geochemistry of some Jordanian oil shales from north Jordan: Journal of Petroleum Geology, v. 5, p. 261-274.

Albrecht, P., M. Vandenbroucke, and M. Mandengue, 1976, Geochemical studies on the organic matter from the Douala Basin (Cameroon) I. Evolution of the extractable organic matter and the formation of petroleum: Geochimica et Cosmochimica Acta, v. 40, p. 791-799.

Ayres, M.G., et al, 1982, Hydrocarbon habitat in main producing area, Saudi Arabia: AAPG Bulletin, v. 66, p. 1-9.

Bakhturov, S.F., 1981, Bituminous carbonate rocks of the Tinnovskii Formation of the periphery of the Patom Highlands: Geologiya i Geofiziks, v. 22, p. 132-135.

Barker, C., 1978, Plate tectonics, organic matter and basin evaluation (abs.): AAPG Bulletin, v. 62, p. 493.

——, 1979, Organic geochemistry in petroleum exploration: AAPG Continuing Education Course Note Series 10, 159 p.

Bass, N.W., 1963, Composition of crude oils in northwestern Colorado and northeastern Utah suggests local sources: AAPG Bulletin, v. 47, p. 2039-2064.

Belfield, J.M., et al, 1983, Monterey fractured reservoir, Santa Barbara Channel, California (abs.): AAPG Bulletin, v. 67, p. 421-422.

Bitterli, P., 1963a, Aspects of the genesis of bituminous rock sequences: Geologie en Mijnbouw, v. 42, no. 6, p. 183-201.

——, 1963b, On the classification of bituminous rocks from Western Europe: Sixth World Petroleum Congress Proceedings, v. 2, p. 155-165.

Bordovskiy, O.K., and N.I. Takh, 1978, Organic matter in the Recent carbonate sediments of the Caspian Sea: Oceanology, v. 18, p. 673-678.

Bordovskiy, O.K., et al, 1974, Evaluation of the role of bottom fauna in the transformation of organic matter in sediments (with specific reference to the deep-sea detritus feeders in the Kuril-Kamchatka Trench): Oceanology, v. 14, p. 128-132.

Botz, R., and G. Müller, 1981, Mineralogie, Petrographie, anorganische Geochemie und Isotopen-Geochemie der Karbionatgesteine des Zechstein 2: Geologisches Jahrbuch, Reihe D, v. 47, p. 3-112.

Botz, R., et al, 1981, Kriterien und Bewertung des Zechstein-Stinkschie-fers im Hinblick auf sein Erdöl- und Erdgaspotential: Geologisches Jahrbuch, Reihe D, v. 47, p. 113-132.

Breger, I.A., 1980, Carbonate sedimentary rocks and the origin of heavy crude oils (abs.): Geological Society of America Abstracts with Programs, v. 12, p. 391.

—— and A. Brown, 1963, Distribution and types of organic matter in a barred marine basin: New York Academy of Science Transactions, v. 25, p. 741-755.

Brognon, G.P., and C.R. Verrier, 1966, Oil and geology in the Cuanza basin of Angola: AAPG Bulletin, v. 50, p. 108-158.

Brukner, A., and I. Vetö, 1983, Extracts from open and closed pores of an Upper Triassic sequence from W. Hungary: a contribution to studies of primary migration, in M. Bjorøy et al, eds., Advances in organic geochemistry 1981: Chichester, John Wiley, p. 175-182.

Calvert, S.E., 1983, Organic matter accumulation in the ocean: implication for the origin of sapropels, black shales, and petroleum source beds (abs.): Canadian Society of Petroleum Geologists News Letter, v. 10, no. 2, p. 1-2.

—— and N.B. Price, 1971, Recent sediments of the South-West African Shelf, in F.M. Delany, ed., Atlantic continental margins: London Institute of Geological Science, p. 175-185.

Cardwell, A.L., 1977, Petroleum source rock potential of Arbuckle and Ellenburger Groups, southern Mid-Continent, United States: Golden, Colorado, Colorado School of Mines Quarterly, v. 72, no. 3, 134 p.

Chave, K.E., 1965, Carbonates: association with organic matter in surface seawater: Science, v. 148, p. 1723-1724.

Cornelius, C., 1978, The role of source rock facies in the origin of petroleum: Erdoel-Erdgas-Zeitschrift, v. 94, no. 3, p. 90-94. (In German.)

Davis, J.B., 1982, Catalytic effects of smectitic clays in hydrocarbon generation (abs.): AAPG Bulletin, v. 66, p. 1442.

—— and J.P. Stanley, 1983, Catalytic effect of smectite clays in hydrocarbon generation revealed by pyrolysis-gas chromatography: Journal of Analytical and Applied Pyrolysis, p. 227-240.

Dean, W.E., J.V. Gardner, and P. Cepek, 1981, Tertiary carbonate-dissolution cycles on the Sierra Leone Rise, eastern equatorial Atlantic Ocean: Marine Geology, v. 39, p. 81-101.

—— G.E. Claypool, and J. Thiede, 1981, Origin of organic carbon-rich Mid-Cretaceous limestones, mid-Pacific Mountains and southern Hess Rise: Deep Sea Drilling Project, Initial Report, v. 62, p. 877-890.

Degens, E.T., 1974, Cellular processes in Black Sea sediments, in E.T. Degens and A.A. Ross, eds., The Black Sea—geology, chemistry and biology: AAPG Memoir 20, p. 296-307.

—— and P. Stoffers, 1976, Stratified waters as a key to the past: Nature, v. 263, p. 22-27.

—— et al, 1978, Varve chronology: estimated rates of sedimentation in the Black Sea deep basin: Deep Sea Drilling Project, Initial Report, v. 42, pt. 2, p. 499-508.

De Groot, K., 1980, Origin, migration and accumulation of hydrocarbons, in Panel discussion: Heyden, London,

10th World Petroleum Congress Proceedings, v. 2, p. 50.

Demaison, G.J., and G.T. Moore, 1980, Anoxic environments and oil source bed genesis: AAPG Bulletin, v. 64, p. 1179–1209.

—— and F.T. Bourgeois, 1984, Environment of deposition of middle Miocene (Alcanar) carbonate source beds, Casablanca field, Tarragona basin, offshore Spain, in J.G. Palacas, ed., Petroleum geochemistry and source-rock potential of carbonate rocks: AAPG Studies in Geology 18, this volume.

Dembicki, H., B. Horsfield, and T.T.Y. Ho, 1983, Source rock evaluation by pyrolysis gas chromatography: AAPG Bulletin, v. 67, p. 1094–1103.

Dow, W.G., 1977, Kerogen studies and geological interpretations: Journal of Geochemical Exploration, v. 7, p. 79–99.

Durand, B., and J.L. Oudin, 1980, Exemple de migration des hydrocarbunes dans une séries deltaique: Le delta de la Mahakam, Indonésie: 10th World Petroleum Congress Proceedings, v. 2, p. 1–12.

Eldridge, G.H., 1901, The asphalt and bituminous rock deposits of the United States: U.S. Geological Survey 22d Annual Report, 339 p.

Espitalié, J., M. Madec, and B. Tissot, 1980, Role of mineral matrix in kerogen pyrolysis: influence on petroleum generation and migration: AAPG Bulletin, v. 64, p. 59–66.

—— et al, 1977, Méthode rapide de charactérisation des roches mères, de leur potential pétrolier et de leur degré d'évolution: Revue de l'Institut Français du Pétrole, v. 32, p. 23–42.

Fouch, T.D., 1983a, Character of ancient petroliferous lake basins of the world (abs.): Canadian Society of Petroleum Geologists News Letter, v. 10, no. 3, p. 1–3.

——, 1983b, Lacustrine siliciclastic rocks and hydrocarbons (abs.): AAPG Bulletin, v. 67, p. 462.

Gardner, W.C., and E.E. Bray, 1984, Oils and source rocks of Niagaran reefs in the Michigan basin, in J.G. Palacas, ed., Petroleum geochemistry and source-rock potential of carbonate rocks: AAPG Studies in Geology 18, this volume.

Garrison, R.E., and R.G. Douglas, eds., 1981, The Monterey Formation and related siliceous rocks of California: SEPM, Pacific Section, Special Publication, 327 p.

Gehman, H.M., 1962, Organic matter in limestones: Geochimica et Cosmochimica Acta, v. 26, p. 885–894.

Grabowski, G.J., 1981, Source-rock potential of the Austin Chalk, Upper Cretaceous, southeastern Texas: Gulf Coast Association of Geological Societies Transactions, v. 31, p. 105–113.

——, 1984, Generation and migration of hydrocarbons in Upper Cretaceous Austin Chalk, south-central Texas, in J.G. Palacas, ed., Petroleum geochemistry and source-rock potential of carbonate rocks: AAPG Studies in Geology 18, this volume.

Gransch, J.A., and J. Posthuma, 1974, On the origin of sulfur in crudes, in B. Tissot and F. Bienner, eds., Advances in organic geochemistry 1973: Paris, Éditions Technip, p. 727–739.

Halpern, H.I., and I.R. Kaplan, 1982, Mineral-kerogen interactions in laboratory experiments—significance for petroleum genesis (abs.): AAPG Bulletin, v. 66, p. 1690.

Hassan, F., and S. El-Dashlouty, 1970, Miocene evaporites of Gulf of Suez region: AAPG Bulletin, v. 54, p. 1686–1696.

Hedberg, H.D., 1968, Significance of high-wax oils with respect to the genesis of petroleum: AAPG Bulletin, v. 52, p. 736–750.

Heybroek, F., 1965, The Red Sea Miocene evaporite basin, in Salt basins around Africa: London Institute of Petroleum, p. 17–40.

Hood, A., and J.R. Castaño, 1974, Organic metamorphism: Its relationship to petroleum generation and application to studies of authigenic minerals: United Nations ESCAP, CCOP Technical Bulletin 8, p. 85–118.

—— C.C.M. Gutjahr, and R.L. Heacock, 1975, Organic metamorphism and the generation of petroleum: AAPG Bulletin, v. 59, p. 986–996.

Hunt, J.M., 1963, Composition and origin of the Uinta Basin bitumens, in A.L. Crawford, ed., Oil and gas possibilities of Utah, re-evaluated: Utah Geological and Mineralogical Survey Bulletin 54, p. 249–273.

——, 1967, The origin of petroleum in carbonate rocks, in G.V. Chilingar, H.J. Bissell, and R.W. Fairbridge, eds., Carbonate rocks: New York, Elsevier, p. 225–251.

——, 1979, Petroleum geochemistry and geology: San Francisco, W.H. Freeman, 617 p.

——, F. Stewart, and P.A. Dickey, 1954, Origin of hydrocarbons of Uinta basin, Utah: AAPG Bulletin, v. 38, p. 1671–1698.

Hutton, A.C., A.J. Kantsler, and A.C. Cook, 1980, Organic matter in oil shales: Australian Petroleum Association Journal, v. 20, pt. 1, p. 44–68.

Isaacs, C.M., 1980, Lithostratigraphy of the Monterey Formation, Goleta to Point Conception, Santa Barbara coast, California: AAPG Field Guide 4, p. 9–24.

——, 1983, Compositional variation and sequence in the Miocene Monterey Formation, Santa Barbara coastal area, California, in D.K. Larue and R.J. Steel, eds., Cenozoic marine sedimentation, Pacific margin, U.S.A.: SEPM, Pacific Section, p. 117–132.

Jacob, H., and K. Kuckelhorn, 1977, The carbonization profile of the Miesback 1 well and its geological interpretation: Erdoel-Erdgas-Zeitschrift, v. 93, p. 115–124. (In German.)

Johns, W.D., 1982, Clay mineral catalysis and petroleum generation (abs.): AAPG Bulletin, v. 69, p. 1445.

Jones, R.W., 1981, Some mass balance and geological constraints on migration mechanisms: AAPG Bulletin, v. 65, p. 103–122.

——, 1983, Organic matter characteristics near the shelf-slope boundary, in D.J. Stanley and G.T. Moore, eds., The shelfbreak: critical interface on continental margins: SEPM Special Publication 33, p. 391–408.

—— and G.J. Demaison, 1982, Organic facies—stratigraphic concept and exploration tool, in A. Saldivar-Sali, ed., Proceedings of the Second ASCOPE Conference and Exhibition: Manila, p. 51–68.

Kohm, J.A., and R.O. Louden, 1979, Ordovician Red River (2)—Exploration success depends upon finding Ordovician Red River structures: Oil and Gas Journal, v. 77, no. 29, p. 89–94.

Krejci-Graf, K., 1963, Origin of oil: Geophysical Prospecting, v. 11, p. 244–275.

——, 1964a, Organic geochemistry: Naturwissenschaftliche Rundschau, v. 16, p. 175–186.

——, 1964b, Geochemical diagenesis of facies: Yorkshire Geological Society Proceedings, v. 34, pt. 4, no. 23, p. 469–521.

Larskaia, E.S., 1977, Disseminated organic matter of carbonate rocks and the oil forming process: Akademiya Nauk SSSR, Seriya Geologicheskaya, no. 12, p. 90–98. (In Russian.)

———, 1978, Distribution, balance, and type of disseminated organic matter in the Paleozoic deposits of the Russian Platform, depending on conditions of sedimentation: Lithology and Mineral Resources, v. 12, p. 323–333.

Leythaeuser, D., et al, 1980, Hydrocarbon generation in source beds as a function of type and maturation of their organic matter: a mass balance approach: 10th World Petroleum Congress Proceedings, v. 2, p. 31–42.

Lindbloom, G.P., and M.D. Lupton, 1961, Microbial aspects of organic geochemistry: Developments in Industrial Microbiology, v. 2, p. 9–22.

Maksimov, S.P., et al, 1976, On aspects of facies-genetic types of dispersed organic matter and its role in process of oil and gas generation: Moscow, Nauka, Issledovaniya organicheskogo veshchestva sovremennyki i iskopaemykh osadkov, p. 168–175. (In Russian.)

Malek-Aslani, M., 1980, Environmental and diagenetic controls of carbonate and evaporite source rocks: Gulf Coast Association of Geological Societies Transactions, v. 30, p. 445–458.

McCames, J.G., and L.S. Griffith, 1967, Middle Devonian facies relationships, Zama area, Alberta: Bulletin of Canadian Petroleum Geology, v. 15, p. 434–467.

Momper, J.A., 1978, Oil migration limitations suggested by geological and geochemical consideration, *in* Physical and chemical constraints on petroleum migration: AAPG Short Course Notes, v. 1, p. B1–B60.

Müller, P.J., and E. Suess, 1979, Productivity, sedimentation rate, and sedimentary organic matter in the oceans—I. Organic carbon preservation: Deep-Sea Research, v. 22A, p. 1347–1362.

——— and A. Mangini, 1980, Organic carbon deposition rates in sediments of the Pacific manganese nodule belt dated by 230Th and 231Pa: Earth and Planetary Science Letters, v. 51, p. 94–114.

Murris, R.J., 1980, Middle East: stratigraphic evolution and oil habitat: AAPG Bulletin, v. 64, p. 597–618.

Neglia, S., 1980, Migration of fluids in sedimentary basins: Reply to R.E. Chapman: AAPG Bulletin, v. 64, p. 1543–1547.

Oehler, J.H., 1984, Carbonate source rocks in the Jurassic Smackover trend of Mississippi, Alabama, and Florida, *in* J.G. Palacas, ed., Petroleum geochemistry and source-rock potential of carbonate rocks: AAPG Studies in Geology 18, this volume.

Ozimic, S., 1982, Depositional environment of the oil shale–bearing Cretaceous Toolebuc Formation and its equivalents, Eromanga basin, Australia, *in* J.H. Gary, ed., 15th Oil Shale Symposium Proceedings, p. 137–148.

Palacas, J.G., 1983, Carbonate rocks as sources of petroleum: Geological and chemical characteristics and oil–source correlations: World Petroleum Congress Proceedings, PD1, Origin, migration and accumulation of hydrocarbons, preprint, p. 1–13.

———, 1984, South Florida basin, a prime example of carbonate source rocks of petroleum, *in* J.G. Palacas, ed., Petroleum geochemistry and source-rock potential of carbonate rocks: AAPG Studies in Geology 18, this volume.

Palmer, S.E., and J.H. Zumberge, 1981, Organic geochemistry of upper Miocene evaporite deposits in Sicilian basin, Sicily, *in* J. Brooks, ed., Organic maturation studies and fossil fuel exploration: New York, Academic Press, p. 393–426.

Parparova, G.M., and S.G. Neruchev, 1977, Foundations of the genetic classification of dispersed organic matter in rock: Geologiya: Geofizika, v. 18, no. 5, p. 45–51.

Peterson, J.A., and R.J. Hite, 1969, Pennsylvanian evaporite–carbonate cycles and their relationship to petroleum occurrence, southern Rocky Mountains: AAPG Bulletin, v. 53, p. 884–908.

Powell, T.G., 1984, Some aspects of the hydrocarbon geochemistry of a Middle Devonian barrier-reef complex, western Canada, *in* J.G. Palacas, ed., Petroleum geochemistry and source-rock potential of carbonate rocks: AAPG Studies in Geology 18, this volume.

——— and L.R. Snowden, 1983, A composite hydrocarbon generation model: Erdöl und Kohle, Erdgas, Petrochemie, v. 36, no. 4, p. 167–175.

Pratt, L.M., 1982, A paleo-oceanographic interpretation of the sedimentary structures, clay minerals, and organic matter in a core of the Middle Cretaceous Greenhorn Formation drilled near Pueblo, Colorado: Princeton, New Jersey, Princeton University unpublished Ph.D. dissertation, 176 p.

Price, L.C., J.L. Clayton, and L.L. Rumen, 1981, Organic geochemistry of the 9.6 km Bertha Rogers No. 1 well, Oklahoma: Organic Geochemistry, v. 3, p. 59–77.

Redwine, L.E., 1981, Hypothesis combining dilation, natural hydraulic fracturing, and dolomitization to explain petroleum reservoirs in Monterey Shale, Santa Maria area, California, *in* R.E. Garrison and R.G. Douglas, eds., The Monterey Formation and related siliceous rocks of California: SEPM, Pacific Section, Special Publication, p. 221–249.

Roehl, P.O., 1981, Dilation breccia—a proposed mechanism of fracturing, petroleum expulsion and dolomitization in the Monterey Formation, California, *in* R.E. Garrison and R.G. Douglas, eds., The Monterey Formation and related siliceous rocks of California: SEPM, Pacific Section, Special Publication, p. 285–316.

Rogers, M.A., 1980, Application of organic facies concepts to hydrocarbon source rock evaluation: 10th World Petroleum Congress Proceedings, v. 2, p. 425–440.

——— and C.B. Koons, 1971, Generation of light hydrocarbons and establishment of normal paraffin preferences in crude oils: Origin and refining of petroleum: American Chemical Society, Advances in Geochemistry Series, p. 67–80.

Savin, S.M., 1982, Oxygen isotopic studies of diagenetic clay minerals: Implications for geothermometry, diagenetic reaction mechanisms, and fluid migration (abs.): AAPG Bulletin, v. 66, p. 1447.

Shearman, D.J., and P.A.d'E. Skipwith, 1965, Organic matter in recent and ancient limestones, and its role in their diagenesis: Nature, v. 208, p. 1310–1311.

Shimoyama, A., and W.D. Johns, 1972, Formation of alkanes from fatty acids in the presence of $CaCO_3$: Geochimica et Cosmochimica Acta, v. 36, p. 87–91.

Smith, J.W., 1983, The chemistry which created Green River Formation oil shale: American Chemical Society,

Symposium on Geochemistry and Chemistry of Oil Shales, preprint, p. 76–84.

—— and K.K. Lee, 1982, Chemistry and physical paleolimnology of Piceance Creek oil shales, in J.H. Gary, ed., 15th Oil Shale Symposium Proceedings: Golden, Colorado, p. 101–114.

Snowdon, L.R., and T.G. Powell, 1982, Immature oil and condensate—modification of hydrocarbon generation model for terrestrial organic matter: AAPG Bulletin, v. 66, p. 775–788.

Spillers, J.P., 1965, Distribution and hydrocarbons in south Louisiana by types of traps and trends (abs.): AAPG Bulletin, v. 49, p. 1749–1751.

Stach, E., et al, 1982, Stach's textbook of coal petrology, 3rd edition: Berlin, Gebrüder Borntraeger, 535 p.

Suess, E., 1970, Interaction of organic compounds with calcium carbonate—I. Association phenomena and geochemical implications: Geochimica et Cosmochimica Acta, v. 34, p. 157–168.

——, 1973, Interaction of organic compounds with calcium carbonate—II. Organo-carbonate association in Recent sediments: Geochimica et Cosmochimica Acta, v. 37, p. 2435–2447.

Summerhayes, C.P., 1981, Oceanographic controls on organic matter in the Miocene Monterey Formation, offshore California, in R.E. Garrison and R.G. Douglas, eds., The Monterey Formation and related siliceous rocks of California: SEPM, Pacific Section, Special Publication, p. 213–220.

Surdam, R.C., and K.O. Stanley, 1981, Diagenesis and migration of hydrocarbons in the Monterey Formation, in R.E. Garrison and R.G. Douglas, eds., The Monterey Formation and related siliceous rocks of California: SEPM, Pacific Section, Special Publication, p. 317–327.

Tissot, B., 1979, Effects on prolific source rocks and major coal deposits caused by sea-level changes: Nature, v. 277, p. 463–465.

—— and D.H. Welte, 1978, Petroleum formation and occurrence: Berlin, Springer-Verlag, 538 p.

—— and J. Roucaché, 1980, Principles of a genetic classification of crude oils, in A.D. Miall, ed., Facts and principles of world petroleum occurrence: Canadian Society of Petroleum Geologists Memoir 6, p. 209–217.

—— et al, 1974, Influence of nature and diagenesis of organic matter in formation of petroleum: AAPG Bulletin, v. 58, p. 499–506.

Toth, D.J., and A. Lerman, 1977, Organic matter reactivity and sedimentation rates in the ocean: American Journal of Science, v. 277, p. 465–485.

Uspenskii, V.A., and A.S. Chernysheva, 1951, The composition of the organic material from Lower Silurian limestones in the region of Chudovo City, in Contributions to geochemistry: Israel Program for Scientific Translations, 1965, p. 103–114.

Veber, V.V., and A.I. Gorskaya, 1965, Bitumen formation in sediments of carbonate facies: International Geology Review, v. 7, p. 816–825.

Waples, D.W., 1980, Time and temperature in petroleum formation: application of Lopatin's method to petroleum exploration: AAPG Bulletin, v. 64, p. 916–926.

Weber, K.J., and E. Daukow, 1975, Petroleum geology of the Niger Delta: World Petroleum Congress Proceedings, v. 2, p. 209–221.

Zhuze, N.G., 1972, Bitumen content of carbonate deposits of Upper Devonian and Lower Carboniferous of the Dnepr-Donets basin: International Geology Review, v. 15, p. 1212–1219.

Zumberge, J.E., 1984, Source rocks of the La Luna Formation (Upper Cretaceous) in the middle Magdalena Valley, Colombia, in J.G. Palacas, ed., Petroleum geochemistry and source-rock potential of carbonate rocks: AAPG Studies in Geology 18, this volume.

Anoxic Environments and Oil Source Bed Genesis[1]

G. J. DEMAISON[2] and G. T. MOORE[3]

Abstract The anoxic aquatic environment is a mass of water so depleted in oxygen that virtually all aerobic biologic activity has ceased. Anoxic conditions occur where the demand for oxygen in the water column exceeds the supply. Oxygen demand relates to surface biologic productivity, whereas oxygen supply largely depends on water circulation, which is governed by global climatic patterns and the Coriolis force.

Organic matter in sediments below anoxic water is commonly more abundant and more lipid-rich than under oxygenated water mainly because of the absence of benthonic scavenging. The specific cause for preferential lipid enrichment probably relates to the biochemistry of anaerobic bacterial activity. Geochemical-sedimentologic evidence suggests that potential oil source beds are and have been deposited in the geologic past in four main anoxic settings as follows.

1. *Large anoxic lakes*—Permanent stratification promotes development of anoxic bottom water, particularly in large lakes which are not subject to seasonal overturn, such as Lake Tanganyika. Warm equable climatic conditions favor lacustrine anoxia and nonmarine oil source bed deposition. Conversely, lakes in temperate climates tend to be well oxygenated.

2. *Anoxic silled basins*—Only those landlocked silled basins with positive water balance tend to become anoxic. Typical are the Baltic and Black Seas. In arid-region seas (Red and Mediterranean Seas), evaporation exceeds river inflow, causing negative water balance and well-oxygenated bottom waters. Anoxic conditions in silled basins on oceanic shelves also depend upon overall climatic and water-circulation patterns. Silled basins should be prone to oil source bed deposition at times of worldwide transgression, both at high and low paleolatitudes. Silled-basin geometry, however, does not automatically imply the presence of oil source beds.

3. *Anoxic layers caused by upwelling*—These develop only when the oxygen supply in deep water cannot match demand owing to high surface biologic productivity. Examples are the Benguela Current and Peru coastal upwelling. No systematic correlation exists between upwelling and anoxic conditions because deep oxygen supply is often sufficient to match strongest demand. Oil source beds and phosphorites resulting from upwelling are present preferentially at low paleolatitudes and at times of worldwide transgression.

4. *Open-ocean anoxic layers*—These are present in the oxygen-minimum layers of the northeastern Pacific and northern Indian Oceans, far from deep, oxygenated polar water sources. They are analogous, on a reduced scale, to worldwide "oceanic anoxic events" which occurred at global climatic warmups and major transgressions, as in Late Jurassic and middle Cretaceous times. Known marine oil source bed systems are not randomly distributed in time but tend to coincide with periods of worldwide transgression and oceanic anoxia.

Geochemistry, assisted by paleogeography, can greatly help petroleum exploration by identifying paleoanoxic events and therefore widespread oil source bed systems in the stratigraphic record. Recognition of the proposed anoxic models in ancient sedimentary basins should help in regional stratigraphic mapping of oil shale and oil source beds.

INTRODUCTION

The most significant progress made in petroleum geology in the last 10 years has been the attainment of a satisfactory understanding of the processes of oil and gas generation and destruction in sedimentary basins. Geochemical techniques now routinely identify oil source beds by analysis of rock samples retrieved from deep wells. However, geologists often question the validity of projecting the presence of source beds identified from a few wells and a few grams of rocks, to entire sedimentary basins. Until the last few years these hesitations were legitimate; it was difficult, if not inconceivable, to map oil source beds without a clear understanding of their depositional environment.

Fortunately, recent oceanographic and geochemical observations now make it possible to comprehend many formerly obscure aspects of the genesis and stratigraphic distribution of petroleum source beds. It is timely, thus, to review depositional environments of oil source beds considering these recent findings and address ourselves to the following questions.

1. What modern sediments are precursors to oil source beds?

©Copyright 1980. The American Association of Petroleum Geologists. All rights reserved.

AAPG grants permission for a *single* photocopy of this article for research purposes. Other photocopying not allowed by the 1978 Copyright Law is prohibited. For more than one photocopy of this article, users should send request, article identification number (see below), and $3.00 per copy to Copyright Clearance Center, Inc., 21 Congress Street, Salem, MA 01970.

[1]The original version of this paper was read at the 10th International Congress on Sedimentology, Jerusalem, 1978, and published in *Organic Geochemistry*. Manuscript received, August 24, 1979; accepted, January 17, 1980.

[2]Chevron Overseas Petroleum Inc., San Francisco, California 94105.

[3]Chevron Oil Field Research Co., La Habra, California 90631.

The writers are indebted to R. R. Hammes, R. W. Jones, R. A. Lagaay, R. S. Oremland, W. D. Redfield, S. R. Silverman, and J. A. Sutherland for critical reviews of the manuscript.

Published with permission of Chevron Overseas Petroleum Inc. and Chevron Oil Field Research Co.

Article Identification Number
0149-1423/80/B008-0002$03.00/0

2. What factors affect the preservation of organic matter in aquatic environments?

3. Why are anoxic conditions more favorable than oxic conditions for organic-matter preservation, both in quality and quantity?

4. What causes favor anoxic conditions in lakes, seas, and oceans?

5. Can natural anoxic settings be scientifically classified to help oceanographers, stratigraphers, and petroleum explorationists?

PRECURSOR SEDIMENTS TO OIL SOURCE BEDS

Potential oil source beds are organic-rich sediments containing a kerogen type that is sufficiently hydrogen-rich (type I or type II; Tissot et al, 1974) to convert mainly to oil during thermal maturation. Kerogen type and thus oil source character in ancient sediments is identified through such approaches as elemental analysis of kerogen and whole-rock pyrolysis with additional support from microscopic organic analysis (Tissot and Welte, 1978, p. 81-91; Hunt, 1979, p. 454-472).

Evaluation of organic content, which can be a gross quantitative index of oil generative potential *if kerogen is oil-prone,* is measured by the amount of organic carbon present in sediments. Documented oil source beds and oil shales around the world always contain hydrogen-rich kerogen and fall into a range of organic carbon content between about 1% and over 20% by weight. The boundary between very rich oil source beds and oil shales is determined by mining and processing economics.

Rich to very rich marine oil source beds and oil shales commonly contain higher than average uranium, copper, and nickel concentrations. Uranium content in most marine oil shales commonly shows a positive correlation with oil yield upon rock pyrolysis (Swanson, 1960).

The measurement of organic carbon in sediments alone is insufficient to identify potential oil source beds. Transported terrestrial organic matter, oxidized aquatic organic matter, and reworked organic matter from a previous sedimentary cycle can create levels of organic carbon in marine sediments up to about 4%. Yet this abnormally concentrated organic matter is hydrogen-poor, gas-prone, and without significant oil generating potential (Tissot et al, 1974; Demaison and Shibaoka, 1975; Dow, 1977). This is essentially the organic facies that has been described in middle Cretaceous marine black shales encountered by several Deep Sea Drilling Project holes in the northwestern Atlantic Basin (Tissot et al, 1979). An identical situation is present today on the Arctic Shelf of the USSR where high organic carbon concentrations in marine sediments result from the influx of large amounts of terrestrial organic matter brought in by fluvial discharge (Bezrukov et al, 1977).

In summary, a high organic carbon content in sediments is not necessarily an indication of oil source rock precursor character. Additional geochemical evidence, such as measurement of hydrogen richness of humic substances and kerogen, pyrolysis yield of whole sediment, and overall soluble-lipid content of the sediment, is needed to ascertain a possible oil source precursor character.

FACTORS INFLUENCING ORGANIC MATTER ACCUMULATION IN AQUATIC ENVIRONMENTS

Factors capable of influencing organic-matter accumulation in sediments are both biologic and physical. Biologic factors include primary biologic productivity of the surface-water layers and of adjoining landmasses and biochemical degradation of dead organic matter by metazoan and microbial scavengers. Physical factors include modes of transit of organic matter to depositional sites, sediment particle size, and sedimentation rates. These factors interact to determine qualitative and quantitative preservation of organic matter in sediments.

Primary Biologic Productivity

The principal source of aquatic organic matter is the phytoplankton (Bordovsky, 1965) composed largely of single-cell microscopic algae residing in the uppermost layers of water illuminated by sunlight, the euphotic zone. The main limiting factor to planktonic productivity, in addition to light, is the availability of mineral nutrients, particularly nitrates and phosphates, which are in short supply in the euphotic zone. Phytoplankton are intensively grazed by zooplankton. Both phytoplankton and zooplankton are then consumed by large invertebrates and fish.

The other source of organic matter in the aquatic environment is transported terrestrial organic matter from streams and rivers. Land-plant productivity is largely dependent on the amount of rainfall on supporting landmasses. Because terrestrial organic matter has undergone considerable degradation in subaerial soils prior to its transport, it is usually hydrogen depleted and refractory in nature.

The traditional view is that fields of high surface productivity in the ocean should be associated with high organic enrichment of underlying sediments; however, after exhaustive investigation, we could not find a systematic correlation between primary biologic productivity (Fig. 1)

Anoxic Environments

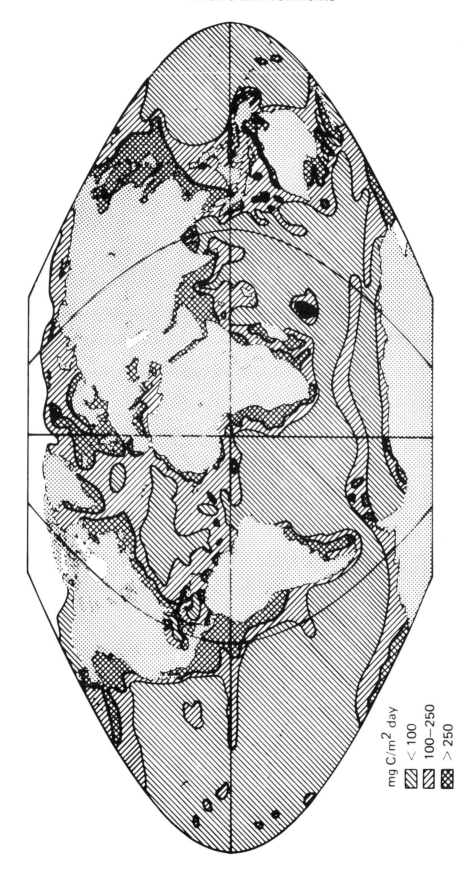

FIG. 1.—Primary biologic production in world ocean, expressed in milligrams of organic carbon per square meter per day (modified from Koblentz-Mishke et al, 1970). Unicellular, microscopic, planktonic algae (phytoplankton) are principal primary producers of organic matter in sea.

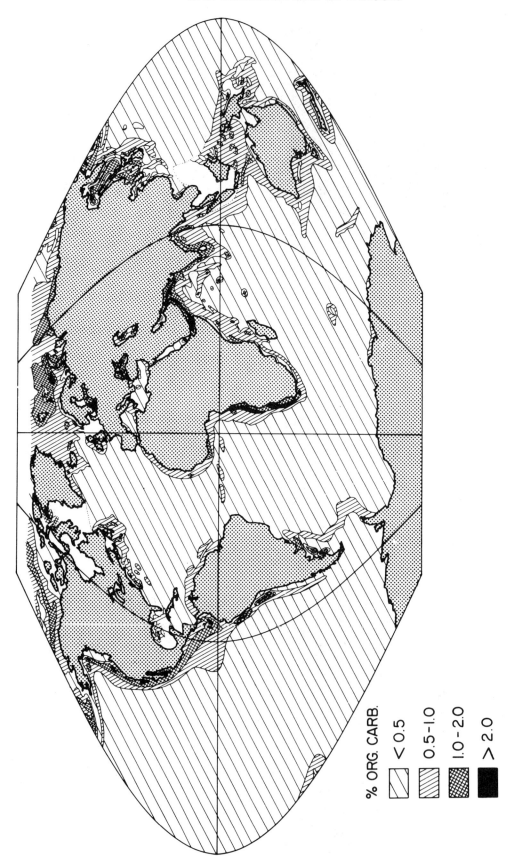

FIG. 2—Percentage of organic carbon in surface layer of sediments in world ocean (data mainly from Bezrukov et al, 1977). Because of biologic sediment mixing under oxic water, reported values may be averages representing time spans in excess of several hundred years.

and organic carbon content of bottom sediments in the oceans (Fig. 2). Some areas of high productivity like the Peruvian Shelf and Southwest Shelf of Africa were enriched; others of equally high productivity like the Grand Banks of Newfoundland, the northeastern Brazilian Shelf, or the Northwest Shelf of Africa were not. Clearly, factors other than surface productivity had to be evaluated.

Biochemical Degradation of Dead Organic Matter

Dead organic matter is inherently unstable thermodynamically and seeks its lowest level of free energy in any given environment. Above all, it serves as a source of energy and nutrients for living organisms. Heterotrophic microorganisms, mainly bacteria, play a critical role in decomposing organic matter in the water column, in the interstitial water of sediments, and in the digestive tracts of animal scavengers. In marine basins bacterial biomass approaches that of the phytoplankton (Bordovsky, 1965). Bacterial degradation proceeds quickly and efficiently in aerobic (oxygen-rich) water. The overall oxidation processes by aerobic bacteria are illustrated by:

$$(CH_2O) + O_2 \rightarrow CO_2 + H_2O.$$

When the oxygen supply becomes exhausted, oxidation of organic matter by anaerobic bacteria using nitrates as a source of oxygen (electron acceptor) takes over according to the generalized scheme:

$$(CH_2O) + 4NO_3 \rightarrow 6CO_2 + 6H_2O + 2N_2.$$

Once nitrates have been exhausted, degradation of organic matter continues by anaerobic bacteria using sulfates as the oxidant by the generalized reaction scheme:

$$CH_2O + SO_4 \rightarrow CO_2 + H_2O + H_2S.$$

The last and least efficient step in anaerobic metabolism is fermentation. Here carboxyl (CO_2) groups and organic acids of the organic matter itself, or from bacterial breakdown, are employed as electron acceptors. A special type of anaerobic fermentation is bacterial methanogenesis (Claypool and Kaplan, 1974). It is a complex process carried out in several steps by types of bacteria which degrade one another's by-products (Manheim, 1976). The methane-generating (methanogenic) bacteria occupy the terminal niche of the anaerobic food web and produce methane by CO_2 reduction by H_2 or by attacking acetate, formate, or methanol (Wolfe, 1971). Higher hydrocarbons, at least up to butane, are always present together with bacterial methane in anoxic water, although in much smaller amounts (Deuser, 1975). Anaerobic methane fermentation is common below the sulfate reducing zone in recent marine sediments and prevails in anoxic waters where both nitrate and sulfate levels are low (e.g., in freshwater lakes). Methane fermentation and sulfate reduction, however, are not mutually exclusive (Oremland and Taylor, 1978). The processes mentioned are presented in a simplified form. They represent an important developing area of research in geomicrobiology.

Anaerobic degradation is thermodynamically less efficient than aerobic decomposition (Claypool and Kaplan, 1974) and results in a more lipid-rich and more reduced (hydrogen-rich) organic residue than does aerobic degradation (Foree and McCarty, 1970; Beliaeva and Romankevich, 1976; Gerschanovich et al, 1976; Pelet and Debyser, 1977; Didyk et al, 1978). Moreover, under such conditions, a significant fraction of the preserved organic matter consists of remains of the bacterial biomass itself (Lijmbach, 1975). The mechanisms leading to this preferential enrichment in lipids are little understood and in need of fundamental research. The enhanced preservation of hydrogen-rich and lipid-rich organic matter in sediments deposited under anoxic water is critical for the genesis of oil source beds.

Explanations proposed for the higher than normal organic matter concentrations observed under anoxic water (Richards, 1970; Deuser, 1971) include the suggestions that anaerobic degradation is inherently slower (Lijmbach, 1975) or that enhanced preservation might be a consequence of high sedimentation rates (Richards, 1970). (The latter is discussed under the effects of other physical parameters.)

As far as biologic factors are concerned, Foree and McCarty (1970) have shown by laboratory experiments that rates of anaerobic degradation of algae by sulfate reduction are identical to those of aerobic degradation. Observations in the natural environment by Orr and Gaines (1974) led to parallel conclusions. This suggests that most of the organic matter consumption at oxygen-rich ("oxic") sediment/water interfaces is by the metazoan, not by the microbial population (Degens and Mopper, 1976).

Under an oxic water column (Fig. 3) the benthic fauna actively scavenges and reworks the "organic rain" falling through the water column. Except in shallow waters where sunlight penetrates to the bottom, there is very little primary production and mainly consumption of organic matter at the benthic boundary. Bottom muds under an oxygen-rich water column are commonly anoxic

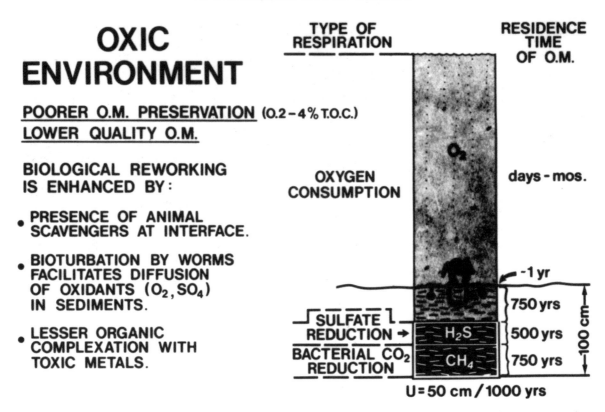

FIG. 3—Degradation of organic matter under oxygenated (oxic) water column. Organic carbon concentration under oxygen-rich water is closely related to sedimentation rate, without regard to surface primary productivity.

but nevertheless extensively disrupted by oxygen-respiring invertebrates such as polychaetes (worms), holothurians, and bivalves. The most common modes of feeding are on particulate organic matter and bacteria in water above the sediment (suspension feeders), and on organic matter and bacteria present within the sediment itself (deposit feeders). Mobile deposit feeders (burrowers) cause mixing and transport of particles, as well as irrigation of sediments, thereby accelerating geochemical processes at or below the sediment-water interface (Rhoads, 1974; Aller, 1978). Many who have studied deep-sea sediments believe that the sediments are mixed to a depth of 5 to 30 cm by biologic activity (Peng et al, 1977). Bioturbation is ubiquitous under oxic water where it has been observed at all water depths, including the deep-sea sediments of the abyssal realm. The range of mixing (bioturbation) rates in the deep sea is comparable to that in nearshore regions, with no apparent correlation between the mixing rate and the sediment type or sedimentation rate (Turekian et al, 1978). This suggests that mixing rate may be almost solely related to benthic biomass variability, and thus to productivity in surface waters and subsequent delivery of edible organic material to the ocean floor.

Under an anoxic water column (Fig. 4), however, oxygen depletion (even in the absence of hydrogen sulfide) depresses and eventually eliminates benthonic metazoan life (Theede et al, 1969). The benthic metazoan biomass is unaffected by oxygen concentrations as low as about 1 ml/l (Rhoads, 1974), but between 0.7 ml/l and 0.3 ml/l it sharply decreases by about a factor of 5 (Rosenberg, 1977). Below 0.3 ml/l, deposit feeders become rare, less active, and soft bodied only. Eventually bioturbation ceases. Below 0.1 ml/l, the suspension feeders disappear, leaving anaerobic bacteria as the only effective reworkers of organic matter.

Once organic matter is incorporated in the anoxic sediment itself, the lack of bioturbation acts as a limiting factor to diffusion of oxidants into the sediment—hence bacterial sulfate reduction is slowed down if not completely arrested. A classic observation is that the pore fluids of anoxic sediments are depleted in sulfates (Manheim, 1976, p. 115-118). Lack of bioturbation under anoxic water results in laminated and organic-rich sediments. Reducing conditions under anoxic water make certain toxic metals, like lead, available for chelation with organic matter (Jones, 1973). Degens and Mopper (1976) have suggested that metal complexation of organic matter makes it less susceptible to microbial at-

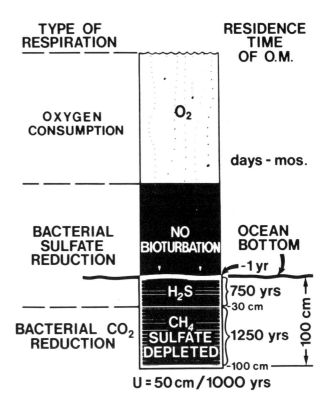

FIG. 4—Degradation of organic matter under anoxic water column. Bioturbation becomes minimal at oxygen concentrations below 0.5 ml/l. This concentration, rather than total absence of oxygen, is effective "biochemical fence" between poor and good organic matter preservation.

tack.

Comparison of Figures 3 and 4 shows that, given identical sedimentation rates, dead organic matter in oxygenated environments is exposed to oxidants much longer than in anoxic environments, owing to bioturbation. For a given sedimentation rate, bioturbation under oxic water prolongs the exposure of organic matter to oxidation by hundreds of years, which strongly enhances degradation. Added to this effect is the grazing of the sediment bacterial biomass by mobile deposit feeders responsible for bioturbation.

Modes of Transit of Organic Matter to Depositional Sites

Dead zooplankton, unconsumed algal cells, fecal pellets, and fish carcasses originating in the shallow euphotic zone continuously sink to the bottom. The speed of fall of particulate organic matter is exceedingly slow and ranges from 0.10 to 5 m per day, according to their shape and size, the smallest particles being the slowest to reach bottom. The transit time for fecal pellets is much faster (Spencer et al, 1978). It is estimated that their passage through a water column 4 km deep will not exceed 15 days.

If organic debris falls through a well-oxygenated water column in which animal life is abundant, it is consumed by scavenging animals, until much of the final organic "rain" which reaches the bottom consists of fecal pellets (Spencer et al, 1978). These are reworked biologically at the sediment interface, thereby losing their original morphologies.

Long residence of organic matter in the water column before sedimentation adversely affects its preservation (Degens and Mopper, 1976). Thus the depth of the water column as well as the size of organic particles affects the quantity and quality of the organic input to the sediments.

Influence of Sediment-Particle Size

In addition to being affected by bioturbation, bacterial activity in recent sediments is also influenced by granulometry. Fine-grained sediments, where diffusibility of oxidizing agents is restrained, have lower levels of bacterial activity than coarse-grained sediments (Bordovsky, 1965), which helps explain the broad correlation between particle size and organic carbon content in modern and ancient sediments. Besides being deposited in high-energy environments, coarse clastic sediments, such as sands, permit easy diffusion of free oxygen and oxidizing salts dissolved in water (Tissot and Welte, 1978, p. 81-91). They are always very low in organic matter.

Influence of Sedimentation Rate

The rate of accumulation of organic carbon in marine sediments is closely related to the bulk accumulation rate (Heath et al, 1977). This regularity is well demonstrated under oxic water in areas such as the Argentine basin (Stevenson and Cheng, 1972), the northwest African continental margin (Hartmann et al, 1976), and the deep regions of the North Pacific and North Atlantic Oceans (Heath et al, 1977). In all these areas there is a positive correlation between sedimentation rate and organic carbon content in sediments. The latter usually ranges between 0.3 and about 4%. Under anoxic water this correlation is less evident, as will be discussed later. The overall range of organic carbon under anoxic water in modern sediments is significantly wider and higher (about 1 to 20+ %), regardless of sedimentation rate.

Furthermore, insufficient documentation exists for modern sediments with regard to possible dilution effects created by very high sedimentation rates (above 500 cm/10^3 year). There is evidence that the positive correlation between organic carbon content and sedimentation rate at low and intermediate sedimentation rates may not extend indefinitely into the realm of very high rates. If it did, prodelta muds should be the richest organic sediments in the marine realm, yet organic carbon contents are systematically below 1% in modern prodelta sediments of the Louisiana Gulf Coast (Dow and Pearson, 1975), Niger delta (Klingebiel, 1976), and Amazon delta (Bezrukov et al, 1977) pointing to a dilution effect. Primary biologic productivities offshore from the Mississippi and Amazon deltas are exceptionally high (Fig. 1) but, as expected in such oxic environments, are not reflected by the sedimentary organic carbon content.

Sediments under anoxic water tend to be organically richer than those under oxic water, largely due to lack of benthonic scavenging and absence of bioturbation at the seafloor. Of even greater importance to the genesis of oil source beds, sedimentary organic matter is more reduced and lipid-rich under anoxic water than under oxic water. The cause for this preferential lipid enrichment probably relates to the biochemistry of anaerobic bacterial activity and requires further research.

Under oxygenated ("oxic") water, fluctuations in organic carbon content are clearly related to sedimentation rates (up to the point where dilution becomes significant), with little influence of surface productivity. Variations in surface productivity appear systematically, compensated by quantitative variations of the benthic biomass responsible for organic matter consumption and bioturbation at the seafloor. Under anoxic water, both sedimentation rates and perhaps surface productivity are factors which explain fluctuation in organic carbon content within the observed range. The respective inputs of these two factors, however, are yet to be satisfactorily elucidated.

DEVELOPMENT OF ANOXIC CONDITIONS IN WATER MASSES

Maximum oxygen saturation in sea water is about 6 to 8.5 ml/l, depending on water temperature and salinity. For this study we define as "anoxic" any water containing less than 0.5 milliliters of oxygen per liter of water (0.5 ml/l), which is the threshold below which the metazoan benthic biomass and, more specifically, bioturbation by deposit feeders, becomes significantly depressed. Therefore, it is proposed as the effective "biochemical fence" between potentially poor or good qualitative and quantitative preservation of organic matter in sediments.

There are two "end member" causes to anoxia in natural waters: excessive oxygen demand and deficient oxygen supply. Anoxic conditions occur where the natural demand for oxygen in water exceeds the supply.

Oxygen Demand in Natural Water

Oxygen consumption in water is essentially a biochemical process resulting from the degradation of organic matter produced in the shallow layers of the euphotic zone. At least 80% of this oxidation occurs in these shallow layers and decreases sharply with depth. However, dead organic matter that has escaped total degradation and has sunk to the bottom continues to create a demand for oxygen which, however weak in relation to that prevailing in the euphotic zone, remains to be satisfied by a matching supply. If this continued demand is not replenished by deep-water circulation, the water column becomes anoxic.

Even water columns with a normal oxygen supply can become anoxic. This occurs in areas of very high primary productivity wherever the oxygen supply near sea bottom cannot cope with the load of descending dead organic matter.

Oxygen Supply in Natural Waters

Oxygen is supplied to water masses by two physical processes: (1) downward movement of oxygen-saturated water from the well-aerated surface layers (oxygen is supplied to the surface layers by exchange with the atmosphere and by photosynthetic oxygen production) and (2) upward movement of oxygen-rich, colder, denser bottom water into intermediate water zones.

Three physical properties of water govern oxygenation of bottom waters: water density increases with increasing salinity; water density increases with decreasing temperature (to 4°C); and oxygen solubility in water varies inversely with decreasing temperature and salinity.

If oxygen-rich surface water becomes denser because it is saltier or cooler than the surrounding water, it sinks to the bottom and circulates there as an aerating undercurrent and causes multilayered vertical stratification in water bodies of all sizes. The terms, thermocline, pycnocline, and halocline, describe temperature, density, and salinity boundaries, respectively.

Stratification of seas and oceans in terms of oxygen concentration has long been recognized by oceanographers. Modern oceans would be entirely anoxic at depth without aeration of their basins by colder and denser oxygen-rich bottom water derived from the polar regions, mainly from the high southern latitudes.

The circulation of these deep oxygen-rich bottom waters is only partly known. Oxygen supply to the intermediate and deep oceanic water depends on patterns of water circulation due to surface-wind stress, density differences, high-latitude cooling, and the Coriolis force (caused by the earth's rotation). The dynamics of general oceanic circulation are complex, fluctuating, and not yet fully understood quantitatively.

The most common cause of anoxia is the incapacity of the oxygen supply in water to meet the biochemical oxygen demand. Hence lack of vertical mixing and oxygen renewal in deep waters is, perhaps, the most important factor controlling the location of anoxic layers and thus, indirectly, the preferred sites of deposition of oil source bed precursors. In the words of Wyrtki (1962), "Biochemical processes are responsible for the existence of oxygen minima, but circulation is responsible for the position."

CLASSIFICATION OF ANOXIC ENVIRONMENTS

The world map of organic carbon (Fig. 2), based mainly on Bezrukov et al (1977), does not distinguish the areas with organic carbon enrichment over 3%. Thus input of transported terrestrial organic matter as well as the effects of high sedimentation rates under oxic water cannot be separated from enrichment in marine organic matter created by anoxic conditions. In the writers' experience with ancient marine sediments, organic carbon contents rarely exceed 3% when transported humic material of terrestrial origin is predominant. Conversely, ancient marine sediments containing more than 3% organic carbon always contain a significant portion of aquatic organic matter.

Consequently, we reviewed in detail studies of recent sediments reported by previous researchers to investigate the implications of enrichments higher than 3%, potentially caused by anoxic conditions. We found it possible to classify present-day anoxic environments into four main types: (1) large anoxic lakes; (2) anoxic silled basins; (3) anoxic layers caused by upwelling; and (4) the anoxic open ocean.

LARGE ANOXIC LAKES

Oxygen depletion in inland seas and large lakes is determined by the supply-demand balance between free oxygen availability in bottom water and planktonic productivity of organic matter in the shallow layers.

Plant nutrients such as phosphates and nitrates are carried into lakes and inland seas by fluvial drainage systems that transport solutes leached from soils. These nutrients usually limit the planktonic productivity of lakes, which then determines the amount of oxygen needed to recycle dead organic matter. Eutrophic lakes are characterized by an abundance of dissolved plant nutrients and by a seasonal oxygen deficiency in the bottom waters. Oligotrophic lakes are deficient in plant nutrients and contain abundant dissolved oxygen in their bottom water.

Oxygen supply in the bottom waters is usually good in areas of contrasting climate with seasonal overturning of the lake (Swain, 1970, p. 73-111). Also, cold, well-oxygenated stream and river water sinks to the bottom and enhances oxic conditions. Oxygen supply is lower in warm tropical climates because of lack of seasonal overturn and lower oxygen contents due to higher water temperatures.

Examples of Large Anoxic Lakes

The two best studied anoxic lakes in the world are Lake Kivu and Lake Tanganyika, both part of the East African rift-lake system.

Lake Tanganyika (Degens et al, 1971)—Lake Tanganyika, proposed as the type example for "large anoxic lakes" (Fig. 5), measures 650 km by up to 70 km. The maximum water depth is about 1,500 m and anoxic conditions lethal to metazoan life prevail below about 150 m. Some hydrogen sulfide is present in the anoxic water. Sediment cores from both deep and shallow water show considerable vertical variability or varving.

Sediments deposited in the shallow aerated water within the lake contain 1 to 2% organic carbon, whereas the deep-water anoxic sediments range between 7 and 11% in organic carbon content. Carbon isotope ratios of recently deposited organic matter in anoxic sediments of Lake Tan-

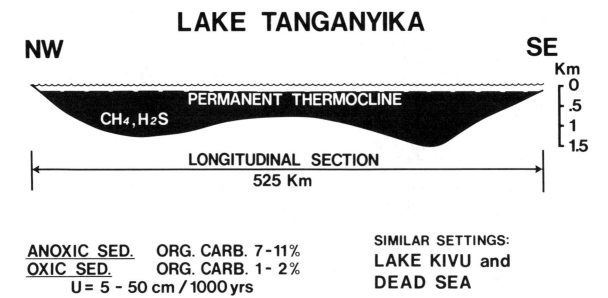

FIG. 5—Type example of "large anoxic lake," Lake Tanganyika. Symbol U in this and following figures expresses sedimentation rates in cm/1,000 years.

ganyika are in a range normally associated with organic matter deposited in marine sediments: −21 to −22 ‰ (Pedee belemnite—PDB). Remains of diatoms are the most abundant fossils in these sediments.

Lake Kivu (Degens et al, 1973)—Lake Kivu is nearly 500 m deep but is entirely anoxic below a depth of about 60 m. The pH of the surface water is around 9 but drops below 7 in the anoxic deep water where hydrogen sulfide is present in detectable quantities. The recent deep-water sediments under the anoxic zone contain up to 15% organic carbon.

The distribution of organic carbon in association with the stratigraphy of carbonate and anhydrite beds, in the past 6,000 years, reveals significant climatic fluctuations. Sapropel-rich beds coincide with humid climate and high water. Periods of dry climate, low water, and evaporites coincide with lower, but still significant, organic enrichments.

Examples of Large Oxic Lakes

Some East African rift lakes are oxic, notable examples being Lake Albert and Lake Victoria. Lake Albert (Talling, 1963) is a thermally stratified lake 50 m deep. There are a few oxygen-depleted pockets down to 0.5 ml/l. The shallow depth and density currents of cooler, oxygenated water are responsible for good ventilation. Lake Victoria is large but shallow (about 100 m) and is affected by seasonal overturns. It is not anoxic (Swain, 1970).

All the very large temperate lakes of the northern hemisphere are well oxygenated. Three examples are typical: (1) Lake Baikal, in Siberia, although the world's deepest lake (1,620 m), is well aerated and no anoxic conditions are present (Swain, 1970); (2) the Great Lakes of North America are oxic (Swain, 1970), although Lake Erie may be partly and intermittently anoxic because of excessive nutrient supply by pollution (phosphates from detergents and nitrates from sewage water); and (3) Great Slave Lake and other large lakes in western Canada are also well oxygenated (Swain, 1970).

Anoxic Environments in Large Lake Systems

Tropical climate, with little seasonal change and moderate to high rainfall year around, promotes permanent water column stratification and therefore anoxia, particularly if water depth is in excess of 100 m. The sediments deposited are commonly both carbonate- and organic-rich. Organic carbon content is higher (commonly over 10%) than in anoxic marine sediments which may be caused by anaerobic degradation in fresh water being largely by fermentation rather than by the more efficient sulfate reduction process prevailing in anoxic marine environments.

Large anoxic lakes are not present in cold and temperate climates. Seasonal overturning of water, the higher capacity of cold water to dissolve oxygen, and density underflows of cold river waters all enhance oxic conditions in bottom lake waters of temperate and cold regions.

Anoxic Environments

Evidence of Large Anoxic Lakes in Past Geologic Time

A spectacular example of past oil shale and oil source bed deposition in large anoxic lake systems is the Eocene Green River Formation in the western interior United States (mainly Colorado and Utah).

The Green River "oil shales" are brown, highly laminated, dolomitic marlstones containing hydrogen-rich kerogen (type I; Tissot et al, 1974). Retorting and conversion of this kerogen, during heat treatment of the whole rock, yields commercial shale oil. The Mahogany zone comprises the richest oil shales of the Green River Formation.

Several chemical and sedimentologic characteristics of these Mahogany zone "oil shales" discussed by Smith and Robb (1973) clearly demonstrate that permanent anoxic conditions existed at lake bottom during deposition. Smith and Robb's conclusions can be summarized as follows.

1. Evidence for complete lack of bioturbation is given by the absence of benthic fossils and the presence of minute seasonal laminations (varves) that can be followed laterally for tens of miles. Such features imply persistent lack of water movement and a lethal environment for benthic fauna in the bottom water.

2. The mineral composition of the "oil shale" implies a very alkaline, sodium carbonate-rich, bottom water whose higher density enhanced permanent chemical stratification of the lake. Eh-pH chemical fences (Krumbein and Garrels, 1952) implied by Mahogany "oil shale" mineral and chemical composition point to high pH (alkaline) and highly reducing (anoxic) conditions in the bottom water. These were lethal to macrolife forms, thus explaining the lack of bioturbation and fine varving of the "oil shales." The upper surface layer of the lake above the density boundary was fully oxygenated and supported planktonic life, for organic matter was continually deposited in the sediment.

The lower part of the Green River Formation, below the Mahogany "oil shales," contains zones of finely laminated or varved, papery, kerogenous shales which usually assay less than 15 gallons per ton of rock (Roehler, 1974). These older anoxic shales differ from the Mahogany "oil shales" in that they were deposited in a freshwater lacustrine environment instead of an alkaline lake.

The lower part of the Green River Formation, mainly below the Mahogany zone "oil shales," has reached the principal stage of oil generation and is responsible for most of the crude oils produced in the Uinta basin (Tissot et al, 1978).

Crude oils generated from anoxic lake beds such as those present in the Green River Formation are highly paraffinic and typically feature high pour points, when undergraded, and very low sulfur contents. The nonmarine Lower Cretaceous oils of China, Brazil, and some coastal west African basins belong in this category.

ANOXIC SILLED BASINS

All silled basins have several physical barriers (or sills) that restrict vertical mixing, thereby enhancing water stratification. Silling alone is not sufficient to create anoxic bottom conditions; certain patterns of water circulation, largely controlled by climate, need also to be present (Grasshoff, 1975).

Basins with a positive water balance have a strong salinity contrast between fresher outflowing surface water and deeper ingoing more saline and nutrient-rich oceanic water. The development of permanently or intermittently anoxic conditions is a general feature of those semi-enclosed seas which have a positive water balance. Basins with a positive water balance also act as nutrient traps, thus enhancing both productivity and preservation of organic matter. Typical examples of anoxic landlocked silled basins with a positive water balance in a humid zone are: the Black Sea (Degens and Ross, 1974); the Baltic Sea (Grosshoff, 1975); Lake Maracaibo (Redfield, 1958); and Saanich Inlet, British Columbia (Nissenbaum et al, 1972).

In basins with a negative water balance, resulting from a hot, arid climate, there is a constant inflow of shallow oceanic water to compensate for high levels of evaporation. As shallow oceanic water enters, it replaces the hypersaline water which sinks and flows out as a density undercurrent into the ocean. Therefore, the basin bottom is both oxygenated and nutrient depleted (Fig. 6). Examples of oxic, negative-water-balance, silled basins without significant bottom organic enrichment are the Red Sea, Mediterranean Sea, and Persian Gulf.

Black Sea

The Black Sea is the largest anoxic landlocked basin in the world. It has been studied in greater detail by oceanographers and geochemists than any other anoxic basin (Degens and Ross, 1974; Usher and Supko, 1978). We propose the Black Sea as the type example of an anoxic silled basin.

The Black Sea has a positive water balance with an excess outflow of fresh water resulting in relatively low salinity of surface water. As a result, a permanent halocline is present which also marks the boundary between oxic and anoxic conditions. Hydrogen sulfide is present in the anoxic zone. Its upper boundary is slightly convex, lying at a depth of about 250 m around the edges of the Black Sea and rising to about 150 m

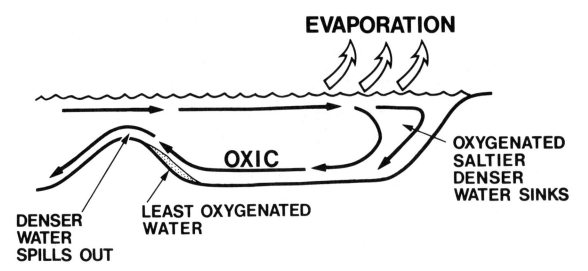

FIG. 6—Schematic model of oxic "negative water balance" basin. Hypersaline water flows out at depth over sill because it is denser than open oceanic water. This underflowing current keeps the basin bottom well ventilated and nutrient depleted.

in the center (Fig. 7). Below those depths the environment is lethal to all fish and invertebrate life.

About 22,000 years ago the Black Sea was a freshwater lake. Approximately 11,000 years ago rising Mediterranean waters began to invade the Black Sea as a consequence of climatic warm-up and ice retreat. About 7,000 years ago, the hydrogen sulfide zone began to form and present conditions were established about 3,000 years ago.

When anoxic conditions began about 7,000 years ago, the maximum organic carbon content of the sediments changed from 0.7 to 20% (Hirst, 1974). The organic-rich microlaminated black layer (7,000 to 3,000 years ago) is about 40 cm thick and is called "the old Black Sea" or "unit 2" (Degens and Ross, 1974). Unit 2 sediments commonly exceed 7% organic carbon.

Unit 1, between 3,000 years ago and the present, is about 30 cm thick and is composed of alternating white coccolith- and black organic-rich microlaminates (50 to 100 laminae per centimeter). Organic carbon content in unit 1 is still significant, but lower (about 1 to 6%) than in the underlying unit 2. In the two deepest basins, organic carbon contents in unit 1 range between 2 and over 5% (Fig. 8).

On the basis of varve chronology (Degens et al, 1978), rates of sedimentation in the deep basin were 10 cm/10^3 years for the sapropel-rich unit 2 and 30 cm/10^3 years for the more recent coccolith-rich unit 1. Thus the organically richer sediments (unit 2) actually correspond to lower sedimentation rates. The correlation is inverse to that found in deep-sea sediments under oxygenated water by Heath et al (1977).

A sedimentation rate map for the whole Black Sea during the last 3,000 years (Ross and Degens, 1974) shows a possible overall negative correlation between sedimentation rate and organic matter concentrations (Shimkus and Trimonis, 1974). Sedimentation rates on the organic-lean upper slope of the Black Sea are higher by a factor of three than in the organically rich deep basins.

Further, Shimkus and Trimonis (1974) observed that "there is no correlation between phytoplankton production and organic-matter content in sediments" of the Black Sea. "Fields of high organic-matter content correspond to areas of low primary production and vice-versa."

Recent geochemical investigations by Pelet and Debyser (1977) on a suite of Black Sea oxic unit 3 and anoxic unit 2 sediments showed that: (1) in both units organic matter is mixed, about ⅓ terrestrial plants and ⅔ marine plankton; and (2) the anoxic sediments (unit 2) have organic carbon contents from 3.85 to 14.95%, whereas the oxic sediments (unit 3) contain 0.65 to 0.69% organic carbon. The humic acid plus fulvic acid contents in relation to organic carbon are twice as high in oxic unit 3 than in anoxic unit 2. Hydrogen/car-

Anoxic Environments

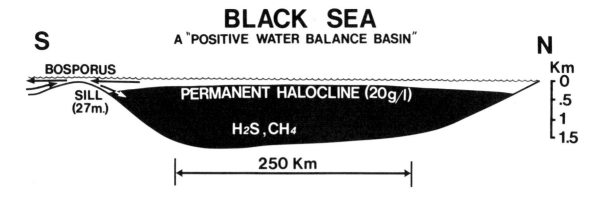

FIG. 7—Type example of "anoxic silled basin," Black Sea. When anoxic conditions began, about 7,000 years ago, average organic carbon content in sediments increased about tenfold. Lipid content in relation to organic carbon also significantly increased in anoxic sediments.

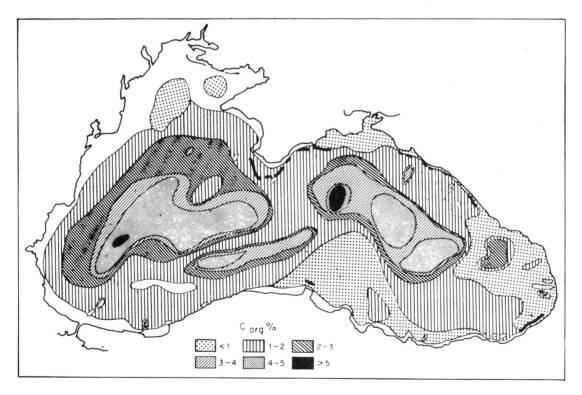

FIG. 8—Content of organic carbon in modern sediments of Black Sea (from Shimkus and Trimonis, 1974). Highest organic carbon concentrations occur in deep abyssal plains. Apparently no correlation is between organic carbon concentration (this figure) and sedimentation rate (Ross and Degens, 1974) in recent anoxic sediments of Black Sea.

bon ratios of humic compounds are higher in sediments from anoxic unit 2 than in the oxic sediments of underlying unit 3. Significantly, the chloroform extractable lipid content, in relation to organic carbon, is five times higher in sediments from the anoxic unit 2 than in the oxic sediments of unit 3.

The organic carbon budget of the Black Sea has been estimated by Deuser (1971). He concluded that about 80% of the organic input is recycled in the top 200 m of water. The remainder falls into the anoxic hydrogen sulfide-poisoned water where approximately half of this organic material is further degraded by sulfate-reducing bacteria, which leaves about 5% of the original input for incorporation into the sediment and 5% solubilized in the anoxic water. Therefore, even in the most favorable model for organic-matter preservation (under anoxic water), 95% of the organic matter escapes fossilization and is eventually recycled.

Baltic Sea

The Baltic Sea, the largest brackish water area in the world, has a positive water balance (Grasshoff, 1975). It is affected by a permanent halocline with a pronounced dip to the east, from the Kattegat into the Gulf of Bothnia. A permanent oxygen deficiency exists below the halocline, with intermittent anoxia and hydrogen sulfide poisoning being present in the lower part of the water column in the Gothland Deep. Increase in bottom-water oxygen depletion during the last 75 years may be due partly to man-made pollution. Organic carbon maps are available for the Baltic Sea (Romankevich, 1977) and show a pattern of organic carbon enrichment above 3% coincident with the areas where the water column is anoxic.

Examples of Oxic Silled Basins

Two of the world's largest silled basins, the Red Sea and the Mediterranean Sea, are well oxygenated at depth and their modern sediments tend to be organic-poor. Both these basins are characterized by a negative water balance, that is, a larger inflow from the ocean than output to the ocean. The negative balance results from loss due to evaporation (Fig. 6).

In the Red Sea, oxygen-rich shallow water flows in from the Gulf of Aden and moves all the way up to the Gulf of Suez (Grasshoff, 1975). As it becomes saltier and denser, the water sinks and returns back at depth into the Gulf of Aden over the shallow sill of Bab el Mandeb. Localized oxygen depletion in the southern Red Sea is developed at the foot of the sill just before the dense, deep bottom water spills over southward into the Gulf of Aden. The northern end of the Red Sea, farthest away from the sill, shows the highest levels of oxygenation in bottom water. Organic matter is low throughout the Red Sea, and sediments are largely biogenic clastic carbonates.

The Mediterranean is the world's largest landlocked silled marine basin, but bottom water is now entirely oxygenated at depth (Fairbridge, 1966). The lack of anoxic conditions in the Mediterranean Sea is due to a negative water balance circulation pattern identical to that observed in the Red Sea. There is, however, evidence from cores taken by DSDP and other oceanographic expeditions of past anoxic events in the eastern Mediterranean during the Pleistocene (Stanley, 1978). Intermittent anoxic conditions prevailed five times during the last 9,000 years. They are possibly the result of large and sudden influxes of fresh water, perhaps due to an increase in precipitations and/or ice retreat, which intermittently turned the eastern Mediterranean into a Black Sea-like, positive water balance basin.

Compared to oxidized Pleistocene sediments which contain less than 0.5% organic carbon, the anoxic muds of the upper Pleistocene of the eastern Mediterranean contain over 10 times more organic carbon (2 to 8%; Fairbridge, 1966). The latter are also laminated and devoid of benthic fauna.

The Persian Gulf is another silled basin with a negative water balance because of arid climate and intense evaporation. Bottom waters are oxygen-rich because of deep hypersaline water outflow and the highest organic carbon values measured (up to 2.5%) are in the silty clays that floor the deepest depressions in the southern part of the Persian Gulf (Hartmann et al, 1971).

Baffin Bay, between Canada and Greenland, is a large silled marine basin where bottom water exhibits mild oxygen depletion (about 3 ml/l) but is nowhere near anoxic conditions. Dense, cold, oxygen-rich arctic water from the northern sill (Nares Strait) slowly sweeps the bottom in pulses and eventually spills out of Davis Strait, the southern sill, and into the Labrador Sea (Palfrey and Day, 1968).

Silled Depressions in Oxic Open Ocean

The world's oceans contain many silled depressions. They are well oxygenated except in rare examples like the Cariaco Trench and the Orca basin.

Cariaco Trench—The Cariaco Trench is a classic example of a relatively small local anoxic depression located on an oxygenated open ocean shelf (Richards and Vaccaro, 1956). This small silled depression, with a maximum depth of 1,400 m, is located on the continental margin north of Venezuela. It is anoxic below 300 m, the sill depth being about 200 m. On the open oxygenated shelf, organic richness of recent sediments averages about 0.8% organic carbon. At the bottom of the trench, where anoxic conditions prevail, organic carbon concentrations increase as much as eight times in relation to the oxic environment, with values reaching 6% (Gormly and Sackett, 1977). Benthic foraminifers are lacking in the laminated anoxic sediments, but fish bones are very abundant. Radiocarbon measurements and studies of pelagic foraminifers indicate that stagnation of the trench was synchronous with an abrupt warming of surface water about 11,000 years ago (Richards, 1976).

Detailed geochemical investigations of the Cariaco Trench anoxic sediments have been reported by Combaz and Pelet (1978) together with identical studies on sediments deposited under oxic water on the Demerara Abyssal Plain and on the outer deltaic cone of the Amazon River.

The anoxic sediments of the Cariaco Trench (1) are organically richer than their sedimentation rate (50 to 75 cm/10^3 years; Combaz and Pelet, 1978) would permit under oxic water; (2) contain insoluble organic matter that is hydrogen-rich, with kerogen in the "oil-prone" type II, category (the Amazon and Demerara kerogens and humic acids are less rich in hydrogen than their Cariaco Trench equivalents; they fit in the type III "gas-prone" category); and (3) contain larger amounts of hydrocarbons and other lipids, in relation to total organic carbon, than the oxic Amazon and Demerara sediments.

Lastly, investigation of fatty-acid content in the interstitial waters of the Cariaco, Demerara, and Amazon sediments, as compared to overlying seawater (Saliot, 1977), shows preferential enrichment in fatty acids of the anoxic sediments in the Cariaco Trench.

Orca basin—The Orca basin is on the lower continental slope in the northwest Gulf of Mexico where complex bathymetry is largely attributed to salt diapirism and slumping of sediments. The basin is a crescent-shaped depression enclosing an area of approximately 400 sq km. The sill depth is approximately 1,900 m on the southeastern flank. The basin has two deeps slightly greater than 2,400 m at the north and south ends of the crescent.

The water in the Orca basin below a depth of 2,200 m, or about 200 m above the basin floor, is a brine. In addition to the pronounced increase in salinity at 2,200 m, there is a reversal of the temperature gradient, and an abrupt decline in oxygen, from 5 ml/l from a sample of 100 m above the halocline, to 0.0 ml/l below the halocline.

Oxygen, as well as nitrate, is depleted by bacterial activity (Shokes et al, 1977). In all probability the brine in the Orca basin came from solution of a salt diapir at or near the surface somewhere around the basin slope but the precise location of this hypothetical feature is not known.

Organic carbon in cored sediments under the anoxic water ranges between 0.8 and 2.9% (Sackett et al, 1978). Some of these values may be affected by dilution from slumped-in sediments from the fairly steep sides of the basin (Sackett et al, 1978).

Evidence of Anoxic Silled Basins in Past Geologic Time

Many anoxic silled basins are suspected in the geologic record but few are well documented in terms of integrated environmental and geochemical studies. Thus, in Great Britain, study of sedimentologic and fossil features in the Lower Jurassic (Toarcian) of Yorkshire by Morris (1979) permitted him to subdivide an apparently monotonous shale sequence into three facies (Fig. 9)—normal, restricted, and bituminous.

The normal shale is a homogeneous, bioturbated sediment with abundant benthic body-fossils and common sideritic nodules. This facies is indicative of well-oxygenated bottom waters. The restricted shale consists of poorly laminated sediments with scattered calcareous concretions, sparse benthic fauna, and thin discrete pyritic burrows; this facies was deposited under oxygen-depleted water, perhaps close to 0.5 ml/l. The bituminous shale is a finely laminated sediment with pyritic concretions, little or no bioturbation, and a benthic fauna which is sparse and does not include burrowing organisms; this facies reflects anoxic conditions in the water column.

The three facies described by Morris (1979) occur in cycles and are indicative of variations in the position of the oxic-anoxic boundary within the water column (Fig. 9), replicating the type settings described for modern sediments in Figures 3 and 4. We applied geochemical kerogen typing by Rock-Eval pyrolysis (Espitalié et al, 1977; Clementz et al, 1979) combined with organic carbon measurements on samples representative of the three facies, kindly supplied by the University of Reading (United Kingdom). The results are summarized on Figure 9; they confirm that the "normal shale" contains only "gas-prone" type III kerogen whereas the restricted shale is a mixed organic facies (type III-type II). Only the anoxic "bituminous" shale contains highly "oil-prone" type II kerogen. Widespread laminated oil shales in the Lower Jurassic (Toarcian) of the Paris basin as documented by Huc (1978) also resulted from anoxic basin conditions.

Alternating coccolith limestones, bioturbated clays, and laminated organic-rich oil shales de-

SHALE FACIES AND KEROGEN TYPE	GEOCHEMICAL DATA	BIVALVE GROUPS	TRACE FOSSILS	CONCRETIONS	ENVIRONMENTAL INTERPRETATION
NORMAL SHALE — TYPE III "GAS-PRONE" KEROGEN	C ORG %: 0.66–3.40 HYDROGEN INDEX: 83–134	EPIFAUNAL & INFAUNAL SUSPENSION, INFAUNAL DEPOSIT	ABUNDANT *CHONDRITES* HORIZONTAL BURROWS	SIDERITIC & CALCAREOUS	Oxic Bottom Water / Oxidizing Conditions / Reducing Conditions
RESTRICTED SHALE — TYPE II-III "MIXED" KEROGEN	C ORG %: 2.59–6.75 HYDROGEN INDEX: 135–216	DOMINANT INFAUNAL DEPOSIT	FEW UNBRANCHED HORIZONTAL BURROWS	CALCAREOUS	Weakly Oxic Bottom Water / Reducing Conditions
BITUMINOUS SHALE — TYPE II "OIL PRONE" KEROGEN	C ORG %: 5.61–11.42 HYDROGEN INDEX: 253–584	DOMINANT EPIFAUNAL SUSPENSION	NONE NO BURROWS	PYRITIC CALCAREOUS	Anoxic Bottom Water / Reducing Conditions

FIG. 9—Interpretation of facies of Liassic (lower Toarcian) of Yorkshire, Great Britain. This format, excepting geochemical data, was adapted from Morris (1979). *Hydrogen Index,* ratio expressing hydrogen richness of kerogen (Espitalié et al, 1977). Investigated rocks are all within same 60-m (197 ft) thick surface section. Degree of maturation of organic matter, as determined from Rock-Eval pyrolysis temperatures, is late immature to early mature.

posited during Late Jurassic time in the Kimmeridgian of southern England (Dorset) show a strong analogy with recent Black Sea sediments. This association reflects cyclic oscillations of an oxic-anoxic boundary within a stratified water column (Tyson et al, 1979) during Late Jurassic time. Kimmeridgian to Volgian organic-rich anoxic shales are also present under parts of the North Sea. Where thermally mature, because of sufficient burial, they are the major source contributors to the North Sea giant oil reserves (Ziegler, 1979).

Anoxic, bituminous shales of very widespread areal extent were also deposited at approximately the same time (Volgian rather than Kimmeridgian) in the Western Siberian basin (Kontorovich, 1971). These Upper Jurassic anoxic shales, where thermally mature, are the source of most of the oil reserves entrapped in the many giant fields of this prolific petroleum province.

The best documented example of a former anoxic silled basin in North America is the Mowry sea which occupied in latest Albian time most of the northwestern interior United States. Byers and Larson (1979) described three distinct sedimentologic facies in the Mowry Shale: (1) laminated mudstone, (2) bioturbated mudstone, and (3) bioturbated sandstone. The three facies represent the following environments: (1) a low-energy, lethal environment, under anoxic water, where bioturbation is absent; (2) a low-energy environment where bioturbation is present because of sufficient oxygenation of the water column; and (3) a high-energy, intensely bioturbated and well-oxygenated environment. These were described by Byers and Larson under the Schaefer (1972) terminology of marine biotopes as (1) lethal isostrate, (2) vital isostrate, and (3) vital heterostrate.

The organically richest zones identified by Nixon (1973) as oil source beds of major significance coincide with the low-energy, unbioturbated, lethal facies, which was deposited under anoxic water in the Mowry sea.

ANOXIC LAYERS CAUSED BY UPWELLING

Upwelling is a process of vertical water motion in the sea wherever subsurface water rises toward the surface. Large-scale upwelling due to wind-stress-induced, Ekman transport occurs along certain coastlines such as those of California, Peru, Chile, South-West Africa, Morocco, and Western Australia. Upwelled water in coastal regions comes from relatively shallow depth, usually less than 200 m.

Ziegler et al (1979) described three distinct geographic situations where favorable conditions are met for coastal upwelling: (1) meridional upwell-

Anoxic Environments

S.W. AFRICAN SHELF

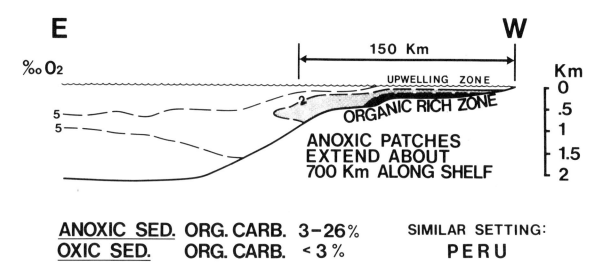

FIG. 10—Type example of "anoxic layers caused by upwelling," South-West African Shelf (Benguela Current). Coastal upwellings are complex current and counter-current systems interacting with offshore winds. Upwelled water, rich in nutrients and low in oxygen, does not originate from great water depths, but from less than 200 m.

ing on north-south coasts between 10 and 40° lat. on the east side of oceans (e.g., California Current); (2) zonal upwelling on east-west coasts of equator-centered continents at about 15° lat. in association with the easterly trade winds (e.g., Caribbean Current); and (3) monsoonal upwellings on diagonal east-facing coasts of equator-centered continents at about 15° lat. in association with easterly trade winds (e.g., Somalia Current).

Upwellings can also occur in the open ocean at water-mass boundaries, and such have been noted around the Antarctic continent, along the equator, and between Iceland and Norway. Upwelled water is rich in nutrients (nitrates and phosphates) and, therefore, promotes high biologic productivity. Recycling of dead organic matter in the water column creates a very high oxygen demand which can trigger anoxic conditions in deeper water layers under the upwelling. A classic example of anoxicity in the water column (essentially created by excessive biochemical oxygen demand due to coastal upwelling) occurs on the shelf offshore South-West Africa (Namibia) in association with the Benguela Current. We propose it as our type example (Fig. 10).

Benguela Current

Considerable oceanographic and geochemical research has been conducted offshore of South-West Africa since the mid-1960s, notably by Calvert and Price (1971a, b). Anoxic conditions occur on the South-West African Shelf (Fig. 10), particularly off Walvis Bay. The oxygen-depleted zone at sea bottom is an elongate area approximately 50 by 340 km (Calvert and Price, 1971a) parallel with and close to the coastline. Beyond the shelf break (about 100 km offshore) normal oxygenated conditions return to the bottom, as the regional oxygen-minimum layer in the Atlantic, south of Walvis Ridge, is only very weakly developed (Bubnov, 1966). Anoxic conditions are created on this narrow shelf by the high oxygen demand from decomposition of large amounts of plankton resulting from the Benguela Current upwelling. The upwelling is due to a combination of a cold coastal current (the Benguela Current) and persistent offshore winds blowing northwest. Shallow surface water is skimmed off by the wind, permitting nutrient-rich subsurface water to ascend from a depth of about 200 m.

In three dimensions the system visualized is one where oxygen-poor, but nutrient-rich water constantly moves up and mixes in the euphotic zone with oxygenated water, causing high biologic productivity along a narrow coastal band. Dead plankton eventually falls to the bottom under the upwelling, and nutrients associated with the organic matter are brought back to the surface by the upwelling instead of being dispersed into the open sea. What is buried and lost to the bottom sediments is replaced by nutrients brought in by the Benguela Current.

Brongersma-Sanders (1972) wrote: "Upwell-

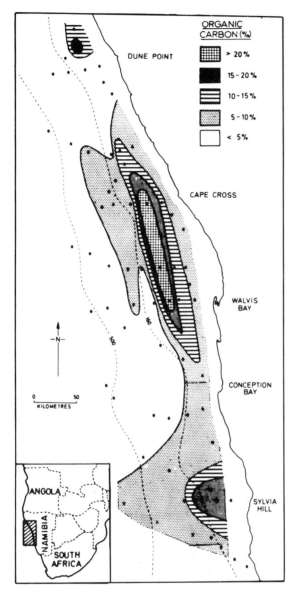

FIG. 11—Content of organic carbon in modern sediments offshore Walvis Bay (Namibia; from Calvert and Price, 1971). Highest values are off Walvis Bay and Cape Cross, where anoxic conditions and bacterial nitrate reduction intermittently prevail in bottom waters.

ings are fertile in the first place because nutrient-rich water is brought to the well-lighted surface layers, but this is not the only or even the main point, which is that upwelling is a kind of counter current system. A counter current system acts as a trap in which nutrients tend to accumulate."

Organic carbon concentrations in sediments under the oxygen-depleted zone range between 5 and 24% (Fig. 11). The highest values (over 20%) are off Walvis Bay and Cape Cross where oxygen-depleted conditions reach the anoxic level (below 0.5 ml/l). Bacterial nitrate reduction occurs in bottom water. Free hydrogen sulfide has been detected in the water of Walvis Bay, causing occasional mass mortality of fish and detectable hydrogen sulfide odor inland. There appears to be a clear positive correlation between level of oxygen depletion in the bottom water and organic carbon enrichment of the sediment.

Most organic matter is planktonic in origin because runoff from the land is intermittent and from a practically rainless desert (Calvert and Price, 1971b). Sedimentation rates were calculated by Veeh et al (1974) for anoxic sediments offshore Walvis Bay. Rates range between 29 and 103 $cm/10^3$ years. Organic carbon values range between 5.65 and 16.36% in the same samples and show no systematic correlation with sedimentation rates.

The anoxic, organic-rich sediments in the Benguela Current contain abnormal concentrations of copper, nickel, uranium, and phosphorus. Copper, molybdenum, and nickel distributions show similarities to that of organic carbon (Calvert and Price, 1971a). Uranium concentrations in sediments of the South-West African Shelf have been investigated by Veeh et al (1974). They reached the following conclusions: (1) the uranium has a positive correlation with organic carbon; (2) the uranium was derived from modern seawater; and (3) fixation of uranium in the sediments was conditioned by the presence of phosphorus together with reducing (anoxic) conditions; uranium is incorporated into apatite and is concentrated in phosphatic nodules and laminae.

When upwellings occur in areas of broad and well-developed intermediate oxygen-depleted layers, the oxygen starvation brought about by the local upwelling reinforces the regional oxygen anomaly. Notable coastal upwellings of this type have been investigated off California and Peru in association with regionally oxygen-depleted intermediate water layers.

Peru Current

The Peru (Humboldt) Current is a well-documented example of an anoxic environment associated with coastal upwelling (Fig. 12; Fairbridge, 1966). The Peru Current refers to a system of relatively shallow currents flowing northward along the west coast of South America. It is a complex system involving two surface currents, an undercurrent, and a countercurrent (Idyll, 1973).

The degree of upwelling offshore of Peru is related to the intensity of the wind stress. Prevailing trade winds off the coasts of northern Chile, Peru, and Ecuador blow principally from the northeast and south. This wind flow pushes the surface wa-

Anoxic Environments

PERUVIAN SHELF

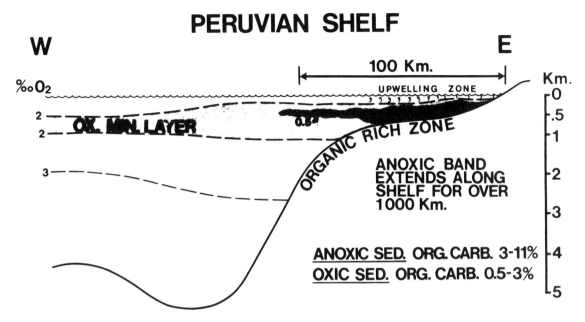

FIG. 12—Example of "anoxic layers caused by upwelling," Peru Current. Geochemical observations on sediments at bottom of Peru Shelf and Trench clearly demonstrate preferential preservation of lipids in those sediments under anoxic water.

ter northward at the same time that the Coriolis force deflects it to the west, thus skimming off the surface layers and letting cold subsurface water well up. Study of the isotherm patterns establishes that the depth of upwelling ranges from about 50 to 240 m (Fairbridge, 1966).

The upwelled water is undersaturated in oxygen but rich in nutrients (phosphates, nitrates). The ample supply of nutrients is associated with an exceedingly high rate of primary productivity. The combination of high productivity in the surface waters and depleted oxygen in the water column generates conditions favorable to enhanced preservation of organic matter in the underlying marine sediments. The concentration is further enhanced by the lack of significant water movement toward the ocean. The Peru Current, like the Benguela Current, is a self-regenerating nutrient trap and an outstanding primary producer of organic matter.

The organic matter in the bottom sediments of the Peruvian Shelf has been investigated by Gershanovich et al (1976). The highest concentrations of organic carbon (average 3.33% with values up to 11%) are in silty clays and laminated diatomaceous oozes under the anoxic zone at water depths between 100 and 500 m (Figs. 12, 13).

Organic-matter quality of its alkali-extractable portion was measured by elemental analysis. Hydrogen/carbon ratios of this humic material reach 1.37 to 1.43, indicating that they are not fulvic and humic acids identical to those of soils or peats but are related to the "sapropelic acids" of Russian researchers. This type of marine humic material is considered as the main precursor of oil-prone kerogens. Hydrogen richness of kerogen precursor material is lower (H/C = 1.23) in the more oxygenated bottom sediments of the trench below 500 m. The same oxygenated sediments contain three times less extractable lipids in relation to organic carbon than those under the anoxic zone. Concurrently, the ratio of total lipid to total organic carbon of the bottom waters versus the sediments was investigated for the anoxic and oxic zones by Beliaeva and Romankevich (1976). Under oxygenated bottom waters in the trench, the soluble lipid to organic carbon ratio decreases by a factor of 22 from the water column into the sediments. On the anoxic shelf, however, the same ratio only decreases by a factor of 7. These observations clearly demonstrate preferential preservation of lipids in sediments under anoxic water.

The sediments of the Peruvian Shelf deposited under anoxic water contain significantly increased quantities of sapropel-type organic matter (70 to 90% of the organic matter); enhanced contents of soluble organic matter (bitumen or total lipids) and significant quantities of hydrocarbon gases including higher hydrocarbons up to hexane (Gershanovich et al, 1976) are also present. Uranium-enriched phosphorite nodules are present along the lower and upper boundaries of the oxygen-minimum layer at the edges of the

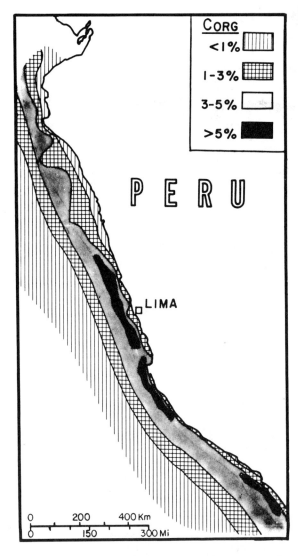

FIG. 13—Content of organic carbon in recent sediments offshore Peru (from Logvinenko and Romankevich, 1973).

ductivity, the circumpolar waters are perhaps the most highly oxygenated on earth, largely because of very low surface temperatures (Fig. 14). Organic carbon values in bottom sediments around Antarctica reported in the *Geological-Geophysical Atlas of the Indian Ocean* (Udintsev, 1975) range between 0.10 and 0.94%.

In the northern Pacific, offshore Japan and the Kuriles, very high productivity caused by upwelling is offset by a constantly renewed supply of oxygen-rich intermediate and deep water coming from the Bering Sea.

Concerning offshore southeastern Brazil, Summerhayes et al (1976) concluded after detailed study, "Beneath the biologically productive upwelled surface waters, there is remarkably little sedimentation of organic matter. Also, although phosphorite deposits are usually associated with upwelling centers, there are no phosphatic sediments off southeastern Brazil . . . this probably results from the high degree of oxygenation of upwelled water which contains 6 ml/l of dissolved oxygen."

Evidence of Anoxic Layers Caused by Upwelling in Past Geologic Time

Paleogeographic reconstructions as applied to occurrence of Phanerozoic marine phosphorites (Ziegler et al, 1979) and petroleum source beds in the Paleozoic (Parrish et al, 1979) clearly point to the following.

1. Phanerozoic phosphorites were deposited in association with upwelling zones at low paleolatitudes.

2. Phanerozoic phosphorites are not randomly distributed in time. Most deposits were formed in upwelling zones during periods of worldwide transgression and expansion of the oxygen-minimum layer.

3. Phosphorites and organic-rich sediments are commonly associated, as in the Permian Phosphoria Formation (Parrish et al, 1979) in North America. The Phosphoria black shale members are major source contributors to Paleozoic oil accumulations in the western interior United States (Claypool et al, 1978; Momper and Williams, 1979), including several giant fields.

Evidence of anoxic sedimentation associated with coastal upwelling zones also is present in the Tertiary of California. Sedimentary suites characteristic of past deposition under the oxygen minimum layer of an upwelling are present at various levels in all the oil producing basins of California. These sediments typically fit the following descriptions: laminated, phosphatic, bituminous, dark-brown shales, and laminated, hard, siliceous, dark-brown, commonly diatomaceous shales. Uranium-bearing phosphatic material is

anoxic zone (Veeh et al, 1973).

Examples of Upwellings Without Anoxic Layers

Not all areas of upwelling and high primary productivity are associated with anoxic layers in the water column and good concomitant organic-matter preservation in bottom sediments. Many areas of high primary productivity are underlain by highly oxygenated water which permits thorough aerobic degradation of the organic matter. Documented examples of such upwelling zones where the oxygen supply at sea bottom exceeds the biochemical oxygen demand are in Antarctica, the northern Pacific, and off southeastern Brazil.

In Antarctica, despite locally high surface pro-

Anoxic Environments

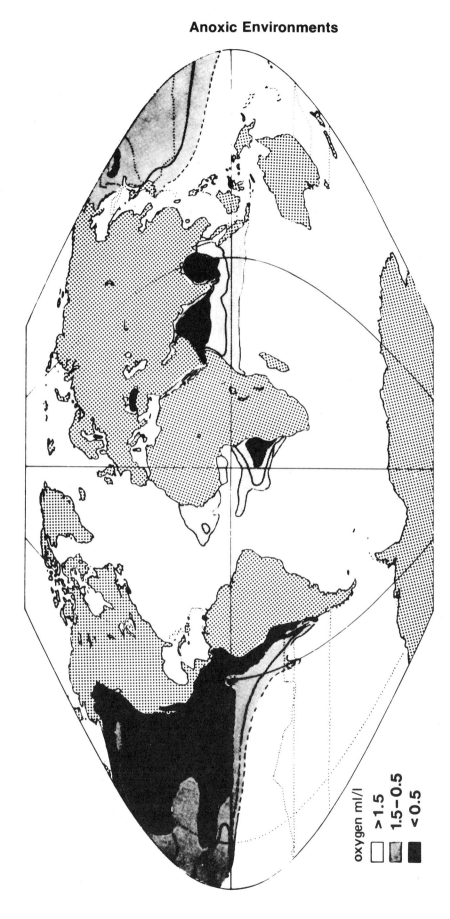

FIG. 14—Maximum extension of oxygen-depleted layers in world ocean. Preferential distribution of oxygen-minimum layers in open ocean is at low latitudes and on eastern sides of basins, largely because of deep oceanic circulation which is controlled by global climate and Coriolis force.

abundant throughout, whether in nodules or finely disseminated, and radioactive background is relatively high. These rocks are generally devoid of macrofossils.

Anoxic organic-rich, bituminous, phosphatic shales are best developed in: the lower Pliocene (base of the Repetto Formation in the Los Angeles basin); the upper Miocene (lower part of the Puente Formation "nodular shale" in the Los Angeles basin, also the McLure and Antelope shales in the San Joaquin Valley); and the middle Miocene (major parts of the Monterey Formation in the Ventura basin, Santa Maria basin, and southern Coast Ranges; middle Miocene, i.e., Luisian, time marks the peak of phosphatic deposition in the Tertiary of California; Dickert, 1966). The formations listed have been geologically recognized as prolific oil sources in the oil-producing California basins, long before the advent of modern petroleum geochemistry (Jenkins, 1943).

The most widespread anoxic sediments caused by past upwelling in California are in the middle Miocene (Luisian Stage) as well as in the lower part of the upper Miocene (Mohnian Stage). These stages, particularly the Mohnian, correspond with highest sea stands and maximum transgression during Miocene time on a global scale (Vail and Mitchum, 1979). Anoxic conditions associated with upwelling were also reinforced locally, in some places, by silled-basin topography. The overall oceanographic setting was clearly similar, however, to that present today in the Peru Current upwelling.

ANOXIC OPEN OCEAN

The intermediate layers of the open ocean both north and south of the equator in the eastern Pacific and northern Indian Ocean have been found to be oxygen depleted over areas of considerable size (Fig. 14). The volumes of anoxic water in today's open oceans greatly exceed the volumes of anoxic water present in the Black Sea.

It has been suggested that the distribution and position of these widespread oxygen-minimum layers result essentially from biochemical oxygen demand created by high planktonic productivity. Oceanographic observations on a global scale, however, do not agree with this hypothesis. Biochemical oxygen demand is indeed the triggering mechanism of anoxicity in general, but deep circulation patterns govern the distribution and position of these regional anomalies.

The Grand Banks of Newfoundland in the northwest Atlantic form one of the world's great fishing grounds with productivities comparable to those offshore California or northwest Africa, yet no oxygen depletion of the intermediate or bottom water has been observed there. The overall water column is well oxygenated (over 6 ml/l) and organic carbon in bottom sediments is present only in low concentrations, averaging 0.5% (Emery and Uchupi, 1972, p. 292, 371-373). The overall lack of correlation between areas of high oceanic productivity and oxygen depletion is also clearly demonstrated in the Pacific and Indian Oceans (Figs. 1, 14). The oxygen-depleted areas in these oceans are cul-de-sacs or are in the lee of sources of cold oxygenated bottom water. Oxygen concentrations decrease very gradually between Antarctica and the northern Indian Ocean and northeast Pacific (Reid, 1965; Gorshkov, 1974) demonstrating very broad regional circulation effects, rather than local anomalies related to overproductive coastal upwellings.

In only two important regions do substantial amounts of cold oxygenated surface water continuously sink to abyssal depth: the North Atlantic and around Antarctica. The deep oxygen concentration in the other oceans diminishes with increasing distance from these polar sources. The deep, cold, oxygen-rich bottom-water underflow from the poles counterbalances the evaporation from shallow waters in the subtropical regions of the oceans.

Strong bottom-current activity is prevalent in the cold, high-latitude regions and on the western side of basins; these are the areas of most turbid bottom water (Hollister et al, 1978). Red pelagic clays also occur on the western floors of oceanic basins (Fairbridge, 1966). These observations are coincident with better overall oxygenation of bottom and intermediate water at high latitudes and on the western sides of the Pacific and Atlantic Oceans (Fig. 14). This unequal latitudinal distribution of the regional oxygen-minimum layer (Fig. 14) indicates a tendency toward stagnation on the eastern sides of oceanic basins. It reflects, according to Stommel (1958), the effects of the Coriolis force upon deep circulation of oxygenated bottom water. It is clearly unrelated to areas of high primary productivity present on *both* sides of the same oceanic basins. Because this asymmetry is created by the earth's rotation, the same effect must have persisted throughout geologic time. Thus it could have had a profound influence on the paleogeographic distribution of oil source beds on continental shelves and slopes.

Significant organic enrichment has been observed around the world wherever the regional oxygen-minimum layer falls at or below 0.5 ml/l and intersects the continental slope and shelf. This correlation has been recognized and described in the northeastern Pacific Basin (Gross et al, 1972), in the Gulf of California (van Andel, 1964), and in the Indian Ocean (Stackelberg, 1972; Konjukhov, 1976).

Anoxic Environments

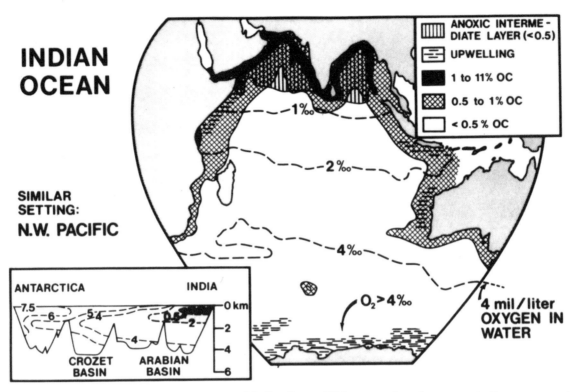

FIG. 15—Type example of "anoxic open ocean," Indian Ocean. Highest organic carbon concentrations are present wherever oxygen minimum layer falls at or below 0.5 ml/l and intersects continental slope and shelf.

Indian Ocean

The Indian Ocean is the type example of the "anoxic open ocean" of this classification.

The upper continental slope of the Indian Ocean from the Gulf of Aden to the Andaman Islands is occupied by a very large and relatively shallow anoxic layer marked in vertical hachures on Figure 15. Wherever this layer impinges on the shelf and slope between 250 and 1,200 m (Fig. 16), abnormally high organic carbon concentrations (between 2 and 10%) have been found by Stackelberg (1972), Konjukhov (1976), and other investigators (Udintsev, 1975). Organic carbon content on the shelf and other parts of the slope under oxic water is lower (between 0.5 and 1%). Similar observations were made in the Gulf of Oman (Hartmann et al, 1971).

Primary productivity varies from high, related to upwelling (marked in horizontal hachures on Fig. 15) on the west side of India, to low in the Gulf of Bengal. Sedimentation rates also vary widely from low, off the Arabian coast and in the Gulf of Oman, to very high off the Indus and Ganges deltas (Lisitzin, 1972, p. 135-171). The strongest correlation which overcomes differences in productivity and sedimentation rates is that between anoxic conditions and organic enrichment.

Pacific Ocean Offshore Washington-Oregon

In the northeast Pacific, offshore Washington and Oregon, a positive correlation is evident between organic carbon content of bottom sediments and oxygen depletion of bottom water (Gross et al, 1972). Where the dissolved-oxygen concentrations fall below 1 ml/l, organic carbon values range between 1 and 3%. Between 1 and 2 ml/l, organic carbon contents of bottom sediments fall to around 1%. Other factors such as grain size and sedimentation rate are significant, but it is statistically clear that the highest organic carbon concentrations occur where the oxygen-minimum layer impinges on the continental slope. In this general area, "there is surprisingly little correlation between known areas of high primary productivity at the ocean surface and the amount of organic matter preserved in the sediment" (Gross et al, 1972).

Gulf of California

The recent sediments of the Gulf of California (Mexico) have been investigated in detail by Calvert (1964), Parker (1964), Phleger (1964), and van Andel (1964) in relation to oxygen content of intermediate and deep waters (Roden, 1964) and primary productivity of surface water. Determi-

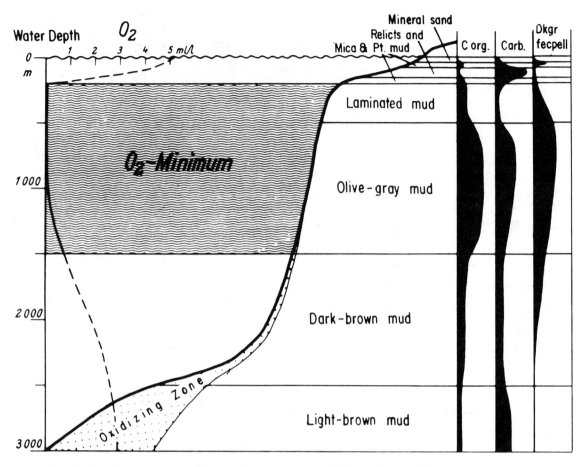

FIG. 16—Section through oxygen minimum layer on western shelf of India (modified from Stackelberg, 1972).

nations of organic carbon were made on approximately 100 core samples throughout this area. The organic carbon content appeared at first to correlate to the amount of clay in the sample. However, inspection of samples with comparable clay contents showed that the largest amounts of carbon (5% and above) are not in the silled basins within the gulf itself, but on the slopes of the southern gulf and in the open sea. In these areas the oxygen-minimum layer in the Pacific Ocean (below 0.5 ml/l) impinges on the slope between about 300 and 1,200 m. These areas of high organic enrichment and anoxic conditions in the water column coincide with finely laminated, richly diatomaceous sediments and the absence or scarcity of burrowing organisms. Conversely, lack of lamination due to homogenization by active bioturbation, as well as lower organic carbon content (below 2%), were demonstrated for those portions of the slope where oxygen concentrations in the water column are higher than 0.5 ml/l (Fig. 17).

Calvert (1964) also observed that oxygen concentrations lower than 0.5 ml/l in the water column did not completely inhibit the development of benthonic foraminifers but virtually suppressed the existence of macroinvertebrates and burrowing organisms.

Atlantic Ocean

Although an oxygen-minimum layer has long been recognized here as well, the Atlantic Ocean in general tends to be oxygenated, for it is an open corridor between the two polar regions where cold, oxygen-rich bottom waters originate. There are only two notable oxygen-depleted intermediate zones, both in the eastern South Atlantic basins (Fig. 14). They are on the west side of Africa, on both sides of the equator, one off Cape Verde (Senegal) and one in the Angola basin (Bubnov, 1966). Only the Angola basin reaches the anoxic level (0.5 ml/l), but detailed geochemical observations on bottom sediments are lacking.

Offshore northwest Africa (Mauritania and Cape Verde), dissolved-oxygen concentrations in the oxygen-minimum layer are still in the oxic range (above 1 ml/l) whereas surface productivity

Anoxic Environments

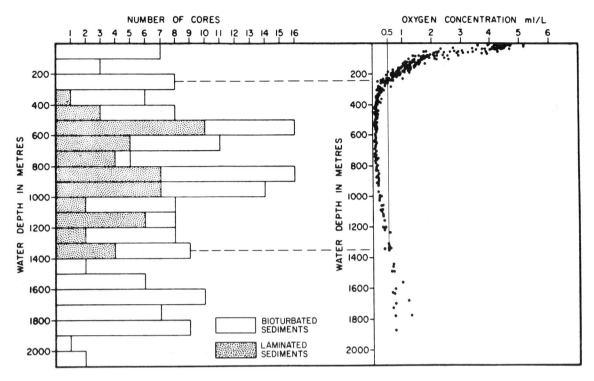

FIG. 17—Degree of bioturbation in relation to oxygen concentration in bottom water observed in Gulf of California (modified from Calvert, 1964). Laminated sediments are present at oxygen concentrations lower than 0.5 ml/l. This level of oxygen depletion, not total absence of oxygen, is effective threshold to lack of bioturbation.

is high owing to local upwelling. Organic carbon measurements by Gaskell et al (1975), Hartmann et al (1976), and Debyser et al (1977) demonstrate that organic preservation is variable in this oxic environment with a range of less than 0.5 to about 4% organic carbon. These organic carbon fluctuations under oxic water are controlled by sedimentation rates and do not reflect present high productivities created by upwelling conditions (Hartmann et al, 1976).

Offshore Ivory Coast and Ghana, oxygen depletion is also present in the intermediate water (about 1.5 ml/l), but does not reach the anoxic level. Organic carbon values range between 3 and 5% but organic matter is largely composed of transported terrestrial plant remains (Klingebiel, 1976). The predominance of transported terrestrial organic matter, originating from rain forests, into an open-marine environment under 1,200 m of water, is an interesting present-day example of a situation common in ancient open-marine sediments. Such a setting implies abundant rainfall on nearby continental masses.

In the Niger delta, in nearly identical climatic surroundings, organic carbon content in prodelta clays is low and ranges between 0.5 and 1.0% (Klingebiel, 1976). Sedimentation rates are very high, suggesting that dilution of organic matter must occur there.

Evidence of Widespread Anoxic Open-Ocean Conditions in Past Geologic Time

Oceanographic observations are made today on a cold earth. Present time represents a relatively brief oscillation within the Pleistocene Ice Age sequence. The modern oceans and seas are relatively well oxygenated at depth because contrasted climatic belts and polar ice caps enhance vigorous convective water mass movements (Berggren and Hollister, 1977). The active bottom circulation that characterizes the present Atlantic, for example, was not initiated until about 3 m.y. ago. Cold-earth conditions as we know them today were heralded by major climatic cooling during late Eocene–Oligocene time followed by late Miocene and Pliocene-Pleistocene global marine regressions and glaciations. The long-term trend since Eocene time is one of global climatic cooling concurrent with a lowering of the mean sea level in the world ocean (Vail and Mitchum, 1979). In the geologic past, deep oceanic circulation was not so active during periods of warm equable global climate which corresponded with high mean sea levels and the absence or extreme reduction of polar ice caps. For example, sedimentary and paleontologic data interpreted within the framework of plate tectonics suggest that the Arctic Ocean and north-polar regions were ice-free in Late Jurassic (Hallam, 1975, p. 52-63) and Late Cretaceous time (Clark, 1977).

Evidence of worldwide oceanic anoxic events (Schlanger and Jenkyns, 1976) in the generally warm Mesozoic Era is indicated by remarkably widespread organic-rich black shales during the Late Jurassic and middle Cretaceous. The global extent and synchroneity of the Cretaceous anoxic events have been shown by the Deep Sea Drilling Project (Degens and Stoffers, 1976; Schlanger and Jenkyns, 1976; Ryan and Cita, 1977; Thiede and van Andel, 1977; Tissot et al, 1979). These paleoanoxic events have also been increasingly recognized on the continental shelves by petroleum exploration drilling.

Identical worldwide oceanic anoxic events are inferred also to have prevailed intermittently in Paleozoic time because the widespread Lower Ordovician, Silurian, and lower Carboniferous organic-rich marine black shales coincide with times of widespread marine transgression and ice cap melting (Berry and Wilde, 1978).

Lastly, there is statistical evidence that the known oil source bed systems present in the stratigraphic record are not randomly distributed in time but coincide with periods of worldwide transgression and oceanic anoxia (Arthur and Schlanger, 1979; Tissot, 1979). Prolific petroleum source beds deposited over part of the Jurassic and Cretaceous account for 70% of the oil in 148 investigated petroleum zones, which, themselves, total about 85% of the world's reserves (Tissot, 1979). Yet this privileged period for oil source bed deposition (Jurassic-Cretaceous) only amounts to 17% of Phanerozoic time. Prolific petroleum regions where parts of the Cretaceous and/or Jurassic play a major oil source role are the Middle East (Kent and Warman, 1972, p. 150), the North Sea (Ziegler, 1979), the West Siberian basin (Kontorovich, 1971), and Mexico (Arthur and Schlanger, 1979).

CONCLUSIONS

The explosive growth in knowledge by recent observations in oceanography, geochemistry, and geomicrobiology has made it necessary to review many traditional concepts relating to oil source bed genesis. The proposed revised concepts follow.

1. Potential oil source beds are organic-rich sediments containing a kerogen type (type I or II) that is sufficiently hydrogen-rich to convert mainly to oil during thermal maturation.

2. Measurement of organic carbon content, alone, is insufficient to identify potential oil source beds. For example, "gas-prone" transported terrestrial organic matter (OM), oxidized aquatic OM deposited at high sedimentation rates, or reworked OM from a previous sedimentary cycle commonly create misleadingly high levels of organic carbon in ancient marine sediments. High organic carbon content does not necessarily equate with oil source potential, hence precise measurement of kerogen type is indispensable to the identification of oil source beds. Insight into the early environmental factors that determine kerogen type, furthermore, is the principal key to understanding the temporal as well as paleogeographic distribution of oil source beds.

2. Organic matter in sediments below anoxic water (defined as water containing dissolved oxygen in amounts less than 0.5 ml/l) is commonly more reduced (hydrogen-rich), more lipid-rich, and more abundant than under oxic water.

3. The organic residues of anaerobic bacterial degradation are more reduced (hydrogen-rich) and more lipid-rich than those resulting from oxidation by aerobic microbes. Hence anaerobic residues are, potentially, the precursors to "oil-prone" kerogens. The latter are essential constituents of potential oil source beds. The mechanisms leading to this preferential enrichment in lipids are little understood and in need of fundamental research in microbial biochemistry.

4. The overall range of organic carbon content under anoxic water is significantly wider and higher (about 1% to over 20% organic carbon) than that in sediments deposited under oxygenated water. It is suggested that variations in sedimentation rate, if perhaps one of the underlying reasons for organic richness fluctuations under anoxic water, are not the main cause of enhanced preservation in reducing environments. Sediments deposited under anoxic marine water also contain abnormally high concentrations of U, Cu, Mo, Ni, P, and S which tend to correlate positively with organic carbon concentrations. Organic matter deposited under anoxic water, wherever investigated, is hydrogen-rich and therefore "oil-prone."

5. Organic carbon content in sediments under oxygenated water is primarily a function of sedimentation rates up to the point (yet ill-defined) where organic matter dilution significantly depresses concentrations. Surface primary productivity appears to have little influence on organic carbon concentrations in sediments under "oxic" water because it is systematically compensated by organic matter consumption at the seafloor by benthic organisms. Organic carbon contents un-

der "oxic" water typically range between less than 0.5% to maxima between 3 and 4%, regardless of surface productivities. Organic matter deposited under oxic water, wherever investigated, is hydrogen depleted and therefore "gas prone."

6. Bioturbation, or biologic sediment mixing by metazoans, is ubiquitous at all water depths under "oxic" water. It causes acceleration of geochemical processes at and below the sediment-water interface. Added to this effect is the grazing of the sediment bacterial biomass by the mobile deposit feeders (burrowers) responsible for bioturbation. Bioturbated sediments under "oxic" water are homogenized whereas sediments deposited under anoxic water are varved or laminated and devoid of burrowing, deposit-feeding infauna.

7. Maximum oxygen saturation in seawater is about 6 to 8.5 ml/l, but benthic metazoans are unaffected by lower oxygen concentrations, down to about 1 ml/l of water. Between 0.1 and 0.5 ml/l, a low diversity, highly stressed, suspension-feeding (non-burrowing) epifauna can still survive *above* the sediment-water interface. Below 0.5 to 0.3 ml/l, bioturbation by deposit feeders (burrowers) virtually ceases and the overall benthic biomass is sharply reduced. Below 0.1 ml/l, all metazoans disappear leaving anaerobic bacteria as the only effective reworkers. Hence 0.5 ml/l (the threshold of arrested bioturbation), not total absence of oxygen in water, is proposed as the effective "biochemical fence" between potentially good or poor qualitative and quantitative organic matter preservation.

8. The tolerance of protozoans, such as benthic foraminifers, to very low oxygen concentrations appears to be higher than that of metazoans. The ecology of benthic foraminifers under oxygen-depleted water is clearly in need of fundamental research.

9. Anoxic water is defined, for this study, as water containing less than 0.5 milliliters oxygen per liter of water. Anoxic conditions occur where the natural demand for oxygen in water exceeds the supply. Oxygen demand relates to surface primary productivity, whereas oxygen supply below the depth of surface mixing depends on water circulation.

10. Lakes are rarely permanently anoxic at depth. It takes special conditions such as overfertilization, lack of overturning of the water in warm climates, and preferably, but not necessarily, deep water to create stable anoxic conditions. When they occur, as in Lake Tanganyika (east Africa), sediments rich in organic matter are deposited. Warm paleoclimatic conditions with lack of seasonal contrast and moderate rainfall should be most favorable to lacustrine oil source bed deposition.

11. At present, the majority of silled basins, whether landlocked or on open shelves, are not anoxic and thus do not favor deposition of oil source bed precursor sediments. Only those silled basins that have a water circulation pattern implying a "positive water balance" become anoxic, for example, the Black Sea or the Baltic Sea. The degree of ventilation in silled depressions on open oceanic shelves follows that one prevailing at comparable depth in the adjoining open oceanic water masses. Hence silled-basin geometry in ancient sedimentary basins does not automatically imply the presence of oil source beds. Silled basins should be prone to anoxia and oil source bed deposition at times of worldwide transgression, both at high and low paleolatitudes. At times of worldwide regression, those landlocked basins in temperate paleoclimatic belts with high rainfall should be more favorable than those in arid zones.

12. Coastal upwellings, which are commonly associated with high primary productivity of the surface layers, do not originate from great water depths. Only upwellings associated with anoxic layers show any abnormal organic preservation as well as phosphorite deposition in bottom sediments (e.g., Benguela and Peru Currents). Extensive oil source beds and phosphorites resulting from upwelling conditions should be sought at low paleolatitudes and preferably at times of worldwide transgression.

13. Large volumes of anoxic water are present in the oxygen-minimum layers of open oceans such as the Indian Ocean without silling or physical barriers. Significant organic enrichment in sediments has been found wherever the oxygen-minimum layer reaches the anoxic level and intersects the continental slope and shelf. The preferential distribution of oxygen-minimum layers in the world ocean is at low latitudes and on the western side of continents, primarily because of deep oceanic circulation spurred and controlled by climatic contrast of the poles and the equator and the Coriolis force. Inferred are similar global patterns, determining the position of oxygen-minimum layers during the geologic past, but with much more widespread extensions of anoxia invading whole oceanic basins. This is suggested by remarkably widespread organic-rich marine black shales at times of global climatic warm-ups and major transgressions as in Late Jurassic and middle Cretaceous time. Known marine oil source bed systems in the stratigraphic record are not randomly distributed in time but coincide with periods of worldwide transgression and oceanic

ANOXIC BASIN TYPE	PALEOGEOGRAPHIC SETTING	STRATIGRAPHIC DISTRIBUTION OF ANOXIC SEDIMENTS
ANOXIC LAKES	EQUABLE, WARM, RAINY. EARLY RIFTS. INTERMOUNTAIN BASINS.	CONTINUOUS WITHIN SAME ANOXIC LAKE SYSTEM.
ANOXIC SILLED BASINS	TEMPERATE TO WARM, RAINY. INTRACRATONIC SEAS. ALSO ANOXIC POCKETS ON SHELVES.	VARIABLE. TENDS TO BE RICHEST AT BOTTOM OF BASIN OR POCKET.
ANOXIC LAYERS W/ UPWELLINGS	OCEANIC SHELVES AT LOW LATITUDES. WEST SIDE OF CONTINENTS.	OFTEN NARROW TRENDS. BUT CAN BE WIDESPREAD. PHOSPHORITES. DIATOMITES.
ANOXIC OPEN OCEAN	BEST DEVELOPED AT TIMES OF GLOBAL WARM-UPS & MAJOR TRANSGRESSIONS.	VERY WIDESPREAD W/LITTLE VARIATION. OFTEN SYNCHRONOUS WORLDWIDE.

FIG. 18—Classification of anoxic depositional models. Certain settings in marine realm may combine two or even three anoxic basin types, for example, local anoxic layers caused by upwelling may reinforce anoxic open-ocean oxygen-minimum layer impinging upon shelf with silled depressions.

anoxia.

14. Geochemistry now provides the means to identify paleoanoxic events in the stratigraphic record. Paleogeographic recognition of the proposed anoxic models—large anoxic lakes, anoxic silled basins, anoxic layers caused by upwelling, and anoxic open ocean—in ancient sedimentary basins should help geologists and geochemists in the regional mapping of oil shales and oil source beds as shown in Figure 18.

REFERENCES CITED

Aller, R. C., 1978, The effects of animal-sediment interactions on geochemical processes near the sediment-water interface, in Estuarine interactions: New York, Academic Press, p. 157-172.

Arthur, M. A., and S. O. Schlanger, 1979, Cretaceous "oceanic anoxic events" as causal factors in development of reef-reservoired giant oil fields: AAPG Bull., v. 63, p. 870-885.

Beliaeva, A. N., and E. A. Romankevich, 1976, Chemical conversion of lipids during oceanic sedimentogenesis, in N. B. Vassoevich, ed., Issledovania Organisheskogo Veshchestva Sovremenyk Iskopaemyk Osadkov: Moscow, Adka. Nauk SSSR, p. 81-102 (in Russian).

Berggren, W. A., and C. D. Hollister, 1977, Plate tectonics and paleo-circulations—commotion in the ocean: Tectonophysics, v. 38, p. 11-48.

Berry, W. B. N., and P. Wilde, 1978, Progressive ventilation of the oceans: Am. Jour. Sci., v. 278, p. 257-275.

Bezrukov, P. L., et al, 1977, Organic carbon in the upper sedimentary layer of the world ocean: Oceanology, v. 17, no. 5, p. 561-564.

Bordovsky, O. K., 1965, Accumulation and transformation of organic substance in marine sediments: Marine Geology, v. 3, p. 3-114.

Brongersma-Sanders, M., 1966, The fertility of the sea and its bearing on the origin of oil: Advancement of Science, v. 23, p. 41-46.

——— 1972, Hydrological conditions leading to the development of bituminous sediments in the pre-evaporite phase, in Geology of saline deposits: Unesco Earth Sci. Ser., no. 7, p. 19-21.

Bubnov, V. A., 1966, The distribution pattern of minimum oxygen concentrations in the Atlantic: Oceanology, v. 6, p. 193-201.

Byers, C. W., and D. W. Larson, 1979, Paleoenvironments of Mowry Shale (Lower Cretaceous), western and central Wyoming: AAPG Bull., v. 63, p. 359-361.

Calvert, S. E., 1964, Factors affecting distribution of laminated diatomaceous sediments in Gulf of California, in Marine geology of the Gulf of California: AAPG Mem. 3, p. 311-330.

——— 1976, The mineralogy and geochemistry of nearshore sediments, in J. P. Riley et al, eds., Chemical oceanography, v. 6, 2d ed.: New York, Academic Press, p. 187-280.

——— and N. B. Price, 1971a, Upwelling and nutrient regeneration in the Benguela Current, October, 1968: Deep-Sea Research, v. 13, p. 505-523.

——— ——— 1971b, Recent sediments of the South-West African Shelf, in Atlantic continental margins: London, Inst. Geol. Sci., p. 175-185.

Clark, D. L., 1977, Climatic factors of the late Mesozoic and Cenozoic Arctic Ocean, in Polar Oceans: Calgary, Alberta, Arctic Inst. North America, p. 603-615.

Claypool, G. E., and I. R. Kaplan, 1974, The origin and distribution of methane in marine sediments, in I. R. Kaplan, ed., Natural gases in marine sediments: New York, Plenum Press, p. 99-140.

——— A. H. Love, and E. K. Maughan, 1978, Organic geochemistry, incipient metamorphism, and oil generation in black shale members of Phosphoria Formation, Western Interior United States: AAPG Bull., v. 62, p. 98-120.

Clementz, D. M., G. J. Demaison, and A. R. Daly, 1979, Well site geochemistry by programmed pyrolysis: Offshore Technology Conf., v. 1, p. 465-470.

Combaz, A., and R. Pelet, eds., 1978, Geochimie organique des sediments marins profonds, ORGON II, Atlantique–N.E. Bresil: Paris, Editions CNRS, 456 p.

Debyser, Y., R. Pelet, and M. Dastillung, 1977, Geochimie organique des sediments marins recents, Mer Noire, Baltique, Atlantique (Mauritanie), in Advances in organic geochemistry, 1975: Madrid, ENADIMSA, p. 289-320.

Degens, E. T., 1974, Cellular processes in Black Sea sediments, in The Black Sea: AAPG Mem. 20, p. 296-307.

—— and K. Mopper, 1976, Factors controlling the distribution and early diagenesis of organic material in marine sediments, in J. P. Riley et al, eds., Chemical oceanography, v. 6, 2d ed.: New York, Academic Press, p. 59-113.

—— and D. A. Ross, eds., 1974, The Black Sea—geology, chemistry and biology: AAPG Mem. 20, 633 p.

—— and P. Stoffers, 1976, Stratified waters as a key to the past: Nature, v. 263, p. 22-27.

—— R. P. Von Herzen, and H.-K. Wong, 1971, Lake Tanganyika; water chemistry, sediments, geological structure: Naturwissenschaften, v. 58, p. 224-291.

—— et al, 1973, Lake Kivu; structure, chemistry and biology of an East African rift lake: Geol. Rundschau, v. 62, p. 245-277.

—— et al, 1978, Varve chronology: estimated rates of sedimentation in the Black Sea deep basin: Initial Rept. Deep Sea Drilling Project, v. 42, pt. 2, p. 499-508.

Demaison, G. J., and M. Shibaoka, 1975, Contribution to the study of hydrocarbon generation from hydrogen-poor kerogens: 9th World Petroleum Cong., Tokyo, Proc., v. 2, p. 195-197.

Deuser, W. G., 1971, Organic-carbon budget of the Black Sea: Deep-Sea Research, v. 18, p. 995-1004.

—— 1975, Reducing environments, in J. P. Riley and R. Chester, eds., Chemical oceanography, v. 3: New York, Academic Press, p. 1-37.

Dickert, P. F., 1966, Neogene phosphatic facies in California: PhD thesis, Stanford Univ.

Didyk, B. M., et al, 1978, Organic geochemical indicators of paleoenvironmental conditions of sedimentation: Nature, v. 272, p. 216-222.

Dow, W. G., 1977, Kerogen studies and geological interpretations: Jour. Geochem. Exploration, v. 7, p. 79-99.

—— and D. B. Pearson, 1975, Organic matter in Gulf Coast sediments: 7th Offshore Technology Conf., Proc., v. 3, OTC 2343, p. 85-94.

Emery, K. O., and E. Uchupi, 1972, Western North Atlantic Ocean: topography, rocks, structures, water, life, and sediments: AAPG Mem. 17, 532 p.

Espitalié, J., et al, 1977, Source rock characterization method for petroleum exploration: Offshore Technology Conf. Proc., v. 3, OTC 2935, p. 439-444.

Fairbridge, R. W., ed., 1966, The encyclopedia of oceanography: New York, Van Nostrand-Reinhold, 1021 p.

Foree, E. G., and P. L. McCarty, 1970, Anaerobic decomposition of algae: Environ. Sci. Technology, v. 4, p. 842-849.

Gaskell, S. J., et al, 1975, The geochemistry of a recent marine sediment off northwest Africa. An assessment of source of input and early diagenesis: Deep-Sea Research, v. 22, p. 777-789.

Gerschanovich, D. E., V. V. Veber, and A. E. Konjukhov, 1976, Organic matter in the bottom sediments of the Peru region of the Pacific Ocean, in N. B. Vassoevich, ed., Issledovania Organisheskogo Sovremenyk Iskopaemyk Osadkov: Moscow, Akad. Nauk, SSSR, p. 121-128 (in Russian).

Gormly, J. R., and W. M. Sackett, 1977, Carbon isotope evidence for the maturation of marine lipids, in Advances in organic geochemistry, 1975: Madrid, ENADIMSA, p. 321-339.

Gorshkov, S. G., 1974, World ocean atlas, v. 1, Pacific Ocean: New York, Pergamon Press, 338 p. (in Russian).

—— 1979, World ocean atlas, v. 2, Atlantic and Indian Oceans: New York Pergamon Press, 350 p. (in Russian).

Grasshoff, K., 1975, The hydrochemistry of landlocked basins and fjords, in J. P. Riley and J. Skirrow, eds., Chemical oceanography, v. 2, 2d ed.: New York, Academic Press, p. 456-597.

Gross, G. M., et al, 1972, Distribution of organic carbon in surface sediment, northeast Pacific Ocean, in The Columbia River Estuary and adjacent ocean waters: Seattle, Univ. Washington Press, p. 254-264.

Hallam, A., 1975, Jurassic environments: Cambridge, U.K., Cambridge Univ. Press, 269 p.

Hartmann, M., et al, 1971: Oberflachensedimente im Persischen Gulf und Gulf von Oman: "Meteor" Forschungsergebnisse, Reihe C, No. 4, 76 p.

—— et al, 1976, Chemistry of late Quaternary sediments and their interstitial waters from the NW African continental margin: "Meteor" Forschungasergebnisse, Reihe C, No. 24, p. 1-67.

Heath, G. R., T. C. Moore, Jr., and J. P. Dauphin, 1977, Organic carbon in deep-sea sediments, in R. N. Andersen, ed., The fate of fossil fuel CO_2 in the oceans: New York, Plenum Press, p. 627-639.

Hirst, D. M., 1974, Geochemistry of sediments from eleven Black Sea cores, in The Black Sea: AAPG Mem. 20, p. 430-455.

Hollister, C. D., R. Flood, and I. W. McCave, 1978, Geological aspects of the benthic boundary layer: Oceanus, v. 21, p. 5-13.

Huc, A-Y., 1978, Geochimie organique des schistes bitumineux du Toarcian du Bassin de Paris: PhD thesis, Univ. Louis Pasteur, Strasbourg, France.

Hunt, J. M., 1979, Petroleum geochemistry and geology: San Francisco, W. H. Freeman and Co., 498 p.

Idyll, C. P., 1973, The anchovy crisis, reprinted 1977, in Ocean science; readings from Scientific American: San Francisco, W. H. Freeman and Co., p. 223-230.

Jenkins, O. P., ed., 1943, Geologic formations and economic development of the oil and gas fields of California: California Div. Mines Bull. 118, 778 p.

Jones, G. E., 1973, An ecological survey of open ocean and estuarine microbial populations. 1, The importance of trace metal ions to microorganisms in the sea, in L. H. Stevenson and R. R. Colwell, eds., Estuarine microbial ecology: Columbia, S.C., Univ. South Carolina Press, p. 233-241.

Kent, P. E., and H. R. Warman, 1972, An environmental review of the world's richest oil-bearing region; the Middle East: 24th Internat. Geol. Cong., Montreal, Proc., sec. 5, p. 142-152.

Klingebiel, A., 1976, Sediments et milieuz sedimentaires dans le Golfe du Benin: Centre Recherche Pau-SNPA, Bull. 10, p. 129-148.

Koblenz-Mishke, O. I., V. V. Volkonsky, and J. G. Kabanova, 1970, Plankton primary production of the world ocean, in W. S. Wooster, ed., Symposium on scientific exploration of the southern Pacific, Scripps Inst. Oceanog., 1968, 9th Mtg. SCOR: Washington, D.C., Natl. Acad. Sci., p. 183-193.

Konjukhov, A. E., 1976, Facies characteristics of contemporary sediments from the western submarine margin of the Indian Peninsula, in N. B. Vassoevich, ed., Issledovania Organisheskogo Veshchestva Sovremenyk Iskopaemyk Osadkov: Moscow, Akad. Nauk SSSR, p. 111-120 (in Russian).

Kontorovich, A. E., ed., 1971, Geochemistry of petroleum bearing Mesozoic formations in the Siberian basins: USSR Ministry of Geology Publication, SNIIG-IMS Bull. 118, 85 p. (in Russian).

Krumbein, W. C., and R. M. Garrels, 1952, Origin and classification of chemical sediments in terms of pH and oxidation-reduction potentials: Jour. Geology, v. 60, p. 1-33.

Lijmbach, G. W. M., 1975, On the origin of petroleum: 9th World Petroleum Cong., Tokyo, Proc., v. 2, p. 357-368.

Lisitzin, A. P., 1972, Sedimentation in the world ocean: SEPM Spec. Pub. 17, 218 p.

Logvinenko, N. V., and Ye A. Romankevich, 1973, Recent sediments of the Pacific Ocean on the coast of Peru and Chile: Lithology and Mineral Resources, v. 8, p. 1-11; also in Litol. Polez. Iskop., no. 1, p. 3-16, 1973 (in Russian).

Manheim, F. T., 1976, Interstitial waters of marine sediments, in J. P. Riley et al, eds., Chemical oceanography, v. 6, 2d ed.: New York, Academic Press, p. 115-186.

Momper, J. A., and J. A. Williams, 1979, Geochemical exploration in Powder River basin, northeastern Wyoming and southeastern Montana (abs.): AAPG Bull., v. 63, p. 497.

Morris, K. A., 1979, A classification of Jurassic marine shale sequences: an example from the Toarcian (Lower Jurassic) of Great Britain: Paleogeography, Paleoclimatology, Paleoecology, v. 26, p. 117-120.

Nissenbaum, A., M. J. Baedecker, and I. R. Kaplan, 1972, Dissolved organic matter from interstitial organic water of a reducing marine fjord, in Advances in organic geochemistry, 1971: Internat. Ser. Mon. Earth Sci., v. 33, p. 427-440.

Nixon, R. P., 1973, Oil source beds in Cretaceous Mowry Shale of Northwestern Interior United States: AAPG Bull., v. 57, p. 136-161.

Oremland, R. S., and B. F. Taylor, 1978, Sulfate reduction and methanogenesis in marine sediments: Geochim. et Cosmochim. Acta, v. 42, p. 209-214.

Orr, W. L., and A. G. Gaines, 1974, Observations on rate of sulfate reduction and organic matter oxidation in the bottom waters of an estuarine basin: the upper basin of the Pettaquamscutt River (Rhode Island), in Advances in organic geochemistry, 1974: Paris, Technip, p. 790-812.

Palfrey, M. K., and G. G. Day, 1968, Oceanography of Baffin Bay and Nares Strait in the summer of 1968: Washington, D.C., Coast Guard Rept. 373-16.

Parker, R. H., 1964, Zoogeography and ecology of macro-invertebrates of Gulf of California and continental slope of western Mexico, in Marine geology of the Gulf of California: AAPG Mem. 3, p. 331-376.

Parrish, J. T., K. S. Hansen, and A. M. Ziegler, 1979, Atmospheric circulation and upwelling in Paleozoic, with reference to petroleum source beds (abs.): AAPG Bull., v. 63, p. 507-508.

Pelet, R., and Y. Debyser, 1977, Organic geochemistry of Black Sea cores: Geochim. et Cosmochim. Acta, v. 41, p. 1575-1586.

Peng, T. H., et al, 1977, Benthic mixing in deep-sea cores as determined by C^{14} dating and its implication regarding climate, stratigraphy and the fate of fossil fuel CO_2, in N. R. Andersen, ed., The fate of fossil fuel CO_2 in the oceans: New York, Plenum Press, p. 355-373.

Phleger, F. B, 1964, Patterns of living benthonic Foraminifera, Gulf of California, in Marine geology of the Gulf of California: AAPG Mem. 3, p. 377-394.

Redfield, A. C., 1958, Preludes to the entrapment of organic matter in the sediments of Lake Maracaibo, in L. G. Weeks, ed., Habitat of Oil: AAPG, p. 968-981.

Reid, J. L., 1965, Intermediate waters of the Pacific Ocean: Baltimore, Johns Hopkins Univ. Press, 85 p.

Rhoads, D. C., 1974, Organism-sediment relationship on the muddy sea floor: Oceanogr. Marine Biology Ann. Rev., v. 12, p. 263-300.

Richards, F. A., 1970, The enhanced preservation of organic matter in anoxic marine environments, in Symposium on organic matter in natural waters: Univ. Alaska Inst. Marine Sci. Occasional Pub. 1, p. 399-411.

―― 1976, The Cariaco basin (trench): Oceanogr. Marine Biology Ann. Rev., v. 13, p. 11-67.

―― and R. F. Vaccaro, 1956, The Cariaco Trench, an anaerobic basin in the Caribbean Sea: Deep-Sea Res., v. 7, p. 163-172.

Roden, G. I., 1964, Oceanographic aspects of Gulf of California, in Marine geology of the Gulf of California: AAPG Mem. 3, p. 30-58.

Roehler, H. W., 1974, Depositional environment of rocks in the Piceance Creek basin, Colorado: Rocky Mtn. Assoc. Geologists Field Conf. Guidebook 25, p. 57-64.

Romankevich, E. A., 1977, Organic geochemistry of oceanic sediments: Moscow, Izd. Nauka, p. 13 (in Russian).

Rosenberg, R., 1977, Benthic macrofaunal dynamics, production and dispersion on an oxygen-deficient estuary of west Sweden: Jour. Exp. Marine Biology Ecology, v. 26, p. 107-133.

Ross, D. A., and E. T. Degens, 1974, Recent sediments of Black Sea, in The Black Sea: AAPG Mem. 20, p.

——— J. T. Parrish, and R. C. Humphreville, 1979, Paleogeography, upwelling and phosphorites (abs.), in P. J. Cook and J. H. Shergold, eds., Proterozoic-Cambrian phosphorites, Project 156 of UNESCO-IUGS: Canberra, Australian Natl. Univ. Press, p. 21. 183-199.

Ryan, W. B. F., and M. B. Cita, 1977, Ignorance concerning episodes of ocean-wide stagnation: Marine Geology, v. 23, p. 197-215.

Sackett, W. M., et al, 1978, A carbon inventory for Orca brines and sediments: Earth and Planetary Sci. Letters, v. 44, p. 73-81.

Saliot, A., 1977, Etude comparative des acides gras de l'eau et de l'eau interstitielle a l'interface eau de mer profonde/sediment: Acad. Sci. Comptes Rendus, ser. D-857, p. 285 (10 Oct. 1977).

Schaefer, W., 1972, Ecology and palaeoecology of marine environments: Chicago, Univ. Chicago Press, 568 p.

Schlanger, S. O., and H. C. Jenkyns, 1976, Cretaceous oceanic anoxic events: causes and consequences: Geologie en Mijnbouw, v. 55, p. 179-184.

Scranton, M. I., and J. W. Farrington, 1977, Methane production in the waters off Walvis Bay: Jour. Geophys. Research, v. 82, p. 4947-4953.

Shimkus, K. M., and E. S. Trimonis, 1974, Modern sedimentation in Black Sea, in The Black Sea: AAPG Mem. 20, p. 249-278.

Shokes, R. F., et al, 1977, Anoxic hypersaline basin in the northern Gulf of Mexico: Science, v. 196, p. 1443-1446.

Smith, J. W., and W. A. Robb, 1973, Aragonite and the genesis of carbonates in Mahogany zone oil shales of Colorado's Green River Formation: U.S. Bur. Mines Rept. 7727, 21 p.

Spencer, D. W., H. Susumu, and D. G. Brewer, 1978, Particles and particle fluxes in the ocean: Oceanus, v. 21, p. 20-25.

Stackelberg, U. V, 1972, Faziesverteilung in Sedimenten des indisch-pakistanischen Kontinentalrandes (Arabisches Meer): "Meteor" Forschungsergebnisse, Reihe C, No. 9, p. 1-173.

Stanley, D. J., 1978, Ionian Sea sapropel distribution and late Quaternary paleo-oceanography in the eastern Mediterranean: Nature, v. 274, p. 149-151.

Stevenson, F. J., and C. N. Cheng, 1972, Organic geochemistry of the Argentine basin sediments: carbon-nitrogen relationships and Quaternary correlations: Geochim. et Cosmochim. Acta, v. 36, p. 653-671.

Stommel, H., 1958, The circulation of the abyss, reprinted, 1977, in Ocean science, readings from Scientific American: San Francisco, W. H. Freeman and Co., p. 139-144.

Summerhayes, C. P., U. de Melo, and H. T. Baretto, 1976, The influence of upwelling on suspended matter and shelf sediments off southeastern Brazil: Jour. Sed. Petrology, v. 46, p. 819-828.

Swain, F. M., 1970, Non-marine organic geochemistry: Cambridge, U.K., Cambridge Univ. Press, 445 p.

Swanson, V. E., 1960, Oil yield and uranium content of black shales: U.S. Geol. Survey Prof. Paper 356-A, p. 1-44.

Talling, J. F., 1963, Origin of stratification in an African rift lake: Limnology and Oceanography, v. 8, p. 68-78.

Theede, H., et al, 1969, Studies on the resistance of marine bottom invertebrates to oxygen deficiency and hydrogen sulfide: Mar. Biology, v. 2, p. 325-337.

Thiede, J., and T. H. van Andel, 1977, The paleoenvironment of anaerobic sediments in the late Mesozoic South Atlantic Ocean: Earth and Planetary Sci. Letters, v. 33, p. 301-309.

Tissot, B. P., 1979, Effects on prolific petroleum source rocks and major coal deposits caused by sea-level changes: Nature, v. 277, p. 377-380.

——— and D. H. Welte, 1978, Petroleum formation and occurrence: New York, Springer-Verlag, 521 p.

——— G. Deroo, and A. Hood, 1978, Geochemical study of the Uinta basin; formation of petroleum from the Green River Formation: Geochim. et Cosmochim. Acta, v. 42, p. 1469-1486.

——— et al, 1974, Influence of nature and diagenesis of organic matter in formation of petroleum: AAPG Bull., v. 58, p. 499-506.

——— et al, 1979, Paleoenvironment and petroleum potential of mid-Cretaceous black shales in Atlantic basins (abs.): AAPG Bull., v. 63, p. 542.

Turekian, K. K., J. K. Cochran, and D. J. De Master, 1978, Bioturbation in deep-sea deposits: rates and consequences: Oceanus, v 21, p. 34-41.

Tyson, R. V., R. C. L. Wilson, and C. Downie, 1979, A stratified water column environmental model for the type Kimmeridge clay: Nature, v. 277, p. 377-380.

Udintsev, Gb., ed., 1975, Geological-geophysical atlas of the Indian Ocean: New York, Pergamon Press, 151 p.

Usher, J. L., and P. Supko, 1978, Initial reports of the Deep Sea Drilling Project, v. 42, pt. 2, 1243 p.

Vail, P. R., and R. M. Mitchum, 1979, Global cycles of sea level change and their role in exploration: 10th World Petroleum Cong., Preprint, Bucharest Panel Discussion 2, Paper 4.

van Andel, T. H., 1964, Recent marine sediments of Gulf of California, in Marine geology of Gulf of California: AAPG Mem. 3, p. 216-310.

Veber, V. V., 1977, Generation of hydrocarbon gases in modern sedimentary formation of the eastern and southern parts of the Atlantic Ocean: Geologyia Nefti i Gaza, no. 1, p. 59-64.

Veeh, H. H., 1979, Modern environment of phosphorite formation and the geochemical balance of phosphorus in the ocean, in P. J. Cook and J. H. Shergold, eds., Proterozoic-Cambrian phosphorites: Project 156 of UNESCO-IUGS: Canberra, Australian Natl. Univ. Press, p. 59.

——— W. C. Burnett, and A. Soutar, 1973, Contemporary phosphorites on the continental margin of Peru: Science, v. 181, p. 844-845.

——— S. E. Calvert, and N. B. Price, 1974, Accumulation of uranium in sediments and phosphorites on the southwest African shelf: Marine Chemistry, v. 2, p. 189-202.

Wolfe, R. S., 1971, Microbial formation of methane: Adv. Microbial Physiology, v. 6, p. 107-146.

Wyrtki, K., 1962, The oxygen minimum in relation to ocean circulations: Deep-Sea Research, v. 9, p. 11-23.

Ziegler, P. A., 1979, Factors controlling North Sea hydrocarbon accumulations: World Oil, v. 189, no. 6, p. 111-124.

Generation, Accumulation, and Resource Potential of Biogenic Gas[1]

DUDLEY D. RICE and GEORGE E. CLAYPOOL[2]

ABSTRACT

Biogenic gas is generated at low temperatures by decomposition of organic matter by anaerobic microorganisms. More than 20% of the world's discovered gas reserves are of biogenic origin. A higher percentage of gases of predominantly biogenic origin will be discovered in the future. Biogenic gas is an important target for exploration because it occurs in geologically predictable circumstances and in areally widespread, large quantities at shallow depths.

In rapidly accumulating marine sediments, a succession of microbial ecosystems leads to the generation of biogenic gas. After oxygen is consumed by aerobic respiration, sulfate reduction becomes the dominant form of respiration. Methane generation and accumulation become dominant only after sulfate in sediment pore water is depleted. The most important mechanism of methane generation in marine sediments is the reduction of CO_2 by hydrogen (electrons) produced by the anaerobic oxidation of organic matter. CO_2 is the product of either metabolic decarboxylation or chemical decarboxylation at slightly higher temperatures. The factors that control the level of methane production after sediment burial are anoxic environment, sulfate-deficient environment, low temperature, availability of organic matter, and sufficient space. The timing of these factors is such that most biogenic gas is generated prior to burial depths of 1,000 m.

In marine sediments, most of the biogenic gas formed can be retained in solution in the interstitial (pore) waters because of higher methane solubility at the higher hydrostatic pressures due to the weight of the overlying water column. Under certain conditions of high pressures and (or) low temperatures, biogenic methane combines with water to form gas hydrates.

Biogenic gas usually can be distinguished from thermogenic gas by chemical and isotopic analyses. The hydrocarbon fraction of biogenic gas consists predominantly of methane. The presence of as much as 2% of heavier hydrocarbons can be attributed to admixture of minor thermogenic gas due to low-temperature degradation of organic matter. The amounts of hydrocarbon components other than methane generally are proportional to temperature, age, and organic-matter content of the sediments. Biogenic methane is enriched in the light isotope ^{12}C ($\delta^{13}C_1$ lighter than -55 ppt) owing to kinetic isotope fractionation by methanogens. The variations in isotopic composition of biogenic methane are controlled primarily by $\delta^{13}C$ of the original CO_2 substrate, which reflects the net isotopic effect of both addition and removal of CO_2. The methane isotopic composition also can be affected by mixing of isotopically heavier thermogenic gas. The possible complicating factors require that geologic, chemical, and isotopic evidence be considered in attempts to interpret the origin of gas accumulations.

Accumulations of biogenic gas have been discovered in Canada, Germany, Italy, Japan, Trinidad, the United States, and USSR in Cretaceous and younger rocks, at less than 3,350 m of burial, and in marine and nonmarine rocks. Other gas accumulations of biogenic origin have undoubtedly been discovered; however, data that permit their recognition are not available.

INTRODUCTION

For effective exploration, a better understanding of processes that lead to the origin, migration, and accumulation of hydrocarbons is necessary. Geologic and geochemical evidence indicates that hydrocarbons are generated from organic matter disseminated in fine-grained sedimentary rocks by a series of complex chemical reactions. The extent of nonbiologic reactions is controlled primarily by temperature and duration of heating (geologic time). The quantity and molecular size of hydrocarbons (liquid or gaseous) generated are influenced by the concentration and type of organic matter preserved in the source rock, and by the stage of thermochemical alteration (maturity).

The generation and occurrence of hydrocarbons can be related to three main stages of thermal maturity of organic matter in sedimentary rocks (Fig. 1):

1. *Immature stage (diagenesis)*—Biologic activity and chemical rearrangement are responsible for converting organic matter to kerogen, an insoluble residue, which is the source of most hydrocarbons. Although small

---

© Copyright 1981. The American Association of Petroleum Geologists. All rights reserved.

[1]Manuscript received, February 15, 1980; accepted, May 19, 1980.
[2]U.S. Geological Survey, Denver, Colorado 80225.

We thank Charles N. Threlkeld of U.S. Geological Survey for chemical and isotopic analyses of natural gases; Branch of Helium Resources of Bureau of Mines, Research Center of Amoco Production Co., and Chevron Oil Field Research Co. for analytical data; and many oil companies for assistance in collecting gas samples. The U.S. Department of Energy funded part of this study.

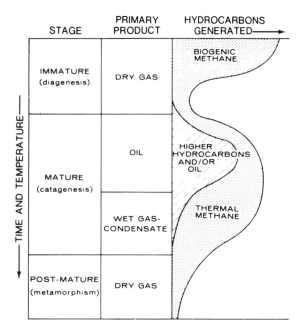

FIG. 1—Diagram showing generation of hydrocarbons with increasing temperature and time.

amounts of hydrocarbons are present which are inherited from or produced by mild degradation of organic tissues, biogenic methane is the only hydrocarbon actually generated in significant volumes during this stage.

2. *Mature stage (catagenesis)*—With increasing temperature and advancing geologic time, the full range of hydrocarbons is produced from kerogen and other nonhydrocarbon precursors by thermal degradation and cracking reactions. Depending on the concentration and type of organic matter, oil generation occurs during this stage accompanied by the production of significant amounts of natural gas. As temperatures increase, lighter hydrocarbons are formed preferentially, owing to breaking of carbon-carbon bonds, affecting both the residual kerogen and the hydrocarbons previously formed. Wet gas and condensate are the primary products of the later part of the stage.

3. *Postmature stage (incipient metamorphism)*—At the end of the mature stage, the kerogen becomes highly polymerized, condensed in structure, and chemically stable. The main hydrocarbon generated is methane which results from the cracking of the existing hydrocarbons. Hydrocarbons larger than methane are destroyed much more rapidly than they are formed.

This paper discusses the methane-rich gas generated in accumulating sediments during the immature stage (diagenesis) by the metabolic activity of anaerobic bacteria. Hereafter, this gas will be referred to as biogenic gas to emphasize that microorganisms are responsible for its formation. We emphasize that biogenic gas not only is generated, but has accumulated in economically significant quantities and warrants consideration in future exploration. Our objectives are to: (1) describe the conditions required for generation of biogenic gas with an emphasis on marine sediments because many ancient accumulations were generated in this environment; (2) discuss the factors favoring the accumulation of biogenic gas; (3) characterize biogenic gas and distinguish it from natural gas of the other stages of hydrocarbon generation; (4) document major ancient occurrences; and (5) estimate potential resources of biogenic gas.

ORIGIN OF BIOGENIC GAS

Biogenic gas is produced during the decomposition of organic matter by microorganisms. Under present-day conditions, biogenic methane formation is controlled by certain physiologic and ecologic restraints. First, methane-producing microorganisms are strict anaerobes and cannot tolerate even traces of oxygen. Second, biogenic methane does not accumulate in significant amounts in the presence of high concentrations of dissolved sulfate. These restrictions confine the activity of methane-producing microorganisms to certain environments such as: dung heaps and anaerobic-sewage digestors (Toerien and Hattingh, 1969); digestive tracts of animals (Bryant, 1965); poorly drained swamps, bays, paddy fields, and anoxic freshwater lake bottoms (Oana and Deevey, 1960; Koyama, 1963; Kim and Douglas, 1972); landfills (Colonna, 1977); glacial drift (Coleman, 1976; Coleman et al, 1977); and marine sediments beneath the zone of active sulfate reduction (Emery and Hoggan, 1958; Atkinson and Richards, 1967; Nissenbaum et al, 1972).

Methane production is readily observable in bogs, swamps, and marshes because it occurs close to the surface and bubbles are released from the sediment. In marine sediments, sulfate reduction is the dominant process of anaerobic bacterial respiration at shallow depths of burial (ZoBell and Rittenberg, 1948). Evidence for significant methane production is usually not detected until dissolved sulfate is almost completely removed from the interstitial water (Nissenbaum et al, 1972). In many recent marine environments, complete removal of sulfate does not occur until sediments have been buried to depths of some tens of meters (Sayles et al, 1973; Waterman et al, 1973; Manheim and Sayles, 1974). Consequently, the geochemical effects of methanogenesis in open marine sediments have been difficult to observe. However, with the advent of deep coring procedures by the Deep Sea Drilling Project (DSDP), information on the nature and distribution of methane generation in marine sediments has become available.

MICROBIAL METABOLISM IN MARINE SEDIMENTS

Microorganisms require energy for growth and cell maintenance which is obtained by metabolizing organic matter by a series of coupled oxidation-reduction reactions. These metabolic processes result in the production of gases including methane. Two general types of metabolic processing by microorganisms in the marine environment are (1) respiration (both aerobic and anaerobic) which utilizes inorganic compounds as electron acceptors, and (2) fermentation in which electron transfer occurs within or between organic compounds.

The metabolic processes yield differing amounts of energy which, along with the environmental constraints, determine the nature of the microbial population. Relative energy yields indicate selective advantage gained by microorganisms capable of catalyzing the various processes. This advantage results in the dominance of certain microbial-population members which obtain the greatest amount of energy when two or more are competing in the same ecologic niche for the same organic substrate. Consequently, an ecologic succession of metabolic processes is established and varies with time and depth in the sediment column, with less efficient organisms being at greater depths. A more detailed account of the succession of diagenetic environments in marine sediments was given by Claypool and Kaplan (1974).

Figure 2 is a cross section through an open-marine, organic-rich sedimentary environment and illustrates the succession of microbial ecosystems. The interactions between sedimentologic and ecologic factors result in three distinct biochemical environments, each of which is characterized by a dominant form of respiratory metabolism. The three resulting zones are: the aerobic zone; the anaerobic sulfate-reducing zone; and the anaerobic carbonate-reducing (methane production) zone. The presence of these zones, which are characterized by successively less efficient modes of respiratory processes, can be inferred in the sediments and interstitial waters by systematic changes in the concentration and isotopic composition of the respiratory metabolites. In each of the zones, the dominant microbial population exploits the environment and eventually creates a new environment that favors a different population. Therefore, the transitions between the various zones are a geochemical consequence of environmental changes induced by microorganisms.

When the ecologic succession in Figure 2 is established, the biochemical zones move upward with time, keeping pace with the addition of new sediments at the sediment-water interface. Likewise, the sediments move downward through the succession of diagenetic environments.

Aerobic respiration using organic matter is the most efficient energy-yielding metabolic process. The aerobic zone in the marine environment is normally developed in the water column and uppermost part of the sediment column. During aerobic respiration, oxygen is rapidly used and the demand often exceeds the rate at which the dissolved gas can be introduced from the atmosphere or overlying water column, particularly in areas of high sedimentation rates. When the oxygen is depleted, obligatory aerobic organisms cannot grow. Facultative anaerobes can switch from aerobic respiration to fermentation or anaerobic respiration using electron acceptors, but these are not abundant in marine sediments and have negligible effect on decomposition of organic matter.

In the marine environment, sulfate reduction becomes the dominant form of respiration after the onset of anaerobic conditions because of the relatively high concentration of sulfate (0.028 M) in normal sea water. In addition, only a few other microbial species can tolerate H_2S which is the end product of sulfate reduction. Sulfate-reducing bacteria are restricted in their range of oxidizable substrates (Postgate, 1965) and these compounds are probably present in limited amounts at a given time. Therefore, active sulfate-reducing bacteria require a symbiotic association with an anaerobic fermenting population to provide a source of oxidizable carbon substrate.

Below the sulfate-reduction zone, CO_2 reduction becomes the dominant process of anaerobic respiration and results in the formation of methane. The extent to which sulfate reduction and methanogenesis are mutually exclusive in marine sediments is currently under investigation (Martens and Berner, 1974, 1977; Oremland and Taylor, 1978; Kosiur and Warford, 1979). Only minor concentrations of methane are present in marine sediments containing dissolved sulfate (Whelan et al, 1975, 1978; Bernard et al, 1978), and rapid methane production appears to begin immediately after (or below the depth where) dissolved sulfate reaches low concentration and sulfate reduction is essentially complete (Nissenbaum et al, 1972; Claypool et al, 1973; Claypool, 1974; Claypool and Kaplan, 1974). However,

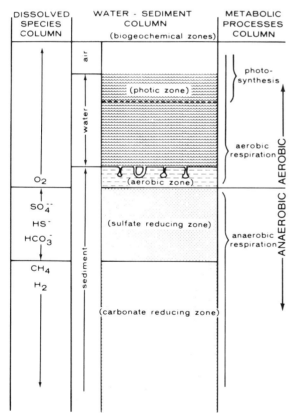

FIG. 2—Diagrammatic cross section of organic-rich, open-marine environment showing succession of microbial ecosystems that lead to methane generation.

several studies suggest that methane is consumed in the overlying sulfate-reducing zone (Barnes and Goldberg, 1976; Reeburgh, 1976; Martens and Berner, 1977; Reeburgh and Heggie, 1977; Kosiur and Warford, 1979). Thus it is likely that the balance between production and consumption of methane is an important factor controlling the distribution of methane in marine sediments (Bernard, 1979).

Methane-producing microorganisms also are even more restricted than sulfate reducers with respect to substrates utilizable for growth. Hydrogen and CO_2 are the preferred substrates studied in pure culture, whereas formate, methanol, and acetate can be used, but the reactions are probably much less energy-efficient (Wolfe, 1971; Zeikus and Wolfe, 1972; Winfrey et al, 1977; Belyaev and Laurinavichus, 1978). CO_2 reduction is energetically the most favorable mechanism of biogenic methane production and probably accounts for most of the methane produced in the marine environment, although other pathways are available.

ISOTOPIC INDICATORS

The progression of metabolic processes that results in methane production in the marine environment can be followed by monitoring the changes in the concentration and stable isotope ratios of the residual reactants or the accumulating products with increasing depth. Where the principle of "steady-state diagenesis" (Berner, 1975) is applicable, the changes with increasing depth are equivalent to the changes with respect to advancing time at the same depth.

The effects of in-situ sulfate reduction by bacteria can be traced in an example from sediments of DSDP hole 148 in the Caribbean Sea (Presley et al, 1973; Claypool, 1974). As illustrated in Figure 3, the sulfate dissolved in the interstitial water is depleted with increasing depth. At the same time, the ^{34}S is relatively enriched in the residual sulfate at any depth as a result of preferential ^{32}S removal in the sulfide with increasing depth.

Sampling techniques commonly have not permitted measurement of the concentration of methane in sediments. Where such measurements are not possible, the effects of methane generation can be examined by the changes in the concentration of $\delta^{13}C$ of bicarbonate dissolved in the interstitial water. Figure 4 is an example of the effects of methane production on the concentration and stable isotope ratio of dissolved bicarbonate from the Gulf of California (Goldhaber, 1974). In these sediments, sulfate concentration goes to zero at a depth of about 2 m. The onset of methane production at this point is reflected in decreased titration alkalinity and increased $\delta^{13}C$ of total dissolved CO_2. However, the interpretation of changes with depth in the concentration and $\delta^{13}C$ of the dissolved bicarbonate due to methane formation is more complex than with sulfate. The complexity results from dissolved bicarbonate at any depth reflecting the net of both addition and removal of CO_2. Metabolic CO_2 added to the interstitial water of anoxic sediments has about the same $\delta^{13}C$ as organic matter, or about -22 ppt (Presley and Kaplan, 1968). The CO_2 removed to form methane is about 70 ppt lighter than the $\delta^{13}C$ of the CO_2 dissolved in the interstitial water at the time of its formation, owing to a large kinetic isotope effect (Rosenfeld and Silverman, 1959).

Possible isotope effects of CO_2 addition and removal during biogenic methane production are illustrated in

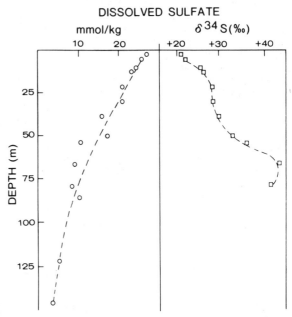

FIG. 3—Changes with depth in concentration and sulfur isotope ratio ($\delta^{34}S$) of dissolved sulfate in interstitial water of Aves Ridge (DSDP Site 148) sediments, caused by bacterial sulfate reduction.

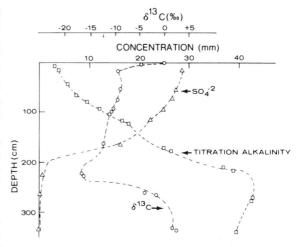

FIG. 4—Changes with depth in concentration of sulfate and titration alkalinity and $\delta^{13}C$ of total dissolved CO_2 in interstitial water of South Guymas basin sediments (Goldhaber, 1974). Shallowest occurrence of methane in these sediments is at depth of 2 m where trends in plotted pore-water constituents reverse.

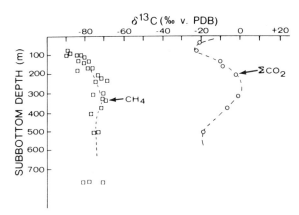

FIG. 5—Parallel changes with depth in $\delta^{13}C$ of dissolved carbonate and methane in sediments of Astoria fan (DSDP Site 174A).

Figure 5 for sediments of the Astoria fan at the mouth of the Columbia River from DSDP hole 174A (Claypool and Kaplan, 1974). Trends of $\delta^{13}C$ of dissolved carbonate and methane are shown with depth. In these sediments, sulfate is present to a depth of about 80 m (Waterman et al, 1973). Just below this depth, biogenic methane with $\delta^{13}C$ of −90 ppt is present. At the onset of methane formation, the concentration of dissolved CO_2 in the interstitial water is relatively high (0.022 M) and has $\delta^{13}C$ of −22 ppt reflecting input of metabolic CO_2 as a result of sulfate reduction. With net removal of isotopically light CO_2 to form methane, the concentration of CO_2 decreases and the $\delta^{13}C$ of the residual CO_2 increases. In such circumstances, methane also becomes isotopically heavier with increasing depth of burial reflecting ^{12}C depletion in the residual dissolved CO_2 from which the methane is being produced. This complementary relation between dissolved bicarbonate and methane is the principal evidence that microbiologic methane production in marine sediments proceeds mainly by the mechanism of CO_2 reduction. These changes also indicate that microbial methane generation can continue for 2 m.y. to depths of 500 m or more in the Astoria fan and probably longer and deeper elsewhere.

In the Astoria fan, the net addition of isotopically light CO_2 causes the $\delta^{13}C$ of both the methane and dissolved CO_2 to shift to more negative values (isotopically lighter) below depths of about 300 m (Fig. 5). Profiles of $\delta^{13}C$ for other sites also reflect a similar trend in which the heaviest values are obtained at intermediate depths with a trend toward lighter values at greater depths (Claypool and Kaplan, 1974). We interpret this to mean that isotopically light CO_2 was being added faster than its removal by methane production. Increases in the concentration of CO_2 over the depth interval in which this trend occurs support this conclusion. Although the early increase in CO_2 content is related to sulfate reduction, this later stage of CO_2 production probably represents a different process. CO_2 production after sulfate reduction has stopped appears to be controlled by the thermal history and age of the sediment as supported by first-order kinetic models (Claypool, 1974). This indicates that low-temperature nonbiologic decarboxylation of organic matter (Johns and Shimoyama, 1972; Tissot et al, 1974; Bray and Foster, 1980) is the most likely mechanism. This later diagenetic CO_2 is isotopically light ($\delta^{13}C$ of −20 to −25 ppt) and serves as a source for continued methane generation as indicated by the shift toward lighter $\delta^{13}C$ values of methane in deeper samples.

REQUIREMENTS FOR METHANE PRODUCTION

Production of biogenic methane is controlled by several critical factors. These requirements must be met not only to commence generation, but also to maintain production over a period of time so that large quantities can accumulate.

1. *Anoxic environment*—Methanogenic microorganisms are obligate anaerobes and cannot tolerate even traces of oxygen.

2. *Sulfate-deficient environment*—In environments where the sulfate concentration of the water is low, such as brackish or fresh water, methane production begins immediately after the oxygen is depleted. However, in the marine environment, sulfate must be reduced almost completely before significant amounts of methane can accumulate.

3. *Temperature*—Methane generation by microorganisms can occur at depths that are equivalent to temperatures between 0 and 75°C (Buswell and Mueller, 1952; Zeikus and Wolfe, 1972). Although methanogens can function over a wide temperature range, the optimum for a specific population generally is confined to several degrees. For example, the optimum temperature for growth of *Methanobacterium thermoautotrophicus*, an extreme thermophile, is 65 to 70°C. In shallow aquatic environments of northern latitudes, seasonal variations in temperature generally confine the maximum methane production to summer months (Koyama, 1963; Barber, 1974; Zeikus and Winfrey, 1976; King and Wiebe, 1978). However, in the deeper marine waters, temperatures and therefore methane-generating rates should remain more constant throughout the year. At the upper end, increasing temperatures lead to denaturation of bacterial tissues and transformation of organic matter to a form (humin/kerogen) which is less susceptible to attack by microorganisms.

4. *Presence of organic matter*—Organic matter is required for methanogenesis and the various metabolic processes that precede and support it. A minimum of metabolizable organic matter equivalent to about 0.5% organic carbon is required to support methane production in marine sediments (Claypool and Kaplan, 1974; Rashid and Vilks, 1977). Organic matter is generally concentrated in finer grained sediments (Hunt, 1972).

5. *Space*—A minimal amount of space is required for bacteria to function, particularly in fine-grained sediments where the organic nutrients are concentrated. Typical shale pores have a median size of 1 to 3 nm (Momper, 1978). However, bacteria have an average size of 1 to 10 μm (Momper, 1978) indicating that they

are unable to function in a compacted shale. However, space would not be a limiting factor for bacterial activity until advanced stages of dewatering and compaction of the muds are achieved at depths of burial of about 2,000 m (Welte, 1972).

These requirements control the timing and depth of methane generation in marine sediments. Physiologic, ecologic, and geologic factors constrain significant levels of biologic methane production to the diagenetic sequence shown in Figure 2. This diagenetic sequence may be incompletely developed, and can be expressed on a depth scale that ranges from as little as a few centimeters to as much as a few hundred meters, depending on relative rates of biologic processes and sediment accumulation. The greatest proportion of biogenic gas probably is generated to burial depths of 1,000 m.

PHYSICAL STATES OF BIOGENIC METHANE IN MARINE SEDIMENTS

The estimated solubility of methane in interstitial water in marine sediments is shown in Figure 6 as a function of depth. The solubility data are taken from Culbertson and McKetta (1951) and have been reduced to 80% of the observed solubility of methane in pure water at a given temperature and depth to adjust for effects of 3.5% salinity. Other factors such as unusually high concentrations of CO_2 may also affect solubility (Bray and Foster, 1980). The diagram shows a series of curves indicating the changes of solubility at different pressure-temperature gradients associated with overlying water depths of 0 to 4,000 m. The pressure gradient is assumed to be hydrostatic (0.1 atm/m) and a geothermal gradient of 0.035° C/m is used. The temperature of sediments with no burial is assumed to be 20°C for a water depth of zero, 5°C for water depths of 500 m, and 2°C for water depths of 1,000 m and greater. In general, methane solubility increases with increasing depth of burial in depths of water less than 1,000 m. However, for deep-sea sediments (greater than 1,000 m) the solubility initially decreases between 1,000 and 2,000 m before increasing at greater depths. This reversal in the trend of methane solubility with increasing depth of burial is due to minimum methane solubility occurring at 82°C. For water depths of less than 800 m, the increase in pressure due to deeper burial is a more significant fraction of the total pressure, and the pressure increase offsets the temperature decrease in methane due to deeper burial.

Figure 6 also shows the conditions under which gas hydrates are stable. Under conditions of high pressure and (or) low temperature, water in a saturated solution of methane can be solidified at temperatures above the freezing point by incorporation of a methane molecule into a water clathrate cage (Hand et al, 1974; Hitchon, 1974; Miller, 1974). Reliable measurements of methane solubility in water in equilibrium with methane hydrate have not been made. The previously reported abrupt decrease in methane solubility in the pressure-temperature region of hydrate stability (Makogon et al, 1972; Claypool and Kaplan, 1974) has been questioned on the grounds that such behavior violates thermodynamic principles (Milton, 1977). Therefore, the dotted lines in Figure 6 are extrapolated from measurements taken outside the region of gas hydrate stability (Culbertson and McKetta, 1951).

Variation of methane solubility under differing conditions of temperature and pressure is important for the accumulation of biogenic methane as will be discussed later. The solubility estimates given in Figure 6 are minimum values and could be increased by less saline waters or higher pressure gradients.

CONDITIONS FAVORING CONCENTRATION AND ACCUMULATION OF BIOGENIC METHANE

The accumulation of biogenic gas depends upon meeting certain physicochemical and sedimentologic conditions for concentration and entrapment.

In nonmarine and (or) brackish-water environments that are generally low in sulfate, methane production begins close to the surface and most of the initially formed gas is lost by aerobic bacterial oxidation or escape to the atmosphere because of low hydrostatic pressures. In marine sediments, methane production begins at a depth below the sediment-water interface because sulfate reduction must precede it. In carbonate-evaporite sequences, biogenic methane production may be inhibited by sulfate. Because of higher hydrostatic pressures due to the overlying water column (Fig. 6), a larger amount of biogenic methane in marine sediments can be retained in solution in the interstitial (pore) waters. This can serve as a holding mechanism until the sediments are compacted, and traps and seals are formed. If conditions are such that methane generation results in

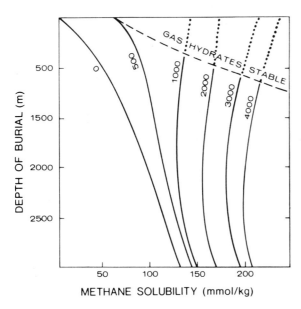

FIG. 6—Estimated methane solubility with depth of burial in marine sediments. Curves are for specified water depths as labeled in meters.

saturation of the waters above the 82°C minimum solubility isotherm, the waters will become supersaturated with further burial and will produce a free gas state. At very shallow depths of burial, free gas will probably bubble to the surface and eventually will enter the atmosphere. This is a situation similar to nonmarine and (or) brackish-water environments.

A free gas phase in marine sediments leading to the accumulation of large quantities of biogenic methane can result from either generation of gas in excess of the solubility, or exsolution of gas brought about by a reduction of hydrostatic pressure. Gas generation in excess of solubility commonly is observed in shallow water sediments. Exsolution by reduction of hydrostatic pressure also may be a frequent cause of a free gas phase available for migration and accumulation, and could result from lowering of sea level, uplift and erosion, or upward migration of gas-bearing waters to zones of lower hydrostatic pressures.

A reservoir, seal, and trap must be present prior to release of gas from solution to insure its retention. These conditions probably are not met until sediments have been buried to depths greater than 500 m. At these depths, the sediments are at least partly dewatered, compacted, and the initial porosity and permeability reduced drastically. A large amount of biogenic methane is probably trapped stratigraphically, at least initially, in rocks of low porosity and permeability. At shallow depths, porous and permeable reservoirs are often flushed with fresh water, and natural gas accumulates only in closed structural traps.

In the marine-shelf environment, the deposition of discontinuous and relatively impermeable silt and sand enveloped by organic-rich mud and clay provides extremely favorable conditions for both in-situ generation and entrapment of biogenic methane. In the Cretaceous section of the northern Great Plains, bentonite beds, which are composed of montmorillonitic clays derived from volcanic ash, are numerous and extensive in marine-shelf sequences and are excellent seals for the widespread gas accumulations.

Another key trapping mechanism of biogenic methane may be the formation of early diagenetic carbonate cements as either layers or concretions; obviously the layers are more effective traps.

Early diagenetic carbonates can be formed in two ways. First, when CO_2 is removed from the dissolved bicarbonate reservoir of interstitial waters by reduction and formation of methane, the pH increases which can result in the precipitation of authigenic carbonates in anoxic sediments. Examples for this type of carbonate from deep-sea sediments (Claypool, 1974) are the Miocene Monterey Shale (Friedman and Murata, 1979), the Upper Cretaceous of Montana (D. L. Gautier, personal commun., 1979), and the Upper Jurassic of England (Irwin et al, 1977).

A second major mechanism for precipitation of isotopically anomalous (-30 ppt) carbonate layers is from CO_2 generated from oxidation of methane (Hathaway and Degens, 1969; Deuser, 1970; Roberts and Whelan, 1975). This process can result from periodic exsolution of methane associated with marine shoreline regressions, escape of the methane upward, and oxidation of methane to CO_2 under aerobic conditions (Hutton and ZoBell, 1949). The isotopically light CO_2 from the oxidation of methane first must be neutralized by reaction with the sediment to form bicarbonate, which is then available for precipitation of carbonate cement.

Under certain conditions of high pressure and low temperature, such as those of deep-sea sediments or areas of permafrost, biogenic methane will combine with water to form hydrates. The hydrates can serve both as a trap for the methane enclosed in the clathrate structure, and also as a seal for hydrocarbons, including biogenic methane generated below the hydrate, or liberated at the base of a gas hydrate zone subsiding into a region of temperature instability. Gas hydrate reservoirs have the potential of trapping greater volumes of gas than free gas trapped in the same space. However, current technology does not permit economic production of the hydrated gas.

During the Pleistocene, large areas of the world were covered with permafrost, and gas hydrates were more widespread because of generally cooler temperatures. Since that time, the areas of permafrost have been retreating and associated hydrates have been decomposing which has resulted in the formation of free gas accumulations. Gas fields in Siberia were cited by Makogon et al (1972) as examples of gas hydrate-involved accumulations.

Finally, most recognizable biogenic gas accumulations have low reservoir pressures because of their shallow depth of burial. However, they are also underpressured in relation to normal hydrostatic pressure gradients. This underpressuring, which is probably related to the removal of overburden and thus may coincide with exsolution of the gas, helps to trap the gas. The subnormal pressures have probably resulted from dilation of pore volume and from a decrease of reservoir temperature associated with uplift and erosion (Barker, 1972; Dickey and Cox, 1977).

CHARACTERISTICS OF BIOGENIC GAS

Gaseous hydrocarbons produced during the three main stages of thermal maturity have distinct chemical compositions and stable carbon isotope ratios (Fig. 7) that enable one, in theory, to distinguish between the products of each stage. However, in most examples, complicating factors require that geologic evidence must also be considered in interpreting the origin of natural gas occurrences.

Biogenic gas consists predominantly of methane, except in nonmarine and (or) brackish-water (low pH) environments where biologically formed CO_2 also may be a major component. Davis and Squires (1954) and Kim and Douglas (1972) demonstrated that trace amounts of the higher alkanes (ethane, propane, butane, and pentane) were generated in laboratory-conducted fermentation studies. Only traces being detected illustrates the usual occurrence of higher hydrocarbons in trace amounts with biogenic methane in recent sediments (Weber and Maximov, 1976).

Accumulations of natural gas of predominantly biogenic origin are usually associated with even greater concentrations of heavier hydrocarbons than are present in laboratory fermentation experiments. Their presence probably indicates a contribution of early thermogenic gas.

Analyses of gas samples from the DSDP indicate that gas of predominantly biogenic origin contains higher hydrocarbons in amounts which are directly proportional to: (1) temperature history, (2) age of the sediments, and (3) organic-matter content of the sediments from which the gas originates (Claypool, 1974; Doose et al, 1978; Whelan and Hunt, 1978). These relations suggest that the higher hydrocarbons are generated in situ by low-temperature degradation of the organic matter. In Figure 8, the ratio of ethane to methane is plotted as a function of depth of burial for a variety of depositional settings. Generally, there is an exponential increase in ethane content with increasing depth of burial. At any given locality, the amount of ethane is proportional to the temperature and age of the sediment. In the Cariaco basin (Fig. 8), the approximate 10-fold increase in the ethane concentration is due to a similar 10-fold enrichment in the organic-matter content as compared with typical sediments (Claypool, 1974). This exponential increase in ethane content with increasing depth of burial continues to depths where gas with composition typical of thermogenic gas associated with liquid petroleum should be generated.

Thus, although methane is the primary hydrocarbon product of anoxic microbial breakdown of organic matter, heavier hydrocarbons amounting to as much as 2% can be expected to be associated with the methane because of early low-temperature degradation. With the cessation of methanogenesis by microorganisms and increasing levels of maturation, the full range of hydrocarbons, typical of the mature stage, including liquid petroleum, are produced. With further increases in temperature, methane-rich gas again becomes the main product in the postmature stage resulting from thermal cracking of the carbon-carbon bonds. However, this gas can be distinguished from biogenic gas by heavier methane carbon isotope ratios.

Two main mechanisms are responsible for carbon isotope fractionation of methane in natural gas biogenic enrichment and thermal cracking. The isotope composition of biogenic methane is the result of the enrichment of the light isotope ^{12}C by microorganisms in the product relative to the substrate (Rosenfeld and Silverman, 1959). In the marine environment where CO_2 reduction is the primary mechanism of methane generation, the isotopic composition of the earliest methane formed is controlled by the $\delta^{13}C$ of the original CO_2 substrate. The isotopic composition of subsequent methane formed reflects the net effect of both addition and removal of CO_2. One of the major sources of CO_2 for reduction is anaerobic oxidation of organic matter during sulfate reduction. The $\delta^{13}C$ of the CO_2 is about -22 ppt (Presley and Kaplan, 1968). Another important source

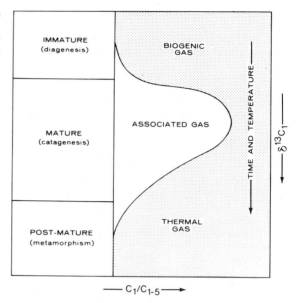

FIG. 7—Diagram showing characteristics of natural gas with increasing temperature and time.

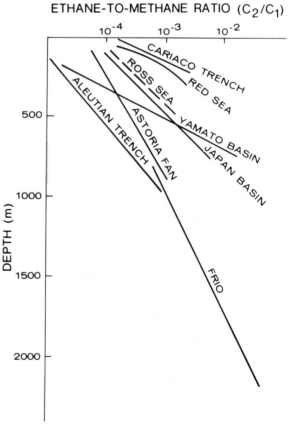

FIG. 8—Ethane-to-methane ratios of natural gas from selected DSDP sites.

of CO_2 at slightly higher temperatures results from decarboxylation which also provides isotopically light CO_2 ($\delta^{13}C = -20$ to -25 ppt).

In general, the $\delta^{13}C$ of the residual dissolved CO_2 becomes isotopically heavier as isotopically light CO_2 is removed from the interstitial waters for methane formation. Consequently the methane produced from this remaining CO_2 also becomes isotopically heavier with depth. This general trend of isotopically heavier methane resulting from removal of isotopically light CO_2 is typical for DSDP sediments buried to depths of 1 km and is illustrated in Figure 5. The $\delta^{13}C$ values over this depth interval range from -90 to -65 ppt. However, if another source of CO_2 is added, such as from the thermogenic decarboxylation, an intermediate trend toward lighter $\delta^{13}C$ values of the methane may result.

An additional mechanism responsible for carbon isotope fractionation of methane is the kinetic isotope effect of thermal cracking. When methane is produced by nonbiologic decomposition or organic matter, ^{12}C-^{12}C bonds are broken at a faster rate than ^{13}C-^{12}C bonds (Stahl, 1974). As a result, gases generated by thermal processes become isotopically heavier with increasing levels of maturation. Gases associated with oil generation have $\delta^{13}C$ values in the range of -50 to -40 ppt; gases in the wet gas-condensate stage have $\delta^{13}C_1$ values of -40 to -35 ppt; and postmature gases are isotopically heaviest with $\delta^{13}C$ values heavier than -35 ppt.

The earliest formed thermogenic methane is isotopically light ($\delta^{13}C = -55$ ppt) because of a kinetic isotope effect (Sackett, 1978; Stahl et al, 1979). However, significant quantities of thermogenic methane are not generated until temperatures of about 100°C are reached (Frank et al, 1974; Hunt, 1979). Minor amounts of thermogenic methane that are generated at low temperatures are mixed with and secondary to biogenic gas previously generated at shallow depths. The addition of minor amounts of thermogenic methane has the effect of gradually increasing the $\delta^{13}C$ of the methane at depths greater than about 1 km to depths of burial where thermogenic processes account for major quantities of gas. Sediments capable of supporting the production of biogenic methane generally contain sufficient quantities of organic matter so that they are also capable of generating thermogenic gas at elevated temperatures. As a result, early formed biogenic gas commonly contains a significant component of later formed thermogenic gas that cannot be distinguished because of compositional changes resulting from mixing.

Because of the effects of the isotopic composition of the initial or residual CO_2 and mixing with thermogenic methane, an absolute lower limit for the isotopic composition ($\delta^{13}C$) of predominantly biogenic methane cannot practically be assigned. However, for classifying ancient accumulations, in which many of these effects are impossible to determine, an arbitrary lower $\delta^{13}C_1$ value of -55 ppt is used, recognizing the fact that biogenic methane can be heavier, although thermogenic methane rarely is lighter.

Table 1 lists the isotopic composition of natural gases of biogenic origin. Although most of the gases are isotopically lighter than -55 ppt, some are as heavy as -45 ppt. Isotopically heavier methane can result from partial microbial oxidation in near-surface sediments. The residual methane after partial oxidation would be enriched in ^{13}C compared with the starting material because of preferential oxidation of $^{12}CH_4$ as discussed by Lebedev et al (1969).

Migration may be a mechanism of carbon isotope fractionation such that methane could either become enriched or depleted with ^{12}C (Galimov, 1967; Columbo et al, 1969; Lebedev and Syngayevski, 1971). Although these effects have been demonstrated in laboratory experiments and may be significant when migrating small quantities of gas, migration has not been shown to produce consistent or significant isotope effects under

Table 1. Isotopic Composition of Natural Gases of Biogenic Origin*

Source and Location	$^{13}C_1(‰)$	References
1. Marsh gas (Delaware, Louisiana, and Minnesota)	-70.2 to -52	Fuex, 1977
2. Lake sediments (Africa)	-45	Deuser et al (1973)
3. Lake sediments (Utah)	-45.8	This report
4. Lake sediments (Connecticut)	-80.2 to -57.2	Oana and Deevey (1960)
5. Marsh gas (USSR)	-69 to -52	Ovsyannikov and Lebedev (1967)
6. Sludge gas (California)	-47.1	Nissenbaum et al (1972)
7. Marsh gas (USSR)	-74 to -52	Lebedev et al (1969)
8. Marine sediments (DSDP, Leg 41)	-73.7 to -51.7	Doose et al (1978)
9. Marine sediments (DSDP, Leg 42B)	-72 to -63	Hunt and Whelan (1978)
10. Marine sediments (Alaska)	-80	Kvenvolden et al (1979)
11. Marine sediments (Gulf of Mexico)	-65.5 to -58	Bernard et al (1976)
12. Marine sediments (DSDP, Leg 15)	-76.3 to -59.6	Lyon (1973)
13. Marine sediments (DSDP, Legs 10, 11, 13, 14, 15, 18, 19)	-88.7 to -47.1	Claypool et al (1973)
14. Marine sediments (British Columbia)	-55.6	Nissenbaum et al (1972)
15. Solution gas (Montana)	-65.32 to -63.94	This report
16. Solution gas (North Dakota)	-71 to -69.2	This report

*Gases isotopically heavier than $-50‰$ may be due to microbial oxidation.

natural conditions (Bernard et al, 1977; Coleman et al, 1977; Stahl et al, 1977).

In addition to the chemical and isotopic characteristics of biogenic methane, the geologic setting must be considered in making the final interpretation of the gas origin. Gases of obvious biogenic origin occur at relatively shallow depths, at low temperatures, and in relatively young sediments and sedimentary rocks. Although most biogenic gases are generated at depths of less than 1 km, subsequently they can be buried as deep as gas of thermal origin. However, deeply buried gas of original biogenic origin invariably will be mixed with gas of later thermogenic origin so that recognition on the basis of chemical and isotopic composition will be difficult. In rare conditions, biogenic gas can be associated with oil if the oil was undersaturated in thermogenic gas and migrated into an immature sedimentary section.

ANCIENT ACCUMULATIONS OF BIOGENIC GAS

Although recent occurrences of biogenic gas are numerous, data that permit recognition of ancient accumulations are available only for those accumulations listed in Table 2. Because a combination of chemical and isotopic composition plus geologic setting has been used to make the interpretations, there are undoubtedly many accumulations that we have not identified.

In general, ancient accumulations of biogenic gas occur in relatively young rock sequences (Cretaceous and younger) that have had a low-temperature history. Although known accumulations occur as deep as 3,350 m, most are at depths of less than 1,800 m. These accumulations are present in both marine and nonmarine rocks. However, the newly documented accumulations occur strictly in marine sequences.

Several known areas containing biogenic gas accumulations listed in Table 2 are reviewed in the following. Detailed data for more than 30 fields are presented in Tables 3, 4, and 5. Locations of newly described fields in the United States and Canada, and fields in northwest Siberia, USSR, are shown in Figures 9 and 12.

Niobrara Formation, Colorado and Kansas

Natural gas was discovered more than 50 years ago in chalks of the Niobrara Formation of Late Cretaceous age in eastern Colorado, but commercial development did not begin until the early 1970s. There are now more than 30 such gas fields in eastern Colorado and northwestern Kansas. Natural gas, much of which is interpreted to be of biogenic origin, is produced from chalk beds that are characterized by high values of porosity, but low values of permeability at depths ranging between about 270 and 850 m. Lockridge and Scholle (1978) indicated that commercial production of hydrocarbons at depths greater than 1,200 m from Niobrara chalk reservoirs will require significant natural fracturing because of greatly reduced porosity and permeability. Thus, the interpreted occurrence of

Table 2. Worldwide Accumulations of Biogenic Gas

Location	Reservoir Age	Depth (m)	$^{13}C_1C^0/_{00}$	Reference
United States				
Cook Inlet, Alaska	Tertiary	910-1,650	−63 to −56	Claypool et al (1980)
Offshore Gulf of Mexico	Pleistocene	460-2,800	−69 to −55	This report
Rocky Mountains (Colorado, Kansas, Montana, Nebraska, New Mexico, South Dakota)	Cretaceous and Tertiary	120- 840	−72 to −55	This report
Illinois	Pleistocene	40	−84 to −72	Wasserberg et al (1963)
Japan	Tertiary	100-1,000	−75 to −65	Nakai (1960)
Italy	Tertiary	400-1,830	−71 to −55	Colombo et al (1966)
Germany	Tertiary	900-1,800	−72 to −64	Schoell (1977)
Canada	Cretaceous	300-1,000	−68 to −60	This report; Fuex (1977)
USSR				
North Aral	Tertiary	320- 350	−72 to −64	Avrov and Galimov (1968)
Siberia	Cretaceous	700-1,300	−68 to −58	Yermakov et al (1970)
Stravapol	Cretaceous and Tertiary	200-1,200	−75 to −57	Alekseyev et al (1972)
North Priaral	Tertiary	300- 500	−72 to −63	Galimov (1969)
Trinidad	Tertiary	980-3,350	−71 to −64	B. D. Carey (personal commun., 1980)

biogenic gas coincides with the development of reservoir properties suitable for development.

Gas samples were collected at seven wells along a southeast-northwest belt that trends as shown in Figure 9. Current depths of burial increase northwestward along this belt from 328 to 842 m. An isopach map of reconstructed thicknesses of the Pierre Shale, the youngest Cretaceous unit exposed at the beginning of the Tertiary, indicates a similar increase in burial depths in an east to west direction. Porosity versus depth plots of the Niobrara and other chalk reservoirs suggest that the Niobrara was probably subjected to greater burial than indicated by present-day burial or that can be accounted for by the thickness of Pierre Shale (Lockridge and Scholle, 1978).

Gases from wells along the trend reflect the increasing depth of burial by becoming isotopically heavier (Fig. 10). We interpret the gases to be of predominantly biogenic origin because of chemical and isotopic composition, reconstructed maximum depths of burial, and source rocks studies by Swetland and Clayton (1976) that indicate that older Cretaceous rocks in this region are immature.

The biogenic gas probably was generated in situ

Table 3. Biogenic Gas Fields in Rocky Mountains, United States

State	Field	Producing Unit[1]	Depth (m)	$^{13}C_1$ (‰)	C_1/C_{1-5}
Colorado	Armel	Niobrara Fm.	482	−62.5	0.981
	Beecher Island	Niobrara Fm.	491-518	−60.8 to −60.1	0.982 to .981
	Republican	Niobrara Fm.	691	−59.7	0.981
	San Luis basin	Alamosa Fm.	300	−70.2 to −69.7	0.999 to .998
	Vernon	Niobrara Fm.	647	−58.8	0.98
	Whisper	Niobrara Fm.	842	−54.7	0.976
Kansas	Wildcat	Niobrara Fm.	328	−65.4	0.993
Montana	Bell Creek	Muddy Ss.	1,387	−65.1	0.98
	Black Coulee	Eagle Ss.	349	−66	0.996
	Bowdoin	Bowdoin and Phillips ss.[2]	224-445	−72.3 to −68.6	0.997 to .995
	Cassady	Eagle Ss.	385	−70	0.998
	Cedar Creek	Eagle Ss.	517	−69.7	0.996
	Guinn	Eagle Ss.	171	−65.2	0.987
	Hardin	Frontier Fm.	253	−65.9	0.989
	Leroy	Eagle Ss.	470	−68.7	0.996
	Liscom Creek	Shannon Ss. Mbr. of Gammon Shale	829	−64.8	0.992
	Lohman	Eagle Ss.	318	−68.1	0.997
	Tiger Ridge	Eagle Ss.	347-432	−65.5 to −63.5	0.997 to .991
Nebraska	Wildcats	Niobrara Fm. and Dakota Ss.	394-655	−66.5 to −62.8	0.956 to .998
New Mexico	Wagon Mound	Dakota Ss.	119-134	−59.8 to −55.3	0.999
South Dakota	West Short Pine Hills	Shannon Ss. Mbr. of Gammon Shale	417-605	−70.0 to −69.7	0.996 to .998

[1]Cretaceous age except Pliocene or Pleistocene Alamosa Fm.
[2]Subsurface usage.

Table 4. Biogenic Gas in Suffield Block, Southeastern Alberta, Canada

Depth (m)	Producing unit	$^{13}C_1$ (‰)	C_1/C_{1-5}
334	Milk River Fm. equivalent	−68	0.997
338	Milk River Fm. equivalent	−67.5	0.997
350	Milk River Fm. equivalent	−68.3	0.993
416	Medicine Hat Ss.	−67.8	0.997
433	Medicine Hat Ss.	−68	0.997
436	Medicine Hat Ss.	−68	0.997
471	Medicine Hat Ss.	−67.1	0.996
563	Second White Specks ss.	−65.7	0.995
564	Second White Specks ss.	−65.3	0.995
566	Second White Specks ss.	−65.4	0.9950
643	Second White Specks ss.	−65.4	0.996
758	Bow Island Formation	−66	0.994
783	Bow Island Formation	−66	0.995
831	Basal Colorado Ss.	−60.2	0.963

Biogenic Gas

Table 5. Biogenic Gas Fields of Pleistocene Age in Offshore Gulf of Mexico

Field	Depth (m)	$\delta^{13}C_1$ (%o)	C_1/C_{1-5}
Ship Shoal 271	1,818 - 1,876	−59.9 to −56.9	0.989 to .987
East Cameron 245	458 - 983	−69.4 to −65.9	0.999
West Cameron 513	2,224 - 2,312	−57.2 to −56.9	0.997 to .972
West Cameron 533	917	−64.3	0.999
West Cameron 543	2,498 - 2,573	−58 to −56.7	0.98 to .978
West Cameron 587	1,355 - 1,562	−58.7 to −58.1	0.981
West Cameron 639	1,414 - 1,512	−63.2 to −57.9	0.996 to .985
West Cameron 643	1,391 - 1,399	−55.4 to −55.3	0.974
High Island A302	2,721 - 2,786	−65.1 to −61.7	0.999 to .997
High Island A309	1,390 - 2,208	−61.5 to −57.4	0.986 to .982
High Island A330	1,456 - 2,430	−66.5 to −64.5	0.997
High Island A343	1,829 - 1,916	−67 to −66.8	0.999 to .998
High Island A370	1,270 - 2,393	−65.3 to −60.6	0.998 to .96

because of low permeability of the chalks that inhibited migration and of the organic-rich laminae within the Niobrara that served as a source for the gas. Organic carbon values generally exceed 1% and some samples contain as much as 5.8%. Additionally, the Niobrara is overlain by a thick section of shale with many bentonite beds in the lower part that served as a seal preventing leakage of the gas from the Niobrara.

Northern Great Plains, Montana and South Dakota

Biogenic gas is being produced from widely spaced fields developed in Lower and Upper Cretaceous marine rocks at depths of less than 800 m (Table 3). However, maximum depths of burial were greater than present-day because most of the Tertiary and some of the Cretaceous section have been removed by erosion. All available evidence, including vitrinite reflectance, thermal alteration index (TAI), pyrolysis, clay mineralogy, and reconstructed depths of burial, suggests that temperatures necessary for thermal generation of significant amounts of hydrocarbons have not been achieved in the organic-rich shales that enclose the reservoirs.

In recent marine environments, biogenic gas is commonly found in sediments accumulating at rates greater than about 50 m/m.y. (Claypool and Kaplan, 1974). The average rate of sedimentation was more than 30 m/m.y. during Late Cretaceous time in the northern Great Plains (Gill and Cobban, 1973). However, the rate of sedimentation for Cretaceous rocks was probably greater than the rate for recent sediments after compensating for compaction, and undoubtedly was sufficiently rapid to insure the maintenance of anoxic conditions necessary for the generation of biogenic gas.

In modern sediments, organic carbon values generally exceed 0.5% where geochemical effects of methane-producing microorganisms are observed (Claypool and Kaplan, 1974; Rashid and Vilks, 1977). Many subsurface samples collected over a 100-m interval in the biogenic gas-productive Bowdoin Dome area and Cedar Creek anticline generally contain more than 0.5% and some samples contain as much as 8.7% organic carbon.

The biogenic gas is trapped in two distinct types of reservoirs and traps as exemplified by productive intervals in the Bearpaw Mountains and Bowdoin Dome. In the Bearpaw Mountains, the primary trapping mechanism is gravity-induced faulting and the reservoirs are porous and permeable marine sandstones enclosed by offshore marine shales (Rice, 1980). Petrographic studies indicate that extensive early mineral diagenesis in the reservoirs has taken place as a result of migrating waters (Gautier, in press). This interpretation suggests that the gas was held in solution in formation waters of the surrounding shales during diagenesis. Subsequent exsolution of the gas during uplift and erosion resulted in a free-gas phase that migrated to and accumulated in the sandstone because of capillary pressure differentials between the shale and sandstone beds.

In the Bowdoin Dome area, biogenic gas is stratigraphically entrapped over a large area (1,500 sq km) in thin (commonly less than 3 cm), discontinuous sandstone and siltstone laminae (Nydegger et al, 1979). Here, laminae are enclosed by organic-rich shales, containing numerous bentonite beds, that served as both a source and a seal for the gas. Because the reservoirs typically exhibit low permeability, stimulation is required to provide flow rates sufficient for economic development. However, this type of reservoir is developed over large areas of the northern Great Plains and structure is not required for trapping the gas.

Suffield Block, Southeastern Alberta

A widespread accumulation of gas covering more than 20,700 sq km was developed in the 1970s in southeastern Alberta. Most of the gas is entrapped in shallow, low-permeability marine reservoirs of Late Cretaceous age similar to those of the Bowdoin Dome area of north-central Montana.

The age occurs at depths of less than 600 m, although maximum depths of burial were greater. Uplift and erosion during the Tertiary probably resulted in the exsolution of the previously formed biogenic gas which was subsequently moved by capillary-pressure differentials from the immature, organic-rich shale to interlaminated

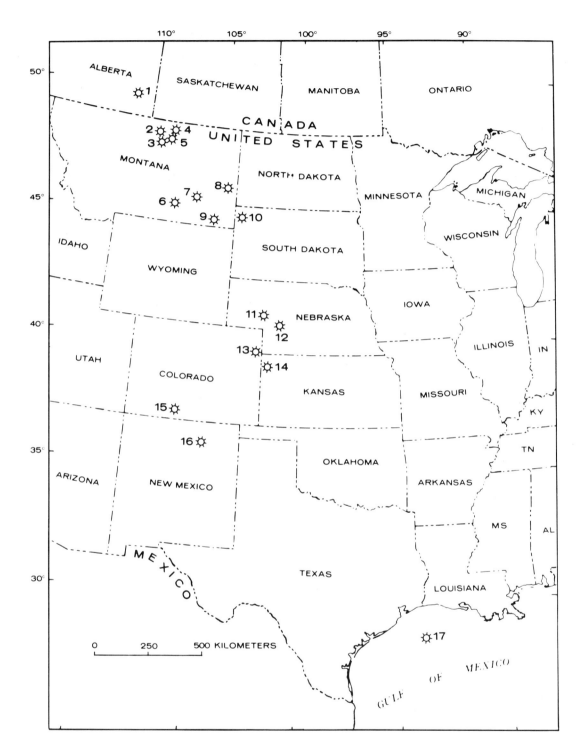

FIG. 9—Map showing location of biogenic gas fields in United States and Canada: (1) Suffield Block; (2) Tiger Ridge, Lohman, Black Coulee, and Cassady; (3) Leroy; (4) Bowdoin; (5) Guinn; (6) Hardin; (7) Liscom Creek; (8) Cedar Creek; (9) Bell Creek; (10) West Short Pine Hills; (11) Wildcat, Sec. 22, T19N, R39W; (12) Wildcat, Sec. 14, T16N, R33W; (13) Armel, Beecher Island, Republican, Vernon, and Whisper; (14) Wildcat, Sec. 10, T95N, R40W; (15) San Luis basin; (16) Wagon Mound; (17) Offshore Gulf of Mexico. Fields are listed in Table 5.

siltstone and sandstone reservoirs.

Gas samples were analyzed from the Suffield Block, a 2,600-sq km area within the large productive area. These gas samples are considered to be typical of the entire productive area. The samples were collected from five stratigraphic units and are arranged in Table 4 in order of descending age and increasing depth of burial. The deeper examples of a particular stratigraphic unit are in the northern part of the block.

The chemical and isotopic compositions, plus the immature stage of the enclosing sediments (Hacquebard, 1977), indicate that the gases of the Suffield Block are of predominantly biogenic origin. Gases in the Milk River Formation equivalent, Medicine Hat Sandstone, Second White Specks sandstone, and Bow Island Formation are distributed over about 400 m of stratigraphic section and have similar chemical and isotopic compositions (Table 4). A sample of gas from the basal Colorado Sandstone of the Colorado Group, which is approximately 120 m below the Bow Island, has a distinctly different chemical and isotopic composition, which may indicate a greater rate of mixing of thermogenic gas than can be explained by the increased depth of burial at the same geothermal gradient.

Offshore Gulf of Mexico

The Gulf of Mexico province lies offshore the states of Louisiana and Texas. It is one of the major hydrocarbon-producing regions of the United States. Many studies, including those of Dow (1978) and Lafayette and New Orleans Geological Societies (1968), have concluded that, because of the young age of the rocks, many of the hydrocarbon accumulations have resulted from extensive vertical migration from deeper, thermally mature source rocks.

Our studies indicate that the generation of biogenic gas at shallow depths in accumulating sediments has made a major contribution to many accumulations, particularly those of Pleistocene age. The gas fields of predominantly biogenic origin (Table 5) occur as deep as 2,800 m, which is much deeper than other documented accumulations (Table 2). In addition, the accumulations are overlain by as much as 200 m of water. However, in this region, temperatures in excess of about 160°C (equivalent to depths of burial of 5,600 m) must be reached before thermal generation of hydrocarbons becomes dominant in sediments of Pleistocene age (Dow, 1978).

Although gases from the fields listed in Table 5 are predominantly of biogenic origin, many other accumulations, some with reservoirs older than Pleistocene, contain gas with biogenic methane as a major component. However, because of localized migration near growth faults and salt domes, these gases are isotopically heavier and (or) contain concentrations of heavier hydrocarbons greater than would be expected from systematic mixing of thermal gas with increasing time and temperature (Rice et al, 1979). Therefore, many accumulations with a significant biogenic gas component are not shown in Table 5.

Buckley et al (1958) conducted a study of water-bearing formations with dissolved hydrocarbon gases in the Gulf of Mexico. They determined that (1) dissolved gas is widespread, (2) the dissolved gas is chiefly methane with concentrations of CO_2 and heavier hydrocarbons ranging from traces to as much as 2% near oil accumulations, and (3) total quantity of dissolved gas probably exceeds known reserves in the area. Although some of the dissolved gas is of thermogenic origin, the dry gases associated with young reservoirs at shallow depths are undoubtedly of predominantly biogenic origin. The dissolved gas is present in quantities ranging from about 30 to 110 mmol/kg with higher concentrations occurring with increasing depth. These concentrations are generally greater than those reported for most deep-sea sediments and probably are analogous to the occurrence of accumulations of free biogenic gas in Gulf Coast Pleistocene reservoirs.

Cook Inlet, Alaska

Two types of natural gas occurrences have been documented in the Cook Inlet. The major reserves are in shallow (generally less than 1,600 m) nonassociated dry gas fields ($C_1/C_{1-5} > 0.99$) that contain methane with $\delta^{13}C$ values in the range of -65 to -56 ppt (Claypool et al, 1980). The gas is trapped in nonmarine sandstones interbedded with coals of Miocene and Pliocene age and is considered to be of biogenic origin. The coals are bituminous or lower in rank (70 to 75% carbon, ash-free). Significant production of thermogenic gas from coal does not occur until higher ranks are attained and the coal-bed gas generated initially is generally wetter than the Cook Inlet nonassociated gas.

Lesser amounts of associated gas and oil are present in nonmarine Oligocene and Paleocene reservoirs. This gas is isotopically heavier ($\delta^{13}C_1$ values in the range of

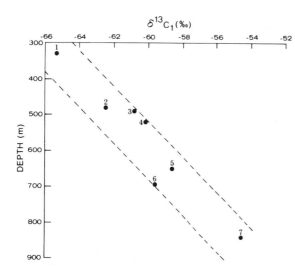

FIG. 10—Changes with depth in $\delta^{13}C_1$ of natural gases from Niobrara Formation in Colorado and Kansas. Numbers and dots correspond to wells located in Figure 11.

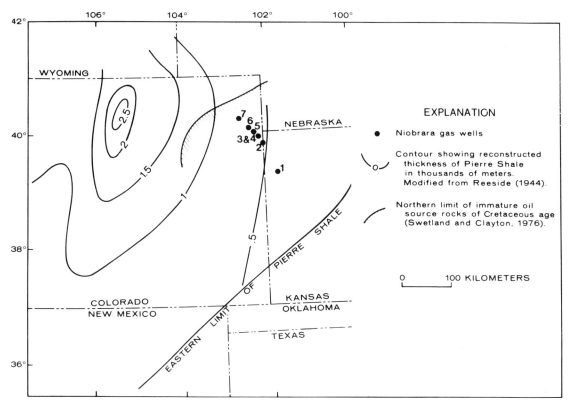

FIG. 11—Map showing location of Niobrara gas wells (Fig. 10), reconstructed thickness of Pierre Shale, and limit of immature oil source rocks.

−49 to −44 ppt) and contains appreciable amounts of heavier hydrocarbons (C_1/C_{1-5} >0.69 to 0.90; Claypool et al, 1980). This gas is interpreted to be of thermogenic origin and to have migrated with the oil from underlying, thermally mature, marine shale source rocks of Jurassic age (Magoon and Claypool, 1979).

Northwest Siberia, USSR

Yermakov et al (1970) studied gases from Jurassic and Cretaceous reservoirs from several giant fields in northwest Siberia. Location and data on these fields are presented in Figure 12 and Table 6. Gases from reservoirs that range in age from Aptian through Cenomanian are extremely dry (C_1/C_{1-5} >0.99). They have carbon isotope values in the range of other ancient biogenic accumulations ($\delta^{13}C$ = −68 to −58 ppt). Nesterov et al (1978) concluded in a study using argon isotopes that these gases were generated in situ.

These Cretaceous gases are interpreted to be dominantly of a biogenic origin. They occur at depths of less than 1,300 m, and are associated with organic-

Table 6. Biogenic Gas Fields in USSR*

Field	Estimated Reserves $10^{12}m^3$ (Tcf)	Depth of Production (m)	$\delta^{13}C_1$ (⁰/₀₀)	C_1/C_{1-5}
Arkticheskoye	0.18 (6.4)	683 - 2,500	−61.7	0.999
Gubkinskoye	0.35 (12.5)	745 - 780	−61.1	0.999
Komsomolskoye	0.46 (16.2)	929 - 985	−64.7	0.998
Medvezhye	1.55 (54.7)	1,057 - 1,207	−58.3	0.999
Severo-Stavropolskoye	0.23 (8.1)	180 - 1,050	−68.7 to −66.5	0.999
Urengoiskoye	5.04 (177.8)	1,043 - 3,160	−59.2 to −59	0.998
Vyngapurskoye	0.29 (10.3)	987 - 1,091	−60	0.998
Zapolyarnoye	2.66 (94.1)	1,120 - 1,300	−60.3	0.999
Total	10.8 (380.1)			

*Data from Alekseyev et al (1972); McCaslin (1977); Yermakov et al (1970). All fields except Severo-Stavropolskoye are in northwest Siberia.

rich rocks at a low level of maturation (lignite to initial cannel coal; Yermakov et al, 1970). During deposition, the area was a swampy lowland which favored the accumulation of organic matter; organic carbon values range between 3 and 6%. This area was along the southern edge of extensive permafrost cover during the Pleistocene. Larger quantities of biogenic gas were initially trapped as hydrates. The permafrost front has since retreated and associated hydrates have decomposed to form free gas (Makogon et al, 1972). Free gas has subsequently accumulated in large structural traps.

In comparison, the gases associated with deeper and older Jurassic reservoirs are different, both chemically and isotopically. The contrasting gases contain more than 6% heavier hydrocarbons and are enriched in ^{13}C ($\delta^{13}C$ values range from -46 to -38‰). The stratigraphic and compositional separation of these two groups of gases, together with the presence of oil in the Jurassic rocks, suggests a thermal origin for the Jurassic gases.

Illinois

Biogenic gas is present both in a free state and in solution in Pleistocene glacial drift of Illinois. Although free gas accumulations are not of commercial significance, many produce sufficient gas for domestic use (Meents, 1960, 1968). Local accumulations are composed primarily of methane, are at an average depth of 40 m, and have $\delta^{13}C_1$ values in the range of -84 to -72 ppt (Wasserburg et al, 1963). Radiocarbon dating of several gas samples indicates that the gas resulted from the microbial degradation of organic matter in the Robein Silt, which dates from 22,000 to 28,000 radiocarbon years B.P. (Coleman, 1976). The sands and gravels just above and below provide the reservoir, and the overlying till serves as the seal.

Coleman et al (1977) analyzed many gas samples in an effort to distinguish gas leaking from underground storage reservoirs from indigenous shallow biogenic gas in solution. They found that migration of gas resulted in chemical fractionation by removal of the heavier hydrocarbons from the gas that leaked out of the injected storage reservoir so that it became methane enriched in a manner similar to biogenic gas. This migration did not significantly affect the isotopic composition. They concluded that the two types of gases can be distinguished by isotopic composition: storage reservoir gas has $\delta^{13}C_1$ values in the range of -46 to -40 ppt whereas biogenic gas in solution is isotopically lighter ($\delta^{13}C_1 = -90$ to -64 ppt).

Japan

Biogenic gas dissolved in formation waters was the main source of natural gas production in Japan from the 1920s to 1960s. The gas is produced from brines confined to the lower part of marine or lagoonal basins by updip meteoric waters. The reservoirs are highly permeable sands and gravels of mostly Pliocene to Pleistocene age. Although the maximum depth of pro-

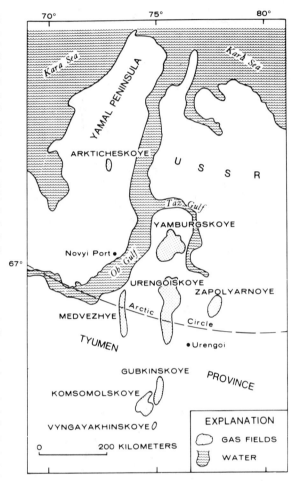

FIG. 12—Map showing location of biogenic gas fields in northwest Siberia. Although chemical and isotopic data are not available for Yamburgskoye and Vyngayakhinskoye fields, their geologic setting suggests that produced gases are of biogenic origin.

duction is 2,000 m, most gas comes from depths of less than 1,000 m.

The initial gas-to-water ratio is close to the maximum solubility of methane in water at reservoir temperatures and pressures (Marsden and Kawaii, 1965). Because the depth of burial controls the pressure, the gas-water ratio is a function of depth. The ratio is about 1.2 m³/m³ at 500 m and 2 m³/m³ at 1,000 m. Most wells flow water naturally with the dissolved gas initially, but need artificial lift before long. The gas generally is released from the water as it flows to the surface and hydrostatic pressure is reduced. However, there are large quantities of water for disposal.

Biogenic origin of the dissolved gas was interpreted by Nakai (1960, 1962) and Sugisaki (1964). Carbon isotope values of gases from several fields range from -75 to -65 ppt (Nakai, 1960). The gas is predominantly methane, although as much as 0.08% ethane has been

Dudley D. Rice and George E. Claypool

Table 7. Estimated Volumes of Biogenic Gas, $10^{12} m^3$ (Tcf)*

Area	Discovered Volumes		Undiscovered Volumes	Reference
	Reserves	Initial In-Place		
Cook Inlet, Alaska	0.21 (7.5)[1]			Blasko (1974); Alaska Div. Oil and Gas (1976)
Gulf of Mexico		0.34 (12)[2]		U.S. Geol. Survey Resource Appraisal Group (unpub. data)
Japan	0.13 (4.7)			Marsden and Kawaii (1965)
Northern Great Plains, United States			2.83 (100)	Rice and Shurr (1980)
Southeast Alberta, Canada		0.42 (15)		Energy Resources Conservation Board (1977)
USSR	13.0 (460)[3]			Alekseyev (1975); McCaslin (1977)

*Total discovered estimate 21.3%; total discovered plus undiscovered estimate 25.5%.
[1] Includes cumulative production of $0.28 \times 10^{12} m^3$ (1 Tcf).
[2] 75% of total estimated volumes because approximately 25% of gas is of thermal origin and has migrated from deeper, mature source rocks.

analyzed in a few rare examples (Marsden and Kawaii, 1965). The isotopic and chemical compositions are typical of biogenic gases. Sugisaki (1964) was able to make a distinction between dissolved biogenic gas and associated gas using nitrogen-argon ratios. Nakai (1962) noted that methane production was preceded by an increase in total CO_2 and concluded that the gas was produced in situ by fermentation in the formation waters.

Italy

Detailed studies on the chemical and isotopic compositions of Italian natural gases, interpreted by us to be of possible biogenic origin, were conducted by Colombo et al (1966, 1969; Table 2). Gases were collected from a thick turbidite sequence of Pliocene age in southern Italy at depths less than 1,600 m below sea level. They found that the $\delta^{13}C$ of the methane fraction became heavier as the C_2/C_1 ratio increased and proposed that this relation could be explained by two mechanisms: (1) mixing of biogenic and thermal gas or (2) migration fractionation of thermal gas from greater depths. They discounted the mixing mechanism because they believed that most biogenic gas is lost to the atmosphere and concluded that the distribution of gases was the result of migration fractionation. However, we believe that the distribution and composition of gases can best be explained by the mixing of predominantly biogenic gas with thermal gas. Early formed biogenic gas was gradually mixed with minor amounts of thermal gas which is generated in amounts proportional to temperature and age of the sediments.

ESTIMATED VOLUMES OF BIOGENIC GAS

Undoubtedly, there are many large areas, both discovered and undiscovered, where conditions were favorable for both the generation and accumulation of biogenic gas. However, in only a few areas are data available for identifying biogenic gas and for estimating volumes, as shown in Table 7. These estimated volumes probably represent only a part of the total resources of biogenic gas. A brief discussion follows on the areas where data are available.

Cook Inlet

Reserves of 0.21×10^{12} (7.5 Tcf) of biogenic gas, which include cumulative production of 0.028×10^{12} m^3 (1 Tcf) have been estimated for the Cook Inlet basin of Alaska (Blasko, 1974; Alaska Div. Oil and Gas, 1976). These reserves are significant because biogenic gas was generated in sediments containing low-rank coal beds of Tertiary age. Minor amounts of associated gas (0.011×10^{12} m^3 or 0.4 Tcf) are entrapped in older deeper Tertiary rocks, but this gas and accompanying oil are interpreted to have migrated from thermally mature Jurassic marine shales (Magoon and Claypool, 1979).

Gulf of Mexico

More than 0.8×10^9 m^3 (5 billion bbl) of oil and condensate, and 1.1×10^{12} m^3 (40 Tcf) of gas have been produced to date from the offshore Gulf of Mexico. Reservoirs of Pleistocene age contain major volumes of predominantly gas. The U. S. Geological Survey Resource Appraisal Group (unpub. data) estimated that rocks of Pleistocene age contain discovered in-place volumes of about 0.45×10^{12} m³ (16 Tcf). Although some oil and thermal gas have migrated from deeper, mature source rocks, we estimate that about 75% of the gas in Pleistocene reservoirs is of predominantly biogenic origin. Therefore, the Pleistocene rocks may contain about 0.34×10^{12} m³ (12 Tcf) of discovered biogenic gas. In addition, accumulations of both associated and nonassociated gas in older Tertiary reservoirs are interpreted to have had a predominantly biogenic origin locally.

Japan

Major quantities of dissolved gas of biogenic origin have been produced in Japan. The gas was adequate for local domestic use and in chemical manufacturing from the 1920s through the 1960s. However, because of production costs, the dissolved gas could not compete with residual oil as a fuel. Marsden and Kawaii (1965) described the three main fields from which biogenic gas is produced (Niigata, Southern Kanto, and Mobara) and estimated the reserves to be about 0.13×10^{12} m³ (4.7 Tcf).

Northern Great Plains (Canada and United States)

The government of Alberta made a detailed evaluation in the early 1970s of the gas reserves of the Suffield Block located in the southeastern part of the province. The block covers an area of approximately 2,600 sq km and is part of a larger productive area of shallow biogenic gas that covers more than 20,700 sq km. The Suffield Block was assigned an in-place gas reserve figure of 0.10×10^{12} m³ (3.7 Tcf) and a recoverable reserve figure of 0.076×10^{12} m³ (2.7 Tcf; Suffield Evaluation Comm., 1974). Most of the gas (94%) is trapped in low-permeability reservoirs of Late Cretaceous age. For the entire productive area of southeastern Alberta, the initial in-place reserves of gas from low-permeability reservoirs of Late Cretaceous age are estimated to be 0.42×10^{12} m³ (15 Tcf; Energy Resources Conservation Board, 1977). This per-section figure is lower than that assessed for the Suffield Block because of the various stages of technology and of different gas prices when the gas was developed. The ultimate size and recoverable reserves for this area in Canada will be controlled by economics and technology.

Similar thin, discontinuous, low-permeability reservoirs are developed over a large part of the northern Great Plains in both Canada and the United States. Rice and Shurr (1980) concluded that their study area of about 300,000 sq km in eastern Montana, western North and South Dakota, and northeastern Wyoming probably contained resources of biogenic gas in excess of 2.8×10^{12} m³ (100 Tcf). They stated that the development of the shallow biogenic gas would depend on gas prices and recovery technology.

USSR

About 40% (26.0×10^{12} m³ or 928 Tcf) of the world's reserves are present in the USSR (McCaslin, 1977). A large part of these gas reserves (10.8×10^{12} m³ or 380 Tcf) has been identified as being of biogenic origin (Table 6). Alekseyev (1975) estimated that about 50% of the total Russian reserves is of biogenic origin; this amounts to 13.0×10^{12} m³ (460 Tcf) on the basis of the figure given by McCaslin (1977). This figure appears accurate because there are several other fields with large reserves which are in the same geologic setting as those identified as being of biogenic origin.

Biogenic gas accounts for more than 20% of the world's gas reserves of 66.4×10^{12} m³ (2,343 Tcf; Table 7). We believe that gas of biogenic origin will make an even greater contribution to future reserves for several reasons. First, natural gas has generally been underpriced as compared with other energy sources. This means that gas, until recently, has not been a major target of hydrocarbon exploration. Next, biogenic gas occurs in shallow reservoirs which are commonly underpressured. Also, much undiscovered biogenic gas is probably trapped in low-permeability marine reservoirs similar to those of the northern Great Plains. Sophisticated evaluation methods and advanced recovery technology are required to detect these accumulations and to provide commercial flow rates. A combination of low pressures and low permeability has resulted in many of these accumulations being uneconomic and unrecoverable in the past. Additionally, large volumes of biogenic gas are either dissolved or in hydrate structures (see Buckley et al, 1958; Claypool and Kaplan, 1974; Trofimuk et al, 1975; Cherskii and Tsarev, 1977; Tucholke et al, 1977; Shipley et al, 1979). Biogenic gas occurring in these phases is a large part of our undeveloped gas resource base. Major technologic advances and increases in gas prices will be required if these potential resources of dissolved and hydrated biogenic gas are to contribute to our energy needs in the future. However, a significant quantity of conventional gas of biologic origin probably has been underestimated and overlooked in past exploration.

SUMMARY AND CONCLUSIONS

Methane-rich gas is a product of low-temperature diagenesis of organic matter in sedimentary rocks. The gas is generated by the degradation of organic matter in rapidly accumulating sediments by anaerobic microorganisms. The gas is referred to as biogenic gas to emphasize that biologic processes are directly responsible for its formation. Biogenic gas is generated in immature sediments and can accumulate in large quantities. Thus it should be considered in future exploration efforts for hydrocarbons. The following conclusions are made about biogenic gas:

1. A succession of sedimentary ecosystems is established in the sediment column with lower energy-yielding metabolic process. In marine sediments, subsequent to oxygen depletion, sulfate reduction becomes the dominant form of respiration. Methane production replaces sulfate reduction only after the high concentrations of sulfate in sea water are reduced. The most important mechanism of methane generation is CO_2 reduction. The CO_2 can be a product of metabolic activity or

later thermal decarboxylation of the organic matter.

2. The succession of metabolic processes that result in methane production can be followed in modern sediments by monitoring the changes in concentration and stable carbon isotope ratios of the residual reactants or accumulating products with depth.

3. Major factors that control methane production after sediment burial are: anoxic environment, sulfate-deficient environment, low temperatures, organic matter, and sufficient space.

4. In marine sediments, most of the biogenic gas is formed in solution in the interstitial (pore) waters because of the limited amount of methane that is generated and the high solubility resulting from hydrostatic pressure due to the weight of the overlying water column. Under conditions of high pressure and (or) low temperature, the methane can combine with water to form gas hydrate compounds.

5. The dissolution of gas in interstitial waters can serve as a holding mechanism until the sediments are compacted, and traps and seals are available. Free gas is formed either when the solubility minimum is exceeded or by exsolution brought about by a reduction in hydrostatic pressure. Possible trapping mechanisms are: structure when the reservoirs are porous and permeable; low-permeability reservoirs; bentonites; early diagenetic carbonate cements, hydrates, and subnormal pressures.

6. The most useful characteristics in distinguishing biogenic gas are the chemical and isotopic compositions. Biogenic gases characteristically are composed almost entirely of methane ($C_1/C_{1-5} > 0.98$). Minor amounts of heavier hydrocarbons can be attributed to low-temperature thermal generation. The most useful criteria for distinguishing biogenic gas are carbon isotope ratios. Biogenic gas is enriched in the light isotope ^{12}C resulting from fractionation during methane generation by anaerobic microorganisms. The $\delta^{13}C_1$ values of biogenic gas are generally lighter than -55 ppt. In making interpretations of gas origin, possible ambiguities in chemical and isotopic evidence are usually resolved by consideration of the geologic setting.

7. Accumulations of biogenic gas are documented from Cretaceous and younger rocks, at depths less than 3,350 m and in both marine and nonmarine rocks. Commercial accumulations are demonstrated from Canada, Germany, Italy, Japan, Trinidad, USSR, and the United States. Other accumulations of biogenic origin are undoubtedly present, but data that permit their recognition are not available.

8. Gas of predominantly biogenic origin has made major contributions to the world's reserves (more than 20%). We predict that biogenic gas will make an even greater contribution to world energy needs in the future.

9. Finally, biogenic gas is an important target for future exploration because it is widespread, shallow, and has been shown to accumulate in commercial quantities.

REFERENCES CITED

Alaska Division of Oil and Gas, 1976, Alaska petroleum production summary by fields for December 1975: Alaska Div. Oil and Gas Bull., February, p. 23.

Alekseyev, F. A., 1975, Zonality in oil and gas formation in the earth's crust based on isotope studies: Petroleum Geology, v. 12, p. 191-193.

_____ V. S. Lebedev, and T. A. Krylova, 1972, Isotope composition of carbon in gaseous hydrocarbons and conditions for accumulations of natural gas: Internat. Geol. Rev., v. 15, p. 300-308.

Atkinson, L. P., and F. A. Richards, 1967, The occurrence and distribution of methane in the marine environment: Deep-Sea Research, v. 14, p. 673-684.

Avrov, V. P., and E. M. Galimov, 1968, Microbiologic nature of a methane pool detected at considerable depth (based on isotope analysis): Akad. Nauk. SSSR Doklady, v. 206, p. 201-202.

Barber, L. E., 1974, Methane production in sediments of Lake Wingra: PhD thesis, Univ. Wisconsin, 145 p.

Barker, C., 1972, Aquathermal pressuring—role of temperature in development of abnormal-pressure zones: AAPG Bull., v. 56, p. 2068-2071.

Barnes, R. O., and E. D. Goldberg, 1976, Methane production and consumption in anoxic marine sediments: Geology, v. 4, p. 297-300.

Belyaev, S. S., and K. S. Laurinavichus, 1978, Microbiological formation of methane in marine sediments, in Environmental biogeochemistry and geomicrobiology: Ann Arbor, Mich., Ann Arbor Science Publishers Inc., v. 1, p. 327-337.

Bernard, B. B., 1979, Methane in marine sediments: Deep Sea Research, v. 26A, p. 429-443.

_____ J. M. Brooks, and W. M. Sackett, 1976, Natural gas seepage in the Gulf of Mexico: Earth and Planetary Sci. Letters, v. 31, p. 48-54.

_____ _____ _____ 1977, A geochemical model for characterization of hydrocarbon gas sources in marine sediments: 9th Offshore Tech. Conf. Proc., v. 3, p. 435-438.

_____ _____ _____ 1978, Light hydrocarbons in recent Texas continental shelf and slope sediments: Jour. Geophys. Research, v. 83, p. 4053-4061.

Berner, R. A., 1975, Diagenetic models of dissolved species in the interstitial waters of compacting sediments: Am. Jour. Sci., v. 275, p. 88-96.

Blasko, D. P., 1974, Natural gas fields—Cook Inlet basin, Alaska: U.S. Bur. Mines Open-File Rept. 35-74, 24 p.

Bray, E. E., and W. R. Foster, 1980, A process for primary migration of petroleum: AAPG Bull., v. 64, p. 107-114.

Bryant, M. P., 1965, Rumen methanogenic bacteria, in Physiology of digestion in the ruminant: Washington, D.C., Butterworths, p. 411-418.

Buckley, S. E., C. R. Hocott, and M.S. Taggart, Jr., 1958, Distribution of dissolved hydrocarbons in subsurface waters, in Habitat of oil: AAPG, p. 850-882.

Buswell, A. M., and H. R. Mueller, 1952, Mechanism of methane fermentation: Ind. Eng. Chemistry, v. 44, p. 550-552.

Cherskii, N. V., and V. P. Tsarev, 1977, Estimating reserves in light of exploration for and extraction of natural gas from world ocean floor sediments: Geol. Geofiz., v. 18, no. 5, p. 15-23.

Claypool, G. E., 1974, Anoxic diagenesis and bacterial methane production in deep sea sediments: PhD thesis, Univ. California, Los Angeles, 276 p.

_____ and I. R. Kaplan, 1974, The origin and distribution of methane in marine sediments, in Natural gases in marine sediments: New York, Plenum Press, p. 99-139.

_____ B. J. Presley, and I. R. Kaplan, 1973, Gas analyses in sediment samples from Legs 10, 11, 13, 14, 15, 18, and 19: Initial Rept. Deep Sea Drilling Project, v. 19, p. 879-884.

_____ C. N. Threlkeld, and L. B. Magoon, 1980, Biogenic and thermogenic origins of natural gas in the Cook Inlet basin: AAPG Bull., v. 64, p. 1131-1139.

Coleman, D. D., 1976, The origin of drift-gas deposits as determined by radioactive dating of methane: Paper presented at the 9th Internat. Radiocarbon Conf., Univ. California, Los Angeles and San Diego.

_____ et al, 1977, Isotopic identification of leakage gas from underground storage reservoirs—a progress report: Illinois Geol. Survey, Illinois Petroleum 111, 10 p.

Colombo, U., et al, 1966, Measurements of C^{13}/C^{12} isotope ratios in Italian natural gases and their geochemical interpretation, in G. D. Hobson and M. C. Louis, eds., Advances in organic geochemistry, 1964: Oxford, Pergamon Press, p. 279-292.

_____ et al, 1969, Carbon isotope study of hydrocarbons in Italian natural gases, in P. A. Schenck and I. Hovenear, eds., Advances in organic geochemistry, 1968: Oxford, Pergamon Press, p. 499-516.

Colonna, R. A., 1977, Methane gas recovery from landfills—a worldwide perspective, in The future supply of nature-made petroleum and gas: New York, Plenum Press, p. 945-962.

Culbertson, O. L., and J. J. McKetta, Jr., 1951, The solubility of methane in water at pressures to 10,000 psia: AIME Petroleum Trans., v. 192, p. 223-226.

Davis, J. B., and R. M. Squires, 1954, Detection of microbially produced gaseous hydrocarbons other than methane: Science, v. 119, p. 381-382.

Deuser, W. G., 1970, Extreme $^{13}C/^{12}C$ in Quaternary dolomites from the continental shelf: Earth and Planetary Sci. Letters, v. 8, p. 118-124.

_____ E. T. Degens, and G. R. Harvey, 1973, Methane in Lake Kivu: new data bearing on its origin: Science, v. 181, p. 51-54.

Dickey, P. A., and W. C. Cox, 1977, Oil and gas in reservoirs with subnormal pressures: AAPG Bull., v. 61, p. 2134-2142.

Doose, P. R., et al, 1978, Interstitial gas analysis of sediment samples from site 368 and hole 369A: Initial Rept. Deep Sea Drilling Project, v. 41, p. 861-863.

Dow, W. G., 1978, Petroleum source rocks on continental slopes and rises: AAPG Bull., v. 62, p. 1584-1606.

Emery, K. O., and D. Hoggan, 1958, Gases in marine sediments: AAPG Bull., v. 42, p. 2174-2188.

Energy Resources Conservation Board, 1977, Alberta's reserves of crude oil, gas, natural gas liquids, and sulphur at 31 December 1977: ERCB Rept. 78-18, 308 p.

Frank, D. J., J. R. Gormly, and W. M. Sackett, 1974, Reevaluation of carbon-isotope compositions of natural methane: AAPG Bull., v. 58, p. 2319-2325.

Friedman, I., and K. J. Murata, 1979, Origin of dolomite in Miocene Monterey Shale and related formations in the Temblor Range, California: Geochim. et Cosmochim. Acta, v. 43, p. 1357-1366.

Fuex, A. N., 1977, The use of stable carbon isotopes in hydrocarbon exploration: Jour. Geochem. Exploration, v. 7, p. 155-188.

Galimov, E. M., 1967, ^{13}C enrichment of methane during passage through rocks: Geochemistry Internat., v. 4, p. 1180-1181.

_____ 1969, Isotopic composition of carbon in gases of the crust: Internat. Geology Rev., v. 11, p. 1092-1104.

Gautier, D. L., in press, Petrology of the Eagle Sandstone, Bearpaw Mountains area, north-central Montana: U.S. Geol. Survey Bull.

Gill, J. R., and W. A. Cobban, 1973, Stratigraphy and geologic history of the Montana Group and equivalent rocks, Montana, Wyoming, and North and South Dakota: U.S. Geol. Survey Prof. Paper 776, 37 p.

Goldhaber, M. B., 1974, Equilibrium and dynamic aspects of the marine geochemistry of sulfur: PhD thesis, Univ. California, Los Angeles.

Hacquebard, P. A., 1977, Rank of coal as an index of organic metamorphism for oil and gas in Alberta, in the origin and migration of petroleum in the western Canadian sedimentary basin, Alberta: Canada Geol. Survey Bull. 262, chap. 3, p. 11-22.

Hathaway, J. C., and E. T. Degens, 1969, Methane-derived marine carbonates of Pleistocene age: Science, v. 165, p. 690-692.

Hand, J. H., D. L. Katz, and V. K. Verma, 1974, Review of gas hydrates with implication for ocean sediments, in Natural gases in marine sediments: New York, Plenum Press, p. 99-139.

Hitchon, B., 1974, Occurrence of natural gas hydrates in sedimentary basins, in Natural gases in marine sediments: New York, Plenum Press, p. 195-225.

Hunt, J. M., 1972, Distribution of carbon in crust of earth: AAPG Bull., v. 56, p. 2273-2277.

_____ 1979, Petroleum geochemistry and geology: San Francisco, W. H. Freeman and Co., 617 p.

_____ and J. K. Whelan, 1978, Dissolved gases in Black Sea sediments: Initial Rept. Deep Sea Drilling Project, v. 42, p. 661-665.

Hutton, W. E., and C. D. ZoBell, 1949, The occurrence and characteristics of methane-oxidizing bacteria in marine sediments: Jour. Bacteriology, v. 58, p. 463-473.

Irwin, H., C. D. Curtis, and M. L. Coleman, 1977, Isotopic evidence for source of diagenetic carbonates formed during burial of organic-rich sediments: Nature, v. 269, p. 209-213.

Johns, W. D., and A. Shimoyama, 1972, Clay minerals and petroleum-forming reactions during burial and diagenesis: AAPG Bull., v. 56, p. 2160-2167.

Kim, A. G., and L. J. Douglas, 1972, Hydrocarbon gases produced in a simulated swamp environment: U.S. Bur. Mines Rept. Inv. 7690, 15 p.

King, G. M., and W. J. Wiebe, 1978, Methane release from soils of a Georgia salt marsh: Geochim. et Cosmochim. Acta, v. 42, p. 343-348.

Kosiur, D. R., and A. L. Warford, 1979, Methane production and oxidation in Santa Barbara basin sediments: Estuar. and Coast. Marine Sci., v. 8, p. 379-385.

Koyama, T., 1963, Gaseous metabolism in lake sediments and paddy soils and the production of atmospheric methane and hydrogen: Jour. Geophys. Research, v. 68, p. 3971-3973.

Kvenvolden, K. A., et al, 1979, Biogenic and thermogenic gas in gas-charged sediment of Norton Sound, Alaska: 11th Offshore Technology Conf. Proc., v. 1, p. 479-486.

Lafayette and New Orleans Geological Societies, 1968, Geology of natural gas in south Louisiana, in Natural gases of North America: AAPG Mem. 9, p. 376-581.

Lebedev, V. S., and E. D. Syngayevski, 1971, Carbon isotope fractionation in sorption processes: Geochemistry Internat., v. 8, p. 460.

_____ V. M. Ovsyannikov, and G. Mogilevskiy, 1969, Separation of carbon isotopes by microbiological processes in the biochemical zone: Geochemistry Internat., v. 69, p. 971-976.

Lockridge, J. P., and P. A. Scholle, 1978, Niobrara gas in eastern Colorado and northwestern Kansas, in Energy resources of the Denver basin: Rocky Mtn. Assoc. Geologists, p. 35-49.

Lyon, G., 1973, Interstitial water studies, Leg 15—chemical and isotopic composition of gases from Cariaco Trench sediments: Initial Rept. Deep Sea Drilling Project, v. 20, p. 773-774.

Magoon, L. B., and G. E. Claypool, 1979, Origin of Cook Inlet oil: 6th Alaska Geol. Soc. Symp., 1977, Proc., p. G1-G17.

Makogon, Y. F., V. I. Tsarev, and N. V. Cherskiy, 1972, Formation of large natural gas fields in zones of permanently low temperatures: Akad. Nauk. SSSR Doklady, v. 205, p. 215-218.

Manheim, F. T., and F. L. Sayles, 1974, Composition and origin of interstitial waters of marine sediments, based on deep-sea drill cores: The Sea, v. 5, p. 527-568.

Marsden, S. S., and K. Kawaii, 1965, "Suiyosei-ten'nengasu," a special type of Japanese natural gas deposit: AAPG Bull., v. 49, p. 286-295.

Martens, C. S., and R. A. Berner, 1974, Methane production in the interstitial waters of sulfate-depleted marine sediments: Science, v. 185, p. 1167-1169.

_____ _____ 1977, Interstitial water chemistry of anoxic Long Island Sound sediments; 1. Dissolved gases: Limnology and Oceanography, v. 22, p. 10-25.

McCaslin, J. C., ed., 1977, International petroleum encyclopedia: Tulsa,, Petroleum Pub. Co., 478 p.

Meents, W. F., 1960, Glacial-drift gas in Illinois: Illinois Geol. Survey Circ. 292, 58 p.

_____ 1968, Illinois glacial-drift gas, in Natural gases of North America: AAPG Mem. 9, v. 2, p. 1754-1758.

Miller, S. L., 1974, The nature and occurrence of clathrate hydrates, in Natural gases in marine sediments: New York, Plenum Press, p. 151-177.

Milton, D. J., 1977, Methane hydrate in the sea floor—a significant resource?, in The future supply of nature-made petroleum and gas: New York, Plenum Press, p. 927-944.

Momper, J. A., 1978, Oil migration limitations suggested by geological and geochemical considerations: AAPG Continuing Education Course Note Ser., no. 8, p. B1-B60.

Nakai, N., 1960, Carbon isotope fractionation of natural gas in Japan: Nagoya Univ. Jour. Earth Sci., v. 8, p. 174-180.

_____ 1962, Geochemical studies on the formation of natural gases: Nagoya Univ. Jour. Earth Sci., v. 10, p. 71-111.

Nesterov, I. I., et al, 1978, Argon isotopes and genesis of natural gas of northwestern Siberia: Akad. Nauk. SSSR Doklady, v. 230, p. 239-240.

Nissenbaum, A., B. J. Presley, and I. R. Kaplan, 1972, Early diagenesis in a reducing fjord, Saanich Inlet, British Columbia—I, Chemical and isotopic changes in major components of interstitial water:

Geochim. et Cosmochim. Acta, v. 36, p. 1007-1027.

Nydegger, G. L., D. D. Rice, and C. A. Brown, 1979, Development of shallow gas reserves in low-permeability reservoirs of Late Cretaceous age, Bowdoin Dome area, north-central Montana: AIME Petroleum Paper no. 7945, 10 p.

Oana, S., and E. S. Deevey, 1960, Carbon-13 in lake waters, and its possible bearing on paleolimnology: Am. Jour. Sci., v. 258, p. 253-272.

Oremland, R. S., and B. F. Taylor, 1978, Sulfate reduction and methanogenesis in marine sediments: Geochim. et Cosmochim. Acta, v. 42, p. 209-214.

Ovsyannikov, V. M., and V. S. Lebedev, 1967, Isotopic composition of carbon in gases of biochemical origin: Geochemistry Internat., v. 42, p. 453-458.

Postgate, J. R., 1965, Recent advances in the study of the sulfate-reducing bacteria: Bacteriol. Rev., v. 29, p. 425.

Presley, B. J., and I. R. Kaplan, 1968, Changes in dissolved sulfate, calcium and carbonate from interstitial water of near-shore sediments: Geochim. et Cosmochim. Acta, v. 32, p. 1037-1048.

_____ et al, 1973, Interstitial water studies, Leg 15, major ions Br, Mn, NH_3, Li, B, Si, and C^{13}: Initial Rept. Deep Sea Drilling Project, v. 20, p. 805-809.

Rashid, M. A., and G. Vilks, 1977, Environmental controls of methane production in Holocene basins in eastern Canada: Organic Geochemistry, v. 1, p. 53-59.

Reeburgh, W. S., 1976, Methane consumption in Cariaco Trench waters and sediments: Earth and Planetary Sci. Letters, v. 28, p. 337-344.

_____ and D. T. Heggie, 1977, Microbial methane consumption reactions and their effect on methane distributions in freshwater and marine systems: Limnology and Oceanography, v. 22, p. 1-9.

Reeside, J. B., Jr., 1944, Map showing thickness and general character of the Cretaceous deposits in the Western Interior of the United States: U.S. Geol. Survey Map OM-10, scale approx. 1 in. to 218 mi.

Rice, D. D., 1980, Coastal and deltaic sedimentation of Upper Cretaceous Eagle Sandstone: relation to shallow gas accumulations, north-central Montana: AAPG Bull., v. 64, p. 316-338.

_____ and G. W. Shurr, 1980, Shallow, low-permeability reservoirs of northern Great Plains—assessment of their natural gas resources: AAPG Bull., v. 64, p. 969-987.

_____ R. B. Powers, and E. W. Scott, 1979, Relation of sedimentary history to natural gas accumulations, western Gulf of Mexico (abs.).: AAPG Bull., v. 63, p. 515.

Roberts, H. H., and T. Whelan, III, 1975, Methane-derived carbonate cements in barrier and beach sands of a tropical delta complex: Geochim. et Cosmochim. Acta, v. 39, p. 1085-1089.

Rosenfeld, W. D., and S. R. Silverman, 1959, Carbon isotope fractionation in bacterial production of methane: Science, v. 130, p. 1658-1659.

Sackett, W. M., 1978, Carbon and hydrocarbons in laboratory simulation experiments: Geochim. et Cosmochim. Acta, v. 42, p. 571-580.

Sayles, F. L., L. S. Waterman, and F. T. Manheim, 1973, Interstitial waters studies on small core samples, Leg 19: Initial Rept. Deep Sea Drilling Project, v. 19, p. 871-874.

Schoell, M., 1977, Die Erdgase der suddeutschen Molasseanwendung von D/H—und $^{13}C/^{12}C$ isot Openanalysen zur Klarung hrer Entstohung: Erdoel Erdgas Zeitschr., v. 93, p. 311-322.

Shipley, T. H., et al, 1979, Seismic evidence for widespread possible gas hydrate horizons on continental slopes and rises: AAPG Bull., v. 63, p. 2204-2213.

Stahl, W., 1974, Carbon isotope fractionations in natural gases: Nature, v. 251, p. 134-135.

_____ E. Faber, and M. Schoell, 1979, Isotopic gas-source rock correlations (abs.): Geol. Soc. America Abs. with Programs, v. 11, p. 522.

_____ et al, 1977, Carbon isotopes in oil and gas exploration, in Internat. Symposium on Nuclear Technology in Exploration, Extraction and Processing of Mineral Resources: Vienna, Internat. Atomic Energy Agency, p. 73-82.

Suffield Evaluation Committee, 1974, Final report relating to evaluation of gas resources of the Suffield Block: Prepared for Province of Alberta, Canada, 31 p.

Sugisaki, R., 1964, Genetic relation of various types of natural gas deposits in Japan: AAPG Bull., v. 48, p. 85-101.

Swetland, P. J., and J. L. Clayton, 1976, Source beds of petroleum in the Denver basin: U.S. Geol. Survey Open-File Rept. 76-572, 23 p.

Tissot, B., et al, 1974, Influence of nature and diagenesis of organic matter in formation of petroleum: AAPG Bull., v. 58, p. 499-506.

Toerien, D. F., and W. H. J. Hattingh, 1969, Anaerobic digestion; I, The microbiology of anaerobic digestion: Water Reseach, v. 3, p.385-416.

Trofimuk, A. A., N. V. Cherskiy, and V. P. Tsarev, 1975, The reserves of biogenic methane in the ocean: Akad. Nauk. SSSR Doklady, v. 225, p. 199-202.

Tucholke, B. E., G. M. Bryan, and J. I. Ewing, 1977, Gas-hydrate horizons detected in seismic-profiler data from western North Atlantic: AAPG Bull., v. 61, p. 698-707.

Wasserburg, G. J., E. Mazor, and R. E. Zartman, 1963, Isotopic and chemical composition of some terrestrial natural gases, in J. Geiss and E. D. Goldber, eds., Earth science and meteoritics: Amsterdam, North Holland Pub. Co., p. 219-240.

Waterman, L. S., F. L. Sayles, and F. T. Manheim, 1973, Appendix II, Interstitial water studies on small core samples, Legs 16, 17, and 18: Initial Rept. Deep Sea Drilling Project, v. 1o, p. 1001-1012.

Weber, V. V., and S. P. Maximov, 1976, Early diagenetic generation of hydrocarbon gases and their variations dependent on initial organic composition: AAPG Bull., v. 60, p. 287-293.

Welte, D. H., 1972, Petroleum exploration and organic geochemistry: Jour. Geochem. Exploration, v. 1, p. 117-136.

Whelan, J. K., and J. M. Hunt, 1978, C_1-C_7 hydrocarbons in holes 378A, 380/380A and 381: Initial Rept. Deep Sea Drilling Project, v. 42, p. 673-677.

Whelan, T., III, J. T. Ishmael, and G. B. Rainey, 1978, Gas-sediment interactions in Mississippi delta sediments: 10th Offshore Tech. Conf. Proc., v. 2, p. 1029-1036.

_____ et al, 1975, The geochemistry of recent Mississippi River delta sediments—gas concentration and sediment stability: 7th Offshore Tech. Conf. Proc., v. 3, p. 71-84.

Winfrey, M. R., et al, 1977, Association of hydrogen metabolism with methanogenesis in Lake Mendota sediments: Applied and Environ. Microbiology, v. 33, p. 312-318.

Wolfe, R. S., 1971, Microbial formation of methane: Adv. Microbial Physiology, v. 6, p. 107-146.

Yermakov, V. I., et al, 1970, Isotopic composition of carbon in natural gases in the northern part of the west Siberian plain in relation to their origin: Akad. Nauk, SSSR Doklady, v. 190, p. 196-199.

Zeikus, J. G., and M. R. Winfrey, 1976, Temperature limitation of methanogenesis in aquatic sediments: Appl. and Environ. Microbiology, v. 31, p. 99-107.

_____ and R. S. Wolfe, 1972, *Methanobacterium thermoautotrophicum* sp. n., an anaerobic, autotrophic, extreme thermophile: Jour. Bacteriology, v. 109, p. 707-713.

ZoBell, C. E., and S. C. Rittenberg, 1948, Sulfate-reducing bacteria in marine sediments: Jour. Marine Research, v. 7, p. 606.

Source Rocks and Hydrocarbons of the North Sea

CHRIS CORNFORD

9.1 Introduction

This chapter has three aims: to review briefly the general characteristics of hydrocarbon source rocks and the generation and migration of oil therefrom; to catalogue, in stratigraphic sequence, the proven and putative source rocks of the North Sea and surrounding areas; and finally to summarise the properties of the reservoired oil and gas in the North Sea area. This text does not attempt an exhaustive treatment of exploration geochemistry. The interested reader is referred, in the first instance, to Tissot and Welte (1984).

Hydrocarbon source rocks can be defined as sediments which are (or were) capable of generating migratable oil or gas. Whether there is sufficient oil or gas to form a commercial accumulation depends largely on the volume and richness of the source rock, its maturity history, the geological framework in which it occurs, and the current economics of exploitation.

A significant abiogenic origin for hydrocarbons (see Gold and Soter, 1982) is not, for a number of serious scientific reasons, considered. The North Sea provides excellent evidence for the association of oil and gas with thick sedimentary sequences, in contrast to the total absence of indigenous hydrocarbons in the surrounding metamorphic shield areas. Overwhelming scientific evidence shows that oil and gas derive from the organic remains incorporated in, and buried with, sedimentary rocks (Tissot and Welte, 1984).

This chapter is concerned mainly with clastic source rocks (mainly shales): the concepts used, and values given to boundary conditions, cannot be uncritically applied to areas such as the Middle East, where chemical sediments (limestones, dolomites, evaporites) comprise major source rock units (Palacas, 1984). Some common depositional environments for clastic source rocks are summarised in Fig. 9.1.

9.2 Recognition of hydrocarbon source rocks

9.2.1 Introduction

The classical hydrocarbon source rock in a clastic environment is an organic-rich, dark olive grey to black, laminated mudstone. The organic matter in the source rock is broadly termed kerogen if solid or insoluble, bitumen if fluid or solvent extractable, and gas if gaseous. The solid kerogen decomposes to bitumen and gas as a result of the increased temperature experienced during burial. The process, termed maturation, gives rise to the generation of oil and gas. However, recent integrated studies of hydrocarbon generation and migration have highlighted a number of geological situations (e.g. Tertiary deltas) where hydrocarbons are sourced from greater thicknesses of relatively organic-lean but well drained sediment. As a result, explorationists now increasingly use source-rock volumetrics and drainage in addition to the amount of oil or gas-prone kerogen, to determine the source potential of a prospect or basin (e.g. Goff, 1983). The concept of source rock drainage is summarised in Fig. 9.2.

Use of this approach requires the recognition of a number of properties of the source rock:

— Quantity of organic matter (TOC or total organic carbon).
— Type of organic matter (oil or gas-prone, or inert).
— Areal extent and lateral variation.
— Thickness of unit and degree of overpressuring.
— Interbedded sands or silts and degree of fracturing.
— Regional maturity boundaries.

These are discussed in more detail later.

Organic-rich shales are fairly uncommon in the geological record, since their deposition requires the coexistence of high bioproductivity and high preservation rates (Muller and Suess, 1979). Recent studies suggest that preservation, rather than productivity *per se*, is the controlling factor. Figure 9.1 summarises schematically some of the environments favouring the deposition of organic rich sediments. High organic preservation is promoted by high sedimentation rates and reduced oxygen concentrations in the water column; see Fig. 9.1 inset (Demaison and Moore, 1980). Reduced oxygen concentrations are found in stratified lakes, in shelf and oceanic basins, in delta swamps, and in oceanic midwater oxygen minima. Sediments associated with these environments are likely to be rich in organic matter. A well-referenced discussion of these processes *vis-à-vis* the Cretaceous black shales of the Atlantic is given by Waples (1983).

The organic matter in a sediment can broadly derive from three sources: higher (land) plants (trees, ferns, etc.), lower (aquatic) plants (planktonic algae, etc.), and bacteria. Volumetrically, animal tissue makes only a minor contribution to kerogen. A land plant

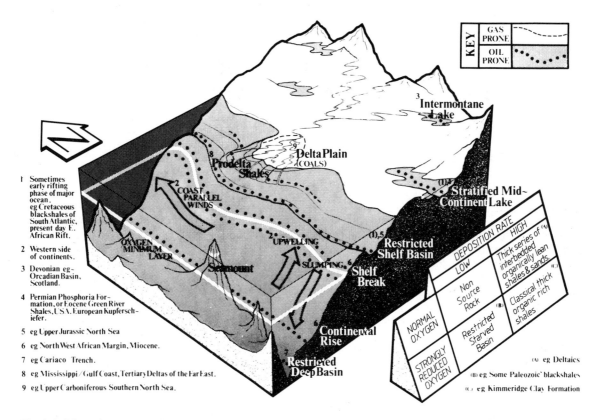

Fig. 9.1. Schematic summary of depositional environments favouring the accumulation of organic-rich oil or gas source-rocks in a clastic regime on a passive continental margin. Note that the gas-prone source rocks of the North Sea (the Westphalian and Middle Jurassic coals) accumulated in a delta plain environment, and the oil-prone black shales of the Upper Jurassic Kimmeridge Clay Formation accumulated in a restricted, shelf-basin environment.

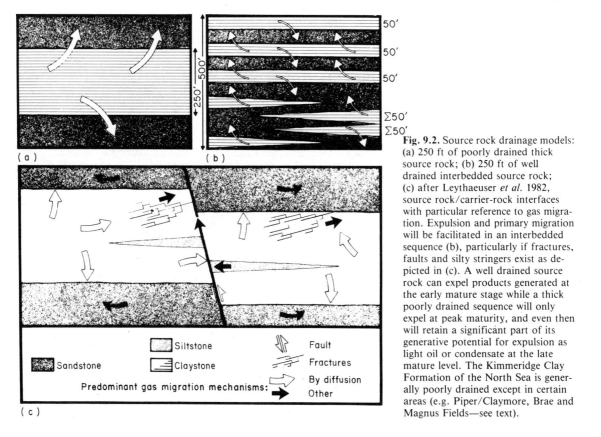

Fig. 9.2. Source rock drainage models: (a) 250 ft of poorly drained thick source rock; (b) 250 ft of well drained interbedded source rock; (c) after Leythaeuser et al. 1982, source rock/carrier-rock interfaces with particular reference to gas migration. Expulsion and primary migration will be facilitated in an interbedded sequence (b), particularly if fractures, faults and silty stringers exist as depicted in (c). A well drained source rock can expel products generated at the early mature stage while a thick poorly drained sequence will only expel at peak maturity, and even then will retain a significant part of its generative potential for expulsion as light oil or condensate at the late mature level. The Kimmeridge Clay Formation of the North Sea is generally poorly drained except in certain areas (e.g. Piper/Claymore, Brae and Magnus Fields—see text).

input to a sediment generally produces a gas-prone, Vitrinitic or Type III kerogen, unless it is altered (e.g. oxidised to charcoal), when it will produce dead carbon, termed Intertinite or Type IV kerogen. However, land plant-derived spores, resin or cuticles can, if present in sufficiently high concentrations in the sediment, generate oil or condensates of a characteristic type (Powell, 1986). This group of land plant tissues is collectively termed Exinite, and comprises part of the Type II kerogen group. An algal and/or bacterial input has the potential to produce oil and associated condensate and gas upon burial. This is termed a Liptinitic kerogen and falls in the Type I or Type II group.

The definition and use of these terms are summarised in Table 9.1 and Figs. 9.3 and 9.4. Using these concepts we can consider the properties of source rocks in more detail.

9.2.2 Quantity of organic matter

An adequate amount of organic matter (measured as organic carbon) is a necessary prerequisite for a sediment to source oil or gas. The quantity of organic matter required for a sediment to be considered a source rock is, like all attempts to define a multi-parameter system by a single variable, a much disputed point. For a typical poorly drained, thick, homogeneous shale (Fig. 9.2a), the following values may be used to rate source rock potential in terms of quantity (but not quality) of kerogen.

<0.5% TOC	very poor
0.5–1.0% TOC	poor
1.0–2.0% TOC	fair
2.0–4.0% TOC	good
4.0–12.0% TOC	very good
>12.0% TOC	oil shale/carbargillite

These ratings should be used to describe the amount of organic matter and not the hydrocarbon source potential, since this will also depend on kerogen type. In addition to kerogen type, drainage must be considered in an assessment of the amount of organic matter required for a sediment to be considered a source rock: in a well drained sequence of interbedded shales and sands (Fig. 9.2b), 1% might be considered good and 2% very good.

A worldwide compilation of TOC data is shown in Fig. 9.5. This shows that high TOC sediments are found in the Cambro-Ordovician and the Lower Carboniferous, with increasing levels from the Jurassic onwards. Claystones have higher TOC contents than sandstones or carbonates, while the size fractions from a single sample show increasing TOC content with decreasing grain size. In addition, petroliferous basins have higher levels than non-petroliferous basins. This is emphasised by the low mean value for the 7253 Deep Sea Drilling Project analyses of non-exploration-related samples.

Table 9.1. Recognition of oil-prone or gas-prone kerogen using various analytical techniques

Hydrocarbon potential	Organic petrography			Pyrolysis[1] (Rock Eval)	$\delta^{13}C$ (pdb)	H/C atomic[2]	Origin (not depositional environment)
Oil- (→condensate →gas) prone	Liptinite	Algal/ Amorphous (sapropel)	Amorphogen[3]	Type I	< −28	1.5–2.0	Aquatic (freshwater or marine) algae, often bacterially degraded to yield amorphous material. Terrigenous spore, pollen, cuticle or resin, can also be degraded.
Condensate- prone	Exinite		Phyrogen	Type II	(Freshwater algae −28 to −32)	1.0–1.5	
Gas-prone	Vitrinite	Herbaceous	Hyalogen	Type III	> −28	0.5–1.0	Terrigenous, ligno-cellulosic tissue, relatively unaltered—present in a particulate or amorphous form.
		Woody					
Inert (dead carbon)	Inertinite	Coaly	Melanogen	Type IV (Type IIIB)		<0.5	Terrigenous as above but altered by oxidation in soil, during transport, or from forest fires, etc.

[1] Rock Eval is a pyrolysis technique carried out on the whole rock, and yields values of Hydrogen and Oxygen Indices as defined in Fig. 9.4. Type II is often a mixture of Type I plus Type III/IV kerogen.
[2] atomic ratio of the immature kerogen.
[3] Bujak, Barss and Williams, 1977.

9.2.3 Type of organic matter

For application to hydrocarbon exploration, sedimentary organic matter can conveniently be divided, like Gaul, into three parts:

— Oil-prone components
— Gas-prone components
— Inert components (or dead carbon)

The kerogen of a typical rock will contain a mixture of all three components. When estimating the oil potential of a sediment, only the oil-prone part of the TOC of that sediment should be considered. For example, a 4% TOC sediment with 50% oil-prone kerogen will have 2% oil-prone organic carbon (2% OPOC). If, in addition, it had 25% gas-prone organic matter it would be said to have 1% gas-prone organic carbon (1% GPOC).

It is convenient to consider the organic carbon content in this way, but it should not be forgotten that hydrogen is the limiting element in all source rocks. Stated another way, once the hydrogen-rich phases of oil and gas (general formulae CH_2 and CH_4 respectively) have been generated from the rock, a carbon-rich graphitic schist almost always remains.

In apportioning gas or oil generative capacity to kerogen, it should be remembered that oil-prone kerogen (or reservoired oil itself) will crack to yield both wet and dry gas if buried, and hence heated, sufficiently. The generation of hydrocarbons from the broad categories of kerogen type is summarised in Fig. 9.3.

Kerogen type is generally determined by microscopy (organic petrography), by pyrolysis (e.g. the Rock Eval method) or by elemental (C,H,O) analysis. Confirmatory evidence can be provided by stable carbon isotope ($\delta\ ^{13}C$), light hydrocarbon, or sediment extract analyses. Because at present, no single technique is totally reliable in all situations, a combination of complementary methods is typically used. Table 9.1 shows a comparison of a number of techniques in terms of their ability to categorise oil and gas potential.

The definition of kerogen Types I, II, III and IV from Rock Eval pyrolysis using Hydrogen and Oxygen indices is shown in Fig. 9.4. The changes of these

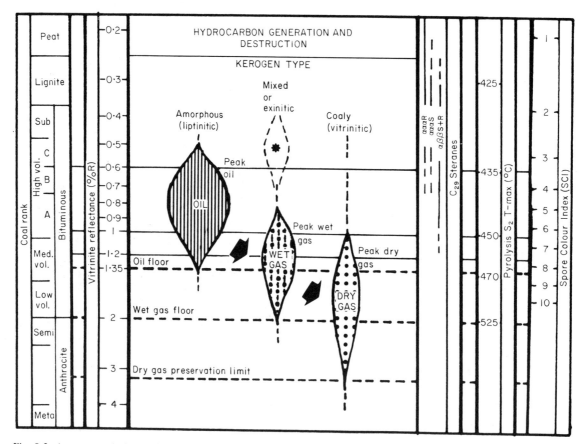

Fig. 9.3. A conceptual view of hydrocarbon generation from different kerogen types (after Dow and O'Connor, 1982). A detailed equivalence of the maturity parameters is not implied since the 'time' factor is not considered—see text and Figs. 9.7 and 9.8 for detailed generation curves. Note that kerogens typically contain a mixture of organic matter types: the pure wet gas-prone kerogen type termed 'mixed' comprises spore/pollen, cuticle, resins and possibly more hydrogen rich vitrinitic debris. The generation of early mature wet-gas/condensate (*) has been described from this type of organic matter at vitrinite reflectance values as low as 0.4%R (e.g. Connan and Cassou, 1977; Snowdon and Powell, 1979, 1982). No generation of early mature condensate has been described in the North Sea area. Sterane stereochemistry designated 5α, 14α, 17α, 20R, etc.

Source Rocks and Hydrocarbons of the North Sea

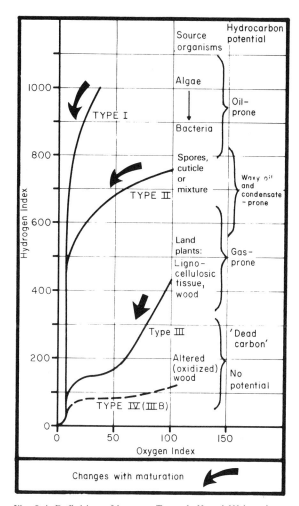

Fig. 9.4. Definition of kerogen Types I, II and III based on a Rock Eval Oxygen Index/Hydrogen Index diagram. Note that Type II kerogen can (occasionally) be a pure spore/pollen/cuticle/resin kerogen, but is generally a mixture of bacterially degraded algal debris of Type I composition, mixed with a terrigenous component of Type III composition. Type IV (sometimes called Type IIIb) comprises altered terrigenous debris (Inertinites).

indices with maturity are indicated by arrows in this figure. As pyrolysis is a bulk determination, a 'Type II' kerogen can be (and in fact generally is) a mixture of Type I algal material, degraded by bacteria and mixed with terrigenous Type III material (Barnard et al., 1981). A pure Type II kerogen comprises spores, pollen, cuticle, etc. The hydrocarbon potential of resin (resinite) is for both naphthenic oil and gas/condensate (Powell, 1986; Mukhopadhyay and Gormley, 1985).

Debate has raged over the possible role of coals as oil or gas/condensate source rocks (Durand and Paratte, 1983): recent publications (Thompson et al., 1985; Powell, 1986) have shown that both selective sedimentation of the more oil-prone tissues of land plants, and their concentration in coals due to the oxidative removal of the humic fraction during sub-aerial exposure at times of low water table, increase the oil potential of the coals.

9.2.4 Recognising source rocks from wireline logs

Organic-rich rocks can often be recognised using a normal suite of wireline logs (Fertl, 1976). Meyer and Nederlof (1984) have recently discussed the identification of source rocks from a statistical evaluation of gamma ray, sonic, density and resistivity log responses, with examples given from the North Sea. Indeed the term 'hot shale' applied to organic-rich sections of the Kimmeridge Clay Formation of the North Sea derives from its high natural radioactivity. The log response of a source rock unit is largely that which would be predicted from its physical properties as summarised in Table 9.2. The use of logs to define source rocks is particularly appropriate for volumetric studies of source rocks since it allows a continuous monitor of source quality over the whole section. Some conventional analyses are, however, always required in order to calibrate the log responses. Typical log responses through some of the Cretaceous and Jurassic rock intervals of the North Sea are shown in Fig. 9.6.

9.3 Maturation and generation

Kerogen matures during burial, that is, it undergoes physical and chemical changes that result largely from the rise in temperature related to the progressive increase in the depth of burial. One result of these changes is the generation of hydrocarbons—first liquids (oil), and then gas. This process is shown schematically in Fig. 9.3.

Quantitative generation from an oil-prone (Type II) kerogen of the type found in the Kimmeridge Clay Formation of the North Sea is shown in Fig. 9.7. The vertical axis is temperature, which can be equated with depth by using an average geothermal gradient (North Sea average 29°C/km (1.6°F/100 ft), range 22–40°C/km (1.2–2.2°F/100 ft); (see Fig. 9.9). North Sea temperature data have been treated in detail by Harper (1971), Cornelius (1975), Cooper et al., (1975), Carstens and Finstad (1981) and Toth et al., (1983). In Fig. 9.7, generation of oil is given in m^3 oil/km^3 of 1% TOC source rock (bbls/acre ft), and associated gas in terms of m^3 gas at standard temperature and pressure/km^3 of 1% TOC source rock (mcf/acre ft).

The progressive generation of dry gas (methane) from coal and coaly shales can be seen in Fig. 9.8, where maturity is measured in terms of vitrinite reflectance and coal rank.

A number of experimentally determined parameters are used to measure kerogen maturity, including vitrinite reflectance, and the properties of coal in general (Teichmüller and Teichmüller, 1979; Bostick, 1979; Robert, 1980). Spore or kerogen colour (Smith, 1983) and, to a lesser extent, fluorescence (van Gijzel, 1981; Smith 1984), are used together with the

Table 9.2. Source rock characterisation from wireline log response

Log (Units)	Response: Change with increase in organic matter	Response: Coals	Values in Kimmeridge Clay Formation	Comments
Natural gamma ray (API units)	Generally very[1] high	Low except in uranium rich coals	Typically 100-200 API units	Response mainly due to uranium[6] (Bjørlykke et al., 1975; Berstad and Dypvik, 1982) associated with planktonic matter. Lower response in gas-prone shale. Very low response in lacustrine organic rich shales.
Formation density (g/cc)	Decreases[2][3]	Very low	Typically 2.2-2.3g/cc	Density contrast can be offset by high pyrite contents.
Resistivity (ohm.m)	Increases	Very high	Immature <5ohm.m Mature >20ohm.m Late mature <2ohm.m (Goff, 1983)	Resistivity low in immature shales due to saline pore water, and high due to oil filled porosity at mature stage. Resistivity falls at late mature stage due to increased pore water content and 'graphitisation' of the organic matter (Meissner, 1978)
Sonic interval velocity (μsec/ft)	Increases[4]	Very high	Variable	Also increases with burial, little affected by pyrite and hole condition.
Neutron (counts/second)	Decreases[5]	Low	—	
Pulsed neutron carbon/oxygen	—	—	—	New logs (Hopkinson et al., 1982): may be able to recognise richness or even, when calibrated, kerogen type in source rocks.

(1) Schmoker (1981) has defined the following relationship between the volume fraction of organic matter (ϕ_0), and γ, the gamma ray log response in API units for the Devonian shales of the Appalachians: $\phi_0 = \gamma_B - \gamma/1.378A$, where γ_B is gamma log response if no organic matter is present and A is the slope of the crossplot of gamma and density logs.
(2) Schmoker (1979) has shown that the volume fraction of the organic matter ϕ_0 is related to log density in the Devonian shales of the Appalachians by $\phi = \rho_B - \rho/1.378$ where ρ_B is the rock density where no organic matter is present.
(3) Tixier and Curtis (1967) have related Fisher Assay pyrolysis yield to log density (ρ) for the Green River oil shales with Type I kerogen as follows: yeild = 154.81 − 59.43ρ gals/ton ($\times 3.8 \approx$ litres/tonne), while Meyer and Nederlof (1984) have found the following relationship for the Posidoniaschiefer of Germany (generally a Type II kerogen): yield = 1.113 + 63.14 $\Delta \rho$ gals/ton where $\Delta \rho$ is the density difference between lean and rich shale.
(4), (5) See Fertl (1976) Figures 10.7 and 10.8 respectively for Fisher Assay pyrolysis yields vs. log response.
(6) The gamma ray log response in the Kimmeridge Clay Formation shales of Amoco's NOCS well 2/11-1 derives 61% from Uranium, 33% from Thorium and 6% from Potassium (Bjørlykke et al., 1975).

temperature of maximum pyrolysis yield (T_{max}) from the Rock Eval pyrolysis technique (Tissot and Welte, 1984), and molecular ratios e.g. sterane/triterpane isomer ratios (Cornford et al., 1983; Mackenzie, 1984; Mackenzie et al., 1980). No one technique can be applied universally, and no simple equivalence can be drawn up between the various techniques. Vitrinite reflectance is the most widely used technique, while the rapidly developing field of molecular ratios has the important advantage that measurements can be made on both source rock and reservoired oil, thus allowing the establishment of a direct genetic link (Cornford et al., 1983). These measurements also have the advantage of reflecting maturation changes within the oil-prone material directly, and do not require empirical correlation with the generation process. An approximate equivalence of some of these maturation parameters is given in Fig. 9.3.

In the absence of measured maturity parameters, a first estimate of the maturity of a source rock can be made from its maximum depth of burial. This initial opinion can be refined by using the geothermal gradient —either past or present—to determine the maximum temperature the strata of interest has experienced

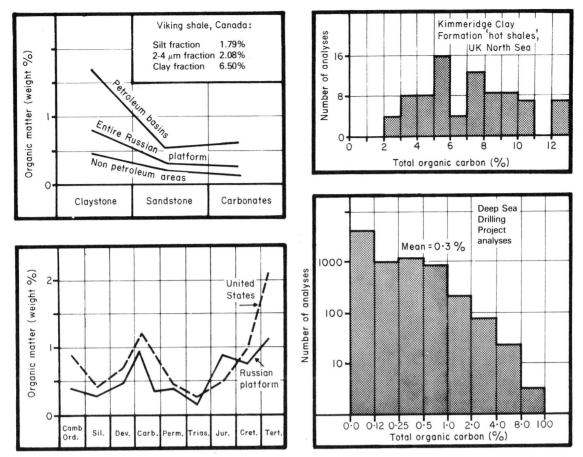

Fig. 9.5. Worldwide compilation of data on Total Organic Carbon (TOC) and organic matter of sediments showing fluctuation with lithology, with grain size within a single sample, and with stratigraphic interval (Bostick, 1974). Histograms are also shown of 7253 analyses from the Deep Sea Drilling Project analyses (McIver, 1975), and the Kimmeridge Clay Formation of the UK North Sea. Note that sampling is probably biased towards higher TOC intervals in all data collections.

during burial. In continuously subsiding basins such as the North Sea Grabens, the maximum temperature experienced is related to depth of burial via the local geothermal gradient (°C/km or °F/100 ft), see Fig. 9.9. The detailed variation of geothermal gradient with geological time, and with depth in the sedimentary section, is a complex function of the interaction of a temporally fluctuating heat flow, the continuously changing thermal conductivity of the progressively compacting strata (e.g. Andrews-Speed et al., 1984), ground water movement, and a surface temperature that changes with the paleo-climate (e.g. Hallam, 1985).

However, the time that a source rock has been exposed to a certain temperature has a major effect on some maturity parameters. For a given temperature, the reflectance of vitrinite is strongly time dependent (Hood et al., 1975; Waples, 1980; Gretener and Curtis, 1982), spore colour is less dependent (Barnard et al., 1981a), whilst some common isomer ratios appear to be variably affected (Mackenzie and McKenzie, 1983; Cornford et al., 1983).

The effect of time on the actual volumetric generation of oil itself is not clear, but it seems to be relatively small (Price, 1983). The quantitative prediction of maturity from subsidence curves and geothermal gradients is a complex matter: computer modelling is the only way to gain a feel for the interactions of heat flow, changes in surface temperatures, sediment compaction, subsidence rates, and uplift and erosion on the maturational process. It is the author's experience that if applying the Lopatin approach, as popularised by Waples (1980), the correlation of Time Temperature Indices (TTI units) with vitrinite reflectance given by Issler (1984) gives better agreement with measured values in the North Sea than the correlation given in the Waples publication. Specific TTI/VR correlations based on North Sea data have been published by Lerche et al. (1984) and Maragna et al. (1984). Those wishing to take this subject further should refer to the publications listed above, and to Waples (1984), Guidish et al. (1985), and Piggot (1985) as a source of references.

9.4 Source rock volumetrics and migration

Attempting a mass-balance between the amount of hydrocarbon generated in the source rock and the amount finally found in place in the reservoir has produced at least three major benefits: firstly, and probably most importantly, it has generated a dialogue between the reservoir engineer, the geophysicist, the exploration geologist, the geochemist, the sedimentologist and the stratigrapher, and has highlighted the interlinking of these disciplines; secondly it constitutes an important part of the quantitative prospect

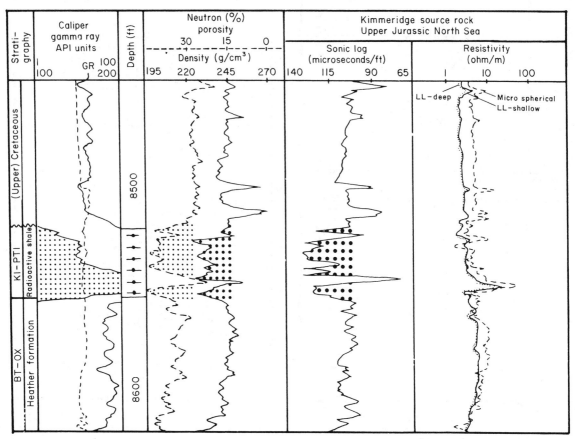

Fig. 9.6. Generalised sketch of log motif through the hot shales of the Kimmeridge Clay Formation source rocks of the North Sea (Meyer and Nederlof, 1984).

evaluation procedure now adopted by many companies; finally it has placed certain important limits on the logistics of the migration process.

9.4.1 Source rock-reservoir volumetric calculations

For exploration in frontier areas it is probably sufficient to know that you have a source rock and that it is likely to be mature; major structures will certainly be drilled even without the comfort of this information. For more marginal structures especially in mature exploration areas, however, a comprehensive prospect evaluation requires evidence that there is sufficient volume of mature source rock of an adequate quality to fill the proposed trap with oil. Also of importance is to determine that the oil was generated at an appropriate time with respect to trap formation. This approach is of great value when rating prospects, that is, deciding which one to drill first, or which acreage to bid for or to relinquish.

The concepts of source rock volumetric estimation have been discussed by Fuller (1975), Tissot and Welte (1978) and Goff (1983). For a simple approach it consists of four steps:

1. Establishing the areal extent and thickness of source rock(s) draining into the structure.

2. Defining those parts of the drainage area that reached maturity after trap/seal formation.
3. Determine the volume of hydrocarbon (oil or gas) generated per unit volume of source rock.
4. Estimate the efficiency of (a) expulsion from the source rock; (b) the migration path to the trap, and (c) of the seal on the trap (Bishop et al., 1984; Downey, 1984).

Although it is beyond the scope of this paper to deal with this topic in detail, the following comments, which are summarised in Fig. 9.10, outline the most important points.

The effective drainage area of a structure may be bounded by one or all of the following features (Fig. 9.10a)

1. The immature/early mature, or early mature/mid-mature boundary, above which no hydrocarbon generation will have occurred. It is of particular importance in areas of marginal maturity to determine this boundary with care: both the sedimentology of the source rock and the type of the kerogen must be considered.
2. The late mature/post-mature boundary for the source rock reconstructed for the time of trap/seal formation. In a continuously subsiding basin, this will generally be stratigraphically deeper than the present day late mature/post-mature boundary.

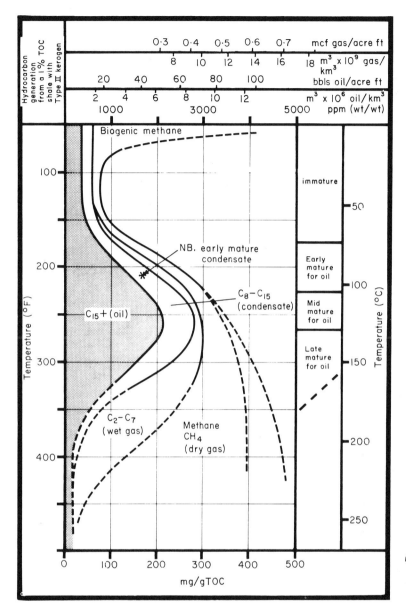

Fig. 9.7. Qualitative generation of oil and condensate: m³ oil/km³ of 1% TOC source rock (bbls/acre ft); and gas: m³CH_4/km³ of 1% TOC source rock (mcf/acre ft), as a function of down-hole temperature for oil-prone kerogen. For details see Brooks, Cornford and Archer (in press). Note the possible generation of larger quantities of early mature condensate from spore/pollen/cuticle/resin rich sediments, and expulsion if drainage is good. To convert to a specific hydrocarbon type of density ρ, use ppm wt/wt × (rock bulk density/ρ) = ppm vol/vol. (-oil ≈ 0.8-0.9; ρ-methane = 7.1×10^{-4} at standard temperature and pressure); lbbl = 5.614 cu ft = 159 litres; 1 acre ft = 1233.5 m³.

3. Sealing faults. The sealing nature of the faults can be evaluated from the nature of the rock on either side of the fault plane, the relative movement along the fault (strike or dip slip), and the timing of the last movement episode.
4. Erosional or regressional sub-crop of the source rock unit.
5. Facies change of the source rock from, say, oil-prone to gas-prone, or from shale to sandstone.

In addition, a maximum migration distance may be a limiting factor, but this is contentious (see next section). Certainly, the regional extent of the proposed migration pathway (e.g. a sheet sand) could place limits on the effective drainage area. The thickness of the source rock unit can be determined from seismic mapping, or from regional compilations. Only that part of the drainage area which reaches maturity after trap formation should be considered. This can be deduced from a subsidence curve for representative parts of the drainage area (Fig. 9.10b). In Fig. 9.10b, part of the source rock in the basin centre (location A) generated its hydrocarbon prior to trap formation at about 110 my bp.

The volume of hydrocarbon generated per unit volume of source rock can be obtained from Figs. 9.7 and 9.8, given an average value for the total organic carbon content of the unit. If the quantity or type of kerogen varies laterally or vertically, then the drainage area should be compartmentalised, and each compartment treated separately.

Finally, source rock studies have now moved into the area of three dimensional modelling of the basic physical and chemical processes (Welte and Yukler,

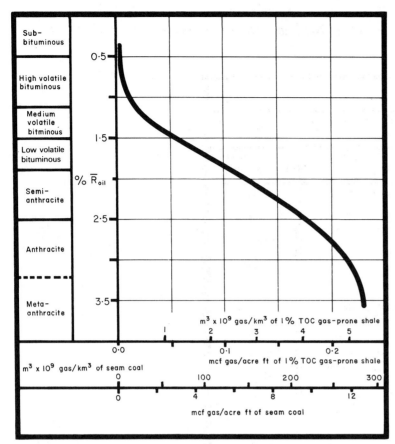

Fig. 9.8. Cumulative methane generation in m³/km³ (millions of cu ft/ acre ft) from 1% TOC gas-prone shale (upper horizontal scale) and coal (lower horizontal scale) as a function of maturity measured in terms of coal rank and vitrinite reflectance. Since methane is highly mobile, the stated volumes of gas will not be present in the coal or shale, but will start to migrate out as soon as generated. The curve is based on laboratory data of Juntgen and Karweil (1966) and stochiometric calculations based on the change of hydrogen content with rank. Strictly, the plot can be used only for vitrinite-rich coals and kerogens, for example, the Westphalian of NW Europe.

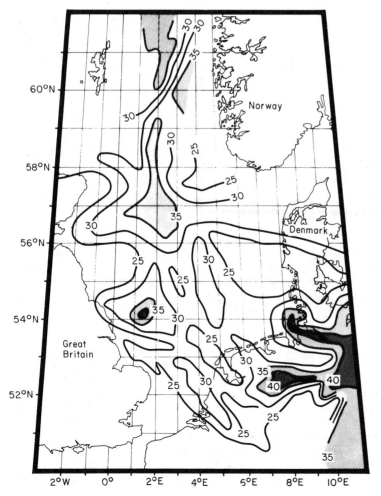

Fig. 9.9. Present day North Sea geothermal gradients (after Cornelius (1975) and Carstens and Finstad (1981); see also Harper, 1971). Considerable errors exist in obtaining geothermal gradients from well-log data—see references above. The gradient plotted is the overall geothermal gradient from sea-floor to the lowest measuring point, normally at terminal depth. Geothermal gradients vary consistently with depth, and with lithology: heat flow is the conserved property at any one location (Oxburgh and Andrews-Speed, 1981).

1981; Ungerer *et al.*, 1984; Yukler and Kokesh, 1984). These large scale computer models are becoming able to predict accurately where oil is generated by thermal modelling, and where it flows by pressure modelling. They do, however, require a large amount of input information. These large, quantitative models force the explorationist to view oil generation and accumulation as a unified process, and have greatly benefitted our understanding by focusing attention on the interrelation of a number of poorly understood but critical areas of hydrocarbon genesis.

9.4.2 Migration of oil and gas

Migration—the movement of hydrocarbons in the subsurface—is controlled by simple physical principles: hydrocarbons diffuse from high to low concentrations; they move along a pressure gradient from high to low pressure; they rise under buoyant forces if the surrounding liquid is of higher density; and, if dissolved or immersed in moving formation water will be transported hydrodynamically.

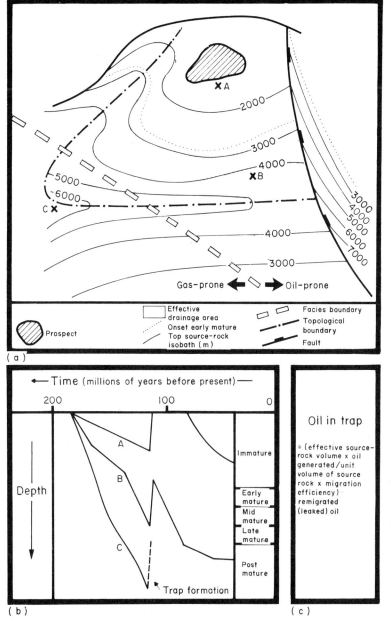

Fig. 9.10. Boundary conditions for source rocks volumetric calculations.
(a) Depth to top source rock on seismic map showing prospect, together with some boundaries for the effective drainage area.
(b) The relationship of generation to trap formation using subsidence plots for three key points A, B and C in the basin (marked with 'x' on map).
(c) A simplistic mass balance—for detailed discussion see Goff (1983), and Fuller (1975).

Migration can usefully be visualised as three consecutive processes

1. Expulsion of the hydrocarbon phase(s) from the kerogen through the fine grained source rock matrix.
2. Primary migration or drainage of the oil or gas through silty stringers, fractures, etc., within the source rock unit.
3. Secondary migration of the oil or gas through regionally extensive more permeable strata to the reservoir.

Uncertainties about migration certainly comprise a significant source of doubt in many prospect evaluations, and contribute the major sources of error in volumetric calculations. Mechanisms of expulsion and migration are currently the topic of animated debate—largely in the absence of sound geological observation (Roberts and Cordell, 1980; Durand, 1983). The movement of light hydrocarbons has, however, been extensively studied (Leythaeuser et al., 1982, and references therein), and the ability of isotopes (James, 1983), gasoline range hydrocarbons (Thompson, 1979, 1983; Kurbskiy et al., 1983), and biomarkers (Cornford et al., 1983) to link reservoired hydrocarbons with their source rocks in terms of maturity, has done much to place limits on the vertical and lateral extent of migration.

Detailed studies of molecular concentration gradients on a decimetre scale (e.g. Leythaeuser et al., 1984), have increased our understanding of the movement of heavier hydrocarbons from a fine grained mudstone source rock into adjacent coarser grained silt or sandstone strata. Diffusive enrichment of the lower molecular weight molecules is seen in the migrated extract. Such fractionation will only be observed during active migration: at equilibrium the higher molecular weight molecules would be expected to have caught up. Lindgreen (1985) has recently discussed migrational processes in the Draupne Formation claystones of the Norwegian Central Graben with respect to micro-porosity and cementation. A long history of overpressuring and fracturing is indicated. An early stage, relating to the loss of fresh, neutral to weakly alkaline compaction water, is followed by synchronous expulsion of hydrocarbons and more alkaline, saline water.

The results of these semi-quantitative studies, summarised in Fig. 9.11. (inset), show that for oil, expulsion of between 10% and 30% of the total generated, is normal in thick, rich source rocks as depicted in Fig. 9.2a. Expulsion efficiencies can rise as high as 50%, or even 80%, in well-drained, interbedded sand-shale sequences (e.g. Fig. 9.2b). The corollary is that well drained source rocks retain less potential for the generation of late mature light oil and gas/condensate, while thick poorly drained source rocks favour such products. Values for migration efficiency derive from mass-balance calculations equating the volume of source rock with the volume of reservoired oils.

Secondary migration efficiencies range from zero (all the oil or gas dissipated within the migration pathway), to as high as 90% for a simple up-dip movement. As shown in the inset of Fig. 9.11, controls include the distance of migration and the effective permeability of the pathway. Inefficient, tortuous, short-distance migration through low permeability rock can be contrasted with efficient, long-distance migration through a high permeability regional sand. Distance affects efficiency due to the residue of oil or gas left in the migration conduit. Experience seems to indicate that oil and gas do not have to saturate the entire rock volume in a secondary migration path. Rather, the hydrocarbon moves in the form of a braided stream along the roof of the conduit, running from high-point to high-point, and favouring the high permeability strata.

Figure 9.11 depicts cartoons of some of the routes for secondary migration invoked to explain the majority of North Sea hydrocarbon occurrences (Cornford et al., 1986). Well drawn detailed seismic-stratigraphic sections, such as those shown in Harding (1984), are necessary to propose a realistic migration pathway. The pathway can be validated by geochemical means, as depicted in Fig. 9.18.

The shortest distance of migration together with the highest oil expulsion efficiencies are to be expected in areas where sand intercalates with the 'hot shales' of the Kimmeridge Clay Formation, as in the Piper, Brae

Fig. 9.11. The logistics of migration in the North Sea (after Cornford, Needham and DeWalque, 1986). A. Short distance migration from interbedded or adjacent sands. Note high efficiency of source rock drainage favours expulsion. B. Migration to the crest of rotated fault blocks. The major source rock areas lie in the back-rotated half-grabens, with the listric faults generally forming a relatively effective barrier to migration. Note paradoxical downward migration of Kimmeridge Clay/Draupne sourced oil, through the Heather into the Brent sands for secondary migration. C. Migration with a major vertical component: a number of explanations have been proposed but unequivocal evidence favouring any one is lacking. D. Upflank migration from the graben axis—not very common but results in super-giant fields. Such a migration route requires a relatively unfaulted graben flank with a regionally extensive conduit (e.g. sandstone). Note that the vast majority of the source rock volume in the graben as depicted, drains to the east (right). E. Remigration on inversion. Flank-reservoired gas remigrates into more axial traps on inversion. This is a relatively inefficient process, requiring an excellent seal (e.g. salt) for the hydrocarbons to survive in significant quantity. F. Unconformity and fault migration. A combination of migrational modes progressively fill a series of unconformity and Tertiary sand traps, with bacterial degradation in the shallower trap. Course clastic lags and weathered surfaces at unconformities provide excellent migration routes: claystone/claystone unconformities like the Late Cimmerian horizon, though well-defined on seismic due to the velocity contrast between the shales, will not provide a plausible migration path (see B and D above).

Inset. Current best estimates of the efficiencies of primary and secondary migration processes based on reservoir/source rock volumetric balances and quantitative investigations of localised redistribution processes (e.g. Leythaeuser et al., 1984).

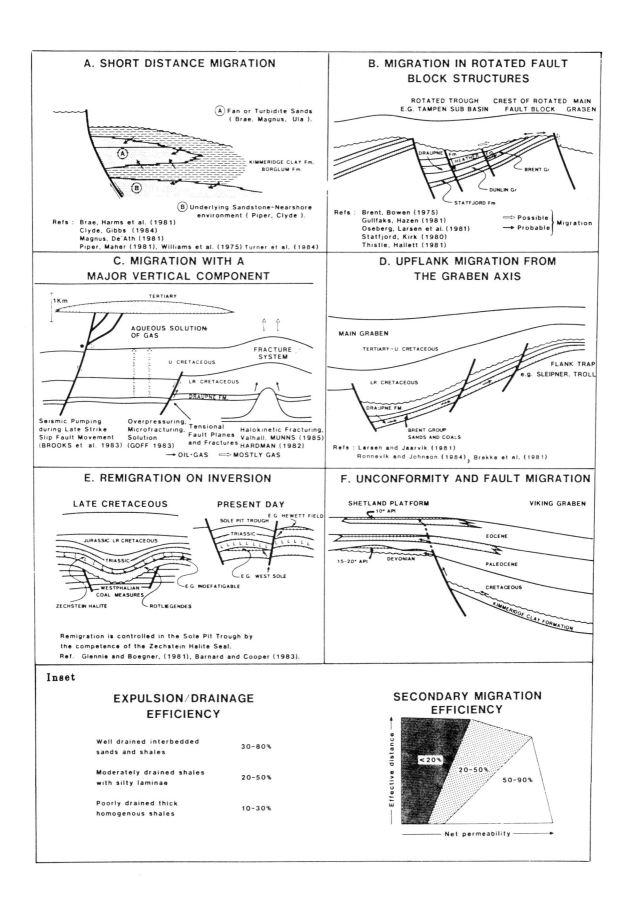

and Magnus Fields (Fig. 9.11A). High migration efficiencies are also associated with upflank movement through regionally extensive, laterally continuous conduits, to a structure at the natural focus (Pratsch, 1983) of the migration path (Fig. 9.11D). Such a simple secondary migration pathway may exist in the Brent Sands of the East Shetland Basin (Fig. 9.11B). In contrast, long distance or devious migration pathways through low permeability conduits (e.g. ratty sands, faults, fractured chalk, etc.) are likely to be much less efficient.

The presence of oil sourced from the Kimmeridge Clay Formation in the Tertiary reservoirs of the North Sea can sometimes pose problems in terms of defining a plausible migration route (Fig. 9.11C). In the Ekofisk Field faults are believed to have played a major role (Van den Bark and Thomas, 1980), while in the case of the Frigg Field overpressure-induced diffusion and hydrodynamic transport in solution have been invoked (Goff, 1984). It is even more difficult to explain the presence of biodegraded Jurassic oils in apparently closed, lensoid, Tertiary sand bodies, with no faulting apparent on seismic lines. In this case, not only must the oil get into the reservoir, but so also must ground water to carry the bacteria and their nutrients (Connan, 1984). In this case, Brooks et al. (1983) have proposed a major role for strike-slip faulting in both the vertical movement of oil and ground water and in the accumulation of the reservoir sands in the area of UK Quadrant 9. Pegrum and Ljones (1984) have also noted the effects of wrench faulting in the Sleipner Field complex, where gas/condensate migration into the Palaeocene sands of the Block 15/9 Gamma structure results.

Gas expulsion is probably nearly 100% efficient but, as Leythaeuser et al. (1982) have pointed out, loss by diffusion during migration and through the reservoir seal can be very significant over geological time. These authors conclude that without continuous topping-up with gas from source rock kitchen areas, most gas fields will have a relatively limited life. In the Ekofisk area, Van den Bark and Thomas (1980) and Munns (1985) have identified 'gas chimneys' on seismic lines, interpreted as the leakage of gas from the Chalk through the Tertiary. Active replenishment is expected in this area (see Section 9.5.8). Other examples of 'gas chimneys' defined on seismic lines are given by Nordberg (1981) and Brewster and Dangerfield (1985), while Hovland and Sommerville (1985) have proved the deep thermal origins of surface gas seeps in the Central and Witch Ground Grabens (Norwegian Block 1/9 and UK Block 15/25). Detailed surface prospecting for vertically migrated gas using stable carbon isotope analyses were reported by Faber and Stahl (1984).

In contrast, the gas in some of the gas fields of the southern North Sea must have been trapped since its generation in the late Cretaceous (Glennie and Boegner, 1981), although in the Broad Fourteens Basin (Oele et al., 1981) and the giant Groningen gas field (van Wijhe et al., 1980) the gas may still be being 'topped-up' to the present day. Remigration to new reservoirs also has occurred as a result of tectonic inversion of former basinal areas (Fig. 9.11E, and see Glennie, chapter 3 this volume). The longevity of the gas fields of the UK southern North Sea may in large part be due to the excellent seal afforded by the Zechstein halite and anhydrite deposits.

9.5 Stratigraphic distribution of recognised and possible source rocks

9.5.1 Introduction

This section comprises a review of the Phanerozoic organic-rich sediments of the North Sea and surrounding areas (Fig. 9.12), giving, where possible, the amount (TOC) and type (oil-prone, gas-prone) of the sedimentary organic matter, and its maturity, where known from published sources. In covering the whole stratigraphic column, it should not be forgotten that the Kimmeridge Clay/Draupne Formation (Upper Jurassic/basal Lower Cretaceous) is overwhelmingly the most important oil source rock, and the Carboniferous Coal Measures (Westphalian) and locally the Middle Jurassic coals are the only well established source rocks for dry gas in the region. Barnard and Cooper (1981) have previously reviewed the hydrocarbon source rocks of the North Sea, whilst more recently, Thomsen et al., (1983) and Rønnevik et al. (1983) have summarised data on Danish and Norwegian acreage respectively. To the north of the 62nd parallel, Mørk and Bjorøy (1984) have reviewed the Mesozoic source rocks of Svalbard, while other papers in Spencer et al. (1984) provide additional source rock information in the northern area.

9.5.2 Cambro-Ordovician

Cambro-Ordovician black shales are found on the shelves and basins of the Iapetus Ocean, within and flanking the Caledonides, from the Kukersite oil shales of Estonia in the east (Duncan and Swanson, 1965) to the Appalachians in the west (Islam et al., 1982). Within the area of interest, the Cambro-Ordovician black shales outcrop in the Southern Uplands of Scotland and in Scandinavia, and could source oil and gas. The Alum shale of southern Sweden contains as much as 17.5% TOC (Andersson et al., 1982; Bitterli, 1963), and is oil-prone (4–8% oil yield upon pyrolysis as an oil shale). At least part of the outcrop currently falls within the oil generation window (Bergstrom, 1980).

In contrast, the Ordovician black shales of the Southern Uplands of Scotland are post-mature for oil and gas generation at Hartfell and Dobb's Linn, and late-mature for gas generation at Mountbenger (average uncorrected graptolite reflectance values of 4.45%R, 3.27%R and 2.20%R respectively; Watson, 1976). Both Watson (1976) and Bergstrom (1980) note a decrease in maturity to the west.

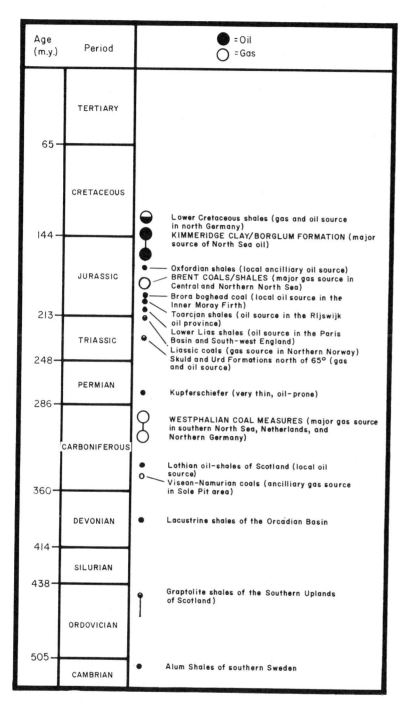

Fig. 9.12. Stratigraphic distribution of source rocks in the North Sea area. Major, proven source rocks are in upper case, less significant or putative source rock intervals are in lower case text. Details are given in Section 9.5 for each stratigraphic interval.

There is no current evidence of a Cambro-Ordovician source for any North Sea hydrocarbon occurrence, but generation from these rocks is known in the palaeo-contiguous Appalachians of North America in the west, and in Estonia to the east.

9.5.3 Devonian

Organic-rich Devonian (Old Red Sandstone) sediments are found in the Orcadian Basin of north-east Scotland and the Shetlands. These lacustrine shales (Donovan, 1980), constitute a minor lithology of an 18 000 ft thick sequence (Donovan et al., 1974), and are often associated with oil staining, bitumens and seeps (Parnell, 1983). Hall and Douglas (1983) obtained TOC values of 0.6 to 5.2% for five Devonian samples. Kerogen type is poorly defined and appears variable from oil-prone to gas-prone, while onshore maturity is probably within the oil window based on estimates from Hall and Douglas' figured sterane and triterpane distributions.

The Orcadian Basin itself extends from eastern Scotland and the Orkneys over to the Hornelen Basin of the west coast of Norway (Ziegler, 1982). The extent to which the lacustrine source rock facies continues under the North Sea is unknown.

9.5.4 Carboniferous

Carboniferous sediments (coals and associated carbonaceous shales) have sourced the major gas fields of the southern North Sea, onshore Netherlands and Germany (Barnard and Cooper, 1983; Bartenstein, 1979; Tissot and Bessereau, 1982), whilst the minor UK onshore oil fields of the East Midlands and Central Valley of Scotland are reputedly sourced from locally developed liptinite-rich sediments within the Carboniferous.

Many of the gas fields sourced by the Carboniferous occur in areas subjected to either early Mesozoic subsidence and inversion (Fig. 9.13), as in the Sole Pit Trough (Glennie and Boegner, 1981) and the Broad Fourteens Basin (Oele *et al.*, 1981), or to areas of higher geothermal gradients (Kettel, 1983).

The Carboniferous source rock story is not so much about kerogen quantity (which is huge), or quality (which is overwhelmingly gas prone), but about maturity and the timing of gas generation relative to the development of reservoir structures and seals.

Dinantian (Visean/Tournaisian)

The Visean oil shales of the Central Valley of Scotland constitute a high quality oil-prone source. TOC values of 11.2% (Bitterli, 1963) and oil yields of about 30 gals/ton (approx. 115 l/t) (Duncan and Swanson, 1965), probably represent the richest beds. Algal ('boghead') coals or 'Torbanites' exist which consist of almost pure compressed bodies of the colonial algae *Botryococcus* (Allan *et al.*, 1980). At outcrop, these

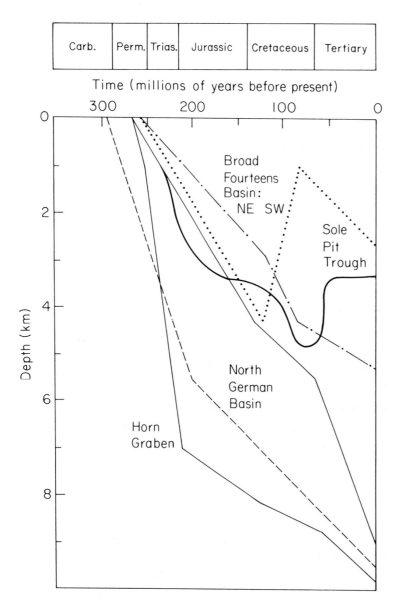

Fig. 9.13. Upper Paleozoic subsidence curves uncorrected for compaction from the deepest parts of the NE and SW Broad Fourteens Basin (Oele *et al.*, 1978), Sole Pit Trough (Glennie and Boegner, 1981), the Horn Graben (Day *et al.*, 1981), the Central Graben (Day *et al.*, 1981) and the North German Basin approximated from Teichmüller *et al.*, 1979; Day *et al.*, 1981; Ziegler, 1982). To estimate the burial of the base of the Westphalian from these curves add about 1200 m for the Stephanian/Westphalian interval. Maturity gradients will differ from location to location, but onset of gas generation is expected after 4–5 km burial.

boghead coals and oil shales appear to be early to mid-mature with respect to oil generation as estimated from the rank of the surrounding coals (high-medium volatile bituminous). They should produce a fairly high-wax oil judging from the n-alkane distribution recorded by Douglas et al., (1969) and Allan et al. (1980) and given the known oil generative capacity of fresh-water algae. The Lothian oil shales are known to have generated mineral hydrocarbons (Parnell, 1984) and may well be the source for the nearby Cousland oil field.

The oil shale facies was probably laterally restricted, but outcrops are known as far apart as Linwood, south of Glasgow in the west, to the Lothians and the Firth of Forth in the east. The extension of this facies further east (offshore) is a matter of speculation (Ziegler, 1982), but its presence has not been reported in any published well information (see also Fig. 2.19).

Coals of Visean age have been recorded over the southern part of the Mid North Sea High, and may constitute a source for relatively small gas accumulations. Elsewhere in the North Sea area, the known Lower Carboniferous does not generally constitute a significant hydrocarbon source, given the high source quality of the overlying strata.

The Namurian

The Namurian rocks may locally possess oil as well as gas potential. The shales of the Visean/Lower Namurian Limestone and Scremerston Coal Group of Northumberland are organic rich (1.08%–3.80% TOC), containing approximately equal quantities of exinite and vitrinite (Powell et al., 1976). These rocks could generate some oil or condensate as well as gas. The bitumens of Windy Knoll in Derbyshire, though deposited in the Visean limestones, are believed to derive from the overlying Namurian Edale Shales (Pering, 1973) possibly under the influence of hydrothermal activity. Namurian shales may also have sourced the oil of the Eakring Field of Nottinghamshire, which is reservoired in the associated Millstone Grit (Kent, 1985).

Namurian marine shales may have oil potential over a limited area to the north of the London-Brabant High. Further north, the Namurian is relatively rich in coal seams and will hence be an additional source for gas in the southern North Sea.

The Coal Measures (Westphalian)

The Coal Measures, comprising coal seams and associated shales, were deposited in a large delta system covering much of north-western Europe (Ziegler, 1982). In the North Sea region, the area extended from about 57°N to the Variscan Deformation Front in the south (Fig. 9.14), although coals are probably restricted to an area south of 55°30′N. The Coal Measures are typically between 1000 m and 2500 m thick, of which in north-west Germany, coal seams comprise about 3% (Lutz et al., 1975). It is still a matter of largely academic debate (e.g. Rigby and Smith, 1982) whether coals or the associated carbonaceous shales constitute the major source of gas: whatever the detailed source, Westphalian strata contain a large thickness of rich, dominantly gas-prone coals and shales.

Carboniferous coal rank in north-western Europe varies from sub-bituminous to meta-anthracite (e.g. Teichmüller et al., 1984). Gas generation occurs between vitrinite reflectance values of 1%R and 3%R (Fig. 9.8), that is, from medium volatile bituminous coal to anthracite. The presence of Kupferschiefer at a sub-bituminous equivalent rank overlying Westphalian anthracites in north Germany and north-east England (Boigk et al., 1971; Gibbons, in press) shows that at least some of the area attained its present maturity by early Zechstein time. In the Variscan tectonic episode, vast amounts of gas and possibly oil must have been generated but then rapidly lost during uplift and erosion. The major Variscan coalification occurred in a series of possibly thrust-soled basins just to the north of the Variscan Front (Fig. 9.14) as evidenced by the anthracites of south Wales, Kent and the Ruhr district.

Many areas survived the Variscan Orogeny with their gas generating potential intact (i.e. having vitrinite reflectance values of less than 2%R), and it is these areas that should be considered as being capable of generating more gas on subsequent burial (Fig. 9.14).

In the centre of the Sole Pit Trough, burial of the top Carboniferous up to about 4000 m occurred progressively from the Triassic to the late Cretaceous (Fig. 9.13; Glennie and Boegner, 1981). Vitrinite reflectance values of 1.0 to >2.8%R are reported by Robert (1980) along the axis of the trough, while values <1.0%R occur on the flanks. Gas generation will thus have occurred along the basin axis during the Jurassic and Cretaceous, with migration to the basin flanks. Late Cretaceous inversion resulted in remigration of gas from the former basin margin structures (e.g. Indefatigable area) into more axial Rotliegend reservoirs (e.g. West Sole and Leman Bank) that had already suffered diagenetic damage (Fig. 9.11E). In the case of the Hewett Field on the south-west flank of the inverted trough, remigration has occurred up into the Triassic. The survival of reservoired hydrocarbons to this day is witness to the exceptional efficiency of the salt seal.

In contrast, Oele et al. (1981) have shown that generation in the Broad Fourteens Basin to the south-east continued from late Jurassic to the present, with a north-easterly migration of the depo-centre. Lee et al. (1985) have placed limits on the timing of the movement of gas in the western fault block of the Broad Fourteens Basin by using K/Ar dating of diagenetic illite. They estimate that gas displaced water some 140 million years ago at the investigated location, in confirmation of the model of Oele et al.

Barnard and Cooper (1983) have noted that the Westphalian can be deeply eroded in areas of inversion

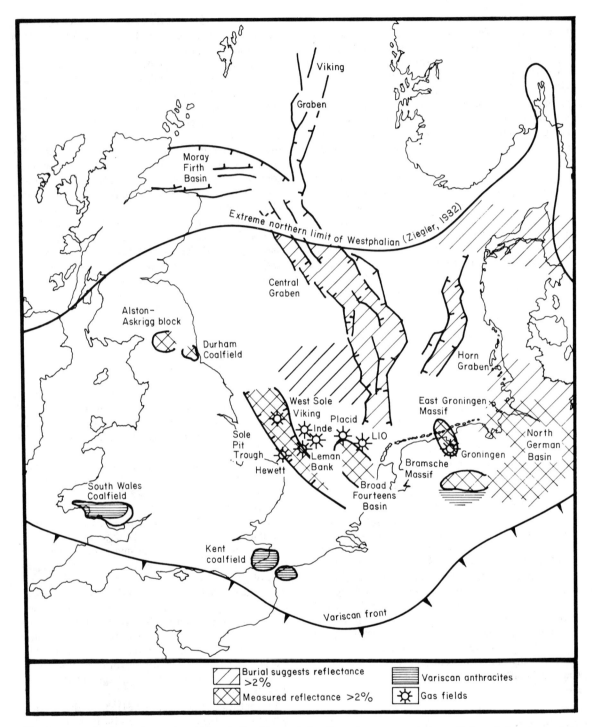

Fig. 9.14. Distribution and maturity of Westphalian (Upper Carboniferous) gas-prone source-rocks bounded to the south by the Variscan Front and to the north by erosion or non-deposition (after Ziegler, 1982, encl. 11). Cross-hatched areas have attained post-Variscan maturity >2%R (Barnard et al., 1983; Teichmüller et al., 1979; Bartenstein, 1979; Kettel, 1983) and diagonals signify areas where base Zechstein burial exceeds 4 km, and hence gas generation may have occurred. Some major gas fields are indicated: note that they are generally associated with areas of inversion (see Fig. 9.12).

as well as on regional highs. It is the areas of inversion, however, that appear to have sourced the major gas fields. They suggest that in some areas only remnants of the Westphalian may be present, in which case the Namurian in Yoredale (coal-shale-limestone cyclothems) or Coal Measures facies may be a possible additional gas source.

A source rock/reservoir mass balance can be attempted for the 693×10^9 m^3 (24.5×10^{12} ft^3) of reserves in the gas fields surrounding the Sole Pit

Trough (Barnard and Cooper, 1983). Using an estimate of the area of the Sole Pit Trough as defined in Glennie and Boegner (1981) of 10.12×10^3 km^2 (2.5×10^6 acres) gives the following values (Fig. 9.8):

Generation of gas from 1% TOC shales
  = 3×10^9 m^3 gas/km^3
  (0.14 mcf/acre ft)
Generation of gas from seam coal
  = 160×10^9 m^3/km^3
  (7 mcf/acre ft)

Multiplication of the area of the trough by the generative potential gives, for shale:
30.6×10^9 m^3 gas/metre (0.364×10^{12} ft^3 of gas/ft) of 1% shale;
and for coals:
1619×10^9 m^3 gas/metre (18×10^{12} ft^3 of gas/ft) thickness of coal seam.

Less than half a metre of seam coal, or 20 m of 1% TOC shale if mature over the whole area of the basin, would be sufficient to yield the gas in place. Clearly there is adequate source potential to account for the reservoired gas even given a relatively incomplete Westphalian section and a low migration/remigration efficiency.

The second area of major post-Variscan burial is in north Germany/southern Denmark which may have sourced the massive Groningen Field in north-east Netherlands. However, Lutz et al. (1975), suggest a source to the south-west of the field in the West Netherlands Basin. A further source area may have been more precisely defined by Kettel (1983) as overlying the East Groningen Massif. Lee et al. (1985) date the first influx of gas into that part of the Groningen Field penetrated by their studied well (on the northern flank of the field) as occurring some 150 million years ago based on the K/Ar dating of diagenetic illite. Younger ages at greater depths within the reservoir section were interpreted as indicating the progressive filling of the structure from 150 to 120 million years before present at the location of the investigated well.

A third area of gas generation from the Westphalian, may exist under the Central Graben and Horn Graben, where Day et al. (1981) show local burial of up to 10 km for the base Zechstein. The Coal Measures, if preserved, are expected on regional grounds to be of bituminous coal rank (Eames, 1975) outside the grabens. The timing of gas generation can be estimated as Triassic-Jurassic from Fig. 9.13.

Finally, some small areas where the Westphalian overlies post-Variscan intrusives (e.g. Bramsche Massif, the Alston granite) could also have generated gas at a time when reservoirs, structures and seals existed. It may be useful to consider in this context that granites, in particular K-feldspar-rich varieties, can give rise to thermal anomalies as a result of the generation of radioactive heat long after the heat of the intrusion itself is dissipated.

No Carboniferous in source rock facies is known from the west of the Shetlands (Ridd, 1981). Carboniferous coals are mined in Ireland (Griffiths, 1983) and the Carboniferous is present in gas-prone Coal Measure facies in the Kish Bank Basin (Jenner, 1981), and possibly in the northern part of the Irish Sea (Barr et al., 1981). Of the two gas finds to the west of England, the Morecambe Field is of unknown source, but compositional similarities with the southern North Sea gas fields (Ebbern, 1981) suggest the Carboniferous—which is known to underlie the area—as the source. In contrast, the Kinsale Head Field is believed by Colley et al. (1981) to have a complex origin from Mesozoic source rocks.

Somewhat beyond the boundaries of the North Sea, but pertinent to current exploration, mature, gas-prone, organic-rich shales and coals of Carboniferous age are present on Bjornoya (Bear Island), offshore northern Norway (Bjorøy et al., 1983). Carboniferous coals and Coal Measures are also present on Svalbard (Spitsbergen). This gas-prone source sequence is part of a separate North Greenland Basin (Rønnevik, 1981).

9.5.5 Permian

The Kupferschiefer/Marl Slate horizon—typically less than 1 m (3 ft) thick—is the only Permian sediment with a high-grade hydrocarbon source potential in the North Sea area. It outcrops in Durham, UK, and north Germany and extends under much of the central and southern North Sea. TOC values are typically 5–8%, and the kerogen type is oil-prone with hydrogen indices above 600 mg/g (Gibbons, in press). Any oil generated would be recognisable by its high porphyrin content. No oil of this type has been reported in the North Sea, where the oils generally have very low porphyrin contents. However, the high Vanadium and Nickel content of Buchan oil (Table 9.6) may suggest a Kupferschiefer contribution. This horizon is present north of the Buchan Field (Taylor, 1981). The absence of effective generation is probably a function of the thinness of this unit, but local oil shows in the Zechstein may derive from the Kupferschiefer. Onshore Netherlands, Van den Bosch (1983) reports oil-prone Zechstein shales and carbonates.

To the north, in the Permian Basin on the east coast of Greenland a probably relatively local facies of source rock is developed (Surlyk et al., 1984). Here in a lagoonal to shallow-marine setting thin, dominantly gas-prone claystones were laid down during the Early Permian, while a thicker more persistent oil-prone shale, with 2–5% TOC in its richest upper part, accumulated in the Upper Permian. Background maturities are in the late-immature to early-mature zone (0.44–0.60% vitrinite reflectance), except where affected by intrusive heating during the Tertiary. To the east of the outcrop the maturity rises as high as 1.83%R.

9.5.6 Triassic

The Triassic sediments have no source potential in the North Sea area. However, in some regions to the north of latitude 62°N, marine sedimentation continued during the Triassic. Knowledge of the source potential is restricted to a limited number of outcrops. The marginal marine Skuld and Urd Formations of Bjornoya Island south of Svalbard (Spitsbergen) contains shales and silty shales with up to 2.0% TOC of dominantly gas-prone kerogen, which is in the lower part of the oil window (Bjorøy et al., 1983). On Svalbard itself, Middle Triassic shales of the Botneheia Formation contain considerable quantities (2.9–5.5% TOC) of oil and gas-prone kerogen, which ranges from late immature to peak oil generating maturity (Forsberg and Bjorøy, 1983; Mørk and Bjorøy, 1984). Small amounts of locally migrated hydrocarbons have been described by Schou et al. (1984). This high quality oil and gas-prone source may extend as far south as about 65°N according to facies maps of Rønnevik (1981) and Rønnevik et al. (1983).

In a totally different facies, the Upper Triassic–Lower Jurassic Hitra coal unit of mid-Norway also has potential for gas/condensate and possibly oil (e.g. Larsen and Skarpnes, 1984; Hollander, 1984).

The Rhaetic (Top Triassic) of southern England may locally develop in an oil or gas-prone facies (Macquaker et al., 1985).

9.5.7 Jurassic

The Jurassic contains the source of the bulk of North Sea oil in the Upper Jurassic-basal Cretaceous claystone unit variously termed the Kimmeridge Clay, Draupne or Borglum Formations, and a major source of gas in the Middle Jurassic coals. Ironically, relatively little detailed work has been published on the best oil source rock, the 'hot shale' of the Kimmeridge Clay/Draupne Formation (Fuller, 1975; Oudin, 1976; Brooks and Thusu, 1977). A typical summary of the Jurassic/Cretaceous source rocks in the North Sea is given in Fig. 9.15, from Kirk's (1980) study of the Statfjord area.

Lower Jurassic (Liassic)

Within the North Sea itself, the Lower Jurassic, where it develops source rock quality, is dominantly gas-prone; only the Upper Lias (Toarcian) is locally developed in an oil-prone facies. The Lower Jurassic is absent, or present only as erosional remnants, in the Central and Witch Ground Grabens and the Outer Moray Firth Basin (Ziegler, 1982). In the extreme south of the Central Graben, in the Broad Fourteens Basin and in the area of the Yorkshire-Sole Pit Trough, the Lower Jurassic is preserved.

In the Yorkshire-Sole Pit Trough area, lean gas-prone shales (0.8% TOC average) and minor coals (jet) occur in the Lower and Middle Liassic sequence. An Upper Lias (Toarcian), rich oil-prone facies is variously known in north-west Europe as the Jet Rock and Bituminous shales of Yorkshire, the Posidoniaschiefer of Germany and Les Schistes Bitumineux in France, Belgium and Luxemberg. These rocks are rich in kerogen of sapropelic type, typically with 2% to 12% TOC (Barnard and Cooper, 1981; Tissot et al., 1971; Huc, 1976; Brand and Hoffman, 1963).

The Hettangian/Sinemurian Blue Lias of Dorset, north Somerset and Glamorgan, UK, is present as cyclic sequences of limestone-marl-shale. The shales, particularly when laminated, contain good yields (up to 18% TOC) of mixed oil and gas-prone kerogen, and are late-immature to early-mature with respect to oil generation (Cornford and Douglas, in preparation). The shales of the Sinemurian Black Venn Marls of Dorset are also rich in mixed oil and gas-prone organic matter but here are immature. The Lias has been suggested as a possible source for the Wytch Farm oil, onshore Dorset (Colter and Havard, 1981).

To the east, in the Danish sub-basin, the open-marine shales of the Fjerritslev Formation have low TOC contents and are probably gas-prone, since they are differentiated by Hamar et al. (1983) from the anaerobic 'hot shales' of the Upper Jurassic/Lower Cretaceous. Apart from some 40 m of Upper Jurassic claystone (Olsen, 1983) the Jurassic is believed to be absent in the Horn Graben (Best et al., 1983).

Since it is absent, or only present as remnants in the graben areas, the Lower Lias will not, without local heating (e.g. Altebaumer et al., 1983), have reached the gas generating stage (vitrinite reflectance levels of about 1.0%R) in the southern part of the North Sea. The Upper Liassic oil-prone shales between Yorkshire and The Netherlands may, however, have reached early maturity for oil generation in the pre-inversion Sole Pit Trough at the end Cretaceous (Glennie and Boegner, 1981), in the Cleveland Basin, Yorkshire (Barnard and Cooper, 1983), and in the Broad Fourteens Basin from the Cretaceous onwards (Oele et al., 1981). The Toarcian is believed to be the source for the oils in the Rijswijk oil province of The Netherlands (Bodenhausen and Ott, 1981).

Minor lean, probably gas-prone, shales of the Statfjord and Dunlin Groups are preserved in certain areas in the northern North Sea such as the East Shetland Basin (e.g. the Magnus, Thistle, and Heather Field areas). In comparison with the overlying Middle Jurassic this is probably not a significant gas-prone source rock. Rønnevik et al. (1983) show mature (for oil) Toarcian marine shales of the Drake Formation, with an average of 2% TOC and oil-prone kerogen in the Norwegian Viking Graben and the Horda Basin, and gas-prone kerogen in the Norwegian sector of the Central Graben.

North of 62°, mature Lower Jurassic-Upper Triassic coals and shales are present in the Haltenbanken and Traenabanken areas as gas and oil-prone source rocks (Rønnevik et al., 1983; Larsen and Skarpnes, 1984; Hollander, 1984). It is these coals and associated

Period/Series Group/FM	Lithology	Thickness range (ft)	TOC[1]	EOM[2]	EOM/TOC	CPI[3]	Ro[4]
Paleocene Rogaland Group	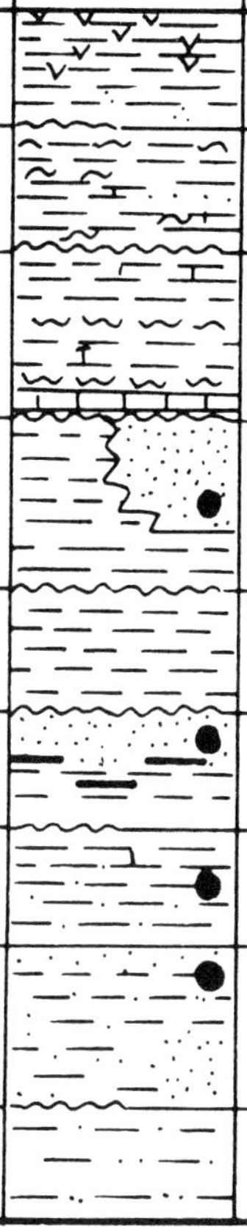	548-820	0.43 (64)	141 (3)	3.3 (3)	1.45 (3)	0.41 (3)
Upper Cretaceous Shetland Group		1758-3998	0.78 (201)	172 (40)	2.64 (40)	1.71 (20)	0.43 (12)
Lower Cretaceous Cromer Knoll Group		0-482	04 (4)	78 (2)	1.41 (2)	–	0.71 (2)
Upper Jurassic Kimmeridge Clay FM (hot shale)		0-298	4.58 (4)	2160 (4)	4.00 (4)	–	–
Middle Jurassic Heather FM		0-994	2.24 (16)	350 (9)	1.71 (9)	1.68 (9)	0.40 (2)
Middle Jurassic Brent FM		0-1023	–	–	–	–	0.44 (5)
Lower Jurassic Dunlin FM		0-1430	1.47 (86)	234 (38)	1.95 (38)	1.68 (36)	0.59 (6)
Lower Jurassic / Triassic Statfjord FM		0-1016	0.44 (19)	–	–	–	0.68 (4)
Triassic Cormorant FM		Unknown. Max. Pen. 6032+ft	0.35 (56)	50 (6)	1.71 (6)	1.37 (3)	0.73 (8)

( ) Number of values
● Oil reservoir

1. Total organic carbon (% rock weight)
2. Extractable organic material (ppm)
3. Carbon Preference Index
4. Vitrinite reflectance

Fig. 9.15. Typical source rock summary for a North Sea well (from Kirk's 1980 study of Statfjord field), identifying the Upper Jurassic Kimmeridge Clay Formation as the best source rock interval (4.58% TOC and 47 mg extract/g TOC), with the Heather and Dunlin being leaner and more gas-prone. The sample density is somewhat puzzling, being highest in the Lower Jurassic, Upper Cretaceous and Paleocene. The gas-prone nature of the Lower Cretaceous Cromer Knoll sediments is highlighted by the presence of altered rather than primary vitrinite, as shown by the off-trend average vitrinite reflectance value of 0.71%R. Otherwise a consistant maturity trend is displayed by the vitrinite reflectance values, with the carbon preference indices being characteristically fickle.

shales, reported to have unusually high Hydrogen Indices (values in the 200–400 mg/gTOC range), that are believed to have generated the gas/condensate in the mid-Norway area. Thompson et al. (1985), in comparing the Norwegian Hitra Formation with the Tertiary coals of South-East Asia, have attempted to account for the oil-prone nature of these coals by considering the palaeo-latitude and delta geometry: they claim that while the delta top rooted coals are gas-prone the peripherally deposited allochtonous coals are enriched in the more oil-prone liptinic tissues. This model yields the opposite distribution of oil-prone kerogen to that proposed by Powell (1986) when discussing Tertiary deltas of the Far East and Canada. His model would predict the most oil-prone coals on the delta top, where a fluctuating water table would allow the preferential oxidation of the humic fraction. No significant source potential is attributed to the Lower Jurassic in Svalbard (Mørk and Bjorøy, 1984).

Middle Jurassic

In the North Sea area, Middle Jurassic deposition is limited (Ziegler, 1982, encl. 19). With the exception of the oil-prone, algal-rich shales and coals of the Inner Moray Firth Basin, the Middle Jurassic generally has gas generating potential in the parallic/deltaic sands, shales and coals of the Yorkshire-Sole Pit Trough area (Hancock and Fisher, 1981), the Moray Firth Basin (e.g. Maher, 1981; Bissada, 1983), the Viking Graben (Eynon, 1981; Pearson et al., 1983), the Unst Basin (Johns and Andrews, 1985), and the Horda, Egersund and North Danish Basins (Koch, 1983; Hamar et al., 1983). Coals are shown as minor components in these sequences (Hancock and Fisher, 1981; Eynon, 1981;

Parry et al., 1981). Goff (1983) estimates a total thickness of 10 m of Brent coals in the drainage area of the Frigg Field, but suggests that the isotopic composition of Frigg gas indicates generation from the shales rather than coals (see Rigby and Smith, 1982). The gas in the giant Sleipner Field is believed to come from Middle Jurassic coals (Larsen and Jaarvik, 1981). The coals and shales frequently contain considerable altered vitrinite and inertinite (Cope, 1980), which somewhat down-grades their gas-generating potential. Indeed, the majority of the major gas fields of the North Sea oil province (e.g. Sleipner, Frigg, Troll) have methane stable carbon isotope compositions incompatible with generation from fully mature seam coals of the type found in the Brent delta. A late mature product of the poorly drained oil-prone Upper Jurassic claystones provides a more plausible source for the bulk of the gas.

The Middle Jurassic shales commonly have TOC values as high as 5% and, like most Coal Measure sequences, the TOC values fluctuate considerably. Their gas-prone nature is shown by the kerogen composition of the claystone facies, which is dominated by vitrinitic material (Parry et al., 1981). Interestingly, minor amounts of oil/condensate-prone algae, resin and cuticle are present in these shales within the Viking Graben. The Callovian shales of the Ninian Field contain up to 4% TOC and oil-prone kerogen (Albright et al., 1980). In addition, Bissada (1983) has reported fairly rich (1.4-2.7% TOC) oil-prone rocks together with coals in the Middle Jurassic of the Moray Firth in the area of the Piper Field. Some of the reports of oil-prone Middle Jurassic may, if not supported by parallel biostratigraphic studies, be the result of downhole caving of drill cuttings from the richer Upper Jurassic section above.

A concentration of the freshwater algae Botryococcus is found in the oil-prone Parrot coal and associated shales of the Brora section of the Moray Firth coast. Barnard and Cooper (1981) have suggested that this might be the source for the high wax Beatrice crude.

In terms of maturity, the Brent coals and shales will be mature for gas generation (>1%R) in the Viking Graben below about 3.9 km (Goff, 1983). Interpolating between the top Trias and base Cretaceous depth maps of Day et al., (1981), gas generation will be confined to the central part of the Viking and Witch Ground Grabens and excluded from the East Shetlands Basin, and the Horda, Egersund and North Danish Basins. Measured onshore reflectance data (up to 0.87%R) from Barnard and Cooper (1983), and offshore subsidence shown by Glennie and Boegner (1981), suggest that the Middle Jurassic of the Yorkshire-Sole Pit Trough area would not have attained gas generating rank even at pre-inversion maximum burial. The Middle Jurassic of the Inner Moray Firth occurs at about 2000-2100 m (6500-6800 ft) in the Beatrice Field area (Linsley et al., 1980), but additional offstructure burial is possible according to the regional seismic maps of Day et al. (1981).

To the west of the North Sea, the Middle Jurassic is present as the arenaceous gas-prone coal-bearing Great Esturine Series in Skye and the Inner Isles, outcrops that may be representative of both of the Minch Basins (Ziegler, 1982). Whilst a similar facies is reported over the Rona Ridge and in the West Shetland Basin, thick dark grey marine shales are present on the north-west flank of the Rona Ridge (Ridd, 1981).

To the north of 62°, the Middle Jurassic Hestberget member of the Andoya Island outlier, contains rich, immature oil-prone (1.0-11.5% TOC) as well as gas-prone (4.6-12.3% TOC) sediments (Bjorøy et al., 1980). A study of an undifferentiated Jurrasic-Cretaceous shale sequence from Spitsbergen (Bjorøy and Vigran, 1980; Mørk and Bjorøy, 1984) indicates good quantities (0.45-16.0% TOC) of oil and gas-prone kerogen.

The Upper Jurassic and basal Lower Cretaceous (Oxfordian to Ryazanian)

The Kimmeridge Clay/Borglum Formation, which falls within this age range, is recognised as the dominant oil (and associated condensate and gas) source rock of the North Sea. It overlies the Oxfordian, which, where developed in an argillaceous facies, is typically a fair to good gas-prone source, but locally also has good oil potential.

Oxfordian

The Oxfordian shales of the North Sea (i.e. the Heather Formation of the Viking Graben, the Haugesund Formation of the Central Graben, and the Egersund Formation of the Fiskebank Sub-basin), are generally fair to rich gas-prone source rocks (Barnard and Cooper, 1981). In the Statfjord Field, Kirk (1980) reports an average of 2.24% TOC for the Heather Formation, which appears gas-prone (Fig. 9.5). Larsen and Jaarvik (1981) have suggested the Heather Formation may be the source for the condensate in the Sleipner Field. Locally, Oxfordian shales can have some oil potential (Bissada, 1983), and where thick, (e.g. in the East Shetland Basin) could augment the Kimmeridge Clay Formation (Goff, 1983). For example Bissada (1983) found the Oxfordian ('non Kimmeridge Clay Upper Jurassic') of UK Quadrants 14 and 15 to be richer (average 1.4-8.8% TOC for eight wells) than, and to contain equally oil-prone kerogen as, the overlying Kimmeridge Clay Formation. Johns and Andrews (1985) report the lower Heather of the Unst Basin to be an excellent oil-prone source rock (4-6% TOC).

Onshore UK, Fuller (1975) reports the Oxford Clay of Yorkshire to be a lean, gas-prone source. In southern England, the Oxford Clay is reported to be a rich but immature potential source for oil in Dorset (Colter and Havard, 1981). Duff (1975) reports mean TOC values of 2.9 and 4.1% for two intervals of the Oxford Clay of central England.

In the Troms area of northern Norway, the Upper Jurassic is argillaceous, with the lower part of this unit being as good as, if not better than, the overlying Kimmeridge Clay equivalent (= Nesna Formation) in terms of oil source potential: it is locally mature on structure in at least some wells (Bjorøy et al., 1983). Larsen and Skarpnes (1984) report a similar situation in the mid-Norway area. On Svalbard, the undifferentiated Jurassic section, presumably containing material representative of the Oxfordian, comprises mature oil-prone (Type II) kerogen (Bjorøy and Vigran, 1980; Mørk and Bjorøy, 1984).

The Kimmeridge Clay/Borglum/Draupne Formation

This formation contains the major oil source rocks of the North Sea (Barnard and Cooper, 1981). They are developed as organic rich 'black' shales in most of the graben areas of the North Sea (Ziegler, 1982; Rønnevik et al., 1983). As previously noted, the high natural gamma ray log response by which the formation is commonly characterised (Fig. 9.6) has earned it the name of 'hot shale'. Typical average TOC values are shown in Table 9.3, which also catalogues studies of the source rock potential of the Kimmeridge Clay Formation. TOC contents vary from area to area, and also with lithology in a given section. In using values from Table 9.3, it should be noted that there is a natural tendency for source rock reports to have analysed the richest lithologies. The richest lithologies are often characterised in well descriptions as 'olive-black' or 'dark olive grey' as opposed to 'dark grey', with sample selection being biased towards intervals with the highest natural gamma log response.

These laminated olive-black claystones were deposited in a marine seaway running from the Barents Sea in the North—down the line of the grabenal system developing during the Upper Jurassic—to the south of the Central Graben (Fig. 9.16a). The Norwegian-Danish and Yorkshire-Dorset Basins were lateral to, and probably atypical of, sedimentation in the major sea-floor furrow extending south from the Boreal Sea (Cornford et al., 1986). The development of Boreal/Tethyan provincialism in at least some groups of organisms (e.g. ammonites), suggests a partial (?climatic) barrier to the south of the Yorkshire-Dorset Basin. In the Viking Graben accumulation occurred during rapid subsidence along the grabenal suture, with major local thickness changes resulting from listric faulting during the break-up of the Brent shelf under tensional stress. In the Central Graben tectonic control of thickness is less marked. Branches of anoxic shale deposition extended into the Moray Firth Basin and to the west of Shetland.

The 'hot shale' facies can be developed locally, and is diachronous: for example, the Tau 'hot shale' Member of the Borglum Formation of the Fiske Sub-basin is of Volgian (uppermost Jurassic) age, whilst the 'hot shale' of the Norwegian sector of the adjacent Central Graben comprises the Ryazanian (basal Cretaceous) Mandal Formation (Hamar et al., 1983). Details of the stratigraphy at the Upper

Table 9.3. TOC values for the Kimmeridge Clay/Borglum Formations of the North Sea

Area	TOC (%)	Reference	Comments
North Sea, unspecified	2.7 av.	Fuller, 1980	= 3.25% organic matter
North Sea, miscellaneous	7.1 av.	Fig. 9.5	2-12% range, richer samples
Unidentified N.Sea well	5.6, 4.9 av.	Brooks and Thusu, 1977	upper and lower intervals
Unst Basin	6-10	Johns and Andrews, 1985	UK well 1/4-1
Ekofisk	1.4-2.6 av.	Van den Bark and Thomas, 1980	NOCS well 2/4-19B
Outer Moray Firth Basin	1.0-3.8 av.	Bissada, 1983	UK Quad. 14 and 15 Piper wells
S. Viking Graben	2.5-4.5	Pearson et al. 1983	UK well 16/22-2
Inner Moray Firth	3-6	Pearson and Watkins, 1983	range of UK Quad. 12 wells
Brae area well	4.29 av.	Reitsema, 1983	UK well 16/7a-19N
E. Shetlands Basin	5.4	Goff, 1983	estimated average
Ninian	6-9	Albright et al. 1980	range
Tern Field	3.4-8.1	Grantham et al. 1980	UK well 210/25-3: 6.8% av.
Statfjord	4.58 av.	Kirk, 1980	Licence 037, av.
Southern Norwegian N.Sea	7	Hamar et al. 1983	7-17.5% in hot shales
S. Norway Shelf	2.1(5.1) av.	Fuller, 1975 (Lindgreen, 1985)	NOCS 2/11-1 core
Danish N. Sea	3.85 av.	Thomsen et al. 1983	NW Central Graben, well I-1
"	1.59 av.	"	SE Central graben, well M-8
Norwegian N. Sea	5 av.	Rønnevik et al. 1983	range 1-5%
Dorset type section (UK)	3.75 av.	Fuller, 1975	bituminous shale
"	1.6 av.	"	grey clay
"	15.3-40.0	Cosgrove, 1970	richest oil shale band
"	0.9-52.7	Farrimond et al. 1984	Selected oil shales and limestones
Yorkshire outcrop	7.95 av.	Fuller, 1975	bituminous shale
"	30.9	"	oil shale
"	3.4	"	dark grey shale
Sutherland outcrop	5.5 av.	"	black shale
Andøya, Svalbard	1.4, 4.3	Dypvik et al. 1979	Northern Norway, Spitsbergen

Jurassic/Lower Cretaceous boundary have been discussed by Rawson and Riley (1982). Thomsen et al. (1983) have shown that while the most oil-prone shales occur at the top of the Upper Jurassic in the deepest (north-west) part of the Danish Central Graben (well I-1), the best source quality in a generally poorer section occurs towards the base of the Upper Jurassic on the south-east flank of the graben (well M-8).

The typical kerogen type of the 'hot shales' (Table 9.4) is a mixture of bacterially degraded algal debris of marine planktonic origin (amorphous liptinite), and degraded humic matter of terrigenous origin (amorphous vitrinite). This amorphous component is mixed with variable amounts of particulate vitrinite (woody debris) and inertinite—highly altered (oxidised or burnt) material of land-plant origin. Land-plant spores, and marine algae such as dinoflagellates and hystrichospheres, are present in minor to trace amounts. Framboidal pyrite is common, attesting to the action of sulphate-reducing bacteria. Under the reflected-light microscope, the kerogen in a polished whole-rock preparation appears as a diffuse fluorescent background (Gutjahr, 1983), while isolated kerogen in transmitted light appears as a clumpy, amorphous, dully or blotchily fluorescent mass (Batten, 1983). Subjected to Rock Eval pyrolysis, this mixture generally comes out as a Type II oil-prone kerogen (Barnard et al., 1981).

A map of the Kimmeridgian-Volgian kerogen type in the North Sea Basin based on 2200 Rock Eval pyrolysis analyses has been published by Demaison et al. (1983). The Chevron group differentiated four essentially arbitary facies: dead carbon (Type IV); gas-prone (Type III); mixed oil and gas-prone (Type II/III); and oil-prone (Type II). This map is reproduced with slight modifications and additions in Fig. 9.16B.

Characterising the kerogens by optical microscopy, Fisher and Miles (1983) have reported that in the Fladen Ground Spur area of the South Viking and Witch Ground Grabens, the kerogen types of the Kimmeridge Clay Formation vary spatially and temporally. Stratigraphically, the Volgian strata contain the best quality source material in this area, while spatially, the most oil-prone kerogen is deposited in the graben centre and away from positive features and the routes of sediment input.

This concept has been generalised by Barnard et al. (1981), using Rock Eval pyrolysis data to illustrate the variation of TOC and four kerogen types down through a single 183 m (600 ft) section of the Kimmeridge Clay Formation, and spatially in terms of the change of dominant kerogen type from platform edge to the graben centre. This model can be briefly summarised as an increase in (gas-prone) terrigenous kerogen, and a decrease in (oil-prone) marine planktonic/bacterial kerogen towards the paleo-coastline.

In a review of the different kerogen types in the Upper Jurassic, Cooper and Barnard (1984) distinguish between an 'algal' and a 'waxy' sapropel. The figured sterane and triterpane data clearly show that the major distinction between the two samples is one of maturity, with the 'algal sapropel' being the least mature, despite the reportedly similar spore colour and vitrinite reflectance values.

It is clear from the above discussion that the characteristics of the 'hot shales' of the Kimmeridge Clay Formation vary considerably. However, for volumetric calculations and broad source rock/oil correlations, some average properties are summarised in Table 9.4. Such a summary is validated by the high compositional uniformity of North Sea crude oils from the South Central Graben to the North Viking Graben (see Section 9.6).

The Upper Jurassic is also developed as a rich oil-prone source beyond the North Sea. North of 62° Rønnevik et al. (1983) predict marine shales with good source quality. More specifically, Bjorøy et al. (1980) report TOC contents of 0.98–6.80% but generally gas/condensate prone kerogen for the Ryazanian to Kimmeridgian of Andoya Island. This section is immature.

In the undifferentiated Jurassic of Svalbard, TOC values are lower but the kerogen is of better quality, and fully mature (Bjorøy and Vigran, 1980; Mørk and Bjorøy, 1984). In the Troms area, the Upper Jurassic contains an organically rich oil and gas-prone claystone (mixed type II and III kerogen), which 'on structure' is late-immature to early-mature with respect to oil generation (Bjorøy et al., 1983; Westre, 1984). Larsen and Scarpnes (1984) have reported mixed Type II and III kerogen with TOC values in the 5 to >10% range in the Traenabanken area of mid-Norway, while similar values (4–13% TOC and Type II kerogen) are reported by Hollander (1984) for the Haltenbanken area.

To the south, the Kimmeridge Clay Formation is not of optimum quality. In the Rijswijk oil province the equivalent Delfland Formation may have been a secondary source to the Toarcian (Bodenhausen and Ott, 1981). Onshore UK, the Kimmeridge Clay contains variable quantities of oil-prone kerogen from Yorkshire to Dorset, being most mature in the North (Williams and Douglas, 1980; Douglas and Williams, 1981; Farimond et al., 1984). Ridd (1981) notes that the Upper Jurassic west of the Shetlands is an organic-rich claystones with a high natural gamma ray response; it thickens to the north-west into the Faeroe Basin.

Depositional environment

The depositional environment of the organic-rich shales of the Kimmeridge Clay Formation has been much disputed. Worldwide, the Upper Jurassic is a time of transgression (Vail and Todd, 1981; Rawson and Riley, 1982), which favours the deposition of organic-rich rocks (Hallam and Bradshaw, 1979), probably due to high bioproductivity and sedimentation rates and low concentrations of dissolved oxygen (Demaison and Moore, 1980 and refs. therein).

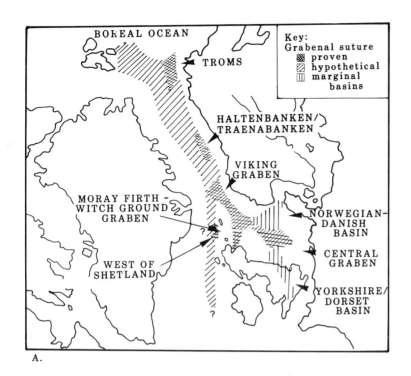

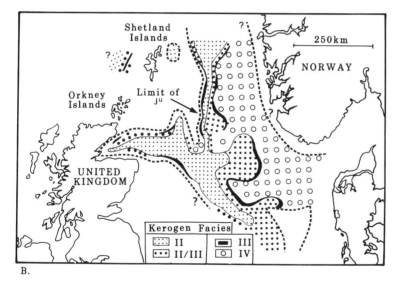

Fig. 9.16. Upper Jurassic Hot shale deposition and kerogen Type: A. Upper Jurassic paleogeography and extent of hot shale deposition (after Cornford et al., 1986) and B. Kerogen type with the North Sea (after Demaison et al., 1983). See Fig. 9.4 for definition of kerogen types from Rock Eval.

Published work on depositional environments is restricted to the UK onshore outcrops where limestones, marls, dolomitic limestones, oil shales, bituminous shale and clay are all recognised (Tyson et al., 1979). Gallois (1976) suggested that blooms of phytoplankton—evidenced by the presence of coccolith limestones interbedded with the oil shales—produced an excess of organic matter over available dissolved oxygen, and hence preservation of organic-rich, anoxic sediments. It was then suggested (Tyson et al., 1979) that the anoxicity was the cause and not the effect of the coccolith blooms. The model of Tyson et al., envisaged a stratified water body (halocline or thermocline), anoxic at depth, which overturned periodically, liberating nutrients and giving rise to algal blooms. A modification of these two models has been suggested by Irwin (1979) who reported bioturbation within the oil shale facies, and noted that biogenic calcite would be rapidly dissolved below any O_2–H_2S interface. Cornford et al. (1980) and Cornford and Douglas (in preparation) have suggested that the influx of terrigeous organic debris may also have reduced the background oxygen levels in extensive shelf seas.

It is clear that deposition occurred below wave base in an oxygen deficient environment with high planktonic bioproductivity (Fig. 9.17). The water depth was probably not great (e.g. approx. 200 m?) since we see condensed sequences on 'highs' in the North Sea area.

Table 9.4. Some average properties of typical Kimmeridge Clay Formation hot shales of the North Sea

Property	Value	Reference	Comment
TOC (%)	5	Table 9.3	Realistic average
H/C ratio (kerogen)	0.9-1.2		Immature kerogen
Rock Eval hydrogen index	450-600	Barnard et al. 1981	Immature kerogen
Kerogen type	II		Rock Eval
$\delta^{13}C$‰ (kerogen) pdb	−27.6 to −28.7 (−25)	Reitsema, 1983; Fuller, 1975	Brae, Moray-Statfjord areas
Organic petrography*			
% amorphous liptinite	30-80	—	Bacterially degraded algae
% particulate liptinite	1-10	—	Dinoflagellates, spores etc.
% vitrinite (am + partic)	20-70	—	(amorphous + particulate)
% inertinite	1-25	—	Fusinite and semi-fusinite
Oil yield (bbls/acre ft per 1% TOC)	50	Fig. 9.7	In the source rock at peak maturity

*Percentage values are visual estimates of area per cent of slide occupied by kerogen components.

Where coarser clastics are associated with the organic-rich facies such as in the Brae area, down-slope transport is indicated, but water depths still need not be great (Stow et al., 1982). The association of these coarse clastics with fault scarps suggests syn-sedimentary fault movement.

The overall picture is one of deposition within a system of linear sea-floor depressions overlying the developing graben system, all within a relatively shallow sea. Highly productive surface waters and anoxic water at depth existed with occasional overturning and mixing. The anoxic Upper Jurassic basin was almost certainly cut off from the Tethys to the south and east by some sort of barrier, but to the north effective exchange of oxygenated oceanic water from the Boreal Ocean may have been restricted merely by the extreme distance (ca.2000 km) and the shallowness of the seaway. The highly organic 'oil shale' facies may have accumulated in deeper sea-floor depressions controlled by fault subsidence, where ponding of anoxic and hydrogen sulphide-saturated water and higher sedimentation rates would favour organic preservation. The surrounding land areas must have been rich in higher-plant vegetation, which contributed the vitrinitic components to the kerogen.

Oil source correlations

The 'hot-shales' of the Kimmeridge Clay Formation have repeatedly been shown (e.g. Oudin, 1976; Mackenzie et al., 1983) to give the best match with North Sea oil properties using oil/source rock correlation techniques (Tissot and Welte, 1978). Considering only the more recent detailed studies, Reitsema (1983) has demonstrated that the Brae oils correlate with the 'hot shales' using n- and isoprenoid alkane, sterane and triterpane distributions, as well as the stable carbon isotope curve-matching technique. He emphasised the need to correlate the properties of the different oils within the Brae complex with specific facies of the hot shale. Mackenzie et al., (1983) show the close correlation of the Brae oils with the interbedded hot shales using gc-ms and isotope data. Interestingly, the oil appears less mature than the interbedded source rock. This is exactly what would be expected if the reservoired oil is a mixture of early, mid- and late-mature hydrocarbons (Fig. 9.18a-c), while the rock extract represents the present day late mature regime commensurate with the burial depth. As suggested by the authors, this interpretation requires that reservoired oil associated with a relatively clean sandstone matrix, may be more stable relative to these types of reaction, than oil in the finer grained, clay-rich source rock.

Fisher and Miles (1983) have noted that in the Southern Viking and Witch Ground Grabens, a correlation can be made between specific kerogen types, maturity, and oil properties.

Oil/source rock correlations have latterly relied heavily on sterane and triterpane fingerprinting using the computerised gas chromatography—mass spectrometry technique (gc-ms). Cornford et al. (1983) using a suite of about 100 North Sea oils and 'hot shale' extracts, have demonstrated a detailed correlation, emphasising that the 'hot shale' generates a recognisable oil type at each maturity stage (Fig. 9.18). It is concluded that the bulk of reservoirs have received a full spectrum of maturity products from the source rock on the flanks of the graben as shown in Fig. 9.18c. Grantham et al. (1981) have suggested an oil/source rock correlation in the Tern area (UK Block 210/25) using the C_{27} and C_{28} triterpanes to identify a specific interval of the 'hot shales'. Fuller (1975) has shown a good correlation between North Sea oils and

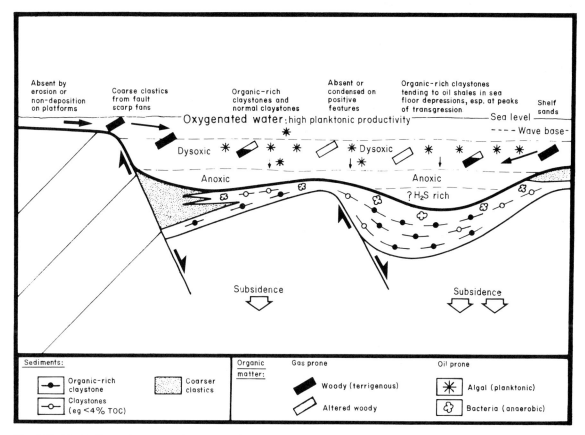

Fig. 9.17. Model illustrating some of the constraints on the depositional environment of the Kimmeridge Clay Formation organic-rich shales of the North Sea. Sedimentation rates are high (typically 20–30 m/m.y. decompacted) and locally can be much higher especially where allochthonous sands are injected into the basin.

the Upper Jurassic source rocks using optical rotation and stable-carbon isotope data.

Data from Mackenzie et al. (1983) were able to differentiate Central Graben oil from Viking and Witch Ground Graben oils on the basis of a lower abundance of C_{28} hopanes and higher abundance of tri-aromatic steroidal molecules in the former. Cornford et al. (1983) have shown that the C_{28} hopane abundance is in part maturity controlled in the source rock, suggesting that this differentiation may merely be the result of the deeper burial of the source rocks in the Central Graben (Fig. 9.19). A similar differentiation appears possible using stable carbon isotope values of whole oils. Cooper and Barnard (1984) quote the following values (see also Table 9.6):

East Shetland Basin oils
 −29.5 to −30.0 per mil.
Viking Graben oils
 −27.5 to −31.0 per mil.
Outer Moray Firth oils
 −29.5 to −30.0 per mil.
Central Graben oils
 −28 approx. per mil.
Ekofisk Complex oils
 −27.0 to −28.5 per mil.
Egersund Sub-basin oils
 −30 approx. per mil.
Beatrice oil (Mackenzie et al.)
 −31.2 per mil.

A more limited tabulation from Mackenzie et al. (1983) also indicates somewhat less negative (isotopically heavier) oils in the Central Graben. Again this difference could be due at least in part to maturity.

Using multi-variate statistics on a number of molecular species, Øygard et al. (1985) have established 3 different oil types together with varying degrees of biodegradation on a suite of 20 oils from 10 closely spaced North Sea wells.

Volumetric modelling

The North Sea oil province, with a well defined single oil source rock and a simple history of maturation, is ideal for developing and testing volumetric (mass balance) models for the generation and accumulation of hydrocarbons (see Section 9.4 and Welte et al., 1983 and refs. therein). Fuller (1975) detailed a mass balance for one basin in the North Viking Graben area. Goff (1983) has published an attempt to account for the oils of the North Viking Graben (East Shetland

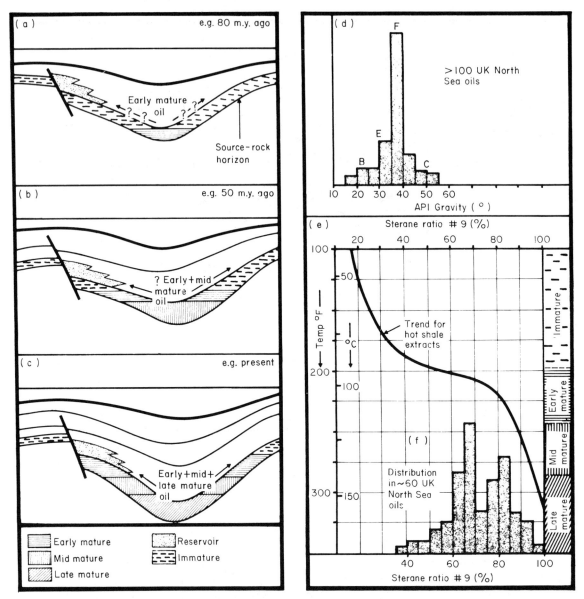

Fig. 9.18. North Sea oil properties controlled by source rock maturity (Cornford et al., 1983).
(a), (b), (c) Progressive oil generation from a single source rock horizon being buried in a graben. Early mature oil, generated first in the graben centre (a), cannot migrate due to poor drainage. Effective expulsion from the graben centre occurs only when this location reaches the mid (b) or even late mature stage (c), by which time early mature generation is occurring on the basin flanks, where improved drainage may facilitate expulsion.
(d), (e), (f) show the result of this process on oil properties. In the distribution of API gravities of UK North Sea oil (d), four oil types are identified B = Biodegraded, C = Condensate/light oil, E = Early mature, F = Full maturity spectrum. (A full maturity spectrum oil is one that contains molecules characteristic of the expelled products of early, mid and late mature source rocks). The maturity designation is derived, amongst other properties, from comparison of a sterane ratio trend measured on a maturity series of source rocks (e), with the same ratio measured on a collection of about 60 oils (f). Early mature oils are rare, and are found on the flanks of the basin, generally in Upper Jurassic reservoirs. The bulk of the oils contain molecules characteristic of a full maturity spectrum, or just a late maturity fraction of source rock products, consistent with the poor drainage of the hot shale source rock. Full maturity spectrum oils are found in reservoirs that have received oil from a range of source rock maturities as in the graben model c.

Basin) in terms of the volumes of mature source rock of known richness. Goff uses the equation:

Oil generated = SRV × OMV × GP × FOIH × TR × VI

where

SRV = Source Rock Volume (m³ or acre ft)

OMV = Organic Matter content by Volume (from TOC, see below)

GP = Genetic Potential (fraction of oil-prone kerogen = 0.7 for the Kimmeridge Clay Formation)

FOIH = Fraction of Oil in Hydrocarbon yield (0.8 according to Goff)

TR = Transformation Ratio (the extent to which the kerogen is converted—depends on the maturity of the source rock)

VI = Volume Increase on oil generation (1.2 according to Goff)

To this must be added a migration and entrapment factor. From these calculations he obtained the overall efficiency of generation—migration and entrapment to be 20–30%, having allowed for some additional generation from the Heather Formation. A major factor not adequately discussed by Goff (1983), and important for prospect evaluation is the definition of the drainage area (see Section 9.4).

Maturity

The maturity of the Kimmeridge Clay/Borglum Formation can be estimated from a base Cretaceous seismic depth map such as that produced by Day et al. (1981) and reproduced in part in Fig. 9.19). Similar, more detailed maps have been produced by Goff (1983) for the East Shetland Basin and Viking Graben, and by Michelsen and Andersen (1983) for the Danish Central Graben area. Depth maps can be used to define the area of mature rocks using a scheme such as shown in Table 9.5, where the temperatures for generation boundaries taken from Cornford et al. (1983) are converted to depth using a range of geothermal gradients (i.e. 25, 30, 35°C/km; 1.4, 1.6 and 1.9°F/100 ft), plus 4°C, (40°F) at sea-floor. This spread covers the majority of mapped variation in geothermal gradient in the North Sea (Harper, 1971; Cornelius, 1975; Carstens and Finstad, 1981) as shown in Fig. 9.9.

Maps of maturity *per se* have been published by Rønnevik *et al.* (1983) for the Norwegian North Sea, and for the Moray Firth–South Viking Graben area by Fisher and Miles (1983). Goff (1983) has produced a series of maps of the East Shetland Basin to show the progressive expansion of the areas of maturity with time from the basin centre to the basin edge (Fig. 9.21). Equating temperature with maturity in this way, however, ignores the effect of the 'cooking' or effective heating time. Within the North Sea region there are areas of both rapid and slow Neogene and recent

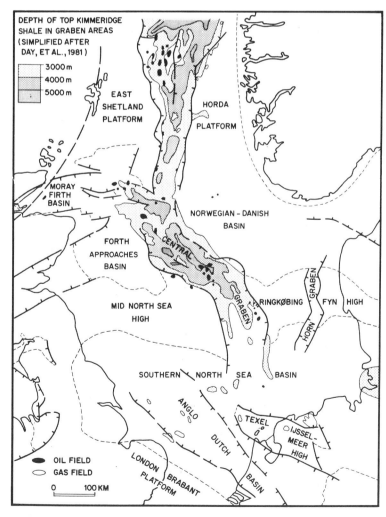

Fig. 9.19. Depth of burial at the top of the Kimmeridge Clay/Borglum Formations (simplified by Glennie from Day *et al.*, 1981). Major burial to oil-generating depths (e.g. >3000 m) occurs in the Central, Witch Ground and Viking Grabens. The relationship between depth of burial of the source rock and hydrocarbon type can be made by comparison with Fig. 9.21.

Table 9.5. Depths to specific maturity levels for areas of high, average and low geothermal gradient

Maturity level for oil (and associated gas)*	Temperature		Depth (km) to boundary at geothermal gradients (°C/km)† of:		
	°C	°F	25	30	35
Immature					
	80	175	3.0	2.5	2.2
Early mature					
	110	225	4.2	3.5	3.0
Mid mature					
	130	265	5.0	4.2	3.6
Late mature					
	155?	310?	6.0	5.0	4.3

* after Cornford *et al.* 1983
† = 1.4, 1.6, 1.9° F/100 ft respectively

subsidence, as well as of Tertiary uplift and erosion (Fig. 9.20).

It is well established that under conditions of rapid burial the increase in reflectance of vitrinite is retarded (e.g. Waples, 1980), but it is not at all clear whether anything other than extremely rapid burial significantly retards the generation of hydrocarbons in terms of m³ (bbls) generated per km³ (acre ft) of source rock (Cornford *et al.*, 1983). Since the majority of the North Sea suffered fairly uniform burial during the Tertiary, and the temperature zones used have been established within the North Sea Basin (Cornford *et al.*, 1983), these maturity intervals can probably be used with some confidence throughout the North Sea region except in areas of uplift and erosion such as the Inner Moray Firth, or areas with very high Neogene burial rates as in certain areas of the Central Graben.

As the North Sea Basin moves into a mature phase of exploration, it is of increasing importance to define the early mature generation phase as precisely as possible, and hence to recognise those geological situations under which the early mature oil will be effectively drained from the source rocks (Fig. 9.11A). This knowledge will optimise the search for oil in the areas of marginal maturity on the flanks of the grabens, areas such as the West Forties Basin, the Egersund Basin, the Inner Moray Firth Basin and the western margin of the East Shetland Basin.

Maturity in the mid-Norway area appears to be controlled by fairly constant Mesozoic and Tertiary burial (Larsen and Skarpnes, 1984; Hollander, 1984) in the basinal areas: structural highs, as usual, reveal unconformities and non-sequences.

9.5.8 The Cretaceous

The Lower Cretaceous (excluding the Ryazanian, see previous section) is developed as a dominantly shaley facies over much of the North Sea (Ziegler, 1982, encl. 21,22; Hesjedal and Hamar, 1983). Major depocentres (>500 m thickness) occurred in the Moray Firth Basin, the Viking and Central Grabens, in the Horda, Egersund and North Danish Basins, and in the Broad Fourteens and Lower Saxony Basins (Ziegler, 1983, encl. 31). In all these basins the section is believed to contain fair to good amounts of gas-prone kerogen (Barnard and Cooper, 1981).

TOC values that average 0.6% to 1.4% were reported for seven wells in the Moray Firth Basin (Bissada, 1983), with the kerogens of most wells being of gas/condensate prone land-plant type except in UK Block 14/20, where the kerogen is of lower-plant origin. In the Statfjord area, the TOC values of the Lower Cretaceous Cromer Knoll Group average 1.04% (Kirk, 1980) and appear to be gas-prone.

Locally, particularly during the Albian and Aptian, a thinly developed oil-prone kerogen facies was deposited in the southern half of the North Sea and may be equivalent to the pyritous, black, Speeton Clay, which outcrops in the east of England.

The Lower Cretaceous is, along with the Jurassic, the source for some of the north-west German gas fields, according to Tissot and Bessereau (1982). The Wealden facies of the Lower Cretaceous, as displayed on the south coast of England, is in a gas-prone facies with thin lignite seams and stringers and carbonaceous shales. The high sand/shale ratios will favour efficient drainage of any generated hydrocarbons.

North of 62°, variable quantities of mature gas- and oil-prone kerogen occur in the uppermost part of the Skarstein Formation (Aptian/Barremian) on Andoya Island (Bjorøy *et al.*, 1980). Rich gas- and oil-prone shales also occur on Svalbard (Bjorøy and Vigran, 1980; Leythaeuser *et al.*, 1983).

The regional maturity of the Lower Cretaceous can be deduced from the base Cretaceous depth map (Fig. 9.19), which indicates that it will be mature for gas (>1%R, >4000 m) in the Central, Viking and Witch Ground Grabens. It also reaches gas generating maturities in the North German Basin, where it is

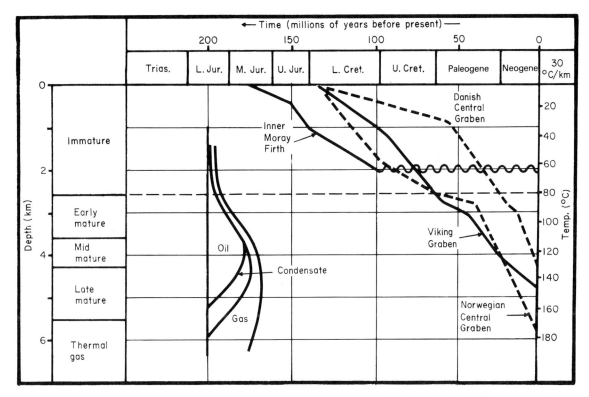

Fig. 2.20. The timing of generation from the Upper Jurassic of the North Sea as deduced from the subsidence curves (uncorrected for compaction) from the central parts of the Danish Central Graben (Hansen and Mikkelson, 1983), the Norwegian Central and Viking Grabens (Rønnevik et al., 1983), and the Inner Moray Firth/Beatrice Field (Linsley et al., 1980). Approximate maturity with respect to oil generation is shown using an average geothermal gradient of 30°C/km (Table 9.5). This shows that generation started in the late Cretaceous/Early Tertiary in the Graben centres, being earlier in the deepest parts of the Central than in the Viking Graben.

believed to have contributed to commercial gas fields. The Cretaceous would be mature within the Møre Basin according to the seismic mapping of Hamar and Hjelle (1984).

The Upper Cretaceous is devoid of source potential in the southern part of the North Sea, where it is developed mainly as chalk.

9.5.9 Tertiary

In the North Sea area, Tertiary sediments are unlikely to be effective source rocks because of their low degree of maturity. Day et al. (1981) and Gowers and Saebøe (1985) show that the top Cretaceous is buried below 3000 m only in the Central Graben. A similar situation exists in the Møre Basin (Hamar and Hjelle, 1984). Paleocene shales, however, are lean and gas and condensate prone. They were suggested by Pennington (1975), as the source for the oil in the Argyll Field, but this is not now thought probable on the grounds of regional maturity and kerogen type. TOC values average 0.43% in the Palaeocene Rogaland Group in the Statfjord area (Kirk, 1980), whilst two Paleocene intervals of Ekofisk well NOCS 2/4–12 gave average TOC values of 1.8 and 2.0% (Van den Bark and Thomas, 1980). Ungerer et al. (1984) report TOC values of 0.5–1.0% and Rock Eval Hydrogen Indices of about 210 mg/gTOC for the Eocene-Palaeocene of the Viking Graben. In the Ekofisk area, the shale has reached maturities of 0.59%–0.62%R at 3030–3070 m (9930–10060 ft). Since this is in the area of deepest Paleocene burial, no significant gas or condensate generation is expected from this formation.

North of 62°, the Tertiary may be a potential source for gas where burial is sufficient. Palaeocene/Eocene coals on Spitsbergen are immature with respect to gas generation (0.4–0.7% vitrinite reflectance), but are interbedded with black marine shales containing abundant amorphous kerogen (Manum and Throndsen, 1977).

9.6 North Sea hydrocarbons

A typical North Sea oil sourced by the Kimmeridge Clay Formation is a low sulphur, medium gravity, naphtheno-paraffinic oil (Table 9.6). Compilations of the composition of the oils have been made by Aalund (1983a, b), Cornford et al. (1983) and Cooper and Barnard (1984).

Gas sourced from the Westphalian Coal Measures of the southern North Sea is sweet and dry in composition (Table 9.7). Little is reported on the properties

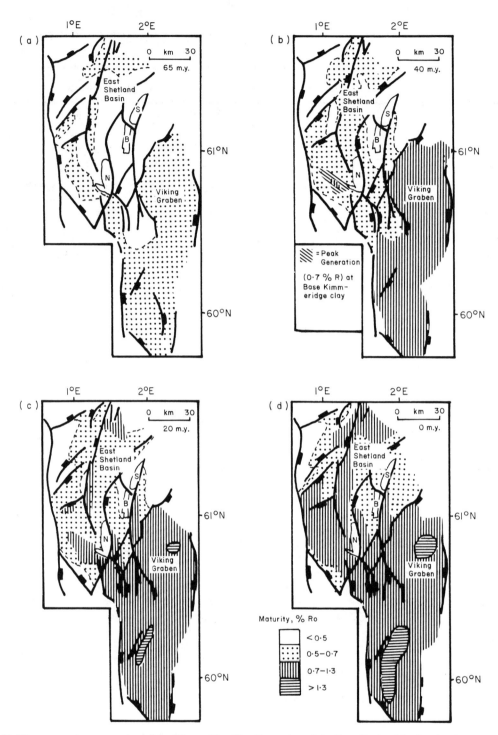

Fig. 2.21. The progressive maturation of the Kimmeridge Clay Formation of the East Shetland Basin with time (reproduced from Goff, 1983 with permission). (a) 65 million years ago, (b) 40 million years ago, (c) 20 million years ago, and (d) present day. Note that generation started in the Viking Graben about 40 m.y. ago, and only in the last 20 m.y. has moved into the half-grabens of the East Shetland Basin. Light oil and condensate generation, initiated about 20 m.y. ago, is restricted to the central portions of the Viking Graben. N = Ninian, B = Brent, S = Statfjord.

of the associated gas in the oil province: Frigg Field gas (questionably sourced from Middle Jurassic coals or shales) is dry and sweet (Heritier et al., 1979; Goff, 1983). From an analysis of the stable carbon isotope ratios of separated gas components, James and Burns (1984) have suggested that dry gases can result from the bacterial degradation of a wet gas or condensate. This mechanism, together with the presence of a thin heavy oil leg at the base of the dry gas leg of the Frigg Field, makes generation from the

Table 9.6. Average properties of North Sea oils sourced from the Kimmeridge Clay Formation

Property	Range	Average (±sd)	Comments
API gravity (°)[1]	17-51	36(±6.5)	Dead oils
Sulphur content (%) (a)[2]	0.13-0.55	0.32(±0.12)	15 oils excl. (b) below
(b)	0.56-1.57	1.0	Piper, Claymore, Tartan Buchan
Gas/Oil ratio (sft^3/bbl) (a)[3]	216-1547	671(±415)	Analytically heterogeneous
(b)	216-952	562(±271)	Excl. Ekofisk
Asphaltenes (%)[1]	0.1-5.1	1.2(±1.2)	Excl. asphaltene-enriched oils up to 35%
Saturate/Aromatics[1]	0.62-8.0	2.02(±1.2)	Topped oils
Pristane/nC_{17} ratio[1]	0.3-1.0	0.63(±0.17)	Excl. biodegraded oils
Phytane/nC_{18} ratio[1]	0.2-1.1	0.56(±0.18)	" "
Pristane/Phytane ratio[1]	0.6-1.9	1.24(±0.25)	" "
$\delta^{13}C$ (‰)[4]	28.4-29.8	–	Brae, Statfjord-Moray oils
	–	∼29	Piper area oils (av.)
V, Ni (ppm, wt)[2] V	0.53-6.0	3.1(±1.8)	Av. of 9 oils (excl.
Ni	0.5-5.0	1.8(±1.3)	Buchan at 26ppm V, 4.5ppmNi)
Wax (%) (a)[2]	4.0-7.7	6.3(±1.1)	5 oils
(b)[3]	–	17	Beatrice[5]

(1) Cornford et al. 1983 (approx. 60 analyses of mainly exploration DST and RFT samples).
(2) Aalund, 1979, 1983 (Production Crudes).
(3) Compiled from field reports in Illing and Hobson, 1981, Woodland, 1975, and Halbouty, 1980.
(4) Reitsema, 1983; Bissada, 1983, Fuller, 1975 (also see text).
(5) Beatrice field may have another source.
   sd = standard deviation

Table 9.7. Properties of gas from the North Sea and adjacent areas

	C_1(%)	C_2+(%)	N_2(%)	CO_2(%)	$\delta^{13}CH_4$‰
U.K. southern N. Sea[1]					
average	91.2	5.2	3.6	0.27	-
range	83.2-95.0	3.7-8.2	1.0-8.4	0.1-0.5	-
Groningen[1]	81.6	2.7	14.8	0.9	−36.6
Kinsale Head[2]	99.1	0.2	0.4	0.3	−45.5 to −48.3
Morcambe[3]	Dry	-	7-8	variable	-
Sleipner[4]	78-80	12-16	-	7-9	-
Frigg[5]	95.5	3.6	0.4	0.3	−43.3
NOCS Block 31/2[6]	92.6	5.4	1.5	0.5	-
NW Germany[7]	65-91	8-30	1	-	−54 to −44
Holland, Waddenzee Fields[8]	77.1-88.7	2.87-6.4	3.1-19.7	0.02-1.26	−31.1
Broad Fourteens (K/13)[9]	85.3	6.7	5.4, 7.1	1.7, 0.1	-

(1) Barnard and Cooper, 1983.
(2) Colley et al., 1981.
(3) Ebbern, 1981.
(4) Larsen and Jaarvik, 1981 (C_2 + figures include N_2).
(5) Héritier et al. 1981.
(6) Brekke et al. 1981.
(7) Tissot and Bessereau, 1982 (Jurassic to Lower Cretaceous source).
(8) Cottençon et al. 1975; Van den Bosch, 1983.
(9) Roos and Smits, 1983 (remigrated Westphalian gas).

underlying late mature Upper Jurassic oil-prone source rocks a more likely prospect. The gas of the Sleipner Field, which also contains significant condensate is wetter, with a high CO_2 content (Larsen and Jaarvik, 1981).

Little information is available for gas (Barnard and Cooper, 1983). Boigk and Stahl (1970) and Boigk *et al.* (1971) have noted regional trends in the contents of methane, nitrogen and carbon dioxide within the southern North Sea and north German offshore and land areas, although a number of possible explanations for these trends were offered. Kettel (1982) presents evidence to show that gases with high nitrogen contents derive from the most mature source rocks. The nitrogen content is generally higher in the Bunter than in the Rotliegend reservoirs of the UK southern North Sea (Barnard and Cooper, 1983), which may suggest enrichment during (re)migration. Some key properties of the oils and gases are summarised in Tables 9.6 and 9.7. In addition to sweet medium gravity oils, condensates are known in the Beryl Embayment, the Witch Ground Graben and the Central Graben (Fig. 9.22) associated with the most deeply buried Kimmeridge Clay Formation (Fig. 9.19), a relationship detailed by Barnard and Cooper (1981).

Heavy oils (API gravity $<20°$) are found over much of the North Sea (Barnard and Cooper, 1981; Cooper and Barnard, 1984), generally in reservoirs with present day temperatures less than 60°C (approx. 2000 m at 30°C/km and hence in Tertiary rocks. There is good evidence to show that they are produced within the reservoir by bacterial degradation of normal oil sourced by the Kimmeridge Clay Formation (e.g. Oudin, 1976). Bacterial degradation of reservoired oil is now readily recognised by its molecular and isotopic composition (Tissot and Welte, 1978; Connan, 1984; Volkman *et al.*, 1983). There is no commercial production of heavy oil in the North Sea.

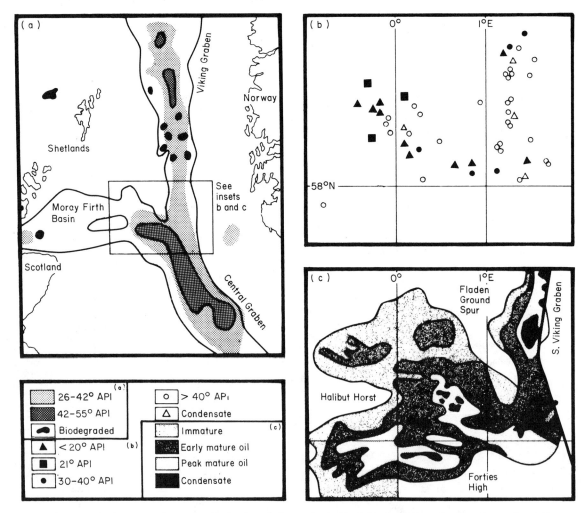

Fig. 9.22. Generalised map showing (a) the distribution of oil types in the North Sea by API gravity (from Barnard and Cooper, 1981) with insets (b) showing detailed distribution of oil according to gravity, and (c) predicted hydrocarbon product based on kerogen type and maturity for the Witch Ground Graben (Fisher and Miles, 1983). The general distribution of oil is clearly controlled by the maturation of the Kimmeridge Clay Formation (Fig. 9.18), while local variation may also be influenced by kerogen type and source rock drainage.

Intermediate gravity oils (20-35° API gravity) are known from the Clair Field, west of Shetland (22-25° API gravity), which Barnard and Cooper (1981) have classified as a biodegraded oil, and from the Troll Field oil. In many cases these intermediate gravity oils result from a mixing of heavy degraded oil (produced when the reservoir was more shallowly buried), together with a more recent influx of fresh oil, which arrived at the reservoir after it had been buried below the limiting depth (temperature) for active biodegradation. The same authors have noted that early mature North Sea oils may fall in the 28-35° API gravity range, in agreement with analytical data presented by Cornford et al. (1983).

Higher sulphur oils are found in some of the Witch Ground Graben Fields (e.g. Piper, Claymore, Tartan and Buchan). Beatrice oil, though light (approx. 38° API) and sweet, has a high wax content (17%) and pour point (65°F; 18°C), and low gas/oil ratio (Linsley et al., 1980). A similar high-wax (16%), low sulphur (0.3%), 35° API, low-GOR oil is produced from the De Lier Field of the Rijswijk oil province (Bodenhausen and Ott, 1981). Both oils are believed to come from non-marine source rocks.

9.7 Acknowledgements

I am pleased to thank Ken Glennie and Gordon Speers for detailed comments on a draft version of the text, and my many former colleagues in the Oil and Organic Geochemistry group at the KFA, Juelich, and at Britoil plc., Glasgow who have influenced my understanding of the genesis of North Sea oil. Douglas Kelso and Sara Cornford deserve special thanks for drafting the majority of the figures and typing the text respectively.

References

Aalund, L.R. (1979) Guide to world crudes—2. *Oil and Gas Journal*, Dec. 3, 99-113.

Aalund, L.R. (1983a) Guide to export crudes of the 80s—3. North Sea now offers 14 export crudes. *Oil and Gas Journal*, May 23, 69-76.

Aalund, L.R. (1983b) Guide to export crudes of the 80s—4. North Sea crudes: Flotta to Thistle. *Oil and Gas Journal*, June 6, 75-79.

Albright, W.A., Turner, W.L. and Williamson, K.R. (1980) Ninian field, U.K. sector, North Sea. In: Halbouty, M.T. (Ed.) *Giant oil and gas fields of the decade: 1968-1978* (AAPG Memoir 30).

Allan, J., Bjorøy, M. and Douglas, A.G. (1980) A geochemical study of the exinite group maceral alginite selected from three Permo-Carboniferous torbanites. In: Douglas, A.G. and Maxwell, J.R. (Eds.) *Advanced in organic geochemistry* 1979. Permagon Press, 599-618.

Altebäumer, F.J., Leythaeuser, D. and Schaefer, R.G. (1983) Effects of geologically rapid heating on maturation and hydrocarbon generation in Lower Jurassic shales from N.W. Germany. In: Bjorøy, M. et al. (Eds.) *Advances in organic geochemistry* 1981. John Wiley, 80-86.

Andersson, A., Dahlman, B. and Gee, D.G. (1982) Kerogen and uranium resources in the Cambrian Alum shales of Billingen-Falbygden and Närke areas, Sweden. *Geol. För. Stockh. Förh.* **104** (3), 197-209.

Andrews-Speed, C.P., Oxburgh, E.R. and Cooper, B.A. (1984) Temperatures and depth dependent heatflow in western North Sea. *AAPG Bull.* **68** (11), 1764-1781.

Asheim, S.M. and Larsen, V. (1984) The Tyrihans discovery —preliminary results from Well 6407/1-2. In: Spencer et al., Ibid, 285-292.

Barnard, P.C. and Cooper, B.S. (1981) Oils and source rock of the North Sea area. In: Illing, L.V. and Hobson, G.D. (Eds.) q.v., 169-175.

Barnard, P.C. and Cooper, B.S. (1983) A review of geochemical data related to the North-west European gas province. In: Brooks, J. (Ed.) q.v., 19-33.

Barnard, P.C., Collins, A.G. and Cooper, B.S. (1981) Identification and distribution of kerogen facies in a source rock horizon—examples from the North Sea basin. In: Brooks, J. (Ed.) *Organic maturation studies and fossil fuel exploration*. Academic Press, 271-282.

Barnard, P.C., Collins, A.G. and Cooper, B.S. (1981a) Generation of hydrocarbons—time, temperature and source rock quality. In: Brooks, J. (Ed.) *Organic maturation studies and fossil fuel exploration*. Academic Press, 337-342.

Barr, K.W., Colter, V.S. and Young, R. (1981) The geology of the Cardigan Bay—St George's Channel basin. In: Illing, L.V. and Hobson, G.D. (Eds.) q.v., 432-443.

Bartenstein, H. (1979) Essay on coalification and hydrocarbon potential of the N.W. European Paleozoic. *Geol. Mijnbouw* **59** (2), 155-168.

Batten, D.J. (1983) Identification of amorphous sedimentary organic matter by transmitted light microscopy. In: Brooks, J. (Ed.) q.v., 275-287.

Bergstrøm, S.M. (1980) Conodants as paleotemperature tools in Ordovician rocks of the Caledonides and adjacent areas in Scandinavia and the British Isles. *Geol. För. Stockh. Förh.* **102** (4), 377-392.

Berstad, S. and Dypvik, J. (1982) Sedimentological evolution and natural radioactivity of Tertiary sediments from the Central North Sea. *J. Petrol. Geol.* **5** (1), 77-88.

Best, G. Kockel, F. and Schöneich, H. (1983) The geological history of the southern Horn Graben. *Geol. Mijnbouw.* **62** (1), 25-33.

Bishop, R.S., Gehman, H.M and Young, A. (1984) Concepts for estimating hydrocarbon accumulation and dispersion. Demaison, G. and Murris, R.J. q.v., 41-52 (Also in *AAPG Bull.* (1983) **67**, 337-348).

Bissada, K.K. (1983) Petroleum generation in Mesozoic sediments of the Moray Firth Basin, North Sea area. In: Bjorøy, M. (Ed.) *Advances in geochemistry* 1981. John Wiley, 7-15.

Bitterli, P. (1963) Aspects of the genesis of bituminous rock sequences. *Geol. Mijnbouw.* **42** (6), 183-201.

Bitterli, P. (1963) On the classification of bituminous rocks from Western Europe. In: *Proceedings of 6th World Petroleum Congress, Frankfurt*, Section 1, 155-165.

Bjørlykke, K., Dypvik, H. and Finstad, K.G. (1975) The Kimmeridgian shale, its compositions and radioactivity. In: Finstad, K.G. and Selley, R.C. (Eds.) (1976) *Jurassic Northern North Sea Symposium*. 12.1-12.20. NPS.

Bjorøy, M. and Vigran, J.O. (1980) Geochemical study of the organic matter in outcrop samples from Agardgfjellet, Spitzbergen. In: Douglas, A.G. and Maxwell, J.R. (Eds.) *Advances in organic geochemistry* 1979. Pergamon Press, 141-147.

Bjorøy, M, Hall, K. and Vigran, J.O. (1980) An organic geochemical study of Mesozoic shales from Andøya, North Norway. In: Douglas, A.G. and Maxwell, J.R. (Eds.) *Advances in organic geochemistry* 1979. Permagon Press, 77-91.

Bjorøy, M., Mork, A. and Vigran, J.O. (1983) Organic geochemical studies of the Devonian to Triassic succession on Bjornøya and the implications for the Barent Shelf. In: Bjorøy, M. *et al.* (Eds.) *Advances in organic geochemistry* 1981. John Wiley, 49-59.

Bodenhausen, J.W.A. and Ott, W.F. (1981) Habitat of the Rijswijk oil province, onshore The Netherlands. In: Illing, L.V. and Hobson, G.D. (Eds.) q.v., 301-309.

Boigk, V.H. and Stahl, W. (1970) Zum Problem der Entstehung nordwestdeutcher Erdgaslagerstätten. In: *Erdöl and Kohle— Erdgas—Petrochemie, 111* Brennstoff Chemie **23** (6), 325-333.

Boigk, H., Stahl, W., Teichmüller, M. and Teichmüller, R. (1971) Metamorphism of coal and natural gas. *Forschr. Geol. Rheinld. u. Westphal.* **19**, 104-111.

Bostick, N.H. (1974) Phytoclasts as indicators of thermal metamorphism, Franciscan assemblage and Great Valley sequence (upper Mesozoic), California. *Geol. Soc. Am. Spec. Paper* **153**, 1-17.

Bostick, N.H. (1979) Microscopic measurement of the level of catagenesis of solid organic matter in sedimentary rocks to aid exploration for petroleum and to determine former burial temperatures: a review. SEPM Special Publication No. 26 Society of Economic Paleontologists and Mineralogists, 17-43.

Brand, E. and Hoffman, K. (1963) Stratigraphy and facies of the North-west German Jurassic and genesis of its oil deposits. In: *Proceedings of 6th World Petroleum Congress, Frankfurt.* Section 1, Paper 7, 223-246.

Brekke, T., Pegrum, R.M. and Watts, P.B. (1981) First exploration results in Block 31/2 offshore Norway. In: *Norwegian Symposium on Exploration, Bergen, Sept. 14-16.* Norwegian Petroleum Society, p. 16.1-16.34.

Brewster, J. and Dangerfield, J.A. (1984) Chalk fields along the Lindenes Ridge, Eldfisk. *Marine and Petroleum Geology* **1** (3), 239-278.

Brooks, J. (Ed.) (1983) *Petroleum geochemistry and exploration of Europe.* Blackwell Scientific Publications.

Brooks, J., Cornford, C., Nicholson, J. and Gibbs, A.D. (1984) Geological controls on occurrence and composition of Tertiary heavy oils, northern North Sea. Fossil Fuels of Europe Conference, Geneva, 1984; Abstract *AAPG Bull.* **68** (6), 793.

Brooks, J. and Thusu, B. (1977) Oil source rock identification and characterisation of the Jurassic sediments in the northern North Sea. *Chem. Geol.* **20**, 283-294.

Brooks, J., Cornford, C. and Archer, R. (in press). The role of marine source rocks in petroleum exploration. In: *Marine petroleum source rocks* meeting, 17-18 May 1983. Petroleum Geochemistry group of the Geological Society.

Brooks, J. and Welte, D. (1984) *Advances in Petroleum Geochemistry* 1. Academic Press, London.

Bujak, J.P., Barss, H.S. and Williams, G.L. (1977) Offshore Canada's organic type and colour and hydrocarbon potential. *Oil and Gas J.* **4**, 198-202.

Carstens, H. and Finstad, K.G. (1981) Geothermal gradients of the northern North Sea Basin, 59-62° N. In: Illing, L.V. and Hobson, G.D. (Eds.) q.v., 152-161.

Colley, M.G., McWilliams A.S.F. and Myres, R.C. (1981) The geology of the Kinsale Head gas field, Celtic Sea, Ireland. In: Illing, L.V. and Hobson, G.D. (Eds.) q.v., 504-510.

Colter, V.S. and Havard, D.J. (1981) The Wytch Farm oil field, Dorset. In: Illing, L.V. and Hobson, G.D. (Eds.) q.v., 494-503.

Connan, J. (1984) Biodegradation of crude oil in reservoirs In: Brooks, J. and Welte, D., q.v., 299-335.

Connan, J. and Cassou, A.M. (1977) Properties of gases and petroleum liquids derived from terrestrial kerogen at various maturation levels. *Proceedings of the International Palynological Congress, Leon, Spain.*

Cooper, B.S. and Barnard, P.C. (1984) Source rocks and oil of the Central and Northern North Sea. In: Demaison, G. and Murris, R.J., q.v., 303-314.

Cooper, B.S., Coleman, S.H., Barnard, P.C. and Butterworth, J.S. (1975) Paleotemperatures in the northern North Sea Basin. In: Woodland, A.W. (Ed.) q.v., 487-492.

Cope, M.J. (1980) Physical and chemical properties of coalified and charcoalified phytoclasts from some British Mesozoic sediments: an organic geochemical approach to paleobotany. In: Douglas, A.G. and Maxwell, J.R. (Eds.) *Advances in organic geochemistry* 1979. Pergamon Press. 663-677.

Cornelius, C.D. (1975) Geothermal aspects of hydrocarbon exploration in the North Sea area. *Norges Geol. Unders.* **316**, 29-67.

Cornford, C. and Douglas, A.G. (In preparation) Maturity, depositional environment and hydrocarbon source potential of Lower Lias limestone/shale sequences of southwest Britain.

Cornford, C., Morrow, J.A., Turrington, A., Miles, J.A. and Brooks, J. (1983) Some geological controls on oil compositions in the U.K. North Sea In: Brooks, J. (Ed.) q.v., 175-194.

Cornford, C., Needham, C.C. and DeWalque, L. (1986) Geochemical habitat of North Sea oils and gases. Paper presented at the Norwegian Petroleum Society Meeting, Stavanger, October 1985, and submitted for publication in conference volume, due 1986.

Cornford, C., Rullkötter, J. and Welte, D. (1980) A synthesis of organic petrographic and geochemical results from DSDP sites in the eastern central North Atlantic. In: Douglas, A.G. and Maxwell, J.R. (Eds.) *Advances in organic geochemistry* 1979. Pergamon Press, 445-453.

Cosgrove, M.E. (1970) Iodine in the bituminous Kimmeridge Shales of the Dorset coast, England. *Geochim, Cosmochim. Acta* **34**, 830-836.

Cottençon, A., Parant, B. and Flacelière, G. (1975) Lower Cretaceous gas fields in Holland. In: Woodland, A.W. (Ed.) q.v., 403-412.

Dahl, B. and Speers, G. (in press) Oseberg geochemistry. Part I. Presented at NPF—Meeting 'Organic Geochemistry in the Exploration of the Norwegian Shelf', Stavanger Oct. 22-24, 1984.

Day, G.A., Cooper, B.A., Anderson, C., Burgers, W.F. Rønnevik, H.C. and Schöneich, H. (1981) Regional seismic structure maps of the North Sea. In: Illing, L.V. and Hobson, G.D. (Eds.) q.v., 76-84.

Demaison, G., Holck, A.J.J., Jones, R.W. and Moore, G.T. (1984) Predictive source bed stratigraphy: a guide to regional petroleum occurence, North Sea Basin and eastern North American continental margin. *Proceedings of the World Petroleum Congress*, London 1983, Paper PD-1 (2). 17-29.

Demaison, G.J. and Moore, G.T. (1980) Anoxic environments and oil source bed genesis. *Org. Geochem.* **2**. 9-31.

Demaison, G., and Murris, R.J. (1984) *Petroleum geochemistry and Basin Evaluation. AAPG Memoir* **35**. AAPG Tulsa.

Donovan, R.N. (1980) Lacustrine cycles, fish ecology and stratigraphic zonation in the Middle Devonian of Caithness *Scott. J. Geol.* **16**, 35-72.

Donovan, R.N., Foster, R.J. and Westoll, T.S. (1974) A stratigraphical revision of the Old Red Sandstone of northeastern Caithness. *Trans. R. Soc. Edin.* **69**, 167-201.

Douglas, A.G. and Williams, P.F.V. (1981) Kimmeridge oil shale: a study of organic maturation. In: Brooks, J. (Ed.) *Organic maturation studies and fossil fuel exploration.* Academic Press, 256-269.

Douglas, A.G. Eglinton, G. and Maxwell, J.R. (1969) The organic geochemistry of certain samples from the Scottish Carboniferous formation. *Geochim. Cosmochim. Acta,* **33**, 579-590.

Dow, W.G. and O'Connor, D.I. (1982) Kerogen maturity and type by reflected light microscopy applied to petroleum exploration. In: Staplin, F.L. *et al.* (Ed.) *How to*

assess maturation and paleotemperature, SEPM Short Course No. 7, 133–157.
Downey, M.W. (1984) Evaluating seals for hydrocarbon accumulations. *AAPG Bull.* **68** (11), 1752–1763.
Duff, K.L. (1975) Paleoecology of a bituminous shale—the Lower Oxford Clay of central England. *Palaeontology* **18**, 443–482.
Duncan, D.C. and Swanson, V.E. (1965) Organic-rich shale of the United States and world land areas. *US Geol. Survey Circ.* No. 523, Washington.
Durand, B. (1980) *Kerogen—Insoluble organic matter from sedimentary rocks.* Editions Technip, Paris.
Durand, B. (1983) Present trends in organic geochemistry in research in migration of hydrocarbons. In: Bjorøy, M. et al. (Eds.) *Advances in organic geochemistry* 1981. John Wiley, 117–128.
Durand, B. and Paratte, M. (1983) Oil potential of coals: a geochemical approach. In: Brooks, J., q.v., 255–265.
Dypvik, H., Rueslätten, H.G. and Thondsen, T. (1979) Composition of organic matter from North Atlantic Kimmeridgian shales. *Am. Assoc. Petrol. Geol. Bull.* **63** (12), 2222–2226.
Eames, T.D. (1975) Coal rank and gas source relationships—Rotliegendes reservoirs. In: Woodland, A.W. (Ed.) q.v., 191–203.
Ebbern, J. (1981) The geology of the Morecombe gas field. In: Illing, L.V. and Hobson, G.D. (Eds.) q.v., 485–493.
Eynon, G. (1981) Basin development and sedimentation in the Middle Jurassic of the northern North Sea. In: Illing, L.V. and Hobson, G.D. (Eds.) q.v., 196–204.
Faber, E. and Stahl, W. (1984) Geochemical surface exploration for hydrocarbons in the North Sea. *AAPG Bull.* **68** (3), 363–386.
Farrimond, P., Comet, P., Eglinton, G., Evershed, R.P., Hall, M.A., Park, D.W., and Wardroper, A.M.K. (1984) Organic geochemical study of the Upper Kimmeridge Clay of the Dorset type area. *Marine and Petroleum Geology* **1** (4), 340–354.
Fertl, W.H. (1976) Elucidation of oil shales using geophsical well logging techniques. In: Yen and Chilingarian (Eds.), *Oil Shale* Elsevier, pp. 199–213.
Fisher, M.J. and Miles J.A. (1983) Kerogen types, organic maturation and hydrocarbon occurrences in the Moray Firth and South Viking Graben, North Sea Basin. In: Brooks, J. (Ed.) q.v., 195–201.
Forsberg, A. and Bjorøy, M. (1983) A sedimentalogical and organic geochemical study of the Botneheia Formation, Svalbard, with special emphasis on the effects of weathering on the organic matter in the shales. In: Bjorøy, M. et al. (Eds.) *Advances in organic geochemistry* 1981. John Wiley, 60–68.
Fuller, J.G.C.M. (1975) Jurassic source rock potential—and hydrocarbon correlation, North Sea. In: *Proceedings of the Symposium on Jurassic—northern North Sea.* Norwegian Petroleum Society meeting.
Fuller, J.G.C.M. (1980) In: Jones, J.M. and Scott, P.W. *Progress report on fossil fuels—exploration and exploitatation.* Proc. Yorks. Geol. Soc. **42**, 581–593.
Galois, R.W. (1976) Coccolith blooms in the Kimmeridge Clay, and origin of North Sea oil. *Nature* **259**, 473–475.
Gibbons, M. (in press) The depositional environment and petroleum geochemistry of the Marl Slate/Kupferschiefer. In: Brooks, J. and Fleet, A. (Eds.) (1984) *Marine petroleum source rocks.* Blackwell Scientific Publications.
Gibbs, A.D. (1984) Clyde Field growth fault secondary detachment above base faults in North Sea. *AAPG Bull.* **68** (8), 1029–1039.
Glennie, K.W. and Boegner, P.L.E. (1981) Sole Pit inversion tectonics. In: Illing, L.V. and Hobson, G.D. (Eds.) q.v., 110–120.
Goff, J.C. (1983) Hydrocarbon generation and migration from Jurassic source rocks in the E. Shetland Basin and Viking Graben of the northern North Sea. *J. Geol. Soc. Lond.* **140**, 445–474.
Goff, J.C. (1984) Hydrocarbon habitat of East Shetland Basin and North Viking Graben of the northern North Sea. Fossil Fuels of Europe Conference, Geneva, 1984, Abstrat: *AAPG Bull.*, 794.
Gold, T. and Soter, S. (1982) Abiogenic methane and the origin of petroleum. *Energy Exploration and Exploitation* **1** (2), 89–104.
Gowers, M.B. and Saebøe, A. (1985) On the structural evolution of the Central Trough in the Norwegian and Danish sectors of the North Sea. *Marine and Petroleum Geology* **2** (2), 298–318.
Grantham, P.J., Posthuma, J. and De Groot, K. (1980) Variation and significance of the C_{27} and C_{28} triterpane content of a North Sea core and various North Sea crude oils. In: Douglas, A.G. and Maxwell, J.R. (Eds.) *Advances in organic geochemistry* 1979. Pergamon Press, 29–38.
Gretener, P.E. and Curtis, C.D. (1982) Role of temperature and time on organic metamorphism. *Am. Assoc. Petrol. Bull.* **66** (8), 1124–1129.
Griffiths, A.E. (1983) The search for petroleum in Northern Ireland. In: Brooks, J. (Ed.) q.v., 213–222.
Guildish, T.M., Kendall, C.G. St. C., Lerche, I., Toth, D.J. and Yarzab, R.F. (1985) Basin evaluation using burial history calculations: an overview. *AAPG Bull.* **69** (1), 92–105.
Gutjahr, C.C.M. (1983) Incident light microscopy of oil and gas source rocks. *Geol. Mijnbouw* **62** (3), 417–425.
Halbouty, M.T. (Ed.) 1980 *Giant oil and gas fields of the decade 1968–1978.* Am. Assoc. Petrol. Geol., Tulsa. Memoir 30.
Hall, P.B. and Douglas, A.G. (1983) The distribution of cyclic alkanes in two lacustrine deposits. In: Bjorøy, M. et al. (Eds.) *Advances in organic geochemistry* 1981. John Wiley, 576–587.
Hallam, A. (1985) A review of Mesozoic climates. *J. Geol. Soc. Lond.* **142** (3), 433–446.
Hallam, A. and Bradshaw, M.J. (1979) Bituminous shales and oolitic ironstones as indicators of transgressions and regressions. *J. Geol. Soc. Lond.* **136**, 157–164.
Hamar, G.P. (1975) A Jurassic structure complex in northern North Sea. In: Finstad, K.G. and Selley, R.C. (Eds.) (975) *Proceedings of the Jurassic Northern North Sea Symposium.* Norwegian Petroleum Society, 17.1–17.18.
Hamar, G.P., Fjaeran, T. and Hesjedal, A. (1983) Jurassic stratigraphy and tectonics of the south-southeastern Norwegian offshore. *Geol. Mijnbouw* **62** (1), 103–114.
Hamar, G.P. and Hjelle, K. (1984) Tectonic framework of the Møre Basin and the northern North Sea. In: Spencer et al. q.v., 349–358.
Hancock, N.J. and Fisher, M.J. (1981) Middle Jurassic North Sea deltas with particular reference to Yorkshire. In: Illing, L.V. and Hobson, G.D. (Eds.) q.v., 186–195.
Hansen, J.M. and Mikkelsen, N. (1983) Hydrocarbon geological aspects of subsidence curves: interpretation based on released wells in the Danish Central Graben. *Bull. Geol. Soc. Denmark* **31**, 159–169.
Harding, T.P. (1984) Graben hydrocarbon occurrences and structural style. *AAPG Bull.* **68** (3) 333–362.
Harper, M.L. (1971) Approximate geothermal gradients in the North Sea Basin. *Nature* **230**, 235–236.
Héritier, F.E., Lossel, P. and Wathne, E. (1979) Frigg field—large submarine fan trap in Lower Eocene rocks of North Sea. *Am. Assoc. Petrol. Geol. Bull.* **63**, 1999–2020.
Héritier, F.E., Lossel, P. and Wathne, E. (1981) The Frigg gas field. In: Illing, L.V. and Hobson, G.D. (Eds.) q.v., 380–391.
Hesjedal, A. and Hamar, G.P. (1983) Lower Cretaceous stratigraphy and tectonics of the south-southeastern Norwegian offshore. *Geol. Mijnbouw* **62** (1), 135–144.

Hollander, N.B. (1984) Geohistory and hydrocarbon evaluation of the Haltenbanken area. In: Spencer, A.M. *et al.*, q.v., 383–388.

Hood, A., Gutjahr, C.C.M. and Heacock, R.L. (1975) Organic metamorphism and the generation of petroleum. *Am. Assoc. Petrol. Geol. Bull.* **59**, 986–996.

Hopkinson, E.C., Fertl, W.H. and Oliver, D.W. (1982) The continuous carbon/oxygen log—basic concepts and recent field experience. *J. Petrol. Technol.*, October 1983, 2441–2448.

Hovland, M. and Sommerville, J.H. (1985) Characteristics of two natural gas seepages in the North Sea. *Marine and Petroleum Geology* **2**, 319–326.

Huc, A. (1976) Mise en evidence de provinces geochimiques dans le Schistes Bitumineux du Toarcien de l'est du Bassin de Paris, *Rev. Inst. Franç. Petrol.*, **31** (6), 933–953.

Illing, L.V. and Hobson, G.D. (Eds.) (1981) *Petroleum geology of the Continental Shelf of north-west Europe.* Heyden & Son.

Irwin, H. (1979) An environmental model for the type Kimmeridge Clay (comment and reply). *Nature* **279**, 819–820.

Islam, S., Hesse, R. and Chagnon, A. (1982) Zonation of diagenesis and low grade metamorphism in Cambro-Ordovician Flysche of Gaspé peninsula, Quebec Appalachians. *Canadian Mineralogist* **20**, 155–167.

Issler, D.R. (1984) Calculation of organic maturation levels for offshore eastern Canada—implications for general application of Lopatin's method. *Canadian Journal of Earth Science* **21** (4), 477–488.

James, A.T. (1983) Correlation of natural gas by the use of carbon isotopic distribution between hydrocarbon components. *AAPG Bull.* **67**, 1176–1191.

James, A.T. and Burns, B.J. (1984) Microbial alteration of subsurface natural gas accumulations. *AAPG Bull.* **68** (8), 957–960.

Jenner, J.K. (1981) The structure and stratigraphy of the Kish Bank basin. In: Illing, L.V. and Hobson, G.D. (Eds.) q.v., 426–431.

Johns, C.R. and Andrews, I.J. (1985) The petroleum geology of the Unst Basin, North Sea. *Marine and Petroleum Geology* **2** (4), 361–372.

Juntgen, J. and Karweil, J. (1966) Gasbildung und Gasspeicherung in Steinkohlflözen. *Erdöl, Kohle, Erdgas, Petrochemie* **19**, 251–258, 339–344.

Kent, P.E. (1985) U.K. onshore oil exploration 1930–1964. *Marine and Petroleum Geology* **2** (1), 56–64.

Kettel, D. (1982) Norddeutsche Erdgase: Stickstoffgehalt und Isotopenvariationen als Reife-und Faziesindikatoren. *Erdöl und Kohle, Erdgas, Petrochemie mit Brenstoffchemie* **35** (12), 557–559.

Kettel, D. (1983) The East Groningen Massif; detection of an intrusive body by means of coalification. *Geol. Mijnbouw* **60** (1), 203–210.

Kirk, R.H. (1980) Statfjord Field, a North Sea giant. In: Halbouty, M.T. (Ed.) *Giant Oil and Gas Fields of the decade 1968–1978.* (AAPG Memoir 30), 95–116.

Koch, J.-O. (1983) Sedimentology of the Middle and Upper Jurassic reservoirs of Denmark, *Geol. Mijnbouw* **62** (1), 115–129.

Kurbskiy, G.P., Bogdanchikov, A.I. and Abuchayeva, V.V. (1983) Changes in hydrocarbon composition of benzine fraction during catagenesis. *Geochemistry International* **20** (1), 97–106.

Larsen, R.M. and Jaarvik, L.J. (1981) The geology of the Sleipner field complex. In: *Norwegian Symposium on Exploration, 1981, Bergen.* Norwegian Petroleum Society, 15.1–15.31.

Larsen, R.M. and Skarpnes, O. (1984) Regional interpretation and hydrocarbon potential of the Traenabanken area. In: Spencer, A.M. *et al.*, q.v., 217–236.

Lee, M., Aronson, J.L. and Savin, S.M. (1985) K/Ar dating of time of gas emplacement in Rotliegendes sandstone, Netherlands *AAPG Bull.* **69** (4) 1381–1385.

Lerch, I., Yarzab, R.F. and Kendall, C.G., St. C. (1984) Determination of paleoheat flux from vitrinite reflectance data. *AAPG Bull.* **68** (11), 704–717.

Leythaeuser, D. Mackenzie, A.S., Schaefer, R.G., Altebäumer, F.J. and Bjorøy, (1983) Recognition of migration and its effects within two coreholes in shale/sandstone sequences from Svalbard, Norway. In: Bjorøy, M. *et al.* (Eds.) *Advances in organic geochemistry* 1981. John Wiley, pp. 136–146.

Leythaeuser, D., Mackenzie, A.S., Schaefer, R.G. and Bjorøy, M. (1984) A novel approach for the recognition and quantitation of hydrocarbon migration effects in sandstone shale sequences. *AAFG Bull.* **68** (2) 196–219.

Leythaeuser, D., Schaefer, R.G. and Yukler, A. (1982) The role of diffusion in primary migration of hydrocarbons. *Am. Assoc. Petrol. Geol. Bull.* **66** (4), 408–429.

Lindgreen, H. (1985) Diagenesis and primary migration in Upper Jurassic claystone source rocks in North Sea. *AAPG Bull.* **69** (4), 525–536.

Linsley, P.N. Potter, H.C., McNab, G. and Racher, D. (1980) The Beatrice Field, Inver Moray Firth, U.K. North Sea. In: Halbouty, M.T. (Ed.) *Giant oil and gas fields of the decade 1968–1978* (AAPG Memoir 30), 117–129.

Lutz, M., Kaasschieter, J.P.H. and van Wijhe, D.H. (1975) Geological factors controlling Rotliegend gas accumulations in the mid-European basin. *Proceedings of 9th World Petroleum Congress.* Applied Science, 93–103.

Mackenzie, A.S. (1984) Applications of biological markers in petroleum geochemistry. In: Brooks, J. and Welte, D., q.v., 115–214.

Mackenzie, A.S., Maxwell, J.R., Coleman, M.L. and Deegan, C.E., (1983) Biological marker and isotope studies of North Sea crude oils and sediments. *Proceedings of the World Petroleum Congress*, London, 1983, Paper **PD-1** (4), 45–56.

Mackenzie, A.S. and McKenzie, D. (1983) Isomerisation and aromatisation of hydrocarbons in sedimentary basins formed by extension. *Geol, Mag,* **120** (5), 417–470.

Mackenzie, A.S., Patience, R.L., Maxwell, J.R., Vandenbroucke, M. and Durand, B. (1980) Molecular parameters of maturation in the Toarcian shales, Paris basin, France. 1. Changes in the configuration of the acyclic isoprenoid alkanes, steranes and triterpanes. *Geochim. Cosmochim. Acta* **44** 1709–1721.

Macquaker, J.H.S., Farrimond, P. and Brassel, S.C. (1985) Rhaetian black shales: potential oil source rocks of S.W. Britain. Paper presented to the *12th International Meeting on Organic Geochemistry*, Jülich, West Germany, Abstract **B37**.

Maher, C.E. (1981) The Piper oilfield. In: Illing, L.V. and Hobson, G.D. (Eds.) q.v., 358–370.

Manum, S.B. and Throndsen, T. (1977) Rank of coal and dispersed organic matter and its geological bearing in the Spitzbergen Tertiary. *Norsk. Polarinst. Arbok* 1977, 159–177, 179–187.

Maragna, B., Zaro, G. and Pessina, P. (1984) Block 33/6 geochemical evaluation. Presented to *Organic Geochemistry in Exploration of the Norwegian Shelf.* Stavanger, Oct. 1984.

McIver, R. (1975) Hydrocarbon occurrences from JOIDES Deep Sea Drilling Project. *World Petroleum Congress, Tokyo,* Panel Discussion, vol. 1, 1.

Meissner, F.F. (1978) Petroleum geology of the Bakken Formation, Williston Basin, North Dakota and Montana. *Williston Basin Symposium*, Montana Geol. Soc., 207–227.

Meyer, B.L. and Nederlof, M.H. (1984) Identification of source rocks on wireline logs by density/resistivity and interval velocity/resistivity cross-plots. *Am. Assoc. Petrol. Geol. Bull.* **68** (2), 121–129.

Michelsen, O. and Andersen, C. (1983) Mesozoic structural and sedimentary development of the Danish Central Graben. *Geol. Mijnbouw* 62 (1), 93-102.

Mørk, A. and Bjorøy, M. (1984) Mesozoic source rocks in Svalbard. In: A.M. Spencer *et al.*, q.v., 371-382.

Mukhopadhyay, P.K. and Gormley, J.R. (1985) Hydrocarbon potential of two types of resinite. In: Schenk, P.A. *et al.*, q.v., 439-454.

Müller, P.J. and Suess, E. (1979) Productivity, sedimentation rate and sedimentary organic matter in the oceans. 1. Organic carbon preservation. *Deep Sea Research* 26A, 1347-1362.

Munns, J.W. (1985) The Valhall Field: a geological review. *Marine and Petroleum Geology* 2 (1), 23-43.

Nordberg, H.E. (1981) Seismic hydrocarbon indicators in the North Sea. *Norwegian Symposium on Exploration, Bergen.* Norwegian Petroleum Society, 8.1-8.40.

Oele, J.A., Hol, A.C.P.J. and Tiemens, J. (1981) Some Rotliegend gas fields of the K and L blocks, Netherlands offshore (1968-1978)—a case history. In: Illing, L.V. and Hobson, G.D. (Eds.) q.v., 289-300.

Olsen, J.C. (1983) The structural outline of the Horn Graben. *Geol. Mijnbouw* 62 (1), 47-50.

Oudin, J-L. (1976) Étude géochimique du Bassin de la Mere du Nord. *Bull. Centre Rech. Pau-SNPA* 10 (1), 339-358.

Oxburgh, E.R. and Andrews-Speed, C.P. (1981) Temperature, thermal gradients and heat flow in the south-west North Sea. In: Illing, L.V. and Hobson, G.D. (Eds.) q.v., 114-151.

Øygard, K., Grahl-Nielsen, O. and Ulvoen, S. (1984) Oil/oil correlation by aid of chemometrics. In: Schenk, P.A. *et al.*, q.v., 561-567.

Palacas, J.G. (1984) Petroleum Geochemistry and Source Rock Potential of Carbonate Rocks. *AAPG Studies in Geology* 18, AAPG, Tulsa.

Parnell, J. (1983) The distribution of hydrocarbon minerals in the Orcadian Basin. *Scott. J. Geol.* 19, 205-213.

Parry, C.C., Whitley, P.K.J. and Simpson, R.D.H. (1981) Integration of palynological and sedimentalogical methods in facies analysis of the Brent Formation. In: Illing, L.V. and Hobson, G.D. (Eds.) q.v., 205-215.

Pearson, M.J. and Watkins, D. (1983) Organofacies and early maturation effects in Upper Jurassic sediments from the Inner Moray Firth Basin, North Sea. In: Brooks, J. (Ed.) q.v., 147-160.

Pearson, M.J., Watkins, D., Pittion, J-L., Caston, D. and Small, J.S. (1983) Aspects of burial diagenesis, organic maturation and paleogeothermal history of an area in the southern Viking Graben, North Sea. In: Brooks, J. (Ed.) q.v., 161-173.

Pegrum, R.M. and Ljones, T.E. (1984) The 15/9 Gamma Gas Field offshore Norway, new trap type for North Sea Basin with regional implications. *AAPG Bull.* 68 (7), 874-902.

Pennington, J.J. (1975) The geology of the Argyll field. In: Woodland, A,W, (Ed.) q.v., 285-291.

Pering, K.L. (1973) Bitumens associated with lead, zinc and fluorite ore minerals in North Derbyshire, England. *Geochim. Cosmochim. Acta.* 37, 401-417.

Piggot, J.D. (1985) Assessing source rock maturity in frontier Basins: Importance of time, temperature and tectonics. *AAPG Bull.* 69 (8), 1269-1274.

Powell, T.G. (1986) Developments in concepts of hydrocarbon generation from terrestrial organic matter. Presented at the Beijing Petroleum Symposium, China, Sept. 1984. *AAPG Bull*, 1986.

Powell, T.G., Douglas, A.G. and Allan, J. (1976) Variations in the type and distribution of organic matter in some Carboniferous sediments from Northern England. *Chemical Geol.* 18, 137-148.

Pratsch, J-C. (1983) Gasfields, N.W. German Basin: secondary gas migration is a major geologic parameter. *J. Petrol. Geol.* 5(3), 229-244.

Price, L.C. (1983) Geologic time as a parameter in organic metamorphism and vitrinite reflectance as an absolute paleogeothermometer. *J. Petrol. Geol.* 6 (1), 5-38.

Rawson, P.F. and Riley, L.A. (1982) Latest Jurassic-Early Cretaceous events and the 'Late Cimmerian Unconformity' in the North Sea area. *AAPG Bull.* 66 (12), 2628-2648.

Reitsema, R.H. (1983) Geochemistry of North and South Brae areas, North Sea. In: Brooks, J. (Ed.) q.v., 203-212.

Ridd, M.F. (1981) Petroleum geology west of the Shetlands. In: Illing, L.V. and Hobson, G.D. (Eds.) q.v., 414-425.

Rigby, D. and Smith, J.W. (1982) A reassessment of stable carbon isotopes in hydrocarbon exploration. *Erdöl, Kohle, Erdgas, Petrochemie* 35 (9), 415-417.

Robert, P. (1980) The optical evolution of kerogen and geothermal histories applied to oil and gas exploration. In: Durand, B. (Ed.), q.v., chapter 11, 340-414.

Roberts III, W.H. and Cordell, R.J. (1980) *Problems of petroleum migration.* AAPG Studies in Geology No. 10. Am. Assoc. Petrol. Geol., Tulsa.

Rønnevik, H.C. (1981) Geology of the Barents Sea. In: Illing, L.V. and Hobson, G.D. (Eds.) q.v., 395-406.

Rønnevik, J., Eggen, S. and Vollset, J. (1983) Exploration of the Norwegian Shelf. In: Brooks, J. (Ed.) q.v., 71-93.

Rønnevik, H. and Johnson, S. (1984) Geology of the Greater Troll Field area. *Oil and Gas Journal* 82 (4), 100-106.

Roos, B.M. and Smits, B.J. (1983) Rotliegend and main Buntsandstein gas fields in Block K/13—a case history. *Geol. Mijnbouw* 62 (1), 75-82.

Schenk, P.A., deLeeuw, J.W. and Lijmbach, G.W.M. (Eds.) (1985) Advances in Organic Geochemistry 1983. Proceedings of 11th International Meeting on Organic Geochemistry, Hague. *Organic Geochemistry special volume* 6, 1-892.

Schmoker, J.W. (1979) Determination of organic content of Appalachian Devonian shales from formation-density logs. *Am. Assoc. Petrol. Bull.* 63 (9), 1504-1509.

Schmoker, J.W. (1981) Determination of organic matter content of Appalachian Devonian shale from gamma ray logs. *Am. Assoc. Petrol. Bull.* 65, 1285-1298.

Schou, L., Mørk, A. and Bjorøy, M. (1984) Correlation of source rocks and migrated hydrocarbons by GC-MS in the Middle Triassic of Svalbard. In: Schenk, P.A., *et al.* (Eds.), q.v., 513-520.

Smith, P.M.R. (1983) Spectral correlation of spore colour standards. In: Brooks, J. (Ed.) q.v., 289-294.

Snowdon, L.R. and Powell, T.G. (1979) Families of crude oils and condensates in the Beaufort-Mackenzie Basin. *Bull. Canadian Petrol. Geol.* 27 (2), 139-162.

Snowdon, L.R. and Powell, T.G. (1982) Immature oil and condensate—modification of hydrocarbon generation model for terrestrial organic matter. *Am. Assoc. Petrol. Geol. Bull.* 66 (6), 775-788.

Spencer, A.M., et al., (1984) *Petroleum Geology of the North European Margin.* Graham and Trotman /Norwegian Petroleum Society (Proceedings of a NPS meeting, Trondheim, May 1983).

Stow, D.A.V., Bishop, C.D. and Mills, S.J. (1982) Sedimentology of the Brae field North Sea: fan models and controls. *J. Petrol. Geol.* 5 (2), 129-148.

Surlyk, F., Piasecki, S., Rolle, F., Stemmerik, L., Thomsen, E. and Wrang, P. (1984) The Permian Basin of East Greenland. In: Spencer, A.M., *et al.* q.v., 303-315.

Taylor, J.C.M. (1981) Zechstein facies and petroleum prospects in the central and northern North Sea. In: Illing, L.V. and Hobson, G.D. (Eds.) q.v. 176-185.

Teichmüller, M. and Teichmüller, R. (1979) Diagenesis of coal (coalification) In: Larsen, G. and Chilingar, G.V.

(Eds.) *Diagenesis in sediments and sedimentary rocks.* Elsevier, 207-246.

Teichmüller, M., Teichmüller, R. and Bartenstein, H. (1979) Inkohlung und Erdgas in Nordwestdeutschland. Eine Inkohlungskarte de Oberflacher des Oberkarbons. *Fortschr. Geol. Rheinld. u. Westf.* **27**, 137-170.

Teichmüller, M., Teichmüller, R. and Bartenstein, H. (1984) Inkohlung und Erdgas—eine neue Inkohlungskarte der Karbon—Oberflache in Norwestdeutschland. *Forschr. Geol. Rheinland und Westf.* **32**, 11-34.

Thompson, K.F.M. (1979) Light hydrocarbons in subsurface sediments. *Geochim. Cosmochim. Acta* **43**, 657-672.

Thompson, K.F.M. (1983) Classification and thermal history of petroleum, based on light hydrocarbons. *Geochim. Cosmochim. Acta* **47**, 303-316.

Thompson, S., Morley, R.J. Barnard, P.C. and Cooper, B.S. (1985) Facies recognition of some Tertiary coals applied to the prediction of oil source rock occurrence. *Marine and Petroleum Geology* **2**(4), 288-297.

Thomsen, E. Lindgreen, H. and Wrang, P. (1983) Investigation on the source rock potential of Denmark. *Geol. Mijnbouw* **62** (1), 221-239.

Tissot, B.P. and Bessereau, G. (1982) Géochimie des gaz naturels et origine des gisements de gaz en Europe occidentale. *Rev. Inst. Français de Petrol.* **37** (1), 63-77.

Tissot, B.P. and Welte, D.H. (1984) *Petroleum Formation and Occurrence.* Springer Verlag. 2nd Edition.

Tissot, B., Chalifet-Debyser, Y., Deroo, G. and Oudin, J.L. (1971) Origin and evolution of hydrocarbons in early Toarcian shales, Paris Basin, France. *Am. Assoc. Petrol Geol. Bull.* **55** (12), 2177-2193.

Tixier, M.P. and Curtis, M.R. (1967) Oil shale yield predicted from well logs. *Proceedings of 7th World Petroleum Congress, Mexico City,* vol. 3, 713-715.

Toth, D.J., Lerche, I., Petroy, D.E., Meyer, R.J. and Kendall, C.G. St. C. (1983) Vitrinite reflectance and the derivation of heat flow changes with time. In: Bjorøy, M. et al. (Eds.) *Advances in organic geochemistry* 1981. John Wiley, 588-596.

Turner, C.C., Richards, P.C., Swallow, J.L. and Grimshaw, S.P. (1984) Upper Jurassic stratigraphy and sedimentary facies in the central Outer Moray Firth Basin. *Marine and Petroleum Geology* **1** (2), 105-117.

Tyson, R.V., Wilson, R.C.L. and Downie, C. (1979) A stratified water column environmental model for the Kimmeridge Clay. *Nature* **277**, 377-380.

Ungerer, P., Bessis, F., Chenet, P.Y., Durand, B., Nogaret, E., Chiarelli, A., Oudin, J.L., Perrin, J.F. (1984) Geological and geochemical models in oil exploration: Principles and practical examples. In: Demaison, G. and Murris, R.J., q.v., 53-77.

Vail, P.R. and Todd, R.G. (1981) Northern North Sea Jurassic unconformities, chronostratigraphy and sea-level changes from seismic stratigraphy. In: Illing, L.V. and Hobson, G.D. (Eds.) q.v., 216-235.

Van den Bark, E. and Thomas, O.D. (1980) Ekofisk: first of the giant oil fields in Western Europe. In: Halbouty, M.T. (Ed.) *Giant oil and gas fields of the decade 1968-1978* (AAPG Memoir 30), 195-224.

Van den Bosch, W.J. (1983) The Harlingen gas field, the only gas field in the Upper Cretaceous chalk of the Netherlands. *Geol. Mijnbouw* **62** (1), 145-156.

van Gijzel, P. (1981) Applications of geomicrophotometry of kerogen, solid hydrocarbons and crude oils to petroleum exploration. In: Brooks, J. *Organic Maturation Studies and Fossil Fuel Exploration*, Academic Press, London, 351-377.

van Wijhe, D.H., Lutz, M. and Kaasschieter, J.P.H. (1980) The Rotliegend in The Netherlands and its gas accumulations. *Geol. Mijnbouw* **59**, 3-24.

Volkman, J.K., Alexander, R., Kagi, R.I. and Rullkötter, J. (1983) Demethylated hopanes in crude oils and their applications in petroleum geochemistry. *Geochim. Cosmochim. Acta* **47**, 1033-1040.

Waples, D. (1980) Time and temperature in petroleum formation; application of Lopatin's method to petroleum exploration. *Am. Assoc. Petrol. Geol. Bull.* **64**, 916-926.

Waples, D.W. (1983) Reappraisal of anoxia and organic richness with emphasis on Cretaceous of North Atlantic. *Am. Assoc. Petrol. Geol. Bull.* **67** (6), 963-978.

Waples, D.W. (1984) Thermal models for oil generation. In: Brooks, J. and Welte, D., q.v., 7-67.

Watson, S. (1976) Sedimentary geochemistry of the Moffat Shales; a carbonaceous sequence in the Southern Uplands of Scotland. Unpublished Ph.D., University of St Andrews, Scotland.

Welte, D.H. and Yukler, M.A. (1981) Petroleum origin and accumulation in basin evolution—a quantitative model *Am. Assoc. Petrol. Geol. Bull.* **65** (8), 1387-1396.

Welte, D.H., Yukler, M.A., Radke, M., Leythaeuser, D., Mann, U. and Ritter, U. (1983) Organic geochemistry and basin modelling—important tools in petroleum exploration. In: Brooks, J. (Ed.) q.v., 237-254.

Westre, S. (1984) The Askeladden gas find—Troms-1. In: Spencer, A.M., et al., q.v., 33-40.

Williams, P.F.V. and Douglas, A.G. (1980) A preliminary organic geochemical investigation of the Kimmeridge oil shales. In: Douglas, A.G. and Maxwell, J.R. (Eds.) *Advances in organic geochemistry* 1979. Pergamon Press, 531-545.

Woodland, A.W. (Ed.) (1975) *Petroleum and the Continental Shelf of North-west Europe,* vol. 1, Geology. Applied Science Publishers.

Yukler, M.A. and Kokesh, F. (1984) A review of models used in petroleum resource estimation and organic geochemistry. In: Brooks, J. and Welte, D., q.v., 69-113.

Ziegler, P.A. (1982) *Geological atlas of Western and Central Europe.* Shell Int. Pet. Maats. B.V.

The proceedings of the Norwegian Petroleum Society Conference on Organic Geochemistry in Exploration of the Norwegian Shelf (Stavanger, Oct. 1984) is now published (Eds. B.M. Thomas et al., Graham & Trottman, 1985). This book contains a wealth of detailed geochemistry concerning the North Sea. Of particular interest are the detailed discussions of migration on a basinal scale (Thomas et al. and Field) and details of the crude oils of the Greater Ekofisk area (Hughes et al.).

SOURCE ROCK EVALUATION

Principles of Geochemical Prospect Appraisal[1]

ANDREW S. MACKENZIE and TOM M. QUIGLEY[2]

ABSTRACT

Our current understanding of the geochemical processes responsible for the creation of commercial accumulations of oil and gas can be reduced to simple rules and equations to allow the petroleum geologist to use geochemistry to estimate—before drilling—the likely volumes and compositions of oil and gas delivered to the trap.

In most sedimentary basins, oil is expelled from source rocks between 120° and 150°C, whereas most gas and gas condensate are released between 150° and 230°C. When the initial potential of source rocks exceeds 10kg/MT, oil expulsion efficiencies are between 60 and 90%.

Expelled oil and gas migrate as petroleum-rich phases driven by fluid potential gradients. Most flow is laterally updip in beds with effective horizontal permeabilities greater than 1 md. In lower permeability rocks, the petroleum fluids move vertically, up or down, along the path of least resistance (i.e., the most permeable or thinnest beds) separating them from a high-permeability lateral carrier bed. The residual saturations of petroleum left behind along the migration pathway are of the order of a few percent of the porosity.

Geochemical analysis of source rocks helps determine the yields and compositions of petroleum fluids expelled in the catchment area of a prospect. These fluid masses are converted to volumes from knowledge of the phase relations and densities of petroleum fluids. Evaluation of the total volume of rock through which the petroleum potentially can migrate and the mean porosity of the rocks that constitute the migration pathway allow migration losses to be assessed. These losses constrain both the range of migration of petroleum fluids away from the mature source rock and the volumes of petroleum remaining for entrapment in the prospect.

INTRODUCTION

The prime function of petroleum geochemistry is to predict, in advance of drilling, the volume and composition of petroleum trapped in a prospect. During the last 20 years, understanding of geological processes that create commercial accumulations of petroleum has improved to the point where quantitative prediction can be made. Much of this understanding has been summarized by Hunt (1979) and Tissot and Welte (1984). Some of the most important advances include Smith's (1971) study of the compaction of sedimentary rocks, Tissot and Espitalié's (1975) work on the kinetics of oil and gas formation, Neglia's (1979) and Ungerer et al's (1983) investigation of phase relations during petroleum migration, Bishop et al's (1983) ideas on the mass balance of source rock processes, and Hubbert's (1953) and Durand et al's (1984) theories of petroleum migration.

Understanding gained from geochemistry is often used in computer models that couple the petroleum-forming processes with parallel developments in the thermal evolution and burial of sedimentary basins (e.g., Welte and Yükler, 1981; Durand et al, 1984). Although such mathematical models are useful tools (and teaching aids), they sometimes are unwieldy in the hands of a busy petroleum geologist, working against deadlines, with a less than complete understanding of the geological area of interest. This paper demonstrates an alternative approach by summarizing the understanding of our own oil company (Goff, 1983; Cooles et al, 1986; England et al, 1987; Mackenzie et al, 1988), in the form of a simplified but practical guide to the geochemical appraisal of a petroleum prospect. The approach we propose can be applied rapidly by the geologist using his own maps, basic geochemical data, and a hand calculator.

Definition of Petroleum

We use "petroleum" as a collective term to mean any natural subsurface material that can be produced as oil or gas, including some associated nonhydrocarbons. Under pressure and temperature conditions at the surface, components of petroleum with six or more carbon atoms (C_{6+}) are found mainly in the liquid phase, whereas components with fewer than six carbon atoms (C_1-C_5) are in the gaseous phase. Reservoir engineers describe these phases as "oil" and "gas," respectively.

Under subsurface conditions of temperature and pressure, petroleum liquids contain dissolved components that are gaseous under surface conditions (i.e., C_1-C_5). For simplicity, this subsurface liquid is often also referred to as "oil" (e.g., oil migration). Similarly, subsurface gaseous petroleum will contain dissolved components that would be liquids under surface conditions (e.g., C_{6+} or reservoir engineer "oil" or "condensate"). Again, for simplicity, this subsurface gaseous phase is often referred to either as "gas" (i.e., lean in C_{6+} components) or "gas condensate" (i.e., rich in C_{6+} components). Thus, the average composition of subsurface

---

©Copyright 1988. The American Association of Petroleum Geologists. All rights reserved.
[1]Manuscript received, February 9, 1987; accepted, December 14, 1987.
[2]BP Exploration Company Limited, Britannic House, Moor Lane, London EC2Y 9BU, England.
We thank G. P. Cooles, W. A. England, J. R. Gray, R. A. James, A. L. Mann, D. M. Mann, and S. W. Richardson for helpful discussions and results of their experiments, and BP for permission to publish.

petroleum is given by its gas/oil ratio (GOR) or condensate/gas ratio (CGR).

BASIC EQUATIONS

All calculations are performed in standard (S.I.) units. Conversion factors for oil-field units are given in Table 1. A list of symbols used in this paper is given in Appendix 1.

Our procedure is based on three equations required for geochemical appraisal of a prospect. Equation 1,

$$M_E = \sum^{SUM} (P_o)(PGI)(PEE_n)(\rho_{ROCK})(y)(AREA) \quad (1)$$

calculates the mass of petroleum fluids (M_E) expelled from slabs of source rocks of approximately equal maturity; these are called "isomaturity slabs" (Figure 1). The slabs contain source rock of mean thickness (y) and bulk subsurface density (ρ_{ROCK}). AREA corresponds to the area of the isomaturity slab in the catchment area; P_o is the average initial petroleum potential (the concentration of organic matter that can be transformed into petroleum fluids at elevated temperatures); PGI is the petroleum generation index (the fraction of petroleum-prone organic matter that has been transformed into petroleum); and PEE_n is the net petroleum expulsion efficiency (the fraction of petroleum fluids formed in the source rock that has been expelled toward the prospect). For example, if the reservoir of the prospect of interest lies above the source rock, only petroleum expelled upward would be included in PEE_n.

The volume of petroleum lost (V_L) along the migration pathway is calculated:

$$V_L = \phi f V_D, \quad (2)$$

where ϕ is porosity, f is residual saturation, and V_D is the volume of rock through which the petroleum flows. The volume of petroleum lost is proportional to the pore volume through which the petroleum flows (MacGregor and Mackenzie, 1987). The volume of rock that constitutes the migration pathway is called the drainage volume (V_D) (Figure 1).

The volume of petroleum charge delivered to the prospect (V_C) is the difference between expelled volume (V_E) and lost volume (V_L). Thus,

$$V_C = V_E - V_L. \quad (3)$$

When lost volume exceeds expelled volume, no petroleum is predicted to reach the prospect (i.e., calculated negative values of V_C imply that the trap is dry).

In this paper, we show how the variables of equations 1-3 can be estimated. We define procedures for determining the bulk composition of the expelled petroleum that allow us to evaluate the petroleum phase and density and hence to convert from mass expelled (equation 1) to volumes expelled (equation 3).

Uncertainties

By simplifying things slightly for practical purposes, some of the uncertainties in the approach are not immediately apparent. The potential errors and uncertainties are discussed individually as they arise in the text. The overall aggregate of these errors means that calculations of petroleum volumes entrapped are unlikely to be more precise than, say, ±50%. Furthermore, other geologic unknowns commonly mean that we can only predict entrapped petroleum volumes to within an order of magnitude. In our experience, however, even this limited precision can improve the estimation of exploration risk.

MASS AND COMPOSITION EXPELLED

The mass and composition of petroleum expelled from source rocks are governed by type of organic matter, concentration of organic matter, and maximum temperature of petroleum expulsion.

Figure 2 summarizes the scheme for classifying organic matter. Initial oil is the oil content of immature source rocks (i.e., those not heated to temperatures greater than 100°C, with vitrinite reflectance less than about 0.5%). Kerogen is divided into inert and reactive portions. Inert kerogen rearranges toward graphitelike structures at high temperatures and pressures, whereas reactive kerogen is transformed at elevated temperatures into petroleum. The concentration of reactive kerogen is given by the pyrolysis yield of source rock samples (the S_2 peak of the Rock-Eval) (Espitalié et al, 1977). Inert kerogen is the difference between the total organic matter content (TOC) of the source rock samples (measured by combustion) and the solvent extractable plus pyrolyzable fractions.

This classification means that in equation 1, initial petroleum potential (P_o) equals initial oil plus initial reactive kerogen. The petroleum generation index (PGI) equals initial oil plus petroleum generated, divided by initial oil plus initial reactive kerogen; this also equals initial oil plus (initial minus remaining) reactive kerogen, divided by initial petroleum potential. Petroleum expulsion efficiency (PEE) equals petroleum expelled divided by initial oil plus petroleum generated, which equals one minus the remaining petroleum divided by initial oil plus (initial minus remaining) reactive kerogen. At this stage, we define only the magnitude of bulk expulsion efficiency (PEE) and not net expulsion in the direction of the prospect (PEE_n). Cooles et al (1986) provide a fuller discussion of Figure 2, the preceding equations, and the laboratory measurements used to estimate oil and kerogen concentrations.

Two major uncertainties are inherent in the scheme described in Figure 2. First, because the hydrogen/carbon ratio of oil is less than that of gas, the conversion of oil to gas can only go to completion if there is an external source of hydrogen. Potential sources of hydrogen include clay minerals and water, as well as inert kerogen. As inert kerogen is transformed to graphite (pure carbon) it loses hydrogen and other noncarbon atoms. When

Andrew S. Mackenzie and Tom M. Quigley

Table 1. Conversion Factors Between Standard International and Oil-Field Units

Quantity	SI and Derived Units	Oil-Field Unit	Conversion Factor
Length	meter (m)	foot	0.3048
	[kilometer (km)]	[mile]	[1.6093]
Area	square meter (m^2)	square foot	0.0929
	[square kilometer (km^2)]	[acre]	[4.047×10^{-3}]
Volume	cubic meter (m^3)	cubic foot	0.0283
		barrel	0.1590
		acre foot	1.234×10^3
GOR*	(kg/kg)	scf/bbl	2.12×10^{-4} (approx.)
CGR+	(kg/kg)	bbl/mmcf	4.55×10^{-3} (approx.)
Mass	kilogram (kg)	long ton	1016
Density	(kg/m^3)	g/cm^3	1000
		°API = 141500/(kg/m^3) − 131.5	
Pressure	Pascal (Pa)	pound per square inch (psi)	6.894×10^3
		bar	10^5
Tension	Newton per meter (N/m)	dyne/cm	10^{-3}
Temperature	[°C]	°F	(°C × 1.8) + 32
Permeability	(m^2)	Darcy	0.9869×10^{-12}

*35°API oil, 1 kg/m^3 gas.
+45°API condensate, 1 kg/m^3 gas.

there is insufficient hydrogen a hydrogen-poor coke is a by-product of the oil to gas reaction; hence, gas yields are reduced. Partly because of circumstantial evidence presented by Cooles et al (1986), all our equations assume that the external supply of hydrogen is adequate and that coke formation is minimal. Therefore, we may overestimate gas yields, and hence gas expulsion efficiencies. The problem of hydrogen balance during oil and gas formation needs more attention; this requires more complete integration of inorganic and organic geochemistry.

The second area of difficulty is with the measurement of oil yields. Yields are based on either the total solvent extract (TSE) or vaporizable organic matter (S_1 peak of Rock-Eval) (Espitalié et al, 1977), suitably corrected for work-up and sampling losses (see Cooles et al, 1986). Deciding whether either corrected TSE or corrected S_1 is the better measure of the oil content of source rocks is difficult. TSE is generally the greater of the two measurements because it includes more heavy components (C_{25}-C_{40}). However, S_1 includes light materials (C_6-C_{14}) that are found in oil, but are absent from TSE due to work-up losses. We normally perform our calculations using corrected TSE as a measure of oil content, then repeat them using corrected S_1 to examine the effect of this uncertainty on the precision of our final answer.

Mass Balance Approach

By comparing geochemical measurements made on mature source rock samples with corresponding measurements on immature equivalents, P_o, PGI, and PEE may be obtained. However, mature samples can be compared directly with, for example, only immature equivalents at the basin margin if organic matter type and initial concentration are identical. Normally, these are not identical. Therefore, Cooles et al (1986) devised a method whereby the concentrations of initial oil and initial reactive kerogen could be determined for mature samples by assuming a starting ratio of initial oil/initial reactive kerogen/inert kerogen. This ratio is governed only by organic matter type and can be measured on immature equivalents, either from the same region or from different regions where source rocks were deposited under similar conditions.

Carbon arising from inert kerogen remains in the source rock throughout the zone of generation and expulsion of petroleum. Therefore, by measuring the inert kerogen content of a mature source rock, the sample's initial concentration of oil and reactive kerogen (and hence P_o, PGI, and PEE) can be calculated using the adopted ratio of the initial concentrations relative to inert kerogen (Figure 3).

Kinetic Approach

Cooles et al (1986) provide a data base of the variation in relative concentrations of reactive kerogen as a function of maximum temperature. This data base suggests that reactive kerogen should be divided into two parts (Figure 2): a labile portion that is transformed between 100° and 150°C into a petroleum that is chiefly oil, and a refractory portion that is transformed at temperatures greater than 150°C into gas. Kerogens derived from algae, bacteria, and some parts of land plants are comprised chiefly of labile kerogen and have small amounts of refractory and inert kerogen. Kerogens derived exclu-

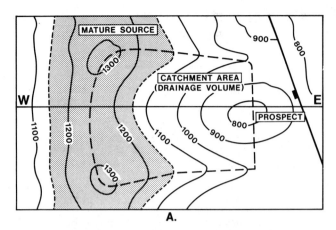

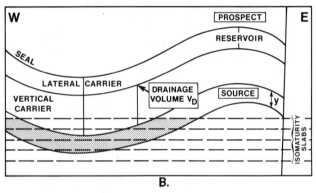

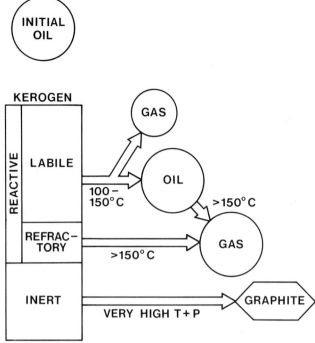

Figure 1—A. Simplified map of catchment area of hypothetical prospect for petroleum fluids. B. East-west section through region and prospect. Contours are drawn on interface between reservoir facies and overlying seals. Migration route from mature source rock to prospect is called "drainage volume." Drainage volume has a component where petroleum fluids move vertically through vertical carriers and a component where petroleum moves laterally through lateral carriers. To solve equation 1, source rock is divided into isomaturity slabs (see lower panel), which are portions of source rock within catchment area of prospect that everywhere have values of P_o, PGI, PEE_n, ρ_{ROCK}, and y close to average value for that slab. Thus equation 1 is solved for each isomaturity slab and results are added together to give total mass of petroleum expelled.

sively from the vascular parts of land plants are mixtures of refractory and inert kerogen. Thus, labile kerogen is chiefly derived from the lipid and resin constituents of living organisms, whereas refractory kerogen corresponds to the C_1-C_3 substituents attached to aromatic nuclei within the lignin biopolymer. These aromatic constituents, together with the condensed remnants of proteins and carbohydrates, form inert kerogen. Our kerogen classification is analogous to that of Tissot et al (1974), but instead of discrete kerogen types we recognize a continuum of changing kerogen composition. Tissot et al's type I kerogen corresponds to a kerogen whose reactive fraction is predominantly labile; their type III kerogen has a reactive fraction that is chiefly refractory. Type II kerogen lies between types I and III, although perhaps closer to type I. Thus, type II reactive kerogen has more labile than refractory material.

Examples of Cooles et al's (1986) results are shown in Figure 4, which plots reactive kerogen against maximum

Figure 2—Classification and fate of organic matter in source rock. Kerogen is divided into reactive and inert portions. Inert kerogen rearranges toward graphitelike structures at very high temperatures (T) and pressures (P). Reactive kerogen is subdivided into a refractory part that yields mainly gas and a labile portion that is transformed into petroleum, which is chiefly oil at the surface. Initial oil corresponds to solvent soluble organic matter normally present in immature source rocks. Relative amounts of initial oil, labile, refractory, and inert kerogen are determined by precursory organisms and depositional setting of host source rock. From Cooles et al (1986).

temperature for two chiefly labile kerogens and one that is predominantly refractory. In the two labile kerogen examples, present temperature is the maximum temperature. For the refractory kerogen, the maximum temperature has been estimated from vitrinite reflectance measurements using our own experience with the technique. The kinetic model used by the writers (Figure 4) assumes that the petroleum-forming reactions are governed by first-order Arrhenius kinetics, the breakdown of each reactive kerogen portion being determined by a Gaussian distribution of activation energies with some mean ($\bar{E}$) and standard deviation (σ) (Pitt, 1962; Campbell et al, 1980). Our scheme is similar to, but not directly analogous to, those used by Tissot and Espitalié (1975), Jüntgen and Klein (1976), Ungerer and Pelet (1987), and Sweeney et al (1986). Appendix 2 gives the relevant Arrhenius parameters. We are encouraged that the kinetic scheme is chemically sensible because we are also able to model the breakdown of reactive kerogen in the laboratory at temperatures up to 400°C hotter and heating rates 10^{12} times faster than natural conditions (Figure 5). Other methods, such as the Lopatin approach (Waples, 1980), perform less well in these heating regimes (McKenzie, 1981). Figure 5 shows the results of two types of pyrolysis experiment, together with the kinetic predic-

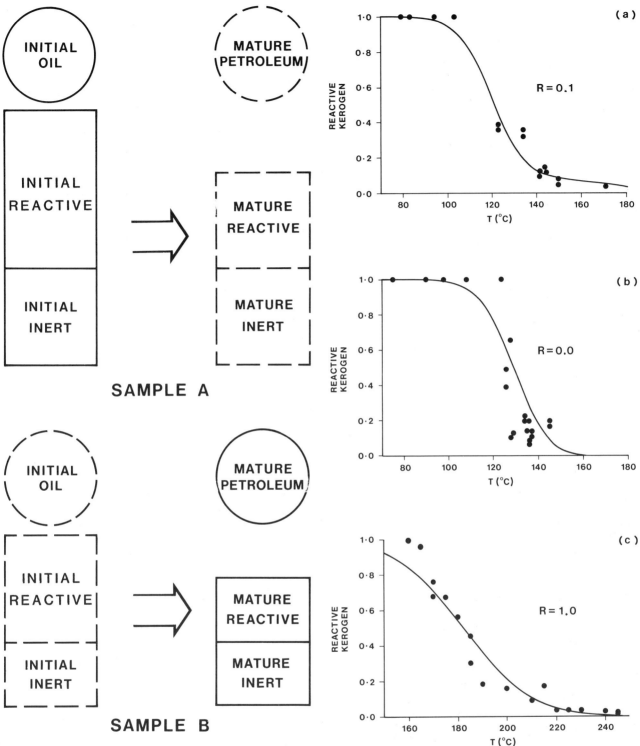

Figure 3—Schematic diagram illustrating mass balance approach for calculation of masses of petroleum generated and expelled, based on analysis of two source rock samples—one immature (sample A), one mature (sample B). Quantities shown as dashed lines are calculated assuming: concentration of inert kerogen is not affected by maturity, and organic matter type present in samples is same (i.e., initial ratio of reactive kerogen/inert kerogen). Thus, sample A is used to determine initial ratio of oil/reactive kerogen/inert kerogen; hence initial state of sample B. By comparing calculated initial state and measured mature state of sample B, masses of petroleum generated and expelled from sample B may be determined.

Figure 4—Results of Cooles et al's (1986) mass balance method showing concentrations of reactive kerogen relative to initial concentrations (solid circles). Examples are (a) Upper Jurassic Kimmeridge Clay Formation, North Sea, (b) Cenomanian Brown Limestone Formation, Gulf of Suez, (c) Westphalian coals, northwest Europe. Reactive kerogens for (a) and (b) are chiefly labile; those of (c) are chiefly refractory. Solid lines are corresponding predictions of writers' kinetic scheme using average geologic heating rates for regions: (a) 1 °C/m.y., (b) 5 °C/m.y., (c) 6 °C/m.y. R is fraction of reactive kerogen that is refractory.

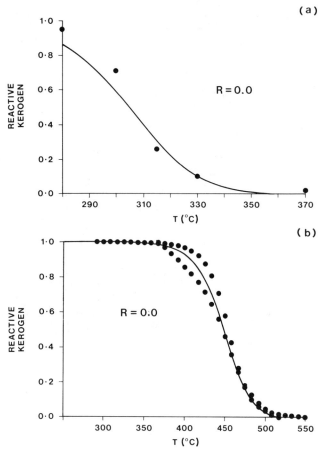
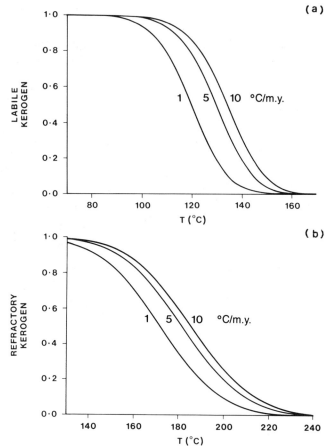

Figure 5—Concentrations of reactive kerogen as function of maximum temperature for two laboratory heating experiments (solid circles): (a) immature Middle Jurassic oil source rock from northeast Scotland heated for three days in platinum capsules at different temperatures; (b) two immature samples of Brown Limestone Formation, Gulf of Suez, heated at 25°C/min in a Rock-Eval pyrolysis apparatus. Solid lines are corresponding predictions of our kinetic scheme for conversion of labile kerogen into petroleum fluids.

Figure 6—Calculated concentrations of reactive labile (a) and refractory (b) kerogens, relative to initial reactive kerogen, as function of maximum temperature for range of heating rates shown. Mean heating rates of about 0.5°C/m.y. occur in old stretched basins such as continental shelf off eastern North America and margin of Russian platform; mean heating rates of about 10°-50°C/m.y. occur in basins produced by extension since Miocene time, such as Los Angeles basin, Pannonian basin, and back-arc basins in Indonesia.

tions. Agreement between theory and experiment is good.

Figure 6 shows calculated labile and refractory kerogen breakdown as a function of maximum temperature for a range of geologically reasonable heating rates (Gretener and Curtis, 1982). The curves may be used, if the heating rate is known, to calculate the petroleum generation index (PGI). First, estimate maximum temperature and geologic heating rate using one of the many models for reconstructing thermal history (e.g., Beaumont, 1978; Royden et al, 1980; Wright, 1980; McKenzie, 1981; Cochran, 1983). Read the ratio (λ) of present labile kerogen to initial labile kerogen from Figure 6a. When the maximum temperature is less than 150°C, PGI may be calculated using:

$$PGI \approx \frac{(1 - \lambda)(1 - R) + \Theta}{(1 + \Theta)}. \quad (4)$$

Θ is the ratio of initial oil to initial reactive (labile plus refractory) kerogen and is constant for similar types of organic matter and may be estimated from analogous source rocks or from the immature equivalents of the same source rock when organic matter type remains constant (Cooles et al, 1986).

We have observed empirically (A. Mann, 1986, personal communication) from pyrolysis gas chromatographic analysis (e.g., Dembicki et al, 1983) of immature kerogens, that the ratio (Γ) of C_1-C_5 molecules produced by pyrolysis divided by the total pyrolysis yield is equal to about 0.41 for reactive kerogen that is predominantly refractory, and equal to approximately 0.15 for reactive kerogen that is predominantly labile. Hence, the approximate refractory fraction of any given reactive kerogen is

$$R \approx 3.84\Gamma - 0.58. \quad (5)$$

When the maximum temperature exceeds 150°C, refractory kerogen is transformed into gas (Figure 6b) and PGI will then be given by

$$PGI \approx \frac{1 - R\rho + \Theta}{1 + \Theta}, \quad (6)$$

236

where ρ is the ratio of present to initial refractory kerogen.

The validity and usefulness of the much simplified kinetic approach, as summarized by equations 4 and 6 and Figure 6, have been confirmed using the full kinetic description of petroleum generation coupled to a mathematical model of sediment thermal history (e.g., see Guidish et al, 1985, for a review of such models). However, for very complicated source-rock thermal histories involving, for example, one or more phases of uplift and reburial when the source rock enters the petroleum generation zone more than once, computer modeling of thermal history coupled to the kinetic equations for kerogen breakdown may be necessary. In our experience, this process is rarely required as most oil-bearing sedimentary sequences have experienced only one period of major burial (i.e., when sediments reach their maximum temperature), perhaps followed by significant uplift. Earlier or subsequent burial and uplift phases are commonly minor in comparison, and may be ignored for the purposes of estimating the maturity of source rocks. Thus, source rock maturity can be estimated from the maximum temperature achieved during the major period of burial. When more than one phase of burial exists, one can consider only the later phases if source rock temperatures are similar to, or greater than, those achieved during the first phase.

Iteration and Extrapolation

Where possible, PGI should be calculated using both the mass balance approach and the kinetic approach. Inconsistencies between the two methods are normally due to an incorrect assumption regarding organic matter type (e.g., one may be wrong to assume that organic matter type is constant within one zone in a basin) or errors in the reconstruction of thermal history. We find that component ratios of organic molecules sensitive to changes in thermal history provide useful additional constraints (Mackenzie et al, 1984). Careful iteration to a unique solution is a powerful tool that can teach much about the organic matter type and thermal history of the source rocks of interest.

Of course, iteration is not possible where no relevant samples are available, a common occurrence when the petroleum geologist wants to extrapolate into a depocenter using samples obtained by drilling structural highs as geochemical control. PGI can then only be estimated using the kinetic approach and the inferred thermal histories of the source rocks in the depocenters. In this case, the initial potential P_o for substitution in equation 1 must be estimated from analogs of the source rocks believed to be present in the depocenter. These may be stratigraphic equivalents that have been drilled on the flanking highs. If petroleum fluids have been discovered in the region of interest, their compositions may be used to constrain the depositional environment of the source that might be active in the depocenter (Moldowan et al, 1985).

Oil-to-Gas Cracking

The transformation of oil to gas at elevated temperature is an important geochemical process (Figure 2). This process can occur in both the source rock and in the reservoir. We cannot yet study oil-to-gas cracking directly under natural conditions. Our approach has therefore been to use the results of closed-vessel pyrolysis to calibrate a kinetic model of oil cracking to gas, and then use subsurface observations as secondary calibrants (i.e., as a check on the validity of the model).

Figure 7 shows the results of one series of experiments at different temperatures. Each experiment was conducted for 3 days. At the end of each experiment, the platinum capsule was burst within the hot injector of a gas chromatograph and the vaporized contents were swept directly onto the chromatography column. The gas/(gas + oil) ratios were determined. In the resulting analysis, C_1-C_5 components are called gas and C_{6+} components are called oil. The bold line on Figure 7 is the corresponding prediction of our kinetic model (see Appendix 2 for relevant Arrhenius parameters).

Figure 8 displays the predicted gas/(gas + oil) ratios as a function of maximum temperature for geologic heating rates. Most cracking occurs between 150° and 180°C (i.e., at temperatures intermediate between labile kerogen breakdown and refractory kerogen breakdown). Figure 8 predicts that major oil accumulations cannot exist at depths below those corresponding to temperatures greater than about 150°-180°C; this prediction agrees with our experience and that of others (e.g., Landes, 1967).

Petroleum Expulsion Efficiency

We distinguish between PEE and PEE_n. The former refers to the overall bulk efficiency with which petroleum fluids migrate out of the source rock, the latter is that fraction expelled in a given direction (i.e. toward a carrier bed system of interest). The two are related thus:

$$PEE_n = \underline{v}PEE \quad (7)$$

The magnitude of $\underline{v}$ (the petroleum that is expelled from a source rock toward a given reservoir) ranges between 0 and 1, and its direction generally is either vertically up or down ($\underline{v}$ and PEE_n are discussed later in this paper).

Petroleum expulsion efficiency (PEE) can be calculated directly by the mass balance approach where both immature and mature samples of similar organic matter type are available. However, the petroleum explorer frequently must compute the masses of petroleum expelled from source rocks in regions where geochemical data are sparse. Figure 9 is taken from Cooles et al (1986), who observed that, between 120° and 150°C and above a critical value of PGI (usually 0.2-0.4), PEE is strongly dependent on initial petroleum potential (P_o). Figure 9 shows

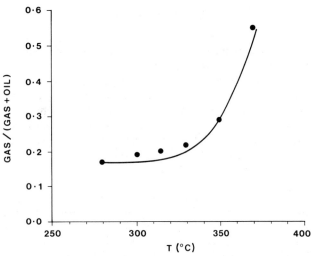

Figure 7—Measured concentration of gas relative to gas + oil (solid dots) during platinum capsule experiments similar to those described in Figure 5a. Each experiment was conducted for 3 days at temperatures shown. Solid line is corresponding predictions of our kinetic scheme of oil cracking to gas.

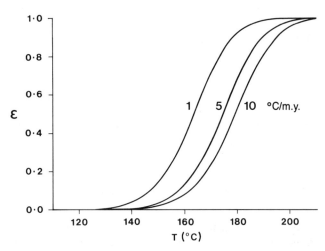

Figure 8—Predicted fraction of generated oil that has been cracked to gas (ϵ = gas/[oil + gas]) as a function of maximum temperature for a range of geological heating rates.

that when P_o is less than 0.005 kg/kg (5 kg/MT) expulsion is inefficient; whereas for P_o greater than 0.01 kg/kg (10 kg/MT) between 60 and 90% of the petroleum fluids generated in the source rock are expelled. Above 150°C, unexpelled oil is rapidly cracked to gas. In our experience (Cooles et al, 1986), this rapid cracking invariably leads to high expulsion efficiencies of gas or gas condensate, irrespective of the value of P_o.

The P_o averages shown in Figure 9 are for intervals between 30–350 m thick. In heterogeneous source rocks, we consider the average P_o values on this length scale, not the P_o values of the richest thin laminae. We have not yet observed any major differences in expulsion efficiency caused by changes in lithology, although we agree with Young and McIver (1977) that the very high absorption of hydrocarbons by kerogen could reduce expulsion efficiencies (especially of oil) from very organic-rich source rocks such as coals.

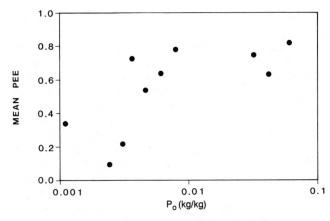

Figure 9—Mean petroleum expulsion efficiency (PEE) between petroleum generation index (PGI) 0.48 and 0.8, as function of initial petroleum potential (P_o) for 10 source rocks. From Cooles et al (1986).

The control of thickness on expulsion efficiency is not straightforward. Mackenzie et al (1988) have shown that capillary forces can increase PEE above the values shown in Figure 9 in source rocks thinner than 5 m interbedded with sandstones. However, the data of Figure 9 are from source rocks between 30 and 350 m thick, and within this range no major control of thickness on PEE was discernible.

Durand et al (1984), England et al (1987), and Mackenzie et al (1988) have argued that most petroleum migration occurs by the pressure-driven flow of discrete petroleum-rich phases. Molecular diffusion may perturb the compositions of the petroleum phases, but is not a major transport process. Our observation that P_o greater than 0.005 kg/kg (5 kg/MT) is necessary for efficient oil expulsion supports this argument; 0.005 kg/kg (5 kg/MT) is equivalent to approximately 0.02 m³ of petroleum per 1 m³ of rock (taking a subsurface petroleum density of 600 kg/m³ and a rock density of 2,400 kg/m³ (England et al, 1987). A typical source rock porosity at 120°–150°C (corresponding to a depth of 3–4 km in most basins studied by us) is about 5%. This measurement implies that petroleum must displace water and saturate about 40% of the pores of a source rock before major expulsion of an oil-rich petroleum fluid can begin.

Source Rock Class and Composition Expelled

Increased understanding of the factors that control the generation of petroleum fluids from kerogen and their subsequent expulsion from source rocks has allowed the derivation of some simple rules to predict the composition of the expelled petroleum. Our rules predict bulk petroleum composition in terms of the gas/oil ratio of expelled petroleum as a function of temperature. Although in prospect appraisal estimating likely API gravity and concentrations of nonhydrocarbon gases such as N_2, H_2S and CO_2 is useful, in practice we recommend that this be done by reference to existing discoveries nearby or to accumulations thought to have arisen

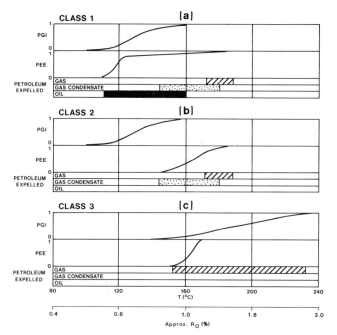

Figure 10—Petroleum generation index (PGI) and petroleum expulsion efficiency (PEE) as function of maximum temperature for three classes of source rock. Principal petroleum phases expelled over relevant temperature ranges are shown. Curves were constructed assuming mean heating rate of 5 °C/m.y.

from a similar source rock. For example, Tissot and Welte (1984) suggested that carbonate source rocks give rise to oils with higher sulfur contents and lower API gravity than aluminosilicate source rocks. One must also ascertain whether in-reservoir biodegradation is causing a reduction in API gravity in traps in the exploration region (e.g., Philippi, 1977).

The gas/oil ratio of expelled petroleum is controlled in part by initial kerogen concentration and kerogen type. This control is illustrated by defining three classes of source rocks, whose variation in PGI and PEE as a function of temperature for a typical geological heating rate of 5 °C/m.y. is shown in Figure 10.

Class 1 is a source rock whose reactive kerogen is mainly labile and has a concentration greater than 0.01 kg/kg rock (10 kg/MT). At temperatures greater than 100 °C, the labile kerogen starts to generate an oil-rich petroleum (Figure 10a). Our pyrolysis experiments and field observations show that the gas/oil ratio of the generated petroleum is typically about 0.2 kg/kg (or about 1,000 scf/bbl). Where pressures are greater than 30 MPa (i.e., pressures normally associated with source rock temperatures greater than 100 °C), published correlations (Glasø, 1980; England et al, 1987) predict that all the gas will be dissolved in the oil and the petroleum will exist as a single liquid phase. Because the initial concentration of labile kerogen is high for class 1 source rocks, the petroleum will rapidly displace water from and saturate much of the source rock's pore volume. Between 60 and 90% of the petroleum fluids will be expelled, as oil with dissolved gas, between 120° and 150 °C. By 150 °C, generation from labile kerogen is almost complete. The small amounts of remaining unexpelled oil will start to be cracked to gas, leading to the expulsion of a gas phase that initially is rich in dissolved condensate. The experiments of Price et al (1983) suggest condensate/gas ratios (CGR) of about 1 kg/kg (or about 200 bbl/mmscf) are possible at pressures normally associated with temperatures of 150 °C (i.e., 40-70 MPa according to England et al, 1987). By 180 °C, oil-to-gas cracking is almost complete, and expulsion of small amounts of dry gas drains the source rock of any remaining petroleum potential by about 200 °C.

Class 2 (Figure 10b) and class 1 have identical mixtures of kerogen types (i.e., predominantly labile), but class 2 has an initial concentration less than 0.005 kg/kg rock (5 kg/MT). Therefore, although PGI variation with temperature is identical to that of class 1, because insufficient oil-rich petroleum is formed between 100° and 150 °C, expulsion of this phase is inefficient. PEE remains low (Figure 10b) and nearly all of the oil remains in the source rock pores until temperatures reach 150 °C, then the oil is cracked to gas and expelled as gas condensate followed by dry gas.

Class 3 (Figure 10C) contains reactive kerogen that is chiefly refractory. No major petroleum generation and expulsion occur until 150 °C, above which relatively dry gas is generated and expelled.

These three source rock classes are end members of most kinds of source rock. By summing the masses and net GORs of expelled petroleum from each isomaturity slab (Figure 1), we can calculate the mean surface GOR of the total petroleum fluids expelled from a source interval in the catchment area of a given prospect.

Marine shales that are oil source rocks, such as the Kimmeridge Clay Formation in the North Sea (Cornford, 1984) or the Bakken Shale in the Williston basin (Meissner, 1984), belong to class 1. Source rocks on the continental shelf off Nova Scotia (Powell, 1982) and in the Nile delta belong to class 2. Paleozoic coals of Europe and North America belong to class 3 (Teichmüller and Durand, 1983). Other coals are mixtures of the different classes: those in the Mahakam delta (Durand and Oudin, 1979) and Gippsland basin (Davidson et al, 1984) are combinations of classes 1 and 3, whereas many other Australasian coals behave as mixtures of classes 2 and 3 (Davidson et al, 1984). Such mixtures can be treated by adding the curves in Figure 10 in the appropriate proportions, and then renormalizing.

MIGRATION OF PETROLEUM

In the preceding section, we discussed all but one of the variables used in equation 1, the exception being PEE_n. Although we have evaluated its bulk magnitude (PEE), we must understand the fractions (v) expelled upward or downward. Determining v and $\overline{PEE}_n$ requires understanding of petroleum migration. This understanding also allows an estimate of losses incurred during migration from source to trap, and hence the solution of equation 2.

The migration of petroleum is determined by gradients in the potential energy of the petroleum phase or phases.

Hubbert (1953) first used this approach to describe the entrapment of petroleum; Durand et al (1984) extended the approach to petroleum migration; and England et al (1987) found it convenient to subdivide the gradient in petroleum potential into three components. (1) A component arising from the gradient in the piezometric water pressure (actual pressure minus hydrostatic pressure). This pressure (known also as overpressure) is caused principally by compaction disequilibrium. Fine-grained low-permeability sediments fail to compact to their equilibrium porosity for an applied overburden stress, because water cannot be squeezed out fast enough. (2) A component arising from the natural buoyancy of less dense petroleum phases surrounded by denser water. (3) A component arising from capillary pressure. Assuming most rocks are water wet, the pressure in the petroleum phase is greater than the pressure in the adjacent water phase. This difference is termed "capillary pressure," and increases with decreasing pore size.

Gradients in water overpressure dominate the petroleum potential gradient (and hence determine the vertical flow of petroleum fluids) in fine-grained rocks more deeply buried than 2-3 km in actively subsiding sedimentary basins (England et al, 1987). The petroleum potential gradient in coarser reservoir facies is normally dominated by buoyancy; that is, petroleum migrates laterally updip in these sediments. However, water overpressure gradients in reservoir facies can be comparable in magnitude to buoyancy forces. Because overpressure generally decreases updip regionally within a basin, migration is still updip but the geometry of trapped petroleum may be altered (Hubbert, 1953).

Capillary pressure gradients are significant only at lithologic boundaries (Mackenzie et al, 1988). The capillary pressure difference between the large pores of a coarse-grained reservoir and the small pores of a fine-grained cap rock causes the cap rock to function as a seal (Berg, 1975).

England et al (1987) have argued that, at least in very general terms, most fluids (oil, gas, or water) in rocks with permeabilities greater than about 10^{-15} m^2 (1 md), such as sandstones or fractured carbonates, move laterally, whereas most fluid movement in rocks with permeabilities less than 10^{-15} m^2, such as shales or siltstones, is vertical. This permeability cutoff arises because the buoyancy-driven updip and lateral flow of petroleum in beds less permeable than 10^{-15} m^2 is not fast enough to travel kilometers in millions of years. Although lateral overpressure potential gradients (which are up to two orders of magnitude greater than the buoyancy gradients) could, in theory, drive petroleum significant distances laterally in beds of 10^{-18}-10^{-16} m^2 permeability (1-100 microdarcys), such gradients are not likely to persist over long periods of geologic time because water can move nearly as quickly as petroleum to eliminate such gradients. If such gradients do persist, they imply lateral barriers to flow, for example, faults with permeabilities less than 10^{-18} m^2 (1 microdarcy). These barriers would stop petroleum flow as well as water flow. We therefore conclude that 10^{-15} m^2 (1 md) can be used as a practical division between predominantly lateral flow and predominantly vertical flow.

These principles predict that petroleum migrates vertically out of a source rock through any immediately overlying or underlying low permeability rocks (i.e., less than 10^{-15} m^2 or 1 md), until it intersects a laterally continuous reservoir-type facies. The petroleum then migrates laterally updip. But what determines the direction of vertical movement (i.e., up or down) to the potential lateral carrier? Case histories tell us that when the lateral carrier directly overlies or underlies the source, most petroleum is expelled vertically upward or downward, respectively. When the source rock is encased in potential vertical carriers (i.e., rocks less permeable than 10^{-15} m^2 or 1 md), the situation is more complicated. We have studied this problem using a one-dimensional model of pore-fluid overpressure (England et al, 1987), which was suitably calibrated by case history.

Figure 11 shows predicted overpressure with depth (England et al, 1987). When the source is separated, above and below, from potential, normally pressured, lateral carriers by shale sequences of similar thickness and permeability, overpressure in the source/shale sequence is a maximum in the center of the source (Figure 11a). If we assume that in fine-grained rocks the gradients in petroleum potential are dominated by the compaction disequilibrium term (England et al, 1987), then petroleum flows in the same direction as water and about half the expelled petroleum flows upward and half downward. If our target reservoir is the deeper sandstone, then $\underline{v}$ in equation 7 has a magnitude of 0.5 and its direction is vertically downward.

If the overlying sequence of potential vertical carriers has a different thickness (Figure 11b) or permeability (Figure 11c) than the lower fine-grained sequence, the overpressure maximum is shifted upward or downward toward the thicker or less permeable sequence. Most expelled petroleum then migrates vertically away from the source rock through the thinner or more permeable sequence. Therefore, for the example shown in Figure 11b, most petroleum moves down through the thin shale; if the underlying vertical carrier is the target reservoir, $\underline{v}$ in equation 7 has a magnitude of approximately 1.0. In the example shown in Figure 11c, most petroleum flow is upward through the more permeable siltstones, hence $\underline{v}$ in equation 7 is approximately zero.

The petroleum stays in the lateral carrier initially encountered unless this carrier is immediately overlain (but not underlain) by a yet more permeably sequence, or juxtaposed with another potential carrier updip at an unconformity or fault. These general rules allow the migration pathway to be evaluated and allow $\underline{v}$ and PEE$_n$ to be estimated together with the drainage volume (V_D) and mean porosity (ϕ) in equation 2.

Additional examples to illustrate our procedure are shown in Figure 12. In Figure 12 the source rock is overlain by siltstones that are thicker than those which underlie it; the petroleum expelled from the mature source rock will migrate vertically downward toward the sandstone, then laterally updip within the sandstone toward the prospect. The lower sandstone will not be reached. The drainage volume will be the sum of (1) the product of the area of the sandstone in the prospective reservoir (which has a focusing geometry toward the prospect) and the

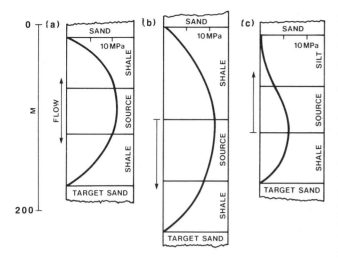

Figure 11—Water overpressure and flow directions as function of depth for three source rocks interbedded with shales and siltstones of varying thickness and permeability, and overlain and underlain by normally pressured sandstones. Sequences have been buried to depths of 3.5 km at rates of 100 m/m.y. Overpressures were calculated using procedures of England et al (1987). In (c), boundary between siltstone and shale lithology occurs in middle of source rock.

mean net sandstone thickness, and (2) the product of the area of mature source rock within the catchment area of the prospect and the mean thickness of the siltstone that separates the source from the sandstone.

The prospect does not receive significant charge from the very deep mature source rock (at the right in Figure 12a), because most flow in the source is vertically downward and lateral flow in the underlying sandstone is updip, away from the fault.

In Figure 12b, the source rock overlaps impervious volcanics; petroleum expelled from mature source rocks will migrate vertically upward through the overlying siltstones, then laterally updip within the sandstone into the prospect. The drainage volume is the product of the catchment area of the prospect (circled, Figure 12b) and the sum of mean siltstone and mean net sandstone thickness minus the volume of siltstone above the area where source rock is absent.

The prospect shown in Figure 12c is more complicated. Petroleum migrates vertically downward from the mature source rock into a sandstone bed because the sandstone is much more permeable than the overlying limestone, then travels laterally updip within that sandstone. An accumulation can form at the updip end of the sandstone. However, because of the inadequate sealing

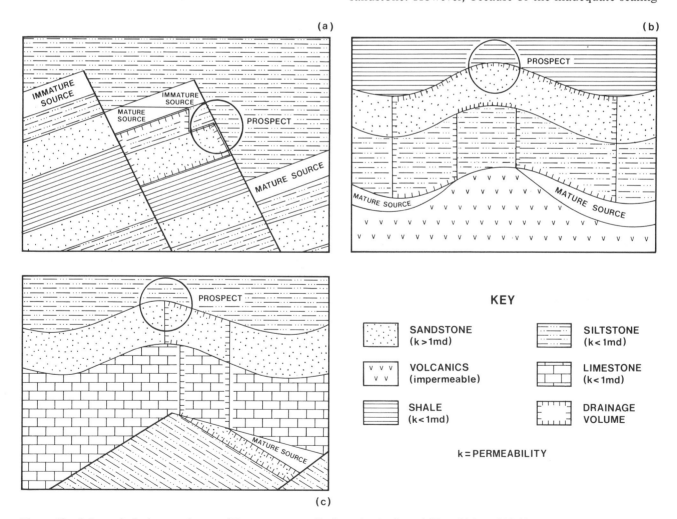

Figure 12—Schematic drainage volumes of three prospects. Each section is about 1-2 km thick and 10-20 km wide. Pre-focusing in lower sandstone in (c) reduces drainage volume, and therefore migration loss, compared to prospect in (b).

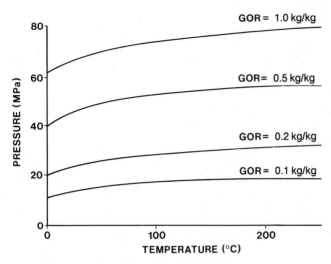

Figure 13—Solubility of gas in a 40° API oil as function of pressure and temperature, as predicted by England et al (1987) using experimental results collected by Glasø (1980).

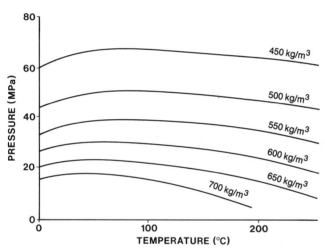

Figure 14—Variation in density of a gas-saturated oil phase as function of pressure and temperature, as predicted by England et al (1987) using experimental results collected by Glasø (1980).

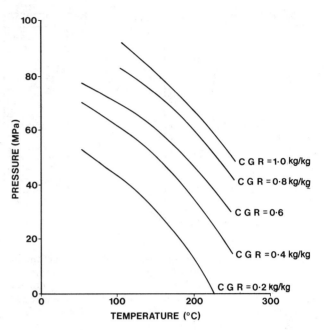

Figure 15—Solubility of 44° API oil (or condensate) in methane as function of pressure and temperature, recalculated by England et al (1987) from experimental results of Price et al (1983).

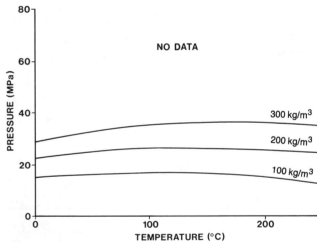

Figure 16—Variation in density of a condensate-saturated gas phase as function of pressure and temperature, as recalculated by England et al (1987) from experimental results of Price et al (1983).

properties of the limestone, petroleum flows vertically upward through the limestone, then laterally updip in the upper sandstone into the prospect. The drainage volume of this migration pathway can be calculated using procedures similar to those of Figure 12a and b.

We have now shown how to calculate all the variables in equations 1 and 2 except the residual saturation factor (f). The best way to evaluate f is using the exploration results from geologically similar regions, where some of the major traps have been drilled and evaluated: (1) calculate the subsurface volumes of accumulated petroleum fluids; (2) calculate the volumes of petroleum fluids expelled from source rocks within the catchment area of the traps using the techniques described in this paper (masses are converted to volumes using average phase densities and GORs. Phase densities and GORs are calculated using the average pressure and temperature of the drainage volume, and Figures 13-16); (3) estimate the drainage volume and its mean porosity; (4) subtract the in-place volumes from the expelled volumes to calculate the lost volumes; and (5) divide the lost volumes by the product of drainage volume and porosity to estimate f.

Our studies (e.g., MacGregor and Mackenzie, 1987; England et al, 1987) suggest typical values of f for sandstones and shales are 0.02 m³/m³ for both oil and gas. But there is a large and somewhat uncertain error bar—perhaps as much as ±0.01 m³/m³. This does not mean that an evenly distributed 2% of the pores of the drainage volume remain petroleum-filled after petroleum has passed through. Oil-staining patterns show that residual saturations as high as 20-50% occur. However, because

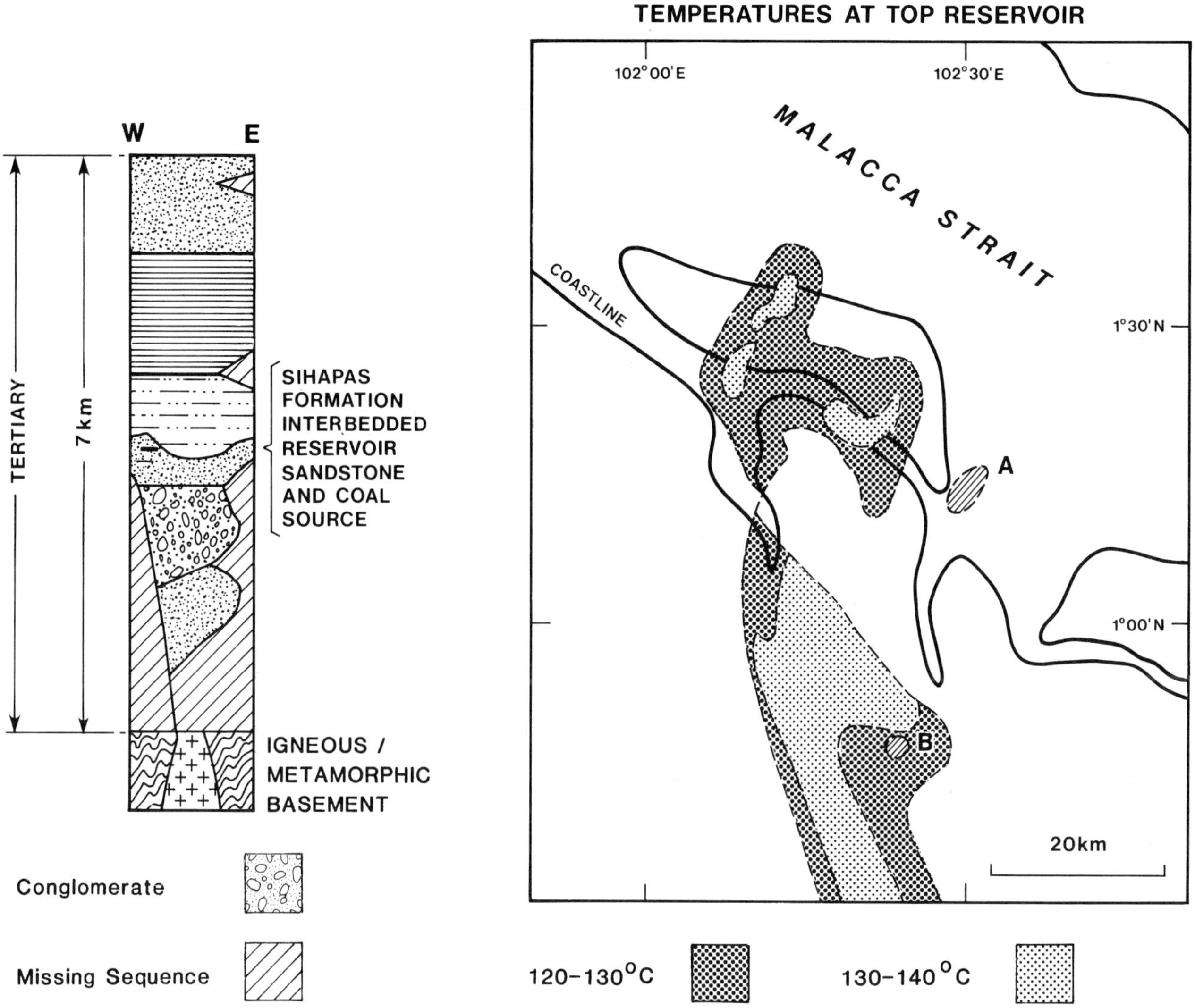

Figure 17—Map with isotherms at top of Sihapas and west to east geologic column for region of Malacca Strait. Key as for Figure 12, except for new symbols shown here. (A) and (B) are prospects considered in case history.

these occupy only 5-10% of the total drainage volume—the remaining 90-95% being unstained—the apparent residual saturation is about 2%. This amount is consistent with capillary forces focusing migrating petroleum through large-pored streaks within the heterogeneous rocks of the drainage volume (Du Rouchet, 1981; England et al, 1987). Our estimates of migration losses are too imprecise to allow us to resolve differences due to variations in the fabric of the carrier rock. Similarly, we are unable to measure the magnitude of gas diffusion from a migrating oil phase into the surrounding carrier rock, although we suspect this may be significant.

The magnitude of secondary migration losses reduces considerably the volume of petroleum available for entrapment and, hence, limits the distance that petroleum can migrate from the mature source rock. The losses must be accounted for when assessing the petroleum budget of a given region.

Table 2. Results of Calculations for Malacca Strait Prospects

	Prospect A	Prospect B
Petroleum mass expelled (M_E)	5×10^{10} kg	7×10^{10} kg
Average petroleum density	700 kg/m^3	700 kg/m^3
Volumes expelled (V_E)	7.1×10^7 m^3	10^8 m^3
Volumes lost (V_L)	10^8 m^3	6.2×10^7 m^3
Volume of charge (V_C)	-2.9×10^7 m^3	3.8×10^7 m^3
		(240 million bbl)

VOLUMES OF PETROLEUM CHARGE

The volume of petroleum (V_C) delivered to a prospect is given by the expelled volume (V_E) minus the lost volume (V_L) (equation 3). We can substitute the appropriate value of V_L, calculated from equation 2, directly into equation 3. However, we must first convert the masses expelled (M_E), calculated from equation 1, into volumes to obtain V_E. To convert from mass to volume we need to know the mean subsurface densities of the petroleum fluids, density being dependent upon pressure, temperature, and composition (England et al, 1987).

Extension to empirical relationships formulated by petroleum engineers for use in the processing side of the industry show that density is governed by petroleum phase, and that petroleum in the subsurface can exist either as a liquid, gas, or a mixture of the two. One must remember that the surface state of a petroleum is a poor guide to its subsurface state; surface gas can dissolve in oil as associated gas and surface petroleum liquids can dissolve in gas and are then called condensates.

Figure 13 shows that the amount of gas that can dissolve in oil increases with increasing pressure and temperature and hence depth, leading to a reduction in the density of the oil (Figure 14). Likewise, Figure 15 demonstrates the amount of petroleum liquid (condensate) that can dissolve in gas increases with increasing pressure and temperature and therefore, with depth, leading, in small part, to an increase in gas density (Figure 16). However, gas density chiefly increases with increasing depth because it is highly compressible. The density of gas at 3 km (30-40 MPa; Figure 16) is over half that of oil (cf. Figure 13). Similar conclusions were reached by Neglia (1979), Ungerer et al (1983), and Heum et al (1986).

The trends of Figures 13-16 show that the properties of oil and gas are similar at depth, but as the oil and gas migrate to shallower depths, they will exsolve gas and oil (condensate) respectively. The uncertainty in Figures 13-16, caused by compositional changes, can be as much as 25%. If greater precision is required, we recommend the use of Glasø's (1980) work directly.

Figures 13-16 can be used to evaluate, very roughly, the density of expelled petroleum as follows. (1) Estimate the GOR of the expelled petroleum in kg/kg using the techniques described in this paper under Mass and Composition Expelled. (2) Compare this with predicted GOR or CGR of the oil and gas phases respectively (for the mean pressure and temperature of the drainage volume) using Figures 13 and 15.

If the expelled GOR is less than the maximum (or saturation) GOR of the oil phase, the petroleum exists entirely in the liquid phase and its density can be read from Figure 14.

If the expelled GOR exceeds the minimum (or saturation) GOR (i.e., 1/CGR) of the gas phase, the petroleum exists entirely in the gas phase and its density can be read from Figure 16.

If the expelled GOR is between the maximum GOR of the oil phase and the minimum GOR (i.e., 1/CGR) of the gas phase, the expelled petroleum is two phase and the mass fraction of the total petroleum (M_E) that is oil or liquid (L) is given by

$$\frac{L}{M_E} = \frac{\Upsilon_G - \Upsilon_F}{\Upsilon_G - \Upsilon_O}, \qquad (8)$$

where Υ_G is the minimum $1/(1 + CGR)$ for a gas phase at the mean pressure and temperature of the drainage volume; Υ_O is maximum $GOR/(1 + GOR)$ for the oil phase; and Υ_F is $GOR/(1 + GOR)$ for the total expelled petroleum.

Equation 8 allows calculation of the weighted average density of the expelled petroleum fluids.

The volume of petroleum expelled (V_E) is calculated by dividing the total expelled mass (M_E) by the appropriate petroleum density. Thus, equation 3 may be used to calculate the petroleum charge volumes (V_C). When the calculated petroleum charge is negative, the petroleum expelled from the source rock runs out before it reaches the prospect of interest, and the prospect is predicted to be dry. If the calculated petroleum charge is positive, the expelled petroleum volumes exceed the volumes lost during migration and petroleum will flow into the prospect. By comparing the petroleum charge with the pore volume available in the trap, one can calculate the degree of fill. If the expelled petroleum is predicted to be two-phase, the petroleum accumulation will consist of an oil leg and gas cap; otherwise, only one phase, oil or gas, is present in the trap.

Two other kinds of losses that occur after accumulation must be accounted for: leakage of petroleum through a seal and in-reservoir biodegradation. We refer to papers by Schowalter (1976) for guidance on seals and Rubinstein et al (1977) for guidance on biodegradation.

The overall calculation may be cross-checked by comparing prediction with observations for a nearby region where petroleum has been found. One must demonstrate that any discovered petroleum fluids correlate chemically with the source rocks that the model predicts gave rise to them (Seifert and Moldowan, 1981).

EXAMPLE CASE HISTORY

The case history is from the Malacca Straits near central Sumatra, Indonesia (Figure 17). Oil-prone coals, similar to those in the Mahakam delta, Kalimantan, Indonesia (e.g., Durand and Oudin, 1979), are the source of oils that are reservoired in the Miocene Sihapas sandstones, within which the coals are interbedded. We do not intend to convince the reader of the correctness of the case history; we only exploit the example to illustrate the type of calculation that can be performed with our approach to geochemical prospect appraisal. For a full discussion of the petroleum geology of the region, including maps and cross sections as well as a justification of the choice of catchment areas of the prospects, refer to MacGregor and Mackenzie (1987).

We consider two prospects (A and B in Figure 17). Their locations relative to the 120° and 130°C isotherms on top of the Sihapas Formation are shown on Figure 17. The values used in equations 1-3 are listed in Appendix 3. For both prospects, the average initial petroleum potential (P_o), measured on over 30 immature samples, is 0.2 kg/kg (200 kg/MT), and average coal thickness (y) and density (ρ_{ROCK}) from electric logs are 5 m and 1,800 kg/m^3, respectively. We divided the source rock into two isomaturity slabs based on present (and maximum) temperature (120°-130°C and 130°-140°C); and PGI and PEE were calculated according to observations for similar coals from the Mahakam delta (Cooles et al, 1986; their Table 2). The appropriate isomaturity slab areas are shown in Appendix 3; the drainage volume for each prospect is simply the product of net Sihapas thickness (150 m) and catchment area. Because the coals are interbedded with a lateral carrier of permeability greater than 10^{-15} m^2 (about 100 md average), migration is purely lateral. Mean Sihapas sandstone porosity (ϕ) is 0.2 (20%) and typical values of migration residual saturation (f) observed in the region are 0.02 m^3/m^3 (2%) (for details, see MacGregor and Mackenzie, 1987).

The results are shown in Table 2. The expelled fluid is predicted to be a single liquid phase of density 700 kg/m^3 at the temperatures and pressures of the drainage volumes. Lost volumes exceed expelled volumes for prospect A, so a dry hole was predicted. Expelled volumes exceed lost volumes in prospect B; therefore, an oil discovery was expected. In practice, both prospects were drilled and the pre-drill prognosis was confirmed. Prospect A was dry and prospect B contained oil-bearing Sihapas sandstone (MacGregor and Mackenzie, 1987).

SUMMARY

Our study of the geological processes responsible for the genesis and migration of petroleum fluids in the subsurface has lead to a set of basic rules.

1. Kerogen may be split into a reactive component that is transformed into petroleum fluids at elevated temperatures, and an inert portion whose carbon remains in the source rock during generation and expulsion of petroleum; inert kerogen slowly rearranges toward graphitelike structures.

2. Reactive kerogen has a labile part that breaks down into oil with small amounts of dissolved gas between 100° and 150°C, and a refractory part that yields gas between 150° and 230°C. Unexpelled oil and oil in the reservoir are rapidly cracked to gas between 150° and 180°C.

3. Labile kerogen is derived from lipids of living organisms, refractory kerogen from C_1-C_3 substituents on the lignin biopolymer, and inert kerogen from lignin and condensed proteins and carbohydrates. The relative amounts of labile, refractory, and inert kerogen define kerogen type and are determined by the nature of the precursory organisms and the depositional setting of the host sediment.

4. Oil expulsion efficiencies are great (>60%) when the initial concentration of labile kerogen exceeds 0.010 kg/kg of rock (10 kg/MT) and temperature exceeds 120°C. Gas expulsion efficiencies are invariably great whatever the initial potential.

5. Petroleum migrates by the bulk flow of petroleum-rich phases. In general, petroleum moves vertically (up or down) in rocks less permeable than 10^{-15} m^2 (1 md) and laterally updip in rocks more permeable than 10^{-15} m^2 (1 md). Petroleum takes the shortest or most permeable vertical migration route to connect with a potential lateral carrier.

6. During migration, an average of 1-3% of the total available pore space along the migration pathway remains petroleum-filled after petroleum migration.

7. Any petroleum fluid discovered in the region of interest should correlate chemically with the source rock the model (rules 1 to 6) predicts gave rise to it.

These general rules may be used together with knowledge of the geology surrounding a given prospect to estimate the volume and composition of petroleum entrapped. The techniques can also be used to evaluate petroleum available for entrapment in whole regions or reservoir sequences. Thus, prospects, regions, or reservoir sequences can be graded and the relative rankings used to influence exploration policy.

APPENDIX 1

List of Symbols

Symbol	Unit	Definition
AREA	(m^2)	Area of isomaturity slab of source rock.
CGR	(kg/kg)	Condensate/gas ratio of gaseous phase.
f	(m^3/m^3)	Residual petroleum saturation, expressed as a fraction of total pore volume of the drainage volume.
GOR	(kg/kg)	Gas/oil ratio of liquid (oil) phase.
L	(kg)	Total mass of expelled liquid (oil) petroleum.
M_E	(kg)	Total mass of expelled petroleum.
P_o	(kg/kg)	Average initial petroleum potential of the source rock.
PEE		Bulk petroleum expulsion efficiency.
PEE_n		Net petroleum expulsion efficiency in direction of a given target reservoir.
PGI		Petroleum generation index.
R	(kg/kg)	Refractory fraction of reactive kerogen.
S_1	(kg/kg)	Vaporizable organic matter.
S_2	(kg/kg)	Total pyrolysis yield.
TSE	(kg/kg C)	Total solvent extract relative to total organic carbon.
TOC	(kg/kg)	Total organic carbon.
V_C	(m^3)	Volume of petroleum charge available for entrapment.
V_D	(m^3)	Total drainage volume.
V_E	(m^3)	Volume of petroleum expelled from mature source rocks.
V_L	(m^3)	Volume of petroleum lost during secondary migration.
y	(m)	Source rock average thickness.
Υ_F	(kg/kg)	(GOR)/(1 + GOR) for total expelled petroleum.
Υ_G	(kg/kg)	Saturation 1/(1 + CGR) for a gas phase at mean pressure and temperature of the drainage volume.
Υ_O	(kg/kg)	Saturation GOR/(1 + GOR) for a liquid (oil) phase at mean pressure and temperature of the drainage volume.
Γ	(kg/kg)	C_1-C_5 molecules produced by pyrolysis expressed as a fraction of total pyrolysis yield.
ϵ		Mass fraction of oil that has cracked to gas [ϵ = gas/(oil + gas)].
θ	(kg/kg)	Ratio of initial oil to initial reactive kerogen.
λ	(kg/kg)	Ratio of present to initial labile kerogen.
$\underline{\nu}$		Mass fraction of petroleum expelled from a source rock in direction of given target reservoir.

ρ	(kg/kg)	Ratio present : initial refractory kerogen.
ρ_{ROCK}	(kg/m^3)	Rock density.
ϕ	(m^3/m^3)	Rock fractional porosity.

APPENDIX 2

Arrhenius Parameters for Petroleum Generation and Oil Cracking to Gas

$$k = Ae^{-\frac{E}{RT}},$$

where E is distributed according to D(E):

$$D(E) = \frac{(2\pi)^{-1/2}}{\sigma}\exp-[(E - \bar{E})^2/2\sigma^2].$$

Labile kerogen: $A = 1.58 \times 10^{13} s^{-1}$; $\bar{E} = 208$ kJ/mol; $\sigma = 5$ kJ/mol.

Refractory kerogen: $A = 1.83 \times 10^{18} s^{-1}$; $\bar{E} = 279$ kJ/mol; $\sigma = 13$ kJ/mol.

Oil-to-gas cracking: $A = 1.0 \times 10^{13} s^{-1}$; $\bar{E} = 230$ kJ/mol; $\sigma = 5$ kJ/mol.

APPENDIX 3

Input to Equations 1-3 for Malacca Strait Prospects

For All Isomaturity Slabs

P_o	0.2 kg/kg (200 kg/MT)
ρ_{ROCK}	1,800 kg/m^3
y	5 m
ϕ	0.2

For Isomaturity Slab 1 (120°-130°C)

PGI	0.35
PEE$_n$	0.80

For Isomaturity Slab 2 (130°-140°C)

PGI	0.50
PEE$_n$	0.80

Catchment Areas

Prospect A	Total	1.67×10^8 m^2
	Isomaturity Slab 1	4.16×10^7 m^2
	Isomaturity Slab 2	4.17×10^7 m^2
Prospect B	Total	1.03×10^8 m^2
	Isomaturity Slab 1	1.94×10^7 m^2
	Isomaturity Slab 2	8.35×10^7 m^2

REFERENCES CITED

Beaumont, C., 1978, Foreland basins: Geophysical Journal of the Royal Astronomical Society, v. 65, p. 291-329.

Berg, R. R., 1985, Capillary pressures in stratigraphic traps: AAPG Bulletin, v. 69, p. 939-956.

Bishop, R. S., M. H. Gehman, and A. Young, 1983, Concepts for estimating hydrocarbon accumulation and dispersion: AAPG Bulletin, v. 67, p. 337-348.

Campbell, J. H., G. Gallegos, and M. Gregg, 1980, Gas evolution during oil shale pyrolysis; 2. kinetic and stoichiometric analysis: Fuel, v. 59, p. 727-732.

Cochran, J. R., 1983, Effects of finite rifting times on the development of sedimentary basins: Earth and Planetary Science Letters, v. 66, p. 289-302.

Cooles, G. P., A. S. Mackenzie, and T. M. Quigley, 1986, Calculation of masses of petroleum generated and expelled from source rocks, in D. Leythaeuser and J. Rullkötter, eds., Advances in organic geochemistry, 1985: Oxford, England, Pergamon Press, p. 235-246.

Cornford, C., 1984, Source rocks and hydrocarbons of the North Sea, in K. W. Glennie, ed., Introduction to petroleum geology of the North Sea: Oxford, England, Blackwell Scientific Publications, p. 171-209.

Davidson, J. K., G. J. Blackburn, and K. C. Morrison, 1984, Bass and Gippsland basins: a comparison: APEA Journal, v. 24, p. 101-109.

Dembicki, H., Jr., B. Horsfield, and T. Y. Ho, 1983, Source rock evaluation by pyrolysis-gas chromatography: AAPG Bulletin, v. 67, p. 1094-1103.

Durand, B., and J. L. Oudin, 1979, Example de migration des hydrocarbures dans une serie deltaique: le delta de la Mahakam, Kalimantan, Indonesie: Proceedings of the 10th World Petroleum Congress: London, Heyden and Son, p. 1-9.

——— P. Ungerer, A. Chiarelli, and J. L. Oudin, 1984, Modélisation de la migration de l'huile: application à deux examples de bassins sedimentaires: Proceedings of the Eleventh World Petroleum Congress, v. 2, geology, exploration, reserves: Chichester, England, Wiley, p. 3-16.

Du Rouchet, J., 1981, Stress fields, a key to oil migration: AAPG Bulletin, v. 65, p. 74-85.

England, W. A., A. S. Mackenzie, D. M. Mann, and T. M. Quigley, 1987, The movement and entrapment of petroleum fluids in the subsurface: Journal of the Geological Society, London, v. 144, p. 327-347.

Espitalié, J., J. L. Laporte, M. Madec, F. Marquis, P. Leplat, J. Paulet, and A. Boutefeu, 1977, Méthode rapide de caracterisation de roches mères de leur potentiel pétrolier et de leur degne d'évolution: Révue L'Institut Francais du Petrole, v. 32, p. 23-42.

Glasø, O., 1980, Generalized pressure-volume-temperature correlations: Journal of Petroleum Technology, v. 32, p. 785-795.

Goff, J. C., 1983, Hydrocarbon generation and migration from Jurassic source rocks in the East Shetland basin and Viking graben of the northern North Sea: Journal of the Geological Society, London, v. 140, p. 445-474.

Gretener, P. E., and C. D. Curtis, 1982, Role of temperature and time on organic metamorphism: AAPG Bulletin, v. 66, p. 1124-1149.

Guidish, T. M., C. G. St. C. Kendall, I. Lerche, D. J. Toth, and R. F. Yarzab, 1985, Basin evaluation using burial history calculation: an overview: AAPG Bulletin, v. 69, p. 92-105.

Heum, O. R., A. Dalland, and K. K. Mesingolt, 1986, Habitat of hydrocarbons at Haltenbanken (PVT-modeling as a predictive tool in hydrocarbon exploration), in A. M. Spencer, ed., Habitat of hydrocarbons on Norwegian continental shelf: London, Graham & Trotman, p. 259-274.

Hubbert, M. K., 1953, Entrapment of petroleum under hydrodynamic conditions: AAPG Bulletin, v. 37, p. 1954-2026.

Hunt, J. M., 1979, Petroleum geochemistry and geology: San Francisco, Freeman, 617 p.

Jüntgen, H., and J. Klein (1976), Entstehung von Erdgas aus kohlingen sedimenten: Erdöl und Kohle-Erdgas-Petrochemie, v. 28, p. 65-75.

Landes, K. K., 1967, Eometamorphism and oil and gas generation in time and space: AAPG Bulletin, v. 51, p. 828-841.

MacGregor, D. S., and A. S. Mackenzie, 1987, Quantification of oil generation and migration in the Malacca Strait region: Proceedings of 15th Annual Convention of Indonesian Petroleum Association, 1986, Jakarta, p. 305-319.

Mackenzie, A. S., 1984, Applications of biological markers in petroleum geochemistry, in J. Brooks and D. H. Welte, eds., Advances in petroleum geochemistry, v. 1, London, Academic Press, p. 115-214.

——— C. Beaumont, and D. P. McKenzie, 1984, Estimation of the kinetics of geochemical reactions with geophysical models of sedimentary basins and applications: Organic Geochemistry, v. 6, p. 875-884.

——— D. Leythaeuser, P. Müller, T. M. Quigley, and M. Radke, 1988, Movement of hydrocarbons in shales: Nature, v. 331, p. 63-65.

McKenzie, D. P., 1981, The variation of temperature with time and hydrocarbon maturation in sedimentary basins formed by extension: Earth and Planetary Science Letters, v. 55, p. 87-98.

Meissner, F. F., 1984, Petroleum geology of the Bakken Formation, Williston basin, North Dakota and Montana, in G. Demaison and

R. J. Murris, eds., Petroleum geochemistry and basin evaluation: AAPG Memoir 35, p. 159-179.

Moldowan, J. M., W. K. Seifert, and E. J. Gallegos, 1985, Relationship between petroleum composition and depositional environment of petroleum source rocks: AAPG Bulletin, v. 69, p. 1255-1268.

Neglia, S., 1979, Migration of fluids in sedimentary basins: AAPG Bulletin, v. 63, p. 573-597.

Philippi, G. T., 1977, On the depth, time, and mechanism of origin of the heavy to medium-gravity naphthenic crude oils: Geochimica et Cosmochimica Acta, v. 41, p. 33-52.

Pitt, G. J., 1962, The kinetics of the evolution of volatile products from coal: Fuel, v. 41, p. 267-274.

Powell, T. G., 1982, Petroleum geochemistry of the Verrill Canyon Formation: a source for Scotian Shelf hydrocarbons: Bulletin of Canadian Petroleum Geology, v. 30, p. 167-179.

Price, L. C., L. M. Wegner, T. Ging, and G. W. Blout, 1983, Solubility of crude oil in methane as a function of pressure and temperature: Organic Geochemistry, v. 4, p. 201-221.

Royden, L., J. G. Sclater, and R. P. Van Herzen, 1980, Continental margin subsidence and heat flow: important parameters in formation of petroleum hydrocarbons: AAPG Bulletin, v. 64, p. 173-187.

Rubinstein, I., O. P. Strausz, C. Spyckerelle, R. S. Crawford, and D. W. S. Westlake, 1977, The origin of the oil sand bitumen of Alberta, a chemical and microbiological simulation study: Geochimica et Cosmochimica Acta, v. 41, p. 1341-1353.

Schowalter, T. T., 1979, The mechanics of secondary hydrocarbon migration and entrapment: AAPG Bulletin, v. 63, p. 723-760.

Seifert, W. K., and J. M. Moldowan, 1981, Paleoreconstruction by biological markers: Geochimica et Cosmochimica Acta, v. 45, p. 783-794.

Smith, J. E., 1971, The dynamics of shale compaction and evolution of pore fluid pressure: Mathematical Geology, v. 3, p. 239-269.

Sweeney, J. J., A. K. Burham, and R. L. Braun, 1986, A model of hydrocarbon generation from type I kerogen: application to the Uinta basin, Utah, *in* J. Burrus, ed., Thermal modeling in sedimentary basins: Paris, Technip, p. 547-561.

Teichmüller, M., and B. Durand, 1983, Fluorescence microscopic rank studies on liptinites and vitrinites in peat and coals, and comparison with results of the Rock-Eval pyrolysis: International Journal of Coal Geology, v. 2, p. 197-230.

Tissot, B., and J. Espitalié, 1975, L'évolution thérmique de la matière organique des sediments: applications d'une simulation mathématique: Révue l'Institut Français du Pétrole, v. 30, p. 743-777.

——— and D. H. Welte, 1984, Petroleum formation and occurrence (2d edition): New York, Springer Verlag, 699 p.

——— B. Durand, J. Espitalié, and A. Combaz, 1974, Influence of nature and diagenesis of organic matter in formation of petroleum: AAPG Bulletin, v. 58, p. 499-506.

Ungerer, P., and R. Pelet, 1987, Extrapolation of the kinetics of oil and gas formation from laboratory experiments to sedimentary basins: Nature, v. 327, p. 52-54.

——— F. Behar, D. Deschamps, 1983, Tentative calculation of the overall volume of organic matter; implications for primary migration, *in* M. Bjorøy et al, eds., Advances in organic geochemistry, 1982: Chichester, England, Wiley, p. 129-135.

Waples, D. W., 1980, Time and temperature in petroleum formation: application of Lopatin's method to petroleum exploration: AAPG Bulletin, v. 64, p. 916-926.

Welte, D. H., and M. A. Yükler, 1981, Petroleum origin and accumulation in basin evolution—a quantitative model: AAPG Bulletin, v. 65, p. 1387-1396.

Wright, N. J. R., 1980, Time, temperature and organic maturation—the evolution of rank within a sedimentary pile: Journal of Petroleum Geology, v. 2, p. 411-425.

Young, A., and R. D. McIver, 1977, Distribution of hydrocarbons between oils and associated fine-grained sedimentary rocks—physical chemistry applied to petroleum geochemistry II: AAPG Bulletin, v. 61, p. 1407-1436.

From
Petroleum Formation and Occurrence
by B.P. Tissot and D.H. Welte

Locating Petroleum Prospects: Application of Principle of Petroleum Generation and Migration — Geological Modeling

In the early days of petroleum exploration, wells were drilled where oil or gas seepages indicated the presence of petroleum. Later, with increasing sophistication of geological knowledge, and especially with the advent of exploration geophysics, the decision to drill a well was additionally taken on the basis of the recognition of suitable structures, such as anticlines, fault traps, unconformity traps and reefs. Frequently, the selection of a structure to be drilled was based on intuition and general experience rather than on pertinent information, because very little information was available whether or not a trap would contain hydrocarbons.

A systematic study, utilizing the new understanding of petroleum generation and migration, can help to decrease the uncertainty in predicting a petroleum-filled structure, and hence the financial risk when drilling a well. This is of special importance in offshore areas and remote, hostile exploration regions where drilling is extremely costly. The consequent application of geochemistry can also define new exploration targets in relatively well-known basins where the rate of discoveries has declined.

The purpose of this chapter is to show how petroleum exploration can benefit, firstly by collecting source rock and maturational information (acquired at relatively low cost) from application of organic geochemical studies, and secondly by relating this knowledge intelligently to the geological framework of a given exploration area. The basic idea of this concept is to identify the source rocks present, the hydrocarbon potential of each source rock in time and space, and relate this information to the geological evolution of the basin. At the end of such a study, most favorable zones for petroleum accumulation are located in the basin. This is done by relating the hydrocarbon generation of a source rock at any given time during basin evolution to the most likely paths of petroleum migration, and to the formation and age of a trap.

The determination of the most favorable zones of petroleum accumulation in a basin consists of a sequence of steps to prepare and combine the necessary geological and geochemical information, and produce maps and sections showing and rating the potential targets. This can be done "by hand" in a conventional manner, or with the help of computers. In Figure V.3.1 a scheme is presented that shows the sequence of steps and, in a general way, the kind of geological information required in order to apply organic geochemical studies to petroleum exploration. This scheme can easily be followed in practice. It is no more difficult than the introduction of facies analyses, isopach maps, or depth contour maps has previously been. In the past, all these techniques have also been assimilated by

A decisive exploration philosophy incorporating geochemical data:
Determination of most favourable zones of petroleum accumulation

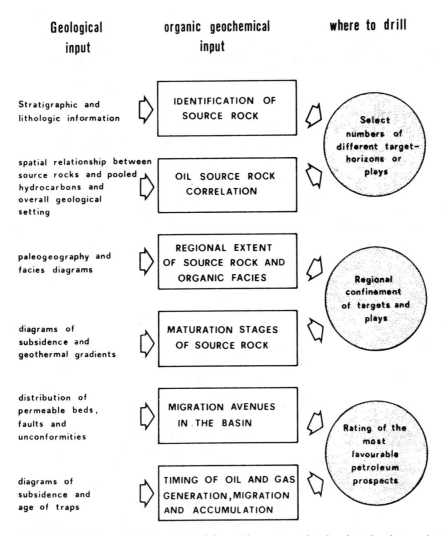

Fig. V.3.1. Schematic summary of the various steps that lead to the determination of most favorable zones of petroleum accumulation in a basin

petroleum geologists, and nowadays geochemical data should be incorporated in the philosophy of exploration.

Geological-geochemical modeling with the help of high speed computers is possibly the most effective, suitable, and certainly the most advanced method for determination of the favorable zones of petroleum accumulation. It integrates and quantifies the newly gained geochemical knowledge on petroleum formation and migration into the geological framework of basin evolution. This computer-aided numerical simulation of geological and geochemical processes has the enormous advantage that through the combination of regionally derived geologi-

3.1 Acquisition of the Geochemical Information

cal data with information about the principal processes of petroleum generation and migration, tailor-made models of a specific exploration area are available for the explorationist. In this way a ranking of exploration targets based on a network of logically consistent data can be established. This diminishes greatly the exploration risk.

The following chapter outlines the acquisition of the geochemical information (Sect. 3.1) and presents a first conceptual model of petroleum generation in a basin (Sect. 3.2). Then geological modeling is presented (Sect. 3.3), whereas geochemical modeling will be introduced in the next Chapter V.4.

3.1 Acquisition of the Geochemical Information

When a new exploration program is designed in a more or less unknown area, only a limited amount of general geological information is normally available. This, of course, applies nowadays only to certain remote areas and offshore regions. Sample material of the most important stratigraphic and lithologic units represented in the basin must be collected, outcrops at the rim of the basin being a readily accessible source. A careful study of the available information on the sedimentary filling of the basin (a facies analysis) might result in a preliminary *identification* of one or more *potential source rocks* or source rock sequences. A source rock-identification program using organic geochemical techniques to analyze available sample material should help to define potential source rocks more clearly. At the same time, neighboring nonsource rock strata should also be checked for porosity and permeability, and their role in offering possible migration avenues. If little or no sample material is available, a limited sampling program may be started to obtain surface samples and complementary unweathered core samples from a few shallow drill holes (up to 100 m depth) at strategic locations. In offshore basins, information from adjacent, comparable strata on land, and sometimes samples taken from the sea bottom by dredging or shallow coring may be helpful.

A different case is presented where basins have been abandoned some years ago, and where fresh interest arises for new exploration. The advent of the geochemical techniques and new knowledge of the principles of petroleum generation and migration frequently justifies a fresh look at old prospects. In such cases, old sample material from early wells can still be used to evaluate the source rock potential. However, in all cases, the first step in a basin study is to identify the number and kind of potential source rocks in the basin. The sample material has to be analyzed for amount and type of kerogen, and its stage of maturity.

In fact, the concept of source rock identification and determination of the main zones of oil and gas formation can be introduced into an exploration campaign at any stage. In particular, before a well is drilled, it is critical to ensure that a maximum of information can be gained with respect to both source and carrier rocks. To obtain this information from the well, an effective sampling program must be prepared. Major lithologies *and* regular depth intervals should be sampled. Cuttings should be collected at intervals of 3 to 30 m, depending on the changes in lithology. Preferably, samples should be canned in small air-tight containers (about 300–500 cc), with ample water to keep the original moist saturation, particularly if light hydrocarbon studies are intended. In addition, some core or side wall core samples should be obtained, in order to support and verify information gained from cuttings. Core sampling is already done routinely for large-scale testing of physical properties. The sample material should be subjected to geochemical analyses in such a way that basic analytical information on source rock properties etc. is available, ideally by completion of the well or, at least, with a minimum time-lag. Therefore, *rapid methods* have to be applied for source rock analyses. Such methods *should be considered as screening techniques*, rather than as a means to obtain detailed analytical results. Methods suitable for screening are the pyrolysis technique and the quantitative determination of light hydrocarbons. Both have been described in Chapter V.1.

These two techniques will provide independent *geochemical logs*, which can be readily interpreted in terms of hydrocarbon potential and level of maturity (main zones of oil and gas formation).

Even information with respect to hydrocarbon migration can be derived from these two logs. Analytical instruments for both kinds of screening methods can be installed and used directly at the well site. Under these circumstances, considerable time is saved because sample processing can be simplified (no canning, packing and transportation etc.), and results can be utilized while drilling is still in progress. The advantages of "real-time" geochemical logging are obvious, and certainly can help to cut costs and gain valuable information. For instance, wellhead geochemical logging by pyrolysis can indicate when a overmature zone is reached, and no more oil is to be expected. Useless continuation of drilling, and hence the waste of money, can thus be prevented. Conversely, total amount of light hydrocarbons and certain hydrocarbon ratios may give an indication for nearby hydrocarbon accumulation, and thus stimulate further drilling and improve the success rate. Knowledge of a nearby accumulation will also stimulate necessary safety measures for further drilling.

In addition to the screening analyses, further sample material has to be analyzed in a well-equipped geochemical laboratory, to verify and complete the information gained from screening. The number of samples to be analyzed in more detail for source rock identification, type of kerogen, and maturity can be reduced considerably by utilizing the screening results to select zones of homogeneous organic facies, in particular those of most promising hydrocarbon potential. Whenever possible, a few core samples (side wall cores or others) should be included to support results obtained from cuttings and/or serve as fixed geochemical reference points in a well. Such detailed geochemical information on selected rock samples, based on various complementary physicochemical and optical microscopic techniques (Chap. V.1), serve to refine results from screening techniques and allow the improvement and updating of the preliminary conceptual model. In addition, analyses of pooled hydrocarbons possibly collected from formation tests (FT samples), can provide information on the source rock–oil relationship: identification of the source by direct correlation and the stage of maturity at the time of hydrocarbon expulsion can be attempted (Chap. V.2). In the case of severe thermal alteration occurring in the reservoir, such information

3.2 First Conceptual Model of Petroleum Generation in a Basin

cannot be obtained. In some cases, where source rocks have not yet been found, the analyses of pooled hydrocarbons may allow prediction of the type of source rock to be expected (Chap. V.2).

3.2 First Conceptual Model of Petroleum Generation in a Basin

The scientific approach for locating petroleum prospects, described below, can be introduced at any stage of exploration. In fact, it is desirable to update this determination of most favorable zones of petroleum accumulation continuously during the entire exploration campaign.

With the help of the geochemical data, a preliminary conceptual model of the basin with respect to source rocks and generation of petroleum is developed. For each source rock identified, a map is produced showing the regional extent of the source bed, its thickness, and depth level. It is also advisable to construct preliminary diagrams of the subsidence history for each source layer at strategic points in the basin (Fig. II.5.11). This probably could be done at a relatively early stage of the exploration campaign in basins situated at a passive continental margin. In the passive margin setting, the depth–time relationship is usually rather simple, as normally there has been minimal erosion. The stratigraphic identification of seismic marker horizons is the basis for the construction of diagrams of subsidence before the drilling of the first exploration wells. By interpretation of the maturity data of the rock samples investigated, the subsidence curves of source rock units help to predict approximate levels of maturity in different parts of the basin. In this way, it is possible to obtain a first idea about the depth level of the main zone of oil formation throughout the basin.

For a determination of the most favorable zones of petroleum accumulation the following series of questions has to be answered:

1. Which beds have a source rock potential? What is their regional extent and their relationship to paleography?
2. During which geological time interval, at what depth, and in which area of the basin were the above-mentioned source rocks mature enough to generate petroleum?
3. During which geological time interval, and along which preferred routes was petroleum migration possible?
4. At what geological time did the formation of traps occur, as compared to the timing of petroleum generation and migration? Where are these traps located, with respect to possible migration routes?

These questions can be answered on the basis of a set of maps and sections prepared "by hand". However, the quantitative approach using geological and geochemical models is a more effective method for answering the questions.

3.3 Numerical Simulation of the Evolution of a Sedimentary Basin — Geological Modeling

The starting point for a quantitative, mathematical-numerical handling of the formation and migration of petroleum is the simulation of the development of a sedimentary basin. The relevant factors of such an undertaking are represented schematically in Figure V.3.2. The idea is to retrace geological steps, which follow each other as a causal sequence beginning from an initial condition at time ($t = T_o$), over intermediate conditions ($t = T_1$), and to the basin configuration observed today ($t = T_x$). A particular difficulty is that only the end result at ($t = T_x$) — today's stratigraphic section, porosities, etc. — and not the initial condition at ($t = T_o$) is known for the observed sedimentary basin as empirical data available for comparison. The evolution of a particular sedimentary basin must be completely computed out, using estimated empirically derived initial data, and then the result must be compared with the real basin to see whether the present condition at time ($t = T_x$) has been reached.

The conceptual model is first set up for a real system, that is, for a real sedimentary basin, which naturally is less well known at the start of an exploration campaign than at the end. The conceptual model is the basis for a mathematical model, which then will be calculated through with estimated initial data. The first iteration will generally not achieve good correspondence with the values that describe the present-day condition of the sedimentary basin. Now the initial input data must be changed and/or the conceptual model modified as long as necessary for the calculations to give the geologically known end result. A flow diagram (Fig. V.3.3) illustrates this procedure schematically.

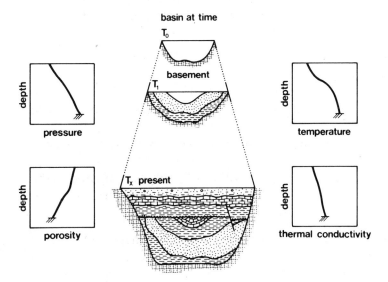

Fig. V.3.2. Evolution of a sedimentary basin from an initial condition at time ($t = T_o$) to the basin configuration observed today ($t = T_x$). Such parameters as pressure, porosity, temperature and thermal conductivity are changing continuously in each sedimentary unit with increasing depth of burial. (After Welte, 1982)

3.3 Numerical Simulation of the Evolution of a Sedimentary Basin

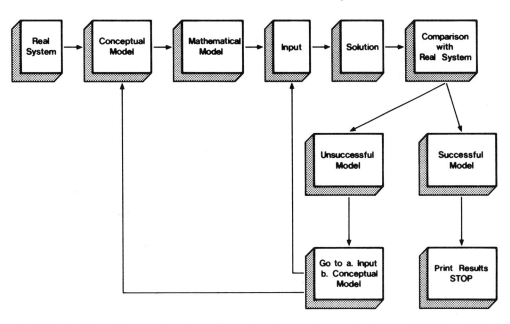

Fig. V.3.3. Flow diagram for three-dimensional quantitative basin model. (After Welte and Yükler, 1981)

The computed values are made to resemble those observed in nature within tolerance by means of successive iterations. In this way the simulation model works its way from a lower level of correspondence to another higher level until the whole simulation is ended. The fact that the most important parameters to be calculated are mutually dependent is of fundamental importance. It means that an inherent logic is followed through the iterations toward the most important parameters for the model, such as compaction and sediment thickness, or temperature and maturity. The core for the three-dimensional deterministic dynamic basin model is a pair of non-linear partial differential equations of energy and mass balance that describe the course of the respective physicochemical processes. The backbone of this approach was presented by Welte and Yükler (1981). A geological, a hydrodynamic, a thermodynamic, and a geochemical part of the program can be distinguished among the input data (Fig. V.3.4). The sedimentary basin to be investigated covering an area of many thousand km^2 and a depth of several km is divided up into a three-dimensional grid, rather, into elementary partial volumes.

The essential parameters, such as pressure, temperature, porosity, degree of maturity of the kerogen, etc., are then calculated with their variations over geological time spans for the numerous grid points, usually in the range of 100,000 of the three-dimensional grid. The necessity of repeating the programmed calculations at all grid points several times in a sequence of iterations was already pointed out. The computed values, which are mutually dependent, are adapted to each other, and they are brought into agreement with those measured in nature by this iteration process (Figs. V.3.3 and V.3.4). It is obvious that only the fastest and most capable of large computers can be used for such calculations.

Locating Petroleum Prospects: Application of Organic Geochemistry

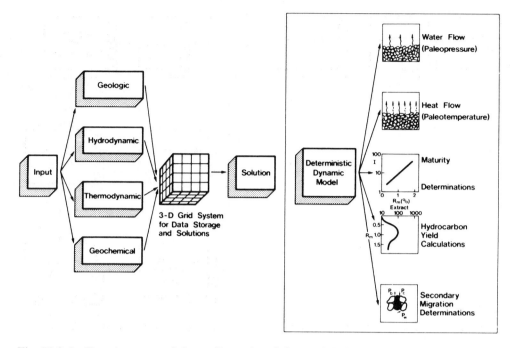

Fig. V.3.4. Development of three-dimensional deterministic dynamic model. Groups of input parameters and results after solution are schematically shown. (After Welte and Yükler, 1981)

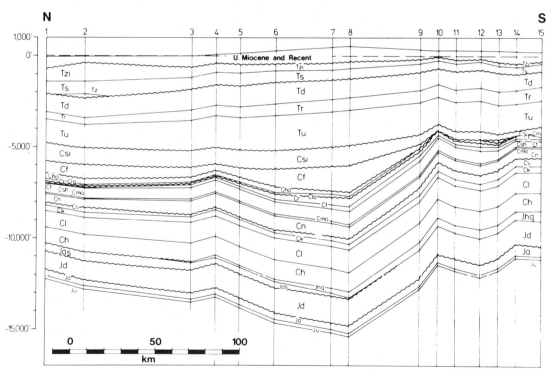

Fig. V.3:5. Computed cross-section through a real basin as based on quantitative basin modeling. Grid points in the horizontal plain are shown with numbers *1–15* (N–S). (After Welte, 1982)

3.3 Numerical Simulation of the Evolution of a Sedimentary Basin

The first important stage in the simulation of a basin's history from a geological standpoint is the calculation of cross sections (Fig. V.3.5) of the sedimentary basin that is under investigation. The comparison of the computed sedimentary thickness on the cross-sections with thickness actually measured in wells or seismic profiles is one of the first control possibilities for bringing concept and reality into agreement through purposeful iterations. How finely the cross-section is divided into discrete units in the horizontal and vertical dimensions is governed by both the complexity of the geology and by the level of geological information. The integration over time of the individual processes of the system leads to the computation of sedimentary thickness in the corresponding geological time spans.

One of the most important goals of integrated basin studies for petroleum exploration is the computation of the hydrocarbon potential of a particular source rock at every given point in time during the history of a sedimentary basin. This is possible with a relatively high level of confidence with the help of the numerical simulation of the basin's development. The maturity mapping of a source rock in space and time is an essential part when determining the hydrocarbon generating zone. In a first step, it is acceptable for an exploration campaign to establish a valid ranking of exploration targets in a given basin rather than to present a truly quantitative prediction of amounts of hydrocarbons. Whilst a truly quantitative evaluation necessitates the use of an elaborate kinetic model of kerogen degradation (presented in the next Chap. V.4), it is sufficient for a first evaluation of exploration targets to calculate in a semi-empirical manner vitrinite reflectance values (i. e., maturity). A relatively simple method to arrive at calculated vitrinite reflectance values using time–temperature relationships was, for instance, proposed by Lopatin (1971) and later modified by Waples (1980). This method is further discussed in the next chapter. Following this line of thought a present-day maturity map based on vitrinite reflectance values is presented in Figure V.3.6 for a source rock in an area of active exploration of about 165,000 km^2. These values can be converted into amounts of hydrocarbon generated per rock unit, if the quality and thickness of the source rock is known. In this example, the computed basin model typically contains about 350 data points for one slice of source bed extending over an area of 165,000 km^2. Maps of this kind can be drawn by the computer at any desired geological time for every parameter that is contained in the set of differential equations of the simulation program.

When the hydrocarbon potential of a source rock has been computed in space and time, this information can be combined in appropriate form with information about the presence of porous carrier and reservoir rocks and their geological position in the developing basin. Structural maps of such reservoir rocks or of any sealing rock can also be called up from the computer for any given point in time (Fig. V.3.7).

The numerical simulation of the geological development of a sedimentary basin makes possible an uncommon density of geological information and geological data of a completely new quality. The great advantage of the numerical simulation described here lies not only in the quantifying of geological parameters, which normally are only available through wells, but more importantly in the fact that a continuous net of geological information covering the whole area and

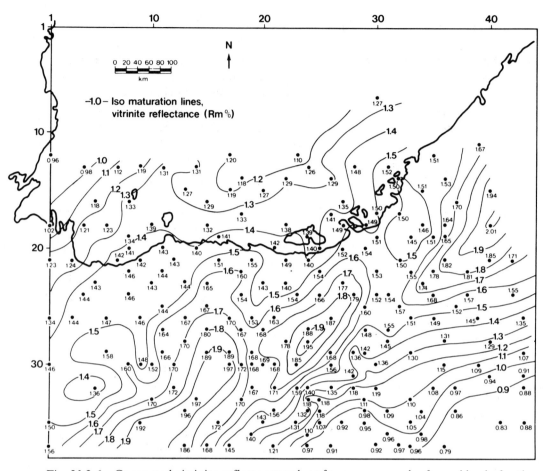

Fig. V.3.6. Computed vitrinite reflectance values for a source rock of a real basin for the present day as based on quantitative basin modeling. (After Welte, 1982)

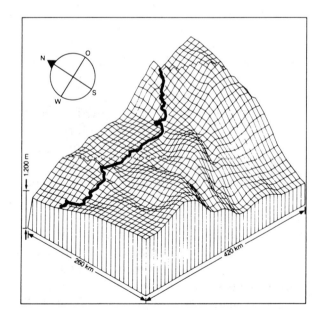

Fig. V.3.7. Three-dimensional subsurface contour map of the top of a reservoir formation in a real basin as based on quantitative basin modeling. (After Welte, 1982)

volume is made available in dependence with time. The computed maps have a considerably higher "reality content" than conventional geological subsurface maps through the mutual influence of individual parameters on each other and through the cross-checks with real geological control points.

It is quite obvious that the application of computer-aided modeling for geological and geochemical processes is only in its initial stages. Furthermore, it is realized that a truly quantitative prediction of hydrocarbon potentials and ultimate oil and gas reservees of sedimentary basins and not only a ranking of exploration targets is only possible if the most important geological and geochemical processes responsible for the formation of hydrocarbon accumulations are understood in greater detail than at present. Next to the geological modeling the geochemical modeling, especially of the kinetics of kerogen degradation and hydrocarbon generation, is of great importance in this connection. This will be discussed in the next chapter.

Summary and Conclusion

A systematic utilization of the new understanding of petroleum generation, migration, and accumulation can help to decrease the uncertainty in predicting petroleum-filled traps. This is done by identifying the source rocks, the hydrocarbon potential of each source rock in time and space, and by relating this information with the geological evolution of the basin. This determination of most favorable zones of petroleum accumulation in a basin can be done "by hand" in a conventional manner or with the help of computers.

In the initial stage of an exploration program, even with (limited) information on potential source rocks, a first conceptual model of petroleum generation in a basin is developed. In this model the relative position of potential source rocks, their presumed state of maturity and their extension throughout the basin are depicted in maps and profiles.

Such a conceptual model may provide valuable information for the selection of the site of exploratory wells. Rapid geochemical screening techniques can be applied at the well site to provide real-time geochemical logs which can be readily interpreted in terms of hydrocarbon potential of source rocks and level of maturity. Further sample material has to be analyzed in a laboratory, to verify and complete the information gained from screening.

A numerical simulation of geological and geochemical processes that lead to the formation of hydrocarbon accumulations is the prerequisite to make quantitative predictions about oil and gas reserves in a basin. Simulation of geochemical processes necessitates a kinetic model of hydrocarbon formation, and also a model of migration and accumulation mechanisms which are not yet known in sufficient detail. However, a ranking of exploration targets can be achieved by simulating the evolution of a sedimentary basin, ie., by geological modeling and a semi-empirical incorporation of relevant geochemical processes. This diminishes greatly the exploration risk.

The backbone of geological modeling is a set of non-linear partial differential equations of energy and mass balance that describe the course of respective geological processes. Main results of the model calculations are, for instance, computer maps showing porosity data, temperatures, maturities or sedimentary thicknesses after compaction for any given sedimentary unit during evolution of the basin. A typical three-dimensional grid system of an exploration area is in the range of 10,000 to 100,000 grid points. It is thus obvious that geological modeling makes possible a formerly uncommon density of geological information and geological data of a completely new quality.

From
Petroleum Formation and Occurrence
by B.P. Tissot and D.H. Welte

Geochemical Modeling: A Quantitative Approach to the Evaluation of Oil and Gas Prospects

4.1 Necessity of a Quantitative Approach to Petroleum Potential of Sedimentary Basins

The purpose of the evaluation of a sedimentary basin in terms of petroleum exploration is to know the amount of oil and gas that has been generated and accumulated and then to locate it.

In this respect, the following information is desirable:
a) Amount of oil and gas generated (per km^2) in each source rock and in every part of the basin.
b) Time of hydrocarbon formation to compare it with the age of deposition of impermeable seals and with the age of folding and faulting, i.e., with respect to formation of traps.
c) Amount of oil and gas which has been expelled out of the source rock into the porous reservoirs, amount and location of hydrocarbons accumulated in traps.
d) Evaluation of the ultimate oil and gas reserves of a sedimentary basin. This parameter will be required increasingly when exploration reaches an advanced stage. Then the problem arises whether to continue or stop exploration, depending on the possible reserves remaining to be discovered.

A quantitative approach allowing a computation of the amount of oil and gas generated and migrated in any place of the basin as a function of time is necessary to answer the above questions. As petroleum and natural gas result from kerogen degradation, various indices have been proposed to characterize the quality and evolution stage of the organic matter. These have been reviewed in Chapter V.1. They can be used as a first approach, in a semi-empirical manner, for ranking of exploration targets (see above, Chap. V.3). However, these indices do not allow a truly quantitative evaluation of hydrocarbon generation as a function of time. In particular, most of them do not account for the respective kinetics of degradation of the various types of organic matter and also for complex burial histories.

Mathematical models based on kinetics of kerogen degradation and polyphasic fluid flows and using computer simulation allow a truly quantitative approach. The problem seems to be adequately expressed in respect to formation of petroleum and gas as a function of time, but additional research is still required on migration of hydrocarbons to achieve the evaluation of the ultimate reserves.

Kinetics of kerogen degradation was first introduced through the consideration that temperature and time might, to some extent, compensate for each other. This view was known for a long time, as Maier and Zimmerley (1924) investigated the kinetics of bitumen generation from the Green River shales. Along that line, McNab et al. (1952) and Abelson (1963) investigated experimentally various degradation schemes (decarboxylation, cracking) and showed that by using an Arrhenius plot, such reactions may occur at the relatively low temperature in the sedimentary basins over geological periods of time. Another aspect is the relation between age or heating time of source rocks and temperature threshold of the main zone of oil generation. For a given type of organic matter, the threshold generally decreases with increasing age or heating time (Connan, 1974, 1976) of the source rock, according to an Arrhenius-type relationship.

On the other hand, coal researchers also considered the kinetics of the progressive carbonization, either natural or artificial during cokefaction, and the related elimination of volatile matter and increase of carbon content and reflectance. Kinetics were introduced by using nomograms, or an approximate computation (Huck and Karweil, 1955; van Krevelen, 1961; Lopatin, 1971; Lopatin and Bostick, 1973). In respect to pyrolysis of oil shales, experimental data on kinetics of kerogen degradation were obtained by Hubbard and Robinson (1950) and mathematical models were subsequently proposed by Allred (1966), Fausett (1974), Braun and Rothman (1975).

Lopatin (1971), followed by Waples (1980), made an attempt to calculate in a simple manner the combined effect of time and temperature on maturation of the organic matter, without using the actual kinetic equations, which require a numerical simulation and the use of computers. It is assumed that the rates of chemical reactions approximately double for every 10°C rise in temperature (Lopatin, 1971; Waples, 1980). Upon progressive burial they calculate an interval maturity by multiplying the time of residence ΔT in a temperature interval of 10°C by a factor 2^n, which itself changes to 2^{n+1}, 2^{n+2}, etc. every 10°C. The total maturity is expressed by TTI (time–temperature index) which is the sum of the interval maturities:

$$TTI = \sum_n 2^n (\Delta T_n).$$

Then a table is provided to calibrate TTI versus vitrinite reflectance and, through it, to interpret this index in terms of stages of oil and gas generation.

Although this calculation procedure is attractive, because it does not require the use of computers, the basic assumption that the rate of reactions doubles for every 10°C rise is only valid for reactions with an activation energy in the range of $E = 10$ to 25 kcal mol^{-1} (at temperatures from 20° to 160°C which include the stages of oil and gas generation). This order of magnitude is acceptable at the onset of oil generation, when relatively weak chemical bonds are broken in kerogen (Tissot, 1969; Connan, 1974).

However, it has been recognized from experiments (Weitkamp and Gutberlet, 1968) and observations in sedimentary basins (Tissot and Espitalié, 1975) that the activation energies involved in kerogen degradation progressively increase, with increasing maturation, as stronger bonds are successively broken. For instance, at the peak of oil generation the average activation energy is E = 40 to 60 kcal mol^{-1}, and in the gas zone it reaches E = 50 to 70 kcal mol^{-1}. It should be remembered that an activation energy E = 50 kcal mol^{-1} corresponds approximately to reaction rates multiplied by 5 every 10°C and E = 65 kcal mol^{-1} to rates multiplied by 10, in the temperature ranges corresponding to the peak of oil generation and the gas zone, respectively. In that interval of maturation, over a temperature increase of 30°C, the TTI will account for a rate multiplied by $2^3 = 8$, whereas the actual rate will be multiplied by 125 to 1000, if the average E = 50 to 65 kcal mol^{-1}. More information about the role and significance of activation energies may be found in Sections V.4.2 and V.4.5. In particular, there are differences in the kinetics of degradation due to differences in chemical composition of the respective kerogen types. Thus it is difficult to account with a single parameter for hydrocarbon generation from the various types of source rocks.

4.2 Mathematical Model of Kerogen Degradation and Hydrocarbon Generation

A mathematical model of petroleum generation, accounting explicitly for geological time, was first introduced by Tissot (1969) and is fully discussed in Tissot and Espitalié (1975). The model is based on kinetics of kerogen degradation and uses the general scheme of evolution, described in Part II of this book. Kerogen is a macromolecule composed of polycondensed nuclei bearing alkyl chains and functional groups, the links between nuclei being heteroatomic bonds or carbon chains. As the burial depth and temperature increase, the bonds are successively broken, roughly in the order of increasing rupture energy. The products generated are first heavy heteroatomic compounds, carbon dioxide, and water; then progressively smaller molecules; and finally hydrocarbons. At the same time, the residual kerogen becomes progressively more aromatic and evolves towards a carbon residue. All these mechanisms have been described in Part II and are summarized in Figure V.4.1. The purpose of the mathematical model is to represent the kinetics of the parallel and successive reactions shown in this figure.

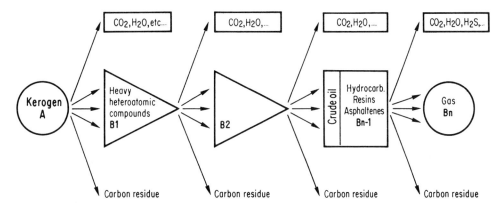

Fig. V.4.1. General scheme of kerogen degradation (Tissot and Espitalié, 1975)

Reactions are considered here as not reversible. In fact, when source rocks have been buried to a certain depth, then brought again to surface by subsequent folding and erosion, the organic content keeps the composition and physicochemical properties corresponding to the maximum burial depth. Furthermore, some of the by-products of kerogen evolution, such as water and carbon dioxide, are highly mobile in sub-surface conditions and could not be available for recombination, should it be the case.

For simulating the system of reactions, shown in Figure V.4.1, it is advisable to consider only a limited number of steps and particularly:

- A, B_1, B_{n-1}, which was the first version proposed by Tissot (1969),
- A, B_{n-1}, B_n, which appears from experience to be sufficient to account for the successive oil and gas formation (Tissot and Espitalié, 1975), and will be used here.

The last formulation corresponds to:

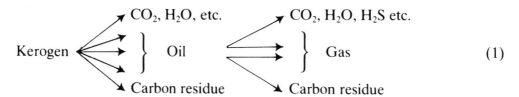

$$(1)$$

The following symbols will be used. A is kerogen, comprising n_i bonds of a given type i, at the time t; x_i is the amount of organic matter reacting in the ith reaction (breaking of i-type bond). B_{11} to B_{1m} are the products of the first step of reactions (formation of oil), their respective amount at time t being y_1 to y_m. B_{21} to B_{2n} are the products of the second step of reactions (cracking), their respective amount at time t being u_1 to u_n.

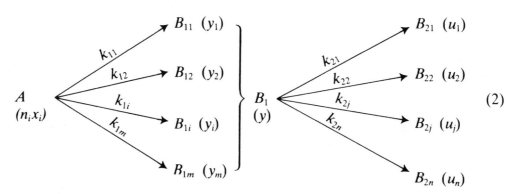

$$(2)$$

The first set of transformations represents a certain number of parallel and/or successive reactions. It is assumed that the probability of breaking a bond of type i is independent of the abundance of bonds of the other types and also independent of the time elapsed. Then the breaking of i-type bond obeys the Poisson law:

$$-\frac{dn_i}{n_i} = k_{1i}dt,$$

4.2 Mathematical Model of Kerogen Degradation and Hydrocarbon Generation

k_{1i} being a constant at a given temperature. If the structure is homogenous, i.e., the density of bonds is statistically the same in the kerogen volume, we have:

$$x_i = \mu n_i,$$

where μ is a constant, and, with regard to parallel reactions:

$$x_{i0} = x_0 P_i,$$

P_i being the frequency of i-type bond at the origin $t = 0$, and $x_0 = \Sigma x_{i0}$ being the total amount of labile — or pyrolyzable — organic matter. Then, it becomes:

$$-\frac{dx_i}{dt} = k_{1i} x_i.$$

Therefore, we obtain for each of the i reactions a kinetic equation similar to that of the first-order reactions. It has been shown that such formulation is able to account for kerogen degradation in geological conditions (Tissot, 1969).

If we make the additional hypothesis that the B_{1i} constituents of oil have the same general behavior in respect to cracking, we can consider for the second step of reactions $B_1 = \Sigma B_{1i}$. The respective amounts x_i, y_i and u_j can be obtained from the following system of differential equations [Eq. (3)]:

$$\left. \begin{aligned} -\frac{dx_i}{dt} &= k_{1i} x_i \\ \frac{du_j}{dt} &= k_{2j} y \\ y &= \sum_i y_i \\ \sum_i x_{i0} + \sum_i y_{i0} + \sum_j u_{j0} &= \sum_i x_i + \sum_i y_i + \sum_j u_j \end{aligned} \right\} \quad (3)$$

The first two equations express the kinetics of the system, whereas the last two express the mass balance. The variation of k_{1i} and k_{2j} with temperature may be described by using the Arrhenius formula:

$$k_{1i} = A_{1i} e^{-\frac{E_{1i}}{RT}}$$

which is basically valid for fast laboratory or industrial reactions, but may be extended to slow reactions occurring under geological situations (Tissot, 1969; Connan, 1974). E_{1i} is the activation energy of the ith reaction, A_{1i} is a constant and T is the absolute temperature (Kelvin). In geological situations, temperature is a function of time through subsidence and related burial. As burial is reconstructed by geologists on the basis of successive time intervals (e.g., Lower Cretaceous, Upper Cretaceous, Paleocene, etc.) depth is considered to be a linear function of time during each interval. Thus temperature also is approximated as a succession of linear functions of time during the successive time intervals. Under these conditions the system [Eq. (3)] cannot be easily integrated, as k_{1i} becomes a complex function of the time t. The easiest way of solving the system is by numerical integration, with successive Δt increments. This is done by using the computer, and the program includes the adjustment of the various parameters x_{i0}, E_{1i}, A_{1i} and y_0, which is in fact the calibration of the model to the particular type of organic matter.

For this calibration, only the first set of reactions (formation of oil) is considered. At the beginning, values measured on comparable recent sediments (e.g., Black Sea sediments for type-II kerogen) are used for y_0, and approximate values for A_{1i} (Tissot, 1969). In respect to the E_{1i}, it is necessary to consider all activation energies from a few kcal mol^{-1}, corresponding to the rupture of weak bonds, as in adsorption, to about 80 kcal mol^{-1}, corresponding to breaking carbon–carbon bonds. The objective of the adjustment should be to determine the frequency, P_i, of each i-type of bonds, or the amount x_{i0} of labile organic material involved in the ith reaction. The calibration is based on values of extractable organic matter (hydrocarbons + resins + asphaltenes) naturally present in cores taken at different depths in the sedimentary basin and/or on comparable figures resulting from laboratory experiments at higher temperature. The adjustment is based on the method of the least squares. The results concerning the three main types of kerogen I, II and III, are shown in Table V.4.1 and Figure V.4.2.

These figures are based on the data obtained from the source rocks of the Uinta, Paris and Douala basins which were originally used to define the three main types of kerogen. They offer a first approach to calculate the amount of oil

Table V.4.1. Distribution of activation energies and genetic potential of the principal kerogen types

Activation energies		Kerogen types					
Class	Average value (kcal mol^{-1})	Type I		Type II		Type III	
		x_{i0}	A	x_{i0}	A	x_{i0}	A
E_{11}	10	0.024	4.75 10^5	0.022	1.27 10^5	0.023	5.20 10^3
E_{12}	30	0.064	3.04 10^{16}	0.034	7.47 10^{16}	0.053	4.20 10^{16}
E_{13}	50	0.136	2.28 10^{25}	0.251	1.48 10^{27}	0.072	4.33 10^{25}
E_{14}	60	0.152	3.98 10^{30}	0.152	5.52 10^{29}	0.091	1.97 10^{32}
E_{15}	70	0.347	4.47 10^{32}	0.116	2.04 10^{35}	0.049	1.20 10^{33}
E_{16}	80	0.172	1.10 10^{34}	0.120	3.80 10^{35}	0.027	7.56 10^{31}
Genetic potential of kerogen $x_0 = \sum_i x_{i0}$		0.895		0.695		0.313	

A is expressed as 10^6 yr^{-1}

value of y_0
- Type I : 0.051
- Type II : 0.035
- Type III: 0.018

4.3 Genetic Potential of Source Rocks. Transformation Ratio

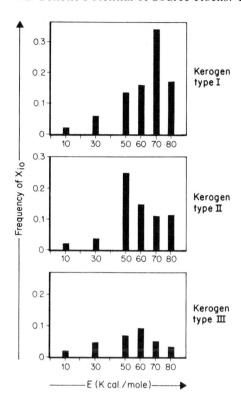

Fig. V.4.2. Distribution of the activation energies involved in the degradation of the three main types of kerogen (Tissot and Espitalié, 1975)

and gas generated from other source rocks of comparable type. However, due to the variability of kerogen composition, a specific adjustment based on laboratory assays, which provide a direct calibration of the model for each particular source rock, is generally preferable.

Calibration of the second set of reactions — formation of gas — has been made from laboratory experiments on cracking, including the results of McNab et al. (1952), and Johns and Shimoyama (1972). Consideration of a single reaction with an activation energy of 50 kcal mol^{-1} seems convenient to account for gas generation in the deep parts of the sedimentary basins.

4.3 Genetic Potential of Source Rocks. Transformation Ratio

The total amount of hydrocarbons which can be produced by a certain kerogen, provided it is heated to a sufficient temperature over a sufficient time, is the quantity

$$x_0 = \sum_i x_{i0}$$

that appears in Table V.4.1. This quantity is equivalent to the *genetic potential of the kerogen*, defined in Chapter II.7.2. The value depends on the type of kerogen, i.e., on its original chemical composition. A *source rock* could be defined as a rock whose genetic potential is above a threshold value, e. g., 0.25 or 0.30 related to the unit weight of kerogen, or (if we consider an average value of 1% organic carbon in rock) 0.25 to 0.30% related to unit weight of rock.

At any time, the stage of evolution is measured by using the *transformation ratio r*, defined in Chapter II.7.2, which is the ratio of the kerogen already transformed to the genetic potential:

$$r = \frac{\Sigma x_{i0} - \Sigma x_i}{\Sigma x_{i0}} = \frac{x_0 - \Sigma x_i}{x_0}.$$

The transformation ratio is zero at shallow depths and progressively increases to 1, which is reached when all labile organic material has been expelled leaving a carbonaceous residue.

4.4 Validity of the Model

In various sedimentary basins, comparison between the figures computed by using the model and the corresponding amounts of petroleum generated through burial shows excellent agreement, with a quadratic deviation lower than 10^{-2}, and a correlation coefficient better than 0.9. The model has been used to simulate experimental heating — either isothermal or with a regular rate of temperature increase — during various times from an hour to one year, again with satisfactory agreement. Furthermore, the same set of constants A_i and E_i, shown in Table V.4.1, is sufficient to account for all conditions of kerogen degradation (Tissot and Espitalié, 1975) including (a) natural evolution in sedimentary basins at relatively low temperature (50–150°C) over a period of 10 to 400 million years; (b) artificial evolution through laboratory experiments (180–250°C) (c) high-temperature (400–500°C) retorting of oil shales. The fact that a single model with the same set of constants is able to simulate such different situations is a confirmation of the validity of the hypothesis made on kinetics (statistics of bonds and activation energies, first-order reactions, etc.).

The timing of oil generation provided by the model may also in some cases be checked against geological data. An example from northern Sahara (Algeria) has been shown in Chapter II.7.6.

4.5 Significance of the Activation Energies in Relation to the Type of Organic Matter

The parameters E_{1i} used in the model are called activation energies for the sake of simplification. They have a role similar to that of activation energy, but they are not strictly activation energies, as the latter are normally defined in respect to a particular and single reaction. This is not the case in the model, where we consider the "formation of petroleum" from numerous parallel and/or successive reactions.

4.5 Significance of the Activation Energies

Many types of bonds are originally present in kerogen, with distinct rupture energies, and particularly:

- weak bonds corresponding to physical or chemical adsorption (hydrogen bonds, etc.),
- carbonyl and carboxyl bonds,
- ether and sulfur bonds,
- carbon–carbon bonds.

Furthermore, the rupture energy of most types of bonding may vary according to neighboring functional groups or substituents, length of chains, etc. Consequently, consideration of a distribution of the activation energies E_{1i} from 0 to 80 kcal mol^{-1}, as we did in Section 2.2, is probably closer to the effective mechanisms than a hypothetical measurement of each individual type of bond. Therefore, the best representation of kerogen composition may be the histogram of activation energies, derived from Table V.4.1, presented in Figure V.4.2 for type-I, -II and -III kerogens. With increasing burial and temperature (and decreasing $\frac{1}{T}$) the various bonds corresponding to the successive activation energies E_i are progressively broken, roughly in order of increasing E_i. This is suggested by the temperature dependence of the reaction parameters k_i in Figure V.4.3.

The genetic potential $x_0 = \Sigma x_{i0}$, and the distribution of the activation energies change according to the type of kerogen (Tissot and Espitalié, 1975):

a) Type-I kerogen contains a large proportion of labile organic material x_0, and thus its genetic potential is high. The distribution of the activation energies include few with low values corresponding to weak bonds. Most values are

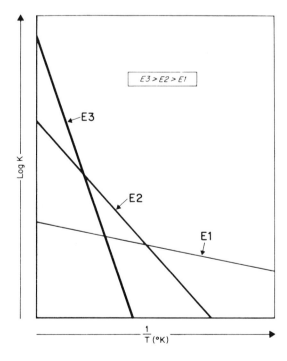

Fig. V.4.3. Temperature dependence of the reaction parameters k_i involved in kerogen degradation. The figure shows that the various bonds corresponding to the successive activation energies are progressively broken, in order of increasing activation energy (Tissot and Espitalié, 1975)

grouped around 70 kcal mol^{-1}, and may correspond to carbon–carbon bonds. Therefore, this particular kerogen requires higher temperatures for generating oil than the other types.

b) Type-II kerogen contains slightly less of the labile organic material resulting in a genetic potential slightly lower than in type I. However, the distribution of activation energies is wide and includes lower values than type I with a mode at 50 kcal mol^{-1}. Thus, the main formation of petroleum starts at somewhat lower temperature than in the previous type.

c) Type-III kerogen contains still less of the labile organic material than the others, and its genetic potential is low. As a result, the total amount of oil generated is comparatively small. The distribution of the activation energies is smooth with a maximum frequency at 60 kcal mol^{-1}.

A high concentration of organic matter related to type I or II results in a rich oil shale with a high oil yield. The Green River shales belong to type I and their values of x_0 is 0.8 to 0.9, i.e., 80 to 90% of the organic matter is able to be converted to oil. The Toarcian shales of Western Europe belong to type II, and the corresponding value of x_0 is 0.6, i.e., 60% of the organic matter can be converted to oil. Type-III organic matter, on the other hand, may be concentrated to form certain coals or carbonaceous shales, but with a low oil yield ($x_0 = 0.25$) and no commercial oil shales.

The relationship between the three main types of kerogen, as defined by their chemical properties, and the respective distribution of activation energies, may be used for a quick calibration of the model. If one knows the specific kerogen type (I, II or III) from optical examination or elemental analysis, the appropriate values of A_i and E_i from Table V.4.1 can be used in the model.

A review of the apparent activation energies proposed in the literature for petroleum generation, carbonization of coal, or oil-shale retorting shows that the values range from 8 to 65 kcal mol^{-1}. These apparent activation energies were computed on the basis of a single reaction, although it deals in fact with a set of successive and/or parallel reactions. As the true activation energies E_i, corresponding to the breaking of various bonds, are affected roughly in order of ascending values with increasing temperature, the apparent activation energy is close to the lowest E_i at low temperature and close to the highest E_i at high temperature. This behavior is confirmed by experimental data obtained from laboratory pyrolysis of oil shales by Weitkamp and Gutberlet (1968): they observed a progressive increase of the apparent activation energy from 20 to 60 kcal mol^{-1}, when the conversion rate of kerogen increased from 0 to 80%. This consideration is sufficient to explain the conclusion that apparent activation energies related to the beginning of oil formation are on the average 10–15 kcal mol^{-1} (Tissot, 1969; Connan, 1974) whereas apparent activation energies related to oil shale pyrolysis or carbonization of coal are about 50 to 65 kcal mol^{-1} (Abelson, 1963; Hanbaba and Jüntgen, 1969, etc.).

4.6 Application of the Mathematical Model to Petroleum Exploration

The mathematical model of kerogen degradation provides quantitative value of oil and gas generated as a function of time. Therefore, it is directly applicable for evaluation of the oil and gas potential of a sedimentary basin, and for determination of the timing of petroleum formation for comparison with the age of structural or stratigraphic traps.

The data requested for use of the model are presented in Figure V.4.4: (a) a direct calibration of the E_i distribution on laboratory assays, which in fact represents the chemical composition of the organic matter, or alternatively identification by optical or chemical methods of the kerogen type and subsequent use of the corresponding distribution shown in Table V.4.1; (b) the burial curve, i.e., the depth versus time relationship, from the time of source rock deposition to the present time, and (c) the geothermal gradient, for computation of the temperature versus time relationship. The gradient may vary with geological periods according to geotectonic conditions, and its determination will be discussed in the following paragraphs.

Provided such data are available, the amount of oil and gas per ton of rock generated in any place of the basin can be computed as a function of time (Fig. V.4.4). In order to cover the whole sedimentary basin, it is convenient to divide up the volume of the basin into a three-dimensional grid (as shown in the previous chapter, Fig. V.3.4) and to make the computation for each unit. The elementary units may be defined in different ways: in platform basins with gentle folding, a grid based on latitude and longitude is generally adequate; in mobile areas where folding is fairly strong, elementary units delineated by isobath curves are more nearly homogeneous in respect to depth and temperature history. The most

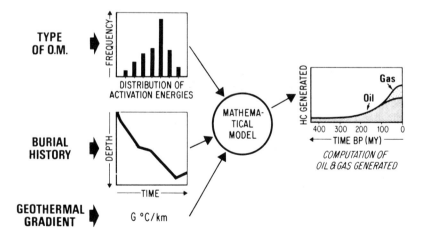

Fig. V.4.4. Principle of use of the kinetic model of oil and gas generation. The data provided are the type of organic matter (or the related distribution of activation energies), the burial history and the geothermal gradient. The model calculates the quantities of oil and gas generated, as a function of time (Tissot et al., 1980)

Geochemical Modeling: A Quantitative Approach to the Evaluation of Prospects

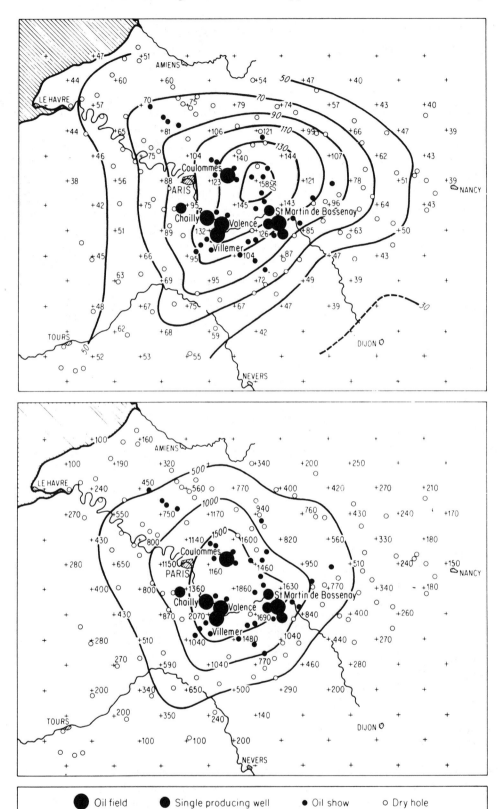

4.6 Application of the Mathematical Model to Petroleum Exploration

effective way to use this geochemical modeling is obviously to introduce this simulation into the basin geological modeling presented in the previous chapter.

The application of the mathematical model to oil generation in Jurassic source rocks of the Paris Basin is shown in Figure V.4.5. It can be expressed by the fraction of kerogen, which has been converted to petroleum, i.e., $r\,x_0$ in grams petroleum per kilogram of organic matter. Alternatively, if the distribution of organic carbon across the basin is also known, it can be related to the whole rock and expressed as grams petroleum per ton of rock. From the figure, it is clear that all known oil fields fall inside the area, where more than 110 g petroleum has been generated per kilogram of organic carbon, and inside the area, where more than 1500 g petroleum has been generated per ton of rock. Regarding the timing of oil generation, most petroleum has been formed in this basin during Cretaceous and Tertiary.

The example of oil and gas generation in Paleozoic source rocks of the Illizi basin, Algeria, is shown in Figure V.4.6 and deals with a more advanced stage of thermal evolution. Geological history of the basin includes two periods of sedimentation and subsidence separated by Hercynian folding and erosion. During Paleozoic subsidence, the Silurian and Devonian source rocks were buried to a great depth (about 3000 m) in the southwestern part of the basin and the kerogen was deeply degraded (transformation ratio $r \cong 1$) to generate gas. This gas subsequently migrated into anticlines formed by Hercynian folding. However, a deep post-Hercynian erosion reached the lower Paleozoic reservoirs and precluded conservation of gas. In the other parts of the basin, Paleozoic subsidence was moderate and only initiated oil generation. Folding and erosion were followed by an important Mesozoic subsidence, particularly in the northeastern part of the basin. The new burial exceeded the maximum pre-Hercynian depths in all places of the basin, except the southwestern part, where kerogen was definitely unable to evolve any more. In all other areas, kerogen degradation continued generating abundant oil and also gas in the northern and eastern parts of the basin where Cretaceous sedimentation was very thick. The amount of petroleum formed has been calculated, with consideration of the respective timing of generation. Thus, oil and gas potential may be described as: (a) low for both oil and gas in the southwest, due to erosion after generation; (b) good for oil in the central and eastern parts of the basin with some gas associated in the east; (c) good for gas in the northeast, with some oil associated in the less buried or colder areas.

The example of the Hassi Messaoud area in northern Sahara, Algeria, has been presented in Chapter II.7.6. There, purely geological considerations allow an independent confirmation of the timing of petroleum generation (Figs. II.7.5 and II.7.6). The mathematical model shows that the main phase of oil generation from Silurian source rocks is not reached until Cretaceous time, and extends

Fig. V.4.5. Application of the mathematical model to the generation of oil from Lower Jurassic source rocks of the Paris Basin. *Top:* the oil generated is expressed in g petroleum per kg organic carbon. *Bottom:* the oil generated is expressed in g petroleum per t rock (Deroo et al., 1969; Tissot and Pelet, 1971)

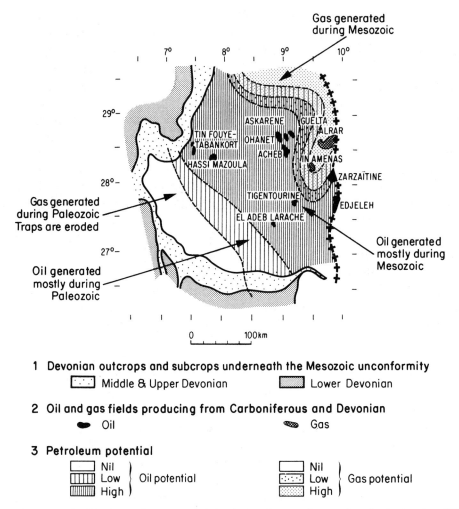

Fig. V.4.6. Hydrocarbon potential of the Lower Devonian beds in the Illizi Basin (Algeria), as determined by use of the mathematical model (Tissot and Espitalié, 1975)

mostly over Cretaceous and Tertiary. In this area all traps are equally prospective, whatever age they are, from Paleozoic to Cretaceous.

4.7 Reconstruction of the Ancient Geothermal Gradient

4.7.1 Heat Flow and Geothermal Gradient

The temperature increase with increasing depth results from a heat flow from the inside to the outside of the earth. The temperature T is a function of depth z and the geothermal gradient is $G = \dfrac{dT}{dz}$. At a given place of the earth's crust the heat flow Φ and the geothermal gradient G satisfy the relation:

$$G = \xi \cdot \Phi,$$

4.7 Reconstruction of the Ancient Geothermal Gradient

where $1/\xi$ is the thermal conductivity. G is expressed as the temperature increase per unit of depth, e. g., degrees Celsius km^{-1}. It may range from 10 to 80°C km^{-1}, with an average value of about 30°C km^{-1}. Φ is expressed in $\mu cal\ cm^{-2}\ s^{-1}$ (10^{-6} calories per square centimeter per second). It may range from less than 1 to 8 $\mu cal\ cm^{-2}\ s^{-1}$, with an average value approximately 1–2 $\mu cal\ cm^{-2}\ s^{-1}$.

Thermal conductivity varies with the composition of rocks. For instance, the values are quite high in salt (Fig. III.2.3). The variations cause a variability of geothermal gradient, whereas values of heat flow are less scattered. Table V.4.2 shows the average values of heat flow in continental and oceanic basins according to geotectonic conditions.

Heat flow is low in Precambrian shields (0.9 $\mu cal\ cm^{-2}\ s^{-1}$) and in areas where orogeny and/or magmatism are of Paleozoic age or older (1.2 to 1.3 $\mu cal\ cm^{-2}\ s^{-1}$). It is somewhat higher in areas where orogeny and/or magmatism is of Mesozoic or Tertiary age (1.9 $\mu cal\ cm^{-2}\ s^{-1}$); also in Tertiary volcanic areas (2.2 $\mu cal\ cm^{-2}\ s^{-1}$); and moreover in thermal areas where the flow may exceed 3 $\mu cal\ cm^{-2}\ s^{-1}$. Thus high values of heat flow and geothermal gradient are reported from synorogenic or postorogenic basins of the Alpine belt (Pannonian basin extending over Hungary, Yugoslavia and Romania) and of the circum-Pacific belt: Okhotsk and Japan seas, Sakhalin, North Sumatra, Fiji, etc.

Oceanic areas also show a large range of heat-flow variation, from oceanic ridges to oceanic basins and troughs. Furthermore, the heat flow is quite high along the axial zone of the ridges, i. e., along the crest and rift valleys, where values as high as 8 $\mu cal\ cm^{-2}\ s^{-1}$ are measured. It decreases progressively away from the crest. Within 150 km on both sides of the crest of the East Pacific ridge, half of the recorded values are still in the range of 3 to 8 $\mu cal\ cm^{-2}\ s^{-1}$; beyond that distance, the heat flow is always lower than 3 $\mu cal\ cm^{-2}\ s^{-1}$. A similar observation has been made along the mid-Atlantic ridge in regard to a 50-km zone on both sides of the crest. The present oceanic expansion being approximately 3 to 6 cm yr^{-1} in the East Pacific and 1 to 2 cm yr^{-1} in the Atlantic, the belt with a high heat flow corresponds approximately to an oceanic basement younger than 5 million years.

More precisely, Le Pichon and Langseth (1969) have computed the average values of heat flow as a function of distance to the axis of oceanic ridges (Table V.4.3). To measure the distance D they used, instead of kilometers, a unit corresponding to magnetic anomaly No. 5, which is definitely dated as slightly less than 10 million years. Thus, distances are in fact computed in terms of age of the oceanic crust, and oceans with different rates of expansion can be compared. From Table V.4.3, it is clear that high heat flow (2 or 3 times the average value in oceanic basins) are related to a belt with an age lower than 5 million years. Then the heat flow decreases with increasing age and is no longer distinct from the average oceanic basin where the age of the basement reaches 15 to 30 million years, depending on the area.

This view is confirmed by heat-flow values recorded in narrow elongated basins corresponding to Upper Tertiary and/or Recent expansion (in the Aden Gulf, the Red Sea, and the California Gulf; 3.9, 3.4 and 3.4 $\mu cal\ cm^{-2}\ s^{-1}$ respectively) and also in continental grabens of similar origin (Baikal Lake and Rhine-Graben), where flows as high as 2.5 and 4 $\mu cal\ cm^{-2}\ s^{-1}$, respectively, have been measured.

Table V.4.2. Average values of geothermal flux. (After Lee and Uyeda, 1965)

Geotectonic Provinces	Average geothermal flux (10^{-6} cal cm^{-2} s^{-1})	Number of measurements
Precambrian shields	0.92	26
Areas stable since Precambrian[a]	1.32	16
Areas with Paleozoic orogenies	1.23	21
Areas with Mesozoic or Cenozoic orogenies	1.92	19
Areas of Cenozoic volcanism	2.16	11
Oceanic basins	1.28	273
Oceanic ridges	1.82	338
Oceanic troughs	0.99	21

[a] Excluding Australia (Cenozoic volcanism).

Table V.4.3. Geothermal flux on oceanic ridges as a function of normalized distance to the axes of the ridge (Le Pichon and Langseth, 1969)

A. East Pacific ridge

Normalized distance (D)	Flux	
	Average value (μcal cm^{-2} s^{-1})	Standard deviation
$0 < D < 0.54$	3.31	1.94
$0.54 < D < 1.8$	2.00	1.29
$1.8 < D < 3.0$	1.47	1.03

B. Atlantic Indian ridge

Normalized distance (D)	Flux	
	Average value (μcal cm^{-2} s^{-1})	Standard deviation
$0 < D < 0.46$	2.72	2.33
$0.46 < D < 1.4$	1.45	0.94
$1.4 < D < 3.1$	1.10	1.14

4.7 Reconstruction of the Ancient Geothermal Gradient

4.7.2 General Rules for Reconstitution of Ancient Geothermal Gradient

From the observed distribution of geothermal data, some general rules may be deduced for the reconstitution of paleogeothermy, according to the geotectonic setting of the sedimentary basins.

4.7.2.1 Stable Platforms

Present and ancient geothermal gradients can be about the same in sedimentary basins where no tectonic or magmatic activity has been known since deposition of the potential source rocks. This normally occurs in areas which, after possible folding and magmatic intrusions, are incorporated into a stable platform. Then, subsequent sedimentation of potential source and reservoir rocks results in a new sedimentary basin, now generally flat-lying or gently folded. In this case, the present geothermal gradient is generally applicable to source rock evolution.

This situation occurs in undisturbed Paleozoic basins resting on folded or even flat-lying pre-Cambrian basement, such as several Paleozoic basins of Australia, and in the northern and eastern Sahara in Algeria; in Mesozoic undisturbed basins resting on Paleozoic or older folded basement, like most of the West Canadian basin, the Paris Basin, etc. In such cases the values of geothermal gradient are often 25 to 35°C km^{-1}.

4.7.2.2 Orogenic Basins

The similarity of past and present geothermal gradients is not acceptable in basins where the last orogeny and/or magmatic intrusion is contemporaneous with, or younger than, source rock deposition. This situation occurs in synorogenic or postorogenic intramontane basins such as the Pannonian basin in Central Europe, associated with the Carpathes alpine folding, or in Permo-Carboniferous basins associated with the Hercynian orogeny in Western Europe and Eastern Australia.

In these basins, geothermal gradient and flux may be high and quite variable with respect to location during the periods of orogenic activity or immediately following it. For instance, values as high as 50°C km^{-1} are now observed in several parts of the Pannonian Tertiary basin, where Pliocene source rocks have been brought to sufficiently high temperature for generating oil, despite the short period of time.

4.7.2.3 Stable Continental Margins

The basins of deposition located along the stable continental margins resulting from oceanic opening and expansion are worth a special consideration. From the data reported in Section 4.7.1 and Table V.4.3, it can be assumed that during the first 5 million years after creation of a new oceanic basement, the first sediments deposited on the new basement have been subjected to a heat flow two or three times higher than the average. Then the flow progressively decreases, to reach the average value after 15 or 20 million years. Under such circumstances, moderately buried source rocks may reach the principal zone of oil formation within 5 or 10 million years after deposition.

This situation occurs mainly at the beginning of ocean spreading, when continental masses are still close to the hot axial zone, and provide abundant material for a thick sedimentation over this zone followed by quick burial. In later stages of oceanic evolution, when the oceanic spreading has already moved the continents far away from the axial zone, this hot axial zone generally receives a reduced and slow sedimentation. Examples of thick sedimentation associated with a high heat flow may occur in Cretaceous coastal basins of West Africa and South America as a result of the opening of the South Atlantic Ocean; in Jurassic and Cretaceous basins resulting from the opening of the North Atlantic Ocean; and basins of Western and Northwestern Australia; and in Mio-pliocene sediments of the Red Sea, Aden Gulf, etc.

These considerations, deduced from the present distribution of heat flows, are in agreement with observations made on the kerogen of ancient sediments of the Aquitaine basin, whose opening is linked to the expansion of the Atlantic Ocean. Robert (1971) observes the reflectance of the organic matter as a function of depth in bore holes, and finds an abrupt increase correlated with a lower Cretaceous stage. Such a shift is explained by the influence of a considerable heat flow over the period of time associated with the opening of the Aquitaine basin.

4.7.2.4 Evolution of Heat Flow

The heat flow through undisturbed sedimentary basins is usually either constant or decreasing as a function of geological time. This is also true with respect to orogenic and magmatic activity: heat flows are high when this activity is young and decrease when it becomes older. After a sufficient time, it remains constant, as present values in Precambrian or Paleozoic folded basins are comparable. The same general scheme — high values, then decrease to reach a rather constant state — is, for other reasons, valid for stable margins related to oceanic opening, as discussed in Section 4.7.2.3.

However, the reverse may happen, and present geothermal gradients may be, in certain cases, higher than the ancient ones. In particular, continental grabens are associated with rather complex mechanisms, including possible rifting, rising of thermal waters, etc. In the Rhine-Graben, heat flows and geothermal gradients are generally high and variable with the location. They may reach peak values like $80°C\,km^{-1}$ in the Landau bore hole. However, the stage of kerogen degradation, as well as vitrinite reflectance, is not consistent with such a high value (Doebl et al., 1974). Mathematical simulation shows that geothermal gradient was moderate until late Tertiary, and then increased substantially up to the present value, probably as a result of faulting and rising of thermal waters.

4.7 Reconstruction of the Ancient Geothermal Gradient

4.7.3 Evaluation of the Present Geothermal Gradient

In many basins of stable platforms, the present value of geothermal gradient is a satisfactory evaluation of the paleo-gradient (Sect. 4.7.2.1). Determination of the present gradient requires a sufficient number of temperature measurements in bore holes. Electric, acoustic or nuclear logging devices are usually equipped with a maximum reading thermometer. However, the time between drilling and logging is usually restrained to a minimum, so that thermal equilibrium with the formation is not established, and the formation temperatures are generally underestimated, resulting in too low gradient values.

Better results are obtained if extended testing is made on reservoir beds containing oil or gas, followed by long pressure build-up. Then the equilibrium is reached, resulting in accurate formation-temperature measurements and computation of gradient.

4.7.4 Evaluation of the Ancient Geothermal Gradient

McKenzie (1978, 1981) studied the time-temperature history in sedimentary basins formed by extension, followed by subsidence. The stretching model calculates the subsidence and the temperature gradient, within a subsiding basin with compaction. The principal results may also apply to some other basins produced by certain thermal disturbances of lithosphere.

Reconstitution of ancient geothermal data may also be attempted by measuring physical or chemical properties of kerogen and simulating its evolution by use of a mathematical model until satisfactory adjustment is made with the measured value. This technique requires the knowledge of the burial versus time curve, based on geological reconstitution. It is best achieved by integrating this calculation into the geological modeling presented in the previous chapter.

The parameters which may be used include vitrinite reflectance, electron spin resonance of kerogen (ESR), hydrocarbon compositions, etc. Some authors have tried the direct use of such measurements as "maximum reading thermometers". The variations of these properties all result from degradation of kerogen, which is governed by kinetic laws, and thus is a function of both time and temperature. Thus, it is not possible to relate any of these properties of kerogen to maximum temperature only, as they result from the whole thermal history. Furthermore, the different parameters may be linked to different aspects of the evolution of kerogen structure, and thus time and temperature may influence the various indices in different ways.

Among the physical properties of kerogen, which can be used for reconstitution of thermal history, one of them, vitrinite reflectance, is discussed here.

The qualitative aspects of the increase of vitrinite reflectance with burial depth are discussed in Chapter V.1. A model of vitrinite degradation has been developed on the basis of a sequence of Logbaba shales in the Douala basin (Cameroon). There, the organic matter consists mainly of vitrinite and of amorphous organic cement with the same physical properties (elemental composition, IR spectra, reflectance etc.). Thus it has been possible to calibrate a model

Geochemical Modeling: A Quantitative Approach to the Evaluation of Prospects

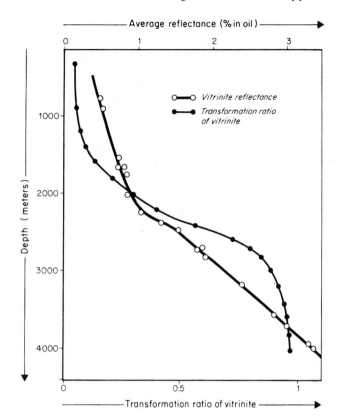

Fig. V.4.7. Changes of the transformation ratio of vitrinite and of its reflectance in the Logbaba wells, Douala Basin (Tissot and Espitalié, 1975)

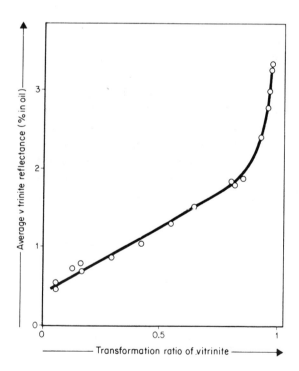

Fig. V.4.8. Relationship between the transformation ratio of vitrinite and its reflectance (Tissot and Espitalié, 1975)

4.7 Reconstruction of the Ancient Geothermal Gradient

(adjusted along the methods outlined in Sect. 4.2) to simulate the degradation of the organic matter of vitrinite type. Furthermore, it is possible to establish a relationship between reflectance and transformation ratio of vitrinite, by measuring both parameters on the same samples. Figure V.4.7 shows both parameters as a function of depth in the sedimentary sequence of the Douala basin, and Figure V.4.8 shows the relationship between reflectance and degradation of vitrinite.

Thus, it becomes possible to make a quantitative interpretation, in terms of thermal history, of a well located in any basin, provided the rocks contain some vitrinite. The interpretation is based on measurements made at different depths in a bore hole, and reconstitution of burial for each sample, made on purely geological information. Each reflectance measurement is converted into the corresponding transformation ratio, according to Figure V.4.8. Independently, the model of vitrinite evolution is used to calculate a transformation ratio, based on the burial history, as a function of geothermal gradient G. Then, an adjustment is made on variable G, until minimizing the quadratic deviation between the computed values of the transformation ratio and the values deduced from observations through Figure V.4.8. This procedure can be used either to adjust a simple law expressing G as a function of time, or to calculate an average value of G which is the best evaluation of the past geothermal gradient over the period of maximum burial.

This method has been used with success in several types of basins, with ancient gradients equal, higher or lower than the present one (Table V.4.4). In the Rhine Graben, the calculated values are in agreement with the observations made by Doebl et al. (1974) that the present geothermal gradient could not have lasted for

Table V.4.4. Present and ancient geothermal gradients

Basin	Well	Geothermal gradients ($°C\ km^{-1}$)		Period of time
		presently	computed	
Paris Basin (France)	Essises	33	31.0	From middle Jurassic
Aquitaine Basin (France)	LA 104	31	33.0	From middle Cretaceous
	PTS I	33	35.4	
	RSE I	31	32.0	
	Nat I	26	25.4	
Illizi Basin (Algeria)	Irarraren	33	32.9	Paleozoic
	Oudoume	25	26.4	Paleozoic to Cretaceous
Rhine-Graben (W. Germany)	Landau 2	77	58	Tertiary
	Sandhausen 1	41	30	Tertiary
Douala Basin (Cameroon)	Logbaba	32	47	Cretaceous Paleocene

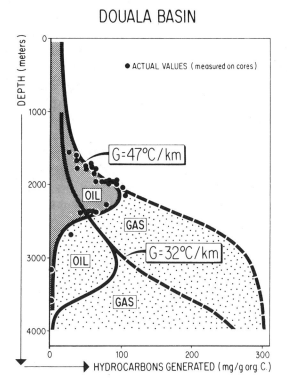

Fig. V.4.9. Comparison of the calculated amount of oil and gas generated in the Douala Basin, by using two different values of the geothermal gradient G. Calculation with the present gradient $G = 32°C$ km^{-1} differ markedly from the observed values, whereas calculation with the computed past gradient $G = 47°C$ km^{-1} shows a good agreement with actual values measured on cores

a very long time. The high values presently observed are probably a result of rising of thermal waters through a fault system. This interpretation would also explain the variability of the observed geothermal gradient over relatively short horizontal distances.

In the Douala Basin, the ancient geothermal gradient was rather high, probably due to the vicinity of the midoceanic ridge in the early stage of the basin. During the first 5 million years heat flow and geothermal gradient were probably well over average, progressively decreasing to reach a standard value after ca. 20 million years. Figure V.4.9 shows the amount of oil and gas generated at Logbaba, computed by the model of hydrocarbon generation using the burial versus depth curve and two different values of the geothermal gradient G: the present $G = 32°C$ km^{-1} and the computed past value $G = 47°C$ km^{-1}. It is obvious that the observed quantities of oil generated (measured on core material) are in agreement with the simulation using the calculated value of average past geothermal gradient ($G = 47°C$ km^{-1}) and differ markedly from the quantities based on the present gradient ($G = 32°C$ km^{-1}).

4.8 Migration Modeling

The other major aspect of geochemical modeling is migration from source rocks to reservoirs and traps, and possibly to the surface. Although the mechanisms of migration are less completely understood than those of generation of hydrocarbons, a first attempt to simulate shale compaction and water expulsion by using a

4.8 Migration Modeling

numerical model was made by Smith (1971). More recently, Ungerer et al. (1983) and Durand et al. (1983) presented an integrated model of oil generation and migration, where the generation part is the kinetic model described above. The migration part is based on the conclusion, expressed in Part III, that the principal mode of fluid migration in sedimentary basins is a polyphasic mode with separate hydrocarbon and water phases.

The model presented by Ungerer et al. (1983) and Durand et al. (1983) is diphasic: fluid movements are governed by the Darcy's law applied to polyphasic flows with the help of the relative permeability concept. The main driving force is compaction due to the sedimentary load, which is supported partly by the solid mineral phase and partly by the pore fluids, water and hydrocarbons, thus increasing internal pressure. Another cause of pressure rise in source rocks is the conversion of solid kerogen into fluid hydrocarbons with a related increase of the average specific volume of the organic phases. Capillary pressures, and the change of oil viscosity with temperature are also accounted for. As presented by these authors, the model is two-dimensional, i.e., it works along a geological section. This section is represented by a grid of finite elements, each element having a specific lithology and organic content. The model calculates for each element and each interval of time: porosity, hydrocarbon saturation, pore fluid pressure, and velocity of fluid flows.

Durand et al. (1983) applied this scheme to part of the Viking Graben in the North Sea, and to the Mahakam delta in eastern Kalimantan, Indonesia. The results from the Viking Graben are presented in Figures V.4.10 and V.4.11. The section represented is bordered by two major faults, on both eastern and western sides, and comprises a tilted block of Jurassic beds underneath the major unconformity. The source rocks belong to Kimmeridge Clay, Heather and Dunlin formations, and also to Brent coals, all of Jurassic age. Sand reservoirs belong to Brent and Statfjord formations. The two major faults bordering the section are considered as impermeable barriers to fluid flow. The oil saturations are presented in Figure V.4.10 at 65 m.y.b.p. (Maestrichtian), when an accumulation is beginning to form in the Upper Brent sands, and also at the present time, when accumulations are located in Jurassic sandstones on the structural high. In source rocks, high saturations are observed in the syncline. It is noted that the onset of oil expulsion from source rocks occurs beyond 20% oil saturation. The model also calculates the excess pressure versus hydrostatic pressure, shown in Figure V.4.11 for the same intervals of time. Important excess pressure is predicted beyond 3000 m depth, and this has been effectively observed in wells. At a given depth, this excess pressure is more important on the structural high, but the maximum values are computed where burial of the Heather formation is the deepest.

The same model used in a very different situation, i.e., the Mahakam delta in Indonesia, has been able to calculate the location of the oil pools, the hydrocarbon reserves in place and the distribution of abnormal pressures, all values being in agreement with observed data. The fact that the model is able to account for very different geological histories, such as the North Sea and the Mahakam delta, suggests that the hypotheses and approximations used in the model are reasonably satisfactory.

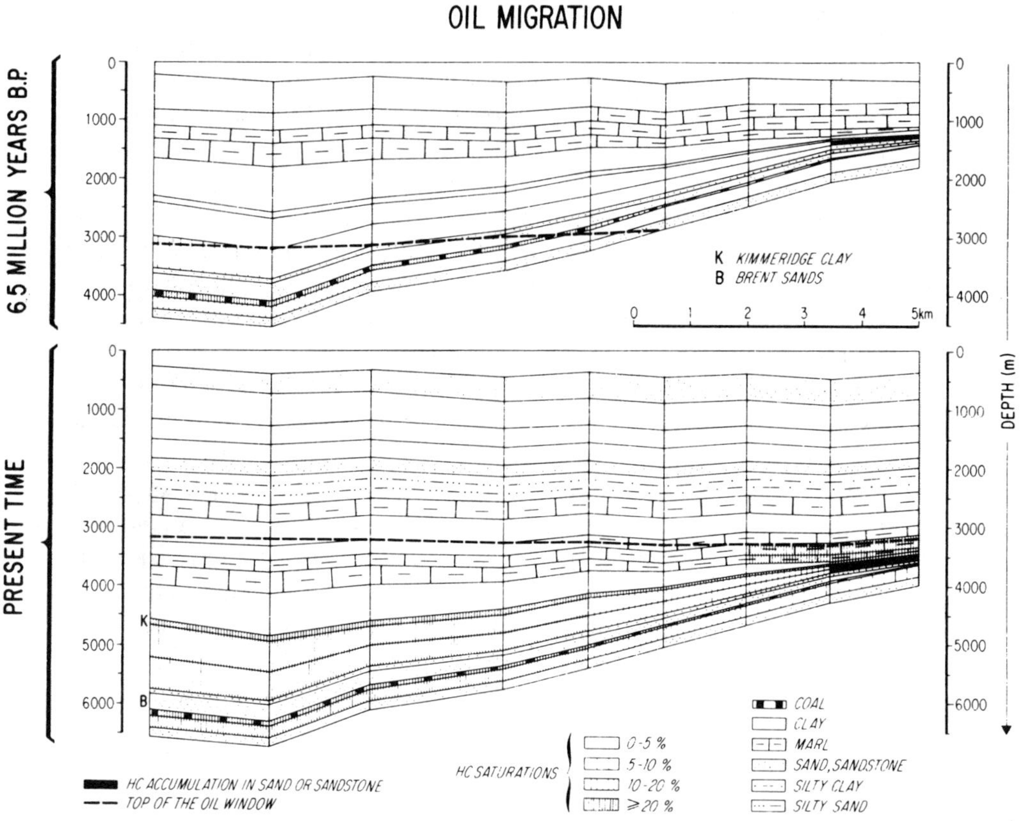

Fig. V.4.10. Hydrocarbons generation and migration modeling in part of the Viking Graben, North Sea. The section is bordered by two major faults on the eastern *(right)* and western *(left)* sides, and comprises a tilted block of Jurassic beds underneath the major unconformity. The model calculates oil saturation in all beds for each interval of time: *(top)* situation at 65 m.y.b.p. (Maestrichtian) which is the beginning of oil expulsion; *(bottom)* situation at the present time, showing accumulation in Jurassic sands (Durand et al., 1983)

Polyphasic models of oil migration certainly represent a significant advance, as most fundamental studies have now concluded that this is the principal mode of hydrocarbon migration. There are, however, specific difficulties inherent to this type of model. Darcy's law is widely used in reservoir engineering, but the conditions in source rocks are very different with respect to heterogeneity, and also extension of the relative permeability concept. Thus, the results presented here should be considered as a first approach. They have shown that, in certain geological situations, the model is able to generate data in agreement with observed facts: location of petroleum accumulations, distribution of pressure including overpressured zones. A present restriction is the two-dimensional character which is acceptable in grabens, such as the Viking Graben, or elongated basins. This should soon be overcome by the increasing availability of high speed computers, which will make the use of three-dimensional models possible.

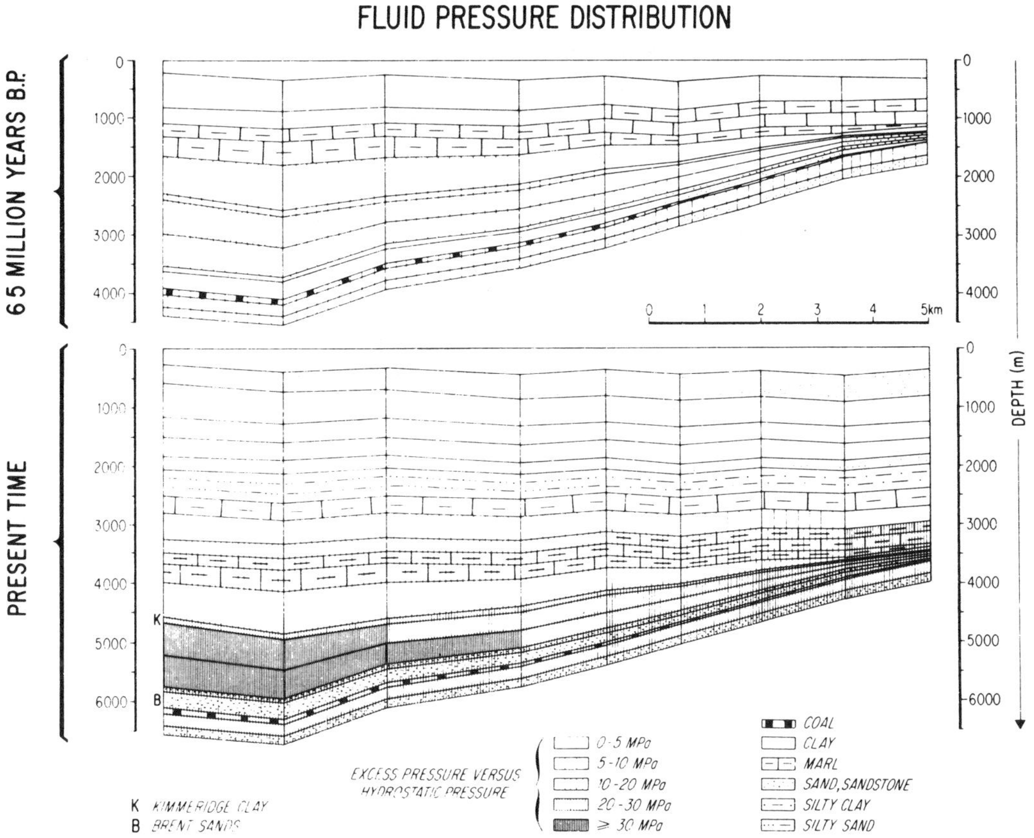

4.9 Conclusion

FLUID PRESSURE DISTRIBUTION

Fig. V.4.11. Excess pressure in part of the Viking Graben. The section is the same as shown in Figure V.4.10. The model calculates the excess pressure versus hydrostatic pressure for the same intervals of time (65 m.y.b.p. and 0 m.y.) (Durand et al., 1983)

4.9 Conclusion

It is now possible to simulate thermal degradation of the main types of kerogen by using a mathematical model. The model is calibrated against each type of organic matter by using a certain distribution of activation energies. A single model is able to simulate the whole thermal history from low temperatures in sedimentary basins to high temperatures in the oil shale industry.

Although migration mechanisms are less completely understood than kinetics of kerogen degradation, it is already possible to simulate migration and calculate hydrocarbon saturation and fluid pressure in each individual bed, as a function of time and location in the basin.

It is realized that modeling of geological processes is still in an initial phase and will rapidly develop, especially with respect to truly three-dimensional simulation procedures. However, some present and future applications of modeling can be listed as follows.

a) determination of prospective areas for oil and gas in sedimentary basins;
b) timing of oil and gas generation and migration, for comparison with the age of formation of traps and impermeable seals;
c) evaluation of the ultimate reserves of a sedimentary basin;
d) distribution of abnormal pressures; history of water flow in the basin;
e) reconstitution of ancient geothermals gradients and thermal history of sediments; this application is not limited to petroleum exploration but concerns geology in general;
f) simulation of retorting conditions of oil shales.

The number of geological/geochemical models used in petroleum exploration is certainly growing fast. Chenet et al. (1983) and Ungerer et al. (1983) discussed the influence of the various subsidiary models (subsidence and compaction, thermal history, hydrocarbon generation, fluid movements, thermodynamics of phase separation and exchange) in a comprehensive model of basin evolution. Bishop et al. (1983) estimated successively the quantities of oil and gas provided by the source rock within the drainage area, the quantities of gas lost by diffusion and dissolution and the trap volume. Then, by comparing these values, they made a probabilistic estimation of trapped hydrocarbon volume and nature (only oil, oil and gas, or only gas). A different type of model, mainly on a statistical, probabilistic basis was presented by Nederlof (1980) and Sluijk and Nederlof (1983). It is designed to calculate the chances that a given trap contains more than a certain volume of oil and/or gas. It draws from a world-wide collection of geological and geochemical data which form the statistical basis and calibration set for a prospect appraisal.

Summary and Conclusion

Geochemical models of petroleum generation offer a quantitative approach to exploration problems. They provide the amount of oil and gas generated (as kg per ton of rock) from the different source rocks in the various parts of the basin. Furthermore, these results are expressed as a function of time, and the timing of petroleum generation can be compared with the age of traps.

The data required as input to the model are the type of kerogen, the burial history (depth versus time curve) and the geothermal gradient. On stable platforms, the ancient gradient can frequently be approximated by the present gradient. In other basins, the ancient geothermal gradient can be calculated by using a model based on vitrinite reflectance.

Migration models provide information on hydrocarbon and water movement. They define the areas and sedimentary beds of hydrocarbon accumulation, and the occurrence of abnormal pressures. They also offer a clue to the problem of ultimate reserves of a sedimentary basin.

Recent Advances in Petroleum Geochemistry Applied to Hydrocarbon Exploration[1]

B. P. TISSOT[2]

ABSTRACT

This paper presents a critical review of recent advances in the study of generation, maturation, and migration of petroleum. The different source potentials for generating hydrocarbons depend on the type of kerogen. In addition to the classic types I, II, and III, a residual, oxidized organic matter is also observed, which has no potential for hydrocarbon generation. Coal is widely considered as a potential source for gas, but some coals may generate sizeable amounts of oil as well. The development of routine analytical tools, particularly pyrolysis, makes possible the preparation of geochemical logs that are the basis for screening samples and for interpreting other analyses. Among optical techniques, the major subjects of discussion concern the nature of amorphous organic matter, which may encompass different chemical types of kerogen, and the applicability of vitrinite reflectance techniques to type I and II kerogens. Following important advances in identification of biological markers in sediments and crude oils, these markers are used for oil-source rock correlation, and proposed for reconstruction of depositional environment and subsequent thermal evolution. Migration of hydrocarbons is now better understood, and the importance of a hydrocarbon-phase migration is widely recognized. Overpressuring of pore fluids is mainly responsible for expulsion of petroleum. Geologic models of basin evolution (subsidence, compaction, thermal history) and geochemical models (hydrocarbon generation and migration) become progressively available. Their integration into a comprehensive model should be one of the major developments in future petroleum exploration.

INTRODUCTION

The emergence of a general scheme of origin and migration of petroleum during the late 1960s and 1970s has now resulted in practical concepts which can be integrated in any scientifically based exploration program (Tissot and Welte, 1978; Hunt, 1979). These concepts use markers or indices which can be routinely determined in the laboratory or even at the well site on a large number of samples from exploration wells. In addition, sophisticated equipment, such as gas chromatograph–mass spectrometers, have become widely available and allow a systematic determination of specific parameters in the laboratory. Based on these concepts and using these tools, organic geochemistry today plays an important role in the strategy of petroleum exploration, as well as in the day-to-day monitoring of exploration drilling.

Among the recent advances in organic geochemistry, various subjects have had a major impact on exploration. A list of these may include the following (they are not listed in order of importance but cover successively aspects of generation, maturation, migration, alteration, and geologic distribution of petroleum): (1) kerogen composition and its role in petroleum generation; (2) coal as a potential oil source rock; (3) development of routine tools (pyrolysis, optical examination) to characterize the hydrocarbon potential of source rocks; (4) identification of biological markers (geochemical fossils) and their use in understanding petroleum source material, diagenetic and catagenetic evolution, and in oil–source rock correlation; (5) primary migration of petroleum, and the importance of hydrocarbon-phase migration; (6) the composition and properties of heavy oils, their geologic/geochemical history, and the nature and importance of the heavy end of crude oil (resins and asphaltenes); (7) basin-scale geologic and geochemical modeling which allows a quantitative approach to the generation, migration, and entrapment of hydrocarbons; and (8) integration of geologic/geochemical data on world distribution and composition of hydrocarbons into the framework of global geologic history.

These various aspects of petroleum geochemistry and geology are reviewed successively. The problem of gas generation is not dealt with in this paper.

KEROGEN AND ITS ROLE IN HYDROCARBON GENERATION

The classification of kerogen, as defined by elemental analysis and evolution paths on the Van Krevelen diagram (Figure 1), is still the most reliable method for characterizing the nature and petroleum potential of organic matter. It also offers a framework for the interpretation of other physical, chemical, or optical methods. Type I kerogen, mostly deposited in lacustrine environments, and type II kerogen, generally related to marine reducing environments, are derived from planktonic and other material more or less extensively reworked by microorganisms living in the sediment (Demaison and Moore, 1980). They offer a high generation potential for oil and gas, depending on the stage of thermal evolution. Type III kerogen is

---

© Copyright 1984. The American Association of Petroleum Geologists. All rights reserved.
[1]Manuscript received, April 7, 1983; accepted, October 25, 1983.
[2]Institut Français du Pétrole, BP 311, 92506 Rueil-Malmaison, France.
I thank G. Demaison, W. G. Dow, B. Durand, and R. Pelet for discussion and comments in preparing this review, and I am greatly indebted to W. G. Dow for help in editing the paper.

mainly derived from terrestrial plants transported to a marine or nonmarine environment of deposition, and which undergo only a moderate level of degradation before burial. Hydrogen content and consequently hydrocarbon source potential are comparatively lower in type III than in the other types. Coal is a special variety of this terrestrially derived organic matter, and is discussed in the next section.

In addition to these basic types, intermediate kerogens are common, particularly between types II and III on the Van Krevelen diagram (e.g., some middle Cretaceous black shales cored by the DSDP program in the North Atlantic; Tissot et al, 1980). Intermediate elemental composition may result from a mixture of marine and terrestrially derived organic matter, although partial alteration contemporaneous with sedimentation may also play a role.

A fourth basic type of kerogen is residual organic matter (Figure 1), which is characterized by an abnormally high atomic O/C ratio (0.25 or more), associated with a low atomic H/C ratio (0.5 or 0.6). This material (sometimes called type IV kerogen) either may be recycled from older sediments by erosion, or deeply altered by subaerial weathering, combustion, or biological oxidation in swamps and soils prior to redeposition. Examples have been reported in Cretaceous shales of the northeastern Atlantic (DSDP Leg 48) (Tissot et al, 1979), and also in some recent peats where high-reflectance particles are observed. Type IV kerogen is a form of "dead carbon" and has no potential for hydrocarbon generation. Another form of dead carbon is material that has lost its hydrogen through the thermal-maturation process. In such cases both atomic H/C and O/C ratios are very low.

Many experimental simulations of kerogen evolution by heating a source rock or an isolated kerogen have been performed in the laboratory (Ishiwatari et al, 1976, 1978; Harwood, 1977; Peters, 1978; Lewan et al, 1979; Larter and Douglas, 1980; Monin et al, 1980; Peters et al, 1981; Tissot and Vandenbroucke, 1983; Winters et al, 1983; a bibliography on the subject was prepared by Barker, 1978). The compositional changes of kerogen in the laboratory compare well with those observed along natural evolution paths provided the original organic matter did not contain a large proportion of oxygen (i.e., the sample belongs to type I or II and the diagenesis stage is almost completed). Type III kerogen and coal, especially from the diagenesis stage, show a different behavior; their artificial evolution does not follow the natural evolution path and presents a shift due to higher oxygen content. A possible interpretation is the preferential elimination of oxygen as water (a hydrogen-consuming process) during experiments instead of a preferential elimination as carbon dioxide in the geologic situation (Monin et al, 1980; Tissot and Vandenbroucke, 1983).

Two important aspects of the influence of water and the mineral matrix on the nature of hydrocarbons generated in pyrolysis experiments were investigated by heating kerogen in conditions closer to subsurface conditions. Lewan et al (1979) and Winters et al (1983) developed a "hydrous pyrolysis" technique. They observed that, in the presence of water, no olefins were generated and the general composition of the hydrocarbon fraction formed was comparable to compositions observed in geologic situations. Espitalié et al (1980) investigated the role of the mineral matrix. During pyrolysis, some minerals, such as illite, are responsible for the retention of heavy hydrocarbons (beyond C_{15}) that are subsequently cracked. Thus, the effluent is richer in light hydrocarbons, compared to pyrolysis of pure kerogen without a mineral fraction. Horsfield and Douglas (1980) obtained comparable results on various coal macerals. These observations obviously have a direct consequence for the interpretation of source rock analysis by pyrolytic methods. However, the possible implications for generation and migration of petroleum in geologic conditions are not easily inferred.

An important feature observed in the effluents of kerogen pyrolysis experiments is the occurrence of recognizable biological markers (geochemical fossils) previously linked by chemical bonds to the kerogen network. Seifert (1978) used this procedure to correlate crude oils with the pyrolysate of their respective source rocks. A comparable observation was made on asphaltenes, the heavy constituents of crude oils. Rubinstein et al (1979)

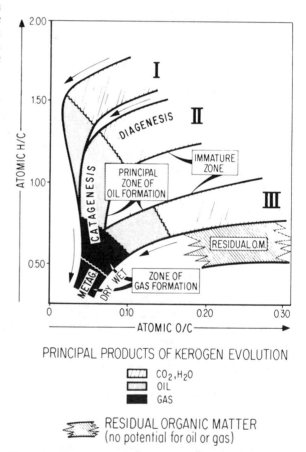

Figure 1—Van Krevelen diagram showing types of kerogen, and principal products generated during thermal evolution of types I, II, and III kerogen. Residual organic matter has no potential for petroleum and no real evolution path (Tissot et al, 1980).

showed that the products from pyrolysis of Prudhoe Bay asphaltenes are comparable to the hydrocarbon constituents of the crude oil. Aref'yev et al (1980) made a similar observation on biodegraded crude oils from USSR, and showed that pyrolysis of the asphaltene fraction can reproduce the degraded paraffinic fraction, which is no longer present. Chappe (1982) submitted kerogens and asphaltenes to a selective cleavage of ether bonds by chemical reactives. Both materials yielded comparable biological markers of the isoprenoid hydrocarbon class, showing that the same type of structures are present in both kerogen and asphaltenes.

Based on these data, asphaltenes can be thought of as small fragments of kerogen, having a comparable structure. Thus kerogen, asphaltenes, resins, and hydrocarbons might be considered to form a continuum, with decreasing size, heteroatoms content, and polarity.

COAL AS SOURCE ROCK

Durand and Paratte (1983) discussed the relationship between type III kerogen and coal, and the possible role of coal as a source rock for oil. Coal is now widely recognized as an important source for gas, as exemplified by the numerous gas fields of northwestern Europe (North Sea, Netherlands, West Germany) (Lutz et al, 1975; Tissot and Bessereau, 1982). The study of related gases is well documented by carbon and hydrogen isotope studies (Stahl, 1978; Schoell, 1980).

The similarity of chemical composition between the upper Tertiary coals of the Mahakam delta (Kalimantan, Indonesia) and the disseminated kerogen present in the interbedded shales was shown by Durand and Oudin (1979) and Vandenbroucke et al (1983) (Figure 2). They proved that both the oil and gas of Handil and neighboring fields are derived from these deep-seated coals and shales. The close association of waxy oils and coal previously was recognized in the Officina area of Venezuela and the Midland coal measures in Great Britain (Banks, 1959; Hedberg, 1968).

Certain types of coal, therefore, are considered to be oil source rocks, provided the coals have attained the adequate rank or maturation. Their potential, expressed in kg/MT of organic material, is comparable to that of disseminated type III kerogen (i.e., 3 or 4 times less than that of type I or II kerogen). To some extent, the abundance of organic matter in coal can compensate for that comparatively low potential. This consideration, however, does not apply to all types of coal. For example, European coals of Westphalian age yielded huge amounts of gas in the North Sea, Netherlands, and northern Germany, but no significant oil accumulation, even in limited areas where vitrinite reflectivity of 1% or less was measured by Teichmüller et al (1979) or Robert (1980). In East Midlands, however, the

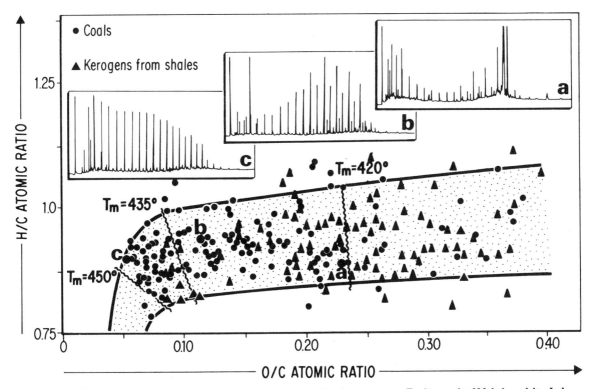

Figure 2—Van Krevelen diagram showing identity of chemical composition between upper Tertiary coals of Mahakam delta, Indonesia, and disseminated kerogen present in interbedded shales. Three insets (a, b, c) present chromatograms of hydrocarbons extracted from corresponding samples (adapted from Vandenbroucke et al, 1983, with additions by Vandenbroucke et al).

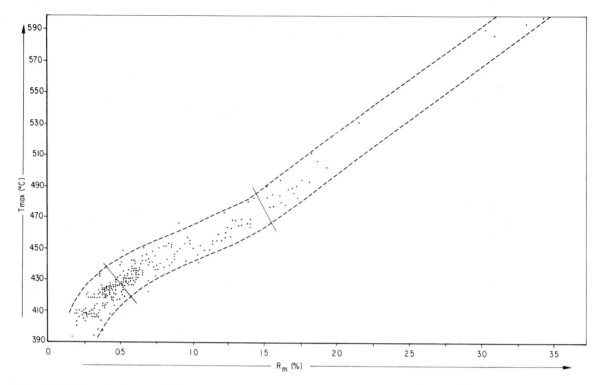

Figure 3—Relationship between vitrinite reflectance (R_m%) and temperature index T_{max} from Rock-Eval pyrolysis for coals and type III kerogens (Teichmüller and Durand, 1983).

coal measures are the source for several small fields of high-wax oil. In Australia, both situations seem to be represented, with Mesozoic and Tertiary coals generating oil, and Permian coals yielding mainly gas (Thomas, 1981). An explanation for the different generation potential of coals might be the different proportion of hydrogen-rich organic matter (e.g., resinite and other macerals of the liptinite/exinite group) included in coal particles (Snowdon, 1980; Thomas, 1981; Powell et al, 1982), and perhaps also their degree of alteration prior to or during sedimentation.

METHODS FOR CHARACTERIZING KEROGEN AND RELATED SOURCE ROCK POTENTIAL

The most reliable methods for characterization of source rocks are based on kerogen, which is solid matter and therefore definitely in place in the rock sample. The bitumen fraction (extractable organic matter) is often affected by fluid displacements (drainage or accumulation) or by contamination during drilling or handling the sample. There are, however, some interesting methods recently proposed using bitumen composition, which are reviewed here.

Chemical and physicochemical methods using elemental analysis, infrared and other spectroscopic analyses, and electron microscopy of kerogen remain the basis for characterization of this insoluble organic material at the level of fundamental research. Recently, new possibilities of applying nuclear magnetic resonance (NMR) to kerogen studies have been offered by the Cross Polarization/Magic Angle Spinning (CP/MAS) ^{13}C NMR operating on the solid phase. Dennis et al (1982) applied it to kerogen from Cretaceous black shales of the Atlantic, whereas Miknis et al (1982) evaluated the genetic potential of various oil shales.

Pyrolysis techniques performed directly on a large number of rock samples, however, have made a major contribution to systematic analysis of kerogen. In particular, the Rock-Eval pyrolyzer (Espitalié et al, 1977; Clementz et al, 1979) provides a fast determination of the type and evolution stage of kerogen, together with a direct evaluation of the hydrocarbon source potential. The type and quality of the kerogen are usually interpreted on a graph derived from the traditional Van Krevelen diagram by replacing H/C and O/C ratios with the hydrogen (HI) and oxygen (OI) indices. Maturation stage is obtained from T_{max}, the temperature corresponding to the peak of kerogen pyrolysis. This parameter is measured on the bulk of the kerogen, and has proved to be a reliable method for characterizing thermal evolution (Figure 3). It is particularly valuable in the case of planktonic kerogen (type I or II) where vitrinite is often scarce or absent. Alternatively, a composite HI-T_{max} diagram allows a visualization of the type and maturation of kerogen.

Recently, Orr (1983) made an evaluation of Rock-Eval-type pyrolysis with respect to qualitative and quantitative significance of hydrocarbon yields. He showed that the carbon in the evolved products (which in turn determines the hydrogen index) usually agrees with the value calculated from elemental analysis within the experimental error (i.e., ±5-10%). He also observed that if an independent maturity parameter (e.g., vitrinite reflectance) is known, elemental composition (H/C and O/C ratios) can be inferred within a narrow range. Comparison of pyrolysis of whole-rock and isolated kerogens shows that results may be affected by 2 factors in whole-rock pyrolysis. The most important is the retention effect of the mineral matrix (first reported by Espitalié et al, 1980) when the organic content is low. In such situations, it is advisable to use pyrolysis as a screening technique and to verify the data with kerogen elemental analysis of a small number of samples. The second factor is the contribution of heavy extractable bitumen (mainly asphaltene) to the pyrolysis peak. This effect was first reported by Clementz (1979), and is significant only when the rock contains a large amount of heavy bitumen. Solvent extraction can eliminate that contribution. Furthermore, both pyrolytic hydrocarbons and total organic carbon are increased by that effect, with the consequence that the hydrogen index is little affected.

Being rapid and inexpensive, pyrolytic analyses may be performed in either the laboratory or at the well site, at regular depth intervals, and the results plotted in the form of a geochemical log showing hydrocarbon source potential and maturation. This log can also be used as a screening technique for selection of the samples submitted to more sophisticated analyses. The log provides excellent geologic/geochemical correlations between different wells located in the same area, and it also offers a specific approach to the identification of organic facies changes across a basin. Alternative versions of rock pyrolysis allow trapping and subsequent gas chromatographic analysis of either free hydrocarbons or hydrocarbons generated by pyrolysis or both.

Optical techniques are also widely used to characterize the type and evolution stage of the organic matter. Studies in transmitted light resulted in the definition of palynofacies (Combaz, 1980) and improved classification of the types of organic particles present (Rogers, 1979; Masran and Pocock, 1981). Algal, amorphous, herbaceous, woody, and coaly are the terms most frequently used to qualify the various organic constituents. Their relationship with the kerogen types defined on the basis of chemical analyses is generally clear (Tissot and Welte, 1978) except for the amorphous facies, which is a special problem. Although an amorphous kerogen may be an alteration product of any primary type of organic matter (Rogers, 1979), it is commonly considered to be dominantly algal in origin, and thus equated with type II or type I kerogen. For this reason, "amorphous" kerogen is considered to have a good source potential for generating oil. In fact, systematic elemental analysis performed on a set of amorphous kerogens from various origins has shown that, although some of them belong to type II, the chemical composition of the amorphous kerogen may spread over the entire Van Krevelen diagram (Figure 4) (Durand and Monin, 1980). In particular, some of the atomic H/C and O/C ratios are typical of type III kerogen (i.e., humic-

type material derived from terrestrial plants); one is even located within the field of residual organic matter. Powell et al (1982), using 58 samples from the Beaufort-Mackenzie basin and the Canadian Arctic Islands, showed a low level of correlation between the content of amorphous material and the chemical analysis of kerogen. Thus, optical studies in transmitted light should be interpreted along with data obtained by chemical analysis or

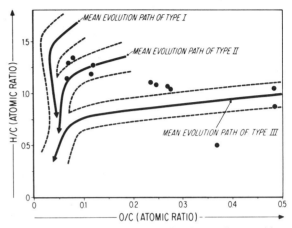

Figure 4—Van Krevelen diagram showing elemental composition of 12 samples of various origins, identified by optical examination (transmitted light) as amorphous organic matter (Durand and Monin, 1980).

pyrolysis. In this framework, they can provide useful additional information, particularly on heterogeneity of organic matter, occurrence of recycled kerogen, etc. Fluorescence studies may also help to separate type I or II kerogens (which are generally fluorescent up to the end of the oil generation stage) from type III kerogen and residual organic matter (which are generally nonfluorescent or weakly fluorescent beyond the very low-rank stages) (Teichmüller and Durand, 1983).

Optical studies in reflected light are commonly oriented toward characterization of the maturation stage of the organic matter, but they also include kerogen identification (Stach et al, 1975; Dow and O'Connor, 1982). The main groups of macerals observed in coal petrography are used for kerogen description: huminite/vitrinite, liptinite/exinite, and inertinite. This analysis can be applied across the entire maturation range (Dow and O'Connor, 1982), whereas type identification by transmitted light may become difficult at higher ranks. The finely divided organic matter, however, is hardly identifiable. Vitrinite and inertinite obviously have a relationship with type III and residual (or type IV) kerogen, respectively. However, the situation is more complex with the liptinite/exinite group where attempts to correlate the abundance of this maceral in coals or kerogens with the hydrogen content (atomic H/C ratio or hydrogen index) sometimes showed a low level of correlation. Once again, results from optical examination should be interpreted along with data from chemical analyses and pyrolysis.

Vitrinite reflectance was used for many years by coal petrographers; later it became one of the most widely used parameters for measuring source rock maturity (Dow, 1977, Alpern, 1980). Among the aspects of vitrinite reflectance frequently discussed are the limitations due to occurrence or proper identification of vitrinite, and the extent of the "oil window" in the reflectance scale. On the first subject, Alpern (1980) pointed out that a limitation for using vitrinite reflectance may be the scarcity or the absence of syngenetic vitrinite in type I or type II kerogens. In addition, the mode of selection of the particles used for measurement is of great consequence when different populations of subordinate macerals are present in those marine or lacustrine kerogens. Alpern (1980) observed 5 populations in the same shallow bed of the lower Toarcian shales of the Paris basin, including a predominant maceral of the liptinite group (bituminite with very low reflectance, 0.10-0.15%) and 3 gelified macerals: (1) granular gels (reflectance R_1 = 0.2-0.3%), (2) vitrodetrinite (R_2 = 0.4-0.7%), and (3) reworked vitrinite (R_3 = 0.8-1.3%). On the north, close to the Ardennes-Rhine massif, which was the source of the detrital material, R_2 and R_3 are predominant. Moving southward (i.e., far from the coast line), R_1 is increasingly abundant and becomes predominant. Thus, careful consideration of the histograms of vitrinite reflectance based on a sufficient number of measurements, and their interpretation in the light of geologic and geochemical data, is always to be preferred to the simple use of an average reflectance value.

Over the years, many views have been expressed about the extent of the oil window in terms of vitrinite reflectance (Dow, 1977; Héroux et al, 1979; Alpern, 1980; Robert, 1980; Teichmüller, 1982). In fact, there are no sharp boundaries of the oil zone. Kerogen may vary in composition according to its original type, and the relative abundance of the various chemical bonds vary accordingly (Tissot and Welte, 1978). Thus, the onset of oil generation may occur relatively early (about 0.5% R_m) when some weaker heteroatomic bonds are abundant (type II); it may occur relatively late (about 0.7% R_m) where mainly stronger C-C bonds are involved (type I). Also, it should be remembered that different thermal histories (fast heating over a short period of time, or slow heating over a long period of time) may result in the same vitrinite reflectance but in different maturation levels of the nonvitrinitic constituents of the organic matter predominant in type I and type II kerogens. This phenomenon is due to the different kinetics of degradation. This applies also to other optical indices of maturation, such as Thermal Alteration Index (TAI) and fluorescence.

Vitrinite reflectance is certainly one of the most valuable tools for measuring the maturation stage of type III kerogen and coal. Other marine or lacustrine kerogens (types I and II) may also contain particles resembling vitrinite, although their composition and the evolution of optical properties may be different (Alpern et al, 1978). In this case, it is advisable to interpret reflectances in the framework of kerogen type and to associate this optical index with a chemically derived parameter. For example, Dow (1977) used a plot of H/C ratio versus vitrinite reflectance. Alternatively, the indices obtained from pyrolysis may be correlated with vitrinite reflectance. The principal maturation parameter from Rock-Eval pyrolysis (temperature, T_{max}, corresponding to the maximum yield of pyrolysis) can also be compared to vitrinite reflectance. Figure 3 (Teichmüller and Durand, 1983), for example, shows an excellent relationship between T_{max} and R_m for coals and type III kerogens. The relation is not strictly a linear correlation due to the somewhat different physicochemical phenomena involved—mainly the cracking of side chains and the increasing size and aromaticity of the remaining kerogen, for T_{max} and R_m, respectively—but the general relationship is very good, as both phenomena are interdependent. The graph would be somewhat different for type I and type II kerogens in the low maturity levels.

Fluorescence under ultraviolet excitation has been studied recently by Teichmüller and Ottenjann (1977), Ottenjann et al (1981), and Teichmüller and Durand (1983). On a series of coals, these authors showed that the coalification of liptinite macerals (sporinite, cutinite, and fluorinite) is characterized by a decrease of fluorescence intensity and a shift of the fluorescence color toward red. These changes vary quantitatively for different liptinite macerals, but remain the same qualitatively. The good relationship presented by Teichmüller and Durand (1983) among fluorescence parameters, vitrinite reflectance, and Rock-Eval pyrolysis tends to confirm that fluorescence is a useful tool for measuring the rank of coal and the evolution stage of source rocks at relatively low maturity levels. A parameter frequently used by these authors is the red shift measured on sporinite by the ratio of fluorescent intensity at 2 different wavelengths (Q = intensity at 650 nm/intensity at 500 nm).

A secondary liptinite maceral, exsudatinite, is also observed in certain coals, commonly filling cells, approximately over the main stage of oil generation (0.4-1.0% vitrinite reflectance). It is markedly fluorescent and is interpreted as an oil-type material. This indicates that microscopic observation of source rocks with the help of fluorescence may also be of interest in migration studies.

A certain fluorescence is also observed in some huminites/vitrinites. At very low maturation ranks, it decreases drastically, and it is completely lost at about 0.5% reflectance. Beyond that stage, another, weaker, type of fluorescence is attributed to the generation of liquid bitumen (Teichmüller and Durand, 1983). The latter phenomenon is accompanied by the occurrence of exsudatinite.

Recently, other maturation parameters have been proposed based on the concentrations of free hydrocarbons which can be measured in source rock samples and, in some cases, in pooled oils. Leythaeuser et al (1979) studied the generation of light hydrocarbons (C_2–C_7) in source rocks by using a hydrogen-stripping technique. Schaefer and Leythaeuser (1980, 1983) combined this technique with thermovaporization, and showed on a selected sequence of source beds that the total concentration of the light hydrocarbons increases by approximately 2 orders of magnitude when vitrinite reflectance increases from 0.4 to 0.9%. Radke et al (1982) investigated the distribution of individual aromatic hydrocarbons, and particularly the phenanthrene (3 aromatic rings) derivatives. Radke et al introduced a methylphenanthrene index (MPI) based on the ratio of 2- and 3- methylphenanthrenes (considered as products of thermal degradation) to 1- and 9- methylphenanthrenes (frequently observed in immature samples). A strong correlation between MPI and vitrinite reflectance was observed in coals and associated type III kerogen of the Deep Basin, western Canada, and a general relationship is proposed—the MPI increases regularly from 0.5 to 1.5 over the main zone of oil generation.

BIOLOGICAL MARKERS (GEOCHEMICAL FOSSILS) AND THEIR APPLICATION TO PETROLEUM EXPLORATION

Biological markers (or geochemical fossils) are molecules inherited from the organisms living at the time of sediment deposition which have been preserved without subsequent alteration, or with only minor changes, so that they keep the main features of their chemical structure. Many advances in this field have taken place in recent years. In fact, progress in identifying fossil molecules in recent or ancient sediments and progress in the chemistry of natural products and their distribution in living organisms have been simultaneous and interrelated. In some instances (hopane group of triterpenes), fossil molecules were identified in all types of sediments before their widespread occurrence in presently living prokaryotes was recognized.

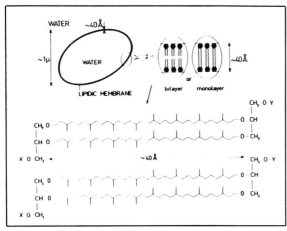

Figure 5—Glycerol alkyl-diethers and alkyl-tetraethers comprising isoprenoid chains are basic constituents of archaebacterial membranes (Chappe, 1982). X, Y = hydrophilic groups (phosphorylated glycerol or other polyols).

Archaebacteria represent ancestral bacterial life forms presently confined in ecological niches corresponding to extreme environments. They include thermoacidophilic, halophilic, and methanogenic bacteria. The latter are anaerobic, methane-producing bacteria usually living in young sediments. Characteristic differences between Archaebacteria and other organisms (eubacteria and eukaryotes) include genetic differences (nucleic acids) and specific compositions of membrane lipids (mainly glycerol ethers in Archaebacteria instead of esters as in other organisms) (de Rosa et al, 1977, 1980). The hydrocarbon chains bound to glycerol are generally branched, and particularly include C_{40} isoprenoid chains with a "head-to-head" junction of the C_{20} subunits (Figure 5). Michaelis and Albrecht (1979) and Chappe et al (1979, 1980, 1982) identified these typical glycerol tetraethers in kerogen, in the asphaltene fraction of ancient sediments and crude oils, and also in the polar fraction of recent sediments. The related free isoprenoid hydrocarbons with the head-to-head junction, which originated from degradation of the ethers during catagenesis, were reported by Moldowan and Seifert (1979) and Albaiges (1980).

Tetracyclic steroids and pentacyclic triterpenes are widely distributed in living organisms. Occurrences of these molecules or their alteration products have been reported in recent sediments, source rocks, and in crude oils (Corbet et al, 1980; A. S. Mackenzie et al, 1981; Schmitter et al, 1981; Ludwig, 1982; Trendel et al, 1982).

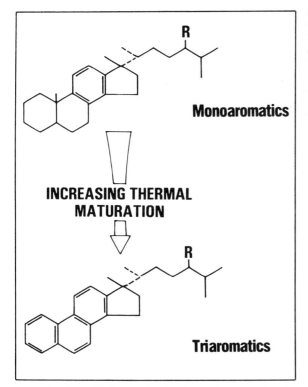

Figure 6—Aromatization of steroids: increasing thermal maturation changes monoaromatic steroids into triaromatics.

The fate of these biological markers during diagenesis and catagenesis has been widely investigated. Elimination of functional groups, isomerization, and aromatization are among the major steps of diagenesis. The process of aromatization of steroids frequently starts in ring C from sterenes precursors, and progressively evolves toward triaromatic steroids (Figure 6) (Ludwig et al, 1981; A. S. Mackenzie et al, 1981; Ludwig, 1982). Less commonly it may start from ring A, as reported by Hussler et al (1981) from Cretaceous black shales cored by the Deep Sea Drilling Project in the Atlantic. This second type of aromatiza-

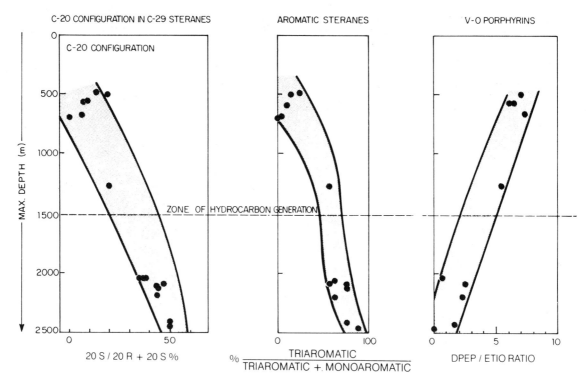

Figure 7—Changes of chemical structure of biological markers with increasing burial depth in lower Toarcian shales of Paris basin: stereochemistry of steranes, aromatization of steroids, conversion of vanadyl porphyrins. Adapted from Mackenzie et al (1980b, 1981, 1980a), respectively.

tion may be an indication of an anoxic depositional environment. More generally, there is a tendency for aromatization to increase with burial from monoaromatic to triaromatic steroids. This situation has been observed in several basins, particularly by A. S. Mackenzie et al (1981) in the lower Toarcian shales of the Paris basin (Figure 7).

The evolution of pentacyclic triterpenes during diagenesis and catagenesis is also documented. In particular, tetracyclic compounds resulting from the opening or the loss of ring A, possibly controlled by microorganisms, have been reported in deltaic and lacustrine sediments (Corbet et al, 1980; Schmitter et al, 1981). A progressive aromatization was also observed in sediments and crude oils by Spyckerelle et al (1977a, b). The most frequent triterpenoid molecules reported in sediments are the hopane series. The parent molecules are rather ubiquitous compounds in prokaryotes (bacteria and blue-green algae), where pentacyclic polyalcohols (bacteriohopanes) are important constituents of the membrane lipids. According to the environmental conditions of deposition and diagenesis, the hopanes could generate the related alkanes and acids (Ourisson et al, 1979).

The geochemistry of porphyrins has been reviewed by Baker (1980) and Maxwell et al (1980). In addition to the most common nickel or vanadyl porphyrins, copper porphyrins were reported as a possible indicator of oxidized terrestrial organic matter by Palmer and Baker (1978), whereas aluminium porphyrins and gallium complexes were found in bituminous coal by Bonnett and Czechowski (1980). The occurrence of a distribution of molecular weights, suggesting a series of alkyl-substituted porphyrins, has been the subject of many discussions. A. S. Mackenzie et al (1980a) showed that C_{35}-C_{39} porphyrins are generated below 2,000 m (6,600 ft) in the Toarcian shales of the Paris basin, and interpreted them as resulting from thermal cracking of kerogen during catagenesis. Comparable conclusions were reached by Quirke et al (1980).

The recent discovery of a natural nickel complex with a porphinoid skeleton in a methanogenic Archaebacteria (Pfalz et al, 1982) offers an alternative to the classical interpretation of Treibs (1934) which related porphyrins to the chlorophyll or hemin molecules. More work is certainly needed before the significance of this discovery can be estimated.

Finally, kerogen (the insoluble organic constituent of source rocks) and asphaltenes (the heaviest polar fraction of crude oils and bitumens) are now recognized to contain biological markers. Pyrolysis of kerogen or asphaltenes yields isoprenoids, steroids, triterpenoids, and porphyrins (Huc, 1978; Rubinstein et al, 1979; Samman et al, 1981). Selective cleavage of chemical bonds in kerogen or asphaltenes by appropriate reactant may also yield some of these products (Chappe et al, 1982). In addition, the biological markers may have been protected, to some extent, from degradation by trapping in the kerogen or asphaltene matrix.

The use of biological markers for oil-oil and oil-source

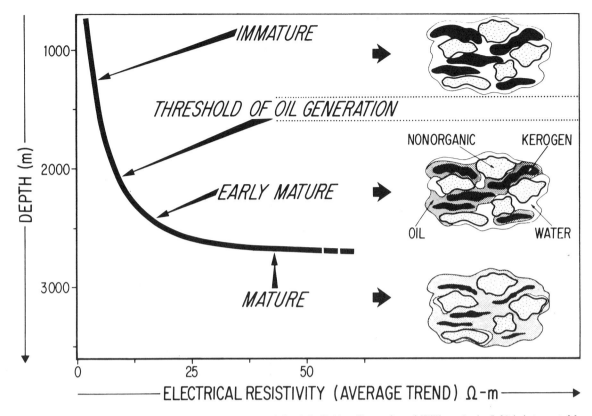

Figure 8—Evolution of electrical resistivity as function of depth in Bakken Formation of Williston basin (left) is interpreted by increase of oil saturation in pore space (right) forming a continuous oil phase. Adapted from Meissner (1978) by Durand (1983).

rock correlation is now widely acknowledged. Correlation problems were reviewed by Welte et al (1975), Deroo (1976), Seifert and Moldowan (1978, 1981), and Seifert et al (1980). The combination of gas chromatography and mass spectrometry has made possible the production of mass fragmentograms, corresponding to specific fragments of fossil molecules, which can be compared from one crude oil to another or to an extract from a source rock. Even biodegraded oils can be correlated. In moderately degraded crude oils, steranes and triterpanes are often found unaltered. In more degraded oils, the related aromatic compounds may be used (Tissot and Welte, 1978). Finally the most degraded heavy oil can provide significant biological markers through pyrolysis of the asphaltene fraction (Rubinstein et al, 1979; Aref'yev et al, 1980).

Other proposed applications of geochemical fossils are in reconstruction of environments of deposition and evaluation of conditions of diagenesis and catagenesis. Brassell et al (1980) discussed the contribution of the various classes of lipid components as potential indicators of environments. Huang and Meinschein (1979) considered the significance of sterols in that respect. A possible limitation to such developments exists, however, because of the predominant input of biological markers inherited from bacteria, including those living in young sediments. In fact, the most widespread biological markers originate in the most resistant part (membranes) of bacteria, which are the last member of the chain of living organisms in the cycle of organic carbon. A consequence of this is that remnants of organisms living in the basin of deposition may be obliterated by the effect of microbial activity. Occurrence of specific biological markers may be a witness to the presence of certain types of organisms, but absence of markers cannot be taken as a proof of absence of the organisms.

Minor, but significant, alterations of biological markers (which preserve the general structure of the molecule) serve as evidence of diagenesis and catagenesis. Stereochemistry of isoprenoids, steroids, and hopanes, aromatization of steroids (from monoaromatics to triaromatics) and conversion of DPEP-porphyrins to etio-porphyrins (Figure 7) have been proposed by Seifert and Moldowan (1978) and A. S. Mackenzie et al (1980a, b, 1981) for evaluation of thermal maturation in the subsurface. However, standard scales, valid in different geologic situations, probably cannot be developed, and specific calibrations will have to be made in each basin. In fact, different types of chemical reactions (isomerization, aromatization, etc) have different kinetic constants, and thus some of them will be more influenced than others by different heating rates caused by different rates of subsidence and geothermal gradients. This conclusion was reached by A. S. Mackenzie et al (1982) when comparing the situation in the Pannonian basin of central Europe (heating rate 15°C/

m.y., 27°F/m.y.), Mahakam delta of Indonesia (7°C/m.y., 12.6°F/m.y.), and Paris basin (0.5°C/m.y., 0.9°F/m.y.)

This observation could be used to help reconstruction of past sedimentation rates and thermal events (A. S. Mackenzie et al, in press). For this purpose, it is advisable to use single, well-established chemical processes, without competitive or side reactions. For example, the progressive decrease of the DPEP-porphyrin/etio-porphyrin ratio observed during burial is commonly considered to be due to the conversion of DPEP-porphyrins to etio-porphyrins. However, Barwise and Roberts (in press) believe that the conversion occurs in the diagenetic stage. The decrease of the DPEP/etio ratio observed during catagenesis would be due, in fact, to a faster rate of destruction of DPEP-porphyrins compared with etio-porphyrins. Elucidation of the controlling process is obviously a prerequisite to using kinetic considerations for geologic reconstructions.

MIGRATION OF PETROLEUM FROM SOURCE ROCK TO RESERVOIR

Of the 2 steps of petroleum migration—from source rocks to reservoir (primary migration) and within the reservoir toward a trap (secondary migration)—the second step is reasonably understood, with the difference of density between water and hydrocarbons (buoyancy) being the major driving force. Much research is still devoted to primary migration. The current views on the subject have been summarized by Durand (1983).

The possible modes of primary migration may require a movement of the water phase, such as transportation of petroleum as molecular or micellar solution, or they may be independent of water movement, such as hydrocarbon phase migration or diffusion. Several observations tend to restrict the feasibility of a water-driven expulsion of hydrocarbons. Additional data have been provided on solubility of pure hydrocarbons in water and also on solubility of whole crude oils in saline waters (McAuliffe, 1980). The influence of pressure on methane solubility has been investigated by Bonham (1978). It is difficult, however, to explain the movement of large quantities of crude oil by molecular or micellar solution in water. Jones (1980) used mass balance calculations in the Williston and Los Angeles basins to show that solution migration would be ineffective. In general terms, hydrocarbon generation occurs at depths where most of the original water has already been expelled and crude oil micellar or molecular solubility is too low. Comparison between chemical compositions of source rock bitumen and pooled crude oil shows that observed changes are almost the contrary of what would be expected from transport in water solution (Tissot and Welte, 1978).

Hydrocarbon phase migration seems to be the most likely way of expulsion from source rock. Meissner (1978) presented an example of oil phase migration in the Bakken Formation of the Williston basin. With increasing burial, most of the water was expelled from this organic-rich shale before hydrocarbons were generated. Then, at peak generation stage, relative saturation of the pore space by hydrocarbons became high, and a continuous oil phase formed, as shown by the very high electrical resistivity (Figure 8). An oil phase expulsion then became possible. McAuliffe (1980) considered a 3-dimensional kerogen network wettable with oil and acting as transport avenue for a separate oil phase.

In a more general situation, overpressuring of pore fluids could create the necessary avenues for hydrocarbon phase expulsion, in particular by microcracking (Snarsky, 1962; Tissot and Pelet, 1971; Momper, 1978). The causes of the important increase of internal pore pressure are discussed by Durand (1983), and include thermal expansion of water, overall volume expansion of organic matter by generation of gaseous and liquid hydrocarbons from kerogen (Ungerer et al, 1983), and partial transfer of geostatic stress from the solid rock matrix to the enclosed pore fluid (Meissner, 1978) The internal pressure buildup would overcome the capillary pressures (if these are not too high) or exceed the mechanical strength of the rock and cause microfracturing, thus offering avenues for hydrocarbon expulsion.

Leythaeuser et al (1984) and A. S. Mackenzie et al (1983) have studied migration effects in Cretaceous and Paleocene sand and shale series of Spitzbergen, Norway. The sediments were buried to 2,500-2,800 m (8,200-9,200 ft)—equivalent to 0.85% vitrinite reflectance—during the Tertiary, then rapidly uplifted to shallow depth (0-240 m, 0-800 ft) since the early Holocene, thus freezing the process of hydrocarbon generation. In addition, increased viscosities resulting from that temperature drop and also from introduction of permafrost may have slowed the migration process as well. Those authors observed that thin shales interbedded in sands and the edges of thick shale units are depleted in hydrocarbons to a much higher degree than the centers of thick shale units. Migration has also a compositional fractionation effect: n-alkanes of lower molecular weight are expelled preferentially, and isoprenoids (pristane, phytane) are expelled less than the related straight-chain alkanes. Calculation of expulsion efficiencies showed a strong decrease with increasing chain length of n-alkanes. The composition of hydrocarbons in reservoir sands is in agreement with that fractionation. This effect may have some consequences for the use of certain geochemical indices. For instance, the pristane/n-C_{17} ratio, which is sometimes used as a maturity parameter, is also controlled by the degree of fractionation during migration.

Leythaeuser et al (1980, 1982) considered that in special geologic conditions diffusion of hydrocarbons through the water-saturated pore space may represent an effective process for moving light hydrocarbons within source rock. They determined diffusion coefficients and calculated the cumulative amount of hydrocarbons escaping with time from several source rocks, including a gas-prone Mesozoic source rock of western Canada. Diffusion is thought to be effective within the source bed over relatively short distances, and for gas only. Diffusion of light hydrocarbons in the subsurface may also have adverse effects and destroy gas accumulations by dissipation through the cap rock.

A few case histories of primary migration have been reported, for example, the Bakken shales of the Williston basin (Meissner, 1978), the upper Tertiary of the Mahakam delta (Durand and Oudin, 1979), and the Mesozoic Deep Basin of Alberta (Welte et al, 1982).

RESINS, ASPHALTENES, AND HEAVY OILS

In recent years, much interest has been focused on heavy oils and tar sands because of very large reserves reported in western Canada, eastern Venezuela, and elsewhere (Demaison, 1977). Resins and asphaltenes, the heavy ends of normal crude oils, are major constituents of heavy oils. They are largely responsible for the main difficulties in producing heavy oils (particularly oils with very high viscosity), and also for other problems in enhanced oil recovery (such as oil wettability) and in refining. For these reasons, much work has been devoted to resins and asphaltenes, and this work in turn may make clearer their role in petroleum generation and migration.

Heavy oils commonly contain more than 40% resins plus asphaltenes (Tissot and Welte, 1978). Sulfur content is generally higher than in normal crude oils. For the Western Canada basin, Deroo et al (1977) showed a direct relationship between the abundance of sulfur and the content of aromatics plus resins and asphaltenes. In fact, sulfur is distributed between benzothiophene derivatives (i.e., the sulfur-aromatic fraction), resins, and asphaltenes. There are, however, heavy oils derived from low-sulfur paraffinic-naphthenic crude oils which contain less than 1% sulfur (Emeraude field, Congo; Bemolanga tar sands, Malagasy) (Claret et al, 1977). Some Utah tar sands fall in the same category.

The origin and occurrence of heavy oils and tar sands were discussed by Demaison (1977), particularly for examples from the Western Canada and Eastern Venezuela basins. He showed that the area of source rock maturation and hydrocarbon generation is located basinward and well outside the area of heavy oil occurrence. Thus an extensive lateral migration occurred over distances of tens to more than 100 km (62 mi). Source-bed drainage was favorable because of stratigraphic features, such as sandstone distribution, and structural factors, such as the Peace River arch in western Canada. The oils were subsequently degraded (including biodegradation) in the reservoirs. Bailey et al (1974) and Deroo et al (1977) showed the successive stages of degradation, and also the relationship between degradation and the access of meteoric fresh waters into the reservoir beds.

Resins and asphaltenes are also normal, usually subordinate, constituents of all crude oils. Their structure was described by Yen (1979) as resulting from the stacking of several polyaromatic sheets bearing aliphatic chains and functional groups. Heteroatoms seem to play an important role in cross-linking by sulfur or oxygen bridges. Nitrogen and the rest of the sulfur are engaged in heterocycles of the polyaromatic nuclei (Moschopedis et al, 1978; Speight et al, 1979).

Pyrolysis of asphaltenes shows that their thermal degradation follows the same major steps observed in pyrolysis of kerogen, that is, elimination of carbonyl or carboxyle groups, followed by hydrocarbon generation, and finally by leaving a highly condensed carbonaceous residue (Moschopedis et al, 1978). Rubinstein et al (1979) analyzed the hydrocarbons generated by pyrolysis of Prudhoe Bay asphaltenes, and found them comparable to those naturally occurring in this particular crude oil. Chappe (1982) used a chemical process of asphaltene degradation, and identified constituents comparable to those already observed in kerogen. Behar et al (in press) pyrolyzed both the asphaltenes present in source rocks and the related kerogen. The gross composition of the products was similar, as well as the n-alkane distribution. A similar result was obtained by comparing pyrolysis products from crude oil asphaltenes with the constituents of the parent oil.

From these observations, it is inferred that the asphaltene constituents of source rock bitumen and crude oils are similar to fragments of kerogen. Most of them may result from early breakdown of heteroatomic or other bonds in kerogen liberating smaller fragments, which dissolve in solvents (including crude oil), but keep their original structure. Alternatively, other asphaltene constituents (being composed of less condensed fragments although comparable in nature to kerogen) may form together with kerogen (Behar et al, in press). In both cases, asphaltenes are able to generate molecules of smaller size, such as resins and hydrocarbons. Thus H/C, O/C, and S/C ratios of asphaltenes progressively decrease with increasing thermal maturation. This suggests the existence of evolution paths of asphaltene comparable to the evolution paths defined for kerogen (Tissot, 1981). In that sense, asphaltenes may be intermediate products along some of the pathways leading progressively from kerogen to hydrocarbons.

A question remains as to what is the exact role of resins and asphaltenes in the migration of petroleum. It has been observed that in source rock bitumen, the average ratio of resins plus asphaltenes to hydrocarbons is about 1 to 1. The same ratio is dramatically reduced in normal reservoired crude oils, where it averages only 1 to 6 (Tissot, 1981). An interpretation is that asphaltenes are solubilized by an "atmosphere" of resins, which, in turn, are surrounded by aromatic hydrocarbons. Based on this model, originally presented by Pfeiffer and Saal (1940), it is proposed that each petroleum is able to carry out of the source rock a quantity of asphaltenes in relation to the available concentration of resins and aromatic hydrocarbons. This interpretation would also be valid for the original composition of the present heavy oils, but subsequent degradation in the reservoir would result in a different composition where the relative proportion of hydrocarbons is greatly decreased, whereas the proportion of resins and asphaltenes is increased (Tissot, 1981).

GEOCHEMICAL MODELING OF SEDIMENTARY BASINS

The objective of geochemical and geologic modeling is to calculate, with the help of computers, the quantities of petroleum that have been generated, migrated, and finally accumulated. This quantitative evaluation is performed as a function of geologic time for each location in the basin.

Geochemical modeling is perhaps an example of a scientific development that appeared too early to be used by industry. The first papers using mathematical-numerical handling appeared in the late 1960s and the early 1970s (Allred, 1966, on oil shale pyrolysis; Tissot, 1969, and Tissot and Pelet, 1971, on generation of hydrocarbons in the subsurface; and Smith, 1971, on shale compaction and water expulsion), but they did not generate much interest. A semiempirical calculation of the thermal maturity of organic matter in rocks presented by Lopatin (1971) aroused more interest because it did not involve complex mathematical handling of reaction kinetics. During the late 1970s and early 1980s, however, a growing concern by the petroleum industry to integrate geochemical information into a global exploration program led to the development of several geochemical and geologic models, some of which have been published. Both geochemical modeling, which calculates the amount of hydrocarbons generated, and geologic modeling, which integrates basin evolution and migration as well, are reviewed here.

Geochemical modeling calculates the amount of oil and gas generated as a function of geologic time in each location of the basin. Waples (1980) calibrated the time-temperature index (TTI) of maturity, introduced originally by Lopatin (1971). This index is calculated on the basis of the time of residence in a given temperature range. Therefore, knowledge of burial history and of present and past geothermal gradients is necessary. Waples (1980) correlated TTI with other geochemical indices and defined the thresholds of the various maturation stages. This semiempirical procedure (Figure 9) does not require numerical simulation or the use of computers. The method, however, is not able to account for different types of organic matter with their different generative potential and kinetics of degradation, or a complex burial history with 2 cycles of sedimentation separated by folding and erosion (e.g., the Paleozoic source rocks of the Algerian Sahara). Also, the numerical calculation involved is based on the hypothesis that the rates of chemical reactions approximately double for every 10°C (18°F) rise in temperature. Based on Arrhenius law, this assumption is acceptable only for reactions with an activation energy in the range from 10 to 20 kcal/mole. Such values may be expected at the beginning of oil generation, when rather weak bonds are broken, but they are questionable when stronger bonds, such as carbon-carbon bonds, are involved with much higher activation energies. For example, activation energies in the range from 40 to 60 or 70 kcal/mole seem to be common at the peak of oil generation or later, when gas is generated by cracking (Weitkamp and Gutberlet, 1968; Tissot and Espitalié, 1975). These activation energies would correspond to reaction rates multiplied by 4 to 10 (instead of 2) for a 10°C (18°F) rise in the corresponding temperature range.

A kinetic model, where time, temperature, and the type of organic matter are explicitly introduced, is presented by Tissot and Espitalié (1975), Tissot and Welte (1978), and Tissot et al (1980). This computerized numerical model can account for any burial and temperature history, including complex situations with 2 or more cycles of burial. The data input includes type of organic matter (represented by a range of activation energies), burial history, and geothermal gradient (Figure 10). Past geothermal history, if different from the present, may be approached by

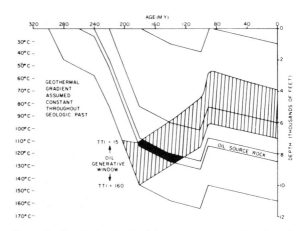

Figure 9—Graph showing burial reconstruction as function of time for succession of formations in 1 location. Depth of oil-generation window is based on time-temperature index (TTI) introduced by Lopatin (1971) and later calibrated by Waples (1980). This semiempirical calculation is valid for reactions with activation energy in range from 10 to 20 kcal/mole (illustration reproduced from Waples, 1980).

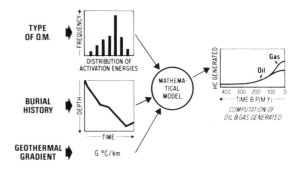

Figure 10—Principle of calculation of oil and gas generation as function of time, by use of kinetic model accounting for various activation energies actually involved in oil and gas generation (Tissot et al, 1980).

using a subordinate model based on vitrinite reflectance or comparable data. Geothermal models of basin evolution calculating temperature as a function of time have been presented and can be used to provide input to geochemical models (D. Mackenzie, 1981).

A special point of interest is the timing of oil and gas generation (Tissot et al, 1980). This information can be compared with the time of trap formation or seal deposition. Most existing oil and gas fields were generated during the Cretaceous and Tertiary. These relatively young accumulations may amount to more than 90% of petroleum reserves. They include not only hydrocarbons generated from Cretaceous and Tertiary source rocks, but also hydrocarbons generated from Paleozoic or Mesozoic source rocks which were never sufficiently buried until the Cretaceous or Tertiary. Such rocks were deposited in platform basins where subsidence was moderate during the Paleozoic, for example the Silurian source rocks of the northern Algerian Sahara (Figure 11), and the Devonian

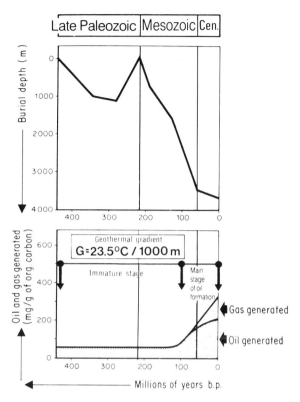

Figure 11—**Burial history of Silurian source rocks in Hassi Messaoud area of Algerian Sahara (top) and generation of petroleum computed by kinetic model (Tissot and Welte, 1978; Tissot et al, 1980).**

source rocks of the Leduc area, Alberta. Additional burial generally occurred in the Cretaceous or Tertiary in response to post-Early Cretaceous global tectonics.

Geologic modeling integrates geochemical information into the framework of basin evolution with its geologic, hydrodynamic, and thermodynamic aspects (Yükler et al, 1978). A 3-dimensional model presented by Welte and Yükler (1981) is based on a set of nonlinear partial differential equations of energy and mass balance that describes the course of the various geologic processes. Space is divided into a regular grid (e.g., vertical interval of 10 m or 33 ft and lateral interval of 1,000 m or 3,300 ft) with a time step (e.g., 100,000 years). A set of initial physical (e.g., porosity and permeability) and thermal (e.g., thermal conductivity) parameters is chosen depending on the lithologic data. The values are subsequently adjusted by means of successive iterations based on presently measurable parameters. The model then calculates pore pressure, temperature, physical and thermal properties, maturation of organic matter, and amount of hydrocarbons at each grid point as a function of time. These results are used to determine possible secondary migration directions which are, in turn, related to the occurrence of traps.

Durand et al (1983) presented a model of petroleum generation and migration in which hydrocarbon maturation and generation are calculated by the kinetic model of kerogen degradation (Tissot and Espitalié, 1975). Durand et al emphasized migration modeling with calculation of the relative hydrocarbon and water saturation as a function of time and the distribution of abnormal pore pressure. On the basis of 2 examples (North Sea and Mahakam delta), they trace the progressive accumulation of hydrocarbons in structural and stratigraphic traps and eventually their destruction by dismigration to the surface. They observe that several parameters (e.g., content of organic matter in synclines where no well has been drilled, past geothermal reconstruction, and viscosity of hydrocarbon phases) are rather sensitive. Triphasic models (water, oil, and gas) require a better knowledge of oil/gas solubility for modeling gas phase migration, and these models raise some numerical problems.

The number of geologic/geochemical models used in petroleum exploration is growing rapidly. Chenet et al (1983) and Ungerer et al (1983) discussed the influence of various subsidiary models (subsidence and compaction, thermal history, hydrocarbon generation, and fluid movements) in a comprehensive model of basin evolution. Nederlof (1979) and Sluijk and Nederlof (in press) presented a model for prospect appraisal calibrated on worldwide geologic and geochemical data. Bishop et al (1983) made a probabilistic estimation of trapped hydrocarbon volume and nature by estimating the historical evolving effects of each factor (kerogen amount, type, and maturity; migration; trap growth; etc).

The major advantage of geologic/geochemical modeling is that it integrates observed regional geologic data and measured geochemical parameters into a framework of fundamental knowledge of basin evolution and petroleum generation and migration. This approach is quantitative and it accounts for geologic time. Future developments should allow predictions to be made of possible reserves and their locations to be pinpointed.

The validity of each of the proposed models will result from comparison with reality, once they are applied to a large number of basins. Some encouraging features already exist. For example, the kinetic models of kerogen degradation are able to account for, with the same set of constants, very different conditions of hydrocarbon generation, ranging from high temperatures and short periods of time (minutes and hours) in the laboratory and oil shale retorts to low temperature and long periods of time (up to 500 m.y.) in some geologic situations. This result is a proof of the validity of the kinetic equations used in the model. The quantitative results from basin modeling are compatible with observed data on porosities, pressure distribution (including overpressured zones), and location of oil and gas accumulations.

GEOLOGIC HISTORY AND WORLD RESERVES

Advances in understanding the fundamentals of petroleum generation, migration, and accumulation make possible interpretation of the observed distribution of world petroleum reserves within the framework of geologic history. This approach has been followed by relatively few authors from North America, western Europe, and the USSR (Olenin, 1977; North, 1979; Tissot, 1979; Bally and

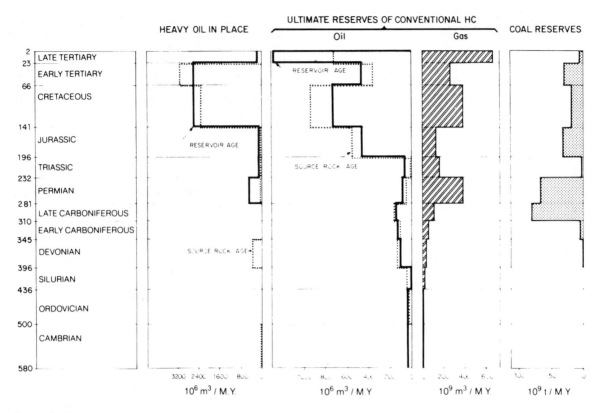

Figure 12—Distribution of oil, gas, and coal reserves according to reservoir and source rock ages. Heavy oil in place in tar sands is also shown (Bois et al, 1982).

Snelson, 1980; Huff, 1980; Klemme, 1980, 1981; Sokolov, 1980; Bois et al, 1982).

Most of these authors attempted to classify petroleum occurrences with respect to geotectonic characters of basins. Bois et al (1982) pointed out that a general trend can be observed with decreasing geologic age. Most Paleozoic oil reservoirs are in cratonic and pericratonic areas. The relative importance of platform reservoirs generally decrease through Mesozoic and Tertiary rocks, whereas the number of accumulations that formed in rapidly subsiding basins increases. Petroleum reservoirs in this last type of basin are particularly important above the lower Tertiary. Bally and Snelson (1980), however, pointed out that very rich and very lean basins—or even basins with no production at all—are present in practically every type of basin in their classification. These authors finally concluded that "the classification of basins does little to improve our hydrocarbon volume forecasting ability." On the other hand, Klemme (1981) believed that "basins analysis and classification permit the discernment of many helpful analogies that may be applied to new or partially explored basins."

Age distribution of oil and gas reserves (Figure 12) offers several characteristics which tentatively may be interpreted in terms of generation and migration of petroleum. A general decrease of oil reserves with increasing age was observed by Bois et al (1982), who explained that trend by thermal evolution (which, with increasing time and temperature, transforms oil into gas) and also by progressive destruction of oil fields through leakage to the surface, faulting, and erosion which destroy the seals.

Peaks of hydrocarbons reserves, and particularly the peak occurring in the Cretaceous, seem to result from a widespread distribution of rich source rocks containing a large amount of predominantly marine organic matter. A high primary production associated with a low rate of destruction is a prerequisite for development of such source rocks. Deposition of planktonic, relatively labile, organic matter requires an oxygen-depleted water column (Demaison and Moore, 1980). Under these circumstances, particularly in anoxic or subanoxic basins (Demaison and Moore, 1980), organic matter is deposited, and the final result is a reducing chemical environment (Bois et al, 1982).

Shallow epicontinental seas transgressing over platform depressions offer such favorable conditions of landlocked basins rich in nutrients with algal blooms and intermittent or permanent anoxia (Gallois, 1976). In fact, there seems to be a relationship between the major transgression phases and a wide distribution of rich source rocks in relatively shallow basins. Tissot (1979) observed that worldwide transgressions, including those associated with the high sea level in the Cretaceous (Vail et al, 1977), are associated with the major occurrences of prolific source rocks. Other examples may be found in the Jurassic (Toarcian and Kimmeridgian of western Europe), Permo-

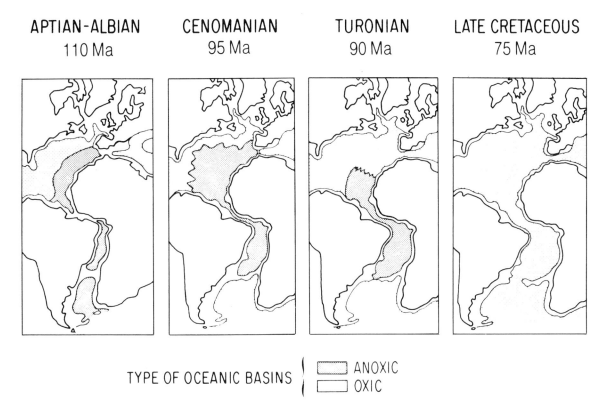

Figure 13—Distribution of depositional environments through part of Cretaceous time in Atlantic basins, reconstructed by using type and origin of organic matter. In places, anoxic conditions may have been intermittent (Tissot et al, 1980).

Carboniferous, Late Devonian, and Silurian.

These circumstances, however, are not sufficient to explain all occurrences of organic-rich source rocks. Schlanger and Jenkyns (1976) and Arthur and Schlanger (1979) reported occurrences of organic-rich sediments in the Upper Jurassic and Lower to middle Cretaceous from very different environments including oceanic basins. They named the occurrences "oceanic anoxic events." Tissot et al (1979, 1980) carried out a geochemical study of the Cretaceous black shales of the Atlantic basins and attempted to reconstruct the depositional environments based on the type and origin of the organic matter (Figure 13). These concepts were subsequently enlarged by de Graciansky et al (1982). The result of these investigations led Bois et al (1982) to the conclusion that, in addition to the global transgressions governing the extension of shallow seas, other factors may have influenced Cretaceous sedimentation in oceanic basins. The lack of communication between the Mesogean and the polar ocean prevented the global circulation which presently brings to low latitudes the deep, oxygen-rich, polar waters. In addition, a warm equable climate with high seawater temperatures (Fischer and Arthur, 1977; Arthur and Schlanger, 1979) decreased the rate of oxygen dissolution. Such events, although they create general conditions favorable for source-bed deposition, have to be associated with local or regional factors (silled basins, upwellings, etc) for widespread deposition of rich source rocks containing marine organic matter.

All these considerations suggest that paleogeography, interpreted in the framework of plate tectonics, will provide the rationale for the distribution of source rock and reservoirs in continental margins and deep offshore basins.

CONCLUSIONS

During recent years, the general scheme of petroleum generation by thermal breakdown of kerogen has been confirmed. Different source rock potentials depend on the original composition of the kerogen in the source rock. The Van Krevelen diagram (atomic H/C ratio versus atomic O/C ratio) is the framework for interpretation of other physical, chemical, and optical analyses. Two special cases are noted. Residual organic matter that has suffered subaerial weathering or biological oxidation has little or no potential for hydrocarbons. Coal is now widely considered as a potential source for gas, but some coals may also generate sizeable amounts of oil as well.

Development of routine analytical tools has favored systematic introduction of organic geochemistry into exploration programs. Pyrolysis offers a wide range of possibilities. It is particularly efficient as a screening technique and for the preparation of geochemical logs, which are subsequently interpreted in terms of correlation and facies changes, source rock potential, and maturation. Optical studies may be interpreted profitably on the basis of geochemical logs and provide useful additional information. The main subjects of discussion in optical studies concern the nature of amorphous organic matter in transmitted light, which may encompass very different chemical compositions of kerogen, and the identification of vitrinite particles for measuring reflectance in marine or lacustrine type I and II kerogens. Fluorescence is becoming a valuable tool for characterizing type and maturation of organic matter.

Many advances have been made in the identification of biological markers in sediments and crude oils, and also in the chemistry and distribution of natural products, which are the ancestors of these fossil molecules. In turn, biological markers are used for oil–source rock correlation, and have been proposed for reconstruction of the environment of deposition and early diagenesis, as well as for evaluation of thermal maturation.

The importance of a hydrocarbon-phase migration from source rock to reservoir has now been clearly established. Overpressuring of pore fluids can create avenues for this expulsion, either by overcoming capillary pressures or by exceeding the mechanical strength of the rock and opening microfractures.

Geochemical and geologic model of sedimentary basins has received much attention. Kinetic models are able to calculate the quantities of oil and gas generated as a function of time by each source bed in any location in a basin. Migration models have become progressively available. Already, they are able to account for the observed distribution of hydrocarbons in several geologic situations, and further progress is expected. Integration of these geochemical models into a comprehensive geologic model of sedimentary basins (including subsidence, compaction, thermal history, etc) should be one of the major developments in future petroleum exploration.

SELECTED REFERENCES

Albaiges, J., 1980, Identification and geochemical significance of long chain acyclic isoprenoid hydrocarbons in crude oils, in A. G. Douglas and J. R. Maxwell, eds., Advances in organic geochemistry 1979: Physics and Chemistry of the Earth, Oxford, England, Pergamon Press, v. 12, p.19-28.

Allred, V. D., 1966, Kinetics of oil shale pyrolysis: Chemical Engineering Progress, v. 62, p. 55-60.

Alpern, B., 1980, Pétrographie du kérogène, in B. Durand, ed., Kerogen: Paris, Éditions Technip, p. 339-371.

———— B. Durand, and C. Durand-Souron, 1978, Propriétés optiques de résidus de la pyrolyse des kérogènes: Revue de l'Institut Français du Pétrole, v. 33, p. 867-890.

Aref'yev, O. A., V. M. Makushina, and A. A. Petrov, 1980, Asphaltenes as indicators of geochemical history of crude oils: Akademiya Nauk SSSR Izvestiya, Seriya Geologicheskaya, v. 1980, n. 4, p. 124-130 (in Russian).

Arthur, M. A., and S. O. Schlanger, 1979, Cretaceous "oceanic anoxic events" as causal factors in development of reef-reservoired giant oil fields: AAPG Bulletin, v. 63, p. 870-885.

Bailey, N. J. L., C. R. Evans, and C. W. D. Milner, 1974, Applying petroleum geochemistry to search for oil: examples from Western Canada basin: AAPG Bulletin, v. 58, p. 2284-2294.

Baker, E. W., 1980, Evolution of the tetrapyrrole diagenesis model, in A. A. Prashnowsky, ed., Internationales Alfred Treibs Symposium, Munich 1979: Munich, West Germany, Universität Würzburg, p. 1-11.

Bally, A. W., and S. Snelson, 1980, Realms of subsidence, in A. D. Miall, ed., Facts and principles of world petroleum occurrence: Canadian Society of Petroleum Geologists Memoir 6, p. 9-75.

Bandurski, E., 1982, Structural similarities between oil-generating kerogens and petroleum asphaltenes: Energy Sources, v. 6, p. 47-66.

Banks, L. M., 1959, Oil-coal association in central Anzoátegui, Venezuela: AAPG Bulletin, v. 43, p. 1998-2003.

Barker, C., 1978, Pyrolysis techniques in organic geochemistry: a bibliography: Tulsa, Oklahoma, University of Tulsa Earth Sciences Publication 78-1.

Barwise, A. J. G., and I. Roberts, in press, Diagenetic and catagenic pathways for porphyrins in sediments, in P. A. Schenck, ed., Advances in organic geochemistry 1983.

Behar, F., R. Pelet, and J. Roucaché, in press, Geochemistry of asphaltenes, in P. A. Schenck, ed., Advances in organic geochemistry 1983.

Bishop, R. S., H. M. Gehman, and A. Young, 1983, Concepts for estimating hydrocarbon accumulation and dispersion: AAPG Bulletin, v. 67, p. 337-348.

Bois, C., P. Bouché, and R. Pelet, 1982, Global geological history and the distribution of hydrocarbon reserves: AAPG Bulletin, v. 66, p. 1248-1270.

Bonham, L. C., 1978, Solubility of methane in water at elevated temperature and pressure: AAPG Bulletin, v. 62, p. 2478-2481.

Bonnet, R., and F. Czechowski, 1980, Gallium porphyrins in bituminous coal: Nature v. 283, p. 465-467.

Brassell, S. C., P. A. Comet, G. Eglinton, P. J. Isaacson, J. McEvoy, J. R. Maxwell, I. D. Thomson, P. J. C. Tibbetts, and J. K. Volkman, 1980, The origin and fate of lipids in the Japan Trench, in A. G. Douglas and J. R. Maxwell, eds., Advances in organic geochemistry 1979: Physics and Chemistry of the Earth, Oxford, England, Pergamon Press, v. 12, p. 375-392.

Chappe, B., 1982, Fossiles moléculaires d'Archébactéries: PhD thesis, Université Louis Pasteur, Strasbourg, France, 164 p.

———— P. Albrecht, and W. Michaelis, 1982, Polar lipids of Archaebacteria in sediments and petroleums: Science, v. 217, p. 65-66.

———— W. Michaelis, and P. Albrecht, 1980, Molecular fossils of Archaebacteria as selective degradation products of kerogen, in A. G. Douglas and J. R. Maxwell, eds., Advances in organic geochemistry 1979: Physics and Chemistry of the Earth, Oxford, England, Pergamon Press, v. 12, p. 265-274.

———— ———— and G. Ourisson, 1979, Fossil evidence for a novel series of archaebacterial lipids: Naturwissenschaften, v. 66, p.522-523.

Chenet, P. Y., F. G. Bessis, P. M. Ungerer, and E. Nogaret, 1983, How mathematical geological models can reduce exploration risks: 11th World Petroleum Congress, London, Special Paper 7, 19 p.

Claret, J., J. B. Tchikaya, B. Tissot, G. Deroo, and A. Van Dorsselaer, 1977, Un exemple d'huile biodégradée à basse teneur en soufre: le gisement d'Emeraude (Congo), in R. Campos and J. Goni, eds., Advances in organic geochemistry 1975, Madrid, Enadimsa, p. 509-522.

Clementz, D. M., 1979, Effect of oil and bitumen saturation on source rock pyrolysis: AAPG Bulletin, v. 63, p. 2227-2232.

———— G. J. Demaison, and A. R. Daly, 1979, Well site geochemistry by programmed pyrolysis: 11th Offshore Technology Conference, OTC 3410, v. 1, p. 465-470.

Combaz, A., 1980, Les kérogènes vus au microscope, in B. Durand, ed., Kerogen; insoluble organic matter from sedimentary rocks: Paris, Éditions Technip, p. 55-111.

Corbet, B., P. Albrecht, and G. Ourisson, 1980, Photochemical or photomimetic fossil triterpenoids in sediments and petroleum: Journal of the American Chemical Society, v. 102, p. 1171-1173.

de Graciansky, P. C., E. Brosse, G. Deroo, J. P. Herbin, L. Montadert, C. Müller, A. Schaaf, and J. Sigal, 1982, The Cretaceous Series in the northern Atlantic and their organic matter: an attempt to reconstruction of paleoenvironments: Revue de l'Institut Français du Pétrole, v. 37, p. 275-335.

Demaison, G. J., 1977, Tar sands and supergiant oil fields: AAPG Bulletin, v. 61, p. 1950-1961.

———— and G. T. Moore, 1980, Anoxic environments and oil source bed genesis: AAPG Bulletin, v. 64, p. 1179-1209.

Dennis, L. W., G. E. Maciel, P. G. Hatcher, and B. R. T. Simoneit, 1982, ^{13}C nuclear resonance studies of kerogen from Cretaceous black shales thermally altered by basaltic intrusions and laboratory simulations: Geochimica et Cosmochimica Acta, v. 46, p. 901-907.

Deroo, G., 1976, Corrélations huiles brutes-roches mères à l'échelle des bassins sédimentaires: Bulletin du Centre de Recherches de Pau, v. 10, 317-335.

―――― T. G. Powell, B. Tissot, and R. G. McCrossan, 1977, The origin and migration of petroleum in the Western Canadian sedimentary basin, Alberta; a geochemical and thermal maturation study: Geological Survey of Canada Bulletin 262, 136 p.

de Rosa, M., S. de Rosa, A. Gambacorta, and J. D. Bu'Lock, 1980, Structure of calditol, a new branched chain nonitol, and of the derived tetraether lipids in thermoacidophile Archaebacteria of the Caldariella group: Phytochemistry, v. 19, p. 249-254.

―――― ―――― L. Minale, and J. D. Bu'Lock, 1977, Chemical structure of the ether lipids of thermophilic acidophilic bacteria of the Caldariella group: Phytochemistry, v. 16, p. 1961-1965.

Dow, W. G., 1977, Kerogen studies and geological interpretations: Journal of Geochemical Exploration, v. 7, p.79-99.

―――― 1978, Petroleum source beds on continental slopes and rises: AAPG Bulletin, v. 62, p. 1584-1606.

―――― and D. I. O'Connor, 1982, Kerogen maturity and type by reflected light microscopy applied to petroleum exploration, in How to assess maturation and paleotemperatures: SEPM Short Course 7, p. 133-157.

Durand, B., 1983, Present trends in organic geochemistry in research on migration of hydrocarbons, in M. Bjorøy, C. Albrecht, C. Corning, K. de Groot, G. Eglinton, and G. Speers, eds., Advances in organic geochemistry 1981: Chichester, John Wiley, p. 117-128.

―――― and J. L. Oudin, 1979, Exemple de migration des hydrocarbures dans une série deltaïque: le delta de la Mahakam, Kalimantan, Indonésie, in Exploration, supply and demand: 10th World Petroleum Congress, Bucharest, Proceedings, v. 2, p. 3-11.

―――― and J. C. Monin, 1980, Elemental analysis of kerogens (C, H, O, N, S, Fe), in B. Durand, ed., Kerogen; insoluble organic matter from sedimentary rocks: Paris, Éditions Technip, p. 113-142.

―――― and M. Paratte, 1983, Oil potential of coals; a geochemical approach, in J. Brooks, ed., Petroleum geochemistry and exploration of Europe: Geological Society Special Publication 12, Oxford, Endland, Blackwell, p. 255-265.

―――― P. Ungerer, A. Chiarelli, and J. L. Oudin, 1983, Modelisation de la migration de l'huile. Application à deux exemples de bassins sédimentaires: 11th World Petroleum Congress, London, Proceedings, Paper 1, p. 1-13.

Espitalié, J., M. Madec, and B. Tissot, 1980, Role of mineral matrix in kerogen pyrolysis: influence on petroleum generation and migration: AAPG Bulletin, v. 64, p. 59-66.

―――― ―――― ―――― J. J. Mennig, and P. Leplat, 1977, Source rock characterization method for petroleum exploration: 9th Offshore Technology Conference, OTC 2935, v. 3, p. 439-444.

Fischer, A. G., and M. A. Arthur, 1977, Secular variations in the pelagic realm, in Deep-water carbonate environments: SEPM Special Publication 25, p. 19-50.

Gallois, R. W., 1976, Coccolith blooms in the Kimmeridge Clay and origin of North Sea oil: Nature, v. 259, p. 473-475.

Harwood, R. J., 1977, Oil and gas generation by laboratory pyrolysis of kerogen: AAPG Bulletin, v. 61, p. 2082-2102.

Hedberg, H. D., 1968, Significance of high-wax oils with respect to genesis of petroleum: AAPG Bulletin, v. 52, p. 736-750.

Héroux, Y., A. Chagnon, and R. Bertrand, 1979, Compilation and correlation of major thermal maturation parameters: AAPG Bulletin, v. 63, p. 2128-2144.

Horsfield, B., and A. G. Douglas, 1980, The influence of minerals on the pyrolysis of kerogen: Geochimica et Cosmochimica Acta, v. 44, p. 1119-1131.

Huang, W. Y., and W. G. Meinschein, 1979, Sterols as ecological indicators: Geochimica et Cosmochimica Acta, v. 43, p. 739-745.

Huc, A. Y., 1978, Geochimie organique des schistes bitumineux du Toarcien du Bassin de Paris: PhD thesis, Université Louis Pasteur, Strasbourg, France.

Huff, K. F., 1980, Frontiers of world exploration, in A. D. Miall, ed., Facts and principles of world petroleum occurrence: Canadian Society of Petroleum Geologists Memoir 6, p. 343-362.

Hunt, J. M., 1979, Petroleum geochemistry and geology: San Francisco, W. H. Freeman and Co., 617 p.

Hussler, G., B. Chappe, P. Wehrung, and P. Albrecht, 1981, C_{27}-C_{29} ring A monoaromatic sterioids in Cretaceous black shales: Nature, v. 294, p. 556-558.

Ishiwatari, R., M. Ishiwatari, I. R. Kaplan, and B. G. Rohrback, 1976, Thermal alteration of young kerogen in relation to petroleum genesis: Nature, v. 264, p. 347-349.

―――― B. G. Rohrback, and I. R. Kaplan, 1978, Hydrocarbon generation by thermal alteration of kerogen from different sediments: AAPG Bulletin, v. 62, p. 687-692.

Jones, R. W., 1980, Some mass balance and geological constraints on migration mechanisms, in Problems of petroleum migration: AAPG Studies in Geology 10, p. 47-68.

Klemme, D. H., 1980, The geology of future petroleum resources: Revue de l'Institut Français du Pétrole, v. 25, p. 337-349.

―――― 1981, Types of petroliferous basins, in J. F. Mason, ed., Petroleum geology of China: Tulsa, Oklahoma, Pennwell Publishing Co., p. 101-115.

Larter, S. R., and A. G. Douglas, 1980, A pyrolysis-gas chromatographic method for kerogen typing, in A. G. Douglas and J. R. Maxwell, eds., Advances in organic geochemistry 1979: Oxford, Pergamon Press, p. 579-583.

Lewan, M. D., J. C. Winters, and J. H. McDonald, 1979, Generation of oil-like pyrolyzates from organic rich shales: Science, v. 203, p. 897-899.

Leythaeuser, D., R. G. Schaefer, and A. Yükler, 1980, Diffusion of light hydrocarbon through near-surface rocks: Nature, v. 284, p. 522-525.

―――― ―――― ―――― 1982, Role of diffusion in primary migration of hydrocarbons: AAPG Bulletin, v. 66, p. 408-429.

―――― A. Mackenzie, R. G. Schaefer, and M. Bjorøy, 1984, A novel approach for recognition and quantitation of hydrocarbon migration effects in shale-sandstone sequences: AAPG Bulletin, v. 68, p. 196-219.

―――― R. G. Schaefer, C. Cornford, and B. Weiner, 1979, Generation and migration of light hydrocarbons (C_2-C_7) in sedimentary basins: Organic Geochemistry, v.1, p. 191-204.

Lopatin, N. V., 1971, Temperature and geologic time as factors in coalification: Akademiya Nauk SSSR Izvestiya, Seriya Geologicheskaya, n. 3, p. 95-106 (in Russian).

Ludwig, B., 1982, Steroides aromatiques de sédiments et pétroles: PhD thesis, Université Louis Pasteur, Strasbourg, France, 143 p.

―――― G. Hussler, P. Wehrung, and P. Albrecht, 1981, C_{26}-C_{29} triaromatic steroid derivatives in sediments and petroleum: Tetrahedron Letters, v. 22, p. 3313-3316.

Lutz, M., J. P. H. Kaasschieter, and D. H. van Wijke, 1975, Geological factors controlling Rotliegend gas accumulations in the Mid-European basin: 9th World Petroleum Congress Proceedings, v. 2, p. 93-103.

Mackenzie, A. S., C. Beaumont, and D. P. Mackenzie, in press, Estimation of kinetics of chemical reactions with geophysical basin models and applications, in P. A. Schenck, ed.: Advances in organic geochemistry 1983.

―――― C. F. Hoffman, and J. R. Maxwell, 1981, Molecular parameters of maturation in the Toarcian shales, Paris basin, France. III. Changes in the aromatic steroid hydrocarbons: Geochimica et Cosmochimica Acta, v. 45, p. 1345-1355.

―――― N. A. Lamb, and J. R. Maxwell, 1982, Steroid hydrocarbons and the thermal history of sediments: Nature, v. 295, p. 223-226.

―――― J. M. E. Quirke, and J. R. Maxwell, 1980a, Molecular parameters of maturation in the Toarcian shales, Paris basin, France; II, evolution of metalloporphyrins, in A. G. Douglas and J. R. Maxwell, eds., Advances in organic geochemistry 1979: Physics and Chemistry of the Earth, Oxford, England, Pergamon Press, v. 12, p. 239-248.

―――― D. Leythaeuser, R. G. Schaefer, and M. Bjorøy, 1983, Expulsion of petroleum hydrocarbons from shale source rocks: Nature, v. 301, p. 506-509.

―――― R. L. Patience, J. R. Maxwell, M. Vandenbroucke, and B. Durand, 1980b, Molecular parameters of maturation in the Toarcian shales, Paris basin, France; I, changes in the configuration of acyclic isoprenoid alkanes, steranes, and triterpanes: Geochimica et Cosmichimica Acta, v. 44, p. 1709-1721.

Mackenzie, D., 1981, The variation of temperature with time and hydrocarbon maturation in sedimentary basins formed by extension: Earth and Planetary Science Letters, v. 55, p. 87-98.

Masran, Th. C., and S. A. Pocock, 1981, The classification of plant-derived particulate organic matter, in J. Brooks, ed., 5th International Palynological Conference, Proceedings: London, Academic Press, p. 145-175.

Maxwell, J. R., J. M. E. Quirke, and G. Eglinton, 1980, Aspects of modern porphyrin geochemistry and the Treibs hypothesis, in A. A. Prashnowsky, ed., Internationales Alfred Treibs Symposium, Munich 1979: Munich, West Germany,Universität Wurzburg, p. 37-55.

McAuliffe, C. D., 1980, Oil and gas migration: chemical and physical constraints, in Problems of petroleum migration: AAPG Studies in Geology 10, p. 89-107.

Meissner, F. F., 1978, Petroleum geology of the Bakken formation, Williston basin, North Dakota and Montana, in The economic geology of the Williston basin; Montana, North Dakota, South Dakota, Saskatchewan, Manitoba: Williston Basin Symposium, Montana Geological Society, p. 207-227.

Michaelis, W., and P. Albrecht, 1979, Molecular fossils or Archaebacteria in kerogen: Naturwissenschaften, v. 66, p. 420-421.

Miknis, F. P., D. A. Netzel, J. W. Smith, M. A. Mast, and G. E. Maciel, 1982, ^{13}C NMR measurements of the genetic potential of oil shales: Geochimica et Cosmochimica Acta, v. 46, p. 977-984.

Moldowan, J. M., and W. K. Seifert, 1979, Head-to-head linked isoprenoid hydrocarbons in petroleum: Science, v. 204, p. 169-171.

Momper, J. A., 1978, Oil migration limitations suggested by geological and geochemical considerations, in Physical and chemical constraints on petroleum migration: AAPG Continuing Education Short Course Note Series 8, p. B1-B60.

Monin, J. C., B. Durand, M. Vandenbroucke, and A. Y. Huc, 1980, Experimental simulation of the natural transformation of kerogen, in A. G. Douglas and J. R. Maxwell, eds., Advances in organic geochemistry 1979: Physics and Chemistry of the Earth, Oxford, England, Pergamon Press, v. 12, p. 517-530.

Moschopedis, S. E., S. Parkash, and J. G. Speight, 1978, Thermal decomposition of asphaltenes: Fuel, v. 57, p. 431-434.

Nederlof, M. H., 1979, The use of habitat of oil models in exploration prospect appraisal, in Exploration, supply and demand: 10th World Petroleum Congress, Bucharest, Proceedings, v. 2, p. 13-21.

North, F. K., 1979, Episodes of source-sediment deposition; 1: Journal of Petroleum Geology, v. 2, p. 199-218.

Olenin, V. B., 1977, Petroleum geology terrain classification using the criterion of genesis: Moscow, Izdatel'stvo Nedra, 223 p. (in Russian).

Orr, W. L., 1983, Comments on pyrolytic hydrocarbon yields in source rock evaluation, in M. Bjorøy, C. Albrecht, C. Corning, K. de Groot, G. Eglinton, and G. Speers, eds., Advances in organic geochemistry 1981: Chichester, John Wiley, p. 775-787.

Ottenjann, K., M. Wolf, and E. Wolff-Fischer, 1981, Beziehungen zwischen der Fluoreszenz von Vitriniten und den technologischen Eigenschaften von Kohlen, in International Conference on Coal Science, Proceedings: Essen, Glückauf, p. 86-91.

Ourisson, G., P. Albrecht, and M. Rohmer, 1979, The hopanoids; paleochemistry and biochemistry of a group of natural products: Pure and Applied Chemistry, v. 51, p. 709-729.

Palmer, S. E., and E. W. Baker, 1978, Copper porphyrins in deep-sea sediments: a possible indicator of oxidized terrestrial organic matter: Science, v. 201, p. 49-51.

Peters, K. E., 1978, Effects on sapropelic and humic proto-kerogen during laboratory-simulated geothermal maturation experiments: PhD thesis, University of California at Los Angeles, Los Angeles, California, 172 p.

——— G. Rohrback, and I. R. Kaplan, 1981, Geochemistry of artificially heated humic and sapropelic sediments—I: protokerogen: AAPG Bulletin, v. 65, p. 688-705.

Pfaltz, A., B. Jaun, A. Fässler, A. Eschenmoser, R. Jaenchen, H. H. Gilles, G. Diekert, and R. K. Thauer, 1982, Zur Kenntnis des Faktors F 430 aus methanogenen Bakterien; Struktur des porphinoiden Ligandsystems: Helvetica Chimica Acta, v. 65, p. 828-865.

Pfeiffer, J. P., and R. N. J. Saal, 1940, Asphaltic bitumen as colloid system: Journal of Physical Chemistry, v. 44, p. 139-149.

Powell, T. G., S. Creaney, and L. R. Snowdon, 1982, Limitations of use of organic petrographic techniques for identification of petroleum source rocks: AAPG Bulletin, v. 66, p. 430-435.

Quirke, J. M. E., G. J. Shaw, P. D. Soper, and J. R. Maxwell, 1980, Petroporphyrins: II the presence of porphyrins with extended alkyl side chains: Tetrahedron, v. 36, p. 3261-3267.

Radke, M., D. H. Welte, and H. Willsch, 1982, Geochemical study on a well in the Western Canada basin: relation of the aromatic distribution pattern to maturity of organic matter: Geochimica et Cosmochimica Acta, v. 46, p. 1-10.

Ritter, U., in press, The influence of time and temperature on vitrinite reflectance, in P. A. Schenck, ed., Advances in organic geochemistry 1983.

Robert, P., 1980, The optical evolution of kerogen and geothermal histories applied to oil and gas exploration, in B. Durand, ed., Kerogen; insoluble organic matter from sedimentary rocks: Paris, Éditions Technip, p. 385-414.

Rogers, M. A., 1979, Application of organic facies concept to hydrocarbon source evaluation, in Exploration, supply and demand: 10th World Petroleum Congress, Bucharest, Proceedings, v. 2, p. 23-30.

Rubinstein, I., C. Spyckerelle, and O. P. Strausz, 1979, Pyrolysis of asphaltenes: a source of geochemical information: Geochimica et Cosmochimica Acta, v. 43, p. 1-6.

Samman, N., T. Ignasiak, C. J. Chen, O. P. Strausz, and D. S. Montgomery, 1981, Squalene in petroleum asphaltenes: Science, v. 213, p. 1381-1383.

Schaefer, R. G., and D. Leythaeuser, 1980, Analysis of trace amounts of hydrocarbons (C_2-C_8) from rock and crude oil samples and its application in petroleum geochemistry, in A. G. Douglas and J. R. Maxwell, eds., Advances in organic geochemistry 1979: Physics and Chemistry of the Earth, Oxford, England, Pergamon Press, v. 12, p. 149-156.

——— ——— 1983, Generation and migration of low molecular weight hydrocarbons in sediments in Site 511 of DSDP/IPOD Leg 71, Falkland Plateau, South Atlantic, in M. Bjorøy, C. Albrecht, C. Corning, K. de Groot, G. Eglinton, and G. Speers, eds., Advances in organic geochemistry 1981: Chichester, John Wiley, p. 164-174.

Schlanger S. O., and H. C. Jenkyns, 1976, Cretaceous anoxic events: causes and consequences: Geologie en Mijnbouw, v. 55, p. 179-184.

Schmitter, J. M., P. J. Arpino, and G. Guiochon, 1981, Isolation of degraded pentacyclic triterpenoids acids in a Nigerian crude oil and their identification as tetracyclic carboxylic acids resulting from ring A cleavage: Geochimica et Cosmochimica Acta, v. 45, p. 1951-1955.

Schoell, M., 1980, The hydrogen and carbon isotopic composition of methane from natural gases of various regions: Geochimica et Cosmochimica Acta, v. 44, p. 649-661.

Seifert, W. K., 1978, Steranes and terpanes in kerogen pyrolysis for correlation of oils and source rocks: Geochimica et Cosmochimica Acta, v. 42, p.473-484.

——— and J. M. Moldowan, 1978, Applications of steranes, terpanes and monoaromatics to the maturation, migration and source of crude oils: Geochimica et Cosmochimica Acta, v. 42, p. 77-95.

——— ——— 1981, Paleoreconstruction by biological markers: Geochimica et Comoschimica Acta, v. 45, p. 783-794.

——— ——— and R. W. Jones, 1980, Application of biological marker chemistry to petroleum exploration, in Exploration, supply and demand: 10th World Petroleum Congress, Bucharest, Proceedings, v. 2, p. 425-438.

Sluijk, D., and M. H. Nederlof, in press, Worldwide geological and geochemical experience as a systematic basis for prospect appraisal, in Basin geochemistry: AAPG Memoir 35.

Smith, J. E., J. G. Erdman, and D. A. Morris, 1971, Migration accumulation and retention of petroleum in the earth: 8th World Petroleum Congress Proceedings, v. 2, p. 13-26.

Snarsky, A. N., 1962, Die primäre migration des Erdöls: Freiberger Forschungshefte, v. C123, p. 63-73.

Snowdon, L. R., 1980, Resinite—a potential petroleum source in the Upper Cretaceous/Tertiary of the Beaufort-Mackenzie sedimentary basin, in A. D. Miall, ed., Facts and principles of world petroleum occurrence: Canadian Society of Petroleum Geologists Memoir 6, p. 509-521.

Sokolov, B. A., 1980, Evolution and hydrocarbon potential of sedimentary basins: Moscow, Nauka (in Russian).

Speight, J. G., and S. E. Moschopedis, 1979, Some observations on the molecular "nature" of petroleum asphaltenes: Division of Petroleum Chemistry, American Chemical Society Preprints, v. 24, p. 910-923.

Spyckerelle, C., A. Greiner, P. Albrecht, and G. Ourisson, 1977a, Hydrocarbures aromatiques d'origine geologique; III, Un tetrahydrochrysene, derive de triterpenes, dans les sediments recents et anciens; 3,3,7-trimethyl 1,2,3,4-tetrahydrochrysene: Journal of Chemical Research, (summary) p. 330-331; (full manuscript) p. 3746-3777.

——— ——— ——— ——— 1977b, Hydrocarbures aromatiques d'origine geologique; IV, Un octahydrochrysene, derive de triterpenes, dans un schiste bitumineux; 3,3,7,13-tetramethyl 1,2,3,4,11,12,13,14-octahydrochrysene: Journal of Chemical Research (summary), p. 332-333; (full manuscript), p. 3801-3828.

Stach, E., M. T. Mackowsky, M. Teichmüller, G. H. Taylor, D. Chandra, and R. Teichmüller, eds., 1975, Stach's textbook of coal petrology, 3d edition: Berlin and Stuttgart, Gebrüder Borntraeger, 428 p.

Stahl, W., 1978, Reifeabhängigkeit der Kohlenstoff-Isotopenverhältnisse des Methans von Erdölgasen aus Norddeutschland: Erdöl und Kohle, Erdgas, Petrochemie, v. 31, p. 515-518.

Teichmüller, M., 1982, The importance of coal petrology in prospecting for oil and natural gas, in E. Stach, M. T. Mackowsky, M. Teichmüller, G. H. Taylor, D. Chandra, and R. Teichmüller, eds., Textbook of coal petrology, 3d edition: Berlin and Stuttgart, Gebrüder Borntraeger, p. 399-412.

———— and B. Durand, 1983, Fluorescence microscopical rank studies on liptinites and vitrinites in peat and coals, and comparison with the results of the Rock Eval pyrolysis: International Journal of Coal Geology, v. 2, p. 197-230.

———— and K. Ottenjann, 1977, Art und Diagenese von Liptiniten und lipoiden Stoffen in einem Erdölmuttergestein auf Grund fluoreszenzmikroskopischer Untersuchungen: Erdöl Kohle, v. 30, p. 387-398.

———— R. Teichmüller, and H. Bartenstein, 1979, Inkohlung und Erdgas in Nordwestdeutschland. Eine Inkohlungskarte der Oberfläche des Oberkarbons: Fortschritte in der Geologie von Rheinland und Westfalen, v. 27, p. 137-170.

Thomas, B. M., 1981, Land plant source rocks for oil and their significance in Australian basins: APEA Journal, v. 22, p. 164-178.

Tissot, B., 1969, Premières données sur les mécanismes et la cinétique de la formation du pétrole dans les sédiments; simulation d'un schéma réactionnel sur ordinateur: Revue de l'Institut Français du Pétrole, v. 24, p. 470-501.

———— 1979, Effect on prolific petroleum source rocks and major coal deposits caused by sea-level changes: Nature, v. 277, p. 462-465.

———— 1981, Connaissances actuelles sur les produits lourds du pétrole: Revue de l'Institut Français du Pétrole, v. 36, p. 429-446.

———— and G. Bessereau, 1982, Géochimie des gaz naturels et origine des gisements de gaz en Europe occidentale: Revue de l'Institut Français du Pétrole, v. 37, p.63-77.

———— and J. Espitalié, 1975, L'évolution thermique de la matière organique des sédiments; applications d'une simulation mathématique potentiel petrolier des bassins sedimentaires et reconstitution de l'histoire thermique des sediments: Revue de l'Institut Français du Pétrole, v. 30, p. 743-777.

———— and R. Pelet, 1971, Nouvelles données sur les mécanismes de genèse et de migration du pétrole, simulation mathématique et application à la prospection: 8th World Petroleum Congress Proceedings, v. 2, p. 35-46.

———— and M. Vandenbroucke, 1983, Geochemistry and pyrolysis of oil shales: Preprints, Division of Fuel Chemistry, American Chemical Society, v. 28, p. 92-99.

———— and D. H. Welte, 1978, Petroleum formation and occurrence; a new approach to oil and gas exploration: Berlin, Heidelberg, New York, Springer-Verlag, 538 p.

———— J. F. Bard, and J. Espitalié, 1980, Principal factors controlling the timing of petroleum generation, in A. D. Miall, ed., Facts and principles of world petroleum occurrence: Canadian Society of Petroleum Geologists Memoir 6, p. 143-152.

———— G. Deroo, and J. P. Herbin, 1979, Organic matter in Cretaceous sediments of the North Atlantic: contribution to sedimentology and paleogeography, in M. Talwani, C. G. Harrison, and D. E. Hayes, eds., Deep drilling results in the Atlantic Ocean: continental margins and paleoenvironment: Washington, American Geophysical Union, Maurice Ewing Series 3, p. 362-401.

———— G. Démaison, P. Masson, J. R. Delteil, and A. Combaz, 1980, Paleoenvironment and petroleum potential of middle Cretaceous black shales in Atlantic basins: AAPG Bulletin, v. 64, p. 2051-2063.

Treibs, A., 1934, Chlorophyll und Häminderivate in bituminösen Gesteinen: Erdölen und Erdwachsen und Asphalten: Annalen der Chemie, v. 510, p. 42-62.

Trendel, J. M., A. Restle, J. Connan, and P. Albrecht, 1982, Identification of a novel series of tetracyclic terpene hydrocarbons (C_{24}-C_{27}) in sediments and petroleum: Journal of the Chemical Society, Chemical Communications, p. 304-306.

Ungerer, P., F. Bessis, P. Y. Chenet, J. M. Ngokwey, E. Nogaret, and J. F. Perrin, 1983, Geological deterministic models and oil exploration: principals and practical examples (abs.): AAPG Bulletin, v. 67, p. 185-186.

Vail, P. R., R. M. Mitchum, Jr., and S. Thompson, III, 1977, Seismic stratigraphy and global changes of sea level, part 4: global cycles of relative changes of sea level, in Seismic stratigraphy—applications to hydrocarbon exploration: AAPG Memoir 26, p. 83-97.

Vandenbroucke, M., B. Durand, and J. L. Oudin, 1983, Detecting migration phenomena in a geological series by means of C_1-C_{35} hydrocarbon amounts and distribution, in M. Bjorøy, C. Albrecht, C. Corning, K. de Groot, G. Eglinton, and G. Speers, eds., Advances in organic geochemistry 1981: Chichester, John Wiley, p. 147-155.

Waples, D. W., 1980, Time and temperature in petroleum formation—application of Lopatin's method to petroleum exploration: AAPG Bulletin, v. 64, p. 916-926.

Weitkamp, A. W., and L. C. Gutberlet, 1968, Application of a microretort to problems in shale pyrolysis: Division of Petroleum Chemistry, American Chemical Society Preprints, v. 13, p. F71-F85.

Welte, D. H., and A. Yükler, 1981, Petroleum origin and accumulation in basin evolution—a quantitative model: AAPG Bulletin, v. 65, p. 1387-1396.

———— H. W. Hagemann, A. Hollerbach, D. Leythaeuser, and W. Stahl, 1975, Correlation between petroleum and source rock: 9th World Petroleum Congress Proceedings, v. 2, p. 179-191.

———— R. G. Schaefer, M. Radke, and H. M. Weiss, 1982, Origin, migration, and entrapment of natural gas in Alberta Deep Basin (abs.): AAPG Bulletin, v. 66, p. 642.

Winters, J. C., J. A. Williams, and M. D. Lewan, 1983, A laboratory study of petroleum generation by hydrous pyrolysis, in M. Bjorøy, C. Albrecht, C. Corning, K. de Groot, G. Eglinton, and G. Speers, eds., Advances in organic geochemistry 1981: Chichester, John Wiley, p. 524-533.

Yen, T. F., 1979, Structural difference between petroleum and coalderived asphaltenes: Division of Petroleum Chemistry, American Chemical Society Preprints, v. 24, p. 901-909.

Yükler, M. A., C. Cornford, and D. H. Welte, 1978, One-dimensional model to simulate geologic, hydrodynamic and thermodynamic development of a sedimentary basin: Geologisches Rundschau, v. 67, p. 960-979.

MODERN APPROACHES IN SOURCE-ROCK EVALUATION
DOUGLAS W. WAPLES[1]

ABSTRACT

It is well established that oil and gas are formed by the decomposition of organic matter (kerogen) at elevated temperatures (above approximately 70°C). Numerous techniques have been developed for evaluating the hydrocarbon-source potential of a possible source rock. Quantity, type, and thermal maturity of its kerogen can all be determined by a variety of analytical methods. These measurements present occasional difficulties, however, and should always be checked for possible errors.

In the last few years, the traditional analysis-based approach to petroleum geochemistry is gradually being replaced by a model-based approach. Conceptual models are constructed first; then measured data serve as tests of the model, which is continuously revised to reflect new knowledge. The use of a model provides a much better framework for interpolating among sparse data points, or for extrapolating beyond data. In addition, the model can be used even in frontier or explorationally immature areas where data are unavailable.

Models currently in use predict organic facies, thermal maturity, migration and volumes of hydrocarbons generated, migrated, and trapped.

INTRODUCTION

In the last few years several important technical and philosophical advances have greatly increased the role of organic geochemistry in petroleum exploration. Virtually every major oil company invests hundreds of thousands or millions of dollars annually in geochemical studies. Some of the larger independent exploration companies have followed suit, as have a few small independents. In short, in the last five years, there has been an increased awareness throughout much of the petroleum industry of the value and utility of modern geochemical techniques in exploration.

To what can we attribute this new attitude? Certainly the persistence of geochemists is in part responsible, for organic-geochemical programs have existed within some companies for 30 years or more. Secondly, geochemistry has become more user-friendly as exploration geologists have learned about it. Thirdly, geochemists are finally developing predictive models of great value in frontier areas or in inadequately tested parts of mature provinces. Finally, geochemists are learning to speak the language of the explorationist and to answer those questions of most pressing concern to exploration programs. Communication between geochemists and geologists is blossoming as never before, and should continue to do so in the coming years.

MECHANISMS OF HYDROCARBON GENERATION

In the last 25 years, it has become generally accepted that both oil and natural gas are produced by the thermal breakdown of large, complex, organic molecules as a result of elevated temperatures acting over extended periods of time in subsurface environments. Although the chemical reactions involved in hydrocarbon generation are manifold and the details of the mechanisms involved not yet completely understood, the basic facts seem incontrovertible. Plant and algal debris deposited under reducing conditions is preserved in sediments, and begins to be transformed to kerogen during early burial. These kerogen molecules are reasonably stable at low temperatures, but at temperatures exceeding about 70 or 80°C (160 to 180°F) the rate of kerogen decomposition becomes significant if adequate time spans are available.

During thermal decomposition of kerogen, both liquid and gaseous hydrocarbons are produced (Figure 1). The exact sequence of liquid and gas generation is still a matter of some debate, but most workers agree that the main phase of oil generation precedes the main phase of gas generation. Thus, most kerogens will first yield liquid hydrocarbons, with a small proportion of wet gas. Later, as oil generation approaches completion, gas generation increases, both in absolute quantity and in comparison with oil generation. Furthermore, as gas generation proceeds the gaseous products become drier and drier; that is, the proportion of methane in-

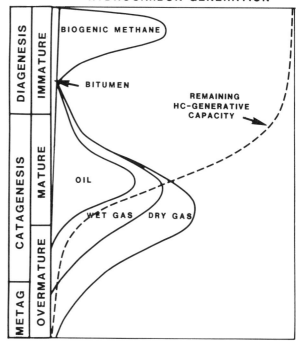

Figure 1. Generalized diagram of hydrocarbon generation.

---
[1] Geochemical Consultant, 1717 Place One Lane, Garland TX 75042 and Vice-president for Exploration Science, Resource Reconnaissance, Inc., 5055 Keller Springs Road, Suite 404, Dallas, TX 75248.

creases. The last products are dry gases, whose hydrocarbon components consist almost exclusively of methane. Dry gases also often contain appreciable amounts of carbon dioxide and nitrogen that were generated concurrently with the methane.

The basic sequence of hydrocarbon-generation processes depicted in Figure 1 is valid for the various types of kerogen. Some kerogens yield mainly liquids, with far less gas than kerogen shown in Figure 1. Conversely, many other kerogens will yield much lower quantities of oil compared to gas.

There can also be some modification to the sequence if resinite is present in the kerogen. Snowdon and Powell (1982) and Teichmuller and Durand (1983) have shown that at least some resinites (from plant resin) decompose very early to yield liquid hydrocarbons more similar to those of condensates than to those present in normal oils. Cases where resinite is a dominant component of kerogen appear to be exceptional, but in some Tertiary shales and coals resinites may represent 5-15% of the organic matter. In such cases the generation sequence would be condensate-oil-wet gas-dry gas, with overlap between adjacent stages.

The early stage of organic transformation occurring at low temperatures is often referred to as *diagenesis*. The stage during which liquid and wet-gas hydrocarbons are produced by thermal decomposition of kerogen is called *catagenesis*, whereas the late stage involving dry gas formation is called *metagenesis*. During late catagenesis and metagenesis, the rates at which petroleum and wet gas hydrocarbons are cracked exceed the rate at which they are formed from kerogen. Thus oil itself becomes a source for methane, ethane, and other small hydrocarbon molecules.

In recent years it has become apparent that much gas found in shallow reservoirs is not produced by the processes outlined above. These shallow gases, which consist almost exclusively of methane, are generated by bacteria (Schoell, 1982, 1983; Rice, 1983). Bacterially produced methane is formed in sediments at low temperatures during diagenesis as a by-product of bacterial metabolism of certain small organic molecules. The existence of bacterial methane has been known for a long time, but its common occurrence in reservoirs is a fact which has only recently been appreciated. In some provinces, in fact, at least half the economically recoverable methane is attributed to bacteria (Claypool and others, 1980; Rice and Claypool, 1981; Mattavelli, and others, 1983.

One major problem has been to distinguish bacterial methane produced during diagenesis from thermal gases produced during catagenesis and metagenesis. Because methane (CH_4) is a very simple molecule, it does not bring with it much information about its origin. One clue lies in a property already mentioned: bacterial gas is very dry, containing only trace amounts of hydrocarbons other than methane. This property distinguishes it from catagenetic gases, which are much wetter (Rice, 1983). But how can we distinguish bacterial gas from metagenetic gas, which also can be very dry?

To answer this question, we must rely on the two pieces of information each methane molecule carries: the isotopic compositions of its 1) carbon and 2) hydrogen atoms. The carbon atoms in bacterial methane are isotopically light, having $\delta^{13}C$ values between about -55 and $-90°/°°$ versus the PDB standard (Rice, 1983). The limit for biogenic gas at the less negative end of the range is not perfectly known. Schoell, (1980), for example, has used $-60°/°°$ (1982, 1983) and $-65°/°°$ (1980). Recently it has been suggested that $\delta^{13}C$ values of biogenic methane can be as positive as $50°/°°$ (Coleman, 1984). It is probably wise to consider values between -55 and $-65°/°°$ as falling near the limits of our present knowledge.

Metagenetic methane, on the other hand, has a heavier carbon-isotope composition (-20 to $-35°/°°$ or so: Rice, 1983). For comparison, catagenetic methane falls in the range of about -35 to $-55°/°°$. Bernard and others (1976) have combined the wetness and carbon-isotope parameters into a single convenient diagram for distinguishing among biogenic, catagenetic, and metagenetic hydrocarbon gases (Figure 2). $\delta^{13}C$ values of methane become progressively less negative as source maturity increases (Schoell, 1980, 1982; Reitsema and others, 1981; Rice, 1983; James, 1983).

A second, less common way of determining the source of hydrocarbon gas is to measure the isotopic composition of ethane, propane, and heavier hydrocarbons (Stahl and Carey, 1975; James, 1983; Schoell, 1983). This technique is not applicable to extremely mature (dry) gases, because it requires careful separation of adequate amounts of each gas for isotopic determination.

A third method for determining the origin of natural gases involves measuring both the hydrogen- and carbon-isotopic composition of methane, and plotting D versus $\delta^{13}C$ (D represents deuterium, the heavy stable isotope of hydrogen). Schoell (1980, 1982, 1983) and Rice (1983) both find that inclusion of deuterium data helps answer questions about mixing, and may help narrow the "gray" area between biogenic and catagenetic methane.

In summary, the overwhelming majority of geologists and geochemists believe that oil is formed by thermal transformation of kerogen at temperatures that range from 70°C to 150°C, depending on the "cooking" time. Hydrocarbon gases can have a wider range of origins: methane can be formed biogenically at low temperatures

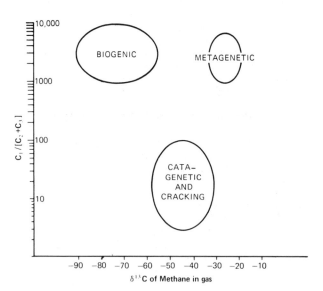

Figure 2. **Determining the origin of natural gas.**

(less than about 70°C); wet gas is formed in small proportion during oil generation, and in cracking of oil. Dry gas is produced during the late stages of metagenesis by decomposition of a variety of larger organic molecules, including Type I, II, and III kerogens, oil, condensate, and wet gas.

A knowledge of the process by which hydrocarbons are formed is important in applying organic geochemistry in an exploration program. It is the basis for source-rock evaluation, by which we determine whether rocks have or ever had the potential to generate hydrocarbons.

METHODS OF SOURCE-ROCK EVALUATION
Definitions

Much of modern petroleum geochemistry depends upon accurate assessment of hydrocarbon-source capacities of sedimentary rocks. Although the term "source rock" is generally employed to describe such rocks, the present usage of that term is too broad. Therefore the following distinctions are made in this paper:

Effective source rock - any sedimentary rock which has already generated and expelled hydrocarbons.

Possible source rock - any sedimentary rock that has not yet been evaluated, but which may have generated and expelled hydrocarbons.

Potential source rock - any immature sedimentary rock known to be capable of generating and expelling hydrocarbons if the level of thermal maturity were higher.

It follows from these definitions that the same stratum could be an effective source rock in one place; a potential source rock in another, less mature area; a possible source rock in another, unstudied region; and conceivably might be known to have no source potential at all in a fourth area where important facies changes had resulted in a drastically content of organic matter. The Phosphoria Formation is a good example of a formation that belongs to each of these classifications in different areas (Claypool and others, 1978).

It is also evident from the preceding discussion that the term "effective source rock" encompasses a wide range of generative histories. When we analyze a rock sample, we actually measure its remaining (untapped) source capacity, G. This quantity is most meaningful if we can compare it to the original source capacity, G_O. G_O must either be estimated or obtained from a sample of the same facies in an immature state; that is, where it is a potential source rock. The difference between G_O and G represents the hydrocarbons already generated in the effective source rock.

Table 1. Categories of Source Rocks and Their Source Capacities.

Category of Source Rock	Original Source Capacity*	Remaining Source Capacity#	Hydrocarbons Generated
Possible	G_O	unmeasured	unmeasured
Potential	G_O	G_O	0
Effective	G_O	G	$G_O - G$

*G_O is not necessarily the same for all rocks.
#Measured in the laboratory.

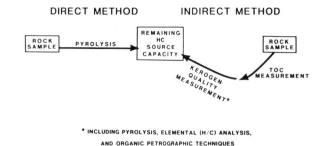

* INCLUDING PYROLYSIS, ELEMENTAL (H/C) ANALYSIS, AND ORGANIC PETROGRAPHIC TECHNIQUES

Figure 3. Direct and indirect methods of determining remaining hydrocarbon-generative capacities of possible source rocks.

Remaining Source Capacity

We noted above that the quantity actually measured in the laboratory is always G, the remaining source capacity. There are currently two different approaches to the problem of determining G. One method measures G directly; the other calculates G based on indirect indicators (Figure 3). In the direct method the rock sample is pyrolyzed (heated rapidly to a high temperature in the absence of oxygen, thus preventing combustion), a process which mimics on a greatly accelerated time scale the natural process of hydrocarbon formation in the subsurface. The quantity of hydrocarbons generated and released from the kerogen during pyrolysis is a direct measure of G. The instrument most frequently used at the present time for such measurements is the Rock-Eval (Espitalie and others, 1977). In the direct method, the Rock-Eval S_2 value is equal to G.

Bernard Durand of the French Petroleum Institute, probably the world's foremost expert on kerogen, has noted some severe problems with Rock-Eval data for shales, however (Durand, 1983). He and other workers (e.g., Katz, 1981; Davis and Stanley, 1982; Espitalie and others, 1980; Dembicki and others 1983) feel that during pyrolysis a strong and variable mineral-matrix effect resulting from interactions of organic molecules with certain clays can yield anomalously low S_2 values in some cases. If the pyrolysis products are then run through a gas-chromatographic column, very different product spectra are obtained with and without the mineral matrix. The extent of the effect in shales is not predictable, but is most important in samples having low TOC values. Neither carbonates nor coals seem to give problems. Durand (1983) therefore recommends that Rock-Eval analyses for shales to be regarded as minimum values, and that in order to obtain accurate data, one should first isolate a kerogen concentrate by dissolving the inorganic matrix in acids, and then pyrolyze the kerogen concentrate.

Kerogen isolation is expensive and time-consuming and cannot be recommended as part of a routine screening program the way whole-rock Rock-Eval pyrolysis has functioned in the last few years. Perhaps the best compromise is to carry out whole-rock pyrolysis to screen for samples that should be studied further as kerogen concentrates. No matter what course the explorationist decides to take in his geochemical studies, the warning from Durand (1983) is clear: don't trust whole-rock pyrolysis data for shales unless there is independent confirmation from pyrolysis of the isolated kerogen.

A new technique for evaluating G employing nuclear magnetic resonance (nmr) on ground whole-rock samples has been proposed by Miknis and others (1982). Further work is necessary to determine its utility, accuracy, and cost-effectiveness over a wide range of samples.

There are three indirect methods for calculating G (Figure 3). All of these approaches break the quantity G down into two components: quantity and type (quality) of organic matter. In all three approaches quantity of organic matter is measured as "total organic carbon" (TOC) content of the rock, expressed in weight percent. The method for measuring TOC is well documented and presents few problems. TOC can also be estimated from gamma-ray logs (e.g., Schmoker, 1981), formation-density logs (Schmoker and Hester, 1983), or rock color (e.g. Charpentier and Schmoker, 1982), but such cases require local calibration and probably will not be of value in most cases.

The three approaches differ in the ways in which they determine the type of organic matter present. The first of these again utilizes the Rock-Eval instrument. Rock-Eval pyrolysis is carried out as before, but the resultant S_2 value is first divided by the TOC value and then multiplied by 100 to give what is called the "hydrogen index" (Espitalie and others, 1977). Dividing S_2 by TOC content normalizes the pyrolysis yield, so that comparisons between rocks are based on the yield per unit of organic carbon rather than per unit of rock. The hydrogen index is thus an indicator of kerogen type, whereas S_2 is an indicator of the combined effects of quantity and type. The only advantage in using this indirect approach as opposed to the direct Rock-Eval method is that the relative effects of TOC and kerogen type on G have been separated. In the direct method there is no way to identify the individual influences of each.

The second approach is elemental analysis (that is, to measure the atomic H/C ratio of the isolated kerogen). It is generally accepted that because hydrogen is the limiting reagent for the reaction that converts kerogen to hydrocarbons, measurement of the H/C ratio of a kerogen appraises indirectly the remaining hydrocarbon-source potential of that kerogen. Saxby (1980) has published an equation relating H/C ratio to oil yield during pyrolysis. Because one must first isolate the kerogen by acid treatment, this method is slow and relatively expensive.

The third approach is to utilize organic-petrographic techniques to estimate the relative abundances of the different kerogen macerals within each sample. Both reflected and transmitted-light microscopy are employed for this purpose, and many workers now also use fluorescence microscopy as an auxiliary tool.

Microscopic organic analysis has had a turbulent history. One problem recognized long ago is that many organic particles are hard to identify, especially when finely divided. More recently, however, Powell and others (1982) have shown that there is a poor correlation between maceral type (identified microscopically) and both hydrogen content and G. Both vitrinite and amorphous material have caused great difficulties. For many years vitrinite has been systematically excluded from the list of macerals having oil-source capacity. However, it has been demonstrated that although most vitrinites follow the conventional wisdom and generate mainly gas, some vitrinites do generate large amounts of liquid hydrocarbons when pyrolyzed. Durand concludes that each vitrinite specimen is chemically unique, even though microscopically they may appear to be the same. He rejects transmitted-light microscopy as an adequate means of appraising kerogen type, and prefers instead to use either elemental analysis or pyrolysis of kerogen concentrates.

The problem of evaluating G is therefore not as simple as it appeared to be a few years ago. Whole-rock pyrolysis has severe limitations in some shales. Pyrolysis or elemental analysis of isolated kerogens is much more expensive and thus is not as well suited for routine sample screening. Perhaps the most cost-effective compromise will prove to be the use of TOC measurements as a preliminary screen, whole-rock pyrolysis as a second screen, and kerogen pyrolysis or elemental analysis, coupled if desired with visual kerogen analysis, for more detailed and accurate studies of the possible source-rock intervals.

Maturity

Knowing a rock's remaining source capacity G solves only one part of the puzzle; it is also necessary to know what level of thermal maturity is represented by G. For example, if G is very low, is it because the rock never had a high initial source capacity, or because the rock is "burned out" (i.e., overmature, in which case virtually all the initial hydrocarbon-source capacity has already been used up)? The exploration implications of these two scenarios are, of course, very different.

In the last two decades, several techniques have been utilized in an effort to assess accurately thermal maturity levels of kerogen (Table 2). These were reviewed and discussed recently by Heroux and others (1979) and Staplin and others (1982). All of the methods have deficiencies, however, and Durand (1983) cautions strongly that care be exercised in interpreting maturity data. The need for concern stems from several sources:

(1) inherent inexactness in the measurements themselves as a result of technical limitations and human error
(2) inappropriateness of many of the measured parameters as indicators of hydrocarbon generation (most measure some unrelated or marginally related change that may not always be concurrent with hydrocarbon generation)
(3) uncertainty about the actual threshold for hydrocarbon generation and expulsion, and about how organic richness and kerogen type affect the threshold.

Table 2. Maturity indicators commonly used in source-rock evaluation

COMMON MATURITY INDICATORS

Vitrinite reflectance (R_O)
Thermal Alteration Index (TAI)
Pyrolysis temperature (T_{max})
Fluorescence
Conodont coloration
Biomarker ratios (GC/MS)

The most popular method today for estimating kerogen maturity is by measuring the reflectance of vitrinite particles. This method was originally developed for measuring rank of coals, where vitrinite is usually very common. This method is based on the fact that as kerogen maturity increases, reflectance values also increase.

In many kerogens, however, vitrinite is rare or absent. Furthermore, reworked vitrinite is much more common in kerogen than in coals. Durand (1983) has emphasized that microscopically indistinguishable but chemically distinct vitrinites may increase in reflectance at different rates. Moreover, vitrinite is not generally thought to be an important oil-generating maceral, and thus any information derived from it is only an indirect indication of oil generation. Finally, the change in vitrinite reflectance might have nothing at all to do with oil generation, even if vitrinite were important in oil generation, because reflectance is a property of the residual kerogen, rather than of the oil itself. In fact, kinetic analyses by Toth and others (1983) indicate that the energetics of reflectance increases are completely different from those of hydrocarbon generation. If their conclusions are correct, the usefulness of vitrinite reflectance is based on a fortuitous coincidence that in some cases may not be even approximately valid.

Thermal Alteration Index (TAI) has been used for about 15 years to determine kerogen maturity. Using transmitted light, a microscopist assesses the darkening of certain kerogen particles, usually recognizable microfossils. The technique is probably more subjective than vitrinite reflectance, but makes maximum use of the human eye, a very sensitive instrument. The microfossils analyzed are more likely to be important hydrocarbon-source materials than vitrinite. On the other hand, the changes measured, like those for vitrinite, are in the residual material rather than in the hydrocarbons, and thus may be unrelated to hydrocarbon generation, or else may have quite different kinetic parameters.

Pyrolysis temperatures are commonly used as maturity indicators, the reasoning being that with increasing maturity the remaining kerogen becomes more and more resistant to pyrolysis. The parameter T_{max} (temperature at which the maximum rate of pyrolysis is occurring) has become a standard part of Rock-Eval data output, and is increasing in popularity. Espitalié and others (1982) and Dembicki and others (1983) have noted, however, that for kerogens of equal maturity, T_{max} is somewhat dependent upon kerogen type. Furthermore, the presence of migrated bitumen in a sample may lower T_{max}. Thus, at the present time it is not possible to set an exact range of T_{max} values that delineate the hydrocarbon-generation window for all kerogens.

Fluorescence has been used by a few workers to indicate maturity, but its applications thus far are somewhat limited. It is an expensive and time-consuming technique when applied as Robert (1981) recommends. Because there are a number of rather subtle difficulties that have not yet been overcome, fluorescence should be regarded as a special technique that is still under development.

Two final difficulties apply to all these techniques. Geochemists have not yet succeeded in defining perfectly either the hydrocarbon-generation or hydrocarbon-expulsion windows. The multiplicity of possible source materials and products makes this a very difficult task, and there is little hope that an exact solution will ever be forthcoming.

A few years ago, an answer to the problems inherent in estimating the progress of hydrocarbon generation by measuring kerogen maturity seemed to have arrived. As a result of pioneering work by Wolf Seifert and his colleagues at Chevron, methods for estimating thermal maturity of bitumen and oils were developed (Seifert and Moldowan, 1978, 1980, 1981; Seifert and others, 1980). When perfected, these techniques were expected to replace to a large extent the kerogen maturity indicators.

Seifert's method involves running hydrocarbons separated from the whole bitumen or crude oil through a combined gas chromatograph-mass spectrometer (GC/MS) system. GC/MS allows one to separate and identify individual compounds among the high-molecular-weight, structurally complex biomarkers (mainly having 20 to 35 carbon atoms and three to five rings). The ratios of concentrations of certain closely related biomarker compounds were reported by Seifert to change systematically with maturity. Mackenzie and his co-workers (1980a, b; 1981a, b) and Radke and others (1982) have extended Seifert's original ideas to other biomarkers.

Unfortunately actual application of GC/MS data to maturity questions has proven more difficult than most workers originally anticipated. First of all, GC/MS can be rather costly because the equipment is expensive. Maintaining costs at a reasonable level requires running many samples per day and developing liberal amortization policies. Secondly, the various biomarker maturity parameters for a single sample are often not internally consistent, and do not change as regularly as was originally thought. Some of these problems may disappear as our experience increases, but there may be more fundamental difficulties as well. For example, complete separation of compounds on the gas chromatograph may not be achieved, leading to inaccurate estimates of compound concentrations. Migration into and out of the rock may distort the original ratios. Various mineral matrices may have different catalytic effects, and the several organic reactions being monitored may respond in different ways to those catalysts.

GC/MS is being used with mixed success at the present time in direct determination of hydrocarbon maturities. It is a unique method that holds promise for the future, but should be regarded as somewhat experimental at the present.

The feeling of most workers today, therefore, is that there is no single maturity indicator that tells the whole story accurately all the time. All of the techniques discussed above are useful and probably reasonably accurate if the analytical work is carefully done. The key to using maturity parameters wisely lies mainly in looking carefully (and sometimes skeptically) at the data, and whenever possible, in obtaining more than one maturity parameter so that problems can more easily be discovered.

A final important question in the evaluation of possible source rocks is what type of hydrocarbons the kerogen will generate. Much progress has been made in this area in the last few years, but delicate questions about gas versus oil remain under debate. The basic statements that can be made with some assurance at the present time are given in Table 3.

Table 3. Kerogen types and their hydrocarbon products.

Kerogen type	Product of Catagenesis	Confidence level
Resinite	Condensate	Accepted by many workers
High sulfur, hydrogen-rich	Heavy, high-sulfur oils	Established
Low-sulfur, Hydrogen-rich, Marine organic matter	Normal oils	Established
Low-sulfur, Hydrogen-rich Terrestrial organics important, although most organics may be algal	Waxy oils	Established
Hydrogen-poor including humic coals	Gas or Gas/oil mix; oil depends on hydrogen content, and may be waxy	Accepted for gas; oils as products are debated, especially from coals.

The most hotly debated point is whether humic coals can generate and expel important amounts of liquid hydrocarbons. Two workers who have considered the issue very carefully have come to diametrically opposed conclusions. Bernard Durand states on the basis of laboratory studies and work on the Mahakam Delta that coals have in some cases been effective source rocks for oil, and that future exploration should take this fact strongly into consideration (Durand and Parratte, 1982). Detlev Leythaeuser (personal communication, 1983) in West Germany is just as emphatically opposed to any significant role for humic coals as oil-source rocks. The issue remains unresolved, but presents a problem of importance in the Rocky Mountain region.

Regional Implications

Even if we assume that in many cases we can adequately define G and the corresponding level of thermal maturity for all possible source rocks in an outcrop section or in a single well profile, we are still missing much informtion that would be of use in completing our source-rock analysis. What we lack is the ability to extrapolate our data through time and space. The data we have were derived from a certain suite of rocks as they now exist in a certain small part of the earths crust. If these rocks are presently beyond the maturity levels at which oil generation and expulsion occurred, they provide no clues as to when the generation, expulsion, and migration took place. If on the other hand they are immature, they give us no real idea about where we might find mature rocks (the answer "deeper" is too qualitative to be very useful). Furthermore, we often obtain from the rocks themselves little evidence regarding the areal extent of these facies, nor a firm basis for predicting what kind of rocks we might find juxtaposed, either vertically or laterally.

At this point we see revealed the main weakness of the analytically based geochemistry that evolved through the late 1970's: in relying almost exclusively on analytical data we constrain ourselves far too much by sample availability. In theory, in order for an analytical approach to be valid, we should have enough data coverage of the area of interest to allow interpolation between points, rather than extrapolation beyond control. In many regions, particularly frontier areas, adequate data are simply not available. As a result, in using an analytical approach we are forced to extrapolate what data we do have across large unsampled voids. Furthermore, drilling is naturally biased in most areas toward settings along basin margins, whose stratigraphy may not be representative of that of the depocenter (Figure 4). We need to overcome the biases and inadequacies of analytical data in order to evaluate lithologically inhomogeneous volumes of rock accurately. Modeling is the best aid in this endeavor.

MODELING

The year 1980 represented a crucial turning point for the successful application of organic geochemistry in hydrocarbon exploration. Two papers were published that brought the concepts of modeling and its potential for extrapolating limited analytical data to the attention of explorationists. One was my discussion of Lopatin's method of predicting thermal maturity, and the other was Demaison and Moore's (1980) analysis of depositional environments conducive to the preservation of organic matter. Both papers drew on years of research by the authors

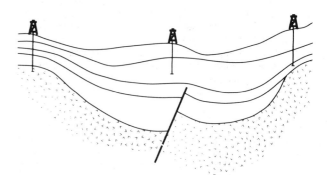

Figure 4. Typical drilling pattern, showing how rocks in the deepest parts of a basin are seldom available for analysis.

and other scientists (for example, Lopatin's paper was originally published in 1971) but prior to 1980 those ideas had mainly been hidden in research laboratories rather than disseminated to exploration personnel.

Thermal Models

Lopatin's method is similar to the Shell LOM method published earlier (Hood and others, 1975), but has been much more successful in catching the interest of explorationists. Lopatin's method and other current models for predicting maturity have been compared recently (Waples, 1983a, 1984). Briefly, Lopatin's method uses the temperature history of a sedimentary rock during its entire burial history as a means of calculating its thermal maturity (Figure 5). Time and temperature are assumed to be the only two factors that materially influence maturity. Maturity calculations can be carried out at any point on the earth's surface provided we can estimate the temperature history of the rocks involved. Such calculations are thus extremely useful in extrapolating beyond data control, especially into depocenters or in frontier areas. If some measured maturity data are available nearby to serve as a check on the accuracy of the high thermal history that we put into our model, very satisfactory results can be obtained. A calibrated model may also be useful in assessing and rectifying assumptions on burial and thermal history assumed in a model. Even where no data are available for checking, the results, while less certain, can often provide important information that will lead to highgrading of exploration prospects.

At the present time, we do not have sufficient data to select a "best" thermal-maturation model among those currently in use. All published models seem to work rather well for their developers within the time-temperature ranges found in most hydrocarbon-source rocks, a comforting result that suggests we can extract much useful information from maturity models even if they are not precisely correct (Waples, 1983a, 1984).

In settings where temperatures are much higher than those normally encountered during hydrocarbon generation (e.g., in geothermal regimes), the effect of temperature becomes so dominant because of its exponential influence on reaction rates that the effect of time may appear to be negligible (Barker, 1983). Barker's results should not be interpreted as indicating that time is insignificant for hydrocarbon generation.

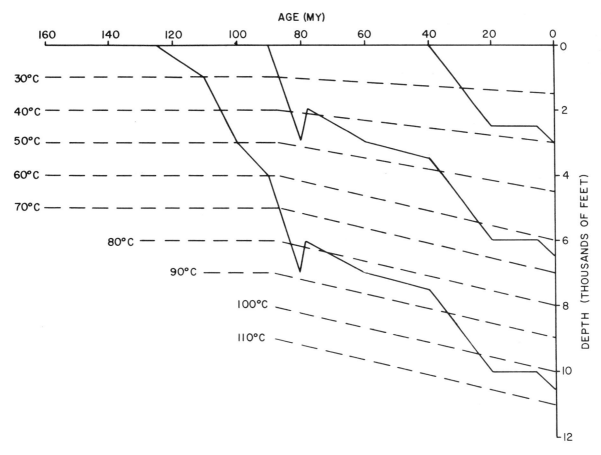

Figure 5. Burial-history curves and thermal histories for calculating maturity by Lopatin's method. After Waples, 1980; reprinted by permission.

Lopatin's method in particular has frequently been critized for being simple-minded (see Waples, 1983a, 1984) but its beauty lies precisely in its simplicity and in the plain fact that the results of many applications to exploration have shown that it generally works well (e.g., Falvey and Deighton, 1982; Magoon and Claypool, 1983; Middleton and Falvey, 1983; Ibrahim, 1983; Bachman and others, 1983; Moshier and Waples, in preparation). Future work will undoubtedly refine these models, but should not drastically change the conclusions being drawn today.

My synopsis in 1980 of Lopatin's method gave some suggestions for its appliction to petroleum exploration. Specifically mentioned were its usefulness in assessing timing of hydrocarbon generation; in predicting preservation deadlines; and in contouring maturity values in plan view. The utility and importance of comparing timing of hydrocarbon generation with trap-forming events have been well established (e.g., Magoon and Claypool, 1983; Tissot and others, 1975; McMillan, 1980; Ibrahim and others, 1981; Ziegler and Spotts, 1978; Leonard, 1983; Angevine and Turcotte, 1983; Ibrahim, 1983; Middleton and Falvey, 1983; Zielinski and Bruchhausen, 1983). Use of calculated maturity values to predict the depth limits for hydrocarbons of any particular composition (especially the oil deadline and the liquid deadline) has proven feasible, although because Lopatin's method was developed to predict generation, its correct application in establishing hydrocarbon deadlines requires some modifications (Waples, 1983a, 1984). Furthermore, the paucity of data in the high-maturity range in my original (1980) calibration of Lopatin's method prevented an accurate correlation between TTI values and gas deadlines (Katz and others, 1982).

Knowledge of depth limits of hydrocarbon stability in untested areas is of great value in evaluating deep plays, and is of particular utility in mature areas where available data from shallow drilling provide a good check for the accuracy of the thermal histories employed. Published applications of thermal modeling to predict the cracking of oil to gas have not been common. The work of Tissot and others (1975), Angevine and Turcotte (1983), and Magoon and Claypool (1983) are of interest in this regard.

Several papers have presented the data from thermal modeling in plan view, as maturity contour maps (e.g., Ayres and others, 1982; Leonard, 1983; Bachman and others, 1983; Ibrahim, 1983). Others, however, have chosen to place predicted iso-maturity lines on cross-sections (e.g., Falvey and Deighton, 1982; Middleton and Falvey, 1983; Zielinski and Bruchhausen, 1983; Moshier and Waples, in preparation) a technique first employed many years ago by coal workers (e.g., Bottcher and others, 1949). Both techniques are useful.

Cross-sections are particularly valuable for understanding the maturity status of an entire basin (Figure 6), whereas plan view usually emphasizes a single stratum or formation. A series of cross-sections representing different times in the past, with the appropriate maturity data superimposed, is an extremely efficient means of discussing the hydrocarbon generation and migration history of an area.

Other interesting and important applications of thermal models have emerged. Magoon and Claypool (1983) noted that predicted and measured maturity values could be brought into agreement by adjusting the paleogeothermal gradient. Thus, use of the thermal models led to an improved understanding of past geologic events. In other cases, discrepancies between measured and predicted data can lead to a reassessment of the magnitude of erosional unconformities.

Of particular importance to geologists working in thrusted areas are current re-evaluations of how thrusting affects the temperatures (and hence thermal maturity) of both the over- and underthrust sheets. Angevine and Turcotte (1983) studied heating of footwall sequences in overthrust belts. Furlong and Edman (1983) have discussed thermal effects of thrusting on both the hanging-wall and footwall sequences. According to those workers, the importance of thermal perturbations will depend on slab thickness, thermal conductivities of the rocks, convective heating, thrusting rates, and erosion rates. Angevine and Turcotte applied their model to the Fossil Syncline of the Wyoming Overthrust Belt. Erdman and Surdam (1983) studied effects of thrusting on the Phosphoria Formation using the model of Furlong and Edman. Further work will be necessary to evaluate the importance of thermal perturbation in other overthrust areas.

Thermal modeling has become an integral part of exploration programs in many oil companies, and is virtually certain to become even more popular. Many major oil companies have developed their own unpublished algorithms for relating time and temperature to hydrocarbon generation; other groups have adopted one of the published models. My own studies suggest that all these models are likely to give satisfactory results in most cases (Waples, 1983a, 1984), so the main problems lie not in the models themselves, but rather in finding creative ways to apply and interpret them.

Organic-facies models

Demaison and Moore (1980) first called attention to ways that explorationists could use sedimentological principles to understand and predict the occurrence of organic-rich rocks having high original source capacities (G_O). Since then other researchers have sought a more detailed knowledge of the most important factors influencing hydrocarbon-source capacities of sedimentary rocks. Most of the debate has centered on the question of the relative importances of productivity and preservation.

It is well known that productivity is unusually high in oceanic regions where upwelling of subsurface nutrient-laden waters occurs. High organic productivity often results in high organic-carbon contents in the underlying sediments. It therefore is logical both to look in the geologic record for evidence of ancient upwelling regimes as possible indicators of the presence of hydrocarbon-source rocks, and to attempt to predict the loci of such upwellings from more basic principles. These approaches have been promoted by Parrish (1982), who superimposed climate patterns on plate-tectonics reconstructions of the Paleozoic in an effort to predict where upwellings would have occurred.

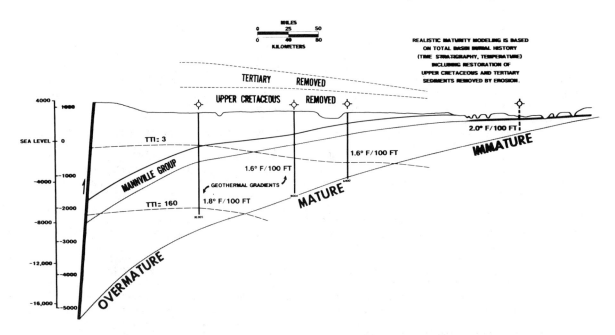

Figure 6. Iso-maturity (TTI) lines on a cross section of the Western Canada Basin. From Mosher and Waples, in preparation.

Although Parrish, Summerhayes (1981a, b; 1983), and some other workers consider upwelling and consequent high productivity to be the most important factor in deposition of many organic-rich sediments, still others prefer to attribute most instances of organic richness to unusually favorable conditions for preservation. As Demaison and Moore (1980), Tissot and others (1979), and others have noted, anoxia or near-anoxia in bottom waters can strongly influence preservation of organic matter. Demaison and Moore (1980) and Kirkland and Evans (1981) catalogued environments in which anoxia can occur.

There are numerous mechanisms capable of producing anoxic waters in these environments. Several of these involve restricting movement of bottom waters, so that oxygen replenishment cannot occur easily; others mechanisms lower oxygen levels through rapid consumption. In evaporite settings high salinities lower the number and diversity of scavengers and predators, thus inhibiting consumption of organic debris. Deciding which mechanism or mechanisms are dominant in any particular case can be rather difficult, because it requires a detailed sedimentologic (and sometimes tectonic) study of the surrounding area. Pisciotto and Garrison (1981), for example, showed that several different mechanisms for creating anoxic depositional environments were operative in the Monterey Formation of California (Figure 7).

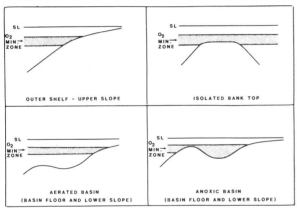

Figure 7. Four different depositional environments capable of preserving organic-rich sediments in the Miocene-age Monterey Formation of California. After Pisciotto and Garrison, 1981; reprinted by permission.

The origin of the common mid-Cretaceous "black shales" of the North and South Atlantic has been a matter of much concern. Numerous ideas, including oceanic or global anoxia, local anoxia caused by either high productivity or local restriction, and global or oceanic oxygen depletion that facilitated local anoxia have been espoused by various workers (see Waples, 1983b, and references therein), but the problem has still not been completely solved.

The combination of local variations in the causes of anoxia and the difficulty we have in distinguishing among these possible causes leads to some problems in using organic-facies models as predictive tools in exploration settings. Nevertheless, if we are to use maturity modeling effectively, we must also be able to model organic facies, for any maturity predictions will only be meaningful if we can also identify the source rocks. It is tempting, of course, to take the path of least resistance, and to assume that the organic characteristics of a certain rock do not change as we move out beyond the area of data control (e.g., Ayres and others, 1982). In areas such as Saudi Arabia, where the source rock is part of an extensive layer-cake package, such assumptions may be reasonably good, but in other less homogeneous areas, they may be disastrous. Even where an organic facies is reasonably homogeneous over large areas there must be limits to its extent.

The Permian complex of eastern Idaho, Wyoming, Utah, and Montana (Phosphoria, Park City, Shedhorn, and Goose Egg formations: Figure 8) is particularly relevant in this respect, changing as it does from an organic-rich phosphatic shale and chert to phosphorite, to non-phosphatic carbonates and shales, then to sands, and finally to red beds (Claypool and others, 1978; McKelvey, 1959). Even within a single member of the Phosphoria organic-geochemical characteristics vary greatly (Claypool and others, 1978). The factors which influence both sedimentology and organic facies are simply too subtle to permit one to make assumptions about homogeneity in most cases.

Nevertheless, it is useful to make whatever statements and predictions we can about organic facies by combining measured data with our facies models. For example, I proposed a model for the development of various phosphatic facies, including phosphatic shales and phosphorites, and attempted to predict the occurrence of phosphatic, organic-rich rocks on the basis of phosphorite occurrences (Waples, 1982). Parrish (1983) has broadened the scope of the phosphorite model to include other types of upwelling deposits as well. These models could be applied to the Phosphoria complex to develop a clearer picture of its source-rock capacities.

Organic-facies modeling is also useful in explaining the origin and predicting the occurrence of high-sulfur oils. It has been established that high-sulfur oils are derived from high-sulfur kerogens. The sulfur in high-sulfur kerogens comes mainly from inorganic sulfate dissolved in bottom waters and pore water. Certain types of anaerobic bacteria reduce sulfate to elemental sulfur and various sulfide species. If iron and other heavy-metal ions are absent, these reduced-sulfur species can react with organic matter to produce high-sulfur kerogens. Most high-sulfur kerogens occur in non-ferruginous carbonates and biogenic siliceous sediments (Gransch and Posthuma, 1974). Fresh-waters generally contain little dissolved sulfate (unless extensive evaporite erosion is occurring in the vicinity), and therefore do not give rise to high-sulfur kerogens. High-sulfur oils are therefore sourced from nonclastic marine sediments deposited under euxinic or other anoxic conditions. Work on the origin of Phosphoria (Powell and others, 1975), Monterey (Orr, 1984), Saudi Arabian (Ayres and others, 1982), and Venezuelan oils (Gransch and Posthuma, 1974) has thus far confirmed the sulfur model.

Another type of organic-facies modeling involves tracing a basin's evolution from birth to cessation of sediment accumulation. World-wide studies of the process of basin evolution have given us some ideas of "normal" sedimen-

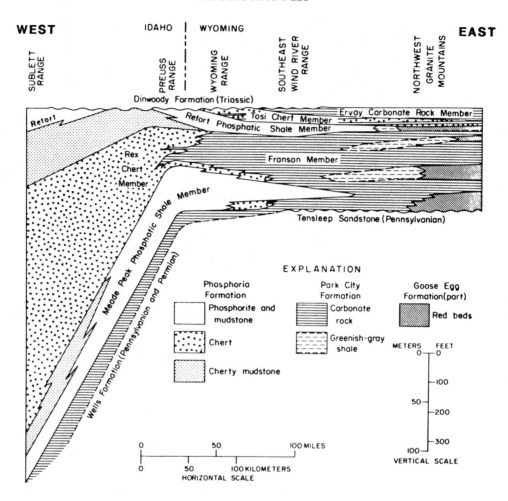

Figure 8. West-East cross section of Phosphoria complex, Idaho and Wyoming. From Claypool and others, 1978.

tary sequences that we can anticipate even in untested basins (e.g., White, 1980; Kingston and others, 1983a, b.) For example, it is well known that rift basins commence accumulating sediment with coarse terrestrial clastics around the basin margin. If marine incursion can be delayed, such as in the present-day African Rift Valleys, fresh-water lakes will develop locally in the middle of the graben. Given adequate water depth (not usually a problem in rift basins) and warm climates, the waters in these lakes may become permanently stratified, and hence anoxic at the bottom. Because subsidence rates are rapid in rift basins, great thicknesses of organic-rich lacustrine muds may accumulate in these lakes, and may eventually become oil-source beds.

The earliest marine incursions into rift basins are likely to be rather tentative. In many such basins, particularly those with warm, arid climates, extensive evaporite sequences (like those of the Aptian in the South Atlantic) will form. Development of hypersalinity is often accompanied by preservation of large amounts of organic matter in black shales juxtaposed in time and space with the salts (Kirkland and Evans, 1981).

As subsidence and basin widening continue, the marine environment becomes less restricted. Even during this period, however, organic-rich sediments may be laid down in areas where productivity is high, where local restrictions occur, or where the oxygen-minimum layer meets the sediment-water interface.

Our general model for the evolution of rift basins thus gives us important clues about possible source rocks. We should consider contributions from lake beds, salt-associated black shales, and marine shales. Integration of our rift-basin models will enhance its effectiveness. Other types of basins are equally amenable to this kind of modeling.

Modeling is of course useful in frontier areas, but is also valuable in mature areas, where enough geochemical and sedimentological data may be available to construct detailed facies maps. I believe that organic-facies modeling can and should be used (in different ways, of course) at every stage of exploration. Facies-modeling capabilities are far from perfected, but they are advanced enough to be of value to explorationists.

Migration

Our understanding of migration is far from complete at the present time, despite many years of effort to elucidate its mechanisms and clarify their implications for exploration.

Nevertheless, even with the gaps in our knowledge of migration we are able to make several important statements that can be integrated into exploration models.

1. Hydrocarbons are mainly expelled from their source rocks (primary migration) as a separate non-aqueous phase (Jones, 1981), as a consequence of potential energy gradients between source rock and adjacent rocks (du Rouchet, 1981; Vandenbroucke and others, 1981; Durand and Oudin, 1981).
2. Hydrocarbon expulsion is almost certainly not 100% efficient. Momper (1978) has suggested a threshold value of hydrocarbon concentration below which no expulsion occurs. Momper's ideas and postulated threshold value of 15 barrels per acre-foot seem reasonable, but have not yet been substantiated.
3. Hydrocarbon migration through porous conduits (secondary migration) also occurs mainly in a discrete, non-aqueous phase. The driving mechanism is principally bouyancy (Gies, 1982a). Movement of subsurface waters may enhance or retard movement caused by buoyancy.
4. The details of primary and secondary migration need to be fully understood in order for explorationists to employ migration models.
5. Lateral (intrastratal) migration is usually much easier and more efficient than vertical (interstratal) migration. By "efficient" I mean that hydrocarbon movement is focused and drainage is from a large volume of source rock. There is thus a greater chance of creating giant or supergiant accumulations.
6. Vertical migration occurs mainly when lateral migration is inhibited by lateral discontinuity of carrier beds. Discontinuity may be caused by faulting or by changes in stratigraphy. Vertical migration is usually more dispersive than is lateral migration, leading to smaller accumulations and probably more loss of hydrocarbons to the earth's surface. Instances in which migration has both lateral and vertical components are also possible. Diffusion is a viable mechanism for vertical movement of gases (not oil), but is by definition a dispersive rather than accumulative mechanism (Leythaeuser and others, 1982).
7. Distance *per se* is not a problem for either lateral or vertical migration providing there is sufficient charge to overcome residual capillary saturation requirements along migration pulls. Rates of subsurface fluid movement are high enough to allow extensive migration in the millions or tens of millions of years normally available. In many cases, however, lateral or vertical migration is limited by the imposition of an effective barrier to migration. Few basins have sufficient size, stratigraphic continuity of carrier beds over large areas, and tectonic stability (lack of disruptive faulting) to permit accumulation of supergiant hydrocarbon deposits. Where these conditions exist, however (e.g., Saudi Arabia, Eastern Venezuela, Alberta), we find the largest accumulations in the world (Demaison, 1977).

    Migration over great vertical distances can occur if continuous fault or fracture conduits exist. Perhaps the best examples of such conduits are intermittently active faults, such as the growth faults in the Gulf Coast or the Niger Delta (du Rouchet, 1981). These faults are dispersive conduits, however, because they allow the moving hydrocarbons to bleed off into any porous strata through which the faults pass. Vertical migration distance is limited by the extent of the vertical conduit system.

8. Given the proper geologic setting lateral migration distances of 200 miles (300 km) or more are easily achieveable. Very few settings will provide lateral continuity over such distances, however. More typical lateral migration distances will not exceed a few tens of miles or kilometers. In cases where lateral continuity is very poor, as in some rift basins, lateral fluid movement may be negligible.
9. Vertical migration is limited by the distance between the earth's surface and the site of hydrocarbon generation. At typical geothermal gradients in sediments of Cenozoic or Mesozoic age oil generation usually occurs somewhere between about 5,000 and 15,000 feet (1,500 and 4,500 meters). Generation and expulsion of gas can occur somewhat deeper — perhaps down to 25,000 feet (7500 meters) in some cases. These distances therefore become our upper limits for vertical migration. Most actual cases will be much more modest, however. Extensive vertical migration involving movement of a few thousand feet through several formations will occur mainly in basins where tensile or generation-induced overpressured fracture systems is active during the phase of hydrocarbon generation (Meissner, 1981). Stacked sand bodies in delta systems can also be important vertical conduits (Galloway and others, 1982).
10. All hydrocarbon accumulations are in a sense kinetic accumulations; that is, they persist because the rate of loss of hydrocarbons through fractured or otherwise permeable "seals" and from biodegradation, oxidation, cracking, and water-washing either has been small during the accumulation's existence, or has been compensated by an influx of newly generated hydrocarbons. The Alberta deep basin gas accumulation has been shown to be a true kinetic accumulation of the latter type (Welte and others, 1982; Gies, 1982b), and it is likely that many of the young Monterey-sourced oil accumulations in Southern California are being replenished faster than they can leak away through their generally poor cap rocks (see McCulloh, 1969). Other examples of true kinetic accumulations, which exist only transiently, are probably more common than we realize.

    Once we recognize that all hydrocarbon accumulations have finite lifetimes because of the many forces conspiring to destroy them, we can at least qualitatively model the effect of time and probabilities for preservation. It is logical to assume that, other things being equal, the more recently an accumulation was formed, the more likely it will be preserved today. It is thus important once again to be able to predict the timing of hydrocarbon generation. A simple comparison of source-rock ages is inadequate, because, for example, Kimmeridgian-age source rocks may have generated hydrocarbons during the Late Cretaceous in one area, and during the Neogene in another.

11. Where intrastratal migration is dominant, it is possible to define drainage areas of prospects if the time of hydrocarbon generation is known, and if paleostructure maps can be created for that time. Pratsch (1982) has illustrated how drainage areas can be calculated by assuming that fluid movement is perpendicular to structural contours. Knowledge of drainage areas is impor-

tant in carrying out volumetric calculations (see next section).

There remain many unanswered questions about migration, some of which are very important in appraising the hydrocarbon potential of a region. For example, what are the relative efficiencies of migration of oil and gas? How efficient are primary and secondary migration? At the present time we are just beginning to answer these questions by carrying out mass-balance calculations on closed systems where both source-rock capacity and in-place hydrocarbons are known. There are relatively few examples that meet all those criteria, however, and those that do may not be typical of hydrocarbon systems world-wide.

Nevertheless, although our use of migration models is in an early learning phase, if we attempt to compare areas rather than derive absolute quantities of hydrocarbons, many of our uncertainties and errors will cancel, and migration models may become valuable.

Volumetric Calculations

It is useful to be able to predict organic richness, thermal maturity, and migration parameters, because these are critical regulators of the hydrocarbon potential of any area. For an explorationist, however, it is of far greater importance to know how much hydrocarbon is trapped within a given prospect, play, or basin. To answer this question one must not only determine or predict source-rock richness, maturity, hydrocarbon type, and migration direction and efficiency, but also integrate these parameters into a single model from which volumetric calculations can be made.

McDowell (1975) presented an early volumetric model in which he calculated the quantity of hydrocarbons generated in the source rock and then partitioned those hydrocarbons among the reservoir, the migration conduit, the source rock, and the surface. Although the partitioning coefficients he used have been improved somewhat as our knowledge of migration systems has increased, the fundamental soundness of his approach has remained unchallenged.

All volumetric approaches share one common philosophical basis, but split into two or three groups on a second important point. The common basis is that in order for hydrocarbons to have accumulated in a reservoir, a series of simultaneous conditions must have been met: the hydrocarbons must have been generated in and expelled from the source rock and then migrated to and been trapped in the reservoir. If the quantity of hydrocarbons that successfully negotiates each of these stages can be estimated, we can carry out an overall volumetric calculation.

From this common basis the various models diverge from each other in two ways. Some models (e.g. Nakayama and Van Siclen, 1981; Welte and Yukler, 1981; Bishop and others, 1983) are deterministic, in that they assume that they provide a complete and accurate description of the hydrocarbon system. Sluijk and Nederlof (in press) have criticized this approach for two reasons: data input may be inaccurate, and the model itself may be incomplete or otherwise deficient. They recommend instead that a calibrated approach be followed, through which the importance of each of the factors in the model can be tested. Using the calibrated approach they found that in fact some of the factors in their original model were unimportant, and thus were able to build a stronger model.

The second way in which the various models differ is in the form of input and output data. The deterministic models of Welte and Yukler (1981) and Nakayama and Van Siclen (1981) use unique values for each piece of input data. Output data thus consist of a single number, which is the most probable value. Probablistic approaches, in contrast, enter data as probability distributions (for example, TOC can vary from 0.5% (p = .00) to 10% (p = 1.00), with a median value of 3% (p = .50)), and carry out all mathematical manipulations on the entire probability distribution using Monte-Carlo procedures (Sluijk and Nederlof, in press; Bishop and others, 1983).

Output data emerge as probability curves as well. The user of such a model is thus not constrained to interpreting a single value, but rather can base his decision on the entire probability curve. For example, consider the output data for two models, one probabilistic and the other non-probabilistic. Suppose the non-probabilistic model yields a single value of 50 million barrels of oil for Prospect A. The probabilistic model, in contrast, might give a result like that in Figure 9.

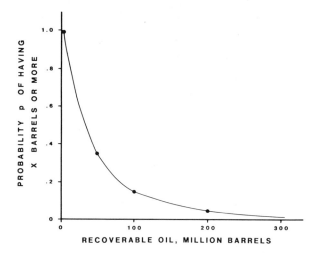

Figure 9. Probabilistic output curve for oil recoverable in Prospect A.

We can extract much more information from Figure 9 than we can from the single value of 50 million barrels. For example, we see that the probability of finding 50 million barrels or more is about 0.35 (35%). The curve also tells us that it is virtually certain (p = .99) that we will find at least 3 million barrels, and that our chances of finding 200 million barrels or more are only 5% (p = .05). If this were an offshore play in an ice-bound area where 100 million barrels was the minimum economic cut off, we would be able to say that there is only a 15% probability of finding that much.

When output data are expressed as probability curves, the data can be used for many creative purposes, because far more information is contained in a probability curve than in a single average value. Furthermore, using probability curves for input data allows one to take into consideration uncertainties in the quality of the data. These uncertainties are automatically translated to the output data.

Sluijk and Nederlof's (in press) criticism of deterministic models is particularly valid *vis-a-vis* the problem of migration. Buoyancy is generally the dominant driving force for migration, but it is likely that other mechanisms (e.g., solution, diffusion, micells) play some role in most cases, and may even be dominant at times. It is beyond our capability to design a volumetric approach that could consider adequately all migration mechanisms. Hence Sluijk and Nederlof's calibration of their non-deterministic model using case studies is sensible. "Migration" is thus treated more as a single variable, rather than being broken down into its components as the deterministic models recommend. Expulsion is also best dealt with by a non-deterministic model. Hydrocarbon generation, in contrast, is somewhat simpler and better understood, and may be amenable to a deterministic approach.

One further improvement to published models could be recommended. While Sluijk and Nederlof (in press) and Bishop and others, (1983) have considered losses of hydrocarbons from traps by spillage and by leakage through the seal, no one has included cracking of oil to gas, oxidation of methane to carbon dioxide, or biodegradation of oil in a model. These are well-known, amply documented phenomena that have a large impact on exploration success. Furthermore, they should be predictable on the basis of thermal maturity of the reservoirs and local hydrodynamics. Future models should include these factors as perturbations of the generation-accumulation system.

CONCLUSIONS

There remain some uncertainties in our capabilities for analyzing and interpreting data from possible source rocks, particularly with regard to determining original and remaining source capacities and present levels of thermal maturity. If proper care is taken, however, reliable results can be obtained.

By combining predictive models for organic facies, thermal maturity, and migration, we can greatly increase the exploration value of our measured data. Thermal models presently in use seem to be adequate, although some refinements would certainly be welcome. Organic-facies models are for the first time being applied in a predictive mode in exploration, and should prove extremely valuable.

Finally, integration of models for generation, expulsion, migration, and preservation of hydrocarbons into a single model that will permit an integrated appraisal of the hydrocarbon potential of an area is underway. Much of the early work in this field will necessarily be somewhat experimental, and will help form the learning set for future studies. Nevertheless, volumetric models are already good enough to be of great utility to explorationists, especially when they can be used qualitatively or semi-quantitatively to compare and contrast areas.

As model-based geochemical approaches replace analysis-based ones, organic geochemistry will become ever more linked with geology and geophysics. The most advanced exploration programs will utilize approaches that integrate these disciplines with each other.

REFERENCES

Angevine, C.L. and D.L. Tourcotte, 1983, Oil generation in overthrust belts: AAPG Bulletin, v. 67, p. 235-241.

Ayres, M.G., M. Bilal, R.W. Jones, L.W. Slentz, M. Tartir, and A.O. Wilson, 1982, Hydrocarbon habitat in main producing areas, Saudi Arabia: AAPG Bulletin, v. 66, p. 1-9.

Bachman, S.B., S.D. Lewis, and W.J. Schweller, 1983, Evolution of a forearc basin, Luzon Central Valley, Philippines: AAPG Bulletin, v. 67, p. 1143-1162.

Barker, C.E., 1983, Influence of time on metamorphism of sedimentary organic matter in liquid-dominated geothermal systems, western North America: Geology, v. 11, p. 384-388.

Bernard, B.B., J.M. Brooks, and W.M. Sackett, 1976, Natural gas seepage in the Gulf of Mexico: Earth and Planetary Science Letters, v. 31, p. 48-54.

Bishop, R.S., H.M. Gehman, Jr., and A. Young, 1983, Concepts for estimating hydrocarbon accumulation and dispersion: AAPG Bulletin, v. 67, p. 337-348.

Bottcher, H., M. Teichmuller, and R. Teichmuller, 1949, Hilt's Law in the Bochem Trough of the Ruhr: Gluckauf, v. 85, p. 81-92 (in German).

Charpentier, R.R. and J.W. Schmoker, 1982, Volume of organic-rich Devonian shale in Appalachian Basin relating "Black" to organic-matter content: AAPG Bulletin, v. 66, p. 375-378.

Claypool, G.E., A.H. Love, and E.K. Maughan, 1978, Organic geochemistry, incipient metamorphism, and oil generation in black shale members of Phosphoria Formation, Western interior United States: AAPG Bulletin, v. 62, p. 98-120.

Claypool, G.E., C.N. Threlkeld, and L.B. Magoon, 1980, Biogenic and thermogenic origins of natural gas in Cook Inlet Basin, Alaska: AAPG Bulletin, v. 64, p. 1131-1139.

Coleman, D.D., 1984, The isotope geochemistry of microbial methane [abs.] in M. Schoell and G. Demaison, eds., Geochemistry of Natural Gases; Program and Abstracts: AAPG Research Conference, 1984, San Antonio, Texas.

Davis, J.B. and J.P. Stanley, 1982, Catalytic effect of smectite clays in hydrocarbon generation revealed by pyrolysis-gas chromatography: Journal of Analytical and Applied Pyrolysis, v. 4, p. 227-240.

Demaison, G.J., 1977, Tar sands and supergiant oil fields: AAPG Bulletin, v. 61, p. 1950-1961.

Demaison, G.J. and G.T. Moore, 1980, Anoxic environments and oil source bed genesis: AAPG Bulletin, v. 64, p. 1179-1209.

Dembicki, H., Jr., B. Horsfield, and T.T.Y. Ho., 1983, Source rock evaluation by pyrolysis gas chromatography: AAPG Bulletin, v. 67, p. 1094-1103.

Durand, B., 1983, Oral Presentations at National Conference on Earth Science, Banff, Sept., 1983, unpublished proceedings.

Durand, B. and J.L. Oudin, 1981, Example of hydrocarbon migration in a deltaic series: Mahakam, Kalimantan, Indonesia: Institut Francais du Petrole Ref. 29092, 16 p.

Durand, B. and M. Parratte, 1982, Oil potential of coals, a geochemical approach: Institut Francais du Petrole, Ref. 30560, 16 p.

du Rochet, J., 1981, Stress fields, a key to oil migration: AAPG Bulletin, v. 65, p. 74-85.

Edman, J.D. and R.C. Surdam, 1983, Influence of overthrusting on maturation of hydrocarbons in Phosphoria Formation, Idaho-Wyoming Overthrust Belt: AAPG Bulletin, v. 67, p. 454-455.

Espitalie, J., J.L. Laporte, M. Madec, F. Marquis, P. Leplat, J. Poulet, and A. Boutefeu, 1977, Rapid method of characterizing source rocks and their petroleum potential and degree of maturity: Revue de l'Institut Francais du Petrole, v. 32, p. 23-42 (in French).

Espitalie, J., M. Madec, and B. Tissot, 1980, Role of mineral matrix in kerogen pyrolysis; Influence on petroleum generation and migration: AAPG Bulletin, v. 64, p. 59-66.

Espitalie, J., F. Marquis, and I. Barsony, 1982, Geochemical logging: Institut Francais du Petrole Ref. 30820, 29 pp.

Falvey, D.A. and I. Deighton, 1982, Recent advances in burial and thermal geohistory analysis: Australian Petroleum Exploration Association Journal, v. 22, pt. 1, p. 65-81.

Furlong, K., and J.D. Edman, 1983, Geophysical approach to determination of hydrocarbon maturation in overthrust terrains: AAPG Bulletin, v. 67, p. 465.

Galloway, W.E., D.K. Hobday, and K. Magara, 1982, Frio Formation of Texas Gulf Coastal plain: depositional systems, structural framework, and hydrocarbon distribution: AAPG Bulletin, v. 66, p. 649-688.

Gies, R.M., 1982a, Basic physical principles of conventional and deep basin gas entrapments [abs.]: AAPG Bulletin, v. 66, p. 572.

_____, 1982b, Origin, migration, and entrapment of natural gas in Alberta deep basin - Part 2 [abs.]: AAPG Bulletin, v. 66, p. 572.

Gransch, J.A., and J. Posthuma, 1974, On the origin of sulfur in crudes; *in* B. Tissot and F. Bienner, eds., Advances in organic geochemistry 1973: Paris, Editions Technip., p. 727-739.

Heroux, Y., A. Chagnon, and R. Bertrand, 1979, Compilation and correlation of major thermal maturation indicators: AAPG Bulletin, v. 63, p. 2128-2144.

Hood, A., C.C.M. Gutjahr, and R.L. Heacock, 1975, Organic metamorphism and the generation of petroleum: AAPG Bulletin, v. 59, p. 986-996.

Ibrahim, M.W., 1983, Petroleum geology of southern Iraq: AAPG Bulletin, v. 67, p. 97-130.

Ibrahim, M.W., M.S. Khan, and H. Katib, 1981, Structural evolution of Harmaliyah oil field, eastern Saudi Arabia: AAPG Bulletin, v. 65, p. 2403-2416.

James, A.T., 1983, Correlation of natural gas by use of carbon isotopic distribution between hydrocarbon components: AAPG Bulletin, v. 67, p. 1176-1191.

Jones, R.W., 1981, Some mass balance and geological constraints on migration mechanisms: AAPG Bulletin, v. 65, p. 103-122.

Katz, B.J., 1981, Limitations of "Rock-Eval" pyrolysis for typing organic matter: AAPG Bulletin, v. 65, p. 944.

Katz, B.J., L.M. Liro, J.E. Lacey, H.W. White, and D.W. Waples, 1982, Time and temperature in petroleum formation: application of Lopatin's method to petroleum exploration: AAPG Bulletin, v. 66, p. 1150-1152.

Kingston, D.R., C.P. Dishroon, and P.A. Williams, 1983a, Global basin classification system: AAPG Bulletin, v. 67, p. 2175-2193.

Kingston, D.R., C.P. Dighroon, and P.A. Williams, 1983b, Hydrocarbon plays and global basin classification: AAPG Bulletin, v. 67, p. 2194-2198.

Kirkland, D.W. and R. Evans, 1981, Source-rock potential of evaporitic environment: AAPG Bulletin, v. 65, p. 181-190.

Leonard, R., 1983, Geology and hydrocarbon accumulations, Columbus basin, offshore Trinidad: AAPG Bulletin, v. 67, p. 1081-1093.

Leythaeuser, D., R.G. Schaefer, and A. Yukler, 1982, Role of diffusion in primary migration of hydrocarbons: AAPG Bulletin, v. 66, p. 408-429.

Lopatin, N.V., 1971, Temperature and geologic time as factors in coalification: Akademiia Nauk SSSR. Izvestiia Seriia Geologicheskalu, no. 3, p. 95-106, (in Russian).

Mackenzie, A.S., R.L. Patience, J.R. Maxwell, M. Vandenbroucke, and B. Durand, 1980a, Molecular parameters of maturation in the Toarcian shales, Paris basin, France — I. changes in the configuration of acyclic isoprenoid alkanes, steranes, and triterpanes: Geochimica et Cosmochimica Acta, v. 44, p. 1709-1721.

Mackenzie, A.S., J.M.E. Quirke, and J.R. Maxwell, 1980b, Molecular parameters of maturation in the Toarcian shales, Paris basin, France - II. Evolution of metalloporphyrius, *in* A.G. Douglas and J.R. Maxwell, eds., Advances in Organic Geochemistry, 1979: Oxford, Pergamon, p. 239-248.

Mackenzie, A.S., C.F. Hoffham, and J.R. Maxwell, 1981a, Molecular parameters of maturation in the Toarcian shales, Paris basin, France - III. Changes in the aromatic steroid hydrocarbons: Geochimica et Cosmochimica Acta, v. 45, p. 1345-1355.

Mackenzie, A.S., C.A. Lewis, and J.R. Maxwell, 1981b, Molecular parameters of maturation in the Toarcian shales, Paris basin, Francis - IV. Labaoratory thermal studies. Geochimica et Cosmochimica Acta, v. 45, p. 2369-2376.

Magoon, L.B. and G.E. Claypool, 1983, Petroleum geochemistry of the North Slope of Alaska: time and degree of thermal maturity; *in* M. Bjoroy ed., Advances in Organic Geochemistry 1981: New York, John Wiley, p. 28-38.

Mattavelli, L., T. Ricchiuto, D. Grignani, and M. Schoell, 1983, Geochemistry and habitat of natural gases in Po Basin, northern Italy: AAPG Bulletin, v. 67, p. 2239-2254.

McCulloh, T.H., 1969, Geologic characteristics of the Dos Cuadras offshore oil field: USGS Professional Paper 679-C, p. 29-46.

McDowell, A.N., 1975, What are the problems in estimating the oil potential of a basin?: The Oil and Gas Journal, v. 73, no. 23, p. 85-90.

McKelvey, V.E., J.S. Williams, R.P. Sheldon, E.R. Cressman, T.M. Cheney, and R.W. Swanson, 1959, The Phosphoria, Park City and Shedhorn formations in the Western Phosphate field: USGS Professional Paper 313A, 47 pp.

McMillan, L., 1980, Oil and gas of Colorado: a conceptual view; *in* H.C. Kent and K.W. Porter, eds., Colorado Geology: Denver, Rocky Mountain Association of Geologists, p. 191-197.

Meissner, F.F., 1981, Abnormal pressures produced by hydrocarbon generation and maturation and their relation to processes of migration and accumulation. AAPG Bulletin, v. 65, p. 2467.

Middleton, M.F. and D.A. Falvey, 1983, Maturation modeling in Otway basin, Australia: AAPG Bulletin, v. 67, p. 271-279.

Miknis, F.P., J.W. Smith, E.K. Maughan, and G.E. Maciel, 1982, Nuclear magnetic resonance: a technique for direct nondestructive evaluation of source-rock potential: AAPG Bulletin, v. 66, p. 1396-1401.

Momper, J.A., 1978, Oil migration limitations suggested by geological and geochemical considerations; *in* Physical and chemical constraints on petroleum migration, Tulsa, AAPG Course Note Series 8, p. B1-B60.

Moshier, S.O., and D.W. Waples, in preparation, Quantitative evaluation of the Lower Cretaceous Mannville Group as the source for Alberta's heavy oils; AAPG Bulletin.

Nakayama, K. and D.C. Van Siclen, 1981, Simulation model for petroleum exploration: AAPG Bulletin, v. 65, p. 1230-1255.

Orr, W.L., 1984, Geochemistry of asphaltic Monterey oils from the Santa Maria basin and Santa Barbara Channel area offshore: American Chemical Society Meeting, 187th, 1984, St. Louis, Book of Abstracts, Geochemical Division, Paper no. 50.

Parrish, J.T., 1982, Upwelling and petroleum source beds, with reference to Paleozoic: AAPG Bulletin, v. 66, p. 750-774.

Parrish, J.T., 1983, Upwelling deposits: nature of association of organic-rich rock, chert, chalk, phosphorite, and glauconite: AAPG Bulletin, v. 67, p. 529.

Pisciotto, K.A. and R.E. Garrison, 1981, Lithofacies and depositional environments of the Monterey Formation, California; *in* R.E. Garrison and R.G. Douglas, eds., The Monterey Formation and related siliceous rocks of California; Proceedings of SEPM symposium dedicated to examine the paleontology, sedimentology, depositional environments and diagenesis of the Monterey formation: Los Angeles, SEPM, Pacific Section, p. 97-122.

Powell, T.G., P.J. Cook, and D.M. McKirdy, 1975, Organic geochemistry of phosphorites - relevance to petroleum genesis: AAPG Bulletin, v. 59, p. 618-632.

Powell, T.G., S. Creaney, and L.R. Snowdon, 1982, Limitations of use of organic petrographic techniques for identification of petroleum source rocks: AAPG Bulletin, v. 66, p. 430-435.

Pratsch, J.C., 1982, Focused gas migration and concentration of deep-gas accumulations: Erdol and Kohle-Erdgas-Petrochemie vereinigit Brennstoff-Chemie, v. 35, p. 59-65.

Radke, M., D.H. Welte, and H. Willsch, 1982, Geochemical study on a well in the Western Canada basin: relation of the aromatic distribution pattern to maturity of organic matter. Geochimica et Cosmochimica Acta, v. 46, p. 1-10.

Reitsema, R.H., A.J. Kaltenback, and F.A. Lindberg, 1981, Source and migration of light hydrocarbons indicated by carbon isotopic ratios: AAPG Bulletin, v. 65, p. 1536-1542.

Rice, D.D., 1983, Relation of natural gas composition to thermal maturity and source rock type in San Juan basin, northwestern New Mexico and southwestern Colorado: AAPG Bulletin, v. 67, p.1199-1218.

Rice, D.D. and G.E. Claypool, 1981, Generation, accumulation, and resource potential of biogenic gas: AAPG Bulletin, v. 65, p. 5-15.

Robert, P., 1981, Classification of organic matter by means of fluorescence; application to hydrocarbon source rocks: International Journal of Coal Geology, v. 1, p. 101-137.

Saxby, J.D., 1980, Atomic H/C ratios and the generation of oil from coals and kerogens: Fuel, v. 59, p. 305-307.

Schmoker, J.W., 1981, Determination of organic-matter content of Appalachian Devonian shales from gamma-ray logs: AAPG Bulletin, v. 65, p. 1285-1298.

Schmoker, J.W. and T.C. Hester, 1983, Organic carbon in Bakken Formation, United States portion of Williston basin: AAPG Bulletin, v. 67, p. 2165-2174.

Schoell, M., 1980, The hydrogen and carbon isotopic composition of methane from natural gases of various origins: Geochimica et Cosmochimica Acta, v. 44, p. 649-662.

_____, 1982, Applications of isotope analysis to petroleum and natural gas research: Spectra, v. 8, p. 32-41.

_____, 1983, Genetic characterization of natural gases: AAPG Bulletin, v. 67, p. 2225-2238.

Seifert, W.K. and J.M. Moldowan, 1978, Applications of steranes, terpanes, and monoaromatics to the maturation, migration, and source of crude oils: Geochimica et Cosmochimica Acta, v. 42, p. 77-92.

_____, 1980, The effect of thermal stress on source rock quality as measured by hopane stereochemistry, in A.G. Douglas and J.R. Maxwell, eds., Advances in Organic Geochemistry, 1979. Oxford, Pergamon, p. 229-237.

_____, 1981, Paleoreconstruction by biological markers. Geochimica et Cosmochimica Acta, v. 45, p. 783-794.

Seifert, W.K., J.M. Moldowan, and R.W. Jones, 1980, Application of biological marker chemistry to petroleum exploration, in World Petroleum Congress, 10th, 1980, Bucharest, Proceedings, v. 2: Heyden, Applied Science, p. 425-438.

Sluijk, D. and M.H. Nederlof, in press, Geochemical evaluation of sedimentary basins: in G.J. Demaison and R.J. Morris, eds., AAPG Memoir, Tulsa.

Snowden, L.R. and T.G. Powell, 1982, Immature oil and condensate — modification of hydrocarbon generation model for terrestrial organic matter. AAPG Bulletin, v. 66, p. 775-788.

Stahl, W.J. and B.D. Carey, 1975, Source-rock identification by isotope analysis of natural gases from fields in the Val Verde and Delaware basins, West Texas: Chemical Geology, v. 16, p. 257-267.

Staplin, F.L., W.G. Dow, C.W.D. Milner, D.I. O'Connor, S.A.J. Pocock, P. van Gijzel, D.H. Welte, and M.A. Yukler, 1982, How to assess maturation and paleotemperatures; Tulsa, SEPM Short Course Number 7.

Summerhayes, C.P., 1981a, Oceanographic controls on organic matter in the Miocene Monterey Formation, offshore California, in R.E. Garrison and R.G. Douglas, eds., The Monterey Formation and related siliceous rocks of California, Proceedings of on SEPM symposium dedicated to examine the paleontology sedimentology, depositional environments and diagenesis of the Monterey Formation: Los Angeles, SEPM, Pacific Section, p. 213-219.

_____, 1981b, Organic facies of middle Cretaceous black shales in deep North Atlantic: AAPG Bulletin, v. 65, p. 2364-2380.

_____, 1983, Sedimentation of organic matter in upwelling regimes; in J. Thiede and E. Suess, eds., Coastal Upwelling; Part B, New York, Plenum, p. 29-72.

Teichmuller, M. and B. Durand, 1983, Fluorescence microscopical rank studies on liptinites and vitrinites in peat and coals, and comparison with results of the Rock-Eval pyrolysis: International Journal of Geology, v. 2, p. 197-230.

Tissot, B., G. Deroo, and J. Espitalie, 1975, Comparative study of the duration of formation and expulsion of oil in several geological provinces; in World Petroleum Congress, 9th, 1975, Tokyo, Proceedings, v. 2: London, Applied Science Publishers, p. 159-169 (in French).

Tissot, B., G. Deroo, and J.P. Herbin, 1979, Organic matter in Cretaceous sediments of the North Atlantic: contribution to sedimentology and paleogeography; in M. Talwani, W. Hay, and W.B.F. Ryan, eds., Deep drilling results in the Atlantic Ocean continental margins and paleoenvironment: Washington, American Geophysical Union, p. 362-374.

Toth, D.J., I. Lerche, D.E. Petroy, R.J. Meyer, and C.G. St. C. Kendall, 1983, Vitrinite reflectance and the derivation of heat flow changes with time; in M. Bjoroy, ed., Advances in Organic Geochemistry, 1981: Chichester, John Wiley, p. 588-598.

Vandenbroucke, M., B. Durand, and J.L. Oudin, 1981, Detecting migration phenomena in a geological series by means of $C_1 - C_{35}$ hydrocarbon amounts and distributions: Institut Francais du Petrole Ref. 29439, 21 p.

Waples, D.W., 1980, Time and temperature in petroleum formation: application of Lopatin's method to petroleum exploration: AAPG Bulletin, v. 64, p. 916-926.

_____, 1982, Phosphate-rich sedimentary rocks: significance for organic facies and petroleum exploration. Journal of Geochemical Exploration, v. 16, p. 135-160.

_____, 1983a, Physical-chemical models for oil generation: Colorado School of Mines Quarterly, v. 78, no. 4, p. 15-30.

_____, 1983b, Reappraisal of anoxia and organic richness, with emphasis on Cretaceous of North Atlantic. AAPG Bulletin, v. 67, p. 963-978.

_____, 1984, Thermal models for oil generation, in J.W. Brooks and D.H. Welte, eds., Advances in Petroleum Geochemistry: London, Academic Press, in press.

Welte, D.H. and M.A. Yukler, 1981, Petroleum origin and accumulation in basin evolution - a quantitative model: AAPG Bulletin, v. 65, p. 1387-1396.

Welte, D.H., R.G. Schaefer, M. Radke, and H.M. Weiss, 1982, Origin, migration, and entrapment of natural gas in Alberta deep basin, - Part 1: AAPG Bulletin, v. 66, p. 642.

White, D.A., 1980, Assessing oil and gas plays in facies-cycle wedges: AAPG Bulletin, v. 64, p. 1158-1178.

Zieglar, D.L. and J.H. Spotts, 1978, Reservoir and source bed history of Great Valley, California: AAPG Bulletin, v. 62, p. 813-826.

Zielinski, G.W. and P.M. Bruchhausen, 1983, Shallow temperatures and thermal regime in the hydrocarbon province of Tierra del Fuego: AAPG Bulletin, v. 67, p. 166-177.

PREDICTIVE SOURCE BED STRATIGRAPHY; A GUIDE TO REGIONAL PETROLEUM OCCURRENCE

NORTH SEA BASIN AND EASTERN NORTH AMERICAN CONTINENTAL MARGIN

Gerard Demaison and A. J. J. Holck, *Chevron Overseas Petroleum Inc., San Francisco, CA, USA*; R. W. Jones and G. T. Moore, *Chevron Oil Field Research Company, La Habra, CA, USA.*

Abstract. Recent progress in understanding the palaeo-oceanographic aspects of petroleum source rock deposition has made it feasible to implement the stratigraphy of source beds in sedimentary basins, mainly by the mapping of organic facies.

Basin geochemical studies carried out in the North Sea Basin accurately identify and outline that segment of the basin which holds over 95% of oil and gas reserves: a central fairway where the favourable organic facies in the Kimmeridgian–Volgian source beds coincides with adequate thermal maturation.

In the eastern continental margin of North America, which is as yet sparsely drilled, an excellent correlation also exists between the type and maturity of the organic facies and petroleum occurrences so far. Whether and where the thick Jurassic section under parts of the present continental slope, is mature and under a favourable organic facies is the key issue to the prediction of petroleum reserves in this Atlantic margin.

The reviewed cases demonstrate that it is feasible, through these integrated studies, to predictively rank the various segments of sedimentary basins in terms of their potential petroleum yield. Thus a critical aspect of exploration risk, the uncertainty relevant to hydrocarbon charge of potential traps detected by geophysical methods, may be reduced.

Résumé. Les progrès recents accomplis dans la compréhension des aspect paléo-oceanographiques de la déposition des sédiments riches en matière organique, permettent de mettre en oeuvre une stratigraphie des roches mères dans les bassins sédimentaires, basée principalement sur la cartographie rationnelle des faciès organiques.

Les études géochimiques de bassin effectuées dans la Mer du Nord ont permis de circonscrire la partie du bassin qui contient plus de 95% de l'huile et du gaz: une zone centrale où les faciès organiques favorables dans les roches-mères du Kimméridgien–Volgien coincident avec une maturation suffisamment élevée pour la genèse du pétrole.

Sur la marge Atlantique du continent Nord-Américain, on a pu également mettre en évidence une excellente correlation entre les types et maturités des faciès organiques d'une part, et les resultats d'exploration, d'autre part. La question posée la plus importante pour l'avenir de l'exploration a trait aux faciès organiques et maturations des sédiments Jurassiques qui existent sous le talus continental actuel.

Nos études démontrent qu'il est possible, par ces méthodes géochimiques intégrées à la géophysique et la géologie, de classifier les différentes portions des bassins sédimentaires en fonction de leur potentiel régional de genèse du pétrole.

De cette manière, un aspect critique du risque en exploration pétrolière, peut être évalué de façon objective, grâce à la géochimie: l'incertitude concernant la présence d'hydrocarbures en quantité suffisante pour remplir les pièges détectés par la géophysique.

1. INTRODUCTION

Economic factors dictate that oil and gas fields in offshore basins and other hostile environments must be large. Such fields commonly result from the occurrence of large structural traps draining prolific oil-generative depressions located within relatively short migration distances. Predicting the locations and geometry of oil-generative depressions in sedimentary basins (so-called 'hydrocarbon kitchens') can be achieved by applying calibrated geothermal modelling to seismic structural compilations.[1,2] However this procedure also requires establishing the actual presence of source beds in the subsurface and locating their precise stratigraphic placement. This requirement is met by analysing a large number of rock samples retrieved from all available deep wells and compiling the obtained data into geochemical logs.

A fundamental problem in basin geochemistry is the integration of these source rock data, into a mapping system capable of predicting regional source bed occurrence and variability. The problem faced is conceptually similar to that of stratigraphic mapping of reservoir trends, an exercise which has been made possible by long standing research into clastic and carbonate sedimentation.

Similarly, the organic stratigraphic principles proposed and applied in this paper would not have evolved without the advances made in oceanography, sedimentology, geomicrobiology, and organic geochemistry[4,8] during the last decade. These advances concur to demonstrate that the critical aspect in the deposition of oil source beds is not how much organic matter is produced but rather how much is preserved from the destructive processes acting in bottom sediments.

The most useful stratigraphic principle applicable to petroleum source bed mapping is the Organic Facies concept. It is a stratigraphic concept similar to that of lithofacies or biofacies: we have previously defined[3] organic facies as a mappable subdivision of a sedimentary sequence, identifiable solely from the character of its organic matter. The organic facies concept is derived from the following fundamental observations:

Kerogen types I, II and III,[4] which have different hydrocarbon source potentials, are strongly correlated to the past degree of oxygenation of the water above the sediment surface (Fig. 1).

Kerogen types are also influenced by past sedimentation rates, which determined the residence time of the dead organic matter in zones of microbial reworking (Fig. 1).

Lastly, kerogen types are influenced by the dominant type of plant debris (whether aquatic or terrestrial) which was deposited in the sediment.

Three main environments are recognizable (Fig. 1) in ancient and modern sediments, in terms of their degree of oxygenation:

– oxic (or aerobic)
– sub-oxic (or dysaerobic)
– anoxic (or anaerobic).

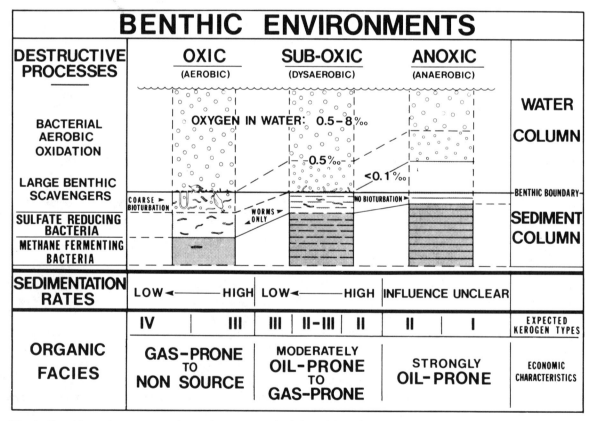

Fig. 1. Benthic environments and petroleum source bed deposition. Organic facies mapping depends upon a clear understanding of the relationship between early depositional factors and kerogen types.

Planktonic and terrestrial organic matter deposited in anoxic or sub-oxic environments are spared much of the aerobic biodegradation and are almost exclusively reworked by less efficient anaerobic bacterial processes. The latter enhance the preservation of fatty or waxy, hydrogen-rich lipidic plant substances, which constitute the precursor material to Type I and Type II kerogens.[3] Furthermore, anaerobic bacterial activity transforms planktonic remains into an amorphous residue that owes much of its composition to bacterial lipids. These are also important precursors to oil and gas.[5]

In oxygenated environments, planktonic organic matter, which is more susceptible to biodegradation, is preferentially destroyed. As a result, marine or lacustrine Type III kerogens commonly contain, by subtraction, much land plant debris at various degrees of reworking, mainly by aerobic microbes. Their hydrocarbon potential is, in most cases, mainly for gas with condensate rather than oil. However, there are documented cases,[6,7] notably in the Tertiary of the Far East, where paralic and coal-rich, rather than marine, organic facies III source beds are responsible for high-wax crude accumulations.

Due to the interaction of oxygenation and sedimentation rate, four organic facies, accurately identifiable by chemical kerogen typing, are recognizable in marine and lacustrine sediments.

1. Organic Facies I (strongly oil-prone) are typical of highly reducing (deeply anoxic) environments. Eventually, organic carbon contents can reach values as high as 30%. Organic Facies I is commonly but not uniquely found in anoxic lacustrine environments, but has also been identified in marine rocks deposited in highly reducing environments. The sediments are laminated and free from benthic fauna.

2. Organic Facies II (oil-prone) is typical of anoxic environments, and also of sub-oxic environments with high sedimentation rates. This facies can be very widespread in both marine and lacustrine environments, with organic carbon contents commonly ranging between about 1% and 10%. The sediments are partly to continuously laminated and can contain sparse benthic foraminifera and worm burrows, when the environment is sub-oxic (dysaerobic) rather than deeply anoxic.

3. Organic Facies III (gas-prone) is typical of mildly oxic conditions at sea or lake bottom, in combination with average to fairly high sedimentation rates. Organic carbon contents commonly range between 1% and 5.0% in such sediments. Where terrestrial plant input is high, it contributes to most of the kerogen composition. The sediments, when marine, are commonly bioturbated by large metazoans and contain fairly abundant benthic foraminifera. Organic Facies III, with organic carbon contents in excess of 20% can also evolve from continental or marginal marine 'coal marsh' to lagoonal environments.

4. Organic Facies IV (non-source) is typical of any environment characterized by very long residence times of the organic matter in zones of aerobic microbial degradation. Ventilated oceanic conditions with low sedimentation rates, such as those encountered at abyssal depths as well as shallow, high-energy, ventilated shelves, lead to widespread occurrence of Organic Facies IV. Organic carbon contents in the non-source (IV) facies have been observed to commonly range between 0.2% and 3%. The sediments have usually been homogenized by vigorous bioturbation.

Organic facies mapping depends upon a clear understanding of the relationship between depositional factors and kerogen type, and on a perception of the various oxygen-depleted natural settings that favour source bed deposition in the marine and lacustrine realms.

Fundamentally, widespread occurrence of oil-prone Organic Facies I and II is caused by the equally widespread existence of permanently stratified sub-oxic (0.5‰ to 0.1‰ oxygen) to anoxic (less than 0.1‰ oxygen) water masses in the depositional basin.[8]

These conditions are met in lakes and silled basins (Fig. 2) that are permanently stratified because of near-horizontal density boundaries within the water masses, due to stable temperature or salinity contrasts.

Oxygen-deficient settings prone to oil source bed deposition can also exist in open marine waters, without sills or physical barriers (Fig. 3). Such settings occur where the oxygen minimum layer contains a sub-oxic or anoxic core, which impinges on a continental slope or shelf. Such conditions tend to occur in zones of weaker oceanic mixing, such as the east side of oceanic basins and in those zones sheltered or removed from the influence of deep oxygenated water. Impingement by an oxygen minimum layer can also be reinforced locally by coastal upwellings which foster inordinately high biologic productivity, hence increased oxygen demand.

Lastly there is statistical evidence that certain periods in the Earth's history were more prone to widespread Organic Facies II deposition than others. These were the periods of so-called 'oceanic anoxic events.'[9] They were generally near-synchronous with past high sea stands and major transgressions over the continental plates, equable

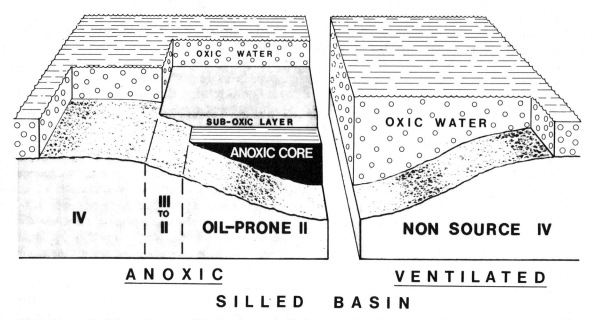

Fig. 2. Source bed depositional models in lakes and silled basins. Anoxic or sub-oxic environments favourable to oil source bed deposition tend to occur when the water becomes stagnant because of permanent density stratification. Well-ventilated silled basins or lakes characterized by active mixing of the water masses tend to remain wholly oxic.

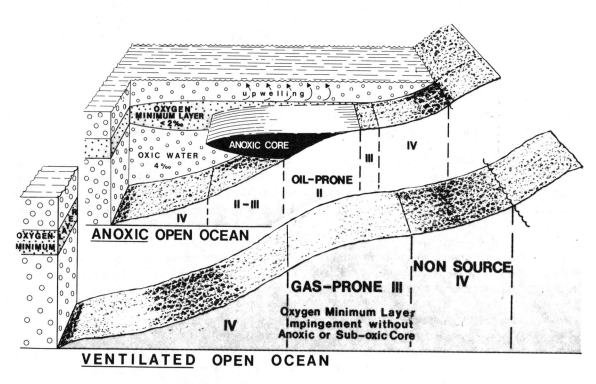

Fig. 3. Source bed deposition in open marine waters. These settings occur where the oxygen minimum layer contains an anoxic, or sub-oxic core which impinges upon a continental slope or shelf. Such settings, when reinforced by coastal upwellings, can lead to prolific source bed deposition.

global climate, warmer oceanic waters, and the absence of polar ice caps. Well-documented examples of these events are found in late Jurassic and middle to late Cretaceous time.

Periods of 'oceanic anoxic events' should not be interpreted as universally leading to synchronous and ubiquitous oil source bed deposition. Anoxic events simply reflect a general tendency to stagnation and inefficient mixing in the world ocean. Widespread oil source-bed deposition only took place when this general tendency was locally or regionally reinforced by other additive factors (silled basins, oceanic culs-de-sac, zones of upwelling, etc.).

Application of these principles and depositional models permit us to understand and predict the location and distribution of potential oil source beds in sedimentary basins.

The studies that follow examine two historically different cases: (1) the North Sea Basin, where a very large data base permits explanation, in hindsight, of regional patterns of oil occurrence and (2) the eastern North American continental margin, which is as yet sparsely drilled and whose ultimate potential is still highly uncertain.

2. THE NORTH SEA BASIN

The heavily explored portion of the North Sea between the 55th and 62nd parallels is a major oil and gas province with estimated reserves of over 24 billion barrels of oil and over 55 tcf of gas.[10] Reservoirs range in age from Devonian to Eocene typically producing on tilted fault blocks and, to a lesser degree, on salt swells. The pre-eminent source rock is the Kimmeridgian–Volgian shale.

As much as 9 km of sediment of Devonian to Recent age is preserved in the deepest parts of the North Sea graben system. More typically, the sediments total 4 to 6 km. Distributed within this sedimentary sequence is a diverse spectrum of organic facies (Fig. 4).

On the basis of a data bank of over 2200 analyses by Rock-Eval pyrolysis with organic carbon, the kerogen types and, therefore, the organic facies of the Mesozoic–Tertiary sequence in the intensely explored area are now well understood. Only rocks of Jurassic age, and particularly the Kimmeridgian–Volgian, display significant concentrations of oil and gas prone kerogen.

On Fig. 5 there is a hydrogen index–oxygen index cross-plot, illustrating the various Kimmeridgian–Volgian kerogen types. This plot demonstrates that, in aggregate, the changes in kerogen types are almost infinitely and subtly variable. As discussed earlier, this is caused by the interplay between oxygen concentrations in the bottom sea water, type of plant input and rates of sedimentation. However, when the vertical and horizontal distribution of the kerogens is examined, two aspects become evident.

Firstly, in the vertical dimension, major kerogen type units (and resulting major organic facies), commonly a few hundred to several thousand feet thick,

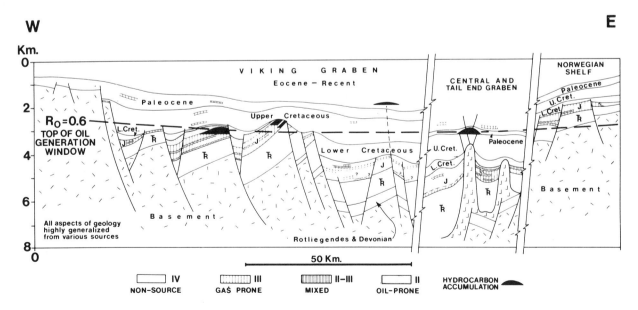

Fig. 4. Composite organic stratigraphic model of the North Sea Basin. Except for parts of the Kimmeridgian–Volgian, the bulk of the sedimentary fill is either non-source or gas-prone in character.

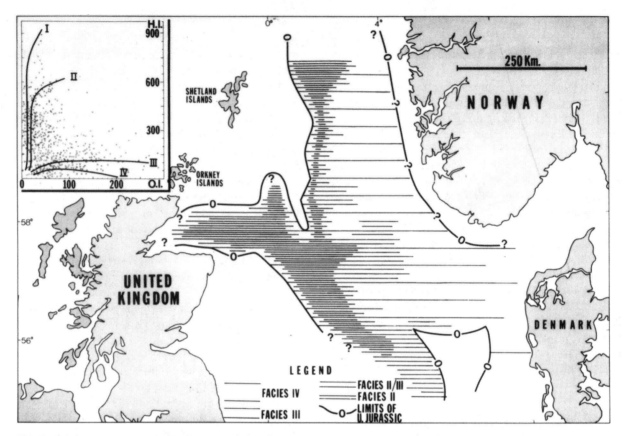

Fig. 5. Regional map of the distribution of the Kimmeridgian–Volgian organic facies. This document is based on a data bank of 2200 analyses by Rock–Eval pyrolysis with organic carbon.

can be recognized in each well. Within each major organic facies unit, short intervals of improved or, alternatively, deteriorated facies can usually be observed. For example, a major gas-prone facies unit may have thin interbeds of inert material or minor zones of mixed facies.

Secondly, boundaries between the major organic facies units can be traced laterally for considerable distances despite local variability caused by onlap, non-deposition, or truncation on structures. These time–space considerations, alone, indicate that the overriding control for organic facies distribution was of basinwide extent and is related to past oxygen concentrations in the bottom water during sedimentation.

Oxic waters prevailed most of the time over the shallower shelf areas, resulting in predominantly inert kerogen (Organic Facies IV) and very low preservation of plankton. In structural lows and sub-basins with greater water depths, permanent stagnancy of the bottom water resulted in widespread sub-oxic or anoxic conditions leading to the formation of hydrogen-rich (oil-prone) kerogens. This proposed North Sea Kimmeridgian–Volgian anoxic depositional model is similar to that already documented by Hallam[11] and Morris[12] for Liassic and late Jurassic shales of the British Isles. Furthermore, detailed sedimentologic observations made in Kimmeridgian outcrop sections by Tyson et al.[13] support and concur with our proposed environmental interpretation.

Regardless of variability caused by onlap, non-deposition or truncation on local structures, a regional map of the distribution of the Kimmeridgian–Volgian organic facies can be developed, as shown in Fig. 5.

The flanking shelves and areas adjacent to some of the major positive elements are characterized by Facies IV, the non-source inert facies deposited under oxic water. Gradationally (in a regional sense), the facies improve through types III, II–III, and II, towards the central parts of the graben system. This testifies to sub-oxic and then permanent anoxic conditions, due to increasing water depth. This trend was reinforced basinwards by higher sedimentation rates, and, possibly, less influence from terrigenous sources of organic matter. On the basis of this observed organic facies distribution alone, a favourable exploration trend can be outlined. This map agrees closely, in a regional sense, with the kerogen facies patterns previously mapped by Barnard and Cooper.[14]

It should be noted that, in detail, there is no smooth gradation between the facies, since tongues and lenses of low quality organic facies extend grabenward and, conversely, oil prone and mixed facies occur in stratigraphically updip positions. Vertical and lateral alternations of facies are characteristic. Very few wells encounter only a single facies.

This interfingering of the various organic facies within the Kimmeridgian is attributed to vertical fluctuations of the sub-oxic/oxic boundaries throughout the water column, periodically sweeping across the basin slopes and shelves. We propose that the approximate limits between each facies (Fig. 5) reflect the mean past intersects of oxygen concentration boundaries in the water column with the slopes and shelves of the North Sea Basin, during late Jurassic time.

The occurrence of favourable organic facies is a necessary but not sufficient cause for regional oil occurrence; the role of maturation is equally critical. In much of the North Sea, and particularly on the Norwegian shelf, the Kimmeridgian–Volgian is immature (Fig. 6) as measured by vitrinite reflectance. On the other hand, in the deep portions of the graben system these rocks are totally within the generative window for gas. As with organic facies, a favourable trend can be delineated on maturation grounds, alone (Fig. 6).

The combination of favourable organic facies and favourable maturation fairway outlines that portion of the North Sea Basin containing over 95% of the oil and gas fields. The correlation between regional petroleum occurrence and favourable Kimmeridgian organic facies and maturation fairway is virtually one to one. Migration distances are relatively short, because of tectonic style, and are limited by the size of the drainage areas that surround the structures.

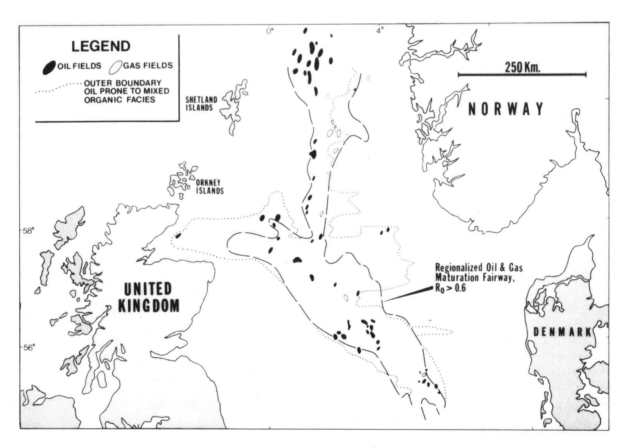

Fig. 6. Kimmeridgian–Volgian organic facies and maturation in relation to petroleum occurrence in the North Sea Basin. The favourable organic facies and maturation fairway outlines that portion of the basin containing over 95% of the petroleum reserves.

3. EASTERN CONTINENTAL MARGIN OF NORTH AMERICA

The eastern continental margin of North America is a passive margin in the terminology of plate tectonics. Rifting began in Late Triassic in offshore southeastern Canada, perhaps somewhat later (Early Jurassic) off the eastern United States and not until Early Cretaceous time in offshore Labrador.

The continental margin contains several depocentres superimposed on thick sedimentary prisms whose sediments range from Triassic to Quaternary in age. They are still sparsely drilled and include areas of both severe disappointment (Baltimore Canyon) and conspicuous success (Hibernia).

The first sediments deposited on this developing passive margin were rift valley sediments that are highly variable in nature. The oldest sediments are Late Triassic–Early Jurassic in age (Fig. 7). Offshore Newfoundland and Nova Scotia they are represented by thick marine evaporites that were deposited in hypersaline restricted basins which developed during the ingress of the western Tethys Sea into the rift system. Farther south, the Triassic rift valleys formed westward of the location of continental separation and were not invaded by the sea: their sediments are dominantly continental red beds.

Later, marginal marine to marine sedimentation became dominant in Late Jurassic time and has continued to this day as the Atlantic Ocean widened (Fig. 8). However, volumetrically, the sediments off the east coast of the North American continent are dominated by shelf deposits whose organic matter is oxidized in varying degrees (Organic Facies III and IV). In addition, the organic-rich sediments deposited during the Cretaceous 'anoxic events' of the North Atlantic are also dominated by terrestrially derived, oxidized or recycled organic matter.[15,20,21] This is due to the somewhat pervasive oxic sedimentation that prevailed on this side of the Cretaceous proto-Atlantic, regardless of water depth.[21] Thus, moving away from the continent does not necessarily lead to organic facies improvements. Despite these general problems, areally limited oil source rock environments do exist (mainly in the Jurassic) within an overall unfavourable situation, and others can be rationally postulated. We have synthesized our observations into the Lower Cretaceous and Upper Jurassic maps shown on Fig. 8. Organic facies distributions are based on available well data and are attempting to predict those more favourable areas, keeping in perspective that these predictions are only broadly regional in nature. Despite the fact that detailed data are available from many wells, most of the control points are grouped in widely spaced clusters. Projections between these control points are based on the implications of Late Jurassic and Early Cretaceous palaeo-oceanography, according to the general models developed earlier in this paper and plate tectonic configurations. Moreover, the substantial thicknesses of both mapped units imply that the

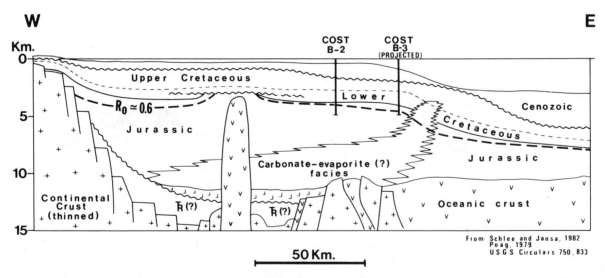

Fig. 7. Geologic section through the Northern Baltimore Canyon Basin area, eastern continental margin of North America. Note the carbonate 'reef' trend, a feature that extends from the Gulf of Mexico and Florida up to the Georges Bank Basin area. The organic facies of the Jurassic section in front of the 'reef' is unknown. This is a key issue to the prediction of petroleum potential in the deep-water offshore portions of this Atlantic Margin.

organic facies within each unit have been averaged, except where a substantial oil proneness is either proved or reasonable. Despite these rather severe limitations, it is possible to demonstrate that a satisfactory correlation exists between the different organic facies shown on the Upper Jurassic and Lower Cretaceous maps and the outcome of exploratory drilling.

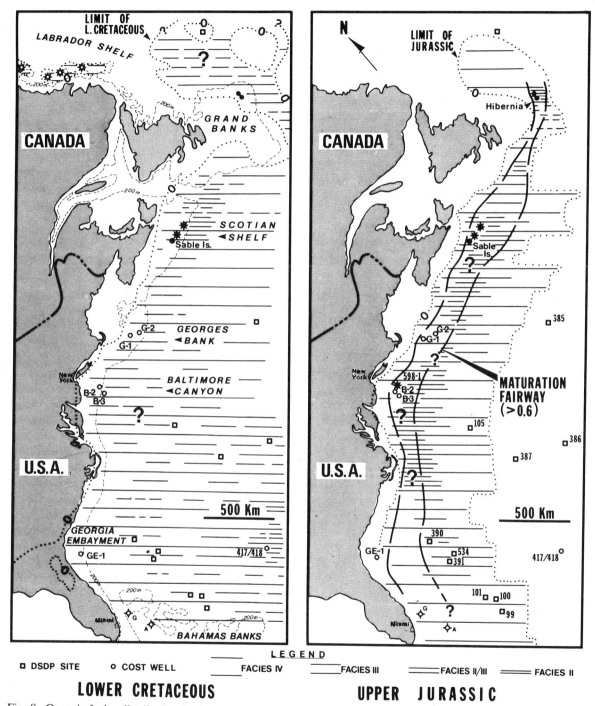

Fig. 8. Organic facies distribution in the Lower Cretaceous and Upper Jurassic of the eastern continental margin of North America. Note the excellent correlation between the type and maturity of the organic facies penetrated in wells and the presence or lack of potentially commercial accumulations of petroleum. The areas underlain by mature Upper Jurassic source rocks are the most prospective and include the significant discoveries at Hibernia, Sable Island and Baltimore Canyon.

For instance, the Lower Cretaceous-hosted oil in the Hibernia field (northeastern Grand Banks) is related to a geographically limited area of mature Upper Jurassic shales (Mic-Mac Formation) present under a rich Organic Facies II. These rocks are synchronous with the event that led to the formation of several prolific petroleum zones in the Northern Hemisphere, including the North Sea and Western Siberia.

On the Scotian Shelf most of the production has also been attributed to the Upper Jurassic to Lower Cretaceous Verrill Canyon Formation which is present as a mixed Organic Facies II-III. The Verrill Canyon Formation overlapped on to the shelf over and between the banks located at the northern end of the Jurassic–Lower Cretaceous carbonate 'reef' trend (Fig. 8) which extends along the entire eastern coast of the United States, from the Gulf of Mexico. Farther south, off the United States, there is no direct evidence to date that rocks characterized by Organic Facies II or II–III were ever deposited on the back side of this carbonate 'reef' barrier.

The Jurassic and, locally, basal Lower Cretaceous part of the 'back-reef' section, however, is in the thermal window for wet-gas generation and within the maturation fairway shown on Fig. 7. The gas discovered in the Baltimore Canyon area is likely to be related to this particular gas-prone source sequence.

Under most of the present continental slope, to the east of the Jurassic 'reef' trend, the sedimentary overburden is inadequate to depress any significant volume of Cretaceous rocks into the generative window. However, due to deeper burial, the Jurassic under the continental slope is likely to be thermally mature at least in certain segments of this margin, but its organic facies is yet unknown (Fig. 8). Part of the hope that exploration results will improve as drilling extends eastward into deeper water, beyond the present shelf, is the existence of seismically defined Jurassic grabens that could, perhaps, contain marine sediments with Organic Facies II source beds. This postulate is based on the palaeo-oceanographic possibility of a relatively narrow and possibly oxygen-poor seaway, in Late Jurassic time, immediately to the east of the carbonate bank-edge. The Jurassic and Lower Cretaceous sediments present under unfavourable organic facies, farther to the east, in the Deep Sea Drilling Project core-holes[15] may have been deposited in shallower oxygenated water.

The following is a rapid review of the major areas of exploratory interest.

Labrador[22-28]

Largely terrestrial-derived organic matter (gas-condensate-prone Organic Facies III) was deposited offshore Labrador from the beginning of rifting in the Early Cretaceous well into the Tertiary. Lack of Organic Facies II deposition in these marine sediments is explained by widespread and persistent oxic conditions which did not permit adequate preservation of plankton, in spite of fairly high sedimentation rates. Because of maturity constraints related to burial depth, most of the gas-condensate deposits found to date on the horst blocks within the rift system (Bjarni, Snorri, Gudrid) must have originated in mature Organic Facies III in the Lower Cretaceous deposited in the intervening grabens. Mature younger sediments of a similar organic facies may have also contributed to some accumulations.

Grand Banks[29-34]

Over 30 unsuccessful exploratory tests were drilled in this area of generally unfavourable organic facies, prior to the discovery of the Hibernia Field by the Chevron *et al.* Hibernia P-15 well in 1979.[31,32] The major reservoirs hosting the oil at Hibernia are in Lower Cretaceous and Upper Jurassic sandstones. The productive interval is underlain by a mature oil-prone organic facies of Upper Jurassic age, a facies of limited distribution that was recognized earlier in the immature state in a well drilled a few miles to the south.[31] No other mature oil-prone source rock system is known in the Grand Banks area and it is a reasonable presumption that future exploration success will depend on the mapping and correct prediction of the distribution of the mature Upper Jurassic Facies II source beds. It appears as if Organic Facies II deposition, in Late Jurassic time, preferentially took place in the central portions of depocentres where anoxic or sub-oxic conditions must have prevailed.

Scotian Shelf[35-39]

Potentially commercial oil and gas condensate discoveries have been made in predominantly Lower Cretaceous sandstone reservoirs associated with salt-related structures.[37,41] The source is probably the transgressive Verrill Canyon Formation of Upper Jurassic and Lower Cretaceous age, which contains a mature, mixed-oil-gas-prone organic facies.[37] Most of the discoveries have been made near the present shelf edge in the vicinity of Sable Island. There is also evidence for the existence of a thick Jurassic–Lower

Cretaceous section under the continental slope. As in the Grand Banks, it could improve in quality (Organic Facies II?) where deposition took place in deeper, perhaps more oxygen-depleted water. The possible improvement in the source potential of the Verrill Canyon Formation, under the continental slope, and the existence of both salt and basement-controlled structures makes this area one of the more prospective offshore eastern North America.

Georges Bank Basin[40–44]

No significant shows of oil or gas have been recorded in either the COST G-1 and G-2 wells or the oil industry exploratory tests, all of which were drilled in shelf sediments lying to the west of the Jurassic 'reef'.[41] Organic facies throughout these Upper Jurassic and Lower Cretaceous shelf sediments, in Georges Bank are unfavourable, because they are dominated by residual and oxidized organic matter of Organic Facies IV.[40] The sediments deposited east of the reef are considerably thinner than on the Scotian Shelf.[40] If a favourable organic facies was deposited there, its maturity remains in question. On the standpoint of potential regional petroleum yield alone, Georges Bank is ranked among the highest risk segments of the North American Atlantic Margin.

Baltimore Canyon Trough[45–51]

This area has been unrewarding, so far, to the oil exploration industry. Only one potentially economic gas-condensate discovery (in the Jurassic) has been made. Other shows, including a small amount of oil, have been reported in the Jurassic section. In wells contributing organic geochemical data, the organic facies are poor (either III or IV) in both the Lower Cretaceous and Upper Jurassic non-marine and shelf sediments. All of the wells drilled to date, including the COST B-3 well, drilled in over 3000 feet of water, lie shelfward to the west of the Upper Jurassic–Lower Cretaceous 'reef' trend.

A major disappointment resulted from drilling the 'Great Stone Dome', a large domal structure caused by a Lower Cretaceous mafic pluton (Fig. 8). It was found void of hydrocarbons, despite adequate reservoirs and seals. Mature organic facies penetrated within the drainage area of the dome consisted mostly of non-source Organic Facies IV, but included some gas-prone Organic Facies III. The 'reef' trend itself, off Baltimore Canyon, yet remains to be tested, but lies in very deep water. Its potential depends on the yet undetermined organic facies and maturity of Jurassic sediments present on the slope side of the 'reef' trend.

Southeast Georgia Embayment, Blake Plateau Trough, and the Bahamas[52–56]

The unsuccessful tests in the Bahamas and adjacent areas to the north (Georgia Embayment) were drilled in shelf sediments where oil-prone organic facies have not been identified, as in the COST GE-1 well,[42,44] and are highly unlikely. This prediction is based on the widespread oxic conditions of sedimentation which prevailed in the 'back-reef' and carbonate platform sediments. The question mark on Fig. 8 (Jurassic Map) indicates that there is a remote chance for narrowly localized favourable organic facies in the Upper Jurassic in grabens under the present-day channels.[56] Adequate maturation of such source rocks, provided they exist, is also in question due to the abnormally low geothermal gradients that prevail in the Bahama Banks and southeast Florida. For the above reasons, this area is considered to be a high risk segment of the North American Atlantic margin.

In summary, exploration results to date, on the eastern North American continental margin, permit some generalizations and speculations based on both the known and possible distributions of organic facies:

- An excellent correlation exists between the type and maturity of the organic facies penetrated in wells and the presence or lack of potentially commercial accumulations of petroleum. In particular, the areas underlain by mature Upper Jurassic source rocks are the most prospective (Fig. 7) and include the significant discoveries at Hibernia, Sable Island and Baltimore Canyon.
- Migration distances for known accumulations have been short and are restricted to migration from mature oil- or gas-condensate-prone organic facies within the drainage areas of individual structures.
- Some unexplored areas of high potential interest exist, but lie mostly under rather deep water. Most worthy of interest are potential entrapment conditions associated with the thick, thermally mature Jurassic and (locally) Lower Cretaceous sections that occur rather extensively under the present continental slope. Juxtaposition of a large, well-sealed, closure on the Lower Cretaceous–Jurassic carbonate 'reef' trend next to a mature oil-prone Jurassic generative depression under the slope represents, perhaps, one of the most desirable exploration models. Thus, whether and where the thick Jurassic section close to the carbonate 'reef' trend contains a mature oil-prone organic facies is the

key issue to the prediction of major petroleum occurrence on this Atlantic margin. Uncertainty will remain very high as long as these 'fore-reef' sediments are not drilled and neighbouring traps remain untested.

- Some discoveries may occur in association with zones of local anoxic deposition in carbonate platforms, in otherwise broad areas of very widespread unfavourable organic facies. The probability that they are significant enough to contribute to the formation of large, economic, accumulations, by offshore standards, is considered to be low.

4. CONCLUSIONS

The application of fundamental stratigraphic principles to the distribution of organic matter types in sedimentary rocks has made it rational to map the occurrence and variability of oil and gas source beds in sedimentary basins. The construction of these predictive mapping systems, however, requires that a reasonable amount of regional control has already been acquired through geophysics, drilling, and subsurface geochemistry. It should be underlined that much of the organic facies mapping offshore North America could not have been carried out without results from the Deep Sea Drilling Project.

Basin geochemical studies in the North Sea Basin and Eastern Continental Margin of North America demonstrate the feasibility of explaining, in hindsight, the fundamental factors responsible for geographic patterns of regional petroleum occurrence. Furthermore, these studies permit a high degree of predictivity in identifying those basins or parts of basins carrying the highest expectations for future exploratory success.

REFERENCES

1. Welte, D. H., Yukler, M. A. and Leythauser. In G. Atkinson and J. J. Zukerman (eds.), *Origin and Chemistry of Petroleum*, Pergamon Press, Elmsford, N.Y., 1981, pp. 67–88.
2. Waples, D. W. *Bull. Am. Assoc. Petrol. Geol.*, 1980, **64**, 916–926.
3. Jones, R. W. and Demaison, G. J. *Bull. Am. Assoc. Petrol. Geol. (Abstract)*, 1980, **6415**, 729.
4. Tissot, B. P. and Welte, D. H. *Petroleum Formation and Occurrence*, Springer Verlag, New York, 1978.
5. Lijmbach, G. W. M. Proc. 9th World Petroleum Congress, Tokyo, 1975, **2**, 357–368.
6. Durand, B. and Oudin, J. L. Proc. 10th World Petroleum Congress, Bucharest, 1979, **2**, 3–11.
7. Thomas, B. M. *The Australian Petroleum Exploration Association Journal*, 1982, **22**, Part 1, 164–178.
8. Demaison, G. J. and Moore, G. T. *AAPG Bull.*, 1980, **64** (8), 1979–2109.
9. Jenkyns, H. C. *Geol. en Mijnbow*, 1980, **55**, 179–184.
10. Ziegler, P. A. *World Oil*, Nov. 1979, 111–124.
11. Hallam, A. *Facies Interpretation and the Stratigraphic Record*, W. H. Freeman and Co., Oxford and San Francisco, 1981, pp. 95–101.
12. Morris, K. A. *J. Geol. Soc. London*, 1980, **137**, 157–170.
13. Tyson, R. V., Wilson, R. C. L. and Donnie, C. *Nature*, 1979, **277**, 377–380.
14. Barnard, P. C. and Cooper, B. C. *Petroleum Geology of the Continental Shelf of North-West Europe*, Illing and Hobson (eds.), Inst. Pet., London, 1981, 169–175.
15. De Graciansky, P. C., Degree, G., Schaaf, A. *et al.*, *Revue de l'Institut Français du Pétrole*, 1982, **37** (3), 275–336.
16. Dow, W. G. *Bull. Am. Assoc. Petrol. Geol.*, 1978, **62** (9), 1584–1606.
17. Schlee, J. S., Dillon, W. P. and Crow, J. A. In 'Geology of Continental Slopes', 1979, SEPM Spec. Pub. No. 27, pp. 95–118.
18. Tucholke, B. E., Houtz, R. E. and Ludwig, W. J. *Bull. Am. Assoc. Petrol. Geol.*, 1982, **66** (9), 1384–1395.
19. Deep Drilling Results in the Atlantic Ocean: Continental Margins and Palaeoenvironment, Am. Geoph. Union, Marine Ewing Series, 3, 1979.
20. Herbin, J. P. and Deroo, G. *Bull. Soc. Geol. France*, 1982, **XXIV** (3), 497–510.
21. Tissot, B. *et al. Bull. Am. Assoc. Petrol. Geol.*, 1980, **64** (12), 2051–2063.
22. Cassou, A. M., Cannon, J. and Porthault, B. *Bull. Can. Petrol. Geol.*, 1977, **25** (1), 174–194.
23. Heroux, Y. *et al. Can. Journ. Earth Sci.*, 1981, **18** (2), 1856–1878.
24. McWhae, J. R. H., Elie, R., Laughton, K. C. and Gunther, P. R. *Bull. Can. Petrol. Geol.*, 1980, **28** (4), 460–488.
25. Powell, R. E. *Geol. Sur. Can.*, Paper 79-1C, 1979, 91–95.
26. Rashid, M. A., Prucell, L. P. and Hardy, I. A. In 'Facts and Principles of the World's Petroleum Occurrence', A. D. Miall (ed.), Can. Soc. Petrol. Geol., Memoir 6, Calgary, 1979.
27. Snowdon, L. R. and Powell, T. G. *Bull. Am. Assoc. Petrol. Geol.*, 1982, **66** (6), 775–788.
28. Umpleby, D. C. *Geol. Surv. Can.*, Paper 79-13, 1979, 1–34.
29. Cutt, B. J. and Laving, J. G. *Bull. Can. Petrol. Geol.*, 1977, **25** (5), 1037–1058.
30. McKenzie, R. M. *Oil Gas Journ.*, Sept. 21, 1981, 240–246.
31. Swift, J. H. and Williams, J. A. In 'Facts and Principles of the World's Petroleum Occurrence', A. D. Miall (ed.), Can. Soc. Petrol. Geol., Memoir 6, Calgary, 1979, pp. 567–588.
32. Henderson, G. G. L., Arthur, K. R. and Kushnir, D. W. *Bull. Am. Assoc. Petrol. Geol. (Abstract)*, 1981, **65** (5), 937.
33. Given, M. M. *Bull. Can. Petrol. Geol.*, 1977, **25** (1), 63–91.
34. Jansa, L. F. and Wade, J. A. *Geol. Surv. Can.*, Paper 74-30, **2**, 1975, 51–105.
35. Powell, T. G. and Snowdon, L. R. *Bull. Can. Petrol. Geol.*, 1979, **27** (4), 453–466.

36. Powell, T. G. and Snowdon, L. R. In 'Facts and Principles of the World's Petroleum Occurrence', A. D. Miall (ed.), Can. Soc. Petrol. Geol., Memoir 6, Calgary, 1979, pp. 421–436.
37. Powell, T. G. *Bull. Can. Petrol. Geol.*, 1982, **30** (2), 167–179.
38. Purcell, L. P., Rashid, M. A. and Hardy, I. A. *Bull. Am. Assoc. Petrol. Geol.*, 1979, **63** (1), 87–105.
39. Purcell, L. P., Umpleby, D. C. and Wade, J. A. In 'Facts and Principles of the World's Petroleum Occurrence', A. D. Miall (ed.), Can. Soc. Petrol. Geol., Memoir 6, Calgary, 1979, pp. 551–566.
40. Austin, J. A., Uchupi, E., Shaughnessy III, D. R. and Ballard, R. D. *Bull. Am. Assoc. Petrol. Geol.*, 1980, **64** (4), 501–526.
41. Poag, C. W. *Bull. Am. Assoc. Petrol. Geol.*, 1982, **66** (8), 1021–1041.
42. 'Geologic Studies of the COST G-1 and G-2 Wells, United States North Atlantic Outer Continental Shelf', P. A. Schole and C. R. Wenkham (eds.), USGS Geol. Circ. 861, 1982.
43. 'Geologic and Operational Summary, COST No. G-2 Well, Georges Bank Area, North Atlantic OCS', US Geol. Sur. Open File Report 80-268, 1980.
44. 'Geologic and Operational Summary, COST No. G-2 Well, Georges Bank Area, North Atlantic OCS', US Geol. Sur. Open File Report 80-269, 1980.
45. French, J. L. *Bull. Am. Assoc. Petrol. Geol.*, 1981, **65** (8), 1476–1484.
46. Poag, C. W. *Bull. Am. Assoc. Petrol. Geol.*, 1979, **63** (9), 1452–1466.
47. Schlee, J. S. *Bull. Am. Assoc. Petrol. Geol.*, 1981, **65** (1), 26–53.

Biological markers in fossil fuel production

R. Paul Philp
School of Geology and Geophysics, University of Oklahoma, Norman, Oklahoma 37019

I. Introduction	1
II. Biological markers	2
III. Sources, significance, and utilization of biological markers	5
A. *n*-Alkanes	5
B. Isoprenoids and branched hydrocarbons	12
C. Bicyclic sesquiterpenoids	17
D. Diterpenoids	19
E. Extended tricyclic terpanes	22
F. Tetracyclic terpanes	24
G. Triterpanes	25
H. Steranes	39
I. Aromatic hydrocarbons	45
K. Porphyrins	46
IV. Summary	47
V. References	48

I. INTRODUCTION

In the past 15–20 years, organic compounds known as *biological markers* have played an increasingly important role in the exploration for and utilization of fossil fuels. The purpose of this article is to review the use of biological markers in these areas and illustrate the type of information that is produced from their presence in fossil fuels. The scope of the review will be governed by the definitions of fossil fuel production and biological markers.

The term "fossil fuel production" is broad and far-reaching and the aspects covered in this review need to be delineated. Fossil fuel production includes all aspects of exploration for crude oil, coal and oil shale, production of liquid fuels from these sources, and characterization of both the source materials and the products. The two areas in which biological markers have had the greatest impact are the exploration for and characterization of new fossil fuel resources, in particular crude oil and, to a lesser extent, coal and oil shale. The major emphasis of this article is therefore directed at the exploration for and characterization of natural fossil fuel resources. Wherever possible, examples of the utilization of biological markers in production of synthetic liquid fuels from coal and oil shales are also given. A more detailed definition of biological markers will be given below but first it is necessary to explain why an article of this nature is appearing in a journal concerned with mass spectrometry.

In most solvent extracts from fossil fuel sources, biological markers are present in relatively low concentrations and as part of complex mixtures containing several hundred other compounds. The complexity of these mixtures means that the use of mass spectrometry (MS) has played a major role in the detection and identification of biological markers in these extracts. In many cases it is necessary to combine MS with an analytical separation technique, such as gas chromatography (GC) or high-performance liquid chromatography (HPLC), to resolve the complex mixtures before any identifications can be made. More recently techniques such as MS/MS have been used in an attempt to reduce the amount of sample fractionation required prior to analysis. Alternative methods of ionization such as fast atom bombardment (FAB) have also started to play an important role in identifying biological markers such as porphyrins. No background information will be given on any instrumental techniques in this article since it has to be assumed that readers of this journal are familiar with many aspects of MS or can refer to the original papers cited in this article for the additional information.

Before discussing the applications of biological markers to the various fossil fuel production problems, a brief explanation of biological markers and the history of their development is presented.

II. BIOLOGICAL MARKERS

Biological markers are organic compounds present in the geosphere whose structures can be unambiguously linked to the structures of precursor compounds occurring in original source material. Any alteration that may occur to the carbon skeleton of the biological marker during deposition and burial of the organic material into the sedimentary record should be minimal and wherever possible limited to stereochemical changes. For many classes of biological markers, definite precursor/product relationships have been established. For example, the pathways by which sterols are converted to steranes have been studied in great detail and many of the conversion reactions are well documented (1). In other cases the situation is not as clear, and tentative conversion pathways have been inferred. Further work is required to elucidate the reactions that can occur to the biological markers after deposition.

The term "biological markers" has evolved from another term, "chemical fossils," introduced in early papers by Eglinton and Calvin (2) and Eglinton (3). In the original work the term was introduced to describe organic compounds found in ancient rocks and oils thought to be derived from organic material deposited in the sediments. In the past few years, the term "chemical fossils" has not been widely used but has been replaced by the term "biological markers" or the abbreviated form "biomarkers." The emphasis in many geochemical studies is now being placed on establishing precursor/product relationships rather than just identifying the biological markers. This shift has probably resulted from a significant increase in studies of recent sediments and source materials, such as algae and higher plants, in the 1970s (4,5). These studies enabled the source of many biological markers to be established with some certainty.

Although the terminology and definitions are relatively new, the basic concept of biological markers is not. Indeed it was in the 1930s that Treibs (6,7) isolated and identified porphyrins in crude oils. From these classical studies he was able to imply a biogenic origin for crude oils and in particular a specific input of organic material to the oil. Many of the early MS studies of biological markers were limited to the isolation and/or identification of individual components. In 1956 de Mayo and Reed (8) reported the determination of molecular weights for a number of steranes and triterpanes by mass spectrometry. However, at that time the precision of the mass spectrometer was such that the weight of the side chains of these molecules could be estimated only to within about three mass units.

Another early example of a biological marker study was the isolation by Jarolim et al. (9) of a number of crystalline triterpenes from Montan Wax (north Bohemian brown coal) and their identification on the basis of their melting points and comparison with authentic standards. The same compounds had been isolated earlier by Ruhemann and Raud in 1932 (10) but not identified at that time. Jarolim et al. (9) commented on the remarkable presence of triterpenes in brown coal and suggested that further research would be of interest both from a paleobotanical point of view and for purposes of an exact classification of different types of coals. Carruthers and Watkins (11) used chemical techniques and infrared spectroscopy to identify 1,2,3,4-tetrahydro-2,2,9-trimethylpicene in an American crude oil after initially isolating the compound in 1954 (12). In another attempt to relate products and precursors, Carruthers and Watkins (11) suggested that this compound either arose directly from triterpenes of the β-amyrin class or from isoprenoids by stepwise cyclization and dehydrogenation.

In the late 1950s and early 1960s MS was being used in the petroleum industry for characterizing the composition of boiling point fractions and identifying specific components in these fractions (13,14). By the time Eglinton and Calvin (2) published their article on chemical fossils in May 1967, the combined technique of GC/MS had been developed and was being used to analyze complex organic mixtures such as those obtained from oils, coals, and other geochemical samples. Thus, the amount of painstaking work and laborious effort of early workers in the field, such as Treibs (6), Carruthers and Watkins (11), Jarolim et al. (9), and many others, required to identify individual biological markers was dramatically reduced for subsequent researchers following the development of GC/MS. The commercial availability of GC/MS systems in the 1970s and their associated data systems, led to an exponential increase in the number of biological markers whose structures were confirmed. In addition, many more biological markers have been ten-

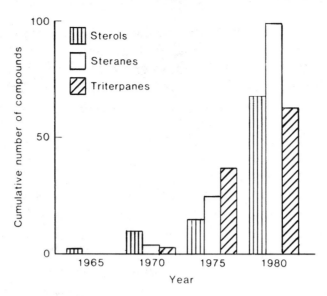

Figure 1. The number of polycyclic biomarkers that have been identified in geological samples has increased exponentially as shown in this diagram taken from a recent paper by Mackenzie et al. (1). [Reprinted with permission. Copyright 1982 by the AAAS.]

tatively identified on the basis of their MS data. Mackenzie et al. (1) published an interesting summary of the number of sterane and triterpane biological markers that had been unambiguously identified over the last ten years (Fig. 1). A similar exponential expansion has probably occurred in other classes of biological markers.

This tremendous increase has been brought about by a number of factors, the first of these being the advances that have occurred in analytical capability. In the early stages of development, GC/MS systems were equipped with packed GC columns. Many advances have been made in gas chromatography and column technology that have led to high-resolution, low-activity, fused-silica capillary columns which are now routinely used by most workers analyzing geochemical samples to determine the distribution of biological markers. In parallel with the increase in resolution available from chromatography columns, fast-scanning quadrupole or magnetic mass spectrometers have been developed, with a consequent increase in the number of components that can be identified in each mixture. A second reason for the increase in the number of biological markers identified is a better understanding of the mechanisms involved in the transformation of organic matter following its deposition. This has provided clues as to the types of compounds expected to be produced from a particular type of organic material or precursor compound which has in many cases assisted in identification of the products. A third reason is an increase in the number of research groups examining an ever-increasing variety of geochemical samples. Furthermore, the use of biological markers, not only in fossil fuel studies but also in other areas such as environmental studies, has led to a greater awareness of their value and potential to solve problems that cannot be approached by more conventional methodologies.

In the following sections the occurrence and utility of the major classes of biological markers that are of importance in fossil fuel production are discussed. Information on the origin of the compounds, distributions observed in geochemical samples, and the significance of these distributions is included. Most of the classes discussed are hydrocarbons since the biological markers of greatest use in fossil fuel studies have been converted from their oxygenated or unsaturated precursors into hydrocarbons as a result of diagenesis and maturation changes. Nitrogen and sulfur compounds will be mentioned only very briefly as they are covered in more detail in other reviews in this volume. Oxygenated compounds such as fatty acids and phenols are mentioned briefly since they occur in many fossil fuels and in some cases are quite useful as biological markers.

III. SOURCES, SIGNIFICANCE, AND UTILIZATION OF BIOLOGICAL MARKERS

A. *n*-Alkanes

n-Alkanes (**1**) have a simple structure but have been used as biological markers quite successfully for several decades as a result of their abundance and ease of detection by GC alone. The most common application of the *n*-alkane fingerprint is to infer the source of the organic material in a sample.

1

For example, it has been clearly demonstrated (15–17) that leaf wax alkanes have a much higher concentration of odd-numbered alkanes (n-C_{27}, C_{29}, and C_{31}) than even-numbered alkanes (n-C_{28} and n-C_{30}). This characteristic leaf wax *n*-alkane fingerprint can be observed in sediments or petroleum source rocks of low maturity that have a high input of terrestrial source material. Another example of the use of *n*-alkane fingerprints as a source indicator is the case of phytoplankton which are characterized by the predominance of the n-C_{17} alkane (18–20). A predominance of *n*-alkanes in the n-C_{16}–C_{24} region of a chromatogram with no marked odd/even predominance is indicative of a significant contribution of bacterial organic material to a sediment that can be confirmed by the presence of branched hydrocarbons (21). Additional examples of the relationship between *n*-alkane signatures and source materials have been described by Brassell et al. (22).

As the maturity of the sediment or fossil fuel precursor material increases, the distributions of the *n*-alkanes change and much of their source specificity is lost. *n*-Alkane distributions with a marked odd/even predominance found in samples of low maturity are converted into a distribution with little or no odd/even preference. As the maturity level of the sample increases further, the position of the *n*-alkane maximum tends to shift toward lower-carbon-number alkanes. Ultimately only condensates of light hydrocarbons or methane are produced. At such high levels of maturity n-alkane distributions are of little use as source of maturity indicators. The mechanisms by which the n-alkane distributions change have been the subject of many experimental and simulation studies. Cooper and Bray (23) initially postulated that odd-numbered alkanes were formed from decarboxylation of even-numbered fatty acids. Kvenvolden (24) also suggested that the normal fatty acids may be precursors for many normal alkanes of intermediate and high molecular weights on the basis of evidence available in the literature at that time. Henderson et al. (25) performed pyrolysis experiments on n-octacosane ($C_{28}H_{58}$) in the presence of clay minerals and showed that the n-alkane thermally degrades to a complete homologous series of n-alkanes and n-alkenes. Each n-alkane generated in this way can undergo further degradation and thus remove any odd/even preference from samples at higher levels of maturity. Pyrolysis of even-numbered n-alcohols commonly found in higher plants has also been shown to produce both even- and odd-numbered alkanes and alkenes (26–28).

In addition to variations in source and maturity, biodegradation can also influence the distribution of n-alkanes is fossil fuels and thus their utility as biological markers. Many articles have been published on biodegradation of fossil fuels (29–31) and it will not be discussed in detail in this article. In summary, the first stages of bacterial degradation of a crude oil are characterized by removal of the low-molecular-weight alkanes, followed by alkanes in the n-C_{16} to n-C_{25} range and finally by those above n-C_{25}. Biodegradation is sometimes accompanied by water washing which can also influence the distribution of hydocarbons in crude oils (32) and produce distributions similar to those resulting from biodegradation. Figure 2 shows gas chromatograms of two Australian oils, presumed to be from the same source, one of which has been biodegraded and had the n-alkanes removed and the other unaltered. Removal of the n-alkanes in this way reduces their usefulness as a correlation tool for altered oils.

The fourth factor that can affect n-alkane distributions specifically in crude oils is migration. Recently, Mackenzie et al. (33) investigated preferential migration of n-C_{15}–C_{19} alkanes in the Rundalen sequence from Spitsbergen, Norway. They found that thin shale sequences tend to expel the n-C_{15}–C_{19} alkanes more rapidly than thick shales as a result of compaction and expulsion. On this basis, it was concluded that up to 90% of lower-carbon-number n-alkanes may be expelled in favorable conditions although only 10% of the total extract is expelled. Chromatographic effects can also influence the distributions of the expelled n-alkanes with the higher-molecular-weight n-alkanes moving more slowly than the lower members of the series during migration.

The four major factors that can affect the n-alkane distributions in fossil fuels, namely source, maturity, biodegradation, and migration, have been briefly summarized. In the following sections a number of studies in which n-alkane distributions have played a role in exploration or production are described.

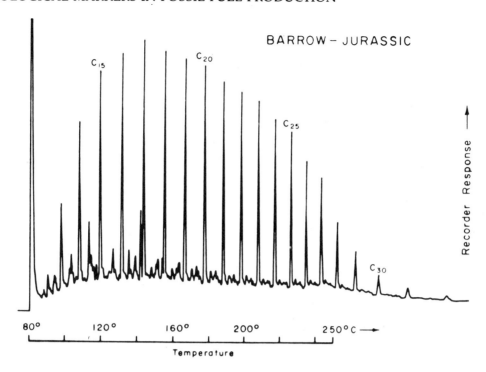

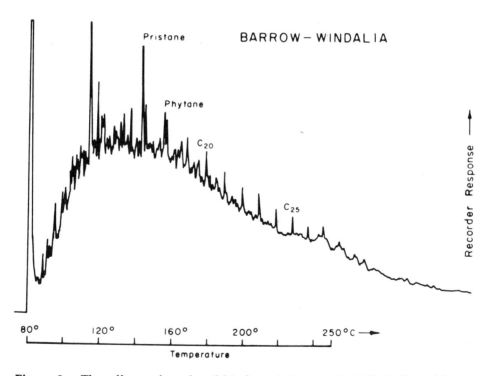

Figure 2. The effects of crude oil biodegradation are initially indicated by removal of the *n*-alkanes as illustrated by the two Australian oils whose chromatograms are shown in this diagram. These oils are assumed to be derived from the same or similar sources but the Windalia oil is biodegraded (34). [Reprinted with permission from the Director, Bureau of Mineral Resources.]

1. Distribution of n-*alkanes in crude oils and source rocks*

There have been many reports in the literature on the characterization of crude oils, source rock extracts, shales, and coals from their *n*-alkane distributions. As mentioned previously, the ease of determination of the *n*-alkanes by using GC alone has played an important role in many of these studies. The majority of references to *n*-alkane utilization given in this article are taken from the last five or six years but many of these recent papers contain citations of earlier papers involved with *n*-alkane distributions in fossil fuel sources. The references given are not meant to be exhaustive but to demonstrate the wide range of samples that have been totally or partially characterized from their *n*-alkane distributions.

Powell and McKirdy (34,35) characterized a number of Australian crude oils on the basis of boiling point and *n*-alkane distributions and concluded that many of these oils were derived from terrestrial source material and differed in maturity as indicated by their *n*-alkane fingerprints. Variations in microbial activity during early diagenesis of the organic source material were also thought to contribute to differences observed in crude oil compositions. In a more detailed study of oil occurrence in the Dampier Sub-Basin, Western Australia, Powell (36) used the *n*-alkane distributions to differentiate oils in this basin into two groups, one being a paraffinic naphthenic type and the other a napthenic type of oil. Subsequent work using GC and GC/MS has shown that the oils in this basin probably have the same source but differ as a result of biodegradation (37).

Illich, Haney, and Jackson (38) analyzed 15 oils from the Maranon Basin, Peru, and distinguished two major types of oils on the basis of *n*-alkane and isoprenoid distributions. In another study by Illich, Haney, and Mendoza (39), oils from the Santa Cruz Basin, Bolivia, were examined and variations in the *n*-alkane distributions and other hydrocarbons were rationalized on the basis of solubility variations for the various hydrocarbons in aqueous solution and fractionation occurring during migration. Kovacev et al. (40) showed that the *n*-alkane distributions from a Bulgarian crude oil maximized at n-C_{16} over the range of n-C_{12} to n-C_{33}. A genetic classification scheme for crude oils and condensates from the Beaufort–Mackenzie Basin was established on the basis of their *n*-alkane and isoprenoid distributions (41), which were determined by GC using an inorganic salt eutectic column of the type previously described by Snowdon and Peake (42). The hydrocarbons of an unnamed terrestrial crude oil were determined by Richardson and Müller (43,44) and the *n*-alkanes were found to maximize in the n-C_{23}–C_{27} region as expected for an oil sourced from terrestrial material. A crude oil from Bradford, Pennsylvania, was analyzed by Hoering (45) and it was found that in addition to *n*-alkanes and other hydrocarbons, 2% of the oil occurred as olefins. Olefins are not commonly reported constituents of crude oils or ancient sediments since they are chemically reactive and do not survive maturation reactions. However, Simoneit and Lonsdale (46) reported *n*-alkanes and olefins in samples of hydrothermal petroleum from the Guaymas Basin, Gulf of California. It was proposed that this unusual "immature" petroleum had been derived from biological detritus by thermal alteration and rapid quenching by hydrothermal removal followed by condensation at the sea bed.

The distribution of n-alkanes in suspected source rock samples is often used to give an indication of the maturity of the rock, but, as mentioned above, the odd/even predominance of the n-alkanes common in immature rocks disappears with increasing maturity. With the discovery of more sophisticated maturity parameters, such as sterane and triterpane distributions described below, the n-alkane parameter is not as widely used as it was a few years ago due to its lack of specificity. A recent example was given by Szucs and Wein-Brukner (47) in an examination of core samples from a Hungarian oil well in which the n-alkane distributions were used to obtain information on both the type and the maturity of the organic matter. A survey of n-alkane distributions throughout the geological record was provided by Tissot et al. (48) who statistically reviewed n-alkane distributions from 1300 analyses of rocks and crude oils. It was found that n-alkanes of high molecular weight, without odd or even preference, or with a slight preference, appeared in rocks from the lower Paleozoic (Table I) and are probably derived from bacterial lipids. On the contrary, n-alkane distributions with a strong odd predominance in the n-C_{27} to n-C_{31} range occurred only in the Mesozoic and mostly in lower Cretaceous rocks (see Table I). These distributions are related to the occurrence of higher plants and particularly to the expansion of angiosperms. Distributions of n-alkanes with an even-number preference from C_{16} to C_{30} occurred essentially in carbonates and evaporite series of the Cretaceous or the Tertiary age (Table I) and were probably determined by the nature of the depositional environment. The distribution of medium-molecular-weight n-alkanes sometimes showed an odd predominance from C_{11} to C_{19} in rocks from the lower or middle Paleozoic series; they were probably derived from saturated fatty acids. However, a strong predominance of C_{15} and C_{17} alkanes was found in some shallow sediments from Devonian to Recent (Table I) which would have been derived from lipids synthesized by phytoplankton and benthic algae.

The occurrence of n-alkane distributions with an even/odd predominance is not as widespread as the odd/even preference. As mentioned above, Tissot et al. (48) noted their occurrence in carbonates and evaporite series. Albaiges and Torradas (49) found an even/odd predominance of n-alkanes in the Amposta marine crude oil from Spain, although most of the Spanish oils had a smooth n-alkane distribution (50). Although the even/odd predominance of n-alkanes in crude oils is not common, a few examples of fossil fuel deposits in which even-carbon-numbered n-alkanes predominate over part of the homology are known. These include Recent sediments (51,52), borehole samples of shales (53), carbonate-rich sediments (54) and wurtzilites (55), bituminous shales from the Ghareb Formation from Nelu Musa (Jordan Valley) (56), and an organic-rich rock of Mississippian age containing algal-like remnants (57).

In addition to characterizing oils and source rocks, hydrocarbon compo-

Table I. Summary of the geological time scale.

		Approximate oldest age in millions of years before present
Cainozoic[a]		
	Quaternary	
	Pleistocene	2
	Tertiary	
	Pliocene	7
	Miocene	26
	Oligocene	38
	Eocene	54
	Paleocene	65
Mesozoic[a]		
	Cretaceous	
	Upper	100
	Lower	162
	Jurassic	
	Upper	162
	Middle	172
	Lower	195
	Triassic	
	Upper	205
	Middle	215
	Lower	225
Paleozoic[a]		
	Permian	
	Upper	240
	Lower	280
	Carboniferous	
	Upper (Silesian)	325
	Lower (Dinantian)	345
	Devonian	
	Upper	359
	Middle	370
	Lower	395
	Silurian	440
	Ordovician	
	Upper	445
	Lower	500
	Cambrian	
	Upper	515
	Middle	540
	Lower	570

[a] "Era"; the subheadings, e.g., Quaternary, correspond to "Periods" and the remainder to "Epochs."

sitions, including n-alkanes, of seawater or surface sediments have been used to determine whether seepage is occurring and to identify previously unknown petroleum sources in marine environments (58,59). It is possible to differentiate between marine sediments receiving petroleum contributions from those sediments in which only biogenic hydrocarbons were present on the basis of the n-alkane distributions.

2. Distribution of n-alkanes in coals, oil shales, and synthetic fuels

The production of liquid hydrocarbons from coals and oil shales has led to a number of reports on the nature of products formed in these reactions, including n-alkanes. Samples examined include syncrudes from Western Kentucky Coal (60), solvent-refined coal produced from Kentucky bituminous coals (61), and jet fuels produced from coal, oil shale and tar sands (62).

As well as being produced during liquefaction reactions, n-alkanes are relatively abundant in the soluble extracts from coals as a consequence of plant wax contributions. Variations in source material and maceral type lead to variations in n-alkane distributions as was shown in a study of a series of carboniferous vitrinites and sporinites of bituminous rank (63). n-Alkanes extractable from coal by supercritical gas extraction have been analyzed by using GC and GC/MS (64) as were the fatty acids to determine whether some of the alkanes were formed by decarboxylation of the fatty acids during the extraction process (65). When extracts from seven Hungarian coals were examined by GC to determine both the n-alkane and aromatic hydrocarbon distributions and the carbon preference index (CPI)* as calculated for each coal, it was found that coals previously thought to be related had similar n-alkane distributions (66). A series of 26 German coals were characterized in detail by Radke et al. (67) to determine relationships between composition of the soluble organic matter in coals and rank and liptinite fluorescence. At a reflectance of about 0.9%, it was found that there was a gradient change in the odd/even predominance of long-chain n-alkanes, a bimodal n-alkane distribution appeared, and there was a sharp drop in concentration of individual n-alkanes. It was proposed that changes observed in n-alkane distributions at this level of maturity could be explained by a shift in the prevailing type of reaction experienced by the coals from hydrocarbon generation to fragmentation. For further information on aliphatic hydrocarbons in coal and coal-related products, readers are referred to a review published in 1978 (68) and another article in 1982 (69) which provide a more detailed coverage of n-alkane distributions than can be given here.

Renewed interest in oil shale liquefaction in recent years has led to the appearance of several papers on the components of shale oils and oil shale extracts. Robinson and Cook (70) examined in some detail the n-alkane distributions from samples of Uinta Basin (Utah) oil shale relative to stratigraphic position within the Green River Formation. Variations observed in these distributions were related to differences in source material, burial depth, or environmental differences. Young and Yen (71) oxidized Green River kerogen and examined the straight-chain aliphatic structures released in this way to obtain information on kerogen structure. The n-alkanes from Bulgarian oil shales have been examined by Kovachev (72) who found a homologous series in the range C_{13}–C_{34}. The bimodal n-alkane distribution had maxima at C_{17} and C_{27} indicative of both algal (C_{17}) and higher plant (C_{27})

---

*The CPI for hydrocarbons is the sum of the odd-carbon-number homologs over a specified range divided by the sum of the even-carbon-number homologs over the same range (23).

contributions. The major components in the solvent extracts from Posidonomia shales from the Dotternhausen Formation (West Germany) were identified as n-alkanes by using GC and GC/MS (73). The n-alkanes ranged from C_{10} to C_{26} with a maximum between C_{15} and C_{19}. The distribution and position of the n-alkane maxima confirmed the marine origin of the Dotternhausen shales. The absence of any odd/even preferences in the n-alkane distributions indicated that extensive diagenesis of the samples had also occurred. Australian oil shales and shale oils have been studied extensively in the past few years. Regtop, Crisp, and Ellis (74) have identified over 500 compounds in shale oil from the Rundle deposit. The bimodal distribution of the n-alkane distributions indicated organic material derived from both algal and higher plant sources with algal sources predominating. In another study Ingram et al. (75) compared extracts of oil shales and shale oils from the Mahogany Zone, Green River Formation, and Kerosene Creek seam, Rundle formation, Australia. Distributions of various classes of compounds including the n-alkanes were used to determine source and maturity differences between the samples. Finally, it is of interest to mention that retorting of oil shales produces, among other products, homologous series of 1-alkenes. In a recent paper Riley et al. (76) have suggested the possibility of using these compounds as markers to identify sediments that have been in contact with shale oils or their refined products.

To summarize, it should be clear that n-alkanes are ubiquitous in fossil fuels. Many studies have been undertaken to determine their distributions in source materials and also in oils, source rocks, coals, and shales. Furthermore, it has been shown that n-alkane fingerprints can be used to provide information on source and maturity of a sample and, in the case of oils, whether or not biodegradation or water washing has occurred. Evidence is now becoming available to show differential rates of migration for individual n-alkanes and a potential use for n-alkanes in petroleum migration studies. The utility of n-alkane distributions in fossil fuel studies is somewhat limited due to their ubiquity, lack of specificity, and differences in many of the samples examined. In many cases the n-alkanes can be used to give an overall impression on the origin and history of a sample, and distributions of specific biomarkers such as steranes and triterpanes can be determined to provide more detailed information on the samples. The next class of biological markers to be discussed, the isoprenoids, are structurally more complex than the n-alkanes and the information obtainable from them also more specific. It will be seen throughout this article that as the complexity of the structures of the biological markers increases so does the amount of information that can be obtained from the distributions.

B. Isoprenoids and branched hydrocarbons

In the preceding section it was noted that the distribution of n-alkanes in fossil fuel samples could be determined by using GC alone without the aid of mass spectrometry. The isoprenoids, with the exception of the regular isoprenoids in the C_{13}–C_{20} range, are generally present in much lower concentrations than the *n*-alkanes and cannot always be detected by using GC alone. The use of GC/MS and single-ion monitoring (SIM) or multiple-ion detection (MID) techniques therefore is necessary. However, before discussing the detection and utilization of isoprenoids in fossil fuel samples, it is appropriate to describe briefly the different types of isoprenoids found in fossil fuels.

Isoprenoids are formed from various combinations of C_5 isoprene units, and three main types of linkages have been discovered for these isoprene units in fossil fuel samples. These are the regular *head*-to-*tail* linkage; *tail*-to-*tail* linkage, and *head*-to-*head* linkage. Head-to-tail linkages are most abundant and include pristane (2), phytane (3), and other members of a homologous series up to C_{40} and tentatively identified up to C_{45} (77). Head-to-tail anteisoprenoids, where the branching points are at the 3, 7, 11, 15 positions rather than the 2, 6, 10, 14 positions of the regular isoprenoids, have also been detected in a number of fossil fuel samples and used to monitor marine oil pollutants (78). Isoprenoids with a single tail-to-tail linkage found in fossil fuels include squalane (4) (79), perhydro-β-carotane (5) (80), and lycopane (6) (81), all three compounds having naturally occurring unsaturated precursors. Head-to-head-linked isoprenoids have been found recently in crude oil (77,82), kerogen degradation products (83), and in cell wall membranes of thermoacidophilic bacteria of the Calderiella series (84).

The regular C_{13}–C_{20} isoprenoids have received much attention in the past since they are present in oils and coals in relatively high concentrations and are easily detectable by using GC. It is commonly assumed that the major source of these compounds is the phytol side chain of chlorophyll (7). Brooks, Gould, and Smith (85) suggested the preferential formation of pristane in an oxidizing-type depositional environment (coal-swamp), whereas phytane is preferentially formed from chlorophyll in a more reducing deltaic or marine-type environment. This idea was expounded by Powell and McKirdy (86) who determined that variations in the pristane/phytane ratio could be used to indicate different types of environments. However, the degradation of chlorophyll and phytol in the environment is extremely complex and many

7

intermediates are involved in the formation of pristane and phytane as summarized in a review by Didyk et al. (87).

Pristane/phytane ratios and also pristane/C_{17} and phytane/C_{18} ratios have been used in many fossil fuel studies to obtain information on the environment of deposition and, to a lesser extent, as correlation parameters in crude oil and source rock studies. For example Vogler, Meyers, and Moore (88) compared a number of oils from the Michigan Basin and used these ratios, and other parameters, to differentiate Devonian, Silurian, and Ordovician oils. The Windalia oil and suspected source rocks from the Barrow Sub-Basin in Western Australia were examined by Alexander, Kagi, and Woodhouse (89), and the ratio of pristane + C_{17}/phytane + C_{18} was used as one of many correlation parameters. Rashid (90) determined pristane/phytane ratios for Labrador Shelf sediments and noted that the ratio increased as the amount of terrestrial material and burial depth increased. Gibert et al. (91) analyzed a shale extract from the Permian Irati Formation of Brazil and observed an excess of regular isoprenoids over n-alkanes as a result of diagenesis and not biodegradation as is normally the case for this unusual distribution. These are only a few examples of studies where pristane/phytane ratios were used to obtain geochemical information. Since the ratio is determined easily from GC data, virtually all of the papers described in the previous section on n-alkanes also contain mention of the distribution of the regular isoprenoids.

In addition to pristane/phytane ratios, use can also be made of the fact that pristane has two chiral centers and three possible stereoisomers. This has made it possible to show that the stereochemistry of pristane in geochemical samples can be related to the enzymatically controlled stereospecificity of the presumed precursor natural products (92,93). Sequential changes have been observed to occur after isomerization at the chiral centers with increasing sample depth in samples from the Paris Basin (94,95). Variations of these ratios, along with those determined from other biomarkers, have been used to obtain information on relative maturities of the samples including those in the middle of the oil generation zone.

The ease of detection of the regular isoprenoids up to C_{20} by using GC has led to much information on their distribution and consequently to explanations for variations in these distributions. The use of GC/MS and SIM and MID has extended the range and type of isoprenoids identified in fossil fuels. In some early work, Han and Calvin (96) identified C_{22}–C_{25} isoprenoids in Bell Creek crude oil by using GC/MS. Similarly, Haug and Curry (97) tentatively identified regular isoprenoids up to C_{30} in a Costa Rican oil seep.

Waples, Haug, and Welte (98) identified a C_{25} regular isoprenoid from highly saline Tertiary sediments and proposed that it could be used as a biological marker for lagoonal-type saline environments. The presence of regular head-to-tail isoprenoids up to C_{40} in Spanish crude oils was established by Albaiges and Torrados (99) and Albaiges, Borbon, and Salagre (100) by using GC/MS. The long-chain regular isoprenoids have been used along with head-to-head isoprenoids in correlation studies of Spanish crude oils with their suspected source rocks (77). Moldowan and Seifert (82) also reported the occurrence of head-to-head isoprenoids up to C_{40} in a California Miocene crude following GC/MS analysis of the sample (Fig. 3). A similar distribution of long-chain isoprenoids has been observed in a new Zealand oil that is almost certainly derived from terrestrial source material. The origin of these long-chain isoprenoids still remains to be determined although Han and Calvin (96) proposed a possible origin from long-chain oligoterpenyl alcohols present in many naturally occurring materials. A similarity has been noted between the distribution of C_{16}–C_{30} saturated isoprenoids in the neutral lipid fraction of thermoacidophilic bacteria and the same compounds in ancient sediments and petroleum (101). This similarity in isoprenoid distribution has led to the suggestion that the thermophilic bacteria are an alternative source for the isoprenoids in crude oils and source rocks. The isolation of C_{40} head-to-head-linked isoprenoids in the degradation products of a kerogen led Michaelis and Albrecht (102) to propose that the mild geological history of the Messel shale used in the study precluded the origin of the head-to-head isoprenoids

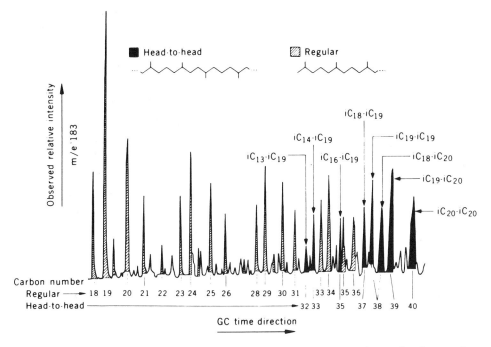

Figure 3. Mixtures of regular and head-to-head isoprenoids can be detected in crude oils by SIM of the ion at m/z 183. This example is from a California Miocene crude oil (82). [Reprinted with permission. Copyright 1979 by the AAAS.]

from thermophilic members of the *Archaebacteria* and alternatively to suggest that they could have been generated from the methanogenic members of the *Archaebacteria*. Further evidence that a significant part of geological organic matter in sediments and petroleum is derived from the lipids of membranes of microorganisms was provided by the identification of glycerol tetraethers with C_{40} head-to-head isoprenoids.

Isoprenoids with tail-to-tail linkages are limited to a much smaller number of compounds than those with head-to-head or head-to-tail linkages, namely squalane **(4)**, perhydro-β-carotane **(5)**, and lycopane **(6)**, all of which are derived from their unsaturated counterparts that occur as natural products. These compounds have not been widely used directly as biological markers in fossil fuel studies although the cyclization of squalene to various sterols and triterpanes ensures its importance in such studies. The role of tail-to-tail isoprenoids in the formation of C_{17} and C_{22} regular isoprenoids has also been discussed in some detail by Albaiges, Borbon, and Salagre (100). Squalane has also been tentatively identified in Libyan (103) and Nigerian crude oils (79). It has also been reported in *Archaebacteria* (101), and these organisms can, therefore, represent a direct input of the saturated hydrocarbon into the sediments and crude oils and consequently an alternative source for squalane in fossil fuels. Furthermore, the C_{25} isoprenoid (2,6,10,15,19-pentamethyleicosane, **(8)** with a tail-to-tail linkage has been found in methanogenic *Archaebacteria* only and has been proposed as a marker for methanogen contributions to marine sediments (104).

A number of other branched hydrocarbons have also played an important role in fossil fuel studies. The first of two noteworthy examples is botryococcane **(9)** which is presumably derived from botryococcene **(10)** which has been observed only in the fresh or brackish water alga, *Botryococcus braunii* (105). The subsequent discovery of botryococcane **(9)** in Indonesian crude oils led Moldowan and Seifert (106) to propose that these oils were generated principally from the organic material of prehistoric fresh or brackish waters. The second example is 2,6,10-trimethyl-7-(3-methylbutyl)-dodecane **(11)** which has recently been found to occur in both Recent sediments and Rozel Point crude oil (107). Although its exact origin still remains to be determined, the limited occurrence of this compound in geological samples means that it could be a useful input marker.

From the preceding discussion it should be clear that isoprenoids are utilized in many aspects of petroleum exploration studies including source identification, maturity measurements, and correlation studies. Isoprenoids have not been used so widely in other areas of fossil fuel production but one occurrence that should be reported concerns the production of prist-1-ene and prist-2-ene from pyrolysis of coal or oil shale. Prist-1-ene was observed in the pyrolysis products of all kerogen types by Larter et al. (108), and van de Meent et al. (109) observed them in pyrolysis products of Messel shale and a marine kerogen from the Cariaco trench. DiSanzo, Uden, and Siggia (110) reported the occurrence of prist-1-ene (12) in shale oils from the TOSCO process (The Oil Shale Corporation, Denver, Colorado) and Burnham et al. (111) studied pyrolysis of Green River kerogen under different conditions and observed variations in rates of formation and relative concentrations of isoprenoids. Baset, Pancirov, and Ashe (112) and Philp and Saxby (113) reported the occurrence of isoprenoid alkenes from the pyrolysis of a variety of American and Australian coal samples, respectively. It has been proposed that the precursor of the prist-1-ene formed during pyrolysis is also the precursor of the pristane and phytane formed under geological conditions (108).

C. Bicyclic sesquiterpenoids

Bicyclic sesquiterpenoids (13) have not been investigated extensively in fossil fuel research, and relatively little effort has been devoted to determining their distributions in fossil fuel samples. One of the earliest reports of these compounds was their occurrence in a degraded Gulf Coast oil (114). Two bicyclic sesquiterpenoids which were predominant in this oil were partially identified on the basis of data acquired by using GC/MS but the precise nature of the substituent pattern could not be determined on the basis of the mass spectral data alone. Bendoraitis (114) proposed a possible origin for these compounds from degradation of pentacyclic triterpenes such as β-amyrin (14) during the maturation process. One of the C_{14} bicycloalkanes identified in this study had a mass spectrum very similar to a dialkylsubstituted hexahydrindane previously observed (115) in extracts from the Green River Shale. Anders and Robinson (115) proposed that the substituted hexahydrindane was formed via a steroid fragmentation pathway.

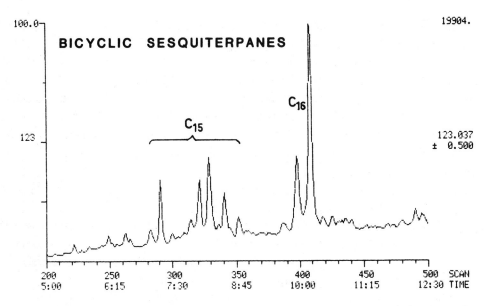

Figure 4. Bicyclic sesquiterpenoids can be detected in crude oils by SIM of the ion at m/z 123. The compounds appear to be particularly widespread in the terrestrially sourced oils from Australia and New Zealand as shown by this New Zealand oil.

Philp, Gilbert, and Friedrich (116) reported the distribution of bicyclic sesquiterpenoids in a number of Australian crude oils determined by using GC/MS and MID. A complex mixture of the C_{14}–C_{16} compounds found in the oils (Fig. 4) was virtually identical in all samples examined but quite different from those observed by Bendoraitis (114). Richardson and Miiller (43) used GC/MS to analyze the saturated fraction of an unnamed oil derived from terrestrial source material and identified a number of isomeric C_{15} and C_{16} bicyclic sesquiterpenoids. The identity of two bicyclic sesquiterpanes found in Australian crude oils was recently confirmed by synthesizing two of the standards, namely drimane **(15)** and eudesmane **(16)** (117). Sesquiterpenes based on eudesmane are widely distributed in higher plants and appear to provide unambiguous evidence for a terrestrial contribution to crude oils. The presence of drimane was confirmed in a number of crude oils including Cambrian–Ordovician samples where land plant input was not significant, and it was therefore proposed that the drimane is most likely of microbial origin (117). Recent studies of Alaskan oils (118) have shown that they contain a number of the bicyclic sesquiterpanes, as do New Zealand oils (Philp, unpublished results). All of these oils are known to be at least partially derived from terrestrial material, and it would appear that further work should establish the usefulness of these compounds as terrestrial source indicators.

The reports described above have been concerned with the use of bicyclic sesquiterpanes as source indicators in petroleum studies. It is also possible that many of the isomers present in these mixtures are formed following an increase in maturity of the samples, which in turn leads to the possibility that ratios of certain bicyclic sesquiterpane isomers may be useful as maturity parameters. In view of the fact that these compounds do not appear to be affected by biodegradation (116), their use as maturity parameters would strongly complement parameters derived from the steranes whose distributions are affected by biodegradation.

The presence of bicyclic sesquiterpanes in coal has not been widely reported, although in a recent report Baset, Pancirov, and Ashe (112) noted the occurrence of at least nine C_{15} compounds in the aliphatic fraction of Wyodak coal. Baset, Pancirov, and Ashe (112) also examined the pyrolysate of Wyodak coal by using GC/MS and noted the absence of sesquiterpenes from the pyrolysate but the presence of a number of alkylnaphthalenes including cadalene (17) which has been proposed to form as a result of dehydrogenation of bicyclic sesquiterpenes (119). Grantham and Douglas (120) identified a number of sesquiterpenoids including cedrane (18) and cuparane (19) in fossil resins and lignites and proposed that these compounds could be useful markers of such a contribution to coals.

D. Diterpenoids

The main role of diterpenoids in fossil fuel studies has been as an indicator of terrestrial source material, particularly resins, to oils, source rocks, and coals. The precursors of diterpanes are generally accepted to be diterpenoid acids based on the abietane (20), pimarane (21), and labdane (22) skeletons, which occur in plant resins as complex mixtures of di- and trienoic acids. In the few reports of diterpenoids in fossil fuels, GC/MS has generally played an important role in their detection and identification. As with the bicyclic sesquiterpenoids described above, authentic standards for many of the diterpanes are not available and thus identification is often tentative and based on mass spectral interpretation alone. The majority of reported diterpenoid occurrences have been concerned with coal and not with oils or source rocks as is the case with many other classes of biological markers.

Hagemann and Hollerbach (121) examined a number of lithotypes of soft brown coal from the Rhenisch area, West Germany, by using a variety of geochemical and petrographical techniques to identify a number of diterpanes including phyllocladane **(23)**, kaurane **(24)**, norabietane **(25)**, and norpimarane **(26)**. The first two compounds are common in Recent phyllocladus

and kauriconifers (119) and the remaining compounds have been identified as common constituents of fossil resins (122) and lignites (55). A similar study has been undertaken of the lithotypes in the various brown coal deposits located in southeastern Australia by Chaffee (123). Extensive use was made of GC/MS to determine variations of many classes of biomarkers, including diterpanes, in the lithotypes of this region. Many of the diterpanes found were similar to those described by Hagemann and Hollerbach (121). In a study of oils from the offshore Gippsland Basin, which is adjacent to the brown coal fields of Victoria, Philp, Gilbert, and Friedrich (116) found a mixture of nine C_{19} and C_{20} diterpenoids in all of the oils examined by using GC/MS and SIM of the characteristic ion at m/z 123 (Fig. 5). Certain similarities were noticed between the diterpanes in the oils and those previously observed in the brown coals, and an attempt was made to investigate any correlation between the diterpanes in brown coal resins and the Gippsland oils (124). This study proved inconclusive although an earlier study along the same lines of the Mackenzie Delta Basin, Canada, by Snowdon (125,126) had been far more successful. Oils from the Mackenzie Delta Basin contained high concentrations of diterpanes, one of which was tentatively identified as sandarocopimarane, and examination of resinite from this basin also showed a high concentration of the same diterpanes. Further work using additional geochemical information confirmed that resinite is indeed a major source contributor to the oils from this basin (127).

Diterpanes are found in coals at the lignite stage of coalification but are virtually absent at the subbituminous or higher levels of coalification (128). This observation can be explained by the fact that in the early stages of coalification, defunctionalization occurs to form the saturated hydrocarbon diterpanes whereas at higher stages of coalification, aromatization occurs

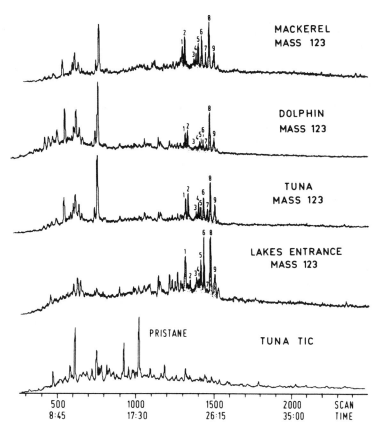

Figure 5. Diterpenoids are abundant in crude oils from the Gippsland Basin, Australia, as illustrated in this example (116). Peaks 1 and 2 are C_{19} diterpenoids and peaks 3–9 are C_{20} diterpenoids.

and the diterpenoid structures are subsequently converted into methyl-substituted phenanthrenes and other aromatic components. Hollerbach and Hagemann (128) proposed that changes in distribution of various classes of biomarkers, such as the diterpanes, could be used as additional organic geochemical coalification parameters besides the petrologically determined values such as vitrinite reflectance. A detailed GC/MS study by Hayatsu et al. (129) of trapped organic compounds and aromatic units in three coals (lignite, bituminous, anthracite) also showed that aromatization of the diterpenoids increased as coalification proceeded.

Diterpenoid hydrocarbons have been reported in coal liquefaction products, although, under the liquefaction conditions commonly used, aromatization of the diterpenoids will normally occur to form substituted phenanthrenes, anthracenes, and naphthalenes. However, Jones, Pakdel, and Bartle (130) used GC/MS to examine liquefaction products obtained by low-temperature conversion of a number of coals and lignites, and tentative evidence was found for the presence of tricyclic terpanes in the C_{17}–C_{26} range. Philp, Russell, and Gilbert (131) studied liquid products formed by hydrogenation of *Tasmanites* sp. concentrates at a number of temperatures by using a combined thermal distillation GC/MS technique and showed that a complex mixture of diterpenoids was formed at low temperatures, the diterpenoids being progressively aromatized as the hydrogenation temperature was increased. The diterpenoids were not related to the abietane or pimarane skeletons and probably represented a series of diterpenoids associated with algal-type material. In a subsequent study of the insoluble residues formed during the same series of reactions, pyrolysis–GC/MS was used to show that the diterpenoids could be detected from the residues of the low-temperature reactions and unaltered *Tasmanites* (132).

E. Extended tricyclic terpanes

In early MS and GC/MS studies of the extracts from the Green River Shale, Anders and Robinson (115) and Gallegos (133) identified many cyclic components including a series of extended tricyclic terpanes that Reed (134) tentatively identified in a sample of weathered petroleum obtained from the same basin. Subsequent studies have suggested that these extended tricyclic terpanes **(27)**, unlike the C_{19} and C_{20} diterpenoids, are not derived from terrestrial material but from marine sources (135,136). The extended tricyclic terpanes are readily detected in sediments and crude oils by using GC/MS and MID, particularly of the abundant characteristic fragment ion at *m/z* 191, and from this it appears that the series extends from C_{19} to C_{30} [137; Fig. 6(a)]. Aquino Neto et al. (138) reported the synthesis of several short-chain tricyclic terpanes and identified two C_{19} and the C_{20} members of the major

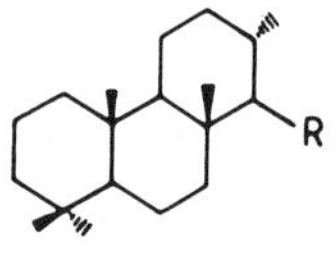

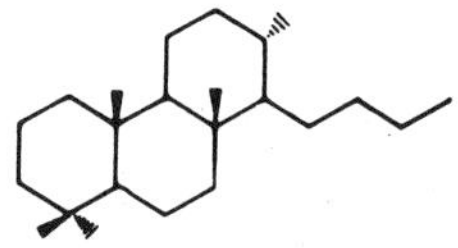

tricyclic series in sediments and petroleum. They proposed a C_{30} precursor for this series and suggested that it could be the tricyclohexaprenol formed anaerobically from a universal cell constituent, hexaprenol. The saturated counterpart of hexaprenol has been reported in petroleum (100) and in the lipids of *Archaebacteria* (101). Ekweozor and Strausz (139,140) observed the tricyclic series in the Athabasca oil sand bitumen and isolated the C_{23} member of the series by preparative GC and subsequently proposed a structure for the compound by using ^{1}H-NMR shown below **(28)**. It was suggested that such a compound could be derived from a C_{30} hexaisoprenoid precursor synthesized with either a head-to-tail coupling of six isoprene units or head-to-tail dimerization of two C_{15} units. A recent study has shown the presence of C_{20}–C_{26} tricyclic carboxylic acids in the Alberta oil sands (141). Decarboxylation of the acids produced a series of alkanes identical to the native tricyclic alkanes in the C_{21}–C_{26} range. However, decarboxylation of the acids is not sufficient to explain the presence of the other tricyclic alkanes in the oil

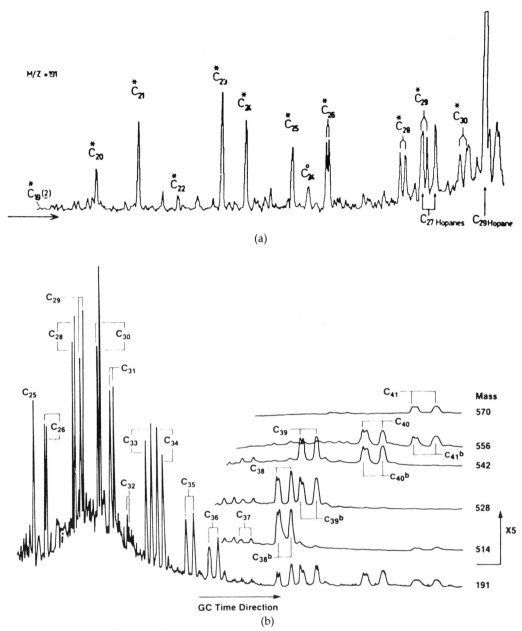

Figure 6. (a) Mass fragmentation (m/z 191) of the tricyclic (*) and tetracyclic (○) terpanes in a crude oil from the Emeraude oil field (Congo) (137). [Figure 6 from ref. 137. Copyright 1983 John Wiley & Sons, Ltd. Reprinted by permission of John Wiley & Sons, Ltd.] (b) Recent work by Moldowan and Seifert (142) has now extended the series of tricyclic terpanes past C_{40} as shown in this diagram for a California crude oil.

sands. The series of tricyclic hydrocarbons has recently been extended from the range C_{19}–C_{30} to C_{19}–C_{45}, and it has been proposed that this supports the prediction of the biogenic origin of the tricyclic terpanes from cyclization of regular polyprenols in bacterial membranes [142, Fig. 6(b)]. The precursor of the C_{45} tricyclic terpane is probably a C_{45} or larger unsaturated isoprenoid alcohol known to occur in plants (96). A novel series of cyclic terpenoid sulfoxides (29) and sulfides (30) in the Athabasca bitumen is present in similar concentrations and distribution as the tricyclic terpanes, and this has led to the suggestion of a possible product/precursor relationship between these classes of compounds and another possible source for the tricyclic terpanes (143). Thermal alteration of polar fractions of oils, such as resins and asphaltenes, has also produced tricyclic terpanes and led Aquino Neto et al. (137) to suggest that precursors of the tricyclic terpanes were present in the resin and asphaltene fractions. A similar theory was also proposed by Ekweozer and Strausz (139) to account for the presence of the tricyclic terpanes in Athabasca tar sands. Variations in the distributions of individual members of this homologous series relative to the ubiquitous hopanes makes the tricyclic terpanes a valuable source correlation parameter (144).

F. Tetracyclic terpanes

Tetracyclic terpanes (31), like the tricyclic terpanes described above, are a relatively new class of biomarkers which have not as yet been studied in great detail. As with the tricyclic terpanes (above) and pentacyclic hopanes described below, the tetracyclic terpanes are detected in fossil fuel samples by using GC/MS and MID of the characteristic ion at m/z 191. Tetracyclic terpanes were reported in the early studies on Green River Shale by Anders and Robinson (115) and Gallegos (133), and the weathered petroleum from the Mahogany Zone in the Green River Shale by Reed (134). Scholefield and Whitehurst (145) analyzed some Georgia–South Carolina clays and partially elucidated the structure of a tetracyclic component present in the extracts. The presence of similar structures in sponges growing in the Bay of Naples led to the suggestion that a possible origin for the tetracyclic hydrocarbon might be diagenesis of terpenes contained in sponges contemporaneous with the deposition of clay sediments (145).

Four tetracyclic terpanes in the C_{24}–C_{27} range have been synthesized by Trendel et al. (146) from hop-17(21)-ene, or trisnorhopan-21-one and have been shown to occur in various geological samples on the basis of mass spectra and GC retention times. Theories advanced for the origin of these compounds include thermocatalytic degradation of hopane precursors, microbial opening of ring E of hopanoids, or cyclization of the precursor squalene stopping at ring D to produce tetracyclic precursors which could be further reduced by geochemical processes (146). The co-occurrence of the tetracyclic terpanes over the same range as the hopanes and their absence in Recent or immature sediments would tend to favor one of the first two hypotheses for their origin. Tetracyclic terpanes have not been widely reported in crude oils or source rocks, but their proposed origin by thermal evolution from pentacyclic triterpanes would suggest a role as thermal maturity indicators in petroleum exploration studies. Ekweozor et al. (147) used GC/MS to detect the C_{24}–C_{27} tetracyclic terpanes in crude oils from various oil fields of the Tertiary Niger delta and also proposed that these compounds were formed from sequential cleavages of the terminal rings of pentacyclic precursors during thermal evolution of the corresponding petroleum. A C_{25} tricyclic alkane present in the oil was thought to be formed by thermal cleavage of the tetracyclic terpanes.

G. Triterpanes

It was mentioned above that an increase in structural complexity will generally lead to an increase in the utility of a biological marker in the solution of fossil fuel production or exploration problems. Triterpanes and steranes, which are discussed below, are classic examples of this statement. In many cases the structural complexity of these molecules means that the standards necessary to confirm unambiguously the structure of the biological markers are not available. However, in certain applications where the biological marker distributions are used for correlation purposes, it is not always necessary to know the absolute structure of the compounds since it is only the distributions, or fingerprints, in the various samples that are compared.

The triterpanes are far more diversified in their applications than many of the biological markers discussed in the preceding sections and can be used as source, maturity, migration, and biodegradation indicators. In order to discuss the triterpanes it is best to divide them into two groups. The pentacyclic hopanes and structurally related moretanes represent the most extensively used group of triterpanes in fossil fuel studies whereas the triterpanes that do not have the hopane-type structure but are found in fossil fuels have been used mainly as source indicators.

1. Pentacyclic triterpanes based on the hopane skeleton

Hopane-type triterpanes (**32**) are ubiquitous biological markers in fossil fuels, and their precursors are widely distributed among bacteria and the cyanobacteria (blue-green algae) (148,149) and have also been found in tropical trees, some grasses and lichens, and several ferns. Two commonly accepted precusor structures for the hydrocarbons are diploptene (33), present in several contemporary organisms, and the C_{35} tetrahydroxyhopane (34), which has been found in various microorganisms and Recent sediments. The discovery of extended hopanes up to C_{40} introduces the possibility of other, yet to be discovered, precursors for hopanes present in fossil fuels (150).

The saturated hopanes are relatively easy to detect in oils and extracts from shales and coals by using GC/MS and MID since two major fragment ions, m/z 191 (see 32) and m/z 148 + R, are formed from the parent ion in the ion source of the mass spectrometer.

Variations in the abundance of these two ions are diagnostic in determining the stereochemistry at the C-17 and C-21 positions of the parent molecule. The naturally occurring precursor compounds have the 17β(H), 21β(H) stereochemistry which is thermodynamically less stable than the 17α(H), 21β(H) stereochemistry (148). Diagenesis and maturation of the organic material containing the precursors leads to defunctionalization and formation of saturated hopanes with the thermodynamically more stable 17α(H), 21β(H) configuration possessed by the majority of hopanes in crude oils and mature source rocks. At the same time, formation of another series of hopane analogs, known as moretanes, with the 17β(H), 21α(H) configuration occurs. The mass spectra of the three C_{29} isomers with the 17β(H), 21β(H); 17β(H), 21α(H); and 17α(H), 21β(H) configurations are shown in Figure 7. From these spectra it can be observed that there are significant variations in the relative abundance of the ions of masses 191 and 148 + R (for a C_{29} isomer this is m/z 177). To generalize, it can be said that for the 17β(H), 21β(H) configuration 148 + R ≫ 191; 17β(H), 21α(H) 148 + R ≃ > 191; and for 17α(H), 21β(H) 148 + R ≪ 191. In most fossil fuel applications, the distribution of the hopanes can be monitored by using SIM of the ion m/z 191. An example of a typical distribution of hopanes in a crude oil is shown in Figure 8 and the identities of the various hopanes are given in Table II.

Before discussing some of the uses of hopanes in exploration problems, it is informative to illustrate changes that occur with increasing maturity in the hopane distributions. Figure 9 shows m/z 191 chromatograms for two samples at different levels of maturity. It can be seen from examining these

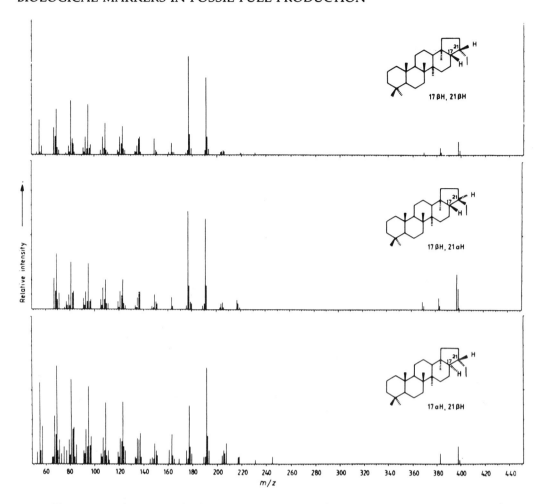

Figure 7. Spectra of the three isomers of the C_{29} hopane structure [17β(H),21β(H); 17β(H),21α(H);17α(H),21β(H)].

chromatograms that the less mature sample contains relatively high concentrations of the 17β(H), 21β(H) hopanes (i.e., naturally occurring stereochemistry) and, as the maturity level increases, the 17α(H), 21β(H) isomers predominate and the 17β(H), 21α(H) moretanes appear. It should also be noted that for the C_{31} and higher homologs, 22S and 22R epimers are formed. The naturally occurring precursor compounds have the 22R configuration, but increase in maturation leads to a mixture of 22R and 22S epimers whose equilibrium ratio is approximately 60 : 40. Van Dorsselaer, Albrecht, and Connan (151) studied these stereochemical changes in a sample of Yallourn lignite which was subjected to thermal alteration in the laboratory, while Ensminger et al. (152) studied the evolution of the hopanes under the effect of burial in early Toarcian shales in the Paris Basin. Mackenzie, Lewis, and Maxwell (153) have also performed laboratory maturation studies with samples of the Toarcian shales and examined changes in configurational isomerization of a number of biological markers including the hopanes.

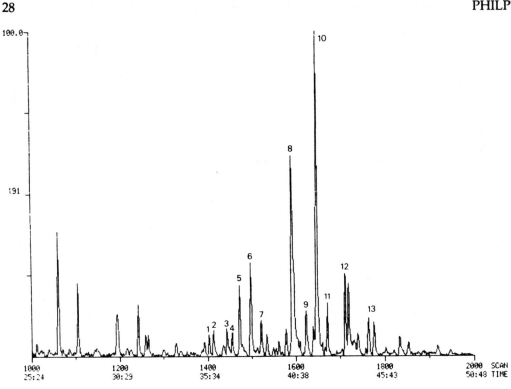

Figure 8. Typical distribution of hopane-type triterpanes in an Australian crude oil. Peak identities are listed in Table II.

Changes in the hopane distributions as a result of increasing maturity have been used to indicate whether a source rock is mature enough to have generated crude oil. Virtually all crude oils that have been analyzed by using GC/MS contain only the 17α(H)-hopane series plus minor concentrations of the moretane [17β(H), 21α(H)] series and the 17β(H)–C_{27} isomer, and the 22S and 22R epimers for C_{31} and higher homologs in a 60 : 40 ratio. In order for a crude oil to have been generated from a particular rock, the hopane

Table II. Identification of triterpanes present in the *m/z* 191 chromatogram shown in Fig. 8.

Peak number	Compound
1	
2	Tricyclic terpanes
3	
4	
5	18α(H)-22,29,30-trisnorhopane
6	17α(H)-22,29,30-trisnorhopane
7	17β(H)-22,29,30-trisnorhopane
8	17α(H),21β(H)-30-norhopane
9	17β(H),21α(H)-30-normoretane
10	17α(H),21β(H)-hopane
11	17β(H),21α(H)-moretane
12	22S and R-17α(H),21β(H)-30-homohopanes
13	22S and R-17α(H),21β(H)-30,31-bishomohopanes

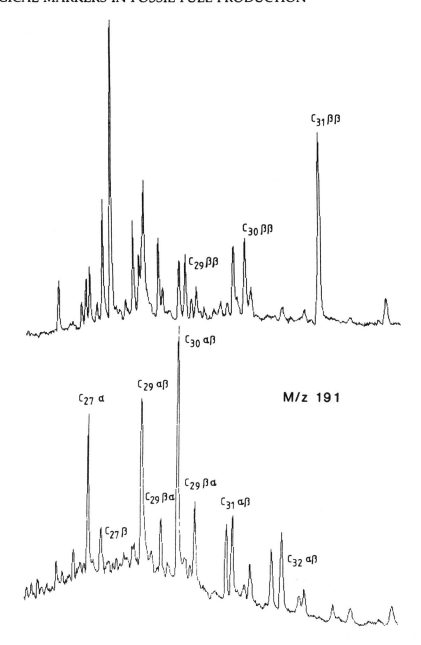

Figure 9. Hopanes in extracts of immature oil shales are complex mixtures of both the 17β(H),21β(H) and 17α(H),21β(H) isomers as shown in the upper chromatogram for an Australian oil shale. As the level of maturity increases (lower chromatogram), the chromatograms are dominated by the 17α(H),21β(H) series observed in crude oils as well as mature source rocks.

distributions will need to be similar to those present in the oil. In other words, if the source rock extracts contain large amounts of the 17β(H) hopanes and excessive amounts of the 22R epimer, it is unlikely that the rock has reached sufficient maturity to have generated petroleum (154). An exception to this statement has been observed in the case of oils from the Gippsland Basin (Australia), where an oil was found to have a 9 : 1 predominance of the naturally occurring C_{31} 22R epimer. Philp and Gilbert (155) proposed that this was possibly due to the oils coming into contact with immature coals in the basin, similar to Yallourn lignites, which are distinguished by very high concentrations of the 22R epimer of the $17\alpha(H)$–C_{31} hopane relative to 22S marking the relatively low level of maturity of the lignite.

The hopane fingerprint as determined by using SIM of *m/z* 191 has been used in many studies for the correlation of oils with suspected source rocks or families of oils thought to be derived from the same source. Any correlation study will also need to include a number of other biological marker parameters to increase the reliability of the results, but the hopane fingerprint provides a good starting point. Seifert and co-workers have published a large number of examples of hopane distributions in oils and source rocks from many different basins including the McKittrick Field, California (156); the Rocky Mountain Overthrust Belt of Wyoming and Utah (157); Sumatra, Indonesia (157); and Prudhoe Bay Field, Northern Alaska (144). Volkman et al. (158) described hopanes in oils from the Barrow Sub-Basin, Western Australia, and Philp, Gilbert, and Friedrich (159), using GC/MS, examined hopane distributions from a number of Australia basins. Hopane fingerprints were among the many properties determined in a study of oils and shales from the Shengli Oilfield, China, by Ji-Yang et al. (160). Welte et al. (161) analyzed a representative collection of oils and Neogene (Miocene–Pliocene) rock samples from the Vienna Basin and used hopane fingerprints in an attempt to determine the origin of the oils and to correlate them with suspected source rocks. In a similar study of oils and source rocks in the German Molasse Basin, Wehner and Teschner (162) used hopane fingerprints determined by using GC/MS and MID including *m/z* 191 to divide the oils into various groups and to correlate with possible source rocks. Ekweozor et al. (163) found that hopanes predominated in oils from both the western and eastern side of the Niger delta, but variations in concentrations of oleanane-type triterpanes, representative of higher plant material, were present in higher concentrations in oils from the eastern part of the delta. Philp (164) analysed 60 oils from the San Jorges Basin, Argentina, by using GC/MS and found that hopane distributions, including demethylated hopanes, could be used to divide the oils into four different groups.

In the majority of oils and source rocks studied, the regular hopanes observed range from C_{27} to C_{35} but do not include the C_{28} member of the series. The formation of the C_{28} hopane (**32**; R = CH_3) from a higher homolog in the series requires cleavage of two carbon–carbon bonds in the side chain rather than one as required for the other members of the series. However, in a few notable exceptions, a C_{28} hopane has been reported as the predominant triterpane. Petrov et al. (165) noted high concentrations of this compound in a crude from the Northern Volga Ural (Siva) Region (USSR), and Seifert et al. (166) isolated the compound from the Monterey shale extract and proved its structure to be $17\alpha(H), 18\alpha(H), 21\beta(H)$-28,30-bisnorhopane

(**35**; R = CH₃) by using a number of techniques including MS and x-ray crystallography. Grantham, Posthuma, and DeGroot (167) detected significant concentrations of the C$_{28}$ hopane, by using GC/MS, in North Sea oils and cores. The origin of this compound has provoked much discussion; Seifert et al. (166) initially suggested a possible origin from certain constituents of ferns, but Grantham, Posthuma, and DeGroot (167) noted the absence of fern detritus, including spores, in cores containing the compound from the North Sea. They suggested, as an alternative, that the presence of the C$_{28}$ hopane may reflect a specific environment of deposition, since there was a correlation between C$_{28}$ hopane concentrations and sulfur content in several of the samples examined.

In the first section of this article it was mentioned that the first effect of crude oil biodegradation was removal of *n*-alkanes. More extensive biodegradation will also affect the hopanes and lead to the formation of a series of demethylated hopanes. Seifert and Moldowan (168) proposed that under these circumstances the methyl group is removed from the A or B ring of the hopane system, and the ion of *m/z* 177, rather than *m/z* 191, can be used for monitoring the presence of the demethylated hopanes. In 1982 Rullkötter and Wendisch (169) isolated a nuclear demethylated hopane from a biodegraded asphalt from Madagascar and used high field proton NMR and MS to show that the methyl group was actually removed from the C-10 position in the ring system to produce the 17α(H)-25-norhopanes (**36**). There is a close correlation between the distribution of regular hopanes and 25-norhopanes, since all the regular hopanes appear to be degraded at approximately the same rate. A comparison between the *m/z* 191 and *m/z* 177 plots in a biodegraded oil clearly shows the relationship between the two series (Figure 10). Philp (164) observed the 25-norhopanes in the oils from San Jorges Basin, Argentina, and Volkman et al. (158), and Alexander et al. (170) have determined their presence in a number of Australian crudes by using GC/MS. In severely degraded crude oils, it has been proposed that the demethylated hopanes can be used as maturity parameters in the same way as the regular hopanes are used in nondegraded oils (158). A study by Connan, Restle, and Albrecht (171) showed that biodegradation does not always appear to cause alteration to the sterane and triterpane distribution. *In vitro* degradation of two oils from the Aquitaine Basin (southwest France) by using *Pseudomonas oleovorans* was allowed to occur for periods of 5 days to 3 months and, although the linear, branched, and isoprenoid hydrocarbons were rapidly removed as a result of bacterial action, no changes had occurred in the hopane distributions.

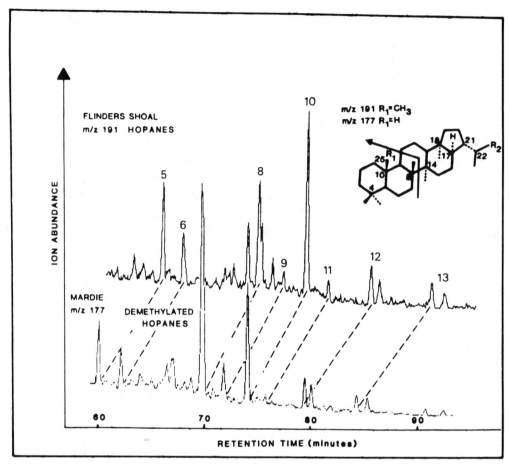

Figure 10. Biodegradation of regular hopanes in crude oils will produce a corresponding series of 25-norhopanes. This diagram taken from the work of Alexander et al. (170) [Copyright 1972. Reprinted by permission of APEA.] shows the relationship between the two series of hopanes for Flinders Shoal oil from the Carnarvon-Barrow Basin (western Australia).

Goodwin, Park, and Rawlinson (172) collected oils from a refinery spill and performed laboratory experiments designed to enhance the naturally occurring biological activity. After several months, substantial alteration of the sterane and triterpane distributions was noted. Specific changes that were observed included reduction in relative concentration of the C_{27} regular and rearranged steranes plus a reduction of the C_{35}-$\alpha\beta$ hopanes. Evidence was also obtained to suggest that (22R) epimers of naturally occurring extended hopanes are degraded at a slightly faster rate than the related (22S) epimers. Thus, contrary to the observations of Connan, Restle, and Albrecht (171) described above, these experiments (172) have demonstrated that sterane and triterpane alteration can proceed by *in vitro* biodegradation.

An unidentified C_{27} triterpane was detected in a number of sediments and oils from the North Sea and Norwegian Continental Shelf by several groups and its structure proved to be very elusive (173,174). Bjorøy and Rullkötter (175) initially and tentatively suggested from NMR data that this compound was a 25,28,30-trisnormoretane. However, Volkman et al. (176), using GC retention time and mass spectral data, have shown that this unusual triterpane is in fact the C-10 demethylated analog of the C_{28} bisnorhopane described above, that is the 17α(H), 18α(H), 21β(H)-25,28,30-trisnorhopane (**35**; R = H).* Rullkötter (177) has recently summarized the efforts to detect nuclear demethylated hopanes by using GC/MS and the utility of these norhopanes in providing a record of microbial action in oils and sediments.

In addition to the use of regular or demethylated hopane fingerprints as a correlation tool in petroleum exploration, in the past few years a number of parameters have been established that provide information on source, maturity, and migration of crude oils. Many of these parameters have been proposed by Seifert and Moldowan and have been described in detail in their papers. The most significant parameters involving the hopanes are summarized in Table III.

In certain sediments and crude oils a number of degraded hopanes or secohopanes and hopanoic acids have been found that can also be used to give information on maturity and changes resulting from biodegradation. Schmitter, Sucrow, and Arpino (178) identified, by the use of GC/MS and authentic standards, a series of 8,14-secohopanes (**37**) ranging from C_{27} to

37

C_{31} in a Nigerian crude oil. The distributions of the 8,14-secohopanes paralleled the distribution of the regular hopanes and supported the theory that the latter are the precursors of the secohopanes. [The formation of 17,21-secohopanes in the C_{24}–C_{27} range and their distribution in sediments and crude oils have been described in the tetracyclic terpane section above (146).] Seifert (179) published a review in 1974 on the distributions of naturally occurring carboxylic acids in crude oils and since that time only a small number of reports have appeared on the topic. Schmitter, Arpino, and Guiochon (180) observed 3,4-secohopanoic acids in the Nigerian crude oil, which were previously thought to form by photochemical decay of pentacyclic ketoterpenoids (181). They also investigated the distribution of hopanoic acids, including the moretane series of acids, in the same oil (182). It was shown that the distribution of hopanoic acids could be used for correlation purposes and maturity determinations in the same manner as their hydrocarbon analogs.

Aromatization of the hopanes can also occur in sediments and crude oils, and the presence of aromatic hopanes in Recent sediments suggests that they can form as a result of microbial reactions as well as maturation. Extensive studies by the geochemistry group at the University of Strasbourg (183–185) have led to the identification of aromatized hopanes, phenan-

*A recent paper by Moldowan et al. (233) has proposed that the major 25,28,30-trisnorhopane is most probably a peak containing two compounds, 17α(H), 21β(H) plus 17β(H), 21α(H) and the later eluting minor component is the 17β(H), 21β(H)-25,28,30-trisnorhopane.

Table III. Summary of biomarker parameters based on steranes and triterpanes that have been used in petroleum studies.

Biomarker parameter	Application	Reference
Tricyclic terpanes (%)	Source	156
Change in paraffin concentration plus increase in sterane and triterpane concentration	Migration and/or maturation specific if common source established	156
$\dfrac{C_{29} + C_{30} \text{ primary terpanes}}{C_{27} + C_{28} \text{ secondary terpanes}}$	Source and maturation	156
$\dfrac{17\alpha(H)\text{-}22,29,30\text{-trisnorhopane }(T_m)}{18\alpha(H)\text{-}22,29,30\text{-trisnorhopane }(T_s)}$	Source and maturation	156
5β(H) steranes/17α(H) hopanes	Migration if common source established	156
$\dfrac{17\beta(H),21\beta(H)}{17\beta(H),21\beta(H) + 17\beta(H),21\alpha(H) + 17\alpha(H),21\beta(H)}$ hopanes	Maturation	94
$\dfrac{22R}{22R + 22S}$ 17α(H),21β(H)-homohopanes	Maturation	94
$\dfrac{20S}{20S + 20R}$ 13β(H),17α(H)-diacholestane	Maturation	94
$\dfrac{24S}{24S + 24R}$ (20R)-24-methyl-ααα-cholestane[a]	Maturation	94
$\dfrac{20R}{20R + 20S}$ -24-ethyl-ααα-cholestane[a]	Maturation	94
$\dfrac{\alpha\beta\beta}{\alpha\beta\beta + \alpha\alpha\alpha}$ (20R + 20S) 24-ethylcholestane[a]		
ααα-C_{28} steranes[a] / ααα-C_{29} steranes	Source	198

$\dfrac{\alpha\alpha\alpha\text{-}C_{27} \text{ steranes}^a}{\alpha\alpha\alpha\text{-}C_{29} \text{ steranes}}$	Source	207
$\dfrac{\beta\alpha\alpha\ (20R) + \alpha\beta\beta\ (20R)}{\alpha\alpha\alpha\ (20R)} C_{29}$ steranes	Migration	144
$\dfrac{13\beta(H),17\alpha(H)(20S)}{\alpha\alpha\alpha\ (20R)} C_{27} + C_{28} + C_{29}$ steranes	Migration	144
$\dfrac{\beta\alpha\alpha\ (20R)\text{-}C_{28} + \alpha\beta\beta\ (20R + 20S)\text{-}C_{28} + \beta\beta\beta\ (20R + 20S)\text{-}C_{29}}{\alpha\alpha\alpha\ (20R)\text{-}C_{28}}$ steranes	Source	144
$\dfrac{C_{27} \text{ triaromatic steranes}}{C_{27} \text{ triaromatic} + C_{28} \text{ monoaromatic steranes}}$	Maturation and aromatization	95
$\dfrac{C_{21} \text{ monoaromatic steranes}}{C_{21} \text{ triaromatic} + C_{28} \text{ monoaromatic steranes}}$	Maturation and bond breaking	95
$\dfrac{C_{26} \text{ triaromatic steranes}}{C_{20} \text{ triaromatic} + C_{27} \text{ triaromatic steranes}}$	Maturation and bond breaking	95
Preferential removal of C_{27}–20S diasterane over 20R epimer	Biodegradation	170
Preferential removal of regular steranes over diasteranes	Biodegradation	170
$17\alpha(H)\text{-}C_{30}$ hopane	Differentiation of source rock stratigraphies by pyrolysis	229
$17\beta(H)\text{-}(C_{29} + C_{30})$ moretanes		
$17\alpha(H)/17\beta(H)$—trisnorhopanes	Differentiation of source rock stratigraphies by pyrolysis	229

$^a\alpha\alpha\alpha$, $\alpha\beta\beta$, $\beta\beta\beta$, $\beta\alpha\alpha$ refer to the stereochemistry of the H atom at the 5, 14, and 17 positions, respectively, of the regular steranes.

threne, and chrysene derivatives based on the hopane structure (**38–41**). These studies have provided useful information on the geochemical pathways leading to the formation of the compounds in fossil fuels as well as their origin and the stages of sedimentation at which the reactions occur.

The distributions of hopanes have been used in a number of studies on various other fossil fuel resources. The oil sands of Northern Alberta represent a vast source of fossil fuel whose exploration and exploitation is of major economic importance. Several theories for the origin of this material have been proposed, such as its formation from organisms living in the environment of the original sand deposits, or evaporation of older Devonian pooled oils after exposure on the surface. Rubinstein et al. (186) attempted to determine whether the oil sands were related to the Lower Cretaceous pooled oils found elsewhere in Alberta by either water washing or biodegradation. As part of this study they pyrolyzed asphaltenes isolated from Prudhoe Bay conventional crude oil before and after biodegradation to ascertain the effects of biodegradation (187). Pyrolysis of the asphaltenes produced homologous series of the 17α(H), 21β(H) hopanes and lesser amounts of the 17β(H), 21β(H) series with the distribution of the 17α(H), 21β(H) series being identical to that observed in the whole oils. From the results of these pyrolysis experiments, it was concluded that asphaltene molecules were produced in an early period during the formation of the crude oil and compaction of the original sediment, and that the information locked into the asphaltene is representative of the organic matter present at that time. The presence of the 17β(H)-hopane isomers may result from the early incorporation and stabilization of these isomers in the asphaltene fractions. The asphaltenes will also contain geochemical information from a period prior to the complete degradation of the oil.

Oil shales are generally at low levels of maturity, and as a result the hopanes that occur in the shales generally have the 17β(H), 21β(H) configuration. Kimble et al. (81) used GC/MS to identify the triterpanes in the Messel oil shale and first suggested the possibility that the relative amounts of 17β(H)/17α(H) hopanes could be used to provide an indication of the relative maturities or thermal histories of shales. More recently several papers on the composition of soluble extracts from Australian oil shales have appeared. Ingram et al. (75) compared oil shales and shale oils from the Rundle formation (Australia) and the Green River Formation, USA. The presence of triterpenoids, including hopanes, in the shale oils has been taken to indicate the feasibility of relating the composition of a shale oil to the petrology and chemical structure of the parent organic matter and mineralogy of the inorganic matrix. A more detailed analysis of the extractable components from the Rundle shale (188) revealed the presence of the thermodynamically unstable 17β(H) hopanes. The presence of the 17β(H) hopanes plus a number of unsaturated hopenes demonstrated that the oil shale was relatively immature. DiSanzo, Uden, and Siggia (110) observed steranes and triterpanes in shale oils from the TOSCO process but did not identify any of them in detail.

Although hopane-type triterpanes are also present in coals, they have not received a great deal of attention, which is somewhat surprising since there is a possibility that they could be used for producing stratigraphic correlations of coal seams in analogous fashion to the oil/source rock correlations described above. Allan, Bjorøy, and Douglas (189) used GC/MS to study variations in triterpane distributions, including hopanes, in a series of coal macerals of different ranks. In a study of five U.S. coals, Gallegos (190) demonstrated the production of the C_{27} and C_{29}–C_{30} hopanes by pyrolysis of the coals. In addition, he showed that the concentration ratio of the 17β(H) isomer to the 17α(H) isomer decreased with the geothermal stress experienced by the coal deposit. Chaffee, Perry, and Johns examined the hopanes produced from the pyrolysis of Australian brown coals (191) and bituminous coals (192). The relative immaturity of the brown coals was reflected in the presence of the 17β(H) series of hopanes as well as minor amounts of the 17α(H) series. The bituminous coals completely lacked the 17β(H) series and any other triterpanes apart from the 17α(H) hopanes, which is in accord with their higher level of maturity. Naturally occurring triterpenoid skeletons are found most commonly as natural products in angiosperms, and their absence from Australian Permian coals is consistent with the fact that angiosperms did not exist in Permian times. Baset, Pancirov, and Ashe (112) pyrolyzed a sample of Wyodak coal in a hydrogen atmosphere and characterized the products by using a number of analytical techniques. The lack of thermal stress experienced by the coal was demonstrated by the presence of the 17β(H), 21β(H) hopanes.

2. Pentacylic triterpanes other than hopanes

The distribution of triterpanes among angiosperms, gymnosperms, algae, bacteria, and fungi has been reviewed by Whitehead (193,194). Although a large number of compounds were cited, only a small number of the hydrocarbon analogs have been identified in fossil fuels. The major classes that have been detected and found to be of some use as markers are lupanes (42), oleananes (43), fernanes (44), and ursanes (45). In most reports on these

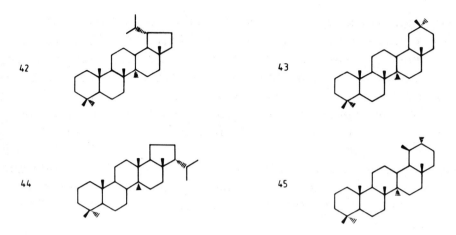

compounds, their use as source indicators has been emphasized. Unlike the hopanes, none has been used extensively for correlation or maturity purposes. Furthermore, these compounds are found most frequently in coals or in oils derived from terrestrial source material, because their precursors are widely distributed in higher plants.

(a) *Lupanes:* Lupanes **(42)** are structurally similar to hopanes although the isopropyl substituent is at C-19 and not at C-21. Lupane was tentatively identified in Nigerian crude oil by Hills and Whitehead (195), but the results were later revised by Whitehead (193). It was not until a report by Rullkötter, Leythauser, and Wendisch (196) appeared that lupanes were proven to be present in any sediment or crude oil. Rullkötter, Leythauser, and Wendisch (196) used MS, GC/MS, and NMR to identify the presence of 17α(H) and 17β(H)-23,28-bis-norlupane in Tertiary sediments from west Greenland and the Gulf of Suez. The biogenic precursors are supposed to be derived from terrestrial higher plants and altered by early diagenetic processes. However, the unusual geographical distribution of shales of Paleocene age bearing the bisnorlupanes in high concentrations in Tunorssuaq Valley on the Nugssuaq peninsula (west Greenland) indicates early diagenetic formation of the bisnorlupanes under specific conditions, possibly by microbial action. Lupanes have also been reported in the Rhenish lignites by Hagemann and Hollerbach (121).

(b) *Oleananes, fernanes, and ursanes:* The major compounds identified in oils from the Niger Delta, Nigeria, were hopanes and various oleananes (163). Oils from the eastern part of the delta were found to contain a higher concentration of oleananes **(43)** than oils from the western part. This difference in concentration reflected a higher concentration of higher plant material in the eastern oils than in the western oils. The variation in the concentration of oleananes in Nigerian petroleum with source facies was large enough to permit reliable oil–source correlations to be made based on triterpane fingerprints. A subsequent examination of extracts from the immature shales from the Agbada formation of the Tertiary Niger Delta revealed low concentrations of the hopanes with the dominant triterpanes being olean-13(18)-ene, a pair of C_{28} isomeric pentacyclic triterpanes of unknown structure, and

18α(H)-oleanane (197). In simulation experiments using samples of the shale, the disappearance of olean-13(18)-ene, and generation of oleananes were observed, and this provided strong evidence to support the current hypothesis on the origin of oleananes in Nigerian petroleum.

Richardson and Miiller (44) showed that the predominant pentacyclic terpanes of an oil derived from terrestrial organic material occurring in an unnamed basin between Malaysia and New Zealand were hopanes, oleananes, and possibly lupanes. The presence of oleananes indicated the terrestrial contribution to the oil and the possible presence of the lupanes led to the conclusion that the oil was of Tertiary age. The precursors of the lupanes are lupeols generated exclusively by angiosperms, which, in the region under investigation in this study, are probably Indonesian sapotaceous trees. Since these and other angiosperms did not evolve until the Late Cretaceous and flourished during the Tertiary, it was concluded that the oil was Tertiary since there were no Cretaceous sediments in the area where oil occurs. This is a good example of biological markers providing useful exploration information on both the source and the age of a crude oil.

A study of Victorian brown coal lithotypes revealed the presence of oleanane. A number of unsaturated fernanes (44) and ursanes (45) (123) were tentatively identified. Their presence in brown coals is consistent with the higher plant material present in the brown coals. A similar distribution of ursene- and oleanene-type triterpenes was observed in the Rhenish lignites (121).

To summarize, it can be seen that triterpanes play an important role in fossil fuel studies. The hopanes and their various analogs are particularly useful in obtaining information on source, maturity, and extent of biodegradation of crude oils and related source materials. Triterpanes not related to the hopanes are important as source indicators, particularly for terrestrially derived samples. A combination of the information obtainable from the distributions of triterpanes and steranes, to be described in the following section, provides a powerful tool for use in any petroleum exploration study.

H. Steranes

Steranes (46) are as important as the hopane-type triterpanes in any biological marker study of fossil fuels. Steranes are derived from sterols that are widely dispersed in plants and microorganisms, with the C_{27} and C_{28} sterols most abundant in marine organisms and the C_{29} sterols in higher plants. Variations in side chain substituent pattern and stereochemistry of the sterols can be used to obtain a more specific indication of source organisms. This aspect of sterol geochemimstry and the reactions by which sterols are converted to steranes need not be discussed in detail here since these topics are covered in a recent review article by Mackenzie et al. (1). The major use of steranes in fossil fuel studies centers on the stereochemical complexity of the basic skeleton, the details of which have been elucidated previously (168). The stereochemical changes that can occur to the molecule after its formation from the precursor sterol are summarized in the following scheme. Sterols generally contain a Δ5,6 double bond and occur naturally as the 20R epimer with the 14α(H), 17α(H) configuration. As diagenesis begins, the double bond is hydrogenated, and a new epimeric center with a mixture of 5α(H) and 5β(H) stereochemistries is formed with 5α(H) predominant. As the level of maturity increases the 14β(H), 17β(H) isomers, which predominate at high maturity levels in petroleum (157), are formed as a mixture of the 20R and 20S epimers (cf. **46a–c**). Because of its greater

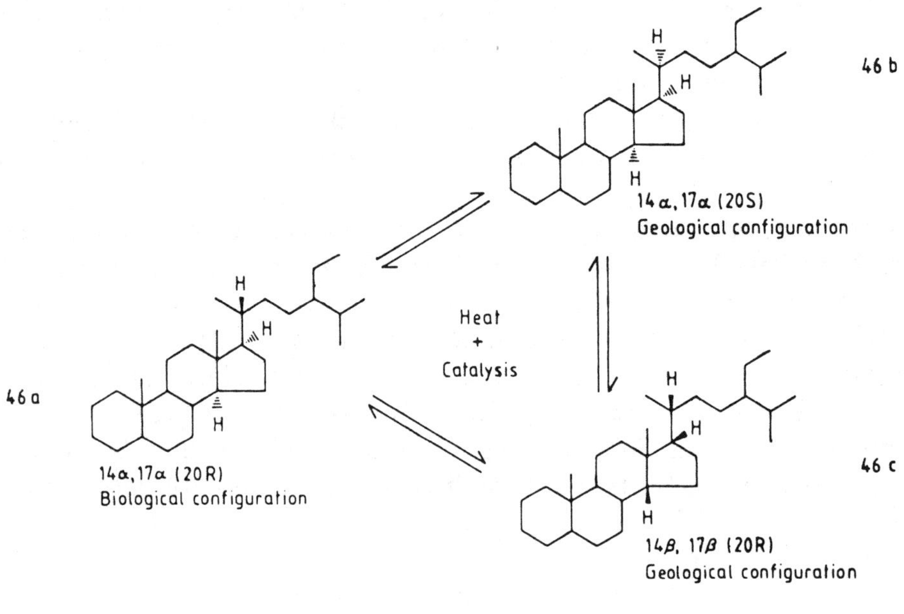

thermal stability, the 5α(H) stereochemistry predominates. The 5β(H) epimers are present in lesser amounts but are generally masked by the more abundant 5α(H) components (198). The stereochemical interconversions result in a complex sterane fingerprint that can be used for making source correlations and maturity determinations, and, as will be seen below, observing the effects of biodegradation on crude oils. Some results have also been presented on the effects of migration on sterane distributions, although much more work is obviously required in this area. The complexity of the sterane fingerprint is further complicated by the fact that diasteranes **(47)**, or rearranged steranes, formed by acid-clay-catalyzed backbone rearrangements of regular sterenes (199), are also found in crude oils and source rocks

47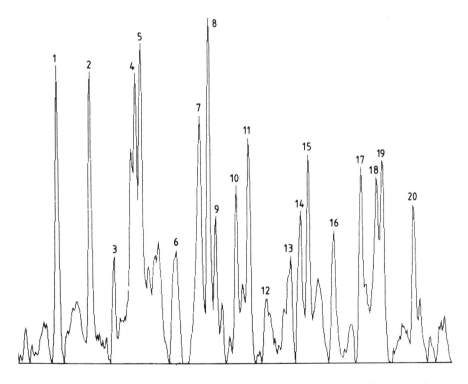

(156,200). An example of the sterane distribution in a crude oil determined by using GC/MS and SIM of the key ion of m/z 217 is given in Figure 11 and the identities of the various components are listed in Table IV.

In virtually all of the literature reports describing the occurrence of steranes in fossil fuels, GC/MS has played an important role in the determination of their distributions because steranes, like triterpanes, are present in relatively low concentrations in most oils and source rock extracts. Furthermore, many fractionation steps would be required to isolate sterane concentrates from individual samples. Hence, the preferred method of analysis has been to use GC/MS and SIM or MID, although an alternative approach to the MID method was recently published (201). In this alternative approach, the specificity of the sterane distributions is enhanced by monitoring the spontaneous fragmentation of sterane parent ions occurring in the first field-free region of a double-focusing mass spectrometer. In this way the sterane metastable parent ion transitions can be monitored during a single GC/MS run to enhance both sensitivity and specificity.

Figure 11. Mass fragmentogram (m/z 217) showing the distribution of steranes in an Alaska crude oil. Peak identifications are given in Table IV.

Table IV. Identification of steranes in m/z 217 chromatogram shown in Fig. 11.

Peak number	Compound
1	13β,17α-Diacholestane (20S)
2	13β,17α-Diacholestane (20R)
3	13α,17β-Diacholestane (20S)
4	13α,17β-Diacholestane (20R)
5	24-Methyl-13β,17α-diacholestane (20R)
6	24-Methyl-13β,17α-diacholestane (20R)
7	24-Methyl-13α,17β-diacholestane (20S) + 14α-cholestane (20S)
8	24-Methyl-13β,17α-diacholestane (20S) + 14β,17β-cholestane (20R)
9	14β,17β-Cholestane (20S) + 24-methyl-13α,17β-diacholestane (20R)
10	14α,17α-Cholestane (20R)
11	24-Ethyl-13β,17α-diacholestane (20R)
12	24-Ethyl-13α,17β-diacholestane (20S)
13	24-Methyl-14α,17α-cholestane (20S)
14	24-Ethyl-13α,17β-diacholestane (20R) + 24-methyl-14β,17β-cholestane (20R)
15	24-Methyl-14β,17β-cholestane (20S)
16	24-Methyl-14α,17α-cholestane (20R)
17	24-Ethyl-14α,17α-cholestane (20S)
18	24-Ethyl-14β,17β-cholestane (20R)
19	24-Ethyl-14β,17β-cholestane (20S)
20	24-Ethyl-14α,17α-cholestane (20R)

In studies involving some elegant chromatography, Mulheirn and Ryback (202) isolated a number of steranes from Rozel Point, Utah, and determined their stereochemical configurations by using a number of techniques. In a second paper, the same authors (203) described mass spectral details of a number of synthetic sterane standards and showed that many of them had very similar spectra, although relative abundances of certain key ions were found to vary with stereochemical changes.

Subsequently, Seifert and co-workers published several papers in which they described the use of sterane distributions in crude oils and source rock extracts to obtain information valuable from an exploration point of view. In one of their early papers, Seifert and Moldowan (156) described a series of novel biological marker parameters for use in the analysis of crude oils from the McKittrick Field, California, which they then applied to a wide range of crude oils and source rocks. Table III represents a summary of biological marker parameters that have been used in exploration studies, as reported in the literature. It can be seen that many of these parameters are taken from the work of Seifert and Moldowan and include ratios based on both the sterane and triterpane distributions.

Since the early work of Seifert and Moldowan, many papers have appeared on the correlation of oils and source rocks by using similar parameters. Several oils from the German Molasse Basin were examined and divided into three groups on the basis of their n-alkane and aromatic hydrocarbon distributions (204). A rather crude separation of the oils into the three groups was initially made on the basis of their saturated hydrocarbon content. The groups consisted of oils in reservoirs of Triassic age in the western part of the basin, oils in reservoirs of Tertiary age, and finally oils from the northeastern part of the Molasse Basin. Following the use of GC/MS and SIM of the ion of m/z 217 for steranes, it became clear that the third type of oil was actually different as a result of biodegradation. This is one of several examples in the literature that show the value of using SIM techniques to correlate biodegraded and nonbiodegraded oils, which is extremely difficult using the classical correlation techniques. Richardson and Müller (44) noted the predominance of the C_{29} steranes in oils with a high input of terrestrial organic matter. Philp, Gilbert, and Friedrich (159) used sterane distributions for correlating oils from a number of basins in Australia. Oils previously thought to be derived from different sources in the Carnarvon Basin were shown (on the basis of their sterane fingerprints) to be from the same source but to differ as a result of biodegradation. Welte et al. (161) also used sterane fingerprints in conjunction with hopane distributions for correlation purposes with their oils from the Vienna Basin.

Biodegradation has been shown to have an effect on the sterane fingerprint of crude oils, and care needs to be taken when making maturity determinations based on ratios of certain sterane isomers. The first epimer to be preferentially removed is the C_{29}-2OR epimer (158,170) and, as the extent of biodegradation increases, the regular steranes are removed, leaving only the rearranged steranes in the most extensively biodegraded oils (168). Curiale, Harrison, and Smith (205) examined some solid bitumens from southeastern Oklahoma by using pyrolysis–GC/MS. The ratio of rearranged to regular steranes was found to be higher in the pyrolysates of the bitumens than in oils in the geographical proximity which is consistent with a bitumen origin due to biodegradation of oil. Rullkötter and Wendisch (169) in their study of the degraded Madagascar asphalts noted that degradation of the steranes started from the lower-molecular-weight end (C_{27}). For the C_{28} and C_{29} sterane isomers, they observed preferential degradation of the 20R diastereomers.

In addition to regular steranes, a number of aromatized steroid hydrocarbons have been found in crude oils and source rocks. Tissot et al. (206) first recognized the occurrence and potential of aromatic steroids for geochemical correlations. Two types of monoaromatized steroid hydrocarbons occur, one type with A-ring aromatization **(48)** and the other with C-ring aromatization **(49)**, with the C-ring compounds being predominant in crude oils. These latter compounds are readily detected by using SIM of the ion of m/z 253; a series of demethylated aromatized steroid hydrocarbons is also present in many samples and can be detected by monitoring the ion of m/z 239. Seifert and Moldowan have used the distribution of monoaromatized steroid hydrocarbons (m/z 253 and 239) for oil–oil source and maturity correlations and also as evidence for a cosourcing concept for the origin of the oil at Prudhoe Bay (144). More recently they have reported the synthesis of monoaromatized steroid hydrocarbon standards and determined new source parameters for oil correlation studies based on these compounds (207). Schaefle et al. (208) also synthesized various monoaromatized steroid hydrocarbons, but Seifert, Carlson, and Moldowan (207) claimed that these standards are not the major petroleum monoaromatized steroid hydrocarbon series. This claim was later supported by Mackenzie, Lamb, and Maxwell (209). The A-ring monoaromatized steroid hydrocarbons are not major constituents of petroleum but have been detected and identified in immature Cretaceous black shales from the southern and eastern North Atlantic (210). C-ring aromatized steroid hydrocarbons occur in two groups, namely C_{20} and C_{21} and C_{27}–C_{29}. The lower members of the series are formed as a result of maturation changes to the higher members, and the ratio of the C_{20}/C_{27} components can be used as a maturity indicator (156).

In laboratory thermal alteration studies, Mackenzie, Lewis, and Maxwell (153) heated samples of an immature Toarcian shale from the Paris Basin under reduced pressure for varying periods of time and measured (a) extent of configurational isomerization at (i) C-6 and C-10 in pristane, (ii) C-20 and C-14 and C-17 in $5\alpha(H),14\alpha(H),17\alpha(H)$-steranes, and (iii) C-17 and C-22 in hopanes; and (b) extent of aromatization in aromatic steroid hydrocarbons. Extensive use was made of GC/MS and SIM to detect the various isomers of interest. It was found that both configurational isomerization and aromatization occurred during heating, but that the latter process was accelerated relative to the former. This compares favorably with the maturation effect experienced by the Toarcian shales as a result of increasing burial depth and the associated temperature increase. On the basis of these results, it was proposed that it might be possible to distinguish the thermal histories of sediment sequences by comparing the extent of isomerization at C-20 in the $5\alpha(H)$, $14\alpha(H)$, $17\alpha(H)$-steranes and of aromatization in the aromatic steroid hydrocarbons.

A recent study by Shi Ji-Yang et al. (160) of petroleums and shales from the Shengli oil field (China) provides valuable evidence to support the results of Mackenzie, Lewis, and Maxwell (153) described above. The shales showed a progressive increase in the extent of aromatization of aromatic steroid hydrocarbons and in the extent of configurational isomerization in the triterpanes which correlates well with depth. Furthermore, the determination of certain maturity measurements such as the %20S-/20R + 20S-24-ethyl-$5\alpha(H),14\alpha(H),17\alpha(H)$-cholestane or %22S-/22R + 22S-$17\alpha(H),21\beta(H)$-bishomohopane proved valuable. However, oils may appear mature from one measurement and immature from another. In other words, reactions based on stereochemical conversions occur at different rates than those based on epimerization, and thus at one level of maturity the two reactions will have proceeded to different extents (209). McKenzie et al. (211) have also shown that sterane isomerization and aromatization reactions occurring in basins formed by stretching of the lithosphere can be reproduced in detail by a simple model of basin formation. They demonstrated that small quantities of biological markers contain a record of the subsidence history of a sedimentary basin and that the progress of the reactions is especially sensitive to uplift which can "freeze" the various isomers if the uplift is sufficiently great (211).

In addition to monoaromatic steranes, a number of triaromatic steroids have been found in crude oils and source rocks (212). The major series of triaromatic steroids (50) in petroleum (207) forms by further aromatization of the monoaromatic compounds (1). It has been shown that in the Paris Basin, aromatization of monoaromatic steranes to triaromatic steranes increases with increasing maximum burial depth or maturity. The extent of aromatization and side chain cracking in the triaromatic components may be used for maturity assessment over a wide range ($R_0 = 0.3$–1.5%) (213).

50

There are few reports concerning steranes in oil shales and coals. Anders and Robinson (115) and Gallegos (133) reported steranes in Green River shale, and, more recently, Ingram et al. (75) and Regtop et al. (188) have detailed the occurrence of steranes in various Australian oil shales and shale oils.

I. Aromatic hydrocarbons

Aromatic compounds are known to occur extensively in fossil fuels. Those that can be related to steroids and triterpenoids from biological material are particularly useful because structural comparison with their precursors is possible. The occurrence and utilization of aromatized hopanes and steranes in petroleum exploration have been discussed above. Radke, Welte, and Willsch (214) used phenanthrenes derived from steroids and triterpenoids and the four isomeric phenanthrenes produced by methylation to devise a methyl phenanthrene index (MPI) that reflects the increasing thermal maturation of organic matter and exhibits a good correlation with vitrinite reflectance. The utility of the index was demonstrated with a set of samples from a well in the Western Canada Deep Basin (214) and a series of 24 German bituminous coals of varying rank (215). George and Beshai (216) examined polyaromatic hydrocarbons in Alberta bitumens and oils and also showed that increasing maturity led to increasing aromaticity and depletion of alkyl substituents. White and Lee (217) identified 78 compounds in the aromatic fraction of a Homestead Kentucky coal extract by using GC/MS, and a number of these aromatic compounds could be related to biological marker precursors known to occur in higher plants such as diterpenoids and sesquiterpenoids. Gallegos (218) presented some early results on the distribution of monoaromatic compounds in Green River shale and more recently discussed the presence of alkylbenzenes in pyrolysis oils from coals (219). He proposed that the alkylbenzenes were derived from carotenoid precursors and observed that their concentration decreased with increasing rank of the coal samples, which raises the possibility of using them as a maturity indicator.

K. Porphyrins

Porphyrins (51) were the first class of compounds used as biological markers by Treibs in the 1930s, and their presence enabled him to conclude that petroleum was derived from plant and animal organic remains (6,7). Baker

and Palmer (220) reviewed the geochemistry of porphyrins to 1977 and discussed identification methods, occurrence, and diagenetic reactions of porphyrins. In the past, porphyrins were not widely used as biological markers because their derivatives were not sufficiently volatile for analysis by using GC/MS. However, in 1980 Alexander et al. (221) showed that it was possible to analyze porphyrins following the preparation of bis(trimethylsiloxy) silicon (IV) derivatives by using fused-silica capillary columns in the gas chromatograph. An extension of this study was the investigation of the TMS derivatives of silicon (IV), aluminum (III), gallium (III), and rhodium (III) alkyl-substituted porphyrins and their analysis. Porphyrins with ester side chains have also been studied as their silicon (IV) derivatives by using GC/MS (222). The development of these GC and GC/MS methods for porphyrin analysis will complement the HPLC/MS methods that have been developed for petroporphyrins, for example, by McFadden et al. (223) who described the HPLC/MS analysis of petroporphyrins from Boscan oil and La Luna shale. Differences in distributions of the petroporphyrins between these samples were thought to reflect geothermal differences between the original samples. Quirke et al. (224) isolated porphyrins from Deep Sea Drilling Project (DSDP) samples by using HPLC prior to their analysis by MS. Similarly, HPLC and MS were used to study demetallated porphyrins in shales and oils from the Shengli oilfield in China as part of an attempt to correlate the oils with suspected source rocks (160). Porphyrins are widely distributed in coals and can originate from plant as well as bacterial chlorophylls as shown by Palmer et al. (225) in a study of 42 U.S. coals ranging from lignite to anthracite. The major porphyrins were found to be of the etio series which is thought to form during early stages of coal formation.

The specificity of porphyrin distributions has long been recognized to have potential for correlation purposes. As more attention is paid to developing separation and analytical techniques, it is clear that porphyrins will indeed play a more important role in this field. With this in mind, a novel method for separating porphyrins by using alkyl sulfonic acid functionalized silica was recently described for separation of Boscan vanadyl porphyrins prior to their MS and HPLC analysis (226). Five metalloporphyrin series were observed ranging from C_{25} to C_{60}, and variations in metalloporphyrin distributions obtained by these techniques provided additional supporting evidence for correlation data based on the sterane and triterpane fingerprints previously obtained (226). Petroporphyrins isolated from Gilsonite from the Uinta Basin, Utah, USA, by using HPLC have been determined by using MS, NMR, and chemical degradation. They provided the first definitive and supporting structural evidence for the Treibs hypothesis that the petroporphyrins were produced directly by defunctionalization of chlorophyll a (227). In their extensive studies of biological markers from the Toarcian shales of the Paris Basin, Mackenzie, Quirke, and Maxwell (228) studied the evolution of metalloporphyrins and observed that major changes occurred to the porphyrins with increasing maximum burial depth. These changes were (i) conversion of DPEP components to etio components, (ii) decrease in chain lengths of β-alkyl substituents, and (iii) a decrease in the ratio of nickel to vanadyl porphyrins. The changes were used to indicate an increase in maturation from north to south across the basin in parallel with changes observed in other classes of biological markers described above (153,213).

In summary, although porphyrins have not been widely studied compared with many other biological markers, they obviously have many important uses in fossil fuel production or exploration. Porphyrin distributions provide information on source material, porphyrin fingerprints can be used for correlation purposes, and finally they can also be used to give relative maturity measurements. Once again it should be emphasized that no one family of biological markers gives the complete picture. Hence the evidence obtained from the porphyrins needs to be considered along with data obtained from all other classes of biological markers. A more detailed discussion of porphyrins can be found in the article of Gallegos in this issue.

IV. SUMMARY

The use of biological markers in fossil fuel production is an area that is rapidly expanding with the development of increasingly sensitive and sophisticated analytical techniques. The range and number of biological markers being discovered continues to expand exponentially. Examination of samples from different environments also leads to the discovery of new marker compounds. Determination of the precursors for these biological markers and the reactions involved in their formation can provide valuable information on depositional environments and possible postdepositional changes. This review summarizes information available on only the most widely exploited groups of biological markers. Other classes of compounds such as those containing nitrogen are reviewed elsewhere in this issue.

What lies in the future for biological markers? There are many other classes of compounds, such as fatty acids, known to be present in fossil fuels but not investigated in great detail in recent years. The development of mass spectrometers with extended mass ranges and alternative ionization techniques will lead to an increase in the discovery of biological markers with higher molecular weights than those currently being observed. Alternative ionization techniques such as FAB have yet to be applied to petroleum exploration problems. Likewise little attention has been given to the use of MS/MS techniques in this field. These techniques could provide new information on the precursors of many of the biological markers currently being used in correlation studies. Studies on the reaction rates of conversion of various stereoisomers, in particular the steranes, should be a particularly fruitful area of research and provide information on the thermal history of basins and basin development (229). At the present time only a limited amount of information is available on the effects of migration on biological marker distributions. Studies in this area should provide an insight into rates of migration and migration pathways.

Finally, it should be realized that the information that has been obtained to date on biological markers has resulted mainly from the analysis of the extractable organic material in fossil fuels. The bulk of organic material in source rocks, coals, and oil shales is present in the insoluble form referred to as kerogen. If a suitable method or methods can be developed, then kerogen has the potential to become the most important biological marker of all. A vast amount of information on source and maturity is stored in kerogen, and a great deal more emphasis should be placed on developing methods to release this information systematically. A certain effort has already been made in this area by using various pyrolysis techniques. This review has not attempted to cover this area but the interested reader is referred to a number of papers dealing with this subject (230–232).

Thus, the field of biological markers is full of challenges. It is still a relatively new area and the surface has only just been scratched. Examination of more samples will lead to the continual discovery of new markers and refinement of current theories. At the same time it is important for geochemists and geologists to collaborate more closely so that the results from each source can be integrated into an overall picture of the formation of fossil fuels.

V. REFERENCES

1. Mackenzie, A. S.; Brassell, S. C.; Eglinton, G.; Maxwell, J. R. *Science* **1982**, *217*, 491–504.
2. Eglinton, G.; Calvin, M. *Sci. Am.* **1967**, *216*, 32–43.
3. Eglinton, G. *Rev. Pure Appl. Chem.* **1968**, *34*, 611–632.
4. Eglinton, G.; Maxwell, J. R.; Philp, R. P. In: "Advances in Organic Geochemistry 1973"; Tissot, B.; Bienner, G., Eds.; Editions Technip: Paris, 1974; pp. 941–961.

5. Philp, R. P.; Maxwell, J.R.; Eglinton, G. *Sci Prog.* **1976**, *63*, 521–545.
6. Treibs, A. *Ann. Chem.* **1934**, *509*, 103–114.
7. Treibs, A. *Angew. Chem.* **1936**, *49*, 682–686.
8. de Mayo, P.; Reed, R. I. *Chem. Ind.* **1956**, 1481–1482.
9. Jarolim, V.; Streibl, M.; Horak, M.; Sorm, F. *Chem. Ind. (London)* **1958**, 1142–1143.
10. Ruhemann, S.; Raud, H. *Brennst. Chem.* **1932**, *13*, 341.
11. Carruthers, W.; Watkins, D. A. M. *Chem. Ind. (London)* **1963**, 1433.
12. Carruthers, W.; Cook, J. W. *J. Chem Soc.* **1954**, 2047.
13. O'Neal, M. J.; Hood, A. *Am. Chem. Soc. Div. Pet. Chem.* **1956**, *1*(4), 127–135.
14. Hood, A.; O'Neal, M. J. *Adv. Mass Spectrom.* **1959**, *1*, 175–192.
15. Eglinton, G.; Hamilton, R. J.; Raphael, R. A.; Gonzalez, A. G. *Nature* **1962**, *193*, 739–742.
16. Eglinton, G.; Hamilton, R. J. In: "Chemical Plant Taxonomy"; Swain, T., Ed.; Academic: New York, 1973; pp. 187–208.
17. Caldicott, A. B.; Eglinton, G. In: "Phytochemistry 3, Inorganic Elements and Special Group of Chemicals"; Miller, L. P., Ed.; Van Nostrand: New York, 1973; pp. 162–185.
18. Oro, J.; Tornabene, T. G.; Nooner, D. W.; Gelpi, E. *J. Bacteriol.* **1977**, *93*, 1811–1818.
19. Blumer, M.; Guillard, R. R. L.; Chase, T. *Mar. Biol.* **1971**, *8*, 183–189.
20. Gelpi, E.; Schneider, H.; Mann, J.; Oro, J. *Phytochemistry* **1970**, *9*, 603–612.
21. Han, J. C-Y., "Chemical studies of terrestrial and extraterrestrial life," Ph.D. Thesis, University of California, Berkeley, 1970, 317 pp.
22. Brassell, S. C.; Eglinton, G.; Maxwell, J. R.; Philp, R. P. In: "Aquatic Pollutants—Transformation and Biological Effects"; Hutzinger, O.; Van Lelyveld, I. H.; Zoeteman, B. C. J., Eds.; Pergamon: Oxford, 1978; pp. 69–86.
23. Cooper, J. E.; Bray, E. E. *Geochim. Cosmochim. Acta* **1963**, *27*, 1113–1127.
24. Kvenvolden, K. A. *J. Am. Oil Chem. Soc.* **1967**, *44*, 628–636.
25. Henderson, W.; Eglinton, G.; Simmonds, P.; Lovelock, J. E. *Nature* **1968**, *219*, 1012–1016.
26. Douglas, A. G.; Coates, R. C.; Bowler, B. F. J.; Hall, K. In: "Advances in Organic Geochemistry 1975"; Campos, R.; Goni, J., Eds.; ENADIMSA: Madrid, 1977; pp. 357–374.
27. Mitterer, R. M.; Hoering, T. C. *Carnegie Inst. Washington Year Book* **1968**, *66*, 510.
28. Connan, J. In: "Advances in Organic Geochemistry 1973"; Tissot, B.; Bienner, F., Eds.; Editions Technip: Paris, 1975; pp. 73–95.
29. Winters, J.; Williams, J. A. In: "Symposium on Petroleum Transformation in Geologic Environments"; Am. Chem. Soc. Div. Pet. Chem.: New York, 1969; Pap. PETR 86, pp. E22–E31.
30. Jobson, A.; Cook, F. D.; Westlake, D. W. S. *Appl. Microbiol.* **1972**, *23*(6), 1082–1089.
31. Bailey, N. J. L.; Jobson, A. M.; Rogers, M. A. *Chem Geol.* **1973**, *11*, 203–221.
32. Milner, C. W. D.; Rogers, M. A.; Evans, C. R. *J. Geochem. Explor.* **1977**, *7*, 101–153.
33. Mackenzie, A. S.; Leythauser, D.; Schaefer, R. G.; Bjoroy, M. *Nature* **1983**, *301*, 506–509.
34. Powell, T. G.; McKirdy, D. M. *APEA J.* **1972**, *12*, 125–131.
35. Powell, T. G.; McKirdy, D. M. *Am. Assoc. Pet. Geol. Bull.* **1975**, *59*, 1176–1197.
36. Powell, T. G. *J. Geochem. Explor.* **1975**, *4*, 441–466.
37. Alexander, R.; Kagi, R. I.; Woodhouse, G. W.; Volkman, J. K. *APEA J.* **1983**, *23*, 53–63.
38. Illich, H. A.; Haney, F. R.; Jackson, T. J. *Am. Assoc. Pet. Geol. Bull.* **1977**, *61*, 2103–2114.
39. Illich, H. A.; Haney, F. R.; Mendoza, M. *Am. Assoc. Pet. Geol. Bull.* **1981**, *65*, 2388–2402.
40. Kovacev, G.; Ubik, K.; Minceva, T.; Nikolov, R. *Collect. Czech. Chem. Commun.* **1975**, *40*, 3728–3730.
41. Snowdon, L. R.; Powell, T. G. *Bull. Can. Pet. Geol.* **1979**, *27*, 139–162.
42. Snowdon, L. R.; Peake, E. *Anal. Chem.* **1978**, *50*, 379–381.
43. Richardson, J. S.; Miiller, D. E. *Anal. Chem.* **1982**, *54*, 765–768.
44. Richardson, J. S.; Miiller, D. E. *Fuel* **1983**, *62*, 524–528.

45. Hoering, T. C. *Chem. Geol.* **1977,** *20,* 1–8.
46. Simoneit, B. R. T.; Lonsdale, P. F. *Nature* **1982,** *295,* 198–202.
47. Szucs, I,; Wein-Brukner, A. *J. Chromatogr.* **1982,** *241,* 113–120.
48. Tissot, B.; Pelet, R.; Roucache, J.; Combaz, A. In: "Advances in Organic Geochemistry 1975"; Campos, R.; Goni, J., Eds.; ENADIMSA: Madrid, 1977; pp. 117–154.
49. Albaiges, J.; Torradas, J. M. *Nature* **1974,** *250,* 567–568.
50. Albaiges, J.; Torradas, J. M. In: "Advances in Organic Geochemistry 1975"; Campos, R.; Goni, J., Eds.; ENADIMSA: Madrid, 1977; pp. 99–115.
51. Welte, D. H.; Ebhardt, G. *Geochim. Cosmochim. Acta* **1968,** *32,* 465–466.
52. Sever, J. R.; Haug, P. *Nature* **1971,** *234,* 447–450.
53. Welte, D. H.; Waples, D. W. *Naturwissenschaften* **1973,** *60,* 516.
54. Le Tran, K.; Connan, J.; Van der Weide, B. *Bull. Centre Rech. Pau* **1974,** *8,* 111–137.
55. Douglas, A. G.; Grantham, P. J. In: "Advances in Organic Geochemistry 1973"; Tissot, B.; Bienner, F., Eds.; Editions Technip: Paris, 1975; pp. 261–276.
56. Spiro, B. *Nature* **1977,** *269,* 235–237.
57. Dembicki, H.; Meinschein, W. G.; Hattin, D. E. *Geochim. Cosmochim. Acta* **1976,** *40,* 203–208.
58. Reed, W. E.; Kaplan, I. R. *J. Geochem. Explor.* **1977,** *7,* 255–293.
59. Venkatesan, M. I.; Kaplan, I. R.; Ruth, E. *Am. Assoc. Pet. Geol. Bull.* **1983,** *67,* 831–840.
60. Bertsch, W.; Anderson, E.; Holzer, G. *J. Chromatogr.* **1976,** *126,* 213–224.
61. Schulz, R. V.; Jorgenson, J. W.; Maskarinec, M. P.; Novotny, M.; Todd, L. J. *Fuel* **1979,** *58,* 783–789.
62. Solash, J.; Hazlett, R. N.; Hall, J. M.; Nowack, C. J. *Fuel* **1978,** *57,* 521–528.
63. Allan, J.; Douglas, A. G. *Geochim. Cosmochim. Acta* **1977,** *41,* 1223–1230.
64. Bartle, K. D.; Jones, D. W.; Pakdel, H.; Snape, C. E.; Calimli, A.; Olcay, A.; Tugrul, T. *Nature* **1979,** *277,* 284–288.
65. Snape, C. E.; Stokes, B. J.; Bartle, K. D. *Fuel* **1981,** *60,* 903–908.
66. Alexander, G.; Hazai, I. *J. Chromatogr.* **1981,** *217,* 19–38.
67. Radke, M.; Schaefer, R. G.; Leythaeuser, D.; Teichmüller, M. *Geochim. Cosmochim. Acta* **1980,** *44,* 1787–1800.
68. Bartle, K. D.; Jones, D. W.; Pakdel, H. In: "Analytical Methods for Coal and Coal Products. Vol II"; Karr, C., Jr., Ed; Academic: New York, 1978; pp. 209–239.
69. Bartle, K. D.; Jones, D. W.; Pakdel, H. In: "Coal and Coal Products: Analytical Characterization Techniques"; Fuller, E. L., Jr., Ed.; *Am. Chem. Soc. Symp. Ser.* **1982,** *205,* 29–45.
70. Robinson, W. E.; Cook G. L. *U.S. Bur. Mines Rep. Invest.* 8017, **1975,** 40 pp.
71. Young, D. K.; Yen, T. F. *Geochim. Cosmochim. Acta* **1977,** *41,* 1411–1417.
72. Kovachev, G. *Acta Chim. Acad. Sci. Hung.* **1980,** *104,* 415–419.
73. Heller, W.; Schallies, M.; Schmidt, K. *J. Chromatogr.* **1979,** *186,* 843–849.
74. Regtop, R. A.; Crisp, P. T.; Ellis, J. *Fuel* **1982,** *61,* 185–192.
75. Ingram, L. L.; Ellis, J.; Crisp, P. T.; Cook, A. C. *Chem. Geol.* **1983,** *38,* 185–212.
76. Riley, R. G.; Garland, T. R.; O'Malley, M. L.; Mann, D. C.; Wildung, R. E. *Environ. Sci. Technol.* **1982,** *16,* 709–713.
77. Albaiges, J. In: "Advances in Organic Geochemistry 1979"; Douglas, A. G.; Maxwell, J. R., Eds.; Pergamon: Oxford, 1980; pp. 19–28.
78. Albaiges, J.; Albrecht, P. *Int. J. Environ. Anal. Chem.* **1979,** *6,* 171–190.
79. Gardner, P. M.; Whitehead, E. V. *Geochim. Cosmochim. Acta* **1972,** *35,* 259–263.
80. Murphy, M. T. J.; McCormick, A.; Eglinton, G. *Science* **1967,** *157,* 1040–1042.
81. Kimble, B. J.; Maxwell, J. R.; Philp, R. P.; Eglinton, G.; Albrecht, P.; Ensminger, A.; Arpino, P.; Ourisson, G. *Geochim. Cosmochim. Acta* **1974,** *38,* 1165–1181.
82. Moldowan, J. M.; Seifert, W. K. *Science* **1979,** *204,* 169–171.
83. Chappe, B..; Albrecht, P.; Michaelis, W. *Science* **1982,** *217,* 65–66.
84. de Rosa, M.; de Rosa, S.; Gambacorta, A.; Bu'lock, J. D. *Chem. Commun.* **1977,** 514.
85. Brooks, J. D.; Gould, K.; Smith, J. W. *Nature* **1969,** *222,* 257–259.
86. Powell, T. G.; McKirdy, D. M. *Nature* **1973,** *243,* 37–39.
87. Didyk, B. M.; Simoneit, B. R. T.; Brassell, S. C.; Eglinton, G. *Nature* **1978,** *272,* 216–222.

88. Vogler, E. A.; Meyers, P. A.; Moore, W. A. *Geochim. Cosmochim. Acta* **1981,** *45,* 2287–2293.
89. Alexander, R.; Kagi, R. I.; Woodhouse, G. W. *Am. Assoc. Pet. Geol. Bull.* **1981,** *65,* 235–250.
90. Rashid, M. A. *Chem. Geol.* **1979,** *25,* 109–122.
91. Gibert, J. M.; De Andrade Brunning, I. M. R.; Nooner, D. W.; Oro, J. *Chem. Geol.* **1975,** *15,* 209–215.
92. Patience, R. L.; Rowland, S. J.; Maxwell, J. R. *Geochim. Cosmochim. Acta* **1978,** *42,* 1871–1875.
93. Shlyakhov, A. F.; Volkava, L. G. *Geochem. Int.* **1977,** *14*(5), 89–94.
94. Mackenzie, A. S.; Patience, R. L.; Maxwell, J. R.; Vandenbroucke, M.; Durand, B. *Geochim. Cosmochim. Acta* **1980,** *44,* 1709–1721.
95. Mackenzie, A. S.; Maxwell, J. R. In: "Organic Maturation Studies and Fossil Fuel Exploration"; Brooks, J., Ed.; Academic: London, 1981; pp. 239–254.
96. Han, J.; Calvin, M. *Geochim. Cosmochim. Acta* **1969,** *33,* 733–742.
97. Haug, P.; Curry, D. J. *Geochim. Cosmochim. Acta* **1974,** *38,* 601–610.
98. Waples, D. W.; Haug, P.; Welte, D. H. *Geochim. Cosmochim. Acta* **1974,** *38,* 381–387.
99. Albaiges, J.; Torradas, J. In: "Advances in Organic Geochemistry 1975"; Campos, R.; Goni, J., Eds.; ENADIMSA: Madrid, 1977; pp. 99–115.
100. Albaiges, J.; Borbon, J.; Salagre, P. *Tetrahedron Lett.* **1978,** *6,* 595–598.
101. Holzer, G.; Oro, J.; Tornabene, T. G. *J. Chromatogr.* **1979,** *186,* 795–809.
102. Michaelis, W.; Albrecht, P. *Naturwissenschaften* **1979,** *66,* 420–422.
103. Hills, I. R., Smith, G. W.; Whitehead, E. V. *J. Inst. Pet. (London)* **1970,** *56,* 127.
104. Brassell, S. C.; Wardroper, A. M. K.; Thomson, I. D.; Maxwell, J. R.; Eglinton, G. *Nature* **1981,** *290,* 693–696.
105. Maxwell, J. R.; Douglas, A. G.; Eglinton, G.; McCormick, A. *Phytochemistry* **1968,** *7,* 2157.
106. Moldowan, J. M.; Seifert, W. K. *Chem. Commun.* **1980,** 912–914.
107. Yon, D. A.; Maxwell, J. R.; Ryback, G. *Tetrahedron Lett.* **1982,** *23,* 2143–2146.
108. Larter, S. R.; Solli, H.; Douglas, A. G.; de Lange, F.; de Leeuw, J. W. *Nature* **1979,** *279,* 405–408.
109. van de Meent, D.; Brown, S. C.; Philp, R. P.; Simoneit, B. R. T. *Geochim. Cosmochim. Acta* **1980,** *44,* 999–1014.
110. DiSanzo, F. P.; Uden, P. C.; Siggia, S. *Anal. Chem.* **1980,** *52,* 906–909.
111. Burnham, A. K.; Clarkson, J. E.; Singleton, M. F.; Wong, C. M.; Crawford, R. W. *Geochim. Cosmochim. Acta* **1982,** *46,* 1243–1251.
112. Baset, Z. H.; Pancirov, R. J.; Ashe, T. R. In: "Advances in Organic Geochemistry 1979"; Douglas, A. G.; Maxwell, J. R., Eds.; Pergamon: Oxford, 1980; pp. 619–631.
113. Philp, R. P.; Saxby, J. R. In: "Advances in Organic Geochemistry 1979"; Douglas, A. G.; Maxwell, J. R., Eds.; Pergamon: Oxford, 1980; pp. 639–653.
114. Bendoraitis, J. G. In: "Advances in Organic Geochemistry 1973"; Tissot, B.; Bienner, F., Eds.; Editions Technip: Paris, 1975; pp. 209–224.
115. Anders, D. E.; Robinson, W. E. *Geochim. Cosmochim. Acta* **1971,** *35,* 661–678.
116. Philp, R. P.; Gilbert, T. D.; Friedrich, J. *Geochim. Cosmochim. Acta* **1981,** *45,* 1173–1180.
117. Alexander, R.; Kagi, R.; Noble, R. *Chem. Commun.* **1983,** 226–228.
118. Williams, J. A.; Dolcater, D. L.; Olson, R. K. *Proc. Am. Assoc. Pet. Geol. Symp. North Slope of Alaska*, in press.
119. Striebl, M.; Herout, V. In: "Organic Geochemistry Methods and Results"; Eglinton, G.; Murphy, M. T. J., Eds.; Springer: Berlin, 1969; pp. 402–424.
120. Grantham, P. J.; Douglas, A. G. *Geochim. Cosmochim. Acta* **1980,** *44,* 1801–1810.
121. Hagemann, H. W.; Hollerbach, A. In: "Advances in Organic Geochemistry 1979"; Douglas, A. G.; Maxwell, J. R., Eds.; Pergamon: Oxford, 1981; pp. 631–638.
122. Maxwell, J. R.; Pillinger, C. T.; Eglinton, G. *Q. Rev. Chem. Soc.* **1971,** *25,* 571–628.
123. Chaffee, A. "The organic geochemistry of Australian coals," Ph.D. thesis, University of Melbourne, Australia, 1983, 399 pp.
124. Philp, R. P.; Simoneit, B. R.; Gilbert, T. D. In: "Advances in Organic Geochemistry 1981"; Bjorøy, M.; et al., Eds.; Wiley: London, 1983; pp. 698–704.

125. Snowdon, L. R. "Organic geochemistry of the Upper Cretaceous Tertiary delta complexes of the Beaufort–Mackenzie Sedimentary Basin, Northern Canada," Ph.D. thesis, Rice University, Houston, 1978, 130 pp.
126. Snowdon, L. R. *Can. Soc. Pet. Geol.* **1980,** *6,* 421–446.
127. Snowdon, L. R.; Powell, T. G. *Am. Assoc. Pet. Geol. Bull.* **1982,** *66*(6), 775–788.
128. Hollerbach, A.; Hagemann, H. W. *Proc. Int. Conf. Coal Sci. Dusseldorf* **1981,** 80–85.
129. Hayatsu, R.; Winans, R. E.; Scott, R. G.; Moore, L. P.; Studier, M. H. *Fuel* **1978,** *57,* 541–548.
130. Jones, D. W.; Pakdel, H.; Bartle, K. D. *Fuel* **1982,** *61,* 44–52.
131. Philp, R. P.; Russell, N. J.; Gilbert T. D. *Fuel* **1981,** *60,* 937–944.
132. Philp, R. P.; Russell, N. J.; Gilbert, T. D. *Fuel* **1982,** *61,* 221–226.
133. Gallegos, E. J. *Anal. Chem.* **1971,** *43,* 1151–1160.
134. Reed, W. E. *Geochim. Cosmochim. Acta* **1977,** *41,* 237–247.
135. Simoneit, B. R. T. *Int. J. Environ. Anal. Chem.* **1982,** *12,* 177–193.
136. Simoneit, B. R. T.; Kaplan, I. R. *Mar. Environ. Res.* **1980,** *3,* 113–128.
137. Aquino Neto, F. R.; Trendel, J. M.; Restle, A.; Connan, J.; Albrecht, P. In: "Advances in Organic Geochemistry 1981"; Bjorøy, M.; et al., Eds.; Wiley: London, 1983; pp. 659–667.
138. Aquino Neto, F. R.; Restle, A.; Albrecht, P.; Ourisson, G.; Connan, J. *Tetrahedron Lett.* **1982,** *23*(19), 2027–2030.
139. Ekweozer, C. M.; Strausz, O. P. In: "Advances in Organic Geochemistry 1981"; Bjorøy, M.; et al., Eds.; Wiley: London, 1983; pp. 746–766.
140. Ekweozer, C. M.; Strausz, O. P. *Tetrahedron Lett.* **1982,** *23*(27), 2711–2714.
141. Cyr, T. D.; Strausz, O. P. *Chem. Commun.* **1983,** 1028–1030.
142. Moldowan, J. M.; Seifert, W. K.; Gallegos, E. J. *Geochim. Cosmochim. Acta* **1983,** *47,* 1531–1534.
143. Payzant, J. D.; Montgomery, D. S.; Strausz, O. P. *Tetrahedron Lett.* **1983,** *24*(7), 651–654.
144. Seifert, W. K.; Moldowan, J. M.; Jones, R. W. *Proc. 10th World Pet. Congr.* **1979,** Paper SP 8, pp. 425–440.
145. Scholefield, D.; Whitehurst, J. S. *Chem. Commun.* **1980,** 135–136.
146. Trendel, J. M.; Restle, A.; Connan, J.; Albrecht, P. *Chem. Commun.* **1982,** 304–306.
147. Ekweozer, C. M.; Okogun, J. I.; Ekong, D. E. V.; Maxwell, J. R. *J. Geochem. Explor.* **1981,** *15,* 653–662.
148. Ourisson, G.; Albrecht, P.; Rohmer, M. *Pure Appl. Chem.* **1979,** *51,* 709–729.
149. Ourisson, G.; Albrecht, P.; Rohmer, M. *Trends Biochem. Sci.* **1982,** *7,* 236–239.
150. Rullkötter, J.; Philp, P. *Nature* **1981,** *292,* 616–618.
151. Van Dorsselaer, A.; Albrecht, P.; Connan, J. In: "Advances in Organic Geochemistry 1975"; Campos, R.; Goni, J., Eds.; ENADISMA: Madrid, 1977; pp. 53–59.
152. Ensminger, A.; Albrecht, P.; Ourisson, G.; Tissot, B. In: "Advances in Organic Geochemistry 1975"; Campos, R.; Goni, J., Eds.; ENADISMA: Madrid, 1977; pp. 45–52.
153. Mackenzie, A. S.; Lewis, C. A.; Maxwell, J. R. *Geochim. Cosmochim. Acta* **1981,** *45,* 2369–2375.
154. Seifert, W. K.; Moldowan, J. M. In: "Advances in Organic Geochemistry 1979"; Douglas, A. G.; Maxwell, J. R., Eds.; Pergamon: Oxford, 1980; pp. 229–237.
155. Philp, R. P.; Gilbert, T. D. *Nature* **1982,** *299,* 245–247.
156. Seifert, W. K.; Moldowan, J. M. *Geochim. Cosmochim. Acta* **1978,** *41,* 77–95.
157. Seifert, W. K.; Moldowan, J. M. *Geochim. Cosmochim. Acta* **1981,** *45,* 783–794.
158. Volkman, J. K.; Alexander, R.; Kagi, R. I.; Woodhouse, G. W. *Geochim. Cosmochim. Acta* **1983,** *47,* 785–794.
159. Philp, R. P.; Gilbert, T. D.; Friedrich, J. *APEA J.* **1982,** 188–199.
160. Ji-Yang, S.; Mackenzie, A. S.; Alexander, R.; Eglinton, G.; Gowar, A. P.; Wolff, G. A.; Maxwell, J. R. *Chem. Geol.* **1982,** *35,* 1–31.
161. Welte, D. H.; Kratochvil, H.; Rullkötter, J.; Ladwein, H.; Schaefer, R. G. *Chem. Geol.* **1982,** *35,* 33–68.
162. Wehner, H.; Teschner, M. *J. Chromatogr.* **1981,** *204,* 481–490.

163. Ekweozer, C. M.; Okogun, J. I.; Ekong, D. E. U.; Maxwell, J. R. *Chem. Geol.* **1979**, 27, 11–28.
164. Philp, R. P. *Geochim. Cosmochim. Acta* **1983**, 47, 267–275.
165. Petrov. A. V.; Pustil'nikova, S. D.; Abriutina, N. N.; Kagramanova, G. R. *Neftekhimiya* **1976**, 16, 411–427.
166. Seifert, W. K.; Moldowan, J. M.; Smith, G. W.; Whitehead, E. V. *Nature* **1978**, 271, 436–437.
167. Grantham, P. J.; Posthuma, J.; DeGroot, K. In: "Advances in Organic Geochemistry 1979"; Douglas, A. G.; Maxwell, J. R., Eds.; Pergamon: Oxford, 1980; pp. 29–39.
168. Seifert, W. K.; Moldowan, J. M. *Geochim. Cosmochim. Acta* **1979**, 43, 111–126.
169. Rullkötter, J.; Wendisch, D. *Geochim. Cosmochim. Acta* **1982**, 43, 1545–1554.
170. Alexander, R.; Kagi, R. I.; Woodhouse, G. W.; Volkmann, J. K. *APEA J.* **1983**, 53–63.
171. Connan, J.; Restle, A.; Albrecht, P. In: "Advances in Organic Geochemistry 1979"; Douglas, A. G.; Maxwell, J. R., Eds.; Pergamon: Oxford, 1980; pp. 1–19.
172. Goodwin, N. S.; Park, P. J. D.; Rawlinson, A. P. In: "Advances in Organic Geochemistry 1981"; Bjorøy, M.; et al., Eds.; Wiley: London, 1983; pp. 650–658.
173. Dastillung, M.; Albrecht, P.; Tissier, M. J. In: "Geochimie organique des sediments marino profonds. Orgon I. Mer de Norvege"; Combaz, A.; Ed.; Editions CNRS: Paris, 1977; pp. 209–228.
174. Bjorøy, M.; Hall, K.; Vigran, J. O. In: "Advances in Organic Geochemistry 1979"; Douglas, A. G.; Maxwell, J. R., Eds.; Pergamon: Oxford, 1980; pp. 77–91.
175. Bjorøy, M.; Rullkötter, J. *Chem. Geol.* **1980**, 30, 27–34.
176. Volkman, J. K.; Alexander, R.; Kagi, R. I.; Rullkötter, J. *Geochim. Cosmochim. Acta* **1983**, 47, 1033–1041.
177. Rullkötter, J. *Int. J. Mass Spectrom. Ion Phys.* **1983**, 48, 39–42.
178. Schmitter, J. M; Sucrow, W.; Arpino, P. J. *Geochim. Cosmochim. Acta* **1982**, 46, 2345–2350.
179. Seifert, W. K. In: "Progress in the Chemistry of Organic Natural Products"; Hertz, W.; Grisebach, H.; Kirby, J. W., Eds.; Springer-Verlag: New York, 1974; Vol. 32, pp. 1–49.
180. Schmitter, J. M.; Arpino, P. J.; Guiochon, G. *Geochim. Cosmochim. Acta* **1981**, 45, 1951–1955.
181. Corbet, B.; Albrecht, P.; Ourisson, G. *J. Am. Chem. Soc.* **1980**, 102(3), 1171–1173.
182. Schmitter, J. M.; Arpino, P.; Guiochon, G. *J. Chromatogr.* **1978**, 167, 149–158.
183. Griener, A. Ch.; Spyckerelle, C.; Albrecht. P. *Tetrahedron* **1976**, 32, 257–260.
184. Griener, A. Ch.; Spyckerelle, C.; Albrecht, P.; Ourisson, G. *J. Chem. Res. (M)* **1977**, 3829–3836.
185. Spykerelle, C.; Griener, A. Ch.; Albrecht, P.; Ourisson, G. *J. Chem. Res. (M)* **1977**, 3801–3828.
186. Rubinstein, I.; Strausz, O. P.; Spyckerelle, C.; Crawford, R. J.; Westlake, D. W. S. *Geochim. Cosmochim. Acta* **1977**, 41, 1341–1353.
187. Rubinstein, I.; Spyckerelle, C.; Strausz, O. P. *Geochim. Cosmochim. Acta* **1979**, 43, 1–6.
188. Regtop, R. A.; Ellis, J.; Crisp, P. T.; Bolton, P. D. *Geochim. Cosmochim. Acta* (submitted).
189. Allan, J.; Bjorøy, M.; Douglas, A. G. In: "Advances in Organic Geochemistry 1975"; Campos, R.; Goni, J., Eds.; ENADIMSA: Madrid, 1977; pp. 633–654.
190. Gallegos, E. J. In: "Analytical Chemistry of Liquid Fuel Sources"; Uden, P. C.; Siggia, S.; Jensen, H. B., Eds.; *Am. Chem. Soc. Adv. Chem. Ser.* **1978**, 170, 2–36.
191. Chaffee, A. L.; Perry, G. J.; Johns, R. B. *Fuel* **1983**, 62, 303–310.
192. Chaffee, A. L.; Perry, G. J.; Johns, R. B. *Fuel* **1983**, 62, 311–316.
193. Whitehead, E. V. In: "Advances in Organic Geochemistry 1973"; Tissot, B.; Bienner, F., Eds., Editions Technip: Paris, 1974; pp. 225–245.
194. Whitehead, E. V. In: "Petroanalysis '81"; Crump, G. B., Ed.; Wiley: London, 1982; pp. 31–76.
195. Hills, I. R.; Whitehead, E. V. *Nature* **1966**, 209, 977.
196. Rullkötter, J.; Leythauser, D.; Wendisch, P. *Geochim. Cosmochim. Acta* **1982**, 46, 2501–2509.

197. Ekweozer, C. M.; Okogun, J. I.; Ekong, D. E. U.; Maxwell, J. R. *Chem. Geol.* **1979,** *27,* 29–37.
198. Seifert, W. K. In: "The Impact of the Treibs of Porphyrin Concept on the Modern Organic Geochemistry"; Prashnosky, A., Ed.; Universitat: Wurzburg, 1980; pp. 13–35.
199. Rubinstein, I.; Sieskind, O.; Albrecht, P. *J. Chem Soc. Perkin I* **1975,** 1833–1836.
200. Ensminger, A.; Joly, G.; Albrecht, P. *Tetrahedron Lett.* **1978,** 1575–1578.
201. Warburton, G. A.; Zumberge, J. E. *Anal. Chem.* **1983,** *55,* 123–126.
202. Mulheirn, L. J.; Ryback, G. *Nature* **1975,** *256,* 301–302.
203. Mulheirn, L. T.; Ryback, G. In: "Advances in Organic Geochemistry 1975"; Campos, R.; Goni, J., Eds.; ENADIMSA: Madrid, 1977; pp. 173–193.
204. Hufnagel, H.; Teschner, M.; Wehner, H. In: "Advances in Organic Geochemistry 1979"; Douglas, A. G.; Maxwell, J. R., Eds.; Pergamon: Oxford, 1981; pp. 51–66.
205. Curiale, J. A.; Harrison, W. E.; Smith, G. *Geochim. Cosmochim. Acta* **1983,** *47,* 517–523.
206. Tissot, B.; Espitalie, J.; Deroo, G.; Tempere, C.; Jonathan, D. In: "Advances in Organic Geochemistry 1973"; Tissot, B.; Bienner, F., Eds.; Editions Technip: Paris, 1974; pp. 315–334.
207. Seifert, W. K.; Carlson, R. M. K.; Moldowan, J. M. In: "Advances in Organic Geochemistry 1981"; Bjorøy, M.; et al., Eds.; Wiley: London, 1983; pp. 710–724.
208. Schaefle, J.; Ludwig, B.; Albrecht, P.; Ourisson, G. *Tetrahedron Lett.* **1978,** 4163–4166.
209. Mackenzie, A. S.; Lamb, N. A.; Maxwell, J. R. *Nature* **1982,** *295,* 223–226.
210. Hussler, G. Chappe, B.; Wehrung, P.; Albrecht, P. *Nature* **1981,** *294,* 556–558.
211. McKenzie, D.; Mackenzie, A. S.; Maxwell, J. R.; Sajgo, C. *Nature* **1983,** *301,* 504–506.
212. Ludwig, B.; Hussler, G.; Wehrung, P.; Albrecht, P. *Tetrahedron Lett.* **1981,** 3313–3316.
213. Mackenzie, A. S.; Hoffmann, C. F.; Maxwell, J. R. *Geochim. Cosmochim. Acta* **1981,** *45,* 1345–1355.
214. Radke, M.; Welte, D. H.; Willsch, H. *Geochim. Cosmochim. Acta* **1982,** *46,* 1–10.
215. Radke, M.; Willsch, H.; Leythauser, D.; Teichmüller, M. *Geochim. Cosmochim. Acta* **1982,** *46,* 1831–1848.
216. George, A. E.; Beshai, J. E. *Fuel* **1983,** *62,* 345–349.
217. White, C. M.; Lee, M. L. *Geochim. Cosmochim. Acta* **1980,** *44,* 1825–1832.
218. Gallegos, E. J. *Anal. Chem.* **1973,** *45,* 1399–1403.
219. Gallegos, E. J. *J. Chromatogr. Sci.* **1981,** *19,* 177–182.
220. Baker, E. W.; Palmer, S. W. In: "The Porphyrins. Vol. 1"; Dolphin, D., Ed.; Academic: New York, 1978; pp. 485–551.
221. Alexander, R.; Eglinton, G.; Gill, J. P.; Volkman, J. K. *J. High Resolut. Chromatogr.* **1980,** *3,* 521–522.
222. Marriott, P. J.; Eglinton, G. *J. Chromatography* **1982,** *249,* 311–321.
223. McFadden, W. H.; Bradford, D. C.; Eglinton, G.; Hajibrahim, S. K.; Nicolaides, N. *J. Chromatogr. Sci.* **1979,** *17,* 518–522.
224. Quirke, J. M. E.; Eglinton, G.; Palmer, S. E.; Baker, E. W. *Chem. Geol.* **1982,** *35,* 69–85.
225. Palmer, S. E.; Baker, E. W.; Charney, L. S.; Louda, J. W. *Geochim. Cosmochim. Acta* **1982,** *46,* 1233–1241.
226. Barwise, A. G.; Whitehead, E. V. In: "Advances in Organic Geochemistry 1979"; Douglas, A. G.; Maxwell, J. R., Eds.; Pergamon: Oxford, 1980; pp. 181–192.
227. Eglinton, G.; Hajibrahim, S. K.; Maxwell, J. R.; Quirke, J. M. E. In: "Advances in Organic Geochemistry 1979"; Douglas, A. G.; Maxwell, J. R., Eds.; Pergamon: Oxford, 1980; pp. 193–203.
228. Mackenzie, A. S.; Quirke, J. M. E.; Maxwell, J. R. In: "Advances in Organic Geochemistry 1979"; Douglas, A. G.; Maxwell, J. R., Eds.; Pergamon: Oxford, 1980; pp. 239–248.
229. Mackenzie, A. S.; McKenzie, D. *Geol. Mag.* **1983,** *120*(5), 417–528.
230. Seifert, W. K. *Geochim. Cosmochim. Acta* **1978,** *42,* 473–484.
231. Larter, S. R. In: "Analytical Pyrolysis, Vol. II"; Voorhees, K. E., Ed.; Butterworths: London, 1984; pp. 212–275.
232. Philp, R. P. *Trends Anal. Chem.* **1982,** *1*(10), 237.
233. Moldowan, J. M.; Seifert, W. K.; Arnold, E.; Clardy, J. *Geochim. Cosmochim. Acta* **1984,** *48,* 1651–1661.

Guidelines for Evaluating Petroleum Source Rock Using Programmed Pyrolysis[1]

K. E. PETERS[2]

ABSTRACT

Rock-Eval pyrolysis is used to rapidly evaluate the petroleum-generative potential and thermal maturity of rocks. Accurate conclusions require pyrograms every 30-60 ft (9-18 m), understanding of interpretive pitfalls, and supporting data, such as visual kerogen, vitrinite reflectance, and elemental analyses.

The generative potential of coals is commonly overestimated by pyrolysis and is best determined by elemental analysis and organic petrography. Most coals show high S2/S3 (>5) and low HI values (<300 mg HC/g TOC). Migrated oil and mud additives, which alter Rock-Eval data, can sometimes be removed by special processing. For immature rocks, bimodal S2 peaks and PI values over 0.2 indicate contamination.

Pyrolysis downgrades organic-poor, clay-rich rocks, which show lower HI and higher T_{max} values than isolated kerogen because of adsorption of pyrolyzate on the clays. T_{max} values for small S2 peaks (<0.2 mg HC/g TOC) are unreliable. T_{max} is affected by maturation, organic matter type, contamination, and the mineral matrix.

S3 is sensitive to inorganic and adsorbed carbon dioxide, and to instrumentation problems. Acidification of carbonate-rich samples and proper maintenance improves S3 measurement.

Constant sample weights (100 mg) are recommended. Below a threshold weight, T_{max} increases by up to 10°C, and other parameters decrease. Organic-rich samples, which overload the detector, can be diluted with carbonate. Detector linearity is determined by pyrolyzing various weights of an organic-rich rock.

INTRODUCTION

A basic problem in petroleum exploration is how to assess the petroleum-generative potentials of prospective source rocks. Rock-Eval pyrolysis of whole rock (Espitalié et al, 1977)

provides information on the quantity, type, and thermal maturity of the associated organic matter (OM). Pyrolysis is defined as the heating of OM in the absence of oxygen, to yield organic compounds. In Rock-Eval pyrolysis, pulverized samples are gradually heated under an inert atmosphere. This heating distills the free organic compounds (bitumen), then cracks pyrolytic products from the insoluble OM (kerogen). In some laboratories, the method has replaced extraction techniques for separating these materials because it is faster, involves less manpower, and requires only small samples.

This paper provides guidelines for understanding and interpreting Rock-Eval pyrolysis data. Guidelines are needed because pyrolysis is widely accepted among exploration geologists as a rapid and effective means of characterizing the quality and thermal maturity of prospective source rocks. Although the method is generally straightforward, the simplicity of the data occasionally can be misleading. Thus, this paper also attempts to define the pitfalls in interpreting pyrolysis results.

DESCRIPTION OF PYROLYSIS METHOD

Instrumentation

Various Rock-Eval pyroanalyzers are commercially available. The Rock-Eval I pyroanalyzer was designed by the Institut Français du Pétrole and built by Girdel, Inc. Two similar versions of the Rock-Eval II were produced by independent companies: Delsi, Inc., and Geocom, Inc. Compared to the Rock-Eval I, both Rock-Eval II units are easier to operate and troubleshoot, and provide more analyses per unit time. The Rock-Eval II (from Delsi, Inc.) combines pyrolysis analysis with the capability of determining total organic carbon. The Rock-Eval III (Oil Show Analyzer) differs from the Delsi Rock-Eval II in that it analyzes gas and oil separately and does not measure carbon dioxide. This paper discusses Rock-Eval II analyses but is generally applicable to the other instruments.

Obviously, an analysis of Rock-Eval data must consider the limitations of the instrument, such as the linear range of detectors and tolerances of the integrators. Because of variations between instruments and accessories, this aspect will not be discussed but should be considered by each interpreter for the particular apparatus used.

Operating Conditions

Operating parameters for the Rock-Eval instruments are similar to those described in Clementz et al (1979). Samples of ground whole rock weighing up to about 100 mg are pyrolyzed

---

©Copyright 1986. The American Association of Petroleum Geologists. All rights reserved.
[1]Manuscript received, May 20, 1985; accepted, December 2, 1985. This paper was presented at the 189th National Meeting of the American Chemical Society, Symposium on Organic Geochemistry of Humic Substances, Kerogen and Coal, Philadelphia, Pennsylvania, August 26-31, 1984.
[2]Chevron Oil Field Research Company, P.O. Box 446, La Habra, California 90631.

I thank G. J. Demaison, W. G. Dow, R. A. Stuart, B. A. Patterson, A. R. Daly, and R. W. Jones for critiques of the manuscript, and Chevron Oil Field Research Company for permission to publish.

at 300°C for 3-4 min, followed by programmed pyrolysis at 25°C/min to 550°C, both in a helium atmosphere. Allowing time to cool the oven, each analysis requires about 20 min.

Measured Parameters

The Rock-Eval II technique provides several measurements (Espitalié et al, 1977) (Figure 1). A flame ionization detector (FID) senses any organic compounds generated during pyrolysis. The first peak (S1) represents milligrams of hydrocarbons that can be thermally distilled from one gram of the rock. The second peak (S2) represents milligrams of hydrocarbons generated by pyrolytic degradation of the kerogen in one gram of rock. (Although the literature expresses S1 and S2 in milligrams of "hydrocarbons" per gram of rock, the FID also senses nonhydrocarbons provided carbon atoms are present.) The third peak (S3) represents milligrams of carbon dioxide generated from a gram of rock during temperature programming up to 390°C, and is analyzed by thermal conductivity detection (TCD). In some United States publications, S1, S2, and S3 are respectively referred to as P1, P2, and P3.

During pyrolysis, the temperature is monitored by a thermocouple. The temperature at which the maximum amount of S2 hydrocarbons is generated is called T_{max}.

The hydrogen index (HI) corresponds to the quantity of pyrolyzable organic compounds or "hydrocarbons" (HC) from S2 relative to the total organic carbon (TOC or C_{org}) in the sample (mg hydrocarbon/g organic carbon = mg HC/g C_{org}). The oxygen index (OI) corresponds to the quantity of carbon dioxide from S3 relative to the TOC (mg CO_2/g C_{org}). The production index (PI) is defined as the ratio S1/(S1 + S2).

Pyrograms

Interpretation of Rock-Eval data requires access to the pyrograms. Figure 2 shows a typical pyrogram from the Rock-Eval II unit. The temperature profile has been offset relative to the FID trace (6 mm on actual pyrograms) so that the dual pens do not interfere. For reference, the beginning of the isothermal period at 300°C (noted by an asterisk in the figure) corresponds temporally with the first sharp increase in FID response to S1. Note that parameters are reported other than the traditional display of S1, S2, and S3 peaks and oven-temperature profile. These include ID number, day and month, time of day, TOC (if known), sample weight (mg), and detector attenuations. The two rows of numbers immediately above the pyrogram are percentage of scale and oven temperature (°C).

Following a sample analysis and entry of such parameters as sample weight and TOC, the system calculates S1, S2, S3, and T_{max}. At the end of each sequence of samples, a histogram and summary report can be printed.

In Figure 2, the first number to the right of S1 and S2 represents mg HC/g rock. The first number to the right of S3 represents mg CO_2/g rock. The second number to the right of S1, S2, and S3 represents the peak area ("area counts") measured by the integrator. For example, the S2 peak consists of 4.273 × 10^5 area counts, equivalent to 2.793 × 10^1 mg HC/g rock.

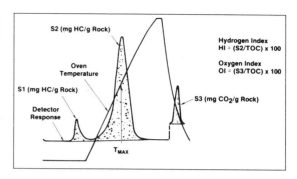

Figure 1—Schematic of pyrogram showing evolution of organic compounds from rock sample during heating (increasing time from left to right). Important measurements include S1, S2, S3, and T_{max}. Hydrogen and oxygen indices are calculated as shown.

"Unknown" indicates that the sample is from an active study and is not a reference standard. The standard used at Chevron is a homogenized composite of various rocks.

APPLICATIONS OF ROCK-EVAL TO GEOLOGIC PROBLEMS

Because Rock-Eval pyrolysis is rapid (20 min) and requires only small samples (100 mg), it is commonly used at the well site to screen large numbers of core and ditch-cutting samples for further study. However, critical interpretations should always be supported by additional geochemical analyses (Peters et al, 1983).

Rock-Eval pyrolysis can be used to describe the petroleum-generative potentials of prospective source rocks by providing information about their OM: (1) quantity, (2) type, and (3)

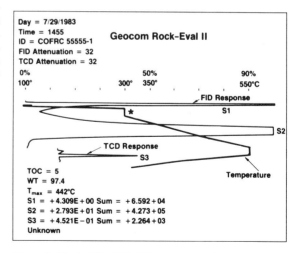

Figure 2—Typical Rock-Eval II pyrogram and report showing oven-temperature profile, FID (S1 and S2), and TCD (S3) responses (increasing time from top to bottom). Asterisk near temperature curve designates isothermal heating period at 300°C. See text for explanation of annotations.

thermal maturity (Tables 1-3). The method can also be used to indicate prospective reservoir rocks and locate the top ("birthline") and bottom ("deadline") of the oil-generative "window." Other applications include correlating reservoirs or source rocks between wells, mapping regional thermal maturity or richness of source rocks, and predicting regions of likely petroleum accumulations within basins.

Quantity of Organic Matter

Two useful measurements related to quantity of OM are the total organic carbon (TOC) and the fraction of TOC that is generated as hydrocarbons and other compounds by pyrolysis (hydrocarbon index or HI). Tables 1 and 2 show how TOC, S2, and HI are used to describe potential source rocks.

Typically, TOC is determined by the direct combustion method: crushed samples are acid treated in a filtering crucible to remove carbonate, combusted at about 1,000°C, and analyzed as carbon dioxide. Comparable results can be obtained by: (1) an indirect TOC method (total carbon minus carbonate carbon), or (2) a modified direct method using nonfiltering crucibles. These methods are discussed further in Peters and Simoneit (1982).

The "Rock-Eval II plus TOC" (Delsi, Inc.) determines TOC by summing the carbon in the pyrolyzate with that obtained by oxidizing the residual OM at 600°C. For small samples (100 mg), this method provides more reliable TOC data than conventional combustion methods, which require about 1-2 g of ground rock. However, mature samples, where vitrinite reflectance (R_o) is more than about 1%, yield poor TOC data when determined by the Rock-Eval III because the temperature is insufficient for complete combustion.

TOC is usually greater than the sum of the pyrolyzable carbon in S1, S2, and S3. For example, a rock containing abundant graphite and no other OM will show a large TOC but essentially no pyrolysis response.

Pyrolysis is useful for predicting the quantity of organic compounds that could be generated from a rock upon further maturation. An example is the excellent correlation of S2 with Fischer assay oil yield (Figure 3). Bienner and Espitalié (1980) have determined linear regressions between pyrolysis (S1 + S2) and Fischer assay yields for various types of OM.

Types of Organic Matter

The most familiar method of classifying OM type is the van Krevelen or atomic H/C vs. O/C diagram (Tissot et al, 1974). In this diagram (Figure 4), different types of kerogens are shown as Type I (very oil prone), Type II (oil prone), and Type III (gas prone). Type IV (inert) kerogens contain very little hydrogen and plot near the bottom of the figure. The thermal maturation of each kerogen type is described by pathways; the most mature samples are near the lower left corner (little hydrogen or oxygen relative to carbon in the kerogen).

Espitalié et al (1977) attempted to use the Rock-Eval method to derive the same type of information conveyed by the H/C vs. O/C diagram. Because pyrolysis does not require kerogen isolation, this method proved more rapid and less expensive than elemental analysis. Espitalié et al showed that oxygen in the kerogen is proportional to the carbon dioxide liberated during pyrolysis (S3) and that the hydrogen content is proportional to the hydrocarbons liberated (S2). They defined the hydrogen index (HI) vs. oxygen index (OI) plot (Figure 5), where HI and OI are (S2/TOC) × 100 and (S3/TOC) × 100, respectively. This plot can be used to describe the type of OM in any sample, but the results should be supported by microscopy, elemental analysis, or both. The good correspondence between data in Figures 4 and 5 is not typical for reasons discussed later. Using the HI or the S2/S3 ratio, Table 2 suggests the types of products (gas or oil) that will be generated from source rocks at a level of thermal maturation equivalent to $R_o = 0.6\%$.

Thermal Maturity of Organic Matter

The level of thermal maturation can be roughly estimated from the HI vs. OI plot described above. Table 3 shows how to use the production index (PI) and T_{max} to estimate maturity. In general, PI and T_{max} values less than about 0.1 and 435°C, respectively, indicate immature OM. A T_{max} greater than 470°C represents the wet-gas zone. The PI reaches about 0.4 at the bottom of the oil window (beginning of the wet-gas zone) and increases to 1.0 when the hydrocarbon-generative capacity of the kerogen has been exhausted. Usually some S1 will remain as adsorbed dry gas, even in highly postmature rocks.

The T_{max} and PI are crude measurements of thermal maturity and are partly dependent on other factors, such as type of OM. Thus, conclusions regarding thermal maturity should be supported by other geochemical analyses, such as vitrinite reflectance or thermal alteration index (TAI).

Table 1. Geochemical Parameters Describing Source Rock Generative Potential

Quantity	TOC (wt. %)	S1*	S2*
Poor	0-5	0-0.5	0-2.5
Fair	0.5-1	0.5-1	2.5-5
Good	1-2	1-2	5-10
Very good	2+	2+	10+

*Nomenclature:
  S1 = mg HC/g rock
  S2 = mg HC/g rock

Table 2. Geochemical Parameters Describing Type of Hydrocarbon Generated

Type	HI (mg HC/g C_{org})*	S2/S3*
Gas	0-150	0-3
Gas and oil	150-300	3-5
Oil	300+	5+

*Assumes a level of thermal maturation equivalent to $R_o = 0.6\%$.

Table 3. Geochemical Parameters Describing Level of Thermal Maturation

Maturation	PI [S1/(S1 + S2)]	T_{max} (°C)	R_o (%)
Top oil window (birthline)	~0.1	~435-445*	~0.6
Bottom oil window (deadline)	~0.4	~470	~1.4

*Many maturation parameters (particularly T_{max}) depend on type of OM.

INTERPRETATION GUIDELINES

Proper interpretation of Rock-Eval and TOC data requires information on lithologies, the relative abundances of OM and mineral matrix, well conditions, the presence or lack of generated hydrocarbons, pyrograms, and geochemical logs. The following list of Rock-Eval interpretive problems describes the more common difficulties.

Large Amounts of Data

The most reliable geochemical interpretations are based on wells for which large amounts of pyrolysis data are available. We recommend one sample every 30-60 ft (9-18 m). Large amounts of Rock-Eval and other data can be readily manipulated and displayed by computer using a geochemical log (Espitalié et al, 1984). By using the geochemical log, trends and anomalous results become readily apparent, and questionable data can then be tested by closer examination of pyrograms, inquiries regarding well conditions or sampling methods, or both. Figure 6 is an example of a Chevron geochemical log.

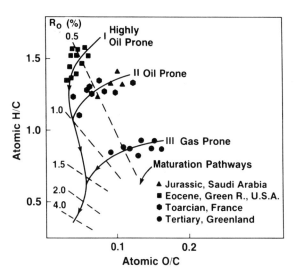

Figure 4—Plot of kerogen atomic H/C vs. O/C showing various hydrocarbon-generative types. Plot is calibrated to vitrinite reflectance (R_o). Thermal maturity is indicated by distance along converging maturation pathways, with the most mature samples in lower left corner. Mixed samples showing compositions between each pathway are common.

Very Immature Sediments

Pyrograms of immature sediments typically show poorly separated S1 and S2 peaks (the "trough" between them is well above the baseline), which can give anomalous S1 and PI results.

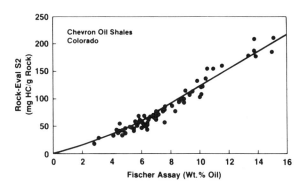

Figure 3—Plot of S2 vs. Fischer assay oil yield for Chevron oil shales, Colorado. Correlation line shows usefulness of pyrolysis in predicting quantity of organic compounds that could be generated on further maturation. Whereas Fischer assay requires about 100 g of rock and 1 hour per analysis, comparable results can be obtained in 20 min by pyrolysis of only 100 mg of rock.

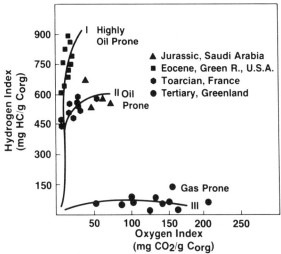

Figure 5—Plot of whole rock HI vs. OI showing various hydrocarbon-generative types. Pyrolysis of whole rock yields results comparable to those obtained by elemental analysis (Figure 4), which requires time-consuming and expensive kerogen isolation from comparatively larger volumes of rock.

Correlations between HI and atomic H/C or between OI and atomic O/C (Orr, 1983) can be particularly poor for immature sediments for various reasons. First, pyrolysis is done using whole rock whereas elemental analysis requires isolated kerogen. OM can be altered while preparing the kerogen (Durand and Nicaise, 1980), and minerals affect pyrolysis results (see below). The Rock-Eval FID does not measure hydrogen or water, which are important pyrolysis products of immature samples. Finally, the pyroproducts of different types of OM respond differently at the detector. The FID responds to C-H bonds and carbon mass. Thus, FID response will be approximately equivalent for one molecule of benzene or six molecules of methane, although the atomic H/C ratios of these molecules differ by a factor of four.

Variation in Types of Organic Matter

Mixtures of types of OM in rocks are common. The Type I, II, and III pathways on the HI vs. OI diagram (Figure 5) are only generalized; many samples plot between the kerogen evolution pathways. Kerogen evolution paths on an HI vs. OI diagram are actually broad bands. Type IV OM shows very low HI values, and plots near the bottom of the figure.

The TOC does not necessarily relate to pyrolysis response. Although higher TOC samples typically generate more pyroproducts (S2), this is not a rule because the types of OM vary between samples. Table 4 shows examples of TOC and pyrolysis results from the same well. The shallowest two samples show nearly identical TOC values, but their S1 and S2 responses differ by about an order of magnitude. As indicated by the HI and OI data, the sample from 1,950 ft (594 m) contains Type II (oil-prone) OM, whereas that from 1,975 ft (602 m) contains Type III (gas-prone) OM. These results were verified by atomic H/C ratios and by microscopy of the macerals in the kerogens. (Macerals are recognizable constituents of kerogen that can be differentiated as to type of OM by their morphology.) The sample from 1,950 ft (594 m) has an atomic H/C of 1.15, and microscopy indicates 90% amorphous OM. The sample from 1,975 ft (602 m) shows an atomic H/C of 0.75, and contains 85% structured and 10% inert OM. Determinations of type of OM can also be supported by carbon monoxide pyrograms (Daly and Peters, 1982).

Like pyrolysis, transmitted-light and reflected-light petrographic identifications of maceral composition are not foolproof. "Amorphous" OM is commonly equated with oil-prone Types I or II. However, not all amorphous OM shows good oil-generative potential (Tissot and Welte, 1984, p. 503). Ultraviolet fluorescence microscopy is particularly useful for identifying amorphous oil-prone OM. For example, the maceral interpretation for the sample from 1,950 ft (594 m) in Table 4 is based on the presence of recognizable algal material and amorphous OM that fluoresces in ultraviolet light.

Preliminary evidence (A. J. Holck and G. J. Demaison, 1982, personal communication) indicates that the Type III pathway on the HI vs. OI diagram of Espitalié et al (1977) in Figure 5 is incorrectly located. Figure 7 shows a contour plot of the percentage of amorphous OM for many samples from the Kimmeridgian-Volgian interval of the North Sea. As discussed, not all amorphous OM is oil prone. The high density of sample points showing a low HI (< 150 mg HC/g C_{org}) defines a gas-prone (Type III) pathway. Type IV OM probably follows a pathway along the base of the figure with very low HI values. From the limited data in Figure 7, one might also modify the pathways for Types I and II. However, further analyses will be required before these changes can be made with confidence.

Coaly Samples

Coals of higher plant origin (Type III) do not generally respond to pyrolysis in the same way as dispersed Type III OM. Therefore, HI vs. OI plots may misrepresent the type of OM. For reasons not fully understood, some coals known from elemental analysis and microscopy to be Type III plot between Type II and Type III kerogens on HI vs. OI diagrams (Figure 8). The result overestimates the liquid-hydrocarbon-generative potential for these types of coals. Typically, coals show HI values below about 300 mg HC/g C_{org} (low compared to Type II OM), with S2/S3 values greater than 5. When coal fragments are evident, OM type is best determined by atomic H/C vs. O/C diagrams and organic petrography.

The reasons for the anomalous behavior of certain Type III coals probably relate to differences in procedure and product detection between Rock-Eval pyrolysis and elemental analysis. The HI from pyrolysis is determined using whole rock, and only those products that are cracked in the range $300°-550°C$ are analyzed. The atomic H/C is determined using kerogen. Because the FID responds only to carbon mass and carbon-hydrogen bonds, common pyroproducts such as water and diatomic hydrogen are not included in the HI but are measured for the H/C. Further, the FID will respond approximately the same to six molecules of methane (H/C = 4) as to one molecule of benzene (H/C = 1).

Differences in the mode of detection may also explain anomalously low OI values for certain coals. Coals of low maturity (R_o less than about 0.6%) generate large amounts of carbon dioxide during pyrolysis, and OI values are generally consistent with the corresponding atomic O/C. However, certain more mature coals show OI values lower than expected based on the atomic O/C. At higher levels of maturity, proportionally more pyrolytic oxygen is released as carbon monoxide, which is not analyzed by the Rock-Eval. This may explain the nonlinear relationship between pyrolytic carbon dioxide and atomic O/C observed by Teichmüller and Durand (1983).

Altered Samples

Oxidized, highly mature, or Type IV kerogens usually show high T_{max} values or lack an S2 peak. Type IV OM can be confirmed by microscopy for maceral composition and vitrinite reflectance. These types of kerogen are common in siltstones and sandstones accessible to oxygenated ground waters. The sample from 2,007 ft (612 m) in Table 4 is an example of an immature rock that is dominated by recycled OM. Although this sample contains 1.04 wt. % TOC, it shows no pyrolysis peak, thus suggesting recycled OM. Organic petrography confirms that it contains over 60% inertinite.

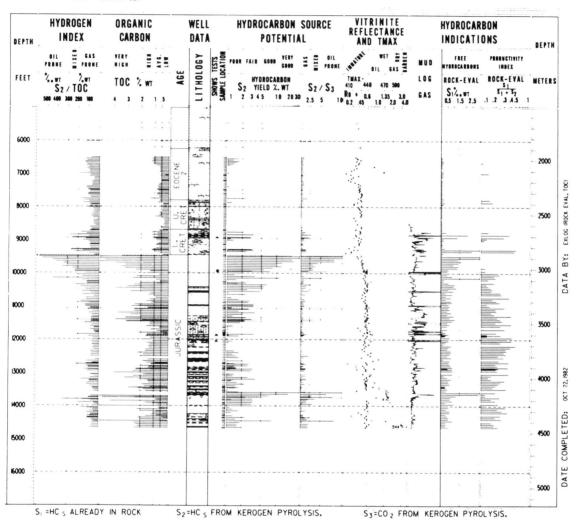

Figure 6—Example of Chevron geochemical log of Rock-Eval and TOC results from well X in North Sea. Log includes option to plot supporting data, such as lithology, vitrinite reflectance, and mud-log gas. Solid T_{max} symbols were determined using cuttings. Open T_{max} symbols were determined using cores; corresponding data in other columns are dashed. (Courtesy of G. J. Demaison.)

All types of OM are subject to oxidation during transport, deposition, and diagenesis. Oxidation tends to remove hydrogen and add oxygen to the kerogen (Durand and Monin, 1980, p. 127), changing the HI vs. OI plot.

Demaison and Bourgeois (1984) showed that the locations of samples along the Type II pathway on the HI vs. OI plot are influenced by thermal maturation and early diagenetic oxidation of the OM (Figure 9). The figure shows more mature ($R_o \approx 0.62\%$) Casablanca 5 samples that plot higher on path II than the less mature ($R_o \approx 0.52\%$) Casablanca 4 samples. The latter points reflect loss of hydrogen from the OM as a result of more oxygenated conditions during deposition and burial diagenesis.

Outcrop samples commonly show depletion in S1 and S2,

Table 4. Total Organic Carbon and Rock-Eval Pyrolysis Results for Selected Core Samples from a Well in Montana*

Depth (ft)	Description	TOC (wt. %)	S1	S2	S3	T_{max} (°C)	PI	HI	OI
1,950	dark-gray, laminated, calcareous shale	3.54	1.77	23.81	1.21	422	0.07	673	34
1,975	dark-gray, massive, calcareous shale	3.56	0.28	2.96	1.21	427	0.09	83	34
2,007	medium-gray, massive shale	1.04	0.04	0	0.55	—	0	0	53
2,073	black, fissile calcareous shale	2.43	0.09	0.56	0.62	432	0.14	23	26
2,076	medium-gray, calcareous shale	0.38	0.05	0.25	0.51	432	0.17	66	134
2,090	brown siltstone (oil stained)	0.61	3.61	4.08	0.12	415	0.47	669	20
2,146	medium-gray, massive shale	0.52	0.04	0.14	0.45	422	0.22	27	87

*Nomenclature:
S1 = mg HC/g rock
S2 = mg HC/g rock
S3 = mg CO_2/g rock
PI = S1/(S1 + S2)
HI = mg HC/g C_{org}
OI = mg CO_2/g C_{org}

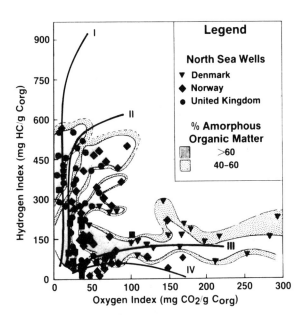

Figure 7—HI vs. OI plot showing contours of percent amorphous OM for Kimmeridgian-Volgian, North Sea. Figure suggests that locations of pathways for Types I, II, and III may require modification. In this figure, Type III pathway has been moved to fit the data, and Type IV pathway has been added (compare with Figure 5). (After A. J. Holck, 1982, personal communication).

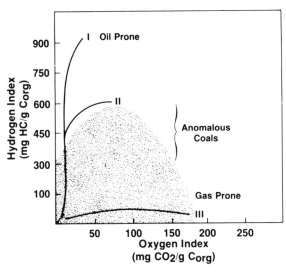

Figure 8—HI vs. OI plot showing range of coal compositions (stippled). Most coals show HI < 300 mg HC/g C_{org} and S2/S3 > 5.

and high S3 values due to weathering. Drying or heating of cuttings samples during storage can reduce the S1, as can solvents such as acetone. Growth of fungi on wet samples can seriously affect pyrolysis results by increasing S1, S2, and S3.

Contamination by Well Additives

Oil-based and water-based mud additives and lubricants can contribute to S1, S2, S3, and T_{max} of the indigenous OM (Figure 10). These types of contaminants generally lower T_{max}. Similar problems result from migrated oil or large amounts of bitumen (Clementz, 1979). Additional information is necessary to distinguish additives from natural, migrated contaminants. The well operator can usually provide samples of

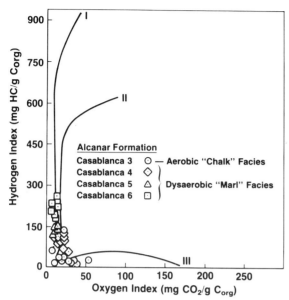

Figure 9—HI vs. OI plot for Alcanar Formation samples from four Spanish wells. The more mature ($R_o \approx 0.62\%$) Casablanca 5 samples plot higher on maturation pathway than less mature ($R_o \approx 0.52\%$) Casablanca 4 samples, which were subjected to more oxic conditions during diagenesis resulting in loss of hydrogen. (After Demaison and Bourgeois, 1984).

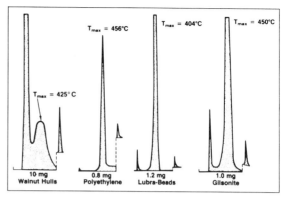

Figure 11—Pyrograms of particulate mud additives show that contaminants may contribute to S1, S2, S3, and T_{max} of rock. T_{max} of contaminants may be low (Lubra-Beads) or high (polyethylene). Note large S1 and S3 peaks generated from walnut hulls.

additives and the depths where they were used. Gas chromatography and other analyses can help resolve the problem.

Particulate contaminants such as walnut hulls, polyethylene, "Lubra-Beads," and gilsonite can raise or lower T_{max} (Figure 11). All organic contaminants tend to increase the HI.

Careful sample preparation is absolutely necessary for useful pyrolysis results. Most particulate additives can be identified and removed using a binocular microscope. Many of the water-soluble contaminants accompanying water-based drilling muds can be washed from samples prior to analysis. Oil-based contaminants are discussed below.

Figure 12 is a simplified geochemical log for the Cathedral River well, Alaska, showing the effects of contamination. A gradual increase in T_{max} with depth ends abruptly at about 8,500 ft (2,590 m). Further investigation showed that a synthetic polymer similar to Lubra-Beads (Figure 11) was injected at this depth, which accounts for the decrease in T_{max} and the increases in TOC and HI (R. W. Jones, 1983, personal communication).

At about 11,000 ft (3,350 m) on the log in Figure 12, T_{max} returns to expected values based on a vitrinite reflectance measurement ($R_o = 1.47\%$). However, at about 12,000 ft (3,660 m), T_{max} again drops sharply while both TOC and HI increase. The anomalous data resulted from a second injection of the polymer.

Contamination by Bitumen and Migrated Oil

Heavy ends of migrated oil and indigenous bitumen can affect the S2 and T_{max} of the kerogen (Clementz, 1979). Oil-based mud contamination or migrated oil is likely for immature rocks (T_{max} less than about 435°C) where PI or S1/TOC is greater than 0.2 or 0.3, respectively. The problem is minor when the ratio of bitumen or oil to kerogen is small. Heavy ends and oil-based additives can usually be eliminated by rinsing suspect samples with an organic solvent prior to analysis; however, the S1 will also be lost or reduced.

Natural contamination is most severe where oil has migrated into coarse-grained or fractured organic-lean rocks. If unrecognized, these rocks might be interpreted as prolific source rocks, based on their high HI values. The problem can be identified from the pyrogram. Pyrograms of contaminated or reservoir rocks, such as that from 2,090 ft (637 m) in Table 4 (Figure 13), commonly show an S1 peak greater than 2 mg HC/g rock, an anomalously high PI and low T_{max} (compared to adjacent samples), and a bimodal S2 peak. If this oil-stained

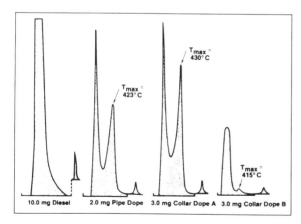

Figure 10—Pyrograms of oil-based mud additives show that contaminants may contribute to S1, S2, S3, and T_{max} of rock (time increases from left to right for each pyrogram). Diesel (left) shows large peak spanning range of both S1 and S2.

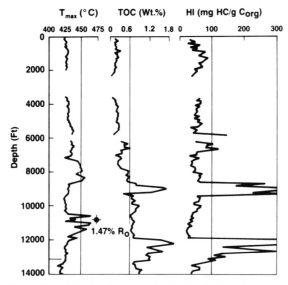

Figure 12—Abbreviated geochemical log for ditch cuttings from Cathedral River well, Alaska. Anomalous results for T_{max}, TOC, and HI at 8,500 and 12,000 ft (2,590 and 3,660 m) are due to injection of particulate mud additive. (Data from Exlog. Plotted by R. G. Huppi, December 28, 1982.)

siltstone was not recognized as containing migrated oil, the high HI (669 mg HC/g C_{org}) could be interpreted as indicating a prospective oil-prone source rock. The example shows the importance of pyrogram examination and information on lithology as a part of interpretation, especially if an anomalous condition exists.

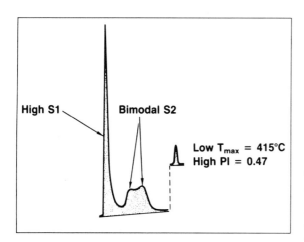

Figure 13—Pyrogram of siltstone contaminated by migrated oil. Pyrolysis data that suggest oil contamination include high S1 peak (>2 mg HC/g rock), bimodal S2 peak, low T_{max}, and high PI. Anomalously low T_{max} (415°C) was measured by Rock-Eval II on first of two S2 peaks. This lower S2 represents heavy ends of contaminating oil. Higher temperature S2 peak (right) is attributable to kerogen.

Determining Bitumen Content

Although S1 is an independent measure of the amount of bitumen in a sample, it does not correspond directly to solvent-extracted bitumen because of procedural differences. S1 represents organic compounds from C_1 to about C_{32} (Tarafa et al, 1983, p. 84) and is based primarily on separation by distillation temperature (up to about 400°C in the Rock-Eval) and the FID response factor of each compound. Extractable bitumen represents the weight percentage of organic compounds removed by an organic solvent. This percentage varies with solvent polarity, extraction method, sample size, and extraction time. Compounds below about C_{15} are lost during evaporation of the solvent. Claypool and Reed (1976) oven-dried pre-extracted samples for 2 hours at 105°C to remove light compounds up to about C_{15}. Using this method, they found a better correlation between extractable bitumen and S1.

Interference by Mineral Matrix

For argillaceous (clay-rich) rocks containing less than 0.5 wt. % TOC, HI values are likely to be too low and T_{max} too high because of adsorption of pyrolytic organic compounds onto the mineral matrix. This adsorption generally decreases in the following order: illite, montmorillonite, calcite, kaolinite (Espitalié et al, 1980). Dembicki et al (1983) showed that for synthetic mixtures of kerogens and minerals, up to 85% of the pyrolyzate (S2) can be retained by an illite matrix. Evidence suggests that expanding three-layer clays are the most active in retaining pyrolyzate and that they may also act as catalysts in converting some compounds to lower molecular weight (Davis and Stanley, 1982). Type III kerogen is most prone to this problem because it generates less pyrolyzate per gram of OM than do Type I or Type II.

Therefore, comparison of pyrolysis results for whole rock and kerogen from the same rock is unwise, particularly if the OM to clay ratio in the rock is low. Partial adsorption on clay in the whole rock delays escape of the pyrolyzate, thus resulting in a high T_{max}. Some of the adsorbed pyrolyzate is thermally converted to nonvolatile char, thus reducing the S2 and HI compared to that for clay-free kerogen pyrolysis. For example, pyrolysis of one clay-rich whole rock (TOC = 0.38%) yields an HI of 21 mg HC/g C_{org} and a T_{max} of 431°C. Pyrolysis of the corresponding kerogen yields an HI of 70 mg HC/g C_{org} and a T_{max} of 423°C. The whole rock and kerogen samples were weighed so that approximately equal amounts of carbon were pyrolyzed in each experiment. As discussed later, sample size can affect pyrolysis results.

The S1 and S2 peak sizes thus depend on (1) type of mineral matrix, (2) type of OM, and (3) the ratio of OM to matrix. If the samples under investigation are broadly uniform in composition, HI values should reflect changes in hydrocarbon-generative potential. The interpreter should always be aware of lithologic and TOC differences among samples.

Large numbers of core and ditch-cutting samples can be rapidly screened for further analyses using pyrolysis. We recommend pyrolysis of whole rock for screening, and pyrolysis of kerogen for detailed work.

Problems Associated with S3

Impurities, solid solution (Katz, 1983), and pyrolytic generation of organic acids generate carbon dioxide from carbonate below the maximum trapping temperature of 390°C. This inorganic carbon dioxide can result in anomalously high OI values, particularly for samples containing less than 0.5 wt. % TOC. Acidification of suspect samples with 2N HCl to remove carbonates improves S3 accuracy. This process should be done at low temperature to minimize loss of S1. One method of avoiding OI problems is to use HI vs. T_{max} (Espitalié et al, 1984) rather than HI vs. OI plots to differentiate maturation pathways.

Anomalous OI values for samples containing less than 0.5 wt. % TOC can result from adsorbed carbon dioxide or oxygen. Each Rock-Eval sample is purged with helium prior to pyrolysis. The observed S3 is sensitive to purge rate and time, probably because of small amounts of carbon dioxide and oxygen adsorbed on the sample, crucible, and plumbing. If the purge is incomplete, the adsorbed carbon dioxide, and carbon dioxide from partial combustion of the OM by adsorbed oxygen, can contribute to S3.

The S3 peak is particularly susceptible to instrumentation problems. Improper maintenance of the carbon dioxide trap can result in the accumulation of contaminants and the elution of contaminant gases that interfere with the integration of S3. The molecular sieve in the carbon dioxide trap must be routinely changed. Poor blanks can result from air leaks in the apparatus or from dirty crucibles. We wash all crucibles with reagent-grade acetone, and dry them at 110°C in a vacuum oven. Crucibles are handled with tongs.

Oxidation of OM, with resulting high S3 values, can occur because of exposure during weathering (outcrops), prolonged storage, or overgrinding during sample preparation. Early diagenetic oxidation was discussed previously.

Problems Associated with T_{max}

T_{max} increases regularly with depth in many wells. However, variations can result due to unconformities, faults, changes in geothermal gradient, and other factors described below.

Accuracy of T_{max} is about 1°–3°C, depending on the instrument, program rate, sample size, and positioning or electronic calibration of the thermistor in the oven. T_{max} values for samples with S2 peaks less than 0.2 mg HC/g rock are often inaccurate and should be rejected. Based on the general trend of increasing T_{max} in Table 4, samples at 2,090 and 2,146 ft (637 and 654 m) are anomalous. The low T_{max} for the sample at 2,090 ft (637 m) results from migrated oil (discussed above). The T_{max} for the sample at 2,146 ft (654 m) should be rejected because of the small S2 peak.

Variations in type of OM affect the T_{max} vs. depth trend, especially in rocks that have not entered the oil-generative window (Figure 14). The most extreme examples are immature rocks dominated by recycled OM that can still generate a small S2 peak. If the total S2 from the recycled OM greatly exceeds that from any accompanying primary OM, T_{max} can be much higher (40°C or more) than expected based on the actual matu-

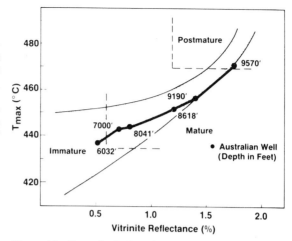

Figure 14—Generalized plot of T_{max} vs. vitrinite reflectance based on large number of analyses conducted at Chevron. Dashed lines separating maturity zones are from data in Table 3. Immature samples can show T_{max} variations up to about 20°C due to differences in type of OM. Data for the Australian well show smooth relationship between T_{max} and R_o because OM in the well is uniformly Type III.

rity of the sediments. The sample from 2,007 ft (612 m) in Table 4 is dominated by recycled OM, but no T_{max} was determined because of insufficient quantities of S2. If the OM had generated a small S2, the T_{max} would probably have been anomalously high.

Typically, samples of equivalent low maturity ($R_o < 0.6\%$) that are dominated by recycled OM show variations in T_{max} up to about 10°C. However, where the kerogen is composed of essentially one maceral, variations occur in T_{max} up to about 20°C (Figure 14). For example, the shallowest two samples in Table 4 are only 25 ft (8 m) apart, yet their T_{max} values differ by 5°C. Microscopy shows that the sample at 1,950 ft (594 m) consists of 90% oil-prone OM, whereas that at 1,975 ft (602 m) is 85% gas-prone OM. These differences decrease as the oil-generative zone is approached. T_{max} differences for mature or postmature OM are generally less than 5°C.

T_{max} values at the threshold of oil generation vary among petroleum source rocks because of differences in OM type. Most rocks enter the oil-generative window at about 435°C. Tissot et al (1978) concluded that the threshold of oil generation for oil-prone Type I kerogens is higher than others. Although generally true, we have observed exceptions to this rule, particularly for kerogens showing high sulfur levels. The resistance of Type I OM to thermal degradation may be due to strong cross-linkage of long, aliphatic chains and a general scarcity of thermally labile heteroatomic bonds. Oil-prone Type II kerogens do not always show higher T_{max} values than Type III kerogens. The samples from 1,950 and 1,975 ft (594 and 602 m) in Table 4 are examples. The simple description of kerogen as Type I, II, III, or IV appears insufficient to predict relative T_{max} values from pyrolysis.

The heavy ends of oil typically appear as S2 rather than S1 (Clementz, 1979), resulting in low T_{max} values and in extreme

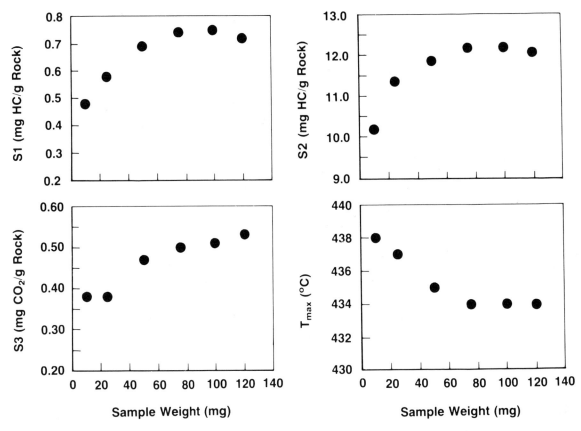

Figure 15—Rock-Eval II pyrolysis of sample COFRC 34212-768 (2.4% TOC) at various sample weights shows reduction of S1, S2, and S3, and increase of T_{max} when sample is less than 75 mg. Magnitude of these effects varies between instruments and is attributed to adsorption of pyroproducts to carbonaceous coating on exposed walls of crucible and plumbing.

cases, a bimodal S2 peak (Figure 13). This result can also be caused by solidified bitumen, such as gilsonite, or certain macerals, such as resinite.

Figure 11 shows the pyrogram of a gilsonite. Unlike most resinites, this material tends to cause high T_{max} values for rocks in which it is found. Resinite is a maceral of the liptinite group and consists of fossil tree resin. It is generally found in immature land-derived sediments ($R_o < 0.6\%$) and has a T_{max} typically in the range of 400°–420°C. Significant amounts of primary (indigenous) or secondary (migrated) resinite in a rock can severely reduce T_{max}. This reduction occurs when the S2 from the kerogen is minor compared to that from the resinite. For example, a marginally mature ($R_o = 0.54\%$) carbonaceous shale from 8,420 ft (2,566 m) in the COST 2 well, Norton Sound, Alaska, contains 6% resinite, and shows a T_{max} of only 422°C. For this type of sample where independent maturity information is available, T_{max} values up to 25°–30°C too low have been observed. For soluble resinite, more accurate T_{max} data can be obtained by solvent extraction prior to pyrolysis. However, most pyrobitumens and some resinites are not soluble in organic solvents.

The presence of pyrobitumens and pure macerals in cuttings is often indicated by the well-site geologist. Samples showing anomalous pyrolysis results should be checked by other analyses. Organic petrography can be used to identify these materials.

When bimodal peaks are well resolved, Rock-Eval assigns T_{max} to the largest S2 peak. If the two peaks are not well resolved, T_{max} is assigned to the first (heavy-ends) peak. T_{max} is shifted by up to 5°C for each peak in the direction of the other. Thus, the T_{max} of the lower temperature peak might be raised 5°C whereas that of the higher temperature peak could be reduced up to 5°C.

Adsorption of pyrolyzate by the mineral matrix is most pronounced in lean (TOC < 0.5 wt. %), argillaceous rocks. For this type of rock, heavy products of pyrolysis are retained and cracked on the clays (Espitalié et al, 1980), thus increasing T_{max} by up to 10°C. Transitions from carbonate-rich to clay-rich sections typically produce a sharp increase in T_{max}, although part of this shift could also be the result of differences in OM.

Organic-Lean Samples

Rocks containing less than about 0.5 wt. % TOC are most likely to be strongly affected by adsorption of pyrolyzate by the mineral matrix, resulting in reduced S2 and HI, and increased T_{max} and OI. Small S2 or S3 peaks can result in unreliable HI or OI values, particularly when the TOC is also low, because calculating S2/TOC and S3/TOC involves dividing one small number by another.

Sample Weight

All samples should be approximately the same weight (100 mg) because results for S2, T_{max}, and other parameters can be affected by sample size. Figure 15 shows an example of reduced S1, S2, and S3, and increased T_{max} when the sample weight is below a threshold. This threshold varies between instruments but typically is less than 75 mg. T_{max} for some small samples can be increased by up to 10°C above that of a 100-mg sample. Only a small part of the reduction in S1 and S2 for small samples is caused by nonlinearity of the FID. These effects appear largely because pyrolyzate is adsorbed on residual carbon that coats exposed surfaces of the crucible and plumbing.

Coals and organic-rich samples are a special problem because a 100-mg sample might overload (saturate) the FID. One solution is to dilute the coal with a comparatively inert matrix of pure carbonate, followed by pyrolysis of 100 mg of the mixture.

Although uncommon, some very rich samples generate enough pyrolyzate to saturate the FID, resulting in low S2 and T_{max} results. The interpreter should be aware of the linear range of the detectors. Linearity can be determined by pyrolyzing various weights of an organic-rich rock. For example, FID is saturated at the point where a plot of integrated S2 area counts vs. weight becomes nonlinear. Our Rock-Eval II becomes nonlinear for FID response at about 300,000 area counts.

CONCLUSIONS

Rock-Eval pyrolysis is a rapid method for screening large numbers of rock samples for further geochemical analyses. Additional information, readily available to the practicing geologist, that can be used to improve interpretation of the data includes: lithology, amount and type of OM, well conditions, and presence of contaminants or migrated oil. Accurate conclusions require an understanding of the pitfalls to interpretation, access to the pyrograms, and geochemical logs.

Pyrolysis was not intended to be used without supporting geochemistry. For critical samples, interpretations from pyrolysis results should be verified by other analyses, such as kerogen elemental composition, vitrinite reflectance, and gas chromatography.

REFERENCES CITED

Bienner, F., and J. Espitalié, 1980, Caracterisation de roches bitumineuses d'origines diverses par essais Fischer et correlations avec les resultats de pyrolyses du type "Rock-Eval": Institut Français du Pétrole, 17 p.

Claypool, G. E., and P. R. Reed, 1976, Thermal-analysis technique for source-rock evaluation: quantitative estimate of organic richness and effects of lithologic variation: AAPG Bulletin, v. 60, p. 608-626.

Clementz, D. M., 1979, Effect of oil and bitumen saturation on source-rock pyrolysis: AAPG Bulletin, v. 63, p. 2227-2232.

—— G. J. Demaison, and A. R. Daly, 1979, Well site geochemistry by programmed pyrolysis: Proceedings of the 11th Annual Offshore Technology Conference, v. 1, p. 465-470.

Daly, A. R., and K. E. Peters, 1982, Continuous detection of pyrolytic carbon monoxide: a rapid method for determining sedimentary organic facies: AAPG Bulletin, v. 66, p. 2672-2681.

Davis, J. B., and J. P. Stanley, 1982, Catalytic effect of smectite clays in hydrocarbon generation revealed by pyrolysis-gas chromatography: Journal of Analytical and Applied Pyrolysis, v. 4, p. 227-240.

Demaison, G., and F. T. Bourgeois, 1984, Environment of deposition of middle Miocene (Alcanar) carbonate source beds, Casablanca field, Tarragona basin, offshore Spain, in Petroleum geochemistry and source rock potential of carbonate rocks: AAPG Studies in Geology 18, p. 151-162.

Dembicki, H., Jr., B. Horsfield, and T. T. Y. Ho, 1983, Source rock evaluation by pyrolysis-gas chromatography: AAPG Bulletin, v. 67, p. 1094-1103.

Durand, B., and J. C. Monin, 1980, Elemental analysis of kerogens (C, H, O, N, S, Fe), in B. Durand ed., Kerogen: Paris, Editions Technip, p. 113-142.

—— and G. Nicaise, 1980, Procedures for kerogen isolation, in B. Durand, ed., Kerogen: Paris, Editions Technip, p. 35-53.

Espitalié, J., M. Madec, and B. Tissot, 1980, Role of mineral matrix in kerogen pyrolysis: influence on petroleum generation and migration: AAPG Bulletin, v. 64, p. 59-66.

—— F. Marquis, and I. Barsony, 1984, Geochemical logging, in K. J. Voorhees, ed., Analytical pyrolysis—techniques and applications: Boston, Butterworth, p. 276-304.

—— M. Madec, B. Tissot, J. J. Mennig, and P. Leplat, 1977, Source rock characterization method for petroleum exploration: Proceedings of the 9th Annual Offshore Technology Conference, v. 3, p. 439-448.

Katz, B. J., 1983, Limitations of "Rock-Eval" pyrolysis for typing organic matter: Organic Geochemistry, v. 4, p. 195-199.

Orr, W. L., 1983, Comments on pyrolytic hydrocarbon yields in source-rock evaluation, in M. Bjorøy et al, eds., Advances in organic geochemistry 1981: New York, Wiley, p. 775-782.

Peters, K. E., and B. R. T. Simoneit, 1982, Rock-Eval pyrolysis of Quaternary sediments from Leg 64, Sites 479 and 480, Gulf of California, in J. R. Curray et al, eds., Initial Reports of the Deep Sea Drilling Project, v. 64, p. 925-931.

—— J. K. Whelan, J. M. Hunt, and M. E. Tarafa, 1983, Programmed pyrolysis of organic matter from thermally altered Cretaceous black shales: AAPG Bulletin, v. 67, p. 2137-2146.

Tarafa, M. E., J. M. Hunt, and I. Ericson, 1983, Effect of hydrocarbon volatility and adsorption on source-rock pyrolysis: Journal of Geochemical Exploration, v. 18, p. 75-85.

Teichmüller, M., and B. Durand, 1983, Fluorescence microscopical rank studies on liptinites and vitrinites in peat and coals, and comparison with results of the Rock-Eval pyrolysis: International Journal of Coal Geology, v. 2, p. 197-230.

Tissot, B. P., and D. H. Welte, 1984, Petroleum formation and occurrence: New York, Springer-Verlag, 699 p.

—— G. Deroo, and A. Hood, 1978, Geochemical study of the Uinta basin: formation of petroleum from the Green River Formation: Geochimica et Cosmochimica Acta, v. 42, p. 1469-1486.

—— B. Durand, J. Espitalié, and A. Combaz, 1974, Influence of nature and diagenesis of organic matter in formation of petroleum: AAPG Bulletin, v. 58, p. 499-506.

Identification of Source Rocks on Wireline Logs by Density/Resistivity and Sonic Transit Time/Resistivity Crossplots[1]

B. L. MEYER[2] and M. H. NEDERLOF[2]

ABSTRACT

Source rock formations generally show a lower density, a lower sonic transit time, and a higher resistivity than other sediments of equal compaction and comparable mineralogy. This phenomenon can be used to identify source rocks on wireline logs provided the source rocks have a minimum thickness within the resolution of the sondes used and are sufficiently rich in organic matter. Classification rules have been established to assist in the recognition of source rocks on a combination of logs.

Because of the low density contrast between water and organic matter, the method becomes inaccurate at high water saturations (i.e., low compaction). Within limits, the amount of organic matter contained in a sediment can be estimated from log anomalies.

When source rocks become mature, free oil is present in addition to kerogen, and the resistivity increases by a factor of 10 or more.

INTRODUCTION

To estimate the hydrocarbon charge (total amount of oil or gas generated) of a given basin or prospect, it is necessary to define the source rock volume, yield, and state of maturity. Source rocks are commonly shales and limestones that contain organic matter. This organic matter may be derived from aquatic organisms and bacteria (Lijmbach, 1975), in which case it is commonly called kerogenous, or from land plants. Such material is referred to as humic (Tissot and Welte, 1978). Various mixtures of these two constituents may occur. In Shell's experience, source rocks containing kerogenous organic matter are usually laminated; humic matter, with the exception of coal, is usually dispersed through the sediment. When heated to a certain temperature, commonly as the result of deep burial, the source rocks will release oil and/or gas and are then considered to be mature for oil and/or gas generation, respectively.

The amount and type of organic matter and its maturity are determined in the laboratory by chemical and microscope analysis of sample sets. For a review of the analytical methods, the reader is referred to Tissot and Welte (1978).

©Copyright 1984. The American Association of Petroleum Geologists. All rights reserved.
[1]Manuscript received, January 10, 1983; accepted, July 13, 1983.
[2]Shell Internationale Petroleum Mij. B.V., P.O. Box 162, 2501 The Hague, Netherlands.
We are indebted to Shell Internationale Petroleum Mij. B.V. for permission to publish this paper. We gratefully acknowledge the assistance received from our colleagues in Shell Research.

At Shell, we use an in-house method of pyrolysis calibrated against the total organic carbon content and oil yield as determined by a modified Fisher assay. Sample sets from wells, however, are generally incomplete or are biased in favor of reservoir formations. Moreover, the chemical analysis of large suites of samples is time consuming, expensive, and not always conclusive, especially for ditch cuttings.

Because the most complete and readily available well data are wireline logs, it seemed desirable to investigate the possibility that source rocks might be recognizable from wireline logs. The results of this investigation are described in this paper.

The data available for this study were standard Schlumberger logs: gamma ray, density/neutron, sonic, and a variety of resistivity logs. From the Shell Research Laboratory, Rijswijk, Netherlands, we had available routine pyrolysis data and, from a considerable number of cores, organic carbon content and oil yield (modified Fisher assay).

HYPOTHESIS

Sediments can be regarded as consisting of heavy and light fractions. The heavy fraction is the mineral matter, and the light fraction is the formation fluids. In source rocks, the contained organic matter is also part of the light fraction. During compaction, water is expelled; consequently, the density increases and the sonic transit time decreases. Because of the presence of organic material, source rocks retain a greater amount of the light fraction than organic-lean sediments, and thus appear on sonic and density logs to be somewhat less compacted. Organic matter, like the mineral matrix, is normally not electrically conductive. Organic-rich sediments, therefore, have a higher resistivity than organic-lean sediments. These principles are illustrated in Figure 1. The overlying and underlying shales have a much higher bulk density than the source rock interval. Furthermore, the formation resistivity increases substantially in the source rock interval as compared with the shales above and below. The maximum organic content in this illustrated source rock is 20% (see also Figure 8), whereas the organic content of the overlying shale is only 0.5%.

GAMMA-RAY LOGS

Organic-rich rocks can be relatively highly radioactive, that is, they can have a higher gamma-ray reading than ordinary shales and limestones (Schmoker, 1981) (Figures 2-4). Correlations of gamma-ray readings with organic content indicate that the higher radioactivity levels are related to the presence of organic matter. Our own investigations in the North Sea (Figure 2) show that this radioactivity is due to uranium enrichment. We postulate that plankton absorb uranium ions that are generally present in sea water together with other trace elements, and that uranium thus becomes concentrated in the source rocks. Lacustrine source rocks (Figures 5-7), in contrast, have no gamma-ray anomaly owing to the scarcity or absence of uranium ions in fresh water.

DENSITY LOG

The density log measures the bulk density (ρ_b) of the formation. This density consists of the combined matrix density (ρ_{ma}) and the fluid density (ρ_f). The more fluid a formation contains, the more porous it is. It follows, therefore, that

$$\phi = \frac{\rho_{ma} - \rho_b}{\rho_{ma} - \rho_f}.$$

In shales with a similar degree of compaction, all other factors being equal, water saturation also should be equal.

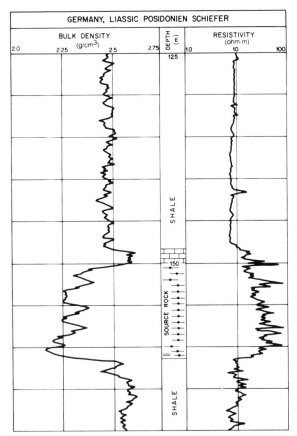

Figure 1—Density and resistivity of the Liassic Posidonien Schiefer in shallow well in south Germany. Resistivity is from Microlaterolog. High resistivity peaks in shale are thin limestone intercalations.

It follows, therefore, that if $\rho_{source\ rocks}$ is smaller than ρ_{shale}, it must be a function of the amount of organic matter present:

$$\text{Vol. \% organic matter} = \frac{\rho_{shale} - \rho_{source\ rock}}{\rho_{shale} - \rho_{organic\ matter}}.$$

The value of $\rho_{organic\ matter}$ is approximately the same as ρ_{water} (i.e., + 1 g/cm³).

The linear relationship between oil yield and bulk density of the source rock has been described by Tixier and Curtis (1967) for the Green River Shale, and by Schmoker (1979) for the Ohio oil shale. Our own investigations arrived at similar results (Figure 8).

Figures 1 and 9 illustrate the reduction in bulk density of source rocks compared with the formations above and below. The vertical resolution of the FDC sonde is approximately 2 ft (60 cm). At shale densities of 2.25 g/cm³ or greater, the minimum concentration of organic matter for the density log to respond is about 1%. Bulk density readings may not be a reliable indication of organic matter in the presence of larger concentrations of heavy minerals (pyrite). Rugose boreholes also render density readings unreliable.

SONIC LOGS

Like density logs, sonic logs also show the difference in compaction between organic-lean sediments and source rocks (Figures 2-7, 9). Sonic logs, therefore, can be used to determine source rocks qualitatively: a relative decrease in sonic transit time and an increase in resistivity indicate an organic-rich layer in nonpermeable sediments. Where density logs are affected by rugosity of the borehole wall or by the presence of pyrite, sonic logs may prove more reliable than density logs. Therefore, it is always useful to make both sonic/resistivity and density/resistivity crossplots (Figures 10, 11) for source rock identification.

Because the interval transit time is affected by the water/organic matter ratio, mineral composition, carbonate/clay content, and grain-to-grain pressure, sonic logs cannot be used alone to estimate the organic content of source rocks. They can be used, however, if correlated with density logs or, even better, if calibrated with cores.

RESISTIVITY LOGS

In principle, any resistivity log may be used to evaluate impermeable formations such as source rocks. Generally, shallow and deep-penetration logs should give the same reading (Figures 3-5, 9). Discrepancies may occur as a result of borehole effects, anisotropy of the formation (e.g., different vertical and horizontal resistivities), and tool characteristics (e.g., depth of investigation, vertical resolution). As source beds are generally thin discrete layers even in thicker source rock formations, richness can vary considerably in a vertical direction, and maximum vertical resolution, therefore, is desirable.

Source rocks are generally laminated and thus are electrically anisotropic. This anisotropy increases the resistivity value measured by spherically focused logs (Figures 2, 7).

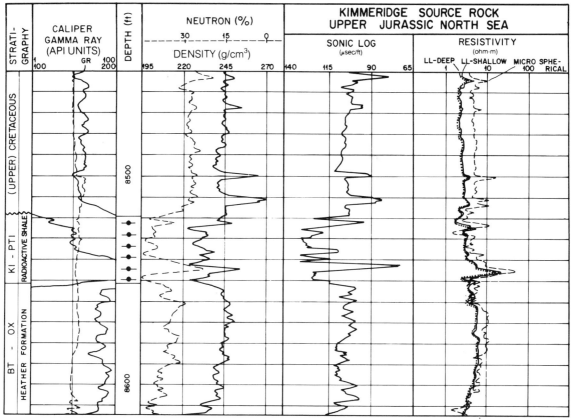

Figure 2—Composite log of Kimmeridge Shale, North Sea. Note high gamma-ray reading. Resistivity is only marginally higher than overlying shale, but density and sonic transit time are considerably lower. Underlying Heather Formation is very silty.

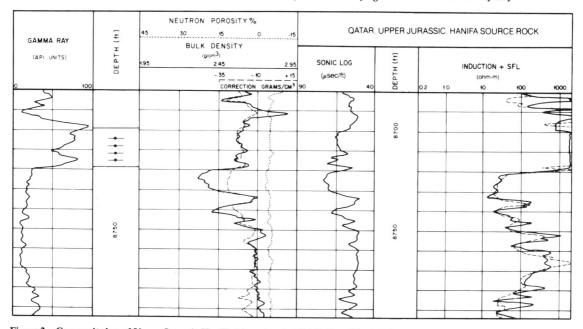

Figure 3—Composite log of Upper Jurassic Hanifa Limestone in Middle East. No shale is present in this section. Gamma-ray anomaly is due to presence of organic matter. Note high resistivity, offscale. At this well, source rock is mature and generating oil as indicated by core analysis.

When source rocks become mature, free oil is present in voids and fractures (Du Rochet, 1981). This phenomenon has been observed in cores of the Hanifa (Figure 3) and other formations. With maturity, the resistivity of a source rock increases by a factor of 10 or more. Thus resistivity can be used as a maturity indicator for a given source rock formation.

Resistivity Correction for Temperature

The resistivity measured is dependent on temperature. In the cases presented, formation temperatures ranged from 70° to 230°F (21° to 110°C). It was necessary, therefore, to calculate the resistivity at one standard temperature (75°F, 24°C). This calculation was done using the Arps

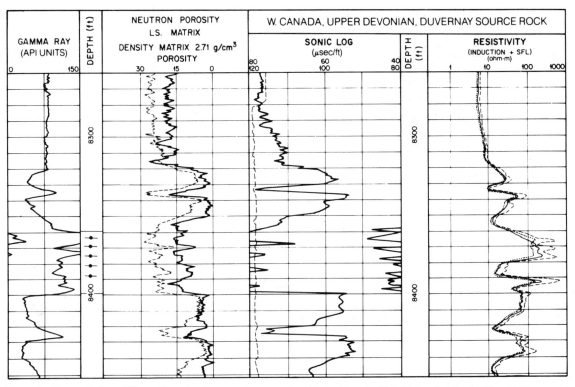

Figure 4—Composite log of Upper Devonian Duvernay source rock, Alberta, Canada. Note high resistivity. In this well, Duvernay is mature and probably has already expelled some oil.

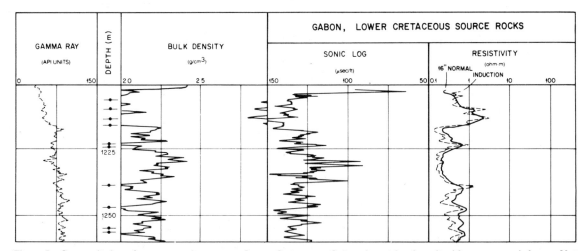

Figure 5—Composite log of a source rock sequence, Lower Cretaceous, Gabon, lacustrine deposit with many organic layers. Note that gamma-ray log does not respond to presence of organic-rich layers.

formula (Schlumberger, 1972):

$$R_T = R_{75} \times 82/(T + 7)$$

or

$$R_{75} = R_T \times (T + 7)/82,$$

where T is the formation temperature (in °F) at the depth concerned. These temperatures are derived from a gradient calculated from the bottom-hole temperatures of the various logging runs.

THIN-BED EFFECT

The recognition of thin source rock layers is dependent

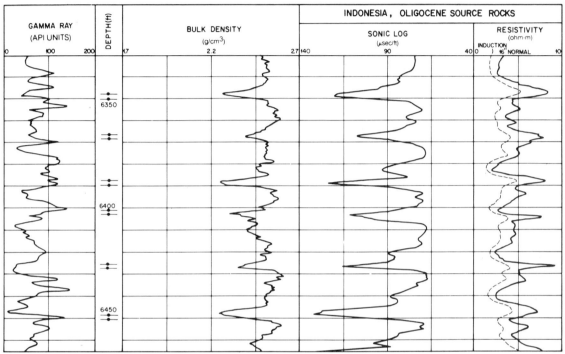

Figure 6—Composite log of an Oligocene source rock sequence, Indonesia lacustrine deposit. Note cyclical occurrence of organic-rich layers and absence of a gamma-ray response.

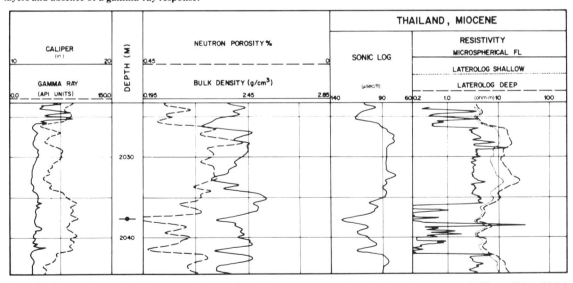

Figure 7—Composite log of a Miocene section in Thailand. Note log response of a thin, lacustrine source rock (30 cm, 12 in., thick in core) intercalated between organic-lean shales.

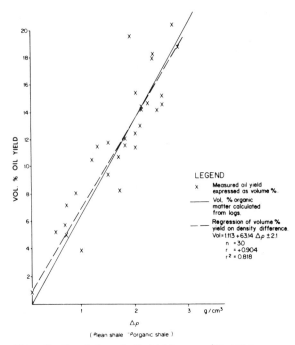

Figure 8—Correlation of oil yield from modified Fisher assay and bulk density. Value of $\Delta\rho$ is density difference between organic-lean shale and organic-rich shale (Liassic, Posidonien Schiefer, Germany; see also Figure 1). Vol. % oil yield is calculated using following parameters: 1 U.S. gal = 3.785 L; 1 short ton = 0.907 metric ton; density = 2.45–2.55 g/cm^3.

on the vertical resolution of the tools used. Among porosity tools, density logs have the best vertical resolution (approximately 2 ft or 60 cm), whereas microresistivity logs have a resolution of a few inches (5 cm).

Figure 7 shows the log response of a 30-cm (12-in.) thick source rock interval in a core. As the formation was soft, the interval is probably somewhat thicker in situ. This example illustrates perhaps the thinnest source rock layers measurable.

STATISTICAL ANALYSIS

The purpose of this statistical analysis was to establish simple classification rules for separating source rocks from non-source rocks on the basis of quantitative wireline log parameters.

The data set consisted of 169 intervals from 15 wells in 9 different countries where, apart from some missing data, gamma-ray, sonic-Δt, density, and resistivity data were available. The source rock data come from the following formations:

Oman	Ara Shale	Precambrian
Canada	Duvernay formation	Devonian
Netherlands Germany	Posidonomya Shale	Lower Jurassic
North Sea	Kimmeridge Shale	Upper Jurassic
Qatar	Hanifa Limestone	Upper Jurassic
Gabon	Melania Shale	Lower Cretaceous
Venezuela	La Luna Limestone	Upper Cretaceous
Indonesia	Talang Akar equivalent	Oligocene
Germany	Ries Crater fill	Upper Miocene

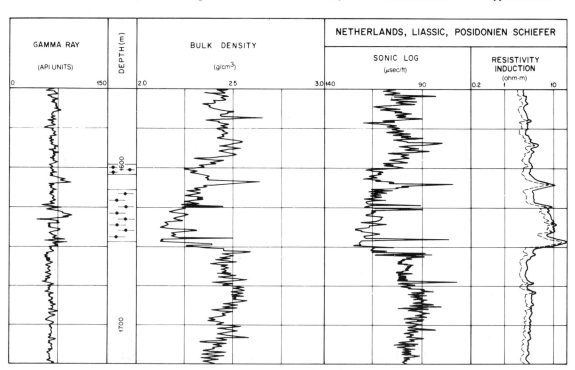

Figure 9—Composite log of Liassic Posidonien Schiefer in Netherlands. Note low radioactivity of this bituminous shale. This shale also has thin limestone intercalations.

Non-source rock data were measured from overlying and underlying strata.

The data points were subdivided into two classes on the basis of geochemical sample analyses: class 1 = source rocks (71 samples) and class 2 = non-source rocks (98 samples). The geochemical analyses were based mainly on pyrolysis, and a cut-off value of approximately 1.5% total organic carbon was used to classify samples into source rock or non-source rock.

Using various combinations of log parameters, a discriminant analysis was performed using the scheme of pseudoregression (Kendall, 1961). Assuming the observations (source rocks and non-source rocks) are plotted on a graph such as Figure 10, the log parameters (density, resistivity) are the coordinates used to locate the observations. For classification, the best possible line separating class 1 from class 2 has to be established. For calculation purposes, it is better to find the equation of a line which is perpendicular to that used for classification. This equation is called the discriminant function. The observations can be projected onto the discriminant line (and hence moved in a direction parallel to the classification line). The chief property of the discriminant function is that the distance between the means of the class 1 and class 2 projections is maximized whereas at the same time the spread of points within the classes is minimized (e.g., see Davis, 1973, p. 442).

Kendall's (1961) method consists of assigning a dummy value of Y to each observation: $N2/(N1 + N2)$ for the class 1 observations and $-N1/(N1 - N2)$ for the class 2 observations (N1 = source rocks, N2 = non-source rock). Then Y is regressed on the X variables (log parameters) using any convenient multiple-regression program. The resulting "pseudoregression" equation is then the discriminant function, where Y is replaced by D.

This analysis results in a linear equation for the discriminant score D. The log parameters used were transformed taking common logarithms. Therefore, a typical equation for the sonic/resistivity combination for shales is:

$$D = -6.906 + 3.186 \log_{10} \Delta t + 0.487 \log_{10} R75,$$

where Δt is in μsec/ft and R_{75} is the resistivity corrected to a standard temperature of 75°F (24°C).

To test for source rock, Δt and the resistivity at 75°F (24°C) have to be entered into the equation to calculate D. If D is positive, the rock is a probable source rock; if D is negative, the rock is probably barren; and if D is zero, the case remains undecided. The various log combinations and the effect of inclusion or exclusion of limestones in the sample gave rise to 8 different results (Table 1).

For the two-variable cases of sonic/resistivity and density/resistivity, crossplots have been made showing the classification boundary based on the discriminant for-

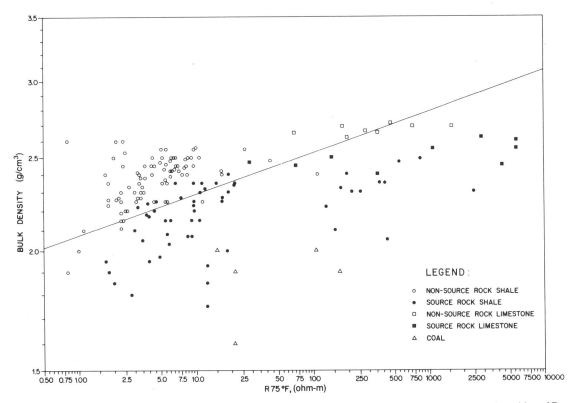

Figure 10—Bulk-density and resistivity crossplot. Density and resistivity plotted on logarithmic scale. Oblique line is position of D = 0 (discriminant analysis). Points below this line (D = positive) = source rocks; points above this line (D = negative) = no source rocks. Misclassification occurs at low shale densities, borderline cases, and high limestone resistivities.

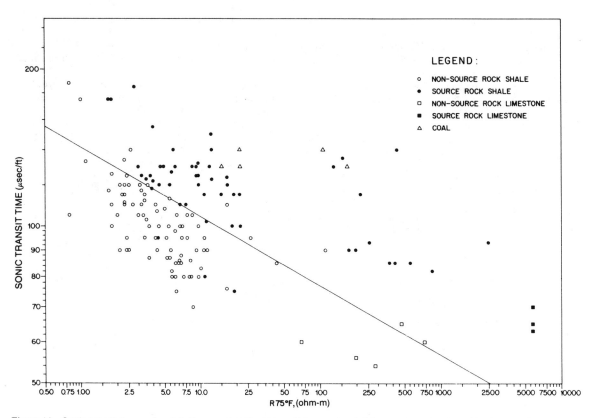

Figure 11—Sonic transit time and resistivity crossplot. Sonic transit time and resistivity are plotted on logarithmic scale. Oblique line is the position of D = O (discriminant analysis). Points below this line (D = negative) = no source rock; points above this line (D = positive) = source rock. Note that most misclassifications occur at low interval velocities, borderline cases, and high limestone resistivities.

Table 1. Results of Discriminant Analysis

Description	Shales Only				All Lithologies			
	G/S/R	G/D/R	S/R	D/R	G/S/R	G/D/R	S/R	D/R
Sample size (N)	142	142	150	151	151	160	159	169
Source rock (N1)	55	55	60	61	58	65	63	71
Non-source rock (N2)	87	87	90	90	93	95	96	98
% Misclassification	7.7	7.7	8.7	8.6	7.9	6.9	10.1	13.0
Separation of class means in terms of pooled standard deviation	2.42	2.63	2.18	2.26	2.51	2.70	2.22	2.18
Class 1 mean	.358	.382	.320	.330	.369	.381	.327	.312
Class 1 standard deviation	.319	.299	.314	.330	.313	.298	.302	.316
Class 2 mean	−.227	−.242	−.214	−.224	−.230	−.260	−.214	−.226
Class 2 standard deviation	.176	.189	.185	.166	.178	.184	−.197	.182
Coefficients:[2]								
Beta 0[3]	−8.094	0.817	−6.906	2.278	−7.758	0.794	−7.335	2.512
Beta gamma ray	0.739	0.856	–	–	0.671	0.825	–	–
Beta sonic	3.121	–	3.186	–	3.015	–	3.410	–
Beta density	–	−7.524	–	−7.324	–	−7.385	–	−7.920
Beta resistivity	0.399	0.292	0.487	0.387	0.412	0.315	0.453	0.339

[1]Results are tabulated for different wireline log combinations:
G = gamma ray (in $\log_{10}$ of API units)
S = sonic (in $\log_{10}$ of μsec/ft)
D = density (in $\log_{10}$ of g/cm^3)
R = resistivity at 75°F (in $\log_{10}$ of ohm-m)
[2]All beta coefficients of the discriminant functions are statistically significant. T-test values for individual beta coefficients range from 4.5 to 11.7.
[3]Beta 0 is the intercept. A typical equation based on the third column is $D = -6.906 + 3.186\log_{10}\Delta t + 0.487\log_{10}R_{75}$. $D < 0$: non-source rock; $D > 0$: source rock; $D = 0$: indeterminate.

mula, which roughly separates the source rocks from the non-source rocks (Figures 10, 11).

The results of the analysis show that such a classification will not always be correct. Where the equation was used to classify the 150 samples in the learning set, five source rocks were classified as barren, and eight lean shales were classified as source rocks. This method gave about 9% misclassifications, or 91% confidence in the ability to correctly classify. Thus the results have to be considered in a probabilistic sense. The D scores calculated for the learning set result in two approximately normal distributions for class 1 and class 2 (each of which has its own mean and standard deviation), but which overlap to some extent. Such distributions can be employed to obtain the probability of having a source rock (Geer, 1971).

The percentage of misclassifications was derived by applying the discriminant function to the learning data set from which it was derived. Cross-validation, although not conducted, would undoubtedly have shown that the classification performance would be somewhat lower for new samples (e.g., see Diaconis and Efron, 1983).

Misclassifications may occur for several reasons.

1. A miscorrelation between chemical and petrophysical data can occur because of the thin-layer effect. An individual layer, from which samples for chemical analysis were taken, may be organically lean even though intercalated in a source rock sequence, and vice versa.

2. Heavy minerals such as pyrite can produce high density readings if sufficiently concentrated.

3. In undercompacted soft formations, the physical contrast between layers that are rich or lean in organic matter becomes minimal, particularly when organic concentrations are low.

4. In high density/high velocity formations such as certain limestones and dolomites, the resistivity correction may be exaggerated.

CONCLUSION

Density/resistivity and sonic/resistivity crossplots can be used to identify source rock intervals on wireline logs. The crossplots are a useful tool in the evaluation of frontier exploration wells as well as in basin evaluation. Only a low possibility exists for error. Even so, it is suggested that, wherever possible, the log interpretation method should be corroborated by geochemical sample analysis.

REFERENCES CITED

Davis, J. C., 1973, Statistics and data analysis in geology: New York, John Wiley & Sons, 550 p.

Diaconis, P., and B. Efron, 1983, Computer-intensive methods in statistics: Scientific American, v. 248, n. 5 (May 1983), p. 116-130.

du Rochet, J., 1981, Stress fields, a key to oil migration: AAPG Bulletin, v. 65, p. 74-85.

Geer, V. D., 1971, Introduction to multivariate analysis for the social sciences: San Francisco, W. H. Freeman & Co., 239 p.

Kendall, M. G., 1961, A course in multivariate analysis: London, Macmillan, 185 p.

Lijmbach, G. W. M., 1975, On the origin of petroleum: 9th World Petroleum Congress, Tokyo, Proceedings, v. 2, p. 357-369.

Schlumberger, 1972, Log interpretation charts: Houston, Schlumberger Limited, 92 p.

Schmoker, J. W., 1979, Determination of organic content of Appalachian Devonian shales from formation density logs: AAPG Bulletin, v. 63, p. 1504-1509.

―――― 1981, Determination of organic-matter content of Appalachian Devonian shales from gamma-ray logs: AAPG Bulletin, v. 65, p. 1285-1298.

Tissot, B. P., and D. H. Welte, 1978, Petroleum formation and occurrence: Berlin, Springer-Verlag, 521 p.

Tixier, M. P., and M. R. Curtis, 1967, Oil shale yield predicted from well logs: 7th World Petroleum Congress, Mexico City, Proceedings, v. 3, p. 713-715.

A geochemical reconstruction of oil generation in the Barrow Sub-basin of Western Australia

JOHN K. VOLKMAN*, ROBERT ALEXANDER†, ROBERT IAN KAGI, ROHINTON A. NOBLE and GARRY WAYNE WOODHOUSE

Petroleum Geochemistry Group, School of Applied Chemistry, Western Australian Institute of Technology, South Bentley, Western Australia, 6102, Australia

(Received January 20, 1983; accepted in revised form August 26, 1983)

Abstract—A suite of crude oils and petroleum source rock extracts from the Barrow Sub-basin of Western Australia have been analysed for biological marker compounds by capillary GC-MS, and for volatile hydrocarbons by whole oil capillary GC. These analyses were used to calculate values for twenty-three biomarker parameters in order to assess aspects of source type, maturity, migration and biodegradation of the hydrocarbons.

The crude oils had a source in the Upper Jurassic Dingo Claystone formation. These hydrocarbons accumulated in the reservoir sands and in some cases were biodegraded. Several accumulation and biodegradation episodes have been recognised while the basin continued to subside, which resulted in a suite of oils showing marked differences in composition.

INTRODUCTION

THERE HAVE BEEN many recent advances in the use of biological marker (biomarker) hydrocarbons, such as steranes, triterpanes and aromatic steroids, for the correlation of crude oils (LEYTHAEUSER et al., 1977; SEIFERT, 1977; SEIFERT and MOLDOWAN, 1978, 1979, 1981; JI-YANG et al., 1982; MCKIRDY et al., 1982; PHILP et al., 1982; WELTE et al., 1982). The relative proportions of some biomarkers, such as pristane and phytane (POWELL and MCKIRDY, 1973a), and C_{27}, C_{28} and C_{29} steranes (MEINSCHEIN and HUANG, 1981) provide information about the type of organic matter from which the petroleum was formed. Other biomarkers provide a record of the severity of thermal maturation from the extent of such reactions as epimerisation at chiral centres, skeletal rearrangement, carbon-carbon bond cleavage and aromatisation. Other biomarkers can be used to determine the extent to which an oil has been biodegraded and water washed in the reservoir (SEIFERT and MOLDOWAN, 1979; WELTE et al., 1982).

Although much information can be obtained from the distribution of polycyclic alkane biomarkers in petroleum, a detailed reconstruction of the history of oil generation and accumulation in a basin often requires information derived from the distributions of structurally less complex, lower molecular weight alkanes. This integrated approach is particularly important when some of the oil accumulations have been biodegraded. Polycyclic alkane biomarkers provide little information in cases of mild to moderate biodegradation, whereas the distributions of low molecular weight alkanes provide a sensitive indicator of this phenomenon (e.g. WELTE et al., 1982). Conversely, the light alkanes provide little information about severe biodegradation, but specific stages in this process can be recognized from changes to the distributions of polycyclic alkanes brought about by preferential biodegradation of some components (e.g. VOLKMAN et al., 1983).

In this paper we describe the application of an integrated approach to crude oil correlation which uses information derived from polycyclic alkane distributions, saturated alkane fractions and whole oil chromatograms. The geological setting of the study area is one in which the reservoir sands have undergone continuous subsidence since deposition. These reservoirs appear to have been charged with petroleum at various times and depths of burial, and each accumulation has been biodegraded to a different extent. As a consequence, the crude oils recovered contain complex mixtures of hydrocarbons which pose a challenging problem for geochemical correlation.

GEOLOGICAL SETTING

The Mesozoic Barrow Sub-basin is one of several geological provinces within the Carnarvon Basin of Western Australia. This basin is an elongate crustal depression which straddles the central Western Australian coastline from 20°S to 29°S. The geology of this area has been reviewed by CRANK (1973), THOMAS and SMITH (1974, 1976) and THOMAS (1978). A structural cross-section, showing major oil drilling sites, is shown in Fig. 1. A generalised stratigraphy is shown in Fig. 2.

During the Early to Middle Jurassic, a northeast to southwest rift was formed by the breaking up of Gondwanaland. A period of progressive sedimentation then followed so that by the Early Cretaceous over 5000 m of deltaic, fluvial and marine sediments had been deposited. The Upper Jurassic Dingo Claystone and Neocomian Barrow Formation were deposited mostly as rift-infill sediments in distal and pro-delta environments. The Dingo Claystone consists mainly of siltstones and claystones, with minor sandstones, dolo-

---

* Present address: CSIRO Marine Laboratories, Division of Oceanography, GPO Box 1538, Hobart, Tasmania, 7001.
† Author to whom correspondence should be addressed.

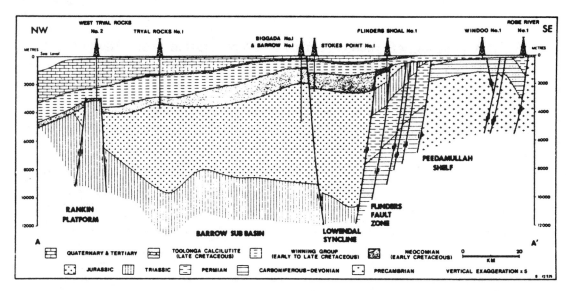

FIG. 1. Structural cross-section of the Barrow Sub-basin from south-east to north-west (after THOMAS and SMITH, 1976).

mites and limestones. The initiation of deltaic sedimentation is marked by the fine-grained sandstones of the Dupuy Member of the Dingo Claystone. These grade into the coarser-grain sandstones and interbedded siltstones of the overlying Barrow Formation.

The Cretaceous Winning Group sediments were deposited on the continental margin during a marine transgression which developed following separation of the continental masses. WISEMAN (1979) has suggested that an unconformity at the base of this Group, marking the change in sedimentation from deltaic to continental shelf, developed due to a fall in sea level of about 200 m during the earliest Neocomian. As marine conditions became established, the Muderong Shale was deposited. Within this marine unit of shale and siltstone, several thin bands of sandstone occur which are present-day reservoirs of oil and gas. The Mardie Greensand Member, which is also referred to as the Muderong Greensand (CRANK, 1973), forms the basal part of the Muderong Shale and consists of glauconitic, silty sandstone. In some areas of the upper part of the Muderong Shale there is a shallow-water marine sandstone, the Windalia Sand Member, which has its maximum development at Barrow Island. The upper sections of the Winning Group consist of the Windalia Radiolarite, which was deposited during a period of high primary productivity, and the Gearle Siltstone. Later Cretaceous and Tertiary sediments consist mainly of carbonate rocks deposited on a regionally subsiding continental margin.

In May 1964, the Barrow 1 well was drilled on Barrow Island and was completed in the Dupuy Member of the Dingo Claystone as a Jurassic '6700 ft. sand' (2040 m) oil discovery in August 1964. Subsequent drilling located small deposits of highly paraffinic oil at depths of 1830 and 1890 m, within the Early Cretaceous Barrow Group and at 2010 m in the Late Jurassic. Highly naphthenic-aromatic oils were also located in the Mardie Greensand (Muderong oil) at 850 m and in the Windalia Sand Member (Windalia oil) at 670 m. The main producing reservoir in the Barrow Island oil field is within the Windalia Sand Member, with minor production from the Mardie Greensand and Gearle Siltstones. Gas accumulations are found throughout the sediment column, and condensates have been recovered from several Middle Jurassic reservoirs.

Most previous studies of oils from the Barrow Sub-basin concluded that at least two different sources must be involved to account for the two very different types of crude oil composition observed (PARRY, 1967; POWELL and McKIRDY, 1972, 1973b; THOMAS and SMITH, 1974, 1976; THOMAS, 1978; ALEXANDER et al., 1980). There has been general consensus, based on a study of the geology of the area (CRANK, 1973), that the deeper paraffinic oils are most likely derived from sediments within the Jurassic Dingo Claystone (see Fig. 2), although geochemical data confirming this have not been presented. Many authors have suggested a Cretaceous source for crude oils recovered from the Winning Group reservoirs (e.g. THOMAS and SMITH, 1974; THOMAS, 1978; ALEXANDER et al., 1980), but Cretaceous sediments at Barrow Island have low vitrinite reflectance values (COOK and KANTSLER, 1980), so any oil produced would be very immature. ALEXANDER et al. (1981) demonstrated from stable carbon isotope data that Cretaceous sediments could be the source of Windalia oil, but the possibility that these oils might have been produced from the Upper Jurassic Dingo Claystone was not examined. Cretaceous sediments downdip, for example within the Barrow Formation in the Tryal Rocks well (Fig. 1), are sufficiently mature to have generated oil (COOK and KANTSLER, 1980),

TERTIARY	MIOCENE	CAPE RANGE GROUP		
	EOCENE	GIRALIA CALCARENITE		
	PALEOCENE	CARDABIA GROUP		
CRETACEOUS	SENONIAN	TOOLONGA CALCILUTITE		
	TURONIAN CENOMANIAN	UPPER GEARLE SILTSTONE		WINNING GROUP
	ALBIAN	LOWER GEARLE SILTSTONE		
	APTIAN	WINDALIA RADIOLARITE		
	LATE NEOCOMIAN EARLY	WINDALIA SAND MEMBER	MUDERONG SHALE	●
		MARDIE GREENSAND MBR		●
		BARROW FORMATION		●
JURASSIC	TITHONIAN	DUPUY MEMBER		
	KIMMERIDGIAN OXFORDIAN	DINGO CLAYSTONE		

FIG. 2. Generalized stratigraphy of the Barrow Sub-basin. (●) Denotes oil accumulations.

but migration through the relatively impermeable Windalia Sandstone is thought to be unlikely (THOMAS and SMITH, 1976). However, migration may have occurred through the Birdrong Sands (COOK and KANTSLER, 1980).

GOULD and SMITH (1978), and later PHILP et al. (1981, 1982), suggested that the compositional differences are due to biodegradation of the shallower oils. Stable carbon isotope data show that there are extensive secondary biogenic carbonate deposits, indicative of bacterial activity, associated with some parts of the Windalia reservoir, but not with the deeper Jurassic reservoirs (GOULD and SMITH, 1978). Although bacterial activity is thought to be very low in the Cretaceous reservoirs at the present time (J. PARRY, pers. commun.), extensive biodegradation could well have occurred until fairly recent times.

EXPERIMENTAL

Samples studies

Crude oils. Three paraffinic oils from the 2040, 2010 and 1980 m reservoirs and two naphthenic/aromatic oils from the Muderong and Windalia reservoirs were obtained from wells drilled on Barrow Island by West Australian Petroleum Limited (WAPET). Two non-commercial biodegraded crude oils, Flinders Shoal and Mardie oils, from wells drilled elsewhere in the sub-basin, were also studied. Detailed chemical data for these latter two oils have been presented previously (VOLKMAN et al., 1983). Geological data for each of the Barrow Sub-basin oils are given in Table 1. The Flinders Shoal oil was obtained from a well drilled midway between Barrow Island and the coast at Robe River. This reservoir is located in the Birdrong Sands at 700 m which is stratigraphically related to the basal section of the Muderong Shale at Barrow Island (THOMAS and SMITH, 1974). This oil was chosen as an example of a severely biodegraded and water-washed crude oil, thought to be from the same source as Windalia oil (THOMAS, 1978). Mardie oil was obtained from a well drilled onshore at Robe River. This more severely biodegraded and water-washed oil occurs in the Mardie Greensand at a present depth of only 77 m.

Source rocks. Hydrocarbon extracts of eight core samples from the Barrow No. 1 well were investigated using GC and GC-MS. These sediments ranged in depth from 307 m to 2835 m and included examples of sediments deposited from the Late Jurassic to Late Cretaceous. These sediments were chosen to provide a representative example of most of the sediment types found in the Sub-basin. Previous work (ALEXANDER et al., 1981) had shown that the Cretaceous sediments have good potential for oil generation, although they are immature, whereas the Jurassic sediments are sufficiently mature to have generated crude oil. Selected data are presented for three samples; more detailed analyses of these and other cores from the Carnarvon Basin will be presented elsewhere. Core 10 is a siltstone from 1035 m within the Neocomian Barrow Group, below the Muderong reservoir. The present temperature at this depth is 64°. Core 34 and Core 37 were obtained from 2402 m and 2835 m respectively, within the Upper Jurassic Depuy Member of the Dingo Claystone. Sediments at depths corresponding to Core 10, Core 34 and Core 37 have vitrinite reflectance values of 0.31, 0.62 and 0.73 respectively (COOK and KANTSLER, 1980).

Compound isolation and identification

Saturated and aromatic hydrocarbon fractions were isolated from 100 mg samples of the crude oils by column chromatography on silicic acid. The saturated hydrocarbons were eluted with 20 ml of redistilled pentane and the aromatic hydrocarbons by further elution with 20 ml of 1:1 pentane: diethyl ether. An aliquot of the 'total saturates' was also treated with activated 5 Å molecular sieves to separate the n-alkanes from branched and cyclic alkanes. Each fraction was analysed by capillary gas chromatography (GC) and capillary gas chromatography-mass spectrometry (GC-MS). An HP 5880 gas chromatograph, fitted with either a 20 m × 0.2 mm ID WCOT fused-silica OV 101 capillary column (Hewlett Packard) or a 50 m × 0.2 mm ID WCOT fused-silica SE 30 capillary column (SGE Australia), was used for all analyses. The sample in pentane was injected in the splitless mode at 40°, and after 5 min the oven was heated to 290° at 4°/min. H_2 at a linear velocity of 28 cm/sec was used as the carrier gas. Injector and FID detectors were 270 and 320° respectively. Chromatograms of the whole oil were obtained using the 50 m SE 30 column by injecting 0.1 to 0.3 μl of the oil in the split mode at 30°. After a 5 min isothermal period the oven was heated to 280° at 4°/min. During this initial isothermal period the chart speed was 1 cm/min, compared with 0.5 cm/min for the rest of the analysis, to facilitate recognition of the closely eluting peaks.

GC-MS analyses of the saturated hydrocarbon fractions were carried out using an HP 5895B capillary GC-quadrupole MS-computer data system. The GC-MS was fitted with a 50 m × 0.2 mm ID WCOT fused-silica OV 101 capillary column (Hewlett Packard) connected directly to the ion source. Each sample was analysed in the multiple ion detection (MID) mode, with a cycle time of 0.57 sec and dwell times of 10 m sec for each of the 20 ions monitored. Typical MS operating conditions were: EM voltage 2400 V; electron energy 70 ev; source temperature 250°. Selected samples were also analysed in the full data acquisition mode by scanning from 50 to 450 amu in 1.3 sec cycles. Compounds were identified by comparison of their retention times and full mass spectra with literature data (WARDROPER et al., 1977; SEIFERT and MOLDOWAN, 1979, 1980; MACKENZIE et al., 1980; RULLKÖTTER and WENDISCH, 1982) and by coinjection with standards.

RESULTS AND DISCUSSION

Previous studies (e.g. POWELL and MCKIRDY, 1972, 1973b) demonstrated that oils from the Barrow Sub-basin can be separated into two major categories depending on their chemical composition. Oils from Late Jurassic or Early Cretaceous (Neocomian) reservoirs, such as the 2040 m, 2010 m and 1890 m sands, are all light (36 to 40° API gravity), highly paraffinic crudes. Those from the Cretaceous Winning Group sands,

Table 1. Geochemical Data for Barrow Sub-basin Oils

Petroleum	Type	Reservoir Depth (m)	Reservoir Age	Reservoir Temperature(°C)*	API** Gravity
Barrow 2040 m	Paraffinic oil	2040	Late Jurassic	103	37.2
Barrow 2010 m	Paraffinic oil	2010	Late Jurassic	102	39.6
Barrow 1890 m	Paraffinic oil	1890	Neocomian	97	36.1
Muderong	Naphthenic oil	850	Aptian	68	33
Windalia	Naphthenic oil	700	Aptian	65	36
Flinders Shoal	Naphthenic oil	700	Aptian	57	23
Mardie	Naphthenic oil	77	Aptian	25	19

* Data from COOK and KANTSLER (1980), GOULD and SMITH (1978) and THOMAS (1978)
**Data from CRANK (1973) and THOMAS (1978)

such as Windalia, Muderong and Flinders Shoal oils, are typically heavy, naphthenic-aromatic crudes almost devoid of *n*-alkanes. The only reliable way of determining whether these oils are related, that is, derived from a similar source rock formation, is by a comparison of their distributions of polycyclic alkanes. Barrow Sub-basin oils contain very low concentrations of these hydrocarbons, suggesting that these oils were generated at high levels of maturity with consequent extensive cracking of hydrocarbons with polycyclic ring systems (ALBRECHT *et al.*, 1976). It was thus necessary to use highly sensitive GC-MS selected ion monitoring techniques to obtain sufficient data for comparisons of the oils to be made. Mass fragmentograms of characteristic ions were used for these comparisons: m/z 217 and 218 for steranes, m/z 259 for diasteranes, m/z 191 for hopanes, moretanes and other triterpanes, and m/z 177 for demethylated triterpanes.

Polycyclic alkane biomarkers for oil correlation

Source bed recognition. In qualitative terms the close similarity of the sterane distribution (m/z 217, Fig. 3) in all of the oils indicates that each was derived from the same source or very similar sources. Unfortunately, there are comparatively few biomarkers for assessing the degree of source correlation

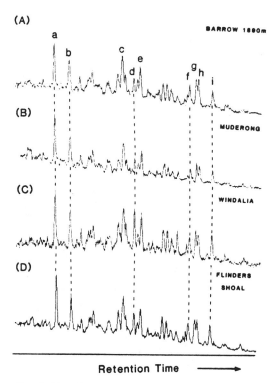

FIG. 3. Mass fragmentograms of m/z 217 (steranes plus diasteranes) for: (A) Barrow 1890 m oil, (B) Muderong oil, (C) Windalia oil and (D) Flinders Shoal oil. Peaks a and b are C_{27}(20S)- and (20R)-13β(H),17α(H)-diasteranes; peaks c and e are C_{29}(20S)- and (20R)-13β(H),17α(H)-diasteranes; peak d is C_{27}(20R)-5α(H),14α(H),17α(H)-sterane; peaks f and i are C_{29}(20S)- and (20R)-5α(H),14α(H),17α(H)-steranes; peaks g and h C_{29}(20R)- and (20S)-5α(H),-14β(H),17β(H)-steranes.

in a quantitative way. The proportion of C_{27}, C_{28} and C_{29} steranes in an oil is related to the type of organic matter in the source rocks and the palaeoenvironment of deposition (MEINSCHEIN and HUANG, 1981). A high proportion of C_{29} steranes is expected to occur in oils derived from sediments containing mostly terrigenous organic matter, reflecting the predominance of precursor C_{29} sterols in higher plants. In most of the Barrow Sub-basin oils, the C_{29} steranes are only slightly more abundant than the C_{27} steranes (A; Table 2) which is consistent with a mixed marine and terrigenous source. An exception to this generality is Windalia oil which contains more C_{27} steranes than C_{29} steranes (Table 2). This implies that marine organic matter comprised a larger proportion of the organic matter in the sediments which produced this oil, but it does not necessarily indicate a different source formation to that which produced the other Barrow Sub-basin oils.

The ratio of C_{27} to C_{29} steranes varies from 0.54 in the Cretaceous Core 10 to values of 1.0 and 1.3 for the two cores from the Upper Jurassic Dingo Claystone (A, Table 2). The latter values are sufficiently close to those of the oils to suggest that the oils could all have been generated from sediments within the Dingo Claystone formation. Further evidence for this is the very close similarity of the entire sterane distribution in the mature Core 37 with the distributions in the oils (compare Figs. 3 and 4B).

Another parameter which should be useful for source correlation is the ratio of C_{27} to C_{29} 13β(H),17α(H)-diasteranes (B, Table 2). This ratio varies from 1.2 to 1.7 in the oils, and from 0.9 to 1.5 in the three core samples. It is significant that the values of the two Dingo Claystone sediments encompass the values calculated for most of the oils (B, Table 2), which provides a further indication that this formation could have been the source of the oils. An interesting feature of these data is that the ratios of C_{27} to C_{29} diasteranes are considerably higher than the ratios of C_{27} to C_{29} steranes in all three cores, and in all the oils with the exception of Windalia. Since the ratio of C_{27} to C_{29} steroidal hydrocarbons is not thought to be significantly affected by maturity differences (MEINSCHEIN and HUANG, 1981), this phenomenon probably reflects some feature of the early diagenetic history of steroidal compounds in the sediment.

The presence or absence of specific triterpenoids, other than the ubiquitous 17α(H)-hopanes and moretanes (Fig. 5), can in some cases be used for source correlations. A C_{28} triterpane, identified as 28,30-bisnorhopane from its retention time and mass spectrum (SEIFERT *et al.*, 1978), is abundant in many of the Cretaceous sediments in the Barrow Sub-basin, as exemplified by the triterpane distribution in the Cretaceous Core 10 (peak b, Fig. 6A). This unusual triterpane does not appear to be common in Jurassic cores from the Sub-basin, and none could be detected in Core 37 from the Dingo Claystone (Fig. 6B). The Barrow Sub-basin oils contain either no 28,30-bisnorhopane or a trace amount (Fig. 5), which is again consistent with this formation being the source of the oils. The close correspondence between the m/z 191 mass fragmentograms for the oils and that of Core 37 offers further support for this proposal. Two other triterpanes, peaks c and d (Figs. 5 and 6), are present in all the oils (Fig. 5) and in both sediments (Fig. 6). These triterpanes have not been identified, but they are conspicuous in the triterpane distributions of Australian coals (unpublished data), suggesting that they may be general indicators of terrigenous organic matter.

Maturity. A large number of biomarker parameters have been derived to measure the thermal maturity of the organic matter at the time when the oil migrated from the source rocks. Two measurements are available from the sterane distributions: the ratios of C_{29}(20S)-5α(H),14α(H),17α(H)-sterane/C_{29}(20R)-

Table 2   Polycyclic Alkane Biomarker Parameters

Parameter:	A	B	C	D	E	F	G	H	I	J	K
Mass Fragmentogram:	217	259	217	217	259	191	191	191	191	259 / 217	191
Type of Measurement*	a	a	b	b,c	b	b	b	b	b	c	d
Oil											
Barrow 2010 m	0.80	1.7	1.1	0.94	1.8	58	0.13	0.39	0.69	0.55	0.02
Barrow 1890 m	0.94	1.2	1.0	0.76	1.8	60	0.11	0.36	0.57	0.68	0.14
Muderong	0.80	1.5	0.80	0.94	1.5	57	0.13	0.36	0.69	0.64	0.40
Windalia	1.4	1.3	0.78	1.2	1.5	57	0.13	0.62	0.69	0.72	0.49
Flinders Shoal	0.95	1.4	0.78	0.80	1.6	59	0.11	0.39	0.53	0.72	0.57
Cores											
10 1036 m	0.54	0.9	0.27	0.33	1.2	32	0.19	1.9	0.15	1.2	0.01
34 2402 m	1.0	1.2	0.81	0.62	1.6	58	0.15	1.5	0.10	0.41	0.02
37 2835 m	1.3	1.5	1.2	0.63	1.4	59	0.10	0.38	0.44	0.31	0.0

* a: Source; b: Maturity; c: Migration; d: Residues

A: C_{27} (20R)-5α(H),14α(H),17α(H)-sterane / C_{29} (20R)-5α(H),14α(H),17α(H)-sterane
B: C_{27} (20R)-13β(H),17α(H)-diasterane / C_{29} (20R)-13β(H),17α(H)-diasterane
C: C_{29} (20S)-5α(H),14α(H),17α(H)-sterane / C_{29} (20R)-5α(H),14α(H),17α(H)-sterane
D: C_{29} (20R)-5α(H),14β(H),17β(H)-sterane / C_{29} (20R)-5α(H),14α(H),17α(H)-sterane
E: C_{27} (20S)-13β(H),17α(H)-diasterane / C_{27} (20R)-13β(H),17α(H)-diasterane
F: C_{31} (22S)-17α(H)-homohopane / (same + C_{31} (22R)-17α(H)-homohopane) x 100
G: C_{30} 17β(H),21α(H)-moretane / C_{30} 17α(H),21β(H)-hopane
H: C_{27} 17α(H)-22,29,30-trisnorhopane / C_{27} 18α(H)-22,29,30-trisnorneohopane (T_m/T_s)
I: T_s / C_{30} 17α(H)-hopane
J: C_{29} (20R + 20S)-13β(H),17α(H)-diasteranes / C_{29} 5α(H)-steranes
K: C_{29} 17α(H)-25-norhopane / C_{29} 17α(H)-30-norhopane.

5α(H),14α(H),17α(H)-sterane (C; Table 2) and C_{29}(20R) - 5α(H),14β(H), - 17β(H) - sterane/C_{29}(20R)-5α(H),14α(H),17α(H)-sterane (D; Table 2). Both ratios increase with increasing thermal maturity, with parameter C tending towards a limiting value between 1.1 and 1.3 corresponding to equilibration between the two isomers (SEIFERT and MOLDOWN, 1979, 1981; MACKENZIE et al., 1981b, 1982). The values for ratios C and D (Table 2), show that all of the Barrow Sub-basin petroleum liquids are mature, although the paraffinic oils appear to be slightly more mature than the naphthenic/aromatic oils. From the values of these ratios it is apparent that the Cretaceous core is much less mature than the oils, and thus it is extremely unlikely that Cretaceous sediments in the vicinity of Barrow Island could have generated the oils accumulated there. Even more mature Cretaceous shales downdip in the Tryal Rocks area have sterane maturity parameters which are still appreciably less than those observed for the oils. The values for Core 34 and Core 37 bracket those of the oils (C and D, Table 2) indicating that the oils were generated at temperatures of about 130°, the present temperature of Core 37 sediment, corresponding to a vitrinite reflectance value of about 0.73. Sediments of Jurassic age which have been subjected to temperatures of 130° for extended periods of time usually have a corresponding vitrinite reflectance in excess of 1.0 (HUNT, 1979). The low value of 0.73 observed in these samples is probably due to high recent heat flow in the basin. In such cases the vitrinite reflectance does not give an accurate measure of the stage of evolution of petroleum from kerogen.

The ratio of 20S to 20R diasteranes (E, Table 2) increased with increasing thermal maturity (MACKENZIE et al., 1980), so this ratio also provided a measure of thermal maturity, complementary to the ratio of 20S to 20R steranes (C, Table 2). A major difference, however, is that parameter E takes on values greater than 1.0, even in immature sediments such as Core 10 which reflects the mechanism by which diasterenes are formed from sterenes (RUBINSTEIN et al., 1975; SIESKIND et al., 1979). Values for the paraffinic oils are marginally higher than the values for the naphthenic/aromatic oils, which is consistent with slightly higher degree of thermal maturity. Values for the two Dingo Claystone sediments are comparable to those in the naphthenic/aromatic oils as would be expected. It is intriguing that the ratio of 20S to 20R epimers in the diasteranes attains much higher values than the comparable ratio in the steranes (compare C and E, Table 2), and yet in both compound classes the epimerisation occurs at the equiv-

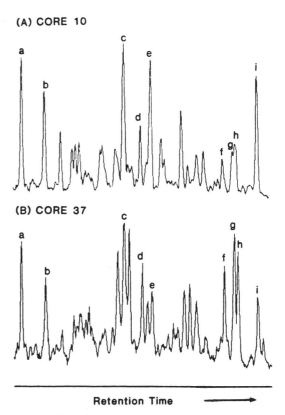

FIG. 4. Mass fragmentograms of m/z 217 (steranes plus diasteranes) for: (A) immature Cretaceous Core 10 and (B) mature Upper Jurassic Core 37. Steranes and diasteranes are identified in the caption to Fig. 3.

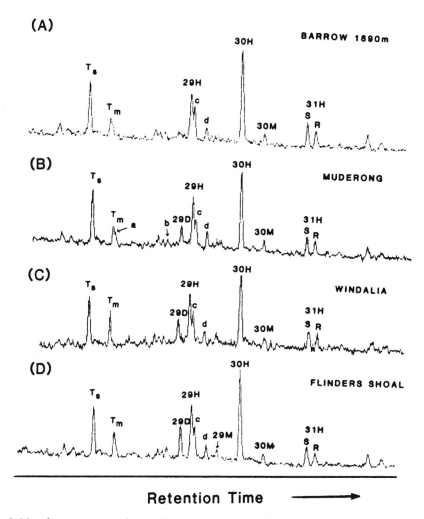

FIG. 5. Mass fragmentograms of m/z 191 (triterpanes) for (A) Barrow 1890 m oil, (B) Muderong oil, (C) Windalia oil and (D) Flinders Shoal oil. $17\alpha(H)$-hopanes are denoted by H, moretanes by M, 25-nor-$17\alpha(H)$-hopanes by D. Numerals indicate carbon number. $T_s = 18\alpha(H)$-22,29,30-trisnorneohopane; $T_m = 17\alpha(H)$-22,29,30-trisnorhopane; a = C_{30} tetracyclic triterpane; b = 28,30-bisnorhopane; c and d were not identified. S and R refer to the configuration at C-22.

alent chiral centre. MACKENZIE et al. (1980) attribute this to steric factors, but this remains unproven.

The close similarity of the triterpane distributions in all of the oils (Fig. 5), provides further support for the contention that all the oils are mature and that they were probably generated from the same source formation. Values for four maturity parameters (F to I) are presented in Table 2.

Parameter F measures the amount of the 22S epimer of the C_{31} extended hopane (homohopane) as a proportion of the total C_{31} 22S and 22R epimers present. Mature source rocks and crude oils have values of about 60% (MACKENZIE et al., 1980), as found for all the Barrow Sub-basin oils. Only very immature oils, such as some from California (SEIFERT and MOLDOWAN, 1979), have values less than 55%. As expected, the immature Cretaceous Core 10 has a low value for this ratio (32%), whereas the Dingo Claystone sediments have values comparable to those for the oils (F, Table 2).

The ratio of $17\beta(H),21\alpha(H)$-moretanes to their corresponding $17\alpha(H),21\beta(H)$-hopanes (G, Table 2), decreases with increasing thermal maturity from a value of about 0.8 in immature sediments to values of less than 0.15 in mature oils and source rocks (MACKENZIE et al., 1980; SEIFERT and MOLDOWAN, 1980). The values for this ratio are virtually identical in all of the oils, and are comparable to values expected for mature oils (SEIFERT and MOLDOWAN, 1978). This ratio shows the expected decrease with sediment depth (G, Table 2), with the oils having values between those for Core 34 and Core 37. A more sensitive maturity indicator for mature oils and source rocks is the ratio of two C_{27} triterpanes, $17\alpha(H)$-22,29,30-trisnorhopane and $18\alpha(H)$-22,29,30-trisnorneohopane (Structures VIII and VII in the Appendix). This ratio is usually labeled T_m/T_s, following the convention introduced by SEIFERT and MOLDOWAN (1978). In some samples a C_{30} tetracyclic triterpane (a, Fig. 5) occurs which coelutes with T_m on some capillary columns (e.g., RULLKÖTTER and WENDISCH, 1982), although it was resolved on the high resolution 50 m capillary column used in the present study. This can lead to anomalous T_m/T_s ratios, so we recommend that the ratio should also be determined from m/z 370 molecular ion mass fragmentograms. Using our GC-MS system, the ratios measured from the m/z 370 mass fragmentograms were a factor of 1.6 greater than the ratios measured from m/z 191 mass fragmentograms, reflecting the different relative intensities of the m/z 191 and 370 ions in the mass spectra of these two hopanes. With the exception of Windalia oil, all of the oils, have very similar maturities, based on the values of this ratio (H, Table 2), which is almost

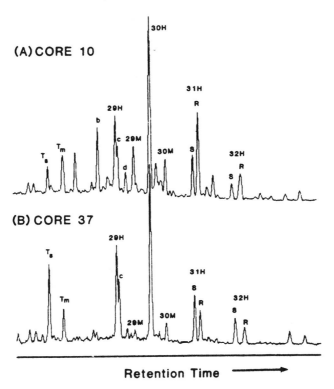

FIG. 6. Mass fragmentograms of m/z 191 for (A) immature Cretaceous Core 10 and (B) mature Upper Jurassic Core 37. Identifications are given in the caption to Fig. 5.

identical with the maturity of the organic matter in Core 37. The value for Windalia oil is about 50% higher than those for the other oils, indicating a slightly lower maturity, but note that this ratio has values less than 1.0 only in very mature oils and source rocks (cf. Core 34, Table 2).

MCKIRDY et al. (1982) suggested that ratios of various hopanes differing in carbon-number, such as C_{29}/C_{30} and C_{34}/C_{35} $17\alpha(H)$-hopanes, can be used as source parameters. In contrast, SEIFERT and MOLDOWAN (1978) proposed that the ratio of "primary terpanes" C_{29} and C_{30} $17\alpha(H)$-hopanes to C_{27} and C_{28} "secondary terpanes" such as T_s, T_m and 28,30-bisnorhopane increases with maturity. Although the proportions of the various hopanoid triterpanes depend to some extent on the type of organic matter, in very mature samples thermal cracking of carbon-carbon bonds will tend to produce triterpanes with lower carbon numbers. Accordingly we propose the use of the ratio of the more stable C_{27} hopane T_s to the C_{30} $17\alpha(H)$-hopane (I, Table 2) as a measure of the maturity of very mature oils and condensates. Our rationale for this proposal is the cleavage of the side-chain of C_{29} and higher carbon number $17\alpha(H)$-hopanes will produce the C_{27} hopane T_s. An extreme example of this phenomenon is afforded by a very mature condensate from the Barrow Sub-basin which has a value for this ratio of 4.9 (unpublished data). The increasing importance of side-chain cracking with increase in thermal maturity has been documented previously in the case of aromatic steroids (MACKENZIE et al., 1981a). All of the oils have similar values for this ratio, ranging from 0.53 to 0.69 which is very much higher than the very immature Cretaceous core and the moderately mature Upper Jurassic Core 34.

Migration. There are few biomarker parameters for assessing whether an oil has migrated over significant distances, and unambiguous measurements are not yet available. SEIFERT and MOLDOWAN (1981) proposed that a plot of the ratio of 20S to 20R $5\alpha(H)$-steranes (C, Table 2) against the ratio of $5\alpha(H),14\beta(H),17\beta(H)$-steranes to $5\alpha(H),14\alpha(H),17\alpha(H)$-steranes (D, Table 2) can be used to assess whether the composition of an oil has been significantly altered by migration. A comparison of the sterane parameters C and D for the Barrow Sub-basin oils (Table 2) suggests that these oils have not migrated over long distances. SEIFERT and MOLDOWAN (1978) contend that diasteranes migrate faster than $5\alpha(H)$-steranes through surface-active sediments such as clays, and thus the ratio of diasteranes to steranes can be used to assess whether an oil is migrated. However, ENSMINGER et al. (1977) have shown that this ratio also increases with increasing maturity.

Since C_{29} diasteranes coelute with C_{27} steranes on apolar capillary columns and so their abundance could not be determined directly from the m/z 217 mass fragmentogram. Therefore an indirect method was used to calculate this ratio (J, Table 2). Thus, their peak areas in the m/z 217 mass fragmentogram were calculated from those of the two well-resolved C_{27} $13\beta(H),17\alpha(H)$-diasteranes using the C_{27}/C_{29} diasterane ratio calculated from the m/z 259 mass fragmentogram (B, Table 2). The proportions of diasteranes to steranes (J, Table 2) is virtually identical in each of the oils, as expected if they were all generated from the same source formation at a similar level of maturity. It is, however, interesting that the values for Core 34 and Core 37 are both lower than the values for the oils, and yet we have established that they have comparable maturities. This could reflect different rates of migration for steranes and diasteranes, but it could simply reflect the natural variation of this ratio in different sediments as illustrated by the fact that the less mature Core 10 actually has a higher diasterane/sterane ratio (J, Table 2).

Demethylated hopanes. Significant amounts of C_{29} triterpane (compound 29D, Fig. 5) are present in the m/z 191 mass fragmentograms of the biodegraded oils, but it is only a trace constituent for the paraffinic oils. This compound has been identified as 17α(H)-25-norhopane (Structure IV Appendix) from its mass spectrum and retention time (RULLKÖTTER and WENDISCH, 1982; VOLKMAN et al., 1983). Mass fragmentograms of m/z 177 (Fig. 7), the A/B ring fragment in the mass spectra, indicates that a series of these compounds are present in the oils. These range in carbon number from C_{26} to C_{34} in severely biodegraded oils, such as Mardie oil (Fig. 7F), but only the C_{28} and C_{29} constituents are sufficiently abundant in the Barrow Sub-basin oils to be readily discernible in the m/z 177 mass fragmentograms. The concentration of demethylated hopanes, relative to hopanes, is very low in the paraffinic oils and becomes progressively greater with decreasing reservoir depth. This is conveniently measured by the ratio of C_{29} 17α(H)-25-norhopane to C_{29} 17α(H)-hopane (K, Table 2), which ranges from a value less than 0.02 in the 2010 m oil to 0.57 in the Flinders Shoal oil.

Although we have identified small amounts of demethylated hopanes in some sediment extracts they are not present in potential source rocks from this location nor are they present in significant concentrations in the undegraded paraffinic oils. All the available evidence suggests that demethylated hopanes are formed from the biotransformation of hopanes (SEIFERT and MOLDOWAN, 1979), at a very late stage of crude oil biodegradation (VOLKMAN et al., 1983). Demethylation of hopanes only occurs when more labile compounds, such as isoprenoid and bicyclic alkanes, have been biodegraded (VOLKMAN et al., 1983). Since these compounds are still present in all the Barrow Sub-basin oils the demethylated hopanes must have been formed from the severe biodegradation of an earlier accumulation of oil which occurred when the reservoir sands were very close to the surface. High concentrations of secondary biogenic carbonates in parts of the Windalia reservoir (GOULD and SMITH, 1978) is additional evidence that severe biodegradation did occur in the Cretaceous. The very low abundance of demethylated hopanes in the deepest Upper Jurassic paraffinic oils indicates that no petroleum accumulated when these reservoir sands were close to the surface. The 1890 m oil however does contain small amounts

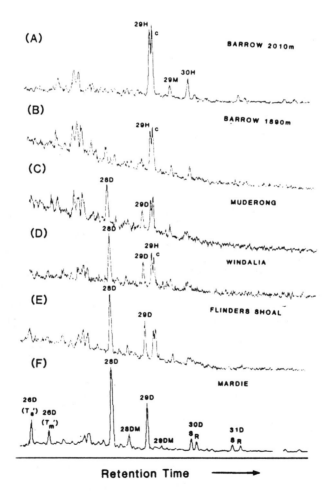

FIG. 7. Mass fragmentograms of m/z 177 (demethylated hopanes) for (A) Barrow 2010 m oil, (B) Barrow 1890 m oil, (C) Muderong oil, (D) Windalia oil, (E) Flinders Shoal oil and (F) Mardie oil. Numerals indicate carbon number.

of demethylated hopanes which indicates that small amounts of oil accumulated when these sands were close to the surface. The presence of demethylated hopanes in this otherwise undegraded oil offers further support for our proposal that the presence of demethylated hopanes in non-biodegraded or moderately biodegraded oils is due to the severely biodegraded residues of an earlier accumulation of oil (VOLKMAN *et al.*, 1983).

Effects of biodegradation

The polycyclic alkane biomarker ratios show that all the oils studied were derived from the same source formation and that they were generated at a high level of thermal maturity. The presence of demethylated hopanes in the naphthenic/aromatic oils indicates that several phases of oil accumulation occurred in these Cretaceous reservoirs, and that some of the oil was severely biodegraded. Unfortunately the polycyclic biomarker parameters provide little information about the less severely biodegraded accumulation, but such information can be deduced from chromatograms of the saturated hydrocarbons.

The chromatogram for the 1890 m oil (Fig. 8) is typical of the paraffinic oils, and it illustrates the high proportion of *n*-alkanes that these oils contain. In contrast, the chromatograms for the shallower oils show very low concentrations of *n*-alkanes, and the unresolved complex mixture (UCM) is very prominent (Fig. 8). Isoprenoid and bicyclic alkanes are the only major hydrocarbons discernible above the UCM, except for Windalia oil which also contains an homologous series of light *n*-alkanes. The major isoprenoid alkanes are pristane, phytane, and the C_{16} isoprenoid 2,6,10-trimethyltridecane. Bicyclic alkanes include *trans*-decalin ($C_{10}H_{18}$), two methyldecalins, and several C_{14}, C_{15} and C_{16} bicyclanes thought to have structures based on the general structure IX (Appendix: KAGRAMANOVA *et al.*, 1976). The latter hydrocarbons are present in many Australian oils (PHILP *et al.*, 1981), and some oils from Texas (BENDORAITIS, 1974; SEIFERT and MOLDOWAN, 1979), but their origin and significance have not been established. Their distributions in each of the Barrow Sub-basin oils are virtually identical (Fig. 8, and unpublished GC-MS data), but this need not imply a common

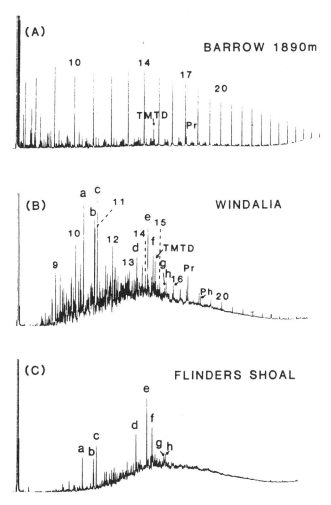

FIG. 8. Capillary gas chromatograms of the saturated hydrocarbon fractions from: (A) Barrow 1890 m oil; (B) Windalia oil and (C) Flinders Shoal oil. Numerals indicate *n*-alkane carbon numbers. Identification of GC peaks a = *trans*-decalin; b and c = methyl decalins; d = C_{14} bicyclane; e and f = C_{15} bicyclanes; g and h = C_{16} bicyclanes. TMTD = 2,6,10-trimethyl-tridecane; Pr = pristane; Ph = phytane.

source for the oils since PHILP et al. (1981) have shown that unrelated oils can have very similar distributions of bicyclanes.

Several parameters have been devised to assess the source and maturity of the oil from the relative abundances of isoprenoid alkanes and n-alkanes. Values for these parameters are given in Table 3. These measurements are less specific than the polycyclic alkane biomarkers ratios but they support the conclusions derived from the latter data. The pristane/phytane ratios are very similar in each of the oils and are consistent with their formation from mixed marine and terrigenous organic matter deposited in a near-shore environment (POWELL and McKIRDY, 1973b). Both the Jurassic Dingo Claystone and Early Cretaceous sediments are of this type. The pristane/nC_{17} and phytane/n-C_{18} ratios (Table 3) are all very low for the paraffinic oils and indicate a high level of thermal maturity (TISSOT et al., 1971; WELTE et al., 1975). Data for the ratio of 2,6,10-trimethyltridecane to pristane are also presented in Table 3. McKIRDY et al. (1982) have used this ratio as a maturity measure, and the sediment data show that it increases with depth. It may also reflect differences in depositional environment, in common with the pristane/phytane ratio.

The very low abundance of n-alkanes, simple branched alkanes and alkylcyclohexanes in the Muderong, Windalia and Flinders Shoal is consistent with the effects of biodegradation of a paraffinic oil. These compounds are more readily degraded by bacteria than are isoprenoid or bicyclic alkanes (EVANS et al., 1971; BAILEY et al., 1973). The high pristane/nC_{17} and phytane/nC_{18} ratios found for the Windalia and Muderong oils (N and O, Table 3) are thus consistent with the preferential biodegradation of n-alkanes compared with isoprenoid alkanes. Such high values are rarely found in undegraded oils or source rocks, even in immature sediments such as Core 10 (Table 3).

The extent to which an oil has been biodegraded can be assessed from the relative proportions of classes of hydrocarbons which differ in their susceptibility to biodegradation. From the fact that Flinders Shoal oil does not contain isoprenoid alkanes, whereas they are present in Muderong and Windalia oils, we conclude that the Flinders Shoal oil is the most severely biodegraded of these three oils. A convenient measure of the extent to which Muderong and Windalia oils are biodegraded is the ratio of the concentration of the first eluting of the two major C_{15} bicyclic alkanes, (Peak e, Fig. 8) to the concentration of pristane (P, Table 3). This value is essentially identical in each of the paraffinic oils and only marginally higher in Muderong oil (Table 3), implying that the relative abundance of isoprenoid and bicyclic alkanes in Muderong oil has not been affected by biodegradation. Their prominence in the saturated hydrocarbon fraction (Fig. 8) is thus solely due to biodegradation of more labile compounds. Windalia oil has a ratio of 2.5 (P, Table 3), which suggests that about 70% of the isoprenoid alkanes have been removed by bacterial degradation, if we assume that a ratio of 0.6 is typical for an undegraded Barrow Sub-basin oil. Windalia oil is thus more biodegraded than Muderong oil, although neither is so biodegraded that the polycyclic alkane distributions are likely to have been altered. According to criteria established by VOLKMAN et al., (1983), both Windalia and Muderong oils would be classified as level 4 biodegraded oils, and Flinders Shoal as level 5. The more degraded Mardie oil, in which the hopane and sterane distributions have been altered by biodegradation, was classified as level 7 (VOLKMAN et al., 1983).

Evidence for multiple accumulation of oil in the Windalia reservoir

Although both Windalia and Muderong oils are moderately biodegraded, they still contain significant amounts of n-alkanes. These are particularly abundant in Windalia oil and yet these n-alkanes should not be present in an oil in which 70% of the isoprenoid alkanes have been removed by biodegradation (BAILEY et al., 1973). The high proportion of short-chain ($<C_{20}$) n-alkanes in Windalia oil is particularly unusual since these are more rapidly degraded by bacteria than are longer-chain alkanes. One explanation for these results is that these n-alkanes must have accumulated in the reservoir at a late stage when bacterial activity was considerably less. That is at least two phases of oil generations are involved. If this is so, then Windalia oil should contain significant concentrations of other light hydrocarbons, and these should have a similar distribution to those in the paraffinic oils.

Whole oil chromatograms (Fig. 9) show that Windalia oil does contain high concentrations of these light hydrocarbons whereas the much more biodegraded Flinders Shoal oil does not. Muderong oil also contains some light hydrocarbons but these are relatively minor components compared with undegraded hydrocarbons, such as pristane and phytane, remaining from the earlier phase of oil accumulation. Note that these light hydrocarbons are lost during the procedures used to produce the saturated hydrocarbon fractions, so chromatograms of the latter (Fig. 8) give the false impression that the n-alkane distributions maximize at n-C_9 or higher. The whole oil chromatograms (Fig. 9) show that n-C_6 is the major n-alkane in all of the oils, and the major hydrocarbon is methylcylohexane.

A number of parameters based on the relative concentrations of specific gasoline-range hydrocarbons have been proposed by assessing the maturity of crude oils (CONNAN and CASSOU, 1980; WELTE et al., 1982, and references therein). PHILIPPI (1981) has shown that the distribution of

Table 3. Acyclic and Bicyclic Alkane Biomarker Ratios

Parameter:	L	M	N	O	P
Oil					
Barrow 2040 m	3.1	1.3	0.19	0.07	0.61
Barrow 2010 m	2.8	1.1	0.19	0.07	0.56
Barrow 1890 m	3.1	1.3	0.21	0.08	0.58
Muderong	2.8	0.9	6.7	2.6	0.65
Windalia	2.8	1.3	2.2	0.81	2.5
Cores					
10 1036 m	2.5	0.07	1.3	0.64	ND*
34 2402 m	2.4	0.19	0.39	0.23	ND*
37 2835 m	2.3	0.85	0.40	0.22	ND*

ND*: Not determined. Ratios for Flinders Shoal oil could not be determined since this oil contains no straight-chain or isoprenoid alkanes.
L - pristane/phytane
M - 2,6,10-trimethyltridecane/pristane
N - pristane/n-heptadecane
O - phytane/n-octadecane
P - C_{15} bicyclic alkane (peak e, Fig. 8)/pristane

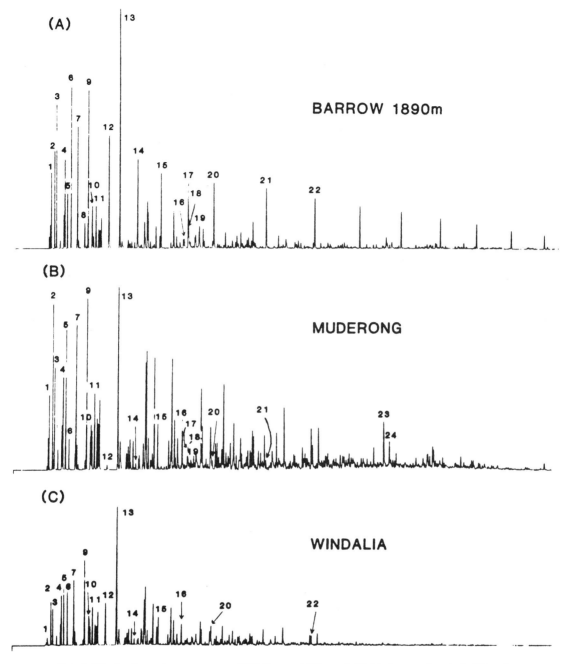

FIG. 9. Whole oil gas chromatograms of (A) Barrow 1890 m oil (B) Muderong oil and (C) Windalia oil. Identification of GC peaks 1: n-butane, 2: 2-methylbutane, 3: n-pentane, 4: 2-methylpentane, 5: 3-methylpentane, 6: n-hexane, 7: methylcyclopentane, 8: benzene, 9: cyclohexane, 10: 2-methylhexane, 11: 3-methylhexane, 12: n-heptane, 13: methylcyclohexane, 14: toluene, 15: n-octane, 16: ethylbenzene, 17: m-xylene, 18: p-xylene, 19: o-xylene, 20: n-nonane, 21: n-decane, 22: n-undecane, 23: 2-methylnaphthalene; 24: 1-methylnaphthalene.

C_6 and C_7 hydrocarbons can be used to correlate an oil with its source formation. On this basis, the similarity of the distribution of these hydrocarbons in Windalia, Muderong and the paraffinic oils (Fig. 9), indicates that each oil was probably derived from the same source rock formation.

Recently, WELTE et al. (1982) demonstrated that six parameters, based on measurement from whole oil chromatograms, can be used to assess the maturity of a crude oil and the extent to which it has been biodegraded or water washed. Although many more parameters could be devised, we have adopted the simpler approach of WELTE et al. (1982) in our study. It must be appreciated however, that conclusions based on these parameters refer only to the light hydrocarbons and may not reflect the history of the oil as a whole, particularly when more than one phase of oil accumulation is involved. Values for these parameters, calculated from the whole oil chromatograms (Fig. 9), are presented in Table 4. These ratios could not be calculated for Flinders Shoal oil due to the absence of light hydrocarbons, which necessarily implies that this oil is extremely biodegraded and water washed.

Table 4. Ratios of Selected Light Hydrocarbons in Barrow Sub-basin Oils

Parameter :	Q	R	S	T	U	V	W
Barrow 2040 m	1.3	0.47	2.2	2.7	6.5	0.67	0.34
Barrow 2010 m	1.4	0.46	2.7	3.6	8.5	0.54	0.32
Barrow 1890 m	1.4	0.52	3.2	3.5	8.2	0.68	0.35
Muderong	0.21	0.03	75	28	93	1.6	4.6
Windalia	0.84	0.31	33	21	53	1.2	1.0

Q : n-hexane / methylcyclopentane
R : n-heptane / methylcyclohexane
S : 3-methylpentane / benzene
T : methylcyclohexane / toluene
U : cyclohexane / benzene
V : 2-methylbutane / n-pentane
W : 3-methylpentane / n-hexane

Ratios were calculated from peak areas in whole oil chromatograms (Fig. 9).

The ratios of n-hexane/methylcyclopentane and n-heptane/methylcyclohexane (parameters Q and R, Table 4) provide a measure of the maturity of crude oils since with increasing temperature, thermal cleavage reactions tend to produce more n-alkanes relative to branched and cyclic alkanes (PHILIPPI, 1975). Unaltered oils give values of 0.7 to 2.8 and 0.43 to 1.5 for parameters Q and R (WELTE et al., 1982), with higher values associated with more mature oils. The Barrow Island paraffinic oils give very similar values for these parameters (Table 4), within the ranges quoted, implying that these oils are unaltered and were generated at similarly high levels of maturity, from the same source rock formation. Muderong and Windalia oils give significantly lower values which might suggest that these oils were generated at lower maturity levels, but lower values are also expected in biodegraded oils (WELTE et al., 1982).

The fact that aromatic hydrocarbons are much more soluble in water than saturated hydrocarbons (MCAULIFFE, 1966) forms the basis for using the ratios 3-methylpentane/benzene and methylcyclohexane/toluene (S and T, Table 4) to assess the extent of water washing (WELTE et al., 1982). Undegraded oils give a wide range of values depending on their source and maturity so it is necessary to establish typical values for related oils for each sedimentary basin studied. There is some scatter in the values of these ratios for the paraffinic oils (S and T, Table 4), but they are sufficiently close to suggest a similar source and maturity. Both Windalia and Muderong oils have very high values indicating extensive water washing. In Table 4, we have included data for an additional parameter, the ratio of cyclohexane to benzene (parameter U), which also seems to be a good measure of water washing. Cyclohexane is usually more abundant than 3-methylpentane in oils (Fig. 9) and it seems more appropriate to compare its abundance to that of benzene to permit a direct comparison with the methylcyclohexane/toluene ratio (parameter T, Table 4).

Two convenient measures of the extent of biodegradation are the ratios 2-methylbutane/n-pentane and 3-methylpentane/n-hexane (V and W, Table 4). These parameters reflect the fact that bacteria degrade straight-chain alkanes more rapidly than branched alkanes (BAILEY et al., 1973). Undegraded oils give very similar values for these parameters: data presented by WELTE et al. (1982) suggest ranges of 0.43 to 0.87 and 0.16 to 0.35 for parameters V and W respectively. Values for the Barrow Island paraffinic oils also fall within these ranges (Table 4), implying that these petroleum accumulations are not biodegraded. The values of ratios V and W for Windalia oil are from 2.0 and 3.1 times higher than averages of the respective ratios for the undegraded oils, indicating that approximately 50 to 70% of the n-alkanes have been removed by biodegradation. Values for Muderong oil are considerably higher reflecting the fact that most of the n-alkanes have been selectively removed during the biodegradation of the oil.

In summary, data from the whole oil chromatograms imply that the paraffinic oils are very closely related, and that they were probably formed from the same source rock formation at a moderate to high level of thermal maturation. There is no evidence that these accumulations have been biodegraded or water washed. There have been at least two major phases of oil accumulation in the Windalia and Muderong reservoirs. The earlier accumulation was extensively biodegraded, implying that the reservoir must have been fairly close to the surface, probably less than 300 m, when this oil accumulated. The more recent accumulation was only lightly biodegraded and water washed and thus probably occurred when these reservoirs were close to their present depths.

TIMING OF OIL GENERATION AND ACCUMULATION IN THE BARROW SUB-BASIN

From the combined information obtained from the acyclic and polycyclic alkane biomarker parameters, and whole oil chromatograms it is now possible to reconstruct the major features of the complex history of oil generation and accumulation in the Barrow Sub-basin. It is apparent that all the oils were produced from sediments within the Jurassic Dingo Claystone, although several different strata are involved. All of the oils are mature and their marked differences in composition reflect the fact that those oils in the Cretaceous reservoirs are very biodegraded. The latter oils represent several accumulations of oil, each of which was biodegraded to a different extent. At least four major phases of oil accumulation can be recognized, and these are discussed in detail below. An unknown amount of oil may have been produced from Triassic sediments and basal sections of the Lower Jurassic, but these sediments are now so deep at Barrow Island that any oil accumulated has been lost as gas and condensate. From the close correspondence of the biomarker maturity parameters for the Upper Jurassic Core 37 and all of the oils (Table 2), we estimate that peak oil generation occurred at about 130°C. In this basin approximately 2800 m of overlying sediment are required to produce this degree of thermal maturity, based on a temperature gradient of 38.6°C/km (COOK and KANTSLER, 1980).

Accumulation of paraffinic Upper Jurassic oils

The absence of demethylated hopanes from the 2040 m and 2010 m oils indicates that substantial accumulation did not occur until these reservoir sands were buried to a depth where bacterial activity was non-existent. If we assume that this depth was at least 800 m, and that about 2800 m of overlying sediment is required to produce oil of this high maturity, we calculate that oil was largely generated during the period of high sedimentation and rapid basin subsidence which occurred in the early Tertiary.

The source formation is undoubtedly the Jurassic Dingo Claystone and it is probable that thick sections laid down in the late Middle Jurassic to late Jurassic were the actual source rocks. The major phase of oil generation probably occurred towards the end of the Neocomian period of the Lower Cretaceous, although small amounts of demethylated hopanes in the 1890 m oil suggest that a small quantity of oil was generated earlier as the Barrow Group was deposited. We cannot rule out that some light oil may have accumulated in these reservoirs in much more recent times but the very close similarity of the biomarker parameters for all 3 paraffinic oils (Tables 2 and 3) suggests that only one major phase of oil accumulation is involved.

Accumulation of oils in Cretaceous reservoirs

The high concentrations of demethylated hopanes in Muderong, Windalia and Flinders Shoal oils indicates that substantial quantities of oil were generated and accumulated when these reservoir sands were close to the surface during the Aptian and Albian of the Lower Cretaceous. This oil was probably generated from sections of the Dingo Claystone laid down during the Upper Jurassic. Continued deposition and basin subsidence in the Upper Cretaceous brought younger sections of the Dingo Claystone into the zone of peak oil generation, with consequent formation of more oil. This oil was not as severely biodegraded as previous accumulations of oil since the reservoir sands were then deeper, and bacterial activity would have diminished. Hydrocarbons from this phase predominate in the gas chromatograms of the saturated hydrocarbon fraction of Muderong and Windalia oils. Different strata within the Dingo Claystone were probably involved in the generation of the two oils since some of the polycyclic alkane biomarker parameters, particularly the ratio of C_{27} to C_{29} steranes (A; Table 2), are not identical.

Continued deposition in the late Tertiary led to further oil generation from upper sections of the Dingo Claystone, but the amount produced was probably less than previously. However, this oil was only subjected to quite mild biodegradation and hence a far greater proportion of it survived to be accumulated in the reservoir sands. Most of this oil is found in the Windalia reservoir, although a small quantity occurs in the Muderong oil.

The Flinders Shoal oil has been described as a biodegraded and water washed Windalia-type oil (THOMAS, 1978), but the polycyclic alkane distributions (Table 2) show a closer affinity to the Muderong and paraffinic Upper Jurassic oils. The high proportion of demethylated hopanes indicates that much of the oil was accumulated during Lower Cretaceous, with another later significant accumulation. Thus, most of the oil was probably generated from Upper Jurassic source rocks probably during the Cretaceous. This oil accumulated in the Birdrong Sands and migrated to its present position where it was extensively biodegraded and water washed. There is little evidence to suggest that there has been any significant accumulation during the Tertiary, which is to be expected since there is very little Upper Jurassic sediment underlying this reservoir (Fig. 2).

CONCLUSIONS

This study has shown that petroleum formation must be considered as a dynamic process which may extend over many millions of years, particularly in a subsiding basin, such as the Barrow Sub-basin. We have identified at least four major phases of accumulation of a light, paraffinic oil generated from the Jurassic Dingo Claystone. Our work confirms and extends earlier speculations by CRANK (1973) that migration began in the Upper Jurassic, but became most important in the late Cretaceous and Tertiary following major arching in the Sub-basin. As a consequence of the hydrogen-poor nature of the mixed marine and terrigenous organic matter, most of this oil was generated at moderate to high levels of thermal maturity. There is no evidence that Cretaceous sediments have been the source of any of the oils at Barrow island, but this does not preclude that there may be such accumulations elsewhere in the basin. Light oil probably first accumulated in the shallower Cretaceous reservoirs during the Aptian but these accumulations were severely biodegraded. Subsequent major generation episodes in the late Cretaceous and Tertiary were less biodegraded giving rise to accumulations of medium API gravity biodegraded oils which still contain abundant light hydrocarbons more typical of undegraded oils.

Our reconstruction of the petroleum geochemistry of the Barrow Sub-basin has made use of compositional data from saturated hydrocarbon fractions, whole oil chromatograms and polycyclic alkane distributions. None of these sets of data used alone would have provided sufficient information to assess the source of these oils, which partly accounts for the divergent views expressed in earlier publications. Twenty-three biomarker parameters have been discussed, and we have shown how these can be used to assess the source and maturity of an oil, and determine the extent to which the oil has been biodegraded and water washed. It needs to be emphasized that these parameters provide information on the specific hydrocarbons from which the parameters were calculated, and any inferences drawn from them about the source or maturity of the oil as a whole need to be made with caution.

Acknowledgements—We thank N. Burns for laboratory assistance, G. Chidlow for some GC-MS analyses, and J. C.

Parry, B. M. Thomas, M. C. Johnstone, and E. L. Horstman for stimulating discussions. Dr. T. G. Powell and J. Stinnet are thanked for their constructive criticism of the manuscript. This work was carried out as part of the National Energy Research Development Demonstration Program of Australia.

REFERENCES

ALBRECHT P., VANDENBROUCKE M. and MANDENGUE M. (1976) Geochemical studies on the organic matter from the Douala Basin (Cameroon). I. Evolution of the extractable organic matter and the formation of petroleum. *Geochim. Cosmochim. Acta* **40**, 791-799.

ALEXANDER R., KAGI R. and WOODHOUSE G. (1980) Origin of the Windalia oil, Barrow Island, Western Australia. *APEA J.* **20**, 250-256.

ALEXANDER R., KAGI R. I. and WOODHOUSE G. W. (1981) Geochemical correlation of Windalia oil and extracts of Winning Group (Cretaceous) potential source rocks, Barrow Sub-basin, Western Australia. *Bull. Amer. Assoc. Petrol. Geol.* **65**, 235-250.

BAILEY N. J. L., JOBSON A. M. and ROGERS M. A. (1973) Bacterial degradation of crude oil: comparison of field and experimental data. *Chem. Geol.* **11**, 203-221.

BENDORAITIS J. G. (1974) Hydrocarbons of biogenic origin in petroleum-aromatic triterpenes and bicyclic sesquiterpanes. In *Advances in Organic Geochemistry—1973* (eds. B. TISSOT and F. BIENNER), pp. 209-224. Editions Technip.

CONNAN J. and CASSOU A. M. (1980) Properties of gases and petroleum liquids derived from terrestrial kerogen at various maturation levels. *Geochim. Cosmochim. Acta* **44**, 1-23.

COOK A. C. and KANTSLER A. J. (1980) The maturation history of the epicontinental basins of Western Australia. *UN ESCAP, CCOP/SOPAC Tech. Bull.* **3**, 171-195.

CRANK K. (1973) Geology of Barrow Island oil field. *APEA J.* **13**, 49-57.

ENSMINGER A., ALBRECHT P., OURISSON G. and TISSOT B. (1977) Evolution of polycyclic alkanes under the effect of burial (Early Toarcian shales, Paris Basin). In *Advances in Organic Geochemistry, 1975* (eds. R. CAMPOS and J. GONI) pp. 45-52. ENADIMSA.

ENSMINGER A., JOLY G. and ALBRECHT P. (1978) Rearranged steranes in sediments and crude oils. *Tetrahedron Letters* pp. 1575-1578.

EVANS C. R., ROGERS M. A. and BAILEY N. J. L. (1971) Evolution and alteration of petroleum in Western Canada. *Chem. Geol.* **8**, 147-170.

GOULD K. W. and SMITH J. W. (1978) Isotopic evidence for microbiologic role in genesis of crude oil from Barrow Island, Western Australia. *Bull. Amer. Assoc. Petrol. Geol.* **62**, 455-462.

HUNT J. M. (1979) *Petroleum Geochemistry and Geology.* Freeman, San Francisco, pp. 335.

JI-YANG S., MACKENZIE A. S., ALEXANDER R., EGLINTON G., GOWAR A. P., WOLFF G. A. and MAXWELL J. R. (1982) A biological marker investigation of petroleums and shales from the Shengli oilfield, the People's Republic of China. *Chem. Geol.* **35**, 1-31.

KAGRAMANOVA G. R., PUSTIL'NIKOVA S. D., PEHK T., DENISOV YU. V. and PETROV A. A. (1976) Sesquiterpane hydrocarbons of petroleums. *Neftekhimiya* **16**, 18-22.

LEYTHAEUSER D., HOLLERBACH A. and HAGEMANN H. W. (1977) Source rock/crude oil correlation based on distribution of C_{29} + cyclic hydrocarbons. In *Advances in Organic Geochemistry, 1975* (eds. R. CAMPOS and J. GONI) pp. 3-20. ENADIMSA.

MACKENZIE A. S., PATIENCE R. L., MAXWELL J. R., VANDENBROUCKE M. and DURAND B. (1980) Molecular parameters of maturation in the Toarcian shales, Paris Basin, France. I. Changes in configurations of acyclic isoprenoid alkanes, steranes and triterpanes. *Geochim. Cosmochim. Acta* **44**, 1709-1721.

MACKENZIE A. S., HOFFMAN C. F. and MAXWELL J. R. (1981a) Molecular parameters of maturation in the Toarcian shales, Paris Basin, France—III. Changes in aromatic steroid hydrocarbons. *Geochim. Cosmochim. Acta* **45**, 1345-1355.

MACKENZIE A. S., LEWIS C. A. and MAXWELL J. R. (1981b) Molecular parameters of maturation in the Toarcian shales, Paris Basis, France—IV. Laboratory thermal alteration studies. *Geochim. Cosmochim. Acta* **45**, 2369-2376.

MACKENZIE A. S., LAMB N. A. and MAXWELL J. R. (1982) Steroid hydrocarbons and the thermal history of sediments. *Nature* **295**, 223-226.

MCAULIFFE C. (1966) Solubility in water of paraffin, cycloparaffin, olefin, acetylene, cycloolefin, and aromatic hydrocarbons. *J. Phys. Chem.* **70**, 1267-1275.

MCKIRDY D. M., ALDRIDGE A. K. and YPMA P. J. M. (1982) A geochemical comparison of some crude oils from Pre-Ordovician carbonate rocks. In *Advances in Organic Geochemistry, 1981* (ed. M. BJORØY). Wiley, pp. 99-107.

MEINSCHEIN W. G. and HUANG W.-Y. (1981) Stenols, stanols, steranes, and the origins of natural gas and petroleum. In *Origin and Chemistry of Petroleum.* (eds. G. ATKINSON and J. J. ZUCKERMAN) pp. 35-55. Pergamon Press.

PARRY J. C. (1967) The Barrow Island Oilfield *APEA J.* **1**, 130-133.

PHILIPPI G. T. (1975) The deep subsurface temperature controlled origin of gaseous and gasoline-range hydrocarbons of petroleum. *Geochim. Cosmochim. Acta* **39**, 1353-1373.

PHILIPPI G. T. (1981) Correlation of crude oils with their oil source formation, using high resolution GLC C_6-C_7 component analyses. *Geochim. Cosmochim. Acta* **45**, 1495-1513.

PHILP R. P., GILBERT T. D. and FRIEDRICH J. (1981) Bicyclic sesquiterpenoids and diterpenoids in Australian crude oils. *Geochim. Cosmochim. Acta* **45**, 1173-1180.

PHILP R. P., GILBERT T. D. and FRIEDRICH J. (1982) Geochemical correlation of Australian crude oils. *APEA J.* **22**, 188-199.

POWELL T. G. and MCKIRDY D. M. (1972) The geochemical characterization of Australian crude oils. *APEA J.* **12**, 125-131.

POWELL T. G. and MCKIRDY D. M. (1973a) Relationship between ratio of pristane to phytane, crude oil composition and geological environment in Australia. *Nature Phys. Sci.* **243**, 37-39.

POWELL T. G. and MCKIRDY D. M. (1973b) Crude oil correlations in the Perth and Carnarvon Basins. *APEA J.* **13**, 81-85.

RUBINSTEIN I., SIESKIND O. and ALBRECHT P. (1975) Rearranged steranes in a shale: occurrence and simulated formation. *J. Chem. Soc. Perkin I.* 1833-1836.

RULLKÖTTER J. and WENDISCH D. (1982) Microbial alteration of 17α(H)-hopanes in Madagascar asphalts: removal of C-10 methyl group and ring opening. *Geochim. Cosmochim. Acta* **46**, 1545-1553.

SEIFERT W. (1977) Source rock/oil correlations by C_{27}-C_{30} biological marker hydrocarbons. In *Advances in Organic Geochemistry, 1975* (eds. R. CAMPOS and J. GONI) pp. 21-44. ENADIMSA.

SEIFERT W. K. and MOLDOWAN J. M. (1978) Applications of steranes, terpanes and monoaromatics to the maturation, migration and source of crude oils. *Geochim. Cosmochim. Acta* **42**, 77-95.

SEIFERT W. K. and MOLDOWAN J. M. (1979) The effect of biodegradation on steranes and terpanes in crude oils. *Geochim. Cosmochim. Acta* **43**, 111-126.

SEIFERT W. K. and MOLDOWAN J. M. (1980) The effect of thermal stress on sourcerock quality as measured by hopane stereochemistry. In *Advances in Organic Geochemistry, 1979* (eds. A. G. DOUGLAS and J. R. MAXWELL) pp. 229-237. Pergamon Press.

SEIFERT W. K. and MOLDOWAN J. M. (1981) Paleoreconstruction by biological markers. *Geochim. Cosmochim. Acta* **45**, 783-794.

Seifert W. K., Moldowan J. M., Smith G. W. and Whitehead E. V. (1978) First proof of structure of a C_{28}-pentacyclic triterpane in petroleum. *Nature* **271**, 436–437.
Sieskind O., Joly G. and Albrecht P. (1979) Simulation of the geochemical transformations of sterols: superacid effect of clay minerals. *Geochim. Cosmochim. Acta* **43**, 1675–1679.
Thomas B. M. (1978) Robe River—an onshore shallow oil accumulation. *APEA J.* **18**, 3–12.
Thomas B. M. and Smith D. N. (1974) A summary of the petroleum geology of the Carnarvon Basin. *APEA J.* **14**, 66–76.
Thomas B. M. and Smith D. N. (1976) Carnarvon Basin. In *Economic Geology of Australia and Papua New Guinea. Vol. 3. Petroleum* (eds. R. B. Leslie *et al.*) pp. 126–155. Australian Institute of Mining and Metallurgy.
Tissot B., Califet-Debyser Y., Deroo G. and Oudin J. L. (1971) Origin and evolution of hydrocarbons in Early Toarcian shales, Paris Basin, France. *Bull. Amer. Assoc. Petrol. Geol.* **55**, 2177–2193.
Volkman J. K., Alexander R., Kagi R. I. and Woodhouse G. W. (1983) Demethylated hopanes in crude oils and their applications in petroleum geochemistry. *Geochim. Cosmochim. Acta* **47**, 785–794.
Wardroper A. M. K., Brooks P. W., Humberston M. J. and Maxwell J. R. (1977) Analysis of steranes and triterpanes in geolipid extracts by automatic classification of mass spectra. *Geochim. Cosmochim. Acta* **41**, 499–510.
Welte D. H., Hagemann H. W., Hollerbach A., Leythaeuser D. and Stahl W. (1975) Correlation between petroleum and source rock. *Proc. 9th World Petrol. Congr.* **2**, 179–191.
Welte D. H., Kratochvil H., Rullkötter J., Ladwein H. and Schaefer R. G. (1982) Organic geochemistry of crude oils from the Vienna Basin and an assessment of their origin. *Chem. Geol.* **35**, 33–68.
Wiseman J. F. (1979) Neocomian eustatic changes—biostratigraphic evidence from the Carnarvon Basin. *APEA J.* **19**, 66–73.

APPENDIX

General structures for compounds discussed in the text.

I: Steranes
II: Diasteranes
III: Hopanes
IV: 25-Norhopanes
V: Moretanes
VI: 25-Normoretanes
VII: T_s (18α(H)-22,29,30-trisnorneohopane)
VIII: T_m (17α(H)-22,29,30-trisnorhopane)
IX: C_{15} Bicyclanes

The American Association of Petroleum Geologists Bulletin
V. 63, No. 12 (December 1979), P. 2128-2144, 2 Figs.

Compilation and Correlation of Major Thermal Maturation Indicators[1]

YVON HÉROUX, ANDRÉ CHAGNON, and RUDOLF BERTRAND[2]

Abstract It is important in hydrocarbon exploration to interpret accurately the thermal maturation stage of sedimentary rocks. A compilation and correlation chart shows the relations of the most commonly used organic and mineral thermal indicators with respect to the degree of maturation. The chart and the discussion of the limitations of each technique are based on the results published by other workers and on the writers' observations.

INTRODUCTION

There is no infallible analytic technique to quantify the degree of transformation of organic matter into hydrocarbons. This is particularly true for lower and middle Paleozoic materials. The absence of, and urgent need for, such an analytic tool for hydrocarbon exploration have led to a proliferation of both thermal maturation parameters and publications dealing with their correlation. Despite all of the research, correlations of equivalent numerical values for the various parameters are still uncertain (Hood et al, 1975, p. 988).

This work provides a short guide to the correlation and significance of some of the most important thermal maturity indicators, based on dispersed organic matter (D.O.M: see Appendix-Glossary) and on mineralogy. This guide should allow petroleum geologists to interpret hydrocarbon generation stages without a time-consuming bibliographic search. Some of the curves for hydrocarbon gas scales shown on Figure 1 are from our work. All other parameters shown on this chart have been discussed in the literature (Fig. 2).

Numerical values of organic geochemical parameters are dependent on time, thermal energy, and type of organic matter (McIver, 1967; Tissot et al, 1971; Van Dorsselaer and Albrecht, 1976; Weber and Maximov, 1976). The values indicated on the chart are, unless otherwise noted, relevant to sapropelic-type organic matter (kerogen type II of Van Krevelen, 1961).

The naphthabitumen parameter values are influenced by the effects of migration or the presence of alteration products such as impregnations and surface indexes (Connan, 1973; Van Dorsselaer et al, 1975). Thus, an idealized source rock system is represented on the chart.

The "birth zone" and "dead zone" of gaseous and liquid hydrocarbons should be considered as gradational. The apparent widths of these zones depend on the analytic techniques as well as on the type of organic matter. For these reasons, lower and upper boundaries of these major vertical divisions must be considered as variable. To avoid ambiguity, selected references are given for each scale on the chart. The chart scale values and boundaries are a consensus of the most recent publications among these selected references (Fig. 2).

The evolution of clays and other index minerals is controlled mainly by the temperature and by chemical and petrologic properties. The patterns shown on the chart apply to fine-grained (pelitic) detrital sequences.

To minimize semantic problems, a glossary of terms is presented in the Appendix. This glossary also serves as a legend for the chart.

MATURATION RANK

As defined by the International Committee for Coal Petrology, "rank" means degree of coalification. This term is herein extended to hydrocarbon generation and premetamorphic transformation stages. Thus, it applies to the thermal maturation attained by a given organic or mineral phase in the rock.

Coalification and Hydrocarbon Generation Stages

Alpern (1976a, p. 178) presented a correlation of coalification scales used by several authors in

---

©Copyright 1978. The American Association of Petroleum Geologists. All rights reserved.

AAPG grants permission for a *single* photocopy of this article for research purposes. Other photocopying not allowed by the 1978 Copyright Law is prohibited. For more than one photocopy of this article, users should send request, article identification number (see below), and $3.00 per copy to Copyright Clearance Center, Inc., One Park Ave., New York, NY 10006.

[1]Manuscript received, November 10, 1977; accepted, December 27, 1978.

[2]Institut National de la Recherche Scientifique (INRS-Pétrole), Université du Québec, Province de Québec, Canada.

The work leading to this paper was financed by the National Research Council of Canada (grant A-4251), by the Ministère de l' Education du Québec (FCAC individual grant EQ-1124) and by the Institut National de la Recherche Scientifique. The writers thank B. Kübler who stimulated this project at INRS-Pétrole, J. Connan who kindly read and criticized a first draft of this paper, and M. Desjardins for in-house translation of the manuscript (RD-149).

Article Identification Number
0149-1423/79/B012-0001$03.00/0

different countries. The A.S.T.M. coalification scale (U.S.A. D-388-66) is used for our chart. Parameters used for the coalification scale also served to establish our hydrocarbon generation scale. The relations of coal rank, hydrocarbon generation, standard parameters, and reflectance of vitrinite herein are based on generally accepted conclusions concerning important economic petroleum occurrences, as reported by Stach et al (1975, p. 326-331). Certain discrepancies in the values of parameters given by the different authors for coal rank and hydrocarbon generation stages can be attributed to differences in kerogen types and source rock facies (Stach et al, 1975).

The wet (early mature condensate) gas substage is not often mentioned in the literature but has been noted by Gunther (1976), Powell et al (1977), and Connan (personal commun., August 1977). In our study of the Labrador offshore (Karlsefni H-13), the presence of such an early wet gas substage has also been detected by studying gases in sealed tin cans particularly by the iso-butane to normal butane ratio (discussed later).

Premetamorphic Zones

The terms diagenesis, catagenesis, and anchimetamorphism are indicated in the chart only for reference and we will not elaborate on them because of the semantic problem they create. Dozens of terms are used to classify the thermal maturation of organic matter and clay minerals (Philippi, 1965; Landes, 1966, 1967; Baker and Claypool, 1970; Kaplan, 1971; Burgess, 1974; Tissot et al, 1974; Hood et al, 1975; Foscolos et al, 1976), and others are applied to more general major changes or processes (Bailey et al, 1971; Bostick, 1971; Hunt, 1973; Leythaeuser, 1973).

Temperature and Time Relation

Most authors (Karweil, 1956; Philippi, 1965; Louis and Tissot, 1967; Vassoyevich et al, 1970; Bostick, 1973, 1974; Castaño and Sparks, 1974; Connan, 1974; Waples and Connan, 1976; Deroo et al, 1977) have stated that the temperature and the durations of exposure to various heat intensities are the most important factors controlling thermal maturation of organic matter. Temperature is not the correct term. In reality, heat or thermal energy, as indicated by measured subsurface temperatures, are more appropriate.

If the maximum temperature reached by Cenozoic sediments can be determined by numerous thermometric measurements, it is hazardous to extrapolate it to Mesozoic and Paleozoic sedimentary rocks, or to any rock sequence that had been involved in an orogenic event. To do so, the main orogenic and tectonic events must be reconstructed. Generally, the authors cited, in their examples, indicated only the original or actual depths but very rarely discussed the assumed T°C reached within Mesozoic or Paleozoic rocks. However, a few authors have suggested temperatures for the liquid hydrocarbon window in Paleozoic sequences. Pusey (1973), on the basis of the work of several authors, listed a temperature range of 65 to 150°C (150 to 300°F). Hood et al (1975) indicated a temperature of 105°C (220°F) for Paleozoic rocks of western Texas. In a recent study, Deroo et al (1977) showed that 88% of initial oil reserves in Alberta have a degree of organic metamorphism comparable with a coal rank of 0.5 to 0.9% Ro. The corresponding paleotemperatures range from 68 to 116°C. No oil was found above 1.3% Ro, or above 143°C.

In our chart, the indicated temperatures of ~65°C and of ±120 to 135°C, for Paleozoic sequences (230 to 570 m.y.) come from Karweil (1956), Castaño and Sparks (1974), and Connan (1974). They applied a modified Arrhenius equation (see Appendix–Glossary) to 12 different oil fields and deduced time and temperature values in agreement with the maximum rate of generation of liquid hydrocarbons. However, these values are only a guide for Paleozoic rocks.

MAJOR THERMAL MATURATION INDICATORS

The following indicators are the most easily obtainable or workable. Maturation indicators obtained from flame photometry detectors (F.P.D.), source rock analyzers with flame ionization detectors (F.I.D.), electron spin resonance (E.S.R.), infrared spectroscopy and others are not included in Figure 1. This applies also for pyrochromatography, oxidation, ash content, coking power, and other properties such as elasticity, grindability, hardness, and plasticity, which require large samples of coal for their determinations and are seldom used for hydrocarbon exploration. Finally, "diagenetic" indicators such as K_2O percentage in mixed-layer minerals, molar ratio (SiO_2 + Fe_2O_3 + MgO/Al_2O_3), cation exchange capacity (C.E.C.), percentage of 2M polymorphs, and sharpness ratio are not shown in the chart. However, good documentation is available for these techniques (Van Krevelen, 1961; Barrabé and Feys, 1965; Espitalié et al, 1973; Foscolos and Kodama, 1974; Durand et al, 1976; Marchand, 1976).

Standard (STD.) Parameters

As previously mentioned, because of the small quantity of dispersed organic matter generally available, many analytic parameters are difficult

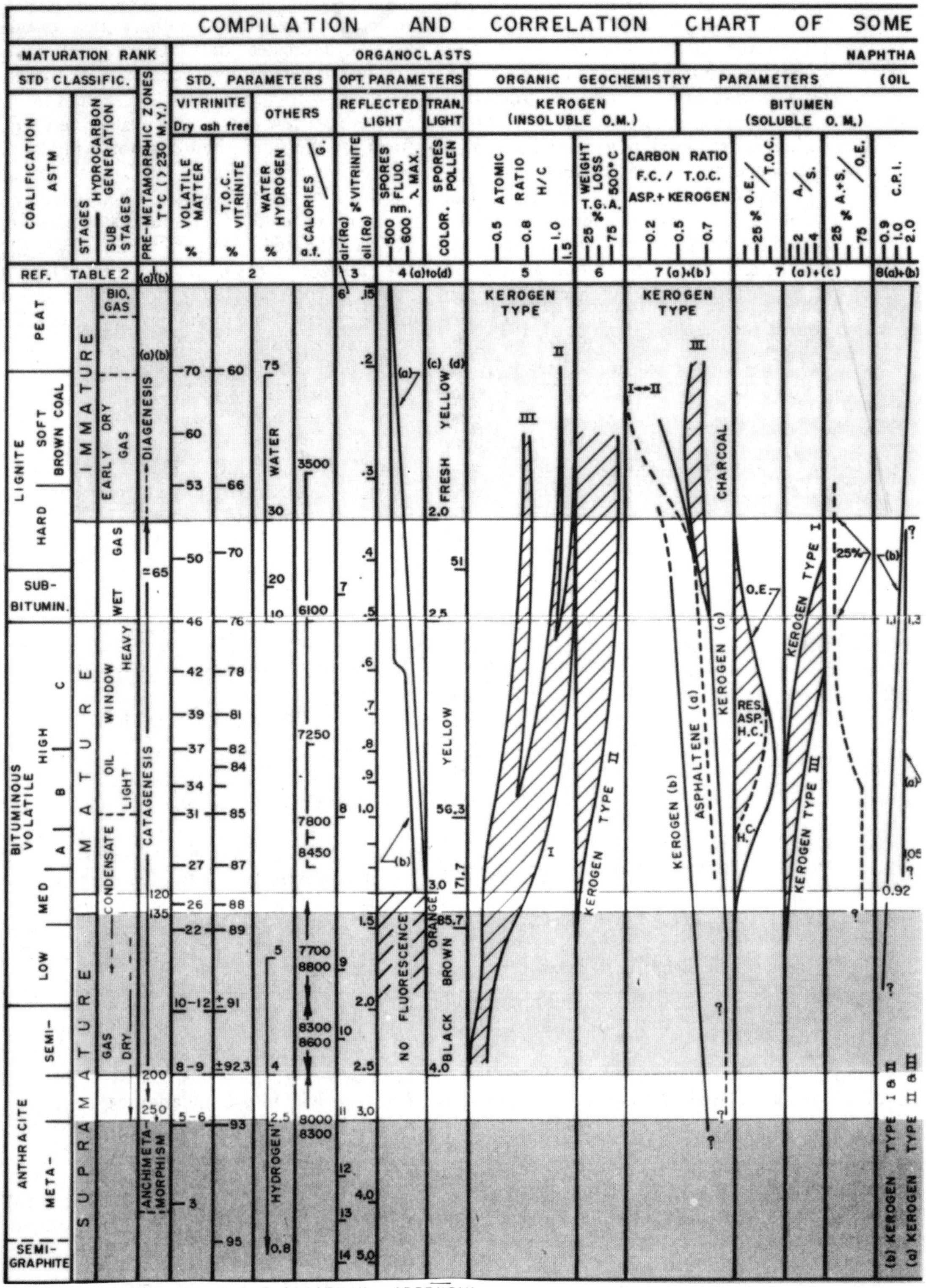

FIG. 1—Compilation and correlation chart of major thermal maturation indicators. For abbreviations, see Appendix-Glossary.

Major Thermal Maturation Indicators

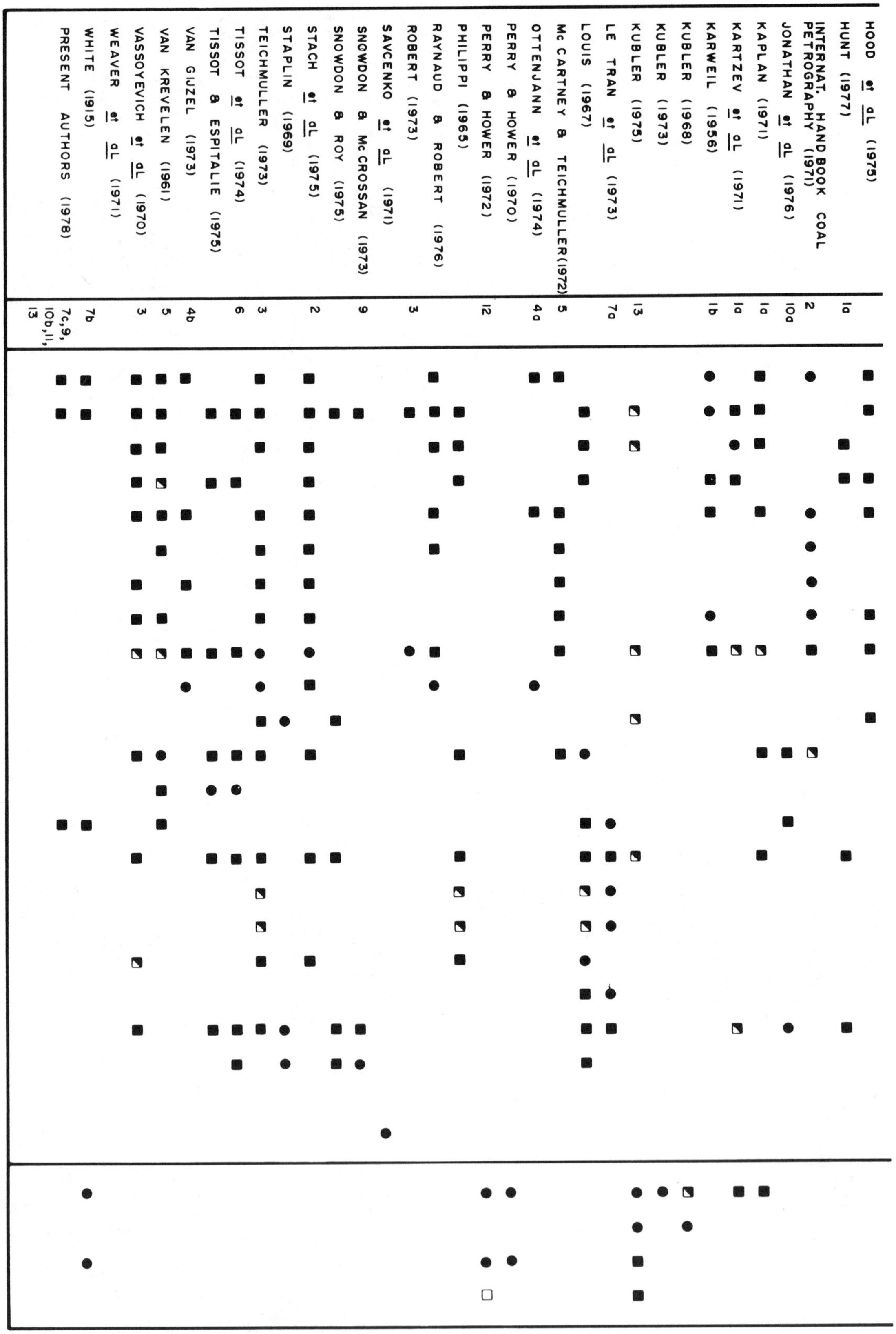

FIG. 2—Compilation of recent publications which discuss and/or correlate thermal indicators and serve as references for Figure 1. For abbreviations, see appendix–glossary.

Major Thermal Maturation Indicators

Authors	Chart # (Fig. 1)
ALLAN & DOUGLAS (1977)	8a
ALPERN (1972)	
ALPERN (1973)	
ALPERN (1976)	
ALPERN et al. (1971)	
BAKER & CLAYPOOL (1970)	
BARKER (1972)	
BARRABE & FEYS (1965)	
BRAY & EVANS (1961)	8b
BURST (1969)	1b
CASTAÑO & SPARKS (1974)	5,7
CONNAN (1970)	
CORREIA (1967)	4c
DUNOYER DE SEGONZAC (1969)	
FEUGÈRE & GÉRARD (1970)	
FOSCOLOS & KODAMA (1974)	12
FOSCOLOS & STOTT (1975)	
FOSCOLOS et al. (1976)	11,12
GUNTHER (1976)	1a
HACQUEBAR & DONALDSON (1970)	4d

Indicators (rows):
- COALIFICATION
- HYDROCARBON GENERATION
- PRE-METAMORPHIC ZONES
- T °C vs TIME
- % V.M.] VITRINITE (d.a.f.)
- % T.O.C.]
- % WATER ; % HYDROGEN
- CALORIES
- REFLECTANCE (VITRINITE)
- FLUORESCENCE] SPORES
- COLOR]
- H/C ATOMIC RATIO
- % WEIGHT LOSS
- F.C./T.O.C.
- O.E./T.O.C.
- A./S.
- A.+S./O.E.
- C.P.I.
- H_2S/T.O.C.
- H.C.G./T.O.C.
- C_1/H.C.G.
- C_2/C_1 vs C_3+/C_1
- $i-C_4/n-C_4$
- INDEX MINERALS
- ILLITE CRYSTALLINITY
- % ILLITE IN ILLITE–2:1 EXPANDABLES
- d(001) OF SMECTITE

● Discussed in more detail than in this report;
■ General discussion of parameters;
◪ Discussed in less detail than in this report;
□ Thermal indicators discussed in different manners than in this report.

to measure, as for the analytic parameters grouped in the chart under the headline "Std. Parameters." They are standard parameters for determining coal rank, together with reflectance and the ultimate or elemental analysis (atomic ratio H/C) of vitrinite (International Committee for Coal Petrology, 1963). Patteisky and Teichmüller (1958) discussed the limitations of analytic techniques and the range of application of the standard parameters. The numerical values shown in the chart are from the International Committee for Coal Petrology (1963, Figs. 3-5), and from Stach et al (1975, p. 34).

Vitrinite reflectance and elemental analysis currently being used in petroleum exploration usually serve as the basic scales for establishing equivalence between coal ranks and hydrocarbon generation stages. Although they could be grouped under the "Std. Parameters" headline in the chart, they are, like other indicators, arbitrarily categorized to outline the type of organic materials (soluble or not) or the analytic methods used (physical: optical, combustion, etc; chemical: chromatography, extract, etc).

Optical (OPT.) Parameters

From these parameters, three scales are established: vitrinite reflectance in oil immersion (Ro), sporinite microspectrofluorescence (Fluo.) and spores and pollen coloration (color) scales. The vitrinite reflectance in air (Ra) scale is shown on the chart for comparison with results obtained before 1970 and is correlated with that of Ro according to Vassoyevich et al (1970). It will not be discussed further.

Ro percentage—Most scientists, working to establish a hydrocarbon maturation scale, use this parameter as a correlation base. The Ro scale in our chart is logarithmic to emphasize the most interesting range (0.35 to 2.50%). The values for that scale are from Kötter's curves (1960, *in* International Committee for Coal Petrology, 1971) and from data reported in Stach et al (1975, p. 326-331).

This technique has the following limitations: (a) lack of references on macerals other than vitrinite; (b) absence of vitrinite in pre-Carboniferous series; (c) low reliability of the vitrinite reflectance technique for values less than 0.3% Ro; (d) identification of vitrinite from other macerals having intermediate characteristics (morphology and reflectance) between vitrinite and inertinite; (e) reworked organic matter that affects reflectance (% Ro) histograms; (f) influence of lithology on reflectance values (at the same maturation stage, reflectance of vitrinite increases from sandstone to siltstone to shale to coal); (g) the reflectance (% Ro) increases with the thickness of coal beds; (h) oxidation reduces vitrinite reflectance, but has a smaller effect than the preceding factors.

Limitations d to h were discussed by Kübler et al (1979).

Some authors (Robert, 1973; Sikander and Pittion, 1976, 1978) tried to establish a Ro scale on asphaltic pyrobitumen. They assumed that the evolution of the reflectance values from pyrobitumen and from vitrinite was the same with increasing temperature and time, although the values from pyrobitumen show a greater scatter than those from vitrinite of equal rank. We consider these results as preliminary because they do not show the influence of each asphaltic pyrobitumen type (elaterite, wurtzilite, albertite, impsomite, and anthraxolite) or "asphaltic pyrobitumen-like" fragments (acritarchs, chitinozoans, etc) on the parameter Ro behavior (research in progress at INRS).

In low-maturity sequences, where vitrinite reflectance values are less than 0.3% Ro, fluorescence and coloration measurements are better indicators of thermal alteration.

Fluorescence (Fluo.)—The fluorescence curves shown on the chart are from Van Gijzel (1973) and Ottenjann et al (1974). The spectral maximum numerical values are in nanometer units of wave length (1 nm = 1 mμ = 10 A).

Fluorescence and reflectance are complementary: fluorescence intensity and reflectance are inversely proportional (Jacob, 1963, *in* International Committee for Coal Petrology, 1975, p. 1). The fluorescence intensity is one of the most useful parameters when vitrinite reflectance values are low (less than 0.3% Ro) and less reliable. However, it cannot be used in the supramature zone (below 25% volatile matter), for sporinite is not fluorescent at this maturation stage.

The fluorescence intensity depends on both the type of organic matter and palynomorph species. The discrepancy between curves a and b shown on the chart is probably the result of measurements done on different species of angiosperm pollen grains and microspores. Another but less significant source of variation may be the use of different methods in calculating λ max. For these reasons spectrofluorometry values must be carefully standardized to obtain valuable diagnosis.

Finally, quoting from the International Committee for Coal Petrology (1975, p. 2): "The fluorescence of organic materials in particular can change as a result of the illumination; this effect is called 'fading effect' . . . or 'alteration' . . . Changes of both the spectrum and the intensity are observed. The possibility of such changes should be kept in mind in all studies of fluorescence." For this reason, the determination of all characteristics of the entire spectrum is recommended (Van Gijzel, 1973).

Color—The coloration of spores and pollens has been described in several publications and several types of scales have been proposed, as pointed out by Hood et al (1975, p. 988): (1) the "yellow through brown to black color scale" of Gutjahr (1966); (2) the "thermal alteration index" of Staplin (1969), based on microscopic observations of both color and structural alteration of organic debris; (3) the "state of preservation" of palynomorphs reported by Correia (1967).

The divisions shown on the chart for that parameter are from Gutjahr (1966) and Correia (1967). The numerical values from Correia (1967), scale C on the chart, are ordinal divisions of a qualitative evaluation of the light adsorption. Those from Gutjahr (1966) are given in 10^{-3} μA on psilate trilete spores.

Organic Geochemistry Parameters of Kerogen

H/C, T.G.A., F.C./T.O.C.—The atomic ratio (H/C), elemental analysis, the percent weight loss or thermogravimetric analysis (T.G.A. at 500°C), and the carbon ratio (Cr/Ct = F.C./T.O.C.) on kerogen are global techniques as are proximate analysis, electron spin resonance (ESR), and infrared spectroscopy. For a better understanding of the behavior of the H/C ratio by combustion and F.C./T.O.C. by pyrolysis with increasing temperature, the carbon in the H/C parameter, which is equal to T.O.C. on kerogen, must be considered as the summation of F.C. plus volatile carbon. Thus, these parameters indicate the quantity of volatile matter (cf. fixed carbon) remaining in the organic material. This loss in volatile components from organic matter is confirmed by the decrease in the H/C ratio, the percent weight loss (T.G.A.), and the increase in carbon ratio (F.C./T.O.C.).

However, the mentioned parameters depend also on the type of dispersed organic matter. Indeed, organic matter derived from invertebrates, plankton, spores, pollen, and microscopic algae originally contains fewer aromatic components than lignin from continental plant material (Jonathan et al, 1976; Tissot, 1976; Van Dorsselaer and Albrecht, 1976; Welte, 1976). Moreover, sporinites show greater overall yields of total alkanes and straight chain alkanes than do the corresponding vitrinites at all ranks (Allan and Douglas, 1977). Hence, kerogen types I and II will give Cr/Ct ratios lower than 0.4 to 0.5 for freshly sedimented organic matter, indicating low aromaticity (high T.G.A. and H/C values). For freshly sedimented kerogen type III, the Cr/Ct ratios are in the range of 0.6 to 0.7 with low T.G.A. and H/C values. Usually, varied proportions of all these types of dispersed organic matter are found.

For these reasons a single scale cannot be established for these parameters. Thus, in the chart, we have indicated for each scale the kerogen type for which the curve is valid. Moreover, considering that this chart is mainly for petroleum geologists, the O/C atomic ratio has been omitted because it undergoes minor changes with increasing temperature for kerogen types I and II in the mature zone.

The foregoing parameters cannot be used alone because they are dependent upon temperature and type of organic matter. When enough organic extract is available, the Cr/Ct ratio can give indications of both the predominant type of dispersed organic matter and its maturation. Therefore, this ratio must be measured on both the asphaltene and kerogen fractions in the same sample (Connan, 1972, *in* Jonathan et al, 1976, p. 98, Fig. 5). Finally, according to Hunt (1978, p. 302): "By plotting the H/C ratio, which indicates differences in maturation, versus the (N+S)/O ratio, which shows differences in organic starting material, it is possible to distinguish most bitumens from coals."

Organic Geochemistry Parameters of Bitumen

Each of the following parameters was discussed in great detail by Le Tran et al (1973).

O.E./T.O.C.—This ratio, expressed in percent of the organic extract (O.E.) over the total organic carbon (T.O.C.), allows the evaluation of a rock "yield." This "yield" is very weak for immature organic matter. It starts increasing at the intense thermal cracking level (birth zone of oil) and then decreases when the organic matter gasification stage is reached (condensate gas or ~ 1.0% Ro) with a maximum yield between 0.8 to 1.0% Ro.

This ratio is also a function of the quantity and type of organic detritus and the alteration or migration effects of the accumulated extracts. Consequently, for type II kerogen, the numerical values shown in the chart are only an indication of a good source rock. Thus, for a mixture of type II (sporinite) and type III (vitrinite) organic matter, the maximum yield of n-alkanes might be at 1% Ro (Allan and Douglas, 1977). So, for the diagnosis of the "oil window" the general trend of the curves is more significant than the numerical values.

A./S. and A.+S./O.E.—The extract composition is generally expressed in percent of hydrocarbons (H.C.), resins (res.) and asphaltene (asp.). The hydrocarbons are subdivided into aromatic (A.) and saturated (S.) components. From these percentages, the ratio A./S. and the hydrocarbons "yield" A.+S./O.E. are calculated.

In general, for a sedimentary sequence where the organic matter is homogeneous, the A./S. ratio tends to decrease with increasing thermal cracking. However, in certain cases, the decrease is hardly detectable because an inverse behavior may occur, as in biologic alterations of liquid hydrocarbons associated with incoming meteoric waters (Connan et al, 1975). Unfortunately, homogeneous organic detritus in sedimentary sequences is not common. So, the nature of extracts is also dependent on the type organic detritus, the environment of deposition, the level of microbial activity (Connan et al, 1975), and the subsequent geologic history (Allan and Douglas, 1977).

Conversely, the A.+S./O.E. and especially the S./O.E. ratio tend to increase with thermal maturation, particularly at thermal maturation stages compatible with the presence of liquid hydrocarbons.

These two parameters, being only a more detailed expression of the O.E./T.O.C. ratio, have the same limitations. However, the loss of hydrocarbons during migration does not affect the A./S. and A.+S./O.E. ratios of remaining "in situ" hydrocarbons, although the migrated product shows some variations in these ratios. With some exceptions, the numerical values for these two parameters are more significant than those of the O.E./T.O.C. ratio in evaluating the maturation degree reached by the sequence. Thus, for a sequence where the organic fraction is sapropelic, a hydrocarbon yield (A.+S./O.E.) greater than 25% strongly suggests that the "oil window" has been reached.

C.P.I.—The carbon preference index (C.P.I.) is one of the parameters relevant to gas chromatographic (GC) analysis of total alkanes, and gives the nature of the n-alkanes extracted from a given rock sample. The n-alkanes distribution has been found to change with thermal maturation of a sedimentary basin and hence has long been used as a clue to crude oil source beds (Bray and Evans, 1961, *in* Allan and Douglas, 1977).

According to Bray and Evans (1961), C.P.I. values (scale b on the chart) range from 2.5 to 5.3 for recent muds, from 0.98 to 2.3 for sedimentary rocks, and from 0.92 to 1.13 for crude oil. Thus, values greater than 1.13 indicate immature sequences, from 0.92 to 1.13 mature zones, and less than 0.92 supramature zones. However, scale a from Allan and Douglas (1977, Fig. 4) shows the mature zone between 1 and 1.35. This last scale was derived from C.P.I. determinations on vitrinite and sporinite (kerogen types II and III) for subbituminous to high-volatile bituminous b coal stages ($0.4 < Ro\% < 1.20$).

The C.P.I. scale from Bray and Evans (b on the chart) seems to cover a wider range of coalification stages and results of measurements on sapropelic (kerogen type I or II or a mixture of both types) organic matter and also on organic matter from recent sediments including freshwater and saline-lake sediments and soils.

Scales a and b on the chart show decreasing C.P.I. with increasing thermal maturation. However, there are discrepancies between them. The different C.P.I. values at same coalification stages between scales a and b on the chart might be explained by variability both in the source of organic matter and in lithologies, as reported by Allan and Douglas (1977). Differences in slopes of curves are explained by Allan and Douglas (1977, p. 1228) by the following process: "fragmentation of the already soluble longer chain compounds to shorter chain constituents occurred at a greater rate than the rate of release of new longer chain structures from the parent material."

Organic Geochemistry Parameters of Gas

Mechanically extractable (adsorbed) gases and chemically extractable (sorbed: occluded) gases are discussed herein. Experimental methods are described for the former in Snowdon and McCrossan (1973, p. 3, 4) and for the latter in Le Tran (1971, p. 324).

$H_2S/T.O.C.$—The hydrogen sulfide parameter shown on the chart is restricted to sorbed and adsorbed occurrences of H_2S in sediments or sedimentary rocks; free or dissolved H_2S is excluded.

The yield of H_2S from organic matter increases with thermal maturation. There are other sources of H_2S, for example, bacterial reduction of sulfates and catalytic reaction of sulfate with hydrocarbons (Orr, 1974). However, for H_2S trapped in ancient rocks, when the trapping is not due to bacterial activities, Le Tran (1971) proposed two reasons: (1) the nature of the organic matter (e.g., algae, annelids, coelenterates, and mollusks constitute an original organic matter rich in sulfur; thus, there is a relation between the sediment facies and the sulfur content because the components are more concentrated in confined pelitic and micritic environments, such as swamps or lagoons); (2) the maturation degree of organic matter.

H.C.G./T.O.C.—The total content of mechanically and/or chemically extractable gases permits the calculation of the "yield" in gaseous hydrocarbons (H.C.G). This "yield" (H.C.G./T.O.C.) varies with the degree of thermal maturation and is a way to visualize the evolution from an early dry gas to the oil zone and finally to a late dry gas zone. Moreover, according to Snowdon and Mc-

Crossan (1973, p. 2), the epigenetic or syngenetic nature of the H.C.G. can be evaluated by the slope of the regression curve of H.C.G./T.O.C. versus burial depth. If the slope of the regression line is high, migration has occurred; if it is low, the H.C.G. have been retained in their source rocks.

Curve a in the chart is from Le Tran et al (1973). Data used to establish this curve were obtained from gases in 65 core samples of Upper Triassic dolomites and Lower Cretaceous limestones crushed and dissolved in acid solutions. Curve c in the chart is from our work on the Labrador offshore basin (in progress). Data compiled for curve c came from 1,000 measurements of adsorbed gases performed by the analytic method described by Snowdon and McCrossan (1973). The cuttings are mainly fine-grained, terrigenous lower Paleocene through Holocene rocks, and the sampling is fairly dense (30 to 60 ft or 9 to 18 m intervals).

The numerical values of curve a range from zero to 200 mL of H.C.G. per gram of rock. On curve c, this ratio, the percent of gas over the percent of carbon ranges from 0.01 to 0.25.

So far, no comparison can be made between scales a and c in the chart. They need to be extended up and down within the hydrocarbon generation stages using the same analytic method, on the same lithologic sequences.

$C_1/H.C.G.$—The generation of methane either biochemically during early diagenesis or physicochemically (by thermal cracking) during late catagenesis, is a well-known phenomenon.

The curve shown on the chart is from Snowdon and McCrossan (1973) and from our study of seven boreholes of the Labrador offshore. For thermal maturation interpretations, this scale should be used together with curve c of H.C.G./T.O.C., curve b of C_2/C_3+, and the $i-C_4/n-C_4$ curve.

C_2/C_3+—The presence of the ethane-butane series in the commonly called "early" and "late" condensate stages is very controversial. The low sensitivity of the analytic technique (Weber and Maximov, 1976, p. 288), together with our poor knowledge of the relation of condensate or wet gas constituents to the type of organic matter, considerably reduces the reliability of interpretations concerning gas evolution and especially the isomers ($i-C_4/n-C_4$). Curve b of C_2/C_3+ column on the chart should be used with caution because the content of adsorbed light hydrocarbon gases is sensitive to the lithology of sedimentary rocks (Feugère and Gérard, 1970, p. 38-39); this is corroborated by our Labrador offshore basin study.

Curve a for this scale is modified from Figure 3 of Feugère and Gérard (1970) and from Figure 15 of Jonathan et al (1976). Curve b of C_2/C_3+ on the chart is from our study of the Labrador offshore. The positions of the C_2/C_3+ and the $i-C_4/n-C_4$ curves with respect to the hydrocarbon generation stage are supported by exhaustive studies of major thermal maturation indicators (work in prep.).

Discrepancies between curves a and b occur only in the upper part of the condensate gas (early wet gas) zone. These curves coincide in the early wet gas zone and the uppermost part of the oil window. Unfortunately, we have no data from the lower part of the condensate and late dry gas zones, and it would be hazardous to make a downward extension of curve b with curve a for C_2/C_3+ in the chart.

$i-C_4/n-C_4$—The only publication that we have seen concerning these light hydrocarbon isomers is by Savčenko et al (1971), who mention the work of several authors. Savčenko et al (1971) concluded that this ratio is dependent on the oil-gas contact in time and space. The ratio is <0.8 if gas is dissolved in oil, extracted from oil, or is or was in contact with oil. The ratio is >0.9 if gas was never in contact with oil. Savčenko et al (1971) also stated that the ratio is independent of stratigraphic level, burial depth, and temperature.

In our Labrador offshore study, with a very dense sampling (30 to 60 ft or 9 to 18 m interval), the isomer concentration ratio log ($i-C_4/n-C_4$) shows the following variations: the beginning of the early wet gas zone (Ro=0.35%) is characterized by relatively high values (>1.0) of the ratio, which thereafter decreases to a relatively constant value (0.7 to 0.8) down to the beginning of the oil window. There, the $i-C_4/n-C_4$ ratio is not affected by T.O.C. variations.

In potential source rocks from offshore Labrador, we observed that the $i-C_4/n-C_4$ ratio reaches values of 0.8 in the upper wet gas zone but higher values for the ratio were obtained from more immature sedimentary rocks at shallower depths. As both the type and the quantity of organic matter are favorable for oil generation and migration effects are not significant, the question arises: for sequences above the oil window, are $i-C_4/n-C_4$ ratio values greater than 0.9 owing to the absence of oil or to insufficient maturation to generate the oil?

Although the basic principle ruling the butane isomer and its thermal maturation relation is still relatively unknown, our work on the Labrador offshore shows that there is a relation between the isomer concentration ratio curve shape and thermal maturation.

Mineralogic Parameters

The temperature and the chemical properties, in the diagenetic environments, control mineral evolution during early and late diagenesis. On the chart (first draft by Kübler, 1976, unpublished, copyright INRS), the mineralogic variations with depth and the index minerals are shown for three types of lithofacies: (a) normal shaly mudstones, (b) fine-grained volcanic detritus, (c) sodium-rich dark shales.

Data from carbonates and sandstones were not included. In these later facies, the significance of index minerals might not be the same as in pelitic sequences.

Normal shaly mudstones are the best documented lithofacies. The diagenetic indicators shown on the chart are the mineral assemblages, the position of the (001) reflection of smectite, the percent of illite layers in the mixed-layer illite −2:1 expandable (smectite of low or high charges), and the illite crystallinity index.

Clay mineralogy—The evolution of smectite toward well-crystallized mica and chlorite and the behavior of kaolinite and dickite, as illustrated on the chart, come from our own observations and from the work of such authors as Weaver (1960), Kübler (1964, 1975), Powers (1967), Burst (1969), Dunoyer de Segonzac (1969), Perry and Hower (1970, 1972), Weaver et al (1971), Johns and Shimoyama (1972), Foscolos and Kodama (1974), and Foscolos et al (1976). The development and transformation of chloritic minerals are less documented and the correlation of this series with other mineralogic and organic parameters is not very reliable at present. Therefore, data pertaining to these minerals are not on the chart.

D (001) of smectite—In the Cenozoic and Mesozoic sedimentary rocks from the Labrador and Nova Scotia Shelves, we observed that the (001) reflection at 14.5 A of smectite moved to 12.5 A to 13 A within the first diagenetic stages. Foscolos (1977, personal commun.) also reported the same observation. This peak displacement corresponds to the first loss of interlayer water in smectites and is correlated with the beginning of stage II as defined by Perry and Hower (1972).

Percent illite in illite −2:1 expandable—There are two methods for calculating the percent of expandable layers in interstratified minerals: the MacEwan et al (1961) method and the Reynolds and Hower (1970) method. The second one should give better results, but each has its advantages.

The illitic to expandable layers ratio was correlated with temperature by Burst (1969) and Foscolos and Kodama (1974). In the chart, this parameter was correlated with other thermal indicators on the basis of temperatures. We also took into account the work of Powers (1967), Perry and Hower (1970, 1972), Johns and Shomoyama (1972), and Foscolos et al (1976). On the chart, the upper inflection point of the "% illite in illite −2:1 expandables" curve corresponds to the beginning of Perry and Hower's (1972) stage II and the lower inflection point to stage III of the same authors.

Illite crystallinity index—The term illite crystallinity index, as used herein, is the width at half height of the peak located around 10 A on a diffractogram. As stated by Kübler (1968), the index values are comparable between laboratories only if instruments are calibrated with the same standards.

As already mentioned, the temperature and the chemical conditions (Kübler, 1968), of the burial environment, control the evolution of this mineral. Moreover, in this situation, a great difficulty arises in data interpretation because of the possible inherited characteristics of illite. Thus, these characteristics must be differentiated from the ones acquired after burial. Mineralogic studies on several granulometric fractions ($< 0.2\ \mu$, 0.2-$2.0\ \mu$ and 2-$20\ \mu$) must be performed routinely as a first step in solving the inherited or acquired characteristic problem. It is also important to consider the vertical trends based on systematic dense sampling on a sufficiently thick sequence (thousands of feet), and not only the numerical values as such.

On the chart, the I.A.G. ("indice d'aigüe glycolé," Kübler, 1968) curve represents index values obtained from glycolated fractions smaller than 2μ, whereas the I.A.N. ("indice d'aigüe naturel") is for unglycolated ones. The I.A.N. values may represent the total width of overlapping peaks. The I.A.G. curve deviation toward higher index (lower crystallinity) results from the appearance of disordered illite, which has wider diffraction peaks and does not swell with glycol.

The curves derived from coarser fractions treated with glycol ($> 2\mu$; not on the chart) show few variations and are used to confirm changes that occurred in incoming material at the time of sedimentation.

For fine-grained volcanic detritus, the index minerals shown in the chart are zeolites such as sodium (Na) and calcium (Ca) clinoptilolite, analcite, and laumontite together with other minerals such as pumpellyite, prehnite, smectite, and corrensite. In this facies, smectite evolves diagenetically to corrensite and then toward chlorite. The minerals listed also may be present in evaporitic sequences, limestones, or low-pressure hydrothermal deposits. However, their diagenetic significance may not be the same as in the fine-grained volcanic detritus shown in the chart. The zonation shown is mainly from the work of Kübler et al (1974) and Kübler (1975).

In the sodium (Na) dark shale facies, where the sodium and organic matter concentrations are both higher than in "normal shale," the main indicators are kaolinite, allevardite (rectorite), pyrophyllite, phengite, and paragonite. The data used here are from Dunoyer de Segonzac (1969) and Kübler (1975).

CONCLUSIONS

To use the accompanying charts presented here it should be kept in mind that, for most parameters, the curve shape is more significant than the numerical values. Moreover, at present, no known thermal parameter is self-sufficient in hydrocarbon exploration.

The confidence level of some parameters needs to be improved both in terms of analytic precision and relations between the measured values and hydrocarbon generation stages.

Other parameters such as those mentioned in the introduction could eventually be added to the chart to improve its usefulness and reliability.

APPENDIX–GLOSSARY

This glossary defines and/or discusses the terms and abbreviations which have been used in the text and figures to characterize thermal maturation in sedimentary rocks and in their analytic methods and which do not appear in the *Glossary of Geology* (Gary et al, 1973) or for which there are differences of usage. The listing of terms and abbreviations is alphabetic.

A.: aromatic.
A./S.: aromatic to saturate ratio.
A.+S./O.E.: (aromatic + saturate) to organic extract ratio; yield in hydrocarbons.
Alkane: see Saturate.
Anchimetamorphism: indicates changes in mineral and organic matter content of rocks under physico-chemical conditions prevailing between catagenesis and true metamorphism (greenschist facies). This term, introduced by Harrassowitz in 1927 (Gary et al, 1973) had an initial definition which encompassed both diagenetic and catagenetic zones, and was restricted to changes in the mineral content of rocks.
Aromatic (A.): aromatic hydrocarbons.
Arrhenius equation: $K = Ae^{-E/RT}$, equation used to establish a relation between chemical reaction rate (K) and temperature (T). The equation can be slightly modified (Connan, 1974) to establish temperature and time relation for O.M. maturation: $\ln(t) = (E/RT) - A$, where E = activation energy (Kcal/mole); A = frequency factor; R = gas constant; T = absolute temperature; and t = time in million years.
Asp.: asphaltene.
Biomonomers: originally formed from bipolymers which are converted to sugars, amino acids, fatty acids, and phenols while incorporated in the sediments under the action of microbes through the use of enzymes. This biodegradation process is confined to the first meter of sediments (Hunt, 1973).
Biopolymers: principal components of living organisms which group carbohydrates, proteins, lipids, lignin, and subgroups such as chitin, glycosides, pigments, fats, and essential oils (Hunt, 1973).

Birth zone of oil: designates the range in which physico-chemical processes begin to generate naphthabitumen from thermal cracking of organic matter. It is the upper boundary of petroleum generation (Stach et al, 1975, p. 327).
Bitumen: herein restricted to liquid phase of naphthabitumen, synonymous with O.E. as opposed to pyrobitumen or kerabitumen which are insoluble in CS_2.
$C_1/H.C.G.$: methane to total hydrocarbon gas ratio.
C_2/C_3+: ethane to propane plus butane isomers ratio.
$i-C_4/n-C_4$: iso-butane to normal butane ratio.
Cal./gr. (Cal./g.): calorific value per gram (= 1.8 Btu/lb).
Carbon preference index (C.P.I.): expression of the odd-even predominance in the n-alkanes spectrum (C.P.I.>1) = odd predominance, (C.P.I.<1) = even predominance, (C.P.I.=1) = no predominance. It is calculated by taking the intensity ratio of the odd to even peaks (area under the curves) for n-paraffins of 25 to 32 carbon atoms. Three ways of calculating this C.P.I. are:

1) Bray and Evans (1961) C.P.I. =
$$\frac{\sum_{25}^{33} \text{n-C odd}}{\sum_{24}^{32} \text{n-C even}} + \frac{\sum_{25}^{33} \text{n-C odd}}{\sum_{26}^{34} \text{n-C even}}.$$

2) Hunt (1973) C.P.I. =
$$\frac{\sum_{25}^{31} \text{n-C odd}}{\sum_{24}^{30} \text{n-C even}} + \frac{\sum_{25}^{31} \text{n-C odd}}{\sum_{26}^{32} \text{n-C even}}.$$

3) approximation (Louis, 1967), C.P.I. =
$$\frac{2\text{n-}C_{29}}{\text{n-}C_{28} + \text{n-}C_{30}}.$$

Carbon ratio (Cr/Ct) or (F.C./T.O.C.): parameter applied to kerogen characterization after pyrolysis at 500 or 900°C. It is the ratio of residual carbon (Cr) or fixed carbon (F.C.) to the total organic carbon (T.O.C. or Ct); the carbon ratio is never established from total organic matter, but only from either kerogen or asphaltene.
Catagenesis (katagenesis): in this report, the catagenetic zone is considered equivalent to the stage of generation and destruction of hydrocarbons by thermal cracking and, therefore, follows definition given by Fersman in 1922 (Gary et al, 1973). In the modern interpretation of Soviet authors, catagenesis includes stages of primary change in rock whereas many American scientists consider it to be equivalent to "late diagenesis" (after Weber and Maximov, 1976). In terms of O.M. transformation, this stage corresponds to geomonomers formation or generation of naphthabitumen or bituminous coal from thermal transformation of kerogen.
Cation exchange capacity: (C.E.C.) the amount of negative charges compensated for by exchangeable cations (Fripiat et al, 1971); expressed in milliequivalent per gram, or per 100 grams.
Caustobiolith (kaustobiolite): rocks which contain an important quantity of organic compounds, or even pure carbon if the latter is of organic origin.
Coloration (color): color of spore and pollen observed in transmitted light microscopy. This parameter gives a clue to the degree of carbonization of the least altered constituents of a palynologic assemblage in a given rock (Allan and Douglas, 1977).
C.P.I.: see carbon preference index.
Death zone of oil: limit of occurrence of in-situ-generated liquid hydrocarbons. It is the lower boundary of petroleum occurrence (Stach et al, 1975, p. 326).
Degree of organic metamorphism: equivalent to many other ex-

pressions to indicate stages of organic transformation or alteration. This scale is based on the rank of humic coal.
Diagenesis: this term as employed in this report and by Soviet authors, is equivalent to early diagenesis (sediment diagenesis) of American scientists. In terms of organic-matter transformation it corresponds to the formation of kerogen from freshly sedimented organic matter, a process which is most effective from surface to 1,000 m (50°C; Hunt, 1973).
Dilatation: analytic parameter for testing the degree of carbonization of coal (coalification) and then the coking power of that coal. Dilatation, in percent, indicates the remaining amount of volatile matter in coal. For more details, see Stach et al (1975, p. 342-344).
Dispersed organic matter: this term is equivalent to organoclast and is sometimes extended to include kerogen.
D.O.M.: this abbreviation should be avoided because it could bring confusion of "dispersed organic matter," "degree of organic metamorphism," and "diagenesis of organic matter" (Tissot et al, 1974), these last two expressions being quite similar.
Exudatinite: this term was defined by Teichmüller (1974) as a secondary bitumen which is a maceral of the liptinite group.
F.C.: fixed carbon.
F.C./T.O.C.: see carbon ratio.
F.I.D.: see flame ionization detector.
Flame ionization detector (F.I.D.): specific detector used in gas chromatography techniques to record hydrocarbons (Barker, 1972; Claypool and Reed, 1976).
Flame photometry detector (F.P.D.): specific detector used in gas chromatography techniques to record sulfur (especially informative in aromatic hydrocarbon analysis to examine compounds containing sulfur thiophenic compounds).
FLUO.: fluorescence.
Fluorinite: liptinite group maceral formed by oil secretion of plants. This maceral is highly transparent in transmitted light with very weak Ro and is the most fluorescent among macerals of the liptinite group.
F.P.D.: see flame photometry detector.
FUS.: fusinite.
Geomonomers: low molecular weight hydrocarbons and organic compounds that are formed by a thermal-alteration-dependent process involving cracking of large molecules to form small compounds; they occur mainly in the 50°C (1,000 m) to 175°C (6,000 m) temperature interval.
Geopolymers: derived from condensation of biomonomers in the first meter of sediment to form organic matter complexes (Hunt, 1973).
H/C: hydrogen to carbon atomic ratio, see ultimate analysis.
H.C.: hydrocarbons (aromatic + saturate).
H.C.G.: low molecular hydrocarbon gases ranging from methane (C_1) to butane (C_4).
I.A.G. ("Indice d'Aigüe des illites Glycolées"): the width at half height of the ± 10 A reflection of illite, on ethylene glycol saturated samples (Kübler, 1968).
I.A.N. ("Indice d'Aigüe des illites Naturelles"): same as the I.A.G., but on untreated samples (Kübler, 1968).
Insoluble organic matter: see organoclast.
Kerogen: equivalent to organoclast or geopolymers of Hunt (1973).
Kerabitumen: equivalent to organoclast.
Liquid window concept: see oil window.
Metamorphism: to avoid confusion with the definition of metamorphism accepted by hard-rock petrologists, we suggest that use of this term be avoided when qualifying the transformation degree of dispersed organic matter in sedimentary rocks and recommend the use of the term be restricted to the definition given by Turner and Verhoogen (1960, *in* Gary et al, 1973).
Naphthabitumen: the natural gas, liquid, and solid mixture of hydrocarbons and hydrocarbonlike compounds derived from caustobioliths. They are soluble in CS_2 and include principally gas, crude oil (O.E.), maltha, asphalt, asphaltite, ozocerite, asphaltoide (Louis, 1967).
O/C: oxygen to carbon atomic ratio. See ultimate analysis.

O.E.: organic extract. Liquid phase of naphthabitumen. Commonly used in oil exploration geochemistry to indicate oily components (hydrocarbons sensu stricto: A. + S.), asphaltenes, resin, carbenes, and carbenoids. These two last components, present in sediments, quickly disappear during diagenesis. Moreover, they are insoluble in CCl_4. The solubility of the different phases in some conventional organic solvents was indicated by Louis (1967).
O.E./T.O.C.: organic extract to total organic carbon ratio. This analytic parameter indicates the organic extract yield of a sedimentary rock.
Oil window: stage of hydrocarbon generation and preservation where crude oil exists. Also see birth and death zones of oil.
O.M.: organic matter.
Organoclast: the fraction of the dispersed organic matter in the sediment or sedimentary rocks that is insoluble in organic solvents (CCl_4, $CHCl_3$, C_6H_6, and CH_3OH, etc). This insoluble organic matter is called organolite or organoclast (Alpern, 1970), kerogen (Brown, 1912, *in* Burgess, 1974, p. 21), or kerabitumen (Louis, 1967).
Peaks (001/001) 10/17 A, (002/003) 10/17 A, etc: on a diffractogram, a peak resulting from a combination of the first reflection (001) of illitic layers at 10 A and the first one of the glycolated smectitic layers at 17 A (001/001) 10/17 A, or of the second reflection (002) of illite and the third (003) of smectite (002/002) 10/17 A, etc. The position of these peaks is a function of the relative proportion of each type of layer in a mixed-layer mineral.
Phytol: an isoprenoid unsaturated alcohol with 20 carbon atoms. During its transformation, chlorophyll loses the phytol group as isoprenoid carbides (pristane-phytane). These isoprenoids are biologic markers (Louis, 1967, p. 20).
Polymorph 2M: 2 is for 2 layers in the unit cell and M means monoclinic. One way of stacking ideal mica layers in an ordered manner in which the rotation is alternatively + 120° and − 120°.
Pyrobitumen: as opposed to bitumen (see this last term). For asphaltic and nonasphaltic pyrobitumen classifications, refer to Abraham (1960), Bell and Hunt (1963), and Hunt et al (1954).
Ra%: reflectance in air.
Reflectance: "The ratio of reflected radiant flux to incident radiant flux" (Gary et al, 1973). The reflectance (R) of a particle is related to its refraction index (n), the adsorption index (K), and the refraction index (N) of the immersion field (air = a and oil = O). The relation is given by the Fresned-Beer equation: $R = (n-N)^2 + n^2K^2/(n+N)^2 + n^2K^2$.
Res.: resinite: exinite group maceral.
Ro%: reflectance in oil.
S.: saturated hydrocarbons: open chain (straight = n-alkane or paraffinic; branched = iso-alkane) or cyclic-type (cyclo-alkanes = naphthenic) organic structure that contains only single bonds between carbon atoms.
Sharpness ratio: the ratio of the total height of the 10 A reflection over the height of the flank of the same peak at 10.5 A (Weaver, 1960).
T.G.A.: thermogravimetric analysis.
Thermal alteration index (T.A.I.): analytic parameter commonly used to characterize the degree of alteration of organic matter by measuring pollen and spore color in transmitted light.
T.O.C.: total organic carbon.
Ultimate analysis: elemental analytic techniques that record parameters such as carbon, hydrogen, nitrogen, sulfur, and iron content of kerogen. The difference from $(100-[C+H+N])$ is reported as oxygen if the analysis is performed on dry, ash-free organic matter (usually vitrinite). From ultimate (elemental) analysis, one may establish the H/C and O/C atomic ratio. H/C is less than 1.0 for ligneous organic matter. O/C is less than 0.1 for sapropelic organic matter. H/C versus N+S/O atomic ratios of kerogen should differentiate many coals from bitumens (Hunt, 1978).
Vi.: vitrinite.
V.M.: volatile matter.

Major Thermal Maturation Indicators

REFERENCES CITED

Abraham, H., 1960, Asphalts and allied substances, v. 1: New York, D. Van Nostrand, 326 p.

Allan, J., and A. G. Douglas, 1977, Variations in the content and distribution of n-alkanes in a series of Carboniferous vitrinites and sporinites of bituminous rank: Geochim. et Cosmochim. Acta, v. 41, p. 1223-1230.

Alpern, B., 1970, Classification pétrographique des constituants organiques fossiles des roches sédimentaires: Inst. Francais du Pétrole Rev., v. 29, p. 1233-1267.

─── 1972, Pétrographie des charbons—bilan des progrès acquis de 1967 à 1971: 7th Cong. Internat. Stratigraphie et Géologie Carbonifère, Krefeld, 1971, Compte Rendu, v. 1, p. 91-126.

─── 1973, Introduction—indices optiques de la matière organique des sédiments, relations avec la paléotempérature et le potentiel pétrolier, in B. Alpern, ed., Pétrographie organique et potentiel pétrolier: Paris, Editions Centre Natl. Recherche Sci., p. 191-193.

─── 1976a, Les combustibles fossiles: Rev. Sciences Encyclopédie Alpha, fasc. 106, p. 169-198.

─── 1976b, Fluorescence et réflectance de la matière organique dispersée et évolution des sédiments (fluorescence and reflectance of dispersed organic matter and evolution of the sediments): Centre Recherche Pau Bull., v. 10, p. 201-220.

─── et al, 1971, Localisation, caractérisation et classification pétrographique des substances organiques sédimentaires fossiles, in Advances in organic geochemistry, 1971: Internat. Ser. Mon. Earth Sci., v. 33, p. 1-28.

Bailey, N. J. L., C. R. Evans, and M. A. Rogers, 1971, Evolution and alteration of petroleum in western Canada: Chem. Geology, v. 8, p. 147-170.

Baker, D. R., and G. E. Claypool, 1970, Effects of incipient metamorphism on organic matter in mudrock: AAPG Bull., v. 54, p. 456-468.

Barker, C., 1972, Pyrolysis techniques for source rock evaluation (abs.): Geol. Soc. America Abs., v. 4, p. 443.

Barrabé, L., and R. Feys, 1965, Géologie du charbon et des bassins houillers: Paris, Masson et Cie, 229 p.

Bell, K. G., and J. M. Hunt, 1963, Native bitumens associated with oil shales: Internat. Ser. Mon. Earth Sci., v. 16, p. 333-366.

Bostick, N. H., 1971, Thermal alteration of clastic organic particles as an indicator of contact and burial metamorphism in sedimentary rocks: Geoscience and Man, v. 3, p. 83-92.

─── 1973, Time as a factor in thermal metamorphism of phytoclasts (coaly particles): 7th Cong. Internat. Stratigraphie et Géologie Carbonifère, Krefeld, 1971, Compte Rendu, v. 2, p. 183-193.

─── 1974, Phytoclasts as indicators of thermal metamorphism, Franciscan assemblage and Great Valley sequence (upper Mesozoic), California, in Carbonaceous materials as indicators of metamorphism: Geol. Soc. America Spec. Paper 153, p. 1-17.

Bray, E. E., and E. D. Evans, 1961, Distribution of n-paraffins as a clue to recognition of source beds: Geochim. et Cosmochim Acta, v. 22, p. 2-15.

Burgess, J. D., 1974, Microscopic examination of kerogen (dispersed organic matter) in petroleum exploration, in Carbonaceous materials as indicators of metamorphism: Geol. Soc. America Spec. Paper 153, p. 19-30.

Burst, J. F., 1969, Diagenesis of Gulf Coast clayey sediments and its possible relation to petroleum migration: AAPG Bull., v. 53, p. 73-93.

Castaño, J. R., and D. M. Sparks, 1974, Interpretation of vitrinite reflectance measurements in sedimentary rocks and determination of burial history using vitrinite reflectance and authigenic minerals, in Carbonaceous materials as indicators of metamorphism: Geol. Soc. America Spec. Paper 153, p. 31-52.

Claypool, G. E., and P. R. Reed, 1976, Thermal-analysis technique for source-rock evaluation; quantitative estimate of organic richness and effects of lithologic variation: AAPG Bull., v. 60, p. 608-612.

Connan, J., 1970, Contribution à la connaissance de la matière organique de quelques sédiments marins et lacustres: PhD thesis, Univ. Strasbourg, 75 p.

─── 1973, Diagenèse naturelle et diagenèse artificielle de la matière organique à éléments végétaux prédominants (abs.), in Advances in organic geochemistry, 1973: Internat. Mtg. Organic Geochemistry Program Abs., no. 6, p. 73-94.

─── 1974, Time-temperature relation in oil genesis: AAPG Bull., v. 58, p. 2516-2521.

─── K. Le Tran, and B. Van Der Weide, 1975, Alteration of petroleum in reservoirs: 9th World Petroleum Cong. Tokyo, Proc., v. 2, p. 171-178.

Correia, M., 1967, Relations possibles entre l'état de conservation des éléments figurés de la matière organique (microfossiles palynoplanctonologiques) et l'existence de gisements d'hydrocarbures: Inst. Français Pétrole Rev., v. 22, p. 1285-1306.

Deroo, G., et al, 1977, The origin and migration of petroleum in the Western Canadian sedimentary basin, Alberta. A geochemical and thermal maturation study: Canada Geol. Survey Bull. 262, 136 p.

Dunoyer de Segonzac, G., 1969, Les minéraux argileux dans la diagenèse. Passage au métamorphisme: Thesis, Univ. Strasbourg, 339 p.

Durand, B., et al, 1976, Etude de kérogènes par résonnance paramagnétique électronique (Study of kerogens by means of electron paramagnetic resonance; abs.): Centre Recherche Pau Bull., v. 10, p. 267-269.

Espitalié, J., et al, 1973, Etude de la matière organique insoluble (kérogène) des argiles du Toarcien du Bassin de Paris. Pt. 2, Etudes en spectroscopie infrarouge, en analyse thermique différentielle et en analyse thermogravimétrique: Inst. Français Pétrole Rev., v. 28, p. 37-67.

Feugère, G., and R. E. Gérard, 1970, Geochemical logging—a new exploration tool: World Oil, v. 170, no. 2, p. 37-40.

Foscolos, A. E., and H. Kodama, 1974, Diagenesis of clay minerals from Lower Cretaceous shales of north eastern British Columbia: Clays and Clay Minerals, v. 22, p. 319-335.

────── and D. F. Stott, 1975, Degree of diagenesis, stratigraphic correlations and potential sediment sources of Lower Cretaceous shale of northeastern British Columbia: Canada Geol. Survey Bull. 250, 46 p.

────── T. G. Powell, and P. R. Gunther, 1976, The use of clay minerals and inorganic and organic geochemical indicators for evaluating the degree of diagenesis and oil generating potential of shales: Geochim. et Cosmochim. Acta, v. 40, p. 953-966.

Fripiat, J., J. Chaussidon, and A. Jelli, 1971, Chimie-physique des phénomènes de surface. Application aux oxydes et aux silicates: Paris, Masson et Cie, 387 p.

Gary, M., R. McAfee, Jr., and C. L. Wolf, eds., 1973, Glossary of geology, 2d printing: Am. Geol. Inst., 805 p.

Gunther, P. R., 1976, Palynomorph color and dispersed coal particle reflectance from three MacKenzie delta boreholes: Geoscience and Man, v. 15, p. 35-39.

Gutjahr, C. C. M., 1966, Carbonization of pollen grains and spores and their application: Leidse Geol. Meded., v. 38, p. 1-30.

Hacquebard, P. A., and J. R. Donaldson, 1970, Coal metamorphism and hydrocarbon potential in the upper Paleozoic of the Atlantic provinces, Canada: Canadian Jour. Earth Sci., v. 7, p. 1139-1163.

Hood, A., C. C. M. Gutjahr, and R. L. Heacock, 1975, Organic metamorphism and the generation of petroleum: AAPG Bull., v. 59, p. 986-996.

Hunt, J. M., 1973, Organic geochemistry of the marine environment (abs.): in Advances in organic geochemistry, 1973: Internat. Mtg. Organic Geochemistry Program Abs., no. 6, p. 593-605.

────── 1977, Distribution of carbon as hydrocarbons and asphaltic compounds in sedimentary rocks: AAPG Bull., v. 61, p. 100-104.

────── 1978, Characterization of bitumens and coals: AAPG Bull., v. 62, p. 301-302.

────── F. Stewart, and P. A. Dickey, 1954, Origin of hydrocarbons of Uinta basin, Utah: AAPG Bull., v. 38, p. 1671-1698.

International Committee for Coal Petrology, 1963, 1971, 1975, International handbook of coal petrography, 2d ed. (1963); supp. to 2d ed. (1971): Paris, Centre Natl. Recherche Sci., 1 vol. (variously paged); suppl. 1975: Internat. Carboniferous Cong., Moscow, 1975.

Johns, W. D., and A. Shimoyama, 1972, Clay minerals and petroleum-forming reactions during burial and diagenesis: AAPG Bull., v. 56, p. 2160-2167.

Jonathan, D., et al, 1976, Les méthodes d'étude physico-chimiques de la matière organique (physico-chemical study methods of the organic matter; abs.): Centre Recherche Pau Bull., v. 10, p. 89-108.

Kaplan, M. Ye., 1971, Criteria for determining zones of catagenesis in terrigenous deposits: Internat. Geology Rev., v. 13, p. 1365-1376.

Kartzev, A. A., et al, 1971, The principal stage in the formation of petroleum: 8th World Petroleum Cong., Moscow, Proc., v. 2, p. 3-11.

Karweil, J., 1956, Die Metamorphose der Kohlen vom Standpunkt der physikalischen Chemie: Deutsch Geol. Gesell. Zeitschr., v. 107, p. 132-139.

Kübler, B., 1964, Les argiles indicateurs de métamorphisme: Inst. Français Pétrole Rev., v. 19, p. 1093-1112.

────── 1968, Evaluation quantitative du métamorphisme par la cristallinité de l'illite. Etat des progrès réalisés ces dernières années: Centre Recherche Pau Bull., v. 2, p. 385-397.

────── 1973, La corrensite, indicateur possible de milieu de sédimentation et du degré de transformation d'un sédiment (Corrensite, a possible guide to the environment of sedimentation and degree of transformation of a sediment): Centre Recherche Pau Bull., v. 7, p. 543-556.

────── 1975, Diagenèse-anchimétamorphisme et métamorphisme: Quebec, Inst. Natl. Recherche Sci., unpublished course notes and tables, 84 p.

────── J. Martini, and M. Vuagnat, 1974, Very low grade metamorphism in the western Alps: Schweizer. Mineral. u. Petrog. Mitt., v. 54, p. 461-469.

────── et al, 1979, Sur le pouvoir réflecteur de la vitrinite dans quelques roches du Jura, de la molasse et des nappes préalpines, helvétiques et penniques (Suisse occidentale et Haute-Savoie): Eclogae Geol. Helvetiae (in press).

Landes, K. K., 1966, Eometamorphism can determine oil floor: Oil and Gas Jour., v. 64, no. 18, p. 172-177.

────── 1967, Eometamorphism, and oil and gas in time and space: AAPG Bull., v. 51, p. 828-841.

Le Tran, K., 1971, Etude géochimique de l'hydrogène sulfuré adsorbé dans les sédiments (Geochemical study of hydrogen sulfide adsorbed in sediments): Centre Recherche Pau Bull., v. 5, p. 321-332.

────── J. Connan, and B. Van Der Weide, 1973, Problèmes relatifs à la formation d'hydrocarbures et d'hydrogène sulfuré dans le bassin sud-ouest aquitain (abs,), in Advances in organic geochemistry, 1973: Internat. Mtg. Organic Geochemistry Program Abs., no. 6, p. 761-789.

Leythaeuser, D., 1973, Effects of weathering on organic matter in shales: Geochim. et Cosmochim. Acta, v. 37, p. 113-120.

Louis, M., 1967, Cours de géochimie du pétrole; cours de l'Ecole Nationale Supérieure du Pétrole et des Moteurs: Paris, Inst. Français Pétrole, Editions Technip, 295 p.

────── and B. Tissot, 1967, Influence de la température et de la pression sur la formation des hydrocarbures dans les argiles à kérogène: 7th World Petroleum Cong., Mexico City, Proc., v. 2, p. 47-60.

MacEwan, D. M. C., A. Ruiz Amil, and G. Brown, 1961, Interstratified clay minerals, in G. Brown, ed., X-ray identification and crystal structure of clay minerals: Mineral. Soc. London, p. 393-424.

Marchand, A., 1976, La résonnance paramagnétique électronique (R.P.E.). Sa mise en oeuvre pour l'étude des kérogènes (Electron spin resonance [E.S.R.] and its utilization in the study of kerogen): Centre Recherche Pau Bull., v. 10, p. 253-266.

McCartney, J. T., and M. Teichmüller, 1972, Classification of coals according to degree of coalification by reflectance of the vitrinite component: Fuel, v. 51, p. 64-68.

McIver, R. D., 1967, Composition of kerogen—clue to its role in the origin of petroleum: 7th World Petro-

leum Cong., Mexico City, Proc., v. 2, p. 25-36.
Orr, W. L., 1974, Changes in sulfur content and isotopic ratios of sulfur during petroleum maturation—Study of Big Horn basin Paleozoic oils: AAPG Bull., v. 58, p. 2295-2318.
Ottenjann, K., M. Teichmüller, and M. Wolf, 1974, Spectral fluorescence measurements of sporinites in reflected light and their applicability for coalification studies, in B. Alpern, 1975, ed., Pétrographie de la matiere organique des sediments, relations avec la paleotemperature et le potentiel petrolier: Paris, Centre National Recherche Sci., p. 49-65.
Patteisky, K., and M. Teichmüller, 1958, Examen des possibilités d' emploi de diverses échelles pour la mesure du rang des charbons et propositions pour la délimitation des principaux stades de houillification: Rev. Industrie Minérale, no. spéc., p. 121-137.
Perry, E. A., Jr., and J. Hower, 1970, Burial diagenesis in Gulf Coast pelitic sediments: Clays and Clay Minerals, v. 18, p. 165-177.
―――― ―――― 1972, Late-stage dehydration in deeply buried pelitic sediments: AAPG Bull., v. 56, p. 2013-2121.
Philippi, G. T., 1965, On the depth, time and mechanism of petroleum generation: Geochim. et Cosmochim. Acta, v. 29, p. 1021-1049.
Powell, T. G., et al, 1977: Paper presented by Foscolos at INRS-Pétrole, March 29, 1977.
Powers, M. C., 1967, Fluid-release mechanisms in compacting marine mudrocks and their importance in oil exploration: AAPG Bull., v. 51, p. 1240-1254.
Pusey, W. C., 1973, The ERS method: a new technique of estimating the organic maturity of sedimentary rocks: Petroleum Times, January 12, p. 21-24, 32.
Raynaud, J. F., and P. Robert, 1976, Les méthodes d'étude optique de la matière organique (Optical methods of studying organic matter): Centre Recherche Pau Bull., v. 10, p. 109-127.
Reynolds, R. C., and J. Hower, 1970, the nature of interlayering in mixed-layer illite-montmorillonites: Clays and Clay Minerals, v. 18, p. 25-36.
Robert, P., 1973, Analyse microscopique des charbons et des bitumes dispersés dans les roches et mesure de leur pouvoir réflecteur. Application à l'étude de la paléogéothermie des bassins sédimentaires et de la genèse des hydrocarbures (abs.), in Advances in organic geochemistry, 1973: Internat. Mtg. Organic Geochemistry Program Abs., no. 6, p. 549-569.
Savčenko, J. P., A. K. Karpar, and Ja. A. Bereto, 1971, De l'intérêt en prospection, de la valeur du rapport de l'isobutane au butane normal dans les gaz d'hydrocarbures: Centre Recherche Pau, Gas. Prom., no. 4, p. 2-5 (French transl.).
Sikander, A. H., and J. L. Pittion, 1976, Reflectance studies on organic matter in lower Paleozoic sediments: Comm. 4th Mtg. Canadian Coal Petrographers Group (CCPG), 12 p.
―――― ―――― 1978, Reflectance studies on organic matter in Lower Paleozoic sediments of Quebec: Bull. Canadian Petroleum Geology, v. 26, p. 132-151.
Snowdon, L. R., and R. G. McCrossan, 1973, Identification of petroleum source rocks using hydrocarbon gas and organic carbon content: Canada Geol. Survey Paper 72-36, 12 p.

―――― and K. H. Roy, 1975, Regional organic metamorphism in the Mesozoic strata of the Sverdrup basin: Bull. Canadian Petroleum Geology, v. 23, p. 131-148.
Stach, E., et al, 1975, Stach's textbook of coal petrology, 2d ed.: Berlin, Gebrüder Borntraeger, 428 p.
Staplin, F. L., 1969, Sedimentary organic matter, organic metamorphism, and oil and gas occurrence: Bull. Canadian Petroleum Geology, v. 17, p. 47-66.
Teichmüller, M., 1973, Generation of petroleum-like substances in coal seams as seen under the microscope (abs.), in Advances in organic geochemistry, 1973: Internat. Mtg. Organic Geochemistry Program Abs., no. 6, p. 379-393.
―――― 1974, Formation et transformation des matières bitumineuses dans les charbons en relation avec la genèse et l'évolution des hydrocarbures: Fortschr. Geol. Rheinland Westfalen, v. 24, p. 65-112 (in German); French transl. by P. Robert, 1977, in Houillification et pétrole—contribution de la pétrologie des charbons à l'exploration de l'huile et du gaz naturel: BRGM, no. 5477, p. 79-138.
Tissot, B., 1976, La transformation de la matière organique (transformation of organic matter; abs.): Centre Recherche Pau Bull., v. 10, p. 87.
―――― J. Espitalié, 1975, L'évolution thermique de la matière organique des sédiments: application d'une simulation mathématique—potentiel pétrolier des bassins sédimentaires et reconstitution de l'histoire thermique des sédiments: Inst. Français Pétrole Rev., v. 30, p. 743-775 (English abs.).
―――― J. L. Oudin, and R. Pelet, 1971, Critères d'origine et d'évolution des pétroles. Application à l'étude géochimique des bassins sédimentaires, in Advances in organic geochemistry (1971): Internat. Ser. Mon. Earth Sci., v. 33, p. 113-134.
―――― et al, 1974, Influence of nature and diagenesis of organic matter in formation of petroleum: AAPG Bull., v. 58, p. 499-506.
Van Dorsselaer, A., and P. Albrecht, 1976, Marquers biologiques: origine, évolution et applications (Biological markers: origin,, evolution and applications): Centre Recherche Pau Bull., v. 10, p. 193-200.
―――― ―――― and J. Connan, 1975, Changes in composition of polycyclic alkanes by thermal maturation (Yallourn lignite, Australia): Advances in Organic Geochemistry, p. 53-59.
Van Gijzel, P., 1973, Polychromatic UV-fluorescence microphotometry of fresh and fossil plant substances with special references to the location and identification of dispersed organic matter in rocks, in B. Alpern, ed., Pétrographie de la matière organique des sédiments, relations avec la paléotempérature et le potentiel pétrolier: Editions Centre Natl. Recherche Sci., p. 67-91.
Van Krevelen, D. W., 1961, Coal-typology, chemistry, physics, constitution: New York, Elsevier Pub. Co., 514 p.
Vassoyevich, N. B., et al, 1970, Principal phase of oil formation: Moskov Univ. Vestnik 1969, no. 6, p. 3-27 (in Russian); English transl.: Internat. Geology Rev., v. 12, p. 1276-1296.
Waples, D., and J. Connan, 1976, Time-temperature re-

lation in oil genesis: discussion and reply: AAPG Bull., v. 60, p. 884-887.

Weaver, C. E., 1960, Possible uses of clay minerals in search for oil: AAPG Bull., v. 44, p. 1505-1518.

——— K. C. Beck, and C. O. Pollard, 1971, Clay water diagenesis during burial: How mud becomes gneiss: Geol. Soc. America Spec. Paper 134, 96 p.

Weber, V. V., and S. P. Maximov, 1976, Early diagenetic generation of hydrocarbon gases and their variations dependent on initial organic composition: AAPG Bull., v. 60, p. 287-293.

Welte, D. H., 1976, The nature of organic matter in sediments (abs.): Centre Recherche Pau Bull., v. 10, p. 85.

White, D., 1915, Some relations in the origin between coal and petroleum: Jour. Wash. Acad. Sci., v. 5, p. 189-213.

MIGRATION

From
Dynamics of Oil and Gas Accumulations
by Alain Perrodon

1.3. PRIMARY MIGRATIONS AND THE SOURCE ROCK CONCEPT

The study of hydrocarbon genesis cannot be separated from that of their expulsion from the source rock towards the reservoirs. A close relationship is generally observed between the oil genesis peak and the onset of primary migration (Jones, 1981), and one may well ask whether the 'manufacture' of oil or gas in the source rock is not facilitated by the more or less simultaneous departure of the resulting products. Moreover, hydrocarbon expulsion mechanisms appear to exert an influence on the very nature of these bodies.

It is customary, and logical, to distinguish between primary and secondary migrations, not for reasons of mechanisms, but due to the difference in characteristics of the pore volumes. The first occurs immediately after hydrocarbon genesis, from fine-grained rocks to a reservoir, and only involves short or very short distances, from around a meter to about ten meters; the second occurs within the permeable rocks and may continue and extend over distances and periods which are much longer (Roberts & Corpel, 1980). Primary migration processes may be divided into two main phases:
— expulsion of hydrocarbons from the kerogen and the source rock,
— transfer of these hydrocarbons through the source rock to its boundaries, in contact with more porous and permeable series.

1.3.1. Different modes of primary migration

It is first necessary to specify the dimensions of the objects to be moved and to compare them with the void configuration of the solid medium through which they are to pass. Migrating hydrocarbons include methane, a gas whose molecules are about 4 Å in diameter (the water molecule has a diameter of 3.2 A), and heavy, solid compounds of the asphaltene type, with molecular diameters up to 50 to 100 A. As for the solid, the pore or micropore diameters of shales range from 50 to 100 Å at a depth of some 2000 m, and are less than 50 Å deeper down.

Theoretically, hydrocarbons may be considered to migrate in the following forms:
— true molecular solutions,
— colloidal and micellar solutions,
— individualized oil or gas phases.

Transport may be considered in the form of a homogeneous flow or in the form of diffusions through a concentrated 'solution'. It is also important to consider whether the water and hydrocarbons transported form a single- or multi-phase mixture, or whether the solid medium is preferentially wetted by oil or by water. The water/hydrocarbon relationship must also be examined, as these hydrocarbons migrate through a porous medium which is always invaded by water.

Migrations by molecular solutions

An important role may be played by interstitial water and water liberated by diagenesis. At shallow depth in particular, large volumes are expelled, and some authors, including Baker (1962), Cordell (1972, 1973), and Rumeau & Sourisse (1975) have postulated that the hydrocarbons entrained in this flow could possibly continue their maturation and conversion into hydrocarbons within the reservoirs themselves. However, it is worth noting that this protopetroleum and its precursors, representing an intermediate stage in the formation of hydrocarbons, have never yet been identified.

We have seen above (Section 1.1 of Part 2) that hydrocarbons display very slight and variable solubility in water, depending on their chemical composition, and that the most soluble compounds are benzene and toluene. Hobson & Tiratsoo (1975), Dickey (1975), Tissot & Welte (1978), and Jones (1981) showed that in these conditions the water volumes expelled would have to be far greater than the quantities normally produced by compaction, in view of the solubility measured, or the hydrocarbon solubility would have to be increased to values of 10,000 ppm to ensure the accumulation of the known oil reserves of a basin.

The first observation is the great discordance existing between the distribution of the different hydrocarbon species in a crude oil and the distribution that would result in accordance with their solubilities in water. The most soluble substances, such as benzene, always occur in very small proportions. Crude oils are especially rich in saturated hydrocarbons, that are among the least soluble and, conversely, they only contain traces of the more soluble hydrocarbons.

Fine hydrocarbon analyses carried out in a source rock, in the adjacent reservoir that it appears to have irrigated, and in the transition zone between them, reveal a depletion of the source in soluble materials (by organic solvents) and a concomitant relative enrichment in heteropolar compounds. On the other hand, the reservoir crudes are enriched in saturated hydrocarbons and depleted of polar compounds containing N, S and O. These observations show the importance of chromatographic absorption mechanisms throughout migration (Tissot and Welte, 1978).

To compensate for the very low solubility of hydrocarbons, many authors, including Baker (1962) and Cordell (1973), suggested the presence of solubilizing agents in the migration waters, including surfactants, which have the property of substantially raising hydrocarbon solubility. These solubilizers occur in the form of microemulsions of micelles with diameters from 100 to 2000 Å. This theory comes up against the following objections:
— the lack of any observation of these solubilizers, which should occur in significant quantities in order to play an active role,
— the dimensions of the micelles, which would prevent them from circulating in fine-grained rocks
— their negative electrical charge identical to that of the mineral surfaces (McAuliffe, 1979; Jones, 1981).

Migration by colloidal or micellar solution

Considering the very low solubility of hydrocarbons, it is tempting to postulate their dispersion in the form of colloids or micelles. However, the dimensions involved would be of the same order of magnitude and normally larger than the pore spaces, about 60 Å on the average, lending little credibility to this hypothesis. In addition, the electrical charge opposition between micelles and surfaces of argillaceous minerals further reduces the likelihood of this process. At best it may be imagined in very slightly compacted series which have not yet reached the critical burial depth of 2000 m.

Transport in the form of bubbles and droplets involves the respective diameters of the objects transported, constrictions between pores, and capillary pressure mechanisms. Movement is only possible if the droplet diameter is smaller than the constrictions, or capillary forces would have to be overcome for the droplets to be deformed and to pass through. For some authors, the rise in pressure in the liquid phase (water and/or hydrocarbons) during compaction associated with fracturing (geotectonic or hydraulic) would enable this process to occur.

Primary migration in aqueous form is undoubtedly not entirely impossible, but it appears to be limited to a very small number of cases, chiefly in the sand/shale zone. This could be the source, for example, of the small neogene oil and gas accumulations of the Nagaoka Plain in Japan, where the particularly high geothermal gradient could have resulted in rapid maturation of the hydrocarbons (Magara, 1968).

Gases are more soluble and, especially at great depth and in deltaic provinces, dissolved gases could well convey the liquid hydrocarbons that they keep in solution (Hedberg, 1979; Price, 1976, and Jones, 1981).

But water is also the vehicle for oxidizing agents and, in this case, it is an important factor in the transformations and biodegradation of oils. Its composition, or more specifically its salinity, plays a major role in this respect. Water is very probably a transport agent whose mechanical or physicochemical action may be considerable, positive or negative. However, its effectiveness must decrease fairly rapidly with depth.

Migration in individualized oil or gas phase

It is logical to believe that, in the initial phase, as long as the macromolecules making up the kerogen are relatively unaltered by link-splitting due to defunctionalization or to breakages of C-C bonds, expulsion by geostatic pressure is hardly credible. But the transformation of 10 to 20 % of the kerogen mass into hydrocarbons (which appears to occur in the 'oil window') must substantially alter the situation: the chemically-disintegrated micromolecules must be transformed into a complex of smaller molecules, mutually free from each other. Subjected to a sufficient stress field, the kerogen thus broken up and softened is plastically deformed. The most mobile molecules probably move in the direction of the weakest main stress (du Rouchet, 1978, internal report). This liberation property could be related to the distribution of organic matter in the rock. This could explain the very low yield of the oil shales, and, conversely, the high yields of carbonate and phosphate rocks (van der Weide, 1978, internal report).

When the oil saturation reaches 30 to 50 % in a source rock which has reached its mature stage, the interstitial water is structured and fixed to the pore walls, if these are not oil wet. Under the action of the pressure gradient, these conditions allow the movement of the oil in a continuous phase (Dickey, 1975; Tissot & Welte, 1978; McAuliffe, 1979; Jones, 1981; du Rouchet, 1981). Analysis by fluorescence microscopy at high magnification (500 ×) reveals oil veins about a micron in size in a mature source rock. This observation, added to the fluorescence of 'exudatinites' in the oil window, confirms the theory of the direct liberation of hydrocarbons from kerogen (Robert, unpublished report) (Plate 1). From the chemical standpoint, a drop in the H/C ratio, corresponding to the expulsion of the product richest in H from the kerogen, is observed in the source rock at these levels. This observation is confirmed by an increase in resistivity on the electric logs at this level. From the isotopic standpoint, an enrichment in ^{13}C occurs during migration.

Kerogen, which is generally hydrophobic, more or less retains the mobile materials in the matrix of its complex structure. Thus its structural organization directly governs the expulsion of hydrocarbons, which is facilitated by the presence of microfissures In a second stage, the hydrocarbons formed force their way across the barrier posed by the fine-grained sediments of the source rock, by opening or re-opening microfissures above the level at which the thrust of the fluids can overcome the confining pressure. This 'breakthrough' is partly facilitated by the increase in pressure caused by the expansion of the aquifers under the effect of temperature, and the individualization of methane microbubbles when the saturation pressure is reached (Hedberg, 1979). It also appears to depend on the quality of the source rocks. Geomechanical confinement is also significantly weakened by tectonic movements (Segor, 1965; du Rouchet, 1978, 1981).

Magara (1978 a, 1978 b, 1980) stresses the role of the oil-permeability of the shales during compaction, while the water-permeability decreases with compaction. This oil-related permeability increases with greater oil saturation, namely following the expulsion of the water with compaction, at least up to a certain point. Thus at a certain stage of compaction, when the bulk of the water is already expelled, but where the relative permeability is still substantial, permeability and oil-related permeability display values offering especially favorable migration conditions for the continuous oil phases (Dickey, 1975; Magara, 1978 a) (Figure 32).

In these conditions, the expulsion of oil in a continuous phase appears to be the most likely hypothesis in a large number of cases, at least when the source rock is sufficiently rich and concentrated. This is acknowledged today by most investigators, including McAuliffe (1979), Baker (1962), Chiarelli and du Rouchet (1977), Dickey (1975), du Rouchet (1978 c) and Jones (1981). This would explain in particular the formation of fields with dense and compact source rocks, like those of the Cretaceous limestones of La Luna in Venezuela, the Devonian limestones of the Swan Hills in Canada, the Devono-Mississippian 'Bakken Shales' of the Williston Basin (Meissner, 1978), and the Silurian shales of the Algerian Sahara (Tissot & Welte, 1978).

1.3.2. The geological framework of primary migration

In overall terms, hydrocarbons begin to take shape in the depths of the sedimentary basins when most of the water initially present is expelled. In fact, the genesis of liquid hydrocarbons occurs in a temperature interval of about 65 to 150 °C (150 to 300 °F), generally corresponding to an advanced degree of compaction, when the bulk of the interstitial water has already been expelled (around 88 % at 500 m, 95 % at 1500 m, 98 % at 2500 m). Primary migration appears to start only above a certain level of kerogen transformation, when the kerogen begins to react to the stresses no longer as a solid, but as a liquid. It continues at depths around 3000 m and beyond in recent series (Young et al., 1977).

Dickey (1975) postulates that a shale buried between 1500 and 4500 m, substantially corresponding to the 'oil window', loses some 11 % of its porosity, mostly because of the loss of water. Magara (1978) states that the water could continue to migrate very slowly but for a very long time in compacted shales with very low porosity. Everything else remaining equal, the volume of water passing through shales of 10 md permeability during one year is equivalent to the quantity passing through a shale of 10^{-5} md during one million years.

An attempt may be made to determine the most favorable period for this expulsion. As noted by Magara (1978 a), the maximum liquid hydrocarbons present marks the start of their transformation into gas, and consequently the beginning of the decline or oil genesis. Consequently the peak phase of oil genesis is situated above, namely before the oil accumulation peak. And the main migration phase could well correspond to this first oil generation peak (Figure 33).

Some authors, including Magara (1974), also claim that osmotic flows, caused both by compaction and by salinity differences between the shale and sandstone waters, add their effects and stimulate primary migration. By raising the salinity in a closed reservoir, this electro-filtration mechanism could also contribute to the concentration of hydrocarbons, by 'precipitating' those that are present in solution (Hobson & Tiratsoo, 1975; Cordell, 1972, 1973 and 1976). Many authors, notably Neglia (1979) and du Rouchet (1979), place the

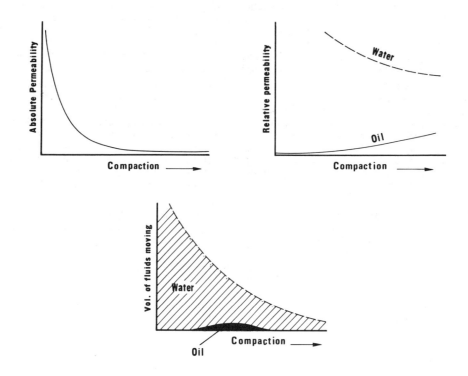

Fig. 32. — Diagrams showing assumed relationship between absolute and relative permeability and oil and water movements as a function of compaction (Magara, 1978).

PLATE 1
Visualization of a source rock by fluorescence microscopy
Essises No. 1 well (France) : Hettangian

Figures 1 and 2 : 2812 m   Figure 3 : 2812.75 m
Catagenesis level : 1 % vitrinite reflectance, end of the 'oil window'
Observations on massive polished rock, without other treatment

Fig. 1-2. — Images of a stratified source rock

     1 : dry front   2 : oil immersion
Shaly rock consisting of alternating
- lamination 20 to 60 μ thick of a fluorescent matrix (yellowish) : this is the productive part of the source rock,
- non fluorescent shale (black), devoid of oil-generating properties.

The fluorescent matrix is sapropelic (its scattered organic fraction is of algal origin) ; it is flecked with small fragments of orange-colored algae.
The exuded oil appears bright green to bright yellow in the fracture fillings (1-2) and in the shale pores (lower quarter Figure 2).

Fig. 3. — Oil identification

Dry-front observation with low magnification.
The low magnification prevents observance of the rock fluorescence, which is less intense, but it shows the geometry of the network of oil-filled fractures (green) over a broad field.
The left-hand part is normal. At the right, the same subject, after cleaning with a solvent pad : the surface oil film has disappeared and the fractures are largely empty.

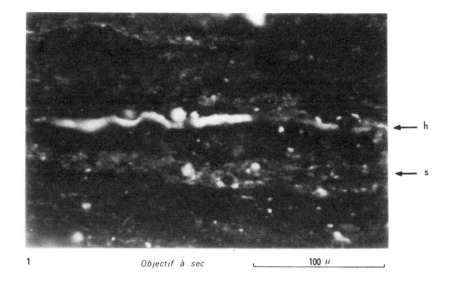

1   *Objectif à sec*   100 μ

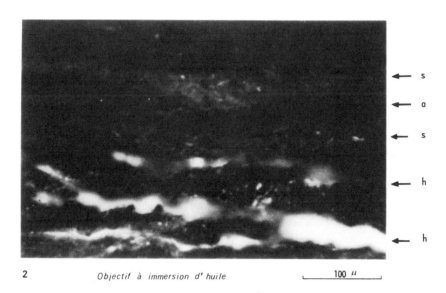

2   *Objectif à immersion d'huile*   100 μ

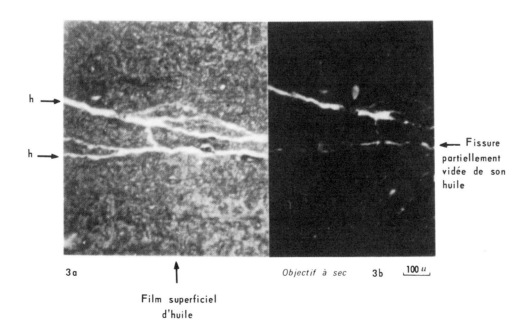

3a   *Objectif à sec*   3b   100 μ

Film superficiel d'huile

Fissure partiellement vidée de son huile

453

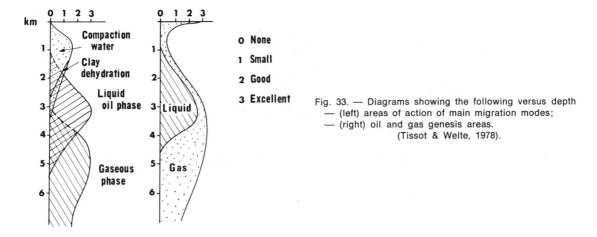

Fig. 33. — Diagrams showing the following versus depth
— (left) areas of action of main migration modes;
— (right) oil and gas genesis areas.
(Tissot & Welte, 1978).

accent on the importance of microfractures affecting the compacted shales and carbonates at these depths, partly due to the rise in pressure caused by the increase in fluid volumes triggered by rising temperatures and hydrocarbon genesis. It also appears that these fractures develop more easily with tighter geomechanical tensile patterns.

While the boundaries of this 'oil kitchen' are fairly well defined especially with respect to the geothermal gradient, considerable uncertainty surrounds the quantity of hydrocarbons expelled. Magara (1978 b) attempted to illustrate this unknown parameter by diagrams.

In the large majority of cases, the yield of this primary migration appears to be low, around 5 to 10 %, and very rarely more, as in the Middle East (Momper, 1978). These movements seem to occur over short distances, around 1 to 10 m. They obviously largely depend on the petrophysical characteristics of the rocks, and are encouraged by the existence of natural migration paths, permeable detrital intervals offering drains for lateral transfers, microfissures and fractures for vertical movements.

It appears that these hydrocarbon expulsion flows occur intermittently throughout geological history, depending on the heat flows and periods of intensified subsidence. In these particular conditions of advanced compaction, in which primary migration appears to occur, special attention is paid to the water produced by catagenetic transformations and to the undercompacted formations.

The role of 'catagenetic water'

In the absence of a good correlation between the expulsion of compaction waters and hydrocarbon genesis, the accent has been placed on the waters expelled during the late compaction phases, and especially the dehydration water of certain minerals like smectite, during genesis and catagenesis (Burst, 1959 and 1969; Weaver, 1959; Powers, 1967; Dunoyer de Segonzac, 1969 and 1970). Magara (1975 and 1978 a) showed that this special release of catagenetic water was a relatively continuous occurrence, and was added to the compaction water, providing a relay in time and guaranteeing a degree of continuity in the water expulsion process. Many researchers have thus stressed the role of smectite, whose transformation into illite in the oil kitchen zone releases water in the proportion of nearly 50 % of its volume (Magara, 1978). A correlation has also been observed between oil pools and smectite.

The case of undercompacted formations

Zones of undercompacted shale, and, in particular, the transition zones above them, play an important role in oil genesis. In fact, these are fairly often beds of sapropelic marine shales with a favorable source rock facies, and in many countries are qualified as 'bituminous zones'. The delay in compaction, which is reflected by abnormally high porosity for the burial depth at which they are found, facilitates the mobilization of the hydrocarbons despite a certain lag in diagenesis/catagenesis. Undercompacted formations thus offer the advantage of delaying the expulsion of large quantities of water at depths where the temperatures are higher and close to those corresponding to hydrocarbon genesis. If these formations are or were more widespread than one is led to believe, this would offer a favorable pattern for certain primary migrations (Chapman, 1972).

The transition zones surrounding them appear to be especially favorable for the expulsion of the hydrocarbons formed, due to their pressure gradient. In the Mackenzie Delta, for example, 90 % of the oil and gas pools lie within this depositional interval (Evans *et al.*, 1975) (Figures 34 and 165). Similar observations have

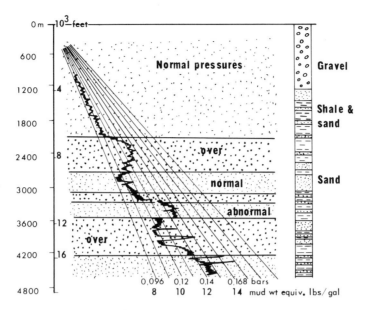

Fig. 34. — Pressure/depth profile in Taglu Field, Mackenzie Delta (Bruce & Parker, 1975).

been reported in the Handil oil zone in East Kalimantan (Durand & Oudin, 1979) (see Chapter 2 of Part 3, page 231).

The migration of gas

Gas migration very probably obeys rules which are different to those of oil, and dissolution in water seems to be a decisive factor.

As Hedberg (1964) states, it may be considered that "overlapping probably occurs between the methane formation depths by biochemical and thermochemical processes, one supplanting the other progressively with depth", the first generally characterized by a higher percentage of isotope ^{12}C than methane of thermochemical origin. The solubility of the methane rises rapidly with pressure, but decreases with higher water salinity. For a fresh water, it increases from 0.7 $m^3 m^{-3}$ at 35 bar pressure to 5 $m^3 m^{-3}$ at 700 bar (Magara, 1978).

At great depths, the quantities of gaseous hydrocarbons dissolved in the aquifers may reach high values, around 5 % by weight, the same order of magnitude as the solubility in oil (Jones, 1978). Considerable quantities are thus accumulated. For instance, it has been estimated that the deep high-pressure aquifers in the onshore areas of Louisiana and Texas could contain more than $6700 \cdot 10^{12}$ m^3, which may be compared with the $80 \cdot 10^{12}$ m^3 of gas discovered to this day in all gas fields worldwide. Some investigators postulate that these high-pressure aquifers release part of their dissolved gas if they undergo pressure drops, as when moving upward under the action of rising gradients. The gases thus released could replenish the traps located in their paths (Hobson, 1978).

Gases could also migrate by diffusion, in particular by a substantial difference between the gas composition in the shales, which contain around 50 % C^1, and those migrating into the neighboring sandstones, where C^1 accounts for 80 to 90 % (van der Weide, 1977; Hinch, 1978). The gas is found to be enriched in ^{13}C during its migration (Stahl, 1977).

The very high molecular diffusion of the gas — especially methane — in water tends to imply that this process is an important mechanism of primary gaseous hydrocarbon migration through the water-impregnated shale series (Leythaueser *et al.*, 1982). Measurements taken on models by these authors show that this process is compatible with the formation of commercial accumulations in a few million years. However, concentration in pools also provides preferential sources of gas diffusion through the seal.

One may thus draw the conclusion that no universal primary migration mechanism exists, but rather several complementary patterns, the relative importance of which depends on the hydrocarbon composition and the burial depth. At shallow depths (1500 m) methane could migrate easily in solution in the water, possibly together with some heavier hydrocarbons. Below 1500 m, in optimal maturation zones, and chiefly in the case of concentrated source rock, liquid hydrocarbons, initially heavy, and then lighter at greater depths, probably migrate mainly in a continuous phase.

In provinces with relatively dispersed source rock and a low geothermal gradient, like the deltaic basins, migration could occur by less effective processes, especially associated with the aquifer phase. Deeper down, where gas and condensates predominate, mechanisms involving gas as a vector of liquid hydrocarbons have been considered. However, "it is doubtful that gas contains the heavier hydrocarbons and N-S-O compounds in high molecular weight crude oil fractions" (McAuliffe, 1979). Moreover, the departure of very large quantities of gas would have to be assumed. Most important is the existence of fairly strong subsidence which, with the burial of several thousand meters of maturing organic matter, guarantees its transformation into hydrocarbons and their expulsion from the rock in which they originated. Despite the name, what is referred to as 'morts terrains' or overburden, plays an active and fundamental role in pool formation.

Once they reach a continuous reservoir horizon, the hydrocarbons continue their travels : this stage is called secondary migration. If a permeability barrier or trap is found in the neighborhood of the source rock, secondary migration does not occur — or only over a very short distance — and the pool may be considered 'in place'. This is referred to as a primary trap or 'trap-reservoir'. This is often the case of reefs, as in Alberta and West Texas, and also in the Jurassic fields of the North Sea. Cormorant, Brent and Magnus, as well as the great Permian (Khuff) gas fields of Iran, Kangan, Nar and Agar.

1.3.3. Characteristics of source rocks

A source rock is a sediment containing a certain quantity of organic matter and having generated appreciable quantities of oil or gas. By extension, this term is applied to facies identical to those of a source rock, but placed in different diagenetic conditions, namely a rock which has produced or could produce hydrocarbons. In this case, one speaks of an exhausted or potential source rock. These rocks seem to be characterized by :
— a percentage of organic matter between 0.5 to 1 and 5 to 6, depending on composition,
— the presence of extractable products (EOM) * denoting at least the possibility of hydrocarbon genesis, with EOM/OM ratios greater than 3, higher for carbonate source rocks than for shales,
— some dispersion of organic matter in a shale phase representing at least 30 % of the rock,
— the existence of laminations, layers and diastems, featuring beds with high concentrations of organic matter and horizontal and/or vertical drainage surfaces.

The composition of the constituents of the organic matter is of vital importance for the petroleum generating potential of a source rock, and hence for the quality of the basins (Rogers, 1979). A relationship has been established between the following two aspects :
— the potential of the Gippsland Basin, southeast of Australia, where some 340 Mt of liquid hydrocarbons have been discovered, and the composition of the shales of the Latrobe/Valley Group, considered to be the main source rock (84 % vitrinite, 12 % exinite and 4 % inertinite) (see Chapter 2 of Part 3, page 228).
— the poverty of the Cooper Basin, which presently contains only 14 Mt of oil and 100 Gm^3 of gas, and the composition of the Permian source rock (Gidgealpa Formation) : 43 % vitrinite, 4 % exinite, 53 % inertinite (from Shibaoka et al., 1978).

The presence of clay minerals in relatively large quantities allows the buildup of high internal pressures, probably causing microfissures, and thus accounting for the expulsion of the fluids formed and especially the hydrocarbons. Carbonate and phosphate source rocks exist, possibly in the form of fine shaly laminations. On the average they contain nearly ten times less organic matter than shales, but substantially the same amount of extractable products, as the organic compounds are generally of the sapropelic type rich in lipids. They also exhibit a more open matrix which probably facilitates transformation into hydrocarbons.

Source rocks exhibit tremendous diversity depending on the composition and evolution of the organic matter (Claret et al., 1981). The hydrocarbons formed do not appear to be extractable except for methane, and large quantities of extractable products remain imprisoned. A source rock is identified by the correlation between the hydrocarbons it is presumed to have contained and the traces that have remained in it. This operation is often difficult due to the differential migrations of the different hydrocarbon species, and the variations recorded by them, especially by chromatographic effect, during their transfer (Figure 35).

(*) EOM : extractable organic matter.

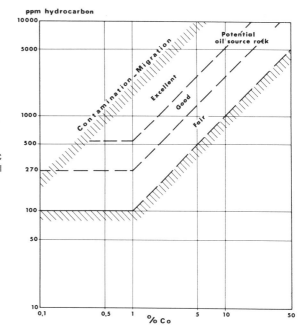

Fig. 35. — Diagram showing percentages of organic C and characteristic hydrocarbon quantities of potential source rocks.

A source rock is characterized by its content of non-migrated extracts, and by the ratio EOM/TOC (*) :

	EOM	EOM/TOC
— very good grade	1500 ppm	5 to 10
— good grade	150 to 1500 ppm	2 to 5
— mediocre or poor	50 to 150 ppm	2

Higher values of the EOM/TOC ratio correspond to carbonate rocks, lower values to shales.

To remove any doubt, it is usually necessary to make correlations based on very fine analyses and reliable markers, the geochemical fossils. Among these are the four- and five-ring saturated hydrocarbons, polyaromatics, isoprenoids and n-alkanes with C_{15}^+, as well as the relationship between the isotopic carbon ratios. All such correlations must be based on several independent parameters which are correctly located in the geological context (Welte et al., 1975). The 'geochemical fossils' that may be found in the source rock and in the crude oils are a very important correlation factor today. The isotopic correlation of kerogen and oils appears to be an increasingly important and reliable criterion for correlation between different crude oils, extending beyond their possible alterations, and also between crude oil and kerogen.

Thus in the Parentis Basin, the oils of the main fields, Cazaux (Albian), Lugos and Parentis (Neocomian), and Lavergne la Teste, display very similar $\delta^{13}C$ values around $-25 ‰$. On the other hand, the deep Parentis and Le Teich 2 crudes appear to be somewhat lighter and seem to belong to another family (Sourisse, Esquevin & Menendez, 1979, unpublished).

Hydorcarbon genesis is observed in the source rock :
— by the existence of sorbed gases, chiefly alkanes from C_1 to C_5 trapped in the mineral matrix, and the C_2/C_1 and C_3^+/C_1 ratios in particular will be examined,
— by the observation of oil microdroplets in microfluorimetry.

Fluorescence shows an intensity peak in immature source rocks close to the oil genesis stage. It declines gradually with advancing maturation. Correlations between the oil of a field and a presumed source rock are always difficult to make and generally require very detailed analyses.

It is first necessary to make sure that the extracts of the probable source rock are correctly in place, namely produced from the organic matter of this source rock and not migrated from another source. The fact that a rock is not a reservoir does not certify that it is not invaded. It should also be noted that the composition of the oils changes during their migration. They grow richer in saturated hydrocarbons, less

---

(*) TOC : total organic carbon.

in aromatics, and lose oxygen, nitrogen and sulfur compounds. In many cases, particularly when the reservoir is found in an immature zone, the hydrocarbons may originate alternatively in:
— older and deeper series, by means of vertical migrations,
— contemporary series buried deeper in neighboring synclines, with lateral migration.

An example of a source rock: the Bakken Formation of the Williston Basin
(Meissner, 1978)

The Bakken Formation is a thin unit that has the special feature of being carbonate as well as rich in organic matter during a major cycle of onlap-offlap sedimentation at the Devonian/Mississippian transition zone, or at the extreme base of the Mississippian, on the Devonian formations. It belongs to a vast sequence of black shales of the same age which covered the North American continent at the time (Figures 36, 165 and 192).

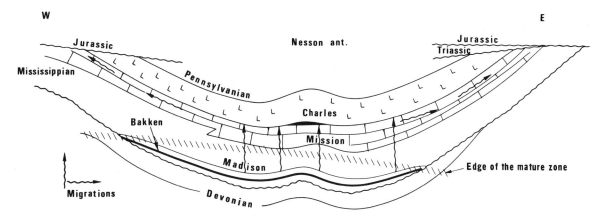

Fig. 36. — Section of Williston Basin showing extension of Bakken source rock and maturation zone (Meissner, 1978).

From the petrographic standpoint, this formation consists of a black dolomitic siltstone, sometimes laminated, bounded by two strongly radioactive more shaly levels, slow on the sonic log and displaying very high or very low resistivity. The central part contains 0.6 to 10 % organic carbon, with a mean of 3.8. When buried deeper than 6300 ft, it functioned as a source rock and produced some 1.4 Ct — 10 billion barrels of oil in the deep part of the basin. The very low speed on the sonic log can be explained by a high percentage of organic matter, and the wide variations in resistivity by the presence or absence of hydrocarbons in the formation, depending on which side of the maturation limit one is placed. This boundary lies between 6200 ft and 8200 ft, in the neighborhood of 70 °C or 160 °F, a critical temperature for a carboniferous source rock. One may in passing note the application of electric logs in detecting the maturation of a particularly rich carbonate source rock, an application which is fairly exceptional but probably offers tremendous possibilities.

Also worth noting is the close correlation between the maturation zones and the zones in which this formation exhibits high internal overpressure, which would imply a direct relationship between hydrocarbon genesis and the presence of high pressures, also observed in the Uinta and Powder River Basins. In these cases the pressure anomalies may be the direct result of the genesis of hydrocarbons, which appear to be the only fluids in excess in this confined environment.

The Bakken Formation is also oil-producing in five fields of the Williston Basin, and especially in the Antelope Field. This pool has produced some 1.4 Mt — 10 million barrels of oil — and about 800 Mm^3 — 29 billion cubic feet — of gas from 44 wells, practically without water inflows. This implies that the hydrocarbons are the only liquids capable of moving in this 'reservoir'. In these overpressure zones, the oil moves through vertical microfissures formed and opened. This offers a visible example of migration (and production) in a fracture system, caused by a high pressure fluid. Wider vertical fractures are very probably the source of vertical migrations on a larger scale, conveying the oil to the main carbonate reservoir of the basin, the Mission Canyon, which is well sealed by the evaporites of the Charles Formation.

Summary and conclusion

Death and transfiguration appear to symbolize the destiny of the organic matter deposited throughout burial. From a complex, decomposing substance, in a few million years, in the depths of the sedimentary basins, temperature creates products with a simpler structure, endowed with powerful energy potential.

Like seeds, organic matter gives way to a new product with great possibilities.

We have seen that this genesis is closely related to the expulsion of the hydrocarbons formed from the rock which gave birth to them, and that these products have excellent chances of leaving in a continuous phase, in a certain sense fully armed. Water probably plays an important role, but it is not quite clear whether its positive effects are greater than its negative effects, which are better known. On the other hand, like the facies distribution, the geochemical context plays an essential and undeniable role.

Throughout these transformations, we have also noted the imprint, initially powerful, of the depositional environment become gradually obliterated during burial under the effect of temperature. Due to their sensitivity and their structural diversity, hydrocarbons thus illustrate perfectly the many aspects assumed by sediments all along their history.

Present Trends in Organic Geochemistry in Research on Migration of Hydrocarbons

B. Durand

The most important recent change in understanding the phenomenon of migration is probably the preference for expulsion mechanisms in which hydrocarbons constitute a separate phasis over those in which hydrocarbons are carried by water. In the new view, hydrocarbon saturation in the source-rock porous space is an essential parameter; it is primarily determined by the organic matter content, which is in turn determined by the sedimentation style of a sedimentary basin. Research in organic sedimentology is thus of major interest to the advance of our knowledge of migration. In this paper, two styles of sedimentary basins are compared: (1) a slowly subsiding basin in which marine organic matter occurs as discrete organic-rich layers, and (2) a quickly subsiding basin in which organic matter is found throughout the sedimentary column, but in low contents. The above considerations are largely theoretical and need to be supported by more observations. There exist tools at the present time, particularly for observing light hydrocarbons the behaviour of which is for the most part unknown, but good samples are still rare. The strong support of oil companies would be invaluable in increasing the number of suitable samples available to researchers. Contrary to widespread opinion, the generation of hydrocarbons, particularly of light hydrocarbons, is not sufficiently understood for the needs of migration studies, as it is seen in the present-day controversy concerning the 'early generation' of methane and light hydrocarbons. Here again, more observation is needed. Since light hydrocarbons are very mobile, however, simulation, usually conducted by pyrolysis, is advisable. Although the validity of simulation by pyrolysis is moot, the field is obviously developing. The handling of correlations between pooled hydrocarbons/source-rock extracts is poor as compared with analytical means, partly because of the length of time required for analysis. This situation will be improved by the development of GC–MS systems. However, the absence of precise information on the fate of biomarkers in migration, due to generally poor cooperation between chemists and geologists, is mainly responsible for this situation. Notable progress in the use of carbon and hydrogen isotopes is due to extensive activity in recent years.

INTRODUCTION

The description and comprehension of the hydrocarbon migration phenomena in sedimentary basins can evidently be achieved only through collaboration among all the disciplines in the field of geology and also need a great deal of experimental and theoretical work.

Therefore isolating the specific contribution of organic geochemistry may appear to be an artificial task. However, themes can be selected, in which organic geochemistry plays the central role and whose progress has immediate impact on the philosophy of oil exploration. They are: the choice of a mechanism of expulsion from the source-rocks, the observation of migration phenomena, and the correlations between pooled hydrocarbons and source rock hydrocarbons. I would like therefore to bring out the progress and present trends of thought in the three above-mentioned fields.

1. SELECTION OF A MECHANISM OF EXPULSION FROM THE SOURCE-ROCK REPERCUSSIONS IN THE FIELD OF OIL EXPLORATION

A great change in the approach to problems of hydrocarbon migration in sedimentary basins is seen in the gradual relinquishing of mechanisms for describing expulsion from source-rock with combined water and hydrocarbons in one phase.

This trend is due to progress in the field of knowledge on hydrocarbon formation in sedimentary basins; we now know that:

(1) Formation essentially takes place at depths at which most of the water has already been expelled from the sediments, and at which the quantities available to propel the hydrocarbons are therefore low;

(2) Differences in distribution of hydrocarbons between pools and source-rocks are difficult to explain in

cases where the hydrocarbons have been transported by water, since they are mostly reverse of solubilities in water.

These observations, combined with the fact that most hydrocarbons have very low solubility in water, even in subsurface conditions (Price, 1976; McAuliffe, 1980), and the difficulty of conceiving a mechanism of exsolution once the hydrocarbons arrive in the reservoirs, make water transportation very unlikely.

Faced with this situation, proponents of this sort of mechanism have made several suggestions:

(1) Creation of water in the hydrocarbon-formation zone by transforming smectites into illites (Powers, 1967; Burst, 1969; Perry and Hower, 1972), so as to increase the quantities of water available;

(2) Formation of micella (Baker, 1962) or emulsions (Cordell, 1973);

(3) Expulsion at shallow depth of a protopetroleum constituted of functionalized molecules, which are thus soluble in water. These molecules defunctionalize afterwards in reservoirs to form petroleum (Dobryansky et al., 1961; Hodgson, 1980).

(4) Dissent with hydrocarbon-formation schemes, by proposing a formation either at shallow depths (Wilson, 1975) where there is still plenty of water, or at great depths, where the higher temperatures and pressures increase the level of solubility (Price, 1976).

The most widespread opinion at the present is that while these mechanisms may sometimes play a role, and it must not be forgotten that methane and light aromatics are fairly soluble in water, the great majority of cases can only be explained by mechanisms in which the hydrocarbons constitute a separate phase — liquid or gas.

The objection that is generally made to the validity of these mechanisms is that the interfacial tension between the water phase and the hydrocarbon phase creates capillary pressures in the source-rock, where the pores are very fine, which are generally deemed to be greater than the pressures developed by compaction. Several additional mechanisms have therefore been proposed.

The ones that evade the obstacle of interfacial tensions by creating a continuous organic phasis, for instance:

(a) The kerogen content is large enough for its transformation into oil and creates a continuous oil phasis (Meissner, 1978a and b),

(b) The kerogen constitutes a three-dimension organic network which is wettable with oil. It is utilized by the hydrocarbons to get out of the source-rock (McAuliffe, 1980),

(c) An organic film exists on the mineral surface (Yariv, 1976); this idea should be seen in conjunction with the preceding one.

The others result in an increase in pore pressure sufficient to exceed the capillary pressure or even microcrack the rock for instance:

(a) Thermal dilatation of the water (aquathermal effect, Barker, 1972),

(b) Increase in specific volume of the organic matter upon transformation of kerogens to hydrocarbons, particularly to gas (Snarskiy 1962, 1970; Sokolov et al., 1964; Vandenbroucke in Tissot and Pelet, 1971; Hedberg, 1974 and 1980; see also Ungerer et al., this congress),

(c) Partial transmission of the geostatic constraints from the solid to the liquid phase when the kerogen is transformed into hydrocarbon (Meissner, 1978). This transmission also might occur to kerogen, due to its plastic behaviour (Du Rouchet, 1978 and 1981), provoking microfracturation even when kerogen is still in a solid state, thus creating ways for further expulsion of hydrocarbons.

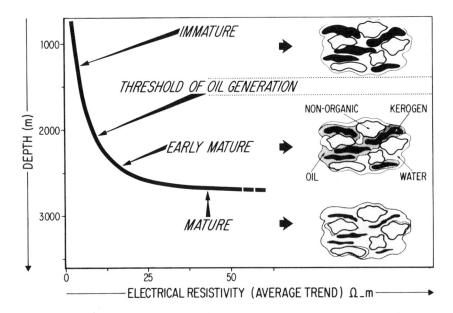

Fig. 1. Bakken formation, after Meissner (1978). *Left*: evolution of resistivity with depth; *Right*: interpretation by increase of pore saturation in oil.

It is possible, however, that the obstacle created by the water/hydrocarbon interfacial tension has been exaggerated. Inaccessible, the source-rock environment is very little understood, and descriptions of its pore space and distribution of organic matter are very abstract. In situ interfacial tension cannot be measured, and the values assumed for calculation are taken from measurements in laboratories on simple systems. Moreover, the source-rock is neither homogeneous nor immutable, and it is modified during compaction. Mineral diagenesis phenomena, some of which occasionally being due to products issued from thermal degradation of kerogen (Moore and Druckam, 1981) may alter the porosity.

To schematize the present trend of thought, let us examine the favourable example of the Bakken formation in the Williston Basin described by Meissner (1978) (see Fig. 1). This formation of lower Mississippian age, is an organic-rich (around 5% wt organic matter) marl. During burial water is expelled from the source-rock before hydrocarbons are formed. If the hydrocarbons are formed from kerogens:

(1) Hydrocarbon saturation of the pores is considerably increased and the hydrocarbons finally constitute in the source-rock a continuous phase which can be expelled without hindrance of water. This phenomenon is shown by the increase in resistivity to very high levels.

(2) The transformation of part of the solid phase (kerogen) to liquid phase (hydrocarbon) results in the partial transmission of the geostatic constraints of the solid phase to the liquid phase. There is thus an increase in pore pressure, which creates in the hydrocarbon phasis pressure gradients favourable to expulsion.

The important fact to retain for oil exploration seems to me to be that unlike the expulsion mechanisms using water transport, separate-phase mechanisms — whatever their proposed variables — depend primarily for their efficacy on the hydrocarbon saturation of the pores of the source-rock.

Hydrocarbon saturation depends initially on the content and distribution of organic matter in the source-rock, and thus on the features of sedimentation in the basin under exploration (the features also determine the gradients of the hydraulic head which define the size and direction of the flows).

By analogy with low permeability reservoirs the flow of hydrocarbons out of the source-rock may be tentatively described by Darcy's law, thanks to the notion of relative permeability.

It should be noted that according to the shape of the relative permeability curves observed for example in the production studies, any increase in hydrocarbon saturation is reflected in a much greater relative permeability to the hydrocarbons. Disregarding, for the moment, the presence of reservoirs capable of receiving the hydrocarbons, the probability of expulsion from the source-rock at a given level of maturity will thus be multiplied, going from a sector which is poor in organic matter to one which is rich, by a factor much greater than the simple ratio of organic matter contents.

Moreover, since the saturation must be above a minimum threshold for the expulsion of the

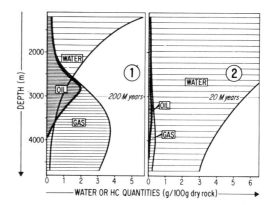

Fig. 2. Comparison of water and hydrocarbons contents in rocks of two sedimentary basins having different sedimentation styles. Basin 1: low sedimentation rate and discrete organic-rich layers. Basin 2: high sedimentation rate and numerous organic-poor layers.

hydrocarbons to be greater than negligible, these mechanisms will not be effective outside certain periods in the history of the source-rock and if its organic-content is too low.

If this line of reasoning is correct, much importance ought to be devoted to organic sedimentological research for the oil exploration industry, particularly research on the source-rock/reservoir systems.* This research is developing (see for example, Demaison and Moore, 1980; the Orgon cruises (CEPM–CNEXO), Tissot et al. (1979, 1980), Deroo and Herbin (1980)); but too little attention is given to the aspects I have just exposed. One instrument in the study of organic sedimentology could be a more extensive development of the use of geochemical logs of the Rock Eval type (Espitalié et al., 1977).

Once expelled from the source-rock, hydrocarbons can mobilize by solubilization along their way hydrocarbons formed in other source-rocks which otherwise could not contribute to migration. This mechanism is cited particularly when the initial expulsion takes place in 'gas' phase.† The experimental demonstration of the possibility of such a mechanism was undertaken by Madame Zhuze's team (Zhuze and Bourova, 1975).

The importance of this effect should be highly variable according to the nature of hydrocarbons pathways towards the reservoir-rock. Good efficiency is obtained when under high pressure conditions, hydrocarbons are forced through relatively low permeabilities series containing mature organic matter. It must be noted that the mobilization capabilities of hydrocarbons which would not otherwise be expelled from the source-rocks is not *a priori* restricted to a gas phase, but may also occur in an oil phase.

For a more complete illustration of the preceding ideas, let us examine the schematic comparison of two

---

* I borrowed this expression from G. Demaison.
† Gas phase is the current expression. However 'supercritical fluid phase' would be preferable, since light hydrocarbons are in this case in a supercritical state.

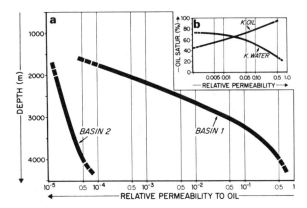

Fig. 3. (a) Evolution of relative permeability to hydrocarbons with depth in Basins 1 and 2 of Fig. 3(b). (b) Curves of relative permeabilities to water and to oil as a function of oil saturation, used to build Fig. 4(a) from water and oil saturation deduced from Fig. 3.

models of sedimentary basins (Fig. 2): the first is a basin of low sedimentation rate (3000 m in 200×10^6 years) with discrete episodes of heavy concentrations of marine organic matter in clay carbonate sediments. Where the porous levels are in sufficient communication with the surface, the slowness of sedimentation and the relative thinness of the fine-grained sedimentation levels make it fairly easy to expel most of the water before the hydrocarbons have begun to form. The second has a high sedimentation rate (3000 m in 20×10^6 years) filled with terrigenous deposits with fairly constant small quantities of organic matter of terrestrial origin throughout the sedimentary column. The sediments still contain a good deal of water when the hydrocarbons form.

In the first model, shortly after their formation the level of saturation of hydrocarbons rapidly rises in the pores rich in organic matter. At the same time, more of the geostatic constraints are transmitted to the liquid phase. The relative permeability of the hydrocarbons and pressure gradients in the hydrocarbons phasis thus rapidly increase, and partial expulsion at the oil stage is likely.

In the second, hydrocarbon saturation remains very low for a long time. Expulsion at the oil stage is unlikely. The oil which is formed is transformed into gas, with a notable increase in pore pressure. Migration, if it takes place, will be later, and will occur preferentially in the 'gas phase'.

Note that in such a basin, the water transport mechanisms can be relatively important, as soon as the hydrocarbons formation reach the gas phase. Organic matter is very dispersed in the sediment, and contact with the water is facilitated. There is a good deal of accessible water compared with the hydrocarbons present. The gaseous hydrocarbons, particularly methane (Bonham, 1978) and the light aromatics are fairly soluble in water under these temperature and pressure conditions.

Figure 3(a) shows the evolution of relative permeabilities to hydrocarbons in basins 1 and 2, estimated from the saturations deduced from curves in Fig. 2 and from the curves of relative permeabilities

versus oil saturation in Fig. 3(b). The values indicated are largely arbitrary, particularly because the curves in Fig. 3(b), extrapolated from a production study (Simlote and Withjack, 1980) have only a distant connection with the actual situations in the sedimentary basins. But it can probably by concluded that, as seen in Fig. 3(b):

1. The relative permeability of source-rocks to hydrocarbons during their expulsion (curves of Fig. 3 do not refer to further drainage and pool formation by secondary migration) is several orders of magnitude lower in basin 2 than in basin 1, although the ratio of concentrations of organic matter is only 1 to 10.

2. This relative permeability increases very rapidly as soon as the hydrocarbons form in basin 1, while it is practically nil to a considerable depth in basin 2.

Basin 1 might be taken as a model for epicontinental platform basins. However the 10% organic carbon content which was used for the making-up of Figs 2 and 3 is rare in such basins. Such a high figure was taken to make a strong contrast with basin 2.

Basin 2 applies for instance to quickly subsiding tertiary basins. Such basins are known for gas occurrence rather than oil occurrence. This feature is widely attributed to the terrestrial nature of organic matter in these basins, which is believed to result in high gas/oil ratios in formation of hydrocarbons during burial (Dow, 1978). Retention of oil and formation of gas by cracking of unexpelled oil seems to me to explain much better the abundance of gas in these basins than do the properties of the organic matter. Indeed, studies of terrestrial kerogens show they form much more oil than gas during burial (Durand et al., in preparation). In this view, the nature of hydrocarbon accumulations (oil or gas) in such a basin should not be very dependent on the nature of organic matter (marine or terrestrial), but much more on organic matter distribution in the sediments. On a large scale, accumulations should be mainly gas ones. However, one should note that in such basins, abnormal pressure zones appear below a certain depth, owing to the high rate of subsidence and the relatively high proportion of clayey sediments in the sedimentary column. Migration of oil will be facilitated, all other things being equal, if hydrocarbon formation takes place in zones where there are high pressure gradients; for example, transition zones near abnormal pressure zones. Also note that percolation in the gas phase across sediments situated above will eventually dissolve part of the hydrocarbon contained therein giving birth to gas-condensate pools or even light oil pools.

Locally, there can be particularly well drained levels, owing to the numerous sand bodies under hydrostatic pressure. There also can be levels with fairly high organic matter-content and even coals (this is the case of many tertiary deltas). These conditions are similar to those of basin 1, and therefore facilitate the precocious expulsion of the hydrocarbons at the oil stage.

Therefore in this view, the nature of hydrocarbon accumulation (oil or gas) in such a basin should not be very dependent on the nature of organic matter (marine or terrestrial), but much more on the organic matter distribution in sediments, and local conditions. On a large-scale, accumulations should be mostly gas ones, owing to a generally poor organic matter content.

After experiments conducted by Bray and Foster (1980), attention was drawn to the role that carbon dioxide might play in the migration of hydrocarbons. These authors showed that slow percolation in an oil bearing rock under the pressure of a water charged with CO_2 permits a greater mobilization of oil than does percolation under the same conditions, but with water free of CO_2. Moreover, the distribution of mobilized hydrocarbons is near that of the oil. It has been concluded by some that the effect of the CO_2 is to increase the solubility of the hydrocarbons in water. In fact, under the experimental conditions, its role appears to be to increase the relative permeability to the oil in the milieu by lowering the interfacial water/oil tension, and the expulsion mechanism is linked up with the mechanisms of migration of the hydrocarbons in separate phase. A qualitatively analogous phenomenon (but quantitatively weaker) would occur by using gaseous hydrocarbons, instead of CO_2. Formation of CO_2 concurrently with that of the hydrocarbons thus appears to be able to facilitate the expulsion of the latter. But CO_2 is probably found only occasionally in great quantities in the source rock at the time when hydrocarbons are formed. Formation of CO_2 from organic matter in fact mostly precedes that of the hydrocarbons. Since it is very soluble in water, most of it is probably expelled with the water, unless it is partially dissolved in the organic matter because of the heavy subsurface pressures. Formation of carbonates is also a sink for CO_2. In some cases, however, the possibility of mobilization of hydrocarbons on the path of the CO_2 from greater depths is not to be excluded (Kvenvolden and Claypool, 1980), in the same way as light hydrocarbons might act (migration in a 'gas' phase).

Let us recall the renewed interests in diffusion mechanisms following the work of Leythaeuser et al. (1980). Without any doubt, diffusion works under subsurface conditions, and its effects reinforce those of all the above-described expulsion mechanisms. It plays a role in equalizing concentrations, and its efficacy, like that of the separate phase expulsion mechanisms, increases with hydrocarbon saturation in the source rock. At the present time, however, its role is usually considered to be minor in the constitution of oil pools owing to very low diffusion coefficients and limitation by low solubilities of hydrocarbons in water and even destructive, given its inherent dispersing effect. It might be more effective in the constitution of gas accumulation, given the higher diffusion coefficients and higher solubilities in water in this case.

2. OBSERVATION OF MIGRATION PHENOMENA

The preceding considerations are largely theoretical, and a realistic discussion of expulsion mechanisms can only be conducted on the basis of precise observation of geological samplings.

The past few years have seen the development of techniques which permit rapid observation of the quantities and distributions of hydrocarbons in sediments. At the forefront of the field are the geochemical logging techniques, such as the Rock Eval (Espitalié et al., 1977) and similar devices (Claypool and Reed, 1976). Thanks to the Rock Eval it is now possible to study the concentrations of free hydrocarbons in rocks, summarily but rapidly — on the well-site for example — and to compare them with the quantities which theoretically should be present according to the characteristics of the kerogen.

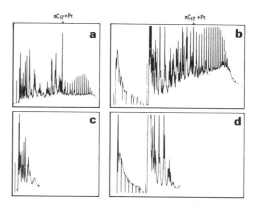

Fig. 4. Comparison of C_1–C_{35} hydrocarbons chromatograms extracted by thermovaporization at 300 °C of: (a) a coal from the Mahakam delta, (c) the same coal mixed at 2% wt with illite. Both chromatograms relate to the same initial weight of organic matter and are recorded with the same shunt. (b) and (d) comparison of hydrocarbons produced by the same samples during a pyrolysis at 470 °C under a nitrogen stream performed after the thermovaporization. The large peak ahead of the chromatograms corresponds to the recording of CH_4.

Over the same period, techniques known as thermovaporization have been developed, by which the distribution of hydrocarbons in sediments is studied in small samples vaporized at temperature < 300 °C and followed by gas-phase chromatography (Bordenave et al., 1970; Jonathan and Lothe, 1975; Schaeffer et al., 1978; Whelan, 1979; Saint Paul et al., 1980).

These techniques are very useful, because observations of migration phenomena on the sedimentary basin scale can now theoretically be conducted by studying a great number of cheaply obtained samples. These techniques also make it possible to study light hydrocarbons, which are not recovered with the usual extraction techniques using solvent, and after which the solvent is evaporated.

Some reservations are required, however, as to the results obtained through these methods at their present level of technology, which apply for the same reasons to results obtained with the Rock Eval. In some cases, particularly where organic matter contents of samples are low and/or samples are very clayey, heaviest hydrocarbons contained in samples are retained in them and a low-temperature cracking may even occur, the effect of which is to produce in the recovered hydrocarbons distributions which are not that of hydrocarbons contained in samples pores. This is the effect of the mineral matrix described by Espitalié et al. (1980). An example — fortunately extreme — is shown in Fig. 4(a) and (c): in this example mineral matrix effect has been simulated by mixing a coal sample to a large excess of illite (2% weight coal in the mixture).

Figure 4(a) is the chromatogram obtained by thermovaporization of pure coal (method of Saint Paul et al., 1980) and Fig. 4(c) is the chromatogram obtained from the coal–illite mixture. In case of Fig. 4(c), only hydrocarbons up to about C_{10} are recovered.

Another important advance is due to progress in fluorescent microscopy applications, thanks to which it is now possible to visualize the hydrocarbons and to study their distribution in pore space (Robert, 1979). At the same time, it has become possible to make a more realistic description of the pore space and distribution of organic matter in a sediment, although at a relatively large scale.

Let us also note the advantages of progress in making electric logs on fine-grained sediments. I have referred above to the use of resistivity logs made by Meissner (1978) to test his hypothesis on hydrocarbon saturation in a porous environment. The same observations were made for the Bazhenov formation in western Siberia (in Bois and Monicard, 1981). The logging study of the compaction phenomenon, the initial expulsion driving force (Chiarelli et al., 1973; Magara, 1978), is also of great value.

An obstacle to observation, however, is the lack of adequate samples, particularly for observing expulsion, and it would be a good idea for the oil companies to exert an effort in this field. The chances of obtaining accumulated samples are greater than those of obtaining drained samples, because drilling is presumably undertaken in accumulation zones. Moreover, the phenomenon of accumulation can be more clearly observed than the phenomenon of drainage because the surface is presumably smaller, and the mobilized quantities per surface unit larger in accumulation zones than in drainage zones. Samples, incidentally, must be taken very carefully, and current drilling practice (turbodrilling; muds of complex composition, often kept secret) complicate or even hinder research in the field.

Because of all these difficulties, interpretative studies of migration phenomena through geochemical observation, are rarely undertaken, and are usually of limited extent when they are. Examples are Vandenbroucke (1972), Barker (1980), Huc and Hunt (1980), Leythaeuser et al. (1980), and during this session, Leythaeuser et al., Schaeffer et al., Schoell et al., and Vandenbroucke et al.

The above-cited study by Meissner (1978) on the Williston basin, using the geochemical results of Dow (1974) and Williams (1974), those of Combaz and de Matharel (1978) and Durand and Oudin (1979) on the Mahakam delta, that of Clayton and Swetland (1980) on the Denver basin, that of Basu et al. (1980) on the Bombay basin are among the studies which have been able to take into account, on the basis of precise geochemical observations, the three-dimensional aspect of the phenomena and the necessity of integration into the geological framework. None of these studies founded on observation has yet been able to establish the nature of expulsion mechanisms definitively.

One fundamental obstacle to interpretation of observations is that migration phenomena can be clearly shown only if the genetic history of the hydrocarbons is well known. And our knowledge of genetic phenomena is

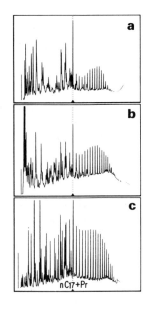
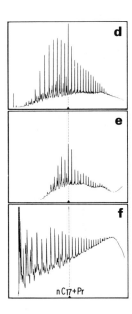

Fig. 5. Comparison of C_1–C_{35} hydrocarbons: (a) of an oil from the Mahakam delta, (b) produced by thermovaporization at 300 °C of a coal from the Mahakam delta at a vitrinite reflectance (in oil) of 0.8%, (c) produced by pyrolysis at 470 °C of a coal from the Mahakam delta at a vitrinite reflectance (in oil) of 0.4% (d), (e), (f) idem for an oil and Lower Toarcian shales from the Paris Basin. Recording of CH_4 produced in cases c and f is not presented here.

still not precise enough, on the quantitative as well as the qualitative level, particularly concerning the formation of light hydrocarbons. I might take as an example that still unresolved controversy on the possible formation processes of light hydrocarbons, notably methane, by thermal degradation of kerogen before the principal zone of oil formation (Powell, 1978; Powell et al., 1978; Snowdon, 1978; Connan and Cassou, 1979; Kubler, 1980). Another example is the very widespread opinion referred to above, which I believe will have to be reexamined, according to which organic matter of terrestrial origin can form almost solely gas.

It will be difficult to establish a very precise scheme of hydrocarbon formation, particularly of light hydrocarbon formation, on the basis of observation alone, precisely because of the element of migration. That is why a correct simulation of the formation phenomenon in the laboratory is needed. This research theme is now highly developed, and pyrolysis in an open milieu under inert atmosphere between 300 and 700 °C is the technique most often used. Many researchers use this technique, in an open or a closed environment, to characterize kerogens (Larter, 1978; Larter et al., 1978; Larter and Douglas, 1980; Philp and Russel, 1980; Van Graas et al., 1980; Van de Meent et al., 1980) but others aim to simulate petroleum formation (Vandenbroucke et al., 1977; Harwood, 1977; Ishiwatari et al., 1978; Peters, 1978; Seifert, 1978; Rohrback, 1979; Peters et al., 1980, Monin et al., 1980).

This technique gives encouraging results, as seen in Fig. 5, which compares for marine and terrestrial organic matters three chromatograms: one is that of an oil (Fig. 5(a) and 5(d)), the other the hydrocarbons obtained by thermovaporization at 300 °C from a sample situated in the principal oil-formation zone Fig. 5(b) and 5(e), and the latter, the hydrocarbons obtained by pyrolysis at 470 °C from an immature kerogen (Fig. 5(c) and 5(f)). Similar patterns are obtained from oil, thermovaporized hydrocarbons and pyrolytic hydrocarbons, however less obviously for marine than for terrestrial organic matter.

Yet the principle of such a simulation is often disputed (Snowdon, 1979). Furthermore, pyrolysis is very sensitive to operating conditions details. The mineral matrix effect is of course observed and may be very important, such as in case of Fig. 5(b) and 5(d) where are compared the hydrocarbons from a pyrolysis at 470 °C of a pure coal and a coal–illite mixture (2% weight of coal in the mixture), so that it is not always wise to pyrolyse organic matter in rock, although pyrolysing isolated organic matter would appear to go against geological logic. In spite of these constraints, it seems to me that the results already obtained encourage the hope of improving the quality of simulations in the near future to a point where the needs of migration studies are satisfied — without, of course, being able to obtain a total similitude of molecular structures in detail. Artificial series will then be close analogous to natural ones and clear tables of correspondence will be established.

In order to do that, experimental techniques will have to be modified. One promising direction seems to be that opened up by Lewan et al. (1979) through pyrolysis of pieces of rock in the presence of water at a temperature of about 350 °C. The distributions observed in the hydrocarbon produced are in fact very close to those observed in the oils. Furthermore the production is limited of n-olefins, which are also present in large quantities in the experiments conducted in open environments without water, and which are scarce in natural oils.

3. CORRELATION BETWEEN POOLS AND SOURCE-ROCKS

These correlations are founded on the comparative study of distributions of hydrocarbons, in particular those of biomakers. They are mainly used in basins studies for an identification of source-rock beds and most often do not refer explicitly to description and explanation of migration. However, the present wide use of such correlations, which means that some satisfaction is obtained, is an implicit argument in favour of migration in a phase separated from the water phase. Indeed, in water phase transportation, distribution of hydrocarbons in source-rocks and in pools would be so different, due to very different solubilities of hydrocarbons in water as a function of molecular weight and molecular structure, that correlations would be much too delicate to be very popular.

However, besides the classical search for source-rock levels, more work on correlation could do much more for solving migration problems and elucidation of mechanisms than is done now. Typically it should be possible to localize more precisely than is done now the sections of source rocks from which the hydrocarbons come and to translate more accurately the differences of distribution in terms of expulsion and transportation mechanisms. The panoply of usable biomarkers is wide, and has recently been expanded to include petroporphyrins, which the development of HPLC has made possible to handle. The contrast between the analytical profusion and the relative paucity of application to understanding migration is all the more visible.

In fact, understanding migration through a careful comparison of distribution in source-rocks and pools comes against serious difficulties, some of which are listed below:

1. The oil represents the average composition of a source-rock over its drainage area, and sometimes is a mixture of different sources. In contrast, a specific sample of the source-rock used for correlation studies may represent a specific kerogen facies of limited extent. Also there may be several possible layers of source-rock in a sedimentary basin having close geochemical characteristics. Therefore a large number of sediment samples should be studied in great detail. The complexity and the time needed by the present analytical methods is difficult to reconcile with this requirement. The development of GC–MS coupling sizeably reduces this obstacle.

2. Available sediment samples are not generally those that the oil comes from, simply because drilling presumably takes place in accumulation zones and not in expulsion zones. At best, in many cases, they represent an immature stage of the effective source-rock.

3. The biomarkers used may be dissolved by the oil on its way, and thus correspond to a rock other than the source-rock.

4. Complex molecules such as biomarkers are easily destroyed by maturation. Therefore their concentration (and consequently their significance) in the hydrocarbons generated at depth has very much decreased when the zone of intense generation (the only significant one for constitution of commercial oil pools) is reached.

5. The behaviour of the biomarkers in migration and in the possible alteration of oils in the reservoir is little known.

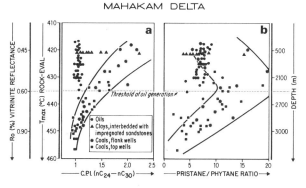

Fig. 6. (a) Evolution of CPI with depth and/or maturation for coals and oils from the Mahakam delta, (b) *idem* for Pristane : Phytane ratio.

As a whole, the effect of migration is previously to lower the concentration of biomarkers in the final oil by 'loss' on the way towards the reservoir-rock, which results in an effect similar to that of maturation and an increased difficulty in interpretation. Moreover, since migration and maturation have presumably similar effects it may be difficult to separate their respective influence on biomarkers distribution.

I might illustrate, with a simple example, certain ambiguities and difficulties relating to the use of markers for migration problems. This example is drawn from the study of the Mahakam delta in Indonesia where the migration phenomena were studied. The organic matter in the delta is of a type which is unique in the sedimentary series. It derives throughout from higher vegetal debris inherited from the continent, and appears either in the dispersed state in clays or in concentration in coal beds (Durand and Oudin, 1979).

Figure 6(a) represents the evolution of the CPI (measured on the n-alkanes in the C_{24}–C_{30} range on oils and $CHCl_3$ extracts of sediments) according to the depth or degree of evolution, which is ascertained by the reflective power of the vitrinite or the T_{max} of the Rock Eval (Espitalié et al., 1977) for: (a) coal samples little affected by migration from the wells situated on the flanks of the Handil and Nilam structures (Vandenbroucke et al., this session), (b) accumulated coal samples from wells situated at the top of the Handil structure, (c) clay samples interbedded between layers of impregnated sandstone, coming from a core taken at shallow depth in the proximity of a reservoir and (d) oil samples taken at different depths.

The CPI of oils is between 0.95 and 1.15. That of coals, whether or not they are accumulated, decreases with evolution and is parallel to that of oils at a degree of evolution corresponding to a T_{max} over 440 °C (or a R_0 of 0.7%). This leads one to believe that the oils were formed at a higher stage of maturity than the one corresponding to this value and that this T_{max} of 440 corresponds to the formation threshold of the oils which may have reached the pools. This result is in agreement with the results obtained through other methods.

The CPI of clays interbedded with impregnated sandstones is highly variable. In the above-described context, the fluctuation in values is interpreted by a greater or lesser impregnation of the samples with oils coming from greater depths.

The CPI of the oil does not appear in the present case to be affected by migration, and the impregnation of the 'accumulated' coals by n-alkanes in the C_{24}–C_{30} range seems minimal, at least at the top of the series (it is impossible to consider in the lower part of the series, because the CPI of the oils and the coals are very close). The values of the Pristane/Phytane ratio (Fig. 6(b)) are too dispersed to be interpretable. A part of this scattering might be due to variations in redox conditions according to variations in the deltaic facies, although the organic matter origin is very constant throughout the whole series. Indeed it is widely accepted that Pristane/Phytane ratio is very sensitive to such variations. Moreover, this ratio — a phenomenon which is now known, at least for organic matter derived from higher plants (Boudou, 1981) — reaches a maximum value in coals little affected by migration (and therefore useful as reference for genetic phenomena) at the end of the immature zone. A low value may thus correspond to two degrees of evolution.

The ratios Pristane nC_{17} and Phytane nC_{18} (Fig. 7(a) and (b)), although they show the disadvantage of reaching maximum values, are less dispersed and more clearly interpretable, once the conclusion is accepted that the oil comes from deep zones:

1. The 'accumulated' coals of the middle and top of the series are there distinguishable from the reference coals, indicating an addition by impregnation by nC_{17}, nC_{18}, Pristane and Phytane from oils coming from the lower depths which add to those generated in situ.

2. The values measured in oils are homogeneous, and correspond to values measured in the deepest coals. An increase in values is observed in the shallowest oils, which suggests that they have dissolved some of the pristane contained in the adjacent rocks during their migration. An interaction, slight but visible, thus takes place with the rocks met on the migration path, which was not the case in the heavy n-alkanes (the behaviour of the oil thus varies according to the fraction under consideration).

These interpretations of markers which are probably the most universally utilized in the geochemical study of sedimentary basins, are here relatively easy, presumably because: (a) it was demonstrated beforehand that the organic matter is of a unique type, (b) the evolution of the markers concentrations with maturation could be defined independently of migration phenomena and (c) the migration phenomena and the geological context were studied elsewhere.

It is evident that the use of these markers in this still simple case, would have been much more delicate if used in interpreting migration phenomena a priori. In more complex cases, for example basins in which organic matter is of several types or of mixed types, difficulties would be even greater, and the instruments for surmounting them even less developed. Here a considerable effort will be required, in a close collaboration of geologists, sedimentologists and geochemists, of a kind which has rarely been carried out heretofore, in three directions:

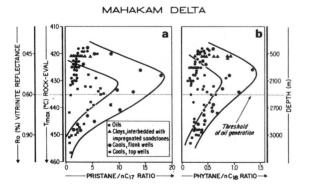

Fig. 7. (a) Evolution of Pristane : n-C17 ratio with depth and/or maturation for coals and oils from the Mahakam delta (b) idem for Pristane : n-C18 ratio.

1. To find easily handled biomarkers, truly specific to the various types of sedimentation. The hopanes exemplify the opposite; having long linked them to higher plants, we are now linking them with bacterial alteration of a ubiquitous character. In this field, the use of n-alkanes derived from cuticle waxes as markers of the continental environment at low stages of evolution, and the abundance of steranes as markers of the marine environment, seem to be the only ones that have withstood the test of time. The work of Baker *et al.* with petroporphyrins, raises some hope; they show (see Baker and Louda, 1981; Louda and Baker, 1981) the great sensitivity of these markers to deposit conditions.

2. To study the evolution of each class of biomarkers with maturation, in homogeneous sedimentary series or by thermal simulation (but thermal simulation might be unrealistic in this particular case since details of molecular structure must be reproduced), so as to be able to extrapolate the results obtained with immature sediments to mature sediments. The following works in the fields of steranes, triterpanes, acyclic isoprenoids and porphyrins are worthy of note: Ensminger (1977), Seifert (1978), Huc (1978), Hajibrahim (1978), Mackenzie (1980), Eglinton *et al.* (1980), Mackenzie *et al.* (1980a and b).

3. To study the fate of the biomarkers in the course of migration or alteration from geological models which have been well studied elsewhere. In this field, we should cite the work of Seifert and Moldowan (1980) on steranes. On the basis of geological examples, these authors indicate that the extent of migration is reflected in the proportions of certain optic isomers in this family of markers. We should also cite the recent work of Connan *et al.* (1980) on the oils of the Aquitaine basin, which confirms more ancient work according to which the distributions of tri-, tetra- and pentacyclic alkanes are not modified by biodegradation unless in extreme cases and might thus serve as correlations in cases of biodegradation.

It would hardly be feasible to follow the paths of gaseous hydrocarbons, particularly methane, with any methods other than isotopic techniques. It is risky, for example, to try to deduce the origin and path of gases from the liquid hydrocarbons which can be associated with them, because, given their vastly different physical properties, it is presumably impossible to be sure that they have come from the same place and at the same time as the gas.

But the isotopic relationships conceal within themselves little information and can thus hardly constitute more than an accessory to the study of migration problems. Also, the factors which determine them, except for the migration phenomena, are numerous, and their respective influence bitterly disputed. These factors are: (a) the nature of the organic matter, (b) the maturity of the organic matter, (c) the genetic process (biogenic gas/thermogenic gas), (d) interaction with the environment (water, CO_2, carbonates in particular).

A good deal of clarification work has been undertaken during the past few years, based on both observation and experimentation. Examples are Schoell (1979), Chung and Sackett (1980), Peters (1980), Redding *et al.* (1980), Fuex (1980), Rice and Claypool (1980). The result is that, in spite of opposing opinions (Neglia, 1979), migration phenomena probably influence the carbon and hydrogen isotopic composition very little. Interactions with the environment also appear to be limited. And the values of $\delta 13 C < -55\%$ are very likely to correspond to biochemical formation processes.

In spite of this simplification, correspondence between the values of isotopic relationships and the nature and stage of maturity of the organic matter, which determine the usefulness of these methods in migration problems by making it possible to localize the gas-formation zones, seem even less sure. As an example, the chart proposed by Stahl (in Schoell, 1979) comparing the degree of maturity of the terrestrial organic matter with the $\delta 13 C$ of the gases that issued from it, does not seem to prove a universal correspondence. One possible explanation, suggested by the work of Redding *et al.* (1980), is that the gas found in the deposits can have different $\delta 13 C$, depending on whether they are directly formed from kerogen (in which case the range of values would be large and the methane issued from overmature samples would be isotopically heavy) or by the succession of two stages: formation of oil then cracking of oil (in which case the range of values would be rather narrow around $-50\%_{00}$). Reservoirs could be supplied with those two types of gas in variable proportions according to nature of organic matter and geological conditions. The values reported by Stahl are those of gases in northern Germany, issued from coal levels of complex geological history (Patijn, 1964). The gas may have been formed at an advanced stage of maturity, during a reburial stage, and thus would have come from a direct degradation of coal without the intermediary of oil formation previously eliminated by migration. This would explain the fact that it is isotopically heavy compared to methane, which also issues from terrestrial organic matter but probably through an oil-formation stage. The Mahakam delta is an example (Schoell *et al.*, this session).

Again, it is through the availability of very carefully prepared geological models that such problems may be resolved.

CONCLUSION

Organic chemistry is only one means for understanding hydrocarbon migration in sedimentary basins. But it is an essential instrument. Observation and prediction of the distribution of organic matter in sediments, as well as the quantities and distributions of hydrocarbons and the reconstitution of the history of their formation, make it possible to establish the initial and final conditions of the system, to make mass-balances and to verify the validity of the proposed mechanisms. Through the three themes I have discussed, some recent important progress in the field is indicated, but there are also some weaknesses. Among the advances, from the conceptual point of view, emergence of expulsion mechanisms in which the hydrocarbons constitute a phasis separated from water seems to me to be a progress as important as the explanation of petroleum formation by the kinetics of organic matter degradation, and from the technological point of view, miniaturization of analytical techniques and increase in analytical speed allow the researchers to study migration in its necessary three dimensions and to realize small and easy to handle experimental systems. Progress in fluorescence microscopy observations offers a possibility of knowing the source-rock milieu in a lesser abstract form.

Among the faults: exploitation for migration studies of those new possibilities is still weak. I also note that the knowledge of hydrocarbon formation, which has advanced little in the past five years, is insufficient for the problems posed. There is also still no clear set of instructions for the use of biomarkers in problems of correlation, taking into account the reality of migration. I believe these faults to be because there is insufficient collaboration among geologists, sedimentologists and geochemists, because the oil companies have placed very few satisfactory geological models at the disposal of researchers, and because the teams are too small to treat problems in their proper prospective.

ACKNOWLEDGEMENTS

This paper greatly benefited of criticism by R. Pelet, T. Powell, P. Ungerer and Mrs M. Vandenbroucke.

I also thank my colleagues at IFP, CFP and SNEA (P), whose support is invaluable in the effort for understanding hydrocarbon migration. I feel particularly indebted to Miss Bessereau and A. Chiarelli and J. L. Oudin.

REFERENCES

Baker, E. G. (1962) Distribution of hydrocarbons in solution. *Bull. Am. Assoc. Pet. Geol.* **46**, 76–84.

Baker, E. W. and Louda, J. W. (1981). Thermal aspects of chlorophyll geochemistry. This volume.

Barker, C. (1972) Aquathermal pressuring: role of temperature on the development of abnormal pressure zones. *Bull. Am. Assoc. Pet. Geol.* **56**, 2068–2071.

Barker, C. (1980) Distribution of organic matter in a shale clast. *Geochim. Cosmochim. Acta* **44**, 1483–1492.

Basu, D. N., Banerjee, A. and Tamhane, D. M. (1980) Source Areas and Migration Trends of Oil and Gas in Bombay offshore basin, India. *Bull Am. Assoc. Pet. Geol.* **64**, 209–220.

Bois, C. and Monicard, R. (1981) Pétrole: peut-on encore découvrir des gisements géants? La Recherche 124, Juillet-Août, pp. 854–865.

Bonham, L. C. (1978) Solubility of methane in water at elevated temperatures and pressures. *Bull. Am. Assoc. Pet. Geol.* **62**, 2478–2481.

Bordenave, M., Combaz, A. and Giraud, A. (1970) Influence de l'origine des matières organiques et de leur degré d'évolution sur les produits de pyrolyse du kérogène. In *Advances in Organic Geochemistry 1966*, ed. by Hobson, G. D. and Speers, Pergamon Press, Oxford, pp. 389–405.

Boudou, J. P. (1981) Diagenèse organique de sédiments deltaiques, delta de la Mahakam, Indonésie, thesis, University of Orléans.

Bray, E. E. and Foster, W. R. (1980) A process for primary migration of petroleum. *Bull. Am. Assoc. Pet. Geol.* **64**, 107–114.

Burst (1969) Diagenesis of Gulf Coast clay sediments and its possible relation to petroleum migration. *Bull. Am. Assoc. Pet. Geol.* **53**, 73–93.

CEPM–CNEXO (Comité d'Etudes Géochimiques Marines) *Géochimie organique des sédiments marins profonds:* Orgon I. (1977) *Mer de Norvège*, ed. by Combaz, A. and Pelet, R.; Orgon II (1979) *Atlantique, NE Brésil*, ed. by Combaz, A. and Pelet, R.; Orgon III (1979) *Mauritanie, Sénégal, îles du Cap Vert*, ed. by Arnould, M. and Pelet, R.; Orgon IV (1981) *Golfe d'Aden, Mer d'Oman*, ed. by Arnould, M. and Pelet, R., Editions du CNRS, Paris.

Chiarelli, A., Serra, O., Gras, C., Masse, P. and Tison, J. (1973) Etude automatique de la sous-compaction des argiles par diagraphies différées. Méthodologie et Applications. *Rev. Inst. Fr. Pét.* **28**, 19–36.

Chung, H. M. and Sackett, W. M. (1979) Carbon isotope effects during pyrolysis of carbonaceous materials. In *Advances in Organic Geochemistry 1979*, ed. by Douglas, A. G. and Maxwell, J. R., Pergamon Press, Oxford, pp. 705–710.

Claypool, G. E. and Reed, P. R. (1976) Thermal analysis technique for source-rock evaluation: quantitative estimate of organic richness and effects of lithologic variation. *Bull. Am. Assoc. Pet. Geol.* **60**, 608–626.

Clayton, J. L. and Swetland, P. J. (1980) Petroleum generation and migration in Denver basin. *Bull. Am. Assoc. Pet. Geol.* **64**, 1613–1633.

Combaz, A. and de Matharel, M. (1978) Organic sedimentation and genesis of petroleum in Mahakam delta, Borneo. *Bull. Am. Assoc. Pet. Geol.* **62**, 1684.

Connan, J. and Cassou, A. M. (1979) Properties of gases and petroleum liquids derived from terrestrial kerogens at various maturation levels. *Geochim. Cosmochim. Acta* **44**, 1–23.

Connan, J., Restle, A. and Albrecht, P. (1980) Biodegradation of crude oil in the Aquitaine basin. In *Advances in Organic Geochemistry 1979*, ed. by Douglas, A. G. and Maxwell, J. R., Pergamon Press, Oxford, pp. 1–18.

Cordell, R. J. (1973) Colloidal soap as proposed primary migration medium for hydrocarbons. *Bull. Am. Assoc. Pet. Geol.* **57**, 1618–1643.

Demaison, G. J. and Moore, G. T. (1980) Anoxic environments and oil source bed genesis. *Bull. Am. Assoc. Pet. Geol.* **64**, 1179–1209.

Deroo, G. and Herbin, J. P. (1980) Bilan des culminations de matière organique pétroligène dans le Crétacé de forages DSDP en Atlantique Nord. *Rev. Inst. Fr. Pet.* **35**, 327–333.

Dobryansky, A. F., Andreyev, P. F. and Bogomolov, A. I. (1961) Certain relations in the composition of crude oils. *Int. Geol. Rev.* **3**, 49–59.

Dow, W. G. (1974) Application of oil-correlation and source rock data to exploration in the Williston Basin. *Bull. Am. Assoc. Pet. Geol.* **58**, 1253–1262.

Dow, W. G. (1978) Petroleum source beds on continental slopes and rises. *Bull. Am. Assoc. Pet. Geol.* **62**, 1584–1606.

Du Rouchet, J. (1978) Elements d'une théorie géomécanique de la migration de l'huile en phase constituée. *Bull. Cent. Rech. Explor. Prod. Elf Aquitaine* **2**, 337–373.

Du Rouchet, J. (1981) Stress Fields, a key to oil migration. *Bull. Am. Assoc. Pet. Geol.* **65**, 74–85.

Durand, B. and Oudin, J. L. (1979) Exemple de migration des hydrocarbures dans une série deltaique: le delta de la Mahakam, Kalimantan, Indonésie. *Proceedings of 10th World Petroleum Congress*, Bucharest, Sept. 1979, Vol. 1, pp. 3–11.

Eglinton, G., Hajibrahim, S. K., Maxwell, J. R. and Quirke, J. M. E. (1980) Petroporphyrins: structural elucidation and the application of HPLC fingerprinting to geochemical problems. In *Advances in Organic Geochemistry 1979*, ed. by Douglas, A. G. and Maxwell, J. R., Pergamon Press, Oxford, pp. 193–203.

Ensminger, A. (1977) Evolution de composés polycycliques sédimentaires. Thesis, University of Strasbourg.

Espitalié, J., Laporte, J. L., Madec, M., Marquis, F., Leplat, P., Paulet, J. and Boutefeu, A. (1977) Méthode rapide de caractérisation des roches-mères, de leur potentiel pétrolier et de leur degré d'évolution. *Rev. Inst. Fr. Pét.* **32**, 23–42.

Espitalié, J., Madec, M. and Tissot, B. (1980) Role of Mineral Matrix in Kerogen Pyrolysis: Influence on Petroleum generation and migration. *Bull. Am. Assoc. Pet. Geol.* **64**, 59–66.

Fuex, A. N. (1980) Experimental evidence against an appreciable isotopic fractionation of methane during migration. In *Advances in Organic Geochemistry 1979*, ed. by Douglas, A. G. and Maxwell, J. R., Pergamon Press, Oxford, pp. 725–732.

Hajibrahim, S. K. (1978) Applications of petroporphyrins to the maturation, migration and origin of crude oils. Ph.D. Thesis, University of Bristol.

Harwood (1977) Oil and gas generation by laboratory pyrolysis of kerogen. *Bull. Am. Assoc. Pet. Geol.* **61**, 2082–2102.

Hedberg, H. D. (1974) Relation of methane generation to undercompacted shales, shale diapirs and mud volcanoes. *Bull. Am. Assoc. Pet. Geol.* **58**, 661–673.

Hedberg, H. D. (1980) Methane generation and petroleum migration. In Problems of Petroleum Migration, AAPG Studies in Geology, No. 10, ed. by Roberts, W. H. III and Cordell, J. R., pp. 179–207.

Hodgson, G. W. (1980) Origin of petroleum: in transit conversion of organic compounds in water. In *Problems of Petroleum Migration*, AAPG Studies in Geology, ed. by Roberts, W. H. III and Cordell, J. R., No. 10, pp. 89–107.

Huc, A. Y. (1978) Géochimie organique des schistes bitumineux du Toarcien du Bassin de Paris. Thesis, University of Strasbourg.

Huc, A. Y. and Hunt, J. M. (1980) Generation and migration of hydrocarbons in offshore South Texas gulf coast sediments. *Geochim. Cosmochim. Acta* **44**, 1981–1989.

Hunt, J. M., Huc, A. Y. and Whelan, J. K. (1980) Generation of light hydrocarbons in sedimentary rocks. *Nature (London)* **288**, 688–690.

Jonathan, D., L'Hote, G. and Du Rouchet, J. (1975) Analyse géochimique des hydrocarbures légers par thermovaporisation. *Rev. Inst. Fr. Pet.* **30**, 65–98.

Ishiwatari, R., Rohrback, B. G. and Kaplan, I. R. (1978) Hydrocarbon generation by thermal alteration of kerogen from different sediments. *Bull. Am. Assoc. Pet. Geol.* **62**, 687–692.

Kubler, B. (1980) Les premiers stades de la diagenèse organique et de la diagenèse minérale. *Bull. Ver. Schweiz. Pet. Geol. Ing.* **45**, 1–22; **46**, 1–22.

Kvenvolden, K. A. and Claypool, G. E. (1980) Origin of gasoline-range hydrocarbons and their migration by solution in carbon dioxide in Norton Basin, Alaska. *Bull. Am. Assoc. Petr. Geol.* **64**, 1078–1086.

Larter, S. R. (1978) A geochemical study of kerogen and related materials. Ph.D. Thesis, University of Newcastle-upon-Tyne.

Larter, S. R. and Douglas, A. G. (1980) Typing of kerogens by pyrolysis capillary G.C. In *Advances in Organic Geochemistry 1979*, ed. by Douglas, A. G. and Maxwell, J. R., Pergamon Press, Oxford, pp. 579–583.

Larter, S. R., Solli, H. and Douglas, A. G. (1978) Analysis of kerogens by pyrolysis-gas chromatography-mass spectrometry using selective ion detection. *J. Chromatogr.* **167**, 421–431.

Lewan, M. D., Winters, J. C. and McDonald, J. H. (1979). Generation of Oil Like Pyrolyzates from Organic Rich Shales. *Science* **203**, 2, March.

Leythaeuser, D., Schaeffer, R. G. and Yükler, A. (1980) Diffusion of light hydrocarbons through near surface rocks. *Nature (London)* **284**, 522–525.

Louda, J. W. and Baker, E. W. (1981) Geochemistry of tetrapyrrole, carotenoid and perylene pigments in sediments from the San Miguel Gap (Site 467) and Baja California borderland (Site 471), DSDP/IPOD Leg 63. Initial Reports of the deep sea drilling project, Vol. 63.

McAuliffe, C. D. (1980) Oil and gas migration: chemical and physical constraints. In *Problems of Petroleum Migration*, ed. by Roberts, W. H. III and Cordell, J. R., AAPG Studies in Geology, No. 10, pp. 89–107.

Mackenzie, A. S. (1980) Application of biological marker compounds to subsurface geological processes. Ph.D. Thesis, University of Bristol.

Mackenzie, A. S., Patience, R. L., Maxwell, J. R., Vandenbroucke, M. and Durand, B. (1980) Molecular parameters of maturation in the Toarcian Shales, Paris Basin, France I. Changes in the configurations of acyclic isoprenoid alkanes, steranes and triterpanes. *Geochim. Cosmochim. Acta* **44**, 1709–1721.

Mackenzie, A. S., Quirke, J. M. E. and Maxwell, J. R. (1980) Molecular parameters of maturation in the Toarcian Shales, Paris Basin, France. II. Evolution of metalloporphyrins. In *Advances in Organic Geochemistry 1979*, ed. by Douglas, A. G. and Maxwell, J. R., Pergamon Press, Oxford, pp. 239–248.

Magara, K. (1978) Compaction and Fluid Migration. Practical Petroleum Geology. Developments in Petroleum Science 9. Elsevier.

Meissner, F. F. (1978a) Compaction and Fluid Migration. Practical Petroleum Geology. Developments in Petroleum Science 9 Elsevier.

Meissner, F. F. (1978b) Petroleum Geology of the Bakken Formation, Williston Basin, North Dakota and Montana. *Proceedings of 1978 Williston Basin Symposium*, Sept. 24–27, 1978. Montana geological society, Billings, pp. 207–227.

Monin, J. C., Durand, B., Vandenbroucke, M. and Huc, A. Y. (1980) Experimental simulation of the natural transformation of kerogen. In *Advances in Organic Geochemistry, 1979*. ed. by Douglas, A. G. and Maxwell, J. R., Pergamon Press, Oxford, pp. 517–530.

Moore, C. H. and Druckman, Y. (1981) Burial Diagenesis and Porosity Evolution, Upper Jurassic Smackover, Arkansas and Louisiana. *Bull. Am. Assoc. Pet. Geol.* **65**, 597–628.

Neglia, S. (1979) Migration of fluids in sedimentary basins. *Bull. Am. Assoc. Petr. Geol.* **63**, 575–597.

Patijn, R. J. H. (1964) Die Entstehung von Erdgas infolge der Nachinkohlung in Nordesten der Niederlande. *Erdöl Kohle* **17**, 2–9.

Perry and Hower (1972) Late stage dehydration in deeply buried sediments. *Bull. Am. Assoc. Petr. Geol.* **56**, 2013–2021.

Peters, K. E. (1978) Effects on sapropelic and humic protokerogen during laboratory simulated geothermal maturation experiments. Ph.D. Thesis, UCLA.

Peters, K. E., Rohrback, B. G. and Kaplan, I. R. (1980) Laboratory simulated thermal maturation of Recent sediments In *Advances in Organic Geochemistry 1979*, ed by Douglas, A. G. and Maxwell, J. R., Pergamon Press, Oxford, pp. 547–555.

Philp, R. P. and Russel, N. J. (1980) Pyrolysis-gas chromatography mass spectrometry of batch autoclave products derived from coal macerals. In *Advances in Organic Geochemistry, 1979*, ed. by Douglas, A. G. and Maxwell, J. R., Pergamon Press, Oxford, pp. 653–661.

Powell, T. G. (1978) An assessment of the hydrocarbons

source-rocks potential of the Canadian arctic islands. *Geological Survey of Canada*, paper 78–12.
Powell, T. G., Foscolos, A. E., Gunther, P. R. and Snowdon, L. R. (1978) Diagenesis of organic matter and fine clay minerals; a comparative study. *Geochim. Cosmochim. Acta* **42**, 1181–1197.
Powers (1967) Fluid-release mechanisms in compacting marine mudrocks and their importance in oil exploration. *Bull. Am. Assoc. Pet. Geol.* **51**, 1240–1254.
Price, L. C. (1976) Aqueous solubility of petroleum as applied to its origin and primary migration. *Bull. Am. Assoc. Pet. Geol.* **60**, 213–224.
Redding, C. E., Schoell, M., Monin, J. C. and Durand, B. (1980) Hydrogen and carbon isotopic composition of coals and kerogens. In *Advances in Organic Geochemistry, 1979*. ed. by Douglas, A. G. and Maxwell, J. R. Pergamon Press, Oxford, pp. 711–723.
Rice, D. D. and Claypool, G. E. (1981) Generation, accumulation and resource potential of biogenic gas. *Bull. Am. Assoc. Pet. Geol.* **65**, 5–25.
Robert, P. (1979) Classification des matières organiques en fluorescence. Application aux roches-mères pétrolières. *Bull. Centr. Rech. Explor. Prod. Elf Aquitaine* **3**, 223–263.
Rohback, B. G. (1979) Analysis of low molecular weight products generated by thermal decomposition of organic matter in recent sedimentary environments. Ph.D. Thesis, UCLA.
Saint Paul, C., Monin, J. C. and Durand, B. (1980) Méthode de caractérisation rapide des hydrocarbures de C_1 à C_{35} contenus dans les roches sédimentaires et dans les huiles. *Rev. Inst. Fr. Pet.* **35**, 1065–1078.
Schaefer, R. G., Leythaeuser, D. and Weiner, B. (1978) Single-step capillary G.C. method for extraction and analysis of sub-ppb quantities of hydrocarbons (C_2–C_8) from rock and crude oil samples and its application in petroleum geochemistry. *J. Chromatogr.* **167**, 355–363.
Schoell, M. (1980) The hydrogen and carbon isotopic composition of methane from natural gases of various origins. *Geochim. Cosmochim. Acta* **44**, 649–662.
Seifert, W. K. and Moldowan, J. M. (1981) Paleoreconstruction by biological markers. *Geochim. Cosmochim. Acta* **45**, 783–794.
Seifert, W. K. (1978) Steranes and terpanes in kerogen pyrolysis for correlation of oils and source-rocks. *Geochim. Cosmochim. Acta* **42**, 473–484.
Simlote, V. N. and Withjack, E. M. (1981) Estimation of tertiary recovery by CO_2 injection, Springer A Sand, Northeast Purdy Unit. *J. Pet. Technol.* May, p. 810.
Snarski, I. (1962) Primary migration of oil. *Geol. Nefti Gaza* **6**, 700–703 (in Russian).
Snarski, I. (1970) Nature of primary migration of petroleum. *Geol. Nefti Gaza* **8**, 11–15 (in Russian).
Snowdon, L. R. (1978) Organic geochemistry of the upper Cretaceous/Tertiary delta complexes of the Beaufort Mackenzie Sedimentary Basin. Thesis, Rice University.
Snowdon, L. R. (1979) Errors in extrapolation of Experimental Kinetic Parameters to organic geochemical systems. *Bull. Am. Assoc. Pet. Geol.* **63**, 1128–1134.
Snowdon, L. R. (1980) Resinite, a potential petroleum source in the upper Cretaceous/Tertiary of the Beaufort Mackenzie Basin. In *Facts and Principles of World Petroleum Occurence*. Canadian Society of Petroleum Geologists, Calgary, Alberta.
Sokolov et al. (1964) Migration processes of gas and oil, their intensity and directionality. *Proceedings of the 6th World Petroleum Congress*. Frankfort 1963, Section 1, pp. 493–505.
Tissot, B. and Pelet, R. (1971) Nouvelles données sur les mécanismes de genèse et de migration. Simulation mathématique et application à la prospection. *Proceedings of the 8th World Petroleum Congress*, Moscow 1971, Vol. 2, pp. 35–46.
Tissot, B., Deroo, G. and Herbin, J. P. (1979) Organic matter in cretaceous sediments of the North Atlantic: contribution to sedimentology and paleogeography. In *Deep Drilling Results in the Atlantic Ocean: Continental Margins and Paleoenvironment*, ed. by Talwain, M., Hay, W. and Ryan, W. B. F. Maurice Ewing Series, 3, Am. Geoph. Union, pp. 362–374.
Tissot, B., Demaison, G., Masson, P., Delteil, J. B. and Combaz, A. (1980) Paleoenvironment and Petroleum Potential of Middle Cretaceous Black Shales in Atlantic Basins. *Bull. Am. Assoc. Petr. Geol.* **64**, 2051–2065.
Vandenbroucke, M. (1972) Etude de la migration primaire: variation de composition des extraits de roche à un passage roche mère/réservoir. In *Advances in Organic Geochemistry 1971*, ed. by Von Gaertner and Wehner, H., Pergamon Press, Oxford–Braunschweig, pp. 547–565.
Vandenbroucke, M., Durand, B. and Hood, A. (1977) Thermal evolution experiments on a kerogen from the Green River Shales formation (Uinta Basin, USA), 8th Int. Meeting on Organic Geochemistry, Moscow.
Van de Meent, D., Brown, S. C., Philp, R. P. and Simoneit, B. R. T. (1980) Pyrolysis-high resolution gas chromatography and pyrolysis–gas chromatography-mass spectrometry of kerogens and kerogen precursors. *Geochim. Cosmochim. Acta* **44**, 999–1014.
Van Graas, G., De Leeuw, J. C. and Schenck, P. A. (1980) Analysis of coals of different rank by Curie pyrolysis/mass spectrometry and Curie point pyrolysis/gas chromatography/mass spectrometry. In *Advances in Organic Geochemistry 1979*, ed. by Douglas, A. G. and Maxwell, J. R., Pergamon Press, Oxford, pp. 485–494.
Whelan, J. K. (1979) C_1 to C_7 hydrocarbons from IPOD Holes 397 and 397 A. *Initial Report DSDP 47*, 531–539.
Williams, J. A. (1974) Characterization of oil types in the Williston Basin. *Bull. Am. Assoc. Pet. Geol.* **58**, 1242–1252.
Wilson, H. H. (1975) Time of Hydrocarbon expulsion, paradox for geologists and geochemists. *Bull. Am. Assoc. Pet. Geol.* **59**, 69–94.
Yariv, S. (1976) Organophilic pores as proposed primary migration media for hydrocarbons in argillaceous rocks. *Clay Sci.* **5**, 19–29.
Zhuse, T. P. and Bourova, E. G. (1977) Influence des différents processus de la migration primaire des hydrocarbures sur la composition des pétroles dans les gisements. In *Advances in Organic Geochemistry 1975*, ed. by Goni, J. and Campos, E., ENADIMSA, Madrid, pp. 493–499.

Detecting Migration Phenomena in a Geological Series by Means of C_1–C_{35} Hydrocarbon Amounts and Distributions

M. Vandenbroucke and B. Durand

I.F.P., Rueil Malmaison, France

J. L. Oudin

TOTAL C.F.P., Talence, France

This paper describes geochemical analyses of samples from various wells in the Mahakam Delta area in Indonesia having different structural positions. One of these wells was partly drilled through an overpressured shale section. Geochemical methods were applied either to coals hand-picked from total cuttings or to kerogens obtained after the acidic destruction of minerals. Analyses include elemental analysis and Rock-Eval pyrolysis of all the samples, as well as more detailed investigation of selected samples, by methods such as extraction and composition of extracts or thermovaporization of free hydrocarbons. The quantitative parameters obtained by these methods are compared for the different wells, and the behaviour of the organic matter is described in relation to the geological environment: formation of hydrocarbons in the source rocks; loss of fractions by migration; and cracking of soluble products if migration does not occur and subsequent formation of both gas and insoluble residues.

GEOLOGICAL SETTING

The Mahakam Delta is situated in Indonesia, on the eastern edge of the Kutei Basin, in south–eastern Borneo (Kalimantan). The exact position of the basement is not known, nor is the age of the first deposits, but deltaic sedimentation is known to have occurred at least since the Middle Miocene, and this eastward prograding series is more than 4000 m thick. The structure of the region is characterized by a succession of NNE–SSW anticlinal axes of decreasing amplitude going from west to east (Fig. 1). Underneath the interbedded clays and sands of the deltaic plain, which are rich in organic matter and coal seams, there are argillaceous platform or prodelta facies, where high-pressure phenomena can occur. Oil fields were discovered along the successive anticlinal axes (Fig. 1). Depending on the case, oil is preferentially found (Handil), which is eventually degraded near the surface (Sanga Sanga) or else gas condensate (Badak). Except for Badak and Nilam, the fields are compartmented by more or less numerous normal faults.

Geochemical analyses in this region have been made on a great many wells (Combaz and de Matharel, 1978; Durand and Oudin, 1979). The results described here have to do with only a portion of these analyses, concerning wells H8, H9bis and H627 of the Handil field and well N25 (formerly Terentang) of the Nilam field.

Their aim is to show how migration phenomena in this type of deltaic sedimentation influence the distribution of hydrocarbons. The position of the wells is shown on the section in Fig. 2. Wells H8 and H9bis, which are next to each other (50 m), are situated at the top of the Handil structure. This structure has a single major fault and mainly contains oil. Well H627 is situated on the eastern flank on the same side of the fault as H8 and H9bis. Well N25 is also situated on the flank of the Nilam structure. This structure is a nose extending towards SW of the Badak structure and having no structural closure above 2200 m. As at Badak, this structure contains mainly gas.

The sedimentary series are characterized from top to bottom by sandy bodies having highly varying sizes and shapes, interbedded with shales (deltaic-plain sedimentation), and then by very fine silty shales (prodelta sedimentation). The hydrocarbons are situated in these sandy bodies. Each reservoir has its own gas/oil and oil/water contact. The Handil field contains hydrocarbons throughout the whole depth where these reservoirs exist, i.e. between 2900 m and almost up to the surface, with a maximum of productive horizons between 1800 and 2450 m. The accumulated oils are lighter and lighter upon approaching the surface.

The prodelta shales are, at the present time, overpressured because subsidence was very fast and the water contained in the porosity was not expelled quickly enough. The transition from well-drained, normal-

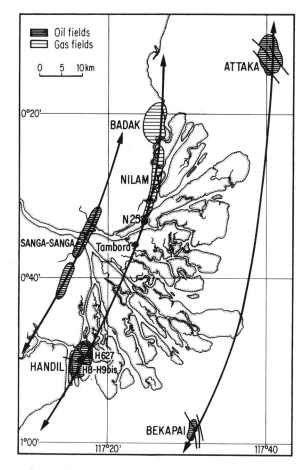

Fig. 1. Regional map of the Mahakam Delta showing the principal oil and gas fields located along parallel anticlinal trends. (Name of the fields in capital letters.)

pressure zones (specific gravity equal to one) to geostatic-pressure zones (specific gravity > 2) occurs in an interval which varies according to the local sedimentological characteristics. It is about 300 m (2850 to 3150 m) at the top of Handil and probably much greater on Nilam and Badak (about 1000 m on the northern side of Badak). The wells examined here penetrated more or less deeply into these zones. For example, well H8 crosses all the way through the transition zone and stops at the top of the high-pressure zone. Well H9bis penetrates 1000 m into this high-pressure zone without reaching the bottom. Well H627 reaches the transition zone at around 3150 m but does not run in very far (less than 100 m). Well N25 penetrates for 180 m into the transition zone (3900 to 4080 m), but the equivalent specific gravity at the bottom of the well is not very high (1.32).

REVIEW OF ANALYTICAL PROCEDURE

This review will be brief. For a detailed description reference is made to the analytical procedure given in various publications dealing with this subject.

Rock–Eval pyrolysis

Samples of 5 to 100 mg are heated quickly (250 to 550 °C at 25°C min^{-1}) in helium (Espitalié et al., 1977). The hydrocarbons which are released are detected by flame ionization and occur in the form of a first peak caused by the free compounds present in the rock, then by a second peak caused by the pyrolysis of the organic matter. The surface area of the pyrolysis peak related to organic

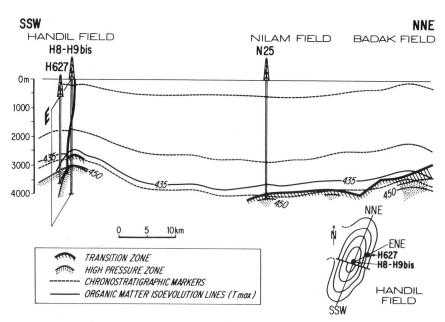

Fig. 2. Section across the Handil–Badak anticlinal trend showing the position of the four wells studied here. Approximate boundaries of transition zone and high-pressure zone are indicated along with isoevolution lines of organic matter (Rock–Eval T_m).

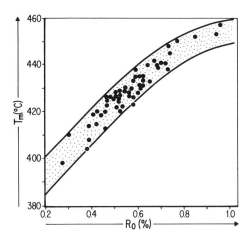

Fig. 3. Correlation between vitrinite reflectance (R_0) and Rock–Eval T_m in Mahakam coals.

carbon, called the hydrogen index, measures the petroleum potential that is still available from the organic matter. The pyrolysis temperature reached at the top of this peak, called T_m, is representative of the maturation state of the organic matter. It gives information equivalent to what is obtained by measuring the reflectance of vitrinite, called R_0, as can be seen in Fig. 3 where the R_0/T_m correlation is given for several Mahakam coals. This parameter T_m, which is obtained faster than R_0, is what we will use hereafter as the common scale of evolution for the wells investigated because the geothermal gradients are relatively variable in the delta. The beginning of oil formation can be roughly situated at about $T_m = 435°C$, the very beginning of gas formation (C_1 to C_5) at around $T_m = 450°C$, and maximum oil formation at around $T_m = 455–460°C$.

Solvent extraction

The ground rock is chloroform extracted (Monin *et al.*, 1978). After the solvent has been filtered and evaporated, the extract is fractionated by thin-layer chromatography (Huc *et al.*, 1976; Huc and Rouchaché, 1981) to obtain saturated hydrocarbons, aromatic compounds and NSO compounds. The hydrocarbons are analysed by gas chromatography and, in some cases, by mass spectrometry (Fabre *et al.*, 1972).

Thermovaporization followed by simulated distillation

The analytical system (Saint-Paul *et al.*, 1980) consists of a sample-holder tube in line with a trap followed by direct introduction into a gas-chromatography column. A sliding oven can be moved from the sample-holder tube to the trap. This system is swept by helium, and the chromatograph has a sub-ambient programming device. The phase used (Dexsil 300) and the programming conditions (-15 to $+370°C$, $8°C$ min^{-1}) enable the total hydrocarbons to be analysed from C_1 to C_{35}, thus renewing interest in this technique which, moreover, has been known for a long time (Bordenave *et al.*, 1970; Jonathan *et al.*, 1975; Schaefer *et al.*, 1978). The sample is quickly heated to 300°C, and the volatilized hydrocarbons are stopped in the trap which is maintained in liquid nitrogen for 15 min. The oven is then moved around the trap. At the same time the temperature programming of the chromatograph is started up. This analysis is thus similar to a chromatographic separation of free hydrocarbons from the first peak in the Rock–Eval method. The distribution of total hydrocarbons (saturated + aromatic) can be calculated along with their amount, between C_1 and C_{35} or in more restricted carbon ranges, by means of a reference standard.

Preparation and analysis of kerogens

The ground and solvent-extracted rock is attacked, under inert gas flow, successively by hydrochloric acid to eliminate the carbonates and then by a hydrochloric-acid/hydrofluoric-acid mixture to eliminate the silicates, and finally by hydrochloric acid once again to destroy the fluorosilicates formed during the preceding stage (Durand and Nicaise, 1980). The pyrite is not attacked during these operations. Between each acid attack, the rock is water washed until it reaches neutrality. Then the organic residue is dried in nitrogen atmosphere. To characterize this kerogen, an elemental analysis is made of C, H, N, O, S and Fe (Durand and Monin, 1980). The organic S is calculated by subtracting from the total S the sulfur belonging to pyrite, this latter being dosed by Fe.

NATURE AND EVOLUTION OF ORGANIC MATTER

Nature of organic matter

Previous publications on this deltaic series (Combaz and de Matharel, 1978; Durand and Oudin, 1979) have already shown that the organic matter it contains comes from higher plants from the continent, that the characteristics of this vegetation have been practically invariable during the entire Tertiary and that there is no difference in either nature or evolution between the organic matter dispersed in the shales and that concentrated in the coals. This is why we have mainly analysed coals here, because analyses can be made on small amounts of these latter samples, easily obtained by hand-picking among cuttings taken at regular intervals in the wells. (Coal is taken here to mean any piece of cutting that can be isolated by tweezers and that can be recognized by its aspect as being mainly made up of pure organic matter. A large proportion of this coal probably comes from the massive coal beds crossed through by the boreholes and spotted by logs.)

These coals were systematically analysed by Rock–Eval pyrolysis and elemental analysis, methods which were also applied to kerogens in the case where coals did not exist in the cuttings. More detailed analyses were made on some coals, in particular the study of distribution of hydrocarbons by thermovaporization or by separation of the chloroform extract. Such analyses are possible only on samples that have not undergone any prior treatment, thus excluding kerogens.

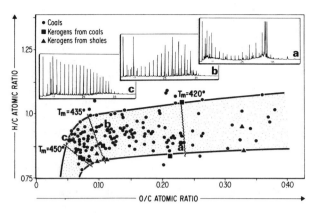

Fig. 4. Evolution of elemental analyses of kerogens and coals in an atomic H/C versus O/C diagram. Some isoevolution lines, indicated by the correspond T_m, are shown. Increasing evolution follows increase of T_m. The chromatograms of saturates from chloroform extracts of coals correspond to increasing evolution from **a** to **c**.

The results of elemental analysis on coals, coal kerogens and shale kerogens coming from the different wells in the Handil field and from well N25 are given in Fig. 4 in the form of a diagram of the H/C versus O/C atomic ratios. The position of these samples is similar to that of other organic matters coming from continental plants and especially with the Type III ones (Tissot et al., 1974; Durand and Espitalié, 1976). The dispersion among coals is in the same range as that between coals and kerogens. Approximate isoevolution lines corresponding to the Rock–Eval T_m of 420, 430 and 450°C are given. Above 450 °C, these isoevolution lines become very close-set on such a diagram.

Formation of hydrocarbons

Figure 4 shows, for three samples having an increasing maturation level and coming from N25, the chromatogram of the $>C_{15}$ saturated hydrocarbons from the chloroform extract. Sample **a**, which is the least evolved, contains very few n-alkanes. Some peaks around the n-C_{15} and n-C_{30} position come from sesquiterpenoids which, in mass spectrometry, give a fragment at $m/e = 204$, and pentacyclic triterpenoids derived from friedelin (Boudou, 1981). Sample **b** contains mainly $>C_{20}$ heavy n-alkanes having a marked odd predominance. The appearance of these alkanes, linked to the decrease in the O/C ratio between stages **a** and **b**, is due to the decarboxylation of vegetal cuticular waxes. In sample **c**, which is the most evolved, the n-alkanes have a regularly decreasing distribution, still with a slight odd predominance. This is the type of distribution that is found in a great many oils in the region. It is due to the thermal cracking of the kerogen and hydrocarbons already formed. Further stages of evolution, that are not obtained here because of the lack of more mature samples, but which are known from other investigations (Albrecht et al., 1976), lead to an alkane distribution whose mode is centred around smaller and smaller carbon numbers and which finally leads to methane.

The formation of hydrocarbons in the coals was also examined by thermovaporization followed by simulated distillation. This method has great advantages compared with chloroform-extract analysis. It is much faster, can be performed on a very small sample and is capable of investigating a wider range of carbon atoms. However, it does not distinguish between saturated hydrocarbons and aromatic hydrocarbons and, until now, was limited to coals for analysis problems.

The quantitative results of this investigation for C_1–C_{35}, C_6–C_9, C_{10}–C_{14} and C_{25}–C_{35} carbon ranges are given in Figs 5 to 8, with the evolution being graduated in Rock–Eval peak temperatures (T_m). Although they were also investigated, the C_{15}–C_{19} and C_{20}–C_{24} distributions are not given here, the representative points of the samples from abnormal-pressure zones being widely scattered because of interference with fuel oil from the drilling mud in these carbon ranges. The amounts of hydrocarbons related to organic carbon are given for the four wells examined, two of which are situated on the top (H8 and H9bis) and two on the flank of the structures (H627 and N25). These amounts of hydrocarbons can be seen to have a regular evolution for the samples from the flank wells, with most of the representative points being within the limit curves traced on each figure, whereas the same is not true for the top wells. This regularity, together with the structural position of the wells and the similarity of the values observed, although these wells are quite far away from one another (Fig. 2), lead us to assume that these curves represent the formation of hydrocarbons by autochthonous organic matter without any influence by migration phenomena. The evolution with T_m of the amounts observed in the different classes effectively corresponds to what can be expected with this hypothesis of hydrocarbon formation, i.e. there is a progressive evolution, when T_m increases, of the carbon number of the hydrocarbons formed towards lighter and lighter fractions, with maximum formation moving from $T_m = 435$–440°C for class C_{25}–C_{35}, $T_m = 450$°C for class C_{10}–C_{14} and T_m far more than 450°C for class C_6–C_9. In this last class, as was shown by Bordenave et al. (1970) for other types of terrestrial organic matter, most of the products formed up to $T_m = 440$°C are represented by light aromatics, benzene, toluene and xylenes, which can easily be spotted on simulated distillation chromatograms (Fig. 9).

The formation curve for total hydrocarbons, both saturated and aromatic in the C_1 to C_{35} range (Fig. 5) has an 'S' shape, with the maximum amount formed being around 40 to 50 mg of hydrocarbons per g C, i.e. 2–2.5 Mt of HC km^{-3} of rock (assuming a rock with a porosity of 10% containing 2% organic carbon), which correlates effectively with the calculation made by Durand and Oudin (1979) using other geochemical parameters independent of the ones used here. The hydrocarbons formed in the diagenesis zone (420 °C $< T_m < 430$°C) by decarboxylation reactions represent about one fifth of the maximum amount, and they are mainly heavy hydrocarbons, as can be seen by comparing the distributions of the C_6–C_9 (Fig. 6), C_{10}–C_{14} (Fig. 7) and C_{25}–C_{35} (Fig. 8) classes. The inflection point of the total-hydrocarbon curve (Fig. 5), situated at the diagenesis/catagenesis transition ($T_m \sim 435$°C), will be interpreted as indicating the transition in the formation mechanisms of hydrocarbons, from decarboxylation to cracking of carbon–carbon bonds. This transition effec-

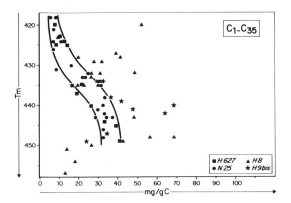

Fig. 5. Variation of the amount of total hydrocarbons in the C_1–C_{35} range with increasing evolution (increasing T_m). Data are obtained by thermovaporization on coals from four wells. H8 and H9bis are located on the top of the Handil structure; H627 and N25 are located respectively on the flank of the Handil and Nilam structures.

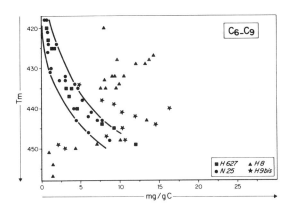

Fig. 6. Variation of the amount of total hydrocarbons in the C_6–C_9 range with increasing evolution (see Fig. 5).

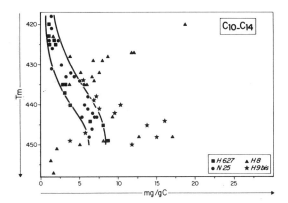

Fig. 7. Variation of the amount of total hydrocarbons in the C_{10}–C_{14} range with increasing evolution (see Fig. 5).

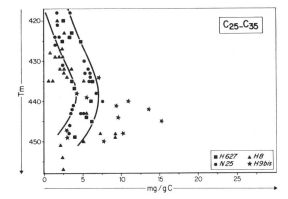

Fig. 8. Variation of the amount of total hydrocarbons in the C_{25}–C_{35} range with increasing evolution (see Fig. 5).

tively corresponds to a maximum value of formation in the C_{25}–C_{35} class, the most frequent carbon range in the higher vegetal waxes.

MIGRATION PHENOMENA

If we return to hydrocarbon distributions by carbon classes, as shown in Figs 5 to 8, for the coals from the wells in the top of the Handil structure (H8 and H9bis), we can see that they do not follow the regular curves defined by the coals from wells H627 and N25. In what follows, we will explain why we attribute this deviation to migration phenomena.

Accumulation phenomena

Let us begin by recalling that a great many reservoirs exist at Handil, and that their filling up with hydrocarbons is all the more frequent as we approach the top of the structure. The positive deviation in total hydrocarbon amounts from C_1 to C_{35} (Fig. 5) among the coals defining the amount formed *in situ* and some of the coals from H8 and H9bis will thus be interpreted as an impregnation of these coals by migrated hydrocarbons. This interpretation is based on two observations.

(1) The depth of the impregnated samples, which is not indicated in Fig. 5, corresponds to that of the zones where hydrocarbon accumulations exist in Handil, i.e. $T_m = 420°C$ corresponding to the shallow zone of the reservoirs (850–1200 m), $T_m = 427$–$430°C$ corresponding to the main zone of the reservoirs (1800–2450 m) and $T_m = 440$–$443°C$ corresponding to the lower zone (2450–2800 m).

(2) The nature of the hydrocarbons impregnating the coals depends, just as for the oils, on the depth at which they are situated in the series. Indeed, if the amounts by carbon classes (Figs 6, 7 and 8) are compared with the curve for total amounts (Fig. 5), we can see that the accumulation in the upper zones (1200 and 2000–2200 m, $T_m = 420$ and 427–$432°C$) is mainly due to light hydrocarbons in classes C_6–C_9 and C_{10}–C_{14}. Class C_{25}–C_{35} is not involved. On the other hand, in the deepest accumulation zone (2600–2800 m, $T_m = 440$–$443°C$), impregnation involves hydrocarbons from all molecular weights, both light and heavy. The curves assumed to represent the genesis of the different hydrocarbon

classes, based on coals from H627 and N25, are justified by arguments of regularity which is not observed in the accumulated samples. Therefore it is not possible to attribute the hydrocarbons in classes C_6–C_9 and C_{10}–C_{14} from these samples entirely to an in-situ origin. Thus, we must assume that the light hydrocarbons, which were formed at greater depth, may have migrated more easily on account of their lower molecular weight and thus have risen higher up in the series.

Depletion phenomena

If we now examine the samples from the transition zone from wells H8 and H9bis (3000–3140 m, $T_m = 448$–457 °C), we find exactly the opposite phenomenon from the preceding one, i.e. the amounts measured are less than the amounts that would be expected from the model of the formation based on the flank wells, both for the total amounts of hydrocarbons (Fig. 5) and for the amounts by carbon classes (Figs 6–8). This cannot, therefore, be the result of a cracking phenomenon, and the systematic loss of hydrocarbons during handling, which may always occur for class C_6–C_9, is not very probable for class C_{10}–C_{14}. Consequently, this phenomenon will be interpreted as a depletion of hydrocarbons resulting from their departure by migration. As was seen previously, this phenomenon is much more appreciable for light hydrocarbons which migrate more easily. Moreover, this loss of the light fraction immediately appears in the comparison of the chromatograms made after the thermovaporization of two coals situated at the top and bottom of the transition zone (Fig. 9). By comparing the amounts remaining in the three deep samples from H8 with the theoretical amounts formed per carbon classes, it is apparent that only one tenth of the C_6–C_9 hydrocarbons and one third of the C_{10}–C_{14} hydrocarbons remain. Class C_{25}–C_{35} is almost unaffected.

Interpretation of the above observations in terms of migration thus reveals:

(1) The importance of working with light hydrocarbons ($< C_{14}$) for investigating migration phenomena, and especially for observing depletion phenomena.

(2) The fact that, in the carbon samples considered, the C_{25}–C_{35} hydrocarbons migrate very little. Therefore they are always representative of organic matter in place or from nearby, hence having a similar degree of evolution. Nonetheless, this comment might not be applicable to shales in which migration may be facilitated by interbedding with sandy drains.

(3) The fact that a differentiation exists in the effects of migration, depending on the carbon classes of the hydrocarbons examined. This may be of help in choosing the most plausible migration mechanism in each case. In our conclusion, we will return to the mechanism considered for explaining the formation of the Handil field.

This interpretation allows the justification, *a posteriori*, of the use of samples coming from the flank wells for modeling the formation of hydrocarbons. Impregnation by migrated hydrocarbons is not very probable on the side of the structure except in the case of stratigraphic trapping. Moreover, these wells stopping near the top of the transition zone, the drainage of this zone is still not very apparent, if it exists.

PHENOMENA SPECIFIC TO THE HIGH-PRESSURE ZONE

Of the two wells from the top of the Handil field described here, one of them, H8, crosses through the entire transition zone (zone in which the pressure of the fluids gradually moves away from hydrostatic pressure and approaches geostatic pressure) and stops at the top of the high-pressure zone. The other well, H9bis, was continued for nearly 1000 m in the high-pressure zone. Samples of cuttings were analysed; unfortunately, the use of an oil-base drilling mud prevented any analysis of coals by thermovaporization and simulated distillation. Whenever possible, the coals were isolated by manual sorting of chloroform-washed cuttings, but a great many samples did not contain any coal. Therefore, kerogens were prepared. Elemental analyses and Rock–Eval pyrolysis characterization were done on both coals and kerogens. An x-ray mineralogical analysis was also made of untreated cuttings and their shaly fraction.

Mineralogical analyses as well as the aspect of the cuttings show that the portion of the high-pressure zone situated below 3250 m is different from the upper part by its richness in silts, and that the kaolinite that is found there and makes up most of the shaly fraction is a product of hydrothermal neogenesis. This shows that it is possible for fluids to circulate inside the high-pressure zone. We will return to this point later.

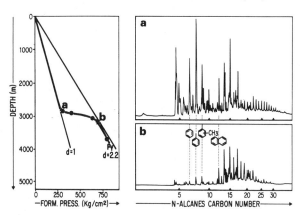

Fig. 9. On the left, evolution of the formation pressure P_f (pressure of the fluids inside the pores) with depth in the top wells of Handil. Equivalent density, d, corresponding to the specific weight of the drilling mud required for equilibrium, is indicated. Normal pressure (hydrostatic pressure) corresponds to $d = 1$. High pressure (geostatic pressure) corresponds to $d = 2.2$. On the right, chromatograms obtained by thermovaporization on coals from Handil top (well H8). Sample **a** (2865 m, 41 mg HC g^{-1} C) located at the normal pressure zone/transition zone boundary is not affected by migration phenomena, whereas sample **b** (3140 m, 13 mg HC g^{-1} C) located at the transition zone/high pressure zone boundary is strongly depleted, mainly in the light carbon range.

478

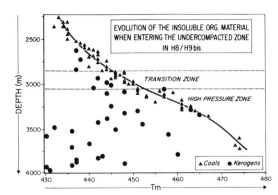

Fig. 10. Evolution of Rock-Eval T_m in the insoluble organic matter (coals and kerogens) from wells H8 and H9bis, when going from normal-pressure zone to high-pressure zone.

The results of Rock–Eval pyrolysis are given in Fig. 10 in the form of a graph showing T_m as a function of depth for coals and kerogens. Whereas in the normal-pressure zone at the top of Handil, including the one not shown here (from 0 to 2250 m), the increase in T_m for both coals and kerogens follows the increase in depth and thus clearly indicates the maturation of the organic matter, a new phenomenon is observed in the abnormal-pressure zone. Coals and some kerogens show a regular and increasing variation of T_m with depth. However, the major part of the kerogens underneath the top of the high-pressure zone (3150 m) does not follow this regular curve, and especially underneath 3250 m. The T_m values may even be lower than 435 °C, i.e. lower than the level of maturation generally considered to be the threshold of the oil generation. Since the succession of the maturation levels of organic matter undergoing increasing burial can, at the most, remain constant but cannot retrograde, this means that there is a change in the nature of the organic matter being analysed. This is confirmed by optical analysis. Beside the humic coals and kerogens showing normal behavior, pyrobitumens appear in the form of finely dispersed amorphous organic particles which are quite recognizable by their optical properties in transmitted and reflected light. These pyrobitumens represent the residual products of thermal and eventually oxidative alteration of a crude oil, besides lighter hydrocarbons which can either migrate toward cooler zones or be again cracked. These pyrobitumens having a less complex and probably less condensed structure than the kerogen are more easily decomposed than this kerogen during Rock–Eval pyrolysis and so have a lower T_m. In fact, in the samples analysed, we observe the result given by the mixture in varying proportions of humic kerogen and pyrobitumens.

The distinction between humic kerogens or coals and pyrobitumens, which appears clearly with methods such as optical observation and Rock–Eval pyrolysis, is much more difficult to detect by elemental analysis. This might be suspected as the result of both the many possible causes of alteration of the measurements (especially for oxygen), and the overall nature of the results obtained. In any case, notwithstanding the many fluctuations, it appears that the H/C ratio of pyrobitumens does not vary and remains around 0.9, whereas the same ratio decreases to 0.7 in coals at the bottom of well H9bis (T_m ~475°C). The maturation level of the autochthonous organic matter at this depth is thus presently at the threshold of the dry-gas formation zone.

The above observations suggest that the presence of the high-pressure zone has a specific influence on the composition of the hydrocarbons produced by the source rock, on account of the delay it causes in the expulsion of water and hydrocarbons. The heavy hydrocarbons formed by the thermal cracking of kerogen inside the high-pressure zone remain in a dispersed state for a longer time inside the source rock because the main cause of primary migration, which is the pressure gradient in the pores caused by compaction, no longer exists. (On the other hand, this cause is particularly effective in the transition zone.) The effect of time and temperature results in lighter and lighter hydrocarbons by cracking, and only the fraction that has been carried by convection or diffusion to the vicinity of the transition zone will be able to escape.

In the light of the above observations, it appears that a high-pressure zone is not necessarily impermeable, and this seems to be the case below 3250 m at Handil. Considering the high silt content and the special composition of the shales in this zone (kaolinite from neogenesis, in large crystals), fluid circulation may possibly be easy there. In fact, there is no communication between this zone and the overlying zone, and the cause of the overall relative impermeability probably lies in the transition zone.

CONCLUSIONS

Formation of the Handil field

In the light of the general results described above, various hypotheses can be formulated to explain the migration mechanisms which have led to the formation of the Handil field. The geochemical examination of hydrocarbon formation shows that the most favorable zone for the formation of liquid hydrocarbons is the one with a T_m of between 440 and 450°C, i.e. mainly the transition zone. The high-pressure zone, on the other hand, gives rise to the creation of gas on account of the delay it causes in fluid flow. This gas percolates through the transition zone and entrains liquid hydrocarbons exactly as a solvent would do during chromatography. This mechanism is probably superposed on the expulsion of oil by compaction in the transition zone, thus improving, and perhaps appreciably so, the efficiency of migration. Most of the oil thus expelled can be found in the main zone of the Handil reservoirs, i.e. 500 to 1000 m above the zone where it was formed. The light fraction continues to rise in the series and, little by little, becomes trapped in successive reservoirs, locally entraining heavy hydrocarbons over shorter distances. A large portion of the methane probably reaches the surface in this way and is released into the atmosphere. This mechanism, which is similar to elution chromatography, explains why the distribution of oil by specific gravity in this field is the opposite from what would have been expected if source rocks of increasing maturation levels had been in immediate contact with the reservoirs, i.e. lighter and lighter oils with increasing depth. The same mechanism also explains why the distributions by carbon atoms in the oils, in the shallow zone and in this zone only, are bimodal: these oils are largely made up of an allochthonous light fraction which has migrated over long distances, the remainder being an autochthonous heavy fraction.

Ubiquity of migration phenomena

This study finally leads to the conclusion that the spatial distribution and composition of the oils in the different reservoirs of the Handil field are mainly controlled by migration phenomena. In many other petroleum basins it can be supposed that the nature and maturation level of the source rocks also influence the distribution of oils; it is nonetheless obvious that migration is a banal phenomenon which always accompanies the formation of oil fields. Therefore, an analysis of possible migration paths and of migration consequences must always be done together with the identification of source rocks, when examining problems of source-rock/reservoir correlation. If the influence of migration phenomena is not regarded in the case described here, we arrive at a hydrocarbon formation scheme in which the light fractions are formed before the heavy ones. This contradicts both laboratory simulations and observations based on various basins in which little migration is known to have occurred because of the absence of fields.

Dynamic nature of migration phenomena

The different oil fields in the Mahakam Delta make up a specially favorable subject for studying migration phenomena. Indeed, since the series are of recent age, the formation and accumulation of hydrocarbons are presently in full evolution and so show the dynamic nature of the phenomena involved. As was previously mentioned in the paper presented at the World Petroleum Congress in Bucharest (Durand and Oudin, 1979), the nature (oil or gas) and amount of the hydrocarbons accumulated in the different fields of the Mahakam delta depend only on the relative position, related to time, of the hydrocarbon formation zone and of the zone from which these hydrocarbons may have been expelled.

So, the existence of a high-pressure zone plays a specific role in this dynamic process. This zone, whose existence is linked to sedimentological phenomena, probably has boundaries that are relatively independent of the rate of burial but, on the other hand, the pressure attained in this zone closely depends on this rate of burial. Therefore, when high-pressure zones exist, the nature of the hydrocarbons found in the reservoirs will depend not only on the conditions under which the petroleum products were formed but also on their migration possibilities. Practically only gaseous hydrocarbons can be expelled from the high-pressure zone. In the example of the Mahakam Delta presented here, where the same potential source rock is found in a high-pressure zone and in an overlying normal-pressure zone as a result of sedimentological conditions and rate of burial, the nature of the hydrocarbons accumulated will change with time, according to the scheme given in Table 1. This scheme is arbitrarily divided into five stages, focusing mainly on the nature, oil or gas, of the hydrocarbons formed in the source rock but, of course, the composition of hydrocarbons during genesis varies continuously from heavy oil to nearly dry methane. Nevertheless, the separation by stages allows to point out the following:

Table 1

	PRESSURE	STAGE 1	STAGE 2	STAGE 3	STAGE 4	STAGE 5
GENESIS IN THE SOURCE ROCK	NP*	Immature	Immature	Oil	Oil + gas	Gas
	HP*	Oil	Oil + gas	Gas	Nothing	Nothing
MIGRATION FROM THE SOURCE ROCK	NP	Nothing	Nothing	Oil	Oil + gas	Gas
	HP	Nothing	Gas	Gas	Nothing	Nothing (or a bit of gas formed in Stage 3)
CONTENTS OF THE RESERVOIRS	NP	Nothing	Gas	Oil + gas	Oil + gas	Gas
EXAMPLE IN MAHAKAM FIELDS		Trend** Dian-Pantuan	Nilam Badak	Handil Attaka		

* NP Normal-pressure zone.
  HP High-pressure zone underneath the NP zone.
** Next trend east of Bekapai - Attaka.

(1) The abundance of gas in this type of situation depends more on migration conditions than on any particular aptitude for gas formation linked to the nature of the source rock.

(2) The chronology for the appearance of hydrocarbons in the reservoirs is also linked to migration conditions and cannot be superposed on the chronology of hydrocarbon formation inside source rocks.

We can thus see to what extent migration conditions can modify in this situation the spatial and temporal distribution in reservoirs of the hydrocarbons produced by a given source rock.

Question

K. de GROOT:

Can you comment on the amount of pseudokerogen, or pyrobitumen found in the overpressured zones?

Answer

M. VANDENBROUCKE:

A distinction between pyrobitumen and autochthonous kerogen is not easy because many geochemical parameters are quite similar. However, optical examination of samples showed that pyrobitumen were abundant, even forming the major fraction of the organic material in some cases. An idea of the amount of pyrobitumen in particular samples, expressed as organic carbon content in the rock, can be deduced from both the lithology and the Rock-Eval data: silts contain mainly allochthonous organic matter, and Rock-Eval T_m allows to distinguish pyrobitumen from kerogens. The carbon content in silts from H9bis below 3500 m and such as $430°C < T_m < 440 °C$, is generally around or slightly higher than 1%. If the concentration of pyrobitumen is homogenized in the lower high-pressure zone by convection and migration phenomena, then the total organic carbon, which varies there around a mean value of 2%, is fairly equally distributed between autochthonous kerogen and pyrobitumen.

At last, the composition of an oil, even when accumulated in a reservoir, evolves with time. Indeed, hydrocarbon migration continues both from the source rock where new hydrocarbons are formed and will in part be added to the ones already in place, and through the cap-rock of the reservoir through which the lightest hydrocarbons escape. Therefore, the size and even the existence of an accumulation are linked to the relative rates of arrival and escape of hydrocarbons, i.e. to migration phenomena.

Acknowledgments

We thank all our colleagues at I.F.P. and TOTAL C.F.P. who have taken part in this project, and especially Mesdames Bessereau and Bernon and Messrs Pittion, Albouy, Buton, Espitalié and Paratte. We also thank C.F.P. for having supplied a great many geological documents and Compagnie Total Indonésie for having sent samples and for having been particularly careful in taking cuttings from well H9bis. Finally, we thank these latter companies together with societies Pertamina and Inpex for having authorized this publication.

REFERENCES

Albrecht, P., Vandenbroucke, M. and Mandengué, M. (1976) Geochemical studies on the organic matter from the Douala Basin (Cameroon) — I Evolution of the extractable organic matter and the formation of petroleum. *Geochim. Cosmochim. Acta* **40**, 791–799.

Bordenave, M., Combaz, A. and Giraud, A. (1970) Influence de l'origine des matières organiques et de leur degré d'évolution sur les produits de pyrolyse du kérogène. In *Advances in Organic Geochemistry 1966*, ed. by Hobson, G. D. and Speers, G. C. Pergamon Press, Oxford, pp. 389–405.

Combaz, A. and de Matharel, M. (1978) Organic sedimentation and genesis of petroleum in Mahakam delta, Borneo. *Am. Assoc. Pet. Geol. Bull.* **62**, 1684–1695.

Durand, B. and Espitalié, J. (1976) Geochemical studies on the organic matter from the Douala basin (Cameroon) — II Evolution of kerogen. *Geochim. Cosmochim. Acta* **40**, 801–808.

Durand, B. and Monin, J. C. (1980) Elemental analysis of kerogens (C, H, O, N, S, Fe). In *Kerogen — Insoluble organic matter from sedimentary rocks*, ed. by Durand, B. Editions Technip, France, pp. 113–142.

Durand, B. and Nicaise, G. (1980) Procedures for kerogen isolation. In *Kerogen — Insoluble organic matter from sedimentary rocks*, ed. Durand, B. Editions Technip, France, pp. 35–52.

Durand, B. and Oudin, J. L. (1979) Exemple de migration des hydrocarbures dans une série deltaïque: le delta de la Mahakam, Kalimantan, Indonesia. *Proc. 10th World Pet. Congr.* **1**, 3.11.

Espitalié, J., Madec, M., Tissot, B., Mennig, J. J. and Leplat, P. (1977) Source rock characterization method for petroleum exploration. OTC paper 2935.

Fabre, M., Leblond, C. and Roucaché, J. (1972) Analyse quantitative par CPG capillaire des n-alcanes de C12 à C32 dans les hydrocarbures saturés d'un pétrole brut ou d'un extrait de roche. *Rev. Inst. Fr. Pét.* **27**, 469–481.

Huc, A. Y. and Roucaché, J. G. (1981) Quantitative thin layer chromatography of sedimentary organic matter. *Anal. Chem.* **53**, 914.

Huc, A. Y., Roucaché, J. G., Bernon, M., Caillet, G. and Da Silva, M. (1976) Application de la chromatographie sur couche mince à l'étude quantitative et qualitative des extraits de roches et des huiles. *Rev. Inst. Fr. Pét.* **31**, 67–98.

Jonathan, D., L'Hote, G. and de Rouchet, J. (1975) Analyse géochimique des hydrocarbures légers par thermovaporisation. *Rev. Inst. Fr. Pét.* **30**, 65–98.

Magnier, P. and Ben Samsu (1975) The Handil oil field in East Kalimantan. *Proceedings of the Indonesian Petroleum Association, 4th Annual Convention*, June 1975.

Monin, J. C., Pelet, R. and Février, A. (1978) Analyse géochimique de la matière organique extraite des roches sédimentaires; IV: Extraction des roches en faibles quantités. *Rev. Inst. Fr. Pét.* **33**, 223–240.

Saint-Paul, C., Monin, J. C. and Durand, B. (1980) Méthode de caractérisation rapide des hydrocarbures de C1 à C35 contenus dans les roches sédimentaires et dans les huiles. *Rev. Inst. Fr. Pét.* **35**, 1065–1078.

Schaefer, R. G., Leythaeuser, D. and Weiner, B. (1978) Single-step capillary GC method for extraction and analysis of sub ppb quantities of hydrocarbons (C2–C8) from rock and crude oil samples and its applications in petroleum geochemistry. *J. Chromatogr.* **167**, 355–363.

Tissot, B., Durand, B., Espitalié, J. and Combaz, A. (1974) Influence of nature and diagenesis of organic matter in formation of petroleum. *Am. Assoc. Pet. Geol. Bull.* **58**, 499. 506.

Trendel in Boudou, J. P. (1981) Diagenèse organique de sédiments deltaïques (Delta de la Mahakam — Indonésie). Thesis, Université d'Orléans, p. 80.

Verdier, A. C., Oki, T. and Suardy, A. (1979) Geology of the Handil field. American Association of Petroleum Geologists' Convention, Houston, March 1979.

Some Mass Balance and Geological Constraints on Migration Mechanisms[1]

R. W. JONES[2]

ABSTRACT

Oil and gas are not at rest in the sedimentary mantle of the earth. They are not in equilibrium, whether they are finely dispersed in a potential source rock or whether they are concentrated in a trap in a reservoir rock. A wide variety of possible escape mechanisms exists; these include diffusion, continuous single phase flow, solution of oil in gas or gas in oil, and solution in water derived from compaction, clay diagenesis, or meteoric sources. The problem is to quantify the possible mechanisms and to rank their relative importance under a given set of physical, chemical, and geologic conditions. The quantitative importance of the various proposed mechanisms can vary by orders of magnitude depending on the physical, chemical, and geologic conditions.

During the past decade, oil-to-source correlations have become reliable and the timing of peak generation and concomitant migration has been sufficiently quantified to allow the geologist/geochemist to make estimates of when and how much petroleum moved from one location to another. Combined with a knowledge of the physical, chemical, and geologic conditions at the time of migration, such quantitative descriptions of subsurface petroleum transfer permit an empirical test of the applicability of the various proposed migration mechanisms. The application of this technique to selected areas suggests that most of the major commercial oil accumulations of the world left their source rock in a continuous oil phase. When bitumen concentrations in the rock are too low for continuous phase flow to exist, other migration mechanisms, which always are operative, will increase in both absolute and relative intensity. Solution of oil in gas may become significant in thick Tertiary delta systems, and meteoric water may be a surprising asset in some very specific geologic settings. However, it is unlikely that solution of oil in water derived from compaction or from dehydration of clay has much to do with the origin of many of the major oil accumulations of the world.

INTRODUCTION

Many mechanisms have been proposed as the primary cause of migration out of the source into the reservoir. Conceptually, most of them fall into two end member types. One depends on water to move the petroleum or petroleum precursors out of the source rock; the other moves the petroleum out of the source rock in a separate phase independently of any associated water movement. In fact, removing or immobilizing most of the water probably is a prerequisite of separate phase movement. Some proposed migration mechanisms and carriers for water-controlled primary migration include: oil in water solution, gas in water solution, other organics in water solution, micellar solution, emulsion, diffusion, convection, meteoric water, compaction water, and clay dehydration water. Some proposed migration mechanisms and carriers for separate phase transport within the pore system include: oil, gas, solution of gas in oil, and solution of oil in gas. Proposed migration mechanisms acting within the kerogen network include diffusion and pressure gradients.

Many of the specific migration mechanisms mentioned were explicitly advocated in the papers on migration presented at the AAPG National Convention in 1978 and included in this volume. It would simplify the problem if we could believe that only one of the authors is correct, and that only one of the proposed mechanisms is "the correct one." However, it is very unlikely that such is the case. All of these mechanisms exist in nature on some scale. Their relative quantitative importance at a given point in space and time is a function of a very large number of factors. Listed below are some of the rock, fluid, and kerogen properties of the source which could be expected to have a significant influence on the migration characteristics of a source rock.

Rock Properties Include: porosity, permeability, pore-size distribution, tortuosity, capillary pressure, wettability, absorption, adsorption, hydrated minerals, particle-size distribution, fracture transmission, and heterogeneity.

Fluid Properties Include: pressure and pressure gradient, fluid potential, temperature and temperature gradient, water salinity, interfacial tension, water bound to clay surfaces, viscosity, saturation, compressibility, density, and buoyancy.

Kerogen Properties Include: quantity, type (generation products), type (absorption characteristics), maturation, and distribution within the source rock.

Even if we could measure all of the rock, fluid, and kerogen proper-

---

© Copyright 1981. The American Association of Petroleum Geologists. All rights reserved.

[1]Manuscript received, August 29, 1978; accepted, August 1, 1979. This manuscript was derived from a paper originally presented in the AAPG Migration Short Course given at the Oklahoma City Annual Meeting, April 4, 1978. Published with permission of Chevron Oil Field Research Company, and reprinted with minor revisions from AAPG *Studies in Geology No. 10* (1980).

[2]Chevron Oil Field Research Company, La Habra, California 90631.

ties of source rocks, the possible permutations and combinations are obviously sufficient to challange a thousand physicists and chemists for a thousand years. In the interim, empirical solutions must be sought. Constraints on possible solutions are provided by what nature has done regarding the subsurface transfer of petroleum. Although it is not yet possible to quantitatively evaluate the large number of combinations of physical and chemical factors that control migration, nature is and has. It might be expected that the amounts and distribution of oil and gas in traps relative to those in the source rock will reflect the migration mechanism that ejected them from the source. Indeed, that is a major thesis of this paper.

To apply mass balance considerations to the problems of evaluating possible migration mechanisms, it is necessary to study basins in which there is in-depth knowledge about: (1) the amount, distribution, and characteristics of the oil and gas, bitumen, and kerogen; (2) the amount, distribution, characteristics, and movement, past and present, of the water; and (3) the structural and stratigraphic history. Our knowledge of the listed factors in several basins is combined with recognized geochemical principles in the next two sections of the paper to argue that (1) continuous phase flow is the mechanism by which petroleum left the source rocks of many of the major oil accumulations of the world; and (2) little of the commercial petroleum of this world left its source rock dissolved in water from compaction or clay diagenesis.

In this paper the concepts of continuous phase flow and aqueous solution in water derived from compaction and clay dehydration are treated as the main adversaries. Although somewhat artificial, the adversary position is convenient owing to the writer's belief in continuous phase flow as the mechanism of primary migration accounting for the majority of the world's large oil fields, and to the strong advocacy in the American literature of aqueous solution in water derived from compaction and clay dehydration as the dominant mechanism of primary migration. Possible contributions from other mechanisms generally are not evaluated or discussed in this paper, except in the last part where two problem areas for any theory of primary migration, the Gulf Coast Tertiary and the Athabasca oil sands, are discussed.

CONTINUOUS PHASE PRIMARY MIGRATION

Introduction

Continuous phase primary migration refers to the movement of oil within and out of the source rock in a continuous oil phase. The movement is believed to be caused by high differential pressures within the rock. The pressure differentials are caused by local supernormal pressure (SNP) generated by normal compaction of a fine-grained rock relative to its more permeable neighbors, and by the continuous increase in volume of the totality of the organic matter (OM) in the rock by generation of organically derived material of lower density. Thus generation is not only the *sine qua non* of migration, it is the cause.

Historically, the combined effects of small pore and capillary size, a water wet pore system, and surface tension effects between water and oil have been thought to preclude the possibility of continuous phase oil flow in and out of a fine-grained source rock (Baker, 1960; McAuliffe, 1979). The physical principles involved are well known and irrefutable; they probably are the primary reason why oil generated in most shales never leaves the shale as oil. However, in organic-rich rocks there are several factors operating which can override the effects of surface tension and capillary pressure. We are not dealing with buoyance effects alone to move the oil, but high differential pressures caused by compaction and hydrocarbon generation. In addition, the pore system need not be water wet owing to the large amount of organic matter in the rock, the tendency for the organic matter to be squeezed into the pore system (Bradley, *in* Hoots et al, 1935), and the fact that the organic matter is oil wet. Perhaps of even greater importance is the heterogeneity of most excellent source rocks. Source rocks are typically varved owing to their low energy depositional environment and the lack of bioturbation required of good source rocks (Demaison and Moore, 1980). Recently Momper (1978) discussed in detail the source rock characteristics such as partings, laminae, various irregular distribution of grains, crystals, kerogen, and microfractures which can contribute to the development of relatively large pores and micropathways which are exploited by the pressure buildup due to hydrocarbon generation. The writer agrees with his evaluation. Because of the thoroughness of Momper's discussion, and other excellent discussions on the possible mechanisms of continuous phase migration published by Snarskii (1970), Dickey (1975), Magara (1978a), and Tissot and Welte (1978), this will not be pursued further here, but instead the writer will turn to some mass balance considerations which are believed to support the continuous oil phase concept independently of our knowledge of the details of the mechanisms.

Numbers and discussions are given for some formations and basins in which primary migration in continuous phase flow probably occurred. No claim of high precision or accuracy is made for the numbers. However, the examples chosen can stand an error by a factor of 2 to 5 and not invalidate the point presented. In addition, some arguments supporting the concept of continuous phase primary migration are presented on the basis of distribution of kerogen and hydrocarbons in source rocks and on the timing of primary migration.

Specific Areas/Formations

Bakken Shale, Williston basin—The Bakken Shale is a thin (0 to 100 ft; 0 to 32 m), organic rich, transgressive marine shale of latest Devonian/earliest Mississippian age which is the source of at least 3×10^9 bbl of in-place oil in the

Williston basin (Dow, 1974). In a perceptive study of the Sanish pool, Antelope field, North Dakota, Murray (1968) pointed out that the visual and log characteristics of the Bakken shales indicated that their pore space was hydrocarbon saturated. Subsequently, Meissner (1976, 1978) described the physical characteristics of the Bakken which support the continuous oil phase concept. These include: (1) no formation water is ever recovered during drillstem tests or initial well completions; (2) where thermally mature, the Bakken is overpressured with respect to both the regional gradient and reservoir beds separated from it by tight strata; and, (3) the electrical resistivity increases abruptly to nearly infinite values where the Bakken enters peak oil generation. Meissner inferred this last relationship from temperature versus resistivity plots, but the proposed relationship has been confirmed by Chevron geochemical data. Meissner used the three facts listed above in combination with Williams' (1974) demonstration that oils in the Bakken, the stratigraphically lower Devonian Nisku Formation, and stratigraphically higher Mississippian Mission Canyon Formation are identical, to argue for continuous phase primary migration in and out of the Bakken, both upward and downward. Meissner further believed that the outward migration of oil takes place through a spontaneously generated fracture system created by high overpressuring caused by hydrocarbon generation. The data in Table 1 support Meissner's conclusions. At low resistivities the hydrogen-to-carbon (H/C) ratios of the organic matter are high, and the ratio of generated to remaining-to-be-generated hydrocarbons is low. At the high resistivities, the amount of generated hydrocarbons is much higher, and the remaining generative potential much lower. In addition, the total hydrocarbon potential per gram of organic carbon in the mature rock is approximately half that of the immature one. Because the type of organic matter is the same in both samples, this suggests that ~167 mg of hydrocarbons per gram of organic carbon have migrated out of the mature rock. For the mature rock described in Table 1, this is ~50,000 ppm (vol.) or ~5% of the rock volume, a figure only compatible with continuous phase migration.

Table 1. Some Contrasts Between Mature and Immature Bakken Shale

	Immature	Mature
Sample Depth (ft)	7,570	11,260
Total Organic Content = TOC (wt. %)	8.8	12.5
Hydrogen/Carbon Ratio (H/C)	1.23	0.83
Resistivity (ohm-meters)	10	> 100
*Hydrocarbons (mg) in bitumen/gm of rock	0.9	6.6
*Hydrocarbons (mg) generated/gm of rock	31.8	18.6
*Hydrocarbons (mg) in bitumen/gm of TOC	10	53
*Hydrocarbons (mg) generated/gm of TOC	360	150
*Hydrocarbon potential = Hydrocarbons (mg)/gm of TOC	370	203

*Measured by programmed temperature pyrolysis with a Rock-Eval pyrolysis device (Espitalie et al, 1977).

Given below are mass balance calculations that demonstrate the impossibility of the Bakken Shale sourced oil in the Williston basin having moved out of the Bakken in aqueous solution. Dow (1974) provided the data for the Bakken Shale in the Williston basin necessary to calculate the minimum solubility required if the Bakken-sourced oils were carried out of the Bakken by compaction water during peak generation. The Bakken is the source of 3×10^9 bbl of in-place oil. The average thickness of mature, organic-rich Bakken is ~40 ft (12.2 m) and the volume of mature Bakken is ~150 cu mi (626 cu km). Thus, minimum oil expelled per cu mi (cu km) of Bakken equals ~20 $\times 10^6$ bbl/cu mi (4.8×10^6 bbl/cu km) or ~770 ppm. If the expulsion of oil occurred while Bakken porosity dropped from 10 to 5%, necessary water solubility would be ~15,000 ppm. If 3.3 times as much oil was expelled from Bakken as was pooled (Dow, 1974), the necessary water solubility would be ~50,000 ppm. Price (1976) presented experimental data which indicated a maximum oil solubility of 50 to 200 ppm at the temperatures at which the Bakken probably reach peak generation (70 to 120°C). Thus the minimum numbers of required solubilities are at least two orders of magnitude too high for the oil to have migrated in aqueous solution in Bakken compaction water at the maximum temperatures reached in the Bakken.

Los Angeles basin, California—A case for the continuous phase migration from a source rock in the Los Angeles basin can be made by using geochemical data and astute microscopic observations provided in a perceptive 43-year-old paper by Hoots et al (1935), which dealt with the origin of the oil in the Playa del Rey field (Figs. 1, 2). The source rock is the "nodular shale," a laminated organic and phosphate-rich rock that directly overlies an oil-filled conglomeratic sandstone resting on a schist basement. Several shale samples yield an average of 1.8 wt. % or ≥4.0 vol. % chloroform extractables. Playa del Rey field is an excellent example of the upward and downward movement of commercial amounts of oil from an organic-rich, bitumen-saturated, oil source rock.

The oil in the upper zone is very similar to the oil below the "nodular shale" but contains more saturates as would be expected from chromatographic effects of migration. The upper zone oil was undoubtedly emplaced from the "nodular shale" along faults that, when cored, are associated with free oil. Hoots et al (1935) argued very convincingly that the types of compositional differences between the extract from the "nodular shale" and the underlying oil indicate that the oil migrated from the overlying shale and that the high chloroform extractable content of the shale is not the result of oil migrating from the reservoir into the shale. In fact, the geochemical data showing a higher percentage of paraffins and naphthenes in the oil than the source reported by Hoots et al probably are the earliest published data which show the differences between oil and source which have frequently been described in recent years (Tissot and Welte, 1978). Perhaps the most fascinating aspects of this study are the microscopic observations of the organic matter by W. H. Bradley, which were included in the Hoots et al (1935) paper. Bradley recognized four types of organic matter microscopically, and described them in detail.

The first...is...spore exines. The second, which is the most abundant ...is deep reddish brown...is essentially structureless...occurs in thin stringers and small irregular flat flakes or sheets and in small irregular flocculent masses...range greatly in size.. .is less altered than most of the organic matter...has undoubtedly undergone some change from its original composition. The third kind of organic matter is light amber in color and is more or less evenly diffused through the rock. It is ...perfectly translucent...has been at some time during its history either a liquid or a very fluid gel...was *forced into all available pores during compaction of the sediment* (italicizing is the writer's). The material of the second sort mentioned appears almost to grade into this pale yellowish material. As the size of the reddish brown flocculent material decreases, it approaches both in color and clarity the perfectly clear yellow stuff of the third class. The fourth kind of organic matter is a dark reddish brown opaque homogeneous stuff ...which occurs rather sparingly, filling cavities...appears to have been either liquid or a quite fluid gel...is quite closely related to the pale yellowish organic matter just described and like the yellowish material was probably derived by decomposition from the reddish brown fragmental organisms.

Probably the first two types of organic matter described by Bradley are kerogen with the exines being a very subordinate component, the third type is the hydrocarbon rich part of the bitumen, and the fourth type is the more asphaltic part of the bitumen which physically was separated in the source due to its less mobile properties. A more modern study of the "nodular shale" would clearly be of interest.

Bradley's work indicates that microscopy never has achieved the importance in source-migration studies which it deserves. Unfortunately, the development of "objective" geochemistry with its sophisticated technology pushed "subjective" microscopy with its 50-year-old technology into the background. Only recently with the development of fluorescence microscopy (see Teichmuller, 1975; Teichmuller and Ottenjann, 1977) and of the scanning electron microscope (SEM) has microscopy started to fulfill its potential in migration studies that was implicit in Bradley's astute observations 40 years ago.

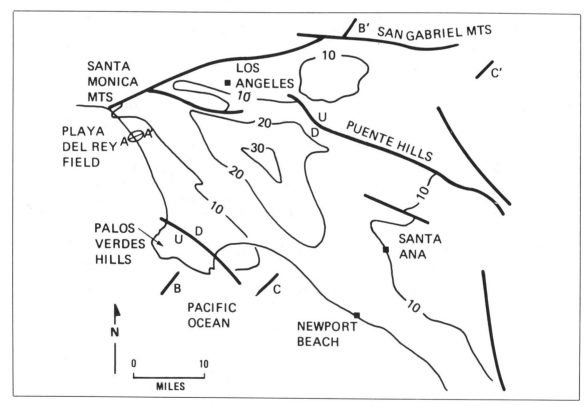

FIG. 1—Generalized basement contours (kilofeet), Los Angeles basin, California. A-A', B-B', C-C' show locations of cross sections depicted in Figs. 2 and 3.

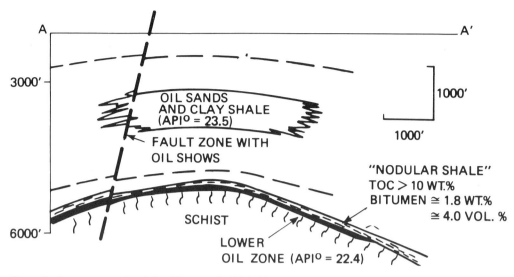

FIG. 2—Generalized structure section (after Hoots et al, 1935), Playa del Rey field, Los Angeles basin, California (A-A' from Fig. 1).

Mass balance calculations confirm the inferences from the geochemical and microscopic observation of Hoots et al (1935), that primary migration of oil in aqueous solution is not important in the Los Angeles basin. Figure 3 contains two northeast-southwest cross sections through the Los Angeles basin which show the basin configuration, location of major oil fields, and the approximate top of peak generation as determined by Philippi (1965) on the west side of the basin and extended across the basin on the basis of other data. The pertinent mass balance data are listed below in a sequential development.

1. Oil in place $\sim 25 \times 10^9$ bbl or ~ 1.0 cu mi (4.2 cu km).
2. Rock volume related to the oil fields $\sim 1,600$ cu mi (6,670 cu km; Barbat, 1958).
3. 600 ppm of pertinent basin volume is oil in traps.
4. Mature and postmature source rock ~ 200 cu mi (835 cu km).
5. 5,000 ppm of source rock volume or $\sim 2,000$ ppm of source rock weight is amount of oil in traps.
6. Assume $\sim 25\%$ of original kerogen converted to oil.
7. Original organic content of source rock ~ 4.0 wt. %.
8. Total oil generated $\sim 1.0\%$ wt. % = 2.5 vol. % of source rock.
9. Total oil generated ~ 5 cu mi (21 cu km).
10. Oil in reservoir/oil generated $\sim 1/5$ (20%).
11. Assume migration occurred during porosity decrease from 10 to 5% in source rock.
12. Fluid loss during oil migration ~ 10 cu mi (42 cu km).
13. Assume all migrated oil is in known traps.
14. Then oil/water ratio during migration from source $\sim 1/9$ (11%).
15. Necessary solubility $\geq 100,000$ ppm.

All of the numbers are impressive. Attention is directed particularly to item 8, which suggests that $\sim 2.5\%$ of the source rock volume was converted to oil; item 10, which demonstrates a high migration and trapping efficiency; and item 15, which demonstrates that migration in aqueous solution did not have much to do with migration from the source rocks in the Los Angeles basin.

Those who support migration of oil in aqueous solution might argue that we are not dependent solely on indigenous water emitted from the source rocks, but that large amounts of water could pass through the source rocks from below and pick up hydrocarbons in the process. This is unlikely. When subsiding and compacting, even a normal nonsource shale with a permeable substratum will be overpressured in its interior and emit fluids both upward and downward (Smith et al, 1971; Evans et al, 1975). When additional overpressuring from the generation of substantial amounts of hydrocarbons is added to the normal overpressuring from compaction, the chances of significant amounts of water moving upward through an organic-rich source rock become remote. The fact that many of the excellent source rocks of the world act as both source and cap for oil sourced within them confirms this interpretation. Interbedded source and reservoir is a perfect combination for draining temporarily overpressured source rocks by continuous phase flow rather than upward moving water and this, of course, is the situation in the Los Angeles basin where turbidite reservoirs are interbedded with pelagic source shales.

There are some additional inferences regarding migration in the Los Angeles basin which can be made from Figure 3 and some additional data on the distribution of oils and their potential sources. The upper Pliocene contains both abundant reservoirs and potential sources by conventional standards (Philippi, 1965), and was structurally deformed during the Pleistocene orogeny which formed the anticlines that contain the large quantities of oil in the lower Pliocene-upper

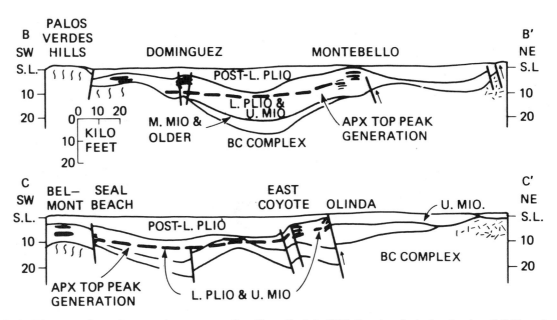

FIG. 3—Diagrammatic northeast-southwest cross sections (from Gardett, 1971), Los Angeles basin, showing oil fields and approximate top peak oil generation (lines B-B' and C-C' from Fig. 1).

Miocene sequence. However, the upper Pliocene contains less than 0.1% of the basin's oil, and this only in association with major faulting in structures containing abundant oil in the deeper section. In addition, the abrupt cessation of reservoired oil near the top of the lower Pliocene section correlates well with the top of peak generation in the basin center for the same stratigraphic units. These facts suggest the following: (1) any compaction and clay diagenesis water moving out of the depocenter of the upper Pliocene carried negligible amounts of liquid hydrocarbons despite the presence of abundant oil-prone source rocks in the early generation stage of thermal diagenesis (Philippi, 1965; Bostick et al, 1978); (2) owing to insufficient generation, continuous phase oil migration did not occur in the post-lower Pliocene section either; and (3) the faults are dominantly sealing rather than leaking. Thus, most of the oil in the lower Pliocene reservoirs probably originated in interlayered, thermally mature, lower Pliocene source rocks deeper in the basin, contrary to Philippi's (1965) hypothesis that the oil in the lower Pliocene reservoirs was emplaced along faults from underlying Miocene source rocks.

Green River Formation, Uinta basin, Utah—Almost everyone who has seen the bitumen dikes in the Green River and Uinta Formations of the Uinta basin is impressed by what nature can do with regard to continuous phase flow of bitumens when the conditions are right. Listed below is a summary of the conditions that probably existed in the Green River Formation and caused the emplacement of the bitumen dikes. The conditions are uncommon, but neither unique nor extreme.

1. 25+ vol. % hydrogen-rich, amorphous organic matter.
2. Varved, heterogeneous rock.
3. Fine-grained, brittle, carbonate matrix.
4. ≥6,000 ft of overburden; the overburden provided the driving force for fluid movement by placing the organic matter under lithostatic pressure, and decreased viscosity of the organic matter, but overburden was not sufficiently thick to raise the temperatures high enough to create conventional thermal maturity.
5. Tectonic tensile fracturing.
6. Release of pressure on the organic matter by intrusion of fractionated organic matter into fractures.

Figure 4 is a diagrammatic section in the Green River basin from Hunt (1963) which shows the gross stratigraphic relationships between the bitumen dikes and oil-saturated sandstones at the updip edge of the basin. Subsequent to the dike formation and the end of tectonic tensile fracturing, additional maturation of the organic matter in the source rocks probably led to continuous phase movement of less viscous bitumens through heterogeneities in the fine-grained source rocks into the laterally equivalent sandstone reservoirs.

One reviewer of this manuscript argued that the demonstrable continuous phase movement from oil shales like the "nodular shale" and the Green River shale has little to do with primary migration from more normal source rocks. The writer rejects this argument on several grounds. (1) Total organic contents (TOCs) of "oil shales" and conventional source rocks overlap. Any division is arbitrary. The Kimmeridgian oil shale of Scotland was famous long before its offshore equivalents became recognized as the source of the oil in the northern North Sea. (2) Both the "nodular shale" and the Green River "oil shale" have demonstrably sourced conventional oil accumulations, although pertinent chemical analyses of both oil and extract have shown them to be relatively immature (Hoots et al, 1935; Tissot et al, 1978). Clearly, further maturation of these rocks would cause them to generate and migrate large amounts of oil. (3) Oil shales are recognized as such because of their immaturity. I am not aware of any published study of former oil shales which are postmature. However, it is reasonable to believe that a total organic content of 4.38 wt. % for the Woodford Shale is approximately 15,000 ft (4,570 m) below the oil deadline at 27,725 ft (8,450 m) in the Anadarko basin in Oklahoma (Price, 1977) and is the remnant of a former oil shale that made a significant contribution to commercial accumulations of hydrocarbons in that basin.

Fractured shale accumulations—There are many commercial accumulations in fractured organic-rich shale in the western United States. The reservoirs are brittle, relatively competent zones which commonly are encased in relatively plastic clay shales that seal the fluids within the fractured zones (Mallory, 1977). Water solubility can have no significance in the origin of these accumulations. The accumulations have low to no water recovery and many depend on gravity drainage. Most of the water either has migrated away or is tightly bound, and it is predominantly the oil that has access to the fractures.

The brittle rocks in which these accumulations are found are not oil shales, but simply good source rocks with total organic content typically ranging from 1 to 4 wt. %. The productive fracturing is due to external forces, not hydrocarbon generation. Development of tension fractures in a brittle rock must have created large pressure gradients between the fractures and nearby pores and literally sucked the oil into the fractures and thereby created a continuous oil phase in the fracture system. It is easy to imagine that where the fractured rock was not enclosed in plastic impermeable shales, the oil moved out in search of more conventional reservoirs and traps. This sequence of events may be important in continuous phase primary migration in certain geologic settings.

Organic Content of Source Rocks

Most of the major oil accumulations of the world originated in source rocks with TOCs ≥ 2.5 wt. % and occasionally with TOCs ranging past 10% by weight. Many examples come to mind: the Kimmeridgian of the northern North Sea, Silurian of Algeria, upper Miocene of the Los Angeles and San Joaquin basins of California, the Cretaceous black shale overlying the major unconformity of the Prudhoe Bay area in Alaska, the Bakken and Woodford Shales of the Paleozoic of the Mid-Continent, the Duvernay of the Devonian of Alberta, Canada, the La Luna limestones of the Cretaceous of Venezuela, and the source rocks of the Middle East.

The most notable exceptions to the previous generalization are the Tertiary delta systems of which the Niger and Mississippi are prime examples. In them, the source rocks underlie the reservoirs and consist of an approximately 10,000 ft (3,048 m) section containing mature-postmature organic matter with TOCs ranging from 0.3 to 1.0 wt. %. The dominant migration process here is unique to these geologic

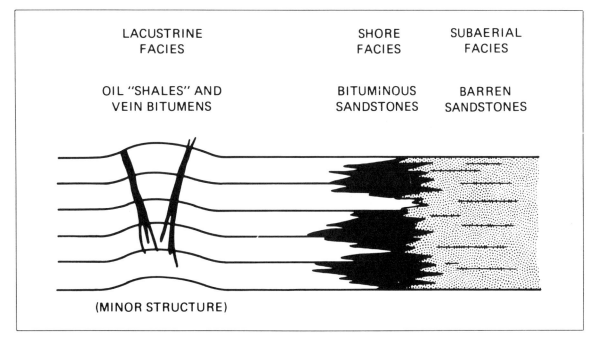

FIG. 4—Association of types of deposits of bitumens with sedimentary facies in the Uinta basin, Utah (from Hunt, 1963).

systems and is discussed later.

Before examining why high total organic content of most source rocks requires continuous phase migration, it is appropriate to examine the question of the minimum TOC necessary for a rock to become an effective source rock (a source of commercial accumulations). An answer to this question is important in understanding migration, but it has been treated in a rather cavalier fashion in most of the literature. Momper's (1978) observations are a clear-cut exception, and are an excellent discussion of how the prevailing number of 0.5 wt. % organic matter became imbedded in the literature without either proof or in-depth discussion. The writer is a bit nonplussed as to why, after his illuminating discussion, Momper appears to fall back on the 0.5 wt. % value as a generally valid criterion. Commercially significant migration probably is unlikely with organic matter contents of 0.5 wt. % unless one is dealing with thick Tertiary delta systems. The reason is simple and straightforward. There are not enough liquid hydrocarbons generated relative to the available pore space to permit a continuous oil phase to form, and nature is forced to turn to less efficient migration mechanisms to move any liquid hydrocarbons out of the source rock.

The minimum amount of organic carbon necessary for a potential source rock to become an effective source is not simply an elusive constant. It is a wide-ranging variable dependent on many other variables. Not only does it depend on both the type of organic matter and its distribution in the rock, but also on the position of the potential source bed with respect to potential carrier beds, the maturation of the organic matter, the physical and chemical characteristics of the inorganic part of the source rock at peak generation, and, of course, the size of the fields which are economic. Thus, most coals with $\geq 90\%$ organic content are not potential oil sources, despite the fact that they generate more heavy hydrocarbons per unit weight than most oil source rocks.

However, vitrinite and inertinite, the dominant organic macerals in most coals, do not become oversaturated and release hydrocarbons until the liquid hydrocarbons are cracked to gas and condensate within the organic matter (Jones, in prep.). In addition, it is easy to visualize that the ease of migration, and hence the amount of organic matter necessary to create an effective source, is partly controlled by whether the organic matter is homogeneously distributed in a potential source bed or distributed in varves of varying organic richness. Most excellent source beds are highly varved. In them the organic matter is likely to be distributed in rich layers which alternate with layers nearly barren of organic matter. This is particularly true of rocks containing an appreciable carbonate or siliceous content like the Green River or Monterey Formations. Such a layered distribution of both organic matter and rock properties facilitates the formation of a continuous oil phase in the rock at peak generation. The organic matter will compress and drive out the pore water and, at peak generation when the kerogen is exuding hydrocarbons, the increased pressure and rock heterogeneities should facilitate the development of microfractures and a continuous oil phase in the existing porosity. On the other hand, an organically rich zone surrounded by lean shales would undoubtedly have difficulty in emitting its liquid hydrocarbons to effective carriers, although if adequately fractured, it might become an effective reservoir and trap. The Sabym field in western Siberia, with at least 10,000 sq km (3,861 sq mi) of potentially productive Upper Jurassic bituminous shales (Auldridge, 1977), is the most notable example of such an accumulation. Thus, the minimum organic matter content necessary to make an effective source depends on a large number of variables which control the ease of primary migration.

A parallel question to the total organic content requirements of a source rock concerns the minimum hydrocarbon content required to eject hydrocarbons from a source rock. This is a legitimate question, but is also as unanswerable a question as what is the minimum TOC necessary for a rock to become an effective source rock. There are too many ifs, ands, and buts. For example: are hydrocarbons in the pore system or in the kerogen; is the rock a lime mud, an illitic shale, or a silty mixed-layer clay; are the hydrocarbons heavy or light; what percentage of the bitumen is hydrocarbons; are the hydrocarbons in thin, organic-rich layers or evenly dispersed through the rock? Momper (1978) handled this question with the statement that "the amount of extractable bitumen needed in a source bed before expulsion begins is about...825 to 850 ppm...*or more.*" (The italics are the writer's.) Momper did not discuss the evidence and geologic and geochemical setting of the 825 ppm minimum figure, although the numbers seem about right for peak generation in the Gulf Coast Tertiary and other major Tertiary delta systems. However, it seems clear from the literature that substantially larger figures exist, they are the norm in organic-rich rocks, and they often are not enough by themselves. Philippi's (1965) data indicate that the extractable $C_{15}+$ hydrocarbons now in the source rocks of the Los Angeles basin reach at least 2,500 ppm; Hunt (1961) gave a 3,000-ppm hydrocarbon content of samples of the Woodford and Duvernay (source) Shales; the Brooks and Thusu (1977) data indicate a hydrocarbon content of several thousand ppm for the Kimmeridgian Shale in the northern North Sea at peak generation; and Tissot et al (1971) measured several thousand ppm $C_{15}+$ hydrocarbons in the Toarcian shales of the Paris basin, despite the fact that the Toarcian apparently was not the source of any of the oil fields in the Paris basin (Poulet and Roucache, 1970). Hoots et al (1935) indicated that the "nodular shale" at Playa del Rey contains ~18,000 ppm chloroform extractable organic matter of which ~50% is probably hydrocarbons (petroleum ether extractable). Thus, minimum bitumen figures, like minimum TOC figures, probably are not very helpful unless they are closely tied to the detailed geology and geochemistry of an area. There are no magic numbers for calculations although there must be magic combinations of organic matter quantity, types, maturity, and distribution in different rock associations. Even with a hydrocarbon content of several thousand ppm the porosity of a mature source rock is several times larger than the hydrocarbon volume. Those who believe in continuous phase primary migration have developed a number of possible explanations of why oil will flow preferentially to water under such circumstances. The explanations involve the concepts of structured versus liquid water (Miller and Low, 1963; Dickey, 1975) and of oil wetness. Dickey (1975) in a perceptive paper, and Magara (1978) in an extension of Dickey's arguments, suggested that in highly compacted source shales the ratio of oil to movable, liquid water will be sufficiently high that oil will preferentially flow out of the shale. In addition, the existence of two or more layers of structured water can be expected to close off the smaller pores to any flow. Because the kerogen particles are, in general, preferentially located in layers and are larger than the associated clay particles (Momper, 1978), the larger connected pores probably are oil wet and contain a high percentage of movable oil. All of the above is somewhat conjectural and considerably more experimental and observational data are needed (Dickey, 1975).

Hydrocarbon Distribution in Source Rocks

At peak generation and slightly beyond, and during primary migration, most of the hydrocarbons are in the pore system. They are not in aqueous solution, not attached to clays, and not in the kerogen. That most of the hydrocarbons in the source rock are not in solution in the pore water at any time is clear from the data in Table 2 which show that even a very low total organic content rock always contains many

Table 2. Hydrocarbon Availability Versus Pore-Water Capacity

Oil-Prone TOC (Wt. %)	Shale Porosity %	C_{15} + Hydrocarbons in Rock (ppm, vol.)	Hydrocarbon Solubility (ppm)	Water Capacity HCs (ppm)	Excess HCs in Rock
0.5	40	30	10	4	8X
0.5	30	70	15	4.5	16X
0.5	20	200	25	5	40X
0.5(5)	10	1,000	100	10	100X (1,000X)
0.5	5	200	500	25	8X

times the hydrocarbons necessary to saturate the pore water. The numbers in Table 2 are hypothetical in the sense that each horizontal line of numbers does not indicate measured values of each parameter. They show the writer's integration of depth versus porosity data summarized by Dickey (1975), generation data from Tissot and Welte (1978), solubility data from Price (1976), and an average geothermal gradient. No claim is made for accuracy in the numbers but, as in the discussion of solubilities in the case of the Bakken Shale and Los Angeles basin, even an error by a factor of five will not invalidate the point the writer is trying to make.

Near peak generation and during primary migration a source rock with a TOC of ~2.5 wt. % contains several hundred times the hydrocarbons required for saturation of the pore water. This in itself is a strong argument against aqueous solution being important in primary migration.

The question of the adsorption of hydrocarbons on clays at the temperatures of peak generation is difficult to answer quantitatively with available data and needs more research. However, the far greater polarities of water and NSO compounds than hydrocarbons, and the decreased adsorption at the temperature of peak generation strongly suggest that most of the hydrocarbons are elsewhere than adsorbed on the clays.

That most of the hydrocarbons in the rock-organic matter-pore system are not in the organic matter at peak generation is strongly indicated by the maturation tracks of unextracted oil-prone organic matter (Fig. 5). The sharp drop in hydrogen/carbon (H/C) ratio of the unextracted oil-prone organic matter with increasing maturity can only mean the ejection from the kerogen—by the kerogen—of the more hydrogen-rich generation products (i.e., the hydrocarbons). In this context the "coalification jumps" of the coal petrographers—which are basically times of rapid emission of volatiles from the oil-prone macerals—are worthy of study by those interested in petroleum migration (see Stach, 1953; Teichmuller, 1975). Figure 5 is more important in the search for an understanding of primary migration than the kerogen (extracted organic matter) H/C versus O/C diagrams which more frequently appear in the petroleum literature. For example, the horizontal nature of the vitrinite curve within the oil generation window from 0.7 to 1.1 vitrinite reflectance clearly shows that vitrinite does not emit significant amounts of hydrocarbons until the condensate-wet gas stage of generation has been reached. During this same maturation interval the unextracted oil-prone organic matter is emitting liquid hydrocarbons into the pore system as shown by its near vertical pathway within the oil generation window.

The presumption from mass balance considerations that at peak generation most of the bitumen is in the pore space (or microfracture) receives confirmation from microscopic observations (Hoots et al, 1935) and resistivity measurements (Murray, 1968; Meissner, 1976, 1978).

Bitumen Redistribution in Source Rocks

The migration implications of the microscopic observations are supported by bitumen analyses of organic-rich, oil-prone shales which indicate that bitumen redistribution begins quite early in their thermal history. For example, in the WOSCO EX-1 core of the Green River Formation, Utah, Robinson and Cook (1975) found greater than a tenfold variation in extraction ratio (bitumen/TOC) for samples with low maturation and similar organic type as indicated by H/C ratios in the 1.45 to 1.55 range. The movement of bitumen implied by the wide variation in extraction ratios is supported by the presence of four bitumen-impregnated tuffs in the same relatively immature section and by the generally inverse correlation between the extraction ratio and the TOC. In less organically rich rocks the redistribution of the bitumen in the source occurs at greater thermal alteration. Deroo et al (1977) noted many "impregnations" in potential source rocks in their study of the Alberta basin and observed that their compositions, as would be expected from migration within the source, are intermediate between the normal bitumen and oils derived from the same source rock. The Russian literature is full of references to "parautochthonous bitumoids" which they interpret as indicators of redistribution of bitumen in the source rock. They have used variations of the composition and amounts of bitumens in source rocks to help quantify the amount of migration (Trofimuk et al, 1974).

Drainage Efficiency

Mature, organic-rich, oil-prone source rocks have a greater hydrocarbon drainage efficiency than less organically rich rocks. For example, in the Los Angeles basin the previous calculations indicated that ~20% of the generated oil is now in the traps in the form of oil. In the Tertiary of the Gulf Coast probably <4.0% of the oil generated is now in traps[3] despite

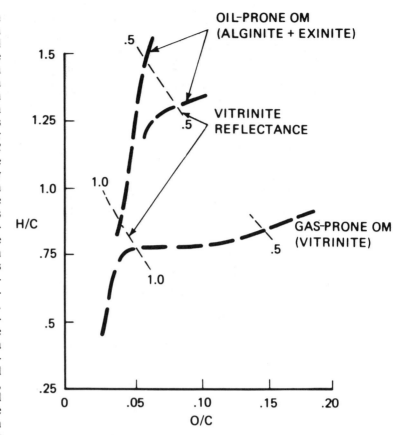

FIG. 5—Diagrammatic maturation tracks of unextracted oil- and gas-prone organic macerals (after Van Krevelen, 1961).

the abundance of traps and oil and the low TOCs. If water movement was more important than separate phase migration, the drainage efficiency of organically lean, oil-prone source rocks should be considerably higher than the organically rich ones. The increased drainage efficiency for the lean rocks would be expected because a given amount of hydrocarbon-saturated water would carry the same amount but a higher percentage of the originally available hydrocarbons. However, as illustrated in the contrast between the Gulf Coast and the Los Angeles basin, such a relationship is not observed.

Time of Primary Migration

At peak generation the bitumens in the source rock show their greatest compositional similarity to the genetically associated, reservoired oils as clearly shown by the various publications of Philippi and the Institut Francais du Petrole. The compositions, of course, are not identical. However, the oils, where unaltered, commonly are composed of lighter, less viscous, less polar material than the bitumens (Tissot and Welte, 1978), a difference very compatible with the hypothesis of continuous phase flow and very much at odds with relative solubilities in aqueous solution (McAuliffe, 1979).

In some productive but rather lean basins the source rocks never reached full peak generation. Examples include the Paris basin, some parts of some California

---

[3]Assuming an average wt. % total organic content of 0.64, a source rock density of 2.5, 50% of organic matter is oil prone, a 25% conversion of organic matter to oil, and that 50% of the basin is mature or postmature source rock, then there was an average of 25×10^6 bbl of oil generated per cu mi of sedimentary rock. Mason (1971) indicated a maximum recovery of 300,000 bbl/cu mi for the richest trends and average recovery for the Gulf Coast Tertiary of 80,000 bbl/cu mi. Assuming a 30 to 35% recovery, the in-place oil would be 1×10^6 and 0.25×10^6 bbl/cu mi respectively. Thus the ratio of oil in traps to oil generated would be 0.04 ($10^6/25 \times 10^6$) for the richest trends and 0.01 ($0.25 \times 10^6/25 \times 10^6$) for the entire basin.

basins, parts of the Cretaceous in the Rocky Mountains, and the upper Paleozoic in parts of the Illinois and Michigan basins. The oil that was emitted prior to full peak generation in these areas came from relatively rich oil-prone rocks, not from rocks with low TOCs.

The first oils emitted from organically rich, oil-prone source rocks (and often most of the oils) commonly are undersaturated with gas, despite the great generating capacity the organic matter has for gas at high levels of thermal maturity. The northern North Sea and the Central Sumatra basin are good examples. This is a clear indication that neither water nor gas had much to do with the migration, and that the migration is triggered when the bitumen saturation in the pore space reaches a critical value.

The time of migration from the source rock of various types of compounds is consistent with the hypothesis of physical ejection from the source rock. The dikes of hydrocarbons and other bitumens in the Tertiary of the Green River basin are not only proof of the existence of separate phase migration from the source rock, but they also indicate that such movement can occur early in the maturation history if certain other conditions are met. Additional thermal maturity created the immature, high-pour-point conventional crudes with a significant carbon preference index (CPI) which characterizes many Green River oils. In the Uinta basin the more mature oils are sourced from the deeper, less organic-rich part of the Green River shale because the richest part never reached full thermal maturity. However, separate phase migration is still indicated by the details of the stratigraphy and production history (Lucas and Drexler, 1975).

The Green River data illustrate several important points which are repeated in many areas of the world where organic-rich, oil-prone source rocks have yielded the major oil accumulations. Most of the oil is emitted during the time of increasing generation, it commonly shows some compositional signs of relative immaturity, can be easily tied to its source by chemical similarities, and commonly is undersaturated with gas unless substantial postmigration subsidence of source and/or reservoir can be demonstrated. Oils and formations in which these characteristics are common include most of the Tertiary oils of California, the Jurassic sourced oils in Cook Inlet, Alaska, Phosphoria oils in Wyoming, and most of the oils in the northern North Sea and in the Middle East. Because these oils moved out of the source rock during peak generation, a severe strain is placed on any proposed mechanism of primary migration which lacks a strong genetic tie between generation and migration.

SOME PROBLEMS WITH MIGRATION OF OIL IN AQUEOUS SOLUTION

Some of the many and varied problems associated with the primary migration of commercial amounts of oil in aqueous solution in water derived from compaction or clay diagenesis are discussed here under four categories: (1) general—all depths; (2) shallow—immature organic matter; (3) peak generation; and (4) super hot (>300°C). Those aspects with minimum previous discussion in the literature are covered here.

General: All Depths

Water movement—In a subsiding and compacting basin, the amount of water which is moving upward from a thermally mature source rock to a cooler reservoir section (the classic solution model for the Gulf Coast) is much less than often envisioned. Figure 6 shows the relative position of two water and one sediment layers as a basin with a relatively nonporous basement and a depositional interface near sea level subsides and compacts without the development of supernormal pressure (SNP). The overall water movement is into the basin and downward rather than out of the basin and upward. The water moves upward stratigraphically, but downward into higher temperatures. Assumptions inferred from Figure 6 are not as restrictive as one might think. For example, if SNP and undercompaction occur for whatever reason, the water will, on average, rise even less stratigraphically than indicated. Bypassing of SNP water along faults will only bring the average water distribution back to that shown on Figure 6. Lateral movement into the section on Figure 6 must be accompanied by lateral movement out elsewhere. Bonham (1978, 1980) discussed the migration implications of such a model for the Gulf Coast Tertiary. He concluded that there are too many reservoired hydrocarbons and too little upward moving water for aqueous solution of hydrocarbons to be a significant factor in the origin of the major oil accumulations in the Gulf Coast. A variation of this model appears to have first been published by Hobson (1961) who explicitly used it as an argument against the aqueous solution theories of Baker (1960).

Coals as source rocks—It is well known that vitrinitic coals are not a source of commercial oil fields, although oily films sometimes are seen in locations of exinite enrichment. Although the generating capacity of vitrinite for liquid hydrocarbons is substantially less than for the more hydrogen-rich macerals, the much greater organic content of coal relative to a normal source rock means that for a given weight of rock, a coal often contains many times the ppm of hydrocarbons that a presumed source rock does. For example, in the Cherokee Formation of southeastern Kansas, Baker (1962) found a mean value of 129 ppm hydrocarbons for 37 samples of the presumed source rock, whereas two associated coals averaged 6,900 ppm. Coals and other vitrinitic organic matter commonly are associated with sandstone aquifers and reservoirs. If water solubility was important in forming most oil accumulations, it is difficult to understand why coal is not a prolific source, and why most coal-bearing sections are devoid of significant oil occurrence unless other types of potential source rocks are present.

Oil composition—The molecular composition of oil bears no relation-

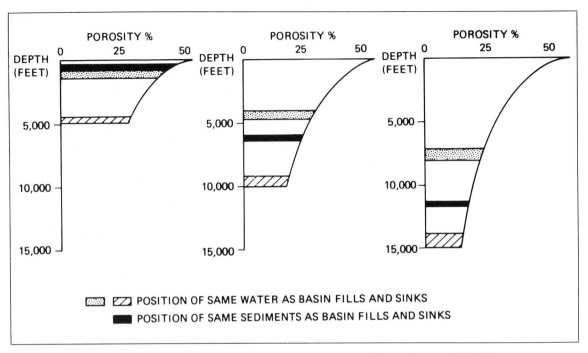

FIG. 6—Vertical movement of water and sediments in a subsiding and compacting basin.

ship to variations in molecular solubility in water. (See McAuliffe [1979] for an in-depth discussion of this important fact.)

Micelles—The problems with micelles as an effective mechanism of primary migration have been thoroughly documented in the literature (most recently by McAuliffe, 1979). The two most condemning problems are the large amount of surfactant needed and the large size of the micelles relative to the pore system at the time of primary migration.

Total organic content of source rocks—With the exception of areas like the Tertiary of the Gulf Coast, which is underlain by at least 10,000 ft (3,048 m) of mature and postmature source rock, rocks with low TOC are not the source of major oil accumulations. Because rocks with TOC >0.2 wt. % nearly always contain more than enough hydrocarbons to saturate the pore water (Table 1), one might expect, if transportation of hydrocarbons in water solution is an important factor regarding oil accumulations, for there to be virtually no such thing as a nonsource rock. At peak generation, where the oil-prone kerogen has emitted most of its hydrocarbons into the pore system, low TOC rocks should saturate their pore water as fully as high TOC rocks. Thus low TOC rocks should be as effective source rocks as high TOC rocks—if solution in water was the primary migration mechanism. But they are not.

The first step in primary migration occurs when generation has proceeded to the point where the absorption capacity of the kerogens is exceeded and bitumen is ejected into the pore system. At this point in the history of an effective source rock there are two to three orders of magnitude more bitumen in the pore system than can dissolve in the water even if we ignore "bound" water (Table 1). It is probably the inability of low TOC rocks to provide high concentrations that precludes significant primary migration from them.

Cap rocks as oil source rocks—Many of the best source rocks in the world are transgressive marine shales which are both cap rock and source rock for oils in reservoirs directly underlying them (see Fig. 2). It is unlikely that enough water could have penetrated these shales from above to form the oil accumulations directly below by a solution mechanism.

Secondary migration—Some writers (Hodgson, 1978; Roberts, 1978) have postulated that hydrocarbons, or other organics, are carried in solution in both the source and carrier rocks and are subsequently precipitated in the trap where the water carrier is presumed to enter the overlying shale. If the mechanism of precipitation in the trap was effective, the distribution of oil and water in a reservoir should not be so completely controlled by capillary effects caused by the pore geometry and the buoyancy of the oil column.

Shallow-Immature Organic Matter

The discovery of enough hydrocarbons in Holocene sediments (Smith, 1954; Kidwell and Hunt, 1958) to form commercial oil fields if the hydrocarbons could be sufficiently concentrated added substantial support to the theory of early and shallow hydrocarbon migration and accumulation. This is, of course, the time in a sediment's history when a maximum amount of water is moving about stratigraphically. The theory of early migration and accumulation that arose from the combination of these two facts is still with us (Wilson, 1975), in part because many notable petroleum geologists (Weeks, 1961; Hedberg, 1964) gave it written support prior to the full advent of modern organic geochemistry. Many explicit arguments have been developed in the last 10 years against the theory of early migration and accumulation. Three are briefly outlined below.

Access to mature source rocks—There are no large oil fields whose reservoirs do not have access to mature source rocks. The Cretaceous of the Williston basin and the upper Pliocene of the Los Angeles basin are clear-cut examples of all the factors needed to make oil fields—except access to thermally mature source rocks.

Bitumen composition—The hydrocarbon composition of immature bitumens is very different from the composition of even the most immature oils. All oils contain a preponderance of thermally formed compounds and the CPIs of oil commonly are not reached in bitumens with vitrinite reflectances of ≤ 0.60. Although it has been argued (Cordell, 1972) that solution of hydrocarbons will not reflect a high CPI in the bitumen source, geochemical evidence does not support this concept. McAuliffe (1979) showed data from a variety of sources which demonstrated that, when water is in equilibrium with crude oils, the concentrations of hydrocarbons found in the water phase are controlled by both the water solubility of each hydrocarbon and its mole fraction concentration in the crude oil. Because n-paraffins which differ by only one carbon number have similar solubilities, the hydrocarbons in water in equilibrium with a bitumen with a high CPI should have essentially the same CPI as the bitumen. Observations of the subsurface support the experimental data. Often slight irregularities in the n-paraffin distribution in an oil will be duplicated in the source (Tissot et al, 1978). In addition, the pristane/phytane ratio commonly is a source indicator which it would not be if primary migration occurred in solution and solution effects smoothed out the original distribution of hydrocarbons with similar solubilities.

Hydrocarbon solubilities—Hydrocarbon solubilities are very low and possible precipitation mechanisms are quantitatively inadequate at shallow depths and low temperatures to account for oil accumulations (Price, 1976; McAuliffe, 1979). In addition, the variations in solubilities with molecular type and weight are not reflected in crude oil composition (McAuliffe, 1979).

Peak Generation

As indicated earlier, many reasons exist for believing that primary oil migration and peak generation occur essentially simultaneously. Because the hydrocarbon compositions of the bitumens at peak generation are most similar to the reservoired oils, some of the arguments against aqueous solution do not apply at peak generation. However, the basic problems of low absolute solubilities and not enough water moving through the source rock remain.

Water availability—Little compaction or clay dehydration water is available at peak generation in many oil-rich areas, particularly those with carbonate source beds.

In several oil-rich late Tertiary basins where peak generation occurs $\geq 150°C$, the shale porosity is low, SNP is minimal, and most of the compaction and clay dehydration water moved out of the source rock prior to peak generation (e.g., Los Angeles basin).

If a section is undercompacted owing to the inability of the water to escape from a thick shale section, the stratigraphically upward movement of water will be even less than depicted in Figure 6, and the absolute downward movement of water in the basin will be greater.

Gulf Coast Tertiary—If solubility in compaction or clay-dehydration water is ever a viable mechanism of primary migration, the Gulf Coast Tertiary is a likely place because there are at least 10,000 ft (3,048 m) of mature-postmature source shales underlying the reservoir section. But in the Gulf Coast Tertiary there is not enough water moving upward from the source, even if we forget about the implications of Figure 6. Outlined below are some approximate mass balance requirements for a highly productive trend in the Gulf Coast Tertiary with an average oil in place estimate of one million bbl per cu mi (Mason, 1971).

Given: (1) Sedimentary prism, 1 mi sq (2.6 sq km) 6 mi (9.7 km) thick.
(2) 6×10^6 bbl of oil in place in traps in the upper part of the 6-cu mi (25 cu km) prism.
(3) Source rock of oil was the entire lower 3 cu mi (12.5 cu km) of sediment where the organic matter is mature or postmature.
(4) Oil left source in aqueous solution.
(5) *All* of the solubilized oil is in the traps.

Table 3 shows the oil solubilities necessary when varied amounts of water move out of the 3 cu mi (12.5 cu km) of source. The required solubility (ppm) approximately equals the oil volume/10^6 water volume.

Even with assumptions which drastically minimize the necessary solubilities, the required solubilities for a realistic water loss are too high. The loss of 5 ϕ units of porosity is roughly equal to the average water loss if a source section of SNP shale 3 cu mi (12.5 cu km) thick was

Table 3. Example of Gulf Coast Tertiary Solubility Requirements

Oil Volume (Barrels)	Water Volume		Average Solubility (ppm)
	(Barrels)	(ϕ Units)	
6×10^6	26×10^9	33.3	230
6×10^6	7.8×10^9	10.0	770
6×10^6	3.9×10^9	5.0	1,540

collapsed to normal pressure and all of the released water driven upward through the overlying reservoirs (Bonham, 1980). Numbers like these and larger forced Price (1976, 1978) to use temperatures of 300 to 350°C (572 to 662°F) to develop the necessary solubility.

Super Hot (≥300°C)

Price (1976, 1977, 1978) has eloquently argued that most of the oil accumulations of the world originated from source rocks that were heated in the subsurface to temperatures in excess of 300°C. Some of the reasons for the general lack of acceptance of his theory are briefly discussed here.

Bypassing water—Essential to Price's hypothesis is the upward movement of considerable water into cooler and more saline regimes. Normal compaction in a subsiding basin does not accomplish this (Fig. 6). It is necessary to postulate water movement along deep-seated faults that extend into the reservoir section and bypass water around the normally compacting section. Such bypassing undoubtedly occurs and specific examples can be spectacular (Price, 1976). The question is whether it is quantitatively significant. I am unaware of any published numbers purporting to quantify the amount of bypassed water. The previously discussed model of Figure 6 suggests it is not very high and, in fact, there is no *a priori* reason for it to be significant.

Barker (1977) argued that: (1) the persistence of SNP limits the amount of bypassed water; (2) the existence of low grade metamorphic minerals in hydrous phases suggests the water staying put rather than moving upward at the temperatures required by Price; and (3) the loss of only 6.6% of the pore water at 300°C and 30,000 ft would reduce a lithostatic pressure to hydrostatic. Although Price (1977) rebutted Barker's statements, both the increasing amounts of chlorite (wt. % water ~12.0%) observed at depth in Gulf Coast well profiles, and the persistence of SNP at depth suggest that we need more definite evidence before the hypothesis of significant amounts of far-ranging, upward-moving, super hot water can be accepted.

Location of exsolution—The Gulf Coast Tertiary is a reasonable test area for Price's migration mechanism because continuous phase migration is difficult to accept owing to the low TOCs, and the fact that the required temperatures of 300° + C exist near the base of part of the sedimentary column. However, the postulated temperatures exist 10,000 to 20,000 ft (3,048 to 6,096 m) below the reservoirs that contain the oil accumulations. What happens in the intervening section? Because the increase in solubility is apparently exponential with temperature, any hot, saturated water moving upward through the shale in the 10,000 to 20,000 ft (3,048 to 6,096 m) below the reservoirs will start to precipitate hydrocarbons within the shale. Most of the hydrocarbons will probably be precipitated within the massive shale long before the reservoir carrier beds are reached (Bonham, 1980) and, in part, in shales that are approximately at the thermal maturity of conventional peak generation. Thus, the basic problem remains of moving the hydrocarbons from the pore system of mature source shales to the reservoir.

Bitumens at high temperatures—To demonstrate bitumen availability at high temperatures, Price (1977, 1978) examined many samples from deep wells and established to his satisfaction that uncommonly high amounts of bitumens occasionally exist at maximum temperatures in excess of 200°C. However, innumerable temperature (depth) versus wt. % bitumen/TOC plots exist in the literature which show severe, abrupt, and permanent drops in the wt. % bitumen/TOC ratio at maximum paleotemperatures well below 200°C. Many company geologists also have made their own unpublished plots of a similar nature which accompanied the demise of the hope for commercial oil accumulations in a given well. In most cases there simply are not enough bitumens in the rocks at >200°C to source commercial oil accumulations. In addition, no evidence has been presented to show that those bitumens which do exist at >200°C ever include the biological markers (e.g., porphyrins, steranes, triterpanes) which are characteristic of most oil accumulations with API° ≤40°.

The writer has no quarrel with Price's observations that substantial amounts of indigenous bitumen occasionally exist in rocks with higher paleotemperatures than normally expected, although others (Barker, 1977) do. However, this phenomenon has always been in very tight rocks, wherever observed by the writer. Thus, the bitumen is there simply because it was not (and is not) available for migration. It is trapped in a minute, high-pressure pocket from which it can neither migrate nor, because of pressure-volume restrictions, further mature by cracking to gas and a carbon-rich residuum. Less dramatic retardation of thermal maturation is occasionally present in very tight calcareous or siliceous rocks at shallower depths, as noted by Bostick and Alpern (1977).

Basin temperatures, past and present—The source rocks in many basins of the world never reached the 300 to 350°C temperatures required by Price's (1978) model. The Cretaceous basins of the Rocky Mountains, many of the Paleozoic basins of the Mid-Continent, and some of the California basins are obvious examples in the United States. Even in the Los Angeles basin where the stratigraphically deepest source rocks may approach 300°C near the center of the basin, the relative immaturity of the oils clearly indicates that migration of the reservoired oils occurred prior to such temperatures being reached. In some basins the paleogeothermal gradient was clearly higher than the present gradient (e.g., Douala basin, Cameroon); however, in most basins the time-temperature-maturation relationships suggest that no significant change in gradient has occurred during the basins' history (Tissot and Welte, 1978).

Price stressed the role of faults and fractures in migration, developed needed solubility data at high pressures and temperatures, and pointed out that we have not adequately evaluated pressure effects in generation. The writer thanks Price, particularly for his provocative assaults on our orthodoxy; however, the writer does not believe that Price has solved the enigma of commercial migration, either in the Gulf Coast Tertiary or elsewhere.

TWO PROBLEM AREAS

Origin of Petroleum in Gulf Coast Tertiary

Despite extensive literature on the subject, the origin of the oil in the Tertiary of the Gulf Coast remains an enigma. It is unlikely that separate phase migration of oil from the pores adjacent to the generating kerogen can be the dominant primary migration mechanism for the Tertiary of the Gulf Coast. There simply appears to be too little kerogen (TOCs ~0.3 to 1 wt. %) and too much pore space (10 to 20%). However, as indicated earlier, if one migration mechanism is inoperative, others will automatically increase in both relative and absolute importance, as nature makes less dense and less viscous hydrocarbons available for migration. Several alternative migration mechanisms which might be applicable to the Gulf Coast Tertiary have been presented (during the migration course and symposium at the 1978 AAPG convention). These included: (1) solution in deep, hot (>300°C) water (Price, 1978); (2) solution near peak generation (Roberts, 1978; Cordell, 1978); (3) diffusion (Hinch, 1980); and (4) transportation with methane (Hedberg, 1980). The reasons for not accepting either Price's thesis that the Gulf Coast petroleums originated at temperatures ≥300°C or the more conventional solution mechanisms, were previously discussed. Hinch's diffusion hypothesis is fascinating, in part because of the use he makes of the concept of bound water, an area where there is a need for more definitive data. However, the writer maintains reservations toward a diffusion hypothesis at this time for the following reasons: (1) diffusion is basically a mode of dispersion rather than of concentration; (2) diffusion coefficients of the heavier petroleum molecules are very low; (3) the multiple reservoirs of the structural traps in the Gulf Coast do not fill from the bottom up; (4) diffusion has low rates of mass transfer over substantial distances, particularly in shales. Thus, one can legitimately question whether diffusion was important in the emplacement of the large oil fields in the Pleistocene sandstones at Eugene Island offshore Louisiana. It is clear from such occurrences as the concentration of the oil in highly faulted intrusive salt domes (Spillers, 1965) and the apparently capricious alternation of oil and water in reservoirs interbedded with shales of similar source characteristics that many of the faults must have been conduits of petroleum at certain times in their history. Insufficient water moved up the faults to form the petroleum by exsolution in the reservoirs (Bonham, 1980). Therefore, the oil probably moved up the fault zones as a continuous oil phase or as the solute in a gas phase. The latter possibility is attractive because recent data on hot, deep, formation fluids in the Gulf Coast Tertiary indicate that they are saturated, and may be over-saturated with methane. In addition, Russian studies (Zhuze et al, 1963; Zhuze and Bourova, 1977) indicate that from both a compositional and quantitative viewpoint, solution of oil in gas is a much more viable migration mechanism than solution in water. However, we are still left with the problem of how the hydrocarbons moved from the source shales to the faults in the volume that they have. More data on the details of the faults and on the distribution of pressure, temperature, fluid, and rock properties near them might help resolve what seems to remain the enigma of primary migration in the Gulf Coast Tertiary. However, it is important to remember that the thick Mesozoic-Tertiary delta systems are unique among the oil basins of the world in many ways, including widespread SNP and at least 10,000 ft (3,048 m) of mature-postmature source rocks beneath the reservoirs, and that one of those unique ways is very apt to be the main mechanism of primary migration.

Origin of Oil Sands of Lower Cretaceous of Alberta

Introduction—Anyone who theorizes on the primary migration of petroleum must address the problem of the origin of the trillion bbl of heavy oil in the Lower Cretaceous of Alberta. After several decades of discussions, it has finally been agreed among most geochemists, and some geologists, that the oil in the Lower Cretaceous sandstones of Alberta originated within the Lower Cretaceous section (Vigrass, 1968; Deroo et al, 1977). Unfortunately, there are no chemical comparisons between the oil sandstones and the bitumens of the organic rich but volumetrically quite limited Jurassic Nordegg shale. Nevertheless, in recent years most of the published discussions have accepted the Lower Cretaceous origin, hotly debated whether the composition of the oil is more dependent on original immaturity or degradation, and completely ignored the compelling question of how this immense amount of oil got to its present location (Montgomery et al, 1974a; Montgomery et al, 1974b; Deroo et al, 1974; Deroo et al, 1977; George et al, 1977). Exceptions to the latter comment are the micellar solution theory of Vigrass (1968) and the molecular solubility theory of Hunt (1977). These will be briefly evaluated later in this paper. It is probable that the virtual ignoring of the question of the mechanism of primary migration of the Lower Cretaceous heavy oils simply reflects the intuitive recognition that all of the more popular migration theories are not applicable.

Continuous phase migration—Continuous phase flow of oil out of the source rock is probably not the answer. To the writer's knowledge, no one has identified an organically rich, highly oil-prone source rock in the Lower Cretaceous. Deroo et al (1977) were puzzled about how a section with TOCs of ~1 to 3%, but with organic matter of dominantly terrestrial origin, could source such large accumulations of oil. Data computed from McIver (1967) and plotted on an H/C-O/C kerogen maturation diagram from Tissot et al (1974) confirm the generally poor quality of the organic matter (Fig. 7). In addition, much of the Lower Cretaceous has not yet reached thermal maturity. If such a source section can generate and migrate a trillion bbl of petroleum by continuous phase flow, the world should be awash in oil.

Compaction water—Solution by compaction waters (Vigrass, 1968; Hunt, 1977) offers no better explanations. Where did all the micelles come from—and go to (McAuliffe, 1979)? Movement by molecular solution means deriving 40 cu mi (167 cu km) of oil from ~50,000 cu mi (209,000 cu km) of shale[4]. This is equivalent to 800 ppm (vol.) of oil emitted from the rock. Allowing a liberal 10% loss of porosity to be the compaction-water loss during oil migration, and assuming all the solubilized oil to be trapped with negligible loss to degradation, an average solubility of approximately 8,000 ppm is needed. Thus, solution in compaction—or clay dehydration water—is not an acceptable answer either.

Meteoric water—Currently, the only migration mechanism that appeals to the writer regarding the origin of the Alberta oil sandstones is the action of the meteoric water over a period of ~50 million years. This proposed solution to the problem of the origin of the heavy oil sandstones of Alberta is implicit in several papers (Hitchon, 1969, 1974; Hodgson et al, 1964), but not explicitly applied with numbers and geology.

The argument for the dominance of meteoric water is indirectly based on the bankruptcy of other possible mechanisms, but is directly based on

---

[4]The volume of shale from which the oil deposits in the Lower Cretaceous were derived is, of course, not known. Hitchon (1968) indicated that ~55,000 cu mi (229,000 cu km) of Lower Cretaceous shale exist between the eastern edge of the disturbed belt and the erosional edge to the east and between 49 and 60°N lat. If we arbitrarily assume that oil-contributing shale in the disturbed belt is slightly less than the shale elsewhere which was not a contributor, a figure of 50,000 cu mi (209,000 cu km) of potentially contributing Lower Cretaceous shale seems a reasonable approximation.

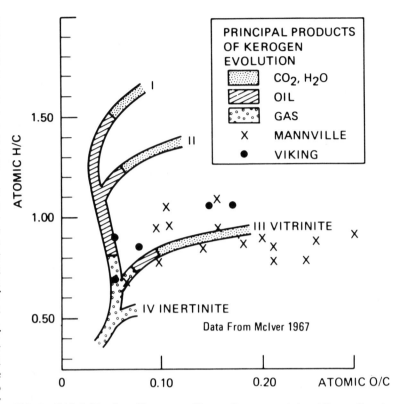

FIG. 7—H/C-O/C ratios of kerogens of Lower Cretaceous shales, Alberta, Canada (after Tissot et al, 1974).

the present surface water flow patterns in Alberta, the reasons for them, and the probable persistence of very similar patterns since Laramide time.

Only a brief summary of the basic geology and arguments can be given here. The reader is referred to Vigrass (1968) for an excellent description of the oil deposits and their geologic setting. Figures 8 and 9 show the areal distribution of the Athabasca and Peace River heavy oil deposits with respect to the hogback of Paleozoic carbonates which strongly influenced the distribution of the thickness and facies of the basal Cretaceous deposits. It is almost certainly not a coincidence that Athabasca lies directly east of a broad sand-filled channel which breached the subdued ridge of Paleozoic carbonates. The possibilities of major long distance, focused migration within the basal Cretaceous are obvious. Undoubtedly much compaction water moved toward and through Athabasca during compaction of the Lower Cretaceous, but that compaction water was only a harbinger of the large volume of water which followed similar paths from early Tertiary time to the present. Figures 10 and 11 are hydraulic head cross sections, through the Athabasca and Peace River deposits by Hitchon (1969, 1974) which show that the two oil accumulations are currently within the ground-water regime. In fact, the unconformity and the basal Cretaceous sandstones are potential sinks, and the break in the buried hogback of Paleozoic strata is essentially a straw which drains the fluids entering the sink into Athabasca. Although the hydraulic head contours and the flow lines generally indicate water flow toward the unconformity area from both above and below, the permeability of the Cretaceous clastic deposits is so much higher than that of the Paleozoic carbonates and evaporites that the eastward-flowing water near the un-

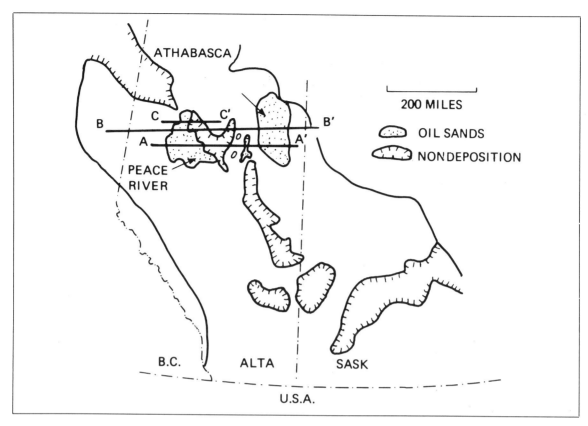

FIG. 8—Lower Mannville (basal Cretaceous) depositional framework, western Canada (from Vigrass, 1968).

conformity can be presumed to be virtually all meteoric water that descended from the present erosion surface. The salinity of < 500 mg per liter for the waters associated with the Peace River deposits (Deroo et al, 1977) is consistent with this hypothesis.

It is very probable that similar water flow patterns have existed since early Tertiary time when the Rocky Mountains began to rise. Certainly, the unconformity and the distribution of the Lower Cretaceous sandstones were identical. The eastern slope of the erosion surface which is an important factor in the persistence of the necessary hydraulic head also has been present in varying degrees since the early Tertiary. Several thousand feet of Cretaceous-Tertiary rocks have been eroded from the plains area west of the oil sandstones since early Tertiary time. Much of this section consisted of coarse clastic rocks which would have easily imbibed rainwater, limited surface runoff, and helped develop the hydraulic head necessary to move meteoric water through the finer grained, underlying Cretaceous clastic sediments. Everything considered, it is difficult not to believe that the gross patterns of subsurface water movement which exist today east of the mountain front have persisted for tens of millions of years. Table 4 shows how the concept of extracting the oil in the oil sandstones from their source by meteoric water flow over a period of approximately 50×10^6 years simplifies the mass balance situation. As is true of most of the numerical examples in this paper, the actual numbers can be varied substantially without

Table 4. Mass Balance Calculations from Meteoric Water Regarding Origin of Oil Sands of Lower Cretaceous of Alberta

1. Oil in place $\cong 10^{12}$ bbls $\cong$ 40 cu mi (167 cu km).
2. Surface drainage area $\cong 330 \times 10^3$ sq mi (777×10^3 sq km).
3. Assume 30 inches (76 cm) of rain for 50×10^6 years.
4. Total rain volume per sq mi (2.59 sq km) $\cong$ 30 inches/year $\times 16 \times 10^{-6}$ mi/in. $\times 50 \times 10^6$ years $\times 1$ sq mi $\cong 24 \times 10^3$ cu mi (99×10^3 cu km).
5. Total rainfall on drainage area $\cong 7.2 \times 10^9$ cu mi (30×10^9 cu km).
6. Assuming 1% of rainfall passes through oil sandstones by subsurface flow, available water $\cong 72 \times 10^6$ cu mi (300×10^6 cu km).
7. ∴ oil/water $\cong 40/72 \times 10^6 \cong 0.56 \times 10^{-6}$ or 0.56 ppm.
8. Could account for oil sands with 3 ppm solubility and 20% precipitation in reservoir.

Migration Mechanisms

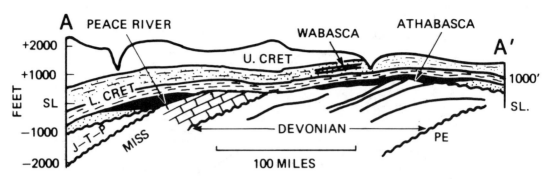

FIG. 9—Diagrammatic east-west cross section showing geologic setting of Lower Cretaceous oil sands, Athabasca-Peace River area (line A-A' from Fig. 8).

changing their implication. Emplacement by meteoric water can handle the mass balance requirements of the origin of the oil sandstones whereas other migration mechanisms cannot.

It is possible that Cretaceous coals might have contributed to the oil in the oil sandstones by means of the meteoric water mechanism. The writer refers this interesting question (and the more general question of the compatibility, or lack thereof, of the chemical composition of the oil with an origin from meteoric water) to the geochemists.

SUMMARY AND CONCLUSIONS

Most major oil accumulations of the world left the source rock in a continuous oil phase. This conclusion is directly based on microscopic observations, log characteistics, and producing characteristics of fractured shale reservoirs; and is indirectly based on the 2 to 3 order of magnitude discrepancy between the amount of hydrocarbons that can be carried in solution by compaction and clay dehydration water and the amount which actually moved from source to traps. Continuous phase primary migration is also supported by bitumen redistribution in source rocks, the bitumen content of the

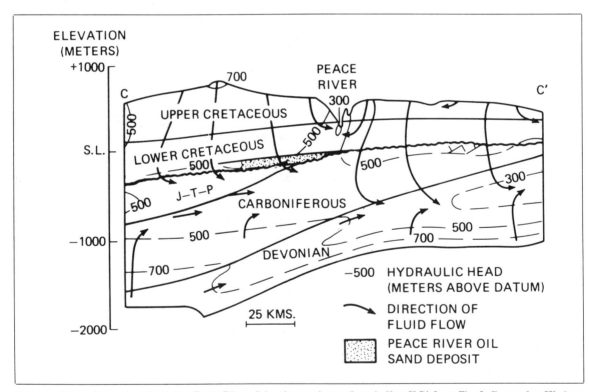

FIG. 10—Hydraulic head cross section, Peace River oil-bearing sandstone deposit (line C-C' from Fig. 8; figure after Hitchon, 1974).

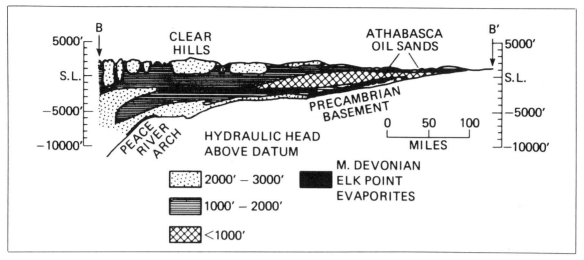

FIG. 11—Hydraulic head distribution, east-west cross section, northern Alberta (line B-B' from Fig. 8; figure after Hitchon, 1969).

pores at peak generation, the high TOCs of most source rocks (2 to 10 wt. %), the usual greater drainage efficiency of organic-rich source rocks, and the genetic tie between peak generation and primary migration.

When bitumen concentrations in the rock are too low for continuous phase flow to exist, other migration mechanisms, which are always operative, will increase in both absolute and relative intensity. Solution of oil in gas may become significant in thick Tertiary deltaic systems and meteoric water may be a surprising contributor in some very specific geologic settings. However, it is unlikely that solution of oil in water from compaction or from clay dehydration has much to do with many of the major oil accumulations of the world.

REFERENCES CITED

Albrecht, P., M. Vanderbroucke, and M. Mandengue, 1976, Geochemical studies on the organic matter from the Douala basin (Cameroon), I. Evaluation of the extractable organic matter and the formation of petroleum: Geochim. et Cosmochim. Acta, v. 40, p. 791-799.
Auldridge, L., 1977, Russia drives to sustain big gains in oil output: Oil and Gas Jour., v. 75, no. 53, p. 69-72.
Bajor, M., M.-H. Roquebert, and B. M. van der Weide, 1969, Transformation de la matiere organique sedimentaire sous l'influence de la temperature: Bull. Centre Recherches Pau - SNPA, v. 3, no. 1, p. 113-124.
Baker, D. R., 1962, Organic geochemistry of Cherokee Group in southeastern Kansas and northeastern Oklahoma: AAPG Bull., v. 46, p. 1621-1642.
Baker, E. G., 1960, A hypothesis concerning the accumulation of sediment hydrocarbons to form crude oil: Geochim. et Cosmochim. Acta, v. 19, p. 309-317.
Barbat, W. F., 1958, The Los Angeles basin area, California, in L. G. Weeks, ed., Habitat of oil: AAPG, p. 62-77.
Barker, C., 1977, Aqueous solubility of petroleum as applied to its origin and primary migration; discussion: AAPG Bull., v. 61, p. 2146-2149.
Bonham, L. C., 1978, Migration of hydrocarbons in compacting basins (abs.): AAPG Bull., v. 62, p. 498-499.
——— 1980, Migration of hydrocarbons in compacting basins: AAPG Studies Geology 10, p. 69-88.
Bostick, N. H., and B. Alpern, 1977, Principles of sampling preparation and constituent selection for microphotometry in measurement of maturation of sedimentary organic matter: Jour. Microscopy, v. 109, pt. 1, p. 41-47.
——— et al, 1978, Gradients of vitrinite reflectance and present temperature in the Los Angeles and Ventura basins, California: Symposium on low temperature metamorphism of kerogen and clay minerals, Pacific Section SEPM, p. 65-96.
Brooks, J., and B. Thusu, 1977, Oil-source rock identification and characterization of the Jurassic sediments in the northern North Sea: Chem. Geology, v. 20, p. 283-294.
Cartmill, J. C., and P. A. Dickey, 1970, Flow of a disperse emulsion of crude oil in water through porous media: AAPG Bull., v. 54, p. 2438-2447.
Cordell, R. J., 1972, Depths of oil origin and primary migration: A review and critique: AAPG Bull., v. 56, p. 2029-2067.
——— 1978, Migration pathways in compacting clastic sediments: paper, AAPG Research Symposium on Problems of Petroleum Migration (Oklahoma City).
Demaison, G. J., and G. T. Moore, 1980, Anoxic environments and oil source bed genesis: AAPG Bull., v. 64, p. 1179-1209.
Deroo, G., et al, 1974, Geochemistry of the heavy oils of Alberta, in Oil sands, fuel of the future: Canadian Soc. Petroleum Geologists, Memoir 3, p. 148-167.
——— et al, 1977, The origin and migration of petroleum in the western Canadian sedimentary basin, Alberta: Canada Geol. Survey Bull. 262, p. 1436.
Dickey, P. A., 1975, Possible primary migration of oil from source rocks in oil phase: AAPG Bull., v. 59, p. 337-345.
Dow, W. G., 1974, Application of oil-correlation and source rock data to exploration in Williston basin: AAPG Bull., v. 58, p. 1253-1262.
Durand, B., and J. Espitalie, 1976, Geochemical studies on the organic matter from the Douala basin (Cameroon), II. Evolution of kerogen: Geochim. et Cosmochim. Acta, v. 40, p. 801-808.
Espitalie, J., et al, 1977, Methode rapide de caracterization des roches meres de leur potential petrolier et de leur degre d'evolution: Inst. Francais Petrole Rev., v. 32, p. 23-42.
Evans, C. R., D. K. McIvor, and K. Magara, 1975, Organic matter, compaction history and hydrocarbon occurrence, Mackenzie Delta, Canada: Proc. 9th World Petroleum Cong., v. 2, p. 149-157.
Gardett, P. H., 1971, Petroleum potential of Los Angeles basin, California, in Future petroleum provinces of the United States: AAPG Memoir 15, v. 1, p. 298-308.
George, A. E., et al, 1977, Simulated geothermal maturation of Athabasca bitumen: Bull. Canadian Petroleum Geology, v. 25, p. 1085-1096.
Hedberg, H. D., 1964, Geologic aspects of origin of petroleum: AAPG Bull., v. 48, p. 1755-1803.
——— 1980, Methane generation and petroleum migration: AAPG Studies Geology 10, p. 179-206.
Hinch, H. H., 1980, The nature of shales and

the dynamics of hydrocarbon expulsion in the Gulf Coast Tertiary section: AAPG Studies Geology 10, p. 1-18.

Hitchon, B., 1968, Rock volume and pore volume data for plains region of western Canada sedimentary basin between latitudes 49°and 60°N: AAPG Bull., v. 52, p. 2318-2323.

―――― 1969, Fluid flow in the western Canada sedimentary basin: I. Effect of topography: Water Resources Research, v. 5, no. 1, p. 186-195.

―――― 1974, Application of geochemistry to the search for crude oil and natural gas, in A. A. Levinson, ed., Introduction to exploration geochemistry: Applied Publishing, Ltd., p. 509-545.

Hobson, G. D., 1961, Problems associated with the migration of oil in "solution": Inst. Petroleum Jour., v. 47, no. 449, p. 170-173.

Hodgson, G. W., 1978, Origin of petroleum—in-transit conversion of organic compounds in water (abs.): AAPG Bull., v. 62, p. 522.

―――― B. Hitchon, and K. Taguchi, 1964, The water and hydrocarbon cycles in the formation of oil accumulations: Recent Research in the Field of Hydrosphere, Atmosphere, and Nuclear Geochemistry, p. 217-242.

Hoots, H. W., A. L. Blount, and P. H. Jones, 1935, Marine oil shale, source of oil in Playa del Rey field, California: AAPG Bull., v. 19, p. 172-205.

Hunt, J. M., 1961, Distribution of hydrocarbons in sedimentary rocks: Geochim. et Cosmochim. Acta, v. 22, p. 37-49.

―――― 1963, Composition and origin of the bitumens of the Uinta basin: oil and gas possibilities of Utah, reevaluated: Utah Geol. and Mineralog. Survey Bull. 54, p. 249-274.

―――― 1977, Ratio of petroleum to water during primary migration in western Canada basin: AAPG Bull., v. 61, p. 434-435.

Kidwell, A. L., and J. M. Hunt, 1958, Migration of oil in recent sediments of Pedernals, Venezuela, in L. G. Weeks, ed., Habitat of oil: AAPG, p. 790-817.

Lucas, P. T., and J. M. Drexler, 1975, Altamont-Bluebell: a major and overpressured stratigraphic trap, Uinta basin, Utah: Rocky Mountain Assoc. Geologists, Symposium (1975), p. 265-273.

Magara, K., 1978a, The significance of the expulsion of water in oil-phase primary migration: Bull. Canadian Petroleum Geology, v. 26, p. 123-131.

―――― 1978b, Primary migration agents (abs.): AAPG Bull., v. 62, p. 538.

Mallory, W. W., 1977, Fractured shale hydrocarbon reservoirs in southern Rocky Mountain basins: Exploration Frontiers of the Central and Southern Rockies (Rocky Mountain Assoc. Geologists Symposium), p. 89-94.

Mason, B. B., 1971, Summary of possible future petroleum potential of Region 6, Western Gulf basin: AAPG Memoir 15, p. 805-812.

McAuliffe, C. D., 1979, Oil and gas migration—chemical and physical constraints: AAPG Bull., v. 63, p. 761-781.

McIver, R. D., 1967, Composition of kerogen—clue to its role in the origin of petroleum: Proc. 7th World Petroleum Cong., v. 2, p. 25-36.

Meinchein, W. G., 1959, Origin of petroleum: AAPG Bull., v. 43, p. 925-943.

Meissner, F. F., 1976, Abnormal electric resistivity and fluid pressure in Bakken Formation, Williston basin, and its relation to petroleum generation, migration, and accumulation (abs.): AAPG Bull., v. 60, p. 1403-1404.

―――― 1978, Petroleum geology of the Bakken Formation, Williston basin, North Dakota and Montana: Williston Basin Symposium, Montana Geological Society, p. 207-227.

Miller, R. J., and P. F. Low, 1963, Threshold gradient for water flow in clay: Soil Sci. Soc. America Proc., v. 27, p. 605-609.

Momper, J. A., 1978, Oil migration limitations suggested by geological and geochemical considerations: AAPG Course Note Series 8, p. B1-B60.

Montgomery, D. S., R. C. Banerjee, and H. Sawatzky, 1974a, Optical activity of the saturated hydrocarbons from the Alberta heavy Cretaceous oils and its relation to thermal maturation: Bull. Canadian Petroleum Geology, v. 22, p. 357-360.

―――― 1974b, Investigation of oils in western Canada tar belt, in Oil sands, fuel of the future: Canadian Soc. Petroleum Geologists, Memoir 3, p. 168-183.

Murray, G. H., 1968, quantitative fracture study—Sanish pool, McKenzie County, North Dakota: AAPG Bull., v. 52, p. 57-65.

Philippi, G. T., 1965, On the depth, time and mechanism of petroleum generation: Geochim. et Cosmochim. Acta, v. 29, p. 1021-1049.

Poulet, M., and J. Roucache, 1970, Etude geochemique des bruits du bassin parisien: Inst. Francais Petrole Rev., v. 25, p. 128-148.

Price, L. C., 1976, Aqueous solubility of petroleum as applied to its origin and primary migration: AAPG Bull., v. 60, p. 213-244.

―――― 1977, Aqueous solubility of petroleum as applied to its origin and primary migration: reply: AAPG Bull., v. 61, p. 2149-2156.

―――― 1978, New evidence for a hot, deep origin and migration of petroleum: AAPG Research Symposium on Problems of Petroleum Migration (Oklahoma City), 1978.

Roberts, W. H., III, 1978, Design and function of oil and gas traps as clues to migration (abs.): AAPG Bull., v. 62, p. 558.

Robinson, W. E., and G. L. Cook, 1975, Compositional variations of organic material from Green River oil shale—WOSCO Ex-1 Core (Utah): U.S. Bur. Mines Rept. Inv. 8017, p. 1-40.

Smith, J. E., J. G. Erdman, and D. A. Morris, 1971, Migration, accumulation and retention of petroleum in the earth: Proc. 8th World Petroleum Cong., v. 2, p. 13-26.

Smith, P. V., 1954, Studies on origin of petroleum: occurrence of hydrocarbons in sediments of Gulf of Mexico: AAPG Bull., v. 38, p. 377-404.

Snarskii, A. N., 1970, The nature of primary oil migration: Neft, Gaz 13, no. 8, p. 11-15 (Translated by Assoc. Technical Serv., Inc.).

Spillers, J. P., 1965, Distribution of hydrocarbons in south Louisiana by types of traps and trends—Frio and younger sediments (abs.): AAPG Bull., v. 49, p. 1749-1751.

Stach, E., 1953, The "Coalification Jump" in the Ruhr Carboniferous: Brennstoff-Chemie, v. 34, p. 353-355.

Teichmuller, M., 1975, Generation of petroleum-like substances as seen under the microscope: Advances in Organic Geochemistry, 1973, p. 378-395.

―――― and K. Ottenjann, 1977, Liptinites and lipoid materials in an oil source rock: type and diagenesis of liptinites and lipoid substances in an oil source rock on the basis of fluorescence microscopic studies: Erdol and Kohle, Erdgas, Petrochemie, v. 30, p. 387-398.

Tissot, B., and D. Welte, 1978, Petroleum formation and occurrence: New York, Springer-Verlag, 530 p.

―――― G. Deroo, and A. Hood, 1978, Geochemical study of the Uinta basin: formation of petroleum from the Green River formation: Geochim. et Cosmochim. Acta, v. 42, p. 1469-1486.

―――― et al, 1971, Origin and evolution of hydrocarbons in early Toarcian shales, Paris basin: AAPG Bull., v. 55, p. 2177-2193.

―――― et al, 1974, Influence of nature and diagenesis of organic matter in formation of petroleum: AAPG Bull., v. 58, p. 499-506.

Toth, J., 1978, Gravity induced cross-formational water flow—possible mechanisms for transport and accumulation of petroleum (abs.): AAPG Bull., v. 62, p. 567-568.

Trofimuk, A. A., et al, 1974, The fractionation of bitumoids in the course of their migration: Geologiya, Geofizika, v. 15, no. 5, p. 122-129.

Van Krevelen, D. W., 1961, Coal: Amsterdam, Elsevier Pub. Co.

Vigrass, L. W., 1968, Geology of Canadian heavy oil sands: AAPG Bull., v. 52, p. 1984-1999.

Weeks, L. G., 1961, Origin, migration and occurrence of petroleum, in Petroleum Exploration Handbook, Chapter 5: New York, McGraw-Hill, p. 1-50.

Williams, J. A., 1974, Characterization of oil types in the Williston basin: AAPG Bull., v. 58, p. 1243-1252.

Wilson, H. H., 1975, Time of hydrocarbon expulsion, paradox for geologists and geochemists: AAPG Bull., v. 59, p. 69-84.

Zhuze, T. P., and E. G. Bourova, 1977, Influence des different processes de la migration primaire des hydrocarbons sur la compositon des petroles dans les gisements: Advances in Organic Geochemistry, 1975, p. 493-500.

―――― G. N. Jushkevich, and G. S. Vshakova, 1963, On the general rules of behavior of gas-oil systems at great depths (in Russian): Dokl. Akad. Nauk. SSSR, v. 152, p. 713-716.

SOME FACTORS IN OIL ACCUMULATION.*

By Professor V. C. Illing, M.A., F.G.S. (Member).

Synopsis.

The main purpose of the paper is to consider the influence of texture and buoyancy in the flow of oil and water mixtures through sands and its bearing on oil accumulation.

In order to drive an oil column continuously forward in a flowing water stream within a sand, a definite excess pressure, the forefront pressure, must be exerted within the oil column. This forefront pressure is inversely proportional to the grade size. When an oil-water stream comes into contact with sands of varying coarseness the low forefront pressure of the coarse sand causes the oil to abandon all further movement in the fine sand and to move only in the coarse. Moreover, when the oil reaches the limit of the coarse sand, it is retained there and cannot enter the fine sand until a sufficient pressure is built up within the oil to attain the forefront pressure of the fine sand. This causes the filtration of the oil at the coarse-fine interface.

With regard to directional movement, the function of buoyancy increases the ease of upward oil flushing. An oil-water column in motion must maintain a certain critical concentration. This is less for upward flow than for horizontal or downward flow. Hence when oil has a choice of alternative paths it selects the most upward one, even though it may mean a movement transverse to the main fluid movement. The result is a differentiation of oil from the main water stream and the production of an oil-pool. The application of this idea to various geological structures is discussed.

In a paper on oil migration read before the Institute some five years ago,[1] the author attempted to clarify the issue by separating the processes of migration into two separate stages. The process of enrichment of the reservoir rock at the expense of the source-rock, which was termed primary migration; and the subsequent readjustment of the oil and gas within the reservoir rock, which was termed secondary migration.

It would be easy to criticize any such attempt to subdivide a set of natural processes into various compartments, for nature is usually complex and her methods interdovetail into one organic whole. These processes of migration are by no means an exception. One might argue, with every justification, that there are cases where the reservoir rock and the source rock are one, and that in such cases primary migration is unnecessary, and also indistinguishable from secondary migration. There would be equal justification in the assertion that both stages of the movement were so intimately interlocked in the filling of a sand lens with oil and gas that it is impossible to draw a hard-and-fast line between them.

In spite, however, of these difficulties, there is a distinct advantage in separating the two stages of migration in order that attention may be directed to fundamental differences in the difficulties involved. In primary migration it is the movement of oil through a normally impervious rock and its retention in the porous rock which is the main problem. In secondary migration our attention is focused on movements of separation between the gas, oil and water in the permeable rock, whereby the oil and gas are concentrated in certain portions of the rock to become commercial oil and gas fields. This is a process of segregation and enrichment, one

---

* Paper presented for discussion at the One Hundred and Eighty-sixth General Meeting of the Institute of Petroleum held on 14th February, 1939.

might almost say a cessation of further movement, quite unlike the first processes of migration.

Many causes have been suggested for the processes of primary migration, such as gravitation, compaction, diastrophism, hydraulic currents, capillarity and surface adsorption. The author has given elsewhere his reasons for considering that most of this primary migration takes place during compaction, and is probably helped by fluid movements associated with diastrophism. It is essentially the outflow of fluids from the compressible rocks, and the oil and gas are entrained in a current of water. The trapping of oil in the reservoir rock is ascribed by the author to the selective filtration of oil and gas from the fluid currents as they pass from the coarse to the fine rocks. This results from the inability of the oil to re-enter the water-wet fine rocks without the development of a considerable pressure differential. This process of selective filtration is essentially associated with interfacial tension, for it depends on the presence of two or more immiscible liquids. It is not, however, a force due to surface tension. One may regard it as Versluys did, as a retention of oil in the coarse sand owing to the excess energy required to develop the much larger surface area of the oil–water interface in the finer rocks.[2] On the other hand, the author prefers to concentrate attention on another aspect of the phenomenon by stressing the required pressure differential. This is a measurable physical phenomenon which definitely prevents oil from entering the water-wet fine rocks. None the less the two views are closely related in so far as they interpret surface tension as a guide, and not as a cause of movement.

The ultimate causes of primary migration are still a fruitful cause of discussion, but as the author's purpose in the present paper is to consider certain aspects of the flow of oil and water in the reservoir rocks, the temptation to linger over the earlier aspects of the phenomena must be avoided.

Physical Relations of Oil and Water in Sands.

It will be sufficient for the present to agree that within the permeable rocks there are fluid movements of water, oil and gas. The extent and velocity of these movements are unknown, but that there is a certain amount of movement will be accepted by everyone save the few who demand a strictly *in situ* origin for oil and gas.[3] Even a cursory study of oil accumulations cannot fail to bring out the importance of their relations to the texture and attitude or structure of the reservoir rocks, and it is therefore reasonable to allow a considerable amount of local segregation even if we decline to accept broad regional migration. For this reason the writer proposes to consider in detail certain aspects of the behaviour of oil–water mixtures in sands, in order to deduce some general principles.

The choice of a two-phase mixture rather than the three phases of gas, oil and water is deliberate, in order to simplify the issue, although the action of gas would have to be considered in a complete review of the whole processes of segregation. This paper is not intended as a complete review, but is rather an attempt to establish principles. Sand is chosen as the typical reservoir rock, as it is convenient and simple to handle. It does not possess all the attributes of every type of reservoir rock, in so far as it is lacking in joints and other fissure planes, but its porosity and permeability make it an ideal medium for holding oil and water.

The pores within such a medium are neither uniform nor are they separate; it is the grains of sand which are separate. The pores represent a continuous interspace of variable dimensions in which each void space between the grains is in communication with contiguous voids. There is thus a great flexibility of choice of path for any fluid moving in this void space. This distinction is important, and must be continuously borne in mind as an essential difference between the pore system of a sand, on the one hand, and a bundle of capillary tubes, on the other. Another important difference between these two forms of capillaries is that the pores in a sand are roughly triangular in cross-section, and of inconstant dimensions, whereas the normal capillary is circular and more or less uniform in cross-section. As indicated above, these differences lead to a much greater flexibility of intercommunication within a sand than could occur in a bundle of capillary tubes, and it is in some respects fallacious to apply the principles of a normal capillary tube to the pore space of a sand. A single instance will suffice to illustrate the difference. The blockage of a capillary at one point would completely eliminate that capillary from all fluid movement along its entire length. The blockage of a single void between sand grains would cut off none of the contiguous voids from general fluid movement, and its effect would be infinitesimal. It is only when blockage becomes widespread that the permeability of a sand suffers.

In spite of this freedom of intercommunication in sand bodies, gas, oil and water interspersed in a water-wet sand will not separate out by gravitation. The surface tensional features at the oil–water interfaces are sufficient to maintain the separate oil bodies in position unless the sand is so coarse that the capillaries are about 2 mm. in diameter—*i.e.*, the sand becomes a grit. The presence of gas tends to help the process of flotation owing to the greater density difference between gas and water and the tendency for gas and oil to become attached to each other where they make contact accidentally. Even this, however, will not produce flotation in normal sands. Experiments which claim to have produced such movement have in fact involved the production of local pressure gradients within the fluids by the generation of gas within the medium.[4] It was these pressure gradients which induced the movement, and the gas globules, being sensitive to pressure changes, were the main medium whereby the pressures tended to equalize.

Most students of oil accumulation agree that there are two fundamental characteristics of a reservoir rock which appear to control the position of the oil. The first is its texture. Many a large oil-pool owes its occurrence entirely to the presence of a zone of coarse sand in fine sand or of sand in clay. The water-bearing finer sand may be just as porous as the coarse sand, the only difference is that the pore spaces are smaller and more numerous. The second important difference is the attitude of the zone of porous rock to the surrounding and more impervious media. This may be summarized by the implications of the term " closure," which really means a porous zone covered by impervious material and from which no escape is possible without downward movement of the enclosed fluid. Most favourable oilfield structures present a condition of closure in one form or another, although in some cases the closure is disguised. The logical conclusion to be reached from this is that gas and oil must seek the highest position they can occupy. It would seem obvious that gravitational separation would be a simple explanation of this condition, but we have noted already that this explanation cannot be accepted. It will therefore be necessary to consider the behaviour of oil in flowing conditions to provide an alternative explanation, and in the following sections the author proposes to consider the effects of texture and the direction of flow on the segregation of oil from water.

Effect of Rock Texture.

It is a well-known principle of oil accumulation that the coarser portions of a reservoir rock tend to contain the oil, whereas the finer portions are more commonly saturated with water. This cannot be a question of porosity in the case of sandy reservoirs, for a fine sand is just as porous as a coarse sand, sometimes even more so. Nor can it be due to selective surface adsorption, for the sands are composed of the same material. Surface tension also cannot be the driving force, for experiments show

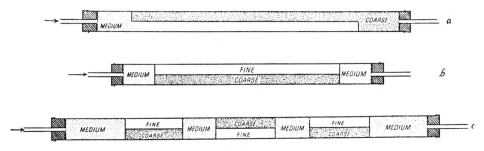

Fig. 1.

that oil globules in water-wet sands tend to resist movement by virtue of surface tension.

If we adopt the principle of selective filtration [1] whereby oil can pass into but not out of a coarse sand which is water-wet, we have a fundamental reason why the coarse streaks should be oil-bearing. One could conceive a set of fluid currents passing along sands of different texture and drifting oil in small quantities with the water. There could be no hindrance to oil passing into the coarse sand, but once it was there, it could not readily pass out again, and so the coarse sands would act as adsorbers of oil from the flowing stream of water and oil. Such was the author's first conception of the process of oil segregation in coarse sand streaks.

In the course of a set of experiments to illustrate this segregation it was at once apparent that this was not the complete explanation. There was clearly a process of path selection in all such movement, whereby the oil, when given the option of a passage through sands of different coarseness, invariably chose the coarser of the two to the complete exclusion of the finer, so that the oil and some of the water passed along the fine sand, while the fine sand contained only flowing water.

In the experiment referred to above, sand was sedimented into a water-filled tube according to the pattern shown in Fig. 1a. The sands chosen were each of one grade, and care was taken in the filling to prevent any streakiness in the texture such as is prone to arise in sedimentation. The two sands were deposited simultaneously, using a moving celluloid strip to guide the deposition, and compaction was ensured by continuous gentle vibration. Gauzes were placed at the ends of the sand column to prevent sand from entering the inlet and outlet tubes.

The tube was placed horizontally and a slow current of water containing about 10–20 per cent. of oil was passed in.

The oil distributed itself fairly uniformly across the section of the tube in the medium sand until it approached the position where the coarse sand had been inserted on half the section. There were now two paths offered to the oil and water, and the oil front immediately began to move more rapidly in the coarse than in the medium sand. This process of selective movement rapidly developed into a complete cessation of oil movement in the medium sand, all the oil passing along the coarse sand streak. When the oil reached a position opposite the end of the medium sand, it rapidly spread to cover the whole of the cross section of the tube, the end of the medium-sand section causing no shadow.

In the first experiment the coarse sand lay above the medium. In order to demonstrate that the result was not due to buoyancy, the experiment was repeated with the position of the sands reversed. The results obtained were identical; the oil chose the path along the lower coarser sand and completely abandoned the upper finer sand after it had advanced less than 1 cm. in the latter. In both experiments movement of water continued in the fine sand throughout the experiment, whereas the fluids moving in the coarse section were a mixture of oil and water.

Experiments were repeated with different arrangements of sand bodies which are illustrated in Figs. 1b and 1c.

In the experiment illustrated by Fig. 1b three sands were used. That of the intermediate grade was placed at the two ends, and the coarse and fine were laid in two contiguous sheets in the middle. The experiment in this case brought into play the function of the coarse–fine interface of the medium to fine sand. The result was that no oil whatever penetrated into the fine sand, but that it all chose the path of the coarse sand. When the oil had reached the end of the coarse sand it was held up at the coarse–medium interface, and the coarse sand was progressively enriched backwards by the displacement of water out of the coarse sand. The water escaped partly at the end of the coarse sand, but also at the junction of the coarse to fine sand along the middle section of the tube.

In the last experiment of the series the sand column was broken up as shown in Fig. 1c. The resulting pattern produced a series of coarse and fine lenses separated by medium sands. The lenses were arranged to alternate in position, and the resulting oil flow and oil enrichment produced the result depicted in the diagram. In each case where the oil reached the end of the coarse lens a temporary halt occurred in its frontal flow, and the coarse sand lens became richer in oil. This process continued until the coarse lens contained about 70–80 per cent. of its pore space filled with oil. The oil then broke across the coarse–medium interface. In no case did oil penetrate the fine sands, although water movement continued in them throughout the experiment.

The previous experiment was repeated with the fluids reversed. The sand was carefully dried and then sedimented into the tube in oil. A fluid stream of oil containing 15 to 20% of water was passed through the tube. It was found that the water in this case chose the path of the coarser sand and became trapped in the coarse lenses. Thus it is clear that neither viscosity nor the absolute surface tension of the two fluids could be the determining factor in the segregation of the two fluids.

The experiments were carried out with distilled water, and oil of a viscosity of 11·5 centipoises at 21° C. The type of oil, however, varied considerably in the different experiments. Generally only moderate velocities were used; if the velocity was greatly increased by the use of high-pressure gradients, it was noted that the clear-cut concentrations of the oil in the coarse chambers did not occur, although the travel path still followed the coarse sands.

In order to study the cause of this path selection and the restriction of the oil current to the coarser sand, it was decided to carry out an experiment in which the pressure gradients of the respective water and oil–water columns could be separately measured. Experiment 1b was repeated using, however, a tube with manometer connections arranged as in Fig. 2, and three sands were introduced as shown in the diagram. Gauze caps were placed at the entry to the manometers to prevent sand flowing into them, and the tubes were kept as small as possible in order to reduce the necessary flow of liquid to cause the registration of pressure changes within the manometers. The manometer leads extended horizontally from the main tube, and the plane of partition between the coarse and medium sands was vertical.

As usual, the tube was filled by sedimentation, so that originally all the sands were water saturated. A current of water with approximately 20 per cent. of oil was slowly passed into the tube, and the pressures in the manometers were noted as the oil column moved forward to the end of the tube.

The observations are plotted in Fig. 2 in terms of the excess pressure of each manometer above the reading of the outlet manometer.

The following points should be noticed :

(a) The rapid rise in pressure occurring in manometers 2, 3, 5, 7, and 9 when the oil reached the position opposite the respective manometer.

(b) The pressure increases of manometers 2 and 9 were similar, as were also those in manometers 3, 5 and 7. The rise in the former was greater than in the latter group.

(c) There was no corresponding rapid increase in manometers 4, 6 and 8 when 3, 5 and 7 rose respectively. The rise in 4, 6 and 8 was more gradual, taking place after the corresponding rise in the opposite capillary.

The experiment suggested that there must be a forefront pressure differential when an oil–water column advanced into a water sand, and that this was an inverse function of the coarseness. This threw a new light on the behaviour of oil in passing through sands of different texture, and it gave a simple explanation of the restriction of oil-flow to the coarse sand.

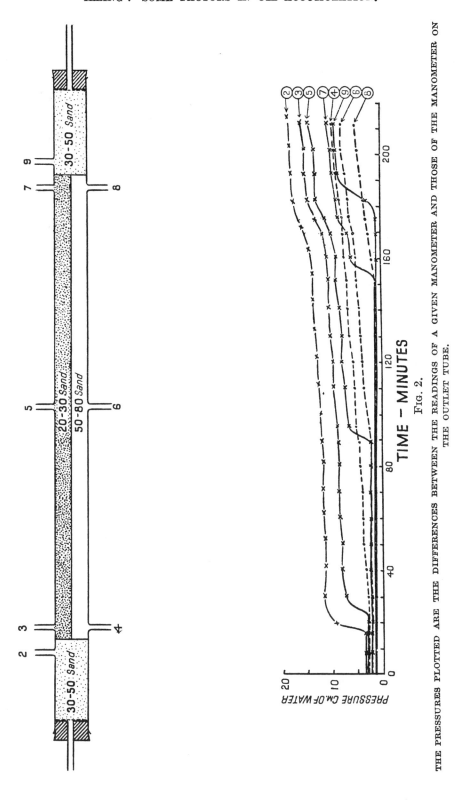

Fig. 2.

THE PRESSURES PLOTTED ARE THE DIFFERENCES BETWEEN THE READINGS OF A GIVEN MANOMETER AND THOSE OF THE MANOMETER ON THE OUTLET TUBE.

To this pressure drop is given the term forefront pressure, and it is clearly a measure of the excess pressure within the oil patches in the advancing fluid measuring the pressure required to continuously deform the fronts of the advancing oil lobes as they penetrate the capillaries. Searching the literature, it was noted that what appeared to be a similar phenomenon had been described by Bartell in 1928, and had been termed by him "Displacement Pressure."[5] In Bartell's discussion of the phenomenon he ascribed it partly to adsorption or to the actual wetting of the mineral matter by the displacing fluid. In the cases which we were examining no such adsorption was taking place, and it was decided to retain for the time being the term forefront pressure, particularly as it was practically self-explanatory. Should, however, it be found later that the two phenomena are identical, Bartell's term would have priority.

Experiments were carried out to study the relations of this forefront pressure differential to the grade of sand and other possible factors (Fig. 3).

The tubes were filled with sand of a uniform grade and a series of manometers spaced along the length of the tube were used to study the pressure conditions in the oil-water column as it travelled past them. The ratio of oil in the inflowing current could be altered at will, and the whole throughput was controlled to maintain a uniform rate of flow of total fluids throughout the experiment.

The pressure rise was shown in each group of manometers, and was found to be the same in all cases, irrespective of the proportional feed of oil to water in the inlet tube. The rate of advance of the oil was, however, naturally dependent on the richness of the inflowing fluid in oil, as well as on the velocity of total fluids.

Experiments were repeated with different rates of flow, and provided that these were not abnormally rapid, the forefront pressures were found to be independent of the general pressure gradient within the sand. At high velocities the forefront pressure was difficult to establish. The forefront pressure was also independent of the direction of flow, whether upwards, downwards or horizontal.

The forefront pressure appears to vary inversely as the diameter of the mean grade size. It is a little difficult to obtain the exact mean size of a sand mass of differing grade composition, but in this case the sands used were of the same origin, and they had been carefully screened to grade dimensions in which the upper and lower limit of the grades were approximately in the ratio of 1·6 to 1 in each sand. The size distribution was therefore likely to be uniform in each case, and errors on this account were reduced to unimportant dimensions. A slight but uniform correction was necessary in all manometer readings on account of capillary effects in the manometers.

Fig. 4 shows the experimental results against a computed curve based on the assumption that the forefront pressure is inversely proportional to the mean grade size. The second curve was obtained by using the results for the grade 30–50 I.M.M. as the basis of computation. The agreement between the observed and computed curves appears sufficiently close to justify the assumption that within these size limits the forefront pressures are inversely proportional to the mean grade size.

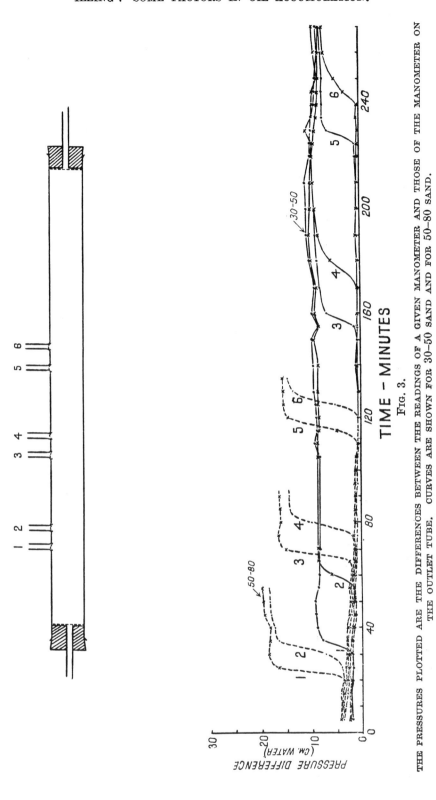

FIG. 3.

THE PRESSURES PLOTTED ARE THE DIFFERENCES BETWEEN THE READINGS OF A GIVEN MANOMETER AND THOSE OF THE MANOMETER ON THE OUTLET TUBE. CURVES ARE SHOWN FOR 30–50 SAND AND FOR 50–80 SAND.

If we assume the validity of this law for silts and clays of finer dimensions, it would appear that the forefront pressure for a silt of 0·1 mm. diameter is 29 cm., and for a particle of 0·01 mm. diameter equivalent to a fine silt the forefront pressure should be about 290 cm. These pressures are not large, but it is probable that the pressure gradients in nature are quite

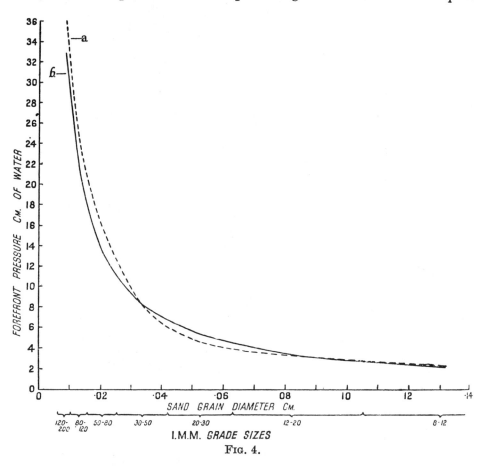

Fig. 4.

a, CURVE OF OBSERVED FOREFRONT PRESSURES, WITH MANOMETER CORRECTION APPLIED USING A VALUE OF 26 DYNES/CM. FOR THE OIL-WATER INTERFACIAL TENSION.

b, CURVE OF FOREFRONT PRESSURES CALCULATED ON THE ASSUMPTION THAT THE FOREFRONT PRESSURE IS INVERSELY PROPORTIONAL TO THE MEAN GRAIN SIZE, AND USING THE EXPERIMENTAL VALUE FOR 30-50 SAND AS BASIS.

small, and such factors as these would be more than adequate to guide oil accumulation.

The existence of this forefront pressure suggested an obvious relation between the selection of the path by the oil and the retention of the oil at a coarse-fine interface. The oil passed into and along the particular sand which allowed it to advance with the least forefront pressure, but it was held up when it reached a fine sand until a sufficient pressure was built up in the oil to reach the necessary forefront pressure of the fine sand. On this assumption the filtration pressure—*i.e.*, the pressure required to force it across the interface—would be equivalent to the difference between the two respective forefront pressures within the two sands.

In order to study these relations, the following experiments were carried out.

A short tube with two manometers was filled in the normal way with coarse and fine sand in two separate segments. Oil was allowed to enter the coarse sand and the forefront pressure noted in the first manometer. The oil was allowed to continue to advance slowly and the pressure build-up was noted in the first manometer immediately preceding the bursting of the coarse–fine interface. The experiment was continued to give the forefront pressure in the fine sand. The filtration pressure was noted to be the difference between the two forefront pressures.

The results of these experiments appear to confirm the relations between the forefront pressure and the filtration pressure, and the following general conclusions may be reached :

(*a*) A definite minimum pressure, the forefront pressure, is required to force oil into and along any water-saturated sand.

(*b*) The pressure is an inverse function of the radius of the sand particles, and for this reason oil will always select the path through the coarse sands.

(*c*) The pressure is required to maintain sufficient pressure within the oil globules to deform their frontal lobes.

(*d*) The process of filtration of oil at the coarse–fine interface is really a question of the need for a build-up of pressure to attain the necessary forefront pressure of the finer sand.

(*e*) This build-up of pressure can be temporarily avoided by the by-passing of water into the finer sand, leading to the enrichment of the oil body. This will continue until much of the water has been driven out, and the interface will then be broken down.

(*f*) Even after the breaking down of the interface the coarser sands will still be richer in oil than the rest unless continuously flushed with water, and the most difficult zone to flush with water is the coarse–fine interface. The fact has an important bearing on the water flushing of sands, and will mean that where there are numerous textural changes in the reservoir rock, it will be increasingly difficult to obtain a high percentage of oil extraction. All such coarse–fine interfaces tend **to** lock away oil in the strata.

The experiments have been carried out at normal atmospheric pressures, and it is not suggested that the same quantitative figures would apply at the pressures and temperatures of oilfield strata. However, it is the principle of the process which is important, and from a qualitative point of view there is no reason to doubt that the same processes will occur at high pressures as at low pressures, so long as the oil and water remain liquids.

The above considerations appear to give a satisfactory explanation of the phenomena associated with the textural accumulation of oil. They will need some modification, in so far as the normal oil reservoir often contains free gas as well as oil and water. This will give an additional complication into which we have not yet fully entered, but the principles ought to remain the same.

Buoyancy.

It would seem at first sight that there need be no difficulty about the upward movement of gas and oil in a reservoir rock. In some cases, where the rocks are fissured or where the cavities are large, there is no doubt that gravitational separation occurs, and the three separate media seek their own hydrostatic levels. The capillaries of the normal sand are, however, too small for such separation, and the upward movement of the gas and oil must be explained in other ways.

In a previous contribution to the *Journal* the author described an experiment, in which, under flowing conditions, the action of buoyancy was clearly indicated.[1] Two separate streams of water and kerosine were partly able to cross each other's path in a tube of sand, the water travelling downwards, whilst the kerosine passed upwards. The separation was not complete, but it was quite definite. The results are also in agreement with the experimental work of many other workers, and there can be no doubt that in flowing currents oil and gas will readily move upwards in a sand body, whereas under static conditions the same oil and gas will not rise.

The behaviour of oil and gas in currents of water within a permeable rock is by no means well understood, and even a cursory review of the literature will reveal many contradictory statements on the subject. Some deny that there can be any downward movement of oil and gas, others argue that there must be some limiting angle of dip below which oil will not migrate up dip; whilst still others deny that oil can be flushed along at all by migrating water currents with the velocities which exist under natural conditions. This confusion of thought is largely because most of the experimental work has been qualitative rather than quantitative, and has therefore failed to reveal the many factors which determine the behaviour of oil and gas. It is, for instance, quite possible to perform experiments which appear to support each misconception referred to above, but only because some factor, such as oil concentration, is strictly limited. The fact is that oil and gas can be carried upwards, downwards or in any other direction, but not with the same facility. It is also possible to have a fast current unable to move oil from a sand at one concentration, and a comparatively slow current able to move oil along in this same sand at another concentration. The factors which determine the behaviour of oil under such conditions include the following: Oil-water concentration, forefront pressure, permeability, velocity, direction of flow, and the attitude of the boundary planes of the porous media and the nature of the contiguous material.

Our experiments on the effects of these factors are still incomplete, but sufficient data have been collected to enable us to reach some conclusions on the question of the buoyancy of oil in currents of water.

To realize adequately the problems in oil-water movement it is, in the first place, necessary to visualize the conditions under which the oil exists in the water-wet sand. The water occurs as a film over all the surface of the sand-grains, and also fills some of the pores. The oil forms irregular bodies which occupy one or more large pore chambers and are completely surrounded by water. Each oil patch is bounded by an oil-water skin of variable shape dependent on the walls of the pores. Its external form under static conditions is therefore dependent upon the constricting pore walls, the oil-water interfacial tension, and the oil's buoyancy. It therefore adopts a form in equilibrium with these three conditions. Should, however, movement be induced in the surrounding water, the pressure gradient set up causes a modification in the form of the oil patch resulting in the reduction of the curvature of the oil-water interfaces on the high-pressure side and increased curvature on the low-pressure side. The extent of this effect will depend on the pore size and the pressure gradient in the water. Since the latter is an inverse function of the permeability and a direct function of the oil-water concentration, it is clear that these two will have a bearing on oil movement.

Should the deformation of the oil cause one of the convex lobes on the low-pressure side to pass through a pore restriction, an immediate change in shape and position of the oil occurs, involving the occupation of a new pore and the retraction from an old one. The movement is spasmodic, not continuous, and is clearly associated with the changing pressures in the surrounding water. The oil moves forward rather like a sinuously-shaped amoeba sending out tongues or lobes into forward chambers and retreating from those behind. In doing so the buoyancy of the oil has full scope to play its part, and it does so in the choice of those lobes which will move forward, the upward ones having the preference if other things are equal. If, however, the pressure gradient be a downward one, or if one of the lower pores be larger than the upper ones, the lower oil lobe will be the one to advance. On this principle the oil movement will always tend to be in the direction of the local pressure gradient, but with a bias towards upward movement in relation to the water.

It also follows that oil movement will depend on the pressure gradient, the permeability (coarseness) and the oil concentration. The latter is important, in that it determines the local pressure gradient in the water. The higher the oil concentration the more restricted is the water passage, and the higher therefore becomes the drag on the oil causing it ultimately to move.

In general, the lower the velocity the more is the oil by-passed by the water, resulting ultimately in a higher oil concentration in the material remaining behind. This concentration increases until the local pressure gradient becomes sufficient to force the oil along, the oil-water concentration remaining constant for this travelling zone provided the pressure gradient remains the same. Thus if oil in small quantities is continuously fed into a slow water stream within a sand, it will produce an oil-water zone within the sand which is more concentrated than the inflowing stream.

The principle that oil flushing depends on concentration as well as on pressure gradient implies that at low concentrations of oil in water currents the oil will remain stationary and be by-passed by the water. This feature is common to all sands, and it must automatically limit the powers of transport of a definite quantity of oil by water currents. The oil, if initially concentrated, will merely be distributed along the line of flow until the concentration is sufficiently diffuse to allow continuous by-passing. This occurs under laboratory conditions when the concentration of oil is reduced to about 10-15 percent of the total pore space.

The above consideration has two important corollaries. First, to produce continuous forward movement of oil there must be a continuous feed of new oil to the flowing stream, and secondly, if this feed of new oil ceases and forward flushing continues, the oil as a whole will be dispersed rather than concentrated. It is, however, true that a portion of it may be locally concentrated by some accident of texture.

Considerations such as these must be given due weight before we allow unlimited scope to the exponents of widespread regional lateral migration, particularly when we realize that in nature changes in texture are numerous and such textural changes are very important in stopping further migration.

There is no lower limit of velocity at which oil movement will not take place provided the oil concentration is sufficient. In fact in an oil-bearing sand which may still contain some 20 per cent. of water in its fluid content there may be no movement of the water at all, but only of the oil.

Turning now to the question of the direction of flow within a sand body, it is obvious that downward movement of oil in a current of water will always be more difficult than upward movement, because the latter is helped by the buoyancy of the oil, whereas in downward flushing the drag of the water has to overcome the buoyancy factor. If now we adopt the principle set forth above, that for a particular grade of sand and a particular concentration there will be a definite velocity at which oil may be moved, it may be deduced that for upward flow the velocity will be lower than for horizontal flow, and for downward flow it will be higher. Alternatively, this will mean that for upward flushing the oil concentration will not be so high as for downward flushing.

This conclusion that for a particular velocity the critical oil concentration for upward flushing is less than for downward flushing, means that whenever an oil–water stream has the option of two courses the oil will invariably choose the upper to the exclusion of the lower. The only condition in which oil would journey in both directions would be when the oil concentration was much higher than the critical oil concentration for that particular pressure gradient.

Furthermore, it may be deduced that where the oil and water are given a certain liberty of choice of path due to variations in the inclination of the boundary surfaces of the sand, the oil will invariably choose the steepest path, provided the texture of the sand remains constant.

A series of experiments were carried out to test the validity of these deductions. The principle adopted in these experiments was to give the passing fluids a choice of path and to examine their adjustment of oil to water which took place under those conditions.

In the initial experiments oil and water were passed into the mid-point of a vertical tube filled with water-saturated sand in a thoroughly compacted condition. The oil and water inflow could be independently controlled, so that the rate of inflow and the proportion of oil to water could be varied in the experiments. The oil, once in the vertical tube, was free to select its path either upwards or downwards, and the factors which governed this selection could be studied. The liquids were free to escape from both the upper and lower end of the tube, and the rate of flow could be adjusted in each case. In the later experiments the single vertical tube with a narrow inlet tube at the side was replaced by a T-tube in which each limb was of the same diameter. The stem of the T was used as the inlet tube, and constituted an improvement on the original narrow inlet tube, in that it was wider and less likely to involve accidental bias in the initial direction of flow of the oil when entering the vertical tube. There was, however, no evidence throughout the experiments that any of the results could have been affected by such accidental bias.

In the first experiment the tube was filled with 12–20 I.M.M. sand and an oil-water current introduced into the side tube. The rate of flow and proportion of oil in the inflowing tube were stabilized as far as possible, and the delivery of both the upper and lower outlet was maintained at the same value. It was noted that after a slight movement of 2 cm. downwards all further downward movement of oil ceased at normal velocities (Fig. 5). On the other hand, the upward movement of the oil was continuous. The experiment was repeated using different sands, altered rates of flow and with different proportions of oil in the inflowing stream.

Comparison of results obtained as shown in Fig. 6 indicate that except at high oil concentrations all the oil normally moves upward and most of the water moves downward. The latter result is due largely to the reduced permeability of the oil-filled section of the tube. Downward flow of oil can generally be achieved by increasing the velocity or by increasing the percentage of oil in the inflowing stream. If the results for sands of different grade size are compared, it is noticeable that it is more difficult to cause downward flow in the coarse than in the fine sands.

In the later experiments with the wide T-tube the exit tubes from both the bottom and top limbs were adjusted initially to give the same rate of flow when water alone was passed into the sand. After this preliminary setting no further adjustments were made in the exit tubes, and the flow along each limb was allowed to adjust itself to the conditions within the tube. Fig. 6 gives the results of this experiment, and indicates that very little downward movement of oil could be induced in these circumstances. It is noticeable that as the oil increases in the upper tube a greater amount of water is forced to flow in a downward direction and less water in an upward direction. In some of these experiments the only way in which oil could be forced downwards was by closing the upper tube entirely. At higher concentrations of oil in the inflow fluid it was more easy to induce downward flow of oil.

The experiments were repeated with the flow tube at an angle of 5° from the horizontal, so that the alternative paths offered to the oil-and-water mixture were 5° downwards and 5° upwards. The flow of the two streams was not regulated after the preliminary setting to give an equal flow of water along each limb. Figs. 7 and 8 show the results and indicate the movement of oil along each limb. The importance of the buoyancy factor is again shown in differentiating the flowing current into two streams, of which the upper contains the main quantity of oil and the lower the main water flow.

One further experiment is an interesting example of the textural factor being utilized in opposition to buoyancy. In one of the first experiments (1a) it was noted that oil would move downwards rather than upwards if it could take advantage of a sand of lower forefront pressure. To explore this condition an experiment was carried out giving alternative paths of

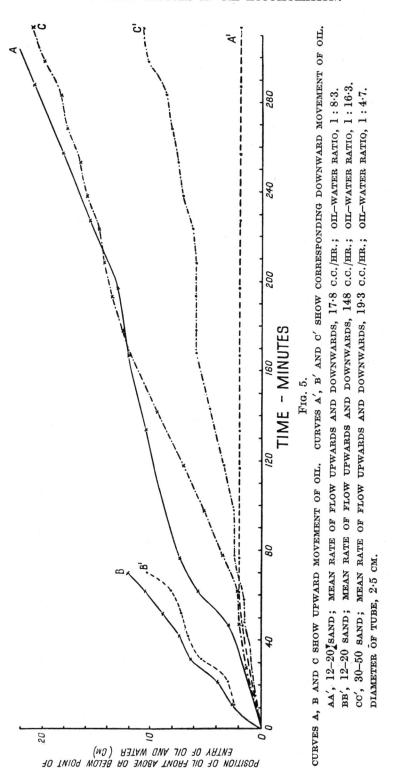

Fig. 5.

CURVES A, B AND C SHOW UPWARD MOVEMENT OF OIL. CURVES A', B' AND C' SHOW CORRESPONDING DOWNWARD MOVEMENT OF OIL.
AA', 12–20 SAND; MEAN RATE OF FLOW UPWARDS AND DOWNWARDS, 17·8 C.C./HR.; OIL–WATER RATIO, 1 : 8·3.
BB', 12–20 SAND; MEAN RATE OF FLOW UPWARDS AND DOWNWARDS, 148 C.C./HR.; OIL–WATER RATIO, 1 : 16·3.
CC', 30–50 SAND; MEAN RATE OF FLOW UPWARDS AND DOWNWARDS, 19·3 C.C./HR.; OIL–WATER RATIO, 1 : 4·7.
DIAMETER OF TUBE, 2·5 CM.

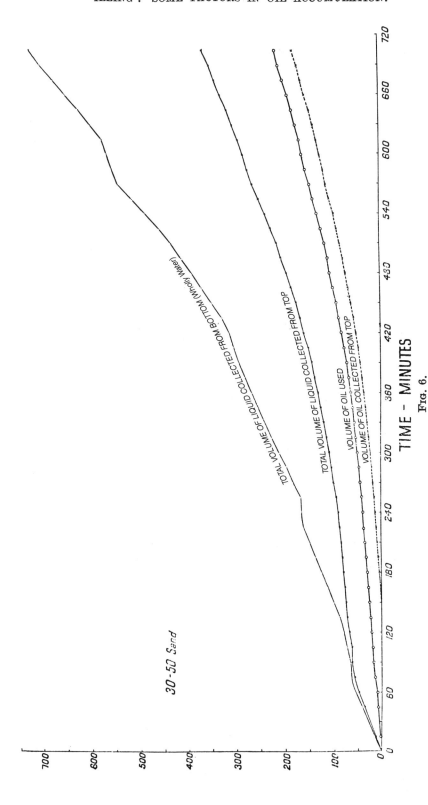

Fig. 6.

DISTRIBUTION OF FLOW OF OIL AND WATER IN SHORT VERTICAL SAND-FILLED TUBE, THE INLET TUBE BEING OF THE SAME DIAMETER AS THE VERTICAL PART. THE FLOWS WERE NOT CONTROLLED AFTER THE INITIAL ADJUSTMENT TO EQUALITY WHEN WATER ONLY WAS BEING INTRODUCED.

flow in a T-tube the lower limb of the tube was filled with 20–30 I.M.M. sand, whilst the upper limb and the inlet tube were filled with 30–50 I.M.M. sand. A current of water containing about 20 per cent. of oil was passed in, and when the oil reached the fork of the T-tube it immediately entered the coarser sand and proceeded downwards against buoyancy. The experiment was continued until flow was stabilized in each limb. A small amount (4 per cent.) of the oil passed upwards into the finer tube, whilst 96 per cent. of the oil took the downward path of the coarser sand.

In order to study the behaviour of oil in a "dead" zone close to a moving column of oil and water, an experiment was carried out in a four-armed tube in the form of a cross. Two of the arms were sealed, and the main limb was placed horizontally with the sealed arms upward and downward. Oil and water were passed along the main stem. It was found that oil tended to segregate out of the main stream in the closed limbs, particularly in

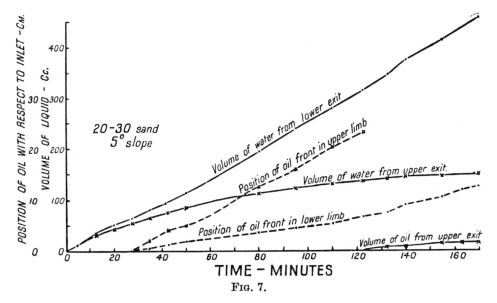

Fig. 7.

the upper one. On revolving the tube in order to reverse the limbs, the main oil concentration was transferred to the new upper limb, which had previously been the lower one.

The following deductions may be made from these experiments:

(a) Where a current of oil and water is given a choice of alternative paths, one upward and the other downward, there is a differentiation of the two streams. Around the critical concentration of oil—i.e., the concentration at which oil flushing is just possible for the particular velocity—all the oil chooses the upward journey. At higher oil concentrations most of the oil continues to move upwards, but a small amount will move downwards.

(b) The differentiation is dependent on the velocity, coarseness, difference in densities of the oil and water, and the concentration of the two fluids. Differentiation is helped by low velocities, low oil concentration, low oil density and coarseness of sand.

(c) Where the oil-water stream is presented with paths of varying gradients, the oil invariably chooses the most upward path possible. This choice is again helped by the same factors as in (b).

(d) Where a dead zone, or zone of low-pressure gradients exists, an oil-water stream tends to feed this zone with oil, and concentration of

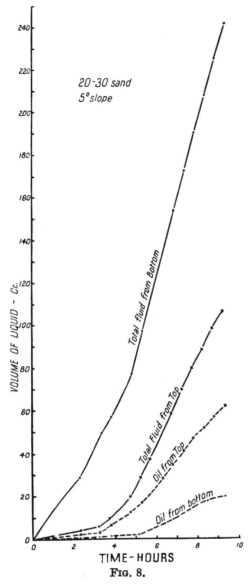

Fig. 8.

the latter takes place. This is particularly true if the " dead " zone occurs in a high rather than a low position.

These general conclusions, based on experimental results, are also in agreement with theoretical deductions, for if it be agreed that oil flows upwards in a water current more easily than downwards, and, secondly, that by-passing is a function of grade, velocity and oil-water concentration, then the differentiation of oil from water in flowing conditions which can be achieved only by by-passing, must also depend on concentration, velocity and the attitude of the sand, as well as on the density difference between the oil and the water.

It appears therefore that the first stage in oil accumulation in a sand body is achieved by the gradual by-passing of oil by the current of water until the oil concentration has been sufficiently enriched to attain a concentration wherein it can move forward without any change in this concentration. This is termed its critical concentration. Such a mass of oil and water can move forward continuously so long as it is fed with oil and water, and will retain its concentration. The degree of concentration will depend on the forefront pressure and the richness of the inflowing oil–water mixture and the velocity at which the fluid is flowing.

The second stage in enrichment is normally produced by a change in texture, and results from the filtration effect at a coarse–fine interface. This can result in an enrichment to a sand containing about 80 per cent. of its pore space filled with oil and the rest filled with water. This is probably as rich a concentration as ever occurs in oil-bearing rocks, save those of very coarse texture or under fissure conditions.

An alternative second stage in enrichment may be produced by changes in the direction of flow of the oil and water. Where the flow is upwards, the critical oil concentration for flow is less, and the oil–water stream becomes less concentrated, but, on the other hand, it moves much faster. Where the flow is downwards, the critical oil concentration for flow is higher, and so the oil stream must become denser to achieve any movement. On the other hand, such movement is slower and much more difficult. Hence when a flowing oil–water stream increases its upward gradient and then diminishes it, there will be an enrichment of oil at the position of diminishing gradient. A similar enrichment is possible where changes occur not in the inclination of the flow, but in the amount of the pressure gradient.

Some Examples of Structural Accumulation Based on Buoyancy.

It will be interesting to consider the application of some of these principles to actual oil-field accumulations.

So far as the textural feature is concerned, little need be said to emphasize its importance, for it is well known in all oil-fields where the reservoir rocks vary in texture, that in general the coarser streaks are those which are oil bearing. Some of our major oil-pools can be described as nothing more than inclined lenses of porous reservoir rock in which the oil and gas occupy the upper part and water fills the rest of the body. Here the accumulation appears to have been brought about by migration into the sand lens and filtration at the coarse–fine interface at its upper end. East Texas and Midway Sunset fields are examples of this type.

Within many fields the oil–water distribution may be found to be extraordinarily complex when the sand bodies are composite. Whilst faulting and fissuring undoubtedly play a part in such complexity, the variation in grade composition is nevertheless a main contributory factor.

It is, however, to some of the structural features that the author wishes to direct attention. The anticline will immediately be called to mind as the striking example of the buoyancy principle. Accumulation of oil in anticlines is, however, a very complex subject. To treat it adequately would need a paper of much greater scope than is intended. The author proposes to consider only a few types of structures of simple form which are recognized as oil-traps, and to discuss the features which in his opinion tend to favour the accumulation of oil. The three examples have been chosen to illustrate the principles set out in the earlier part of the paper.

Structural Terraces.

Perhaps the least definite of all structures which have been claimed as favourable for oil accumulation are the structural terraces. Their value was first noted by Edward Orton, who spoke of them as arrested anticlines. They may be described as local flattenings of the dip on a general monocline, and as they do not have any dip reversal, there is no definite closure to explain the oil concentration. This lack of closure has led many geologists to cast doubt on the validity of the structural terrace as a cause of oil accumulation,[6] and indeed it cannot be claimed that the recent developments in oil discovery have strengthened the claims of the terrace to recognition. Yet we cannot dismiss the terrace and its first cousin the monoclinal nose as of no importance, and their value requires explanation.

The normal explanation of the terrace is that the flattening of the bed provides an area on which the dip is too low for upward oil migration. Hence the oil being carried up the monoclinal dip comes to rest on the flattening, and remains there. This explanation appears to infer that oil cannot be carried along a flat bed, which we know is untrue. The author's chief criticisms, however, of the orthodox explanation of the terrace is that it falls far short of the full story. His conception of accumulation on the structural terrace and monoclinal nose is likewise based on the supposition that they are due to the migration of oil and water up dip. The most important feature is, however, not necessarily the flattening of the dip, but rather the zone of abnormally high dip which must always exist just below a terrace or monoclinal nose. This zone of excess dip—*i.e.*, a dip which is above that of the general average dip of the monocline—is the counterpart of the flattening, and brings the general structure below the terrace into line with the general monocline (Fig. 9a).

The oil–water current which approaches this zone of increased dip is partly differentiated as explained in the previous section. The oil tends to be focused towards the terrace, choosing the path of the maximum gradient. This automatically means that the oil–water mixture which arrives at the edge of the terrace is enriched in oil.

The fluids which flow onwards over the terrace cannot carry out this proportion of oil, and water is bled out in preference to oil, leading to a second cause of enrichment of oil on the terrace. Thus oil is being brought in at abnormally high rates to enrich the concentration, and is being carried out at abnormally low rates. The result is a continuous enrichment, and the creation of a mass of oil, which owing to its viscosity aids as a general block to the migrating fluids. Such a block or dead zone on the

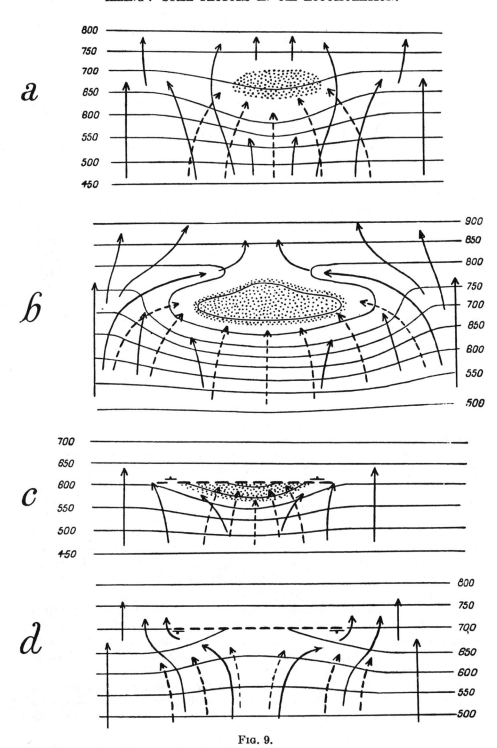

Fig. 9.

FULL ARROWS, CURRENTS CONSISTING MAINLY OF WATER. BROKEN ARROWS, OIL-ENRICHED WATER CURRENTS. DOTTED AREAS SHOW OIL ACCUMULATION.

flattening of structural terrace acts as a back eddy in which further oil accumulation can take place.

Monoclinal Domes.

The simple explanation of the monoclinal dome as a trap for up-dip migration has perhaps required modification in some instances since the discovery of the relations of some of these structures to buried ridges. More particularly is this true of those cases where vertical migration may have played an important part in the oil accumulation. On the other hand, there seems good evidence in most cases to favour at least some amount of regional migration, and in such cases the mode of accumulation would be on the same lines as that of the terrace. Again attention should be directed in the first place to the steepening below the dome and the way the regional stratum contours must bend around the structure focusing the oil enriched stream towards the dome (Fig. 9b). Again, the outflowing current from the dome, being downward, would be deprived of its oil. This would result in the enrichment of the dome in oil and gas at the expense of the migrating fluids. The dome would gradually become a zone of the more viscous oil—in other words, a dead zone in the general pressure gradient, and would then be fed with more oil on its lower side by up-dip migration.

Fault Structures.

Perhaps one of the most important applications of up-dip migration and the trapping of oil by the alteration of the dip of the beds is given in the case of fault structures. It is generally agreed that strike faults are the important trap structures, and that their effect is dependent on closure. On the other hand, it is not sufficiently emphasized that the majority of such fault-traps which have been responsible for oil accumulation are those in which the downthrow is on the up-dip side. Thus along the Balcones fault system it is the supplementary faults with downthrows on the north-west side, not the main fault with the throw in the dip direction, which are the oil accumulators. Even the suggestion that the supplementary faults have screened the oil from the Balcones fault is not satisfactory, for there are gaps in the secondary fault system, and quite considerable feeding-grounds for the main fault.

If, however, we examine the structure contour system, which must in principle occur in association with these two types of strike faults (Figs. 9c and 9d), we note that in the case of the faults which are upthrown on the down-dip side the regional dips near the fault will focus oil towards the fault plane, whereas in the case of the faults with the downthrow in the dip direction the regional dip will have no such focusing tendency. Thus, in spite of the fact that both faults may produce the same actual blockage to fluid migration as a whole, the fluid which accumulates in the case of the favourable fault structure is more highly oil enriched than in the case of the other fault.

There exists, of course, in the case of faults other complications, such as associated warping of the strata, dip faults and zones of vertical migration, and the author does not wish to suggest that the preceding analysis is more than a partial one, but it is useful in its suggestion that the structure approaching a fault may be just as important as the fault itself.

The author wishes to emphasize that these three examples are put forward merely to illustrate the application of the principle that oil always chooses the steepest angle of flow up-dip that it can attain. The treatment is meant to be suggestive and no attempt is made to make it complete.

Conclusions.

In conclusion it will perhaps clarify the position if the broad conclusions of this paper are divided into two sections : (a) the physical principles involved and, (b) their implications with regard to the process of oil accumulation.

(a) *The Physical Principles.*

(1) A flowing mixture of oil and water in a sand body consists of separate lobes of oil completely surrounded by the water medium. The movement of the two liquids is largely independent. That of the water is continuous, whilst that of the individual oil lobes is spasmodic.

(2) A definite forefront pressure is required to force oil into a water-saturated sand. This forefront pressure is greater than the pressure in the contiguous water, the difference being an additive factor due to interfacial tension between the oil and water.

(3) This forefront pressure is inversely proportional to the mean grade size of the sand.

(4) When an oil–water column advances, there appears to be a necessary minimum or critical concentration of oil within the sand before the oil can be moved forward. This critical concentration is dependent on the sand grade and the pressure gradient.

(5) The existence of a critical forefront pressure for each grade of sand involves the consideration that oil will not penetrate from a coarse to a finer sand if both are water-wet, until the pressure in the oil column has been built up to the forefront pressure of the corresponding fine sand. This build-up of pressure corresponds to what the author terms the filtration pressure of the coarse–fine interface.

(6) Oil–water movement in a sand is affected by the forefront pressure, permeability, the oil–water concentration, the pressure gradient and the direction of flow.

(7) The buoyancy factor in oil renders it more easily carried upwards than downwards or horizontally; this results in a predilection for upward flow and causes the oil to separate upwards in a flowing oil–water stream. Under static conditions there would be no separation of the oil and water.

(8) The textural separation of the oil from the water is not due to the fact that water has the higher surface tension and thus goes to the finer sands. If the sands are originally oil-wet, the two liquids can be reversed. The whole process of separation is, however, guided by the forefront pressure conditions, which are the result of the interfacial tension of oil and water.

Application of these Physical Principles to Oil Accumulation.

(1) Oil accumulation is guided very largely by texture and all oil movement tends to be in the coarser rocks.

(2) The oil is trapped, and much of the water is filtered out at the coarse-fine interface—*i.e.*, at the upper surface of the coarse streak. So we have oil sands in clays, oil-bearing coarse sands in water-bearing fine sands.

(3) Whenever the dip of a potential reservoir rock is locally increased oil tends to be focused from the oil–water migrating fluids. Whenever the dip decreases, the flow of oil across such a flattening is hindered. Hence we have the oil accumulation on monoclinal noses, terraces, etc.

(4) A reversal in the regional dip is an excellent trap, because down-dip oil movement is possible only at high velocities or at high concentrations of oil. Since the former are very unlikely in nature, it means that a high oil concentration must be formed on a monoclinal dome before the spilling plane can be approached.

(5) Finally, with reference to fault structures, it is suggested that the reasons why some strike faults form trap structures and others do not may be at least partly explained by the attitude of the contiguous strata. Faults which are upthrown on the down-dip side tend to have a stratum contour system which deflects the oil up-dip towards the fault. The others tend to disperse the oil.

Acknowledgments.

The author would like to express his thanks to Dr. G. D. Hobson, who carried out the experimental work on which this paper is based.

References.

[1] Illing, V. C., "The Migration of Oil and Natural Gas," *J.I.P.T.*, 1933, **19**, p. 229.
[2] Versluys, J., "Compaction an Agent in the Accumulation of Oil at the Anticlines," *Koninklijke Akademie van Wetenschaffen Te Amsterdam, Proc.*, 1930, XXXIII (9), 990.
[3] Howard, W. V., "Accumulation of Oil and Gas in Limestone": Problems of Petroleum Geology, *A.A.P.G.*, 1934, p. 365.
[4] Theil, G. A., "Gas an Important Factor in Oil Accumulation," *Eng. & Min. J.*, 1920, **109**, (15), p. 888.
[5] Bartell, F. E., and Osterhof, H. J., "The Measurement of Adhesion Tension Solid against Liquid," *Colloid Symposium Monograph*, (1928), V, 113.
[6] Wilson, W. B., "Proposed Classification of Oil and Gas Reservoirs": Problems of Petroleum Geology, *A.A.P.G.*, 1934, p. 433.

Stress Fields, A Key to Oil Migration[1]

JEAN du ROUCHET[2]

ABSTRACT

A coherent oil-migration model based on geomechanical considerations includes both the high-molecular kerogen structure and the capillary properties of source rocks. Oil is squeezed from kerogen by compaction following oil generation. This squeezing effect should be created by the differential stress (maximum compressive stress minus least compressive stress) acting on the kerogen which has been chemically broken up by oil formation. In sedimentary bodies whose water is at hydrostatic pressure, the migration of oil seems to involve two processes: (1) lateral transfer, by channeling into the more coarsely microporous layers of the source rocks, from the oil generation site toward the geologic structure or lower pressured zone; and (2) vertical transfer from source rock to reservoir by the opening or reopening of vertical fractures in the few areas, such as structural tops, where the least compressive stress is slightly greater than or equal to the pore pressure, and where the capillary pressure increment ($2\gamma/R$) of oil in the microporosity exceeds the tensile strength of the rock.

In sedimentary bodies whose water is overpressured, the pore pressure should be governed by the least compressive stress and thus migration should begin by oil transfers in a system of small open fractures, and eventually in larger fractures.

The theory demonstrates the impossibility of oil being transferred to the reservoir under true tensile conditions (negative effective compressive stress) and thus explains the large asphaltic veins of southeastern Turkey and the well-known bitumen veins of the Uinta basin.

INTRODUCTION

The migration of hydrocarbons has been a little understood phenomenon, and petroleum geochemists disagree on fundamental points: some acknowledge that migration is accomplished entirely in a hydrocarbon single phase (oil or gas), whereas others feel that part of the transfer occurs in solution with water.

The remarkable chemical analogy between natural petroleums and extracts from their source rock is certainly in favor of oil migration in single phase only and, because the oil in many accumulations is rather immature, it seems necessary to explain migration without referring to pressure increases created by gas formation.

This study gives an explanation of oil migration in a single phase, rendered more consistent by the incorporation of geomechanical data that until now have not been fully exploited. This theory cannot claim to be completely original. Since the early days of oil prospecting, many geologists have believed that oil migrates in a single phase and makes use of fractures to pass from the source rock to the reservoir, or from one reservoir to another.

This concept of migration underlies, for example, several regional studies performed by Landes (1970) and has been brilliantly applied to the genesis of deposits in northern Iraq by Dunnington (1958). These ideas were nevertheless contested by the partisans of oil migration in solution with water during the 1960s and early 1970s and were then defended by geochemists such as Tissot and Pelet (1971), by hydrogeologists such as Dickey (1975), and by exploration geologists such as Hedberg (1964, 1974) and Dow (1974).

The following theory makes direct use of geomechanical ideas first introduced by Hubbert and Willis (1957) and developed by Secor (1965, 1969). Thus the concepts presented here are somewhat similar to the ideas of Schaar (1976) and of Meissner (1978). The concepts also easily fit into the more general work of Gretener (1977) concerning the implications of pore pressure in structural geology.

The treatment is largely theoretical. Most of the "facts" discussed are already well known, such as the existence of dispersed oil in the microporosity of source rocks, the frequent evidence of open fractures in hydrostatic areas, and the often noted relation between pore pressure and fracture pressure (Dickey, 1979, p. 298).

This model strives to understand the entire situation and thus uses concepts originating in petroleum geochemistry, rock mechanics, tectonics, petrophysics, and hydrodynamics. Specialists in these disciplines may be inclined to reproach a lack of precision in the analyses or formulae which directly concern them. The theory is not proposed as a quantitative solution but as an attempt at a global phenomenologic understanding of the problem. Thus, the most simple formulae have been used for convenience and also because of a belief that this is probably the most lucid way of understanding the subject.

This article may also be criticized because it does not consider that loads are transmitted across grain contacts where stress concentrations are produced which are not easily related to the principal stresses supposed to exist in the aggregate (Gallagher et al, 1974). Indeed this

---

© Copyright 1981. The American Association of Petroleum Geologists. All rights reserved.

[1]Manuscript received, January 11, 1980; accepted, June 5, 1980.

[2]Societe Nationale Elf Aquitaine Production, Aquitaine Company of Canada, Calgary, Canada.

The writer thanks the management of SNEA(P) for authorizing the publication of this document.

analysis has been based on the classic "continuum model" which assumes a continuous spatial stress variation, just as if the matter were uniformly distributed throughout the rock. This model is, of course, less realistic than the granular picture but it is simple and efficient for the description of general fracturing. However, we can be certain that stress concentrations must favor compaction of kerogen and microfracturing.

OIL TRANSFER INTO MICROPOROSITY OF SOURCE ROCK

Hypothesis of Kerogen Compaction

Kerogen is a heterogeneous and disordered macromolecular substance. Consideration of this characteristic seems to be an essential condition for the understanding of migration. Any discussion of kerogen structural evolution is rather speculative but Tissot and Welte (1978) best summed up the subject:

Electron diffraction has been used by Oberlin et al. (1974) to show the degree of organization of carbon structures in amorphous kerogen at different scales from 5 to 500 A and their evolution during burial. The method shows the existence of stacks of two to four more or less parallel aromatic sheets in kerogen. The diameter of an aromatic sheet is between 5 and 10 A, corresponding to less than ten condensed aromatic cycles.

As burial, and thus temperature and pressure increase, the structure of the immature kerogen is not any more in equilibrium with the new physicochemical conditions. Rearrangements take place...

Electron microdiffraction has clearly shown that the stacks of aromatic sheets gather to form aggregates or clusters...

Heteroatoms, especially oxygen, are partly removed as volatile products: H_2O, CO_2. The rupture of these bonds liberates smaller structural units made of one or several bound nuclei and aliphatic chains. These fragments are the basic constituents of bitumens.

As temperature continues to increase, the kerogen reaches the stage of catagenesis. *More bonds of various types are broken, like esters and also some carbon-carbon bonds, within the kerogen and within the previously generated bitumen fragments...*

Thus the transformation into hydrocarbons of 10 to 20% of the kerogen, a magnitude generally present in source rocks located in the oil kitchen, should disintegrate the kerogen into a complex of independent molecules. We will assume that, as oil generation progresses, the oil generated passes from a state of molecular dispersion in the kerogen to an organization in thin films between the clusters of aromatic stacks seen in electron microdiffraction. Subjected to a stress field with sufficient differential stress the kerogen in such a state of maturity must be deformed as the relative movement of molecules becomes possible. The smallest and most mobile molecules must be displaced in the direction of the least compressive principal stress.

In a kerogen flake which is large enough in relation to the rock grain, the actual stress system may be that acting on the rock aggregate (du Rouchet, 1978). In a small lump of kerogen squeezed between two mineral grains, the active stress system will be determined by the support between grains (Meissner, 1978).

Let us suppose that the hydrocarbons are squeezed from the kerogen at considerable pressure, on the order of the geostatic load. This pressure may permit the opening of tensile fractures if it is greater than the least compressive principal stress prevailing in the formation (Fig. 1a). It will permit, generally, the injection of oil into the more coarsely microporous zones of the source rock adjacent to the kerogen-bearing zones (Fig. 1b).

The photograph in Figure 1c may represent a natural example of kerogen compaction giving a brittle rupture, but at a shallow depth. It was a migration of premature bitumens in Anatolian Oligocene carbonate rocks. Most numerous small vertical bitumen veins observed in this formation are rooted in kerogen microvarves (Beseme, 1972, p. 896).

In pyrolysis at atmospheric pressure, the oil is released by the kerogen aromatic systems as the kerogen progressively "shrivels up." This process may be considered as a deformation resulting from internal stresses. It is likely that the external stresses of the surrounding rock will also be effective at depth, at least if there is sufficient differential stress.

An example demonstrating pressurization by oil generation without gas generation is the well-known Altamont-Bluebell synclinal accumulation in the Uinta basin (Lucas and Drexler, 1976). If the pressure results from kerogen compaction by the geostatic load, its persistence is only made possible by the low permeability of the source rock.

Summary of Conditions for Development and Maintenance of Open Fractures

In an isotropic solid, the mechanical state at any point is characterized by a spatial distribution of applied stresses in the form of an ellipsoid, whereas in a liquid this distribution is typically spherical (Fig. 2a).

The ellipsoid of stresses is defined by giving the directions of its three orthogonal axes and the corresponding stress values S1, S2, and S3, known as the principal stresses. Assuming that compressive stresses are positive, S1 will be the maximum compressive stress and S3 will be the least compressive stress. Deformations result from a difference in the principal stresses S1 and S3, that is, the differential stress.

In depositional basins, whether tectonically inactive or undergoing extension, the maximum stress S1 will be the geostatic load, the intermediate stress S2 and the minimum stress S3 will be horizontal; Ranalli (1975) wrote: "sedimentary cover rocks show average horizontal stress less than the overburden pressures at depths larger than approximately 500 m."

The least compressive stress S3 (also called the confining stress) can be conveniently expressed as a fraction of the geostatic load S1:

$$S3 = f \cdot S1, \qquad (1)$$

where f can be called the "confining stress coefficient." Even in the absence of any tectonic action the least com-

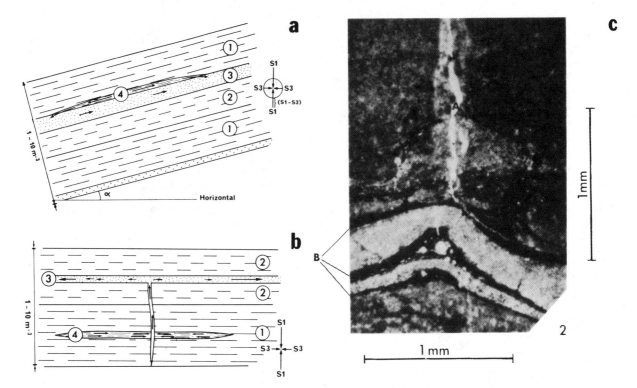

FIG. 1—Compaction of kerogen microflake under differential stress S1-S3. a, Compaction with elastic closure; oil passes into surrounding microporosity. Suggested lithology: 1, montmorillonite clay; 2, silty clay; 3, argillaceous silt; 4, kerogen. Injection pressures into these layers are supposed to be $P_0, 1 > P_0, 2 > P_0, 3$. b, Compaction with brittle failure; initiation of vertical open microfracture permits oil to reach and infiltrate more coarsely microporous layers. c, Example in carbonate rock at shallow depth: bitumen-filled (light) fracture initiated from kerogen flake (dark). Photograph from Beseme (1972); A, bitumen; B, kerogen.

pressive stress S3 varies vertically.

If the rock has only responded elastically to the geostatic load the least compressive stress S3 will be proportional to that load following the standard theory of elasticity:

$$S3 = \frac{v}{1-v} \cdot S1, \qquad (2)$$

where v is Poisson's ratio. Poisson's ratio is strongly dependent on the lithology: in sandstone, v is between 0.20 and 0.33; in marl, v is close to 0.40.

The lithologic effect, in the elastic hypothesis, on the least compressive stress S3 and the differential stress S1-S3 is shown in Table 1. Rocks are flattened primarily by mechanisms other than elastic deformation. In particular the total volume is first sharply reduced by

Table 1. Lithologic Effect on Stress

Lithology	v	S3	S1-S3
Shale	0.40	0.67 S1	0.33 S1
Limestone	0.30	0.43 S1	0.57 S1
Sandstone	0.20	0.25 S1	0.75 S1

"packing" due to compaction; then active flattening is realized by different modes of the pressure-solution process (stylolite formation in limestones, "thermal" quartzite formation in sandstones, recrystallization of clays, etc), which we cannot, at present, quantify.

Measurements of pump pressure during hydraulic fracturing of formations in oil wells show that the lateral compressive stress increases with depth, but that it varies greatly at a given depth (Fig. 2b). Poulet (1976) cited: toward 1,300 m at Zarzaitine (Algeria), S3 = 0.3·S1; toward 2,500 m in 12 Oklahoma wells, S3 = 0.6·S1; toward 3,400 m in the Ordovician of Hassi Messaoud, S3 = 0.85·S1. The confining stress coefficient f seems to be minimal between 1,000 and 2,000 m. It is certain that the lateral least compressive stress is, in many sedimentary basins, much reduced by local or regional tectonic extensions.

By contrast a lateral compressive stress superior to the geostatic load is frequently observed at shallow depths, sometimes down to 1,000 m. This anomaly is probably due to the surface boundary conditions: vertical stress equals atmospheric pressure at 0 m although a tectonic horizontal-stress component may exist.

It is not the object of this paper to describe in detail the theory of the formation and maintenance of open fractures in a state of extension or shearing. This question, geologically of prime importance, has been thoroughly discussed by Secor (1965, 1969).

Leaving aside the poorly known rheologic considerations, the following necessary fluid pressure conditions may be retained.

1. For a fracture to be opened by a fluid (water, oil, or gas), the pressure of the fluid in the porosity P_p must exceed the sum of the least compressive stress S3 and the tensile strength of the rock K at its weakest points of cohesion (Secor, 1965):

$$P_p \geq S3 + K. \tag{3}$$

2. For such a fracture to be propagated the pressure of the fluid in the fracture P_f must continue to fulfill the same conditions: $P_f \geq S3 + K$.

3. For the fracture to remain open without support (e.g., calcite bridges, uneven fracture plane, etc), the following relation is sufficient:

$$P_f = S3. \tag{4}$$

These formulae are approximate, as they do not include stress modifications occurring at crack tips which may lead to continued propagation even when $P_f = S3$. Propagated fractures will close rapidly outside their zone of formation depending on the difference between P_f and P_p.

1. If the pore pressure of the medium surrounding the fracture is equal to the pressure of the fluid within the fracture (in geopressured shales and reservoirs enclosed in these shales, or in porous bodies enclosed in anhydrites or compact limestones), the fractures could remain open for a long period without support.

2. If the pore pressure of the medium surrounding the fracture is lower than the initial pressure of the fluid within the fracture, the fractures will be closed quickly by filtration of the fluid into the porous surroundings; the only fractures which could remain open would be those blocked open by supports in rigid rocks.

Equations 3 and 4 are more generally presented by the introduction of new parameters, the effective stresses: $\sigma1 = S1 - P_p$; $s1 = S2 - P_p$; $s3 = S3 - P_p$. Secor (1965, p. 640), stated, "The failure criterion for tension fracturing can be taken as $s3 = -K$, where $s3$ is the least effective principal stress, and K is the tensile strength. Tension fracturing presumably can occur anywhere in the Earth's crust where this condition is satisfied."

Brace and Martin (1968) have experimentally demonstrated, in rocks of low porosity and permeability, the validity of the effective-stress concept, that is, the major role played by pore pressure in fracturing. Their experiments were carried out under conditions of slow stress establishment comparable with those encountered in geologic processes. In the text which follows, the writer has preferred a notation using the principal stresses S1, S2, and S3 and the fluid pressures P_p and P_f rather than to confuse them in the effective-stress concept. It is sometimes convenient to express the pore pressure as a fraction of the geostatic load, as was done for the least compressive stress S3:

$$P_p = k \cdot S1. \tag{5}$$

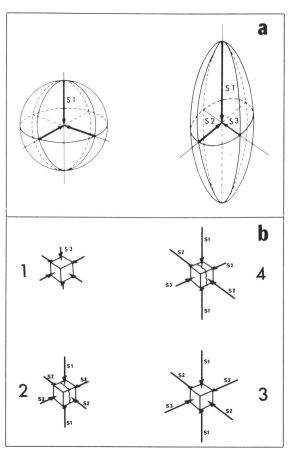

FIG. 2—a, Distribution of stresses: left, spherical as in liquid; right, ellipsoidal as in solid. S1 > S2 > S3; S3 is least compressive stress. b1, Horizontal stresses near surface are generally greater than vertical stress; horizontal open fracturing can only occur if P_p is superior to vertical load. b2, Horizontal stresses between 1,000 and 4,000 m are generally smaller than vertical stress; vertical open fracturing can occur in P_p > S3. b3, 4, Horizontal stress becomes equal to vertical stress below 4,000 m; open fracturing can only occur if S3 is tectontically reduced below pore pressure value S3 < P_p.

Evaluation of coefficient k for normal hydrostatic conditions—If we take 0.23 kg/m·cm² as the average specific weight of the rock, and if H is the depth in meters, S1 ~ 0.23·H in bars: (1) if the average specific weight of water is ϱ_w = 0.11 kg/mg·cm², P_p = 0.11·H, and k = P_p/S1 = 0.48; (2) if the average specific weight of water is ϱ_w = 0.10 kg/m·cm², P_p = 0.10·H, and k = P_p/S1 = 0.43.

In deep argillaceous formations with a delay in clay compaction, the fluid pressure increases and can approach the lateral least compressive stress: k often assumes values between 0.7 and 0.9. It is likely that P_p can exceed S3 only transitorily, for the opening of preexisting or newly induced fissures fixes P_p at the value of S3; that is, that $\sigma3 = S3 - P_p$ is generally zero and not negative, and that effective tensile stresses are only transient.

Fracturing of Source Rock by Oil Released from Kerogen

In a small oil-filled fracture arising from the compressed mature kerogen, equation 3 for fracturing becomes:

$$P_o \geq S3 + K, \quad (3')$$

where P_o, the oil pressure, replaces P_p, the pore pressure.

The propagation or reopening of fractures is made easier by the stress modifications occurring around crack tips. Assuming that the geostatic load is partly transmitted to the oil films of the kerogen during its compaction, the fracturing criterion will be easily satisfied and the fractures will be propagated upward until:

$$P_o = S3. \quad (4')$$

Injection of Oil Released by Kerogen into Microporous Layers of Source Rock

Propagation of the fractures will stop when pressure P_p is equal to or slightly lower than the least compressive stress S3. Because P_o in these fractures will commonly be superior to the pressure required for oil injection into the microporous layers of the source rock crossed by the fractures, the latter will empty their oil into these layers.

The pressure of oil injection into a porous layer depends not only on the pore pressure of the layer P_p but also on the capillary characteristics of the layer. To inject oil into a water-saturated, finely porous layer it is necessary to overcome the force which opposes the development of oil-water interfaces (Leverett, 1941; Hubbert, 1953). These capillary phenomena "arise from the fact that when two immiscible fluids are in contact, molecular attractions between similar molecules in each fluid are greater than the attractions between the different molecules of the two fluids" (Berg, 1979).

The Laplace equation gives the pressure difference P_c existing at equilibrium across an interface between oil and water:

$$P_c = \gamma(1/R1 + 1/R2),$$

where γ is the interfacial tension between oil and water, and R1 and R2 are the two principal radii of the interface.

Because we generally do not know the respective values of R1 and R2, we will write:

$$1/R1 + 1/R2 = 2/R, \quad (5)$$

where R is by definition the mean curvature radius of the interface and may be determined experimentally.

Laplace's equation simplified in this way gives the following criterion for the injection of oil from a fracture into the surrounding porosity:

$$P_o \geq P_p + 2\gamma/R, \quad (6)$$

where R is the mean curvature radius of the interface within the throats between pores (Fig. 3).

As the opening, reopening, or maintenance of a fracture implies at least $P_o = S3$, we can express equation 6 as:

$$S3 \geq P_p + 2\gamma/R,$$
$$\text{or} \quad (6')$$
$$2\gamma/R \leq S3 - P_p.$$

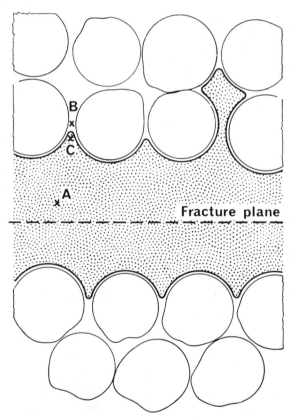

FIG. 3—Condition of oil injection into porous layer from oil-filled fracture. Horizontal section: $P_A = P_B$ in oil. Across oil/water interface: $P_{C(oil)} = P_{B(water)} + 2\gamma/R$, R being mean curvature radius of interface within throats.

Let us examine this criterion for a depth of 3,000 m, corresponding to a geostatic stress $S1 = 690$ b and to a normal pore pressure of 330 b, and by assuming the plausible confining stress coefficient $f = 0.7$ we have $S3 \cong 483$ b.

The application of condition $2\gamma/R < S3 - P_p$ gives $2\gamma/R < 153$ b, which implies that even a very fine micropore space can be injected with oil. However, oil will be preferentially drawn off by the more coarsely microporous layers reached by the fractures. If these injected beds are continuous they will serve as paths for migration toward structural or sedimentary traps. These migration paths will consist of slightly more coarsely porous levels intercalated in the source rocks or in their immediate vicinity.

In equations 6 and 6', if P_p is abnormally high, the terms $P_o - P_p \cong S3 - P_p$ become small and the criterion $2\gamma/R < P_o - P_p$ (approximately $2\gamma/R < S3 - P_p$) will be more difficult to satisfy; the oil may only be injected into the rock if it contains sufficiently coarse porous layers. When $P_p = S3$, the oil transfer will be entirely by interconnected fractures. Under the continuous influx of fluid the fracture networks will, at first, migrate in all directions (but dominantly upward). They will usually meet zones of reduced confining stress and will have a tendency to focus in such zones, for example, anticlines where certain sectors are under extension, or other structural elements such as shear-fault zones.

TRANSFER OF OIL TOWARD RESERVOIRS

The generalization of the concepts introduced in the preceding chapter will permit a better understanding of the processes of oil transfer toward reservoirs.

Conditions for Maintenance of Oil in Open Vertical Fractures

For oil to stay in an open vertical fracture which has crossed a porous bed, its pressure P_o must satisfy the double inequality:

$$S3 = P_o < P_p + 2\gamma/R, \qquad (7)$$

where $S3$ is the least compressive stress, that is, the weaker of the two horizontal or subhorizontal stresses, and P_p is the water pressure in the pores of the porous paths crossed by the fracture.

Equation 7 shows that for the oil to remain in the fracture, P_o must be equal to $S3$, and P_o must be weaker than the pressure of injection into the most coarsely porous bed reached by the fracture. This condition implies $S3 < P_p + 2\gamma/R$, and thus $S3 \cong P_p$. This will only be attained in two types of geologic environment: (1) zones where the water pressure is abnormally high owing to a delay in clay compaction or to pressures created by chemical or thermal changes acting upon rock isolated within a compact environment such as evaporites or impervious carbonates; and (2) sectors where the least compressive stress is drastically reduced by tectonic or structural effects.

Conditions for Oil Transfer by Migration Paths

Under normal conditions $S3$ is greater than the sum of the pore water pressure P_p and the capillary pressure increment $2\gamma/R$ required for the entry and maintenance of oil in the porous bed (eq. 6').

Experimental studies designed to define the retention capacity of argillaceous caprocks in underground gas storage areas (Tek, 1975) suggest that the term $2\gamma/R$ will generally be in the 5 to 30-b range for oil; the clays are probably composed of alternations of layers where injection would be difficult, $2\gamma/R > 30$ b, and layers where injection would be easier, $2\gamma/R > 10$ b.

The most coarsely porous bed reached will probably act as a valve by absorbing all oil present in the fractures and in the more finely porous layers connected with them. If the bed is lenticular there will be immobilization until a possible fracturing occurs owing to modification of the pressure or geomechanical conditions, either increased subsidence or tectonic events. If the bed is continuous there will be lateral oil movement. These two possibilities avoid dispersion of the oil and reduce the effects of "water washing."

The system of initial paths will be in the source rock or in its immediate vicinity; oil in very fine porosity will have a high capillary pressure. However, as the upper and lower zones limiting the path will be even more finely porous, the oil will be unable to disperse and will be strictly channeled. The oil is, in capillary terms, a prisoner of the path which carries it.

In the paths, the displacement will be carried out: (1) under the high pressure of the petroleum expelled from the kerogen; (2) by the effect of differential subsidence—a microporous layer will tend toward geostatic pressure in the deeper part of a syncline; and (3) when sufficiently large hydrocarbon bodies have been accumulated, they will move updip by the effect of differential density. The relative importance of these different processes depends directly on lithology and rate of burial.

Oil Transfer Toward Reservoirs

Let us examine the means or processes which make it possible for hydrocarbons which have undergone this channelization to reach the reservoirs where they have accumulated. Certain geologic conditions greatly favor vertical transfer; the most notable of these is the truncation of migration paths within the source rock, above or below the reservoir, by an unconformity.

The transfer of oil toward reservoirs must be more generally by reopening of the vertical fractures: the criterion $P_o > S3$ should be generally satisfied on parts of the structural forms where $P_p \leq S3$. The geologic frequency of these zones of reduced $S3$ is proven by the commonness of open fractures in structures (Stearns and Friedman, 1972). Because the oil pressure P_o exceeds the water pressure by the capillary increment $2\gamma/R$, a system of open fractures injected with oil may be developed where S3 is slightly superior to P_p.

Equation 7 for the maintenance of oil in open fractures implies that $S3 \leq P_p + 2\gamma/R$, or $2\gamma/R > S3 - P_p$. Thus the capillary pressure $2\gamma/R$ of the oil in the path should be superior to the "effective" confining stress, which implies both a sufficiently fine porosity in the path and similar values of S3 and P_p.

Applying equation 6 in practical units we have:

$$P \text{ (in bars)} = 2.10^{-2} \gamma \text{ (dynes/cm)}/R \text{ (microns)},$$

with R being the mean curvature radius of the interface within the throat.

If depth = 2,000 m, rock density = 0.23 kg/m·cm², water density = 0.11 kg/m·cm², γ = 30 dyn/cm, then S1 = 460 b and P = 220 b; supposing f = 0.5 (a plausible state of extension), $S3 \cong 230$ b, which gives us for $2\gamma/R > S3 - P_p : 2\gamma/R > 10$ b.

This relatively moderate tensile anomaly will give easy vertical fracture initiation for $2\gamma/R > 10$ b which only requires that $R < 6 \mu$.

Equation 7 shows that as soon as the least compressive stress S3 is smaller than the pore pressure P_p plus the capillary pressure increment $2\gamma/R$ of the oil in the path, fractures (usually close to the vertical) will open in the source-rock complex. Some accumulated capillary oil will pass into the fractures and this feeding will permit propagation. As this propagation will be mainly vertical, there will be a maximum chance for the fractures to lead into a continuous reservoir or lens into which they will empty their oil and then close.

Impossibility of Feeding Reservoirs Under Tensile Conditions

When the least compressive stress S3 is lower than P_p the oil can no longer enter into the reservoirs. The injection condition, $S3 > P_p + 2\gamma/R_G$ (eq. 6'), may be written $2\gamma/R_G < S3 - P_p$, which, in tensile conditions, implies that $2\gamma/R_G < 0$, which is impossible.

It is likely that P_p can exceed S3 only transitorily, for the opening of preexisting or newly induced fissures will progressively increase S3 to the value of P_p by the expansion of the rock bulk volume.

When the oil rises in a fracture remaining under tensile conditions it should arrive at the surface without ever being injected into a reservoir. This is probably the explanation of asphalt veins in southeastern Turkey (Lebkuchner et al, 1972) or of the gilsonite veins of the Uinta basin in the United States (Bell and Hunt, 1963). Feeding reservoirs at geostatic pressure is likewise difficult, if not impossible, as indicated by an old empirical rule of petroleum geologists in the Gulf Coast and elsewhere.

Role of Continuing Compaction in Hydrocarbon Transfer Toward Reservoirs and in Opening of Fractures

We have seen that a capillary pressure of at least 10 to 30 b can be considered indispensable for the reopening of vertical fractures by oil. This pressure increment seems all the more attainable when we consider the intervention of a phenomenon we have not yet discussed—the continuation of burial after the positioning of oil in the microporous zones of the source rocks.

The continuation of burial is commonly reflected by an increase in the overburden gradient, thus an increase in the oil capillary pressure gradients in the microporous paths of the source rock. Migration toward high points will be accelerated or, in some areas, will begin.

In microporous lenses, the continuation of burial will make the hydrocarbon pressure P_O rise to the smaller of the two regulating pressures, that of the crossing, by capillary filtration, of argillaceous walls up to the nearest continuous path, or that of open vertical fracture initiation ($P_O \geq S3 + K$), which will also render possible communication with neighboring paths. The supplementary compaction will thus provoke progressive mobilization of the oil immobilized in these sedimentary cul-de-sacs; migration will therefore be continued long after the period of oil production by kerogen has ended.

The effects of continuing burial, and therefore of compaction, also intervene into the zones of oil accumulation. Pore radius will progressively decrease and, hence, capillary pressure will increase. When it is sufficiently high, an open fracture filled with oil will be created in sectors having a sufficiently small least compressive stress S3. Thus the final vertical phase of migration may require minimal compaction of the source-rock carrier paths.

VERTICAL OIL MIGRATION AND TECTONIC REDUCTIONS OF LEAST COMPRESSIVE STRESS S3

Tectonic Reductions of Least Compressive Stress S3

Oil migration processes probably depend only on tectonic events which have occurred since the decomposition of the kerogen. Tectonic events which resulted in only apparently minor deformations were sufficient to reduce S3 enough to permit fracturing. These slight deformations often represent further play over preexisting zones of structural deformation.

Let us consider tectonic extensions. At least three possibilities may be distinguished.

1. Simple extensions may give rise to a system of vertical open fractures. These extensions are often revealed in seismic sections by small, regularly distributed, normal faults.

2. Extensions are also associated with shearing, in particular those active in the evolution of grabens. Sometimes shearing is preceded by an important development of microfissures which "dilate" the volume of rock neighboring the future shear plane; this is the dilatancy phenomenon. More basically, shear zones contain zones under extension, which is to be expected as shearing is characterized by a strong horizontal tectonic action F1 corresponding to S1, and by an abnormally weak horizontal reaction approximately at right angles to the action. Figure 4a gives a sketch of ex-

pected possible tectonic situations resulting from the microstructural analyses of Mattauer (1973).

3. Reductions in the least compressive stress S3 are also linked with structural forms, particularly those ordinarily present in basins undergoing extension. Thus, the numerical models clearly reveal, in anticlines due to basement horsts, sectors where S3 is reduced far below normal or even becomes negative, that is, becomes a tensile stress (Fig. 4). The existence of these sectors with a reduced S3 is proven by open fracturing.

In an anticline created by compressive stress, the open fractures extend parallel with the maximum principal stress S1; they will thus be perpendicular to the direction of the fold and parallel with the thrust as shown by Lebkuchner et al (1972) and by Stearns and Friedman (1972, Fig. 5, p. 89).

Petroleum Case Histories Interpreted with Proposed Model of Migration

Traps over synsedimentary faults of Niger delta—Research in this area by several companies has led to a general belief that oil transfer is along zones neighboring growth faults. The theory outlined in this article explains this transfer process. The source rocks are principally horizons in the undercompacted, overpressured, prodeltaic clays of the Akata Formation. The oil released toward the overlying sandy layers seems to

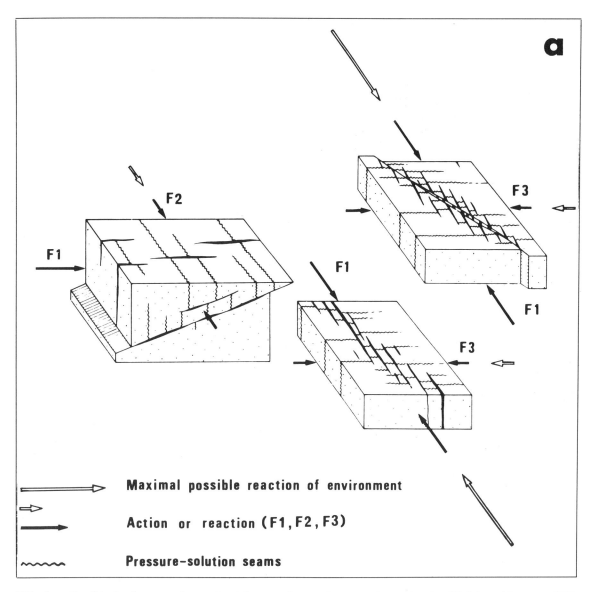

FIG. 4—a, Possible development of extensional fractures in overall compressive setting (modified from Mattauer, 1973).

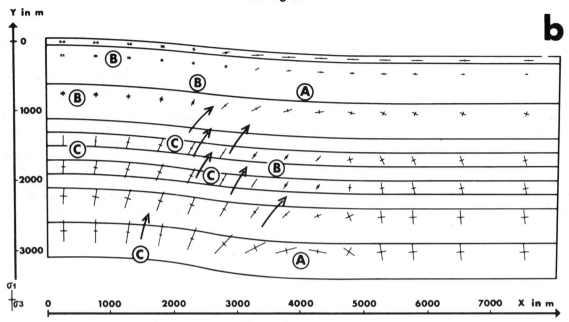

FIG. 4—b, Anomalous distribution of effective stresses due to horst, according to two-dimensional finite-element model computed by Boudon (1976) applying nonlinear elastic behavior and taking joints into account. A, possibility of horizontal open fracturing; B, tensile zones; C, initiation of oil-filled vertical open fractures if oil capillary pressure is sufficient. Tensile stress, x-x. Arrows suggest vertical migration by open fracturing.

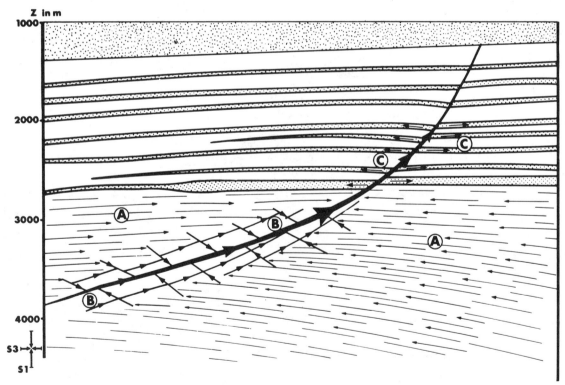

FIG. 5—Interpretation of oil migration toward growth-fault structure in Niger delta. $P_{o,Sh}$, capillary pressure of oil microporous paths of source rock; $P_{o,G}$, capillary pressure of oil in reservoir; S3 is least compressive stress; A, squeezing of oil from kerogen under compression by differential stress S1-S3 and injection into more coarsely microporous layers of source rocks; B, S3 becomes smaller than $P_{o,Sh}$ by structural effect: initiation of vertical open fractures; C, injection of oil from fractures into adjacent reservoir (S3 > $P_{o,G}$).

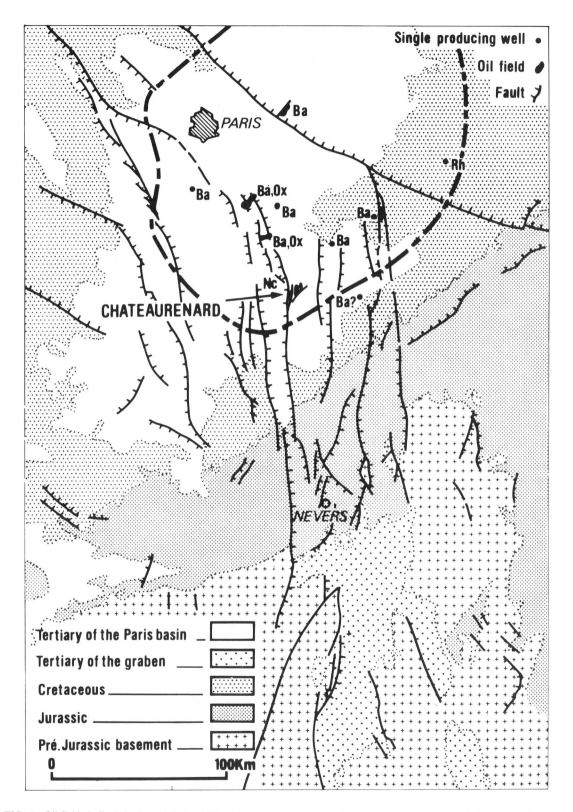

FIG. 6—Oil fields in Paris basin and their relation with mainly tensile tectonic features. Reservoir age: *Rh*, Rhaetian; *Ba*, Bathonian; *Ox*, Oxfordian; *Ne*, Neocomian. Dashed line, outer limit of mature Hettangian source rocks.

have been collected by now extinct faults.

Figure 5, based on a seismic profile through an oil-bearing structure, interprets the migration on the basis of the proposed model. The perceptible prolongation, despite the absence of throw, of the growth fault into the superficial sandy Benin Formation suggests that the fault zone has been recently involved in fracturing. The surroundings of the fault have thus remained under tension, that is, (1) in the injected sands at the upper part of the fault (U), there was at least $S3_U - P_{p,G} \geq 0$, with normal hydrostatic pore pressure $P_{p,G}$ in sandstones; (2) in the Akata, at the base of the fault (L), where pore pressure in shale $P_{p,Sh}$ is geostatic, there was probably $S3_L - P_{p,Sh} = 0$, and because the hydrocarbon pressure $P_{o,Sh} = P_{p,Sh} + 2\gamma/R$, we have $S3_L - P_{o,Sh} < 0$.

The fault must function intermittently.

1. The hydraulic pressuring and opening of the lower part of the fault occur under the influx of hydrocarbons coming from the undercompacted, overpressured, source rocks of the Akata at the pressure $P_{o,Sh} = P_{p,Sh} + 2\gamma/R$.

2. At a stage of this development, relaxation occurs with a rapid propagation of the hydrocarbons contained in the fault toward the upper part and in consequence a sudden opening of the upper part of the fault.

3. Reclosing of the fault occurs when the hydrocarbons become level with a reservoir (or reservoirs), the injection of the hydrocarbon into the pore space sharply reducing the oil pressure in the fault.

4. Repetition of the cycle by repressuring of the lower part of the fault occurs because of hydrocarbon filtration from the undercompacted clays.

Neocomian oil fields of southern Paris basin—These fields have been described in detail by Heuillon (1972). Small oil fields are located in sandstone reservoirs which contain nearly fresh water. The trapping "apparently is in part stratigraphic and in part hydrodynamic." The oil originated in the underlying Hettangian marl, more than 1,000 m deeper, which is shown by geochemical studies to be the only effective source rock in the Paris basin. Migration has been in north-south faults and fractures that obviously correspond to a nonsubsident prolongation of the Limagne Cenozoic graben, that is, a major tensile geostructure. This area under extension is clearly marked on the lineament map of France (France, BRGM, 1980). All the oil fields of the Paris basin seem to have profited from this state of extension which is probably the only element favoring migration in this region of little structure (Fig. 6).

CONCLUSION

The geochemical theory of migration in a single phase as outlined explains the essential aspects of what is known of oil transfer from source rock to accumulation, in particular the lack of dispersion and the usual low volume of light hydrocarbon losses by dissolution in water. It also explains that, apart from the oil in source rocks and reservoir accumulations, oil tends to be localized in fractures.

This model of oil migration can easily be extended to cover gas migration. Most of the results can be directly transposed. Gas would move sooner and farther than oil, as it contains a greater quantity of elastic energy (Hedberg, 1974). It will also migrate faster than oil, because it is less viscous and its interfacial tension with water is considerably reduced at depth (29 dyn/cm at 350°F for 85% methane, Jennings and Newman, 1971). These factors should facilitate the processes of fracture opening, reopening, and entry into reservoirs.

SELECTED REFERENCES

Bell, K. G., and J. M. Hunt, 1963, Native bitumens associated with oil shales, *in* Organic geochemistry: New York, Macmillan Co. (Internat. Ser. Mons. Earth Sci., v. 16), p. 333-366.

Berg, R. R., 1975, Capillary pressures in stratigraphic traps: AAPG Bull., v. 59, p. 939-956.

Beseme, P., 1972, Bitumes figures et deplaces en association dans cinq series a kerogene d'Anatolie (Turquie); description et origine de ces kerabitumes: Inst. Francais Petrole Rev., v. 27, p. 885-912.

Boudon, J., 1976, Application de la methode des elements finis a l'approche mecanique d'un phenomene tectonique: le poinconnement: Theses Fac. Sci. Grenoble, France.

Brace, W. F., 1969, The mechanical effects of pore pressure on the fracturing of a rock, *in* Research in tectonics: Canada Geol. Survey Paper 68-52, p. 113-123.

—— and R. J. Martin, III, 1968, A test of the law of effective stress for crystalline rocks of low porosity: Internat. Jour. Rock Mechanics and Mining Sci., v. 5, p. 415-426.

Chiarelli, A., and J. du Rouchet, 1977, Importance des phenomenes de migration verticale des hydrocarbures: Inst. Francais Petrole Rev., v. 32, p. 189-208.

Dickey, P. A., 1975, Possible primary migration of oil from source rock in oil phase: AAPG Bull., v. 59, p. 337-345.

—— 1979, Petroleum development geology: Tulsa, Oklahoma, Petroleum Publishing Co., 389 p.

Dow, W. G., 1974, Application of oil-correlation and source-rock data to exploration in the Williston basin: AAPG Bull., v. 58, p. 1253-1262.

Dunnington, H. V., 1958, Generation, migration, accumulation, and dissipation of oil in northern Iraq, *in* Habitat of oil: AAPG, p. 1194-1251.

Du Rouchet, J. H., 1978, Elements d'une theorie geomecanique de la migration de l'huile en phase constituee: Elf-Aquitaine Centre Recherche Exploration et Production Bull., v. 2, p. 337-373.

[France] Bureau de Recherches Geologique et Minieres, 1980, Carte des lineaments de la France a 1/1,000,000: Orleans, France.

Gallagher, J. J., Jr., et al, 1974, Experimental studies relating to microfracture in sandstone: Tectonophysics, v. 21, p. 203-247.

Gretener, P. E., 1977, Pore pressure: fundamentals, general ramifications and implications for structural geology: AAPG Continuing Education Course Note Ser. 4, 87 p.

Hedberg, H. D., 1964, Geologic aspects of origin of petroleum: AAPG Bull., v. 48, p. 1755-1803.

—— 1974, Relation of methane generation to undercompacted shales, shale diapirs, and mud volcanoes: AAPG Bull., v. 58, p. 661-673.

Heuillon, B., 1972, Oil fields of Neocomian of Paris basin, France, *in* Stratigraphic oil and gas fields—Classification, Exploration Methods, and Case Histories: AAPG Mem. 16, p. 599-609.

Hubbert, M. K., 1953, Entrapment of petroleum under hydrodynamic conditions: AAPG Bull., v. 37, p. 1954-2026.

—— and D. G. Willis, 1957, Mechanics of hydraulic fracturing: Jour. Petroleum Technology, v. 9, p. 153-168.

Jaeger, J. C., and N. G. W. Cook, 1969, Fundamentals of rock mechanics: London, Methuen, 519 p.

Jennings, H. Y., and G. H. Newman, 1971, The effect of temperature and pressure on the interfacial tension of water against methane-normal decane mixtures: Soc. Petroleum Engineers Jour., June, p. 171-176.

Lajtai, E. Z., 1971, A theoretical and experimental evaluation of the Griffith theory of brittle fracture: Tectonophysics, v. 11, p. 129-158.

Landes, K. K., 1970, Petroleum geology of the United States: New York, Wiley-Interscience, 571 p.

Lebkuchner, R. F., F. Orhun, and M. Wolf, 1972, Asphaltic substances in southeastern Turkey: AAPG Bull., v. 56, p. 1939-1964.

Leverett, M. C., 1941, Capillary behavior in porous solids: AIME Trans., v. 142, p. 152-169.

Lucas, P. L., and J. M. Drexler, 1976, Altamont-Bluebell—a major, naturally fractured stratigraphic trap, Uinta basin, Utah, in North American oil and gas fields: AAPG Mem. 24, p. 121-135.

Mattauer, M., 1973, Les deformations des materiaux de l'ecorce terrestre: Paris, Hermann, 493 p.

Meissner, F. F., 1978, Petroleum geology of the Bakken Formation, Williston basin, North Dakota and Montana, in Williston basin symposium: Montana Geol. Soc. 24th Ann. Conf. Guidebook, p. 207-227.

Phillips, W. J., 1972, Hydraulic fracturing and mineralization: Geol. Soc. London Jour., v. 128, p. 337-359.

Poulet, M., 1976, Apport des experiences de mecanique des roches a la geologie structurale des bassins sedimentaires: Inst. Francais Petrole Rev., v. 31, p. 781-822.

Price, N. J., 1974, The development of stress systems and fracture patterns on underformed sediments, in Advances on rock mechanics: 3d Internat. Cong. Soc. Rock Mechanics, Denver, Proc. v. 1A, p. 497-508.

Ranalli, G., 1975, Geotectonic relevance of rock-stress determinations, Recent crustal movements: Tectonophysics, v. 29, p. 49-58.

Schaar, G., 1976, The occurrence of hydrocarbons in overpressured reservoirs of the Baram delta: Indonesian Petroleum Assoc. 5th Ann. Conv. Proc., p. 163-170.

Secor, D. T., 1965, Role of fluid pressure in jointing: Am. Jour. Sci., v. 263, p. 633-646.

_____ 1969, Mechanics of natural extension of fracturing at depth in the earth's crust, in Research in tectonics: Canada Geol. Survey Paper 68-52, p. 3-48.

Stearns, D. W., and M. Friedman, 1972, Reservoirs in fractured rock, in Stratigraphic oil and gas fields—Classification, Exploration Methods, and Case Histories: AAPG Mem. 16, p. 82-106.

Tek, M. R., 1975, Nouveaux aspects du stockage souterrain du gaz: Inst. Francais Petrole Rev., v. 2, p. 1623-1639.

Tissot, B., and R. Pelet, 1971, Nouvelles donnees sur les mecanismes de genese et de migration du petrole simulation mathematique et application a la prospection: 8th World Petroleum Cong., Moscow, Proc., v. 2, p. 35-46.

_____ and D. H. Welte, 1978, Petroleum formation and occurrence: New York, Springer-Verlag, 521 p.

Williams, J. A., 1974, Characterization of oil types in Williston basin: AAPG Bull., v. 58, p. 1243-1252.

A Novel Approach for Recognition and Quantification of Hydrocarbon Migration Effects in Shale-Sandstone Sequences[1]

DETLEV LEYTHAEUSER,[2] ANDREW MACKENZIE,[3] RAINER G. SCHAEFER,[2] and MALVIN BJORØY[4]

ABSTRACT

A detailed organic geochemical study of over 150 samples from two cores with a total combined length of 320 m (1,050 ft) through sequences of interbedded source rock–type shales (0.84% R_m maturity) and reservoir sandstones allowed recognition and quantitation of a number of migration effects. Detailed gas chromatography–mass spectrometry of steranes and triterpanes was used to insure that samples being compared to investigate migration effects contain organic matter of a similar type. Thin shales interbedded in sands and the edges of thick shale units are depleted in petroleum-range hydrocarbons to a much higher degree than the centers of thick shale units.

For the alkanes, expulsion occurs with pronounced compositional fractionation effects: shorter chain length n-alkanes are expelled preferentially, and isoprenoid alkanes are expelled to a lesser degree than their straight-chain isomers. Based on material balance calculations, expulsion efficiencies were determined and found to be very high in certain instances. For thin interbedded shales, they decrease from about 80% around C_{15} to near zero in the C_{25+} region. There is no evidence for significant redistribution of steranes and triterpanes in the two sequences. Compared to C_{15} to C_{25} n-alkanes, they appear relatively immobile.

The composition of the hydrocarbons impregnating parts of the reservoir sandstones is in agreement with expulsion occurring with pronounced fractionation based on molecular chain length. Hence, consideration of bulk expulsion efficiencies gives an unrealistic picture. Furthermore, the impregnation of a siltstone cap rock from an underlying hydrocarbon accumulation seems to have occurred by bulk-oil migration and without significant fractionation. The degree of hydrocarbon depletion of some of the shales of both sequences appears to be controlled by compaction, and the primary migration process appears to have occurred with chromatographic separation. The migration phenomena observed in both sequences lead us to propose that the main phase of expulsion can be preceded by an earlier stage, during which the edges of thick shales and thin interbedded shales appear to be slowly and continuously depleted by the chromatographic processes.

The composition of the hydrocarbon product accumulating in the reservoirs at this stage appears to be controlled primarily by physical processes rather than by the type and maturity of the organic matter in the generating source rock. By this mechanism, the origin of accumulations of light oils and gas-condensates in low mature sequences bearing predominantly terrestrial-derived organic matter can be explained. Finally, the migration effects documented in this study have some consequences for interpretation of geochemical data (e.g., the pristane/n-C_{17} ratio, a commonly accepted maturity parameter, has been shown to be also controlled by the degree of hydrocarbon expulsion).

INTRODUCTION

Migration of hydrocarbons through a column of sediments in the subsurface profoundly influences the normally water-saturated fluid system of the rock pores and is hence likely to leave permanent traces. The kind and magnitude of these effects in the geologic environment depend primarily on the concentration levels that the migrating hydrocarbons reach in the rock pore systems. Because concentration levels can vary drastically, a wide range of effects, from the visible to the invisible, are known to occur. Hydrocarbon migration at higher concentration levels and through strata of elevated porosities and permeabilities can result in oil or gas "shows" that can be recognized by the geologist during exploratory drilling. Under more ideal conditions, migration can result in the formation of pooled hydrocarbon accumulations. These phenomena usually result from secondary migration,

---

©Copyright 1984. The American Association of Petroleum Geologists. All rights reserved.

[1]Manuscript received, January 17, 1983; accepted, June 3, 1983. An earlier version of this paper was delivered at the Gordon Research Conference on Organic Geochemistry, Holderness, New Hampshire, in August 1982.

[2]Institute for Petroleum and Organic Geochemistry (ICH-5), Kernforschungsanlage Jülich, P.O. Box 1913; 5170 Jülich, F.R. Germany.

[3]Institute for Petroleum and Organic Geochemistry (ICH-5), Kernforschungsanlage Jülich, P.O. Box 1913; 5170 Jülich, F.R. Germany. Present address: British Petroleum Co. Ltd., Exploration & Production Research Division, Research Centre, Chertsey Road, Sunbury-on-Thames, Middlesex TW16 7LN, Great Britain.

[4]Continental Shelf Institute, Postboks 1883, 7001 Trondheim, Norway.

We thank the Store Norske Spitsbergen Kulkompani for permission to sample their core holes for this study and for logistic support during the sampling. We are greatly indebted to the following individuals. F. J. Altebäumer, KFA, for sampling; J. Gormly, KFA, for Rock-Eval pyrolysis data; P. Mukhopadhyay, KFA, for vitrinite reflectance measurements; M. Radke, KFA, for wet chemistry data; J. Rullkötter, KFA, for allowing us to use his mass spectrometric equipment; A. Stensrud, Store Norske Spitsbergen Kulkompani, for information regarding the geology of the core holes; D. H. Welte, KFA, for support and advice.

For technical assistance we thank Mrs. G. Bruch, B. Kammer, B. Schmitz, and B. Winden, as well as W. Benders, U. Disko, F. Leistner, H. Pooch, J. Schnitzler, and H. Willsch.

which has been defined as the migration stage occurring within the carrier and reservoir beds themselves. On the other hand, hydrocarbon migration within the generating source rock formation ("primary migration") rarely leaves visible traces due to a major difference in scale with respect to both concentration levels and volume in the much finer pore size system of source rock shales. Therefore, more sophisticated analytical techniques (e.g., geochemical and microscopic—fluorescence and visible light microscopy, scanning electron microscopy) are required to recognize effects that have arisen from primary migration.

Roberts and Cordell (1980) concluded that an adequate understanding of petroleum migration has not been achieved because of insufficient observations of petroleum moving from one place to another in the subsurface. Numerous attempts have been made by petroleum geochemists and geologists to recognize migration effects, and several geochemical parameters are commonly accepted and applied in studies of exploration wells. A well-documented case history of this kind is the Mahakam delta area, offshore Kalimantan, Indonesia (Durand and Oudin, 1980; Vandenbroucke et al, 1983). Based on a comparison of hydrocarbon compositions between oil pools at a wide depth range in a multiple-pay field and those of the interbedded organic-rich shales and coals, migration was concluded to have occurred in a vertical direction and over a considerable distance from a source interval lying deeper than the deepest oil pool. Drainage effects in source rock-type shales near their contacts with porous reservoir beds have been studied by Vandenbroucke (1972) and Connan and Cassou (1980). The geochemical effects of gas migrating by diffusion from a siltstone into an underlying sandstone unit has been documented by Leythaeuser et al (1983a). Migration phenomena for petroleum-range hydrocarbons moving over very short distances within source rocks have been investigated by attempting a physical separation of the soluble organic matter of mineralogically closed and open pores and comparing their compositions (Beletskaya, 1972; Sajgo et al, 1983).

In a manner analogous to these and other previous investigations, the study reported here intended to search for migration effects by geochemical analysis of a case history, where migration had in all likelihood occurred. This study differs, however, in several respects from most previous ones because: (1) samples from a sequence of interbedded source and reservoir rocks at favorable maturity levels were analyzed, (2) very closely spaced samples (decimeter to meter range) were examined, (3) only good quality conventional core samples were taken, and (4) a variety of modern geochemical techniques was applied. By detailed analysis of more than 150 samples from two deep core holes with a combined length of 320 m (1,050 ft), we have shown that redistribution of petroleum range hydrocarbons from the shales into the sandstones has been significant. Careful analysis of these redistribution effects allowed recognition of major compositional fractionation effects associated with primary migration, calculation of expulsion efficiencies for source rock shales, and examination of cap rock effects in the seal above a small hydrocarbon accumulation.

The consequences of the results of the study reported here and in a short previous publication (Mackenzie et al, 1983) for petroleum geochemistry and geology are considerable. Previously, the amount and composition of solvent-soluble organic matter in organic-rich shales were considered mainly as an expression of their maturity and type of organic matter (e.g., Philippi, 1965; Tissot et al, 1971; Tissot and Welte, 1978; Hunt, 1979). However, as is shown below, primary migration and expulsion can also significantly alter the extract composition. The amount and composition of petroleum-range hydrocarbons from source rocks which have expelled hydrocarbons can be changed to such an extent it is difficult to unambiguously discriminate source- and maturity-related effects from the consequences of migration processes. A result of this study is that the mechanisms of primary migration appear most likely to be understood by assessment of the compositional fractionation effects associated with the expulsion of hydrocarbons from shales into more porous lithologies.

GEOLOGIC AND GEOCHEMICAL BACKGROUND

In order to distinguish migration effects from the consequences of hydrocarbon generation processes, it is necessary (1) that closely spaced samples be available from sequences with source rock–type shales or siltstones, at the maturity required for petroleum generation, and in juxtaposition with good reservoir-type rocks, and (2) that these samples be conventional cores. Interpretation problems—caused by, e.g., cavings or mud contamination—commonly associated with geochemical data from cuttings and sidewall cores are simply too great for the approach to be discussed here. These prerequisites are, to a large extent, met by the sample sets from two deep core holes drilled on Spitsbergen Island, Svalbard, Norway. In the summer of 1979, Store Norske Spitsbergen Kulkompani drilled, in the course of a coal-exploration program, a core hole at Adventdalen and at Reindalen. The Adventdalen site is located close to Longyearbyen and within the area of economic coal mining activities, whereas the Reindalen site is approximately 30 km (19 mi) to the southeast.

Some basic data for the two core holes are listed in Table 1, and their lithologic composition is shown schematically in Figure 1. The 142-m (466-ft) deep Reindalen core hole penetrated the Paleocene Firkanten Formation, which is the main coal-producing sequence on Spitsbergen Island (Harland et al, 1976; Manum and Throndsen, 1978). At Adventdalen, a 239-m (784-ft) succession of Lower Cretaceous sediments was continuously cored. It consisted of the marine Carolinefjellet Formation and the underlying nonmarine Helvetiafjellet Formation (A. Stensrud, oral communication). Both sequences comprise interbedded dark-gray shales and siltstones and light-gray sandstones. The main difference in lithology between the two sequences is shown by the sand to shale ratios: the Adventdalen sequence contains a much greater proportion of sandstones (Table 1, Figure 1). The sandstones of both core holes are fine to medium grained, poorly sorted, and silt rich. This is also apparent in many instances from their elevated organic carbon contents. However, as evident from lower organic carbon contents, the Reindalen sequence sandstones are significantly cleaner than those at Adventdalen. No measurements of porosities and permeabilities of any sandstone samples were made in the

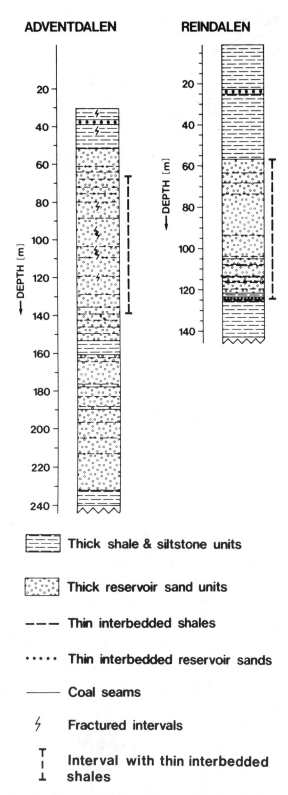

Figure 1—Schematic lithology of two core holes from Svalbard, Norway, analyzed in this study.

course of this study. Several intervals must, however, exhibit reservoir rock properties because gas shows were encountered during drilling of both core holes (A. Stensrud, oral communication). A gas-charged sand was penetrated at the Adventdalen site around 162 m (531 ft) depth, which continued to bleed a methane-rich gas to the surface for a long time after the drill site was abandoned. A comparatively minor gas show was observed in the Reindalen sequence around 141 m (463 ft) depth.

Most shales of both core holes are dark gray to black, partly massive, and partly laminated, with varying proportions of silt. Gradational transitions exist between shales and siltstones, but these are not differentiated in most of the following discussion. Dark, organic-rich shales are present in both sequences in two basic forms: as thick, continuous intervals about 2–10 m (7–33 ft) thick and as very thin layers 5–50 cm (2–20 in) thick. Both forms are interbedded in continuous sandstone intervals (Figure 1). In the lower third of each sequence, coals were encountered: at Adventdalen, two thin layers (5 and 20 cm; 2 and 8 in.) and at Reindalen, a major seam and several thin layers (1.85 m and 25 cm; 6 ft and 10 in.). In the Adventdalen sequence, several distinct intervals of fractured rocks were observed, and samples for geochemical analysis were selected from these intervals.

Both core holes are located on Spitsbergen Island within the area of a large depression with a north-northwest-trending axis, which was rapidly subsiding during the early Tertiary (Birkenmajer, 1972; Kellogg, 1975; Steel et al, 1981). It is filled with up to 2,300 m (7,550 ft) of clastic sediments bearing coal seams in the lowermost and uppermost parts (Paleocene Firkanten Formation and the Oligocene Aspelintoppen Formation; Manum and Throndson, 1978). On the basis of a study of regional variations of rank of the coals in the Firkanten Formation, the same authors concluded that Tertiary sediments were originally deposited in a greater thickness than presently preserved. From their reconstructed-isopach map for the total Tertiary, the maximum burial depths for the Adventdalen and Reindalen sequences can be estimated as about 2,500 m (8,200 ft) and 2,800 m (9,200 ft), respectively. Since the early Holocene, the study area has been subjected to rapid and intense uplift caused by crustal rebound in response to deglaciation. A maximum rate of uplift of approximately 25 m/1,000 years (82 ft/1,000 years) has been estimated for the time 9,000 years B.P. (Boulton, 1979). The locations of both core holes are, owing to their high latitude, within the area of continuous permafrost (Black, 1954). Permafrost thicknesses of 110 m (360 ft) and 80 m (260 ft) were estimated for the Adventdalen and Reindalen sites (A. Stensrud, oral communication), but no downhole geophysical logging measurements, which would have allowed more accurate estimates, were carried out.

The burial history has advanced the rank of the coal seams in the lower parts of each core hole to a similar high-volatile bituminous A stage, equivalent to 0.82 and 0.85% mean vitrinite reflectance (% R_m) at Reindalen and Adventdalen, respectively. Given the shortness of both core holes, these values are believed to represent the true maturity level of the organic matter finely disseminated in

Table 1. Basic Geologic and Geochemical Data for Core Holes

	Adventdalen	Reindalen
Total depth (m)	239	142
Age	Lower Cretaceous	Paleocene
Sand/shale ratio	1.25	0.43
Estimated depth of permafrost (m)	80	110
Depth of gas shows (m)	162	141
Mean organic carbon content (%)		
of shales	2.6	1.7
of sandstones	1.4	0.3
Kerogen type (mean hydrogen index in mgHC/gC$_{org}$)		
of shales	77.6	72.7
of sandstones	69.9	34.4
Maturity (mean vitrinite reflectance, % R$_m$)		
of coal	0.85	0.82
of shales and siltstones		
mean	0.65	0.71
range	0.51-0.91[a]	0.66-0.79[b]

[a]20 samples measured.
[b]7 samples measured.

the shales and siltstones, although their vitrinite reflectance values revealed some scatter and lower means (the mean for 20 samples from Adventdalen is 0.65% R$_m$, whereas for 7 samples from Reindalen it is 0.71% R$_m$). This observation is not unusual. More scatter in mean vitrinite reflectance values measured in shales versus those measured in coal seams interbedded in the same sequences has commonly been observed (e.g., Bostick and Foster, 1975). Regardless of this scatter in vitrinite reflectance values, the conclusion is warranted that the maturity of potential source rocks in the Adventdalen and Reindalen sequences is favorable for hydrocarbon generation. Because of the type III nature of their kerogens, the sequences can be considered to have penetrated the oil generation threshold but not to have reached the peak oil generation interval (Tissot and Welte, 1978).

Many of the shales analyzed from both core holes can be classified as good source rocks, as they meet all three essential requirements for effective hydrocarbon generation: sufficient amounts and an adequate quality of finely disseminated organic matter, as well as the required maturity levels. Mean organic carbon contents of all shales and siltstones analyzed are 2.6 and 1.7% for the Adventdalen and Reindalen sequences, respectively (Table 1). Most of these samples reveal moderate to high extract yields, examples of which are listed in Table 2. The stratigraphic distribution of shales and siltstones with good source rock quality within each core is as follows: thick intervals from 30.4 to 52.5 m (99.7 to 172.2 ft), 153.4 to 160.8 m (503.3 to 527.6 ft), and 232.2 m (761.8 ft) to total depth in the Adventdalen sequence; and 119.8 m (393 ft) to total depth in the Reindalen core and many thin layers interbedded

Table 2. Basic Source Rock Quality Data for Selected Shale Samples

Depth (m)	TOC (%)	C$_{15+}$-Soluble Organic Matter (PPM)	(mg/gTOC)	Hydrogen Index (mg/gTOC)
REINDALEN SITE				
62.5	4.8	1,524	31.8	82
112.8	3.0	766	25.5	81
121.5	2.4	1,824	76.0	120
124.5	5.7	1,436	25.2	135
127.0	5.7	2,310	40.5	97
ADVENTDALEN SITE				
33.5	1.9	828	43.6	72
38.1	1.5	1,311	87.4	64
48.3	2.2	1,877	85.3	92
134.4	2.2	672	24.9	71
137.7	2.8	1,221	43.6	52
155.5	1.6	1,708	106.8	69
162.2	2.9	1,183	40.8	68
164.0	2.8	756	27.0	84
188.4	2.8	1,138	40.6	123
212.5	7.2	3,650	50.7	97
234.0	1.5	680	45.3	54

over most of the continuous sandstone units of both sequences (Figure 1). The quality of the organic matter occurring in both sequences in shales, siltstones, and also in some sandstones remains uniformly of a hydrogen-lean type III kerogen nature, as recognized by the standard Rock-Eval analysis of many samples (Espitalié et al, 1977). In both sequences, there is no systematic change in hydrogen index, neither with depth nor with lithology. Shales and siltstones have almost identical mean hydrogen indices of 77.6 and 72.7 (mgHC/gC$_{org}$) for the Adventdalen and Reindalen sequences, respectively (Table 1), compared to means of 69.6 and 34.4 for the sandstones in the same series.

Based on data from routine geochemical analyses, there are clear indications that major redistribution and migration of light and heavy hydrocarbons have indeed occurred in both sequences (Leythaeuser et al, 1983a, b; Mackenzie et al, 1983). The so-called "production index" (calculated from Rock-Eval pyrolysis data as $PI = S_1/S_1 + S_2$, where S_1 = volatilized organic compounds up to 300°C and S_2 = components produced by the pyrolytic breakdown of kerogen and less volatile heavy heteroatomic components) reveals basic differences between shales and sandstones. As exemplified in Figure 2 for some samples from the Reindalen sequence, most shales have production indices between 0.1 and 0.2, which according to Espitalié et al (1977) is at, or slightly below, the accepted level of indigenous hydrocarbons for this maturity level. Some sandstones have, on the other hand, high S_1 and low S_2 signals resulting in significantly higher-than-normal production indices (maximum of 0.87 at 89.0 m, 292.0 ft, depth in the Reindalen sequence), indicating impregnation by migrated hydrocarbons. In the Reindalen sequence, most sandstone samples below 80 m (260 ft) show this evidence of impregnation. The Rock-Eval traces shown in Figure 3 illustrate the very high production index from an impregnated sandstone below 80 m (260 ft) and, for comparison, an example for the typical response of samples above 80 m (260 ft). The latter example has a production index of 0.25, which is usual for nonimpregnated samples of a maturity of 0.82% R_m, and is hence called "unmodified" (Figure 3). Table 3 includes other values of the production index for organic carbon-poor sandstones from Reindalen. The extract yields (Table 3) are also elevated for those samples that are designated as "impregnated" from Rock-Eval data.

The variation of the production index data in the Adventdalen sequence has been discussed in detail by Leythaeuser et al (1983b), so only a brief summary is given here. Most of the sandstones had indices above the normal values for unmodified sediments of their maturity determined by Espitalié et al (1977), whereas the interbedded shales had indices well below the presumed unmodified value. This difference was interpreted as the result of the shales having been depleted in their volatile components (S_1 peak) by their movement into the adjacent sands, which are consequently impregnated with additional hydrocarbons. Some of the highest production indices were recorded for fractured sandstone intervals at around 100 m (330 ft), and the values for the other sandstones showed a steady decrease in a symmetrical pattern above and below the fractured interval. It was suggested, therefore, that this pattern could be the result of the introduction of migrating hydrocarbons into the sandstone body via the fractures.

In summary, the Adventdalen and Reindalen cores appear to be well suited for a geochemical study of migration effects, because (1) they comprise interbedded sequences of source and reservoir beds, (2) the source beds have reached sufficient maturity levels for hydrocarbon generation as indicated by high extract yields, and (3) closely spaced core samples were available for analysis.

ANALYTICAL METHODS

Since all techniques applied in this study have been described in detail elsewhere, the discussion of analytical

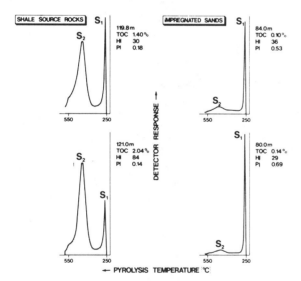

Figure 2—Pyrolysis yields of selected samples from Reindalen sequence. FID-traces from Rock-Eval pyrolysis show S_1 and S_2 peaks for source rock-type shale (left) and reservoir sand samples (right). Designation of "impregnated" samples according to Espitalié et al (1977). TOC = % organic carbon, HI = hydrogen index (mg/g TOC), PI = production index (S_1/S_1+S_2).

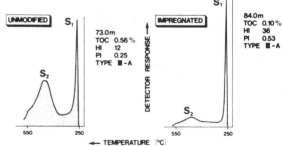

Figure 3—Examples of unmodified and impregnated sandstone samples from Reindalen sequence based on Rock-Eval pyrolysis traces. For explanations see Figure 2. Both samples bear same type of C_{15+} soluble organic matter, designated as type III-A (for explanation see Figure 4).

Table 3. Basic Geochemical Data for Selected Sandstone Samples

Depth (m)	TOC (%)	C_{15+}-Soluble Organic Matter (PPM)	(mg/gTOC)	Production Index $S_1/(S_1+S_2)$
REINDALEN SITE				
23.0	0.1	106	105.8	n.m.
56.5	0.2	253	126.5	0.16
58.0	0.1	160	160.0	0.29
61.0	0.2	149	74.5	0.36
84.0	0.1	228	227.5	0.60
89.0	0.2	174	87.0	0.87
98.9	0.5	152	30.4	0.50
ADVENTDALEN SITE				
52.2	0.5	356	71.2	0.17
93.5	1.9	551	29.0	0.35
109.4	0.9	544	60.4	0.28
139.4	1.3	1,205	92.7	0.29
162.5	0.8	374	46.8	0.36
170.0	0.6	258	43.0	0.11

methods is limited to the minimum level necessary to understand this communication. Organic carbon contents (TOC) were determined by the Leco method. The finely ground rock samples were extracted with dichloromethane, using a slightly modified flow blending technique (Radke et al, 1978), and the saturated and aromatic hydrocarbon fractions were separated by medium-pressure liquid chromatography (Radke et al, 1980). The saturated hydrocarbons were analyzed by glass-capillary gas chromatography. The concentrations of hydrocarbons of certain samples were quantified by the addition during extraction of known amounts of squalane as an internal standard, since this compound is not normally present in significant relative abundance in samples of the maturity studied. In addition, the alkanes of other samples were quantified gas chromatographically by comparison with n-pentadecane as an external standard.

The saturated hydrocarbon fractions were analyzed by combined gas chromatography/mass spectrometry (GC/MS), using methods and instrumentation described by Mackenzie (1983). Four masses were monitored during the analyses: m/z 191, 205, 217, 218, with dwell times of 400 ms and a scan rate of 2 seconds per scan to study the sterane and triterpane patterns (cf. Seifert and Moldowan, 1981).

The samples were also studied by Rock-Eval pyrolysis (Espitalié et al, 1977), and the mean reflectance of vitrinite particles was determined for polished blocks of the seam coals and selected shale and siltstone samples from both core holes. Table 4 shows that a very large number of samples was investigated by each method considering the total combined length of both cores was only 320 m (1,050 ft).

DETERMINATION OF ORGANIC MATTER TYPE

The main approach to recognize and quantify migration effects in the sequences studied was to compare the concentrations and compositions of extractable petroleum hydrocarbons from selected sample pairs representing geologic situations where migration could be expected to have occurred. Three types of sample pairs were chosen: the center portions of thick shale units versus (a) their edges, (b) thin interbedded shales, and (c) reservoir sands. Such comparisons are justified only if it can be ascertained that, as far as can be measured, the samples compared contain a similar type of organic matter of a similar maturity, as hydrocarbon distributions are known to be sensitive to these two factors (e.g., Tissot et al, 1974). In the Adventdalen and Reindalen sequences, it is unlikely that maturity varies much within the cores because of their shortness. Furthermore, mostly samples in close proximity (a few meters) were compared.

Monitoring fluctuations in the nature of the organic matter that was contributed to the sediment during deposition and which has survived diagenesis, was more problematic. It was not believed that the hydrogen index of Rock-Eval pyrolysis would reflect these fluctuations in a sufficiently sensitive manner. The distribution of certain complex molecules known as "biological markers" (because they preserve to a large extent the original carbon skeleton structure synthesized by living organisms) is thought to represent a more sensitive geochemical parameter for the study of fluctuations in the nature of the organic matter input. The distributions of steranes (see Appendix, I–IV, X) and triterpanes (Appendix, V–IX) in

Table 4. Analytical Methods and Samples Analyzed for Migration Study of Adventdalen and Reindalen Core Holes

Methods	No. of Samples
Organic carbon	167
Rock-Eval pyrolysis	135
Solvent extraction and liquid chromatography	100
Capillary gas chromatography of C_{15+} saturated hydrocarbons	
qualitative	100
quantitative	36
GC/MS of biological markers	51
Vitrinite reflectance	29

mature sediments appear to be well suited for that purpose (Mackenzie et al, 1982; Seifert and Moldowan, 1981). Therefore, the saturated hydrocarbon fractions of 51 samples from both core holes were analyzed for sterane and triterpane distributions by combined gas chromatography/mass spectroscopy.

Our knowledge of the sterane and triterpane distributions is not advanced enough to allow a definite assignment of paleoenvironment type for their distribution patterns. It is, however, generally thought that major differences in, for example, the relative abundances of these compounds relate to a difference in organic matter type (Huang and Meinschein, 1979; Mackenzie et al, 1982). Application of this principle allowed the subdivision of the samples into three subtypes in each sequence. These were called III-A, III-B, and III-C in the Reindalen, and III-D, III-E, and III-F in the Adventdalen series, respectively. There are some similarities between the three subtypes of Reindalen and those of Adventdalen rocks, but it was decided to keep the two sequences separate.

In samples bearing type III-A organic matter, for steranes with the same stereochemistry, the concentrations of C_{21}, C_{22}, C_{27}, C_{28}, and C_{29} species are about equal (Figure 4). As well as the ubiquitous pentacyclic $17\alpha(H),21\beta(H)$-hopanoid triterpanes (Van Dorsselaer et al, 1974), a significant amount of tricyclic and tetracyclic terpanes (Aquino et al, 1982; Trendel et al, 1982) is also present in the samples of this organic matter type (Figure 4). Samples bearing type III-C organic matter reveal completely different distributions of biologic markers: the steranes are dominated by C_{29} components and no tricyclic terpanes are present (Figure 5). Of the 37 Reindalen samples analyzed by GC/MS techniques, 33 could be classified as III-A or III-C. The remaining four have C_{27} to C_{29} steranes similar to type III-A, but much lower amounts of C_{21} and C_{22} steranes and of tricyclic terpanes. They were called type III-B. One of the most convenient ways of distinguishing between types III-A and III-B and type III-C in the Reindalen sequence is a triangular diagram (Figure 6) of the relative concentrations of C_{27} to C_{29} $5\alpha(H)$, $14\beta(H)$, $17\beta(H)$, 20R + 20S-steranes (see Appendix, IV), as proposed by Huang and Meinschein (1979).

The main argument as to why these variations in the biologic marker distributions are interpreted to reflect changes in the type of organic matter input, yet cannot be explained by migration effects, is provided by consideration of the stratigraphic distribution of the types III-A, III-B, and III-C in the Reindalen sequence (Figure 7). All samples from the siltstone unit above 55 m (180 ft), including that from the conglomeratic sandstone, belong to type III-A, with one exception from 61 m (200 ft). Likewise, most sandstone samples can be classified as type III-A with the exception of the two siltstone and two sandstone samples closest to the lithologic boundary at 55 m (180 ft), which belong to type III-B. On the other hand, the steranes and triterpanes of all the shales and coals conform to type III-C. Thus, the distribution of organic matter types in the Reindalen sequence appears to be controlled by grain-size variations: coarser grained lithologies—sandstones and siltstones—contain one type of organic material (III-A based on steranes and triterpanes), whereas finer grained lithologies—shales and coals—

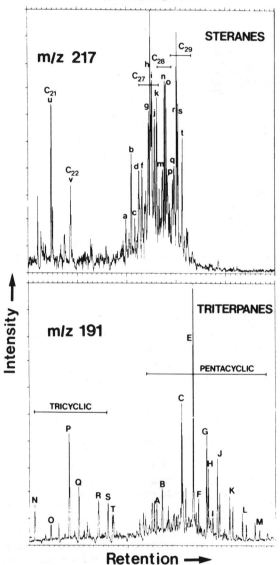

Figure 4—Mass fragmentograms m/z 217 and 191 from GC/MS analysis of siltstone sample from Reindalen sequence, whose organic matter type is classified as type III-A, to show sterane and terpane patterns characteristic for all samples of this type. For explanation of lettering of peaks see Appendix.

contain a quite different type of organic material (type III-C). These differences could not be the result of migration effects, but must mainly reflect differences in the nature of the organic sedimentation and the diagenetic conditions in bottom sediments during deposition.

Steranes derive from sterols (Rhead et al, 1971), and the

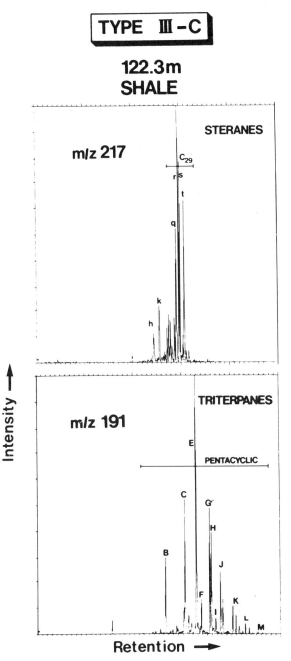

Figure 5—Mass fragmentograms m/z 217 and 191 from GC/MS analysis of shale sample from Reindalen sequence, whose organic matter type is classified as type III-C, to show sterane and terpane patterns characteristic for all samples of this type. For explanation of lettering of peaks, see Appendix.

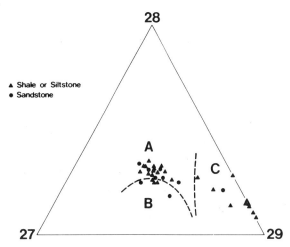

Figure 6—Triangular diagram of relative abundances of C_{27} to C_{29} $5\alpha(H)$, $14\beta(H)$, $17\beta(H)$-steranes for all samples analyzed from Reindalen sequence, with fields for three designated organic matter types (III-A, III-B, III-C) shown.

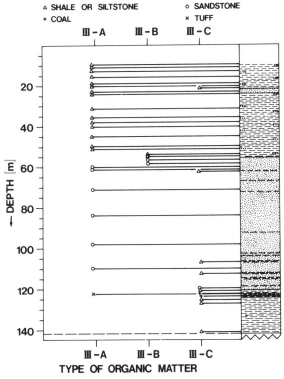

Figure 7—Variation of organic matter type (as defined by composition of biologic markers, see Figures 4-7) as function of depth and lithology for Reindalen sequence.

absence, or low relative abundance, of all carbon numbers other than C_{29} in the shales and coals (type III-C) suggests that the precursor sterols $< C_{29}$ were not deposited in these sediments. Either their biological production ceased during these depositional time periods, or it continued at a greater distance from the Reindalen site and higher energy depositional environments of siltstones and sandstones were required to transport the precursor sterols $< C_{29}$ to the site of deposition in greater amounts. Similar arguments could be advanced to explain the variation of tricyclic terpane concentrations between type III-A and type III-C. The occurrence of shales, without tricyclic terpanes and with steranes < C_{29} in low relative abundance (type III-C), interbedded with sandstones, where the tricyclic terpanes and steranes < C_{29} are present in equivalent amounts to pentacyclic triterpanes and C_{29} steranes respectively (Figure 7), indicates that migration of steranes and terpanes from the shales into the adjacent sands has not occurred. This conclusion is supported by the results of n-alkane analyses discussed later, and suggests that the sterane and terpane distributions can be considered indigenous. As such, their use as indicators of organic matter type is valid. Between the two coal seams of the Reindalen sequence, a volcanic ash deposit is present, whose sterane and terpane distributions are of the "high-energy" pattern (type III-A) in contrast to the type III-C signatures in the coals above and below (Figure 7).

The steranes and triterpanes of the Adventdalen sequence were studied in less detail. Instead of analyzing the whole core, two selected intervals were analyzed by GC/MS techniques. Within the samples analyzed, three subtypes are recognized, although apart from the sterane carbon number distributions, there is little to distinguish the patterns of the different sediments. It should, therefore, be emphasized that the minor differences in sterane carbon number distributions need not always imply significant compositional differences in the rest of the extractable organic matter: the total concentrations of steranes and triterpanes in extracts of sediments with a maturity of 0.8% R_m are, in our experience, well below 1%. The sandstones and shales between 150 and 165 m (492 and 541 ft) plot mainly in two designated areas on the triangular diagram (Figure 8) illustrating the carbon number composition for C_{27}-C_{29} 5α(H), 14β(H),17β(H), 20R + 20S steranes called types III-D (sandstones) and III-E (shales). An exception is the sandstone at 158 m (518 ft), whose steranes plot in a third field, called type III-F, where all the siltstones and shales between 37 and 45 m (121 to 148 ft) depth also occur.

The designations type III-A to type III-F are used to ensure that samples of equal organic matter type are compared. It is reemphasized that the paleoenvironmental significance of these subdivisions is not clear. They are only measures of compositional differences of the biologic markers thought to have been caused by variation in the nature of the depositional environment, and not by maturity differences (see discussion in Mackenzie, 1983) or migration effects.

RECOGNITION OF MIGRATION EFFECTS BASED ON COMPOSITIONAL PATTERNS

Qualitative comparison of compositional patterns of samples from the Reindalen sequence revealed several effects that could best be explained by migration processes (Leythaeuser et al, 1983b). Even depth plots of bulk geochemical parameters, like total extract yields (mg/g C_{org}), showed those effects as exemplified in Figure 9. The uppermost 55-m (180 ft) interval of the Reindalen sequence is an organic carbon–lean, light-gray siltstone (mean TOC = 0.50%) with a 0.20 m (8 in.) thin conglomeratic sandstone unit interbedded in the middle. The carbon-normalized extract yields decrease markedly toward the thin sandstone interbed both above and below it. The latter bears an extract concentration almost three times higher than adjacent siltstone samples at either side do. The composition of the saturated hydrocarbon fraction of these three samples (marked by arrows in Figure 9) is shown in Figure 10. The predominant peaks in these and all following gas chromatograms represent n-alkanes and are marked by their carbon number, whereas the other peaks are due to iso-alkanes, anteiso-alkanes, and acyclic isoprenoid alkanes (among which pristane and phytane, marked A and B in all gas chromatograms, are usually the most abundant). A marked difference in the saturated hydrocarbon composition is obvious from Figure 10. The sandstone sample has a relatively smooth, front-end biased distribution, whereas the siltstones above and below exhibit broad, slightly bimodal distribution patterns. This bimodality becomes gradually more pro-

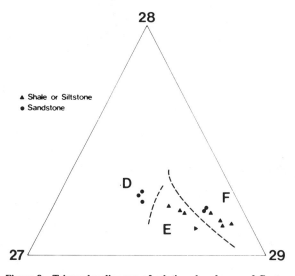

Figure 8—Triangular diagram of relative abundances of C_{27} to C_{29} 5α(H), 14β(H), 17β(H)-steranes for all samples analyzed from Adventdalen sequence, with fields for three designated organic matter types (III-D, III-E, III-F) shown.

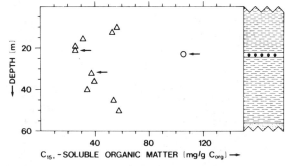

Figure 9—Yield of C_{15+} soluble organic matter as function of depth and lithology (triangles = siltstone: circle = conglomeratic sandstone) for depth interval of 10–60 m (33–197 ft) in Reindalen sequence. Gas chromatograms of samples indicated by arrows are shown in Figure 10.

nounced in this siltstone unit above 21 m (69 ft) and also below 32 m (105 ft) (Leythaeuser et al, 1983b). Thus, it appears that the lower carbon number alkanes have moved preferentially and are hence enriched in the sandstone unit relative to the higher carbon numbers. These compositional differences are likely to reflect migration effects, since all samples shown in Figure 10 bear an identical organic matter type (III-A) and also have an identical maturity level. The movement of hydrocarbons through this siltstone and from it into the sandstone interbed has led to compositional fractionation as a result of differential drainage conditions and possibly related chromatographic-type processes.

Similar effects were seen near the edges of thick source rock–type shale and siltstone units in the Adventdalen sequence (Mackenzie et al, 1983). Changes of saturated hydrocarbon distributions with depth are shown in Figure 11 for an organic-rich siltstone unit (1.5–2.4% TOC) in contact with an underlying sandstone, and in Figure 12 for a 7.4-m (24.2-ft) thick source rock shale (1.2–2.3% TOC) overlain and underlain by sandstone units. From both Figures 11 and 12, regular changes in the alkane distributions are observed to be a function of the lithology of the sample; within the source rock units themselves, they are observed to be a function of the proximity of the samples to the nearest sandstone contact. All samples from the center of the source rock units exhibit broad bimodal alkane distributions (with a valley in the distribution envelope around C_{20}) and slight predominances of the odd carbon numbered n-alkanes above C_{23}. The source rock samples closest to the sandstone contacts have unimodal alkane distributions skewed to higher carbon numbers. There are no differences between the samples from the edges and those from the center of the source rock units, with respect to their organic matter type or their maturity. Therefore, the observed differences could be interpreted as compositional fractionation effects associated with migration. Due to their greater mobility, lower carbon

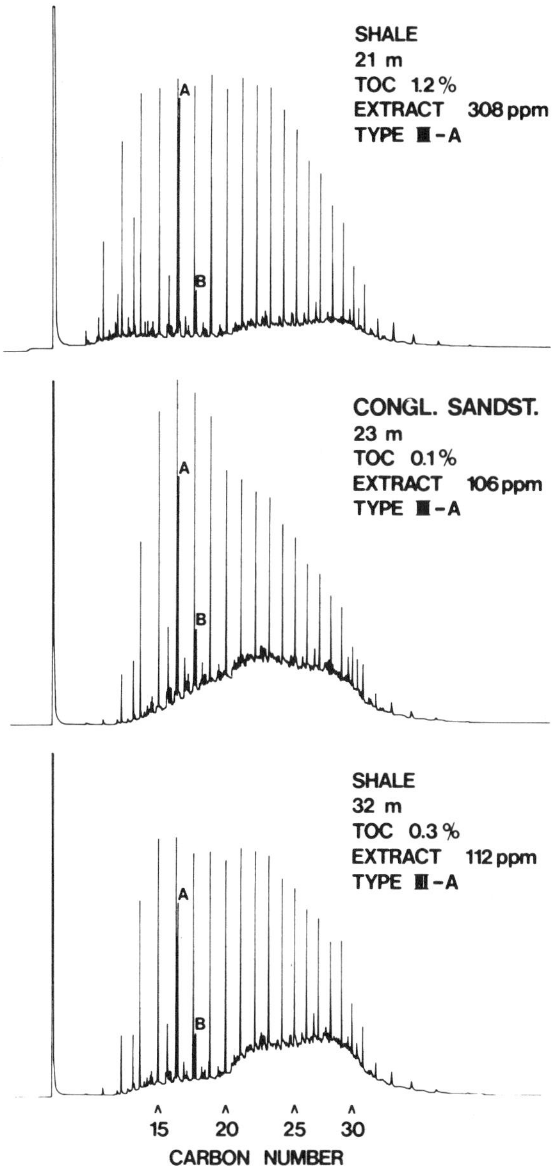

Figure 10—Capillary gas chromatograms showing composition of C_{15+} saturated hydrocarbons for conglomeratic sandstone and two adjacent siltstone samples bearing same type III-A organic matter from Reindalen sequence. Stratigraphic positions of these three samples are shown in Figure 9. Carbon number per molecule for n-alkanes shown at bottom. A = pristane; B = phytane.

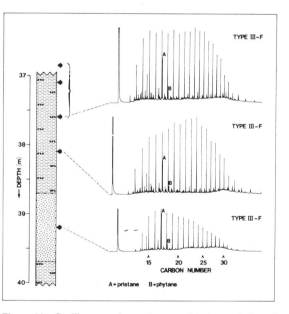

Figure 11—Capillary gas chromatograms showing variation of distributions of C_{15+} saturated hydrocarbons as function of depth and lithology for interval from Adventdalen sequence. All samples analyzed (arrows) bear same type III-F organic matter, as defined by biologic marker composition. Uppermost chromatogram is representative for three samples analyzed between 36.0 and 37.6 m (118 and 123 ft) depth. For explanation of abbreviations, see Figure 10.

number n-alkanes have been expelled preferentially. Consequently, the hydrocarbon distributions in the adjacent sandstone samples are front-end biased and smooth, as clearly shown in Figures 11 and 12.

In summary, the better drained portions of the source rock units close to the sandstones have lost a significant part of their alkanes below about C_{20} and have preferentially retained the higher molecular weight material. In this way, their originally bimodal hydrocarbon distribution became fractionated into two portions of opposite modality. The more mobile front-end portion represents the migrated part that accumulated in the reservoir sandstones, whereas the C_{20+} portion was retained to a greater extent in the depleted source rock.

In the interval from 150 to 163 m (492 to 535 ft) in the Adventdalen sequence (Figure 12), there are some differences in the type of organic matter between the shale and the sandstone samples. However, they do not invalidate the foregoing interpretation of the changes in alkane distributions as migration effects. Within the 7.4-m (24.2-ft) thick shale itself, the organic matter remains uniformly of a type III-E nature and hence the interpretation of preferential expulsion from the edges of this unit remains. Despite a slightly different biologic marker distribution (type III-D) in the sandstone samples analyzed from 150.0, 161.0, and 162.5 m (492, 528, and 533 ft), it is believed that the bulk of the saturated hydrocarbons in these samples has been derived from the shale interval 153.4–160.8 m (503.3–527.6 ft). This reasoning is based on the expected low hydrocarbon generation potential of these sandstone samples due to their low organic carbon contents (0.4, 0.2, and 0.8% in foregoing listed sequence).

Another approach to recognize migration effects was to compare hydrocarbon compositions between the two types of shales (i.e., the very thin shale interbeds in the sandstone intervals versus those from the center of thick shale units). An example of this kind from the Reindalen sequence is shown in Figure 13. Despite equal conditions for hydrocarbon generation in both samples (similar type of organic matter, III-C, and maturity), their alkane distributions differ markedly. The shale at 127.0 m (417 ft) has a broad, bimodal alkane distribution with slight predominances of the odd carbon numbered n-alkanes between C_{23} and C_{29}. Although the shale at 62.5 m (205 ft) also has the latter compositional feature, it has markedly reduced concentrations for the n-alkanes at the front end of the chromatogram. Furthermore, the isoprenoid alkanes pristane and phytane are not reduced in concentration to the same relative extent as their n-alkane isomers. Because variations in organic matter type and maturity can be excluded, the main difference between the two samples shown in Figure 13 must concern their position in the stratigraphic sequence (i.e., their proximity to the nearest permeable reservoir bed). Sample 127.0 m (417 ft) is close to the center of a thicker shale unit and 6.5 m (21 ft) below the nearest sandstone contact, whereas sample 62.5 m (205 ft) is from a thin 5-cm (2-in.) shale interbed in an otherwise continuous reservoir sandstone sequence. The total 5-cm (2-in.) length of this shale core was used for geochemical analysis. Therefore, the compositional differences between both samples shown in Figure 13 is best explained by assuming the thin shale has, owing to its proximity to porous sands and hence better drainage conditions, preferentially expelled the lower carbon number n-alkanes. Due to their lesser degree of depletion, the isoprenoids have been relatively enriched in the thin shale. This is especially obvious in Figure 13 by the high predominance of pristane over the adjacent n-alkanes.

QUANTIFICATION OF MIGRATION EFFECTS LEADS TO DETERMINATION OF EXPULSION EFFICIENCIES

The previous observations concerning migration effects were based on a purely qualitative comparison of hydrocarbon distributions. However, it was thought that the key to a better understanding of migration processes would be to consider absolute concentrations of individual petroleum-range hydrocarbons. We have chosen to calcu-

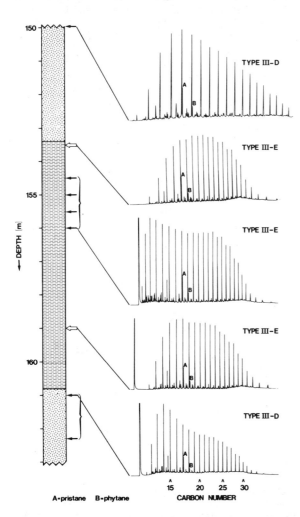

Figure 12—Capillary gas chromatograms showing variation of distributions of C_{15+} saturated hydrocarbons as function of depth and lithology for interval from Adventdalen sequence. All samples analyzed (arrows) bear indicated organic matter type as defined by biologic marker composition.

late concentrations in the rock (μg/g rock) instead of concentrations normalized to organic carbon, which is commonly used in petroleum geochemistry for evaluation of source rock quality. Because the alkanes are considered mobile, the latter concentration units have less meaning due to the inability of the kerogen to migrate.

In Figure 14, the absolute concentrations of alkanes are compared for the same two samples as shown in Figure 13. The concentration difference between both samples must, on the basis of the discussed reasoning, be explained by different degrees of depletion, and is not influenced by differences in amounts of hydrocarbon-generating organic matter. Both shale samples have similar organic carbon contents (62.5 m or 205 ft, 4.8% TOC; 127.0 m or 417 ft, 5.7% TOC). The shaded area in Figure 14, the difference between the thin "depleted" shale and the "unmodified" center of the thick shale, indicates the amount and the composition of the hydrocarbons that have been expelled from the former. Obviously, with increasing molecular size, lesser and lesser quantities of n-alkanes have been expelled from the thin shale. Little or no expulsion of n-alkanes has occurred beyond C_{25+}. As the branched isoprenoid alkanes appear to be expelled to a lesser extent than their nonbranched isomers, the foregoing conclusion of little redistribution of the higher carbon number steroid and triterpenoid hydrocarbons is supported by this finding. The fact that concentration levels in the C_{25+} region are nearly identical (in view of the error limits of the analytical methods) provides additional support for the similarity between both samples with respect to organic matter type and maturity, and hence confirms the validity of our approach for recognition of migration effects.

The amount and composition of the saturated hydrocarbon mixture expelled from the thin shale at 62.5 m (205 ft) into the adjacent reservoir sandstones, as reconstructed by above material balance (Figure 14), are shown as trend A in Figure 15. A very similar alkane distribution envelope was indeed observed in the impregnated sandstone interval of the same sequence (Figure 15, trend B). The similarity between trends A and B is encouraging considering the error limits of the analytical methods. One interesting observation from Figure 15 concerns the distribution envelope of the expelled hydrocarbon mixture, which is smooth, front-end biased, and has a sharp concentration drop from C_{15} to C_{25}. It corresponds to that of a condensate or a very light oil of a presumably high-maturity origin, which is surprising, since these hydrocarbons were generated from shales bearing type-III kerogens. Thus, the observations made here offer a viable explanation for the origin of so-called early diagenetic condensates discovered in many low mature sequences. The fractionation effects associated with expulsion appear to be the main control on the composition of the hydrocarbon product

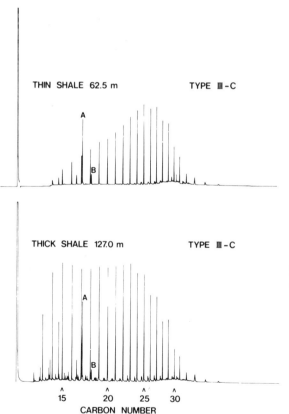

Figure 13—Capillary gas chromatograms showing comparison of distributions of C_{15+} saturated hydrocarbons for two samples from Reindalen sequence. Sample 62.5 m (205 ft) is from thin (5 cm; 2 in.) shale interbedded in sandstone interval. Sample 127.0 m (417 ft) is from the center portion of a thick shale unit. Both samples are from organic-rich, source rock-type shales bearing same type III-C organic matter. For explanation of abbreviations, see Figure 10.

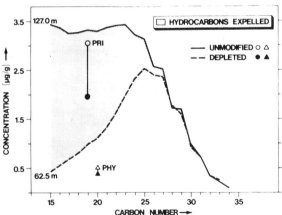

Figure 14—Quantitative comparison of absolute amounts of individual normal and isoprenoid alkanes (μg/g sediment, based on quantitative evaluation of gas chromatograms) for same shale samples as shown in Figure 13. Designation of samples as "unmodified" and "depleted" explained in text. Concentration difference between samples (shaded area) represents hydrocarbon mixture expelled from shale sample 62.5 m (205 ft) during primary migration. PRI = pristane; PHY = phytane.

accumulating in reservoirs, rather than the type of organic matter in the source rock.

When the amounts of individual alkanes presumed to have been expelled (Figure 15, trend A) are expressed as a percentage of the amounts in the unmodified source rock (Figure 14, 127.0 m or 417 ft), a measure for the expulsion efficiency for each compound is obtained (Mackenzie et al, 1983). These values are shown in Figure 16 for the thin shale at 62.5 m (205 ft) depth. Expulsion efficiencies are very high for the lower carbon number range (e.g., reaching over 80% for n-C_{15}), but drop regularly with increasing molecular chain length and approach zero in the C_{25+} region. The isoprenoid hydrocarbons have lower expulsion efficiencies than their straight chain isomers (e.g., pristane has approximately half the expulsion efficiency of n-C_{19}). Thus, expulsion of alkanes from the thin source

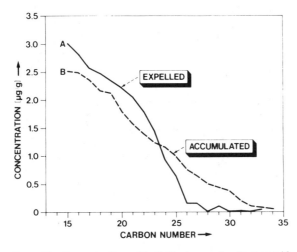

Figure 15—Concentrations of individual normal and isoprenoid alkanes for following examples from Reindalen sequence: A, Hydrocarbon mixture expelled from thin shale 62.5 m (205 ft), based on determination explained in Figure 14.
B, Hydrocarbons impregnating reservoir sandstone at 84.0 m (276 ft) depth.

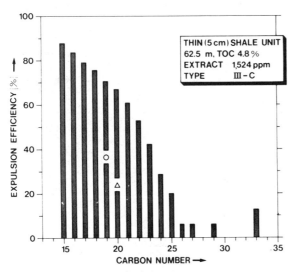

Figure 16—Efficiency of expulsion (%) of individual normal and isoprenoid alkanes from thin shale 62.5 m (205 ft) in the Reindalen sequence. Expulsion efficiencies calculated by expressing concentration values of migrated hydrocarbon mixture shown in Figure 15(A) as percentages of amounts measured for "unmodified" source rock shown in Figure 14. Expulsion efficiencies of n-alkanes (bars), pristane (circle), and phytane (triangle).

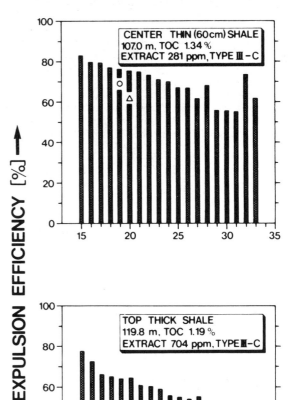

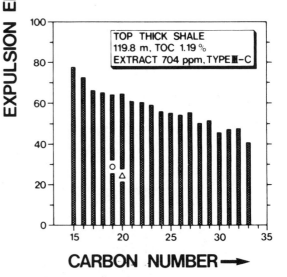

Figure 17—Efficiency of expulsion (%) of individual normal and isoprenoid alkanes from center of a 0.60 m (24 in.) thick shale unit at 107.0 m (351 ft) depth and from uppermost part of thick shale unit extending from 119.8 m (393 ft) downward in Reindalen sequence. Both samples are from organic-rich, source rock-type shales bearing same type III-C organic matter. Expulsion efficiencies were calculated as explained in Figure 16 by hydrocarbon mass balance between "unmodified" shale 127.0 m (417 ft) (Figures 13, 14) and each shale sample at 107.0 and 119.8 m (351 and 393 ft) depth. For explanation of symbols, see Figure 16.

rock shale at 62.5 m (205 ft) was associated with fractionation effects controlled not only by molecular weight but also by molecular structure. This mass balance approach to determine expulsion efficiencies could only be applied to a limited number of shale samples, because of the requirements of similar organic matter type (as defined by sterane and triterpane fingerprints). Figure 17 shows the expulsion efficiencies (determined by quantitative comparison with sample 127.0 m, 417 ft) for a sample from the center of a thin, 0.60-m (2.4-in.) shale interbedded in sandstones and for the topmost part of the thick source rock shale interval in the bottom part of the Reindalen sequence. Similar observations can be made. The compositional fractionation is, however, less pronounced, and for both examples expulsion of C_{30+} alkanes has occurred. These differences are possibly related to differences in pore-size distributions for which no measurements are available. In summary, the expulsion efficiencies for the lower carbon number n-alkanes from thin interbedded shale source rocks, as well as from the edges of a thick shale interval, are generally high and decrease with increasing chain length. Isoprenoids generally have lower expulsion efficiencies than their straight chain isomers (i.e., they are preferentially retained in the source rocks). Thus, the depleted source rocks can have alkane compositions, which suggests a lower maturity than is the case. Extreme cases, like the thin shale at 62.5 m (205 ft) can have hydrocarbon distributions reminiscent of biodegraded oils (Winters and Williams, 1969; Milner et al, 1977).

The same approach has also been applied to determine expulsion efficiencies of total extracts for the same source rock shales from the Reindalen sequence. This application gave a mean value of 23% for these three depleted source rock shales. In view of the previously documented fractionation effects, it appears unrealistic to consider a single whole-oil expulsion efficiency value for a particular source rock. Rather, separate expulsion efficiencies should be considered for individual compounds or compound classes.

It should also be possible to determine expulsion efficiencies for source rock shales based on quantification of alkanes of impregnated sandstones from the Reindalen sequence. Provided these alkanes were derived from the thin interbedded shales and had not experienced much lateral transport within the sandstone intervals themselves, it should be possible by a simple calculation to obtain the alkane concentrations originally present in the generating source rock prior to expulsion (i.e., combining those migrated into the sand with the residual ones in the depleted shales). The alkanes of the sandstone sample from 84.0 m (276 ft), whose production index (PI = 0.53) suggests impregnation, were compared with those of a nearby thin and depleted shale from 62.5 m (205 ft; PI = 0.16). Although this sandstone sample has a different sterane and triterpane distribution from the shale (type III-A versus type III-C, Figure 7), if there has been only minor redistribution of these polycyclic alkanes from shale to sandstone (as suggested) compared with that of the lower carbon number alkanes, then this recombination should still be valid. Besides, consideration of type III-A and type

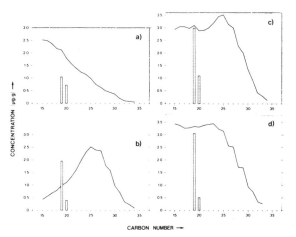

Figure 18—Concentration of individual normal and isoprenoid alkanes for following examples from Reindalen sequence: a—Sandstone sample at 84.0 m (276 ft) depth, impregnated by migrated hydrocarbons. b—Thin (5 cm or 2 in.) shale sample at 62.5 m (205 ft), which is depleted in hydrocarbons as shown in Figures 13 and 14. c—"Reconstituted" hydrocarbon mixture in original source rock, obtained by combining a and b. d—Sample from center portion (127.0 m or 417 ft) of thick shale unit bearing "unmodified" hydrocarbon concentrations. Note that concentration and composition of hydrocarbons in "reconstituted" source rock (c) resembles that of actual source rock (d). Pristane and phytane shown as bars at C_{19} and C_{20}, respectively.

III-C n-alkane distributions from the samples in the Reindalen section, thought to have been unmodified by depletion or impregnation, revealed only minor differences compared to the major differences between "depleted" shales and "impregnated" sandstones.

The quantified C_{15+} hydrocarbon distributions are shown in Figure 18a-d. The impregnated sandstone (Figure 18a) has a unimodal n-alkane distribution pattern biased toward lower carbon numbers, whereas the depleted shale (Figure 18b) contains n-alkanes that are unimodal and biased toward higher homologues. The isoprenoid alkanes—pristane and phytane—are present in higher relative abundance in the depleted shale. Although the sandstone has only 2% of the organic carbon content of the shale (0.09 versus 4.8%) and 20% of its extract concentration (273 versus 1,524 ppm), the absolute amounts of n-alkanes in the two samples are comparable. Indeed, the amounts of C_{15} to C_{20} n-alkanes are higher in the sandstone. The expulsion of n-alkanes and particularly lower carbon numbers is highly favored relative to other solvent extractable organic components. The sum of the "impregnated" and "depleted" distributions (Figure 18c) bears a close resemblance to the n-alkane pattern from the center of the thick shale unit considered previously (Figure 18d).

Expressing the concentrations of the alkanes of the impregnated sandstone (Figure 18c) as percentages of the summed concentrations of the impregnated and depleted sample provides an alternative measure for the expulsion efficiencies of individual alkanes. The values (Figure 19) are very similar to those obtained by comparing thin shales with the center of thick shales (Figure 16). Again, about 80% of the lower carbon number range n-alkanes appear to have been expelled and pristane has about half the expulsion efficiency of n-C_{19}.

LEAKAGE FROM RESERVOIR ACCUMULATION INTO CAP ROCK

The extract and pyrolysis data from the region of the siltstone/sandstone contact at 55.5 m (182 ft) in the Reindalen sequence were, at first, difficult to explain using the ideas developed so far. The upper portion of the sandstone interval below 55.5 m (182 ft) may have contained a small oil accumulation at some time in the past. When the oil column below the siltstone contact reached the length required to achieve the buoyancy necessary to overcome the difference in capillary pressures, it would have forced itself into the lowermost part of the siltstone cap rock. This movement appears to have proceeded to just beyond the siltstone sampled at 55.0 m (182 ft) and not to have reached 54.0 m depth (177 ft) (Figure 20), as indicated by the following evidence: despite similarly lean organic carbon contents, the lowermost siltstone sample at 55.0 m (180 ft) has a drastically higher extract yield than the one farther away from the sandstone contact (Figure 20). A comparison of the pyrolysis data of both samples reveals clear evidence for impregnation of the lower siltstone sample (Figure 21). This is shown by its production index of 0.51, whereas a value of 0.21 (sample 32.0 m, 105 ft) suggests its hydrocarbons are indigenous (Leythaeuser et al, 1983b). Therefore, the sample from 55.0 m (180 ft) is referred to as the "impregnated cap rock" in the following discussion.

The alkanes of the two siltstone samples shown in Figure 20 were quantified in order to examine the mechanism of reservoir leakage and cap-rock enrichment. The absolute amounts of individual hydrocarbons in the impregnated siltstone cap rock are compared in Figure 22 with those of the shallower and unmodified siltstone. Although the shallower sample has twice the organic carbon content of the deeper impregnated sample, this increase is in the direction opposite to the increase in extract yield (Figure 20) and to the increase in individual alkane concentrations (Figure 22). The enrichment factors of individual alkanes, calculated by dividing the differences between the values for the impregnated cap rock (Figure 22b) and those of the unmodified siltstones (Figure 22a) by the unmodified

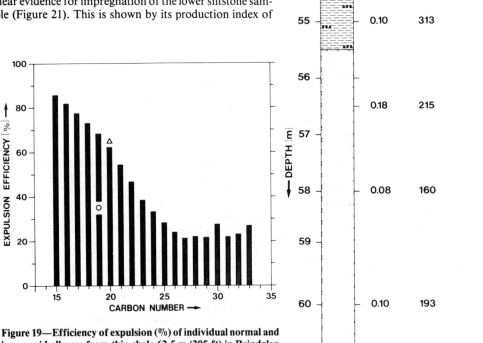

Figure 19—Efficiency of expulsion (%) of individual normal and isoprenoid alkanes from thin shale 62.5 m (205 ft) in Reindalen sequence. Expulsion efficiencies were calculated by expressing concentration values of impregnating hydrocarbon mixture in sandstone sample 84.0 m (276 ft) (Figure 18a) as percentages of the amounts determined for the "reconstituted" source rock (Fig. 18c). For explanation of symbols, see Figure 16. Note that values for expulsion efficiencies of individual compounds determined in the above explained manner for thin shale 62.5 m generally match (with some exceptions) those determined according to different mass balance concept explained in Figure 16.

Figure 20—Schematic lithology of interval 53.5–61.0 m (175.5–200 ft) of Reindalen sequence showing position, basic geochemical data, and interpreted status of extractable hydrocarbons for samples analyzed. Note that lowermost sample from siltstone unit (55.0 m or 180 ft) has, despite lower TOC content, much higher concentration of total extract than sample at 54.0 m (177 ft) from same siltstone unit.

amounts, are shown in Figure 23. All components considered have enrichment factors between about 5 and 6 and there is no difference between normal and isoprenoid alkanes. The corresponding enrichment factor value for the total C_{15+} extract is almost identical (6.1). It appears, therefore, that the impregnation of the siltstone cap rock, occurred without significant fractionation. The mechanism that produced this "cap-rock effect" seems therefore to have involved bulk oil flow. There are no visible traces of the small oil accumulation left today; it can only have existed temporarily.

INTERPRETATION OF MIGRATION PHENOMENA IN TERMS OF MECHANISM

In this section, we attempt to deduce the mechanism and timing of migration events from observations outlined in detail previously. As a basis for this discussion, Figure 24 presents a schematic summary of the stratigraphic occurrences of the major migration phenomena observed in the Reindalen sequence. Down to a depth of 55.5 m (182 ft), the section comprises a relatively organic carbon–lean siltstone unit with a thin (0.20 m, 8 in.) conglomeratic sandstone layer near its center. This porous and permeable layer has functioned as a migration avenue and preferentially depleted the siltstone unit above and below it (Figure 9). The lowermost part of the siltstone unit has functioned as a cap rock, albeit an inefficient one, for a small hydro-

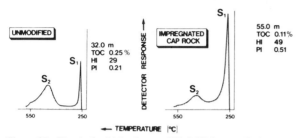

Figure 21—Pyrolysis yields of unmodified (32.0 m or 105 ft) versus impregnated (55.0 m or 180 ft) sample from same siltstone unit in Reindalen sequence. For further explanations see Figure 2.

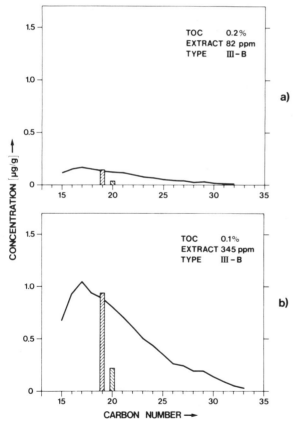

Figure 22—Absolute amounts of individual normal and isoprenoid alkanes (μg/g sediment) for two samples from same siltstone unit in Reindalen sequence: a—Unmodified siltstone at 54.0 m (177 ft) depth. b—Impregnated siltstone due to cap-rock effect at 55.0 m (180 ft) depth. Pristane and phytane shown as bars at C_{19} and C_{20}, respectively.

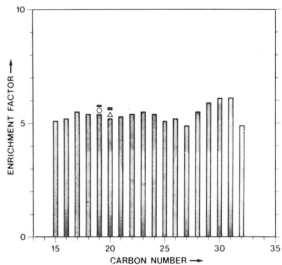

Figure 23—Enrichment factors for individual normal and isoprenoid alkanes of impregnated siltstone cap rock at 55.0 m (180 ft) in Reindalen sequence, calculated by dividing differences between concentration values given in Figure 22 (a) and (b) by those given in Figure 22 (a).

carbon accumulation that existed in the underlying reservoir sandstones (Figures 20–23). The thick sandstone interval extending from 55.5 to 119.8 m (182 to 393 ft) is enriched in migrated hydrocarbons (Figures 2, 3, 18, 19) to varying degrees. This variation is thought to be controlled by both differences in porosity and permeability (for which no measurements are available) and in the route taken by hydrocarbons during the secondary migration process, given that the Reindalen well was not drilled on a structural high. Throughout this thick sandstone unit, sev-

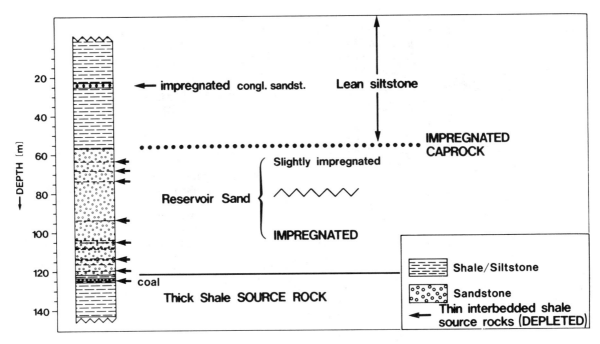

Figure 24—Schematic summary of major migration phenomena observed in Reindalen sequence in relation to depth and lithology.

eral thin shale interbeds are present that are rich in organic carbon and have a similar organic matter type to the underlying thick source rock shales. These thin shale interbeds are highly depleted in hydrocarbons since they have experienced efficient expulsion (Figures 7, 13-16). From 119.8 m (393 ft), a continuous dark-gray shale unit with two coal seams intercalated in its upper part is present. This shale unit can be classified as a hydrocarbon source rock, which has generated hydrocarbons and released them preferentially from its uppermost portion (Figures 7, 13-16).

Consideration of the interpretations presented in this and previous papers concerned with the Adventdalen and Reindalen sequences reveals, at first inspection, several apparent inconsistencies. Their clarification requires that the timing of migration events be discussed. Leythaeuser et al (1983b) proposed that there is active diffusion of light hydrocarbons (C_2–C_8) from the siltstone in the top part of the Reindalen sequence into the sandstone below. Leythaeuser et al (1983b) suggested, however, that petroleum-range hydrocarbons had moved in the opposite direction, from an accumulation into the overlying cap rock. These two interpretations may only be reconciled if the cap-rock impregnation occurred before the diffusion of the light hydrocarbons. The accumulation, which caused the cap rock to be impregnated, could have formed and existed temporarily when the sequence was deeply buried. Later uplift and tilting may have caused the accumulation to either find another trap or to leak to the surface.

Figure 25 attempts to explain schematically how the fractionation phenomena could evolve with time and/or migration distance (represented in vertical sequence by stages I to III). Starting with a bimodal n-alkane mixture, generated in the source rock (top of Figure 25), the lower molecular-weight portion is considered to be expelled preferentially. This leads to an absolute enrichment of the lighter molecules in the migrating fluids and a relative enrichment of the higher molecular compounds in the residual hydrocarbon mixture retained by the source rock (left and right, respectively, of Figure 25). However, the isoprenoids, exemplified in Figure 25 by pristane, are expelled to a lesser degree compared with their straight-chain isomers. Ultimately, the original bimodal n-alkane mixture is split into two quite different distributions, one biased to the lighter components, the other to the heavier species. Figure 25 assumes for the purpose of simplification that the total hydrocarbons are split 50:50 between the migrated and residual hydrocarbon mixtures, but because of fractionation more than 50% of the original hydrocarbons < C_{21} migrate, and less than 50% of those > C_{21} migrate, resulting in the relative enrichment of the migrated fraction in hydrocarbons < C_{21}. This fractionation is developed to an increasing extent moving from stages I to III in Figure 25.

At the most advanced stage of depletion (III in Figure 25), pristane clearly predominates over the n-alkane envelope. Thus, the pristane/n-C_{17} ratio, which has been interpreted in the past as a maturity indicator (Durand and Espitalié, 1973; Connan and Cassou, 1980), could also indicate the degree of expulsion. In the scheme of Figure 25, this ratio increases from 0.8 in the original sample, with increasing expulsion in the residual source rock extracts to 1.1, 1.7, and 2.3 at the depletion stage III. Following the same principle, it is conceivable that also the odd/even predominance of the high molecular weight n-alkanes (Bray and Evans, 1961) is influenced by migration

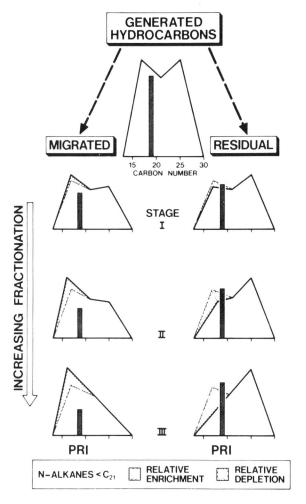

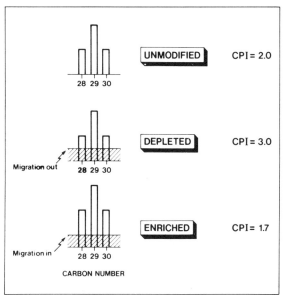

Figure 26—Scheme to illustrate how original predominance of the n-C_{29} alkane (expressed as Carbon Preference Index) of a hypothetical source rock ("unmodified" condition) is changed by expulsion of n-alkanes ("depleted" condition), or by impregnation and admixture of n-alkanes from a different source ("enriched" condition).

Figure 25—Evolution of fractionation effects between expelled hydrocarbon mixture (left) and residual hydrocarbons in source rock (right) with migration distance and/or time represented by stages I-III. Scheme shown assumes that 50% of total n-alkanes are expelled, but that this migrated fraction will contain more than 50% of C_{14}-C_{21} n-alkanes and less than 50% of original C_{22}-C_{29} n-alkanes and pristane to an increasing extent. Only changes in distributions below C_{21} are shown by shading.

processes. This is shown in Figure 26 as a simple numerical exercise for a hypothetical example. If a source rock with an original CPI-value of 2.0 for the n-C_{29} alkane (top of Figure 26) has effectively expelled hydrocarbons, this value increases in the residual n-alkanes extractable from the depleted source rock (middle of Figure 26), provided that the expulsion occurs in ideal, saturated solution. Conversely, if enrichment of the same original source rock has occurred by admixture of n-alkanes with CPI equal to 1.0 from a different source, the predominance of the n-C_{29} is decreased (bottom of Figure 26). Thus, the consideration outlined here introduces depletion and/or enrichment as a new variable for the explanation of downhole variations of CPI values in exploration wells, which in the past have been interpreted only in terms of maturity and source-related effects.

The highly schematic nature of the process outlined in Figures 25 and 26 is emphasized. However, good examples of most stages shown in these figures were seen in the Adventdalen and Reindalen series (Leythaeuser et al, 1983b). Generally, only under certain conditions can true migrated distributions be detected unambiguously. The migrating n-alkanes will in most situations be added to the indigenous components. Thus, true migrated distributions can only be recognized when the absolute amount of the migrated material far exceeds that of the indigenous components.

What is the nature of the process responsible for the observed compositional fractionation effects? The existence of a system in the source rock consisting of a mobile liquid phase (part of the pore water), enriched by various amounts of hydrocarbons of different molecular weight and structure, and a stationary phase (clay minerals) leads to the assumption that chromatographic processes play a role during the initial stages of hydrocarbon migration. In liquid-solid chromatography, a continuous flow of a liquid phase moves through a column usually packed with a porous adsorbent as the stationary phase. If this concept is applied to primary migration of hydrocarbons in source rocks, the expulsion efficiency of a migrating compound is controlled by its retention time, which depends on both its adsorption and desorption rates in this two-phase system. Conditions in nature become more complicated as soon as a third (i.e., hydrocarbon based) liquid phase occurs in the system. This happens when the hydrocarbon concentrations in the pores exceed the solubility limits. Oil globules are then formed. It is suspected, however, that even under these conditions chromatographic effects should occur. The adsorption/desorption equilibrium is controlled by the molecular weight, polarity, and polarizability (by induced dipoles) of the specific compound. Also steric effects may be an important factor during adsorption or "trapping" in the pore structure of the adsorbent if the latter approaches molecular dimensions. The depletion of crude oils in aromatic hydrocarbons, if compared to the residual hydrocarbon content of their source rocks (cf., Hunt, 1979), can be interpreted either by irreversible adsorption during primary migration or by their higher water solubilities. Likewise, Young and McIver (1977) found in experiments that in source rock-type shales the tendency to be adsorbed increases from normal alkanes to branched alkanes to cyclic alkanes to aromatic hydrocarbons for a given carbon number, and that for hydrocarbons of a similar type the adsorption affinity increases with increasing carbon number.

Figure 27 is an attempt to explain the compositional fractionation effects observed in the Adventdalen and Reindalen series by chromatographic processes, for which two scenarios are considered. In both, migration of a hypothetical mixture of compounds of different retention properties (e.g., polarities) occurs from the generating source rock at the bottom in an upward direction through a column of fine-grained rocks. Case 1 shows the chromatographic separation if the solute (i.e., the hydrocarbons expelled from the source rock) is injected instantaneously into the mobile phase. This would be equivalent to elution chromatography, but will probably not be observed under geologic conditions where generation in and expulsion from source rocks occurs over a long time interval. Instead, in the natural process, the input of the solute into the mobile phase occurs continuously over a longer period of geologic time. This appears to be comparable to so-called frontal analysis chromatography and is illustrated in Figure 27 by case 2. As shown in this scheme, evidence for compositional fractionation effects can be recognized only within a certain time interval and/or distance from the source rock. There, a chromatographic front composed of the compounds of lesser retention will emerge only temporarily. With continuing migration, steady-state conditions ideally will be reached, and the chromatographic fractionation effects will disappear. If the mobile phase encounters a permeability barrier (e.g., a cap rock), the components of higher retention will continue to move toward the barrier, and so the fractionation effects will eventually be removed (Figure 27, stage IV). The role of chromatographic processes in hydrocarbon migration has been emphasized by Seifert and Moldowan (1981). They have postulated "geochromatography" as a process for changing crude oil composition by long-distance secondary migration.

The n-alkane compositions of the "impregnated" sandstone (84.0 m, 27.6 ft; Figure 18a) and of the "expelled" hydrocarbons from the thin interbedded shale (62.5 m, 205 ft; Figure 13) in the Reindalen sequence are interpreted to result from chromatographic retention effects. Obviously the migration process did not reach steady-state.

Unexpectedly high is the difference in the retention behavior of the isoprenoid hydrocarbon pristane compared to its straight-chain isomer n-nonadecane. Provided that the interpretation of our results is correct, either minor differences in the molecular polarizabilities or steric effects might cause the elevated retention time of pristane. These effects would necessarily lead to a relative enrichment of pristane in the depleted source rock and to a relative depletion of this compound in the migrating hydrocarbon mixture.

If the fractionation effects seen in the Adventdalen and Reindalen series are the results of a chromatographic process as shown in Figure 27 for case 2, the reason they can be recognized today is that they represent the intermediate stages marked by asterisks in Figure 27. Either migration is

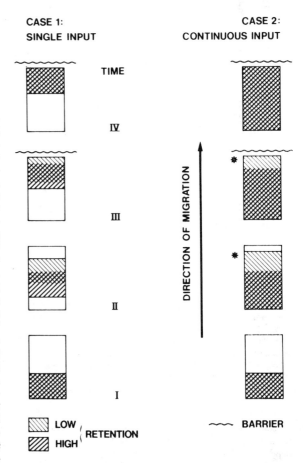

Figure 27—Evolution of chromatographic fractionation effects with migration distance and/or time (stages I-IV) from bottom toward top of illustration. Schemes shown assume migration of a mixture of compounds of different retention properties (e.g., polarities) upward from source rock at stage I and occurrence of permeability barrier between stages III and IV. Case 1—Analogous to elution chromatography (single input). Case 2—Analogous to frontal analysis chromatography (continuous input). Migration effects observed in Adventdalen and Reindalen sequences are interpreted to reflect intermediate stages marked by asterisks.

actively continuing today and the final stage IV in Figure 27 has not been reached, or the migration has slowed drastically—possibly by the development of permafrost or by the rapid and recent uplift of these sequences. The rapid uplift of the study area since the early Holocene could have cooled the sections quickly and viscosities of the migration fluids could hence have risen. In this way, flow rates could have been slowed sufficiently, so as to permit preservation of intermediate stages like II or III in Figure 27, which represent a non-steady state.

It is interesting that the movement of the oil into the cap rock above 55.5 m (182 ft) in the Reindalen sequence occurred without fractionation. When the oil moves as a bulk phase, the interaction with the mineral surfaces is minimized and fractionation effects are not observed. The best chance for interaction with the active sites on the mineral surfaces must be when the oil is still finely disseminated in the source rock. It could be that the main fractionation occurs when the disseminated hydrocarbons, generated throughout the source rock, migrate to form globules of oil (the so-called pore-center network of Barker, 1980), which can be expelled by internal overpressuring. The chances of the components of these globules being significantly fractionated during expulsion are small, since interaction with adsorbing active sites will only occur at the edges of the globule. However, if there is major variation in the compositions of individual globules, caused perhaps by the fractionation associated with movement from a disseminated condition to a pore-center network site (Barker, 1980), then those globules enriched in, for example, lower carbon number n-alkanes will be preferentially expelled because of their lower viscosity. Recent work (Beletskaya, 1978, and references therein; Sajgo et al, 1983) has indeed suggested that the movement of oil disseminated in the mineralogically closed pores to a more accumulated state in more open pore structures is associated with major fractionation, similar to that seen in the Reindalen and Adventdalen sequences.

In earlier papers, it has been proposed that molecular diffusion is responsible for some movement of light hydrocarbons in the subsurface (Leythaeuser et al, 1982, 1983a), but that such a mechanism would not be feasible for the transport of petroleum-range hydrocarbons over longer distances. The rate of molecular diffusion is highly dependent on molecular size and structure (Leythaeuser et al, 1980), and if the movement over the small distance from closed to open pore structures was by such a mechanism, some of the fractionation effects associated with expulsion of hydrocarbons from source rocks could result.

As discussed, an advanced stage of hydrocarbon expulsion has been observed in the Adventdalen and Reindalen sequences for thin shale interbeds as well as for the edges of thick shale units, which leads to the consideration of the role of compaction in explaining the observed migration phenomena. Compaction is one of the driving forces for the movement of fluids from source sediments to reservoir rocks. It is capable of forcing not only water but also hydrocarbons out of the pore system of a source rock (Magara, 1976; Tissot and Welte, 1978; Hunt, 1979). However, one argument against compaction as a driving force is that there is a time gap between stages of main expulsion of formation water from compacting mudstones and onset of effective hydrocarbon generation, which normally occurs at a later stage. Although this is true for sedimentary sequences where compaction proceeds in a normal way (i.e., the density of clays are always in equilibrium with depth), many exceptions are known to exist. In many basins, sequences have been reported with certain intervals that remain in compaction disequilibrium for extended periods of time and where adjustment to normal compaction is delayed (e.g., to beyond the stage when hydrocarbon generation becomes significant). Chapman (1972) proposed that this concept of compaction disequilibrium explains the discrepancy in timing of compaction, migration, and generation events. A well-documented case history reviewed by Hunt (1979) concerns the Maikop Formation in the USSR. It is an interbedded shale/sandstone sequence with regionally variable lithologic composition in terms of thickness and continuity of the interbedded units. Teslenko and Korotkov (1966) have shown that porosities of shales at equal depth levels are highly variable and dependent on their proximity to the nearest interbedded sandstone. Thin shales interbedded with permeable sandstones have much lower porosities than thick shales with fewer intercalated permeability avenues. Due to this compaction disequilibrium, the latter shales have retained a porosity of about 10–15% at a depth interval of 2,000–2,500 m (6,600–8,200 ft), where the threshold of intensive hydrocarbon generation has been documented to occur in the Maikop source rocks (Kartsev et al, 1971). Thus, significant porosity loss with compaction is indeed occurring in this sequence at stages where hydrocarbons have effectively been generated.

In the Adventdalen and Reindalen sequences, the thin shale interbeds and the edges of the thick shale units have, in all likelihood, had a better chance of achieving compaction equilibrium since the expelled fluids could readily escape into the adjacent sandstones. Less material is expected to be expelled from the center of the thicker shale units, which have hence retained, to a large extent, the generated hydrocarbons. In summary, the degree of compaction of the shale samples studied appears, at this maturity level, to control the extent to which petroleum-range components have been expelled. However, to bridge the gap between main stages of compaction and generation, this explanation requires the assumption that both sequences have experienced a retarded compaction, but with differences between thin and thick shales. This explanation is conceivable if the sand intervals encountered in both core holes are of a lenticular nature (i.e., encased updip by shales). Although the presence of lenticular sands appears to be a reasonable assumption, based on the depositional environment of both series, no detailed information is available about the lateral continuity of the encountered sand bodies in the vicinity of both sites. Thus, provided the role of compaction has correctly been assessed in our study, similar migration and expulsion phenomena as observed in the Adventdalen and Reindalen sequences can perhaps not be expected in normally compacted source rock shales.

Although favored here, the role of compaction in the

overall explanation of the migration phenomena observed in the Adventdalen and Reindalen sequences is not unequivocal; an alternative explanation exists. Tissot and Welte (1978) emphasized the role of gas pressure resulting from oil generation in source rocks as the most important mechanism for primary migration at advanced maturity stages. The release of this pressure was likely to have been easier in both sequences for the thin shale interbeds and the margins of thick shale units owing to their proximity to the adjacent sands. Thus, at these places, expulsion of petroleum-range hydrocarbons could have occurred preferentially and more efficiently.

OBSERVED MIGRATION PHENOMENA IN RELATION TO OTHER MIGRATION MODELS

Several migration phenomena documented in the present study, do not, at first glance, fit the currently accepted model for migration (for reviews see Tissot and Welte, 1978; Hunt, 1979). Although both sample series analyzed are coal-bearing sequences and all shales bear a uniformly hydrogen-lean type III kerogen, the source rocks have—as shown above—released an n-alkane mixture, the composition of which is equivalent to that of light oils or condensates. This release of a light oil appears to be in disagreement with the well-documented fact that source strata rich in terrestrial-derived organic matter generate heavy, waxy crude oils (Hedberg, 1968). More recently, the widespread discovery of light oils and gas-condensates in many low-maturity sequences bearing predominantly terrestrial-derived organic matter, as in the Beaufort-Mackenzie basin of northern Canada (Monnier et al, 1983) and the Tertiary of the Gulf Coast (Laplante, 1974), has required explanation. Snowdon and Powell (1982) have explained the "early" oils and condensates in several Canadian frontier basins as derived from resinite particles, which are abundant in those sediments and are known to thermally decompose at lower maturity levels than most other types of plant-derived organic matter (Snowdon, 1978). Based on the results of the Adventdalen-Reindalen study, an additional mechanism is proposed to explain the origin of these light oils and condensates.

An early stage of expulsion, controlled by shale compaction and associated with chromatographic separation effects, appears to precede the main phase of primary migration (oil-phase expulsion, responsible for formation of major oil accumulations). As documented in the present study, "light ends" migrate preferentially and expulsion of hydrocarbons from source rocks occurs at this early stage only from places which are in close proximity with open permeability avenues (short distance migration from thin shales interbedded in carrier beds, the edges of thick shales, and narrow margins along open fractures). Thus, the composition of the hydrocarbon mixture expelled from the source beds at this stage appears to be more controlled by physical processes (chromatographic effects under small-scale drainage conditions) rather than by the composition and structure of the generating source rock itself. Expulsion efficiencies in the molecular range < C_{20} are much higher than previously assumed. Therefore, provided sufficient thicknesses, series of closely interbedded shales rich in terrestrial-derived organic matter and sands would, at early maturity stages, present prolific sequences for light oils and condensates. The centers of unfractured, thick shales are thought to contribute little at this stage to the expelled hydrocarbon fluids; however, they become more effectively depleted during the main phase of expulsion.

Expulsion with chromatographic fractionation at the proposed early migration stage explains also the pronounced compositional discrepancies between crude oils and extracts of their associated source rocks, which have commonly been observed at early stages of maturity (Young and McIver, 1977; Connan and Cassou, 1980). Whether the main phase of expulsion dominated by oil-phase migration will follow this early migration stage, presumably depends on the richness and the type of organic matter in the generating source rocks. This stage may not be reached at all for source strata bearing modest contents of organic matter. This concept was outlined by Hunt (1979, p. 220-221), who wrote, "Primary migration probably involves a combination of different mechanisms operating in varying degrees of intensity depending on the quantity and type of organic matter in the source rock. Rocks with less than 1 percent organic matter, of which an appreciable part is gas generating, would migrate hydrocarbon by diffusion, in solution, or in a gas phase. . . . Source rocks of the offshore U.S. Gulf Coast are in this category. Rocks with more than 5 percent organic matter, most of which is oil generating, would migrate hydrocarbons as an oil or gas phase. The Bakken Shale of the Williston Basin and the basal Green River-Wasatch of the Uinta Basin are examples of this. Source rocks with intermediate organic contents. . .probably involve different migration mechanisms for different types and sizes of hydrocarbon molecules."

SUMMARY

Based on detailed analysis of closely spaced core samples from two interbedded sequences of source rock-type shales (at favorable maturity equivalent to about 0.85% R_m) and from reservoir sandstones from Spitsbergen Island, several geochemical effects associated with primary and secondary migration have been documented. The main observations and conclusions reached can be summarized as follows.

1. A comparison of hydrocarbon compositions from source rock-type shale samples of similar organic matter type reveals that thin shales interbedded in sands and the edges of thick shale units are depleted in petroleum-range hydrocarbons to a higher degree than the centers of thick shale units.

2. For the alkanes, expulsion is associated with pronounced compositional fractionation effects according to molecular chain length and structure. Shorter chain length n-alkanes are expelled more effectively than their higher

homologues and pristane and phytane are expelled to a lesser degree than their straight-chain isomers.

3. The patterns of hydrocarbon composition and certain established migration parameters allowed classification of most samples analyzed into one of these categories: "unmodified" shales and sands, "depleted" shales, or "enriched" sands and siltstones.

4. Absolute quantification of the C_{15+} saturated hydrocarbons of selected sample series allowed calculation of expulsion efficiencies for individual petroleum-range hydrocarbons. The expulsion efficiencies for the n-alkanes from a thin 5-cm (2-in.) source rock-type shale interbedded in a sandstone interval are extremely high at the lower molecular weight range (about 80% around C_{15}), but decrease with increasing carbon number to zero in the C_{25+} region. Pristane and phytane have about one-half the expulsion efficiencies of n-C_{19} and n-C_{20}, respectively. However, among the other examples of thin shale interbeds used for this mass-balance approach, there are considerable variations of the expulsion efficiencies and the degrees of fractionation, which are presently not understood.

5. For the same examples, expulsion efficiencies calculated in a similar manner for the total C_{15+} soluble organic matter contents (comparable to whole oil) are much lower, averaging 23%. Whole-extract expulsion efficiencies are believed, however, to be unrealistic given the observed fractionation effects and the hydrocarbon composition of the impregnated sandstones from the same sequence.

6. Large-size molecules of the so-called "biological-marker type" (steranes and triterpanes) have not been expelled from the same source rock shale samples to any significant extent. This makes the analysis of the biological-marker distributions an excellent tool for accurate characterization of the type of kerogen in the source rock (necessary as a prerequisite for our mass-balance calculations).

7. Impregnation of the lowermost portion of a siltstone unit overlying a sandstone, which was bearing a small hydrocarbon accumulation, was not associated with fractionation effects. It is presumed, therefore, to have occurred by bulk-oil migration.

8. A schematic model is proposed to explain, for the n-alkanes and isoprenoid alkanes, the evolution of fractionation effects between residual and migrated hydrocarbon mixtures with increasing time and/or migration distance (Figure 25).

9. Most of the expulsion phenomena documented in this study can be best explained by a migration mechanism controlled by adsorption/desorption processes leading to chromatographic separation. This separation appears to be according to molecular size and steric effects. It is likely that the evidence for chromatographic fractionation was so clearly preserved in the Advantdalen and Reindalen series due to the correct combination of certain events in their geologic history (generation and migration followed by rapid uplift and development of permafrost).

10. The degree of compaction is thought to control the extent to which petroleum-range hydrocarbons have been expelled from source rock-type shales in the sequences studied. This interpretation is based on the observations summarized under point 1 and on the assumption that conditions of compaction disequilibrium, have existed locally in the Adventdalen and Reindalen sequences.

11. The observed migration phenomena lead us to propose a modification of the currently accepted primary migration model. We propose that the main phase of expulsion is in certain sequences preceded by a stage during which only the edges of thick shale units and very thinly interbedded shales are depleted. This early expulsion stage appears not to be a "pulsed event" but rather a slow and continuous process, which is associated with pronounced compositional fractionation effects.

12. The composition of the hydrocarbon product accumulating in the reservoir trap appears at this stage of the migration process to be primarily controlled by physical processes (short-distance drainage conditions) rather than by the type of organic matter in the generating source rock.

13. This proposed modification can also explain the origin of accumulations of light oils and gas-condensates discovered in many low-maturity sequences bearing predominantly terrestrial-derived organic matter.

14. Based on the results of this study, it appears that several current geochemical concepts need to be modified. The source rock quality of organic-rich shales is commonly assessed by measurement of amount and composition of the solvent-soluble organic matter. For data interpretation, the assumption is then made that only a small portion (in the order of 5-10%) is expelled during primary migration, and hence the material extracted in the laboratory is assumed to be representative for the total hydrocarbons generated by the source rock. However, geochemists should now be aware that there are shales which are highly depleted and where expulsion has occurred with fractionation effects as documented in this study. Application of the above outlined geochemical concept to the data from those shales could result in grossly misleading conclusions about their source rock quality. It may be difficult to discriminate unambiguously source- and maturity-related effects from the consequences of migration processes for certain shale samples. For example, the pristane/n-C_{17} ratio which is commonly used in geochemical data interpretation as a maturity indicator also appears to be influenced by the degree of hydrocarbon expulsion.

15. It is suggested that the results of this study should not lead directly to routine application in petroleum exploration, because it is believed that more research is needed. The complexities of hydrocarbon compositional patterns in the subsurface are great because so many factors are involved. In this study, the recognition of migration effects was restricted to one-dimensional movement (up and down the hole). It is conceivable that samples with n-alkanes showing an "enriched" compositional pattern in a well need not show quantitative enrichment, because the main mass of migrating fluids could have continued to move updip. Likewise, shale samples with "depleted" compositional patterns of the n-alkanes subsequently may have been replenished during later migration stages within the same unit, which may, however, have been associated with less fractionation.

APPENDIX

Sterane Identification and Structures
(m/z 217 fragmentograms) Figures 4, 5

	Compound	Elemental Composition	Structure (Figure 28)
a	13β, 17α-diacholestane (20S)	$C_{27}H_{48}$	I, R = H
b	13β, 17α-diacholestane (20R)	$C_{27}H_{48}$	I, R = H
c	13α, 17β-diacholestane (20S)	$C_{27}H_{48}$	II, R = H
d	13α, 17β-diacholestane (20R)	$C_{27}H_{48}$	II, R = H
e	24-methyl-13β, 17α-diacholestane (20S)	$C_{28}H_{50}$	I, R = CH_3
f	24-methyl-13β, 17α-diacholestane (20R)	$C_{28}H_{50}$	I, R = CH_3
g	24-methyl-13α, 17β-diacholestane (20S)	$C_{28}H_{50}$	II, R = CH_3
	+ 14α, 17α-cholestane (20S)	$C_{27}H_{48}$	III, R = H
h	24-ethyl-13β, 17α-diacholestane (20S)	$C_{29}H_{52}$	I, R = C_2H_5
	+ 14β, 17β-cholestane (20R)	$C_{27}H_{48}$	IV, R = H
i	14β, 17β-cholestane (20S)	$C_{27}H_{48}$	IV, R = H
	+ 24-methyl-13α, 17β-diacholestane (20R)	$C_{28}H_{50}$	II, R = CH_3
j	14α, 17α-cholestane (20R)	$C_{27}H_{48}$	III, R = H
k	24-ethyl-13β, 17α-diacholestane (20R)	$C_{29}H_{52}$	I, R = C_2H_5
l	24-ethyl-14α, 17β-diacholestane (20S)	$C_{29}H_{52}$	II, R = C_2H_5
m	24-methyl-13α, 17α-cholestane (20S)	$C_{28}H_{50}$	III, R = CH_3
n	24-ethyl-13α, 17β-diacholestane (20R)	$C_{29}H_{52}$	II, R = C_2H_5
	+ 24-methyl-14β, 17β-cholestane (20R)	$C_{28}H_{50}$	IV, R = CH_3
o	24-methyl-14β, 17β-cholestane (20S)	$C_{28}H_{50}$	IV, R = CH_3
p	24-methyl-14α, 17α-cholestane (20R)	$C_{28}H_{50}$	III, R = CH_3
q	24-ethyl-14α, 17α-cholestane (20S)	$C_{29}H_{52}$	III, R = C_2H_5
r	24-ethyl-14β, 17β-cholestane (20R)		
	+ unknown sterane	$C_{29}H_{52}$	IV, R = C_2H_5
s	24-ethyl-14β, 17β-cholestane (20S)	$C_{29}H_{52}$	IV, R = C_2H_5
t	24-ethyl-14α, 17α-cholestane (20R)	$C_{29}H_{52}$	III, R = C_2H_5

Triterpane Identification and Structures
(m/z 191 fragmentograms) Figures 4, 5

	Compound	Elemental Composition	Structure (Figure 28)
A	18α(H)-trisnorneohopane	$C_{27}H_{46}$	V
B	17α(H)-trisnorhopane	$C_{27}H_{46}$	VI, R = H
C	17α(H)-norhopane	$C_{29}H_{50}$	VI, R = C_2H_5
D	normoretane	$C_{29}H_{50}$	VII, R = C_2H_5
E	17α(H)-hopane	$C_{30}H_{52}$	VI, R = $CH(CH_3)_2$
F	moretane	$C_{30}H_{52}$	VII, R = $CH(CH_3)_2$
G	17α(H)-homohopane (22S)	$C_{31}H_{54}$	VI, R = $CH(CH_3)C_2H_5$
H	17α(H)-homohopane (22R)	$C_{31}H_{54}$	VI, R = $CH(CH_3)C_2H_5$
	+ unknown triterpane		
I	homomoretane	$C_{31}H_{54}$	VII, R = $CH(CH_3)C_2H_5$
J	17α(H)-bishomohopane (22S and 22R)	$C_{32}H_{56}$	VI, R = $CH(CH_3)C_3H_7$
K	17α(H)-trishomohopane (22S and 22R)	$C_{33}H_{58}$	VI, R = $CH(CH_3)C_4H_9$
L	17α(H)-tetrakishomohopane (22S and 22R)	$C_{34}H_{60}$	VI, R = $CH(CH_3)C_5H_{11}$
M	17α(H)-pentakishomohopane	$C_{35}H_{62}$	VI, R = $CH(CH_3)C_6H_{13}$

Additional Alkane Identification and Structures
Tricyclic and Tetracyclic Terpanes
(m/z 191 fragmentograms) Figures 4, 5

	Carbon Number	Elemental Composition	Structure (Figure 28)
N	21	$C_{21}H_{38}$	VIII, R = C_2H_5
O	22	$C_{22}H_{40}$	VIII, R = C_3H_7
P	23	$C_{23}H_{42}$	VIII, R = C_4H_9
Q	24	$C_{24}H_{44}$	VIII, R = $C_3H_5(CH_3)_2$
R	25	$C_{25}H_{46}$	VIII, R = $C_3H_5(CH_3)C_2H_5$ (17R and 17S)
S	24	$C_{24}H_{42}$	IX
T	26	$C_{26}H_{48}$	VIII, R = $C_3H_5(CH_3)C_3H_7$ (17R and 17S)

Short-Side Chain Steranes
(m/z 217 fragmentograms)

	Compound	Elemental Composition	Structure (Figure 28)
u	5α(H)-pregnane	$C_{21}H_{36}$	X, R = C_2H_5
v	5α(H)-bisnorcholane	$C_{22}H_{38}$	X, R = $CH(CH_3)_2$

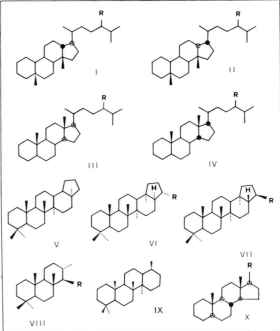

Figure 28—Molecular structures of biological markers referred to in text: A, steranes; B, triterpanes; C, tricyclic and tetracyclic terpanes. Roman numbers refer to identification of molecular types given in Appendix.

REFERENCES CITED

Aquino Neto, F. R., A. Restle, J. Connan, P. Albrecht, and G. Ourisson, 1982, Novel tricyclic terpanes (C_{19}, C_{20}) in sediments and petroleums: Tetrahedron Letters, v. 23, p. 2027-2030.

Barker, C., 1980, Primary migration: The importance of water-mineral-organic matter interactions in the source rock, in W. H. Roberts and R. J. Cordell, eds., Problems of petroleum migration: AAPG Studies in Geology 10, p. 19-31.

Beletskaya, S. N., 1972, A study of distribution of the disseminated bitumens in the pore systems of rocks related to the assessment of the state of migration processes. A comparative study of the gaseous extracts from non-crushed and crushed rocks: Geologija Nefti i Gaza, v. 16, p. 39-45 (in Russian).

Birkenmajer, K., 1972, Tertiary history of Spitsbergen and continental drift: Acta Geologica Polonica, v. 22, p. 193-218.

Black, R. F., 1954, Permafrost—a review: GSA Bulletin, v. 65, p. 839-856.

Bostick, N. H., and J. N. Foster, 1975, Comparison of vitrinite reflectance in coal seams and in kerogen of sandstones, shales, and limestones in the same part of a sedimentary section, in B. Alpern, ed., Pétrographie de la matière organique des sédiments, relations avec la paléotémperature et le potentiel pétrolier: Paris, Editions du C.N.R.S., p. 13-25.

Boulton, G. S., 1979, Glacial history of the Spitsbergen archipelago and the problem of a Barents Sea ice sheet: Boreas, v. 8, p. 31-57.

Bray, E. E., and E. D. Evans, 1961, Distribution of n-paraffins as a clue to recognition of source beds: Geochimica et Cosmochimica Acta, v. 22, p. 2-15.

Chapman, R. E., 1972, Primary migration of petroleum from clay source rocks: AAPG Bulletin, v. 56, p. 2185-2191.

Connan, J., and A. M. Cassou, 1980, Properties of gases and petroleum liquids derived from terrestrial kerogen at various maturation levels: Geochimica et Cosmochimica Acta, v. 44, p. 1-23.

Durand, B., and J. Espitalié, 1973, Evolution de la matiére organique au cours de l'enfouissement des sediments: Comptes Rendues de l'Académie des Sciences, Paris, v. 276, p. 2253-2256.

——— and J. L. Oudin, 1980, Example de migration des hydrocarburs dans une série deltaique: le delta de la Mahakam: 10th World Petroleum Congress Proceedings, v. 2, p. 3-11.

Espitalié, J., M. Madec, and B. Tissot, 1980, Role of mineral matrix in kerogen pyrolysis: influence on petroleum generation and migration: AAPG Bulletin, v. 64, p. 59-66.

——— J. L. Laporte, M. Madec, F. Marquis, P. Leplat, J. Paulet, and A. Boutefeu, 1977, Méthode rapide de caractérisation des roches mères, de leur potentiel pétrolier et de leur degré d'evolution: Revue de l'Institut Francais du Petrole, v. 32, p. 23-42.

Harland, W. B., C. A. G. Pickton, N. J. R. Wright, C. A. Croxton, D. G. Smith, J. L. Cutbill, and W. G. Henderson, 1976, Some coal-bearing strata in Svalbard: Norsk Polarinstitut, Oslo, Skrifter No. 164.

Hedberg, H. D., 1968, Significance of high-wax oils with respect to genesis of petroleum: AAPG Bulletin, v. 52, p. 736-750.

Huang, W. Y., and W. G. Meinschein, 1979, Sterols as ecological indicators: Geochimica et Cosmochimia Acta, v. 43, p. 739-745.

Hunt, J., 1979, Petroleum geochemistry and geology: San Francisco, Freeman, 617 p.

Kartsev, A. A., N. B. Vassoevich, A. A. Geodekian, S. G. Neruchev, and V. A. Sokolov, 1971, The principal stage in the formation of petroleum: 8th World Petroleum Congress Proceedings, v. 2, p. 3-11.

Kellogg, H. E., 1975, Tertiary stratigraphy and tectonism in Svalbard and continental drift: AAPG Bulletin, v. 59, p. 465-485.

Laplante, R. E., 1974, Hydrocarbon generation in Gulf Coast Tertiary sediments: AAPG Bulletin, v. 58, p. 1281-1289.

Leythaeuser, D., R. G. Schaefer, and H. Pooch, 1983a, Diffusion of light hydrocarbons in subsurface sedimentary rocks: AAPG Bulletin, v. 67, p. 889-895.

——— ——— and A. Yükler, 1980, Diffusion of light hydrocarbons through near-surface rocks: Nature, v. 284, p. 522-525.

——— ——— 1982, Role of diffusion in primary migration of hydrocarbons: AAPG Bulletin, v. 66, p. 408-429.

——— M. Bjorøy, A. S. Mackenzie, R. G. Schaefer, and F. J. Altebäumer, 1983b, Recognition of migration and its effects within two core holes in shale/sandstone sequences from Svalbard, Norway, in M. Bjorøy et al, eds., Advances in Organic Geochemistry 1981: Chichester, John Wiley, p. 136-146.

Mackenzie, A. S., 1983, Applications of biological markers in petroleum geochemistry, in J. Brooks and D. H. Welte, eds., Advances in Petroleum Geochemistry Vol. 1: London, Academic Press, p. 115-214.

——— S. C. Brassell, G. Eglinton, and J. R. Maxwell, 1982, Chemical fossils: the geological fate of steroids: Science, v. 217, p. 491-504.

——— D. Leythaeuser, R. G. Schaefer, and M. Bjorøy, 1983, Expulsion of petroleum hydrocarbons from shale source rocks: Nature, v. 301, p. 506-509.

Magara, K., 1976, Water expulsion from clastic sediments during compaction—directions and volumes: AAPG Bulletin, v. 60, p. 543-553.

Manum, S. B., and T. Throndsen, 1978, Rank of coal and dispersed organic matter and its geological bearing on the Spitsbergen Tertiary: Norsk Polarinstitutt Arbok 1977, p. 159-177.

Milner, C. W. D., M. A. Rogers, and C. R. Evans, 1977 Petroleum transformation in reservoirs: Journal of Geochemical Exploration, v. 7, p. 101-153.

Monnier, F., T. G. Powell, and R. L. Snowdon, 1983, Qualitative and quantitative aspects of gas generation during maturation of sedimentary organic matter, examples from Canadian frontier basins, in M. Bjorøy et al, eds., Advances in Organic Geochemistry 1981: Chichester, John Wiley, p. 487-495.

Philippi, G. T., 1965, On the depth, time and mechanism of petroleum generation: Geochimica et Cosmochimica Acta, v. 29, p. 1021-1049.

Radke, M., H. G. Sittardt, and D. H. Welte, 1978, Removal of soluble organic matter from rock samples with a flow-through extraction cell: Analytical Chemistry, v. 50, p. 663-665.

——— H. Willsch, and D. H. Welte, 1980, Preparative hydrocarbon group type determination by automated medium pressure liquid chromatography: Analytical Chemistry, v. 52, p. 406-411.

Rhead, M. M., G. Eglinton, and G. H. Draffan, 1971, Hydrocarbons produced by the thermal alteration of cholesterol under conditions simulating the maturation of sediments: Chemical Geology, v. 8, p. 277-297.

Roberts, W. H., and R. J. Cordell, 1980, Problems of petroleum migration: introduction, in W. H. Roberts and R. J. Cordell, eds., AAPG Series in Geology 10, p. vi-viii.

Sajgo, C., J. R. Maxwell, and A. S. Mackenzie, 1983, Evaluation of fractionation effects during the early stages of primary migration: Organic Geochemistry, v. 5, p. 65-73.

Seifert, W. K., and J. M. Moldowan, 1981, Paleoreconstruction by biological markers: Geochimica et Cosmochimica Acta, v. 45, p. 783-794.

Snowdon, L. R., 1978, Organic geochemistry of the Upper Cretaceous/Tertiary delta complexes of the Beaufort-Mackenzie sedimentary basin, northern Canada: PhD thesis, Rice University, Houston.

——— and T. G. Powell, 1982, Immature oil and condensate—modification of hydrocarbon generation model for terrestrial organic matter: AAPG Bulletin, v. 66, p. 775-788.

Steel, R. J., A. Dalland, K. Kalgraff, and V. Larsen, 1981, The central Tertiary basin of Spitsbergen: sedimentary development of a sheared margin basin: Canadian Society of Petroleum Geologists Memoir 7, p. 647-664.

Teslenko, P. F., and B. S. Korotkov, 1966, Effect of arenaceous intercalations in clays on their composition: International Geology Review, v. 9, p. 699-701.

Tissot, B., and D. H. Welte, 1978, Petroleum formation and occurrence; a new approach to oil and gas exploration: Berlin, Springer-Verlag, 538 p.

——— B. Durand, J. Espitalié, and A. Combaz, 1974, Influence of nature and diagenesis of organic matter in formation of petroleum: AAPG Bulletin, v. 58, p. 499-506.

——— Y. Califet-Debyser, G. Deroo, and J. L. Oudin, 1971, Origin and evolution of hydrocarbons in Early Toarcian shales, Paris basin, France: AAPG Bulletin, v. 55, p. 2177-2193.

Trendel, J. M., A. Restle, J. Connan, and P. Albrecht, 1982, Identification of a novel series of tetracyclic terpene hydrocarbons (C_{24}-C_{27}) in sediments and petroleums: Journal of the Chemical Society Chemical Communications, p. 304-306.

Vandenbroucke, M., 1972, Etude de la migration primaire: variation de composition des extracts à un passage roche mère réservoir, in H. R. Gärtner and H. Wehner, eds., Advances in Organic Geochemistry 1972: Oxford, Pergamon Press, p. 547-565.

——— B. Durand, and J. L. Oudin, 1983, Detection of migration phenomena in a geological series by means of C_1-C_{35} hydrocarbon amounts and distributions, in M. Bjorøy et al, eds., Advances in Organic Geochemistry 1981, Chichester, John Wiley, p. 147-155.

Van Dorsselaer, A., A. Ensminger, C. Spyckerelle, M. Dastillung, O. Sieskind, F. Arpino, P. Albrecht, G. Ourisson, P.W. Brooks, S. J. Gaskell, B. J. Kimble, R. P. Philp, J. R. Maxwell, and G. Eglinton, 1974, Degraded and extended hopane derivatives (C_{27} to C_{35}) as ubiquitous geochemical markers: Tetrahedron Letters, v. 14, p. 1349-1352.

Winters, J. C., and J. A. Williams, 1969, Microbial alteration of crude oil in the reservoir: Symposium on Petroleum Transformations in Geologic Environments, American Chemical Society Meeting Preprint, p. E22-E31.

Young, A., and R. D. McIver, 1977, Distribution of hydrocarbons between oils and associated fine-grained sediments—physical chemistry applied to petroleum geology, II: AAPG Bulletin, v. 61 p. 1407-1436.

Petroleum Geology of the Bakken Formation Williston Basin, North Dakota and Montana

By
FRED F. MEISSNER
Filon Exploration Corporation
Denver, Colorado

INTRODUCTION

The Bakken Formation is a relatively thin unit and is limited in areal distribution to the deeper part of the Williston Basin (Fig. 1). In spite of its insignificant volume when compared with that of the total sedimentary section, the unit is undoubtably one of the most important when considered in relation to the presence of oil and gas. Organic-rich shales in the Bakken have been documented as excellent petroleum source-rocks (Dow, 1974; Williams, 1974) and are believed to have generated the tremendously large volumes of oil found in reservoirs somewhat distantly located above and below the unit. Production has been established within the Bakken itself, and considerable remaining exploration potential may exist within the elusive fracture-type reservoirs which characterize the unit. Since the Bakken is relatively isolated by seemingly impervious overlying and underlying lithologies and is the only source-rock within several thousand feet of vertical stratigraphic section, studies of its hydrocarbon-generation (maturity) pattern, associated physical changes and fluid pressure phenomena, and its relation to known reservoirs and accumulations may be of value in deciphering mechanisms and routes of migration and in adding a factor of predictability to the overall science of petroleum geology.

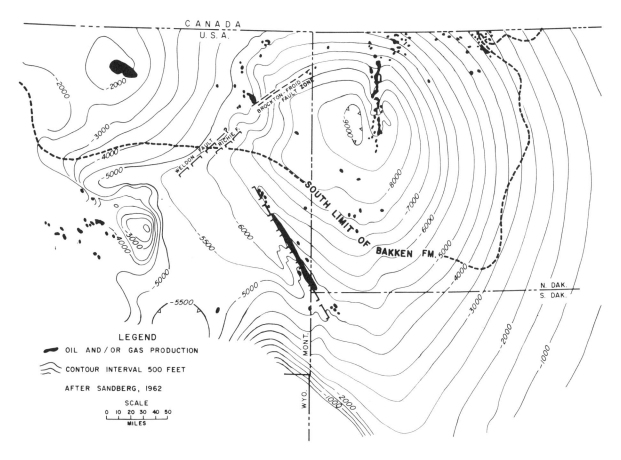

FIGURE 1
Williston basin, United States of America with structure contours on base of Mississippian strata and limit of Bakken Formation.

STRATIGRAPHY OF THE BAKKEN FORMATION AND ASSOCIATED UNITS

The Bakken Formation (Fig. 2) is the relatively thin basal unit of a thick sequence of predominately carbonate rocks (the overlying Madison Group) deposited during a major cycle of onlap-offlap sedimentation which began in uppermost Devonian (?) - lowermost Mississippian time (Bakken transgression) and extended to upper middle-Mississippian (Meremec) time (Charles regression). This onlap-offlap "depositional cycle" has been termed the "Tamaroa sequence" by Wheeler (1963). It can be identified as a rock unit surrounding most of the North American craton and is associated with a major amount of production on a continental scale. The basal Tamaroa transgressive unit equivalent to the Bakken can be recognized as a "black shale unit" known variously as the Exshaw/Banff in the Alberta Basin/Northern Rocky Mountains; the Pilot in the Cordilleran area; the "Lower Missis-

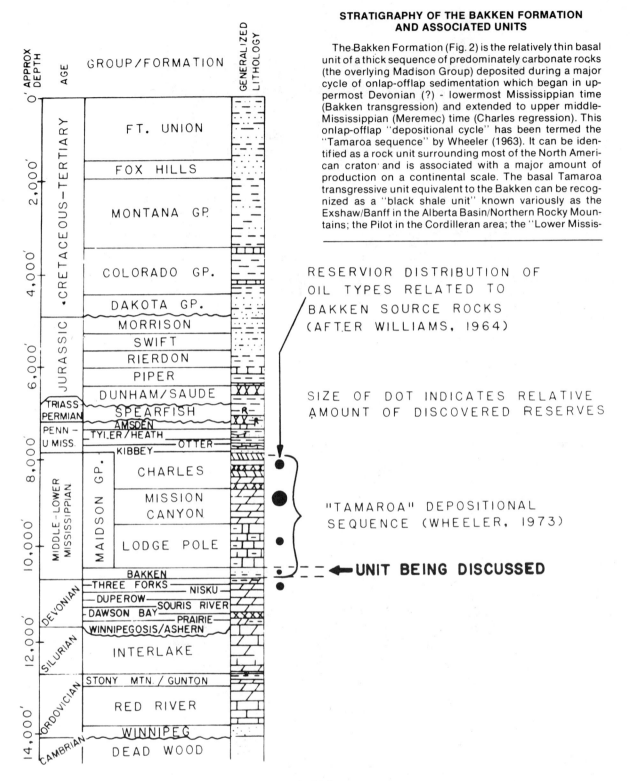

FIGURE 2
Generalized stratigraphic column, central Williston Basin, showing the location of the Bakken Formation and the distribution of oil types related to Bakken source-rocks.

sippian Black Shale" in the Permian Basin; the Woodford in the Anadarko Basin/Arbuckle Mountains; the Chattanooga in the eastern Mid-Continent/Southern Appalachian Basin; the Antrim in the Michigan Basin and the New Albany in the northern Appalachian Basin. Although immediate lateral correlatives of the Bakken are present in outcrops of central and western Montana and adjacent Alberta, the unit as formally defined (Nordquist, 1953) is restricted to the subsurface of the Williston Basin in eastern Montana/western North Dakota in the United States (Fig. 1 and 3) and southern Saskatchewan/southwestern Manitoba in Canada.

The Bakken Formation unconformably overlies the Upper Devonian Three Forks Formation (Fig. 2). The Three Forks (Peale, 1893; Sandberg & Hammond, 1958) averages about 150 feet in thickness, and consists primarily of interbedded yellowish-gray, greenish-gray, orange and red siltstones and shales, generally highly dolomitic. A few of the silty zones contain conglomeratic dolomite clasts and pebbles. Thinly-layered anhydrite beds occur near the base. A sandy zone up to 10 feet thick occurs somewhat erratically at the top of the formation, just beneath the lower Bakken Shale. At Antelope field this sandy zone is productive and is informally referred to as the "Sanish Sand".

At its type locality near the center of the Williston Basin (Fig. 3 and 4), the Bakken may be divided into three members:

1. an Upper Shale Member
2. A Middle Siltstone Member
3. A Lower Shale Member

The total formation ranges in thickness from a maximum of 140 feet near the center of the Basin to a subsurface "0" limit on the eastern, southern and southwestern flanks (Fig. 3). The three members can be correlated regionally, with most thickness changes taking place in the Middle Siltstone Member. Regional correlations indicate that the Bakken overlaps the underlying Devonian unconformity, with each succeedingly younger member overlapping the older member toward the depositional zero limit of the unit (Cross Section, Fig. 4).

The Upper and Lower Shale Members of the Bakken have apparently identical lithologies throughout most of their areal extent and consist of hard brittle often waxy-looking black shale which has a very dark brown color when examined as cuttings with a microscope under strong light. Thin-sections of the shales show them to be composed mostly of indistinct organic material with lesser amounts of clay, silt and dolomite grains. The Lower Bakken Shale appears to become less-organic and more-clayey, -silty, and -dolomitic near its zero limit, particularly on the western flank of the Basin.

The lithology of the Middle Siltstone Member of the Bakken varies somewhat unpredictably from a light-to-medium-gray very-dolomitic fine-grained siltstone to a very-silty fine-crystalline dolomite. Dark carbonaceous mottles and partings are common. The unit is often faintly laminated and occassionally contains fine-scale crossbedding.

The Bakken is conformably overlain by the Lodgepole Formation — the basal unit of the Madison Group (Fig. 2). The lithology of the lowermost Lodgepole, just above the Bakken, generally consists of dark gray dense lime mudstones interbedded with a few dark gray calcareous shales.

RESERVOIR PROPERTIES OF THE BAKKEN FORMATION AND ASSOCIATED UNITS

Effective matrix porosities and permeabilities of the Bakken Formation, the underlying Three Forks and the overlying Lodgepole are very low to non-existant. Only the Sanish Sand of the Upper Three Forks and the Middle Bakken Siltstone have measurable matrix reservoir properties. Core material from the Sanish Sand at Antelope Field indicates porosities averaging between 5 and 6 percent and permeability invariably less than 0.1 md (Murray, 1968). Core analyses from the Middle Bakken Siltstone

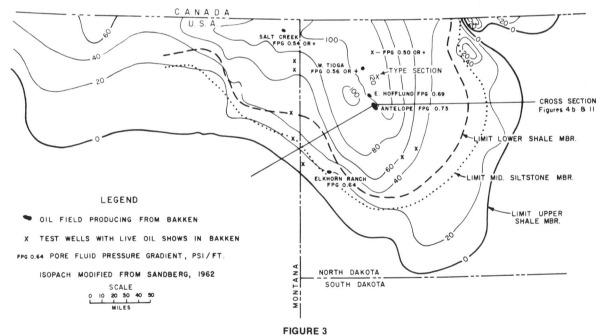

FIGURE 3
Bakken Formation Isopach map with oil fields, shows and fluid pressure gradients.

penetrated in the Sun-Phillips, Dynneson No. 1, Sec. 32, T24N, R58E, Richland Co., Montana, indicate an average porosity of about 5.5 percent and permeability ranging from 0.1 to 57 millidarcies. The lower permeability value is believed to represent matrix properties, while the higher value is believed to represent fracture properties. A two-hour drill-stem test of the Bakken interval in this well recovered 5 feet of 29.4 gravity oil plus 15 feet of oil-and gas-cut mud and 1400 feet of slightly oil-cut water cushion. Matrix porosities and permeabilities in the Sanish Sand and Bakken Siltstone, as documented in these two instances, appear similar to those associated with overpressured fracture-type reservoirs in several recently discovered major oil fields (i.e., Altamont, Uinta Basin and Spearhead Ranch, Powder River Basin).

Although Bakken oil shows are somewhat universal in deeper parts of the Williston Basin, drill-stem tests of most of these shows are usually disappointing. Generally only gas-and oil-cut mud recoveries accompanied by low unstabilized shut-in pressures are obtained by these tests. Significant free-oil recoveries have been obtained in at least 8 widely-spaced wildcat wells in which subsequent completions were not attempted. Bakken production has been established from five fields in the U.S. portions of the Basin (Fig. 3). Four of these fields are small accumulations; the fifth (Antelope field) has about 44 wells which have produced some 10 million barrels of oil and 29 billion cubic feet of gas. A significant fact regarding fluid found in the Bakken is that no formation water is ever recovered during drill-stem tests or initial well completions. Small water cuts have been observed in advanced stages of depletion at Antelope field; however, this water is probably extremely fine-pore capillary water, and is not suggestive of water encroachment or the existance of an oil-water contact. This leads to the conclusions that hydrocarbons are essentially the only movable fluid found in the Bakken.

Most Bakken oil wells have productive rates far in excess of those theoretically possible from the low to nonexistent matrix-reservoir properties present in the inter-

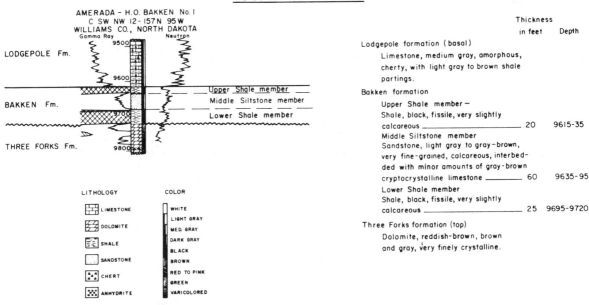

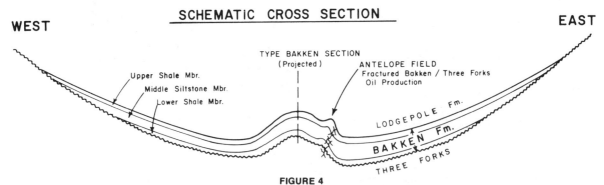

FIGURE 4
Bakken formation type section and east-west schematic cross-section of the Bakken Formation through the Williston Basin (cross-section located in Fig. 3).

val. Murray has made strong argument for tensional (extensional) fracturing being the major cause of reservoir development in the Bakken/Sanish zone at Antelope field. This argument appears to be equally valid wherever oil DST recoveries or actual production has been established in the Bakken. In fact, it would appear that the Bakken will produce oil and gas — and these fluids only — wherever it is found in a fractured state within the deeper part of the Basin.

Although fracturing is apparently of major importance in establishing a viable Bakken reservoir, it appears that the marginal amount of matrix porosity observed in the Middle Bakken Silt and the Sanish Sand are a strong contributing factor in establishing storage volume. Most fracture reservoirs which have been extensively studied have bulk-fracture porosities in the neighborhood of only one-half percent (39 bbls/acre-ft total reservoir volume) or less, even though permeabilities may range to hundreds or thousands of millidarcies. The fact that Bakken/Sanish wells at Antelope field have produced as much as 900,000 barrels of oil per 160 acres from an approximately 120-foot thick gross interval (or 46.4 bbls/acre-ft. actual recovered volume) at initial rates of as much as 1400 bbls per day would seem to indicate that 1) the major contribution to storage volume must come from marginal matrix porosities with high oil saturations in the Middle Bakken Silt and Sanish Sand zones, while 2) the major contribution to statisfactory production rates must come from a pervasive fracture system which renders the section relatively permeable.

Another factor of interest in evaluating Bakken reservoir properties is that all stabilized formation-fluid pressure measurements from wildcat or field wells in which formation fluid has been recovered indicate the Bakken to be anomalously overpressured (Fig. 3). Several unstabilized measurements also indicate overpressure of an unknown maximum magnitude. Documented fluid-pressure gradients are as high as 0.73 psi/foot (Antelope field). These abnormally high pressures are evidently discretely confined to the Bakken/Sanish interval as shown by pressure gradient/depth profile at Antelope field, where overlying and underlying reservoir zones, separated from the Bakken by tight strata, are actually found to be abnormally underpressured (Fig. 5). The reason for the anomalous Bakken fuild overpressure will be discussed in a later section.

GEOCHEMICAL (SOURCE ROCK) PROPERTIES OF THE BAKKEN FORMATION

According to modern theory, most oil and gas is generated from rocks which 1) have a certain critical organic content and which 2) have undergone sufficient thermal alteration (metamorphism) associated with time and burial-temperature to have cracked chemical bonds between complex primary organic materials and released hydrocarbons (Philippi, 1965; Welte, 1965; Landes, 1967; Momper, 1972; LaPlante, 1974). The high organic content of the Upper and Lower Bakken Shales as observed in thin-section studies, together with the fact that small chips of the rock visually generate significant amounts of oil and gas when heated in the test tube (Trask and Patnode, 1942, pg. 62), indicates that they are hydrocarbon source-rocks. Murray (1968), noting the "petroliferous" nature of the shales, their universal association with shows in sample cuttings and certain diagnostic petrophysical properties in their electrical and mechanical log character, was one of the first to speculate on the importance of the Bakken interval as a hydrocarbon source.

Recently, both Dow (1974) and Williams (1974) have expanded knowledge on the source-rock properties of the Bakken. Williams analysed 26 samples of Bakken Shale and found them to contain from 0.65 to 10.33 weight percent organic carbon, with an average of 3.84 percent. On the basis of sophisticated geochemical work on hydrocarbon extracts from the Bakken, he further related the Bakken source rock to the distinctive type of oil found and produced from the underlying Nisku Formation, the overlying Madison group and the Bakken/Sanish interval itself (Fig. 2). Dow's work presented a series of areal geochemical maps of the Bakken and its related oil types. From these maps he was able to speculate on migration paths, accumualtion patterns, etc. He estimates that Bakken source rocks have generated in the neighborhood of 10 billion barrels of oil in the deeper portions of the basin, where they are buried to a postulated "maturity" depth greater than 7000 feet. Considering Dow's evidence and arguments, it seems reasonable that although a large part of the hydrocarbons generated in the Bakken have migrated vertically downward and upward to underlying and overlying Nisku and Madison reservoirs, such migration would probably not have taken place until after all possible matrix-and fracture-reservoir volume in the Bakken had been effectively charged. This would account for the widespread occurrence of Bakken shows and the observation that hydrocarbons are the only mobile fluid phase found within the unit in the deeper part of the basin.

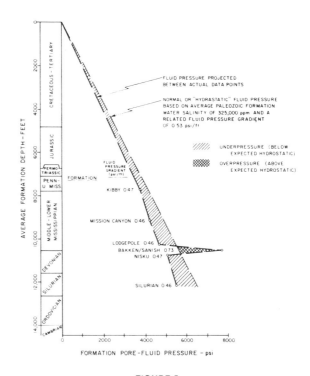

FIGURE 5
Reservoir fluid pressure versus depth, Antelope field, McKenzie Co., No. Dakota.

PETROPHYSICAL PROPERTIES OF THE BAKKEN FORMATION AND THEIR RELATION TO SOURCE-ROCK PRINCIPLES AND RESERVOIR OVERPRESSURING

Murray (1968) noted the peculiar behavior of the Bakken Shale Members are expressed on gamma-ray, sonic, neutron and resistivity logs and was one of the first to relate their petrophysical character to source-rock and overpressured reservoir properties. Using well known log-interpretation principles, the author has expanded considerably on Murray's observations and has utilized resistivity and sonic logs to map 1) source rock maturity and 2) formation fluid overpressure.

The Upper and Lower Bakken Shale Members are regionally characterized by 1) anomalously high gamma-ray radioactivity, 2) anomalously low, but highly variable sonic velocity (high transit time) and 3) either very-high or very-low resistivity. Typical log character of these parameters are shown in Figure 6. The unusually low sound-velocity (high transit-time) indicated for the Bakken shales is believed to be due for the most part to their high content of low-velocity organic material, although additional affects are also contributing factors, as will be discussed shortly.

Most normal shale units are characterized by low electrical resistivity because of 1) the basic conductivity of most shale clays and 2) their relatively high porosity, filled with conductive water. Logs through the Bakken Formation at depths less than about 6500 feet are characterized by these "normal" low shale resistivities; however, logs from deeper penetrations indicate anomalously high re-

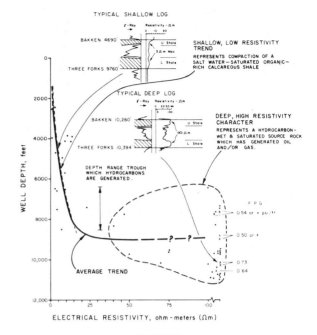

FIGURE 7
Electrical resistivity versus depth for Bakken Shale with "trend" and source-rock "maturity" interpretations indicated.

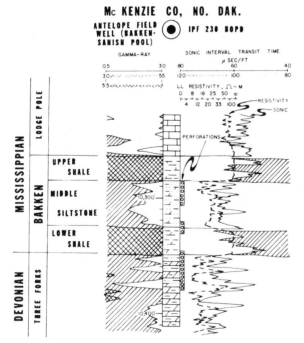

FIGURE 6
Typical electrical and mechanical log behavior, Bakken and adjacent formations.

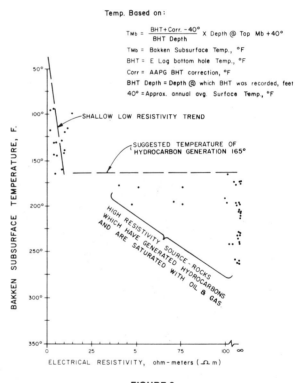

FIGURE 8
Electrical resistivity versus subsurface temperature for Bakken Shale, with "trend" and source-rock "maturity" interprettions indicated.

sistivities. A resistivity vs. depth plot for data obtained in 32 widely-separated control wells is shown in Fig. 7 (index map for these points is shown in Fig. 10). Murray (1968), concluded that the anomalously-high "essentially infinite" resistivity characterizing the Bakken Shale Members throughout the central Williston Basin was a natural consequence of the fact that they were hydrocarbon-saturated source-rocks. In view of this, the relatively rapid depth-related change from low to high resistivity shown in Fig. 7 is believed to represent the onset of "maturity" or hydrocarbon generation and consequent replacement of high-conductive pore water in the organic shales with nonconductive hydrocarbons. Numerous investigators (i.e., Philippi, 1968; Tissot, et al, 1971; Nixon, 1973; Claypool, et al, 1978) have documented the dramatic increase in extractable heavy hydrocarbons from source-rocks in other basins which have gone through the hydrocarbon-generation ("maturity") threshold. Plots of extractable-hydrocarbon to residual organic-carbon ratio vs. depth prepared by these investigators (i.e., Philippi, 1968, Fig. 5, pg. 33 and Fig. 7, pg. 37) have a form similar to the resistivity vs. depth plot of Fig. 7. Although no information is available on the extractable hydrocarbon content vs. depth relation for the Bakken Formation, it is anticipated that such data would show changes in hydrocarbon content which would be correlative with changes in resistivity. If this logic is correct, the resistivity effect might be taken as a measure of the equilibrium saturation required to achieve a continuous hydrocarbon-phase expulsion mechanism from the rock.

As shown in Fig. 7, the depth range through which the shallow low-resistivity/depth trend (which follows a normal shale compaction trend) changes to the high-resistivity character indicates maturity occurs from about 6200 to 8200 feet. The existence of this rather wide depth range suggests that the actual depth of hydrocarbon generations is not uniform throughout the basin - e.g. that a unique surface which may represent "maturity" is non-planar and/or non-horizontal. Since the point at which hydrocarbon generation starts is actually more directly related to temperature than depth, temperatures for the formation resistivities plotted in Fig. 7 were estimated from bottom-hole temperature data and a resistivity versus temperature plot was prepared (Fig. 8). This plot shows an extremely abrupt change from low-resistivity immature Bakken source rocks to abnormally high-resistivity rocks at about 160° F. This temperature is about that of the "critical temperature" for oil generation in Devonian-Mississippian-age rocks according to recent work (Connan, 1974) involving the calibration of thermochemical first order-reaction-rate theory to the concept of hydrocarbon generation (Fig. 9). This is an observation that strongly supports the idea that the change in resistivity of the Bakken source rocks is uniquely related to the onset of hydrocarbon generation and the replacement of conductive pore-water with thermally generated nonconductive hydrocarbons. Figure 10 is a map of Bakken Shale Member electrical resistivities and formation temperatures showing 1) the areas of "mature" high-resistivity and "immature" low-resistivity source rocks and 2) forma-

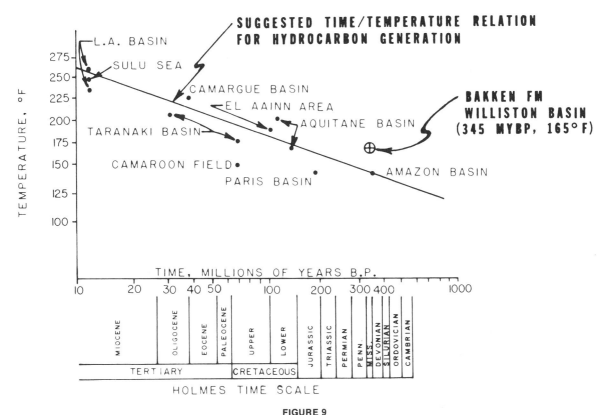

FIGURE 9

Theoretical time versus critical temperature, required to generate hydrocarbons from source-rocks (after Connan, 1974) and its relation to interpreted "maturity" in the Bakken Formation.

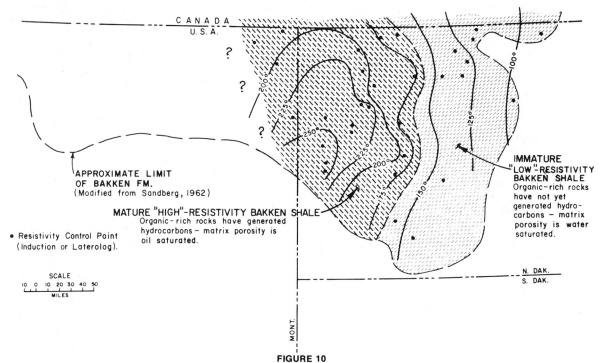

FIGURE 10
Areas of "high" and "low" electrical resistivity in Bakken shales, with subsurface isotherm contours and interpreted area of source-rock "maturity".

tion temperatures related to these maturity/resistivity relms. Figure 11 is a cross section which schematically depicts relations between depth, source rock maturity, resistivity and temperature in the Bakken.

It is a well-known fact that anomalous fluid pressures affect the electrical resistivity of rocks (Hottman and Johnson, 1965; Mac Gregor, 1965); however, the effects of maturity and resulting hydrocarbon saturation which cause high resistivities in the Bakken shales are in the opposite direciton of low resistivities caused by high fluid pressures and far over-shadow them in their total effect. Realizing that sonic velocities are also influenced by formation-fluid pressure, an attempt was made to map the extent of known abnormally-high fluid pressures in the Bakken through the use of sonic-transit-time vs. depth plots as proposed by Hottman and Johnson. As shown in

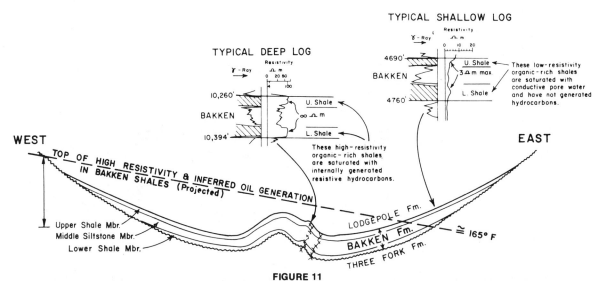

FIGURE 11
Schematic east-west section across the Williston basin showing the occurrence of "high" and "low" resistivity Bakken source-rock shales and their interpreted relationship to hydrocarbon generation (cross-section located in Fig. 3).

Figure 12a (which was prepared from sonic log data in the same 32 wells utilized in construction of Figures 7, 8 and 10), two trends — one "shallow", and one "deep" — can be discerned in the transit time/veolocity vs. depth plot. The "shallow trend" shows little scatter and depicts a normal slight increase of velocity with depth related to the compaction process. The "deep trend" is characterized by considerable scatter and, although within this trend velocities increase with depth as above, the average velocities of this trend are much lower than those predicted by a downward projection of the shallow trend. The shallow trend is further distinguished from the deep trend on the basis of the "low" and "high" resistivity catagories separating water-saturated "immature" and hydrocarbon-saturated "mature" source rocks as described in the preceding paragraph. The fact that the deep trend is known to contain overpressured reservoirs probably accounts for both the general shift to lower velocities and the apparent scatter in the actual values. Because sound velocities in oil are less than those in water, replacement of water by oil in the pore spaces of a source-rock which has become mature leads to a proportional diminishment of sound velocity in the overall rock (Poh-hsi Pan and DeBremaecker, 1970). This effect is shown schematically in the veolicity/transit time vs. depth plot of Fig. 13. The effect of abnormal pressure in lowering the velocity of sound in the rock is shown for a series of Gulf Coast shales in Fig. 14a. The magnitude of abnormality in the velocity of the overpressured shales with respect to that of a normal shale at the same depth is a direct relation to the amount of abnormal fluid pressure. If a series of control points reflecting known pressure conditions in a given rock unit can be obtained, a calibration can be derived and unknown fluid pressures can be determined from sonic-log data (Fig. 14b.). The transit time/velocity vs. depth relation interpreted to show the effects of both oil saturation (maturity) and the magnitude of overpressuring in the Bakken Shale Members is shown in Figure 12b. This interpretation has been extended to the plan map of Fig. 15. This map depicts the extent and magnitude of overpressuring in the Upper and Lower Bakken Shale Members. By innference, the same pressure properties are present in the immediately adjacent Bakken Silt and Sanish Sand (if present).

A comparison between the areas of Bakken overpressuring and the areas of source rock maturity as shown in Figs. 10 and 15 shows that they are essentially identical and suggests that there is a direct relationship between hydrocarbon generation and the occurrence of abnormally high fluid pressure. This phenomenon has been observed elsewhere (i.e., the Uinta and Powder River Basins) and the reason for it is of some importance. Formation-fluid overpressures may be caused by a number of processes (see Houston Geological Society, Abnormal Subsurface Pressure Study Group Report for an excellent resume'). Two processes that seem to be important in the phenomenon being discussed here are 1) undercompaction and 2) metamorphic phase change. It must be realized that in most cases involving undercompaction, fluid overpressuring is a manifestation of excess fluid or excess porosity in a rock when compared to normal fluid pressure (e.g., "hydrostatic" conditions) and porosity at the same depth. The well-known fluid overpressures in deep shales and sands of the Gulf Coast are related to

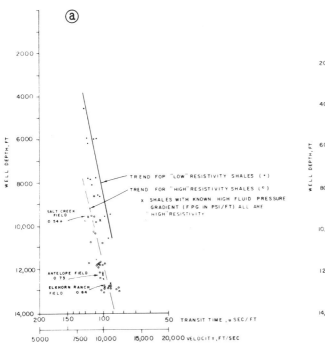
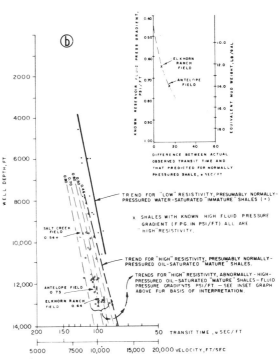

FIGURE 12

Sound velocity and transit-time versus depth for Bakken Shale. a. - data points, trends for "high" (mature) and "low" (immature) resistivity shales, and known occurrence of pore-fluid overpressure. b. - fluid-saturation species and interpreted pore-fluid pressure gradients.

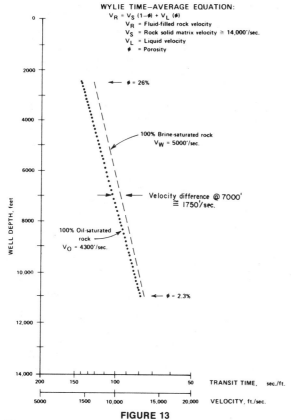

FIGURE 13
Theoretical effects of pore-fluid species on sound velocity in porous rocks based on Wylie time-average equation.

water as the excess fluid, and their origin is most widely attributed to rapid depositional rates not in equilibrium with the normal compaction process. In the case of overpressuring related to hydrocarbon generation, hydrocarbons appear to be the only excess fluid. The author believes that abnormal pressures found associated with mature Bakken source rocks are basically caused by:
1) the inhibited structural collapse of the rock framework as overburden-supporting solid organic material (estimated to be at least 25 volume percent of the rock) is converted to non-overburden supporting hydrocarbon pore fluid (e.g., oil and/or gas).
2) the increased volume occupied by metamorphosed organic residue plus generated hydrocarbon fluids above those occupied by the unaltered organic material.

Anomalous pressures in the Bakken are believed to be maintained by the combination of large hydrocarbon volumes generated at high rates and the relative isolation of the Bakken by extremely tight rocks in the underlying Three Forks and overlying Lodgepole Formations.

THEORY AND CONTROLS FOR LOCALIZING RESERVOIR FRACTURING IN THE BAKKEN

The occurrence of fracture-reservoirs in close association with mature oil-generating source rocks and abnormally-high reservoir-fluid pressures, as found at all fields producing from the Bakken Formation in the Williston Basin, (and some fields elsewhere in the world, as the Uinta and Powder River Basins) suggests a direct cause-and-effect relationship. This relationship is strongly supported by currently accepted basic failure-theory for porous isotropic homogeneous brittle-elastic rocks. This theory (Secor, 1965 and others) is schematically depicted

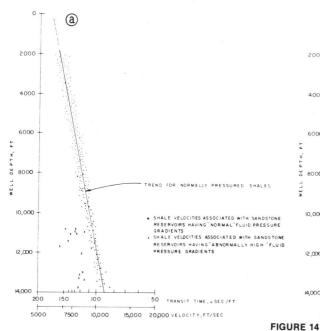
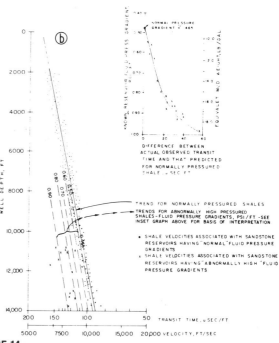

FIGURE 14
Sound velocity and transit-time versus depth for normally- and abnormally-pressured Gulf Coast Oligocene-Miocene shales (after Hottman and Johnson, 1955).
a. - data points, trend for normally-pressure shale, and known occurrence of abnormally- high pore-fluid pressure.
b. - interpreted pore-fluid pressure gradients.

according to standard graphical Mohr's circle and failure envelope representation in Fig. 16a. Shown is a "failure envelope" superimposed on a coordinate system, upon which it is possible to represent any possible stress field characterized by principal and normal stressed (plotted along or parallel to the horizontal abcissa) and shear stressed (plotted along or parallel to the vertical ordinate). Any particular stress field within the coordinate system is represented by a "Mohr's stress circle" which intersects the abcissa at values of the maximum (normally vertical with reference to the earth's surface) and minimum (normally horizontal) principle stresses (e.g. S_1 and S_2 respectively). The shape and dimensions of the failure envelop characterize certain physical properties of the rock, such as tensile strength (intersection of the envelope with the abcissa and normally a negative value), shear strength (intersection of the envelope with the ordinate, with both negative and positive value intersections), and the degree of brittle-elasticity of the material (as measured by the amount of similarity between shape and dimensions of the parabollically-shaped envelope to those predicted by the Griffith and Mohr-Coulomb-Navier failure theory). Stress circles within the concave portion of the failure envelope represent structurally stable non-fracture conditions. Stress circles which, because of an enlarging or laterally shifting stress field circle or a decreasing envelope size, become tangent to the failure envelope represent structurally unstable conditions and indicate fracture failure of the rock (Fig. 16b). If the point of tangency between the stress circle and the failure envelope is on the positive (+) or compressive side of the principle or normal stress coordinate origin, the failure will be in the form of a shear fracture (fault) and will not intrinsically be associated with the formation of reservoir-making porosity and permeability. If the point of tangency is on the negative (-) or tensile side of the origin, the failure will be in the form of an open extension fracture and will be intrinsically associated with the creation of porosity and permeability and hence the creation of fracture-reservoir conditions. Geometric relations between the failure envelope and the stress field coordinate system are such that large stress circles (e.g. large differential stress fields with large differences between maximum and minimum stress values) with compressive principal stresses are required for normal shear fracture failure; smaller stress circles, with at least one of the principal stresses in a negative (-) tensile sense are required for tensile (extension) fracture failure.

The stress fields represented by Mohr's stress circles are actually of two types in porous fluid-filled materials:
1) a "total stress" field, wherein the stress circle is defined by the "total" or externally applied maximum and minimum principal stresses characterized by S_1 and S_2 in Fig. 17a.
2) an "effective stress" field, wherein the stress cirlce is defined by the internal or "effective" maximum and minimum principle stresses characterized by σ_1, σ_2 in Fig. 17a.

According to Hubert and Rubey (1959) the type of stress field which actually causes porous rock failure is that related to internal "effective stresses" rather than externally applied or "total stresses". "Total stress" (S) and "effective stress" (σ) are related by the pressure (p) of fluid filling the pores of the rock according to the following equation:

$$\sigma = S - p$$

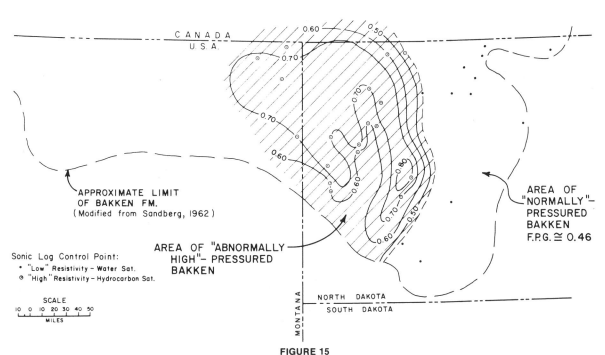

FIGURE 15

Aerial distribution of pore-fluid gradients in the Bakken Formation and areas of "normal" and "abnormally-high" pressure. Note coincidence of "low" electrical resistivity with "normal" pressure and "high" resistivity with "abnormally-high" pressure. Compare area of abnormally-high pressure with area of interpreted source-rock maturity indicated in Fig. 10.

BASIC PRINCIPLES OF ROCK FRACTURE THEORY

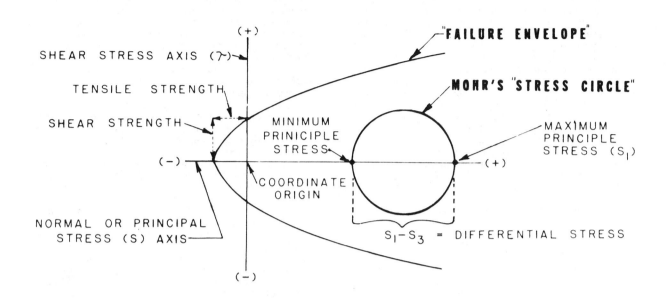

A) GRAPHICAL REPRESENTATION OF A STRESS FIELD (BY A "STRESS CIRCLE") AND A "FAILURE ENVELOPE".

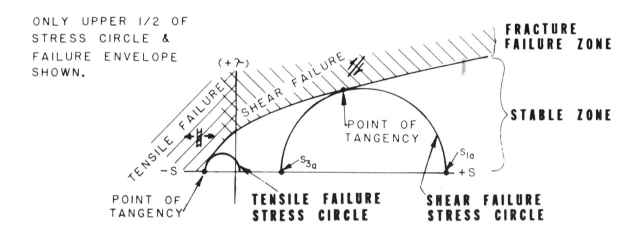

B) GRAPHICAL REPRESENTATION OF STRESS FIELDS PRODUCING SHEAR AND TENSION FRACTURE FAILURE.

FIGURE 16
Basic principles of rock fracture theory as illustrated by "failure envelopes" and "stress circles".

EFFECTS OF PORE FLUID PRESSURE ON ROCK FRACTURE THEORY

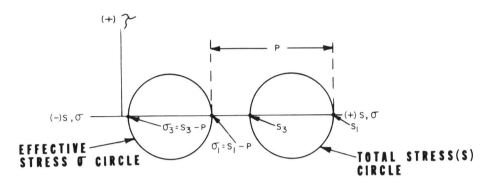

A) GRAPHICAL RELATION BETWEEN TOTAL AND EFFECTIVE STRESSES

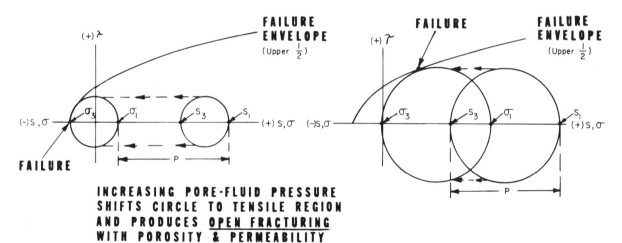

INCREASING PORE-FLUID PRESSURE SHIFTS CIRCLE TO TENSILE REGION AND PRODUCES <u>OPEN FRACTURING</u> WITH POROSITY & PERMEABILITY

INCREASING PORE-FLUID PRESSURE SHIFTS CIRCLE TO LESS-COMPRESSIVE REGION AND PRODUCES <u>FAULTING</u>

B) GRAPHICAL REPRESENTATIONS OF FRACTURE FAILURE PRODUCED BY INCREASING PORE PRESSURE

FIGURE 17
Effects of pore-fluid pressure on rock fracture theory.

As shown in Fig. 17a and b, the position of an effective stress circle with respect to the failure envelope is strongly influenced by any fluid pressure existing within the pores of a rock. The general effect of introducing or increasing pore-fluid pressure in a rock is to shift a stress circle both toward the failure envelope and toward a more tensile direction. Shifting of the stress field due to increasing pore-fluid pressure may cause either closed shear fracture failure (faulting) or open tensile fracture failure. The shifting or large stress field circles which become tangent to the compressive portion of the failure envelope leads to shear failure, while the shifting of critically small circles which become tangent to the tensile portion leads to tension (extension) fracturing. An enlarging stress field

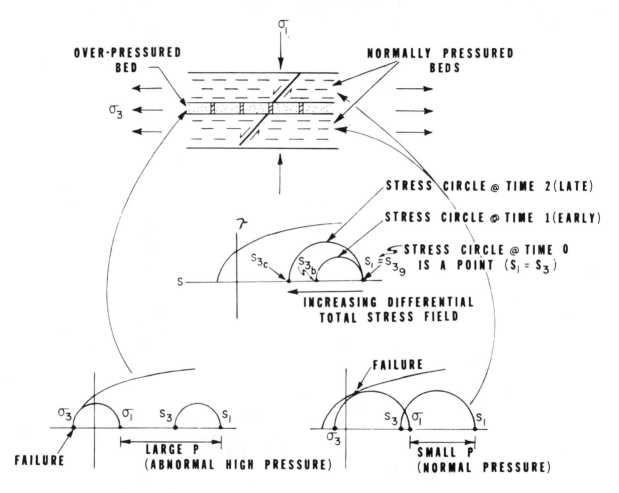

FIGURE 18
Schematic section and fracture-failure diagrams showing that both faulting and open tension (extension) fracturing may be produced in a series of sedimentary beds with a rising stress field affecting units having different pore-fluid pressure.

affecting a thick column of sedimentary rocks may cause failure of different types in individual beds depending on 1) the brittle-elastic or plastic nature of each individual bed, 2) bed strength or 3) bed pore-fluid pressure. An individual overpressured bed may fail in open tension fracturing, usually at an early, low differential stress condition, while overlying and underlying normally-pressured beds fail by faulting at a later and somewhat larger stress condition as shown in Fig. 18.

From the preceding discussion of basic fracture theory, it can be seen that two basic parameters essentially control the type of open tension fractures that create reservoir conditions:
1) Fluid overpressuring
2) Differential stress

Critical fluid overpressures leading to fracture failure can be caused by hydrocarbon generation as discussed in the previous section. Critical differential stress fields can be caused by a number of processes, including:
1) regional tectonic forces
2) burial and uplift
3) diagenetic processes (i.e., stylolitization)
4) secondary "bending" and "stretching" associated with local tectonic features.

Local "bending," which created a critical differential stress field existing in conjunction with regional fluid over-pressuring due to hydrocarbon generation, is believed to have created the fracture reservoir in the Bakken/Sanish interval at Antelope field, McKenzie Co., North Dakota. Murray (1968) and Finch (1969) have written excellent papers describing Antelope field.

The author concurs with the basic data and conclusions of Murray and Finch concerning Antelope fields, with the exception that Finch ascribes fluid overpressuring in the Bakken to hydrocarbon generation rather than to tectonic compression. Further, he believes that the overpressure described by the two investigators is an absolutely essential element in forming the reservoir fracture system, as described in preceding paragraphs. It should be noted that the geometrical factors described as "radius of curvature" and "curvature" are a quantitative measure of the process term referred to as "bending" in this report. Certain additional observations concerning this field can be made. The basic structural configuration of Antelope field as proposed by Finch (Fig. 19) and Murray (Fig. 20) consists of a northwest-southeast trending closed anticline with additional non-closing plunge to the southeast. The structure is strongly asymmetric, with a gentle southwest flank and a much steeper northeast flank. The overall structural configuration of the Antelope field is believed to represent a classic "drape fold" (Stearns, 1971) overlying a vertically uplifted basement block which is faulted at depth beneath the steep limb of the anticline (Fig. 21). As shown by Finch (Fig. 19), "bending" in the shallower, normally-pressured beds over the drape fold has lead to stress conditions producing shear-fracture failure as manifested in a series of compensating "keystone-type" normal faults. In contrast to shallower beds, "bending" (or "curvature") in the deeper overpressured Bakken/Sanish section has led to stress conditions producing open tensile fracture failure, as manifested in the production "fairway" mapped by Murray (Fig. 20). The overall effect of "bending" in producing the two types of secondary fracture failure over the Antelope structure is schematically shown in Fig. 22. Note the similarity between the secondary faulting and tension fracturing pattern shown in this diagram to that depicted in Fig. 18.

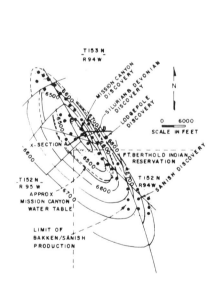

STRUCTURE ON TOP OF MISSION CANYON FM WITH MISSION CANYON & BAKKEN/SANISH PRODUCTIVE LIMITS & RESERVOIR DISCOVERY WELLS.

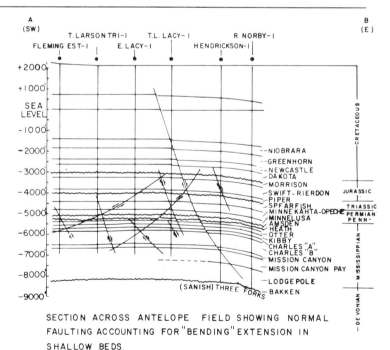

SECTION ACROSS ANTELOPE FIELD SHOWING NORMAL FAULTING ACCOUNTING FOR "BENDING" EXTENSION IN SHALLOW BEDS.

FIGURE 19

Antelope field, McKenzie Co., N. Dakota. Structure-contour map and cross-section interpretation after Finch (1969). Note presence of faulting (shear fracturing) at shallow depths, presumably associated with normal pore-fluid pressures.

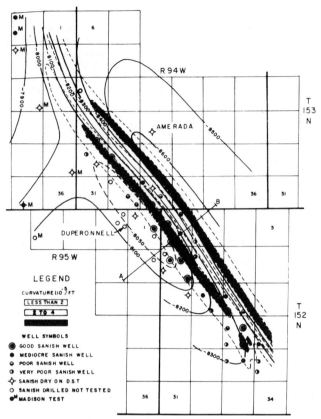
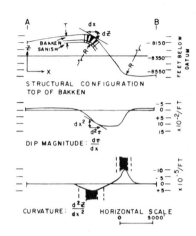

STRUCTURE & "CURVATURE" ON TOP OF BAKKEN FM WITH QUALITY OF RELATED PRODUCTION

FIGURE 20

Antelope field, McKenzie Co., N. Dakota. Structure-contour map and cross-section interpretation after MMurray (1968). Note presence of open tensile fracturing associated with production and "curvature" in the highly over-pressured Bakken Formation.

CONCLUSIONS AND DISCUSSION OF IMPLICATIONS

In the preceding sections an attempt was made to describe the petrophysical behavior the Upper and Lower Bakken Shale Members and to speculate on the reasons for this behavior. Arguments were presented which strongly suggest that the unusual petrophysical behavior is a direct reflection of the source-rock properties of these units. The conclusions drawn from these speculations and arguments — while being of a very preliminary nature and obviously requiring further investigation — have very important implcations regarding 1) petrophysical behavior in general, 2) oil generation and expulsion mechanisms, 3) the identification of "maturity" in source-rocks, and 4) contributions to exploration plays.

An interesting comparison may be drawn between the petrophysical behavior of reservoir rocks into which oil has migrated and accumulated, and sourcerocks from which oil has evidently been generated and expelled. Petrophysical techniques involving well log measurements of such properties as sonic-transit time, electrical resistivity and density have routinely been used to evaluate such paramters as lithology, porosity, hydrocarbon saturation, fluid pore pressure, etc.; however, most of these measurements have been applied to quantify states-of-condition in coarse-grained reservoir rocks. It seems only logical that similar behavior would manifest itself in the lithology (e.g., organic content), porosity, fluid saturation, pore pressure, etc. of fine-grained source rocks and thus be of considerable effort to an exploration concept in which a consideration of source-rocks is important.

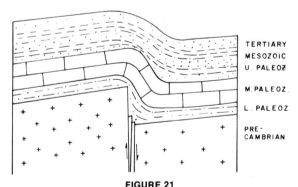

FIGURE 21

Schematic cross section of a typical Rocky Mountain "drape fold" (after Stearns, 1971). Note similarity of overall structure to that at Antelope field.

The petrophysical behavior of the Bakken and its speculative interpretation implies that generated oil is expelled from the rock as a continuous porosity-

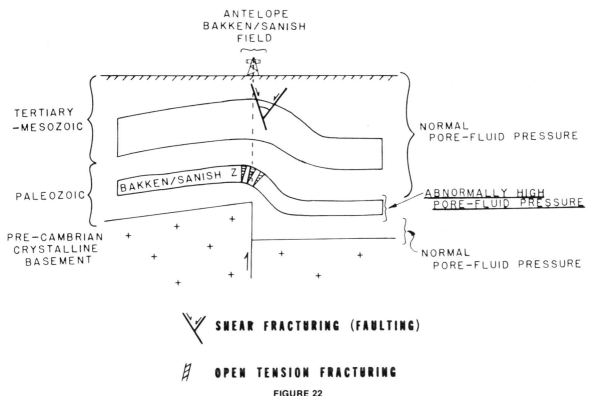

FIGURE 22
Schematic cross section showing fracture patterns in the Antelope field drape fold. Compare with Figure 18.

saturating phase through an oil-wet rock matrix. The inferred mechanism of expulsion appears related to a mechanical-potential energy imbalance caused by anomalous fluid pressure introduced by compactional processes related to rock failure incurred as 1) a result of increasing effective stress due to increasing depth of burial, and also 2) by the conversion of solid overburden-supporting kerogen to liquid, nonoverburden-supporting, mobile and expellable oil. An additional supporting expulsion mechanism could be related to pressure increases produced by the greater volumes occupied by metamorphosed organic material (residual kerogen plus generated hydrocarbons) over those occupied by the original kerogen. A conceptual scheme illustrating the proposed generation, saturation, compaction and expulsion process is shown in Fig. 23.

The petrophysical behavior noted in the Bakken may not be characteristic of all source-rock lithologies; however, some similarities may exist. Analyses suggest the Bakken to be unusually rich in organic carbon and to also contain a considerable amount of inorganic carbonate material. The high weight-percentage of organic carbon in the Bakken may be indicative of extremely large volume percentages of organic material in the rock — perhaps as much as 25 to 50%. Thiss could mean that the majority of mineral matter forming the walls and throat interconnections of the porosity network in the rock consists of preferentially oil-wet organic material. Further, carbonate mineral grains may also be oil-wet due to the absorbtion of organic compounds on their surfaces during deposition in an extremely organic-rich environment.

Petrogrpahic examination of Bakken thin sections shows that organic material is well-distributed throughout the rock and that it is not concentrated into obvious laminations, as is characteristic of many other source-rocks. This may have a great effect on average rock wetability and equilibrium-hydrocarbon saturation, in that less-organic-rich rocks may be mostly water-wet and any hydrocarbon generated or expelled within the porosity network may be confined to a low-saturation insular phase. Similarly, if a continuous network or system of laminations composed or organic oil-wet material exists in the otherwise water-wet rock, hydrocarbon saturation and explusion paths may be limited to only a small portion of the total rock volume. The effects of hydrocarbon saturation in these types of rocks may produce petrophysical changes which are either too subtle to detect or are of a much less obvious magnitude than that typified by the Bakken. However, if these changes are detectable in other source-rocks, a wide field of application to the mapping of mature source-rocks based on existing well control seems possible.

The effect of fluid saturation and pore pressure on the sonic properties of the Bakken shale may make seismic techniques amenable to the delineation of "mature" and geopressured areas.

Speculative conclusions regarding the relationship between source-rock maturity, hydrocarbon generation, geopressuring and fracturing suggest an opportunity in exploration for unrecognized and unlooked-for "unconventional" accumulations of potentially very large retional extent. A logical place to look for oil would naturally

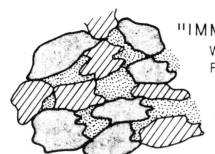

"IMMATURE"
WATER-SATURATED PORES, LOW-RESISTIVITY, NORMAL FLUID-PRESSURE.

///////// OIL GENERATION /////////

"EARLY MATURE"
OIL-WET MATRIX, CONTINUOUS-PHASE OIL- AND WATER-SATURATION; WATER EXPULSION.

MODERATELY-HIGH RESISTIVITY
NO COMPACTION
HIGH FLUID-PRESSURE

"MATURE"
HIGH CONTINUOUS-PHASE OIL-SATURATION AND EXPULSION, LOW DISCONTINUOUS WATER-SATURATION.

VERY HIGH RESISTIVITY
SOME COMPACTION
HIGH FLUID-PRESSURE

LEGEND

MATRIX

 KEROGEN

 NON-ORGANIC

PORE FLUID

 WATER

 OIL

FIGURE 23
Schematic diagram showing changes in pore-fluid volume (porosity) and pore-fluid species which may accompany hydrocarbon-generation (maturity) in source rocks.

be within the source-rock body itself, wherever the generation process has led to the temporary or permanent geopressuring of fluid in the rock and the subsequent creation of economically significant porosity and permeability in a system of extension fractures. Unexplored areas of fractured Bakken are believed to exist in the Williston Basin. The concept may be equally as valid for similar actively generating overpressured source-rocks in other stratgraphic sections and basins.

Occurrences of abnormally high pore-fluid pressure, where hydrocarbons are the only moveable phase present and which may be uniquely related to an actively-generating and well-defined source-rock unit, have been noted in other sedimentary rocks and areas of the world. The author has noted such occurences in the Powder River, Wind River, Green River, Uinta, Paradox, Anadarko, Delaware and San Joaquin Basins.

The relation of oil generation in Bakken source-rock shales to the creation of fluid overpressure and the subsequent formation of vertical open tension fractures may provide a key for discerning migration and accumulation patterns of oils thought to originate from the Bakken. Oils found in extracts of Bakken shale, within fractured Bakken reservoirs and within more conventional reservoirs some distance above the Bakken in the Mission Canyon and some distance below the Bakken in the Nisku are apparently identical and can be presumed to have the same origin (Williams, 1974; Dow, 1974; Fig. 2 this paper). Further, the Bakken is the only source-rock which has been identified for a distance of several hundred feet above or below the reservoir sections. Considering the general impermeability of "tight" rocks in the Lodgepole Formation separating the Bakken source from a Mission Canyon reservoir or of similar "tightness" in rocks of the Three Forks separating the Bakken from a Nisku reservoir, the question may be raised as to how upward and downward vertical migration takes places outward from the Bakken through "tight" intervening rocks to reservoirs in which large volumes of related oil have been found. Dow proposed vertical fractures through the Lodgepole and localized along the Nesson anticlinal axis as being responsible for upward migration into the Mission Canyon and believed lateral migration across faults was necessary to charge the Nisku (Dow, 1974, pgs. 1257-59 and Fig. 12). The author believes vertical fracture paths are much more extensive and are essentially related to the area of oil-generation-caused fluid-overpressure within the area of Bakken source-rock maturity.

The schematic cross-section shown in Figure 24 depicts a scheme of hydrocarbon generation, migration and accumulation relating Bakken source-rocks to the distribution of known types of oil accumulations. Salient features of the scheme are summarized as follows:

1) Hydrocarbon source-rock bodies comprising the upper and lower Bakken Shale units are mature in the deeper portions of the Williston Basin where they have been exposed to present-day maximum burial-related temperatures of about 165° F or more.
2) Hydrocarbon generation within the zone of maturity has caused the creation of a zone of abnormally high fluid pressure within the Bakken and closely adjacent beds.
3) The abnormally high fluid pressures have caused the creation of vertical fractures in the adjacent confining beds comprising the Lodgepole and Three Forks Formations. Fracturing appears to be preferentially upward towards the Mission Canyon regionally-extensive matrix-porosity interval on the eastern flank of the Basin. The Lodgepole on this flank of the Basin must have a lower "fracture gradient" than the Three Forks. On the western flank of the Basin, where the Lodgepole becomes siliceous and thicker, fracturing evidently occurs downward through the Three Forks to the Nisku. Fracturing within the Bakken itself — both shale and siltstone members — also takes place in areas where Bakken is highly stressed, as in the zone of strong "bending" or "curvature" at Antelope field.
4) Outward migration of oil created within the Bakken overpressure "cell" takes place through the spontaneously generated fracture system. Upward vertical migration takes place through Lodgepole fractures and Mission Canyon matrix porosity until the first evaporite unit of the "Charles facies" is encountered. After encountering the Charles "seal" migration occurs laterally undip until structural and stratigraphic traps are encountered and accumulation takes place. Trap types present in the Mission Canyon include those controlled by a) subcrop truncation b) pinchout of porous tidal flat facies into tight sabka evaporite facies c) basement structural traps d) single- and multiple-stage salt solution traps and e) combinations of any of the preceding conditions. Downward vertical migration to matrix porosity in the Nisku takes place where fractures are formed through the Three Forks. Lateral updip migration beneath the Three Forks "seal" then takes place until a trap is encountered. Nisku accumulations appear to be confined to "closed" basement or salt-solution anticlinal features. Based on the known occurrence of Bakken-type oil in a few small fields located primarily on the west flank of the Basin, fracturing and downward migration through the Three Forks must be relatively inefficient when compared to upward fracturing and migration and is essentially limited to the area where the Lodgepole is thick and siliceous and the Mission Canyon has regionally poorer matrix-porosity development.
5) Fractures through the Lodgepole and Three Forks which "confine" the fluid-overpressure cell within the Bakken are believed to "breathe" and act as a safety valve to the development of extremely high fluid-overpressures developed in the Bakken as a result of hydrocarbon generation. Once fracturing occurs at some critical high pressure, generated hydrocarbons bleed-off from the Bakken to the more-normally pressured Mission Canyon or Nisku and relieve pressure within the Bakken. When fluid pressure falls below that of the **minimum principal** tectonic **stress** (e.g., the **effective stress** becomes compressive instead of tensile) the fractures close. The fractures may open again when more oil is generated and the fluid pressure is again raised to a critical value again or the whole system may merely equilibrate to a pressure sufficient to maintain a steady-state of oil generation and outward migration through barely open fractures. The fact that open fractures are present in the overpressured Bakken/Sanish reservoir at Antelope field seems to imply substantial eleastic contrasts between the properties of the Bakken/Sanish lithologies and those of the confining Lodgepole Formation which separates it from a productive Mission Canyon matrix reservoir. Fluid pressure gradients mapped in the Bakken Formation and based on a sonic-behavior

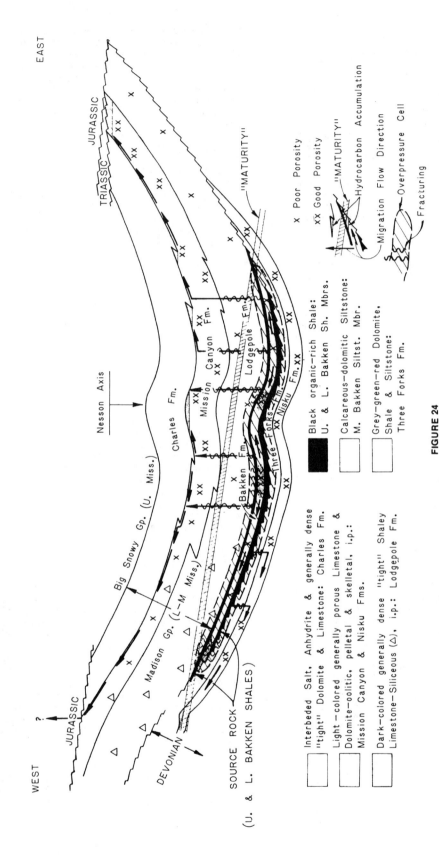

FIGURE 24

Schematic east-west section across the Williston Basin showing source-rock maturity, fluid over-pressure, fracture, migration and hydrocarbon accumulation patterns in the Bakken Formation and adjacent units.

relation (Fig. 15) imply that a fluid pressure ("frac") gradient of about .60 to .80 psi/ft is required to open fractures in confining units and allow outward migration from overpressured and currently hydrocarbon-generating Bakken source-rocks.

The generation/migration scheme postulated for the Bakken source-rock unit and its associated oil-productive reservoir units is believed to be somewhat universal in other units and basins throughout the world. For example — oil in upper Cretaceous sandstone reservoirs of the Powder River Basin in Wyoming are separated from overpressured and actively generating source-rocks in the Niobrara Formation by a thousand feet or so of intervening Pierre shales. A similar scheme of vertical migration through fractured overpressured shale may explain patterns of generation and accumulation on the lower Gulf Coast, where mature source rocks charging Middle and Upper Tertiary reservoirs must be extremely deep and lie within or at the bse of an extremely thick geopressured shale section.

BIBLIOGRAPHY

Anderson, E.M. (1951), The dynamics of faulting: Oliver and Boyd, Edinburg, 206 pp.

Connan, J. (1974), Time-temperature relation in oil genesis: Am. Assoc. Petr. Geol. Bull., v. 58, n. 12, pp. 2516-2521.

Claypool, G.E., Love, A.H. and Maughan, E.K. (1978), Organic geochemistry, incipient metamorphism, and oil generation in black shale members of Phosphoria formation, western interior United States: Am. Assoc. Petr. Geol. Bull., v. 62, pp. 98-120.

Dow, W.G. (1974), Application of oil-correlation and source-rock data to exploration in the Williston Basin: Am. Assoc. Petr. Geol. Bull., v. 58, n. 7, pp. 1253-1262.

Finch, W.C. (1969), Abnormal pressure in the Antelope field, North Dakota: Jour. Petr. Tech., July, pp. 821-826.

Hottman, C.E. and Johnson, R.K. (1965), Estimation of formation pressures from log-derived shale properties: Am. Inst. Min. Metall. Petr. Engr. Trans., v. 234, pp. 717-722.

Houston Geological Society (1971), Abnormal subsurface pressure — A study group report 1969-1971, 92 pp.

Hubbert, M.K. (1951), Mechanical basis for certain familiar geological structures: Geol. Soc. Am. Bull., v. 62, pp. 355-372.

Hubbert, M.K., and Rubey, W.W. (1959), Role of fluid pressure in mechanics of overthrust faulting: Geol. Soc. Am. Bull., v. 70, n. 1, pp. 115-166.

Landes, K.K. (1967), Eometamorphism and oil and gas in time and space: Am. Assoc. Petr. Geol. Bull., v. 51, n. 6, pp. 828-841.

LaPlante, R.E. (1974), Hydrocarbon generation in Gulf Coast Tertiary sediments: Am. Assoc. Petr. Geol. Bull., v. 58, n. 7, pp. 1281-1289.

MacGregor, J.R. (1965), Quantitative determination of reservoir pressures from conductivity log: Am. Assoc. Petr. Geol. Bull., v. 49, n. 9 (Sept.), pp. 1502-1511.

Momper, V.A. (1972), Evaluating source beds for petroleum (abs.): Am. Assoc. Petr. Geol. Bull., v. 56 n. 3, p. 640.

Murray, G.H., Jr. (1968), Quantitative fracture study - Sanish pool, McKenzie County, North Dakota: Am. Assoc. Petr. Geol. Bull., v. 52, n. 1, pp. 57-65.

Nixon, R.P. (1973), Oil Source beds in Cretaceous Mowry Shale of northwestern interior United States: Am. Assoc. Petr. Geol. Bull., v. 57, pp. 136-161.

Nordquist, J.W. (1953), Mississippian stratigraphy of Northern Montana: Billings Geol. Soc., Guidebook 4th Annual Field Conference, pp. 68-82.

North Dakota Geological Survey (1974), Production statistics and engineering data, oil in North Dakota, second half of 1973, 262 pp.

Pan, Poh-hsi and de Bremaeker, J. Cl. (1970), Direct location of oil and gas by the seismic reflection method: Geophysical Prospecting; v. 18 (Dec.) pp. 712-727.

Peale, A.C. (1893), Paleozoic section in the vicinity of Three Forks, Montana: U.S. Geol. Surv. Bull. 110, 56 pp.

Philippi, G.T. (1965), On the depth, time and mechanism of petroleum generation: Geochem. et. Cosmochi. Acta, v. 29, pp. 1021-1049.

_____(1968), Essentials of petroleum formation process are organic source and a subsurface temperature controlled chemical reaction mechanism; in Schenk, P.A. and Havanaar, I. (eds.), Advances in Organic Geochemistry: Oxford, Pergamon, pp. 25-46.

Price, N.J. (1966), Fault and joint development in brittle rock, Pergamon Press, Oxford, 176 pp.

Sandberg, C.A. and Hammond, C.R. (1958), Devonian System in Williston Basin and Central Montana: Am. Assoc. Petr. Geol. Bull., v. 42, n. 10, pp. 2293-2334.

Secor, D.T., Jr. (1965), Role of fluid pressure in jointing: Am. Jour. Sci., v. 263, pp. 633-646.

Stearns, D.W. (1971), Mechanisms of drape folding in the Wyoming province: Wyoming Geol. Assn. Guidebook, 23rd Annual Field Conference pp. 125-143.

Tissot, B., Califet-Debyser, Y., Deroo, G., and Oudin, J.L., (1971), Origin and evolution of hydrocarbons in early Toarcian shales, Paris Basin, France: Am. Assoc. Petr. Geol. Bull., v. 55, pp. 2177-2193.

Trask, P.D. and Patnode, W. (1942), Source beds of petroleum: Am. Assoc. Petr. Geol. Spec. Publ., 556 pp. - See especially p. 62.

Welte, D.H. (1965), Relationship between petroleum and source rock: Am. Assoc. Petr. Geol. Bull., v. 49, n. 12, pp. 2246-2268.

Wheeler, H.E. (1963), Post-Sauk and Pre Absaroka Paleozoic stratigraphic patterns in North America: Am. Assoc. Petr. Geol. Bull., v. 46, n. 8, pp. 1497-1526.

Williams, J.A. (1974), Characterization of oil types in the Williston Basin: Am. Assoc. Petr. Geol. Bull., v. 58, n. 7, pp. 1242-1252.

Geochemical Exploration in the Powder River Basin

James A. Momper
Consultant
Tulsa, Oklahoma

Jack A. Williams
Amoco Production Company
Tulsa, Oklahoma

Geochemical and geological data were used to identify effective source rocks and oil-types, and to determine stratigraphic sequences and areas that are prospective for crude oil and thermal hydrocarbon gas. The source rock volumes and generation-expulsion performance data for each effective source sequence provided the basis for calculating quantities of expelled oil and gas. These quantities readily account for discovered in-place reservoir oil of more than 7 billion barrels and relatively minor amounts of gas, mainly associated.

Lower and Upper Cretaceous source beds expelled most of the indigenous oil. These oils are chemically similar, regardless of their source. Lower Cretaceous Mowry Shale and Upper Cretaceous Niobrara and Carlile formations expelled most of the discovered oil. Oil expulsion from Cretaceous source rocks began during the early Tertiary and continued through much of Miocene time as the expulsion fronts moved up section and updip. Laramide structure controlled directions of migration of Cretaceous oil.

The second major type of oil is nonindigenous to the Powder River Basin and is correlated to the remote Upper Permian Phosphoria Formation source area centered in southeastern Idaho. This oil entered northeastern Wyoming during Late Jurassic time, before the Powder River Basin formed, through carrier beds of Pennsylvanian and Permian age. Phosphoria-type oil is preserved in four separate parts of the basin, primarily in sandstone reservoirs of Early Permian age in the Minnelusa and Tensleep formations.

A minor oil-type found in the southeastern part of the Powder River Basin was expelled from relatively thin, local shales of Pennsylvanian age.

Several giant fields with more than 100 million barrels of recoverable oil and major oil fields of at least 50 million barrels are located on structural positives around the periphery of the Powder River Basin. These salients served as gathering areas to concentrate migrating oil. Other large fields are in stratigraphic traps oriented parallel with structural strike on the eastern flank, this orientation permitting large accumulations to form from a big drainage area in downdip source rocks.

Meteoric water, aerobic bacteria, distillation, and thermal cracking are affecting the quality of preserved oil. Two types of bacterial alteration are common. Much of the gas generated with oil has escaped or dispersed. Oxygenated recharge waters appear to be degrading organic matter in Cretaceous source rocks around the basin perimeter. Both chemical and physical properties of rocks and fluids proved to be useful in defining prospective areas for the various types of oil.

INTRODUCTION

The Powder River Basin in northeastern Wyoming and southeastern Montana is an excellent petroleum province for testing and applying various geochemical techniques and concepts because of its diverse geology and the intensive exploration in the basin for the last 30 years.

A large number of oil, rock, water, and gas samples were analyzed for this study. The 235 oil samples were collected from wells and seeps in widely distributed geographic locations from all of the important reservoirs of the basin and from most of the minor reservoirs. At least 40 discrete productive reservoirs have been discovered in all systems from the Mississippian to the Tertiary inclusively. The majority of the more than 500 rock samples analyzed (cores, cuttings, and outcrops) were from the Cretaceous System. In addition, Bureau of Mines oil sample analytical data and other published geochemical information pertaining mainly to the Cretaceous Mowry Shale were used.

Typical of the several Bureau of Mines studies are those by Wenger and Reid (1958), Biggs and Espach (1960), Wenger and Ball (1961), and Coleman et al (1978). Papers by Hunt (1953), Curtis (1958), Strickland (1958), and McIver (1962) described crude oil and its occurrences in the

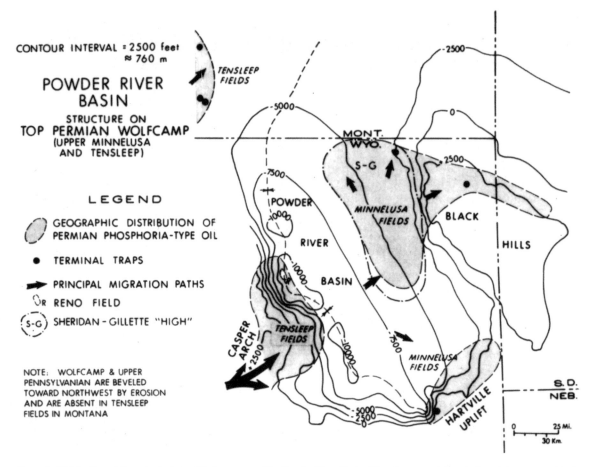

Figure 1. Distribution of Phosphoria-type oil in four areas of the Powder River Basin. Migration preceded evolution of the structural basin. Gaps along migration paths in the basin resulted, in part, from thermal cracking of oil to gas. Isolated shows are present east-southeast of the terminal trap on the northern margin of the Black Hills in South Dakota.

Powder River Basin. Hedberg (1968) discussed the significance of high-wax oils.

Published information on source rocks in the Powder River Basin is quite limited. Four papers on the Mowry Shale (Early Cretaceous) have been published: Schrayer and Zarelia (1963, 1966, 1968) and Nixon (1973). A cursory study of the geochemistry of some Upper Cretaceous shales by Merewether and Claypool was published in 1980.

OIL TYPES

Three basic oil types occur in the basin—a non-indigenous Permian Phosphoria-type (Figure 1), a Cretaceous-type, and a Pennsylvanian-type of minor importance. Phosphoria-type oil was generated in a remote source area external to the basin and migrated into the area during Late Jurassic time. The Powder River Basin formed during Late Cretaceous-early Tertiary time. Phosphoria-type oil occurs in reservoirs of Mississippian through Jurassic age in and around the basin. Expulsion of indigenous Cretaceous and Pennsylvanian oils occurred after the basin formed.

Cretaceous-type oil was derived from multiple source beds in both the Lower and Upper Cretaceous (Figure 2), resulting in two subtypes with only subtle differences in chemical characteristics. The Lower Cretaceous (LK) subtype is principally Mowry oil (Figure 3). Most of the Upper Cretaceous (UK) subtype was derived from the Niobrara and Carlile formations (Figure 4). The two subtypes commonly are commingled and are therefore difficult to distinguish. Identifications of the LK- and UK-subtypes are further complicated by several alteration processes that have affected oil to varied degrees in many of the fields. Cretaceous-type oil is the most abundant and widespread in the basin, having been discovered in reservoirs as old as Pennsylvanian and as young as early Tertiary, and in all but the northernmost quarter of the basin.

Oils expelled from Upper Cretaceous source beds are mainly confined to the southwestern part of the basin and

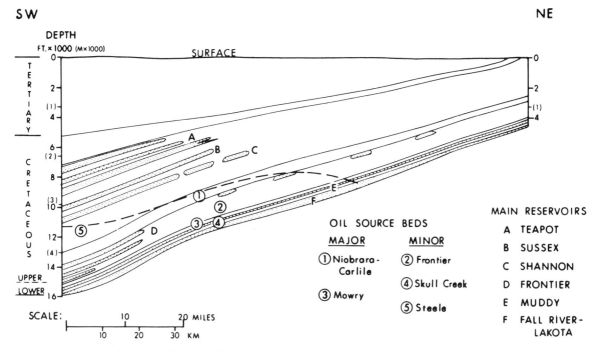

Figure 2. Stratigraphic relationships of oil source beds and main reservoirs within Cretaceous System in the Powder River Basin. Argillaceous rocks with significant oil-generating capability below heavy dashed line expelled oil, and are effective oil source beds. Line of section (southwest-northeast) is on Figure 3.

its margins. Only a few small scattered pools of UK-subtype oil occur updip east of the effective source area (Figure 4). The limiting factor for more entrapment in that updip area appears to be a lack of reservoir beds. Some Upper Cretaceous reservoirs out in the basin are interpreted to contain LK-subtype oil, suggesting that faults or fractures may have permitted transverse migration. Upper Cretaceous reservoirs considerably younger than any of the effective source rocks produce oil (Figure 2), proving that appreciable transverse migration of UK-subtype oil occurred, if not of the LK-subtype. Within the Powder River Basin, most of the Cretaceous-type oil is in stratigraphic traps. Around the basin margin, anticlinal and combination traps dominate.

Minor quantities of Pennsylvanian-type oil are present in Pennsylvanian reservoirs in a few pools in the southeastern sector of the basin. At least one occurrence has been identified in a Permian reservoir and one in a Jurassic reservoir. This type has been difficult to identify because of oil degradation and limited sample availability but we now have sufficient control to conclusively establish the presence of this third type.

SOURCE ROCKS

Effective oil and gas source beds within the basin principally are lower Upper Cretaceous and Lower Cretaceous shales. In the southeastern part of the basin, where Pennsylvanian-type oil occurs, local thin Pennsylvanian shale beds expelled oil. The other systems do not contain thermally mature rocks with significant oil-generating capability. Locally, part of the Jurassic has thermal gas-generating potential and small quantities of recoverable biogenic gas have been generated in the Cretaceous and Tertiary systems.

Several analytical methods were used to evaluate source rocks and to correlate oils to their respective sources. Quantitative methods provided information on the amounts of organic carbon, extractable bitumen, and extractable C_{15+} saturated hydrocarbons in the source rock samples. Oil-rock correlations were established by a variety of analyses on the extracts and oils. Quantities of hydrocarbons already generated and quantities remaining to be generated were obtained from pyrolysis. Qualitative methods were used to determine the convertibility of organic matter to oil and natural gas, and also to determine the extent of organic matter degradation. Thermal carbonization stages of organic matter and the maturity of generated bitumen and hydrocarbons were interpreted from qualitative methods. In addition, several methods indicated whether organic matter was preferentially oil- or gas-generating, or non-generating.

PHOSPHORIA OIL SOURCE

A large quantity, perhaps as much as 2 billion (2×10^9) barrels of Phosphoria-type oil migrated into northeastern Wyoming before the Powder River Basin formed. The rich

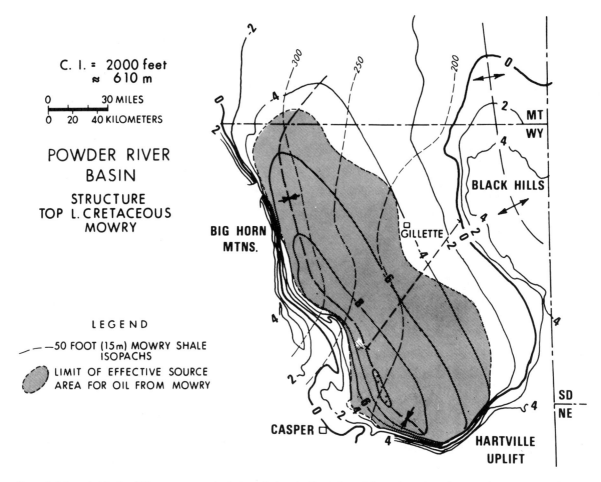

Figure 3. Interpreted limits of Mowry source rocks that expelled crude oil, relative to Mowry isopachs and present-day structure on the top of the Mowry Shale.

Phosphoria source sequence centered in southeastern Idaho expelled oil during Late Jurassic time. An almost ideal, westward-tilted, sealed carrier system comprised mainly of permeable Permian and Pennsylvanian sandstone enabled Phosphoria oil to migrate long distances across Wyoming (Sheldon, 1967) along several migration paths. In western Wyoming, the oil migrated mainly through Pennsylvanian Tensleep sandstone, moving up-section into and updip through Wolfcampian Tensleep-Casper sandstone in eastern Wyoming where that section is preserved.

Phosphoria-type oil has been identified in four separate areas of the basin. It entered northeastern Wyoming via the Casper Arch along one path (Figure 1), and was trapped mainly in Permian Tensleep and Minnelusa reservoirs in three of the areas. Terminal entrapment occurred where permeable sandstone carrier beds change facies into tight evaporites and red beds flanking the Black Hills Uplift. Nevertheless, oil staining and heavy oil residues in outcrops indicate that a large volume of oil escaped and a significant amount was degraded.

A second migration path delivered a considerably smaller amount of oil to Tensleep reservoirs and a Mississippian reservoir in the area now on the northwestern margin of the basin in Montana. This oil migrated through mid-Pennsylvanian-age sandstone and also terminated at an evaporite facies. The truncated subcrop of the Upper Pennsylvanian and Lower Permian is located southeast of this northernmost migration path. Some redistribution of Phosphoria-type oil occurred during basin-subsidence and the Laramide orogeny (Late Cretaceous-early Tertiary).

In the deep, relatively hot southern part of the basin and elsewhere where geothermal gradients are sufficiently high, Phosphoria oil has been cracked to gas. Oil pools have survived in cooler locations as deep as 15,000 ft (4,572 m) in the Reno Field area and more should be discovered along the original migration and remigration paths if shallower, cooler traps are present. Most of the gas accompanying Phosphoria-type oil during its Late Jurassic migration has escaped. Possibly much of the gas resulting from the later thermal cracking is in undiscovered traps.

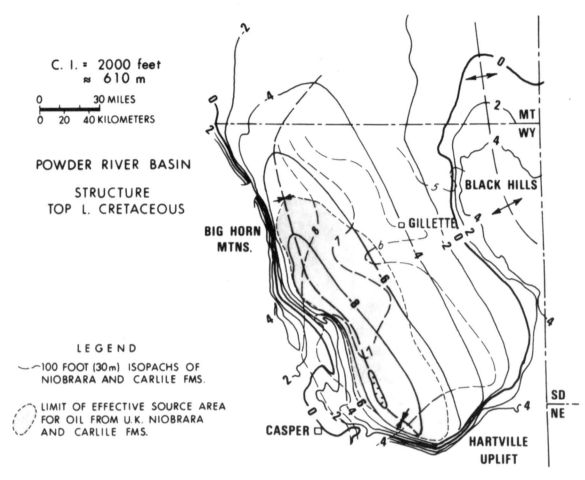

Figure 4. Isopachs of combined Niobrara Formation and Sage Break Shale Member of the Carlile indicate that the interval within the area of effective oil source rocks ranges from about 550 ft (168 m) to more than 800 ft (244 m). However, on the flanks of the basin, the actual thickness of the effective source rocks is less than the total thickness of these units, as shown on Figure 2. This section accounted for most of the UK-subtype oil.

CRETACEOUS SOURCE BEDS

Several effective oil-sources are present in the Cretaceous (Figure 2). The Lower Cretaceous Mowry Shale evidently expelled at least as much oil as all other sources combined. The next most important oil sources were the Upper Cretaceous Niobrara and subjacent Carlile formations. The Frontier, Skull Creek, and Steele shales (and possibly the Fuson Shale) were relatively minor, effective oil source sections.

Mowry Shale

The effective Mowry Shale oil source area is delineated on Figure 3. Superimposed on a structural map of the Lower Cretaceous are 50-ft (15.2-m) isopachs of the Mowry. The Mowry is highly siliceous. It is more than 350 ft (107 m) thick along the western margin of the basin and is beveled by erosion toward the southeast. Geochemical analyses and subsidence profiles (using exposure time and burial temperatures) both indicate that the Mowry in the shaded area expelled oil. Times of expulsion from within this area of effective Mowry source rocks depended on the ages and thicknesses of overburden, as well as on the local geothermal gradients, which today range from about 13 to 20°F per 1,000 ft (305 m) of burial depth. Available evidence suggests that the same relationship of high and low gradients has persisted since the Cretaceous in those deeper parts of the section unaffected by cooling phenomena.

Oil expulsion began in late Paleocene or early Eocene time from the basin deep in the southwest. Elsewhere, expulsion was delayed, persisting through much of Miocene time as the oil-generation front moved updip and up-section. Generation eventually ceased as the shallower section in the basin cooled, owing to deep erosion, uplifting, late Cenozoic climatic cooling, and encroachment of relatively cold recharge water. The Mowry and other

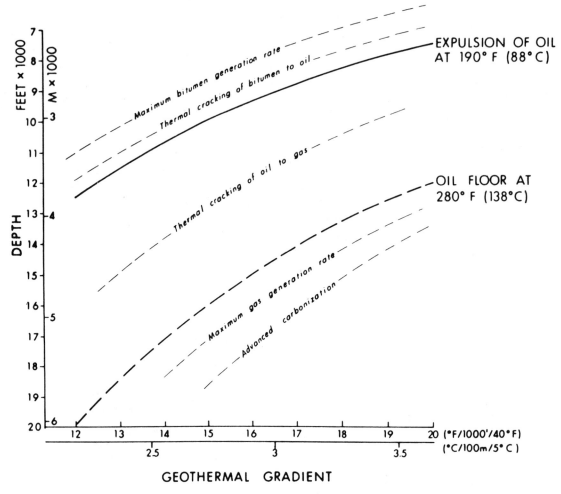

Figure 5. Burial depths needed for oil expulsion from Cretaceous source beds in the Powder River Basin vary with geothermal gradients and exposure times. Expulsion curve shown is for maximum exposure time. Shorter exposure periods required higher temperatures for expulsion. Similarly, the thermal destruction of commercial oil accumulations (oil floor) occurs approximately at the depths indicated for the present range of geothermal gradients and maximum exposure times. In this province, the deepest Cretaceous is less than 15,000 ft (4,572 m) so that the oil floor is encountered only in the highest geothermal gradient settings (that is, generally above 17°F/1,000 ft).

effective source beds remain overpressured today in the deeper part of the basin, although not sufficiently for oil expulsion to occur. The build-up of pressures in these source rock units resulted primarily from fluid-generation during carbonization of organic matter. The shallower parts of the source rocks that had generated substantial quantities of fluids are under-pressured as a result of the cooling trends.

We calculated the Mowry Shale expelled about 11.9 billion barrels of oil (Table 1) from its 10,500 sq mi (27,195 sq km) effective source area, using averages of 3.0 weight percent organic carbon, 240 ft (73.2 m) of shale, and an oil-generating capability of 105 barrels per acre foot (>5,800 ppm).

The expulsion efficiency of 7 percent was obtained by comparing the calculated amount of generated oil with the total of (1) discovered in-place oil, plus (2) an estimate of undiscovered oil, and (3) estimates of the amounts of oil lost by degradation, dispersal, escape, and thermal cracking. Of

Table 1. Total quantity of oil expelled from effective source rocks within Lower Cretaceous Mowry Shale in Powder River Basin.

Organic carbon content	weight percent	3.0
Oil generated	barrels per acre foot	105
Oil expelled at 7% efficiency	barrels per acre foot	7.4
Thickness, average	feet	240
Oil expelled	barrels per acre	1,765
Oil expelled	10^3 barrels per square mile	1,130
Effective source area	square miles	10,500
Total Mowry oil expelled	10^9 barrels	11.9

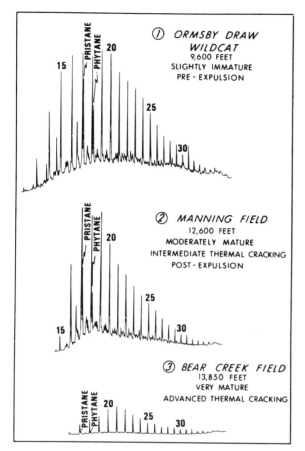

Figure 6. Whole-extract gas chromatograms from Mowry Shale samples at different stages of thermal maturity resulting from progressive thermal cracking with deeper burial (higher temperatures) and long exposure times. Well locations are shown on Figure 7.

Table 2. Total volume of oil expelled from effective source rocks within the combined Niobrara Formation and Sage Breaks Shale Member of Carlile Formation, both Late Cretaceous, in the Powder River Basin.

Organic carbon content	weight percent	2.5
Oil generated	barrels per acre foot	75
Oil expelled at 7% efficiency	barrels per acre foot	5.3
Thickness, average	feet	550
Oil expelled	barrels per acre	2,890
Oil expelled	10^3 barrels per square mile	1,850
Effective source area	square miles	4,100
Total volume oil expelled	10^9 barrels	7.6

these estimates, the least accurate are the values assigned to oil dispersed and escaped. Various combinations of values were tried and usually resulted in efficiencies in the range of 6 to 8 percent.

Niobrara-Carlile Section

The Niobrara and Carlile formations are much thicker than the Mowry. However, they are generally 800 to 1,000 ft (244 to 305 m) shallower than the Mowry and therefore, the areal extent of effective Niobrara-Carlile source rocks is much more limited (Figure 4; Table 2). The Niobrara is dominantly marl and calcareous shale and the Sage Breaks Shale Member of the Carlile is calcareous. In the northwestern sector, where the Niobrara-Carlile section is 900 to 1,000 ft (274 to 305 m) thick, it is still immature. These units accounted for the bulk of the expelled UK-subtype oil.

Undegraded organic matter in the Niobrara-Carlile section seems to be slightly poorer in quality overall than undegraded Mowry organic matter. Also, the effective source volume of the Niobrara-Carlile is more than 10 percent smaller. Consequently, we calculated that about 7.6 billion barrels of Niobrara-Carlile oil were expelled in the Powder River Basin (Table 2), or about one-third less than the yield from the Mowry.

Minor Cretaceous Sources

The continental, somewhat coaly Lower Cretaceous Fuson Shale (between Fall River and Lakota in Inyan Kara Group; Figure 2) expelled gas and speculatively, 0.1 to 0.2 billion barrels of waxy oil. Control is too sparse to definitely list the Fuson as an effective oil source. Gas-expulsion from the Fuson, other than gas that may have been expelled with oil, was restricted to relatively small areas where the geothermal gradient is high and peak gas-generating conditions have been attained. The marine Lower Cretaceous Skull Creek Shale contains Type II organic matter and contributed about 1 billion barrels of the total Cretaceous oil expelled in the basin. Much of this oil is commingled with oil from the Mowry in Muddy Sandstone reservoirs.

The thickness of the net effective oil source rocks within the mixed marine-continental Frontier Formation is relatively small compared to the total thickness of the Frontier. The Frontier is coaly, in part, and much of the shale contains Type III organic matter. Thus, gas would be the main product of thermal generation but peak gas generation has not been attained in the unit. The amount of expelled oil attributed to the Frontier is 0.8 billion barrels but the calculation is based on less sample data than is needed for an accurate assessment.

A few samples from the marine Upper Cretaceous Steele Shale (above the Niobrara) suggest that it may have expelled some oil in the west-central area of the basin. Most of the samples from the Steele contained organic matter that is immature or has little oil-generating capability. Inadequate data permit only a gross estimate of no more than 100 million barrels of oil expelled from the Steele.

Cretaceous Source Rock Performance

The calculated total volume of oil generated in all Cretaceous source beds is more than 300 billion barrels. The volume expelled is 21.5 billion barrels, roughly one-third of which appears to have been trapped in-place. The remainder was dispersed, degraded, escaped, and cracked to gas. Numerous active oil seeps around the basin—more than a score of seeps were mapped on Salt Creek Anticline alone—attest to the considerable loss of Cretaceous-type oil.

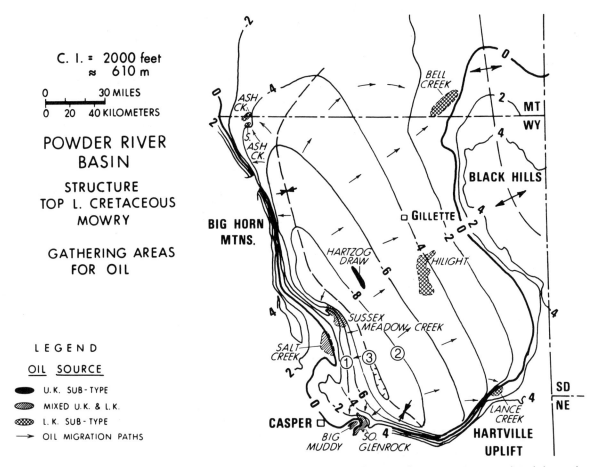

Figure 7. Oil migration essentially at right angles to structural strike enables structurally positive features to concentrate relatively large volumes of oil; four of five giant fields (Bell Creek, Lance Creek, Sussex-Meadow Creek and Salt Creek) are favorably located with respect to structurally controlled gathering areas. A number of important non-giants are similarly situated. Most of the larger stratigraphic accumulations (for example, Hartzog Draw and Hilight) are parallel or subparallel to strike.

Organic Matter Degradation

Many source rock samples from this basin contain kerogen with below-average oil-generating capability. One-third of the samples analyzed by pyrolysis from the Mowry, Niobrara, and Carlile were rated as of non-source or marginal source-rock quality. This is attributed to degradation of the organic matter after expulsion occurred, the degradation being confined to areas of active meteoric water incursion.

OIL FLOOR

Basinwide data indicate that oil expulsion usually occurred above about 190°F (Figure 5). Depending on local geothermal gradients, this minimum temperature is encountered at depths from about 8,000 to 12,000 ft (2,438 to 3,658 m) in Cretaceous source rocks. Oil in reservoirs is being cracked to natural gas liquids and natural gas at temperatures generally above 280°F, or at depths below about 12,500 ft (3,810 m) in the high-gradient areas. Because of comparable exposure times, the oil floor in Paleozoic reservoirs generally is at only slightly higher temperature levels than the floor in the Cretaceous because the Paleozoic reservoirs that contain Phosphoria-type oil were at shallow depths until rapid subsidence occurred during Late Cretaceous time. Pennsylvanian-type oil was not expelled until early Tertiary time and thus had the same exposure time as the older Cretaceous-type oils.

Proximity to the oil-floor in the deepest Cretaceous is shown by gas chromatograms of extracts from Mowry samples collected from three relatively deep wells (Figure 6). Identical sample sizes and instrument settings were used. Extract from well 1 (at 9,600 ft, or 2,926 m, and 175°F), is interpreted as slightly immature. Mowry extract from well 2 (at 12,600 ft, or 3,840 m, and 250°F), remaining after expulsion, is moderately mature. The small quantity of thermally cracked extract from well 3 (at 13,850 ft, or 4,221 m, and 275°F) is very mature. Well locations are shown on

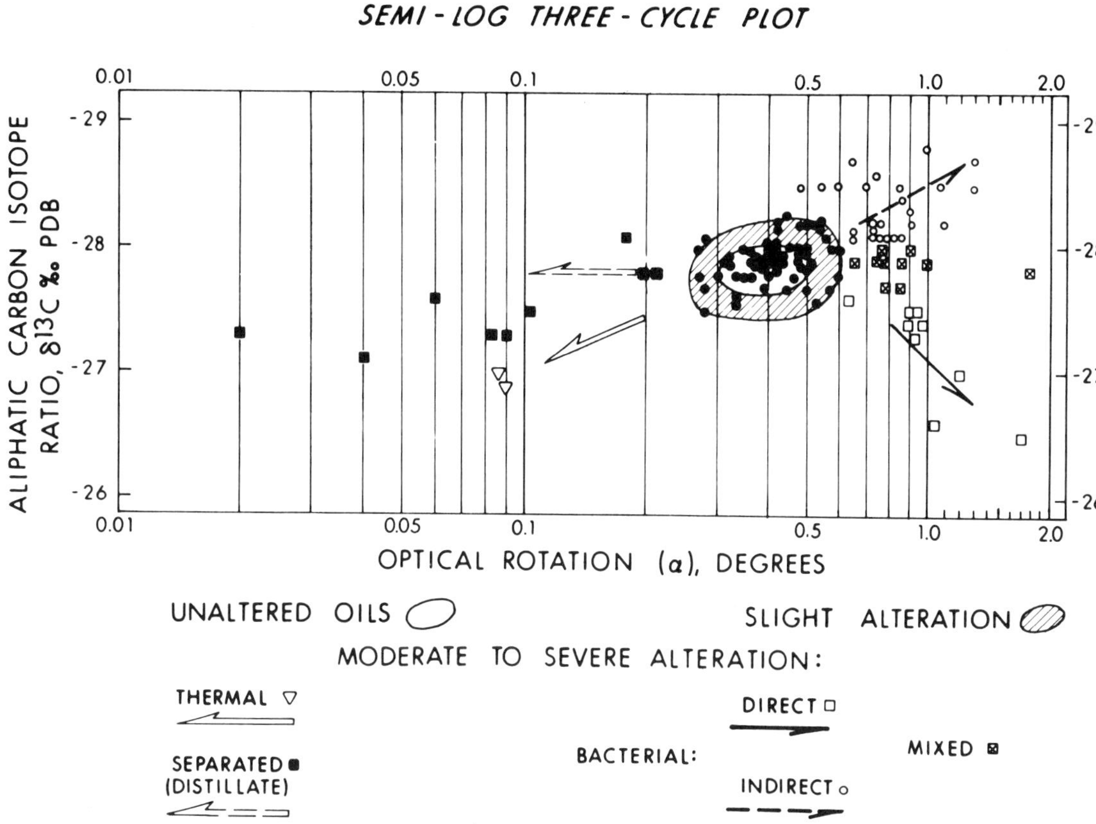

SEMI-LOG THREE-CYCLE PLOT

Figure 8. The stable carbon isotope ratios of the aliphatic fraction of crude oils and the optical activities of the oils provide useful data for determining the types and degree of alteration that affected each sample in its reservoir. All data points represent oil samples from Lower Cretaceous, mainly Muddy, reservoirs and indicate that a high percentage of the samples sustained aerobic bacterial degradation.

Figure 7. Wells 2 and 3 are in an area where the geothermal gradient is moderately high; the gradient at location 1 is significantly lower.

FACTORS CONTROLLING MIGRATION

Migration of Cretaceous-type oil was controlled by the Laramide-induced structural configuration which survives essentially unchanged today (Figure 7). Intrastratal migrations of oil are updip, generally at right angles to structural strike, unless diverted laterally along strike by a structural or stratigraphic barrier. Because migration is a dispersal phenomenon, the most favored accumulation sites tend to be at structurally positive gathering areas that locally concentrate oil.

In this basin, three of the five giant fields (Lance Creek, Salt Creek, Sussex-Meadow Creek) are in sizeable anticlines located on even larger, structurally positive salients, as are a number of major fields. Bell Creek and Hilight are stratigraphic traps that were favored by elongated reservoirs oriented parallel or subparallel with structural strike, another preferred setting for large oil accumulations. Most of the larger stratigraphic accumulations have this orientation. Bell Creek Field also is on a broad anticlinal nose that served as a gathering area.

The oil in Hartzog Draw Shannon reservoir (Figures 2 and 7), a major accumulation, migrated transverse to the bedding. The volume of oil accumulated in Hartzog Draw Field, and in several similar stratigraphic traps located stratigraphically above the shallowest effective source rocks, suggests migration upward along faults or fractures. At Lance Creek, and other fields in its vicinity, cross-fault migration enabled Cretaceous-type oil to be trapped in pre-Cretaceous reservoirs.

ALTERED OILS

Alteration of labile oil begins in most basins soon after the first oil is expelled. The Powder River Basin oils were no exception. One means of recognizing basic oil types and subtypes, and of determining kinds and degrees of alteration, is to plot the stable carbon isotope ratio of the aliphatic fraction of each oil sample versus the optical rotation measurement of the sample. On Figure 8, the large

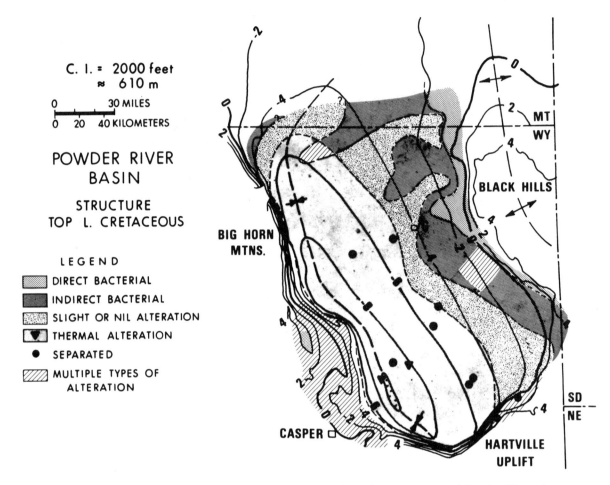

Figure 9. Many of the oil samples from Lower Cretaceous reservoirs show evidence of one or more types of alteration. The areal distribution of unaltered oils and various groups of altered oils are controlled mainly by encroachment of recharge waters and by heat flow.

cluster represents the basic Cretaceous oil type (LK-subtype). The data points in the inner cluster are from oils that evince no alteration. The outer ring of values are from mildly altered oils. To the lower right, the data points represent oils directly degraded by aerobic bacteria in areas and reservoirs invaded by meteoric water. Generally, the farther a point is from the main cluster, the more the oil has been degraded. To the upper right, data points represent oils indirectly altered by the addition of products of biodegradation. They also are from reservoirs in recharge areas, but downdip from directly biodegraded oils. A few of the oils in this group were at reservoir temperatures above 200°F, too high for direct bacterial attack. Intermediate points are for oils that show both types of degradation. The biodegraded products are isotopically light (more negative $\delta^{13}C$), whereas biodegraded oils from which the products were removed are isotopically heavy.

Data points from oils altered by thermal cracking (but not yet condensates) plot to the lower left of the main cluster. Condensates have little or no optically active components. Points scattered almost horizontally to the left represent oils that have separated, or distilled off, from the basic oil. They have properties similar to condensates but retain heavier components and measurable optical rotations. Distillation does not have much effect on carbon isotope ratios whereas condensates and thermally cracked oils are isotopically heavier than the original oil. Because the distillates are also subjected to cracking, they tend to be shifted slightly toward the heavier isotope ratios. We suspect that this separation occurred in the source beds, mainly the Mowry, during the later phase of oil expulsion, the heavy ends having been retained in the source rocks.

Distributions of unaltered and altered oils in Lower Cretaceous reservoirs are mapped on Figure 9. Basinward from the west flank of the Black Hills, an irregular belt encompasses oils that have been directly altered by bacteria and also are water-washed. Downdip, but still in the recharge area, oils have been indirectly affected by the addition of degradation products transported from updip degraded oil accumulations by migrating meteoric water. These oils have been water-washed and many also were affected directly by aerobic bacteria in the shallower pools. In the next deeper belt, the oils are essentially unaltered; recharge waters have not yet penetrated this belt, part of which is underpressured because of cooling, to a significant extent. Farther downdip, seven oils classed as separated were encountered as were two on the southeast flank. The deeper the oil reservoir, the more these separated oils approach a true thermal distillate. The distillates are similar to, but distinguishable from, condensates caused by thermal cracking.

In the deepest basin, where the Cretaceous is appreciably overpressured, Cretaceous oils have been altered by thermal cracking but retain too much of their heavy components to be classed as condensates. However, thermal cracking and gas generation locally have progressed to a level where condensate is to be anticipated in the hottest deep reservoirs. On the flanks of the basin, particularly the southwest and southeast, a variety of altered types are intermingled (cross-hatched areas).

SUMMARY

The Powder River Basin has a large volume of effective oil source rocks of both Early and Late Cretaceous age that evidently expelled more than 20 billion barrels of oil during the Tertiary. Phosphoria oil migrated eastward across Wyoming from its main source area during Late Jurassic time, entering northeastern Wyoming before the Powder River Basin formed. A relatively small volume of oil was generated and expelled from Pennsylvanian source rocks in the southeastern part of the basin. Much of the trapped oil has undergone one or more types of alteration, and most of the gas expelled with the three types of oil has disappeared.

The present-day Powder River Basin structural configuration controlled the migration of Cretaceous-type and Pennsylvanian-type oils, and the redistribution of Phosphoria-type oils.

Gathering areas around the basin controlled the accumulation of several large oil fields. Geochemical data accurately delineate prospective areas and stratigraphic intervals in the Powder River Basin.

ACKNOWLEDGMENTS

This is an updated version of a summary paper published in the Oil and Gas Journal, December 10, 1979. We thank Amoco Production Company for granting clearance.

REFERENCES CITED

Biggs, P., and R.H. Espach, 1960, Petroleum and natural gas fields in Wyoming: U.S. Bureau of Mines Bulletin 582, 538 p.

Coleman, H.J., et al, 1978, Analyses of 800 crude oils from United States oil fields: BETC/RI-78/14, Technical Information Center, U.S. Department of Energy, November, 447 p.

Curtis, B.F., et al, 1958, Patterns of oil occurrences in the Powder River Basin, in Habitat of oil: AAPG Special Symposium Volume, p. 268-292.

Hedberg, H.D., 1968, Significance of high-wax oils with respect to genesis of petroleum: AAPG Bulletin, v. 52, p. 736-750.

Hunt, J.M., 1953, Composition of crude oil and its relation to stratigraphy in Wyoming: AAPG Bulletin, v. 37, p. 1837-1872.

McIver, R.D., 1962, The crude oils of Wyoming—product of depositional environment and alteration: Wyoming Geological Association 17th Annual Field Conference Guidebook, p. 248-251.

Merewether, E.A., and G.E. Claypool, 1980, Organic composition of some Upper Cretaceous shale, Powder River Basin, Wyoming: AAPG Bulletin, v. 64, p. 488-500.

Nixon, R.P., 1973, Oil source beds in Cretaceous Mowry Shale of northwestern interior United States: AAPG Bulletin, v. 57, p. 136-161.

Schrayer, G.J., and W.M. Zarella, 1963, Organic geochemistry of shales, part 1; distribution of organic matter in the siliceous Mowry Shale of Wyoming: Geochimica et Cosmochimica Acta, v. 27, n. 10, p. 1033-1046.

—— and W.M. Zarella, 1966, Organic geochemistry of shales—II; distribution of extractable organic matter in the siliceous Mowry Shale of Wyoming: Geochimica et Cosmochimica Acta, v. 30, p. 415-434.

—— and W.M. Zarella, 1968, Organic carbon in the Mowry Formation and its relation to the occurrence of petroleum in Lower Cretaceous reservoir rocks: Wyoming Geological Association 20th Annual Field Conference Guidebook, p. 35-39.

Sheldon, R.P., 1967, Long-distance migration of oil in Wyoming: The Mountain Geologist, v. 4, n. 2, p. 53-65.

Strickland, J.W., 1958, Habitat of oil in the Powder River Basin: Wyoming Geological Association Guidebook, Powder River Basin, p. 132-147.

Wenger, W.J., and J.S. Ball, 1961, Characteristics of petroleum from the Powder River Basin, Wyoming: U.S. Bureau of Mines, Report of Investigations 5723.

——, and B.W. Reid, 1958, Characteristics of petroleum in the Powder River Basin: Wyoming Geological Association Guidebook, Powder River Basin, p. 148-156.

Gas Generation and Migration in the Deep Basin of Western Canada

D.H. Welte
R.G. Schaefer
W. Stoessinger
M. Radke

Institute of Petroleum and
Organic Geochemistry
Julich, Federal Republic of Germany

The Alberta Deep basin, situated along the northeastern front of the Rocky Mountain belt, is the deepest part of the Alberta synclinal sedimentary basin. This trough-shaped deep basin, extending across northwestern Alberta and into northeastern British Columbia, covers an area of 65,000 sq km (25,000 sq mi).

Enormous volumes of natural gas have been found in recent years within the thick, clastic Mesozoic sediments which partly fill the deep basin. These sediments exceed 3,100 m (10,200 ft) in total thickness.

Based on detailed geochemical analyses of more than 300 rock samples (mainly cutting samples) from several wells in the Elmworth gas field, information was obtained on the hydrocarbon source strata and the generation and redistribution of hydrocarbons.

The clastic Mesozoic rock section contains numerous shaly zones which are very rich in organic matter, and also a suite of coal strata. This section, containing mainly Type III kerogen, is the ideal gas generator. Maturity is defined as 0.5 % vitrinite reflectance to about 2.0 % in the deeper part of the section. Maturity has also been defined in terms of the "Methylphenanthrene Index" which is based on aromatic hydrocarbons. Apparently the mature section is still in an active phase of hydrocarbon generation. Due to the tightness of the rock, the hydrocarbon transport mechanism seems to be dominated by diffusion processes. The light hydrocarbon distribution patterns observed throughout the wells suggest a dynamic trapping mechanism. Light hydrocarbons are lost at the top of the mature hydrocarbon-generating zone and are replenished in the middle part of the section where rich source rocks are found.

Based on this concept numerical treatment of gas diffusion with finite element computation is presented for well 6-28-68-13W6M of the Elmworth area. Using subsidence curves, time-temperature relationships, maturity-related methane generation data for source-rock intervals, and effective diffusion coefficients for methane (2×10^{-5} to 10^{-6} sq cm S^{-1}) concentration/depth curves were calculated as a function of geologic time. The results of the simulation for the present-day status compare remarkably well with the hydrocarbon distribution observed in this well today.

INTRODUCTION

The Deep basin in Western Canada is located east of the tectonically disturbed belt of the Rocky Mountains in the provinces of British Columbia and Alberta (Fig. 1). It represents the deepest part of the huge asymmetric Western Canada basin. The Deep basin is approximately 650-km (404-mi) long and reaches a width of about 130 km (81 mi). A schematic south-southwest to north-northeast cross section shows the principal geological features of the basin (Fig. 2). Paleozoic rock, mainly carbonates, rest unconformably on the Precambrian basement which dips gently to the southwest. They are overlain by a thick Mesozoic sequence largely consisting of Cretaceous dark shales with interlayered sandstones and conglomerates. Coal seams are frequent, particularly in the deeper part of the Lower Cretaceous. The thickness of the Mesozoic increases from approximately 300 m (984 ft) in Eastern Alberta to more than 4,000 m (13,123 ft) near the Foothills and the overthrust belt which forms the border of the basin to the west. The strata in general dip southwestward into the Deep basin where both porosity and permeability of the sandstones decrease significantly due to greater compaction, higher clay content, and more intense diagenesis. Gas occurs only in the deepest part of the basin, over an area of approximately 67,000 sq km, where almost the entire Mesozoic section at a depth level exceeding 1,000 m (3,281 ft) below the surface is gas-saturated as derived from resistivity logs (stippled area; Masters, 1979) and geochemical analyses of several wells (Welte et al, 1980; 1981). These gas-bearing zones correspond to the less porous and less permeable rock situated in a downdip position. The same strata in an updip position east of a transition zone, exhibiting higher porosities and permeabilities, are saturated with water. Thus, the situation is the reverse from a conventional gas field, where above the gas an impermeable seal would be expected; instead there are water-saturated reservoir-type strata. Throughout the Mesozoic, in the Triassic, the Jurassic and first of all in Cretaceous rock some 12 pay zones have been encountered with average porosities of 10 % and permeabilities of about 0.5 md (Masters, 1979). The better parts of these pay zones are often conglomeratic and exhibit permeabilities which range from

50 md to several darcys. To produce the gas the wells generally have to be stimulated by hydraulic fracturing techniques. Recoverable gas in the Deep basin may very well be around 50 tcf (50×10^{12} cu ft equivalent to $1,416 \times 10^9$ cu m, STP) or even more (Masters, 1983, personal communication).

This paper presents some results of combined geological and geochemical research that show that the gas occurrences in the Deep basin can best be explained as a dynamic situation between generation of gas from coals and carbonaceous shales on one hand, and losses to the shallower upper layers of the rock section and going updip on the other hand.

GAS GENERATION AND MIGRATION

The Mesozoic rock section in the area of the Elmworth gas field represents, due to its richness in organic carbon and type of organic matter, probably one of the finest source sections for gas anwhere. A profile monitoring the organic carbon content as derived from cuttings analyses of a typical Elmworth well is shown in Fig. 3. Down to 2,400 m (7,874 ft) depth, the organic carbon content averages around 1%. Then a very rich zone follows with values up to 80% which reflects, in essence, the presence of coal seams in this part of the section (2,400 to 3,200 m, or 7,874 to 10,499 ft). From 3,200 m to 3,555 m (10,499 to 11,663 ft) total depth (T.D.) the organic carbon level is around 2%.

The type of the organic matter was determined by ROCK-EVAL pyrolysis. The diagram showing the hydrogen index plotted versus oxygen index (Fig. 4) corresponds to the van-Krevelen diagram. The data of this well plot fairly close to the Type-III curve indicating a hydrogen-poor kerogen which is mainly derived from higher terrestrial plants. Microscopic investigation revealed low amounts of liptinite macerals, but high contents of vitrinite (up to 70%) which is known to be a good gas generator, and inertinite. Thus, the microscopic data agree with the results from ROCK-EVAL pyrolysis. Therefore, it is concluded that this kerogen is not able to generate large amounts of oil, but is a good source of gas if it is in the right stage of maturation.

Figure 5 shows vitrinite reflectance values (% R_m) for the same Elmworth well.

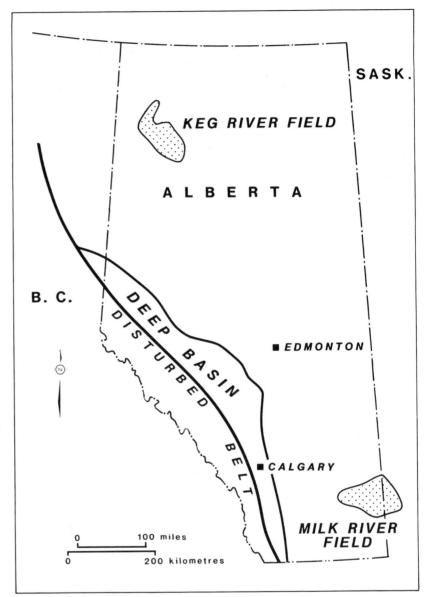

Figure 1. Location map of Western Canada Deep basin.

The maturity of the organic matter is approximately 0.7% mean vitrinite reflectance in the uppermost part of the well and shows only a slight increase down to 1,500 m (4,921 ft) where it reaches 0.8%. At this point there is a progressive change in the reflectance gradient and maturity increases quickly to 2.1% at total depth (T.D.). It is known that Type-III kerogen normally starts to produce liquid hydrocarbons at a maturity level around 0.7% R_m.

Oil generation reaches a maximum around 0.9% R_m and then decreases with further maturation toward the bottom of the "oil window" (at approximately 1.3 to 1.4% R_m). Significant amounts of gas are thought to be generated at maturities exceeding 0.9 to 1.1% R_m. The bottom of the "dry gas window" is not yet reached in this well at T.D.

The organic carbon-normalized concentrations of C_{15+} saturated hydrocarbons

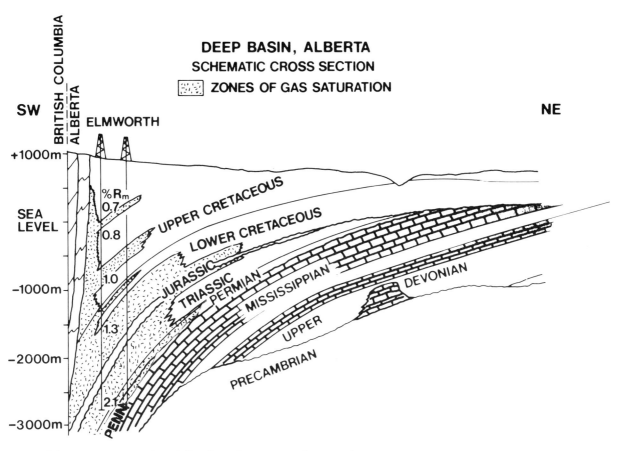

Figure 2. Schematic cross section through Deep basin showing zones of gas saturation.

and C_{15}-C_{35} n-alkanes, as determined for cuttings samples from the above well, when plotted versus depth (Fig. 6) exhibit the typical shape of a generation curve. This indicates that, in general, the C_{15+} hydrocarbons remained at the place where they were generated. There was no major redistribution (that is, migration) in a vertical direction. In this context, it is important to remember that the (liquid) C_{15+} hydrocarbons which were generated during maturation of the source rock at shallower depth are being cracked with further maturation progress and converted to smaller molecules (for example, gas) at greater depth. Therefore, hydrocarbon concentrations of the C_{15+} fraction pass through a maximum as shown in Figure 6. Similar curves exist for both the total extractable aromatics fraction and individual aromatic hydrocarbons. The aqueous solubility of the low-boiling aromatics is larger than that of the corresponding saturates. As these aromatic components also show a typical generation curve, it is suggested that no long-range vertical water flow occurred since the time of intense hydrocarbon generation. Such a water flow which would have blurred the original generation pattern of the low-boiling aromatics is also unlikely because of the very low porosity and permeability of the rocks.

Further evidence against a large vertical migration of the heavier hydrocarbons is derived from the application of the Methylphenanthrene Index (MPI). This chemical parameter permits the definition of the maturity of an oil or a rock extract in terms of calculated vitrinite reflectance (R_c) (Radke and Welte, 1983). In Figure 7 the hydrocarbon internal maturity (R_c) is compared with the mean vitrinite reflectance of the kerogen (% R_m). Obviously, there is an excellent agreement between the two curves. Any invasion of a more mature oil or condensation from greater depths would have resulted in a positive deviation of the R_c values from the vitrinite reflectance curve.

With decreasing carbon number (Fig. 8) the concentration profiles of hydrocarbons in the investigated Elmworth well get broader and eventually lose any similarity to a generation curve. Contrary to a previous figure (Fig. 6) depicting a typical generation curve in the depth profile of C_{15+} hydrocarbons in the shale, the propane concentration profile is rather broad and remarkably constant between 1,000 and 2,400 m (3,281 and 7,874 ft) depth, below which it decreases stepwise (Fig. 9). The propane distribution pattern as presented in this figure should show a definite trend toward lower concentrations going from 2,000 m (6,562 ft) depth upward if it would be influenced by generation rates.

The absolute concentration of total light alkanes (C_2-C_5) reaches a maximum

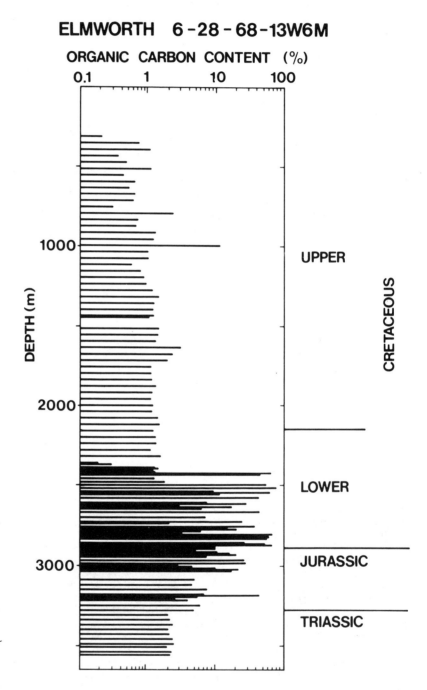

Figure 3. Organic carbon content (note logarithmic scale) plotted against depth for cuttings samples of well 6-28-68-13W6M.

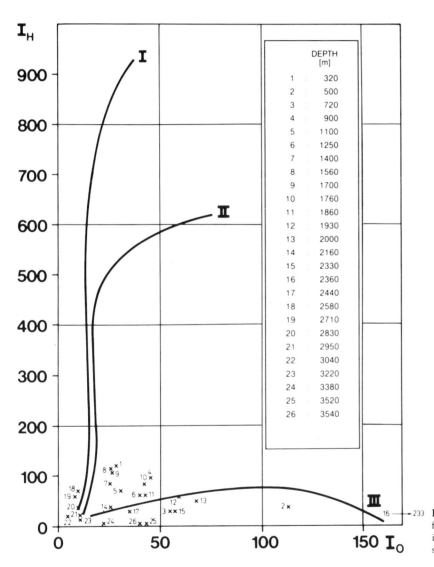

Figure 4. Van-Krevelen-type diagram derived from ROCK-EVAL pyrolysis data (hydrogen index I_H and oxygen index I_O) for selected rock samples of well 6-28-68-13W6M.

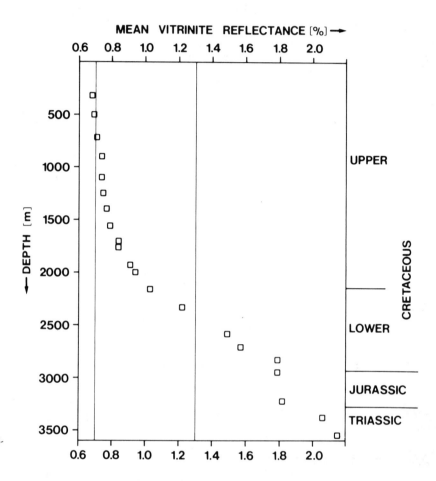

Figure 5. Mean vitrinite reflectance plotted against depth for rock samples of well 6-28-68-13W6M.

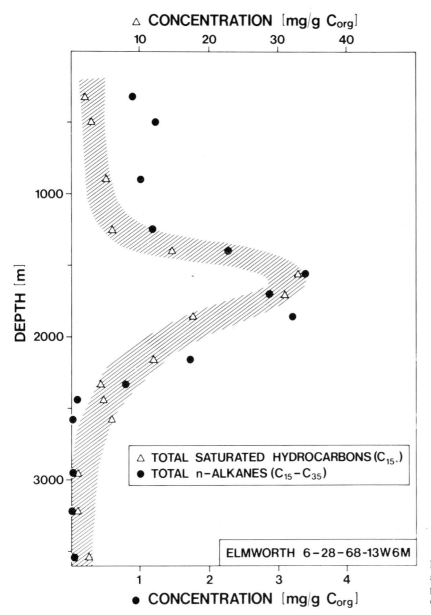

Figure 6. Total saturated hydrocarbon (C_{15+}) and total n-alkane (C_{15}-C_{35}) concentrations plotted against depth for selected rock samples of well 6-28-68-13W6M.

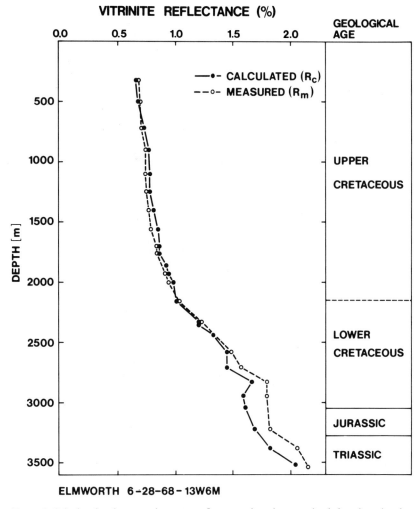

Figure 7. Calculated and measured vitrinite reflectance plotted against depth for selected rock samples of well 6-28-68-13W6M.

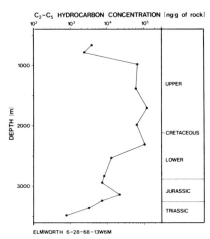

Figure 8. C_2-C_5 hydrocarbon concentration (absolute values obtained by combined thermovaporization/hydrogen stripping) plotted against depth for selected rock samples of well 6-28-68-13W6M.

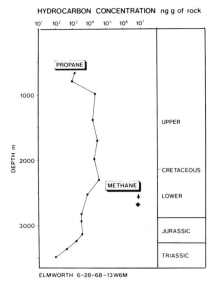

Figure 9. Propane concentration (absolute values obtained by combined thermovaporization/hydrogen stripping) plotted against depth for selected rock samples of well 6-28-68-13W6M. Also shown in this figure is the calculated methane concentration using the results of a pressure core-barrel gas analysis from another Elmworth well at the corresponding maturity level.

of 1.2×10^5 ng/g, or 120 g/metric ton of rock (Fig. 8), which is considered to be a very high value. It becomes even more spectacular when considering the fact that methane is not included. Hence, these geochemical analyses verify the conclusions based on resistivity log interpretations by Masters (1979) concerning the gas saturation. As the bulk of the light hydrocarbons is thought to be generated in maturity levels exceeding 0.9% R_m, corresponding to depth ranges below 2,000 m (6,562 ft), it has to be assumed that a considerable part of the light hydrocarbons (for example, C_1-C_3) has migrated over an appreciable distance into the overlying strata, contrary to the heavy C_{15+} hydrocarbons which apparently remain more or less at the site of their origin.

The concentrations displayed in Figures 8 and 9 were obtained using a combined thermovaporization/hydrogen stripping method developed by Schaefer et al (1978). These data are in good agreement with gas analyses including methane (Fig. 9) of pressurized core-barrel samples taken from another Elmworth well at about 2,600 m (8,530 ft) depth. The measured amount of methane and higher hydrocarbons from the pressurized core-barrel samples makes it possible to extrapolate the absolute methane yield (which cannot be measured from cuttings quantitatively) from the propane concentration at a given maturity. This calculated value is 6.7×10^6 ng/g or about 11 cu m (STP) methane/metric ton of coal, which is only the residual amount of gas present today. Yet, a coal is thought to have produced 8 to 10 times as much methane upon reaching this

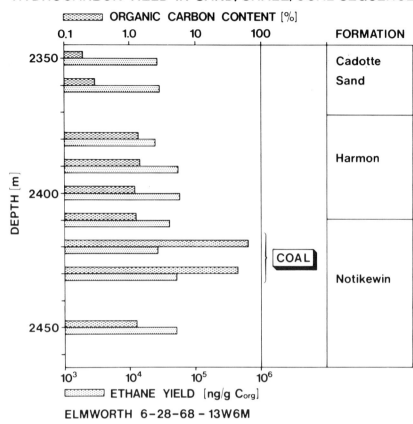

Figure 10. Organic carbon contents and ethane yield (obtained by hydrogen stripping) plotted against depth for a sand/shale/coal sequence of well 6-28-68-13W6M.

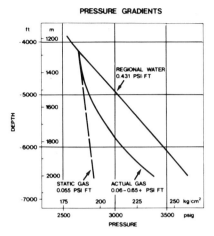

Figure 11. Pressure gradients for regional water, "static gas," and actual gas for the Elmworth area (after Gies, 1982).

maturation stage (approximately 1.3% R_m). This total volume generated cannot be stored by the coal itself and, therefore, migrated into adjacent strata.

Thus, the absolute concentration of methane and other gaseous hydrocarbons like ethane, propane, and butane per rock volume and available pore space of the different sedimentary units becomes an important parameter to understand the problem of gas generation and migration. Furthermore, knowing that this gas is generated from coals and Type-III kerogen, the methane concentrations normalized to content of organic carbon of these rocks inform us about the relative importance of the various gas generators (that is, the sources of the gas). Following this line of thought ethane concentrations were determined in detail in coal containing rock sections. Figure 10 may serve as an example where organic carbon content and ethane yields obtained by hydrogen stripping have been analyzed for a coal-containing 100-m (328-ft) interval.

The figure shows a section of the Lower Cretaceous in the deeper part of the well where the Notikewin coal seam is overlain by Harmon shales grading upward into shaly sands which are followed by the Cadotte sandstone. The organic carbon-normalized ethane content remains fairly constant over the whole depth interval. Therefore, a certain volume of coal contains much more ethane than the same volume of shale as it had to be expected. A similar curve has been found for propane. It can be assumed that methane concentrations follow the same trend (that is, that absolute methane concentration is highest in the coal). The fact that the organic carbon-normalized values for ethane are in the same order of magnitude for different rock sections with varying lithology and also for coals is interpreted as an indication that the gas generation process is still active and that concentration gradients for gas from the coals toward neighboring rocks are maintained. The concentration gradients would have been eliminated by diffusion if the coals had ceased to generate an appreciable amount of methane. These concentration gradients must be the driving force for an active gas diffusion going on today.

Hence, the conclusion is that in the Elmworth gas field there is, even at present, a zone of active gas generation mainly in the deeper part of the rock column exceeding 2,000 m (6,562 ft). In this zone, between about 2,000 and 3,500 m (6,562 and 11,483 ft) depth, temperatures range from about 80 to 120°C and maturities from 0.9 to 2.0% R_m.

Based on detailed pressure studies of the Lower Cretaceous Cadomin sandstone reservoir in the Elmworth gas field, Gies (1982) independently arrived at the same conclusion that the gas accumulation is not static equilibrium but is in a dynamic state with ongoing updip gas migration. The pressure gradients for regional water and the actual gas as determined by Gies (1983) are shown in Figure 11. It can be seen that the actual downdip gas pressures are greater than those predicted by a constructed hypothetical static gas pressure gradient based on a continuous gas saturation over a depth range of 800 m (2,625 ft). The conclusion, derived from these pressure data in the Cadomin sandstone

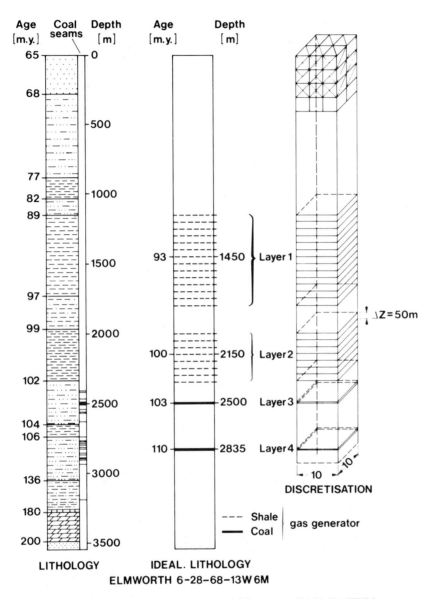

Figure 12. Conceptual model for gas generation and diffusion in well 6-28-68-13W6M.

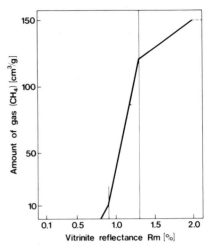

Figure 13. Integral amount of methane generated during coalification (modified after Jüntgen and Klein, 1975).

$$\frac{\partial c}{\partial t} - \frac{\partial}{\partial x_i}\left[(Dd)_{ij}\frac{\partial c}{\partial}x_j\right] = 0$$

where

c = concentration [M/L³]
$(Dd)_{ij}$ = components of the molecular diffusion [L²t]
x_i = components of the coordinates [L]
t = time [t]

The basic system, an idealized sedimentary section, as taken from an Elmworth well and the gas generators in that section, is shown in Figure 12. The left-hand column is a simplified lithologic log of this well, subdivided into individual lithologic units with their respective approximate ages. The column in the middle depicts the idealized system with four gas-generating sources. Layer 1 and layer 2 represent the two shale sequences whereas layer 3 and layer 4 represent the two groups of coal seams interlayered with shaly sands. The thick shales cover a wide maturity range. In order to improve the precision of the gas generation process these shales are subdivided into several "gas generating zones" with a depth interval of 50 m (164 ft). The right-hand column shows the discretized system under

reservoir, is that gas must be migrating upward in response to the pressure drop in an updip direction.

NUMERICAL MODEL OF GAS DIFFUSION

To understand the processes of gas generation and migration in the Elmworth area in a more quantitative manner a numerical model based on finite element discretisation with a moving interface was developed (Stoessinger, 1983). For a first simulation it was assumed that only molecular diffusion occurs within the system. Under these conditions the gas transport for unsteady state can be written in a tensor notation:

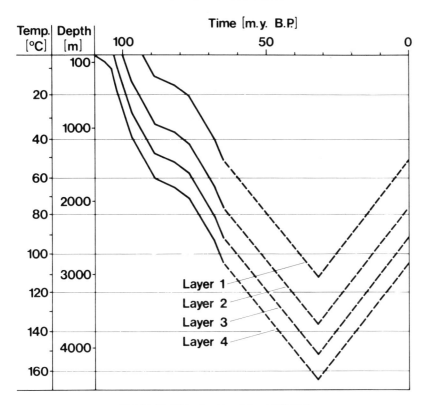

Figure 14. Depositional history for well 6-28-68-13W6M.

consideration. The boundary condition at the top of the system is a zero-concentration surface.

The numerical solution requires the definition of the total methane volume generated until today. The time dependent cumulative methane production can be calculated by using a modified curve from Juntgen and Klein (1975). It shows (Fig. 13) the cumulative amount of methane released from coal as a function of its maturity or degree of coalification. Methane generation begins around 0.8% vitrinite reflectance, then increases strongly between 0.9 and 1.3%; beyond that range it diminishes and levels off around 2.0% in the semianthracite stage.

More precise information on gas generation in natural maturation of organic matter taking into account the kind of source material and the kinetics of the gas generation process is not yet available.

Another prerequisite to calculate methane production is the knowledge of the burial history of the source rocks, which can be transformed into a thermal history by using a certain temperature-depth relationship. A temperature versus depth curve was kindly provided by T. Connolly, Canadian Hunter Exploration Ltd. (1981, personal communication) and was slightly modified (Fig. 14). The evolution of maturity can then be derived using a modified Lopatin (1971) or Waples (1980) approach.

No precise data are available on the thicknesses of the Tertiary and Upper Cretaceous rock which have been eroded since Oligocene (32 m.y.), the time of deepest burial (Hacquebard, 1977). For this reason the maximum burial depth is calculated and illustrated in Figure 14 from the recent mean vitrinite reflectance. In other words, that part of the curial curve corresponding to Tertiary and Quaternary (broken lines) is adjusted in such a way that calculated and measured maturity values (that is, vitrinite reflectance) could be matched.

With the above-mentioned boundary conditions the vitrinite-reflectance and the gas-generation rate as a function of time can be calculated. Both curves are presented in Figure 15. The upper part of the figure shows the evolution of the maturity of each source rock as it has been derived from the thermal history by Waples' relationship. Layer 1 has not yet reached the threshold of gas generation (0.8% R_m).

The time-dependent methane production can be calculated by taking into account the maturation, the thickness of the respective source rock, the organic carbon content and the modified production curve. The lower part of Figure 15 shows the unsteady production curve of the gas sources. The values are normalized to a 1.0 sq m base.

Because little is known about the actual tensor of molecular diffusion D_d and other petrophysical parameters throughout the whole sedimentary column, a sensitivity analysis was made in order to estimate the influence of the molecular diffusion on the concentration distribution. Figure 16 shows the methane concentration versus depth as calculated for present time for different diffusion tensors. The coefficients are assumed constant throughout the entire rock column. It becomes obvious from these curves that the unsteady concentration profiles are strongly influenced by the molecular diffusion. For instance, the bulk of the methane would still be at depths below 1,700 m (5,577 ft) if D is 10^{-7} sq cm S^{-1}. If D is assumed to be 10^{-5} sq cm s^{-1} the gas would have already reached the earth's surface and hence a major portion would have been lost.

Assuming that increasing temperature with mounting depth increases the coefficient of molecular diffusion more than the lower effect of decreasing porosity, a linear variation between $1.0\ 10^{-6} - 2.0\ 10^{-5}$ is considered reasonable. Following this line of thought concentration profiles are computed for different times in order to demonstrate the evolution of the gas concentration as a function of geologic history. The results are plotted in Figure 17. In the first stage, 44 m.y. ago, methane is generated only by the lower group of coal seams (layer 4). Concentration decreases quickly toward shallower depths and at 1,500 m (4,921 ft) above the seams it is virtually zero. The diffusion front has

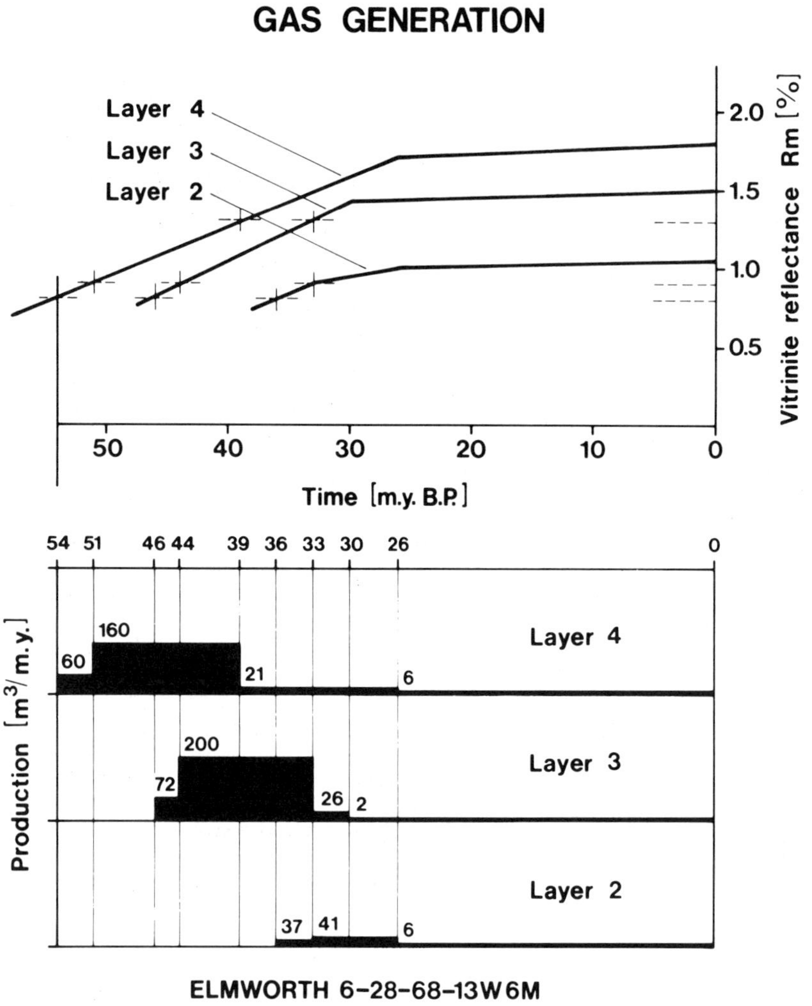

GAS GENERATION

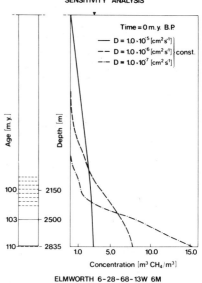

Figure 15. Time-dependent maturation (upper part of figure) and methane generation (lower part of figure) for different layers of well 6-28-68-13W6M (see Fig. 12).

Figure 16. Methane concentration plotted against depth calculated for three different diffusion coefficients in well 6-28-68-13W6M.

migrated upward at 33 m.y., and the curve shows a sharp bend as layer 3 starts to produce large amounts of gas. Seven million years later (26 m.y.) the profile again shows a smooth shape which is caused by both decreasing gas production from layer 3 and beginning of gas generation from the shales (layer 2). Today the methane front has reached the earth's surface according to the model. Although models of this kind are very informative and highly desirable it must not be forgotten that they depend on the assumptions made, as for instance, that only diffusion controls gas migration in our example and other migration processes are negligible.

SUMMARY AND CONCLUSIONS

In summary, the following observations were made in the Elmworth gas field. In the tight, low-porosity center part of the Deep basin nearly the entire Mesozoic rock section is gas saturated. The gas saturation decreases rapidly in the shallower part of the rock column and updip toward the east where more porous and permeable rocks are found. Detailed geochemical analyses show that there is no major redistribution (that is, migration) of heavier hydrocarbons and no flow of water in the tight part of the rock section. However, there is evidence for massive redistribution (that is, migration of gaseous hydrocarbons) and it can be shown that the main transportation mechanism must be diffusion inside the tight rock section.

Using the above concept, calculations show that present dry gas-saturation profiles in the center part of the Elmworth gas field can be simulated when assuming diffusion coefficients for methane in the range of 1.0×10^{-6} to 2.0×10^{-5} sq cm s^{-1} whereas increasingly higher diffusion coefficients have been adopted with increasing depth and temperature.

When compared with the gas analysis of the pressure core-barrel, the calculated concentration at the appropriate depth (2,500 to 3,000 m, or 8,202 to 9,842 ft) is only a factor of two higher than the actual methane concentration. As with any deterministic model the numerical solution requires accurate data. Considering the variation of all parameters which are difficult to measure in the field, this is a remarkably good result. Obviously the diffusional losses and other losses by conventional buoyancy-driven gas transport mainly toward more porous updip situated strata are compensated by a continuing gas generation from coals and organic rich shaly source rocks in the rock section. Temperatures between 80 and 120°C

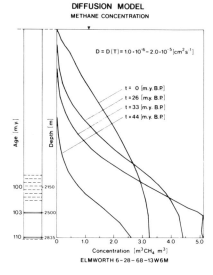

Figure 17. Methane concentration plotted against depth for different time stages assuming a linearly increasing diffusion coefficient with depth for well 6-28-68-13W6M.

seem to be high enough to guarantee an ongoing coalification process and hence gas generation. Balance calculations show the coal measures to be important sources for the gas. Cumulative coal thicknesses of Jurassic to Lower Cretaceous coals in the Elmworth region may range up to 70 m (230 ft) and more (Masters, 1983, personal communication).

All in all the Elmworth gas field is a dynamic situation where gas is continuously being generated in the center part of the Deep basin, and lost toward the surface and the more porous edge. In the inner core of the gas-generating rock column diffusion processes seem to be the predominating mode of transportation.

From the numerical model and its results it can be concluded that given a database similar to the one used in this example, it seems to be feasible in the future to predict gas distributions with much higher accuracy.

ACKNOWLEDGMENTS

We acknowledge the excellent cooperation with Canadian Hunter Exploration Ltd., Calgary, and the stimulating discussions with Mr. John Masters, President of Canadian Hunter Exploration.

We are indebted to Dr. J. Gormly (organic carbon values and ROCK-EVAL data), Dr. P. K. Mukhopadhyay (vitrinite reflectance measurement and maceral analysis), and Mr. H. M. Weiss (subsidence curves, time-temperature calculations), all at KFA-Julich.

Assistance with the experimental work by Mrs. M. Weiner and B. Winden, as well as Messrs. P.W. Benders, U. Disko, W. Laumer, F. Leistner, H. Pooch, F. Schlosser, B. Schmidl, K. Schmitt, H. W. Schnitzler, H. G. Sittardt, R. Weckheuer, and H. Willsch, all at KFA-Julich, is gratefully acknowledged.

REFERENCES

Gies, R.M., 1982, Basic physical principles of conventional and Deep basin gas entrapment (abs.): AAPG Bulletin, v. 66, p. 572.

Hacquebard, P.A., 1977, Rank of coal as an index of organic metamorphism for oil and gas in Alberta, in G. Deroo et al, eds., The origin and migration of petroleum in the western Canadian sedimentary basin, Alberta; a geochemical and thermal maturation study: Geological Society of Canada Bulletin, v. 262, p. 11–22.

Juntgen, H. and J. Klein, 1975, Entstehung von erdgas aus kohligen sedimenten: Erdol und Kohle, Erdgas, Petrochemie, Erganzungsband, v. 1, p. 52-69.

Leythaeuser, D., R. G. Schaefer, and A. Yukler, 1980, Diffusion of light hydrocarbons through near-surface rocks: Nature, v. 284, p. 522-525.

Lopatin, N.V., 1971, Temperature and geologic time as factors in coalification: Akademiya Nauk, Uzb. SSSR, Ser. geologicheskaya, Izvestiya, no. 3, p. 95-106.
(Translation by N.H. Bostick, Illinois State Geological Survey, February 1972).

Masters, J.A., 1979, Deep basin gas trap western Canada: AAPG Bulletin, v. 63, p. 152–181.

Radke, M. and D. H. Welte, 1983, The Methylphenanthrene Index (MPI); a maturity parameter based on aromatic hydrocarbons, in Advances in Organic Geochemistry 1981: Bergen, Proceedings of the 10th International Meeting on Organic Geochemistry, p. 504–512.

Schaefer, R.G., B. Weiner, and D. Leythaeuser, 1978, Determination of sub-nanogram per gram quantities of light hydrocarbons (C_2-C_9) in rock samples by hydrogen stripping in the flow system of a capillary gas chromatograph: Analytical Chemistry, v. 50, p. 1848-1854.

Stoessinger, W., 1983, Numerical treatment of gas diffusion with finite element computation: Internal Report of KFA Julich, unpublished.

Waples, D.W., 1980, Time and temperature in petroleum formation; application of Lopatin's method to petroleum exploration: AAPG Bulletin, v. 64, p. 916–926.

Welte, D.H., et al, 1980, Organic geochemistry of well Canadian Hunter, Elmworth 10-35-71-13W6M, Canada, and its implication for an "unconventional" gas accumulation: Internal Report of KFA-Julich, unpublished.

———, et al, 1981, Organic geochemistry of well Canadian Hunter, Elmworth 6-28-29-13W6M, Western Canada: International Report of KFA-Julich, unpublished.

Johann-Christian Pratsch

Reprinted by permission of Konradin-Industrieverlag GmbH. Published in *Erdoel und Kohle, Erdgas, Petrochemie*, v. 35 (1982), pp. 59-65.

Focused Gas Migration and Concentration of Deep-Gas Accumulations

Northwest German Basin

The observable concentration of the major deep gas accumulation areas in the Northwest German basin is the result of focused secondary migration. Gas generated in several effective depocenters (regional structural lows) from late Carboniferous source beds migrated into adjacent regional structural highs. The preferred migration paths are qualitatively predictable on the basis of present basin geometry. Optimal conditions for gas accumulations exist where the presently trapping regional highs have also been sites of favorable reservoir development. Ongoing and future deep-gas exploration efforts in the basin can be regarded as quite hopeful.

Die sichtbare Konzentration der bedeutenden tiefliegenden Erdgasspeicher auf bestimmte Räume des Nordwestdeutschen Beckens ist das Ergebnis einer gezielten Sekundärmigration. Denn Erdgas, das sich in verschiedenen bedeutenden Depozentren (regionalen Strukturdepressionen) aus spätkarbonen Muttergesteinen gebildet hatte, war dann in benachbarte regionale Strukturaufwölbungen abgewandert. Seine bevorzugten Migrationswege sind ihrer Bedeutung entsprechend auf der Grundlage der heutigen Beckengeometrie vorauszusagen. Optimale Bedingungen für solche Gasansammlungen bestehen dort, wo heutige Gasfallen in regionalen Hochgebieten auch gleichzeitig Raum für eine günstige Speichergesteinsentwicklung zur Verfügung hatten. Laufende und zukünftige Bemühungen zur Exploration tiefliegender Gaslagerstätten innerhalb des Nordwestdeutschen Beckens dürfen deshalb als recht hoffnungsvoll betrachtet werden.

Introduction

Of the 200 to possibly 300 TCF of natural gas reserves discovered so far or to be discovered in the future in Northwest Europe, more than 100 TCF are located in several fields in the Northwest German Basin; included here are the Northeast Netherlands *(Figs. 1, 2)*. Basic geologic parameters that satisfactorily explain these accumulations (their location and size) certainly will apply to the geologic conditions in entire Northwest Europe and elsewhere. Especially interesting is the question to what degree such geologic explanations also have predictive value for future gas exploration in Northwest Germany and thus in Northwest Europe [11, 12, 13].

The gas fields in the Northwest German Basin are not distributed in a randomly statistical order. Rather, they occur in several distinct geographic clusters *(Fig. 2)*. This concentration of gas accumulations is even more pronounced when the quantitative distribution of gas reserves per cluster is shown *(Fig. 3)*. The Groningen and the Oldenburg field areas alone contain the vast majority of gas reserves in the Northwest German Basin (99 TCF or 95% of a basin-wide total of 104 TCF). Inside the German borders, the Oldenburg field area contains by far the largest gas reserves.

The lack of gas fields in the onshore portion of the basin north of a line about between the cities of Bremen and Hamburg is striking. On a first impression, geologic factors like the effect of the northward deepening of the post-Permian basin appear to be the reason. However, the northward increasing lack of reliable geological and geophysical data from deep Permian and pre-Permian beds is a more likely explanation; it offers several positive possibilities for future gas exploration efforts in the basin.

General Concept

The numerous complexities involved in the formation and preservation of any hydrocarbon accumulation prevent quick and easy solutions for presence or absence of fields in a basin. In the case of the Northwest German Basin, too, there are many opinions and data that at first are difficult to place into a coherent explanation scheme. Some explorationists will seek explanations for the observed field concentrations in the presence of favorable structural traps but do not mention the many dry holes drilled in the same basin on presumably equally valid structural anomalies. Others will point to the widespread (yet possibly locally concentrated) mature gas-source beds in late Carboniferous coal and carbonaceous shales *(Fig. 4)*, or at the local concentrations of suitable reservoir beds from Carboniferous and Permian sandstones to late Permian carbonates and Triassic sandstones. Still others will consider the aspects of preservation of trapped gas as most important, related

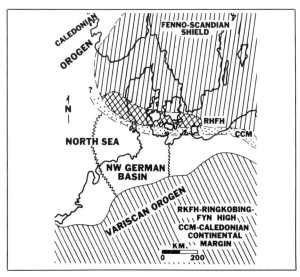

Figure 1: Northwest German Basin—Location Map

Author: Dr. Johann-Christian Pratsch, 6822 Charlmont Circle, Dallas, TX 75248 USA

to the positive influence of capping evaporites of Permian, Triassic, and Jurassic ages. The "right" timing of geologic events may also be seen as controlling the existence and even the location of gas fields.

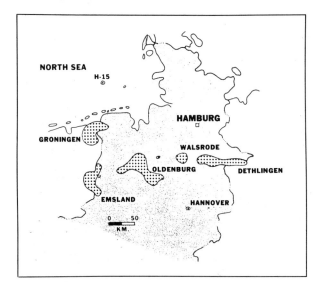

Figure 2: Northwest German Basin—Gas Field Areas

All of these parameters, no doubt, contribute to the observable concentrations of gas reserves in a few small areas in the Northwest German Basin. But none of these parameters alone appears to be more prominent than others. It is the integrated efficiency of many geologic processes that controls the gas distribution. The main problem with a more detailed analysis is the fact that a satisfactory explanation may be found only after all data involved are obtained. This will be late in the development of this or any other basin. And if the parameters cited above indeed were the only ones available, they would have a low predictive value for hydrocarbon exploration. Hydrocarbon exploration would then be controlled to a large degree by statistical approaches [5].

Fortunately, however, there is another geologic parameter that does possess a considerable predictive value in basin evaluation for hydrocarbon exploration and that at the same time allows one to collect and to integrate all available geological data into one coherent synthesis early in an exploration program: this is the geologic parameter of secondary hydrocarbon migration (Fig. 5). In this approach, the common parameters in hydrocarbon exploration—trap, reservoir, cap, and preservation—gain in individual value by their integration into one common process.

Secondary hydrocarbon migration is the process of physical movement of hydrocarbons—oil and gas—from the area of hydrocarbon generation through carrier beds to the area of final entrapment (Fig. 6). In contrast, primary migration is commonly defined as the process of movement of oil and gas from the generating source bed to the carrier bed.

As yet there is no common agreement on all facets of hydrocarbon migration: the physico-chemical processes involved, the necessary physical or geological requirements, the changes and variations in the process [3, 15]. Yet, innumerable geologic observations and the ever increasing laboratory data and concepts permit already to use secondary migration as a decisive geologic parameter in basin evaluations [4, 9, 11]. In such an evaluation three major steps are required. These steps are normally undertaken in any basin evaluation study but are only too often overlooked in the final evaluation phases (Fig. 7):
1. Definition of the effective depocenter(s),
2. Definition of preferred migration paths,
3. Definition of length of migration paths.

These three steps can be used not only to explain existing hydrocarbon accumulations, but also to identify areas of higher-than-normal prospectivity of a certain basin portion, thus of higher-than-normal hydrocarbon concentrations in such preferred

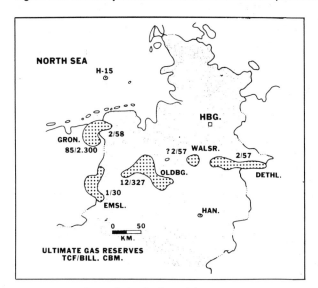

Figure 3: Northwest German Basin—Gas Accumulations

area. Not the number of traps or fields is the answer, but the amount of trapped gas or oil per basin area; thus the distribution of migrated hydrocarbons in a given prospective basin. The results obtained from migration analysis are firstly qualitative. However, when quantitative data are added (like source richness, effectiveness of thermal maturation, of secondary migration, of trapping, remigration or

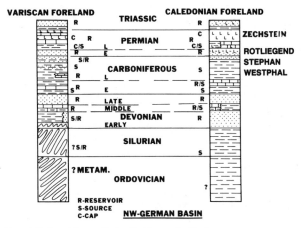

Figure 4: Northwest German Basin—Gas-Prospective Section

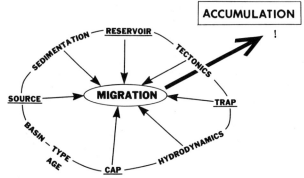

Figure 5: Basic Parameters in Basin Evaluations

preservation), this approach will furnish semi-quantitative or even quantitative data on hydrocarbon distribution in a given basin.

Definition of the Effective Depocenter

A depocenter, as part of a sedimentary basin, contains the maximum thickness of sedimentary section above basement *(Fig. 6)*. An effective depocenter, in petroleum geology, contains the maximum thickness of generative sediments and hence the maximum thickness of thermally mature or overmature hydrocarbon source beds. Thermal maturity of source beds in a depocenter guarantees the presence of hydrocarbons that can migrate. Overmaturity of source beds adds the geological-historical question where the hydrocarbons are that may have migrated away from the area before the onset of overmaturity.

A large number of geological, geochemical and geophysical techniques are available today to define the presence or absence of mature or over-/under-mature source beds. Geochemical data not only explain a portion of the geological history of a basin, they even permit predictions about the occurrence or absence of source beds by age, type and maturity level [15]. Computer-assisted predictive maturation analysis now is routine in this field while quantitative predictions still depend much on the quality and quantity of input data [17].

The geometry (shape) of a depocenter can be obtained routinely through definition of regional structure using gravity, magnetics, refraction or reflection seismic, magneto-tellurics, surface or subsurface geology, or remote data interpretation *(Fig. 8)*. Detailed

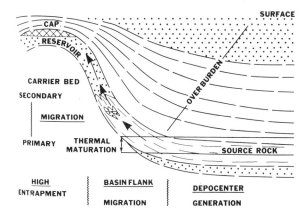

Figure 6: Hydrocarbon System

mapping of the present basin geometry (structure) at the very level of the mature hydrocarbon source bed or regional carrier bed will be of greatest importance. Where data are available only from levels above the mature source beds or the regional carrier bed, such data define at least the minimum subsidence history of the source bed, thus its thermal maturation history and structural history. Quite clearly, all phases of hydrocarbon exploration are involved in the step of depocenter definition. It is from here that the input data for migration prediction are obtained.

Definition of the Preferred Migration Paths

Oil and gas migrate from the areas of their generation to the areas of their final entrapment. Such migration occurs under the

3. **DEFINITION OF LENGTH OF MIGRATION PATHS**

2. **DEFINITION OF PREFERRED MIGRATION PATHS**

1. **DEFINITION OF EFFECTIVE DEPOCENTER (S)**

Figure 7: Steps in Migration Prediction

influence of existing pressure differentials. The natural buoyancy of hydrocarbon particles in a water environment and effective pressure gradients appear to be most important [16]. At this time there is no general agreement in our profession about all the geological

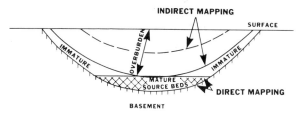

Figure 8: Depocenter Mapping Levels

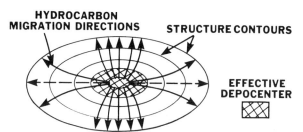

Figure 9: Migration Directions in Oval-Shaped Basins

and physico-chemical parameters that enter into the migration process. There is, however, general agreement on some basic observations that can serve as a basis for migration predictions *(Fig. 6)*:

1. Hydrocarbons migrate up-dip unless extreme pressure differentials prevent this; fresh-water influx from basin flanks or local vertical pressure gradients from high-pressure under-compacted clays to underlying normal-pressure porous layers are two examples of such limiting pressure differentials.

2. Hydrocarbons migrate laterally or vertically depending on geologic conditions. The effectiveness of either one of the two modes is much debated in literature, which is a sign for the complexities and variations from basin to basin [11, 12, 14].

3. Fracturing of rocks critically enhances permeabilities that are required for early, late, lateral and vertical hydrocarbon migration. Fracturing on local and regional scales is much more common and more effective than is commonly assumed as the entire globe is under constant major stress [10].

Ideally, migrating hydrocarbons will seek the shortest possible migration path. In practice, variations in porosities, permeabilities and effective driving forces will cause deviations and variations of migration paths. As a consequence there will always be a statistical component in migration direction definitions related to the quality of data available.

In some basins remigration of once accumulated hydrocarbons is known; here basin geometries changed through time. The present migration paths and hence the present preferred distribution of migrated hydrocarbons are still related to present basin geometry. All evidence indicates that hydrocarbon migration, while lasting over a long period of time where maturation and generation continue, can be quick in geologic terms (a few million years at the most for basin-wide migration patterns).

The most critical parameters derived from basin or depocenter geometry are the planar map-view and the symmetry of the basin *(Fig. 9)*. Hydrocarbons normally will not migrate across the basin axis; each flank of a basin can be considered as a separate unit. Basin symmetry controls the amount of generated hydrocarbons available for migration in each basin portion.

Basin geometries range from circular to elongate and from straight to curved. Simple basins containing only one depocenter will have different migration patterns than complex basins containing more than one depocenter. These distinctions can be used to

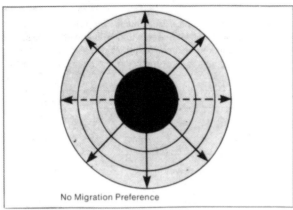

Figure 10: Migration Principles—Circular Symmetrical Basin

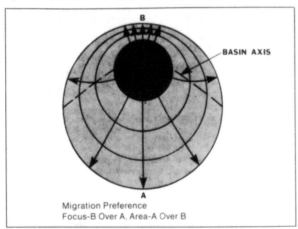

Figure 11: Migration Principles—Circular Asymmetrical Basin

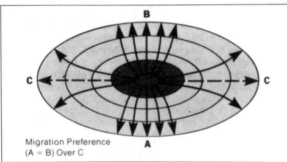

Figure 12: Migration Principles—Elongate Symmetrical Basin

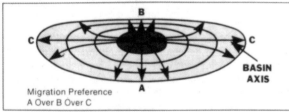

Figure 13: Migration Principles—Elongate Asymmetrical Basin

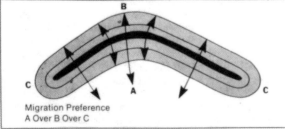

Figure 14: Migration Principles—Elongate Symmetrical Curved Basin

develop a theoretical system of basin geometries and of resulting preferred migration paths, assuming uniform permeability, uniform pressure distribution, uniform source bed richness and uniform kerogen-hydrocarbon conversion rate. When we add data on the variations of these parameters to the primary qualitative/semi-quantitative migration definition, we will obtain an increasingly improved quantitative migration pattern, hence an improved predictive quantitative basin evaluation.

The following basin geometry types can be distinguished:

I. Simple Basins, containing one depocenter

1a. Circular symmetrical (Fig. 10)
No preferred migration direction exists.
Examples: Michigan Basin, USA (parts); single lobes of complicated depocenter patterns.

1b. Circular asymmetrical (Fig. 11)
Crowding of migration rays on the narrow concave side of this geometric form indicates the potential for migration focusing. Dispersion of migration rays on the wide concave side of the "basin" indicates dispersion of migrating hydrocarbons, hence less migration focusing. The degree of basin asymmetry will control the amount of hydrocarbons available for migration in each basin portion.
Examples: No example is known, except possibly again (as in case I. 1a) lobes of complicated depocenters. This form is important, however, to develop the concept of hydrocarbon migration focusing and basin symmetry.

2a. Elongate symmetrical (Fig. 12)
Migration focusing along the long flanks and migration diffusion along the short flanks occur "simultaneously". This and the volumetric differences between long and short flanks result in an accumulation preference along long flanks of such basins over short flanks.
Examples: Depocenters of Rheingraben, Germany, and of Vienna Basin, Austria (12).

2b. Elongate asymmetrical (Fig. 13)
Migration focusing is identical to the previous case (I. 2a). Here the long gentle flank with the larger volume is further preferred due to the increased volume of hydrocarbons generated in a thicker sedimentary section. The degree of asymmetry will determine where the larger volume will be found.
Examples: Great Valley Basin, California; Mid-Magdalena Basin, Colombia [9].

3a. Elongate symmetrical curved (Fig. 14)
The concave long flank A is clearly favored for migration focusing. The width/length ratio of such a basin will determine to what degree the short flanks or the long convex flank B have the largest accumulation (migration focusing potential).
Examples: Szeged Basin, Hungary [12].

3b. Elongate asymmetrical curved (Fig. 15)
The distribution of migration focusing is similar to that of basin form I. 3a—elongate, curved, symmetrical. However, the additional differences in sediment volume between the two flanks of such asymmetrical basins further accentuate the potential of the long concave basin flank A. Another complication can arise when the basin axis lies nearer to the long concave flank A. In such a case, the migration focusing effect of the concave long flank can be offset by the volumetric effects of asymmetry.
Examples: Los Angeles Basin, California; Wind River Basin, Wyoming; Bighorn Basin, Wyoming; Powder River Basin, Wyoming; Reforma Region, Mexico; Eastern depocenter of Po Valley Basin, Italy.

II. Composite basins, containing two or more depocenters

In such basins hydrocarbons migrating from a pair of adjacent depocenters can accumulate in joint trapping areas, or in separate trapping areas. A third or fourth depocenter will merely form more pairs of adjacent depocenters. In composite basins only one such pair has to be analyzed at a time. Conceptually composite basins develop from class I. 3, elongate curved *(Figs. 14 and 15)*.

1. Composite linear (Fig. 16)

Migration focusing will be most pronounced in the center of common flanks; less focusing will occur in individual long flanks, and on short flanks.

Examples: Great Valley Basin, California; Baltimore Canyon Area, USA; Lower Magdalena Basin, Colombia; Mid-Magdalena Basin, Colombia; Reconcavo Basin, Brazil; Gulf of Suez, Egypt; Sirte Basin, Libya; Po Basin, Italy.

2. Composite parallel (Fig. 17)

Migration focusing is identical to the previous case II. 1.

Examples: Mackenzie Delta, Canada; Gippsland Basin, Australia; Pre-Salt Plays, Gabon; Mahakam Delta, Indonesia; Hassi Messaoud Region, Algeria.

3. Composite asymmetrical

In these cases specific hydrocarbon migration patterns and specific hydrocarbon distribution patterns will develop, depending on the degree of symmetry.

Example: Northwest German Basin.

There is no known dependence of migration focusing potential on specific geologic basin types or classes. In other words, basin classifications do not enter into migration prediction. To what degree, quantitatively, hydrocarbons will be generated and can migrate in a given basin depends on specific geologic factors. Of all the regional-geological parameters that have been proposed in attempts to correctly predict hydrocarbon distribution, migration focusing, or preferred migration on the basis of existing basin geometry, appears to be the most powerful predictive concept.

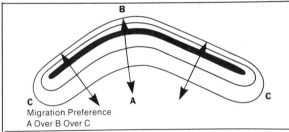

Figure 15: Migration Principles—Elongate Asymmetrical Curved Basin

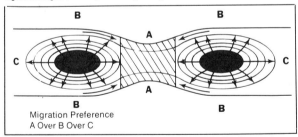

Figure 16: Migration Principles—Composite Linear Symmetrical Basin

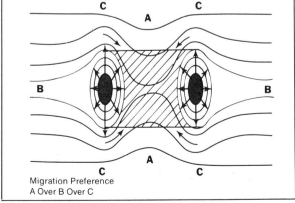

Figure 17: Migration Principles—Composite Parallel Symmetrical Basin

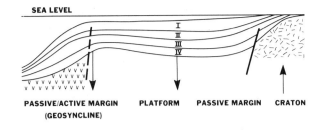

Figure 18: Continental Margin Development—Stage I

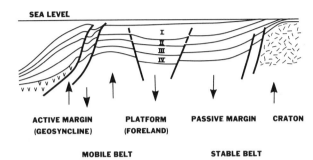

Figure 19: Continental Margin Development—Stage II

Definition of the Length of Migration Paths

Quite commonly, the length of a migration path cannot be predicted. In other words, it is usually not possible to say where in the preferred migration path oil or gas accumulations will be found *(Fig. 6)*. Most detailed exploration techniques like subsurface geology, seismic reflection data or comparative geology will aid in this determination. In most cases potential traps will first be tested in regionally high positions; subsequent tests will be further down-dip until the region of the effective depocenter itself is tested. While interparticle porosity and permeability will decrease with depth, fracturing of originally tight rocks can add new prospects at greater depth.

These limitations of predictive exploration appear negative at first. However, we must realize that in most productive basins 75% or more of hydrocarbon reserves are found on 25% or less of the basinal areas. Even the difficulty to define the length of migration paths cannot be a reason against the extreme usefulness of the concept as an explanatory and predictive hydrocarbon exploration approach.

Application of Migration Focusing in the NW German Gas Region Basin Type

The Northwest German basin north of the outcropping Paleozoics (Variscan mobile belt) possesses all of the characteristics of a long-lived passive continental margin *(Fig. 1)*. The effects of the Variscan orogeny in the mobile belt to the south are visible in the form of deformation style and sedimentary records. Certain additional characteristics reminiscent of back-arc basins are also found. These include shear and tension deformation, post-orogenic volcanic activity, and molasse sediments. This is not surprising because continental passive margins and back-arc basins indeed are products of very similar geologic processes [7]. The development of the Northwest German Basin began with an early Paleozoic passive stage, that, during middle and late Paleozoic time, changed to an active margin in the south *(Figs. 18, 19)* [8]. It is difficult to see how in this case of dynamic progressive geologic development an evaluation of hydrocarbon potentials could be based on a basin classification scheme, as proposed by many, which is always static at best.

The gas productive section in the Northwest German basin, as known today, is largely the result of late Paleozoic post-orogenic events in a quasi-back-arc setting. Northwards, towards the Sveco-Scandian/Russian craton and beneath late Carboniferous sediments, prospects related to the original passive continental margin of early Paleozoic age should be present *(Fig. 4)*.

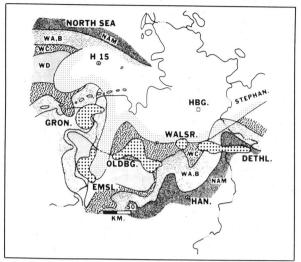

Figure 20: Northwest German Basin—Pre-Permian Paleogeology and Gas Field Areas

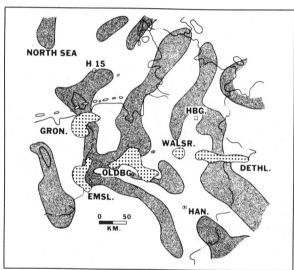

Figure 21: Northwest German Basin—Basement Highs and Gas Field Areas

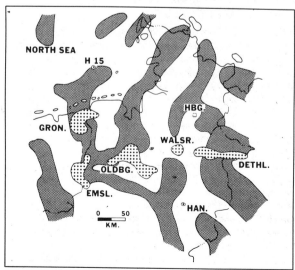

Figure 22: Northwest German Basin—Regional Highs and Gas Field Areas

No detailed regional structure maps of gas productive zones or of mature gas source beds across the NW German basin exist in the public domain. The map that comes closest to the required goal is a regional basement structure map obtained from integrated mag-

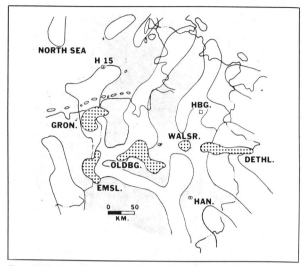

Figure 23: Northwest German Basin—Regional Lows and Gas Field Areas

netic, gravity and subsurface data [8] *(Fig. 21)*.

Comparisons with published regional thickness and facies maps have proven that this basement map is reliable for regional basement geometry. Further verification comes from a basin-wide pre-Perm paleographic map based on well results [1, 8] *(Fig. 20)*. The major gas field areas in the basin superposed upon regional basement structure and upon pre-Permian paleogeology are shown on *Figs. 20 to 23*.

The Northwest German basin is a complex basin with several synsedimentary highs and lows present in linear and parallel arrangements *(Fig. 21)*. The mature gas source bed intervals lie in Late Carboniferous Westphalian coal beds and carbonaceous shales occurring at least across the southern part of the basin [6]. If the present regional lows were synsedimentary lows already during Late Carboniferous time, these lows will be effective depocenters. And even if these present regional lows do not contain thicker or richer gas source beds than adjacent regional highs, the regional lows will still be preferred sites for gas generation because of their pronounced subsidence history since at least Late Carboniferous time. The regional lows in the Northwest German basin thus are seen as effective depocenters from which gas was able to migrate into adjoining regional highs for long geologic times since the onset of major gas generation.

Preferred Gas Migration Paths

The regional low areas in the Northwest German basin appear here as elongate symmetrical depocenters trending in a general north-south direction. Preferred gas migration will be along their individual western and eastern long flanks. In addition, northward plunging regional highs will act as migration focal areas where they plunge into a depocenter at their northern end. A somewhat more simplified view with the same result is to classify the Northwest German basin as complex linear and/or complex parallel. In each case regional highs will be areas of preferred gas migration from the depocenters.

Length of Migration Paths

The longest possible migration path for gas in the Northwest German Basin is less than 100 km and thus not at all extreme. The general problem of length of migration paths, therefore, is here reduced to that of optimum trapping conditions for migrated gas inside the preferred migration paths.

Important is a pronounced special character of synsedimentary regional highs: they act not only as migration focal areas but also as optima for reservoir development (carbonates and clastics) and for trap formation, both stratigraphic and structural *(Fig. 6)*. In fact, in the Northwest German basin all major gas accumulations possess major stratigraphic trap parameters. In Groningen, Rotliegend clastics lie between a northern shale-out and a southern pinch-out zone. The majority of Oldenburg gas fields produce from Zechstein carbonates which shale-out northward and pinch-out southward; Rotliegend sandstone reservoirs in the Dethlingen region appear to undergo diagenetic alterations following a regional pattern possibly related to internal trends of the Dethlingen regional high [2]. And gas reservoirs in early Permian and late Carboniferous sandstones across the basin depend much on pre-existing sedimentation conditions, thus on Carboniferous and Permian subsidence, uplift, and erosion, in addition to secondary diagenesis.

Conclusion

Secondary migration of gas and oil follows simple laws. The preferred direction of hydrocarbons migrating from the generating low effective depocenter to trapping higher basin flanks or adjacent regional highs is qualitatively predictable where meaningful present basin geometry is known. The majority of trapped hydrocarbons will be found along these predictable preferred regional migration paths. This approach allows us to improve regional exploration concepts and programs and helps to define additional exploration plays, even in basins that are at a relatively high degree of exploration development. Where exploration data of all types can be integrated into a coherent basin evaluation, the definition of secondary migration directions can well be utilized as basis for forward-exploration plans.

References

1 *H. Bartenstein:* "Essay on the coalification and hydrocarbon potential of the Northwest European Paleozoic", Geol. en Mijnb, 59, 2, 155-168 (1979)
2 *H. J. Drong:* "Diagenetische Veränderungen in den Rotliegend Sandsteinen im NW-Deutschen Becken", Geol. Rdsch., 68, 1172-1183 (1980)
3 *J. M. Hunt:* "Petroleum Geochemistry and Geology," Freeman, San Franc., 678 p. (1979)
4 *P. L. Lawrence* and *J. C. Pratsch:* 1980; "Regional Analysis of Hydrocarbon Migration Using Geophysics: Gippsland Basin, SE Australia", Abstr., 15 Ann. Mtg. NE Sect. GSA, Philadelphia, 68 (1980)
5 *H. W. Menard:* "Toward a rational strategy for oil exploration," Scient. Amer. 244, 1, 57-64 (1981)
6 *E. Plein:* "Das Deutsche Erdöl und Erdgas", Jh. Ges. Naturk, Württ., 134, 5-33 (1979)
7 *J. C. Pratsch:* "Future Hydrocarbon Exploration on Continental Margins and Plate Tectonics", Petrol. Geol., 1, 2, 95-105 (1978)
8 *J. C. Pratsch:* "Regional Structural Elements in Northwest Germany", J. Petrol. Geol., 2, 2, 159-180 (1978)
9 *J. C. Pratsch:* "Hydrocarbon Concentration Through Preferred Migration—Middle Magdalene Basin, Colombia, South America", Erdöl-Erdgas Ztschrft., in press
10 *J. C. Pratsch:* "Wedge Tectonics Along Continental Margins", 1981 Hedberg AAPG Research Conference, Galveston, AAPG Mem., in press
11 *J. C. Pratsch:* "Focused Migration of Gas and Oil", Abstr., 1981 Annual Mtg., AAPG Bull., 65, 5, 974
12 *J. C. Pratsch:* "Basin Evaluations and Concentrations of Oil and Gas Accumulations", Symp. Complex Oil-Geol. Aspects Offsh. Coastal Adriatic Areas, Split, Yugoslavia, Proc., 53-66
13 *J. C. Pratsch:* "Regional Structural Elements and Major Gas Accumulations in Northwest Europe", Seminar Expl. Gas Fields ECE Region; Proc., Hannover, ECE, 6 p. (1981)
14 *L. C. Price:* "Mobilization and Documentation of Vertical Oil Migration in Deep Basins", J. Petrol. Geol., 2, 4, 353-387 (1981)
15 *B. P. Tissot* and *T. H. Welte:* "Petroleum Formation and Occurrence, A New Approach to Oil and Gas Exploration", Springer, Berlin-New York, 538 p. (1978)
16 *D. H. Welte:* "Neuere Überlegungen und Erkenntnisse über die Erdölmigration", Erdöl-Erdgas-Zeitschrft., 95, 207-208 (1979)
17 *D. H. Welte* and *Yübler:* "Evolution of sedimentary basins from the standpoint of petroleum origin and accumulation—an approach for a quantitative basin study"; Organic Geochem., 2, 1-8 (1980)

Acknowledgement

I thank the Management of Mobil Oil Corporation for the permission to contribute this paper to the "Seminar on the Exploration for Gas Fields in the ECE Region (Geology and Geophysics)," May 18-23, 1981, Hannover, Germany. I also appreciate many discussions on subjects of NW European geology with Dr. H. Bartenstein, Celle, and valuable data contributions by Mr. H. Keshav, Celle.

SURFACE GEOCHEMISTRY

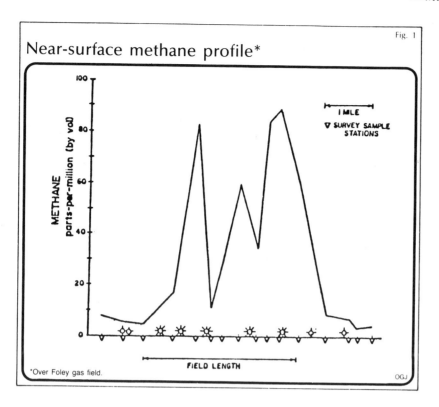

Fig. 1 Near-surface methane profile*
*Over Foley gas field.

Near-surface hydrocarbon surveys in oil and gas exploration

Richard D. McIver
President
McIver Consultants International
Houston

Most petroleum producing regions of the world were first recognized because of oil or gas seeps in their soils, rocks, streams, or lakes.

The first oil and gas wells in many basins were close to such seeps. While easily recognized macroseeps are relatively common, more obscure microseeps must be even more abundant. If these could be identified positively, they would also be extremely important to explorationists. They can!

Unfortunately, pre-1960 methods of identifying hydrocarbons made it difficult and costly to confirm the presence of microseeps. Hydrocarbon determinations were time consuming and insensitive. Therefore, the quest for microseeps led to the development of a variety of "secondary" microseep indicators, i.e., indicators that could be more easily measured. Anomalously high quantities of many of these were found in soils over a few known oil and gas fields, so they were applied to prospecting in general, in spite of the fact that there was little or no chemical or physical basis for their direct or invariable association with seeping hydrocarbons.

In all fairness, however, it must be stated that some of the distinct anomalies formed by these indirect indicators may have been caused by the surface or near-surface outlet of preferred channels for upwelling compaction waters. As these "plumes" reach the water table or the deep soil, their dissolved mineral salts interact with the oxidizing environment and alter the subcrop and soil, thus forming anomalies. There are geological reasons (eg. differential compaction of shales surrounding reefs or thinning over structure) why these preferred channels would coincide with petroleum reservoirs, but an anomaly would not indicate the presence of hydrocarbon in those structures. Measurement of hydrocarbons in the near-surface could.

Methods. Whatever the cause of such anomalies, a host of indirect methods evolved from the

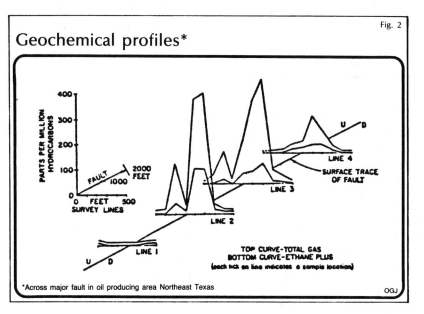

Fig. 2 Geochemical profiles*

*Across major fault in oil producing area Northeast Texas

1930s on. Most involve collection of surface or near-surface soil samples and determining a secondary indicator such as:
- Radiogenic or inert gases
- Radioactivity
- Heavy hydrocarbons, including paraffin dirt
- Methane utilizing bacteria
- Spectral fluorescence or luminescence of soil extracts
- Trace metals in soils and plants
- Inorganic salts or anions
- Carbonates, sometimes analyzed for carbon-13 isotope contents.

Proliferation of these uncertain methods, and extravagant claims for them by their practitioners, were responsible for much of the skepticism and controversy that surrounds near-surface geochemical "prospecting." This was compounded by the fact that surface prospectors 'frequently neglected geological and physical principles, and even common sense, when promoting their services. For example, many of them claimed that hydrocarbons or associated secondary indicators move straight upward from hydrocarbon accumulations so that their surface manifestations are virtually the same size and shape as the oil or gas pools, and precisely above them—regardless of the depth of the accumulations. Others contended that the anomaly is always in the form of a halo wrapped tightly around the surface trace of a pool's oil-water or gas-water contact—again, no matter how deep the accumulation. These untenable claims helped give a black eye to the entire practice.

Methods of sample collection and data handling in such surveys also were often suspect. Some prospectors grabbed soil samples right at the surface or a few inches below, oblivious to the fact that microbes in some soils produce or that sulfate-reducing bacteria in other soils destroy hydrocarbons. So, either the presence or the absence of methane in shallow soils might be completely misleading. Moreover, some "successful" surveys had few or no sample stations beyond the margins of the prospect or pool. Hence, there was no way to establish whether purported "anomalous" values were significantly higher than regional background.

Furthermore, some promoters were apparently unaware that in any single statistical population of data, such as background gas values, there is a predictable number of relatively high values. Thus, many an anomaly "discovered" by a prospector was little more than a cluster of random higher values that had no statistical or exploration significance. There are ways of testing the significance of such data; unfortunately, they are often not used. However, even if satisfactory sampling techniques are employed, and the statistical limitations of the data are taken into account, most secondary indicators are still of dubious value because they have little or no chemical or physical relationship with hydrocarbons.

Development of very sensitive gas chromatographs with flame-ionization detectors that rapidly detect and measure subparts-per-million quantities of hydrocarbons revolutionized near-surface geochemical surveying. Measurement of the full suite of petroleum-like gaseous hydrocarbons eliminated the need to rely on "secondary" indicators such as those listed above. Because of this breakthrough in instrumentation, virtually all of the major oil companies[1,2] and many smaller ones now conduct near-surface hydrocarbon surveys themselves, or contract the work out to service companies. This has become particularly important in helping assess large blocks of offshore acreage in preparation for competitive sales.

The gas chromatograph does more than just measure total hydrocarbon; it separates the individual hydrocarbons first and then measures the amount of each one. Consequently, the geochemist can now determine whether the gas in anomalies is the same as gas in oil and gas fields in the study area.

Composition reveals whether the gas in the near-surface is simply microbially produced marsh gas or is rich in ethane, propane, and heavier hydrocarbons that are only produced in deeply buried rocks, and hence, must be migrating from them. (If the hydrocarbon sought is methane, the carbon-isotope-ratio mass spectrometer can make the same distinction; however, this adds appreciably to costs.) This ability to elucidate the origin of near-surface hydrocarbons has greatly improved the specificity of hydrocarbon surveys.

Of course, microseepages and surface hydrocarbon anomalies are probably produced by more than one process. While diffusion of the lighter, gaseous hydrocarbons may proceed through hundreds of feet of most rocks, even shale[3,4,5], many rocks contain extensive networks of fractures or joints that allow gaseous hydrocarbons to pass more easily. There is disagreement whether faults and fractures are seals or conduits, or both, but many geochemical surveys show a close relationship of near-surface hydrocarbon anomalies with faults and fractures.

There can be no doubt that statistically significant anomalies, i.e., microseeps have been detected in near-surface sediments over many oil and gas fields, particularly those reservoired shallower than about 5,000 or 6,000 ft. The example in Fig. 1 shows a methane profile taken across the Foley gas field in Baldwin County, Ala., shortly after it was discovered. This may be the nearest example we have to a true diffusion anomaly, originating from the shallow (ca 2,000 ft) reservoir.

Cone-shaped plumes. Other things equal, gases migrating up from reservoirs must spread upward and outward into cone-shaped plumes. Therefore, gas originating from deep rocks is greatly dissipated by the time it reaches the surface—but less so, if it is focused by faults, fractures, or joints. This focusing can be seen in profiles taken at close spacing across surface expressions or faults. Fig. 2

shows just such a highly focused hydrocarbon anomaly over a fault in a producing area in Northeast Texas. Fig. 3 shows another example over the prolific Scipio-Albion field in Michigan.

Even with faults, reservoir derived anomalies must spread out and become more patchy or fragmented the further the hydrocarbon has to migrate to reach the surface. Moreover, unless the flux of hydrocarbons is relatively high, all the hydrocarbon can be destroyed by oxygen and or microbes in the upper soil layers, without a trace reaching the shallowest soils.

Therefore, for greatest utility, samples for surveys should be collected from beneath the zone of potential hydrocarbon oxidation, the depth of which varies from area to area. The zone of hydrocarbon destruction usually extends downward 2-10 ft, depending on the porosity of the soil, its moisture content, lithology, etc. Profiles of hydrocarbon concentration in a samples from grass-roots down, in every basin we have worked in so far, show background levels persisting from the surface 2-8 ft, then an abrupt increase, followed by a trend of slowly increasing values downward.

Another extremely important observation is that gas anomalies rapidly disappear over depleted or depressured fields.[6] This means that light hydrocarbon surveys can be used to help detect new pools in mature exploration areas or untapped prospective zones above or below depleted zones in oil fields.

While geochemical surveys are sometimes used as stand-alone exploration projects, we prefer that they be used to evaluate prospects previously located by other geological or geophysical means. This makes it possible to design a sample grid and choose a sample density suitable for the specific problem. For example, a ¼ to ½ mile sample grid used to detect diffusion anomalies (as in Fig. 1) is not suitable for detecting seepage along faults. This may require sampling locations only a few tens of feet apart, as shown in Fig. 2.

However, if geochemistry is used for reconnaissance, and if it should reveal a significant anomaly, the anomalous area should be studied by conventional geological and geophysical means (and probably should be sampled more extensively for near-surface hydrocarbons) to locate potential reservoirs responsible for the near-surface hydrocarbon anomaly.

Conclusion. In conclusion, it must be emphasized strongly that modern near-surface geochemical surveying is another potentially powerful geological tool available to oil and gas ex-

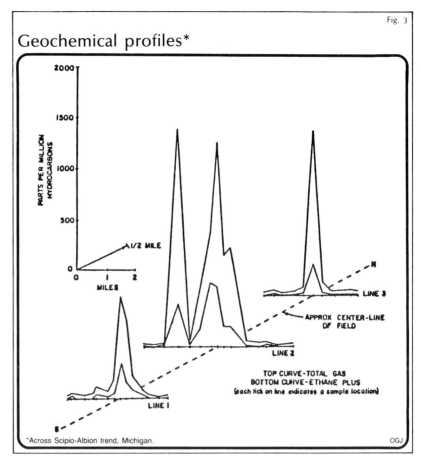

Fig. 3

Geochemical profiles*

*Across Scipio-Albion trend, Michigan.

plorationists. It is not a direct petroleum finding method. However, its judicious use will increase the odds of finding commercial oil and gas. And, in the risky business of oil and gas exploration, the wise operator uses every tool at his disposal.

References
1. "Predictions of oil or gas potential by near-surface geochemistry," 1983, Jones, V.T., and R.J. Drozd (Gulf Research), AAPG Bull., Vol. 67, No. 6, pp. 932-952.
2. "Developments in geochemistry and their contribution to hydrocarbon exploration," 1979, Weisman, T.J. (Gulf Research), Proceedings, Tenth World Petroleum Congress, Review Paper No. 2, September 1979, pp. 369-386.
3. "Diffusion of light hydrocarbons in subsurface sedimentary rocks," 1983, Leythaeuser, D., R.G. Schaefer and H. Pooch, AAPG Bull., Vol. 67, No. 6, pp. 889-895.
4. "Near-surface evidence of migration of natural gas from deep reservoirs and source rocks," 1981, Stahl, W., E. Faber, B.D. Carey and D.L. Kirksey, AAPG Bull., Vol. 65, No. 11, pp. 1543-1550.
5. "Role of diffusion in primary migration of hydrocarbons," 1982, Leythaeuser, D., R. G. Schaeffer and A. Yukler, AAPG Bull., Vol. 66, No. 4, pp. 408-430.
6. "Hydrocarbon geochemical prospecting after thirty years," 1969, Horvitz, L., Proceedings, Conference on Unconventional Methods of Exploration for Oil and Gas, Dallas, pp. 205-218.

Journal of Geochemical Exploration, 17 (1982) 1—34
Elsevier Scientific Publishing Company, Amsterdam — Printed in The Netherlands

SURFACE GEOCHEMICAL METHODS USED FOR OIL AND GAS PROSPECTING — A REVIEW

R.P. PHILP[1] and P.T. CRISP[2]

[1] CSIRO Division of Fossil Fuels, PO Box 136, North Ryde, N.S.W. 2113 (Australia)
[2] Department of Chemistry, The University of Wollongong, Wollongong, N.S.W. 2500 (Australia)

(Received April 30, 1981; revised and accepted December 14, 1981)

ABSTRACT

Philp, R.P. and Crisp, P.T., 1982. Surface geochemical methods used for oil and gas prospecting — a review. J. Geochem. Explor., 17: 1—34.

The majority of the world's oil and gas deposits have been discovered by drilling in the vicinity of natural petroleum seeps, and to date the most successful geochemical prospecting methods still rely upon the surface detection of hydrocarbons. Gas chromatographic techniques are now commonly used in the analysis of hydrocarbon gases for prospecting both onshore (analysis of soils and rocks) and offshore (analysis of near-bottom waters and sediments). Detection of helium fluxes has been partially successful as a geochemical prospecting technique. Many indirect techniques such as the determination of isotope and metal-leaching anomalies in surface rocks and the measurement of radon fluxes have not been widely used.

Onshore geochemical prospecting appears to have more problems associated with it than offshore prospecting due to the more complex migration mechanism of near-surface waters containing dissolved gases. No onshore prospecting studies have been published which thoroughly consider this factor and the success of onshore prospecting remains equivocal. In offshore prospecting "sniffers" have been used to detect hydrocarbon anomalies in near-bottom waters, and coring equipment has been used for the detection of hydrocarbons in near-surface sediments. Success is claimed using these techniques.

Geochemical prospecting methods are complementary to the widely used geophysical methods. Geochemical methods can provide direct evidence for the presence of petroleum accumulations and are relatively cheap and rapid. Failures in prospecting to date are attributable to the simplistic manner in which data have been interpreted; insufficient attention has been paid to the hydrological and geological factors which modify the upward migration of indicator species to the surface. As oil and gas deposits become more difficult to locate, greater attention should be paid to geochemical prospecting techniques, especially as a regional exploration tool.

INTRODUCTION

Surface geochemical prospecting for oil and gas is basically a search for surface chemical phenomena arising from the accumulation of hydrocarbons

0375-6742/82/0000—0000/$11.55 © 1982 Elsevier Scientific Publishing Company

at depth. Surface anomalies arise from the fact that no oil or gas reservoir cap rock is completely impermeable, and so hydrocarbons and other compounds escape from the reservoirs and the more volatile components migrate to the surface where they may be trapped in soils or subsequently diffuse into the atmosphere or oceans. Hydrocarbons that reach the surface may be detected either directly or indirectly through geochemical changes they induce. The bulk of the literature on geochemical prospecting for oil and gas is concerned with the direct detection of hydrocarbon gases in soils, ocean sediments and water. Indirect indications of hydrocarbons include the reductive leaching of metals from surface rocks, microbiological changes in soils, and the surface distributions of helium and radon which accumulate in certain hydrocarbon reservoirs and, due to their volatility, escape to the surface. Geochemical and geophysical prospecting techniques can be viewed as being complementary to each other. Geophysical techniques will delineate regions or structures that might contain hydrocarbon reservoirs, and geochemical methods can be used to determine whether or not there are anomalously high concentrations of hydrocarbons or other indicator species in these regions (Meyer, 1965).

Geochemical methods for oil and gas prospecting were first investigated in the 1930s and early 1940s by Sokolov and coworkers (1933) in Russia, by Laubmeyer (1933) in Germany, and by Horvitz (1939) and Rosaire (1940) in the United States. These early studies were mainly restricted to the analysis of methane and total non-methane hydrocarbons in soil gases, and employed techniques such as condensation, combustion and manometry. Descriptions of the early Russian work were given by Kartsev et al. (1954), Sokolov (1959) and Sokolov et al. (1959). Rosaire (1940) reviewed prospecting techniques involving the analysis of hydrocarbon gases in soils, soil gases and ground waters, and Pirson (1940) compared onshore prospecting methods. Chilingar and Karim (1962) and Karim (1964) described the application of gas surveys to surface sediments, drill fluids and cores and, in reviewing geochemical prospecting for minerals, Boyle and Garrett (1970) also included a section on petroleum. Davis (1967) has also discussed some of the early methods used for geochemical prospecting, including infra-red analysis of soil extracts (Bray, 1956a) and measurement of $^{13}C/^{12}C$ ratio of soil lipid extracts (Bray, 1956b). Horvitz (1939, 1954, 1959, 1969, 1972) has dealt with the basic principles, methods and results obtained from geochemical prospecting studies. Sokolov et al. (1970) have reviewed methods and results obtained from subsurface and deep-gas surveys and discussed the shapes of various anomalies. Kroepelin (1975) considered the physical and chemical conditions required for prospecting and commented on the various methods and results, and Motojima (1975) reported on the state of geochemical prospecting in Japan.*

The idea of prospecting for oil and gas by geochemical means is still viewed with scepticism in many quarters. In most countries, however, the

---

*See "Notes added in proof", p. 33, note 1.

earliest discoveries of petroleum were made by drilling near oil seeps, tar deposits or visible evidence of hydrocarbon leakage. Oil and gas seepage is associated with many oil-producing regions (Link, 1952), and as late as 1940 it was claimed (Degolyer, 1940) that visible evidence of oil was responsible for the discovery of more oil than any other prospecting technique. Modern gas-chromatographic methods are more sensitive detectors of hydrocarbons than the human eye and permit invisible microseepages of volatile hydrocarbons to be located. In the following sections methods used for the detection of surface anomalies of hydrocarbons in onshore and offshore areas are discussed, and a final section deals with prospecting methods not based on the detection of hydrocarbon gases.

ONSHORE PROSPECTING BASED ON HYDROCARBON GASES

Introduction

Liquid or solid hydrocarbons in outcrops are undoubtedly the most valuable direct indicators of petroleum. Hydrocarbon gases, however, are also indicators of petroleum as they are present in all petroleum reservoirs, are sufficiently volatile to diffuse to the surface and are easily detected. Light liquid hydrocarbons experience more difficulty than gases in migrating the distance required to reach the surface and bitumens, due to their low volatility, are rarely able to reach the surface (Sokolov, 1970). The source of liquid and solid hydrocarbons which do reach the surface is unquestionable, whereas the source of the surface gases is often ambiguous.

Methane is not, in general, a reliable indicator of oil or natural gas because it is produced in prodigious quantities by microbiological processes occurring in soils and sediments and methane from this source overwhelms any methane which might have migrated to the surface from underlying petroleum sources. The majority of prospecting techniques are based on the detection of ethane, propane, *n*- and *iso*-butane which are not produced biologically in significant quantities (Davis and Squires, 1954) and are therefore more reliable as indicators of gas seepage from underlying reservoirs. Increasing sensitivity of isotopic ratio mass spectrometers has reduced the sample size required for $\delta^{13}C$ measurements. In the future, it is clear that increasing use will be made of $\delta^{13}C$ ratios to determine the origin of methane in the surface samples (Reitsema et al., 1981; Stahl et al., 1981). Coleman et al. (1977), in a study of leakage gas from underground storage reservoirs, showed that isotopic analysis is a reliable technique for differentiating between methane from thermogenic and biogenic sources. The isotopic composition of the methane was found to be a more reliable indicator of hydrocarbon reservoirs than non-methane hydrocarbon concentrations. Isotopic analysis of methane has in the past been restricted to situations where large quantities (millilitres) of gas could be collected.

Gases generated thermocatalytically during the maturation of a source rock do not usually rise vertically; instead, gas migration is generally thought to follow faults and fractures with the gases being trapped temporarily beneath cap rocks of low permeability. The shape of the surface anomaly depends upon the geology and hydrology of the underlying sediments.

Migration of gases may be caused by differences in concentration (diffusion), pressure (effusion) or a combination of the two (Stegena, 1961). On land the near-surface flow of groundwater causes lateral movement of migrating gases, but these effects have generally been overlooked in geochemical prospecting. Hitchon (1974) stated that there did not appear to be any published studies on surface prospecting which were supported by examination of the flow of near-surface formation water. This statement still appears to be true, although Freeze and Witherspoon (1966, 1967, 1968) have published a theoretical treatment of water flow in basins. Smith et al. (1971) calculated that a 3 m thick aquifer at a depth of 870 m, halfway between the surface and hydrocarbon accumulation and with a lateral water velocity of 1 m/yr, would move the vertically diffusing gases a lateral distance of 190 km. Any application of geochemical prospecting methods to the search for oil or gas therefore needs to study the effects of groundwater movement upon the migration of gases.

Hydrocarbon gases are trapped in soils in several different forms:

(a) as gas in the spaces between soil particles (pore gas);

(b) adsorbed onto particulate material (adsorbed gas);

(c) dissolved in the groundwater (dissolved gas).

In general the three forms of gas will be in equilibrium and no sampling procedure will collect one form exclusively. This theoretical division, however, provides a rough classification for the literature describing onshore prospecting.

Pore-gas analysis

Pore-gas analysis was the earliest method of gas analysis used in geochemical prospecting, and methods for collecting samples of soil air from systematically located bore holes and analysing the samples for traces of methane were described by Laubmeyer (1933). Bore holes were drilled to a depth of 1 to 2 m, sealed for 24 to 48 h, and a sample of air was then extracted for analysis. Sokolov (1933) carried out similar work in Russia. The technique consisted of partially reducing the pressure in the bore hole and then collecting a sample of soil air by displacement of water in a suitable container. The hydrocarbons were separated into two fractions:

(a) a "light" fraction containing methane and ethane (not condensing at liquid-air temperature);

(b) a "heavy" fraction containing a mixture of propane and higher hydrocarbons (condensing at liquid-air temperature).

The results from both Sokolov's study in Russia and Laubmeyer's study in Germany indicated that maximum concentrations of hydrocarbon gases occur in soil air over productive zones, with minimum values over barren areas. It was also suggested that a high ratio of the light fraction to the heavy fraction is indicative of a gas reservoir, while a low ratio was evidence for an oil deposit. This need not necessarily be the case, since later studies show that the composition of gases at the surface is dependent on geological features such as faults and fractures.

Kartsev et al. (1954) provided a description of the common types of pore-gas surveys. In the *ordinary-gas survey*, holes 3 to 4 m deep were drilled and gas was pumped from the lower part of the hole after this section had been isolated from the atmosphere by means of an inflatable rubber collar connected to the upper portion of the sampling tube. The *delayed-gas survey* involved taking samples up to 24 h after hermetically sealing the drill hole. In the *deep-gas survey*, samples were taken along sections of bore holes at depths from 3 to 20 m and the analysis considered not only the absolute concentrations but also variations with depth. In situations where gas concentrations were low a *gas survey adsorbent* (such as silica gel) was used to concentrate the hydrocarbons from the subsoil air. Another type of pore gas survey employed large metal caps pressed into the soil to collect the flux of hydrocarbons migrating to the surface. This method was rarely used and no results were published. Irrespective of the sampling method, the concentrations of methane and non-methane hydrocarbons were determined by adsorption, combustion and manometry.

In the early Russian work, surveys were made using free gases from a depth of 2 to 3 m. On the Aspheron Peninsula, gas anomalies of methane and heavy hydrocarbons were obtained over several fields, and best results were obtained by taking samples from depths of 5 to 10 m. The shapes of the surface anomalies were found to be dependent on geological conditions.

Wood (1980) recently described a gas chromatographic method, suitable for geochemical prospecting, capable of measuring ethane (5 ppb), ethylene (5 ppb) and propane (20 ppb) without preconcentration. Despite the low levels of detection possible and the claim that subtle differences in light hydrocarbon concentrations in the soil gases of petroleum producing areas can be detected, no examples of its use in a complete geochemical survey are given.

In a geochemical prospecting study in the Cooper Basin, Australia, pore gas samples were collected from holes drilled to a depth of 3 m over a grid pattern with holes at 500 m intervals (Philp et al., unpublished results). A probe was inserted into the hole which was subsequently sealed with an inflatable collar. Approximately 500 ml of pore gas were pumped through a trap containing Porapak Q in liquid nitrogen. As soon as practical after collection the gases were analysed in the field by gas chromatography. The results of this preliminary study were encouraging, and anomalously high concentrations of C_2 to C_4 hydrocarbons were detected in an area adjacent to a fault previously determined by a seismic survey. The frustrations of this method of prospecting were made abundantly clear one year later, when another survey was performed in the same region and the anomaly could no longer be detected. The reason for this is not obvious: it could be related to climatic differences such as rainfall and higher surface water movement in the second survey. If this type of occurrence is common, it might explain the paucity of successful reports of onshore geochemical prospecting in the literature.

Adsorbed-gas analysis

Early work in the USA on adsorbed-gas analysis was started by Horvitz and Rosaire in the 1930s and much of this early work is summarized by Horvitz (1939) and Rosaire (1940). In 1939 Horvitz developed an acid treatment method whereby large samples of soil collected in the field from depths of 5 to 10 m were treated with acid and/or heated to release adsorbed hydrocarbons which were then determined by combustion and manometry following the method of Kartsev et al. (1954). Sokolov (1970) reported that in many areas adsorbed-gas analysis gave better results than pore-gas analysis and also found that gaseous hydrocarbon concentrations increased with depth over oil and gas fields as a result of vertical migration from the reservoir. Debnam (1965) and Horvitz (1972) developed more elaborate methods for the isolation and analyses of adsorbed gases which involved the release of pore-gas by gentle heating of the sample under vacuum, followed by the release of adsorbed gases by treatment with acid. Gases were condensed at liquid-nitrogen temperature and analysed by gas chromatography. This method was far more sensitive than the original methods described by Horvitz (1939) due to advances in instrumentation available for analyses of the gases.

Debnam (1969) used a gas chromatographic method, which he had previously developed (Debnam, 1965), to evaluate the application of geochemical prospecting for locating oil and gas fields in Canada. In the areas examined, both halo and solid anomalies were recognized and it was concluded that this geochemical method could be a valuable addition to the usual geological and geophysical exploration methods if used under favourable conditions.

McCrossan et al. (1972) evaluated a commercially available surface geochemical prospecting technique in the western part of Alberta by measuring absorbed gas concentrations and found that the distribution of anomalous points was random and could not be correlated with underlying oil and gas pools. Samples rich in carbonate minerals dissolved in acid to a greater extent than low carbonate samples and released larger amounts of adsorbed hydrocarbons. Corrections based upon the amount of acid-soluble material were not successful and it was concluded that the method had little value in areas of variable carbonate distribution.

Poll (1975) also observed that the amount of gas released depended on the lithological characteristics and type of organic material in the sample, e.g.

calcareous samples gave the highest desorption efficiency. In a geochemical study on the Gippsland Basin, samples were divided into homogeneous sets on the basis of sediment type, coherence, structure, cementation and mineral content. To overcome differences in soil type, the parameter examined was the sum of C_2 to C_5 alkane concentrations divided by the percentage acid-soluble fraction in the soil. This study detected six anomalies in the area, three of which were either weak or incomplete. It was concluded that the three other anomalies, plus one incomplete one, should be subject to a more detailed study to delineate further the anomalies and try to determine whether they were separate entities or formed a pattern related to a common cause.

Devine and Sears (1975, 1977) carried out two soil hydrocarbon surveys in the Cooper Basin, Australia. Soil samples were collected from the areas of the Daralingie and Della gas fields and returned to the laboratory for treatment with acid and analysis of the released gases by gas chromatography. Surface anomalies were detected in both surveys and appeared to be associated with all the known fields traversed and with structures having petroleum potential. The concentration of soil hydrocarbons ranged from 0.2 to 37 ppb for C_2 to C_5 and the values were highest where the hydrocarbon bearing sediments were shallowest and the cap rock was thinnest or absent. In an attempt to eliminate variations in gas concentrations due to changes in mineralogy, the concentrations of C_2 to C_5 divided by the percentage of acid soluble material was also plotted and found to be useful in interpreting some of the data.

Dissolved-gas analysis

Sokolov et al. (1959) pointed out the value of measuring the dissolved hydrocarbon concentrations in formation waters, and Kartsev et al. (1954) briefly described the results of a number of Russian hydrochemical surveys. The most detailed account of a dissolved-hydrocarbon survey given in the literature appears to be that by Stadnik et al. (1977), who studied the near-surface waters above the Pripyat-Dnepr-Donets Basin in Russia, where three different environments were distinguished on the basis of variations in the phase equilibrium of gases between the petroleum deposits (P_g) and the subsurface water (P_w). It was claimed that if $P_g/P_w > 1$ the environment was mainly gaseous, but $P_g/P_w < 0.14$ indicated a poor prospect for either oil or gas, and $0.44 > P_g/P_w > 0.14$ characterized an area with mixed hydrocarbon composition, or an oil and gas prospect. However, this does not seem to be a particularly useful method for performing a fairly rapid geochemical survey over a prospective area.

Zarella et al. (1967) pioneered the use of benzene in subsurface waters from drill holes as a guide to the presence of oil and gas. For a number of hydrocarbon deposits, it was found that benzene concentrations decreased with increasing distance from the deposit and vertical migration of benzene between aquifers was restricted. Shtorgin et al. (1978), following the work of Zarella et al. (1967), determined the benzene content of subsurface waters from Cretaceous and Paleogene rocks in a number of productive, and unproductive, structures of the Crimea. The presence of oil and gas shows was indicated by a benzene concentration of several tenths of a milligram per litre of water. An enhanced benzene concentration was correlated with the presence of hydrocarbon pools. The examples used in this study were all taken from areas that had already been investigated and previously shown to be either productive or dry. As with so many reports in the literature, this method has not been applied to a real exploration problem.

Discussion

Adequate analytical methods are now available for the trace determination of hydrocarbon gases in soils. The problem is how to interpret the analytical results. In any geochemical survey it is usual to collect samples over a grid pattern, plot the results onto the grid and contour the points of similar concentration. Collecting samples over a grid pattern has the trivial advantage of possibly using seismic lines as reference points, but it has a disadvantage in that it assumes blanket emanation of gases from the petroleum accumulation. This does not occur, and it is more practical to use geologic information related to the presence of faults or fractures as a guide in determining the geochemical sampling pattern. The following questions must then be answered. What is the threshold value above which a soil gas concentration is considered anomalous? What is the origin of the anomaly and how is it related by geology and hydrology to an oil or gas reservoir?

The interpretation of the results from geochemical surveys has always been controversial. For example, Smith and Ellis (1963) claimed that results previously obtained by Horvitz (1954) could be explained by gases derived from plant roots, and that the ratio of saturated to unsaturated hydrocarbons was similar for petroleum and plant roots. Some years later Horvitz (1972) replied to these claims by supplying quantitative data, absent from the work of Smith and Ellis (1963), to support his results. These data show that < 1 ppb (by weight) of saturated hydrocarbons is contributed by 1 g of grass or roots to 200 g of soil. This result is much lower than the value normally used for background or for the hydrocarbon content of barren soils. Horvitz (1972) also showed that the saturated to unsaturated hydrocarbon ratios were different in gases from plant roots and from petroleum, and that vegetation in soils does not therefore interfere with the application of hydrocarbon geochemistry to petroleum exploration. However in recent laboratory studies, Hunt et al. (1980) have shown that C_4 to C_7 hydrocarbons can be produced by bacterial degradation of naturally-occurring terpenoids. In the light of these results even greater caution needs to be placed on the interpretation of the origin of surface gases. It is plausible to suggest that microbial degradation of vegetation could produce certain hydrocarbons which in turn would effect the ratios measured by Horvitz (1979).

Of the three types of soil-gas analysis, pore-gas analysis is the most practical and convenient for use directly in the field. The analyses do not depend upon release of the gases from the acid-soluble material in the soil as does adsorbed-gas analysis (Poll, 1975) and the method can be used in regions where the depth of the water table makes dissolved gas analyses impractical.

Avoidance of contamination during collection and storage of samples is of critical importance. Devine and Sears (1977) initially shipped their samples in plastic bags. It was found that on standing for a period of several months, the gas concentrations noticeably decreased. Following this, subsequent soil samples were collected and stored in cans with no noticeable decrease in gas concentration with standing. Oil contamination during mechanical sampling must be avoided; samples should be stored at a low temperature to inhibit microbial growth, and sample containers such as plastic bags should not be used (Smith and Ellis, 1963; Miller, 1976).

Hydrocarbon migration from a petroleum reservoir to the surface may occur by diffusion (due to differences in concentration) or by effusion (due to differences in pressure) or by a combination of both processes (Stegena, 1961; Petukhov, 1977). *Diffusion anomalies* arise from the diffusion of hydrocarbon molecules along the surfaces of mineral crystals and through fluid-filled spaces between minerals. In a uniform geological medium and without hydrological interference this process yields an anomaly centred above the oil or gas deposit in the case of very shallow deposits. Due to the chromatographic effect of the rock, methane may be the only gas from the reservoir capable of reaching the surface. *Effusion anomalies* (filtration anomalies) arise from the migration of gases or water-containing dissolved gases through highly permeable regions such as fractures, cleavages, or fault zones in response to pressure differences. Diffusion of dissolved species along these preferred paths would be much slower than effusion. Effusion is the most plausible method for the migration of significant quantities of hydrocarbons. The problem for geochemical prospecting is that most effusion pathways will involve a large component of lateral movement. In practice, most migration pathways will involve both mechanisms and will be complicated by lateral groundwater movements.

Surface anomalies may consist either of simple maxima or of haloes, i.e. regions with high hydrocarbon concentrations around the periphery and low concentration over the central part. The former are explained by virtually all hypotheses of gas migration (e.g. Hunt, 1979); the latter have puzzled researchers for many years. Horvitz (1939) suggests that mineralization directly over some reservoirs may prevent upward migration of gases. Hedberg (pers. commun. to Hunt, 1979) elaborated on this hypothesis: mineralization of rocks overlying a petroleum reserve might occur due to the upward migration of waters which were originally responsible for transporting petroleum into the trap. An alternative hypothesis is that bacterial activity in the centre of the anomaly leads to a decrease in gas concentration at that point. No combined hydrocarbon/microbiological results have been published to support this view.

A recent paper by Leythauser et al. (1980) has important ramifications for the interpretation of surface prospecting data. A study was made of diffusion of light hydrocarbons through the exposed surfaces of organic-rich shales from West Greenland which, before uplift, had been buried deep enough for hydrocarbon generation. Concentrations of ethane to pentane-range hydrocarbons were found to be depleted within the near-surface 3 m interval due to the diffusion of light hydrocarbons through water saturated pore spaces in the shales.

The reliability of onshore prospecting techniques is still a matter of debate. Early Russian and American workers were optimistic as to the potential of the technique. Devine and Sears (1975, 1977), in two of the few published reports which actually include results from geochemical surveys, were successful in delineating surface anomalies albeit in a known producing area. Other surveys carried out in the Cooper Basin by Philp et al. (in prep.) produced ambiguous results: an anomaly which was apparent one year could not be detected the next. Kvet and Michalicek (1965) concluded, in a survey of geochemical prospecting results in Czechoslovakia, that results did not fulfill expectations and, with one or two exceptions, that only data concerning regional prospects were of any use. This view was shared by Hunt (1979). On the other hand it was recently claimed that geochemical methods outstripped geological and geophysical methods in the discovery of a large gas condensate deposit in Turkmenia in 1976 (Galkin, 1979), and that geochemical methods were of great value in discovering the Veselogorod and Kondrashero deposits in the Ukraine. It therefore appears that the success of onshore geochemical prospecting is highly variable and depends primarily upon the ease with which hydrocarbon migration pathways can be established.

In conclusion, therefore, it has to be said that current opinion on the usefulness of onshore geochemical prospecting is divided. The Russian literature appears to be strongly in favour of this approach and cites various successful examples of this type of prospecting. The remaining literature is not so optimistic, with fewer results, either positive or negative, being published, and with only a few strong supporters of the method. Many of the surveys performed and published to date have been over already proven producing areas and many of the others have probably resulted in failure or produced ambiguous results.

OFFSHORE PROSPECTING BASED ON HYDROCARBON GASES

Introduction

The interpretation of offshore prospecting data is simplified by the fact that there is no near-surface zone of fluid flow in marine sediments. Fluid flow occurs due to the compaction of sediments, but complex fluid movements analogous to the onshore movements of groundwater are absent except in shelf sediments close to shore. Hitchon (1974) described the directions of fluid flows in a thick offshore wedge of sediments and pointed out that, although some deep oil fields may be undetectable due to the downward flowing fluid in a zone of undercompaction, offshore prospecting is, in theory, more likely to be successful than onshore prospecting. Sediment sampling offshore is analogous to soil sampling onshore, and the sampling of near-bottom waters is analogous to the sampling of air at ground level. It is instructive to compare the physical processes affecting offshore and onshore prospecting:

(a) Gaseous diffusion in liquids is several orders of magnitude lower in water than in air. It is likely, therefore, that the surface zone of sediment which is depleted of hydrocarbons by gaseous diffusion will be thinner offshore than onshore.

(b) Temperatures and pressure at the sea bed are more constant than those at ground level on land. Ocean currents at the sea bed are slower than winds and the thermal stratification of the oceans reduces convection currents in ocean waters.

(c) The fluid medium around sediment particles offshore is homogeneous: water containing dissolved materials. Onshore, however, air spaces penetrate to varying depths into the soil until they meet the water table. The air spaces will expand or contract in response to temperature and pressure changes, resulting in bulk movement of air in and out of the soil. This process is absent at the sea floor.

Consideration of these effects indicates:

(a) that sediment sampling offshore should be more reliable as a prospecting guide than soil sampling; and

(b) that sampling of near-bottom water may be a feasible prospecting method, whereas atmospheric air sampling onshore is not.

Factors which may complicate offshore prospecting include the dispersal of dissolved gases by ocean currents, the consumption or generation of hydrocarbons by microbes and the variable effects of bioturbation in producing direct pathways in the sediment for gaseous diffusion. These modifying factors are less severe than those expected for onshore prospecting. In both theory and practice, the most successful offshore techniques have involved the sampling of near-bottom water and the sampling of sediments taken from sub-bottom depths of a few metres.

Offshore geochemical prospecting began around 1960 and is thus a more recent development than onshore prospecting. To date no comprehensive reviews have been published apart from one by Sackett (1977) which concentrated on the use of sniffers in offshore exploration. Offshore prospecting techniques based upon sediment and water analyses show many similarities and are conveniently described together. Methods are discussed in order of increasing complexity of the gas isolation step.

Gas bubble collection

In rare cases, gas bubbles may be collected directly from the water overlying natural seeps or from expansion pockets in sediment cores. Brooks et al. (1974) and Bernard et al. (1976) reported the analysis of gases from 14 natural seeps in the Gulf of Mexico. Samples were collected by snorkel and scuba divers using inverted glass jars. The $C_1/(C_2+C_3)$ ratios of the seep gases ranged from 68 to greater than 1000 and the $\delta^{13}C_{PDB}$ values from -39.9 to -65.5‰. Thermogenic processes produce hydrocarbons with $C_1/(C_2+C_3)$ ratios less than 50 and isotopic values more positive than -50. On this basis 11 of the seeps were ascribed to microbial degradation and three were believed to contain a significant thermocatalytic component. The characterization of hydrocarbon gas sources was discussed further by Bernard et al. (1977). In a review of the chemistry of marine petroleum seeps, Reed and Kaplan (1977) described the analysis of gases from two seeps near Santa Barbara, California. The hydrocarbon compositions and isotope values were used to demonstrate that the gases from these seeps were of thermogenic origin.

Gas expansion pockets from deep cores in the Atlantic and Pacific were sampled by Claypool et al. (1973). The core liner was punctured and a syringe was used to transfer the gas to evacuated glass containers. Simoneit (1978, unpublished results) used a similar procedure to collect gas from shallow cores in the Guaymas Basin, Mexico. Large quantities of ethane and higher hydrocarbon gases were found to result from localized heating of shallow sediments by igneous intrusions (Simoneit et al., 1979).

Kvenvolden et al. (1979) sampled the gases in a submarine seep in Norton Sound, Alaska and found that the major component was carbon dioxide (98% by volume). The isotopic composition of carbon dioxide indicated that it was of igneous origin and it was suggested that the hydrocarbon plume detected previously by Cline and Holmes (1977) arose from hydrocarbon gases swept out of the sediment by the rising flow of carbon dioxide.

Phase equilibration

Hydrocarbon gases are usually present in sea water and marine sediments at $\mu g/l$ concentrations, and more elaborate methods are required for their isolation than gas-bubble collection. A method for the analysis of hydrocarbon gases in aqueous samples was developed by McAuliffe (1969a). The procedure involves equilibration of a water sample with an equal volume of helium and analysis of an aliquot of the gas phase by gas chromatography. The method gives good results for gases such as *n*-alkanes, which partition under these conditions to the extent of 96% or more in the gas phase (McAuliffe, 1966, 1969b). Later, McAuliffe (1971) published a theoretical discussion of the gas chromatographic determination of solutes by multiple-phase equilibration, and described an extension of the method to substances with smaller partition coefficients. The method involves repeated equilibrations of helium with the aqueous sample in a large glass syringe. Gas-chromatographic data on successive equilibrium gas phases are plotted and back-extrapolated to yield the concentration of hydrocarbon in the original aqueous sample. The method is applicable to hydrocarbons containing up to 10 carbon atoms. Williams and Bainbridge (1973) used a procedure based upon that of McAuliffe for the determination of methane, carbon monoxide and hydrogen in surface waters of the South Pacific. A flask was used for the continuous equilibration of air with water. Air was circulated in a closed system through glass frits at the bottom of the equilibrator and sea water was pumped continuously through the equilibrator.

Phase equilibration methods have been applied by a number of authors to the analysis of gases in sediments. McIver (1972; 1973a, b; 1974a, b) reported the analysis of hydrocarbon gases in canned core samples from the Deep Sea Drilling Project. Cans were punctured through a rubber septum clamped to the side. A measured volume of headspace gas was recovered with a syringe and injected into a gas chromatograph. Significant concentrations of non-methane hydrocarbons were found in some surface sediment samples. Rashid et al. (1975) used a similar headspace analysis method for the analysis of methane-rich sediments from the Gulf of St. Lawrence. No thermogenic hydrocarbons were detected and it was concluded that the methane was produced by anaerobic decomposition of the large amount of organic matter in the sediments. Geodekyan et al. (1977) used a phase equilibration method to measure the concentrations of a wide range of gases dissolved in bottom waters of the Indian Ocean and Red Sea. Concentrations of hydrocarbons, hydrogen, helium and carbon dioxide varied with the geological structure of the ocean floor.

Carlisle et al. (1975) reported the results of a geochemical survey based on sediment samples taken from a depth of 1 m. The samples were mixed with helium in a blender on board ship and aliquots of the headspace gases were analysed. Although the location of the survey area was not specified, two thermogenic seeps were found, the larger of which was associated with a fault zone.

Kvenvolden et al. (1981), using a headspace analysis technique, found eight sites in the North Bering Sea where the content of methane rapidly increased with depth in the first 4 m of sediment. The concentration of gases indicated that the interstitial water of the near-surface sediments at these sites were probably gas saturated. The isotopic ratio for methane at four of the sites had a composition clearly indicative of a microbial process (i.e. $\delta^{13}C_{PDB}$ —69 to —80°/oo). At another site the methane had a composition of —36°/oo and was probably derived from thermogenic processes operative at depth in the Norton Basin. High gas concentrations in near-surface sediments can lead to sediment instability and the possibility of sea-floor cratering which in turn could introduce major engineering problems associated with the placement of the drilling structures. Thus the analysis of gases in surface sediments is not only important from a prospecting point of view but also from an engineering standpoint.*

Horvitz (1979), also using a headspace analysis technique, published what appears to be the first report of an offshore hydrocarbon survey which gives data that were collected prior to drilling in an area. This survey involved the collection of 256 sediment samples at a depth of 1.8 m below the sediment surface, offshore Louisiana in the Gulf of Mexico. Methane represented 89% of the light hydrocarbon (C_1 to C_5) gases extracted from the sediment. The $\delta^{13}C_{PDB}$ values for methane lay within the upper range of δ values associated with methane in petroleum gases and it was concluded that these gases were of thermogenic origin. A hydrocarbon anomaly was detected, several exploratory wells were drilled and subsequently a production platform was built in the region.

Vacuum stripping

In 1957, Slobad, Dunlap and Moore applied for a patent, granted in 1959, on "Exploration for Petroliferous Deposits by Locating Oil and Gas Seeps". The patent was issued for an apparatus now called a hydrocarbon "sniffer". The earliest reference to "sniffing" was made by Dunlap et al. (1960). Both these reports described a system utilizing a vacuum pump to continuously extract dissolved gases from a rapidly flowing stream of sea water. Initially, the sea water was collected from a depth of about 3 m and the extracted gas was dried and passed through a non-dispersive infra-red analyser sensitive to methane. Extraction of gas was not quantitative and the analytical procedure could only distinguish methane. Natural seeps were detected up to 9.7 km downcurrent from the source. In a follow-up paper, Dunlap and Hutchinson (1961) described application of the method to prospective drilling locations and demonstrated that the methane sniffing technique could be used to find offshore seeps.

Over the next few years, periodic reports appeared in the commercial literature indicating that "sniffing" was helping to make preliminary evaluations of leases in areas such as the North Sea, off the coasts of Venezuela, Cuba and Libya, and on inland waters of the Northwest Territories in Canada (Anonymous, 1964). Sackett (1977) estimated that in the late 1960s almost every major oil company initiated programs for underwater seep detection. The results of these programs have remained confidential. The only references resulting from this large effort on the part of the oil industry appear to be two printed abstracts of oral presentations (Jeffrey and Zarrella, 1970; Rogers and Edwards, 1975).

During the late 1960s, three improvements were made upon the "sniffer" of Dunlap et al. (1960):

(a) Gas chromatography, rather than infra-red spectrometry, was used to analyse the hydrocarbons stripped from the sea water. This development was

---

*See "Notes added in proof", p. 33, note 2.

important because of the ubiquity of biological methane production and the greater sensitivity of gas-chromatographic techniques.

(b) Surface-water sampling was replaced by near-bottom sampling. Deeper sampling reduced interference from diffusion and from advective currents.

(c) Vacuum stripping was replaced by an air-stripping technique. This change resulted in better than 95% stripping of gases from the water sample (Schink et al., 1971) compared with only 50% with vacuum stripping (Brooks et al., 1973).

Sackett (1977) has reviewed hydrocarbon sniffing in offshore exploration, and Brooks et al. (1973) and Brooks and Sackett (1973) reported extensive analyses in the Gulf of Mexico using a vacuum sniffer coupled to a gas chromatograph. The application of vacuum stripping to the analysis of sediments has been less common. A combination of vacuum stripping and mass spectrometry was applied by Emery and Hogan (1958) to the analysis of methane and several other gases in sediments collected from basins off the coast of southern California. The sediment was mixed with water and transferred to a large evacuated flask; hydrocarbons were collected in a liquid nitrogen trap and analysed by mass spectrometry.

Gérard and Feugère (1969) applied vacuum stripping to the analysis of marine sediment cores (0 to 5 m) collected off the coast of Gabon. The samples were digested at 80°C with hydrochloric acid, hydrocarbon gases were removed by vacuum, purified and analysed by gas chromatography. Some hydrocarbon anomalies were detected and, in contrast to other authors, it was noted that for prospecting purposes the study could have been limited to samples taken from the surface of the sea floor.

Inert-gas stripping

Swinnerton et al. (1962a) described the first application of an inert-gas stripping system to the analysis of dissolved gases in aqueous samples. The essential feature of their method is the use of a sample chamber with a coarse glass frit at the bottom. A small (1 to 2 ml) water sample is introduced onto the frit and a stream of helium gas is allowed to bubble through the frit, rapidly stripping the dissolved gases from the sample and carrying them into a gas chromatograph. The limit of detection (approximately 0.3 $\mu l/l$) is adequate for analysis of gases such as oxygen and nitrogen, which are commonly found at high concentrations, but not for the analysis of hydrocarbons in sea water. A later version of the method (Swinnerton et al., 1962b) using larger (up to 30 ml) samples improved sensitivity by a factor of approximately 15. A modification of this method was used by Atkinson and Richards (1967) for the determination of methane in sulphide-bearing waters from places such as the Cariaco Trench. The method failed to detect methane in open sea water.

The ability of the method of Swinnerton et al. (1962b) to analyse small water samples was applied by Reeburg (1968) to the analysis of gases in the interstitial waters of sediments. Interstitial water was separated from the sediment with a filter press-type sediment squeezer and the dissolved methane, nitrogen and argon concentrations were determined. The same procedure was used by Martens and Berner (1974) for the estimation of methane production in the interstitial waters of sulphate-depleted sediments.

Swinnerton and Linnenbom (1967a) published a more sensitive gas-stripping procedure and also reported the first determination of C_1 to C_4 hydrocarbons in open ocean waters (Swinnerton and Linnenbom, 1967b). The procedure employs a large (1.2 litre) stripping chamber in which hydrocarbons are stripped from solution by means of a stream of helium gas bubbles. The helium is passed through two traps in series. The first, containing activated alumina at $-80°C$, collects all hydrocarbon gases other than methane; the second, containing activated charcoal at $-80°C$, collects methane. The two traps are warmed to $90°C$ and the gases analysed separately by gas chromatography. Separate collection and analysis of the two fractions was necessary to avoid overlap of methane and ethane peaks during chromatography. The technique is capable of determining gaseous hydrocarbons in aqueous solutions at concentrations as low as 1 part in 10^{13} by weight.

Swinnerton and co-workers made extensive application of this procedure to study:

(a) Concentration-depth profiles for methane and ethane plus ethylene (Swinnerton and Linnenbom, 1967b).

(b) Dissolved methane in the surface oceans which was shown to be in equilibrium with the atmosphere in the Western Atlantic (Swinnerton et al., 1969) but not in the North and South Pacific (Lamontagne et al., 1971).

(c) Production of gaseous hydrocarbons from dissolved organic matter (Wilson et al., 1970).

(d) Methane concentrations in a wide range of marine environments (Lamontagne et al., 1973).

(e) C_1 to C_4 hydrocarbons in surface waters of the Pacific Ocean (Lamontagne et al., 1974).

This method was also used for the analysis of hydrocarbons in the Gulf of Mexico (Frank et al., 1970) and southern Beaufort Sea (Macdonald, 1976). Cline and Holmes (1977) used the method for the detection of submarine gas seepage in Norton Sound. Since their chromatographic technique had better resolution than that of Swinnerton and Linnenbom (1967a, b), they were able to employ a single trap of activated alimina at $-196°C$ for collecting hydrocarbons. A simplified version of the original method was used for the analysis of methane in surface waters of the Atlantic Ocean (Scranton and Brewer, 1977).

The most sensitive and elaborate gas-stripping procedure in the literature is that of Grob (1973). Stripping is carried out in a closed system to minimize contamination from the inert gas, and the volatile components are adsorbed onto activated charcoal. Charcoal has the advantage that it selectively adsorbs organic compounds even from moist gas streams. The compounds are removed from the charcoal by extraction with carbon disulphide and analysed by gas chromatography and mass spectrometry. In unpolluted water hundreds of substances up to C_{24} were detected at concentrations down to 1 part in 10^{13} by weight. The method is also applicable to sea water and to sediment-water slurries. The main drawback with the Grob procedure is the difficulty of avoiding contamination. Subsequent papers by Grob and Grob (1974) and Grob et al. (1975) describe the application of the method to potable waters around Zurich and compare the method with a liquid-extraction procedure.

A hydrocarbon sniffer using a gas-stripping system was first described by Schink et al. (1971). The gas-stripping system is the one developed by Williams and Miller (1962) and modified by Schink et al. (1970) for the determination of radon in sea water. The stripping system continuously equilibrates countercurrent streams of sea water and purified air in a small chamber containing a series of rotating discs. Dissolved hydrocarbon gases are transferred to the air stream with greater than 95% efficiency. By comparison, vacuum-stripping systems are only 50% efficient. Introduction of air stripping thus eliminates one of the shortcomings of early sniffers.

The need for near-bottom water sampling when sniffing was emphasized by Schink et al. (1971). On the basis of estimated horizontal diffusion coefficients and vertical eddy diffusion coefficients, it was shown that 1 km downcurrent from a diffusive seep the hydrocarbon concentration 25 m above the sea bottom is 5 orders of magnitude less than at a height of 5 m.

Sigalove and Pearlman (1975) described the application of a "sniffer" incorporating near-bottom sampling and air stripping. The equipment monitors methane, ethane, propane, n- and iso-butane and is offered for service on geophysical prospecting vessels.

Discussion

Hydrocarbon gases in sea water and marine sediments began to receive attention in the late 1960s. This attention was a result of increasing interest in environmental measurements and interest from oil exploration companies who saw the possibility of using marine hydrocarbon measurements as an offshore prospecting guide.

The hydrocarbon sniffers developed by the oil companies were all of similar design. The sniffer consisted of a metal "fish" containing an electric pump and an elaborate sonar system, which was towed behind a ship. Data from the sonar system on the "fish" were compared with shipboard sonar data by means of a small computer which operated an automatic winch for maintaining the fish a few metres above the sea bottom. The water which was continuously pumped to the surface was subsequently analysed for C_1 to C_4 hydrocarbon gases in a number of different ways. One popular method was to warm the water, pass it through a packed column containing a counter-current of air and analyse aliquots of the air at regular intervals by gas chromatography (Sackett, 1977). No results from surveys by these hydrocarbon sniffers have been reported in the literature. The technique is, however, limited to water depths of approximately 300 m due to drag on the cable supporting the "fish". Sniffers do not appear to be in wide use at present since offshore exploration is largely being carried out at water depths in excess of 300 m and petroleum explorers have not developed more sophisticated systems suitable for these depths. It is also difficult to evaluate current usage of sniffer systems since much of this work is proprietary in nature.

The emphasis by exploration companies in water sampling rather than sediment sampling is attributable to the greater ease with which large areas can be surveyed. Two important issues concerning sediment sampling remain to be resolved:

(a) *The optimum sub-bottom sampling depth is subject to debate.* Gérard and Feugère (1969), off the coast of Gabon, found surface sampling as reliable as sampling from a sub-bottom depth of 5 m. Kvenvolden (1978) in a core from Norton Sound, Alaska found anomalous hydrocarbon concentrations below 10 cm. At least one prospecting company (Horvitz, 1979) currently takes samples from a depth of 1 m. The work of Simoneit et al. (1979) in the Guayamas Basin, Mexico showed that gas anomalies did not become obvious in some cases until a sub-bottom depth of approximately 3 m was reached. Optimum sampling depth appears to vary and may be related to the physical properties of the sediment and to factors such as bioturbation. Sampling should be carried out no deeper than is necessary to obtain a reliable indication of the hydrocarbon concentrations but deep enough to minimize surface effects. On that basis an optimum depth of a few metres is indicated.

(b) *The areal extent of gas anomalies is unknown.* Carlisle et al. (1975) indicated that anomalies discovered by them in an unspecified area have a cross-section of approximately up to 2 km, while a typical hydrocarbon anomaly had a cross-section of approximately 100 m. Kvenvolden (1978) found anomalous hydrocarbon levels in only one out of five cores taken in the seep area delineated by Cline and Holmes (1977) on the basis of previous water sample analyses. No comparative study of the sediments and near-bottom water records of any area has yet been published, but it is likely that some sediment anomalies may be too localized to be conveniently and reliably detected by coring. A combination of near-bottom water sampling followed by sediment sampling may prove the most effective technique: water sampling would delineate the area of interest and sediment sampling would pinpoint the seep location.*

PROSPECTING METHODS NOT BASED ON THE DETECTION OF HYDROCARBON GASES

Carbon and oxygen isotope anomalies in surface rocks

Hydrocarbons seeping upwards from petroleum reservoirs are oxidized

---
*See "Notes added in proof", p. 33, note 3.

near the earth's surface. The resulting carbon dioxide reacts with water to produce bicarbonate ions which combine with calcium and magnesium ions in groundwater to produce isotopically distinctive pore-filling cements and surface rocks. If leakage of hydrocarbons is rapid, preferential evaporation of $H_2^{16}O$ from near-surface groundwater may produce changes in oxygen isotopic composition (Nisle, 1941).

Donovan (1974) described striking mineralogic and chemical changes occurring in outcrops of Permian sandstone overlying the oil-productive Cement Anticline in Oklahoma. The sandstone outcrops, which are typically red and friable in the surrounding region, are altered to pink, yellow and white on the flanks of the anticline and to hard carbonate-cemented sandstone above the crest. Calcitized gypsum highly deficient in ^{13}C ($\delta^{13}C_{PDB} = -35^0/oo$) overlies petroleum-productive parts of the anticline near regions where faulting permits vertical migration of fluids. Away from these regions, the influence of hydrocarbons on the isotopic composition of the carbonate cements decreases systematically. Some carbonate cements above the anticline were found to be unusually rich in ^{18}O ($\delta^{18}O_{SMOW} = 35^0/oo$). These cements were interpreted to result from preferential evaporation of $H_2^{16}O$ from pore water into rising bubbles of hydrocarbon gases. Such a process was first suggested by Mills and Wells (1919) and shown to be thermodynamically possible by Nisle (1941).

Donovan and co-workers carried out similar isotope surveys over the Davenport oilfield in Oklahoma and in 1975 over a possible undrilled petroleum deposit near Boulder, Colorado. In dolomitic sandstones over the Davenport oilfield, carbon isotope values ($\delta^{13}C_{PDB} = -5$ to $-11^0/oo$) and oxygen isotope values ($\delta^{18}O_{SMOW} = 29$ to $49^0/oo$) are both heavier than those found over the Cement oilfield. This result suggests more rapid petroleum leakage to the surface with correspondingly less biological oxidation of hydrocarbons and greater evaporation of $H_2^{16}O$ from near-surface groundwaters into the rising gas stream. Near Boulder the carbon and oxygen isotopic anomalies are halo shaped. Both elements become isotopically lighter towards the interior of the halo. $\delta^{13}C_{PDB}$ values range from -3 to $-18^0/oo$ and $\delta^{18}O_{SMOW}$ values from 11 to $24^0/oo$. The halo shape of the anomaly is interpreted to be the result of petroleum seepage around the perimeter of an impermeable cap rock. A review of isotope work by the U.S. Geological Survey is given by Donovan and Dalziel (1977).

Iron and manganese leaching from surface rocks

The passage of hydrocarbons and associated compounds such as hydrogen sulphide through rock strata produces a reducing environment. Consequent reduction and leaching of iron and manganese ions may, therefore, occur in surface rocks overlying petroleum reservoirs. Metal-leaching is most easily detected as a surface decolorization. Changes in colour resulting from the reduction and loss of iron(III) from surface rocks in the presence of petroleum were described in the Apsheron Peninsula and Baku Region of the

USSR (Kartsev et al., 1954, and references therein). Donovan et al. (1974) studied changes in the colour and metal composition of rocks in the same areas where carbon and oxygen isotope studies were conducted. Leaching of iron and manganese from sandstone rocks overlying the Cement oilfield in Oklahoma (Donovan, 1974) was noted visually and later confirmed by metal analyses. There is considerable variation in colour: the normally red sandstones appear pink, yellow or white. Maximum bleaching and metal loss occur along the crestal regions of the anticline. Over the Davenport oilfield (Donovan et al., 1974) examination of rocks for metal-leaching was limited by the availability of permeable metal-bearing rocks. Near Boulder, however, metal-leaching was found to correspond with the halo of anomalous carbon isotope values (Donovan et al., 1975).

Dalziel and Donovan (1980) have discovered anomalously high manganese-to-iron ratios occurring in pine needles and sage leaves over the Recluse oilfield, Wyoming and suggested that they result from the effects of petroleum microseepage on the plants. It was proposed that the magnitude of the plant anomalies, combined with chemical evidence confirming petroleum leakage, made a strong case for the use of plants as a biogeochemical prospecting tool.

Ions associated with petroleum

The following ions have been examined as petroleum indicators, but due to their widespread occurrence are likely to be of little value for prospecting purposes.

Ammonium ions are present at high concentration (100 to 1000 mg/l) in the waters of many oil-bearing sediments (Kartsev et al., 1954; Bogomolov et al., 1970). They are believed to be derived from nitrogeneous compounds originally present in petroleum.

Chloride ions are commonly found at high concentration in the soil overlying a structural uplift (Kartsev et al., 1954). Novosiletskiy et al. (1977) used the ratio of sodium ion concentration to chloride ion concentration as an indicator of petroleum. A value of 1.0 to 0.6 was found to correlate with the presence of an oil or gas accumulation. Interpretation of soil data are, however, complicated by rain and movements of the water table.

Iodide ions in the concentration range 3 to 100 mg/l are usually found in the waters of oil-bearing sediments, whereas fresh groundwaters contain only 10^{-5} to 10^{-3} µg/l and underground saline waters only 10^{-1} to 1 mg/l (Kartsev et al., 1954). Hitchon (1974) reporting a study on the western Canada sedimentary basin found a statistical difference in the iodide concentrations between stratigraphic intervals which were oil- or gas-producing and those which were not.

Naphthenates are surface-active anions of branched cyclic carboxylic acids containing between 10 and 15 carbon atoms. They have been found with phenols in the waters of oil-bearing sediments, but not elsewhere (Kartsev et al., 1954). In model experiments, Norenkova et al. (1978) showed that naphthenic acids can be produced by microorganisms from naphthenic petroleum over a period of 39 months. Naphthenic acids have been observed in the stratal waters of petroleum deposits on previous occasions (Bars and Glezer, 1960). These recent results indicate that the naphthenic acids in groundwaters may be related to the presence of a naphthenic-type petroleum as previously discussed by Davis (1967).

Sulphide ions and other reduced sulphur species are formed by the bacterial reduction of sulphate ions under anaerobic conditions. They are indicators of a reducing environment such as that found near petroleum deposits.

Helium detection

Helium is produced by radioactive decay processes deep within the earth and accumulates with petroleum in reservoirs below rock strata of low permeability. Being inert and of low atomic mass, helium readily diffuses through fluid-filled fissures in rocks and soil and thus the surface distribution of helium can indicate the presence of underground helium reservoirs in which petroleum may also be found.

The use of helium in mineral exploration was reviewed by Dyck (1976). Limited work by Russian scientists and by the Canadian Geological Survey has indicated that:

(a) the helium content of soil gas from a depth of less than 1 m was variable and indistinguishable from that of atmospheric air;

(b) helium concentrations became stable and meaningfuly only below 50 m.

Roberts et al. (1976) reported more encouraging results for the analysis of helium in soil gas near Boulder, Colorado. A helium distribution was found which agreed with the isotope and metal distributions of Donovan et al. (1975) for the same area. The presence of a halo of gas seepage was not, however, as clearly indicated. The results were consistent with seepage of helium through the top of a cap rock as well as around its perimeter.

In the procedure of Roberts et al. (1976), soil samples were collected from a depth of 0.5 m, sealed in aluminium core collectors and transported to the laboratory. Samples of the headspace gas were then analysed by mass spectrometry. Helium concentrations ranged from 5.2 μl/l air (typical in normal atmospheric air) to 8.5 μl/l air.

The same procedure was used by Roberts and co-workers (U.S. Geological Survey, Denver) to demonstrate helium anomalies over the Harley Dome in Utah, the Cement oilfield in Oklahoma, the Redwing Creek field in North Dakota and the Rubelsanto oilfield in Guatemala (Roberts, unpublished results). Current work by this group involves the testing of a soil-gas sampler developed for helium detection during uranium exploration (Reimer, 1976a, b; Reimer et al., 1976, 1979; Reimer and Otton, 1976; Reimer and

Adkisson, 1977; Reimer and Rice, 1977). The sampler consists of a long tube which is driven into the ground. A rod is then inserted to reduce the dead volume of the system. The top of the tube is equipped with a septum through which soil-gas samples are removed by syringe and analysed by a portable mass spectrometer.

Radon-222

Metals such as uranium are believed to accumulate in petroleum deposits due to the presence of a reducing environment and the possibility of chelation by organic molecules. Radon-222 is one of the products in the nuclear decay chain of uranium-235. It has a half life of only 3.8 days and should not be detectable more than 40 to 60 m from its origin if diffusion alone is responsible for its dispersal (Dyck and Smith, 1969). Radon-222 anomalies have, however, been associated with oil and gas deposits for many years (Dyck, 1968) and are presumably due to the bulk migration of groundwater to the surface. Gates and McEldowney (1977) suggested that pressurized reservoirs should cause an increased rate of diffusion of radon to the surface and hence give rise to radon anomalies which could be easily detected by methods originally developed for uranium exploration (Dyck, 1969a, b). Relatively little interest appears to have been taken in the technique.

Microbiological methods

Microbiological prospecting is based upon the detection of microorganisms capable of consuming hydrocarbon gases such as methane, ethane, propane and butane. Davis (1967) summarized the microbial petroleum prospecting methods which had been described in the literature up to that time. The various methods available for the detection of hydrocarbon-utilizing bacteria are discussed in some detail. These methods include plate counting, utilization of radioactive ethane, and determination of rate of oxidation of intermediate alcohols. Although methane-oxidizing bacteria have been used as an index of petroleum-gas seepage, it appears that the determination of the ethane-oxidizing bacteria is far more specific in locating possible petroleum prospects. Brisbane and Ladd (1965, 1972) reviewed the literature on microbiological prospecting for petroleum and concluded that differences in growth associated with different soils introduced so much variability (Brisbane and Ladd, 1968) that the method was unlikely to be commercially successful.

Sealey (1974a, b), in publicizing a microbiological prospecting service, presented a brighter picture. In his results from 196 producing areas, he obtained 163 positive and 33 negative results; from 160 dry holes he obtained 125 negative and 35 positive. When taken together this gives an overall correlation factor of 80% (in a second series he obtained an 86%

correlation). From a series of 89 wildcat locations tested, 24 were positive and 65 negative. Of the 24 positive, 13 produced oil or gas and 11 were abandoned (54% correlation); of the 65 negative, 55 were reported dry and abandoned, 5 showed and 5 produced (82% correlation). No details, however, were given of the culturing technique.

Miller (1976) described a microbial survey, where microorganisms from soil samples were cultured on the basis of their ability to consume gases such as methane, ethane, propane and butane. Different organisms took different times to react to the gases. In one instance, ethane required 23 days compared to 7 days for methane. The number of hydrocarbon consumers in the soil was taken as an indicator of the hydrocarbon flux through the soil. A good correlation was found to exist over the Bell Creek field and the Recluse field in Wyoming.

The main problem with microbial methods is to distinguish between organisms which are normally consuming hydrocarbons for their basic metabolism and those which, under laboratory conditions, merely switch over to hydrocarbon metabolism. Measurement of the time required to start metabolizing hydrocarbons may be a useful indicator here. Other factors which may distort microbiological results are soil type, humus content, moisture and temperature. One of the major disadvantages with microbiological methods of prospecting is the time involved in culturing the bacteria and monitoring the rates of hydrocarbon oxidation. For this reason alone, it seems unlikely that this technique will ever become an important prospecting tool.

Dissolved nitrogen

Zorkin et al. (1976) investigated the geochemistry of nitrogen in the groundwaters of oil- and gas-bearing basins. Nitrogen concentrations were found to increase towards the centres of the basins and also towards ancient intraplatform depressions. Higher nitrogen concentrations were found in the oil-bearing regions.

Getz (1977) pointed out that methane (80%) and ammonia (15%) are the principal gases released during the maturation of coal, and suggested that ammonia may react with iron(III) ions to form nitrogen which may migrate to reservoirs inaccessible to methane. It was therefore proposed that nitrogen-filled reservoirs could indicate the presence of methane-filled reservoirs in the same vicinity.

Very low-frequency electromagnetic induction

Johnson (1978) observed that surface mineralization anomalies for elements such as vanadium, iron, manganese, cobalt, nickel, copper and uranium occur around the perimeter of oil and gas deposits. The measured anomaly to background ratios for these elements vary from 2 to 6. Concentration of these elements alone would result in a bulk mineralization anomaly of 1 to 3%. It was proposed that these anomalies result from horizontal deflection of vertically migrating mineralized water by an oil or gas deposit. As this mineralized water moves vertically towards the surface, physiochemical variations cause the precipitation of much of the transported material. Because the vertically migrating water that is peripheral to the oil or gas deposit is more abundant than either over or considerably away from the deposit, an anomalously mineralized halo is formed. The method proposed for detecting such anomalies is very low-frequency electromagnetic induction. This method is based on the induction of electric currents in buried conductors (alteration zones) by the effects of electromagnetic waves that are initiated either at or above the surface of the earth. Johnson (1978) detected well-defined electromagnetic anomalies for all the 58 producing oil and gas fields examined in the USA.

Remote laser spectrometry

Biryulin et al. (1981) have recently published a preliminary assessment using a laser unit to measure the aureoles of methane diffusing into atmospheric air at ground level above possible oil and gas deposits. It was observed that over a three-day period the surface methane concentrations just before dawn over certain oil- and gas-producing horizons in a region adjoining Slavyansk, Russia, were twice the day-time level. It was also claimed that a methane aureole could be detected over the Anastasiyevka-Troitskoye oilfield above the background value of methane. Although it would appear from this article that methane anomalies are being detected, it has not been proved that biogenic methane is not responsible for the anomalies. As with so many other studies, this survey has been performed over a producing field and untested prospects have not been examined.

Landsat imagery and airborne detection methods

Marrs and Kaminsky (1977) mentioned the use of Landsat imagery for the detection of petroleum-related soil anomalies. They state that efforts by researchers to establish a relationship between tonal and spectral patterns interpreted from Landsat imagery and hydrocarbon reservoirs have proved inconclusive. Field mapping, field spectral measurements, geochemical analyses and statistical analyses of spectral and geochemical values neither confirm the presence of hydrocarbon-related soil anomalies nor define unique characteristics of anomalous areas.

Donovan et al. (1979a) carried out direct detection experiments at the Cement and Garza oilfields, Oklahoma and Texas, respectively, using enhanced Landsat I and II images. The aim of these experiments was to determine whether enhanced Landsat images could discriminate the surface alteration zones in bedrock at the Cement oilfield. It was concluded that although the Cement and Garza oilfields are geologically well suited for a Landsat direct-detection experiment, their climatological and geographic setting combined against marked success because of the spatial and spectral resolution limitations of the imaging system. In a later study Donovan et al. (1979b) detected high wave-number anomalies on profiles from airborne magnetic surveys over the Cement oilfield, Oklahoma. These anomalies appeared to correlate with the near-surface diagenetic formation of magnetite as a direct result of hydrocarbon microseepage from underlying reservoirs (Donovan, 1974).*

Watson et al. (1978) have shown that accurate measurements of the areal extent of oil slicks from selected seeps in the Santa Barbara Channel, California can be achieved using a Fraunhofer Line Discriminator operating in an imaging mode. Although this was developed mainly as a pollution-monitoring technique, it also has potential use as a geochemical prospecting tool in the detection of natural oil seeps.

Halbouty (1980) demonstrated the value of satellite images of 15 giant fields in various parts of the world. Interpretation of the images of each field delineates the principle geological features, and thus areas that should be investigated further by geophysical surveys. In common with most other articles, the work concentrated on areas where fields had already been discovered. It would have been extremely useful to have included prospects where the satellite image might serve as a starting point for further exploration.

CONCLUSION

The literature on geochemical prospecting is frequently contradictory and confused. There is a lack of vigour in many descriptions, much pertinent data from the major oil companies have not been published and some data (usually the most optimistic) are provided by companies offering geochemical prospecting services. On the basis of the inadequate data presently available and the authors' personal experience, the following generalizations are offered:

(a) Onshore prospecting using hydrocarbon gases is a difficult technique due to the likelihood of lateral gas migration, perhaps for many hundred kilometres. The principal reasons for this appear to be the presence of aquifers, faults and fracture zones which influence the upward migration of gas bubbles or water containing dissolved gases. No study has yet been published wich adequately takes these factors into account. It is likely that these hydrological and geological factors will prove too complex to unravel in many situations. The application of onshore prospecting should at present be restricted to areas where the hydrology and geology can be determined, and far more attention should be devoted to these factors.

(b) Offshore prospecting using hydrocarbon gases offers the advantage

---

*See "Notes added in proof", p. 33, note 4.

that the lateral movement of water is much reduced; there is no near-surface zone of fluid flow analogous to the onshore movement of groundwater. Many gaseous seeps have been detected offshore and success is claimed for the hydrocarbon "sniffers" developed by the major oil companies in the 1960s, although little evidence has yet been published to support this. Although offshore sediments have simpler hydrology, geological factors such as faulting must be closely considered. It is not known how successfully observed seeps have been related to hydrocarbon accumulations. Sniffing of near-bottom water appears best for reconnaissance purposes; sediment sampling should be used to confirm sniffing results and pinpoint seep locations. The present lull in offshore sniffing is reportedly due to the inability of sniffers currently available to operate at depths greater than 300 m where most offshore exploration is now occurring.

(c) Other prospecting techniques are in an even more experimental stage of development than those based upon hydrocarbon detection. Measurements of isotope anomalies, helium anomalies and metal leaching in surface rocks give good results over some known oilfields, but their prospecting value remains to be determined and their application will be restricted to areas of suitable geology. Methods based upon the measurement of radon or dissolved nitrogen have barely been examined. Microbiological methods are complicated by the need to consider facultative conversion of bacteria to hydrocarbon metabolism; methods to overcome this are likely to make the technique too tedious for practical use. Methods based on very low-frequency electromagnetic induction, Landsat imagery and airborne techniques probably lack the specificity required for petroleum detection.

In the future it would appear that the use of carbon isotope measurement of surface methane will play an important role in geochemical prospecting. Techniques are becoming available to make such measurements on small quantities of methane, enabling a distinction to be made between biogenic and thermogenic surface gases.

It has been demonstrated that the accumulation of hydrocarbons at depth produces a variety of surface chemical phenomena which can be detected successfully. Failures in prospecting are attributable to the simplistic manner in which data have been interpreted: insufficient attention has been paid to the hydrological and geological factors which modify the upward migration of indicator species to the surface.

ACKNOWLEDGMENTS

The authors wish to thank W.R. Ryall, L. Horvitz, T.J. Donovan and J.M. Hunt for reading this manuscript and making many useful and constructive criticisms of its contents. One of the authors (P.T.C.) was supported during this research by a CSIRO Postdoctoral Research Fellowship and he thanks I.R. Kaplan for his assistance during this period.

REFERENCES

Anonymous, 1964. Searchers "sniffing" for North Sea gas. Oil Gas J., 64: 93.

Atkinson, L.P. and Richards, F.A., 1967. The occurrence and distribution of methane in the marine environment. Deep-Sea Res., 14: 673—684.

Bars, E.A. and Glezer, V.G., 1960. Hydrogeochemical survey in the Archangel region of the Bashir, SSSR. Tr. Geol. Inst., Akad. Nauk SSSR, 1: 314—327.

Bernard, B.B., Brooks, J.M. and Sackett, W.M., 1976. Natural gas seepage in the Gulf of Mexico. Earth Planet. Sci. Lett., 31: 48—54.

Bernard, B.B., Brooks, J.M. and Sackett, W.M., 1977. A geochemical model for characterization of hydrocarbon gas sources in marine sediments. Proc. 9th Offshore Technology Conference, pp. 435—438.

Biryulin, V.P., Golubev, O.A., Mironov, V.D., Popov, A.I., Nazarov, I.M. and Fridman, Sh.D., 1981. Geochemical prospecting for oil and gas by remote laser spectrometry of methane in air at ground level. Int. Geol. Rev., 23: 679—683.

Bogomolov, G.C., Kudel'skiy, A.V. and Kozlov, M.F., 1970. The ammonium ion as an indicator of oil and gas. Dokl. Akad. Nauk SSSR, 195: 938—940.

Boyle, R.W. and Garrett, R.G., 1970. Geochemical prospecting — a review of its status and future. Earth-Sci. Rev., 6: 51—75.

Bray, E.E., 1956a. Geochemical exploration methods. U.S. Pat., 2,742,575.

Bray, E.E., 1956b. Method of geochemical prospecting. U.S. Pat., 2,773,991.

Brisbane, P.G. and Ladd, J.N., 1965. The role of microorganisms in petroleum exploration. Annu. Rev. Microbiol., 19: 351—364.

Brisbane, P.G. and Ladd, J.N., 1968. The utilization of methane, ethane and propane by soil micro-organisms. J. Gen. Appl. Microbiol., 14: 447—450.

Brisbane, P.G. and Ladd, J.N., 1972. Growth of *Mycobacterium paraffinicum* on low concentrations of hydrocarbons. J. Appl. Bacteriol., 35: 659—665.

Brooks, J.M. and Sackett, W.M., 1973. Sources, sinks and concentrations of light hydrocarbons in the Gulf of Mexico. J. Geophys. Res., 78: 5248—5258.

Brooks, J.M., Fredricks, A.D., Sackett, W.M. and Swinnerton, J.W., 1973. Baseline concentrations of light hydrocarbons in the Gulf of Mexico. Environ. Sci. Technol., 7: 639—642.

Brooks, J.M., Gormly, J.R. and Sackett, W.M., 1974. Molecular and isotopic composition of two seep gases from the Gulf of Mexico. Geophys. Res. Lett., 1: 213—216.

Carlisle, C.T., Bayliss, G.S. and van Delinder, D.G., 1975. Distribution of light hydrocarbons in seafloor sediment: correlations between geochemistry, seismic structure and possible reservoired oil and gas. Proc. 7th Offshore Technology Conference, OTC2341, pp. 65—70.

Chilingar, G.V. and Karim, M., 1962. Gaseous survey methods in exploration and prospecting for oil and gas: a review. Alberta Soc. Pet. Geol. J., 10: 610—617.

Claypool, G.E., Presley, B.J. and Kaplan, I.R., 1973. Gas analysis in sediment samples from Legs 10, 11, 13, 14, 15, 18, 19. In: G.S. Creager et al. (Editors), Initial Reports of the Deep Sea Drilling Project. U.S. Government Printing Office, Washington, D.D., Vol. XIX, pp. 879—884.

Cline, J.C. and Holmes, M.L., 1977. Submarine seepage of natural gas in Norton Sound, Alaska. Science, 198: 1149—1153.

Coleman, D.D., Meents, W.F., Liu, C-L. and Keogh, R.A., 1977. Isotopic identification of leakage gas from underground storage reservoirs — a progress report. Ill. Pet., 111: 1—10.

Dalziel, M.C. and Donovan, T.J., 1980. Biogeochemical evidence for subsurface hydrocarbon occurrence, Recluse Oil Field, Wyoming: preliminary results. U.S. Geol. Surv. Circ., 837: 11 pp.

Davis, J.B., 1967. Petroleum Microbiology. Elsevier, Amsterdam, 604 pp.

Davis, J.B. and Squires, R.M., 1954. Detection of microbially produced gaseous hydrocarbons other than methane. Science, 119: 381—382.

Debnam, A.H., 1965. Field and laboratory methods used by the Geological Survey of Canada in geochemical surveys. No. 6. Determination of hydrocarbons in soils by gas chromatography. Geol. Surv. Can., Pap., 64-15: 17 pp.

Debnam, A.H., 1969. Geochemical prospecting for petroleum and natural gas in Canada. Geol. Surv. Can., Bull., 177: 26 pp.

Degoyler, E., 1940. Future position of petroleum geology in the oil industry. Bull. Am. Assoc. Pet. Geol., 24: 1389—1399.

Devine, S.B. and Sears, H.W., 1975. An experiment in soil geochemical prospecting for petroleum, Della Gas Field, Cooper Basin. APEA J., 15: 103—110.

Devine, S.B. and Sears, H.W., 1977. Soil hydrocarbon geochemistry, a potential petroleum exploration tool in the Cooper Basin, Australia. J. Geochem. Explor., 8: 397—414.

Donovan, T.J., 1974. Petroleum microseepage at Cement, Oklahoma: evidence and mechanism. Bull. Am. Assoc. Pet. Geol., 58: 429—446.

Donovan, T.J. and Dalziel, M.C., 1977. Late diagenetic indicators of buried oil and gas. U.S. Geol. Surv., Open-File Rep., 77-817.

Donovan, T.J., Friedman, I. and Gleason, J.D., 1974. Recognition of petroleum-bearing traps by unusual isotopic compositions of carbonate-cemented surface rocks. Geology, 2: 351—354.

Donovan, T.J., Noble, R.L. Friedman, I. and Gleason, J.D., 1975. A possible petroleum-related geochemical anomaly in surface rocks, Boulder and Weld Counties, Colorado, U.S. Geol. Surv., Open-File Rep., 75-47.

Donovan, T.J., Termain, P.A. and Henry, M.E., 1979a. Late diagenetic indicators of buried oil and gas: II. Direct detection experiment at Cement and Garza oil fields, Oklahoma and Texas, using enhanced LANDSAT I and II images. U.S. Geol. Surv., Open-File Rep., 79-243.

Donovan, T.J., Forgey, R.L. and Roberts, A.A., 1979b. Aeromagnetic detection of diagenetic magnetite over oil fields. Bull. Am. Assoc. Pet. Geol., 63: 245—248.

Dunlap, H.F. and Hutchinson, C.A., 1961. Marine seep detection. Offshore, 14: 11—12.

Dunlap, H.F., Bradley, J.S. and Moore, T.F., 1960. Marine seep detection — a new reconnaissance method. Geophysics, 25: 275—282.

Dyck, W., 1968. Radon-222 emanations from a uranium deposit. Econ. Geol., 63: 288—291.

Dyck, W., 1969a. Field and laboratory methods used by the geological survey of Canada in geochemical surveys. No. 10 — Radon determination apparatus for geochemical prospecting for uranium. Geol. Surv. Can., Pap., 68-21.

Dyck, W., 1969b. Development of uranium exploration methods using radon. Geol. Surv. Can., Pap., 69-46.

Dyck, W., 1976. The use of helium in mineral exploration. J. Geochem. Explor., 5: 3—20.

Dyck, W. and Smith, A.Y., 1969. The use of radon-222 in surface waters in geochemical prospecting for uranium. Colo. Sch. Mines, Mag., 64: 223—236.

Emery, K.O. and Hogan, D., 1958. Gases in marine sediments. Bull. Am. Assoc. Pet. Geol., 42, 2174—2185.

Frank, D.J., Sackett, W., Hall, R. and Fredricks, A., 1970. Methane, ethane and propane concentrations in the Gulf of Mexico. Bull. Am. Assoc. Pet. Geol., 54: 1933—1938.

Freeze, R.A. and Witherspoon, P.A., 1966. Theoretical analysis of regional groundwater flow, Part 1. Water Res., 2: 641—656.

Freeze, R.A. and Witherspoon, P.A., 1967. Theoretical analysis of regional groundwater flow, Part 2. Water Res., 3: 623—634.

Freeze, R.A. and Witherspoon, P.A., 1968. Theoretical analysis of regional groundwater flow, Part 3. Water Res., 4: 581—590.

Galkin, Y., 1979. By the smell to mineral deposits. Press release U.S.S.R. Embassy, Canberra, IEN/0 3190, pp. 1—2 (APN).

Gates, T.M. and McEldowney, R.C., 1977. Uranium exploration method may help find gas and oil. World Oil, 184: 55—57.

Geodekyan, A.A., Arulo, V.I. and Bordovskiy, O.K., 1977. Gases in bottom water of the north western part of the Indian Ocean. Dokl. Akad. Nauk SSSR, 237: 1483—1485 (English translation in Geochemistry, 17: 241—243, 1980).

Gérard, R.E. and Feugère, G., 1969. Results of an experimental offshore geochemical prospecting study. In: P.A. Schenck and I. Havenaar (Editors), Advances in Organic Geochemistry 1968. Pergamon, Oxford, pp. 355—371.

Getz, F.A., 1977. Molecular nitrogen: clue in coal-derived methane hunt. Oil Gas J., 75: 220—221.

Grob, K., 1973. Organic substances in potable water and its precursor. Part I: Methods for their determination by gas-liquid chromatography. J. Chromatogr., 84: 255—273.

Grob, K. and Grob, G., 1974. Organic substances in potable water and its precursor. Part II: Applications to the area of Zurich. J. Chromatogr., 90: 303—313.

Grob, K., Grob, K. Jr. and Grob, G., 1975. Organic substances in potable water and its precursor. Part III: The closed loop stripping procedure compared with rapid liquid extraction. J. Chromatogr., 106: 299—315.

Halbouty, M.T., 1980. Geological significance of Landsat data for 15 giant oil and gas fields. Bull. Am. Assoc. Pet. Geol., 64: 8—37.

Hitchon, B., 1974. Application of geochemistry to the search for crude oil. In: A.A. Levinson (Editor), Introduction to Exploration Geochemistry. Applied Publishing Ltd, Wilmette, Ill., Ch. 13.

Horvitz, L., 1939. On geochemical prospecting. Geophysics, 4: 210—225.

Horvitz, L., 1954. Near surface hydrocarbons and petroleum accumulation at depth. Min. Eng., 1205—1209.

Horvitz, L., 1959. Geochemical prospecting for petroleum. 20th Int. Geol. Congr., Symposium de Exploración Geoquimíca, 2: 303—319.

Horvitz, L., 1969. Hydrocarbon geochemical prospecting after 30 years. In: W.B. Heroy (Editor), Unconventional Methods in Exploration for Petroleum and Natural Gas. Southern Methodist University Press, Dallas, Texas, pp. 205—215.

Horvitz, L., 1972. Vegetation and geochemical prospecting for petroleum. Bull. Am. Assoc. Pet. Geol., 56: 925—940.

Horvitz, L., 1979. Near surface evidence of hydrocarbon movement from depth. In: Problems of Petroleum Migration. Am. Assoc. Pet. Geol. Stud. Geol., 10: 241—270.

Hunt, J.M., 1979. Petroleum Geochemistry and Geology. W.H. Freeman and Company, San Francisco, Calif., 617 pp.

Hunt, J.M., Miller, R.J. and Whelan, J.K., 1980. Formation of C_4—C_7 hydrocarbons from bacterial degradation of naturally occurring terpenoids. Nature, 288: 577—578.

Jeffrey, D.A. and Zarella, W.M., 1970. Geochemical prospecting at sea. Bull. Am. Assoc. Pet. Geol., 54: 853—854.

Johnson, A.C., 1978. V.L.F. - electromagenetic induction aids exploration. Oil Gas J., 76: 168—169.

Karim, M.F., 1964. Some Geochemical Methods of Prospecting and Exploration for Oil and Gas. Ph.D. Thesis, University of Southern California, University Microfilms Inc.

Kartsev, A.A., Tabasaranskii, Z.A., Subbota, M.I. and Mogilevskii, G.A., 1954. Geochemical methods of prospecting and exploration for petroleum and natural gas. State Scientific and Technical Publishing House of Petroleum and Mineral Fuel Literature, Moscow, 430 pp. (English translation by Witherspoon, P.A. and Romey, W.D., 1959. University of California Press, 349 pp.).

Kroepelin, H., 1975. Geochemical prospecting. Proc. 7th World Petroleum Congress, 1B: 37—57.

Kvenvolden, K.A., 1978. Geochemical analyses of a possible petroleum gas seep in Norton Basin. 2nd USGS Petroleum Research and Resources Seminar, Golden, Colorado.

Kvenvolden, K.A., Weliky, K., Nelson, H. and Des Marais, D.J., 1979. Submarine seep of carbon dioxide in Norton Sound, Alaska. Science, 205: 1264—1266.

Kvenvolden, K.A., Redden, G.D., Thor, D.R. and Nelson, C.H., 1981. Hydrocarbon gases in near-surface sediment of northern Bering Sea (Norton Sound and Chirikov Basin). In: Eastern Bering Sea Shelf Oceanography and Resources (in press).

Kvet, R. and Michalicek, M., 1965. Results and perspectives of the application of geochemical surface prospecting for bitumen in the CSSR. Geochem. Cesk., 1: 287—294.

Lamontagne, R.A., Swinnerton, J.W. and Linnenbom, V.J., 1971. Nonequilibrium of carbon monoxide and methane at the air-sea interface. J. Geophys. Res., 76: 5117—5121.

Lamontagne, R.A., Swinnerton, J.W., Linnenbom, V.J. and Smith, W.D., 1973. Methane concentrations in various marine environments. J. Geophys. Res., 78: 5317—5324.

Lamontagne, R.A., Swinnerton, J.W. and Linnenbom, V.J., 1974. C_1 to C_4 hydrocarbons in the North and South Pacific. Tellus, 26: 71—77.

Laubmeyer, G., 1933. A new geophysical prospecting method, especially for deposits of hydrocarbons. Petroleum, 29: 1—4.

Leythauser, D., Schaefer, R.G. and Yukler, A., 1980. Diffusion of light hydrocarbons through near-surface rocks. Nature, 284: 522—525.

Link, W.K., 1952. Significance of oil and gas seeps in world oil exploration. Bull. Am. Assoc. Pet. Geol., 36: 1505—1540.

McAuliffe, C., 1966. Solubility in water of paraffin, cycloparaffin, olefin, acetylene, cycloolefin and aromatic hydrocarbons. J. Phys. Chem., 70: 1267—1275.

McAuliffe, C., 1969a. Determination of dissolved hydrocarbons in subsurface brines. Chem. Geol., 4: 225—233.

McAuliffe, C., 1969b. Solubility in water of normal C_9 and C_{10} hydrocarbons. Science, 163: 478—479.

McAuliffe, C., 1971. GC determination of solutes by multiple phase equilibration. Chem. Tech., 1: 46—51.

McCrossan, R.G., Ball, N.L. and Snowdon, L.R., 1972. An evaluation of surface geochemical prospecting for petroleum, Olds — Caroline area, Alberta. Geol. Surv. Can., Pap., 71-31: 101 pp.

MacDonald, R.W., 1976. Distribution of low-molecular-weight hydrocarbons in Southern Beaufort Sea. Environ. Sci. Technol., 10: 1241—1246.

McIver, R.D., 1972. Geochemical significance of gas and gasoline-range hydrocarbons and other organic matter in a Miocene sample from site 134 — Balearia abyssal plain. In: W.B.F. Ryan et al. (Editors), Initial Reports of the Deep Sea Drilling Project. U.S. Government Printing Office, Washington, D.C., 13: 813—817.

McIver, R.D., 1973a. Hydrocarbons in canned muds from sites 185, 186, 198 and 191 — leg 19. In: J.S. Creager et al. (Editors), Initial Reports of the Deep Sea Drilling Project. U.S. Government Printing Office, Washington, D.C., 19: 875—877.

McIver, R.D., 1973b. Hydrocarbon gases from canned core samples, sites 174A, 176 and 180. In: L.D. Kuhn et al. (Editors), Initial Reports of the Deep Sea Drilling Project. U.S. Government Printing Office, Washington, D.C., 18: 1013—1014.

McIver, R.D., 1974a. Residual gas contents of organic-rich canned sediment samples from leg 23. In: R.B. Whitmarsh et al. (Editors), Initial Reports of the Deep Sea Drilling Project. U.S. Government Printing Office, Washington, D.C. 23: 971—972.

McIver, R.D., 1974b. Evidence of migrating liquid hydrocarbons in deep sea drilling project cores. Bull. Am. Assoc. Pet. Geol., 58: 1263—1271.

Marrs, R.W. and Kaminsky, B., 1977. Detection of petroleum-related soil anomalies from LANDSAT. Bull. Am. Assoc. Pet. Geol., 61: 7.

Martens, C.S. and Berner, R.A., 1974. Methane production in the interstitial waters of sulphate-depleted marine sediments. Science, 185: 1167—1169.

Meyer, R.F., 1965. New methods for oil exploration. Anal. Chem. 37: 27A—37A.

Miller, G.H., 1976. Microbial surveys help evaluate geological, geophysical prospects. Oil Gas J., 74: 192—202.

Mills, R.V.A. and Wells, R.C., 1919. The evaporation and concentration of waters associated with petroleum and natural gas. U.S. Geol. Surv., Bull., 693.

Motojima, K., 1975. Geochemical prospecting for petroleum and natural gas deposits. Reg. Min. Res. Dev. Cent., Advis. Text No. 9, March.

Nisle, R.G., 1941. Considerations of the vertical migration of gases. Geophysics, 6: 449—454.

Norenkova, I.K., Arkhangel'Skaya, R.A. and Tarasova, T.G., 1978. Water soluble organic substances produced by microbial oxidation of petroleum. Geokhimiya, 3: 408—414. (English translation Geochem. Int., 15: 54—62, 1978).

Novosiletskiy, R.M., Savka, Ye.P. and Sharum, D.V., 1977. An estimate of prospects for petroleum occurrences in sedimentary rocks based on the metamorphic grade of the formation waters. Int. Geol. Rev., 19: 1396—1398.

Petukhov, A.V., 1977. Principal structural elements of the field of concentrations of hydrocarbon gases. Dokl. Acad. Sci. USSR, Earth Sci. Sect., 233: 189—191.

Pirson, S.J., 1940. Critical survey of recent developments in geochemical prospecting. Bull. Am. Assoc. Pet. Geol., 24: 1464—1474.

Poll, J.J.K., 1975. Onshore Gippsland geochemical survey. A test case for Australia. APEA J., 15: 93—101.

Rashid, M.A., Vilks, G. and Leonard, J.D., 1975. Geological environment of a methane-rich sedimentary basin. Chem. Geol., 15: 83—96.

Reeburg, W.A., 1968. Determination of gases in sediments. Environ. Sci. Technol., 2: 140—141.

Reed, W.E. and Kaplan, I.R., 1977. The chemistry of marine petroleum seeps. J. Geochem. Explor., 7: 255—293.

Reimer, G.M., 1976a. Design and assembly of a portable helium detector for evaluation as a uranium exploration instrument. U.S. Geol. Surv., Open-File Rep., 76-398.

Reimer, G.M., 1976b. Helium detection as a guide for uranium exploration. U.S. Geol. Surv., Open-File Rep., 76-240.

Reimer, G.M. and Adkisson, C.W., 1977. Reconnaissance survey of the helium content of soil gas in Black Hawk, Eldorado Springs, Golden, Morrison, Ralston Buttes and Squaw Pass quadrangles, Colorado. U.S. Geol. Surv., Open-File Rep., 77-464.

Reimer, G.M. and Otton, J.K., 1976. Helium in soil gas and well water in the vicinity of a uranium deposit, Weld County, Colorado. U.S. Geol. Surv., Open-File Rep., 76-699.

Reimer, G.M. and Rice, R.S., 1977. Linear-traverse surveys of helium and radon in soil gas as a guide for uranium exploration, Central Weld County, Colorado. U.S. Geol. Surv., Open-File Rep., 77-589.

Reimer, G.M., Roberts, A.A. and Denton, E.H., 1976. Diurnal effects on the helium concentration in soil gas and near-surface atmosphere. U.S. Geol. Surv., Open-File Rep., 76-715.

Reimer, G.M., Denton, E.H., Friedman, I. and Otton, J.K., 1979. Recent developments in uranium exploration using the U.S. Geological Survey's mobile helium detector. J. Geochem. Explor., 11: 1—12.

Reitsema, R.H., Kaltenback, A.J. and Lindberg, F.A., 1981. Source and migration of light hydrocarbons indicated by carbon isotopic ratios. Bull. Am. Assoc. Pet. Geol., 65: 1536—1542.

Roberts, A.A., Dalziel, M., Pogorski, L.A. and Ouirt, S.G., 1976. A possible helium anomaly in the soil gas, Boulder and Weld Counties, Colorado. U.S. Geol. Surv., Open-File Rep., 76-544.

and Edwards, G.S., 1975. Dissolved hydrocarbon distributions offshore ℓOS, Trans. Am. Geophys. Union, 56: 1000—1001.

ℓ.E., 1940. Symposium on geochemical exploration. Geochemical prospecting petroleum. Bull. Am. Assoc. Pet. Geol., 24: 1400—1433.

ℓkett, W.M., 1977. Use of hydrocarbon sniffing in offshore exploration. J. Geochem. Explor., 7: 243—254.

Schink, D.R., Guinasso, N.L., Charnell, R.L. and Sigalove, J.J., 1970. Radon profiles in the sea: a measure of air—sea exchange. IEEE Trans. Nucl. Sci., NS-17: 184—193.

Schink, D.R., Guinasso, N.L., Sigalove, J.J. and Cima, N.E., 1971. Hydrocarbons under the sea: a new survey technique. Proc. 3rd Offshore Technology Conference, pp. 131—142.

Scranton, M.I. and Brewer, P.G., 1977. Occurrence of methane in near-surface waters of the western subtropical North Atlantic. Deep-Sea Res., 24: 127—138.

Sealy, J.Q., 1974a. A geomicrobiological method of prospecting for petroleum. Part I. Oil Gas J., 72: 142—146.

Sealy, J.Q., 1974b. A geomicrobiological method of prospecting for petroleum. Part II. Oil Gas J., 72: 98—102.

Shtorgin, O.D., Nechina, S.V., Ozernyy, O.M. and Savka, O.M., 1978. Benzene in subsurface waters of the Crimea and its relation to the oil and gas of the interior. Geokhimiya, 12: 1863—1869. (English translation Geochem. Int., 15: 171—177.)

Sigalove, J.J. and Pearlman, M.D., 1975. Geochemical seep detection for offshore oil and gas exploration. Proc. 7th Offshore Technology Conference, 3: 95—102.

Simoneit, B.R.T., Mazurek, M.A., Brenner, S., Crisp, P.T. and Kaplan, I.R., 1979. Organic geochemistry of recent sediments from Guayamas Basin, Gulf of California. Deep-Sea Res., 26A: 879—891.

Slobad, R.L., Dunlap, H.F. and Moore, T.F., 1959. Exploration for petroliferous deposits by locating oil and gas seeps. U.S. Patent No. 2,918,579, December, 22.

Smith, G.H. and Ellis, M.M., 1963. Chromatographic analysis of gases from soil and vegetation, related to geochemical prospecting for petroleum. Bull. Am. Assoc. Pet. Geol., 47: 1897—1903.

Smith, J.E., Erdman, J.G. and Morris, D.A., 1971. Migration, accumulation and retention of petroleum in the earth. In: Proceedings of the 8th World Petroleum Congress. Applied Science Publishers, London, pp. 13—16.

Sokolov, V.A., 1933. New prospecting method for petroleum and gas. Technika, February Bull. NGRI No. 1.

Sokolov, V.A., 1959, Geochemical methods of prospecting for oil and gas deposits (Symposium). Izd. Akad. Nauk SSR, Moscow, 462 pp.

Sokolov, V.A., 1970. The theoretical foundations of geochemical prospecting for petroleum and natural gas and the tendencies of its development. Can. Inst. Min. Metall., 11: 544—549.

Sokolov, V.A., Alexeyev, F.A., Bars, E.A., Geodekyan, A.A., Mogilevsky, G.A., Yurovsky, Y.M. and Yasenev, B.P., 1959. Investigations into direct oil detection methods. Proc. 5th World Petroleum Congress. Section 1, pp. 667—687.

Sokolov, V.A., Geodekyan, A.A., Grigoryev, G.G., Krems, A.Ya., Stroganov, V.A., Zorkin, L.M., Zeiddson, M.I. and Vainbaum, S.Ja., 1970. The new methods of gas surveys, gas investigations of wells and some practical results. Can. Inst. Min. Metall., 11: 538—543.

Stadnik, Ye.V., Mogilevskiy, G.A., Soshnikov, V.K., Yurin, G.A., Gal'chenko, V.A. and Popovich, T.A., 1977. Importance of waters of near surface horizons for petroleum exploration (as in the Pripyat-Dnepr-Donets basin). Int. Geol. Rev., 19: 559—568.

Stahl, W., Faber, E., Carey, B.D. and Kirksey, D.L., 1981. Near-surface evidence of migration of natural from deep reservoirs and source rocks. Bull. Am. Assoc. Pet. Geol., 65: 1543—1550.

Stegena, L., 1961. On the principles of geochemical prospecting. Geophysics, 26: 447–451.

Swinnerton, J.W. and Linnenbom, V.J., 1967a. Determination of C_1 to C_4 hydrocarbons in sea water by gas chromatography. J. Gas Chromatogr., 5: 570–573.

Swinnerton, J.W. and Linnenbom, V.J., 1967b. Gaseous hydrocarbons in sea water: determination. Science, 156: 1119–1120.

Swinnerton, J.W., Linnenbom, V.J. and Cheek, C.H., 1962a. Determination of dissolved gases in aqueous solutions by gas chromatography. Anal. Chem., 34: 483–485.

Swinnerton, J.W., Linnenbom, V.J. and Cheek, C.H., 1962b. Revised sampling procedure for determination of dissolved gases in solution by gas chromatography. Anal. Chem., 34: 1509.

Swinnerton, J.W., Linnenbom, V.J. and Cheek, C.H., 1969. Distribution of methane and carbon monoxide between the atmosphere and natural waters. Environ. Sci. Technol., 3: 836–838.

Watson, R.D., Henry, M.E., Thasen, A.F., Donovan, T.J. and Hemphill, W.R., 1978. Marine monitoring of natural oil slicks and man made wastes utilizing an airborne imaging Fraunhofer Line Discriminator. Proc. 4th Joint Conference on Sensing of Environmental Pollutants, pp. 667–671.

Williams, D.D. and Miller, R.R., 1962. An instrument for on-stream stripping and gas chromatographic determination of dissolved gases in liquids. Anal. Chem., 34: 657–659.

Williams, R.T. and Bainbridge, A.E., 1973. Dissolved CO, CH_4 and H_2 in the Southern Ocean. J. Geophys. Res., 78: 2691–2694.

Wilson, D.F., Swinnerton, J.W. and Lamontagne, R.A., 1970. Production of carbon monoxide and gaseous hydrocarbons in sea water. Relation to dissolved organic carbon. Science, 168: 1577–1579.

Wood, M.B., 1980. An application of gas chromatography to measure concentrations of ethane, propane and ethylene found in interstitial soil gases. J. Chromatogr. Sci., 18: 307–310.

Zorkin, L.M., Stadnik, Ye. V. and Yurin, G.A., 1976. Geochemistry of nitrogen in ground water of oil- and gas-bearing basins. Int. Geol. Rev., 19: 1404–1410.

Zarella, W.M., Mousseau, R.J., Coggeshall, N.D., Morris, M.S. and Schrayer, G.J., 1967. Analysis and significance of hydrocarbons in subsurface brines. Geochim. Cosmochim. Acta, 31: 1155–1166.

NOTES ADDED IN PROOF

[1] In recent articles Duchscherer (1980, 1981) discussed various geochemical prospecting techniques and provided certain case histories.

[2] In another study Kvenvolden and Field (1981) used the same technique to analyse and detect thermogenic hydrocarbons in offshore samples from the Eel River Basin, northern California, and concluded that conditions for petroleum generation had existed within the basin.

[3] More recently, Donovan et al. (1981) have shown, from $\delta^{13}C$ and other parameters, that epigenetic zoning in surface and near-surface rocks in the Velma Oilfield, Oklahoma, results from seepage-induced redox gradients.

[4] Many of these results have been summarized in a recent paper by Donovan (1981).

REFERENCES

Donovan, T.J., 1981. Geochemical prospecting for oil and gas from orbital and suborbital altitudes. In: B.M. Gottlieb (Editor), Unconventional Methods in Exploration for Petroleum and Natural Gas. SMU Press, Dallas, Tex., pp. 95–117.

..., J., Roberts, A.A. and Dalziel, M.C., 1981. Epigenetic zoning in surface and ...ce rocks resulting from seepage-induced redox gradients, Velma Oil Field, ...ma: A synopsis. Shale Shaker, 32: 1—7.
...erer, W., 1980. Geochemical methods of prospecting for hydrocarbons. Oil ...as J., 78: 194—208.
...chscherer, W., 1981. Nongasometric geochemical prospecting for hydrocarbons with case histories. Oil Gas J., 79: 312—327.
Kvenvolden, K.A. and Field, M.E., 1981. Thermogenic hydrocarbons in unconsolidated sediment of Eel River Basin, offshore northern California. Bull. Am. Assoc. Pet. Geol., 65: 1642—1646.